Symbol Table

$\overline{A}$	complement of event A
H_0	null hypothesis
H_1	alternative hypothesis
α	alpha; probability of a type I error or the area of the critical region
β	beta; probability of a type II error
r	sample linear correlation coefficient
ρ	rho; population linear correlation coefficient
r^2	coefficient of determination
r_s	Spearman's rank correlation coefficient
b_1	point estimate of the slope of the regression line
b_0	point estimate of the y-intercept of the regression line
$\hat{y}$	predicted value of y
d	difference between two matched values
$\overline{d}$	mean of the differences d found from matched sample data
s_d	standard deviation of the differences d found from matched sample data
s_e	standard error of estimate
$\mu_{\overline{x}}$	mean of the population of all possible sample means $\overline{x}$
$\sigma_{\overline{x}}$	standard deviation of the population of all possible sample means $\overline{x}$
E	margin of error of the estimate of a population parameter, or expected value
Q_1, Q_2, Q_3	quartiles
$D_1, D_2, \ldots, D_9$	deciles
$P_1, P_2, \ldots, P_{99}$	percentiles
x	data value
f	frequency with which a value occurs
Σ	capital sigma; summation
Σx	sum of the values
Σx^2	sum of the squares of the values
$(\Sigma x)^2$	square of the sum of all values

Σxy	sum of the products of each x value multiplied by the corresponding y value
n	number of values in a sample
$n!$	n factorial
N	number of values in a finite population; also used as the size of all samples combined
k	number of samples or populations or categories
$\overline{x}$	mean of the values in a sample
μ	mu; mean of all values in a population
s	standard deviation of a set of sample values
σ	lowercase sigma; standard deviation of all values in a population
s^2	variance of a set of sample values
σ^2	variance of all values in a population
z	standard score
$z_{\alpha/2}$	critical value of z
t	t distribution
$t_{\alpha/2}$	critical value of t
df	number of degrees of freedom
F	F distribution
χ^2	chi-square distribution
χ^2_R	right-tailed critical value of chi-square
χ^2_L	left-tailed critical value of chi-square
p	probability of an event or the population proportion
q	probability or proportion equal to $1 - p$
$\hat{p}$	sample proportion
$\hat{q}$	sample proportion equal to $1 - \hat{p}$
$\overline{p}$	proportion obtained by pooling two samples
$\overline{q}$	proportion or probability equal to $1 - \overline{p}$
$P(A)$	probability of event A
$P(A \mid B)$	probability of event A, assuming event B has occurred
${}_nP_r$	number of permutations of n items selected r at a time
${}_nC_r$	number of combinations of n items selected r at a time

ANNOTATED INSTRUCTOR'S EDITION

Essentials of Statistics

ANNOTATED INSTRUCTOR'S EDITION

Essentials of Statistics

5TH EDITION

MARIO F. TRIOLA

PEARSON

Boston Columbus Indianapolis New York San Francisco Upper Saddle River
Amsterdam Cape Town Dubai London Madrid Milan Munich Paris Montréal Toronto
Delhi Mexico City São Paulo Sydney Hong Kong Seoul Singapore Taipei Tokyo

Editor in Chief: Deirdre Lynch
Executive Editor: Christopher Cummings
Senior Content Editors: Rachel Reeve and Chere Bemelmans
Assistant Editor: Sonia Ashraf
Senior Managing Editor: Karen Wernholm
Production Project Managers: Tracy Patruno and Mary Sanger
Associate Director of Design: Andrea Nix
Art Director and Cover Designer: Beth Paquin
Digital Assets Manager: Marianne Groth
Media Producer: Vicki Dreyfus
Software Developers: Mary Durnwald and Bob Carroll
Senior Marketing Manager: Erin Lane
Marketing Assistant: Kathleen DeChavez
Senior Author Support/Technology Specialist: Joe Vetere
Image Manager: Rachel Youdelman
Procurement Specialist: Debbie Rossi
Production Coordination, Composition, Illustrations:
Cenveo® Publisher Services
Text Design: Leslie Haimes
Cover Images: (kites) Manuel Fernandes/Shutterstock;
(pencils) Diane Miller/iStockphoto

Credits appear on pages 655–656, which constitute a continuation of the copyright page.

Library of Congress Cataloging-in-Publication Data

Triola, Mario F.
Essentials of statistics Mario F. Triola.--5th ed.
p. cm.
Includes index.
ISBN 0-321-92459-2
1. Statistics. I. Title.
QA276.12.T776 2011
519.5--dc22

2009013574

1 2 3 4 5 6 7 8 9 10—CRK—17 16 15 14 13

Annotated Instructor's Edition
ISBN 10: 0-321-92465-7
ISBN 13: 978-0-321-92465-0

PEARSON

www.pearsonhighered.com

ISBN-10: 0-321-92459-2
ISBN-13: 978-0-321-92459-9

To

Ginny

Marc, Dushana, and Marisa

Scott, Anna, Siena, and Kaia

About the Author

Mario F. Triola is a Professor Emeritus of Mathematics at Dutchess Community College, where he has taught statistics for over 30 years. Marty is the author of *Elementary Statistics*, 12th edition, *Elementary Statistics Using Excel*, 5th edition, *Elementary Statistics Using the TI-83/84 Plus Calculator*, 4th edition, and he is a co-author of *Biostatistics for the Biological and Health Sciences*, *Statistical Reasoning for Everyday Life*, 4th edition, *Business Statistics*, and *Introduction to Technical Mathematics*, 5th edition. *Elementary Statistics* is currently available as an International Edition, and it has been translated into several foreign languages. Marty designed the original STATDISK statistical software, and he has written several manuals and workbooks for technology supporting statistics education. He has been a speaker at many conferences and colleges. Marty's consulting work includes the design of casino slot machines and fishing rods, and he has worked with attorneys in determining probabilities in paternity lawsuits, analyzing data in medical malpractice lawsuits, identifying salary inequities based on gender, and analyzing disputed election results. He has also used statistical methods in analyzing medical school surveys, and analyzing survey results for the New York City Transit Authority. Marty has testified as an expert witness in New York State Supreme Court. The Text and Academic Authors Association has awarded Marty a "Texty" for Excellence for his work on *Elementary Statistics*.

Contents

Preface

This Fifth Edition was written with several goals:

- To provide an abundance of new and interesting data sets, examples, and exercises.
- To foster personal growth of students through critical thinking, use of technology, collaborative work, and development of communication skills.
- To incorporate the latest and best methods used by professional statisticians.
- To include information personally helpful to students, such as the best job search methods and the importance of avoiding mistakes on résumés.
- To provide the largest and best set of supplements to enhance teaching and learning.

GAISE This book reflects recommendations from the American Statistical Association and its *Guidelines for Assessment and Instruction in Statistics Education* (GAISE). Those guidelines suggest the following objectives and strategies.

1. **Emphasize statistical literacy and develop statistical thinking:** Each section exercise set begins with *Statistical Literacy and Critical Thinking* exercises. Many of the book's exercises are designed to encourage statistical thinking rather than the blind use of mechanical procedures.
2. **Use real data:** 92% of the examples and 89% of the exercises use real data.
3. **Stress conceptual understanding rather than mere knowledge of procedures:** Instead of seeking simple numerical answers, exercises and examples involve conceptual understanding through questions that encourage practical interpretations of results. Also, each chapter includes a *Data to Decision* project.
4. **Foster active learning in the classroom:** Each chapter ends with several *Cooperative Group Activities.*
5. **Use technology for developing conceptual understanding and analyzing data:** Computer software displays are included throughout the book. Special *Using Technology* subsections include instruction for using the software. Each chapter includes a *Technology Project.* When there are discrepancies between answers based on tables and answers based on technology, Appendix D provides *both* answers. The CD-ROM included with the book includes instructions for downloading free text-specific software (STATDISK) and data sets formatted for several different technologies, which are also listed in Appendix B.
6. **Use assessments to improve and evaluate student learning:** Assessment tools include an abundance of section exercises, Chapter Quick Quizzes, Chapter Review Exercises, Cumulative Review Exercises, technology projects, "Data to Decision" projects, and Cooperative Group Activity projects.

Audience/Prerequisites

Essentials of Statistics is written for students majoring in any subject. Algebra is used minimally, but students should have completed at least a high school or college elementary algebra course. In many cases, underlying theory is included, but this book does not require the mathematical rigor more suitable for mathematics majors.

Changes in This Edition

As in previous editions, this fifth edition includes a substantial revision of examples, exercises, and Chapter Problems, as shown in the following table:

	Number	New to This Edition	Use Real Data
Exercises	1585	86% (1362)	89% (1411)
Examples	196	85% (166)	92% (181)
Chapter Problems	11	100% (11)	100% (11)

Organization

Combined Sections

- The 4th edition Section 1-2 ("Statistical Thinking") and Section 1-4 ("Critical Thinking") have been combined into one section in this 5th edition:

  Section 1-2 Statistical and Critical Thinking

- The 4th edition Section 2-4 ("Statistical Graphics") and Section 2-5 ("Critical Thinking: Bad Graphs") have been combined into one section in this 5th edition:

  Section 2-4 Graphs That Enlighten and Graphs That Deceive

- The 4th edition Section 7-3 ("Estimating a Population Mean: σ Known") and Section 7-4 ("Estimating a Population Mean: σ Not Known") have been combined into one section in this 5th edition:

  Section 7-3 Estimating a Population Mean

  This change is motivated by two factors: (1) Technology makes use of the t distribution relatively simple, and (2) professional statisticians almost never use the normal distribution when constructing confidence interval estimates of population means.

- The 4th edition Section 8-4 ("Testing a Claim about a Mean: σ Known") and Section 8-5 ("Testing a Claim about a Mean: σ Not Known") have been combined into one section in this 5th edition:

  Section 8-4 Testing a Claim about a Mean

  This change is motivated by two factors: (1) Technology makes use of the t distribution relatively simple, and (2) professional statisticians almost never use the normal distribution when testing claims about a population mean.

Switched Sections Sections 6-6 and 6-7 from the previous edition have been switched so that Section 6-6 is now "Assessing Normality" and Section 6-7 is now "Normal as Approximation to Binomial." This change is motivated by the widespread availability of technology that facilitates methods for assessing normality, while the same technology has diminished the importance of using a normal approximation for a binomial distribution.

Exercises

Many exercises require the *interpretation* of results. Great care has been taken to ensure their usefulness, relevance, and accuracy. Exercises are arranged in order of increasing difficulty and by dividing them into two groups: (1) Basic Skills and Concepts and (2) Beyond the Basics. Beyond the Basics exercises address more difficult concepts or require a stronger mathematical background. In a few cases, these exercises introduce a new concept.

Real data

Hundreds of hours have been devoted to finding data that are real, meaningful, and interesting to students. All of the Chapter Problems are based on real data, 92% of the examples are based on real data, and 89% of the exercises are based on real data. Some exercises refer to the 23 large data sets listed in Appendix B, and 10 of those data sets are new to this edition. Exercises requiring use of the Appendix B data sets are located toward the end of each exercise set, where they are clearly identified.

Flexible Syllabus

This book's organization reflects the preferences of most statistics instructors, but there are two common variations:

- **Early coverage of correlation and regression:** Some instructors prefer to cover the basics of correlation and regression early in the course. *Sections 10-2 ("Correlation") and 10-3 ("Regression") can be covered early.* Simply limit coverage to Part 1 ("Basic Concepts") in each of those two sections.
- **Minimum probability:** Some instructors prefer extensive coverage of probability, while others prefer to include only basic concepts. Instructors preferring minimum coverage can include Section 4-2 while skipping the remaining sections of Chapter 4, as they are not essential for the chapters that follow. Many instructors prefer to cover the fundamentals of probability along with the basics of the addition rule and multiplication rule, and those topics can be covered with Sections 4-1 through 4-4.

Hallmark Features

Great care has been taken to ensure that each chapter of *Essentials of Statistics* will help students understand the concepts presented. The following features are designed to help meet that objective:

Chapter-opening features:

- A list of chapter sections previews the chapter for the student.
- A chapter-opening problem, using real data, motivates the chapter material. Examples and exercises that further explore the ideas presented in the opening problem are marked with an icon.

- The first section is a brief review of relevant earlier concepts and previews the chapter's objectives.

End-of-chapter features:

A chapter **Review** summarizes the key concepts and topics of the chapter.

A **Chapter Quick Quiz** provides 10 review questions that require brief answers.

Review Exercises offer practice on the chapter concepts and procedures.

Cumulative Review Exercises reinforce earlier material.

A **Technology Project** provides an activity for STATDISK, MINITAB®, Excel®, a TI-83/84 Plus calculator, or StatCrunch®.

From Data to Decision is a capstone problem that requires critical thinking and writing.

Cooperative Group Activities encourage active learning in groups.

Other features:

Real Data Sets Appendix B contains printed versions of 23 large data sets referenced throughout the book, including 10 that are new. These data sets are also available on the companion Web site (http://www.pearsonhighered.com/triola), the CD-ROM bound in the back of new copies of the book, and MyStatLab®.

Margin Essays Of 107 margin essays, 20% are new and several others have been updated. New topics include *Statistics for Online Dating, DNA Evidence Misused, Bar Code,* and *How Many People Do You Know?*

Flowcharts The text includes 12 flowcharts that simplify and clarify more complex concepts and procedures. Animated versions of the text's flowcharts are available within MyStatLab and MathXL®.

Top 20 Topics The most important topics in any introductory statistics course are identified in the text with an icon. Students using MyStatLab have access to additional resources for learning these topics with definitions, animations, and video lessons.

Quick-Reference Endpapers Tables A-2 and A-3 (the normal and t distributions) are reproduced on the rear inside cover pages.

Detachable Formula and Table Card This insert, organized by chapter, gives students a quick reference for studying, or for use when taking tests (if allowed by the instructor). It also includes the most commonly used tables.

CD-ROM The CD-ROM was prepared by Mario F. Triola and is bound into the back of every new copy of the book. It contains the data sets from Appendix B available as txt files, MINITAB worksheets, SPSS files, SAS files, JMP files, Excel workbooks, and a TI-83/84 Plus application. The CD also includes sections on *Probabilities Through Simulations* and *Bayes' Theorem*, an index of applications, a symbols table, programs for the TI-83/84 Plus graphing calculator, and instructions for obtaining STATDISK Statistical Software (Version 12).

Technology

New: This edition now includes instructions and displays from the StatCrunch technology, and XLSTAT is now used in Excel screenshots.

As in the preceding edition, there are many displays of screens from technology throughout the book, and some exercises are based on displayed results from technology. Where appropriate, sections end with a *Using Technology* subsection that includes instruction for STATDISK, MINITAB®, Excel®, StatCrunch, or a TI-83/84 Plus calculator. (Throughout this text, "TI-83/84 Plus" is used to identify a TI-83 Plus, TI-84 Plus, or TI-Nspire calculator with the TI-84 Plus keypad installed.) The end-of-chapter features include a *Technology Project.*

The STATDISK (Version 12) statistical software package is designed specifically for this textbook. STATDISK is free to users of this book and instructions for downloading it are included on the CD-ROM.

Supplements

For the Student

Student's Solutions Manual, by James Lapp (Colorado Mesa University), provides detailed, worked-out solutions to all odd-numbered text exercises.
(ISBN-13: 978-0-321-92466-7; ISBN-10: 0-321-92466-5)

Student Workbook for the Triola Statistics Series, by Anne Landry (Florida Community College at Jacksonville) offers additional examples, concept exercises, and vocabulary exercises for each chapter.
(ISBN-13: 978-0-321-89196-9; ISBN-10: 0-321-89196-1)

The following technology manuals include instructions, examples from the main text, and interpretations to complement those given in the text.

Excel Student Laboratory Manual and Workbook for the Triola Statistics Series, by Beverly Dretzke (University of Minnesota).
(ISBN-13: 978-0-321-83799-8; ISBN-10: 0-321-83799-1)

MINITAB Student Laboratory Manual and Workbook for the Triola Statistics Series, by Mario F. Triola.
(ISBN-13: 978-0-321-83379-2; ISBN-10: 0-321-83379-1)

Graphing Calculator Manual for the TI-83 Plus, TI-84 Plus, TI-89 and TI-Nspire, by Kathleen McLaughlin (University of Connecticut) and Dorothy Wakefield (University of Connecticut Health Center).
(ISBN-13: 978-0-321-83803-2; ISBN-10: 0-321-83803-3)

STATDISK Student Laboratory Manual and Workbook for the Triola Statistics Series (Download Only), by Mario F. Triola. These files are available to instructors and students through the Triola Statistics Series Web site, www.pearsonhighered.com/triola, and MyStatLab.

SPSS Student Laboratory Manual and Workbook for the Triola Statistics Series (Download Only), by James J. Ball (Indiana State University). These files are available to instructors and students through the Triola Statistics Series Web site, www.pearsonhighered.com/triola, and MyStatLab.

StatCrunch Manual (Download Only) for the Triola Statistics Series, by Diane Hollister (Reading Area Community College). These files are available to instructors and students through the Triola Statistics Series Web site, www.pearsonhighered.com/triola, and MyStatLab.

For the Instructor

Annotated Instructor's Edition, by Mario F. Triola, contains answers to exercises in the margin, plus recommended assignments, and teaching suggestions.
(ISBN-13: 978-0-321-92465-0; ISBN-10: 0-321-92465-7)

Instructor's Solutions Manual (Download Only), by James Lapp (Colorado Mesa University), contains solutions to all the exercises. These files are available to qualified instructors through Pearson Education's online catalog at www.pearsonhighered.com/irc or within MyStatLab.

Insider's Guide to Teaching with the Triola Statistics Series, by Mario F. Triola, contains sample syllabi and tips for incorporating projects, as well as lesson overviews, extra examples, minimum outcome objectives, and recommended assignments for each chapter.
(ISBN-13: 978-0-321-83372-3; ISBN-10: 0-321-83372-4)

Testing System: TestGen® (www.pearsoned.com/testgen) enables instructors to build, edit, and print, and administer tests using a computerized bank of questions developed to cover all the objectives of the text. TestGen is algorithmically based, allowing instructors to create multiple but equivalent versions of the same question or test with the click of a button. Instructors can also modify test bank questions or add new questions. The software and testbank are available for download from Pearson Education's online catalog.

PowerPoint® Lecture Slides: Free to qualified adopters, this classroom lecture presentation software is geared specifically to the sequence and philosophy of *Essentials of Statistics.* Key graphics from the book are included to help bring the statistical concepts alive in the classroom. These files are available to qualified instructors through Pearson Education's online catalog at www.pearsonhighered.com/irc or within MyStatLab.

Active Learning Questions: Prepared in PowerPoint®, these questions are intended for use with classroom response systems. Multiple-choice questions are available for each chapter of the book, allowing instructors to quickly assess mastery of material in class. The Active Learning Questions are available to download from within MyStat Lab® and from the Pearson Education online catalog.

Technology Resources

On the CD-ROM, Triola Statistics Series Web site (http://www.pearsonhighered.com/triola), and MyStatLab:

- Appendix B data sets formatted for Minitab, SPSS, SAS, Excel, JMP, and as text files. Additionally, these data sets are available as an APP for the TI-83/84 Plus calculators, and supplemental programs for the TI-83/84 Plus calculator are also available.
- **STATDISK** statistical software instructions for download.
- Extra data sets, *Probabilities through Simulations*, *Bayes' Theorem*, and a symbols table.

Triola Stats Visit www.triolastats.com and select Stat Resources for updated links to a variety of statistics resources and data sets recommended by the author.

Video Resources have been expanded and now supplement most sections in the book, with many topics presented by the author. The videos feature technologies found in the book and the worked-out Chapter Review exercises. This is an excellent resource for students who have missed class or wish to review a topic. It is also an excellent resource for instructors involved with distance learning, individual study, or self-paced learning programs. These videos also contain optional English and Spanish captioning. All videos are available through the MyStatLab online course.

MyStatLab™ Online Course (access code required) MyStatLab is a course management system that delivers **proven results** in helping individual students succeed. It provides **engaging experiences** that personalize, stimulate, and measure learning for each student. Tools are embedded to make it easy to integrate statistical software into the course. And, it comes from a **trusted partner** with educational expertise and an eye on the future.

- MyStatLab's comprehensive online gradebook automatically tracks students' results on tests, quizzes, homework, and in the study plan. Instructors can use the gradebook to provide positive feedback or intervene if students have trouble. Gradebook data can be easily exported to a variety of spreadsheet programs, such as Microsoft Excel.

MyStatLab provides **engaging experiences** that personalize, stimulate, and measure learning for each student.

- **Tutorial Exercises with Multimedia Learning Aids:** The homework and practice exercises in MyStatLab align with the exercises in the textbook, and most regenerate algorithmically to give students unlimited opportunity for practice and mastery. Exercises offer immediate helpful feedback, including guided solutions, sample problems, animations, and videos.
- **Adaptive Study Plan:** Pearson now offers an optional focus on adaptive learning in the study plan to allow students to work on just what they need to learn when it makes the most sense to learn it. The adaptive study plan maximizes students' potential for understanding and success.
- **Additional Question Libraries:** In addition to algorithmically regenerated questions that are aligned with your textbook, MyStatLab courses come with two additional question libraries. **450 Getting Ready for Statistics** questions cover the developmental math topics students need for the course. These can be assigned as a prerequisite to other assignments, if desired. The **1000 Conceptual Question Library** requires students to apply their statistical understanding.
- **StatCrunch®:** MyStatLab includes Web-based statistical software, StatCrunch, within the online assessment platform so that students can analyze data sets from exercises and the text. In addition, MyStatLab includes access to **www.StatCrunch.com,** a Web site where users can access thousands of shared data sets, conduct online surveys, perform complex analyses using the powerful statistical software, and generate compelling reports.
- **Integration of Statistical Software:** We make it easy to copy our data sets, both from the ebook and the MyStatLab questions, into software such as StatCrunch, MINITAB, Excel, and more. Students have access to a variety of support tools—Technology Instruction Videos, Technology Study Cards, and Manuals for select titles—to learn how to use statistical software.
- **StatTalk Videos:** Fun-loving statistician Andrew Vickers takes to the streets of Brooklyn, NY, to demonstrate important statistical concepts through interesting stories and real-life events. This series of 24 videos includes available assessment questions and an instructor's guide.
- **Expert Tutoring:** Although many students describe the whole of MyStatLab as "like having your own personal tutor," students also have access to live tutoring from qualified statistics instructors via MyStatLab.

And, MyStatLab comes from a **trusted partner** with educational expertise and an eye on the future.

MyStatLab™ Ready to Go Course (access code required)
These new Ready to Go courses provide students with all the same great MyStatLab features that you're used to, but make it easier for instructors to get started. Each course includes pre-assigned homeworks and quizzes to make creating your course even simpler. Ask your Pearson representative about the details for this particular course or to see a copy of this course.

MathXL® for Statistics Online Course (access code required)
MathXL® is the homework and assessment engine that runs MyStatLab. (MyStatLab is MathXL plus a learning management system.)

With MathXL for Statistics, instructors can:

- Create, edit, and assign online homework and tests using algorithmically generated exercises correlated at the objective level to the textbook.
- Create and assign their own online exercises and import TestGen tests for added flexibility.
- Maintain records of all student work, tracked in MathXL's online gradebook.

With MathXL for Statistics, students can:

- Take chapter tests in MathXL and receive personalized study plans and/or personalized homework assignments based on their test results.
- Use the study plan and/or the homework to link directly to tutorial exercises for the objectives they need to study.
- Access supplemental animations and video clips directly from selected exercises.
- Copy our data sets for use with external statistical software. We make it easy to copy our data sets, both from the ebook and the MyStatLab questions, into software like StatCrunch, Minitab, Excel and more.

MathXL for Statistics is available to qualified adopters. For more information, visit our Web site at www.mathxl.com, or contact your Pearson representative.

StatCrunch®
StatCrunch is powerful Web-based statistical software that allows users to perform complex analyses, share data sets, and generate compelling reports of their data. The vibrant online community offers thousands of data sets for students to analyze.

- **Collect.** Users can upload their own data to StatCrunch or search a large library of publicly shared data sets, spanning almost any topic of interest. Also, an online survey tool allows users to quickly collect data via Web-based surveys.
- **Crunch.** A full range of numerical and graphical methods allow users to analyze and gain insights from any data set. Interactive graphics help users understand statistical concepts, and are available for export to enrich reports with visual representations of data.
- **Communicate.** Reporting options help users create a wide variety of visually appealing representations of their data.

Full access to StatCrunch is available with a MyStatLab kit, and StatCrunch is available by itself to qualified adopters. For more information, visit our Web site at www.StatCrunch.com, or contact your Pearson representative.

The Student Edition of MINITAB is a condensed version of the professional release of MINITAB statistical software. It offers the full range of statistical methods and graphical capabilities, along with worksheets that can include up to 10,000 data points. Individual copies of the software can be bundled with the text (ISBN-13 978-0-13-143661-9; ISBN-10: 0-13-143661-9).

JMP Student Edition is an easy-to-use, streamlined version of JMP desktop statistical discovery software from SAS Institute, Inc. and is available for bundling with the text (ISBN-13: 978-0-321-89164-8; ISBN-10: 0-321-89164-3).

Acknowledgments

I would like to thank the thousands of statistics professors and students who have contributed to the success of this book. I would like to extend special thanks to Mitch Levy, Broward College; Kate Kozak, Coconino Community College; Steve Schwager, Cornell University; Rick Woodmansee, Sacramento City College; Rob Fusco, Broward College; Joe Pick, Palm Beach State College; Richard Weil, Brown College; Donald Burd, Monroe College; James Bryan, Merced College; Richard Herbst, Montgomery County Community College; Diane Hollister, Reading Area Community College; George Jahn, Palm Beach State College; Dan Kumpf, Ventura College; Kim McHale, Heartland Community College; Ken Mulzet, Florida State College at Jacksonville; Sandra Spain, Thomas Nelson Community College; Ellen G. Stutes, Louisiana State University, Eunice; Barbara Ward, Belmont University; Richard Hertz; Chris Vertullo, Marist College; Kelly Smitch, Brevard College; Robert Black, United States Air Force Academy; Michael Huber.

This fifth edition of *Essentials of Statistics* is truly a team effort, and I consider myself fortunate to work with the dedication and commitment of the Pearson Arts and Sciences team. I thank Chris Cummings, Deirdre Lynch, Rachel Reeve, Chere Bemelmans, John Orr (of Cenveo Publisher Services), Tracy Patruno, Mary Sanger, Sonia Ashraf, Christina Lepre, Joe Vetere, and Beth Paquin. I extend special thanks to Marc Triola, M.D., New York University, for his outstanding work on the STATDISK software, and Scott Triola for his great help in creating this new edition.

I thank the following for their help in checking the accuracy of text and answers in this fifth edition: James Lapp, David Lund, and Kimberley Polly.

M.F. T.
Madison, Connecticut
September 2013

ANNOTATED INSTRUCTOR'S EDITION

Essentials of Statistics

1 Introduction to Statistics

1

chapter problem

Survey: Have you ever been hit with a computer virus?

The world in which we live is now saturated with surveys. Surveys are essential tools used in marketing. Surveys determine what television shows we watch. Surveys guide political candidates. Surveys shape business practices and many other aspects of our lives. Surveys provide us with understanding about the thinking of the rest of the world. Let's consider one particular survey dealing with a topic of great concern to all of us who have embraced the use of computer technology. The survey question and responses are given below, and Figure 1-1 graphically depicts the survey results. (Figure 1-1 was generated using Minitab statistical software.)

"Have you ever been hit by a computer virus?"

- Yes: 106,685
- No: 63,378

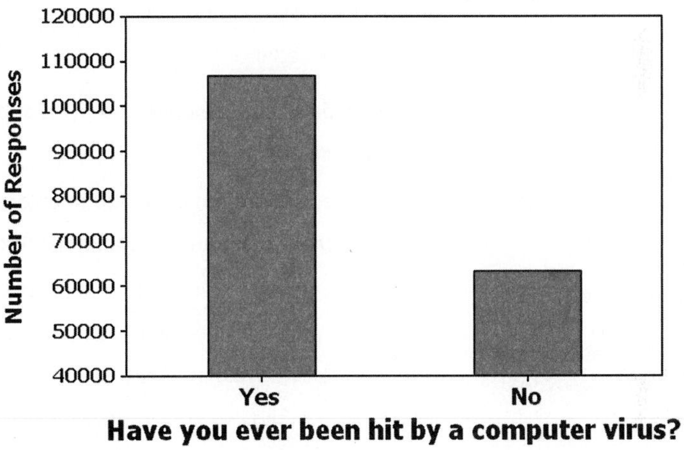

Figure 1-1 **Survey Results**

The results of the survey appear to be quite dramatic. The total number of respondents is 170,063 adults, and that is a very large number of respondents. Many polls have only about one thousand or two thousand respondents. Also, by looking at the bars in Figure 1-1, we see that roughly three times as many respondents have been hit by computer viruses as have not been hit. One important objective of this text is to encourage the use of critical thinking so that such results are not blindly accepted. We might question whether the survey results are valid. Who conducted the survey? How were respondents selected? Does the graph in Figure 1-1 depict the results in a way that is not misleading?

The survey results presented here have two major flaws. Because these two flaws are among the most common, it is especially important to recognize them. Following are brief descriptions of each of the two major flaws.

Flaw 1: Misleading Graph Figure 1-1 is deceptive. Using a vertical scale that does not start at zero exaggerates the difference between the two numbers of responses. Thus Figure 1-1 makes it appear that the "yes" responses are about three times the number of "no" responses, but examination of the actual response counts shows that the "yes" responses are really about 1.7 times the "no" responses. Deceptive graphs are discussed in more detail in Section 2-4.

Flaw 2: Bad Sampling Method The survey responses are from a recent America OnLine survey of Internet users. The survey question was posted on the America OnLine Web site and Internet users decided whether to respond. This is an example of a *voluntary response sample*—a sample in which respondents decide themselves whether to participate. With a voluntary response sample, it often happens that those with a strong interest in the topic are more likely to participate, so the results are very questionable. The large number of respondents does not overcome this flaw of having a voluntary response sample. When we want to use sample data to learn something about a population, it is *extremely* important to obtain sample data that are representative of the population from which the data are drawn. As we proceed through this chapter and discuss types of data and sampling methods, we should focus on these key concepts:

- **Sample data must be collected in an appropriate way, such as through a process of *random* selection.**
- **If sample data are not collected in an appropriate way, the data may be so completely useless that no amount of statistical torturing can salvage them.**

It would be easy to accept the preceding survey results and blindly proceed with calculations and statistical analyses, but if we did so, we would miss the critical two flaws described above. We might then develop conclusions that are fundamentally wrong and misleading. Instead, we should develop skills in statistical thinking and critical thinking so we can understand why the survey is so seriously flawed and why we should not rely on it to yield any valid information.

Note to Instructor

Notes to instructors are included in the margin throughout this text, but they are *not* included in student editions. These notes include comments about changes to this new edition, teaching suggestions, and recommended exercises. Suggested homework assignments can be found in the margin near the beginning of each set of exercises. Also see the *Insider's Guide to Teaching with the Triola Statistics Series,* which is a support manual for adjuncts and full-time instructors.

Organization change to this new edition: Section 1-2 (Statistical Thinking) and Section 1-4 (Critical Thinking) from the 11th edition have been combined into Section 1-2 (Statistical and Critical Thinking) in this edition.

Suggested treatment of Chapter 1: Because introductory statistics courses require careful time management, consider assigning all or part of Chapter 1 as reading to be done independently. Announce that these definitions should be well known and understood: *population, sample, parameter, statistic, discrete data, continuous data, voluntary response sample,* and *simple random sample.*

Most concepts in this chapter are not very difficult and can be easily understood, but these might require some explanation in class: the four levels of measurement in Section 1-3 and the different types of sampling in Section 1-4.

1-1 Review and Preview

The first section of each of Chapter 1 through Chapter 11 begins with a brief review of what preceded the chapter, and a preview of what the chapter includes. This first chapter isn't preceded by much of anything except the Preface, and we won't review that (most people don't even read it). However, we can review and formally define some statistical terms that are commonly used. The Chapter Problem discussed an America OnLine poll that collected sample data. Polls collect data from a small part of a larger group so that we can learn something about the larger group. This is a common and important goal of statistics: Learn about a large group by examining sample data from some of its members. In this context, the terms *sample* and *population* have special meanings. Formal definitions for these and other basic terms are given here.

DEFINITIONS

Data are collections of observations, such as measurements, genders, or survey responses. (A single data value is called a *datum*, a term that does not see very much use.)

Statistics is the science of planning studies and experiments; obtaining data; and then organizing, summarizing, presenting, analyzing, and interpreting those data and then drawing conclusions based on them.

A **population** is the complete collection of *all* measurements or data that are being considered.

A **census** is the collection of data from *every* member of the population.

A **sample** is a *subcollection* of members selected from a population.

Because populations are often very large, a common objective of the use of statistics is to obtain data from a sample and then use those data to form a conclusion about the population. See Example 1.

Example 1 Gallup Poll: Identity Theft

In a poll conducted by the Gallup corporation, 1013 adults in the United States were randomly selected and surveyed about identity theft. Results showed that 66% of the respondents worried about identity theft frequently or occasionally.

Gallup pollsters decided who would be asked to participate in the survey and they used a sound method of randomly selecting adults. The respondents are not a voluntary response sample, and the results are likely to be better than those obtained from the America OnLine survey discussed earlier.

In this case, the population consists of all 241,472,385 adults in the United States, and it is not practical to survey each of them. The sample consists of the 1013 adults who were surveyed. The objective is to use the sample data as a basis for drawing a conclusion about the population of all adults, and methods of statistics are helpful in drawing such conclusions.

Origin of "Statistics"

The word *statistics* is derived from the Latin word *status* (meaning "state"). Early uses of statistics involved compilations of data and graphs describing various aspects of a state or country. In 1662, John Graunt published statistical information about births and deaths. Graunt's work was followed by studies of mortality and disease rates, population sizes, incomes, and unemployment rates. Households, governments, and businesses rely heavily on statistical data for guidance. For example, unemployment rates, inflation rates, consumer indexes, and birth and death rates are carefully compiled on a regular basis, and the resulting data are used by business leaders to make decisions affecting future hiring, production levels, and expansion into new markets.

1-2 Statistical and Critical Thinking

Key Concept This section provides an overview of the process involved in conducting a statistical study. This process consists of "prepare, analyze, and conclude." We begin with a preparation that involves consideration of the context, consideration of the source of data, and consideration of the sampling method. Next, we construct suitable graphs, explore the data, and execute computations required for the statistical method being used. Finally, we form conclusions by determining whether results have statistical significance and practical significance. See Figure 1-2 for a summary of this process.

Figure 1-2 includes key elements in a statistical study. Note that the procedure outlined in Figure 1-2 does not focus on mathematical calculations. Thanks to wonderful developments in technology, we now have tools that effectively do the number crunching so that we can focus on understanding and interpreting results.

Note to Instructor

Section 1-2 combines these two sections from the previous edition: Section 1-2 (Statistical Thinking) and Section 1-4 (Critical Thinking).

Emphasize that statistics is much more than plugging data into a formula. It involves consideration of the factors included in Figure 1-2, which is an excellent summary and overview. Because voluntary response samples are now so common, stress how such samples are totally unsuitable for good statistical methodology.

Prepare

1. **Context**
   What do the data mean?
   What is the goal of study?
2. **Source of the Data**
   Are the data from a source with a special interest so that there is pressure to obtain results that are favorable to the source?
3. **Sampling Method**
   Were the data collected in a way that is unbiased, or were the data collected in a way that is biased (such as a procedure in which respondents volunteer to participate)?

↓

Analyze

1. **Graph the Data**
2. **Explore the Data**
   Are there any outliers (numbers very far away from almost all of the other data)?
   What important statistics summarize the data (such as the mean and standard deviation described later)?
   How are the data distributed?
   Are there missing data?
   Did many selected subjects refuse to respond?
3. **Apply Statistical Methods**
   Use technology to obtain results.

↓

Conclude

1. **Statistical Significance**
   Do the results have statistical significance?
   Do the results have practical significance?

Figure 1-2 Statistical Thinking

Prepare

Context Let's consider the data in Table 1-1. (The data are from Data Set 6 in Appendix B.) The data in Table 1-1 consist of measured IQ scores and measured brain volumes from 10 different subjects. The data are matched in the sense that each individual "IQ/brain volume" pair of values is from the same subject. The first subject had a measured IQ score of 96 and a brain volume of 1005 cm^3. The format of Table 1-1 suggests the following goal: Determine whether there is a relationship between IQ score and brain volume. This goal suggests a possible hypothesis: People with larger brains tend to have higher IQ scores.

Table 1-1 IQ Scores and Brain Volumes (cm^3)

IQ	96	87	101	103	127	96	88	85	97	124
Brain Volume (cm^3)	1005	1035	1281	1051	1034	1079	1104	1439	1029	1160

Source of the Data The data in Table 1-1 were provided by M. J. Tramo, W. C. Loftus, T. A. Stukel, J. B. Weaver, and M. S. Gazziniga, who discuss the data in the article "Brain Size, Head Size, and IQ in Monozygotic Twins," *Neurology*, Vol. 50. The researchers are from reputable medical schools and hospitals, and they would not gain by putting spin on the results. In contrast, Kiwi Brands, a maker of shoe polish, commissioned a study that resulted in this statement, which was printed in some newspapers: "According to a nationwide survey of 250 hiring professionals, scuffed shoes was the most common reason for a male job seeker's failure to make a good first impression." We should be very wary of such a survey in which the sponsor can somehow profit from the results. When physicians who conduct clinical experiments on the efficacy of drugs receive funding from drug companies they have an incentive to obtain favorable results. Some professional journals, such as the *Journal of the American Medical Association*, now require that physicians report such funding in journal articles. We should be skeptical of studies from sources that may be biased.

Sampling Method The data in Table 1-1 were obtained from subjects who were recruited by researchers, and the subjects were paid for their participation. All subjects were between 24 years and 43 years of age, they all had at least a high school education, and the medical histories of subjects were reviewed in an effort to ensure that no subjects had neurologic or psychiatric disease. In this case, the sampling method appears to be sound.

Sampling methods and the use of randomization will be discussed in Section 1-4, but for now, we simply emphasize that a sound sampling method is absolutely essential for good results in a statistical study. It is generally a bad practice to use voluntary response (or self-selected) samples, even though their use is common.

DEFINITION A **voluntary response sample** (or **self-selected sample**) is one in which the respondents themselves decide whether to be included.

The following types of polls are common examples of voluntary response samples. By their very nature, all are seriously flawed because we should not make conclusions about a population on the basis of such a biased sample:

- Internet polls, in which people online can decide whether to respond
- Mail-in polls, in which subjects can decide whether to reply
- Telephone call-in polls, in which newspaper, radio, or television announcements ask that you voluntarily call a special number to register your opinion

With such voluntary response samples, we can draw valid conclusions only about the specific group of people who chose to participate; nevertheless, such samples are often incorrectly used to assert or imply conclusions about a larger population. From a statistical viewpoint, such a sample is fundamentally flawed and should not be used for making general statements about a larger population. The Chapter Problem involves an America OnLine poll with a voluntary response sample. See also Examples 1 and 2, which follow.

Value of a Statistical Life

The *value of a statistical life* (VSL) is a measure routinely calculated and used for making decisions in fields such as medicine, insurance, environmental health, and transportation safety. As of this writing, the value of a statistical life is $6.9 million.

Many people oppose the concept of putting a value on a human life, but the word *statistical* in the "value of a statistical life" is used to ensure that we don't equate it with the true worth of a human life. Some people legitimately argue that every life is priceless, but others argue that there are conditions in which it is impossible or impractical to save every life, so a value must be somehow assigned to a human life in order that sound and rational decisions can be made. Not far from the author's home, a parkway was modified at a cost of about $3 million to improve safety at a location where car occupants had previously died in traffic crashes. In the cost-benefit analysis that led to this improvement in safety, the value of a statistical life was surely considered.

Publication Bias

There is a "publication bias" in professional journals. It is the tendency to publish positive results (such as showing that some treatment is effective) much more often than negative results (such as showing that some treatment has no effect). In the article "Registering Clinical Trials" (*Journal of the American Medical Association*, Vol. 290, No. 4), authors Kay Dickersin and Drummond Rennie state that "the result of not knowing who has performed what (clinical trial) is loss and distortion of the evidence, waste and duplication of trials, inability of funding agencies to plan, and a chaotic system from which only certain sponsors might benefit, and is invariably against the interest of those who offered to participate in trials and of patients in general." They support a process in which *all* clinical trials are registered in one central system, so that future researchers have access to all previous studies, not just the studies that were published.

Example 1 Voluntary Response Sample

Literary Digest magazine conducted a poll for the 1936 presidential election by sending out 10 million ballots. The magazine received 2.3 million responses. The poll results suggested incorrectly that Alf Landon would win the presidency. In a much smaller poll of 50,000 people, George Gallup correctly predicted that Franklin D. Roosevelt would win. The lesson here is that it is not necessarily the *size* of the sample that makes it effective, but the *sampling method*. The *Literary Digest* ballots were sent to magazine subscribers as well as to registered car owners and those who used telephones. On the heels of the Great Depression, this group included disproportionately more wealthy people, who were Republicans. But the real flaw in the *Literary Digest* poll is that it resulted in a voluntary response sample. In contrast, Gallup used an approach in which he obtained a representative sample based on demographic factors. (Gallup modified his methods when he made a wrong prediction in the famous 1948 Dewey/Truman election. Gallup stopped polling too soon, and he failed to detect a late surge in support for Truman.) The *Literary Digest* poll is a classic illustration of the flaws inherent in basing conclusions on a voluntary response sample.

Example 2 Voluntary Response Sample

The ABC television show *Nightline* asked viewers to call with their opinion about whether the United Nations headquarters should remain in the United States. Viewers then decided themselves whether to call with their opinions, and 67% of 186,000 respondents said that the United Nations should be *moved out* of the United States. In a separate poll, 500 respondents were randomly selected and 72% of them wanted the United Nations to *stay* in the United States. The two polls produced dramatically different results. Even though the *Nightline* poll involved 186,000 volunteer respondents, the much smaller poll of 500 randomly selected respondents is more likely to provide better results because of the superior sampling method.

Analyze

Graph and Explore After carefully considering context, source of the data, and sampling method, we can proceed with an analysis that should begin with appropriate graphs and explorations of the data. Graphs are discussed in Chapter 2, and important statistics are discussed in Chapter 3.

Apply Statistical Methods Later chapters describe important statistical methods, but application of these methods is often made easy with calculators and/or statistical software packages. A good statistical analysis does not require strong computational skills. A good statistical analysis does require using common sense and paying careful attention to sound statistical methods.

Conclude

Statistical Significance *Statistical significance* is achieved in a study when we get a result that is very unlikely to occur by chance.

- Getting 98 girls in 100 random births is statistically significant because such an extreme event is not likely to be the result of random chance.
- Getting 52 girls in 100 births is not statistically significant, because that event could easily occur with random chance.

Practical Significance It is possible that some treatment or finding is effective, but common sense might suggest that the treatment or finding does not make enough of a difference to justify its use or to be practical, as illustrated in Example 3.

Example 3 Statistical Significance versus Practical Significance

In a test of the Atkins weight loss program, 40 subjects using that program had a mean weight loss of 2.1 kg (or 4.6 pounds) after one year (based on data from "Comparison of the Atkins, Ornish, Weight Watchers, and Zone Diets for Weight Loss and Heart Disease Risk Reduction," by Dansinger et al., *Journal of the American Medical Association*, Vol. 293, No. 1). Using formal methods of statistical analysis, we can conclude that the mean weight loss of 2.1 kg is statistically significant. That is, based on statistical criteria, the diet appears to be effective. However, using common sense, it does not seem very worthwhile to pursue a weight loss program resulting in such relatively insignificant results. Someone starting a weight loss program would probably want to lose considerably more than 2.1 kg. Although the mean weight loss of 2.1 kg is statistically significant, it does not have practical significance. The statistical analysis suggests that the weight loss program is effective, but *practical* considerations suggest that the program is basically ineffective.

Analyzing Data: Potential Pitfalls

Here are a few more items that could cause problems when analyzing data.

Misleading Conclusions When forming a conclusion based on a statistical analysis, we should make statements that are clear even to those who have no understanding of statistics and its terminology. We should carefully avoid making statements not justified by the statistical analysis. For example, Section 10-2 introduces the concept of a *correlation*, or association between two variables, such as smoking and pulse rate. A statistical analysis might justify the statement that there is a correlation between the number of cigarettes smoked and pulse rate, but it would not justify a statement that the number of cigarettes smoked *causes* a person's pulse rate to change. Such a statement about causality can be justified by physical evidence, not by statistical analysis.

Correlation does not imply causation.

Reported Results When collecting data from people, it is better to take measurements yourself instead of asking subjects to report results. Ask people what they weigh and you are likely to get their *desired* weights, not their actual weights. Accurate weights are collected by using a scale to measure weights, not by asking people to report their weights.

Small Samples Conclusions should not be based on samples that are far too small. The Children's Defense Fund published *Children Out of School in America*, in which it was reported that among secondary school students suspended in one region, 67% were suspended at least three times. But that figure is based on a sample of only *three* students! Media reports failed to mention that this sample size was so small.

Loaded Questions If survey questions are not worded carefully, the results of a study can be misleading. Survey questions can be "loaded" or intentionally worded to

Detecting Phony Data

A class is given the homework assignment of recording the results when a coin is tossed 500 times. One dishonest student decides to save time by just making up the results instead of actually flipping a coin. Because people generally cannot make up results that are really random, we can often identify such phony data. With 500 tosses of an actual coin, it is extremely likely that at some point, you will get a run of six heads or six tails, but people almost never include such a run when they make up results.

Another way to detect fabricated data is to establish that the results violate Benford's law: For many collections of data, the leading digits are not uniformly distributed. Instead, the leading digits of 1, 2, . . . , 9 occur with rates of 30%, 18%, 12%, 10%, 8%, 7%, 6%, 5%, and 5%, respectively. (See "The Difficulty of Faking Data," by Theodore Hill, *Chance*, Vol. 12, No. 3.)

Statistics Is Sexy

CareerCast.com is a job Web site, and its organizers analyzed professions using five criteria: environment, income, employment prospects, physical demands, and stress. Based on that study, here are the top ten jobs: (1) mathematician, (2) actuary, (3) ***statistician*** (author's emphasis), (4) biologist, (5) software engineer, (6) computer system analyst, (7) historian, (8) sociologist, (9) industrial designer, (10) accountant. Lumberjacks are at the bottom of the list with very low pay, dangerous work, and poor employment prospects.

Reporter Steve Lohr wrote the article "For Today's Graduate, Just One Word: Statistics" in the *New York Times*. In that article he quoted the chief economist at Google as saying that "the sexy job in the next 10 years will be statisticians. And I'm not kidding."

elicit a desired response. Here are the actual rates of "yes" responses for the two different wordings of a question:

97% yes: "Should the President have the line item veto to eliminate waste?"

57% yes: "Should the President have the line item veto, or not?"

Order of Questions Sometimes survey questions are unintentionally loaded by such factors as the order of the items being considered. See the following two questions from a poll conducted in Germany, along with the very different response rates:

- "Would you say that traffic contributes more or less to air pollution than industry?" (45% blamed traffic; 27% blamed industry.)
- "Would you say that industry contributes more or less to air pollution than traffic?" (24% blamed traffic; 57% blamed industry.)

Nonresponse A *nonresponse* occurs when someone either refuses to respond to a survey question or is unavailable. When people are asked survey questions, some firmly refuse to answer. The refusal rate has been growing in recent years, partly because many persistent telemarketers try to sell goods or services by beginning with a sales pitch that initially sounds like it is part of an opinion poll. (This "selling under the guise" of a poll is now called *sugging*.) In *Lies, Damn Lies, and Statistics,* author Michael Wheeler makes this very important observation:

> People who refuse to talk to pollsters are likely to be different from those who do not. Some may be fearful of strangers and others jealous of their privacy, but their refusal to talk demonstrates that their view of the world around them is markedly different from that of those people who will let poll-takers into their homes.

Missing Data Results can sometimes be dramatically affected by missing data. Sometimes sample data values are missing because of random factors (such as subjects dropping out of a study for reasons unrelated to the study), but some data are missing because of special factors, such as the tendency of people with low incomes to be less likely to report their incomes. It is well known that the U.S. Census suffers from missing people, and the missing people are often from the homeless or low income groups.

Precise Numbers Example 1 in Section 1-1 included a statement that there are 241,472,385 adults in the United States. Because that figure is very precise, many people incorrectly assume that it is also *accurate*. In this case, that number is an estimate, and it would be better to state that the number of adults in the United States is about 240 million.

Percentages Some studies cite misleading or unclear percentages. Keep in mind that 100% of some quantity is *all* of it, but if there are references made to percentages that exceed 100%, such references are often not justified. In referring to lost baggage, Continental Airlines ran ads claiming that this was "an area where we've already improved 100% in the last six months." In an editorial criticizing this statistic, the *New York Times* correctly interpreted the 100% improvement to mean that no baggage is now being lost—an accomplishment that was not achieved by Continental Airlines.

The following list identifies some key principles to apply when dealing with percentages. These principles all use the basic notion that % or "percent" really means "divided by 100." The first principle is used often in this book.

Percentage of: To find a *percentage of* an amount, drop the % symbol and divide the percentage value by 100, then multiply. This example shows that 6% of 1200 is 72:

$$6\% \text{ of } 1200 \text{ responses} = \frac{6}{100} \times 1200 = 72$$

Fraction→Percentage: To *convert from a fraction to a percentage,* divide the denominator into the numerator to get an equivalent decimal number; then multiply by 100 and affix the % symbol. This example shows that the fraction 3/4 is equivalent to 75%:

$$\frac{3}{4} = 0.75 \rightarrow 0.75 \times 100\% = 75\%$$

Decimal→Percentage: To *convert from a decimal to a percentage,* multiply by 100%. This example shows that 0.25 is equivalent to 25%:

$$0.25 \rightarrow 0.25 \times 100\% = 25\%$$

Percentage→Decimal: To *convert from a percentage to a decimal number,* delete the % symbol and divide by 100. This example shows that 85% is equivalent to 0.85:

$$85\% = \frac{85}{100} = 0.85$$

There are many examples of the misuse of statistics. Books such as Darrell Huff's classic *How to Lie with Statistics,* Robert Reichard's *The Figure Finaglers,* and Cynthia Crossen's *Tainted Truth* describe some of those other cases. Understanding these practices will be extremely helpful in evaluating the statistical data encountered in everyday situations.

What Is Statistical Thinking? Statisticians universally agree that statistical thinking is good, but there are different views of what actually constitutes statistical thinking. If you ask the 18,000 members of the American Statistical Association to define statistical thinking, you will probably get 18,001 different definitions. In this section we have described statistical thinking in terms of the ability to see the big picture; to consider such relevant factors as context, source of data, and sampling method; and to form conclusions and identify practical implications. Statistical thinking involves critical thinking and the ability to make sense of results. Statistical thinking might involve determining whether results are statistically significant and practically significant. Statistical thinking demands so much more than the ability to execute complicated calculations. Through numerous examples, exercises, and discussions, this text will help you develop the statistical thinking skills that are so important in today's world.

Recommended Assignment

Every set of exercises begins with four exercises under the heading *Statistical Literacy and Critical Thinking.* In general, exercises tend to be arranged so that the level of difficulty increases; the earlier exercises are easier than those located near the end. Those exercises under the heading of "Beyond the Basics" are recommended for honors classes or classes with highly motivated and well-prepared students. Also, consider pointing out to students that answers to odd-numbered exercises are found in Appendix D and that there is a *Student Solutions Manual* available for purchase. It includes worked-out solutions to many of the odd-numbered exercises.

Recommended assignment for Section 1-2: Exercises 1–20.

1. Statistical significance is indicated when methods of statistics are used to reach a conclusion that some treatment or finding is effective, but common sense might suggest that the treatment or finding does not make enough of a difference to justify its use or to be practical. Yes, it is possible for a study to have statistical significance but not practical significance.
2. If the source of the data can benefit from the results of the study, it is possible that an element of bias is introduced so that the results are favorable to the source.
3. A voluntary response sample is a sample in which the subjects themselves decide whether to be included in the study. A voluntary response sample is generally not suitable for a statistical study, because the sample may have a bias resulting from participation by those with a special interest in the topic being studied.

1-2 Basic Skills and Concepts

Statistical Literacy and Critical Thinking

1. Statistical Significance versus Practical Significance What is the difference between statistical significance and practical significance? Can a statistical study have statistical significance, but not practical significance?

2. Source of Data In conducting a statistical study, why is it important to consider the source of the data?

3. Voluntary Response Sample What is a voluntary response sample, and why is such a sample generally not suitable for a statistical study?

4. Even if we conduct a study and find that there is a correlation, or association, between two variables, we cannot conclude that one of the variables is the cause of the other.

4. Correlation and Causation What is meant by the statement that "correlation does not imply causation"?

Consider the Source. ***In Exercises 5–8, determine whether the given source has the potential to create a bias in a statistical study.***

5. There does appear to be a potential to create a bias.

5. Physicians Committee for Responsible Medicine The Physicians Committee for Responsible Medicine tends to oppose the use of meat and dairy products in our diets, and that organization has received hundreds of thousands of dollars in funding from the Foundation to Support Animal Protection.

6. There does not appear to be a potential to create a bias.

6. Body Measurements Data Set 1 in Appendix B includes body measurements obtained by the U.S. Department of Health and Human Services, National Center for Health Statistics.

7. There does not appear to be a potential to create a bias.

7. Word Counts Data Set 17 in Appendix B includes word counts obtained by members of the Departments of Psychology at the University of Arizona, Washington University, and the University of Texas at Austin.

8. There does appear to be a potential to create a bias.

8. Chocolate An article in *Journal of Nutrition* (Vol. 130, No. 8) noted that chocolate is rich in flavonoids. The article notes that "regular consumption of foods rich in flavonoids may reduce the risk of coronary heart disease." The study received funding from Mars, Inc., the candy company, and the Chocolate Manufacturers Association.

Sampling Method. ***In Exercises 9–12, determine whether the sampling method appears to be sound or is flawed.***

9. The sample is a voluntary response sample and is therefore flawed.

10. The sample is a voluntary response sample and is therefore flawed.

11. The sampling method appears to be sound.

12. The sampling method appears to be sound.

9. Wi-Fi Security A survey of 721 subjects involved the providing of personal information when using Wi-Fi hotspots. The survey subjects were Internet users who responded to a question that was posted on the electronic edition of *USA Today.*

10. Text Messaging In a survey of 109 subjects, each was asked to indicate how many text messages they send and receive each day. The sample consisted of those who chose to respond to the request posted on the StatCrunch Web site.

11. Applying for a Job In a survey of 514 human resource professionals, each was asked about the importance of the appearance of a job applicant. The survey subjects were randomly selected by Harris Interactive pollsters.

12. Evolution In a survey of beliefs about evolution, Gallup pollsters randomly selected and telephoned 1018 adults in the United States.

13. Because there is a 30% chance of getting such results with a diet that has no effect, it does not appear to have statistical significance, but the average loss of 45 pounds does appear to have practical significance.

14. Because there is only a 1% chance of getting the results by chance, the method appears to have statistical significance. The result of 540 boys in 1000 births is above the approximately 50% rate expected by chance, but it does not appear to be high enough to have practical significance. Not many couples would bother with a procedure that raises the likelihood of a boy from 50% to 54%.

15. Because there is a 23% chance of getting such results with a program that has no effect, the program does not appear to have statistical significance. Because the success rate of 23% is not much better than the 20% rate that is typically expected with random guessing, the program does not appear to have practical significance.

Statistical Significance and Practical Significance. ***In Exercises 13–16, determine whether the results appear to have statistical significance, and also determine whether the results have practical significance.***

13. Waite Diet In a study of the Marisa Waite diet, four subjects lost an average of 45 pounds. It is found that there is about a 30% chance of getting such results with a diet that has no effect.

14. Gender Selection In a study of the Gender Aide method of gender selection, 1000 users of the method gave birth to 540 boys and 460 girls. There is about a 1% chance that such extreme results would occur if the method had no effect.

15. Guessing Technique When making random guesses for difficult multiple-choice test questions with possible answers of a, b, c, d, and e, we expect to get about 20% of the answers correct. The Ashton Prep Program claims to have developed a better method of guessing. In a test of that program, guesses were made for 100 answers, and 23 were found to be correct. There is a 23% chance of getting such results if the program has no effect.

16. IQ Scores Most people have IQ scores between 70 and 130. For $32, you can purchase a computer program from Highiqpro.com that is claimed to increase your IQ score by 10 to 20 points. The program claims to be "the only proven IQ increasing software in the brain training market," but the author of your text could find no data supporting that claim, so let's suppose that these results were obtained: In a study of 12 subjects using the program, the average increase in IQ score is 3 IQ points. There is a 25% chance of getting such results if the program has no effect.

16. Because there is a 25% chance of getting such results with a program that has no effect, the program does not appear to have statistical significance. Because the average increase is only 3 IQ points, the program does not appear to have practical significance.

In Exercises 17–20, refer to the data in the table below. (The pulse rates are from one sample of randomly selected males and a different sample of randomly selected females listed in Data Set 1 in Appendix B.)

Pulse Rate (beats per minute)					
Male	60	64	60	72	64
Female	68	72	88	60	60

17. Context of the Data Refer to the table of pulse rates. Is there some meaningful way in which each male pulse rate is matched with the corresponding female pulse rate? If the male pulse rates and the female pulse rates are not matched, does it make sense to use the difference between any of the pulse rates that are in the same column?

18. Source of the Data The listed pulse rates were obtained for the Third National Health and Nutrition Examination Survey conducted by the U.S. Department of Health and Human Services, National Center for Health Statistics. Is the source of the data likely to be unbiased?

19. Conclusion Given the data in the table, what issue can be addressed by conducting a statistical analysis of the pulse rates?

20. Conclusion If we use the listed pulse rates with suitable methods of statistics, we conclude that when the 64.0 average (mean) of the pulse rates of the five males is compared to the 69.6 average (mean) of the pulse rates of the five females, there is a 36% chance that the difference can be explained by random results obtained from populations of males and females having the same average (mean) pulse rate. Does this prove that the populations of males and females have the same average (mean) pulse rate? Why or why not? Would better results be obtained with larger samples?

17. The male and female pulse rates in the same column are not matched in any meaningful way. It does not make sense to use the difference between any of the pulse rates that are in the same column.

18. Yes, the source of the data is likely to be unbiased.

19. The data can be used to address the issue of whether males and females have pulse rates with the same average (mean) value.

20. The results do not prove that the populations of males and females have the same average (mean) pulse rate. The results are based on a particular sample of five males and five females, and analyzing other samples might lead to a different conclusion. Better results would be obtained with larger samples.

In Exercises 21–24, refer to the data in the table below. The IQ score and brain volume are listed for each of five different subjects. (The values are from Data Set 6 in Appendix B.)

IQ Score and Brain Volume					
Subject	1	2	3	4	5
IQ Score	87	127	101	94	97
Brain Volume (cm^3)	1035	1034	1173	1347	1029

21. Context of the Data Refer to the given table of measurements. Is there some meaningful way in which the IQ scores are matched with the corresponding brain volumes? If they are matched, does it make sense to use the difference between each IQ score and the brain volume that is in the same column? Why or why not?

22. Conclusion Given the context of the data in the table, what issue can be addressed by conducting a statistical analysis of the measurements?

23. Source of the Data The data in the table were obtained by members of departments at Harvard Medical School, Massachusetts General Hospital, Dartmouth College, Dartmouth Medical School, and the University of California at Davis. Funding for the study was provided by awards from the National Institutes of Health, which is an agency of the

21. Yes, each IQ score is matched with the brain volume in the same column, because they are measurements obtained from the same person. It does not make sense to use the difference between each IQ score and the brain volume in the same column, because IQ scores and brain volumes use different units of measurement. It would make no sense to find the difference between an IQ score of 87 and a brain volume of 1035 cm^3.

22. The issue that can be addressed is whether there is a correlation, or association, between IQ score and brain volume.

23. Given that the researchers do not appear to benefit from the results, they are professionals at prestigious institutions, and funding is from a U.S. government agency, the source of the data appears to be unbiased.

U.S. Department of Health and Human Services. Does the source of the data appear to be unbiased?

24. No. Correlation does not imply causation, so a statistical correlation between IQ score and brain volume should not be used to conclude that larger brain volumes cause higher IQ scores.

24. Conclusion If we were to use such data and conclude that there is a correlation or association between IQ score and brain volume, does it follow that larger brains are the cause of higher IQ scores?

What's Wrong? ***In Exercises 25–28, identify what is wrong.***

25. It is questionable that the sponsor is the Idaho Potato Commission and the favorite vegetable is potatoes.

25. Potatoes In a poll sponsored by the Idaho Potato Commission, 1000 adults were asked to select their favorite vegetables, and the favorite choice was potatoes, which were selected by 26% of the respondents.

26. The sample is a voluntary response sample, so there is a good chance that the results are not valid.

26. College Major In a *USA Today* online poll, 728 Internet users chose to respond, and 41% of them said that their college majors prepared them for their chosen careers very well.

27. The correlation, or association, between two variables does not mean that one of the variables is the cause of the other. Correlation does not imply causation.

27. Cell Phones and Pirates In recent years, the numbers of cell phones and the numbers of pirates have both increased, so there is a correlation, or association, between those two variables. Therefore, cell phones cause pirates.

28. The correlation, or association, between two variables does not mean that one of the variables is the cause of the other. Correlation does not imply causation.

28. Storks and Babies In the years following the end of World War II, it was found that there was a strong correlation, or association, between the number of human births and the stork population. It therefore follows that storks cause babies.

Percentages. ***In Exercises 29–36, answer the given questions, which are related to percentages.***

29. a. 397.02 adults
b. No. Because the result is a count of people among the 1018 who were surveyed, the result must be a whole number.
c. 397 adults
d. 25%

29. Evolution A Gallup poll of 1018 adults reported that 39% believe in evolution.

a. What is the exact value that is 39% of 1018?

b. Could the result from part (a) be the actual number of adults who said that they believe in evolution? Why or why not?

c. What is the actual number of adults who said that they believe in evolution?

d. Among the 1018 respondents, 255 said that they did not believe in evolution. What percentage of respondents said that they did not believe in evolution?

30. a. 162.26 women
b. No. Because the result is a count of women among the 427 who were surveyed, the result must be a whole number.
c. 162 women
d. 7%

30. Online Shopping In a *Consumer Reports* Research Center Telephone survey of 427 women, 38% said that they purchased clothing online.

a. What is the exact value that is 38% of 427?

b. Could the result from part (a) be the actual number of women who said that they purchase clothing online? Why or why not?

c. What is the actual number of women who said that they purchase clothing online?

d. Among the 427 respondents, 30 said that they purchase electronics online. What percentage of the women said that they purchase electronics online?

31. a. 322.28 adults
b. No. Because the result is a count of adults among the 2302 who were surveyed, the result must be a whole number.
c. 322 adults
d. 2%

31. Piercings and Tattoos In a Harris poll of 2302 adults, 14% said that they have a tattoo.

a. What is the exact value that is 14% of 2302?

b. Could the result from part (a) be actual number of adults who said that they have a tattoo? Why or why not?

c. What is the actual number of adults who said that they have a tattoo?

d. Among the 2302 respondents, 46 said that they had face piercings only. What percentage of respondents had face piercings only?

32. Dollar for Your Thoughts In a Harris poll of 2513 adults, 76% said that they prefer $1 to be in the form of paper currency.

a. What is the exact value that is 76% of 2513?

b. Could the result from part (a) be the actual number of adults who said that they prefer $1 to be in the form of paper currency? Why or why not?

c. What is the actual number of adults who said that they prefer $1 to be in the form of paper currency?

d. Among the 2513 respondents, 327 said that they prefer $1 to be currency in the form of a coin. What is the percentage of respondents who said that they prefer $1 to be currency in the form of a coin?

33. Percentages in Advertising An ad for Big Skinny wallets included the statement that one of their wallets "reduces your filled wallet size by 50%–200%." What is wrong with this statement?

34. Percentages in Advertising In an ad for the Club, a device used to discourage car thefts, it was stated that "The Club reduces your odds of car theft by 400%." What is wrong with this statement?

35. Percentages in the Media In the *New York Times Magazine,* a report about the decline of Western investment in Kenya included this statement: "After years of daily flights, Lufthansa and Air France had halted passenger service. Foreign investment fell 500 percent during the 1990s." What is wrong with this statement?

36. Percentages in Advertising A *New York Times* editorial criticized a chart caption that described a dental rinse as one that "reduces plaque on teeth by over 300%." What is wrong with this statement?

1-2 Beyond the Basics

37. ATV Accidents The Associated Press provided an article with a headline stating that ATV accidents killed 704 people in the last year. The article noted that this is a new record high and compares it to 617 ATV deaths the year before that. Other data about the frequencies of injuries were included. What important value was not included? Why is it important?

38. Falsifying Data A researcher at the Sloan-Kettering Cancer Research Center was once criticized for falsifying data. Among his data were figures obtained from 6 groups of mice, with 20 individual mice in each group. The following values were given for the percentage of successes in each group: 53%, 58%, 63%, 46%, 48%, 67%. What's wrong with those values?

39. What's Wrong with This Picture? The *Newport Chronicle* ran a survey by asking readers to call in their response to this question: "Do you support the development of atomic weapons that could kill millions of innocent people?" It was reported that 20 readers responded and that 87% said "no," while 13% said "yes." Identify four major flaws in this survey.

1-3 Types of Data

Key Concept A common and important use of statistics involves collecting sample data and using them to make inferences, or conclusions, about the population from which the data were obtained. The terms *sample* and *population* were defined in Section 1-1. We should also know and understand the meanings of the terms *statistic* and *parameter,* as defined below. The terms *statistic* and *parameter* are used to distinguish

32. a. 1909.88 adults
b. No. Because the result is a count of adults among the 2513 who were surveyed, the result must be a whole number.
c. 1910 adults
d. 13%

33. Because a reduction of 100% would eliminate all of the size, it is not possible to reduce the size by 100% or more.

34. If the Club eliminated all car thefts, it would reduce the odds of car theft by 100%, so the 400% figure is misleading.

35. If foreign investment fell by 100%, it would be totally eliminated, so it is not possible for it to fall by more than 100%.

36. Because a reduction of 100% would eliminate all plaque, it is not possible to reduce it by more than 100%.

37. Without our knowing anything about the number of ATVs in use, or the number of ATV drivers, or the amount of ATV usage, the number of 740 fatal accidents has no context. Some information should be given so that the reader can understand the *rate* of ATV fatalities.

38. All percentages of success should be multiples of 5. The given percentages cannot be correct.

39. The wording of the question is biased and tends to encourage negative responses. The sample size of 20 is too small. Survey respondents are self-selected instead of being selected by the newspaper. If 20 readers respond, the percentages should be multiples of 5, so 87% and 13% are not possible results.

Note to Instructor

Consider assigning this section to be read independently. After students have read this section, review Example 3 for the meaning of the term *countable,* as it is used in the definition of *discrete data.* Also discuss the distinction between the interval and ratio levels of measurement, and suggest a "ratio" test: If one number is *twice* the other, is the quantity being measured also twice the other quantity? If yes, the data are at the ratio level. For example, $2 is twice as much money as $1, but 2°F is not twice as hot as 1°F. The money amounts are at the ratio level, but the temperatures are at the interval level.

Six Degrees of Separation

Social psychologists, historians, political scientists, and communications specialists are interested in "The Small World Problem": Given any two people in the world, how many intermediate links are necessary to connect the two original people? In the 1950s and 1960s, social psychologist Stanley Milgram conducted an experiment in which subjects tried to contact other target people by mailing an information folder to an acquaintance who they thought would be closer to the target. Among 160 such chains that were initiated, only 44 were completed, so the failure rate was 73%. Among the successes, the number of intermediate acquaintances varied from 2 to 10, with a median of 6 (hence "six degrees of separation"). The experiment has been criticized for its high failure rate and its disproportionate inclusion of subjects with above-average incomes. A more recent study conducted by Microsoft researcher Eric Horvitz and Stanford Assistant Professor Jure Leskovec involved 30 billion instant messages and 240 million people. This study found that for instant messages that used Microsoft, the mean length of a path between two individuals is 6.6, suggesting "seven degrees of separation." Work continues in this important and interesting field.

between cases in which we have data for a sample, and cases in which we have data for an entire population.

We also need to know the difference between the terms *quantitative data* and *categorical data.* Some numbers, such as those on the shirts of basketball players, are not quantities because they don't measure or count anything, and it would not make sense to perform calculations with such numbers. In this section we describe different types of data. The type of data is one of the key factors that determine the statistical methods we use in our analysis.

Parameter/Statistic

DEFINITIONS

A **parameter** is a numerical measurement describing some characteristic of a *population.*

A **statistic** is a numerical measurement describing some characteristic of a *sample*.

HINT The alliteration in "population parameter" and "sample statistic" helps us remember the meaning of these terms.

Using the foregoing definitions and those given in Section 1-1, we see that the term *statistics* has two possible meanings:

1. Statistics are two or more numerical measurements describing characteristics of samples.
2. Statistics is the science of planning studies and experiments; obtaining data; organizing, summarizing, presenting, analyzing, and interpreting those data; and then drawing conclusions based on them.

We can determine which of these two definitions applies by considering the context in which the term *statistics* is used. The following example uses the first meaning of *statistics* as given above.

Example 1 Parameter/Statistic

In a Harris Poll, 2320 adults in the United States were surveyed about body piercings, and 5% of the respondents said that they had a body piercing, but not on the face. Based on the latest available data at the time of this writing, there are 241,472,385 adults in the United States. The results from the survey are a sample drawn from the population of all adults.

1. **Parameter:** The population size of 241,472,385 is a *parameter*, because it is based on the entire population of all adults in the United States.
2. **Statistic:** The sample size of 2320 surveyed adults is a *statistic*, because it is based on a sample, not the entire population of all adults in the United States. The value of 5% is another statistic, because it is also based on the sample, not on the entire population.

Quantitative/Categorical

Some data are numbers representing counts or measurements (such as a height of 60 inches or an IQ of 135), whereas others are attributes (such as eye color of green or brown) that are not counts or measurements. The terms *quantitative data* and *categorical data* distinguish between these types.

DEFINITIONS

Quantitative (or **numerical**) **data** consist of *numbers* representing counts or measurements.

Categorical (or **qualitative** or **attribute**) **data** consist of names or labels that are not numbers representing counts or measurements.

CAUTION Categorical data are sometimes coded with numbers, but those numbers are actually a different way to express names. Although such numbers might appear to be quantitative, they are actually categorical data. See the third part of Example 2.

Example 2 Quantitative/Categorical

1. **Quantitative Data:** The ages (in years) of survey respondents
2. **Categorical Data as Labels:** The political party affiliations (Democrat, Republican, Independent, other) of survey respondents
3. **Categorical Data as Numbers:** The numbers 12, 74, 77, 76, 73, 78, 88, 19, 9, 23, and 25 were sewn on the jerseys of the starting offense for the New Orleans Saints when they won a recent Super Bowl. Those numbers are substitutes for names. They don't measure or count anything, so they are categorical data.

Include Units of Measurement With quantitative data, it is important to use the appropriate units of measurement, such as dollars, hours, feet, or meters. We should carefully observe information given about the units of measurement, such as "all amounts are in *thousands of dollars*," "all times are in *hundredths of a second*," or "all units are in *kilograms*." Ignoring such units of measurement can be very costly. NASA lost its $125 million Mars Climate Orbiter when the orbiter crashed because the controlling software had acceleration data in *English* units, but they were incorrectly assumed to be in *metric* units.

Discrete/Continuous

Quantitative data can be further described by distinguishing between *discrete* and *continuous* types.

DEFINITIONS

Discrete data result when the data values are quantitative and the number of values is finite or "countable." (If there are infinitely many values, the collection of values is countable if it is possible to count them individually, such as the number of tosses of a coin before getting tails.)

Continuous (numerical) data result from infinitely many possible quantitative values, where the collection of values is not countable. (That is, it is impossible to count the individual items because at least some of them are on a continuous scale, such as the lengths from 0 cm to 12 cm.)

Statistics for Online Dating

The four founders of the online dating site OkCupid are mathematicians who use methods of statistics to analyze results from their Web site. The chief executive officer of OkCupid has been quoted as saying, "We're not psychologists. We're math guys" (from "Looking for a Date? A Site Suggests You Check the Data," by Jenna Worthman, *New York Times*). The OkCupid Web site is unique in its use of methods of statistics to match people more effectively.

By analyzing the photos and responses of 7000 users, analysts at OkCupid found that when creating a profile photo, men should not look directly at the camera, and they should not smile. For women, the appearance of being interesting produces much better results than the appearance of being sexy. They found that brevity is good for the first posted message; the ideal length of the first posted message is 40 words—about what a typical person can type in one minute.

CAUTION The concept of *countable* data plays a key role in the preceding definitions, but it is not a particularly easy concept to understand. Carefully study Example 3.

Example 3 Discrete/Continuous

1. **Discrete Data of the Finite Type:** The numbers of eggs that hens lay in one week are *discrete* data because they are finite numbers, such as 5 and 7 that result from a counting process.
2. **Discrete Data of the Infinite Type:** Consider the number of rolls of a die required to get an outcome of 2. It is possible that you could roll a die forever without ever getting a 2, but you can still *count* the number of rolls as you proceed. The collection of rolls is countable, because you can count them, even though you might go on counting forever.
3. **Continuous Data:** During a year, a cow might yield an amount of milk that can be any value between 0 liters and 7000 liters. There are infinitely many values between 0 liters and 7000 liters, but it is impossible to *count* the number of different possible values on such a continuous scale.

When we are describing smaller amounts, correct grammar dictates that we use "fewer" for discrete amounts and "less" for continuous amounts. It is correct to say that we drank *fewer* cans of cola and that, in the process, we drank *less* cola. The numbers of cans of cola are discrete data, whereas the volume amounts of cola are continuous data.

Levels of Measurement

Another common way of classifying data is to use four levels of measurement: nominal, ordinal, interval, and ratio. When we are applying statistics to real problems, the level of measurement of the data helps us decide which procedure to use. There will be some references to these levels of measurement in this book, but the important point here is based on common sense: *Don't do computations and don't use statistical methods that are not appropriate for the data.* For example, it would not make sense to compute an average (mean) of Social Security numbers, because those numbers are data that are used for identification, and they don't represent measurements or counts of anything.

DEFINITION The **nominal level of measurement** is characterized by data that consist of names, labels, or categories only. The data cannot be arranged in an ordering scheme (such as low to high).

Example 4 Nominal Level

Here are examples of sample data at the nominal level of measurement.

1. **Yes/No/Undecided:** Survey responses of *yes*, *no*, and *undecided*
2. **Political Party:** The political party affiliations of survey respondents (Democrat, Republican, Independent, other)
3. **Social Security Numbers:** Social Security numbers are just substitutes for names; they do not count or measure anything.

Because nominal data lack any ordering or numerical significance, they should not be used for calculations. Numbers such as 1, 2, 3, and 4 are sometimes assigned to the different categories (especially when data are coded for computers), but these numbers have no real computational significance and any average (mean) calculated from them is meaningless.

DEFINITION Data are at the **ordinal level of measurement** if they can be arranged in some order, but differences (obtained by subtraction) between data values either cannot be determined or are meaningless.

Example 5 Ordinal Level

Here are examples of sample data at the ordinal level of measurement.

1. **Course Grades:** A college professor assigns grades of A, B, C, D, or F. These grades can be arranged in order, but we can't determine differences between the grades. For example, we know that A is higher than B (so there is an ordering), but we cannot subtract B from A (so the difference cannot be found).
2. **Ranks:** *U.S. News & World Report* ranks colleges. As of this writing, Harvard was ranked first and Princeton was ranked second. Those ranks of 1 and 2 determine an ordering, but the difference between those ranks is meaningless. The difference of "second minus first" might suggest $2 - 1 = 1$ but this difference of 1 is meaningless because it is not an exact quantity that can be compared to other such differences. The *difference* between Harvard and Princeton cannot be quantitatively compared to the *difference* between Yale and Columbia, the universities ranked third and fourth, respectively.

Ordinal data provide information about relative comparisons, but not the magnitudes of the differences. Usually, ordinal data should not be used for calculations such as an average, but this guideline is sometimes ignored (such as when we use letter grades to calculate a grade-point average).

DEFINITION Data are at the **interval level of measurement** if they can be arranged in order, and differences between data values can be found and are meaningful. Data at this level do not have a *natural* zero starting point at which *none* of the quantity is present.

Measuring Disobedience

How are data collected about something that doesn't seem to be measurable, such as people's level of disobedience? Psychologist Stanley Milgram devised the following experiment: A researcher instructed a volunteer subject to operate a control board that gave increasingly painful "electrical shocks" to a third person. Actually, no real shocks were given, and the third person was an actor. The volunteer began with 15 volts and was instructed to increase the shocks by increments of 15 volts. The disobedience level was the point at which the subject refused to increase the voltage. Surprisingly, two-thirds of the subjects obeyed orders even when the actor screamed and faked a heart attack.

Example 6 **Interval Level**

These examples illustrate the interval level of measurement.

1. **Temperatures:** Outdoor temperatures of 40°F and 90°F are examples of data at this interval level of measurement. Those values are ordered, and we can determine their difference of 50°F. However, there is no natural starting point. The value of 0°F might seem like a starting point, but it is arbitrary and does not represent the total absence of heat.
2. **Years:** The years 1492 and 1776 can be arranged in order, and the difference of 284 years can be found and is meaningful. However, time did not begin in the year 0, so the year 0 is arbitrary instead of being a natural zero starting point representing "no time."

DEFINITION Data are at the **ratio level of measurement** if they can be arranged in order, differences can be found and are meaningful, and there is a natural zero starting point (where zero indicates that *none* of the quantity is present). For data at this level, differences and ratios are both meaningful.

Example 7 **Ratio Level**

The following are examples of data at the ratio level of measurement. Note the presence of the natural zero value, and also note the use of meaningful ratios of "twice" and "three times."

1. **Car Lengths:** Car lengths of 106 in. for a Smart car and 212 in. for a Mercury Grand Marquis (0 in. represents no length, and 212 in. is twice as long as 106 in.)
2. **Class Times:** The times of 50 min and 100 min for a statistics class (0 min represents no class time, and 100 min is twice as long as 50 min.)

HINT This level of measurement is called the ratio level because the zero starting point makes ratios meaningful, so here is an easy test to determine whether values are at the ratio level: Consider two quantities where one number is twice the other, and ask whether "twice" can be used to correctly describe the quantities. Because a person with a height of 6 ft is *twice* as tall as a person with a height of 3 ft, the heights are at the ratio level of measurement. In contrast, 50°F is *not twice* as hot as 25°F, so Fahrenheit temperatures are *not* at the ratio level. See Table 1-2.

Table 1-2 Levels of Measurement

Level of Measurement	Brief Description	Example
Ratio	There is a natural zero starting point and ratios are meaningful.	Heights, lengths, distances, volumes
Interval	Differences are meaningful, but there is no natural zero starting point and ratios are meaningless.	Body temperatures in degrees Fahrenheit or Celsius
Ordinal	Data can be arranged in order, but differences either can't be found or are meaningless.	Ranks of colleges in *U.S. News & World Report*
Nominal	Categories only. Data cannot be arranged in order.	Eye colors

1-3 Basic Skills and Concepts

Recommended Assignment
Exercises 1–8 and even-numbered Exercises 14–28.

Statistical Literacy and Critical Thinking

1. Parameter and Statistic What is a parameter, and what is a statistic?

2. Quantitative/Categorical Data How do quantitative data and categorical data differ?

3. Discrete/Continuous Data Which of the following describe discrete data?

a. The numbers of people surveyed in each of the next several Gallup polls

b. The exact heights of individuals in a sample of several statistics students

c. The number of Super Bowl football games that must be played before one of the teams scores exactly 75 points

4. Identifying the Population In a Gallup poll of 1010 adults in the United States, 55% of the respondents said that they used local TV stations daily as a source of news. Is the 1010 value a statistic or a parameter? Is the 55% value a statistic or a parameter? Describe the population.

1. A parameter is a numerical measurement describing some characteristic of a population, whereas a statistic is a numerical measurement describing some characteristic of a sample.
2. Quantitative data consist of numbers representing counts or measurements, whereas categorical data can be separated into different categories that are distinguished by some characteristic that is not numerical.
3. Parts a and c describe discrete data.
4. The values of 1010 and 55% are both statistics because they are based on the sample. The population consists of all adults in the United States.

In Exercises 5–12, determine whether the given value is a statistic or a parameter.

5. Distracted Driving In a AAA Foundation for Traffic Safety survey, 21% of the respondents said that they recently texted or e-mailed while driving.

5. Statistic

6. States There are 50 state capitols in the United States.

6. Parameter

7. Titanic A study was conducted of all 2223 passengers aboard the *Titanic* when it sank.

7. Parameter

8. Late Flights Among the flights included in the sample of flights in Data Set 15 of Appendix B, 21% arrived late.

8. Statistic

9. IQ Scores The mean IQ score for subjects taking the Wechsler Adult Intelligence Scale IQ test is 100.

9. Parameter

10. Periodic Table The average (mean) atomic weight of all elements in the periodic table is 134.355 unified atomic mass units.

10. Parameter

11. Brain Volume The average (mean) volume of the brains included in Data Set 6 of Appendix B is 1126.0 cm^3.

11. Statistic

12. HDTV Penetration In a random sample of households, it was found that 47% of the sampled households had high-definition TVs.

12. Statistic

In Exercises 13–20, determine whether the given values are from a discrete or continuous data set.

13. Honda Civic Crash Test In Data Set 13 of Appendix B, the measured chest deceleration of a Honda Civic in a crash test is 39 g, where g is a force of gravity.

13. Continuous

14. Honda Civic Cylinders The Honda Civic has 4 cylinders.

14. Discrete

15. Word Count From Data Set 17 in Appendix B we see that a male spoke 13,825 words in one day.

15. Discrete

16. First President George Washington was 188 cm tall.

16. Continuous

17. House of Representatives Currently the House of Representatives has 435 members.

17. Discrete

18. Discrete

18. Crash Test Results Data Set 13 in Appendix B includes crash test results from 21 different cars.

19. Continuous

19. Earthquake From Data Set 16 in Appendix B we see that an earthquake had a measurement of 0.70 on the Richter scale.

20. Continuous

20. Arm Circumference From Data Set 1 in Appendix B we see that a female had an arm circumference of 27.5 cm.

In Exercises 21–28, determine which of the four levels of measurement (nominal, ordinal, interval, ratio) is most appropriate.

21. Nominal

21. Colors of M&Ms (red, orange, yellow, brown, blue, green) listed in Data Set 20 in Appendix B

22. Ratio

22. Depths (km) of earthquakes listed in Data Set 16 of Appendix B

23. Interval

23. Years in which U.S. presidents were inaugurated

24. Ordinal

24. The movie *Avatar* was given a rating of 4 stars on a scale of 5 stars.

25. Ratio

25. Volumes (cm^3) of brains listed in Data Set 6 of Appendix B

26. Nominal

26. Car models (Chevrolet Aveo, Honda Civic, . . . , Buick Lucerne) used for crash testing, as listed in Data Set 13 of Appendix B

27. Ordinal

27. Blood lead levels of low, medium, and high used to describe the subjects in Data Set 5 of Appendix B

28. Interval

28. Body temperatures (in degrees Fahrenheit) listed in Data Set 3 of Appendix B

29. The numbers are not counts or measures of anything, so they are at the nominal level of measurement, and it makes no sense to compute the average (mean) of them.

30. The flight numbers do not count or measure anything. They are at the nominal level of measurement, and it does not make sense to compute the average (mean).

31. The numbers are used as substitutes for the categories of low, medium, and high, so the numbers are at the ordinal level of measurement. It does not make sense to compute the average (mean) of such numbers.

32. The numbers are substitutes for names and are not counts or measures of anything. They are at the nominal level of measurement, and it makes no sense to compute the average (mean) of them.

In Exercises 29–32, identify the level of measurement of the data. Also, explain what is wrong with the given calculation.

29. Political Parties In a preelection survey of likely voters, political parties of respondents are identified as 1 for a Democrat, 2 for a Republican, 3 for an Independent, and 4 for anything else. The average (mean) is calculated for 850 respondents and the result is 1.7.

30. Flight Numbers Data Set 15 in Appendix B lists flight numbers of 48 different flights, and the average (mean) of those flight numbers is 11.0.

31. Lead Levels In Data Set 5 in Appendix B, blood lead levels are represented as 1 for low, 2 for medium, and 3 for high. The average (mean) of the 121 blood lead levels is 1.53.

32. World Series Champs As of this writing, the New York Yankees were the last team to win the World Series, and the numbers of the starting lineup are 2, 18, 25, 13, 20, 55, 24, 33, and 53. The average (mean) of those numbers is 27.0.

1-3 Beyond the Basics

33. a. Continuous, because the number of possible values is infinite and not countable
b. Discrete, because the number of possible values is finite
c. Discrete, because the number of possible values is finite
d. Discrete, because the number of possible values is infinite and countable

33. Countable For each of the following, categorize the nature of the data using one of these three descriptions: (1) Discrete because the number of possible values is finite; (2) discrete because the number of possible values is infinite and countable; (3) continuous because the number of possible values is infinite and not countable.

a. Exact braking distances of cars, measured on a scale from 100 ft to 200 ft

b. Braking distances of cars, measured on a scale from 100 ft to 200 ft and rounded to the nearest foot

c. The numbers of students now in statistics classes

d. The number of attempts required to roll a single die and get an outcome of 7

34. Scale for Rating Food A group of students develops a scale for rating the quality of cafeteria food, with 0 representing "neutral: not good and not bad." Bad meals are given negative numbers and good meals are given positive numbers, with the magnitude of the number corresponding to the degree of badness or goodness. The first three meals are rated as 2, 4, and -5. What is the level of measurement for such ratings? Explain your choice.

35. Interpreting Temperature Increase In the *Born Loser* cartoon strip by Art Sansom, Brutus expresses joy over an increase in temperature from 1° to 2°. When asked what is so good about 2°, he answers that "it's twice as warm as this morning." Explain why Brutus is wrong yet again.

34. Either ordinal or interval is a reasonable answer, but ordinal makes more sense because differences between values are not likely to be meaningful. For example, the difference between a food rated 1 and a food rated 2 is not necessarily the same as the difference between a food rated 9 and a food rated 10.

35. With no natural starting point, temperatures are at the interval level of measurement, so ratios such as "twice" are meaningless.

1-4 Collecting Sample Data

Key Concept An absolutely critical concept in applying methods of statistics is consideration of the method used to collect the sample data. Of particular importance is the method of using a *simple random sample.* We will make frequent use of this sampling method throughout the remainder of this book.

As you read this section, remember this:

If sample data are not collected in an appropriate way, the data may be so utterly useless that no amount of statistical torturing can salvage them.

Part 1 of this section introduces the basics of data collection, and Part 2 describes some common ways in which observational studies and experiments are conducted.

Note to Instructor

This section can be read by students on their own. The previous edition included separate definitions for *simple random sample* and *random sample.* See the note in parentheses included with the definition of simple random sample. At this stage, it is not so important to distinguish between a simple random sample and a random sample. But it is critically important to stress the importance of good sampling techniques. Reinforce the concept that voluntary response samples are unsuitable for sound statistical methods. If the sampling is not done correctly, even very large samples may be totally worthless.

Lurking variable is mentioned in this section. *Nonrandom sampling error* and two other sampling errors are now included in this section.

Part 1: Basics of Collecting Data

Statistical methods are driven by the data that we collect. We typically obtain data from two distinct sources: *observational studies* and *experiments.*

DEFINITIONS

In an **observational study,** we observe and measure specific characteristics, but we don't attempt to *modify* the subjects being studied.

In an **experiment,** we apply some *treatment* and then proceed to observe its effects on the subjects. (Subjects in experiments are called **experimental units.**)

Experiments are often better than observational studies, because experiments typically reduce the chance of having the results affected by some variable that is not part of a study. (A **lurking variable** is one that affects the variables included in the study, but it is not included in the study.) In one classic example, we could use an observational study to incorrectly conclude that ice cream causes drownings based on data showing that increases in ice cream sales are associated with increases in drownings. Our error is to miss the lurking variable of temperature and thus fail to recognize that warmer months result in both increased ice cream sales and increased drownings. If, instead of using data from an observational study, we conducted an *experiment* with one group treated with ice cream while another group got no ice cream, we would see that ice cream consumption has no effect on drownings.

Example 1 Observational Study and Experiment

Observational Study: The typical survey is a good example of an observational study. For example, the Pew Research Center surveyed 2252 adults in the United States and found that 59% of them go online wirelessly. The respondents were asked questions, but they were not given any treatment, so this is an example of an observational study.

Clinical Trials vs. Observational Studies

In a *New York Times* article about hormone therapy for women, reporter Denise Grady wrote about randomized clinical trials that involve subjects who were randomly assigned to a treatment group and another group not given the treatment. Such randomized clinical trials are often referred to as the "gold standard" for medical research. In contrast, observational studies can involve patients who decide themselves to undergo some treatment. Subjects who decide themselves to undergo treatments are often healthier than other subjects, so the treatment group might appear to be more successful simply because it involves healthier subjects, not necessarily because the treatment is effective. Researchers criticized observational studies of hormone therapy for women by saying that results might appear to make the treatment more effective than it really is.

Experiment: In the largest public health experiment ever conducted, 200,745 children were given a treatment consisting of the Salk vaccine, while 201,229 other children were given a placebo. The Salk vaccine injections constitute a treatment that modified the subjects, so this is an example of an experiment.

Whether one is conducting an observational study or an experiment, it is important to select the sample of subjects in such a way that the sample is likely to be representative of the larger population. In Section 1-2 we saw that in a voluntary response sample, the subjects decide themselves whether to respond. Although voluntary response samples are very common, their results are generally useless for making valid inferences about larger populations. The following definition refers to one common and effective way to collect sample data.

DEFINITION A **simple random sample** of *n* subjects is selected in such a way that every possible *sample of the same size n* has the same chance of being chosen. (A simple random sample is often called a random sample, but strictly speaking, a random sample has the weaker requirement that all members of the population have the same chance of being selected. That distinction is not so important in this text.)

Throughout, we will use various statistical procedures, and we often have a requirement that we have collected a *simple random sample*, as defined above.

The definition of a simple random sample requires more than selecting subjects in such a way that each has the same chance of being selected. Consider the selection of three students from the class of six students depicted below. If you use a coin toss to select a row, randomness is used and each student has the same chance of being selected, but the result is not a simple random sample. The coin toss will produce only two possible samples; some samples of three students have *no chance* of being selected, such as a sample consisting of a female and two males. This violates the requirement that all samples of the same size have the same chance of being selected. Instead of the coin toss, you could get a simple random sample of three students by writing each of the six different student names on separate index cards, which could then be placed in a bowl and mixed. The selection of three index cards will yield a simple random sample, because every different possible sample of three students now has the same chance of being selected.

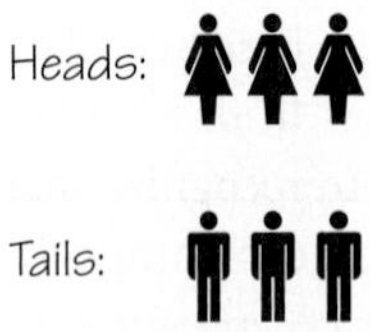

With random sampling we expect all components of the population to be (approximately) proportionately represented. Random samples are selected by many different methods, including the use of computers to generate random numbers. Unlike careless or haphazard sampling, random sampling usually requires very careful planning and execution. Wayne Barber of Chemeketa Community College is quite correct when he tells his students that "randomness needs help."

Other Sampling Methods In addition to simple random sampling, here are some other sampling methods commonly used for surveys. Figure 1-3 illustrates these different sampling methods.

Random Sampling:
Each member of the population has an equal chance of being selected. Computers are often used to generate random telephone numbers.

Simple Random Sampling:
A sample of n subjects is selected in such a way that every possible sample of the same size n has the same chance of being chosen.

Systematic Sampling:
Select some starting point, then select every kth (such as every 50th) element in the population.

Convenience Sampling:
Use results that are easy to get.

Stratified Sampling:
Subdivide the population into at least two different subgroups (or strata) so that subjects within the same subgroup share the same characteristics (such as gender or age bracket), then draw a sample from each subgroup.

Cluster Sampling:
Divide the population into sections (or clusters), then randomly select some of those clusters, and then choose all members from those selected clusters.

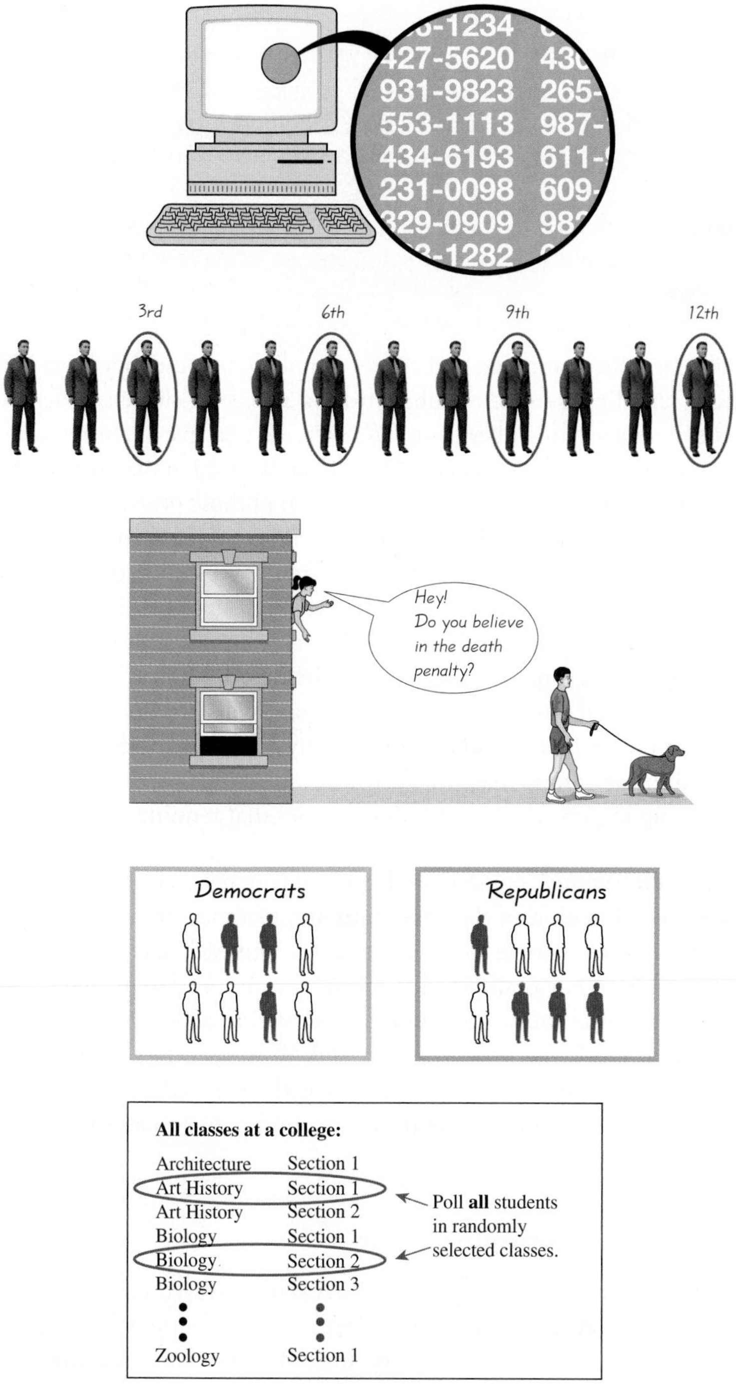

Figure 1-3 Common Sampling Methods

Hawthorne and Experimenter Effects

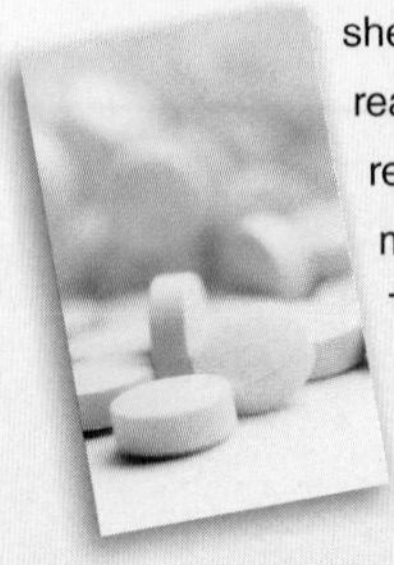

The well-known placebo effect occurs when an untreated subject incorrectly believes that he or she is receiving a real treatment and reports an improvement in symptoms. The Hawthorne effect occurs when treated subjects somehow respond differently, simply because they are part of an experiment. (This phenomenon was called the "Hawthorne effect" because it was first observed in a study of factory workers at Western Electric's Hawthorne plant.) An experimenter effect (sometimes called a Rosenthal effect) occurs when the researcher or experimenter unintentionally influences subjects through such factors as facial expression, tone of voice, or attitude.

DEFINITIONS

In **systematic sampling,** we select some starting point and then select every *k*th (such as every 50th) element in the population.

With **convenience sampling,** we simply use results that are very easy to get.

In **stratified sampling,** we subdivide the population into at least two different subgroups (or strata) so that subjects within the same subgroup share the same characteristics (such as age bracket). Then we draw a sample from each subgroup (or stratum).

In **cluster sampling,** we first divide the population area into sections (or clusters). Then we randomly select some of those clusters and choose *all* the members from those selected clusters.

It is easy to confuse stratified sampling and cluster sampling, because they both use subgroups. But cluster sampling uses *all* members from a *sample* of clusters, whereas stratified sampling uses a *sample* of members from *all* strata. An example of cluster sampling is a preelection poll, in which pollsters randomly select 30 election precincts from a large number of precincts and then survey all voters in each of those precincts. This is faster and much less expensive than selecting one voter from each of the many precincts in the population area. Pollsters can adjust or weight the results of stratified or cluster sampling to correct for any disproportionate representation of groups.

For a fixed sample size, if you randomly select subjects from different strata, you are likely to get more consistent (and less variable) results than by simply selecting a random sample from the general population. For that reason, pollsters often use stratified sampling to reduce the variation in the results. Many of the methods discussed later in this book require that sample data be derived from a *simple random sample*, and neither stratified sampling nor cluster sampling satisfies that requirement.

Multistage Sampling Professional pollsters and government researchers often collect data by using some combination of the basic sampling methods. In a **multistage sample design**, pollsters select a sample in different stages, and each stage might use different methods of sampling. For example, one multistage sample design might involve the random selection of clusters, but instead of surveying all members of the chosen clusters, you might randomly select 50 men and 50 women in each selected cluster; thus you begin with cluster sampling and end with stratified sampling. See Example 2 for an actual multistage sample design that is complex, but effective.

Note to Instructor

The definitions of *stratified sampling* and *cluster sampling* may cause some confusion because they both involve sampling from different groups. Eliminate the confusion by noting that with cluster sampling, we use *all* members from the selected clusters, whereas stratified sampling uses only a *sample* of members from each strata. Associate "cluster" with "all."

Example 2 Multistage Sample Design

The U.S. government's unemployment statistics are based on surveyed households. It is impractical to personally visit each member of a simple random sample, because individual households are spread all over the country. Instead, the U.S. Census Bureau and the Bureau of Labor Statistics collaborate to conduct a survey called the Current Population Survey. This survey obtains data describing such factors as unemployment rates, college enrollments, and weekly earnings amounts. One recent survey incorporates a multistage sample design, roughly following these steps:

1. The entire United States is partitioned into 2025 different regions called *primary sampling units* (PSUs). The primary sampling units are metropolitan areas, large counties, or combinations of smaller counties. These primary sampling units are geographically connected. The 2025 primary sampling units are then grouped into 824 different strata.

2. In each of the 824 different strata, one of the primary sampling units is selected so that the probability of selection is proportional to the size of the population in each primary sampling unit.
3. In each of the 824 selected primary sampling units, census data are used to identify a census *enumeration district,* with each containing about 300 households. Enumeration districts are then randomly selected.
4. In each of the selected enumeration districts, clusters of about four addresses (contiguous whenever possible) are randomly selected.
5. Respondents in the 60,000 selected households are interviewed about the employment status of each household member of age 16 or older.

This multistage sample design includes random, stratified, and cluster sampling at different stages. The end result is a very complicated sampling design, but it is much more practical and less expensive than using a simpler design, such as a simple random sample.

Part 2: Beyond the Basics of Collecting Data

In Part 2 of this section, we refine what we've learned about observational studies and experiments by discussing different types of observational studies and different ways of designing experiments.

There are various types of observational studies in which investigators observe and measure characteristics of subjects. The following definitions identify the standard terminology used in professional journals for different types of observational studies. These definitions are illustrated in Figure 1-4.

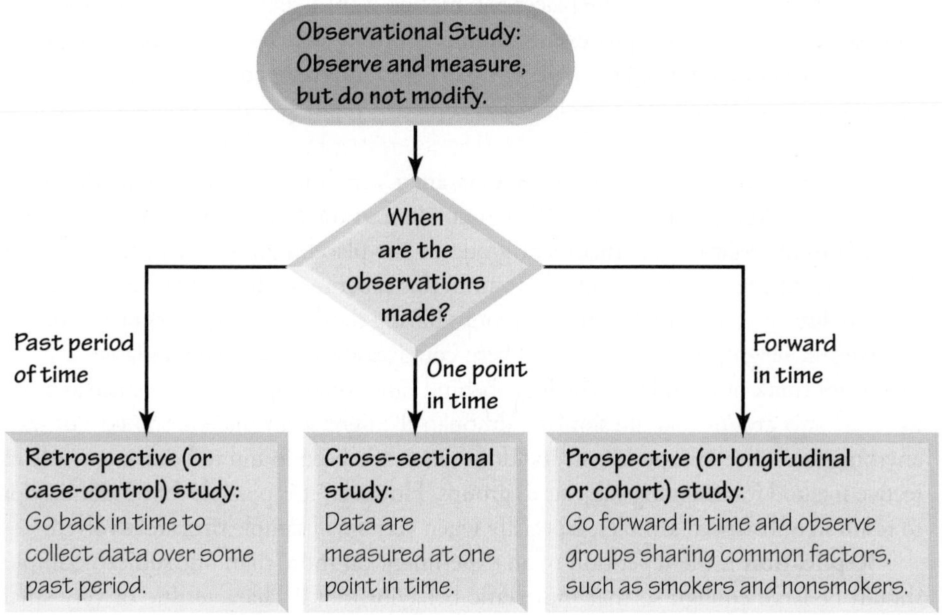

Figure 1-4 Types of Observational Studies

Prospective National Children's Study

A good example of a prospective study is the National Children's Study begun in 2005. It is tracking 100,000 children from birth to age 21. The children are from 96 different geographic regions. The objective is to improve the health of children by identifying the effects of environmental factors, such as diet, chemical exposure, vaccinations, movies, and television. The study will address questions such as these: How do genes and the environment interact to promote or prevent violent behavior in teenagers? Are lack of exercise and poor diet the only reasons why many children are overweight? Do infections impact developmental progress, asthma, obesity, and heart disease? How do city and neighborhood planning and construction encourage or discourage injuries?

This section and several other sections are partitioned into *Part 1* and *Part 2*. The typical one-semester course does not allow enough time to cover all topics in this book, and Part 2 could be omitted from such courses.

National Children's Study

The National Children's Study, launched in 2009, is designed to follow 100,000 U.S. children from before birth to age 21. Subjects have been selected from 105 different counties so that it will include a good mix of characteristics. Because children are observed from birth into the future, this is a good example of a prospective study. The cost of the study was originally expected to be $3 billion, but it has grown to almost $7 billion. Data are being collected from such varied sources as breast milk and carpet dust. The general goal is to collect data that will enhance our understanding of the effects of genetics and the environment on children. It is hoped that this understanding will lead to improvements in children's health. The National Children's Study is the largest, most expensive, and most comprehensive study of its type ever undertaken. It is being sponsored by several federal organizations, including the Department of Health and Human Services, the National Institutes of Health, and the Environmental Protection Agency.

DEFINITIONS

In a **cross-sectional study,** data are observed, measured, and collected at one point in time, not over a period of time.

In a **retrospective** (or **case-control**) **study,** data are collected from a past time period by going back in time (through examination of records, interviews, and so on).

In a **prospective** (or **longitudinal** or **cohort**) **study,** data are collected in the future from groups that share common factors (such groups are called *cohorts*).

The sampling done in retrospective studies differs from that in prospective studies. In retrospective studies we go back in time to collect data about the characteristic that is of interest, such as a group of drivers who died in car crashes and another group of drivers who did not die in car crashes. In prospective studies we go forward in time by following a group with a potentially causative factor and a group without it, such as a group of drivers who use cell phones and a group of drivers who do not use cell phones.

Designs of Experiments

We begin with Example 3, which describes the largest public health experiment ever conducted, and which serves as an example of an experiment having a good design. After describing the experiment in more detail, we describe the characteristics of randomization, replication, and blinding that typify a good design in experiments.

Example 3 **The Salk Vaccine Experiment**

In 1954, a large-scale experiment was designed to test the effectiveness of the Salk vaccine in preventing polio, which had killed or paralyzed thousands of children. In that experiment, 200,745 children were given a treatment consisting of Salk vaccine injections, while a second group of 201,229 children were injected with a placebo that contained no drug. The children being injected did not know whether they were getting the Salk vaccine or the placebo. Children were assigned to the treatment or placebo group through a process of random selection, equivalent to flipping a coin. Among the children given the Salk vaccine, 33 later developed paralytic polio, and among the children given a placebo, 115 later developed paralytic polio.

Randomization is used when subjects are assigned to different groups through a process of random selection. The 401,974 children in the Salk vaccine experiment were assigned to the Salk vaccine treatment group or the placebo group via a process of random selection equivalent to flipping a coin. In this experiment, it would be extremely difficult to directly assign children to two groups having similar characteristics of age, health, sex, weight, height, diet, and so on. There could easily be important variables that we might not think of including. The logic behind randomization is to use chance as a way to create two groups that are similar. Although it might seem that we should not leave anything to chance in experiments, randomization has been found to be an extremely effective method for assigning subjects to groups. However, it is possible for randomization to result in unbalanced samples, especially when very small sample sizes are involved.

Replication is the repetition of an experiment on more than one subject. Samples should be large enough so that the erratic behavior that is characteristic of very small samples will not disguise the true effects of different treatments. Replication is used effectively when we have enough subjects to recognize differences resulting from different treatments. (In another context, *replication* refers to the repetition or duplication of an experiment so that results can be confirmed or verified.) With replication, the large sample sizes increase the chance of recognizing different treatment effects. However, a large sample is not necessarily a good sample. Although it is important to have a sample that is sufficiently large, it is even more important to have a sample in which subjects have been chosen in some appropriate way, such as random selection.

Use a sample size that is large enough to let us see the true nature of any effects, and obtain the sample using an appropriate method, such as one based on *randomness.*

In the experiment designed to test the Salk vaccine, 200,745 children were given the actual Salk vaccine and 201,229 other children were given a placebo. Because the actual experiment used sufficiently large sample sizes, the researchers could observe the effectiveness of the vaccine.

Blinding is in effect when the subject doesn't know whether he or she is receiving a treatment or a placebo. Blinding enables us to determine whether the treatment effect is significantly different from a **placebo effect,** which occurs when an untreated subject reports an improvement in symptoms. (The reported improvement in the placebo group may be real or imagined.) Blinding minimizes the placebo effect or allows investigators to account for it. The polio experiment was **double-blind,** which means that blinding occurred at two levels: (1) The children being injected didn't know whether they were getting the Salk vaccine or a placebo, and (2) the doctors who gave the injections and evaluated the results did not know either. Codes were used so that the researchers could objectively evaluate the effectiveness of the Salk vaccine.

Controlling Effects of Variables Results of experiments are sometimes ruined because of *confounding.*

DEFINITION **Confounding** occurs in an experiment when the investigators are not able to distinguish among the effects of different factors.

Try to design the experiment in such a way that confounding does not occur.

Designs of Experiments See Figure 1-5(a), where confounding can occur when the treatment group of women shows strong positive results. Here the treatment group consists of women and the placebo group consists of men. Confounding has occurred because we cannot determine whether the treatment or the gender of the subjects caused the positive results. It is important to design experiments in such a way as to control and understand the effects of the variables (such as treatments). The Salk vaccine experiment in Example 3 illustrates one method for controlling the effect of the treatment variable: Use a *completely randomized experimental design,* whereby randomness is used to assign subjects to the treatment group and the placebo group. A completely randomized experimental design is one of the following methods that are used to control effects of variables.

Completely Randomized Experimental Design: Assign subjects to different treatment groups through a process of *random selection,* as illustrated in Example 3 and Figure 1-5(b).

Randomized Block Design: A **block** is a group of subjects that are similar, but blocks differ in ways that might affect the outcome of the experiment. Use the following procedure, as illustrated in Figure 1-5(c):

1. Form blocks (or groups) of subjects with similar characteristics.
2. Randomly assign treatments to the subjects within each block.

Survey Pitfalls

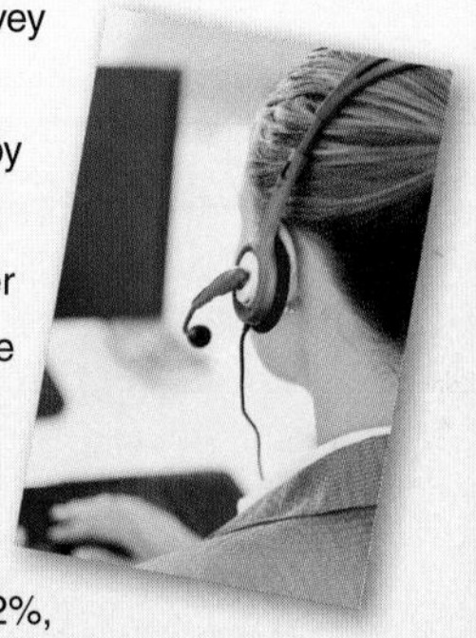

Surveys constitute a huge and growing business in the United States, but survey results can be compromised by many factors. A growing number of people refuse to respond; the average response rate is now about 22%, compared to 36% around the year 2000. A growing number of people are more difficult to reach because they use cell phones (no directories); about 15% of adults now have cell phones and no landlines, and they tend to be younger than average. There are obvious problems associated with surveys that ask respondents about drug use, theft, or sexual behavior, and a *social desirability bias* occurs when survey respondents are not honest because they don't want to be viewed negatively by the person conducting the interview.

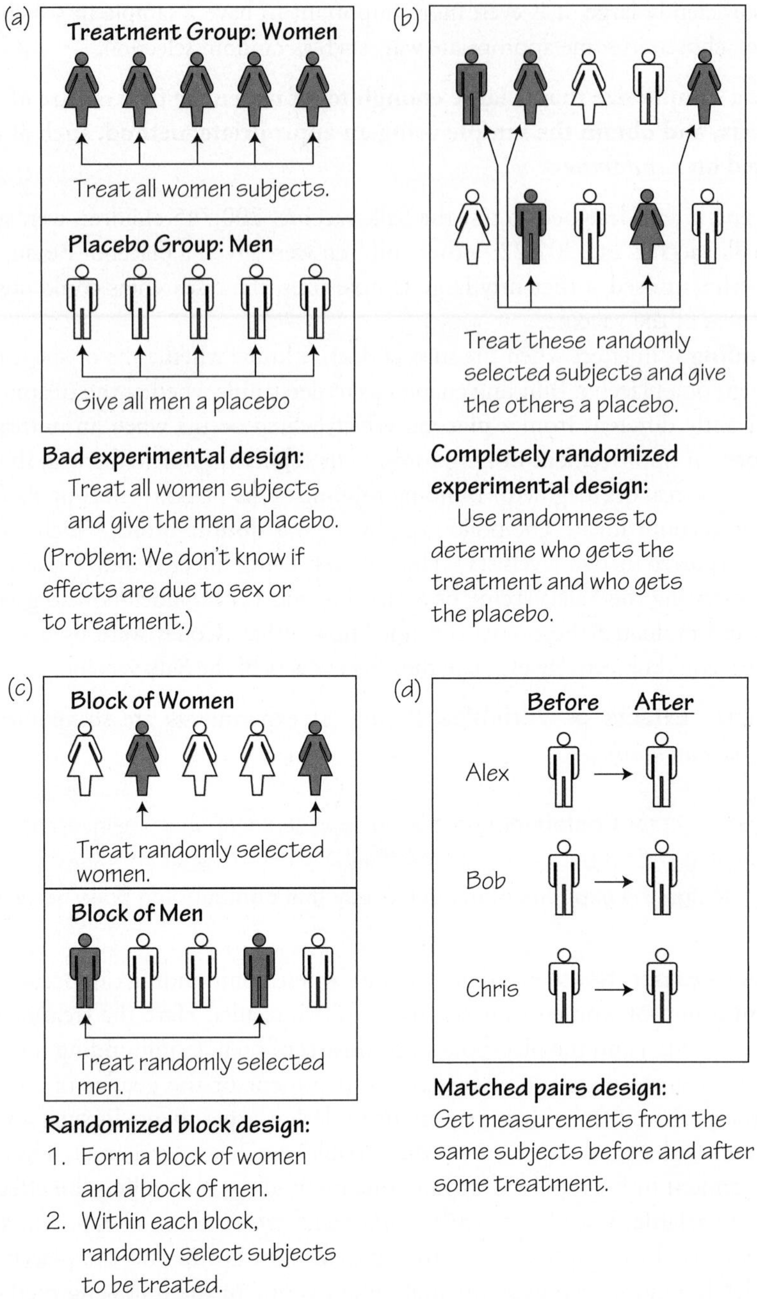
(a)
Treatment Group: Women
Treat all women subjects.
Placebo Group: Men
Give all men a placebo
Bad experimental design:
Treat all women subjects and give the men a placebo.
(Problem: We don't know if effects are due to sex or to treatment.)
(b)
Treat these randomly selected subjects and give the others a placebo.
Completely randomized experimental design:
Use randomness to determine who gets the treatment and who gets the placebo.
(c)
Block of Women
Treat randomly selected women.
Block of Men
Treat randomly selected men.
Randomized block design:
1. Form a block of women and a block of men.
2. Within each block, randomly select subjects to be treated.
(d)
Before
After
Alex
Bob
Chris
Matched pairs design:
Get measurements from the same subjects before and after some treatment.

Figure 1-5 Designs of Experiments

For example, in designing an experiment to test the effectiveness of aspirin treatments on heart disease, we might form a block of men and a block of women, because it is known that the hearts of men and women can behave differently. By controlling for gender, this randomized block design eliminates gender as a possible source of confounding.

A randomized block design uses the same basic idea as stratified sampling, but randomized block designs are used when designing experiments, whereas stratified sampling is used for surveys.

Matched Pairs Design: Compare two treatment groups (such as treatment and placebo) by using subjects matched in pairs that are somehow related or have similar characteristics, as in the following cases.

- Before/After: Matched pairs might consist of measurements from subjects before and after some treatment, as illustrated in Figure 1-5(d). Each subject yields a "before" measurement and an "after" measurement, and each before/after pair of measurements is a matched pair.
- Twins: A test of Crest toothpaste used matched pairs of twins, where one twin used Crest and the other used another toothpaste.

Rigorously Controlled Design: Carefully assign subjects to different treatment groups, so that those given each treatment are similar in the ways that are important to the experiment. In an experiment testing the effectiveness of aspirin on heart disease, if the placebo group includes a 27-year-old male smoker who drinks heavily and consumes an abundance of salt and fat, the treatment group should also include a person with these characteristics (such a person would be easy to find). This approach can be extremely difficult to implement, and often we can never be sure that we have accounted for all of the relevant factors.

Sampling Errors In an algebra course, you will get the correct result if you use the correct methods and apply them correctly. In statistics, you could use a good sampling method and do everything correctly, and yet it is possible for the result to be wrong. No matter how well you plan and execute the sample collection process, there is likely to be some error in the results. Suppose that you randomly select 1000 adults, ask them whether they use a cell phone while driving, and record the sample percentage of "yes" responses. If you randomly select another sample of 1000 adults, it is likely that you will obtain a *different* sample percentage. The different types of sampling errors are described here.

DEFINITIONS

A **sampling error** (or **random sampling error**) occurs when the sample has been selected with a random method, but there is a discrepancy between a sample result and the true population result; such an error results from chance sample fluctuations.

A **nonsampling error** is the result of human error, including such factors as wrong data entries, computing errors, questions with biased wording, false data provided by respondents, forming biased conclusions, or applying statistical methods that are not appropriate for the circumstances.

A **nonrandom sampling error** is the result of using a sampling method that is not random, such as using a convenience sample or a voluntary response sample.

If we carefully collect a random sample so that it is representative of the population, we can use methods in this book to analyze the sampling error, but we must exercise great care to minimize nonsampling error.

Experimental design requires much more thought and care than we can describe in this relatively brief section. Taking a complete course in the design of experiments is a good start in learning so much more about this important topic.

Misleading Statistics in Journalism

New York Times reporter Daniel Okrant wrote that although every sentence in his newspaper is copy-edited for clarity and good writing, "numbers, so alien to so many, don't get nearly this respect. The paper requires no specific training to enhance numeracy and [employs] no specialists whose sole job is to foster it." He cites an example of the *New York Times* reporting about an estimate of more than $23 billion that New Yorkers spend for counterfeit goods each year. Okrant writes that "quick arithmetic would have demonstrated that $23 billion would work out to roughly $8000 per city household, a number ludicrous on its face."

Recommended Assignment:

Exercises 1–8 and even-numbered Exercises 10–26.

1. No. Not every sample of the same size has the same chance of being selected. For example, the sample with the first two names has no chance of being selected. A simple random sample of *n* items is selected in such a way that every sample of the same size has the same chance of being selected.
2. In an observational study, you would examine subjects who consume fruit and those who do not. In the observational study, you run a greater risk of having a lurking variable that affects weight. For example, people who consume more fruit might be more likely to maintain generally better eating habits, and they might be more likely to exercise, so their lower weights might be due to these better eating and exercise habits, and perhaps fruit consumption does not explain lower weights. An experiment would be better, because you can randomly assign subjects to the fruit treatment group and the group that does not get the fruit treatment, so lurking variables are less likely to affect the results.
3. The population consists of the adult friends on the list. The simple random sample is selected from the population of adult friends on the list, so the results are not likely to be representative of the much larger general population of adults in the United States.
4. Because there is nothing about left-handedness or right-handedness that would affect being in the author's classes, the results are likely to be typical of the population. The results are likely to be good, but convenience samples in general are not likely to be so good.
5. Because the subjects are subjected to anger and confrontation, they are given a form of treatment, so this is an experiment, not an observational study.
6. Because the subjects were given a treatment consisting of Lipitor, this is an experiment.
7. This is an observational study because the therapists were not given any treatment. Their responses were observed.

1-4 Basic Skills and Concepts

Statistical Literacy and Critical Thinking

1. Simple Random Sample At a national conference of the American Appliances Association, a market researcher plans to conduct a survey of conference attendees. She uses the list of attendee names and selects every 20th name. Is the result a simple random sample? Why or why not? In general, what is a simple random sample?

2. Observational Study and Experiment You want to conduct a study to determine whether fruit consumption leads to reduced weight. Why would an experiment be better than an observational study?

3. Simple Random Convenience Sample A student of the author listed his adult friends, and then he surveyed a simple random sample of them. What is the population from which the simple random sample was selected? Are the results likely to be representative of the general population of adults in the United States? Why or why not?

4. Convenience Sample The author conducted a survey of the students in all of his classes. He asked the students to indicate whether they are left-handed or right-handed. Is this convenience sample likely to provide results that are typical of the population? Are the results likely to be good or bad? Does the quality of the results in this survey reflect the quality of convenience samples in general?

In Exercises 5–8, determine whether the given description corresponds to an observational study or an experiment. In each case, give a brief explanation of your choice.

5. Contentious Survey The Milgram Research Company wants to study reactions to stress, so it administers surveys in which the person asking the questions pretends to become very angry with the survey subject. At one point, the surveyor screams at the subject and asks how anyone could have such "stupid" opinions.

6. Clinical Trial In a clinical trial of the cholesterol drug Lipitor, 188 subjects were given 20-mg doses of the drug, and 3.7% of them experienced nausea (based on data from Pfizer, Inc.).

7. Touch Therapy Nine-year-old Emily Rosa was an author of an article in the *Journal of the American Medical Association* after she tested professional touch therapists. Using a cardboard partition, she held her hand above one of the therapist's hands, and the therapist was asked to identify the hand that Emily chose.

8. Happiness Survey In a study sponsored by Coca-Cola, 12,500 people were asked what contributes most to their happiness, and 77% of the respondents said that it was their family or partner.

In Exercises 9–20, identify which of these types of sampling is used: **random, systematic, convenience, stratified,** *or* **cluster.**

9. Harry Potter The author collected sample data by randomly selecting 12 different pages from *Harry Potter and the Sorcerer's Stone* and then finding the number of words in each sentence on each of those pages.

10. Sexuality of Women The sexuality of women was discussed in Shere Hite's book *Women and Love: A Cultural Revolution.* Her conclusions were based on sample data that consisted of 4500 mailed responses from 100,000 questionnaires that were sent to women.

11. Twitter Poll In a Pew Research Center poll, 1007 adults were called after their telephone numbers were randomly generated by a computer, and 85% of the respondents were able to correctly identify what Twitter is.

12. Ecology When collecting data from different sample locations in a lake, a researcher uses the "line transect method" by stretching a rope across the lake and collecting samples at every interval of 5 meters.

13. CBS News The CBS News station in New York City often obtains opinions by interviewing neighbors of a person who is the focus of a news story.

14. Acupuncture Study In a study of treatments for back pain, 641 subjects were randomly assigned to the four different treatment groups of individualized acupuncture, standardized acupuncture, simulated acupuncture, and usual care (based on data from "A Randomized Trial Comparing Acupuncture, Simulated Acupuncture, and Usual Care for Chronic Low Back Pain," by Cherkin et al., *Archives of Internal Medicine,* Vol. 169, No. 9).

15. Deforestation Rates Satellites are used to collect sample data used to estimate deforestation rates. The Forest Resources Assessment of the UN Food and Agriculture Organization uses a method of selecting a sample of a 10-km-wide square at every 1° intersection of latitude and longitude.

16. Dictionary The author collected sample data by randomly selecting 20 different pages from a printed version of the *Merriam-Webster Dictionary* and then counting the number of defined words on each of those pages.

17. Testing Lipitor In a clinical trial of the cholesterol drug Lipitor, subjects were partitioned into groups given a placebo or Lipitor doses of 10 mg, 20 mg, 40 mg, or 80 mg. The subjects were randomly assigned to the different treatment groups (based on data from Pfizer, Inc.).

18. Exit Polls On the day of the last presidential election, ABC News organized an exit poll in which specific polling stations were randomly selected and all voters were surveyed as they left the premises.

19. Literary Digest Poll In 1936, *Literary Digest* magazine mailed questionnaires to 10 million people and obtained 2,266,566 responses. The responses indicated that Alf Landon would win the presidential election, but Franklin D. Roosevelt actually won the election.

20. Highway Strength The New York State Department of Transportation evaluated the quality of the New York State Throughway by testing core samples collected at regular intervals of one mile.

Simple Random Samples. ***In Exercises 21–26, determine whether the sample is a simple random sample. Give a brief explanation of your choice.***

21. Bayer Aspirin Bayer HealthCare LLC produces low-dose aspirin pills designed to contain 81 mg of aspirin. Because each pill contains other ingredients, including corn starch, talc, and propylene glycol, it is difficult to check whether manufactured pills contain 81 mg of aspirin. A quality control plan is to select every 1000th pill, which is then tested for the correct amount of aspirin.

22. Market Research In order to test for a gender gap in the way that men and women purchase cars, the Grant Survey Company polls exactly 750 adult men and 750 adult women randomly selected from adults in the United States.

23. Post-Election Survey In the last general election, 132,312 adults voted in Dutchess County, New York. You plan to conduct a post-election survey of 500 of those voters. After obtaining a list of those who voted, you number the list from 1 to 132,312, and then you use a computer to randomly generate 500 numbers between 1 and 132,312. Your sample consists of the voters corresponding to the selected numbers.

8. This is an observational study because the survey subjects were not given any treatment. Their responses were observed.
9. Cluster
10. Convenience
11. Random
12. Systematic
13. Convenience
14. Random
15. Systematic
16. Cluster
17. Random
18. Cluster
19. Convenience
20. Systematic
21. The sample is not a simple random sample. Because every 1000th pill is selected, some samples have no chance of being selected. For example, a sample consisting of two consecutive pills has no chance of being selected, and this violates the requirement of a simple random sample.
22. The sample is not a simple random sample. Not every sample of 1500 adults has the same chance of being selected. For example, a sample of 1500 women has no chance of being selected.
23. The sample is a simple random sample. Every sample of size 500 has the same chance of being selected.

24. Jury Selection According to the State of New York Unified Court System, names of potential jurors are selected from a variety of different sources. When a trial requires a jury, names from the list are randomly selected in a way that is equivalent to writing the names on slips of paper, mixing them in a bowl, and selecting the required number of potential jurors.

25. Mall Research Mall managers commonly research how customers use the malls. The author was approached by a pollster at the Galleria Mall in Dutchess County, New York. The pollster was obviously selecting subjects who appeared to be approachable.

26. Car Crash Tests The National Highway Traffic Safety Administration (NHTSA) conducts crash tests of cars. One car of each different model is randomly selected for testing.

1-4 Beyond the Basics

***In Exercises 27–30, indicate whether the observational study used is* cross-sectional, retrospective, *or* prospective.**

27. Nurses' Health Study The Nurses' Health Study was started in 1976 with 121,700 female registered nurses who were between the ages of 30 and 55. The subjects were surveyed in 1976 and every two years thereafter. The study is ongoing.

28. Drinking and Driving Study In order to study the seriousness of drinking and driving, a researcher obtains records from past car crashes. Drivers are partitioned into a group that had no alcohol consumption and another group that did have evidence of alcohol consumption at the time of the crash.

29. Smoking Study Researchers from the National Institutes of Health want to determine the current rates of smoking among adult males and adult females. They conduct a survey of 500 adults of each gender.

30. Meat and Mortality Researchers at the National Cancer Institute studied meat consumption and its relationship to mortality. Approximately one-half million people were surveyed, and they were then followed for a period of 10 years.

In Exercises 31–34, identify which of these designs is most appropriate for the given experiment: completely randomized design, randomized block design, or matched pairs design.

31. Lisinipril Lisinipril is a drug designed to lower blood pressure. In a clinical trial of Lisinipril, blood pressure levels of subjects are measured before and after they have been treated with the drug.

32. Aspirin A clinical trial of aspirin treatments is being planned to determine whether the rate of myocardial infarctions (heart attacks) is different for men and women.

33. West Nile Vaccine Currently, there is no approved vaccine for the prevention of West Nile virus. A clinical trial of a possible vaccine is being planned to include subjects treated with the vaccine while other subjects are given a placebo.

34. HIV Vaccine The HIV Trials Network is conducting a study to test the effectiveness of two different experimental HIV vaccines. Subjects will consist of 80 pairs of twins. For each pair of twins, one of the subjects will be treated with the DNA vaccine and the other twin will be treated with the adenoviral vector vaccine.

35. Blinding For the study described in Exercise 34, blinding will be used. What is blinding, and why was it important in this experiment?

36. Sample Design Literacy In "Cardiovascular Effects of Intravenous Triiodothyronine in Patients Undergoing Coronary Artery Bypass Graft Surgery" (*Journal of the American Medical Association,* Vol. 275, No. 9), the authors explain that patients were assigned to one of three groups: (1) a group treated with triidothyronine, (2) a group treated with normal saline bolus and dopamine, and (3) a placebo group given normal saline. The authors summarize the sample design as a "prospective, randomized, double-blind, placebo-controlled trial." Describe the meaning of each of those terms in the context of this study.

24. The sample is a simple random sample. Every sample of the same size has the same chance of being selected.
25. The sample is not a simple random sample. Not every sample has the same chance of being selected. For example, a sample that includes people who do not appear to be approachable has no chance of being selected.
26. The sample is not a simple random sample. Not all samples of the same size have the same chance of being selected. For example, a sample that includes only Honda cars has no chance of being selected.
27. Prospective study
28. Retrospective study
29. Cross-sectional study
30. Prospective study
31. Matched pairs design
32. Randomized block design
33. Completely randomized design
34. Matched pairs design
35. Blinding is a method whereby a subject (or a person who evaluates results) in an experiment does not know whether the subject is treated with the DNA vaccine or the adenoviral vector vaccine. It is important to use blinding so that results are not somehow distorted by knowledge of the particular treatment used.
36. Prospective: The experiment was begun and results were followed forward in time. Randomized: Subjects were assigned to the different groups through a process of random selection, whereby they had the same chance of belonging to each group. Double-blind: The subjects did not know which of the three groups they were in, and the people who evaluated results did not know either. Placebo-controlled: There was a group of subjects who were given a placebo; by comparing the placebo group to the two treatment groups, the effects of the treatments might be better understood.

Chapter 1 Review

The single most important concept presented in this chapter is to recognize that when one is using methods of statistics with sample data to form conclusions about a population, it is absolutely essential to collect sample data in a way that is appropriate. Using data from a voluntary response (self-selected) sample is a really bad idea, because such a sample could very easily be biased in that it might not be at all representative of the population. One common and effective method for collecting data is to use a *simple random sample.* With a simple random sample of n items, all possible samples of n items have the same chance of being selected.

Statistical literacy includes a clear understanding of such important terms as *sample, population, statistic, parameter, quantitative data, categorical data, voluntary response sample, observational study, experiment,* and *simple random sample.* Section 1-2 introduced statistical thinking, and Figure 1-2 summarized important issues to consider in preparation for analysis, conducting the analysis, and forming conclusions. Section 1-3 discussed different types of data, and it is crucial to understand the distinction between quantitative data and categorical data. After completing this chapter, you should be able to do the following:

- Distinguish between a population and a sample, and distinguish between a parameter and a statistic.
- Recognize the importance of good sampling methods in general, and recognize the importance of a *simple random sample* in particular. Understand that even though voluntary response samples are common, they should not be used for a statistical analysis.

Chapter Quick Quiz

1. Chicago Bulls The numbers of the current players for the Chicago Bulls basketball team are 1, 2, 3, 5, 6, 9, 11, 13, 16, 20, 22, 26, 32, and 40. Does it make sense to calculate the average (mean) of these numbers?

2. Chicago Bulls Which of the following best describes the level of measurement of the data listed in Exercise 1: nominal, ordinal, interval, ratio?

3. Earthquake Depths Data Set 16 includes depths (km) of the sources of earthquakes. Are these values discrete or continuous?

4. Earthquake Depths Are the earthquake depths described in Exercise 3 quantitative data or categorical data?

5. Earthquake Depths Which of the following best describes the level of measurement of the earthquake depths described in Exercise 3: nominal, ordinal, interval, ratio?

6. Earthquake Depths True or false: If you construct a sample by selecting every sixth earthquake depth from the list given in Data Set 16, the result is a simple random sample.

7. Gallup Poll In a recent Gallup poll, pollsters randomly selected adults and asked them whether they smoke. Because the subjects agreed to respond, is the sample a voluntary response sample?

8. Parameter and Statistic In a recent Gallup poll, pollsters randomly selected adults and asked them whether they smoke. Among the adults who responded to the survey question, 21% said that they did smoke. Is that value of 21% an example of a statistic or an example of a parameter?

9. Observational Study or Experiment Are the data described in Exercise 8 the result of an observational study or an experiment?

10. Statistical Significance and Practical Significance True or false: If data lead to a conclusion with statistical significance, then the results also have practical significance.

Each Chapter Review is followed by a Chapter Quick Quiz, Review Exercises, and Cumulative Review Exercises. Student editions of the text include answers to

- Odd-numbered section exercises
- All Chapter Quick Quiz questions
- All Review Exercises
- All Cumulative Review Exercises

1. No. The numbers do not measure or count anything.
2. Nominal
3. Continuous
4. Quantitative data
5. Ratio
6. False
7. No
8. Statistic
9. Observational study
10. False

Review Exercises

1. Walmart Stores Currently, there are 4227 Walmart stores in the United States and another 3210 stores outside of the United States.

a. Are the numbers of Walmart stores discrete or continuous?

b. What is the level of measurement for the numbers of Walmart stores in different years? (nominal, ordinal, interval, ratio)

c. If a survey is conducted by randomly selecting 10 customers in every Walmart store, what type of sampling is used? (random, systematic, convenience, stratified, cluster)

d. If a survey is conducted by randomly selecting 20 Walmart stores and interviewing all of the employees at the selected stores, what type of sampling is used? (random, systematic, convenience, stratified, cluster)

e. What is wrong with surveying customer satisfaction by mailing questionnaires to 10,000 randomly selected customers?

2. What's Wrong? A survey sponsored by the American Laser Centers included responses from 575 adults, and 24% of the respondents said that the face is their favorite body part (based on data from *USA Today*). What is wrong with this survey?

3. What's Wrong? A survey included 4230 responses from Internet users who decided to respond to a question posted by America OnLine (AOL). Here is the question: How often do you use credit cards for purchases? Among the respondents, 67% said that they used credit cards frequently. What is wrong with this survey?

4. Sampling Seventy-two percent of Americans squeeze their toothpaste tube from the top. This and other not-so-serious findings are included in *The First Really Important Survey of American Habits*. Those results are based on 7000 responses from the 25,000 questionnaires that were mailed.

a. What is wrong with this survey?

b. As stated, the value of 72% refers to all Americans, so is that 72% a statistic or a parameter? Explain.

c. Does the survey constitute an observational study or an experiment?

5. Percentages

a. The labels on U-Turn protein energy bars include the statement that these bars contain "125% less fat than the leading chocolate candy brands" (based on data from *Consumer Reports magazine*). What is wrong with that claim?

b. In a Pew Research Center poll on driving, 58% of the 1182 respondents said that they like to drive. What is the actual number of respondents who said that they like to drive?

c. In a Pew Research Center poll on driving, 331 of the 1182 respondents said that driving is a chore. What percentage of respondents said that driving is a chore?

6. Why the Discrepancy? A Gallup poll was taken two years before a presidential election, and it showed that Hillary Clinton was preferred by about 50% more voters than Barack Obama. The subjects in the Gallup poll were randomly selected and surveyed by telephone. An America OnLine (AOL) poll was conducted at the same time as the Gallup poll, and it showed that Barack Obama was preferred by about twice as many respondents as Hillary Clinton. In the AOL poll, Internet users responded to voting choices that

1. a. Discrete
   b. Ratio
   c. Stratified
   d. Cluster
   e. The mailed responses would be a voluntary response sample, so those with strong opinions are more likely to respond. It is very possible that the results do not reflect the true opinions of the population of all customers.
2. The survey was sponsored by the American Laser Centers, and 24% said that the favorite body part is the face, which happens to be a body part often chosen for some type of laser treatment. The source is therefore questionable.
3. The sample is a voluntary response sample, so the results are questionable.
4. a. It uses a voluntary response sample, and those with special interests are more likely to respond, so it is very possible that the sample is not representative of the population.
   b. Because the statement refers to 72% of all Americans, it is a parameter (but it is probably based on a 72% rate from the sample, and the sample percentage is a statistic).
   c. Observational study
5. a. If they have no fat at all, they have 100% less than any other amount with fat, so the 125% figure cannot be correct.
   b. 686
   c. 28%
6. The Gallup poll used randomly selected respondents, but the AOL poll used a voluntary response sample. Respondents in the AOL poll are more likely to participate if they have strong feelings about the candidates, and this group is not necessarily representative of the population. The results from the Gallup poll are more likely to reflect the true opinions of American voters.

were posted on the AOL site. How can the large discrepancy between the two polls be explained? Which poll is more likely to reflect the true opinions of American voters at the time of the poll?

7. Statistical Significance and Practical Significance The Gengene Research Group has developed a procedure designed to increase the likelihood that a baby will be born a girl. In a clinical trial of their procedure, 112 girls were born to 200 different couples. If the method has no effect, there is about a 4% chance that such extreme results would occur. Does the procedure appear to have statistical significance? Does the procedure appear to have practical significance?

8. Marijuana Survey In a recent Pew poll of 1500 adults, 52% of the respondents said that the use of marijuana should not be made legal. In the same poll, 23% of the respondents said that the use of marijuana for medical purposes should not be legal.

a. The sample of 1500 adults was selected from the population of all adults in the United States. The method used to select the sample was equivalent to placing the names of all adults in a giant bowl, mixing the names, and then drawing 1500 names. What type of sampling is this? (random, systematic, convenience, stratified, cluster)

b. If the sampling method consisted of a random selection of 30 adults from each of the 50 states, what type of sampling would this be? (random, systematic, convenience, stratified, cluster)

c. What is the level of measurement of the responses of yes, no, don't know, and refused to respond?

d. Is the given value of 52% a statistic or a parameter? Why?

e. What would be wrong with conducting the survey by mailing a questionnaire that respondents could complete and mail back?

9. Marijuana Survey Identify the type of sampling (random, systematic, convenience, stratified, cluster) used when a sample of the 1500 survey responses is obtained as described. Then determine whether the sampling scheme is likely to result in a sample that is representative of the population of all adults.

a. A complete list of all 241,472,385 adults in the United States is compiled and every 150,000th name is selected until the sample size of 1500 is reached.

b. A complete list of all 241,472,385 adults in the United States is compiled and 1500 adults are randomly selected from that list.

c. The United States is partitioned into regions with 100 adults in each region. Then 15 of those regions are randomly selected, and all 100 people in each of those regions are surveyed.

d. The United States is partitioned into 150 regions with approximately the same number of adults in each region, then 10 people are randomly selected from each of the 150 regions.

e. A survey is mailed to 10,000 randomly selected adults, and the 1500 responses are used.

10. Marijuana Survey Exercise 8 referred to a Pew poll of 1500 adults, and 52% of the respondents said that the use of marijuana should not be made legal.

a. Among the 1500 adults who responded, what is the number of respondents who said that the use of marijuana should not be made legal?

b. In the same poll of 1500 adults, 345 of the respondents said that the use of marijuana for medical purposes should not be legal. What is the percentage of respondents who said that the use of marijuana for medical purposes should not be legal?

c. In this survey of 1500 adults, 727 are men and 773 are women. Find the percentage of respondents who are men, and then find the percentage of respondents who are women.

7. Because there is only a 4% chance of getting the results by chance, the method appears to have statistical significance. The result of 112 girls in 200 births is above the approximately 50% rate expected by chance, but it does not appear to be high enough to have practical significance. Not many couples would bother with a procedure that raises the likelihood of a girl from 50% to 56%.

8. a. Random
b. Stratified
c. Nominal
d. Statistic, because it is based on a sample.
e. The mailed responses would be a voluntary response sample. Those with strong opinions about the topic would be more likely to respond, so it is very possible that the results would not reflect the true opinions of the population of all adults.

9. a. Systematic
b. Random
c. Cluster
d. Stratified
e. Convenience

10. a. 780 adults
b. 23%
c. Men: 48.5%; women: 51.5%

d. No, although this is a subjective judgment.

e. No, although this is a subjective judgment.

d. Does the difference between the two percentages from part (c) appear to have statistical significance?

e. Does the difference between the two percentages from part (c) appear to have practical significance?

The Cumulative Review Exercises normally involve material from earlier chapters, a feature that is obviously impossible in this first chapter. Instead, these exercises focus on calculator usage. Exercises 9–12 involve the use of *scientific notation.*

1. The mean is 11. Because the flight numbers are not measures or counts of anything, the result does not have meaning.

2. The mean is 101, and it is reasonably close to the population mean of 100.

3. 11.83 is an unusually high value.

4. 0.46

5. 1067

6. 0.0037

7. 28.0

8. 5.3

Cumulative Review Exercises

For Chapter 2 through Chapter 11, the Cumulative Review Exercises include topics from preceding chapters. For this chapter, we present a few *calculator warm-up exercises*, with expressions similar to those found throughout this book. Use your calculator to find the indicated values.

1. Flights Refer to the flight numbers listed in the first column of Data Set 15 in Appendix B. What value is obtained when those 48 numbers are added and the total is then divided by 48? (This result, called the *mean*, is discussed in Chapter 3.) Does the result have any meaning?

2. IQ Scores Refer to the IQ scores listed in Data Set 6 in Appendix B. What value is obtained when those 20 IQ scores are added and the total is then divided by 20? (This result, called the *mean*, is discussed in Chapter 3.) Is the result reasonably close to 100, which is the mean IQ score for the population?

3. Height of Tallest Man Sultan Kosen is the tallest man, and the expression below converts his height of 247 cm to a standardized score. Find this value and round the result to two decimal places. Such standardized scores are considered to be unusually high if they are greater than 2 or 3. Is the result unusually high?

$$\frac{247 - 176}{6}$$

4. Transportation Safety The given expression is used for determining the likelihood that a water taxi will have a total passenger weight that exceeds the maximum safe weight of 3500 lb. Find the given value and round the result to two decimal places.

$$\frac{175 - 172}{\frac{29}{\sqrt{20}}}$$

5. Determining Sample Size The given expression is used to determine the size of the sample necessary to estimate the proportion of college students who have the profound wisdom to take a statistics course. Find the value and round the result to the nearest whole number.

$$\frac{1.96^2 \cdot 0.25}{0.03^2}$$

6. Testing the Effectiveness of Echinacea The given expression is part of a calculation used to study the effectiveness of Echinacea in treating colds. Round the result to four decimal places.

$$\frac{(88 - 88.570)^2}{88.570}$$

7. Variation in Body Temperatures The given expression is used to compute a measure of the variation (variance) of three IQ scores.

$$\frac{(96 - 100)^2 + (106 - 100)^2 + (98 - 100)^2}{3 - 1}$$

8. Standard Deviation The given expression is used to compute the standard deviation of three IQ scores. (The standard deviation is introduced in Section 3-3.) Round the result to one decimal place.

$$\sqrt{\frac{(96 - 100)^2 + (106 - 100)^2 + (98 - 100)^2}{3 - 1}}$$

Scientific Notation. ***In Exercises 9–12, the given expressions are designed to yield results expressed in a form of scientific notation. For example, the calculator-displayed result of 1.23E5 can be expressed as 123,000, and the result of 4.56E-4 can be expressed as 0.000456. Perform the indicated operation and express the result as an ordinary number that is not in scientific notation.***

9. 0.6^{14} **10.** 8^{12} **11.** 7^{14} **12.** 0.3^{10}

9. 0.00078364164
10. 68,719,476,736 (or about 68,719,476,000)
11. 678,223,072,849 (or about 678,223,070,000)
12. 0.0000059049

Technology Project

Simple Random Sample In this project, we will use technology and randomness to identify a simple random sample. Let's assume that we want to conduct a survey of people randomly selected from a population. Instead of using a very large population, we will refer to Data Set 12 in Appendix B for the list of 38 names of the presidents of the United States. (That list does not include presidents who took office as the result of a resignation or assassination.) We will use a process that results in a simple random sample of five of those names. The basic idea is to consider the list of 38 names to be numbered from 1 through 38. We will use technology to randomly generate five numbers between 1 and 38, and then we will identify the five names corresponding to those five random numbers.

STATDISK: Click on **Data** at the top of the screen, then select **Uniform Generator.** In the window that appears, enter a sample size of 5, enter 1 for the minimum value, enter 38 for the maximum value, and enter 0 for the number of decimal places (because we want whole numbers). Click on **Generate** and you will get five **random numbers** between 1 and 38. If any numbers are duplicates, repeat the process to get five *different* numbers between 1 and 38. Identify the five presidents from Data Set 12 that correspond to the five different random numbers.

Minitab: Click on **Calc,** select **Random Data,** then select **Integer.** In the window that appears, enter 5 for the "Number of rows to generate," enter C1 for the column that will contain the random numbers, enter 1 for the minimum value, enter 38 for the maximum value, then click on **OK.** If any numbers are duplicates, repeat the process to get five *different* numbers between 1 and 38. Identify the five presidents from Data Set 12 that correspond to the five different random numbers.

Excel: Click on the toolbar entry of f_x. Select the "Category" item of **Math & Trig.** Select the "function" of **RANDBETWEEN**. Click **OK.** In the dialog box that appears, enter 1 for the "Bottom" and enter 38 for the "top." Click **OK** and a random number will be generated. Repeat this process until five different random numbers between 1 and 38 are obtained.

TI-83/84 Plus: Press MATH and then use the ⟩ key to scroll to **PRB.** In the PRB menu, select **randInt** and press ENTER. The format of the randInt command is to enter the minimum value, the maximum value, and the number of values to be generated. Those entries should be separated by commas so that the command is **randInt(1, 38, 5)**. Press ENTER and you will get five random numbers between 1 and 38 inclusive. If any numbers are duplicates, repeat the process to get five *different* numbers between 1 and 38. Identify the five presidents from Data Set 12 that correspond to the five different random numbers.

StatCrunch Select **Data** from the top menu bar, then select **Simulate data.** Choose the option of **Discrete uniform.** Make the required entries in the dialog box. (For this project, use 5 rows, 1 column, a minimum of 1, and a maximum of 38; store the values in a stacked column; and use a single dynamic seed.) Click on **Simulate.** If any numbers are duplicates, repeat the process to get five *different* numbers between 1 and 38. Identify the five presidents from Data Set 12 that correspond to the five different random numbers.

Note to Instructor

You can require a particular technology, such as the TI-83/84 Plus calculator, or you can allow students to use a variety of different inexpensive scientific calculators. This text is flexible in that it does not require a particular technology. After making your decision, it is important to inform students as soon as possible, preferably during the first class. Be sure to notify the class about any calculators that are prohibited on tests. (Many instructors prohibit any calculator with a "QWERTY" keyboard, such as the TI-92.) Comment that calculators capable of handling two-variable statistics simplify some pretty messy calculations that may come up later.

STATDISK is free to colleges that adopt this text. Students using this text may make their own personal copy of STATDISK. Comment to the class that even if no statistical software package is used, there is value in interpreting the computer displays that appear herein.

The preceding procedure results in a simple random sample of five names selected from a list of 38 different names. The same procedure can be used with a much larger population.

Pollsters often use the above procedure to randomly generate telephone numbers. Suppose that you want to randomly generate phone numbers beginning with an area code of 347 and an exchange of 489 (as in the song "Diary" by Alicia Keys). You can use the above procedure to randomly generate numbers between 0 and 9999, and the result will be a complete phone number that could be called for a survey. Now randomly generate 10 such numbers, and combine them with the area code of 347 and the exchange of 489 so that you have a list of 10 people who can be reached at phone numbers with the format of (347) 489–xxxx.

from data TO DECISION

Critical Thinking

The concept of "six degrees of separation" grew from a 1967 study conducted by psychologist Stanley Milgram. His original finding was that two random residents in the United States are connected by an average of six intermediaries. In his first experiment, he sent 60 letters to subjects in Wichita, Kansas, and asked them to forward the letters to a specific woman in Cambridge, Massachusetts. The subjects were instructed to hand-deliver the letters to acquaintances who they believed could reach the target person either directly or through other acquaintances. Of the 60 subjects who were solicited, 50 participated, and three of the letters reached the target. Two subsequent experiments also had low completion rates, but Milgram eventually reached a 35% completion rate, and he found that for completed chains, the mean number of intermediaries was around six. Consequently, Milgram's original data led to the concept referred to as "six degrees of separation."

Analyzing the Results

1. Did Stanley Milgram's original experiment have a good design, or was it flawed? Explain.

2. Do Milgram's original data justify the concept of "six degrees of separation?"

3. Describe a sound experiment for determining whether the concept of six degrees of separation is valid.

Cooperative Group Activities

1. In-class activity From the cafeteria, obtain 18 straws. Cut 6 of them in half, cut 6 of them into quarters, and leave the other 6 as they are. There should now be 42 straws of three different lengths. Put them in a bag, mix them up, then select one straw, find its length, and then replace it. Repeat this until 20 straws have been selected. (*Important:* Select the straws without looking into the bag, and select the first straw that you touch.) Find the average (mean) of the lengths of the sample of 20 straws. Now remove all of the straws and find the mean of the lengths of the population. Did the sample provide an average that was close to the true population average (mean)? Why or why not?

2. In-class activity In mid-December of a recent year, the Internet service provider America OnLine (AOL) ran a survey of its users. This question was asked about Christmas trees: "Which do you prefer?" The response could be "a real tree" or "a fake tree." Among the 7073 responses received by the Internet users, 4650 indicated a real tree, and 2423 indicated a fake tree. We have already noted that because the sample is a voluntary response sample, no conclusions can be made about a population larger than the 7073 people who responded. Identify other problems with this survey question.

3. In-class activity Identify the problems with the following:

- A recent report on *CNN Headline News* included a comment that crime in the United States fell in the 1980s because of the growth of abortions in the 1970s, which resulted in fewer unwanted children.
- *Consumer Reports* magazine mailed an Annual Questionnaire about cars and other consumer products. Also included were a request for a voluntary contribution of money and a ballot for the Board of Directors. Responses were to be mailed back in envelopes that required postage stamps.

4. Out-of-class activity Find a report of a survey that is based on a voluntary response sample. Describe how it is quite possible that the results do not accurately reflect the population.

5. Out-of-class activity Find a professional journal with an article that includes a statistical analysis of an experiment. Describe and comment on the design of the experiment. Identify one particular issue, and determine whether the result was found to be statistically significant. Determine whether that same result has practical significance.

Note to Instructor

One of the six recommendations included among the *Guidelines for Assessment and Instruction in Statistics Education* is this: Foster active learning in the classroom. The author and many notable leaders in statistics education enthusiastically encourage active involvement in introductory statistics courses, and the Cooperative Group Activities provide some ideas for such involvement. If your course currently has no such involvement, consider beginning with one or two simple activities. These Cooperative Group Activities are located at the end of each chapter.

2 Summarizing and Graphing Data

2

chapter problem

Does exposure to lead affect IQ score?

Data Set 5 in Appendix B includes full IQ scores from three groups of children who lived near a lead smelter. The children in Group 1 had *low* levels of measured lead in their blood (with blood levels less than 40 micrograms/100 mL in each of two years). Group 2 had *medium* levels of measured lead in their blood (with blood levels of at least 40 micrograms/100 mL in exactly one of two years). Group 3 had *high* levels of measured lead in their blood (with blood levels of at least 40 micrograms/100 mL in each of two years).

Let's consider the measured full IQ scores from Group 1 (low lead level) and Group 3 (high lead level), as listed in Table 2-1. It is an exceptionally rare person who can look at both lists of IQ scores and form meaningful conclusions. Almost all of us must work at describing, exploring, and comparing the two sets of data. ("Describing, Exploring, and Comparing Data" would be a great title for a chapter in a statistics book—see Chapter 3.) In this chapter we present methods that focus on summarizing the data and using graphs that enable us to understand important characteristics of the data, especially the *distribution* of the data. These methods will help us compare the two sets of data so that we can determine whether IQ scores of the *low* lead group are somehow different from the IQ scores of the *high* lead group. Such comparisons will be helpful as we try to address this important and key issue: Does exposure to lead have an effect on IQ score?

Table 2-1 Full IQ Scores of Low Lead Group and High Lead Group

Low Lead Level (Group 1)															
70	85	86	76	84	96	94	56	115	97	77	128	99	80	118	86
141	88	96	96	107	86	80	107	101	91	125	96	99	99	115	106
105	96	50	99	85	88	120	93	87	98	78	100	105	87	94	89
80	111	104	85	94	75	73	76	107	88	89	96	72	97	76	107
104	85	76	95	86	89	76	96	101	108	102	77	74	92		
High Lead Level (Group 3)															
82	93	85	75	85	80	101	89	80	94	88	104	88	88	83	104
96	76	80	79	75											

No Phones or Bathtubs

Many statistical analyses must consider changing characteristics of populations over time. Here are some observations of life in the United States from 100 years ago:

- 8% of homes had a telephone.
- 14% of homes had a bathtub.
- The mean life expectancy was 47 years.
- The mean hourly wage was 22 cents.
- There were approximately 230 murders in the entire United States.

Although these observations from 100 years ago are in stark contrast to the United States of today, statistical analyses should always consider changing population characteristics that might have more subtle effects.

Note to Instructor

Organizational change in this edition: Section 2-4 (Statistical Graphics) and Section 2-5 (Critical Thinking: Bad Graphs) have been combined into one section: Section 2-4 (Graphs That Enlighten and Graphs That Deceive).

2-1 Review and Preview

Chapter 1 presented some critically important concepts, including context of data, source of data, sampling method, conclusions, and practical implications. Like the data in Table 2-1, many samples of data are large, so understanding them requires that we organize, summarize, and represent the data in a way that allows us to gain insight. We can organize and summarize data numerically in tables or visually in graphs, as described in this chapter. Of course, our ultimate goal is not the mere generation of tables or graphs; instead, we want to use tables and graphs as keys that unlock the hidden and important characteristics of data. In this chapter we are mainly concerned with the *distribution* of a data set, which is one of the following five characteristics that are typically most important. This chapter focuses mainly on the distribution of data. Chapter 3 presents methods for investigating the other characteristics.

Characteristics of Data

1. **Center:** A representative value that indicates where the middle of the data set is located.
2. **Variation:** A measure of the amount that the data values vary.
3. **Distribution:** The nature or shape of the spread of the data over the range of values (such as bell-shaped).
4. **Outliers:** Sample values that lie very far away from the vast majority of the other sample values.
5. **Time:** Any change in the characteristics of the data over time.

Study Hint: Blind memorization is not effective in remembering information. To remember the above characteristics of data, it may be helpful to use a memory device (or mnemonic) for the first five letters **CVDOT.** Remembering the sentence "**C**omputer **V**iruses **D**estroy **O**r **T**erminate" is an easy way to help us remember the five key characteristics of data.

Critical Thinking and Interpretation: Going Beyond Formulas and Manual Calculations

In the modern statistics course, it is not so important to memorize formulas or manually perform complex arithmetic calculations. Instead, we get results by using technology (a calculator or computer software), and then we focus on making practical sense of results through critical thinking. This chapter includes detailed steps for important procedures, but it is not necessary to master those steps in all cases. However, we recommend that in each case you perform a few manual calculations before using technology. This will enhance your understanding and help you acquire a better appreciation of the results obtained from the technology.

2-2 Frequency Distributions

Key Concept When one is working with large data sets, a *frequency distribution* (or *frequency table*) is often helpful in organizing and summarizing data. A frequency distribution helps us to understand the nature of the *distribution* of a data set.

DEFINITION A **frequency distribution** (or **frequency table**) shows how data are partitioned among several categories (or *classes*) by listing the categories along with the number (frequency) of data values in each of them.

Consider the IQ scores of the low lead group listed in Table 2-1. Table 2-2 is a frequency distribution summarizing those IQ scores. The **frequency** for a particular class is the number of original values that fall into that class. For example, the first class in Table 2-2 has a frequency of 2, so 2 of the IQ scores are between 50 and 69 inclusive.

The following standard terms are sometimes used in constructing frequency distributions and graphs.

Table 2-2
IQ Scores of Low Lead Group

IQ Score	Frequency
50–69	2
70–89	33
90–109	35
110–129	7
130–149	1

Note to Instructor

Recommendation: Cover Section 2-2 (Frequency Distributions) and Section 2-3 (Histograms) in one class. Do not use too much valuable class time describing the details of manually constructing a frequency distribution and histogram. Instead, demonstrate with a class activity, such as this: Ask each student to record his or her pulse rate by counting the number of heart beats in one minute. Collect the anonymous counts, and show how to construct a frequency distribution. After constructing a frequency distribution, identify the class limits, class midpoints, class boundaries, and class width. Then construct the corresponding histogram, and show how the graph is much easier to understand than the table of numbers.

Students typically have difficulty with three items: (1) definition of class width (believing incorrectly that it's the difference between the upper and lower *limits* of a class), (2) identifying class boundaries, and (3) getting a lower class boundary that is negative when negative numbers don't make sense in the context of the data.

DEFINITIONS

Lower class limits are the smallest numbers that can belong to the different classes. (Table 2-2 has lower class limits of 50, 70, 90, 110, and 130.)

Upper class limits are the largest numbers that can belong to the different classes. (Table 2-2 has upper class limits of 69, 89, 109, 129, and 149.)

Class boundaries are the numbers used to separate the classes, but without the gaps created by class limits. Figure 2-1 shows the gaps created by the class limits from Table 2-2. In Figure 2-1 we see that the values of 69.5, 89.5, 109.5, and 129.5 are in the centers of those gaps, and following the pattern of those class boundaries, we see that the lowest class boundary is 49.5 and the highest class boundary is 149.5. Thus the complete list of class boundaries is 49.5, 69.5, 89.5, 109.5, 129.5, and 149.5.

Class midpoints are the values in the middle of the classes. Table 2-2 has class midpoints of 59.5, 79.5, 99.5, 119.5, and 139.5. Each class midpoint is computed by adding the lower class limit to the upper class limit and dividing the sum by 2.

Class width is the difference between two consecutive lower class limits (or two consecutive lower class boundaries) in a frequency distribution. Table 2-2 uses a class width of 20.

CAUTION Finding the correct class width and class boundaries can be tricky. For class width, don't make the most common mistake of using the difference between a lower class limit and an upper class limit. See Table 2-2 and note that the class width is 20, not 19.

For class boundaries, remember that they split the difference between the end of one class and the beginning of the next class, as shown in Figure 2-1.

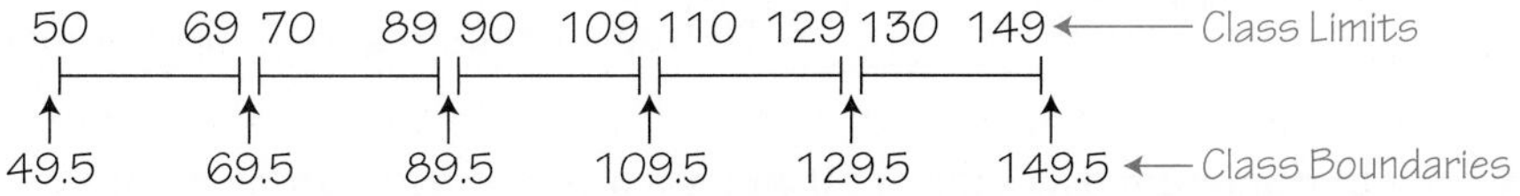

Figure 2-1 Finding Class Boundaries from Class Limits in Table 2-2

Procedure for Constructing a Frequency Distribution

We construct frequency distributions (1) so that we can summarize large data sets, (2) so that we can analyze the data to see the distribution and identify outliers, and (3) so that we have a basis for constructing graphs (such as *histograms,* introduced in the next section). Although technology can generate frequency distributions, the steps for manually constructing them are as follows:

1. Select the number of classes, usually between 5 and 20. The number of classes might be affected by the convenience of using round numbers.

2. Calculate the class width.

$$\text{Class width} \approx \frac{(\text{maximum data value}) - (\text{minimum data value})}{\text{number of classes}}$$

Round this result to get a convenient number. (It's usually best to round *up*.) Using a specific number of classes is not too important, and it's usually wise to change the number of classes so that they use convenient values for the class limits.

3. Choose the value for the first lower class limit by using either the minimum value or a convenient value below the minimum.
4. Using the first lower class limit and the class width, list the other lower class limits. (Add the class width to the first lower class limit to get the second lower class limit. Add the class width to the second lower class limit to get the third lower class limit, and so on.)
5. List the lower class limits in a vertical column and then determine and enter the upper class limits.
6. Take each individual data value and put a tally mark in the appropriate class. Add the tally marks to find the total frequency for each class.

When constructing a frequency distribution, be sure the classes do not overlap. Each of the original values must belong to exactly one class. Include all classes, even those with a frequency of zero. Try to use the same width for all classes, although it is sometimes impossible to avoid open-ended intervals, such as "65 years or older."

Authors Identified

In 1787–88 Alexander Hamilton, John Jay, and James Madison anonymously published the famous *Federalist Papers* in an attempt to convince New Yorkers that they should ratify the Constitution. The identity of most of the papers' authors became known, but the authorship of 12 of the papers was contested. Through statistical analysis of the frequencies of various words, we can now conclude that James Madison is *likely* to have been the author of these 12 papers. For many of the disputed papers, the evidence in favor of Madison's authorship is overwhelming to the degree that we can be almost certain of being correct.

Example 1 IQ Scores of Low Lead Group

Using the IQ scores of the low lead group in Table 2-1, follow the above procedure to construct the frequency distribution shown in Table 2-2. Use five classes.

Solution

Step 1: Select 5 as the number of desired classes.

Step 2: Calculate the class width. Note that we round 18.2 up to 20, which is a much more convenient number.

$$\begin{aligned}\text{Class width} &\approx \frac{(\text{maximum data value}) - (\text{minimum data value})}{\text{number of classes}}\\ &= \frac{141 - 50}{5} = 18.2 \approx 20 \text{ (rounded up to a convenient number)}\end{aligned}$$

Step 3: The minimum data value is 50 and it is a convenient starting point, so use 50 as the first lower class limit. (If the minimum value had been 52 or 53, we would have rounded down to the more convenient starting point of 50.)

Step 4: Add the class width of 20 to 50 to get the second lower class limit of 70. Continue to add the class width of 20 until we have five lower class limits. The lower class limits are therefore 50, 70, 90, 110, and 130.

Step 5: List the lower class limits vertically as shown in the margin. From this list, we identify the corresponding upper class limits as 69, 89, 109, 129, and 149.

50–
70–
90–
110–
130–

Step 6: Enter a tally mark for each data value in the appropriate class. Then add the tally marks to find the frequencies shown in Table 2-2.

So far we have discussed frequency distributions using only quantitative data sets, but frequency distributions can also be used to summarize categorical (or qualitative or attribute) data, as illustrated in Example 2.

Example 2 **East Haven Police Department Traffic Tickets**

Table 2-3 summarizes the race/ethnic classifications recorded on traffic tickets issued by Connecticut's East Haven Police Department during a recent nine-month period. Here is an interesting and revealing fact about the data: Table 2-3 shows that 18 of those given tickets were classified by police as being Hispanic, but in fact, 209 of those given tickets had Hispanic names!

Table 2-3 East Haven Traffic Tickets

Race	Frequency
White	329
Black	15
Asian	0
Hispanic	18
White/Hispanic	4
Blank (no indication)	5

Relative Frequency Distribution

A variation of the basic frequency distribution is a **relative frequency distribution** or **percentage frequency distribution,** in which each class frequency is replaced by a relative frequency (or proportion) or a percentage. In this text we use the term "relative frequency distribution" whether we use relative frequencies or percentages. Relative frequencies and percentages are calculated as follows:

$$\text{Relative frequency for a class} = \frac{\text{frequency for a class}}{\text{sum of all frequencies}}$$

$$\text{Percentage for a class} = \frac{\text{frequency for a class}}{\text{sum of all frequencies}} \times 100\%$$

Table 2-4 is an example of a relative frequency distribution. It is a variation of Table 2-2 in which each class frequency is replaced by the corresponding percentage

Table 2-4 Relative Frequency Distribution of IQ Scores of Low Lead Group

IQ Score	Frequency
50–69	2.6%
70–89	42.3%
90–109	44.9%
110–129	9.0%
130–149	1.3%

Growth Charts Updated

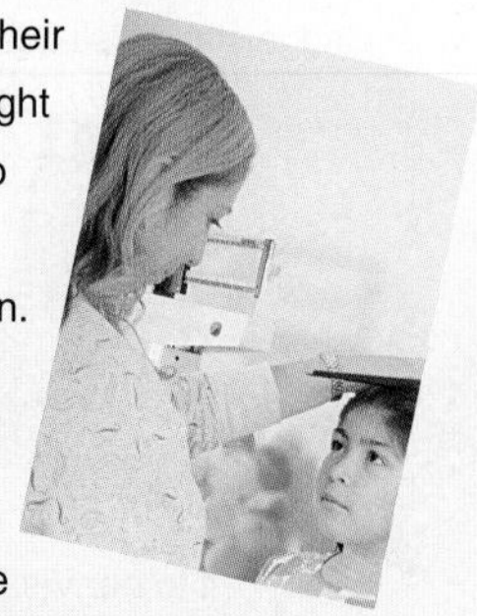

Pediatricians typically use standardized growth charts to compare their patient's weight and height to a sample of other children. Children are considered to be in the normal range if their weight and height fall between the 5th and 95th percentiles. If they fall outside of that range, they are often given tests to ensure that there are no serious medical problems. Pediatricians became increasingly aware of a major problem with the charts: Because they were based on children living between 1929 and 1975, the growth charts had become inaccurate. To rectify this problem, the charts were updated in 2000 to reflect the current measurements of millions of children. The weights and heights of children are good examples of populations that change over time. This is the reason for including changing characteristics of data over time as an important consideration for a population.

value. Because there are 78 data values, divide each class frequency by 78, and then multiply by 100%. The first class of Table 2-2 has a frequency of 2, so divide 2 by 78 to get 0.0256, and then multiply by 100% to get 2.56%, which we rounded to 2.6%. The sum of the percentages should be 100%, with a small discrepancy allowed for rounding errors, so a sum such as 99% or 101% is acceptable. The sum of the percentages in Table 2-4 is 100.1%.

The sum of the percentages in a relative frequency distribution must be very close to 100%.

Cumulative Frequency Distribution

Another variation of a frequency distribution is a **cumulative frequency distribution** in which the frequency for each class is the sum of the frequencies for that class and all previous classes. Table 2-5 is a cumulative frequency distribution based on Table 2-2. Using the original frequencies of 2, 33, 35, 7, and 1, we add 2 + 33 to get the second cumulative frequency of 35, then we add 2 + 33 + 35 to get the third, and so on. See Table 2-5, and note that in addition to the use of cumulative frequencies, the class limits are replaced by "less than" expressions that describe the new ranges of values.

Table 2-5 Cumulative Frequency Distribution of IQ Scores of Low Lead Group

IQ Score	Cumulative Frequency
Less than 70	2
Less than 90	35
Less than 110	70
Less than 130	77
Less than 150	78

Critical Thinking: Using Frequency Distributions to Understand Data

Earlier, we noted that a frequency distribution can help us understand the *distribution* of a data set, which is the nature or shape of the spread of the data over the range of values (such as bell-shaped). In statistics we are often interested in determining whether the data have a *normal distribution.* (Normal distributions are discussed extensively in Chapter 6.) Data that have an approximately normal distribution are characterized by a frequency distribution with the following features:

Normal Distribution

1. The frequencies start low, then increase to one or two high frequencies, and then decrease to a low frequency.
2. The distribution is approximately symmetric, with frequencies preceding the maximum being roughly a mirror image of those that follow the maximum.

Table 2-6 satisfies these two conditions. The frequencies start low, increase to the maximum of 56, then decrease to a low frequency. Also, the frequencies of 1 and 10 that precede the maximum are a mirror image of the frequencies 10 and 1 that follow the maximum. Real data sets are usually not so perfect as Table 2-6, and judgment must be used to determine whether the distribution comes "close enough" to satisfying those two conditions.

Outlier Ticket Price

The cost of a Metro-North train ride from Grand Central Station in New York City to the Goldens Bridge station is usually $10.

Johns Hopkins student Lydia Alcock purchased a ticket and charged it to her Visa credit card. She was surprised when she later received her Visa bill and saw that the train ride had a price of $23,148,855,308,184,500, or $23 quadrillion. She was also charged another $20 for exceeding her account limit.

The $23 quadrillion train ticket price is an outlier because it is substantially higher than all other train ticket prices. In this case, the amount is clearly an error. If we were analyzing a sample of train ticket prices and the $23 quadrillion price was included as a sample value, we should delete that value because it is an obvious error. Another outlier might be very far away from the vast majority of other sample values, but we should not necessarily exclude it unless we are sure that it is an error. If we think that the outlier might be a valid sample value, it would be wise to analyze the sample data with the outlier included, and then again with the outlier excluded, so that we can tell what effect the outlier has on the results.

Note to Instructor

Construction of a frequency distribution is not the ultimate goal. The frequency distribution should be used to better *understand* the data. Explain how Examples 3, 4, and 5 illustrate the role of frequency distributions in describing, exploring, and comparing data sets.

Table 2-6 Frequency Distribution Showing a Normal Distribution

Score	Frequency	Normal Distribution
50–69	1	← Frequencies start low, . . .
70–89	10	
90–109	56	← Increase to a maximum, . . .
110–129	10	
130–149	1	← Decrease to become low again.

The following examples illustrate how frequency distributions are used to describe, explore, and compare data sets.

Example 3 Describing Data: How Were the Weights Obtained in California?

When collecting weights of people, it's better to actually weigh people than to ask them what they weigh. People often tend to round *way* down, so that a weight of 196 lb might be reported as 170 lb. Table 2-7 summarizes the *last digits* of the weights of 100 people used in the California Health Interview Survey. If people are actually weighed on a scale, the last digits of weights tend to have frequencies that are approximately the same, but Table 2-6 shows that the vast majority of weights have last digits of 0 or 5, and this is strong evidence that people reported their weights and were not physically weighed. (Also, the word "interview" in the title of the California Health Interview Survey reveals that people were interviewed and were not physically measured.)

Table 2-7 Last Digits of Weights from the California Health Interview Survey

Last Digit of Weight	Frequency
0	46
1	1
2	2
3	3
4	3
5	30
6	4
7	0
8	8
9	3

Example 4 Exploring Data: What Does a Gap Tell Us?

Table 2-8 is a frequency distribution of the weights (grams) of randomly selected pennies. Examination of the frequencies reveals a large *gap* between the lightest pennies and the heaviest pennies. This suggests that we have two different populations: Pennies made before 1983 are 95% copper and 5% zinc, but pennies made

Missing Data

Samples are commonly missing some data. Missing data fall into two general categories: (1) missing values that result from random causes unrelated to the data values, and (2) missing values resulting from causes that are not random. Random causes include factors such as the incorrect entry of sample values or lost survey results. Such missing values can often be ignored because they do not systematically hide some characteristic that might significantly affect results. It's trickier to deal with values missing because of factors that are not random. For example, results of an income analysis might be seriously flawed if people with very high incomes refuse to provide those values because they fear income tax audits. Those missing high incomes should not be ignored, and further research would be needed to identify them.

after 1983 are 2.5% copper and 97.5% zinc, which explains the large gap between the lightest pennies and the heaviest pennies in Table 2-8.

Table 2-8 Randomly Selected Pennies

Weight (grams) of Penny	Frequency
2.40–2.49	18
2.50–2.59	19
2.60–2.69	0
2.70–2.79	0
2.80–2.89	0
2.90–2.99	2
3.00–3.09	25
3.10–3.19	8

Gaps Example 4 illustrates this principle:

The presence of gaps can suggest that the data are from two or more different populations.

The converse of this principle is not true, because data from different populations do not necessarily result in gaps.

Example 5 Comparing IQ Scores of the Low Lead Group and the High Lead Group

Table 2-1, which is given with the Chapter Problem at the beginning of this chapter, lists IQ scores from the low lead group and the high lead group. Because the sample sizes of 78 and 21 are so different, a comparison of frequency distributions is not easy, but Table 2-9 shows the relative frequency distributions for those two groups. By comparing those relative frequencies, we see that the majority of children in the low lead group had IQ scores of 90 or higher, but the majority of children in the high lead group had IQ scores below 90. This suggests that perhaps high lead exposure has a detrimental effect on IQ scores.

Table 2-9 IQ Scores from the Low Lead Group and the High Lead Group

IQ Score	Low Lead Group	High Lead Group
50–69	2.6%	
70–89	42.3%	71.4%
90–109	44.9%	28.6%
110–129	9.0%	
130–149	1.3%	

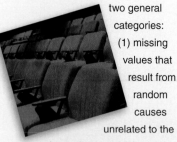

2-2 Basic Skills and Concepts

Statistical Literacy and Critical Thinking

1. Frequency Distribution Table 2-2 on page 45 is a frequency distribution summarizing the IQ scores of a group of children with low lead exposure. Is it possible to identify the original list of the 78 individual IQ scores from Table 2-2? Why or why not?

2. Relative Frequency Distribution After construction of a relative frequency distribution summarizing the times that males spend each day thinking about females, what should be the sum of the relative frequencies?

3. Do You Believe? In a Harris Interactive survey, 2303 adults were asked whether they believe in five different things, and the accompanying table summarizes the results. Does this table describe a relative frequency distribution? Why or why not?

4. Analyzing a Frequency Distribution The accompanying frequency distribution summarizes the heights of a sample of people in Vassar Road Elementary School. What can you conclude about the people included in the sample?

In Exercises 5–10, identify the class width, class midpoints, and class boundaries for the given frequency distribution. The frequency distributions are based on real data from Appendix B.

5.

Age (years) of Best Actress When Oscar Was Won	Frequency
20–29	27
30–39	34
40–49	13
50–59	2
60–69	4
70–79	1
80–89	1

6.

Age (years) of Best Actor When Oscar Was Won	Frequency
20–29	1
30–39	26
40–49	35
50–59	13
60–69	6
70–79	1

7.

Verbal IQ Score of Subject Exposed to Lead	Frequency
50–59	4
60–69	10
70–79	25
80–89	43
90–99	26
100–109	8
110–119	3
120–129	2

8.

Years President Lived after First Inauguration	Frequency
0–4	8
5–9	2
10–14	5
15–19	7
20–24	4
25–29	6
30–34	0
35–39	1

Recommended Assignment
Exercises 1–6, 11–14, 20, 22, 24.
Due to limited space, answers for odd-numbered exercises are not included here, but they are included in Appendix D.

Table for Exercise 3

Believe in the devil	60%
Believe in hell	61%
Believe in UFOs	32%
Believe in astrology	26%
Believe in reincarnation	20%

Table for Exercise 4

Height (in.)	Frequency
35–39	6
40–44	31
45–49	67
50–54	21
55–59	0
60–64	0
65–69	6
70–74	10

2. If percentages are used, the sum should be 100%. If proportions are used, the sum should be 1.

4. The gap in the frequencies suggests that the table includes heights of two different populations: students and faculty/staff.

6. Class width: 10. Class midpoints: 24.5, 34.5, 44.5, 54.5, 64.5, 74.5. Class boundaries: 19.5, 29.5, 39.5, 49.5, 59.5, 69.5, 79.5.

8. Class width: 5. Class midpoints: 2, 7, 12, 17, 22, 27, 32, 37. Class boundaries: −0.5, 4.5, 9.5, 14.5, 19.5, 24.5, 29.5, 34.5, 39.5.

10. Class width: 2.0. Class midpoints: 3.95, 5.95, 7.95, 9.95, 11.95, 13.95. Class boundaries: 2.95, 4.95, 6.95, 8.95, 10.95, 12.95, 14.95.

12. Yes. The frequencies start low, increase to the maximum frequency of 43, and then decrease. Also, the frequencies are approximately symmetric about the maximum frequency of 43.

14. 12, 12, 6, 2

16. The differences are not substantial. Based on the given data, males and females appear to have about the same distribution of white blood cell counts.

White Blood Cell Count	Relative Frequency (Males)	Relative Frequency (Females)
3.0–4.9	20.0%	15.0%
5.0–6.9	37.5%	40.0%
7.0–8.9	27.5%	22.5%
9.0–10.9	12.5%	17.5%
11.0–12.9	2.5%	0.0%
13.0–14.9	0.0%	5.0%

18.

Age (years) of Best Actor When Oscar Was Won	Cumulative Frequency
Less than 30	1
Less than 40	27
Less than 50	62
Less than 60	75
Less than 70	81
Less than 80	82

20. Because there are disproportionately more 0s and 5s, it appears that the weights were reported instead of measured. Consequently, it is likely that the results are not very accurate.

Last Digit	Frequency
0	26
1	1
2	1
3	2
4	2
5	12
6	1
7	0
8	4
9	1

9.

White Blood Cell Count of Males	Frequency
3.0–4.9	8
5.0–6.9	15
7.0–8.9	11
9.0–10.9	5
11.0–12.9	1

10.

White Blood Cell Count of Females	Frequency
3.0–4.9	6
5.0–6.9	16
7.0–8.9	9
9.0–10.9	7
11.0–12.9	0
13.0–14.9	2

Normal Distributions. ***In Exercises 11–14, answer the given questions which are related to normal distributions.***

11. Identifying the Distribution Using a reasonably strict interpretation of the relevant criteria, does the frequency distribution given in Exercise 5 appear to have a normal distribution? Explain.

12. Identifying the Distribution Does the frequency distribution given in Exercise 7 appear to have a normal distribution? Explain.

13. Normal Distribution Refer to the frequency distribution given in Exercise 9 and ignore the given frequencies. Assume that the first two frequencies are 4 and 7, respectively. Assuming that the distribution of the 40 sample values is a normal distribution, identify the remaining three frequencies.

14. Normal Distribution Refer to the frequency distribution given in Exercise 10 and ignore the given frequencies. Assume that the first two frequencies are 2 and 6, respectively. Assuming that the distribution of the 40 sample values is a normal distribution, identify the remaining four frequencies.

Relative Frequencies for Comparisons. ***In Exercises 15 and 16, construct the relative frequencies and answer the given questions.***

15. Oscar Winners Construct one table (similar to Table 2-9 on page 50) that includes relative frequencies based on the frequency distributions from Exercises 5 and 6, and then compare the ages of Oscar-winning actresses and actors. Are there notable differences?

16. White Blood Cell Counts Construct one table (similar to Table 2-9 on page 50) that includes relative frequencies based on the frequency distributions from Exercises 9 and 10, and then compare the white blood cell counts of females and males. Are there notable differences?

Cumulative Frequency Distributions. ***In Exercises 17 and 18, construct the cumulative frequency distribution that corresponds to the frequency distribution in the exercise indicated.***

17. Exercise 5

18. Exercise 6

Constructing Frequency Distributions. ***In Exercises 19–28, use the indicated data and construct the frequency distribution.***

19. Analysis of Last Digits Heights of statistics students were obtained by the author as part of an experiment conducted for class. The last digits of those heights are listed below. Construct a frequency distribution with 10 classes. Based on the distribution, do the heights appear to be reported or actually measured? What do you know about the accuracy of the results?

0 0 0 0 0 0 0 0 0 1 1 2 3 3 3 4 5 5 5 5 5 5 5 5 5 5 5 5 5 5 5 5 6 6 8 8 8 9

20. Analysis of Last Digits Weights of respondents were recorded as part of the California Health Interview Survey. The last digits of weights from 50 randomly selected respondents are listed below. Construct a frequency distribution with 10 classes. Based on the distribution, do the weights appear to be reported or actually measured? What do you know about the accuracy of the results?

5 0 1 0 2 0 5 0 5 0 3 8 5 0 5 0 5 6 0 0 0 0 0 0 8
5 5 0 4 5 0 0 4 0 0 0 0 0 8 0 9 5 3 0 5 0 0 0 5 8

21. Pulse Rates of Males Refer to Data Set 1 in Appendix B and use the pulse rates (beats per minute) of males. Begin with a lower class limit of 40 and use a class width of 10. Do the pulse rates of males appear to have a normal distribution?

22. Pulse Rates of Females Refer to Data Set 1 in Appendix B and use the pulse rates (beats per minute) of females. Begin with a lower class limit of 50 and use a class width of 10. Compare the frequency distribution to the one found in the preceding exercise. Is there a notable difference between pulse rates of males and females?

23. Earthquake Magnitudes Refer to the earthquake magnitudes listed in Data Set 16 of Appendix B. Begin with a lower class limit of 0.00 and use a class width of 0.50. Using a very strict interpretation of the requirements for a normal distribution, do the magnitudes appear to be normally distributed?

24. Earthquake Depths Refer to the earthquake depths listed in Data Set 16 in Appendix B. Begin with a lower class limit of 1.00 km and use a class width of 4.00 km. Using a very strict interpretation of the requirements for a normal distribution, do the depths appear to be normally distributed?

25. Male Red Blood Cell Counts Refer to Data Set 1 in Appendix B and use the red blood cell counts (million cells/μL) for males. Begin with a lower class limit of 4.00 and use a class width of 0.40. Using a very loose interpretation of the requirements for a normal distribution, do the red blood cell counts appear to be normally distributed?

26. Female Red Blood Cell Counts Refer to Data Set 1 in Appendix B and use the red blood cell counts (million cells/μL) for females. Begin with a lower class limit of 3.60 and use a class width of 0.40. Using a very loose interpretation of the requirements for a normal distribution, do the red blood cell counts appear to be normally distributed?

27. Flight Arrival Times Refer to Data Set 15 in Appendix B and use the times of the arrival delays. Begin with a lower class limit of –60 min and use a class width of 30 min. Based on the result, does it appear that most of the American Airline flights from JFK to LAX are close to arriving in Los Angeles without too much delay?

28. Flight Taxi-Out Times Refer to Data Set 15 in Appendix B and use the times required to taxi out for takeoff. Begin with a lower class limit of 10 min and use a class width of 5 min. Based on the result, does it appear that the time required to taxi out can be predicted with reasonable accuracy?

Categorical Data. ***In Exercises 29–32, use the given categorical data to construct the relative frequency distribution.***

29. Titanic Survivors The 2223 people aboard the *Titanic* include 361 male survivors, 1395 males who died, 345 female survivors, and 122 females who died.

30. Train Derailments An analysis of 50 train derailment incidents identified the main causes listed below, where T denotes bad track, E denotes faulty equipment, H denotes

22. Yes. The pulse rates of males appear to be generally lower than the pulse rates of females.

Pulse Rate (Female)	Frequency
50–59	1
60–69	8
70–79	18
80–89	5
90–99	6
100–109	2

24. No, the distribution does not appear to be a normal distribution.

Depth (km)	Frequency
1.00–4.99	7
5.00–8.99	21
9.00–12.99	4
13.00–16.99	12
17.00–20.99	6

26. Yes, the distribution appears to be roughly a normal distribution.

Red Blood Cell Count	Frequency
3.60–3.99	2
4.00–4.39	13
4.40–4.79	15
4.80–5.19	7
5.20–5.59	2
5.60–5.99	1

28. No. The times vary from a low of 12 min to a high of 49 min. It appears that many flights taxi out quickly, but many other flights require much longer times, so it would be difficult to predict the taxi-out time with reasonable accuracy.

Taxi-Out Time (min)	Frequency
10–14	10
15–19	20
20–24	9
25–29	1
30–34	2
35–39	2
40–44	2
45–49	2

30.

Cause	Relative Frequency
Bad Track	46%
Faulty Equipment	18%
Human Error	24%
Other	12%

human error, and O denotes other causes (based on data from the Federal Railroad Administration).

T T T E E H H H H H O O H H H E E T T T E T H O T

T T T T T T H T T H E E T T E E T T T H T T O O O

31. Fatal Plane Crashes Among fatal plane crashes that occurred during the past 60 years, 650 were due to pilot error, 78 were due to other human error, 156 were due to weather, 286 were due to mechanical problems, and 117 were due to sabotage. (These results are based on data from PlaneCrashInfo.com, and they do not include private aircraft, military aircraft, aircraft carrying 10 or fewer people, or helicopters.) What is the most serious threat to aviation safety, and can anything be done about it?

32. California Lottery The digits drawn in one month for the California Daily 4 lottery were recorded. The digits 0 through 9 had these frequencies: 20, 10, 12, 12, 8 11, 9, 10, 9, 19. Do the digits appear to be selected with a process that is functioning correctly? Why are these digits *categorical* data?

2-2 Beyond the Basics

33. Interpreting Effects of Outliers Refer to Data Set 22 in Appendix B for the axial loads of aluminum cans that are 0.0111 in. thick. The load of 504 lb is an *outlier* because it is very far away from all of the other values. Construct a frequency distribution that includes the value of 504 lb, and then construct another frequency distribution with the value of 504 lb excluded. In both cases, start the first class at 200 lb and use a class width of 20 lb. State a generalization about the effect of an outlier on a frequency distribution.

34. Number of Classes According to what is known as Sturges' guideline, the ideal number of classes for a frequency distribution can be approximated by $1 + (\log n)/(\log 2)$ where n is the number of data values. Use this guideline to complete the table in the margin.

2-3 Histograms

Key Concept While a frequency distribution is a useful tool for summarizing data and investigating the distribution of data, an even better tool is a *histogram*, which consists of a graph that is easier to interpret than a table of numbers.

DEFINITION A **histogram** is a graph consisting of bars of equal width drawn adjacent to each other (unless there are gaps in the data). The horizontal scale represents classes of quantitative data values and the vertical scale represents frequencies. The heights of the bars correspond to the frequency values.

A histogram is basically a graph of a frequency distribution. For example, Figure 2-2 shows the Minitab-generated histogram corresponding to the frequency distribution given in Table 2-2 on page 45.

Class frequencies should be used for the vertical scale and that scale should be labeled as in Figure 2-2. The bar locations on the horizontal scale are usually labeled with one of the following: (1) class boundaries (as shown in Figure 2-2), (2) class midpoints, or (3) lower class limits. The first and second options are technically correct, while the third option introduces a small error. It is often easier for us mere mortals to use class midpoints for the horizontal scale. Histograms can usually be generated using technology.

32. The digit 0 appears to have occurred with a higher frequency than expected, but in general the differences are not very substantial, so the selection process appears to be functioning correctly. The digits are qualitative data because they do not represent measures or counts of anything. The digits could be replaced by the first 10 letters of the alphabet, and the lottery would be essentially the same.

Digit	Frequency
0	16.7%
1	8.3%
2	10.0%
3	10.0%
4	6.7%
5	9.2%
6	7.5%
7	8.3%
8	7.5%
9	15.8%

34. 46–90, 91–181, 182–362, 363–724, 725–1448, 1449–2896

Table for Exercise 34

Number of Data Values	Ideal Number of Classes
16–22	5
23–45	6
?	7
?	8
?	9
?	10
?	11
?	12

Note to Instructor

This section introduces *histograms* and *relative frequency histograms.* Other graphs are introduced in Section 2-4. Don't spend too much class time describing the detailed mechanics of constructing those graphs. Emphasize the *interpretation* of the graphs. On a test, don't ask students to construct a histogram from a large list of data. Instead, provide some histograms and ask questions about them.

New to this section: Using normal quantile plots to determine whether sample data appear to be from a population with a normal distribution.

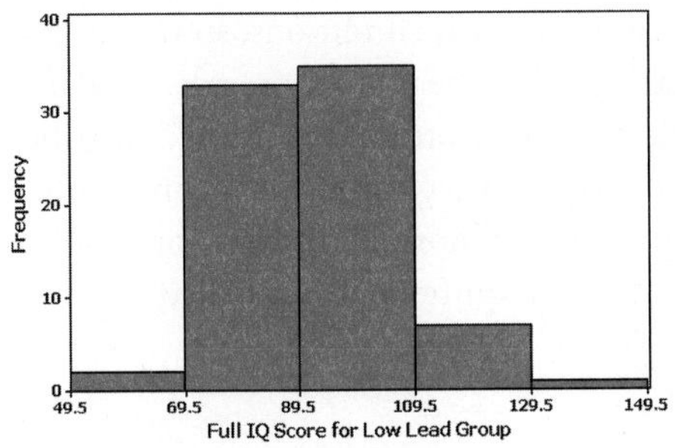

Figure 2-2 Histogram

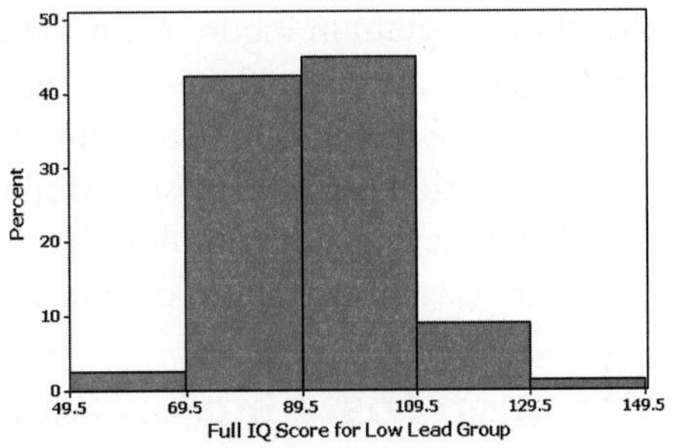

Figure 2-3 Relative Frequency Histogram

Relative Frequency Histogram

A **relative frequency histogram** has the same shape and horizontal scale as a histogram, but the vertical scale uses relative frequencies (as percentages or proportions) instead of actual frequencies. Figure 2-3 is the relative frequency histogram corresponding to Figure 2-2.

Critical Thinking: Interpreting Histograms

Even though creating histograms is more fun than human beings should be allowed to have, the ultimate objective is not creating a histogram, but rather *understanding* something about the data. Analyze the histogram to see what can be learned about "CVDOT": the center of the data, the variation (which will be discussed at length in Section 3-3), the distribution, and whether there are any outliers (values far away from the other values). Examining Figure 2-2, we see that the histogram is centered close to 90, the values vary from around 50 to 150, and the distribution is roughly bell-shaped.

Normal Distribution

When graphed as a histogram, a normal distribution has a "bell" shape similar to the one superimposed in Figure 2-4. In a normal distribution, (1) the frequencies increase to a maximum and then decrease, and (2) the graph has symmetry, with the left half of the histogram being roughly a mirror image of the right half. Figure 2-4 shows

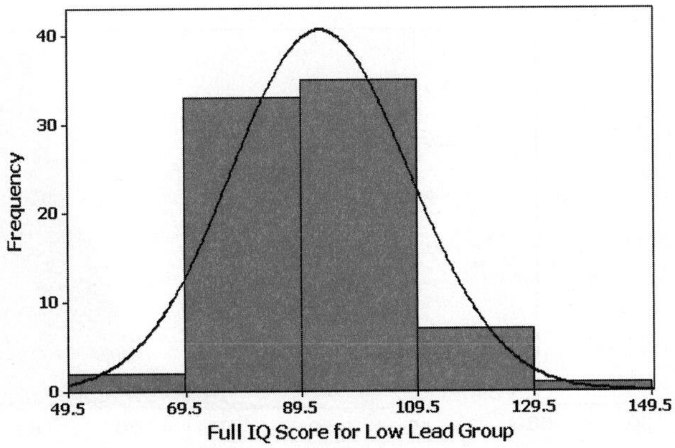

Figure 2-4 Bell-Shaped Distribution
Because this histogram is roughly bell-shaped, we say that the data have a *normal distribution*.

that the histogram in Figure 2-2 roughly satisfies those two conditions, so we say that the IQ scores are approximately normally distributed. (There are more advanced and less subjective methods for determining whether the distribution is a normal distribution; see Section 6-6.) Many statistical methods require that sample data come from a population having a distribution that is approximately a normal distribution, and we can often use a histogram to determine whether this requirement is satisfied.

Common Distribution Shapes

The histograms shown in Figure 2-5 depict four common distribution shapes. We have already discussed the characteristics of the normal distribution. With a **uniform distribution,** the different possible values occur with approximately the same frequency, so the heights of the bars in the histogram are approximately uniform, as in Figure 2-5(b). Figure 2-5(b) depicts outcomes of digits from state lotteries.

Skewness

A distribution of data is **skewed** if it is not symmetric and extends more to one side than to the other. Data **skewed to the right** (also called *positively skewed*) have a longer right tail, as in Figure 2-5(c), which depicts annual incomes (in thousands of dollars) of adult Americans. Data **skewed to the left** (also called *negatively skewed*) have a longer left tail, as in Figure 2-5(d). Distributions skewed to the right are more common than those skewed to the left because it's often easier to get exceptionally

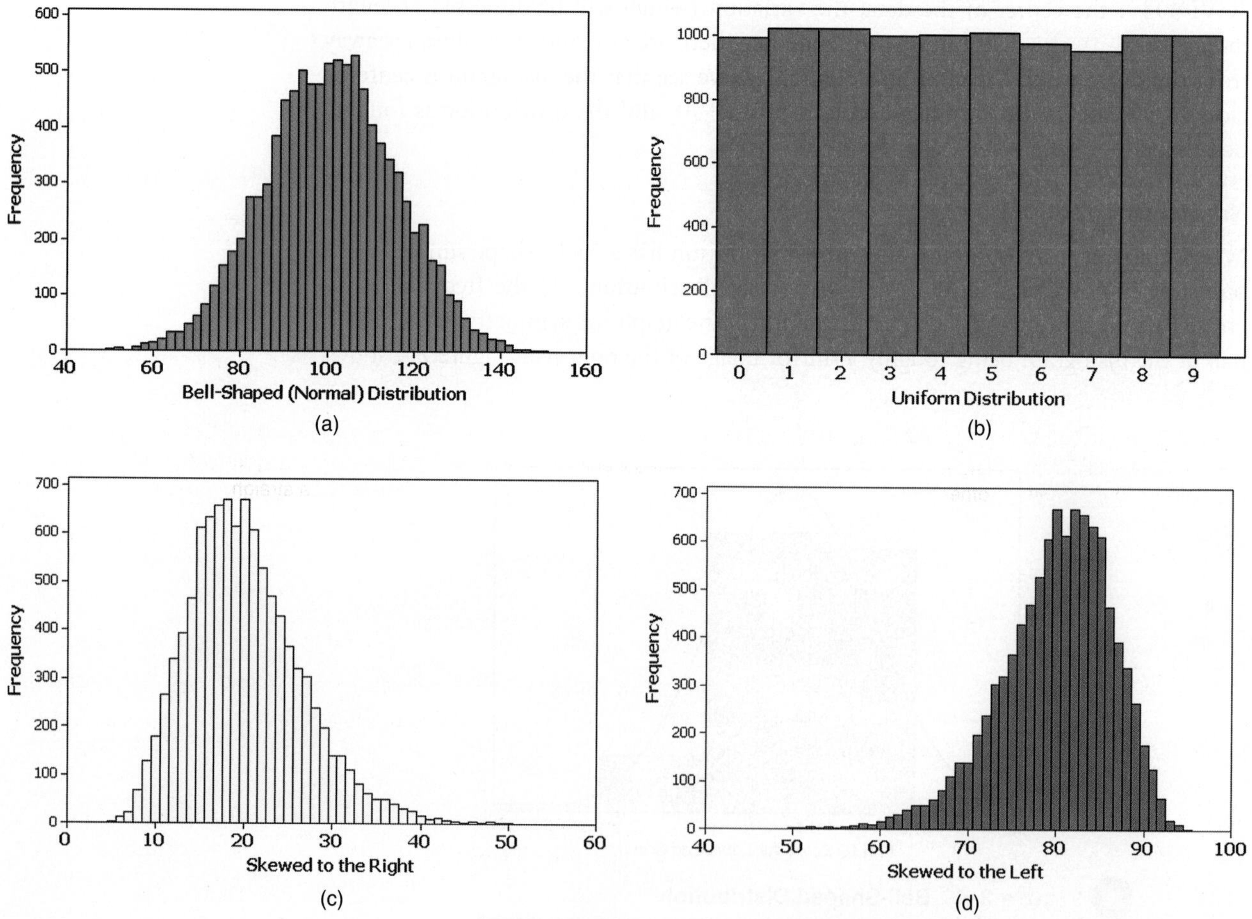

Figure 2-5 Common Distributions

large values than values that are exceptionally small. With annual incomes, for example, it's impossible to get values below zero, but there are a few people who earn millions or billions of dollars in a year. Annual incomes therefore tend to be skewed to the right.

Assessing Normality: Normal Quantile Plot Some really important methods presented in later chapters have a requirement that sample data must be from a population having a normal distribution. We can see that a histogram is often helpful in determining whether the normality requirement is satisfied. However, histograms are not very helpful with small data sets. Section 6-6 discusses methods for *assessing normality*—that is, determining whether the sample data are from a normally distributed population. Section 6-6 includes a procedure for constructing *normal quantile plots,* which involve plotting transformed sample values. Normal quantile plots are easy to generate using technology such as STATDISK, Minitab, XLSTAT, StatCrunch, or a TI-83/84 Plus calculator. Interpretation of a normal quantile plot is based on the following criteria:

Note to Instructor

This section now includes histograms and normal quantile plots as tools for determining whether sample data appear to be from a population with a normal distribution. The procedure for constructing a normal quantile plot is included in Section 6-6, but this section briefly describes the *interpretation* of a normal quantile plot. Consider just mentioning this interpretation now.

Criteria for Assessing Normality with a Normal Quantile Plot

Normal Distribution: The population distribution is normal if the pattern of the points in the normal quantile plot is reasonably close to a straight line, and the points do not show some systematic pattern that is not a straight-line pattern.

Not a Normal Distribution: The population distribution is *not* normal if the normal quantile plot has either or both of these two conditions:

- The points do not lie reasonably close to a straight line.
- The points show some *systematic pattern* that is not a straight-line pattern.

The following are examples of normal quantile plots. Procedures for creating such plots are described in Section 6-6.

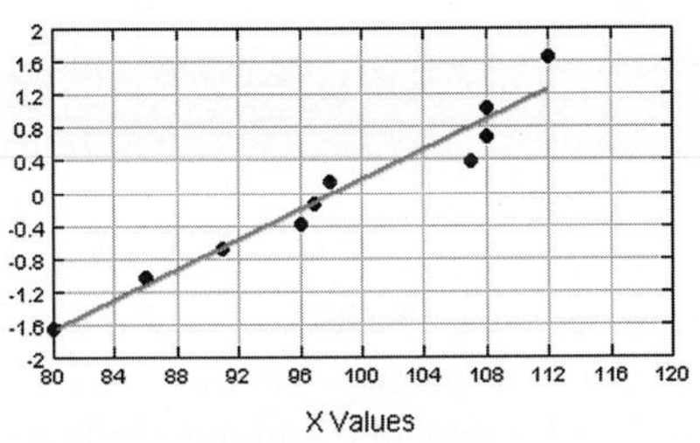

Normal Distribution: The points are reasonably close to a straight-line pattern, and there is no other systematic pattern that is not a straight-line pattern.

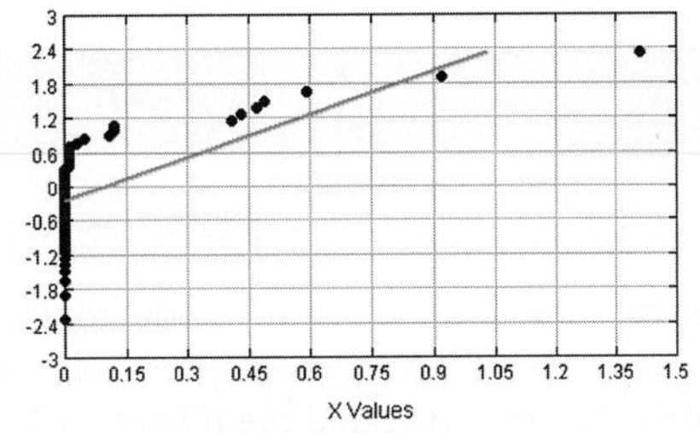

Not a Normal Distribution: The points do not lie reasonably close to a straight line.

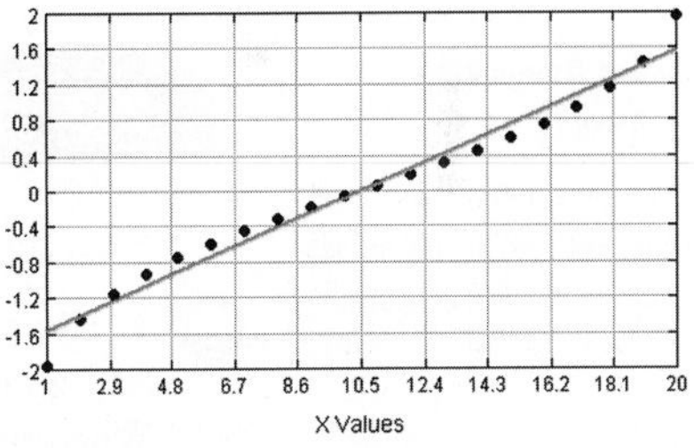

Not a Normal Distribution: The points show a systematic pattern that is not a straight-line pattern.

using TECHNOLOGY

Statistical software packages are effective for generating histograms. Throughout this text, we make frequent reference to STATDISK, Minitab, Excel, the TI-83/84 Plus calculator, and StatCrunch, and all of these technologies can generate histograms. The detailed instructions can vary from easy to complex, so we provide some relevant comments below. For detailed instructions, see the STATDISK, Minitab, Excel, SPSS, SAS, and TI-83/84 Plus manuals that are supplements to this book.

STATDISK Enter or open the data in the STATDISK Data Window, click **Data,** click **Histogram,** and then click on the **Plot** button. To use your own class width and starting point, click on the "User defined" button before clicking on Plot. Click on the **Turn labels on** button to see the frequency for each class. (In addition to generating a histogram, this is also an easy way to identify the entries in a frequency distribution.)

continued

MINITAB **Minitab 15 and earlier versions:** Enter or open the data in a column, click on **Graph,** then **Histogram.** Select the "Simple" histogram. Enter the column in the "Graph variables" window and click **OK. Minitab 16:** Click on **Assistant** and select **Graphical Assistant.** Click on **Histogram,** select the column to be used, and then click **OK.**

Minitab uses default settings for the class width and starting point, but those defaults can be changed as follows: Right-click on the horizontal axis and select **Edit X scale,** then use the **Scale** tab to enter the locations of the tick marks (class boundaries or class midpoints), and use the **Binning** tab to enter the midpoints of the classes. See Figures 2-2, 2-3, and 2-4 for examples of Minitab-generated histograms.

TI-83/84 PLUS Enter a list of data in L1 or use a list of values assigned to a name. Select the **STAT PLOT** function by pressing 2ND Y=. Press ENTER and use the arrow keys to turn Plot1 to "On" and select the graph with bars. The screen display should be as shown here.

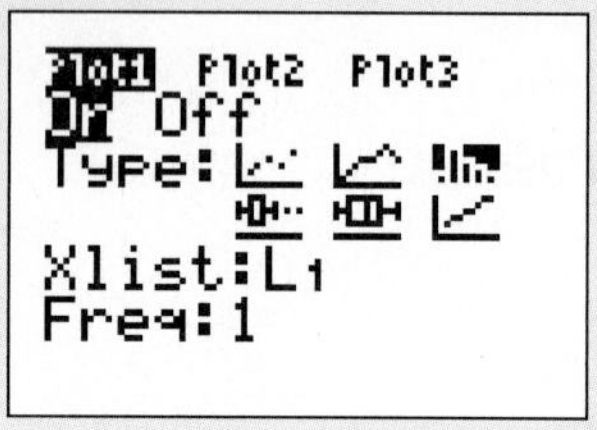

If you want to let the calculator determine the class width and starting point, press ZOOM 9 to get a histogram with default settings. (To enter your own class width and class boundaries, press WINDOW and enter the maximum and minimum values. The Xscl value will be the class width. Press GRAPH to obtain the graph.)

EXCEL Excel can generate histograms, but it is *extremely* difficult. To generate a histogram easily, use XLSTAT. After loading XLSTAT with Excel included, click on **XLSTAT** at the top. Select **Visualizing Data,** then select **Histograms.** In the Data box, enter the range of cells containing the data, such as A1:A78 for 78 values in column A. Click on the "Sample labels" box only if the first cell contains the name of your data. You can click **OK** to get a histogram with default settings, or you can click on the **Options** tab. One of the options is "User defined," which allows you to enter the range of cells containing your desired class boundaries that you must enter in another column.

STATCRUNCH Click on **Open StatCrunch,** then enter or open a data set. Click on **Graphics,** then click on **Histogram.** Select the column containing the data. Click on **Next** to enter a desired starting point and class width. Click **Create Graph.**

Recommended Assignment

Exercises 4–8 and 10, 12, 14.

Due to limited space, answers for some odd-numbered exercises are not included here, but they are included in Appendix D.

2-3 Basic Skills and Concepts

Statistical Literacy and Critical Thinking

1. Histogram Table 2-2 is a frequency distribution summarizing the IQ scores of the low lead group listed in Table 2-1, and Figure 2-2 is a histogram depicting that same data set. When trying to better understand the IQ data, what is the advantage of examining the histogram instead of the frequency distribution?

2. Voluntary Response Sample The histogram in Figure 2-2 on page 55 is constructed from a *simple random sample* of children. If you construct a histogram with data collected from a *voluntary response sample,* will the distribution depicted in the histogram reflect the true distribution of the population? Why or why not?

2. Not necessarily. Because those with special interests are more likely to respond, the voluntary response sample is likely to consist of a group having characteristics that are fundamentally different from those of the population.

3. Small Data NASA provides these duration times (in minutes) of all flights of the space shuttle *Challenger:* 7224, 8784, 8709, 11,476, 10,060, 11,844, 10,089, 11,445, 10,125, 1. Why does it not make sense to construct a histogram for this data set? What is notable about this data set?

4. Normal Distribution When it refers to a normal distribution, does the term "normal" have the same meaning as in ordinary language? What criterion can be used to determine whether the data depicted in a histogram have a distribution that is approximately a normal distribution? Is this criterion totally objective, or does it involve subjective judgment?

4. When referring to a normal distribution, the term *normal* has a meaning that is different from its meaning in ordinary language. A normal distribution is characterized by a histogram that is approximately bell-shaped. Determination of whether a histogram is approximately bell-shaped does require subjective judgment.

Interpreting a Histogram. *In Exercises 5–8, answer the questions by referring to the following Minitab-generated histogram, which represents the heights (inches) of people randomly selected from those who entered New York City's Museum of Natural History during a recent Friday morning.*

MINITAB

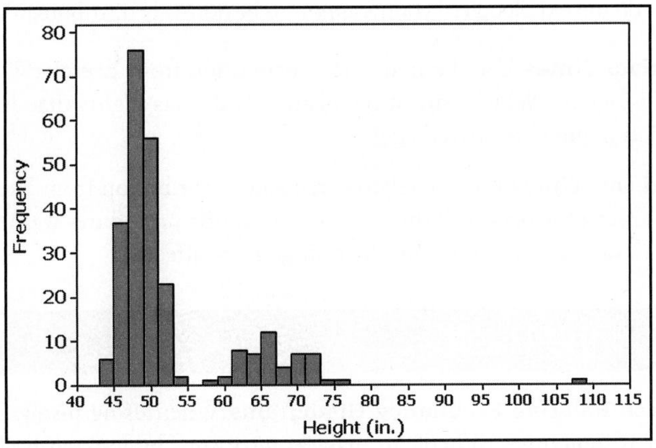

5. Sample Size What is the approximate number of people with heights of 55 in. or less?

6. Class Width and Class Limits What is the class width? What are the approximate lower and upper class limits of the first class?

7. Outlier? What is the height of the tallest person included in the histogram? Where on the histogram is that height depicted? Is that height an outlier? Could that height be an exceptional value that is correct, or is it an error? Explain.

8. Gap What is a reasonable explanation for the gap between the group of people with heights between 43 in. and 55 in., and the group of people with heights between 57 in. and 77 in.?

Constructing Histograms. *In Exercises 9–18, construct the histograms and answer the given questions.*

9. Analysis of Last Digits Use the frequency distribution from Exercise 19 in Section 2-2 to construct a histogram. What can you conclude from the distribution of the digits? Specifically, do the heights appear to be reported or actually measured?

10. Analysis of Last Digits Use the frequency distribution from Exercise 20 in Section 2-2 to construct a histogram. What can you conclude from the distribution of the digits? Specifically, do the weights appear to be reported or actually measured?

11. Pulse Rates of Males Use the frequency distribution from Exercise 21 in Section 2-2 to construct a histogram. Does the histogram appear to depict data that have a normal distribution? Why or why not?

12. Pulse Rates of Females Use the frequency distribution from Exercise 22 in Section 2-2 to construct a histogram. Does the histogram appear to depict data that have a normal distribution? Why or why not?

13. Earthquake Magnitudes Use the frequency distribution from Exercise 23 in Section 2-2 to construct a histogram. Using a loose interpretation of the requirements for a normal distribution, do the magnitudes appear to be normally distributed? Why or why not?

14. Earthquake Depths Use the frequency distribution from Exercise 24 in Section 2-2 to construct a histogram. Using a strict interpretation of the requirements for a normal distribution, do the depths appear to be normally distributed? Why or why not?

6. Class width: 2 in. Approximate lower limit of first class: 43 in. Approximate upper limit of first class: 45 in.

8. The first group appears to be children, and the second group appears to be adults. Knowing that the people entered a museum on a Friday morning, we can reasonably assume that there were many school children on a field trip and that they were accompanied by a smaller group of teachers and adult chaperones and other adults visiting the museum by themselves.

10. The digits 0 and 5 seem to occur much more often than the other digits, so it appears that the heights were reported and not actually measured. This suggests that the results might not be very accurate.

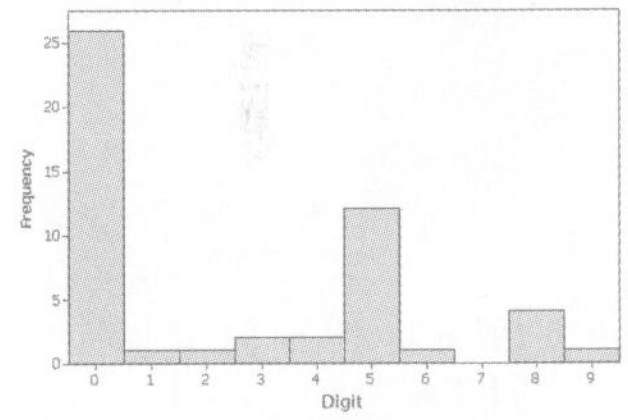

12. The histogram appears to roughly approximate a normal distribution. The frequencies generally increase to a maximum and then decrease, and the histogram is symmetric with the left half being roughly a mirror image of the right half.

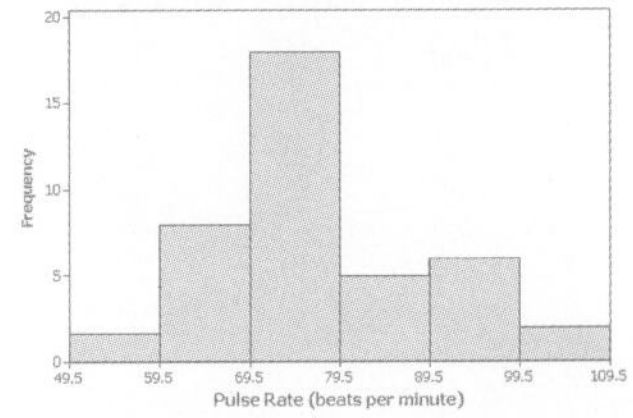

14. No, the histogram does not appear to approximate a normal distribution. The frequencies do not increase to a maximum and then decrease, and the histogram is not symmetric with the left half being a mirror image of the right half.

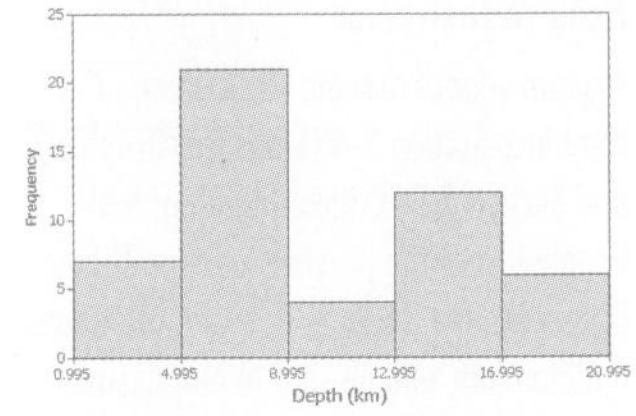

15. Male Red Blood Cell Counts Use the frequency distribution from Exercise 25 in Section 2-2 to construct a histogram. Using a very loose interpretation of the requirements for a normal distribution, do the red blood cell counts appear to be normally distributed? Why or why not?

16. Female Red Blood Cell Counts Use the frequency distribution from Exercise 26 in Section 2-2 to construct a histogram. Using a very loose interpretation of the requirements for a normal distribution, do the red blood cell counts appear to be normally distributed? Why or why not?

17. Flight Arrival Times Use the frequency distribution from Exercise 27 in Section 2-2 to construct a histogram. Which part of the histogram depicts flights that arrived early, and which part depicts flights that arrived late?

18. Flight Taxi-Out Times Use the relative frequency distribution from Exercise 28 in Section 2-2 to construct a histogram. If the quality of air traffic procedures was improved so that the taxi-out times vary much less, would the histogram be affected?

16. The histogram appears to roughly approximate a normal distribution. The frequencies increase to a maximum and then decrease, and the histogram is symmetric with the left half being roughly a mirror image of the right half.

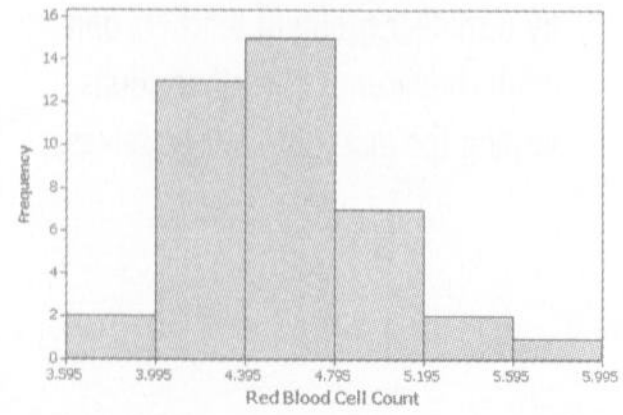

18. Yes, the entire distribution would be more concentrated with less spread.

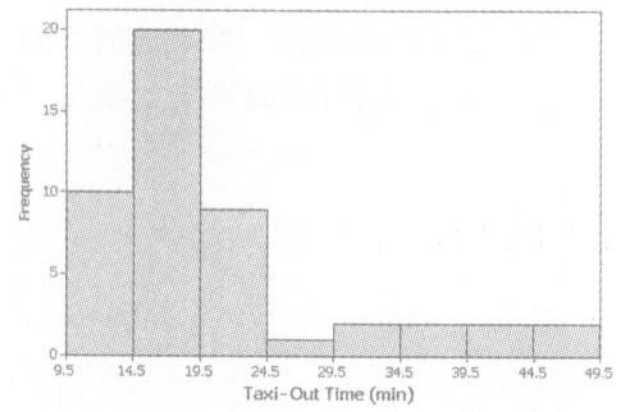

2-3 Beyond the Basics

19. Back-to-Back Relative Frequency Histograms When using histograms to compare two data sets, it is sometimes difficult to make comparisons by looking back and forth between the two histograms. A *back-to-back relative frequency histogram* has a format that makes the comparison much easier. Instead of frequencies, we should use relative frequencies (percentages or proportions) so that the comparisons are not difficult when there are different sample sizes. Use the relative frequency distributions of the ages of Oscar-winning actresses and actors from Exercise 15 in Section 2-2, and complete the back-to-back relative frequency histograms shown below. Then use the result to compare the two data sets.

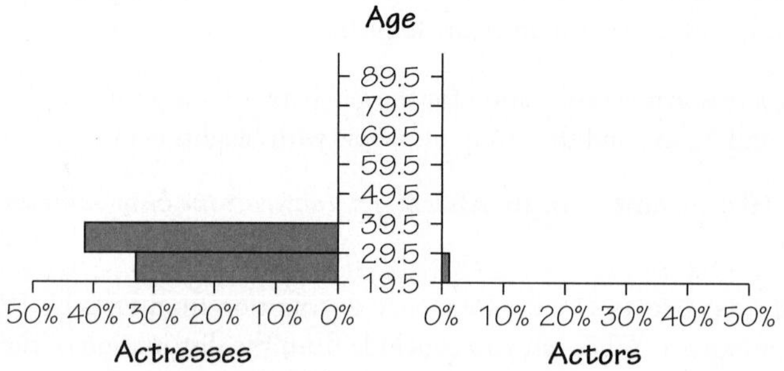

20. Interpreting a Histogram Refer to the histogram given for Exercises 5–8 and answer the following questions:

a. What are the possible values (rounded to the nearest inch) of the largest height included in the histogram? What are those values expressed in feet and inches?

b. Why is it wrong to say that the people with heights between 43 in. and 55 in. are the tallest people because they have the tallest bars in the histogram?

20. a. 107 in. to 109 in.; 8 ft 11 in. to 9 ft 1 in.

b. The heights of the bars represent numbers of people, not heights. Because there are many more people between 43 in. tall and 55 in. tall, they have the tallest bars in the histogram, but they have the lowest actual heights. They have the tallest bars because there are more of them.

2-4 Graphs That Enlighten and Graphs That Deceive

Note to Instructor

Organizational change: This section combines Section 2-4 (Statistical Graphics) and Section 2-5 (Critical Thinking: Bad Graphs) from the previous edition. If time is very limited, include at least scatterplots and dotplots, and be sure to emphasize the ways in which graphs can be used to deceive.

Key Concept Section 2-3 presented the histogram as a graph that is helpful in learning about the shape of the distribution of data. The histogram is a graph that enlightens in the sense that it gives us better understanding of data. In this section we introduce other commonly used graphs that enlighten. We also discuss some graphs that deceive in the sense that they tend to create impressions about data that are somehow misleading or wrong.

The days of charming and primitive hand-drawn graphs are well behind us, and technology now provides us with powerful tools for generating a wide variety of different graphs. In this section, all figures except Figure 2-19 were generated using technology.

Graphs That Enlighten

Scatterplots

A **scatterplot** (or **scatter diagram**) is a plot of paired (x, y) quantitative data with a horizontal x-axis and a vertical y-axis. The horizontal axis is used for the first (x) variable, and the vertical axis is used for the second variable. The pattern of the plotted points is often helpful in determining whether there is a correlation (or relationship) between the two variables. (This issue is discussed at length when the topic of correlation is considered in Section 10-2.)

Example 1 Correlation: Waist and Arm Circumference

Data Set 1 in Appendix B includes the waist circumferences (cm) and arm circumferences (cm) of randomly selected males. Figure 2-6 is a scatterplot of the paired waist/arm measurements. The points show a pattern of increasing values from left to right. This pattern suggests that there is a correlation, or relationship, between waist circumference and arm circumference in males.

Example 2 No Correlation: Weight and Pulse Rate

Data Set 1 in Appendix B includes weights (kg) and pulse rates (beats per minute) of randomly selected males. Figure 2-7 is a scatterplot of the paired weight/pulse rate measurements. The points in Figure 2-7 do not show any obvious pattern, and this lack of a pattern suggests that there is no correlation, or relationship, between the weight and pulse rate of males.

Correlation Coefficient Examples 1 and 2 involve making decisions about a correlation based on subjective judgments of scatterplots, but Chapter 10 introduces more objective methods. Those methods involve calculating a value of a *linear correlation coefficient* r, which is a value between -1 and 1. If r is close to -1 or close to 1, there appears to be a correlation, but if r is close to 0, there does not appear to be a correlation. For the data depicted in the scatterplot of Figure 2-6, $r = 0.788$, and the data in the scatterplot of Figure 2-7 result in $r = -0.213$. Section 10-2 describes the calculation and interpretation of a value of r, so that a decision about correlation is much more objective. Even though the methods based on calculations of r are much more objective than the subjective interpretation of a scatterplot, it is always wise to construct a scatterplot first, so that we can see characteristics that cannot be seen by examining the list of paired data values.

The Power of a Graph

With annual sales around $13 billion and with roughly 50 million people using it, Pfizer's prescription drug Lipitor has become the most profitable and most widely used prescription drug ever marketed. In the early stages of its development, Lipitor was compared to other drugs (Zocor, Mevacor, Lescol, and Pravachol) in a process that involved controlled trials. The summary report included a graph showing a Lipitor curve that had a steeper rise than the curves for the other drugs, visually showing that Lipitor was more effective in reducing cholesterol than the other drugs. Pat Kelly, who was then a senior marketing executive for Pfizer, said "I will never forget seeing that chart. . . . It was like 'Aha!' Now I know what this is about. We can communicate this!" The Food and Drug Administration approved Lipitor and allowed Pfizer to include the graph with each prescription. Pfizer sales personnel also distributed the graph to physicians.

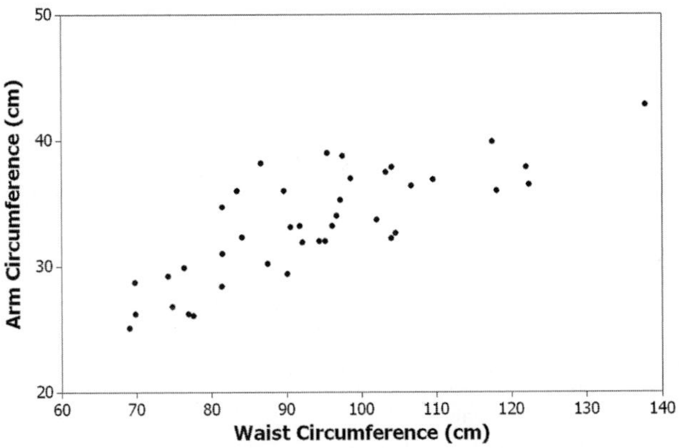

Figure 2-6 Waist Circumference and Arm Circumference in Males

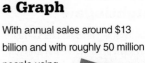

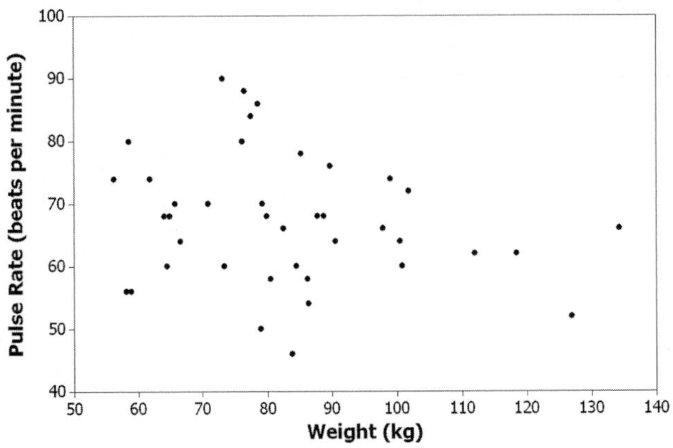

Figure 2-7 Weight and Pulse Rate in Males

Florence Nightingale

Florence Nightingale (1820–1910) is known to many as the founder of the nursing profession, but she also saved thousands of lives by using statistics. When she encountered an unsanitary and undersupplied hospital, she improved those conditions and then used statistics to convince others of the need for more widespread medical reform. She developed original graphs to illustrate that during the Crimean War, more soldiers died as a result of unsanitary conditions than were killed in combat. Florence Nightingale pioneered the use of social statistics as well as graphics techniques.

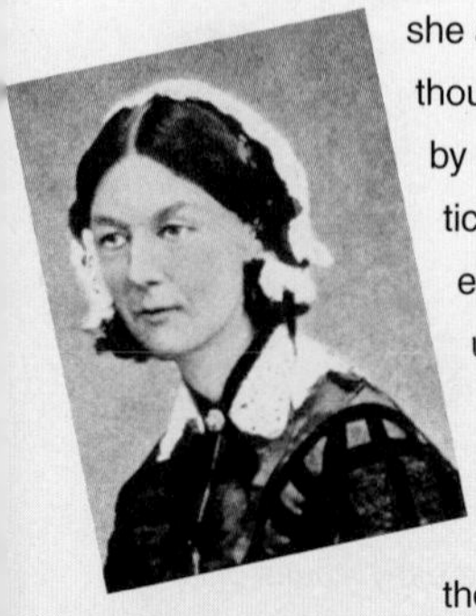

Note to Instructor

Comment in class that scatterplots are discussed again in Chapter 10. Also, discuss the issue of *causation* by asking these questions: "Suppose grades and hours studied are recorded for a sample of students. What would the scatterplot look like? If there is a clear pattern showing a relationship, can we conclude from the scatterplot that more studying *causes* higher grades?" Section 10-2 makes the very important point that correlation does *not* imply causation.

Example 3 Clusters and a Gap

Consider the scatterplot in Figure 2-8. It consists of paired data consisting of the weight (grams) and year of manufacture for each of 72 pennies. This scatterplot shows two very distinct clusters separated by a gap, which can be explained by the inclusion of two different populations: pre-1983 pennies are 97% copper and 3% zinc, whereas post-1983 pennies are 3% copper and 97% zinc. If we ignored the characteristic of the clusters, we might incorrectly think that there is a relationship between the weight of a penny and the year it was made. If we examine the two groups separately, we see that there does *not* appear to be a relationship between the weights of pennies and the years in which they were produced.

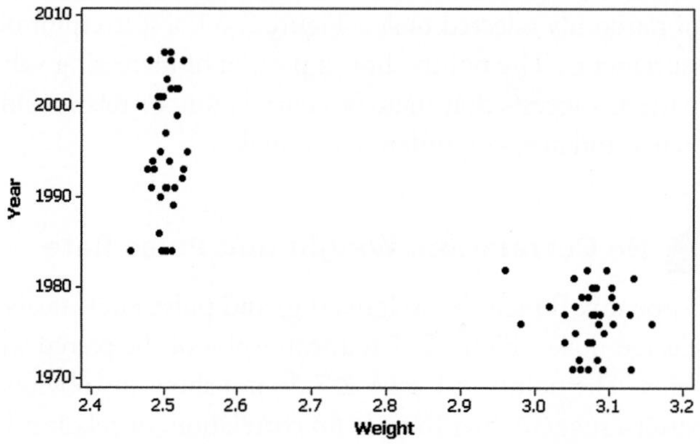

Figure 2-8 Weights (g) of Pennies and Years of Production

Time-Series Graph

A **time-series graph** is a graph of *time-series data,* which are quantitative data that have been collected at different points in time, such as monthly or yearly.

Example 4 Time-Series Graph: Dow Jones Industrial Average

The time-series graph shown in Figure 2-9 depicts the yearly high values of the Dow Jones Industrial Average (DJIA) for the New York Stock Exchange. This graph shows a fairly consistent pattern of increases from 1980 to 1999, but the DJIA high values have been much more erratic in recent years.

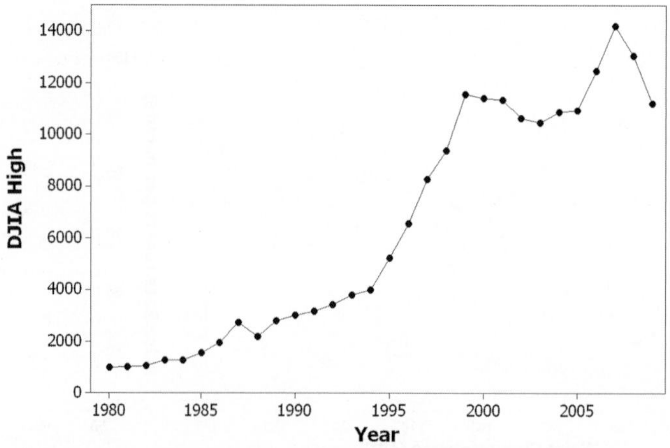

Figure 2-9 Dow Jones Industrial Average

Dotplots

A **dotplot** consists of a graph in which each data value is plotted as a point (or dot) along a horizontal scale of values. Dots representing equal values are stacked.

Example 5 Dotplot: IQ Scores of Low Lead Group

Figure 2-10 shows a dotplot of the IQ scores of the low lead group from Table 2-1 included with the Chapter Problem at the beginning of this chapter. The five stacked dots above the position at 76 indicate that five of the IQ scores are 76. There are three dots stacked above 80, so three of the IQ scores are 80. This dotplot reveals the distribution of the IQ scores. It is possible to recreate the original list of data values, because each data value is represented by a single point.

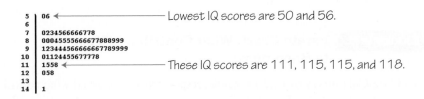

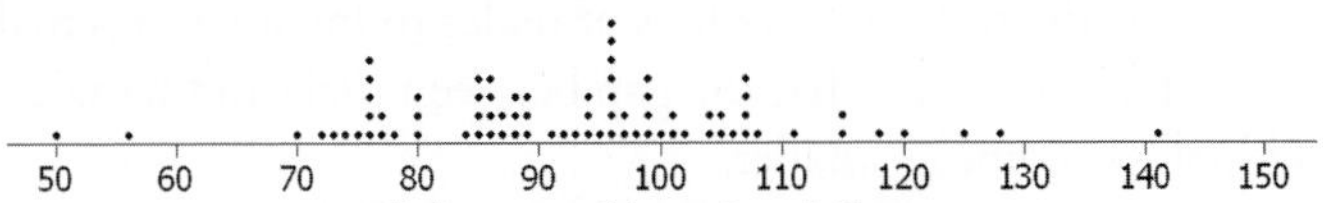

Figure 2-10 Dotplot: IQ Scores of Low Lead Group

Go Figure

Among U. S. men between 20 and 40 years of age and taller than 7 feet, 17% of them play basketball in the NBA.

Stemplots

A **stemplot** (or **stem-and-leaf plot**) represents quantitative data by separating each value into two parts: the stem (such as the leftmost digit) and the leaf (such as the rightmost digit). Better stemplots are often obtained by first rounding the original data values. Also, stemplots can be *expanded* to include more rows and can be *condensed* to include fewer rows, as in Exercise 26.

One advantage of the stemplot is that we can see the distribution of data while keeping the original data values. Another advantage is that constructing a stemplot is a quick way to *sort* data (arrange them in order), which is required for some statistical procedures (such as finding a median, or finding percentiles as described later in this book).

Example 6 Stemplot: IQ Scores of Low Lead Group

The following stemplot displays the IQ scores of the low lead group in Table 2-1 given with the Chapter Problem. The lowest IQ score of 50 is separated into its stem of 5 and its leaf of 0, and each of the remaining values is separated in a similar way. The stems and leaves are arranged in increasing order, not the order in which they occur in the original list. Note that if you turn the stemplot on its side, you can see distribution of the IQ scores in the same way you would see it in a histogram.

```
 5 | 06                      <- Lowest IQ scores are 50 and 56.
 6 |
 7 | 0234566666778
 8 | 00045555666677888999
 9 | 123444566666667789999
10 | 01124455677778
11 | 1558                    <- These IQ scores are 111, 115, 115, and 118.
12 | 058
13 |
14 | 1
```

Bar Graphs

A **bar graph** uses bars of equal width to show frequencies of categories of *categorical* (or qualitative) data. The vertical scale represents frequencies or relative frequencies. The horizontal scale identifies the different categories of qualitative data. The bars may or may not be separated by small gaps. A **multiple bar graph** has two or more sets of bars and is used to compare two or more data sets.

Example 7 Multiple Bar Graph of Income by Gender

See Figure 2-11 for a multiple bar graph of the median incomes of males and females in different years. The data are from the U.S. Census Bureau, and the values for 2010 are projected. From this graph we see that males consistently have much higher median incomes than females, and that both males and females have steadily increasing incomes over time. Comparing the heights of the bars from left to right reveals that the ratios of incomes of males to incomes of females appear to be decreasing, which indicates that the gap between male and female median incomes is gradually becoming smaller.

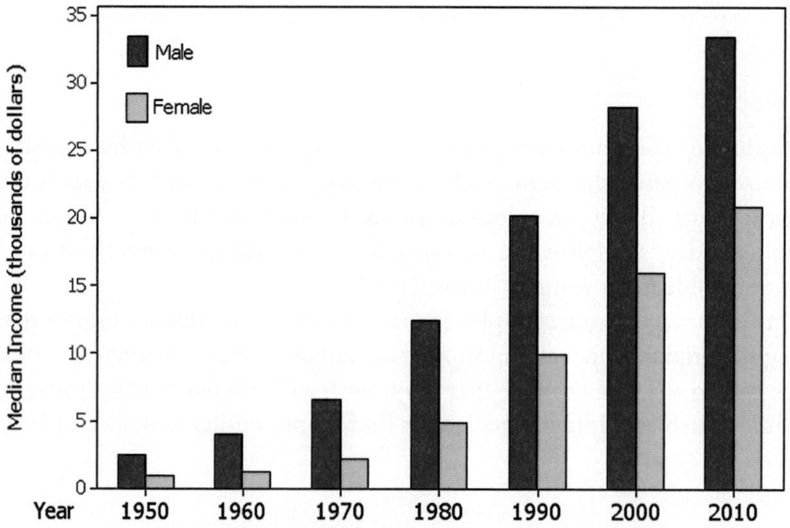

Figure 2-11 Multiple Bar Graph: Median Income by Gender

Pareto Charts

When we want a bar graph to draw attention to the more important categories, we can use a **Pareto chart**, which is a bar graph for categorical data, with the added stipulation that the bars are arranged in descending order according to frequencies. The vertical scale in a Pareto chart represents frequencies or relative frequencies. The horizontal scale identifies the different categories of qualitative data. The bars decrease in height from left to right.

Example 8 Pareto Chart: What Contributes Most to Happiness?

In a Coca-Cola survey of 12,500 people, respondents were asked what contributes most to their happiness. Figure 2-12 is a Pareto chart summarizing the results. We see that family or partner is by far the most frequently selected choice.

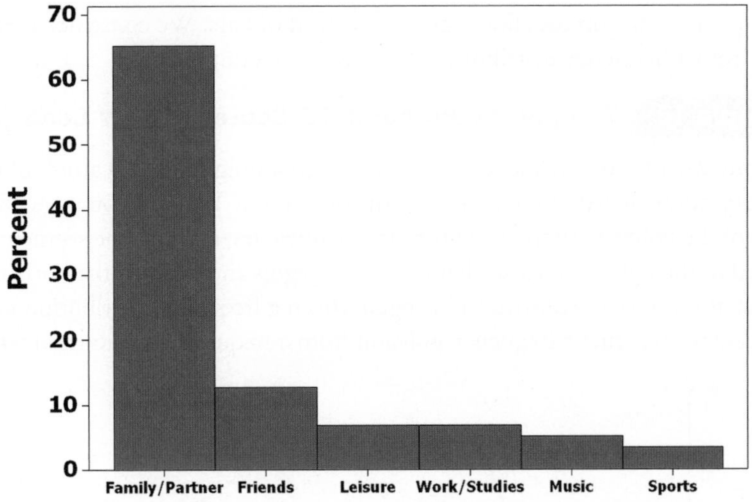

Figure 2-12 Pareto Chart: What Contributes Most to Happiness?

Pie Charts

A **pie chart** is a graph that depicts categorical data as slices of a circle, in which the size of each slice is proportional to the frequency count for the category.

Example 9 Pie Chart: What Contributes Most to Happiness?

Figure 2-13 is a pie chart corresponding to the same data from Example 8. Construction of a pie chart involves slicing up the circle into the proper proportions that represent relative frequencies. For example, the category of friends accounts for 13% of the total, so the slice representing friends should be 13% of the total (with a central angle of $0.13 \times 360^\circ = 47^\circ$).

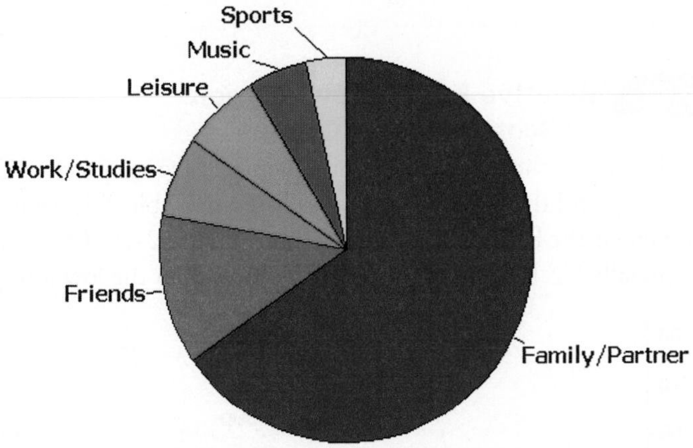

Figure 2-13 Pie Chart: What Contributes Most to Happiness?

The Pareto chart in Figure 2-12 and the pie chart in Figure 2-13 depict the same data in different ways, but the Pareto chart does a better job of showing the relative sizes of the different components. Graphics expert Edwin Tufte makes the following suggestion:

> **Never use pie charts because they waste ink on components that are not data, and they lack an appropriate scale.**

Frequency Polygon

A **frequency polygon** uses line segments connected to points located directly above class midpoint values. A frequency polygon is very similar to a histogram, but a frequency polygon uses line segments instead of bars. We construct a frequency polygon from a frequency distribution as shown in Example 10.

Example 10 Frequency Polygon: IQ Scores of Low Lead Group

See Figure 2-14 for the frequency polygon corresponding to the IQ scores of the low lead group summarized in the frequency distribution of Table 2-2 on page 45. The heights of the points correspond to the class frequencies, and the line segments are extended to the right and left so that the graph begins and ends on the horizontal axis. Just as it is easy to construct a histogram from a frequency distribution table, it is also easy to construct a frequency polygon from a frequency distribution table.

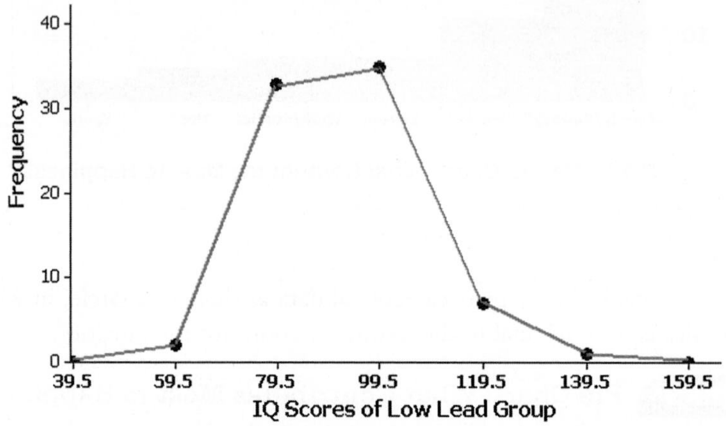

Figure 2-14 Frequency Polygon: IQ Scores of Low Lead Group

A variation of the basic frequency polygon is the **relative frequency polygon,** which uses relative frequencies (proportions or percentages) for the vertical scale. When one is trying to compare two data sets, it is often very helpful to graph two relative frequency polygons on the same axes.

Example 11 Relative Frequency Polygon: IQ Scores of Lead Groups

See Figure 2-15, which shows the relative frequency polygons for the IQ scores of the low lead group and the high lead group as listed in Table 2-1 given with the Chapter Problem at the beginning of this chapter. Figure 2-15 shows that the high lead group generally has lower (farther left) IQ scores than the low lead group.

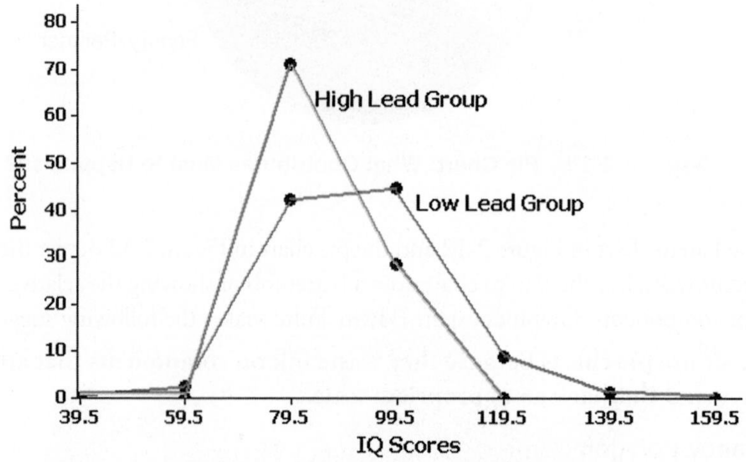

Figure 2-15 Relative Frequency Polygons: IQ Scores

It appears that the greater exposure to lead tends to be associated with lower IQ scores. Figure 2-15 enables us to understand data in a way that is not possible with visual examination of the lists of data in Table 2-1.

Ogive

Another type of statistical graph is an **ogive** (pronounced "oh-jive"), which depicts *cumulative* frequencies. Ogives are useful for determining the number of values below some particular value, as illustrated in Example 3. An ogive uses class boundaries along the horizontal scale and uses cumulative frequencies along the vertical scale.

Example 12 **Ogive: IQ Scores of Low Lead Group**

Figure 2-16 shows an ogive corresponding to the cumulative frequency distribution table (Table 2-5) on page 48. From Figure 2-16, we see that for the low lead group, 35 of the IQ scores are less than 89.5.

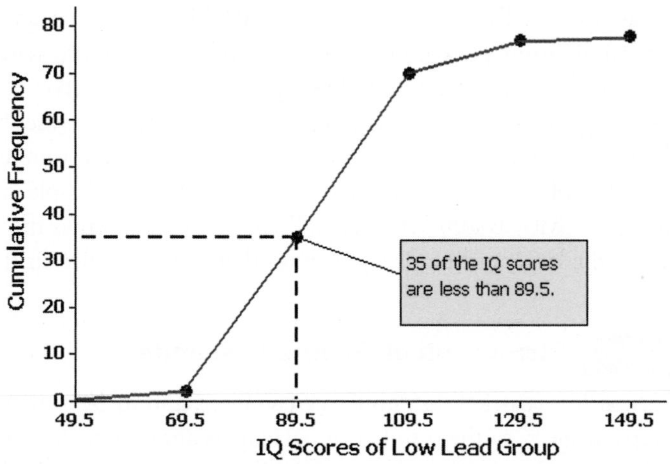

Figure 2-16 **Ogive: IQ Scores of Low Lead Group**

Graphs That Deceive

Some graphs deceive because they contain errors, and some deceive because they are technically correct but misleading. It is important to develop the ability to recognize deceptive graphs. Here we present two of the ways in which graphs are commonly used to deceive.

Example 13 **Nonzero Axis**

Figure 2-17 and Figure 2-18 are based on the same data from Data Set 14 in Appendix B. By using a vertical scale starting at 30 mi/gal instead of at 0 mi/gal, Figure 2-17 exaggerates the differences and creates the false impression that the Honda Civic gets mileage that is substantially better than the mileage ratings found for the Chevrolet Aveo and the Toyota Camry. Figure 2-18 shows that the differences among the three mileage ratings are actually small.

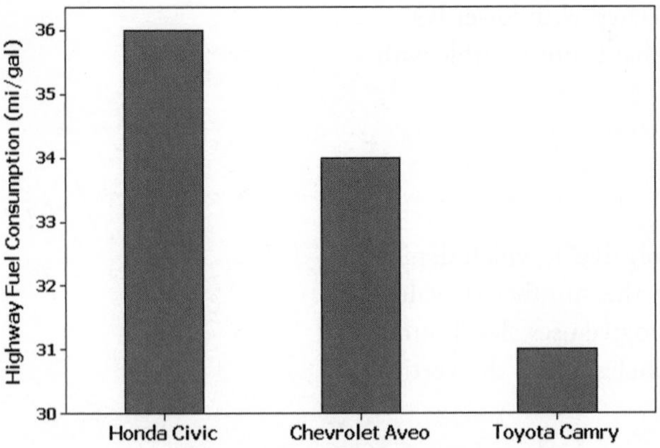

Figure 2-17 Highway Fuel Consumption with Vertical Scale Not Starting at Zero

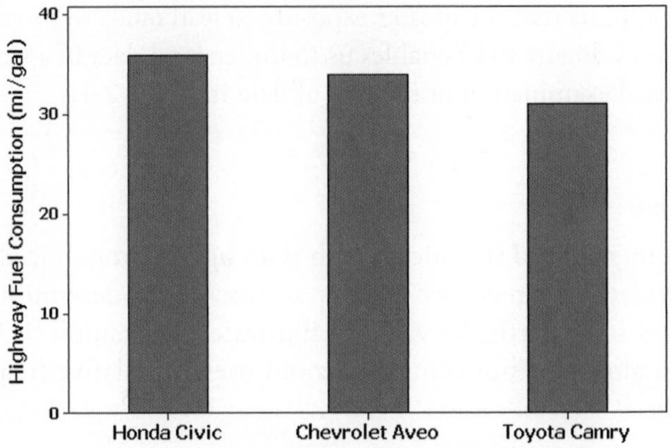

Figure 2-18 Highway Fuel Consumption with Vertical Scale Starting at Zero

Pictographs

Drawings of objects, called *pictographs,* are often misleading. Data that are one-dimensional in nature (such as budget amounts) are often depicted with two-dimensional objects (such as dollar bills) or three-dimensional objects (such as stacks of coins, homes, or barrels). By using pictographs, artists can create false impressions that grossly distort differences by using these simple principles of basic geometry: (1) When you double each side of a square, the area doesn't merely double; it increases by a factor of *four.* (2) When you double each side of a cube, the volume doesn't merely double; it increases by a factor of *eight.* See Figure 2-19 in the following example, and note that the larger airliner is twice as long, twice as tall, and twice as deep as the first airliner, so the volume of the larger airliner is eight times that of the smaller airliner.

Example 14 Pictograph of Airline Passengers

In 1984, U.S. airlines carried 345 million passengers, and in 2010 they carried 706 million passengers, so the number of passengers approximately doubled from 1984 to 2010. The pictograph in Figure 2-19 illustrates these data with images of airliners that are objects of volume. Readers can have a variety of perceptions. Some might think that the numbers of passengers are the same in both images, because the same numbers of seats are included. Others might look at the different sizes of the airliners and see objects of volume, the larger aircraft being roughly eight times the size of the smaller one. Even though Figure 2-19 includes attractive images, it does a very poor job of accurately and unambiguously depicting the data. In contrast, Figure 2-20 is a simple bar graph that does a good job of depicting the data accurately.

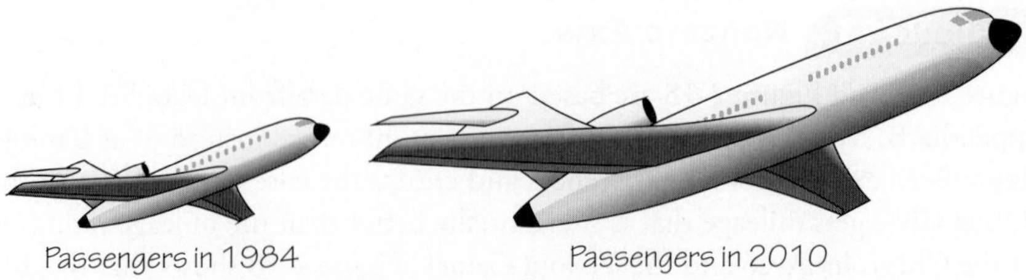

Figure 2-19 Passengers Carried by U.S. Airlines

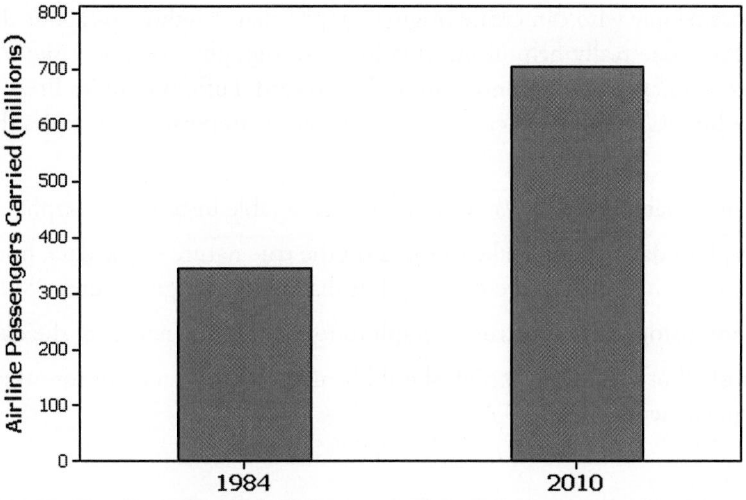

Figure 2-20 Passengers Carried by U.S. Airlines

Examples 13 and 14 illustrate the following principles related to misleading graphs:

- **Nonzero axis:** Always examine a graph to see whether an axis begins at some point other than zero so that differences are exaggerated.
- **Pictographs:** When examining data depicted with a pictograph, determine whether the graph is misleading because objects of area or volume are used to depict amounts that are actually one-dimensional. (Histograms and bar charts represent one-dimensional data with two-dimensional bars, but they use bars with the same width so that the graph is not misleading.)

Conclusion

In this section we saw that graphs are excellent tools for describing, exploring, and comparing data.

Describing data: In a histogram, for example, consider the distribution, center, variation, and outliers (values that are very far away from almost all of the other data values). What is the approximate value of the center of the distribution, and what is the approximate range of values? Consider the overall shape of the distribution. Are the values evenly distributed? Is the distribution skewed (lopsided) to the right or left? Does the distribution peak in the middle? Is there a large gap, suggesting that the data might come from different populations? Identify any extreme values and any other notable characteristics.

Exploring data: Look for features of the graph that reveal some useful and/or interesting characteristics of the data set. For example, the scatterplot included with Example 1 shows that there appears to be a relationship between the waist circumferences and arm circumferences of males.

Comparing data: Construct similar graphs to compare data sets. For example, Figure 2-14 shows a frequency polygon for the IQ scores of a group with low lead exposure and another frequency polygon for a group with high lead exposure, and both polygons are shown on the same set of axes. Figure 2-14 makes the comparison relatively easy.

In addition to the graphs we have discussed in this section, there are many other useful graphs—some of which have not yet been created. The world desperately needs more people who can create original graphs that enlighten us about the nature of data. For some really helpful information about graphs, see *The Visual Display of Quantitative Information,* second edition, by Edward Tufte (Graphics Press, PO Box 430, Cheshire, CT 06410). Here are just a few of the important principles that Tufte suggests:

- For small data sets of 20 values or fewer, use a table instead of a graph.
- A graph of data should make us focus on the true nature of the data, not on other elements, such as eye-catching but distracting design features.
- Do not distort data; construct a graph to reveal the true nature of the data.
- Almost all of the ink in a graph should be used for the data, not for other design elements.

using TECHNOLOGY

Here we list the graphs that can be generated by various technologies. (Detailed instructions can range from quite simple to extremely complex, so see the individual manuals that are supplements to this book.)

STATDISK Histograms, scatterplots, and pie charts

MINITAB Histograms, frequency polygons, dotplots, stemplots, bar graphs, multiple bar graphs, Pareto charts, pie charts, scatterplots, and time-series graphs

EXCEL Histograms and scatterplots

TI-83/84 PLUS Histograms and scatterplots

STATCRUNCH Histograms, scatterplots, pie charts, bar charts, stemplots, and dotplots

Recommended Assignment

Exercises 1–4, 6, 8, 11, 12, and 21–24. Other exercises should be included to correspond to the topics that you covered.

Due to limited space, answers for some exercises are not included here, but they are included in Appendix D.

2-4 Basic Skills and Concepts

Statistical Literacy and Critical Thinking

1. Bar Chart and Pareto Chart A bar chart and a Pareto chart both use bars to show frequencies of categories of categorical data. What characteristic distinguishes a Pareto chart from a bar chart, and how does that characteristic help us in understanding the data?

2. Scatterplot What is a scatterplot? What type of data is required for a scatterplot? What characteristic of the data can be better understood by looking at a scatterplot?

2. A scatterplot is a plot of paired quantitative data, and each pair of data is plotted as a single point. The scatterplot requires paired quantitative data. The configuration of the plotted points can help us determine whether there is some relationship between the two variables.

3. SAT Scores Listed below are SAT scores from a sample of students (based on data from www.talk.collegeconfidential.com). Why is it that a graph of these data will not be very effective in helping us understand the data?

2400 2200 2150 2040 2230 1890 2100 2090

4. SAT Scores Given that the data in Exercise 3 were obtained from students who made a decision to submit their SAT scores to a Web site, what type of sample is given in that exercise? If we had a much larger sample of that type, would a graph help us understand some characteristics of the population?

4. The sample is a voluntary response sample. Because the sample is a voluntary response sample, it is very possible that it is not representative of the population, even if the sample is very large. Any graph based on the voluntary response sample would have a high chance of showing characteristics that are not actual characteristics of the population.

Scatterplots. ***In Exercises 5–8, use the given paired data from Appendix B to construct a scatterplot.***

5. President's Heights Refer to Data Set 12 in Appendix B, and use the heights of U.S. presidents and the heights of their main opponents in the election campaign. Does there appear to be a correlation?

6. Brain Volume and IQ Refer to Data Set 6 in Appendix B, and use the brain volumes (cm^3) and IQ scores. A simple hypothesis is that people with larger brains are more intelligent and thus have higher IQ scores. Does the scatterplot support that hypothesis?

7. Bear Chest Size and Weight Refer to Data Set 7 in Appendix B, and use the measured chest sizes and weights of bears. Does there appear to be a correlation between those two variables?

8. Coke Volume and Weight Refer to Data Set 19 in Appendix B, and use the volumes and weights of regular Coke. Does there appear to be a correlation between volume and weight? What else is notable about the arrangement of the points, and how can it be explained?

Time-Series Graphs. ***In Exercises 9 and 10, construct the time-series graph.***

9. Harry Potter Listed below are the gross amounts (in millions of dollars) earned from box office receipts for the movie *Harry Potter and the Half-Blood Prince.* The movie opened on a Wednesday, and the amounts are listed in order for the first 14 days of the movie's release. Suggest an explanation for the fact that the three highest amounts are the first, third, and fourth values listed.

58 22 27 29 21 10 10 8 7 9 11 9 4 4

10. Home Runs Listed below are the numbers of home runs in major league baseball for each year beginning with 1990 (listed in order by row). Is there a trend?

3317	3383	3038	4030	3306	4081	4962	4640	5064	5528
5693	5458	5059	5207	5451	5017	5386	4957	4878	4655

Dotplots. ***In Exercises 11 and 12, construct the dotplot.***

11. Coke Volumes Refer to Data Set 19 in Appendix B, and use the volumes of regular Coke. Does the configuration of the points appear to suggest that the volumes are from a population with a normal distribution? Why or why not? Are there any outliers?

12. Car Pollution Refer to Data Set 14 in Appendix B, and use the greenhouse gas (GHG) emissions from the sample of cars. Does the configuration of the points appear to suggest that the amounts are from a population with a normal distribution? Why or why not?

Stemplots. ***In Exercises 13 and 14, construct the stemplot.***

13. Car Crash Tests Refer to Data Set 13 in Appendix B and use the 21 pelvis (PLVS) deceleration measurements from the car crash tests. Is there strong evidence suggesting that the data are *not* from a population having a normal distribution?

14. Car Braking Distances Refer to Data Set 14 in Appendix B and use the 21 braking distances (ft). Are there any outliers? Is there strong evidence suggesting that the data are *not* from a population having a normal distribution?

Pareto Charts. ***In Exercises 15 and 16, construct the Pareto chart.***

15. Awful Sounds In a survey, 1004 adults were asked to identify the most frustrating sound that they hear in a day. In response 279 chose jackhammers, 388 chose car alarms, 128 chose barking dogs, and 209 chose crying babies (based on data from Kelton Research).

16. School Day Here are weekly instruction times for school children in different countries: 23.8 hours (Japan), 26.9 hours (China), 22.2 hours (U.S.), 24.6 hours (U.K.), 24.8 hours (France). What do these results suggest about education in the United States?

Pie Charts. ***In Exercises 17 and 18, construct the pie chart.***

17. Awful Sounds Use the data from Exercise 15.

18. School Day Use the data from Exercise 16. Does it make sense to use a pie chart for the given data?

6. The configuration of the points does not support the hypothesis that people with larger brains have higher IQ scores.

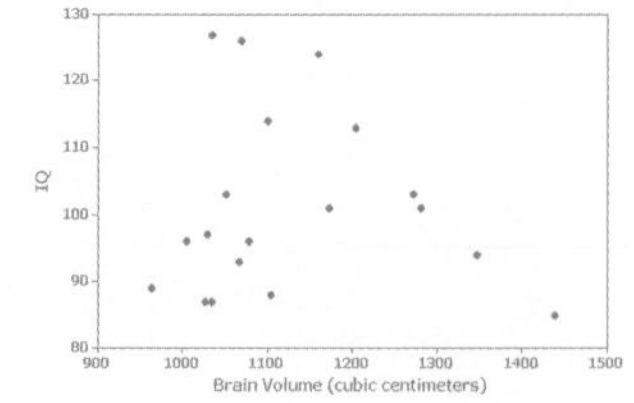

8. Yes. There is a very distinct pattern showing that cans of Coke with larger volumes tend to weigh more. Another notable feature of the scatterplot is that there are five groups of points that are stacked above each other. This is due to the fact that the measured volumes were rounded to one decimal place, so the different volume amounts are often duplicated, with the result that the points are stacked vertically.

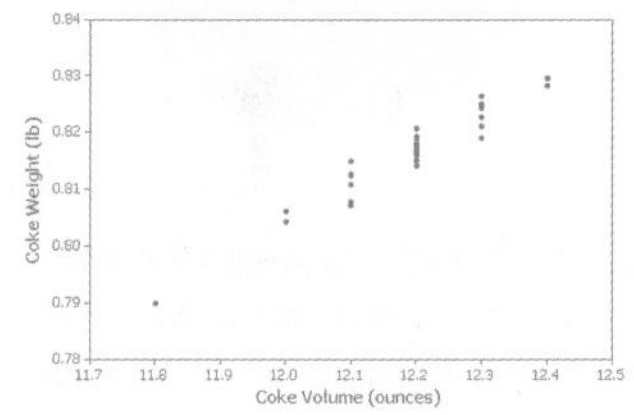

10. The numbers of home runs rose from 1990 to 2000, but after 2000 there was a very gradual decline.

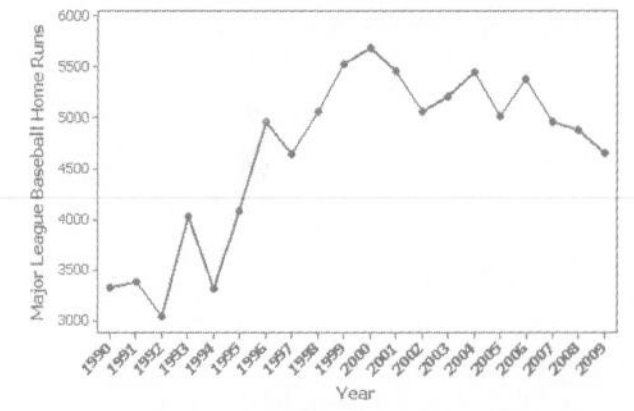

12. No, because the configuration of points is not at all a bell shape, the amounts do not appear to be from a normally distributed population.

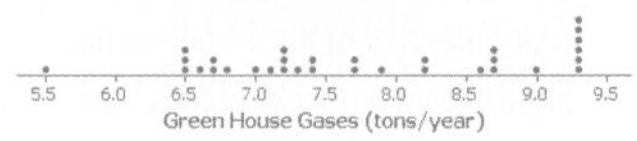

14. There are no outliers. The distribution is not dramatically far from being a normal distribution with a bell shape, so there is not strong evidence against a normal distribution.

12	68
13	1234556667789
14	000335

16. To remain competitive in the world, the United States should require more weekly instruction time.

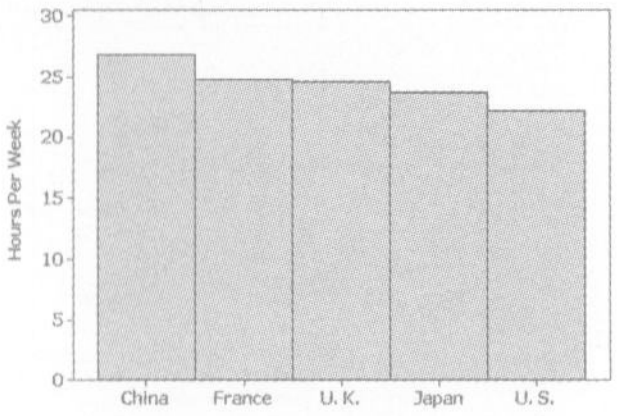

18. Because there is not a single total number of hours of instruction time that is partitioned among the five countries, it does not make sense to use a pie chart for the given data.

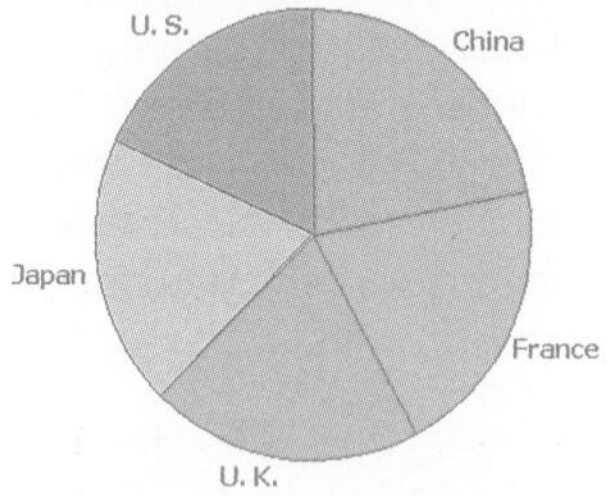

20. No, the frequency polygon does not appear to approximate a normal distribution. The frequencies do not increase to a maximum and then decrease, and the graph is not symmetric with the left half being a mirror image of the right half.

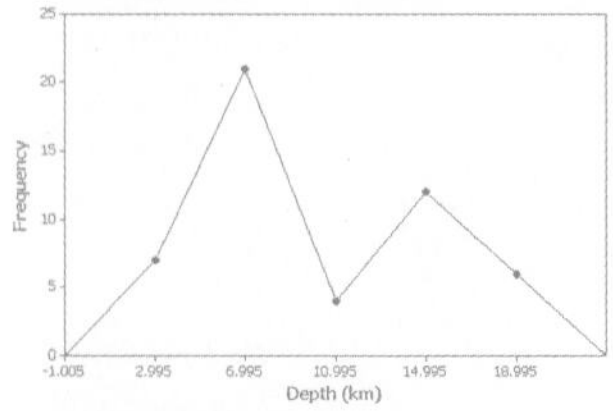

22. The fare doubled from $1 to $2, but when the $2 bill is shown with twice the width and twice the height of the $1 bill, the $2 bill has an area that is *four times* that of the $1 bill, so the illustration greatly exaggerates the increase in the fare.

Frequency Polygon. ***In Exercises 19 and 20, construct the frequency polygon.***

19. Earthquake Magnitudes Use the frequency distribution from Exercise 23 in Section 2-2 to construct a frequency polygon. Applying a loose interpretation of the requirements for a normal distribution, do the magnitudes appear to be normally distributed? Why or why not?

20. Earthquake Depths Use the frequency distribution from Exercise 24 in Section 2-2 to construct a frequency polygon. Applying a strict interpretation of the requirements for a normal distribution, do the depths appear to be normally distributed? Why or why not?

Deceptive Graphs. ***In Exercises 21–24, identify the characteristic that causes the graph to be deceptive.***

21. Election Results The accompanying graph depicts the numbers of votes (in millions) in the 2008 U.S. presidential election.

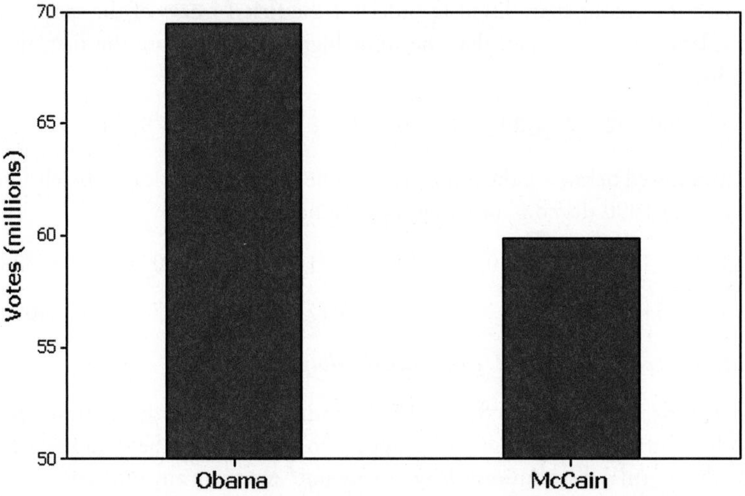

22. Subway Fare In 1986, the New York City subway fare cost $1, and in 2003 the cost was raised to $2, so the price doubled. In the accompanying graph, the $2 bill is twice as long and twice as tall as the $1 bill.

1986 Subway Fare Current Subway Fare

23. Oil Consumption China currently consumes 7.6 million barrels of oil per day, compared to the United States oil consumption of 20.7 million barrels of oil per day. In the accompanying illustration, the larger barrel is about three times as wide and three times as tall as the smaller barrel.

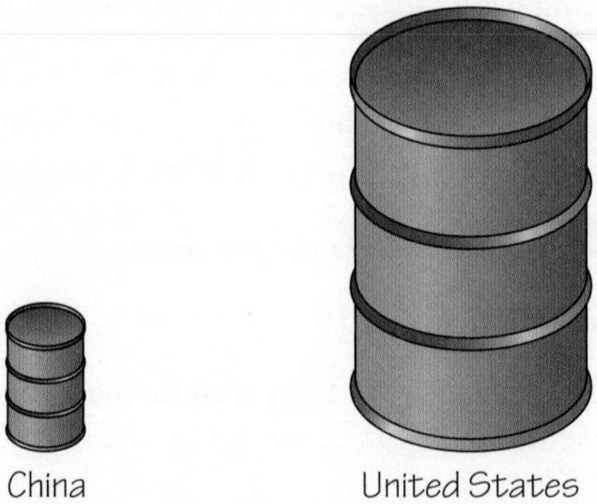

China United States

24. Braking Distance Data Set 14 in Appendix B lists braking distances (ft) of different cars, and the braking distances of three of those cars are shown in the accompanying illustration.

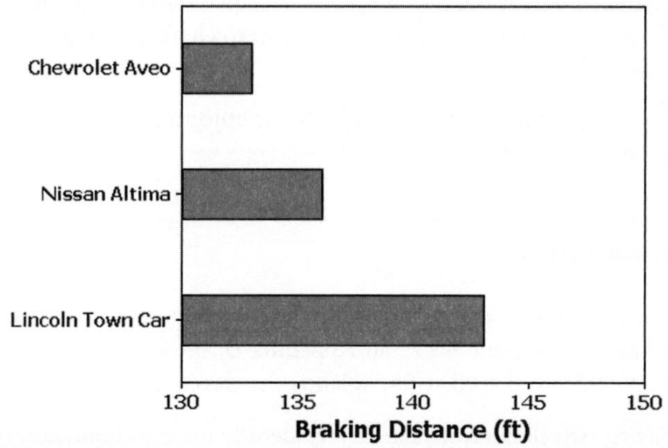

24. The actual braking distances are 133 ft, 136 ft, and 143 ft, so the differences are relatively small, but the illustration has a scale that begins at 130 ft, so the differences are grossly exaggerated.

2-4 Beyond the Basics

25. Back-to-Back Stemplots Exercise 19 in Section 2-3 used back-to-back relative frequency histograms for the ages of actresses and actors that are listed in Data Set 11 of Appendix B. Use the same method to construct back-to-back stemplots of the ages of actresses and actors, and then use the results to compare the two data sets.

26. Expanded and Condensed Stemplots

a. A stemplot can be *expanded* by subdividing rows into those with leaves having digits of 0 through 4 and those with leaves having digits 5 through 9. Using the body temperatures from 12 AM on Day 2 listed in Data Set 3 of Appendix B, the first three rows of an expanded stemplot have stems of 96 (for leaves between 5 and 9 inclusive), 97 (for leaves between 0 and 4 inclusive), and 97 (for leaves between 5 and 9 inclusive). Construct the complete expanded stemplot for the body temperatures from 12 AM on Day 2 listed in Data Set 3 of Appendix B.

b. A stemplot can be *condensed* by combining adjacent rows. Using the LDL cholesterol measurements from males in Data Set 1 of Appendix B, we obtain the first two rows of the condensed stemplot as shown below. Note that we insert an asterisk to separate digits in the leaves associated with the numbers in each stem. Every row in the condensed plot must include exactly one asterisk so that the shape of the condensed stemplot is not distorted. Complete the condensed stemplot. What is an advantage of using a condensed stemplot instead of one that is not condensed?

```
6-7 | 79*778
8-9 | 45678*049
```

26. a.

```
96. | 59
97. | 0001112333444
97. | 5566666678888899
98. | 0000000000000022222334444444444444
98. | 5555666666666666666677777888888899
99. | 001244
99. | 56
```

b. The condensed stemplot reduces the number of rows so that the stemplot is not too large to be understandable.

```
6-7   | 79*778
8-9   | 45678*049
10-11 | 348*234477
12-13 | 01234*5
14-15 | 05*4569
16-17 | *049
18-19 | *6
20-21 | 1*3
```

Chapter 2 Review

This chapter presented methods for organizing, summarizing, and graphing data sets. When one is investigating a data set, the characteristics of center, variation, distribution, outliers, and changing pattern over time are generally very important, and this chapter includes a variety of tools for investigating the distribution of the data. After completing this chapter, you should be able to do the following:

- Construct a frequency distribution or relative frequency distribution to summarize data (Section 2-2).
- Construct a histogram or relative frequency histogram to show the distribution of data (Section 2-3).

Note to Instructor

Reminder: The student edition of this book includes answers to *all* Chapter Quick Quiz exercises, *all* Chapter Review Exercises, and *all* Cumulative Review Exercises. For the regular chapter sections, the student edition continues to include answers for the odd-numbered exercises only.

- Examine a histogram or normal quantile plot to determine whether sample data appear to be from a population having a normal distribution (Section 2-3).
- Construct graphs of data using a scatterplot (for paired data), frequency polygon, dotplot, stemplot, bar graph, multiple bar graph, Pareto chart, pie chart, or time-series graph (Section 2-4).
- Critically analyze a graph to determine whether it objectively depicts data or is somehow misleading or incorrect (Section 2-4).

Chapter Quick Quiz

1. When one is constructing a table representing the frequency distribution of weights (lb) of discarded textile items from Data Set 23 in Appendix B, the first two classes of a frequency distribution are 0.00–0.99 and 1.00–1.99. What is the class width?

2. Using the same first two classes from Exercise 1, identify the class boundaries of the first class.

3. The first class described in Exercise 1 has a frequency of 51. If you know only the class limits given in Exercise 1 and the frequency of 51, can you identify the original 51 data values?

4. A stemplot is created from the intervals (min) between eruptions of the Old Faithful geyser in Yellowstone National Park, and one row of that stemplot is 6 | 1222279. Identify the values represented by that row.

5. In the California Daily 4 lottery, four digits between 0 and 9 inclusive are randomly selected each day. We normally expect that each of the ten different digits will occur about 1/10 of the time, and an analysis of last year's results shows that this did happen. Because the results are what we normally expect, is it correct to say that the distribution of selected digits is a normal distribution?

6. In an investigation of the travel costs of college students, which of the following does not belong: center; variation; distribution; bar graph; outliers; changing patterns over time?

7. In an investigation of the relationship between SAT scores and grade point averages (GPA) of college students, which of the following graphs is most helpful: histogram; pie chart; scatterplot; stemplot; dotplot?

8. As a quality control manager at Sony, you find that defective CDs have various causes, including worn machinery, human error, bad supplies, and packaging mistreatment. Which of the following graphs would be best for describing the causes of defects: histogram; scatterplot; Pareto chart; dotplot; pie chart?

9. What characteristic of a data set can be better understood by constructing a histogram?

10. A histogram is to be constructed from the brain sizes listed in Data Set 6 of Appendix B. Without actually constructing that histogram, simply identify two key features of the histogram that would suggest that the data have a *normal distribution.*

Review Exercises

1. Frequency Distribution of Brain Volumes Construct a frequency distribution of the 20 brain volumes (cm^3) listed below. (These volumes are from Data Set 6 of Appendix B.) Use the classes 900–999, 1000–1099, and so on.

1005	963	1035	1027	1281	1272	1051	1079	1034	1070
1173	1079	1067	1104	1347	1439	1029	1100	1204	1160

1. 1.00
2. −0.005 and 0.995
3. No
4. 61 min, 62 min, 62 min, 62 min, 62 min, 67 min, 69 min
5. No
6. Bar graph
7. Scatterplot
8. Pareto chart
9. The distribution of the data
10. The bars of the histogram start relatively low, increase to some maximum, and then decrease. Also, the histogram is symmetric with the left half being roughly a mirror image of the right half.

Review Exercises

1.

Volume (cm^3)	Frequency
900–999	1
1000–1099	10
1100–1199	4
1200–1299	3
1300–1399	1
1400–1499	1

2. No, the distribution does not appear to be normal because the graph is not symmetric.

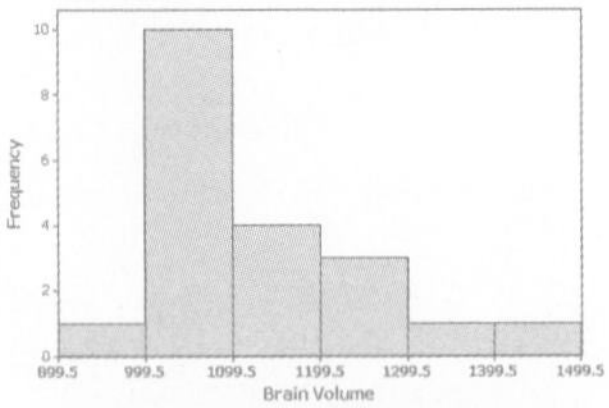

3. Although there are differences among the frequencies of the digits, the differences are not too extreme given the relatively small sample size, so the lottery appears to be fair.

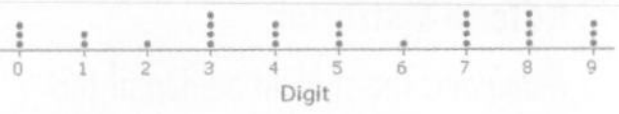

4. The sample size is not large enough to reveal the true nature of the distribution of IQ scores for the population from which the sample is obtained.

8	779
9	66
10	133

2. Histogram of Brain Volumes Construct the histogram that corresponds to the frequency distribution from Exercise 1. Applying a very strict interpretation of the requirements for a normal distribution, does the histogram suggest that the data are from a population having a normal distribution? Why or why not?

3. Dotplot of California Lottery In the California Daily 4 lottery, four digits are randomly selected each day. Listed below are the digits that were selected in one recent week. Construct a dotplot. Does the dotplot suggest that the lottery is fair?

5 3 8 9 2 9 1 1 3 0 9 7 3 8 7 4 7 4 8 5 6 8 0 0 4 7 5 3

4. Stemplot of IQ Scores Listed below are the first eight IQ scores from Data Set 6 in Appendix B. Construct a stemplot of these eight values. Is this data set large enough to reveal the true nature of the distribution of IQ scores for the population from which the sample is obtained?

96 89 87 87 101 103 103 96

5. CO Emissions Listed below are the amounts (million metric tons) of carbon monoxide emissions in the United States for each year of a recent ten-year period. The data are listed in order. Construct the graph that is most appropriate for these data. What type of graph is best? What does the graph suggest?

5638 5708 5893 5807 5881 5939 6024 6032 5946 6022

6. CO and NO Emissions Exercise 5 lists the amounts of carbon monoxide emissions, and listed below are the amounts (million metric tons) of nitrous oxide emissions in the United States for the same ten-year period as in Exercise 5. What graph is best for exploring the relationship between carbon monoxide emissions and nitrous oxide emissions? Construct that graph. Does the graph suggest that there is a relationship between carbon monoxide emissions and nitrous oxide emissions?

351 349 345 339 335 335 362 371 376 384

7. Sports Equipment According to *USA Today*, the largest categories of sports equipment sales are as follows: fishing ($2.0 billion); firearms and hunting ($3.1 billion); camping ($1.7 billion); golf ($2.5 billion). Construct the graph that best depicts these different categories and their relative amounts. What type of graph is best?

5. A time-series graph is best. It suggests that the amounts of carbon monoxide emissions in the United States are increasing.

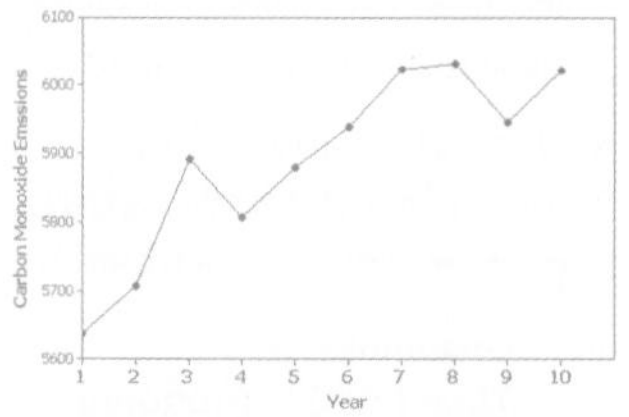

6. A scatterplot is best. The scatterplot does not suggest that there is a relationship.

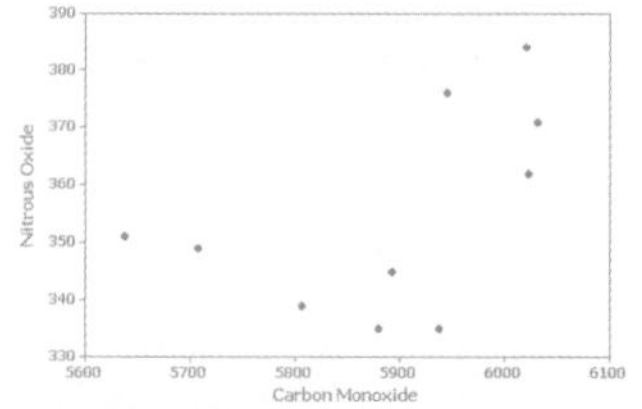

7. A Pareto chart is best.

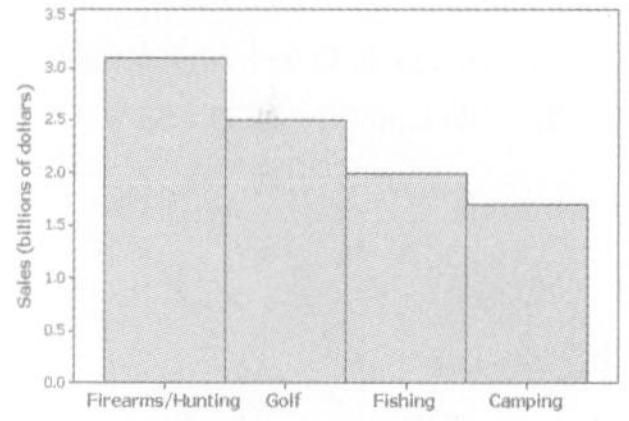

Cumulative Review Exercises

In Exercises 1–5, refer to the table in the margin, which summarizes results from 641 people who responded to a* USA Today *survey. Participants responded to this question: "Who do you most like to get compliments from at work?"

Response	Frequency
Co-workers	260
Boss	241
Strangers	82
People who report to me	58

1. Graph Which of the following graphs would be best for visually illustrating the data in the table: histogram; dotplot; scatterplot; Pareto chart; stemplot?

2. Level of Measurement Is the level of measurement of the 641 individual responses nominal, ordinal, interval, or ratio? Why?

3. Sampling The results in the table were obtained by posting the question on a Web site, and readers of *USA Today* could respond to the question if they chose to. What is this type of sampling called? Is this type of sample likely to be representative of the population of all workers? Why or why not?

4. Misleading Graph How is the accompanying graph misleading? How could it be modified so that it would not be misleading? (The graph is on the top of the next page.)

1. Pareto chart

2. Nominal, because the responses consist of names only. The responses do not measure or count anything, and they cannot be arranged in order according to some quantitative scale.

3. Voluntary response sample (or self-selected sample). The voluntary response sample is not likely to be representative of the population, because those with special interests or strong feelings about the topic are more likely than others to respond, and their views might be very different from those of the general population.

4. By using a vertical scale that does not begin at 0, the graph exaggerates the differences in the numbers of responses. The graph could be modified by starting the vertical scale at 0 instead of 50.
5. 37.6% chose the category of boss. Because it is based on a sample (not on the population), that percentage is a statistic.
6.

Grooming Time (min)	Frequency
0–9	2
10–19	3
20–29	9
30–39	4
40–49	2

7. Because the frequencies increase to a maximum and then decrease, and the left half of the histogram is roughly a mirror image of the right half, the data do appear to be from a population with a normal distribution.

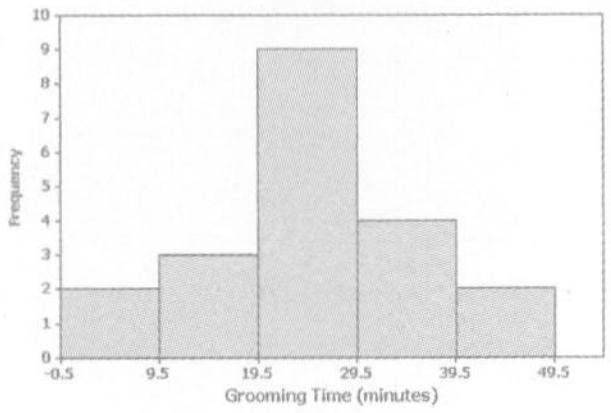

8. Stemplot.

0	05
1	255
2	024555778
3	0055
4	05

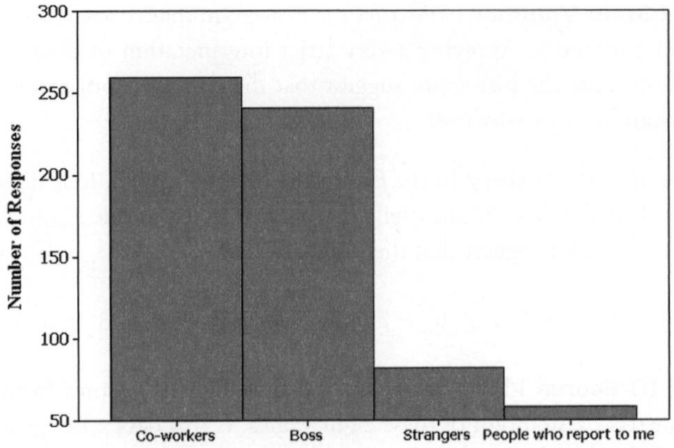

5. Statistic or Parameter? Among the 641 people who responded to the *USA Today* survey, what is the percentage of respondents who chose the category of boss? When considered in the context of the population of all workers, is that percentage a statistic or a parameter? Explain.

6. Grooming Time Listed below are times (minutes) spent on hygiene and grooming in the morning (by randomly selected subjects) (based on data from a Svenska Cellulosa Aktiebolaget survey). Construct a table representing the frequency distribution. Use the classes 0–9, 10–19, and so on.

0 5 12 15 15 20 22 24 25 25 25 27 27 28 30 30 35 35 40 45

7. Histogram of Grooming Times Use the frequency distribution from Exercise 6 to construct a histogram. Based on the result, do the data appear to be from a population with a normal distribution? Explain.

8. Stemplot of Grooming Times Use the data from Exercise 6 to construct a stemplot.

Technology Project

It was noted in this section that the days of charming and primitive hand-drawn graphs are well behind us, and technology now provides us with powerful tools for generating a wide variety of different graphs. The data sets in Appendix B are available as files that can be opened by statistical software packages, such as STATDISK, Minitab, Excel, SPSS, and SAS. Use a statistical software package to open the male and female body measurements from Data Set 1 in Appendix B. Use the statistical software with the methods of this chapter to describe, explore, and compare the blood platelet measurements of males and females. Does there appear to be a gender difference in blood platelet counts? When analyzing blood platelet counts of patients, should physicians take the gender of patients into account? Support your conclusions with printouts of suitable graphs. (Later chapters will present more formal methods for making such comparisons.)

from data TO DECISION

Flight Planning

Data Set 15 in Appendix B includes data about American Airline flights from New York (JFK airport) to Los Angeles (LAX airport). The data are from the Bureau of Transportation.

Critical Thinking

Use the methods from this chapter to address the following questions.

1. Is there a relationship between taxi-out times at JFK and taxi-in times at LAX? Explain.

2. Is there a relationship between departure delay times at JFK and arrival delay times at LAX? Explain.

3. Arrival delay times are important because they can affect the plans of passengers. Explore the arrival delay times by using the methods of this chapter and comment on the results. Is there very small variation among the arrival delay times? Are there any outliers? What is the nature of the distribution of arrival delay times? Based on the results, are arrival delay times very predictable?

Cooperative Group Activities

1. In-class activity Using a package of purchased chocolate chip cookies, each student should be given two or three cookies. Proceed to count the number of chocolate chips in each cookie. Not all of the chocolate chips are visible, so "destructive testing" must be used through a process involving consumption. Record the numbers of chocolate chips for each cookie and combine all results. Construct a frequency distribution, histogram, dotplot, and stemplot of the results. Given that the cookies were made through a process of mass production, we might expect that the numbers of chips per cookie would not vary much. Is that indicated by the results? Explain. (See "Chocolate Chip Cookies as a Teaching Aid" by Herbert K. H. Lee, *American Statistician,* Vol. 61, No. 4.)

2. In-class activity In class, each student should record two pulse rates by counting the number of her or his heartbeats in one minute. The first pulse rate should be measured while seated, and the second pulse rate should be measured while standing. Using the pulse rates measured while seated, construct a frequency distribution and histogram for the pulse rates of males, and then construct another frequency distribution and histogram for the pulse rates of females. Using the pulse rates measured while standing, construct a frequency distribution and histogram for the pulse rates of males, and then construct another frequency distribution and histogram for the pulse rates of females. Compare the results. Do males and females appear to have different pulse rates? Do pulse rates measured while seated appear to be different from pulse rates measured while standing? Use an appropriate graph to determine whether there is a relationship between sitting pulse rate and standing pulse rate.

3. In-class activity Given below are recent measurements from the Old Faithful geyser in Yellowstone National Park. The time intervals between eruptions are matched with the corresponding times of duration of the geyser; thus the interval of 76 min is paired with the duration of 4.53 min, the interval of 84 min is paired with the duration of 3.83 min, and so on. Use the methods of this chapter to summarize and explore each of the two sets of data separately, and then investigate whether there is some relationship between them. Describe the methods used and the conclusions reached.

Intervals (min) between eruptions

76	84	76	103	92	47	98	54	80	91	69	86	83	75	93
89	96	65	94	85	94	60	94	86	93	88	61	96	52	98

Durations (min) of eruptions

4.53	3.83	3.83	4.23	4.70	1.83	4.00	2.00	3.57	4.25	2.75	4.47	3.35	3.27	4.30
4.25	4.05	2.12	4.63	4.18	4.05	2.13	4.60	4.53	3.70	4.17	1.87	4.68	1.83	4.10

4. Out-of-class activity Search newspapers and magazines to find an example of a graph that is misleading. (See Examples 13 and 14 in Section 2-4.) Describe how the graph is misleading. Redraw the graph so that it depicts the information correctly.

5. Out-of-class activity Obtain a copy of *The Visual Display of Quantitative Information,* second edition, by Edward Tufte (Graphics Press, PO Box 430, Cheshire, CT 06410). Find the graph describing Napoleon's march to Moscow and back, and explain why Tufte says that "it may well be the best graphic ever drawn."

6. Out-of-class activity Obtain a copy of *The Visual Display of Quantitative Information,* second edition, by Edward Tufte (Graphics Press, PO Box 430, Cheshire, CT 06410). Find the graph that appeared in *American Education*, and explain why Tufte says that "this may well be the worst graphic ever to find its way into print." Construct a graph that is effective in depicting the same data.

3 Statistics for Describing, Exploring, and Comparing Data

3

chapter problem

How many chips are in a chocolate chip cookie?

This edition of *Elementary Statistics* and previous editions have included data obtained from M&M plain candies. This Chapter Problem continues the legacy of using snack foods for statistical purposes, and the choice here is chocolate chip cookies as suggested by the article "Chocolate Chip Cookies as a Teaching Aid," by Herbert Lee (*The American Statistician,* Vol. 61, No. 4).

Table 3-1 lists the numbers of chocolate chips counted in different brands. The counts were obtained by the author, who found that the counting process was not as simple as it might seem. What do you do with loose chocolate chips that were found in each package? Some chocolate chips were stuck together, so they had to be counted with great care. Care also had to be taken to not count nut particles as chocolate chips. There were some small fragments that were not counted after the author made an arbitrary decision about the minimum size required to be counted as an official chocolate chip.

The counts in Table 3-1 do not include weights of the chocolate chips, and the Hannaford brand had many that were substantially larger than any of the chocolate chips in the other brands. Also, there is an issue with the sampling method. The author used all of the cookies in one package from each of the different brands. A better sampling method would involve randomly selecting cookies from different packages obtained throughout the country, and this would have required extensive travel by the author—a prospect with some appeal. In developing this Chapter Problem, the author learned much about chocolate chip cookies, including the observation that a huge pile of crushed chocolate chip cookies has absolutely none of the appeal of a single cookie untouched by human hands.

Table 3-1 Numbers of Chocolate Chips in Different Brands of Cookies

Chips Ahoy (regular)																			
22	22	26	24	23	27	25	20	24	26	25	25	19	24	20	22	24	25	25	20
23	30	26	20	25	28	19	26	26	23	25	23	23	23	22	26	27	23	28	24
Chips Ahoy (chewy)																			
21	20	16	17	16	17	20	22	14	20	19	17	20	21	21	18				
20	20	21	19	22	20	20	19	16	19	16	15	24	23	14	24				
Chips Ahoy (reduced fat)																			
13	24	18	16	21	20	14	20	18	12	24	23	28	18	18	19	22	21	22	16
13	20	20	23	24	20	17	20	19	21	27	16	24	19	23	25	14	18	15	19
Keebler																			
29	31	25	32	27	31	30	29	31	26	32	33	32	30	33	29	30			
28	32	35	37	31	24	30	30	34	29	27	24	38	37	32	26	30			
Hannaford																			
13	15	16	21	15	14	14	15	13	13	16	11								
14	12	13	12	14	12	16	17	14	16	14	15								

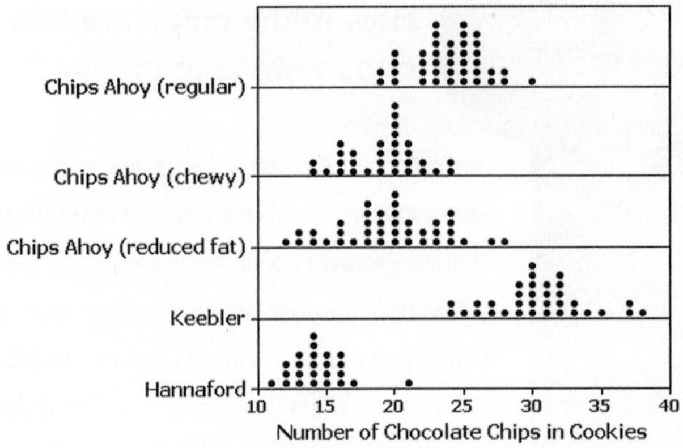

Figure 3-1 Dotplot of Numbers of Chocolate Chips in Cookies

Figure 3-1 is a dotplot (described in Section 2-4) that includes all of the cookies from Table 3-1. Figure 3-1 shows some obvious differences. Instead of relying solely on subjective interpretations of a graph like Figure 3-1, this chapter introduces measures that are essential to any study of statistics. The mean, median, standard deviation, and variance are among the most important statistics presented in this chapter, and those statistics will be used in our description, exploration, and comparison of the counts in Table 3-1.

3-1 Review and Preview

Note to Instructor

Recommendation: Don't require memorization of formulas and don't require that students master the mechanics of manual calculations. Allow the use of a computer or calculator for obtaining results, and really stress the *interpretation* of results. But don't *totally* dismiss manual calculations, which have some merit in some cases. When introducing standard deviation, for example, demonstrate the manual calculations for a small data set and show how the formula accomplishes the objective of measuring variation. On tests, allow students to use a computer or calculator, then ask questions designed to test an *understanding* of the concepts.

Chapter 1 discussed methods of collecting sample data. Chapter 2 presented frequency distributions and a variety of different graphs that help us summarize and visualize data. In Chapter 2 we noted that when describing, exploring, and comparing data sets, these characteristics are usually extremely important: (1) center; (2) variation; (3) distribution; (4) outliers; and (5) changing characteristics of data over time. In this chapter we introduce important statistics, including the mean, median, and standard deviation. Upon completing this chapter, you should be able to find the mean, median, standard deviation, and variance from a data set, and you should be able to clearly understand and interpret such values. It is especially important to understand values of standard deviation by using tools such as the range rule of thumb described in Section 3-3.

Critical Thinking and Interpretation: Beyond Formulas

This chapter includes several formulas used to compute basic statistics. Because many of these statistics can be easily calculated by using technology, it is not so important for us to memorize formulas and manually perform complex calculations. Instead, we should focus on understanding and interpreting the values we obtain from them.

The methods and tools presented in Chapter 2 and in this chapter are often called methods of **descriptive statistics,** because they summarize or describe relevant characteristics of data. Later in this book, we will use **inferential statistics** to make inferences, or generalizations, about a population.

Note to Instructor

Section 3.2 is partitioned into two parts: Part 1 (Basic Concepts of Measures of Center) and Part 2 (Beyond the Basics of Measures of Center). Part 1 introduces the basic definitions of mean, median, mode, and midrange. Part 2 includes frequency distributions, weighted mean, and skewness.

Recommendation: For a one-semester introductory statistics course, omit Part 2.

3-2 Measures of Center

Key Concept The focus of this section is the characteristic of center of a data set. In particular, we present measures of center, including *mean* and *median,* as tools for analyzing data. Our objective here is not only to find the value of each measure of center, but also to interpret those values. Part 1 of this section includes core concepts that should be understood before considering Part 2.

Part 1: Basic Concepts of Measures of Center

In this Part 1, we introduce the mean, median, mode, and midrange as different measures of center.

DEFINITION A **measure of center** is a value at the center or middle of a data set.

There are several different ways to determine the center, so we have different definitions of measures of center, including the mean, median, mode, and midrange. We begin with the mean.

Mean

The mean (or arithmetic mean) is generally the most important of all numerical measurements used to describe data, and it is what most people call an *average*.

DEFINITION The **arithmetic mean,** or the **mean,** of a set of data is the measure of center found by adding the data values and dividing the total by the number of data values.

Important Properties of the Mean

- Sample means drawn from the same population tend to vary less than other measures of center.
- The mean of a data set uses every data value.
- A disadvantage of the mean is that just one extreme value (outlier) can change the value of the mean substantially. (Since the mean cannot resist substantial changes caused by extreme values, we say that the mean is not a *resistant* measure of center.)

Calculation and Notation of the Mean

The definition of the mean can be expressed as Formula 3-1, in which the Greek letter Σ (uppercase sigma) indicates that the data values should be added, so Σx represents the sum of all data values. The symbol n denotes the **sample size,** which is the number of data values.

Formula 3-1

$$\text{mean} = \frac{\Sigma x}{n} \quad \frac{\leftarrow \text{sum of all data values}}{\leftarrow \text{number of data values}}$$

If the data are a *sample* from a population, the mean is denoted by $\bar{x}$ (pronounced "x-bar"); if the data are the entire population, the mean is denoted by μ (lowercase Greek mu). (Sample statistics are usually represented by English letters, such as $\bar{x}$, and population parameters are usually represented by Greek letters, such as μ.)

Average Bob

According to Kevin O'Keefe, author of *The Average American: The Extraordinary Search for the Nation's Most Ordinary Citizen*, Bob Burns is the most average person in the United States.

O'Keefe spent 2 years using 140 criteria to identify the single American who is most average. He identified statistics revealing preferences of the majority and applied them to the many people he encountered. Bob Burns is the only person who satisfied all of the 140 criteria. Bob Burns is 5 ft 8 in. tall, weighs 190 pounds, is 54 years of age, married, has three children, wears glasses, works 40 hours per week, drives an eight-year-old car, has an outdoor grill, mows his own lawn, drinks coffee each day, and walks his dog each evening.

Class Size Paradox

There are at least two ways to obtain the mean class size, and they can have very different results. At one college, if we take the numbers of students in 737 classes, we get a mean of 40 students. But if we were to compile a list of the class sizes for each student and use this list, we would get a mean class size of 147. This large discrepancy is due to the fact that there are many students in large classes, while there are few students in small classes. Without changing the number of classes or faculty, we could reduce the mean class size experienced by students by making all classes about the same size. This would also improve attendance, which is better in smaller classes.

Notation

Σ denotes the *sum* of a set of data values.

x is the *variable* usually used to represent the individual data values.

n represents the *number of data values* in a *sample.*

N represents the *number of data values* in a *population.*

$\bar{x} = \frac{\Sigma x}{n}$ is the mean of a set of *sample* values.

$\mu = \frac{\Sigma x}{N}$ is the mean of all values in a *population.*

Example 1 Mean

Table 3-1 includes counts of chocolate chips in different cookies. Find the mean of the first five counts for Chips Ahoy regular cookies: 22 chips, 22 chips, 26 chips, 24 chips, and 23 chips.

Solution

The mean is computed by using Formula 3-1. First add the data values, then divide by the number of data values:

$$\bar{x} = \frac{\Sigma x}{n} = \frac{22 + 22 + 26 + 24 + 23}{5} = \frac{117}{5}$$
$$= 23.4 \text{ chips}$$

The mean of the first five chip counts is 23.4 chips.

Median

The median can be thought of loosely as a "middle value" in the sense that about half of the values in a data set are less than the median and half are greater than the median. The following definition is more precise.

DEFINITION The **median** of a data set is the measure of center that is the *middle value* when the original data values are arranged in order of increasing (or decreasing) magnitude.

Calculation and Notation of the Median The median is often denoted by $\tilde{x}$ (pronounced "*x*-tilde"). To find the median, first *sort* the values (arrange them in order), and then follow one of these two procedures:

1. If the number of data values is odd, the median is the number located in the exact middle of the sorted list.
2. If the number of data values is even, the median is found by computing the mean of the two middle numbers in the sorted list.

Important Properties of the Median

- The median does not change by large amounts when we include just a few extreme values (so the median is a *resistant* measure of center).
- The median does not use every data value.

Example 2 Median

Find the median of the five sample values used in Example 1: 22 chips, 22 chips, 26 chips, 24 chips, and 23 chips.

Solution

First sort the data values by arranging them in order, as shown below:

22 22 **23** 24 26

Because the number of data values is an odd number (5), the median is the number located in the exact middle of the sorted list, which is 23. The median is therefore 23 chips. Note that the median of 23 chips is different from the mean of 23.4 chips found in Example 1.

Example 3 Median

Repeat Example 2 after including the sixth count of 27 chips. That is, find the median of these counts: 22 chips, 22 chips, 26 chips, 24 chips, 23 chips, and 27 chips.

Solution

First arrange the values in order:

22 22 **23** **24** 26 27

Because the number of data values is an even number (6), the median is found by computing the mean of the two middle numbers, which are 23 and 24.

$$\text{Median} = \frac{23 + 24}{2} = \frac{47}{2} = 23.5$$

The median is 23.5 chips.

CAUTION Never use the term *average* when referring to a measure of center. The word *average* is often used for the mean, but it is sometimes used for other measures of center. To avoid any confusion, we use the correct and specific term, such as *mean* or *median*; *average* is not used by statisticians and it will not be used throughout the remainder of this book when referring to a specific measure of center.

Mode

The mode is another measure of center.

DEFINITION The **mode** of a data set is the value that occurs with the greatest frequency.

Finding the Mode: A data set can have one mode, more than one mode, or no mode.

- When two data values occur with the same greatest frequency, each one is a mode and the data set is **bimodal.**
- When more than two data values occur with the same greatest frequency, each is a mode and the data set is said to be **multimodal.**
- When no data value is repeated, we say that there is **no mode.**

Example 4 **Mode**

Find the mode of these same values used in Example 1: 22 chips, 22 chips, 26 chips, 24 chips, and 23 chips.

Solution

The mode is 22 chips, because it is the data value with the greatest frequency.

In Example 4 the mode is a single value. Here are two other possible circumstances:

Two modes: The values of 22, 22, 22, 23, 23, 23, 24, 24, 26, and 27 have two modes: 22 and 23.

No mode: The values of 22, 23, 24, 26, and 27 have no mode because no value is repeated.

In reality, the mode isn't used much with numerical data. However, the mode is the only measure of center that can be used with data at the nominal level of measurement. (Remember, the nominal level of measurement applies to data that consist of names, labels, or categories only.)

Note to Instructor
The midrange can be easily omitted. It is included mainly to emphasize the point that there are different ways of determining the "center" of a set of data.

Midrange

Another measure of center is the midrange.

DEFINITION The **midrange** of a data set is the measure of center that is the value midway between the maximum and minimum values in the original data set. It is found by adding the maximum data value to the minimum data value and then dividing the sum by 2, as in the following formula:

$$\text{midrange} = \frac{\text{maximum data value} + \text{minimum data value}}{2}$$

Important Properties of the Midrange

- Because the midrange uses only the maximum and minimum values, it is very sensitive to those extremes.
- In practice, the midrange is rarely used, but it has three redeeming features:
  1. The midrange is very easy to compute.
  2. The midrange helps reinforce the very important point that there are several different ways to define the center of a data set.
  3. The value of the midrange is sometimes used incorrectly for the median, so confusion can be reduced by clearly defining the midrange along with the median.

Example 5 **Midrange**

Find the midrange of these values from Example 1: 22 chips, 22 chips, 26 chips, 24 chips, and 23 chips.

Solution

The midrange is found as follows:

$$\text{Midrange} = \frac{\text{maximum data value} + \text{minimum data value}}{2}$$

$$= \frac{26 + 22}{2} = 24$$

The midrange is 24.0 chips.

When calculating measures of center, we often need to round the result. We use the following rule.

Round-Off Rules:

For the mean, median, and midrange, carry one more decimal place than is present in the original set of values.

For the mode, leave the value as is without rounding (because values of the mode are the same as some of the original data values).

Note to Instructor
Try to encourage students to not round too much in the middle of a calculation; instead, carry as many places as the calculator will handle. Only round at the end. Large errors sometimes occur when intermediate results are rounded too much. Round-off rules in the introductory statistics course can cause some confusion, because there are several of them.
Recommendation: Don't be too critical of violations of round-off rules.

When applying this rule, round only the final answer, *not intermediate values that occur during calculations.* For example, the mean of 2, 3, and 5 is 3.333333 . . . , which is rounded to 3.3, which has one more decimal place than the original values of 2, 3, and 5. As another example, the mean of 80.4 and 80.6 is 80.50 (one more decimal place than was used for the original values). Because the mode is one or more of the original data values, we do not round values of the mode; we simply use the same original values.

Critical Thinking

Although we can calculate measures of center for a set of sample data, we should always think about whether the results are reasonable. In Section 1-3 we noted that it does not make sense to do numerical calculations with data at the nominal level of measurement, because those data consist of names, labels, or categories only, so statistics such as the mean and median are meaningless. We should also think about the sampling method used to collect the data. If the sampling method is not sound, the statistics we obtain may be very misleading.

Example 6 Critical Thinking and Measures of Center

For each of the following, identify a major reason why the mean and median are *not* meaningful statistics.

a. Zip codes of the author, White House, Air Force division of the Pentagon, Empire State Building, and Statue of Liberty: 12590, 20500, 20330, 10118, 10004.

b. Rank (by sales) of selected statistics textbooks: 1, 4, 5, 3, 2, 15.

c. Numbers on the jerseys of the starting offense for the New Orleans Saints when they won Super Bowl XLIV: 12, 74, 77, 76, 73, 78, 88, 19, 9, 23, 25.

Solution

a. The zip codes don't measure or count anything. The numbers are actually labels for geographic locations.

b. The ranks reflect an ordering, but they don't measure or count anything. The rank of 1 is from a book with sales substantially greater then the book with rank of 2, so the different numbers don't correspond to the magnitudes of the sales.

c. The numbers on the football jerseys don't measure or count anything. Those numbers are simply substitutes for names.

Example 6 involves data that do not justify the use of statistics such as the mean or median. Example 7 involves a more subtle issue.

Example 7 Class Size

It is well known that smaller classes are generally more effective. According to the National Center for Education Statistics, California has a mean of 20.9 students per teacher, and Alaska has a mean of 16.8 students per teacher. (These values are based on elementary and secondary schools only.) If we combine the two states, we might find the mean number of students per teacher to be 18.85, as in the calculation shown below, but is this result correct? Why or why not?

$$\bar{x} = \frac{20.9 + 16.8}{2} = 18.85$$

Solution

The combined states of California and Alaska do *not* have a mean of 18.85 students per teacher. The issue here is that California has substantially more students and teachers than Alaska, and those different numbers should be taken into account. Combining California and Alaska, we get a total of 6,539,429 students and 315,013 teachers, so the student/teacher ratio is 6,539,429/315,013 = 20.8 (not the value of 18.85 from the above calculation). When using values from different sample sizes, consider whether those sample sizes should be taken into account.

As another illustration of the principle in Example 7, if you have a list of the 50 state per capita income amounts and you find the mean of those 50 values, the result is not the mean per capita income for the entire United States. The population sizes of the 50 different states must be taken into account. (See the *weighted mean* in the following subsection.)

Note to Instructor
Omit Part 2 if you find that Chapters 2 and 3 are taking longer than you had planned.

Part 2: Beyond the Basics of Measures of Center

Calculating the Mean from a Frequency Distribution

The first two columns of Table 3-2 shown here are the same as the frequency distribution of Table 2-2 from Chapter 2. When working with data summarized in a frequency distribution, we don't know the exact values falling in a particular class, so we make calculations possible by pretending that all sample values in each class are equal to the class midpoint. For example, consider the first class interval of 50–69 with a

Table 3-2 IQ Scores of Low Lead Group

IQ Score	Frequency f	Class Midpoint x	$f \cdot x$
50–69	2	59.5	119.0
70–89	33	79.5	2623.5
90–109	35	99.5	3482.5
110–129	7	119.5	836.5
130–149	1	139.5	139.5
Totals:	$\Sigma f = 78$		$\Sigma(f \cdot x) = 7201.0$

frequency of 2. We pretend that each of the two IQ scores is 59.5 (the class midpoint). With the IQ score of 59.5 repeated twice, we have a total of $59.5 \cdot 2 = 119$ as shown in the last column of Table 3-2. We can then add such products to find the total of all sample values, which we then divide by the sum of the frequencies, Σf. Formula 3-2 is used to compute the mean when the sample data are summarized in a frequency distribution. Formula 3-2 is not really a new concept; it is simply a variation of Formula 3-1.

Formula 3-2

First multiply each frequency and class midpoint; then add the products.

↓

Mean from frequency distribution: $$\bar{x} = \frac{\Sigma(f \cdot x)}{\Sigma f}$$

↑

Sum of frequencies

The following example illustrates the procedure for finding the mean from a frequency distribution.

Example 8 Computing Mean from a Frequency Distribution

The first two columns of Table 3-2 constitute a frequency distribution summarizing the full IQ scores of the low lead group in Data Set 5 from Appendix B. Use that frequency distribution to find the mean.

Solution

Table 3-2 illustrates the procedure for using Formula 3-2 when calculating a mean from data summarized in a frequency distribution. The class midpoint values are shown in the third column, and the products $f \cdot x$ are shown in the last column. The bottom row of Table 3-2 shows the two components we need for the calculation of the mean (as in Formula 3-2): $\Sigma f = 78$ and $\Sigma(f \cdot x) = 7201.0$. The calculation using Formula 3-2 is as follows:

$$\bar{x} = \frac{\Sigma(f \cdot x)}{\Sigma f} = \frac{7201.0}{78} = 92.3$$

The result of $\bar{x} = 92.3$ is an approximation because it is based on the use of class midpoint values instead of the original list of full IQ scores. The mean of 92.9 found by using all of the original full IQ scores is a better result.

Calculating a Weighted Mean

When different x data values are assigned different weights w, we can compute a **weighted mean.** Formula 3-3 can be used to compute the weighted mean.

Formula 3-3

$$\text{weighted mean: } \bar{x} = \frac{\Sigma(w \cdot x)}{\Sigma w}$$

Formula 3-3 tells us to first multiply each weight w by the corresponding value x, then to add the products, and then finally to divide that total by the sum of the weights, Σw.

Example 9 Computing Grade Point Average

In her first semester of college, a student of the author took five courses. Her final grades along with the number of credits for each course were A (3 credits), A (4 credits), B (3 credits), C (3 credits), and F (1 credit). The grading system assigns quality points to letter grades as follows: A = 4; B = 3; C = 2; D = 1; F = 0. Compute her grade point average.

Solution

Use the numbers of credits as weights: $w = 3, 4, 3, 3, 1$. Replace the letter grades of A, A, B, C, and F with the corresponding quality points: $x = 4, 4, 3, 2, 0$. We now use Formula 3-3 as shown below. The result is a first-semester grade point average of 3.07. (Using the preceding round-off rule, the result should be rounded to 3.1, but it is common to round grade point averages to two decimal places.)

$$\bar{x} = \frac{\Sigma(w \cdot x)}{\Sigma w}$$

$$= \frac{(3 \times 4) + (4 \times 4) + (3 \times 3) + (3 \times 2) + (1 \times 0)}{3 + 4 + 3 + 3 + 1}$$

$$= \frac{43}{14} = 3.07$$

using TECHNOLOGY

The calculations of this section are fairly simple, but some of the calculations in the following sections are not so simple. Many computer software programs allow you to enter a data set and use one operation to get several different sample statistics, referred to as *descriptive statistics*. Here are some of the procedures for obtaining such displays. (The accompanying displays are based on the numbers of chocolate chips in Chips Ahoy regular cookies, as listed in Table 3-1.)

STATDISK Enter the data in the Data Window or open an existing data set. Click on **Data,** select **Descriptive Statistics,** and enter the column number for the desired data set. Now click on **Evaluate** to get the various descriptive statistics, including the mean, median, midrange, and other statistics to be discussed in the following sections. (Click on **Data** and use the **Explore Data** option to display descriptive statistics along with a histogram and other items discussed later.)

STATDISK

```
Descriptive Statistics
Column 1

Sample Size, n:  40
Mean:            23.95
Median:          24
Midrange:        24.5
RMS:             24.08215
Variance, s^2:   6.510256
St Dev, s:       2.55152
Mean Abs Dev:    2.0075
Range:           11
Coeff. Of Var.   10.65%

Minimum:         19
1st Quartile:    22.5
2nd Quartile:    24
3rd Quartile:    26
Maximum:         30

Sum:             958
Sum Sq:          23198
```

MINITAB Enter the data in the column with the heading C1 (or open an existing data set). Click on **Stat,** select **Basic Statistics,** then select **Descriptive Statistics.** Double-click on C1 or another column so that it appears in the box labeled "Variables." (Optional: Click on the box labeled "Statistics" to check or uncheck the statistics that you want.) Click **OK.** The results will include the mean and median as well as other statistics.

MINITAB

Descriptive Statistics: Chips

Variable	N	Mean	StDev	Minimum	Q1	Median	Q3	Maximum
Chips	40	23.950	2.552	19.000	22.250	24.000	26.000	30.000

EXCEL **XLSTAT:** If XLSTAT is available, click on **XLSTAT** at the top, select **Describing Data,** then select **Descriptive Statistics.** Enter the range of cells (such as A1:A40) in the "Quantitative Data" box. Check the "Sample labels" box only if the first cell contains the name of the data set. Click **OK** to continue. The result will be a list of 33 different statistics, including the mean and median. The accompanying display shows some of the more important results.

XLSTAT

Minimum	19.0000
Maximum	30.0000
Freq. of minimum	2
Freq. of maximum	1
Range	11.0000
1st Quartile	22.7500
Median	24.0000
3rd Quartile	26.0000
Sum	958.0000
Mean	23.9500
Variance (n)	6.3475
Variance (n-1)	6.5103
Standard deviation (n)	2.5194
Standard deviation (n-1)	2.5515

DATA ANALYSIS: If XLSTAT is not available, the Data Analysis add-in must be installed. (If the Data Analysis add-in is not yet installed, install it using the **Help** feature: Search for "Data Analysis," select "Load the Analysis Tool Pak," and follow the instructions.) In Excel 2003, select **Tools,** then **Data Analysis,** then select **Descriptive Statistics** and click **OK.** In Excel 2013, 2010, or 2007, click on **Data,** select **Data Analysis,** then select **Descriptive Statistics** in the pop-up window, and click **OK.** In the dialog box, enter the input range (such as A1:A40 for 40 values in column A), click on **Summary Statistics,** then click **OK.**

If it is necessary to widen the columns to see all of the results, select **Format,** then select the **AutoFit Column Width** option. (In Excel 2003, use **Format, Column, AutoFit Selection**).

Caution: If Excel finds more than one mode, it provides only the first one that it finds, so there may be modes other than the one identified by Excel.

EXCEL (DATA ANALYSIS ADD-IN)

Column1	
Mean	23.95
Standard Error	0.4034308
Median	24
Mode	23
Standard Deviation	2.551520411
Sample Variance	6.51025641
Kurtosis	-0.138051213
Skewness	-0.08759594
Range	11
Minimum	19
Maximum	30
Sum	958
Count	40

TI-83/84 PLUS Enter a list of data in L1 or use a list of values already assigned to a name. (To enter data, press STAT, then select **Edit** and press the ENTER key.) After the data values have been entered or opened, press STAT and select **CALC,** then select **1-Var Stats** and press ENTER twice. The display will include the mean $\bar{x}$, the median, the minimum value, and the maximum value. Press the down-arrow key to view the results that don't fit on the initial display.

TI-83/84 PLUS

```
1-Var Stats
 x̄=23.95
 Σx=958
 Σx²=23198
 Sx=2.551520411
 σx=2.519424537
↓n=40
```

```
1-Var Stats
↑n=40
 minX=19
 Q1=22.5
 Med=24
 Q3=26
 maxX=30
```

STATCRUNCH Click on **Open StatCrunch,** then enter or open a data set. Click on **Stats,** then click on **Summary Stats,** then **Columns.** Select the column containing the data. Click on **Next** to select the desired statistics, then click on **Calculate.**

STATCRUNCH

Summary statistics:

Column	n	Mean	Variance	Std. Dev.	Std. Err.	Median	Range	Min	Max	Q1	Q3
var1	40	23.95	6.5102563	2.5515203	0.4034308	24	11	19	30	22.5	26

Other Technologies

To see results from some other technologies, the displays from SPSS and JMP are shown in the following displays.

continued

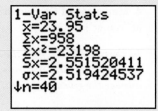

SPSS

	N	Range	Minimum	Maximum	Mean	Std. Deviation	Variance
Chips	40	11.00	19.00	30.00	23.9500	2.55152	6.510
Valid N (listwise)	40						

JMP

Mean	23.95
Std Dev	2.5515204
Std Err Mean	0.4034308
Upper 95% Mean	24.766016
Lower 95% Mean	23.133984

Recommended Assignment
1–4 and even-numbered Exercises 6–24. (If frequency distributions have been covered, also include Exercises 29–32.)

1. No. The numbers do not measure or count anything, so the mean would be a meaningless statistic.
2. The term *average* is not used in statistics. The term *mean* should be used for the value obtained when data values are added, then the sum is divided by the number of data values.
3. No. The price exactly in between the highest and lowest is the midrange, not the median.
4. They use different approaches for providing a value (or values) of the center or middle of a set of data values.
5. $\bar{x}$ = \$159.8 million; median = \$95.0 million; mode: none; midrange = \$199.5 million. Apart from the obvious and trivial fact that the mean annual earnings of all celebrities is less than \$332 million, nothing meaningful can be known about the mean of the population.
6. $\bar{x}$ = \$51,596.6; median = \$51,193.0; mode: none; midrange = \$52,642.5. Apart from the obvious and trivial fact that all other colleges have tuition amounts less than those listed, nothing meaningful can be known about the mean of the population.
7. $\bar{x}$ = 430.1 hic; median = 393.0 hic; mode: none; midrange = 435.0 hic. The safest of these cars appears to be the Hyundai Elantra. Because the measurements appear to vary substantially from a low of 326 hic to a high of 544 hic, it appears that some small cars are considerably safer than others.
8. $\bar{x}$ = 703.7 hic; median = 630.5 hic; mode: none; midrange = 820.5 hic. All of the measures of center are less than 1000 hic, but that does not indicate that all of the individual booster seats satisfy the requirement. One of the booster seats has a measurement of 1210 hic, which does not satisfy the specified requirement of being less than 1000 hic.

3-2 Basic Skills and Concepts

Statistical Literacy and Critical Thinking

1. Employment Data Listed below are results from the National Health and Nutrition Examination.

1 3 3 2 1 1 1 7 9 1 1 1 4 4

These results indicate the working status of the person being surveyed, and the listed values represent these responses: 1 = Working; 2 = With a job but not at work; 3 = Looking for work; 4 = Not working; 7 = Refused to answer; 9 = Don't know. Does it make sense to calculate the mean of these numbers? Why or why not?

2. Average The Web site IncomeTaxList.com lists the "average" annual income for California as \$44,400. What is the role of the term *average* in statistics? Should another term be used in place of *average*?

3. Median In an editorial, the *Poughkeepsie Journal* printed this statement: "The median price—the price exactly in between the highest and lowest— . . . ." Does that statement correctly describe the median? Why or why not?

4. Measures of Center In what sense are the mean, median, mode, and midrange measures of "center"?

In Exercises 5–20, find the (a)* mean, *(b)* median, *(c)* mode, *and (d)* midrange *for the given sample data. Express answers with the appropriate units of measurement. Then answer the given questions.

5. Top 10 Celebrity Incomes Listed below are the earnings (in millions of dollars) of the celebrities with the 10 highest incomes in a recent year. The celebrities in order are Steven Spielberg, Howard Stern, George Lucas, Oprah Winfrey, Jerry Seinfeld, Tiger Woods, Dan Brown, Jerry Bruckheimer, J. K. Rowling, and Tom Cruise. Can this "Top 10" list be used to learn anything about the mean annual earnings of all celebrities?

332 302 235 225 100 90 88 84 75 67

6. Top 10 Most Expensive Colleges Listed below are the annual tuition amounts of the 10 most expensive colleges in the United States for a recent year. The colleges listed in order are Sarah Lawrence, NYU, George Washington, Bates, Skidmore, Johns Hopkins, Georgetown, Connecticut College, Harvey Mudd, and Vassar. What does this "Top 10" list tell us about the population of all U.S. college tuitions?

\$54,410 \$51,991 \$51,730 \$51,300 \$51,196
\$51,190 \$51,122 \$51,115 \$51,037 \$50,875

7. Car Crash Test Measurements Listed below are head injury measurements from small cars that were tested in crashes. The measurements are in "hic," which is a measurement of a standard "head injury criterion." The data are from Data Set 13 in Appendix B, which is based on data from the National Highway Traffic Safety Administration. The listed values correspond to these cars: Chevrolet Aveo, Honda Civic, Volvo S40, VW Jetta, Hyundai Elantra,

Kia Rio, and Kia Spectra. Which car appears to be safest? Based on these limited results, do small cars appear to have about the same risk of head injury in a crash?

371 356 393 544 326 520 501

8. Tests of Child Booster Seats The National Highway Traffic Safety Administration conducted crash tests of child booster seats for cars. Listed below are results from those tests, with the measurements given in "hic," which is a measurement of a standard "head injury criterion." According to the safety requirement, the hic measurement should be less than 1000 hic. Do the results suggest that all child booster seats meet the specified requirement?

774 649 1210 546 431 612

9. Harry Potter Listed below are the gross amounts (in millions of dollars) earned in box office receipts for the movie *Harry Potter and the Half-Blood Prince.* The movie opened on a Wednesday, and the amounts are listed in order for the first 14 days of the movie's release. What important feature of the data is not revealed through the different measures of center? What is an explanation for the fact that the three highest amounts are the first, third, and fourth values listed?

58 22 27 29 21 10 10 8 7 9 11 9 4 4

10. Florida Manatee Deaths Listed below are the numbers of manatee deaths caused each year by collisions with watercraft. Manatees, also called "sea cows," are large mammals that live underwater, often in or near waterways. The data are listed in order for each year of the past decade. The data are from the Florida Fish and Wildlife Conservation Commission. What important feature of the data is not revealed through the different measures of center?

78 81 95 73 69 79 92 73 90 97

11. Ghost Prices Listed below are the prices listed for Norton Ghost 14.0 software from these vendors: Newegg, Dell, Buycheapsoftware.com, PC Connection, Walmart, and Overstock.com. For this collection of data values, are any of the measures of center the most important statistic? Is there a different statistic that is most relevant here? If so, which one?

\$55.99 \$69.99 \$48.95 \$48.92 \$71.77 \$59.68

12. CEO Compensation Listed below are recent annual compensation amounts for these chief executive officers: Mulally (Ford), Jobs (Apple), Kent (Coca-Cola), Otellini (Intel), and McNerney (Boeing). The data are from the Associated Press. What is particularly notable about these compensation amounts?

\$17,688,241 \$1 \$19,628,585 \$12,407,800 \$14,765,410

13. Lead in Medicine Listed below are the lead concentrations (in μg/g) measured in different Ayurveda medicines. Ayurveda is a traditional medical system commonly used in India. The lead concentrations listed here are from medicines manufactured in the United States. The data are based on the article "Lead, Mercury, and Arsenic in US and Indian Manufactured Ayurvedic Medicines Sold via the Internet," by Saper et al., *Journal of the American Medical Association,* Vol. 300, No. 8. What do the results suggest about the safety of these medicines? What do the decimal values of the listed amounts suggest about the precision of the measurements?

3.0 6.5 6.0 5.5 20.5 7.5 12.0 20.5 11.5 17.5

14. Mercury in Sushi Listed below are the amounts of mercury (in parts per million, or ppm) found in tuna sushi sampled at different stores in New York City. The study was sponsored by the *New York Times* and the stores (in order) are D'Agostino, Eli's Manhattan, Fairway, Food Emporium, Gourmet Garage, Grace's Marketplace, and Whole Foods. Which store has the healthiest tuna? Does it appear that the different stores get their tuna from the same supplier?

0.56 0.75 0.10 0.95 1.25 0.54 0.88

9. $\bar{x}$ = \$16.4 million; median = \$10.0 million; mode: \$4 million, \$9 million, and \$10 million; midrange = \$31 million. The measures of center do not reveal anything about the pattern of the data over time, and that pattern is a key component of a movie's success. The first amount is highest for the opening day when many Harry Potter fans are most eager to see the movie; the third and fourth values are from the first Friday and the first Saturday, which are the popular weekend days when movie attendance tends to spike.

10. $\bar{x}$ = 82.7 manatees; median = 80.0 manatees; mode: 73 manatees; midrange = 83.0 manatees. The measures of center do not reveal anything about the pattern of the data over time, and it is important to monitor the numbers of manatee deaths caused by collisions with watercraft, so that corrective action might be taken.

11. $\bar{x}$ = \$59.217; median = \$57.835; mode: none; midrange = \$60.345. None of the measures of center are most important here. The most relevant statistic in this case is the minimum value of \$48.92, because that is the lowest price for the software. Here, we generally care about the lowest price, not the mean price or median price.

12 $\bar{x}$ = \$12,898,007.4; median = \$14,765,410.0; mode: none; midrange = \$9,814,293.0. The compensation amount of \$1 for Jobs is an outlier because it is very far from all of the other values.

13. $\bar{x}$ = 11.05 μg/g; median = 9.50 μg/g; mode: 20.5 μg/g; midrange = 11.75 μg/g. There is not enough information given here to assess the true danger of these drugs, but ingestion of any lead is generally detrimental to good health. All of the decimal values are either 0 or 5, so it appears that the lead concentrations were rounded to the nearest one-half unit of measurement (μg/g).

14. $\bar{x}$ = 0.719 ppm; median = 0.750 ppm; mode: none; midrange = 0.675 ppm. Fairway has the tuna with the lowest level of mercury, so it has the healthiest tuna. Because of the large range of values, it does not appear that the different stores are getting their tuna from the same supplier.

15. Years to Earn Bachelor's Degree Listed below are the lengths of time (in years) it took for a random sample of college students to earn bachelor's degrees (based on data from the National Center for Education Statistics). Based on these results, does it appear that it is common to earn a bachelor's degree in 4 years?

4 4 4 4 4 4 4.5 4.5 4.5 4.5 4.5 4.5 6 6 8 9 9 13 13 15

16. Cell Phone Radiation Listed below are the measured radiation emissions (in W/kg) corresponding to these cell phones: Samsung SGH-tss9, Blackberry Storm, Blackberry Curve, Motorola Moto, T-Mobile Sidekick, Sanyo Katana Eclipse, Palm Pre, Sony Ericsson, Nokia 6085, Apple iPhone 3G S, and Kyocera Neo E1100. The data are from the Environmental Working Group. The media often present reports about the dangers of cell phone radiation as a cause of cancer. The Federal Communications Commission has a standard that cell phone radiation must be 1.6 W/kg or less. If you are planning to purchase a cell phone, are any of the measures of center the most important statistic? Is there another statistic that is most relevant? If so, which one?

0.38 0.55 1.54 1.55 0.50 0.60 0.92 0.96 1.00 0.86 1.46

17. JFK to LAX Flight Delays Listed below are the arrival delay times (in minutes) of randomly selected American Airline flights from New York's JFK airport to Los Angeles (LAX). Negative values correspond to flights that arrived early before the scheduled arrival time, and positive values represent lengths of delays. Based on these very limited results, what do you conclude about the on-time performance of American Airlines? (The data are from the Bureau of Transportation, and more data are listed in Data Set 15 in Appendix B.)

−15 −18 −32 −21 −9 −32 11 2

18. Freshman 15 According to the "freshman 15" legend, college freshmen gain 15 pounds (or 6.8 kilograms) during their freshman year. Listed below are the amounts of weight change (in kilograms) for a simple random sample of freshmen included in a study ("Changes in Body Weight and Fat Mass of Men and Women in the First Year of College: A Study of the 'Freshman 15'" by Hoffman, Policastro, Quick, and Lee, *Journal of American College Health,* Vol. 55, No. 1). Positive values correspond to students who gained weight and negative values correspond to students who lost weight. Do these values appear to support the legend that college students gain 15 pounds (or 6.8 kilograms) during their freshman year? Why or why not?

11 3 0 −2 3 −2 −2 5 −2 7 2 4 1 8 1 0 −5 2

19. Saints in Super Bowl Listed below are the numbers on the jerseys of the starting lineup for the New Orleans Saints when they recently won their first Super Bowl football game. What do the measures of center tell us about the team? Does it make sense to compute the measures of center for these data?

9 23 25 88 12 19 74 77 76 73 78

20. Phenotypes of Peas Biologists conducted experiments to determine whether a deficiency of carbon dioxide in the soil affects the phenotypes of peas. Listed below are the phenotype codes, where 1 = smooth-yellow, 2 = smooth-green, 3 = wrinkled-yellow, and 4 = wrinkled-green. Can the measures of center be obtained for these values? Do the results make sense?

2 1 1 1 1 1 1 4 1 2 2 1 2 3 3 2 3 1 3 1 3 1 3 2 2

In Exercises 21–24, find the* mean *and* median *for each of the two samples; then compare the two sets of results.

21. Speeding and Race Listed below are speeds (in mi/h) of cars on the New Jersey Turnpike. All cars are going in the same direction, and all of the cars are from New Jersey. The speeds were measured with a radar gun and the researchers observed the races of the drivers.

15. $\bar{x}$ = 6.5 years; median = 4.5 years; mode: 4 and 4.5 years; midrange = 9.5 years. It is common to earn a bachelor's degree in four years, but the typical college student requires more than four years.

16. $\bar{x}$ = 0.938 W/kg; median = 0.920 W/kg; mode: none; midrange = 0.965 W/kg. If purchasing a cell phone with concern about radiation emissions, you might be more interested in the fact that the maximum emission is 1.55 W/kg, which is less than the FCC standard of 1.6 W/kg. You might also be interested in the radiation emission for the particular cell phone you are considering.

17. $\bar{x}$ = −14.3 min; median = −16.5 min; mode: −32 min; midrange = −10.5 min. Because the measures of center are all negative values, it appears that the flights tend to arrive early *before* the scheduled arrival times, so the on-time performance appears to be very good.

18. $\bar{x}$ = 1.9 kg; median = 1.5 kg; mode = −2 kg; midrange = 3.0 kg. No, because the mean weight gain is only 1.9 kg, which is far below the 6.8 kg weight gain given in the legend.

19. $\bar{x}$ = 50.4; median = 73.0; mode: none; midrange = 48.5. The numbers do not measure or count anything; they are simply replacements for names. The data are at the nominal level of measurement, and it makes no sense to compute the measures of center for these data.

20. $\bar{x}$ = 1.9; median = 2.0; mode = 1; midrange = 2.5. The mode of 1 correctly indicates that the smooth-yellow peas occur more than any other phenotype, but the other measures of center don't make sense with these data at the nominal level of measurement.

21. White drivers: $\bar{x}$ = 73.0 mi/h; median = 73.0 mi/h. African American drivers: $\bar{x}$ = 74.0 mi/h; median = 74.0 mi/h. Although the African American drivers have a mean speed greater than the white drivers, the difference is very small, so it appears that drivers of both races appear to speed about the same amount.

The data are from Statlib and the authors are Joseph Kadane and John Lamberth. Does it appear that drivers of either race speed more than drivers of the other race?

White drivers:	74	77	69	71	77	69	72	75	74	72
African American drivers:	79	70	71	76	76	74	71	75	74	74

22. Parking Meter Theft Listed below are amounts (in millions of dollars) collected from parking meters by Brinks and others in New York City during similar time periods. A larger data set was used to convict five Brinks employees of grand larceny. The data were provided by the attorney for New York City, and they are listed on the DASL Web site. Do the limited data listed here show evidence of stealing by Brinks employees?

Collection contractor was Brinks:	1.3	1.5	1.3	1.5	1.4	1.7	1.8	1.7	1.7	1.6
Collection contractor was not Brinks:	2.2	1.9	1.5	1.6	1.5	1.7	1.9	1.6	1.6	1.8

23. Political Contributions Listed below are contributions (in dollars) made to the two presidential candidates in a recent election. All contributions are from the same zip code as the author, and the data are from the *Huffington Post.* Do the contributions appear to favor either candidate? What do you conclude after learning that there were 66 contributions to Obama and 20 contributions to McCain?

Obama: \$275 \$452 \$300 \$1000 \$1000 \$500 \$100 \$1061
\$1200 \$235 \$875 \$2000 \$350 \$210 \$250

McCain: \$50 \$75 \$240 \$302 \$250 \$700 \$350 \$500
\$1250 \$1500 \$500 \$500 \$40 \$221 \$400

24. Customer Waiting Times Waiting times (in minutes) of customers at the Jefferson Valley Bank (where all customers enter a single waiting line) and the Bank of Providence (where customers wait in individual lines at three different teller windows) are listed below. Determine whether there is a difference between the two data sets that is not apparent from a comparison of the measures of center. If so, what is it?

Jefferson Valley (single line):	6.5	6.6	6.7	6.8	7.1	7.3	7.4	7.7	7.7	7.7
Providence (individual lines):	4.2	5.4	5.8	6.2	6.7	7.7	7.7	8.5	9.3	10.0

Large Data Sets from Appendix B. *In Exercises 25–28, refer to the indicated data set in Appendix B. Use computer software or a calculator to find the* **means** *and* **medians.**

25. Earthquakes Use the magnitudes (Richter scale) of the earthquakes listed in Data Set 16 in Appendix B. In 1989, the San Francisco Bay Area was struck with an earthquake that measured 7.0 on the Richter scale. That earthquake occurred during the warm-up period for the third game of the baseball World Series. Is the magnitude of that World Series earthquake an *outlier* (data value that is very far away from the others) when considered in the context of the sample data given in Data Set 16? Explain.

26. Flight Data Refer to Data Set 15 in Appendix B and use the times required to taxi out for takeoff. For American Airlines, how is it helpful to find the mean?

27. Presidential Longevity Refer to Data Set 12 in Appendix B and use the numbers of years that U.S. presidents have lived after their first inauguration. What is the value of finding the mean of those numbers?

28. IQ Scores Refer to Data Set 6 in Appendix B and use the listed IQ scores. IQ tests are designed so that the mean IQ of the population is 100. Does the sample mean suggest that the sample is consistent with the population?

22. Collection contractor was Brinks: $\bar{x}$ = \$1.55 million; median = \$1.55 million. Collection contractor was not Brinks: $\bar{x}$ = \$1.73 million; median = \$1.65 million. The data do suggest that collections were considerably lower when Brinks was the collection contractor.

23. Obama: $\bar{x}$ = \$653.9; median = \$452.0. McCain: $\bar{x}$ = \$458.5; median = \$350.0. The contributions appear to favor Obama because his mean and median are substantially higher. With 66 contributions to Obama and 20 contributions to McCain, Obama collected substantially more in total contributions.

24. Jefferson Valley: $\bar{x}$ = 7.15 min; median = 7.20 min. Providence: same results as Jefferson Valley. Although the measures of center are the same, the Providence times are much more varied than the Jefferson Valley times.

25. $\bar{x}$ = 1.184; median = 1.235. Yes, it is an outlier because it is a value that is very far away from all of the other sample values.

26. $\bar{x}$ = 21.0 min; median = 18.5 min. The mean taxi-out time is important for calculating and scheduling the arrival time.

27. $\bar{x}$ = 15.0 years; median = 16.0 years. Presidents receive Secret Service protection after they leave office, so the mean is helpful in planning for the cost and resources used for that protection.

28. $\bar{x}$ = 101.0; median = 96.5. The mean of 101.0 does not differ from the population mean of 100 by an amount that is substantial, so it appears that the sample is consistent with the population.

In Exercises 29–32, find the* mean *of the data summarized in the given frequency distribution. Also, compare the* computed means *to the* actual means *obtained by using the original list of data values, which are as follows: (Exercise 29) 35.9 years; (Exercise 30) 44.1 years; (Exercise 31) 84.4; (Exercise 32) 15.0 years.

29. $\bar{x} = 35.8$ years. This result is quite close to the mean of 35.9 years found by using the original list of data values.

30. $\bar{x} = 44.5$ years. This result is not substantially different from the mean of 44.1 years found by using the original list of data values.

29.

Age of Best Actress When Oscar Was Won	Frequency
20–29	27
30–39	34
40–49	13
50–59	2
60–69	4
70–79	1
80–89	1

30.

Age of Best Actor When Oscar Was Won	Frequency
20–29	1
30–39	26
40–49	35
50–59	13
60–69	6
70–79	1

31. $\bar{x} = 84.7$. This result is close to the mean of 84.4 found by using the original list of data values.

32. $\bar{x} = 15.0$ years. When rounded, this result is the same mean of 15.0 years found by using the original list of data values.

31.

Verbal IQ Score of Subject Exposed to Lead	Frequency
50–59	4
60–69	10
70–79	25
80–89	43
90–99	26
100–109	8
110–119	3
120–129	2

32.

Years President Lived after First Inauguration	Frequency
0–4	8
5–9	2
10–14	5
15–19	7
20–24	4
25–29	6
30–34	0
35–39	1

3-2 Beyond the Basics

33. a. 0.6 parts per million b. $n - 1$

33. Degrees of Freedom Carbon monoxide is measured in San Francisco on five different days, and the mean of those five values is 0.62 parts per million. Four of the values (in parts per million) are 0.3, 0.4, 1.1, and 0.7. (The data are from the California Environmental Protection Agency.)

a. Find the missing value.

b. We need to create a list of n values that have a specific known mean. We are free to select any values we desire for some of the n values. How many of the n values can be freely assigned before the remaining values are determined? (The result is referred to as the *number of degrees of freedom.*)

34. 15.0 years; 15.2 years or greater. The results do not differ by much.

34. Censored Data Data Set 12 in Appendix B lists the numbers of years that U.S. presidents lived after their first inauguration. As of this writing, five of the presidents are still alive and after their first inauguration they have lived 33 years, 21 years, 17 years, 9 years, and 1 year so far. We might use the values of 33+, 21+, 17+, 9+, and 1+, where the plus signs indicate that the actual value is equal to or greater than the current value. (These values were said to be *censored* at the time that this list was compiled.) If you use the values in Data Set 12 and ignore the presidents who are still alive, what is the mean? If you use the values given in Data Set 12 along with the additional values of 33+, 21+, 17+, 9+, and 1+, what do we know about the mean? Do the two results differ by much?

35. Trimmed Mean Because the mean is very sensitive to extreme values, we say that it is not a *resistant* measure of center. By deleting some low values and high values, the **trimmed mean** is more resistant. To find the 10% trimmed mean for a data set, first arrange the data in order, then delete the bottom 10% of the values and delete the top 10% of the values, and then calculate the mean of the remaining values. Refer to the BMI values for females in Data Set 1 in Appendix B, and change the highest value from 47.24 to 472.4, so the value of 472.4 is an outlier. Find (a) the mean; (b) the 10% trimmed mean; (c) the 20% trimmed mean. How do the results compare?

35. $\bar{x} = 39.070$; 10% trimmed mean: 27.677; 20% trimmed mean: 27.176. By deleting the outlier of 472.4, the trimmed means are substantially different from the untrimmed mean.

36. Harmonic Mean The **harmonic mean** is often used as a measure of center for data sets consisting of rates of change, such as speeds. It is found by dividing the number of values n by the sum of the *reciprocals* of all values, expressed as

$$\frac{n}{\Sigma \frac{1}{x}}$$

(No value can be zero.) The author drove 1163 miles to a conference in Orlando, Florida. For the trip to the conference, the author stopped overnight, and the mean speed from start to finish was 38 mi/h. For the return trip, the author stopped only for food and fuel, and the mean speed from start to finish was 56 mi/h. Is the actual "average" speed for the round-trip the mean of 38 mi/h and 56 mi/h? Why or why not? What is the harmonic mean of 38 mi/h and 56 mi/h, and does this represent the true "average" speed?

36. The mean of 47 mi/h is not the actual average speed, because more time was spent at the lower speed. The harmonic mean is 45.3 mi/h, and it does represent the true "average" speed.

37. Geometric Mean The **geometric mean** is often used in business and economics for finding average rates of change, average rates of growth, or average ratios. Given n values (all of which are positive), the geometric mean is the nth root of their product. Find the *average growth factor* for money deposited in annual certificates of deposit for the past 5 years (as of this writing) with annual interest rates of 1.7%, 3.7%, 5.2%, 5.1%, and 2.7% by computing the geometric mean of 1.017, 1.037, 1.052, 1.051, and 1.027. What single percentage growth rate would be the same as having the five given successive growth rates? Is that result the same as the mean of the five annual interest rates?

37. Geometric mean: 1.036711036, or 1.0367 when rounded. Single percentage growth rate: 3.67%. The result is not exactly the same as the mean, which is 3.68%.

38. Quadratic Mean The **quadratic mean** (or **root mean square,** or **R.M.S.**) is usually used in physical applications. In power distribution systems, for example, voltages and currents are usually referred to in terms of their R.M.S. values. The quadratic mean of a set of values is obtained by squaring each value, adding those squares, dividing the sum by the number of values n, and then taking the square root of that result, as indicated below:

$$\text{Quadratic mean} = \sqrt{\frac{\Sigma x^2}{n}}$$

Find the R.M.S. of these voltages measured from household current: 0, 100, 162, 162, 100, 0, −100, −162, −162, −100, 0. How does the result compare to the mean?

38. R.M.S.: 114.8 volts, which is very different from the mean of 0 volts.

39. Median When data are summarized in a frequency distribution, the median can be found by first identifying the *median class,* which is the class that contains the median. We then assume that the values in that class are evenly distributed and we interpolate. Letting n denote the sum of all class frequencies, and letting m denote the sum of the class frequencies that *precede* the median class, the median can be estimated as shown here:

$$(\text{lower limit of median class}) + (\text{class width})\left(\frac{\left(\frac{n+1}{2}\right) - (m+1)}{\text{frequency of median class}}\right)$$

Use this procedure to find the median of the frequency distribution given in Exercise 29. How does the result compare to the median found from the original list of data, which is 33.0 years? Which value of the median is better: the value computed using the frequency table or the value of 33.0 years?

39. 34.0 years (rounded from 33.970588 years); the value of 33.0 years is better because it is based on the original data and does not involve interpolation.

3-3 Measures of Variation

Key Concept The topic of variation discussed in this section is the single most important topic in statistics. This section presents three important measures of variation: *range, standard deviation,* and *variance.* Being measures of variation, these statistics result in numbers, but it is essential to know that our focus is not only finding numerical values but developing the ability to *interpret* and *understand* the numbers that we get. This section is not really about arithmetic; it is about conceptual understanding, and the concept is one of paramount importance.

Study Hint: Part 1 of this section presents basic concepts of variation and Part 2 presents additional concepts related to the standard deviation. Although both parts contain several formulas for computation, do not spend too much time memorizing those formulas and doing arithmetic calculations. Instead, make it a priority to *understand* and *interpret* values of standard deviation.

Got a Second?

The time unit of 1 second is defined to be "the duration of 9,192,631,770 periods of the radiation corresponding to the transition between the two hyperfine levels of the ground state of the cesium-133 atom." That definition redefines time to be based on the behavior of atoms instead of the earth's motion. It results in accuracy of ±1 second in 10,000,000 years, which is the most accurate measurement we use. Because it is so accurate, the second is being used to define other quantities, such as the meter. The meter was once defined as 1/10,000,000 of the distance along the surface of the earth between the North Pole and the equator (passing through Paris). The meter is now defined as the length of the distance traveled by light in a vacuum during a time interval of 1/299,792,458 sec.

When dealing with time measurement devices, the traditional standard deviation has been found to be poor because of a trend in which the mean changes over time. Instead, other special measures of variation are used, such as Allan variance, total variance, and TheoH.

Unrelated to statistics but nonetheless interesting is the fact that ads for watches usually show a watch with a time close to 10:10. That time allows the brand name to be visible, and it creates a subliminal image of a happy face. The time of 10:10 has been the industry standard since the 1940s.

Part 1: Basic Concepts of Variation

To visualize the effect of variation, see Figure 3-2 and verify this important observation: The numbers of chocolate chips in Chips Ahoy cookies (top dotplot) have more *variation* than those in the Triola cookies (bottom dotplot). The top dotplot shows more spread than the bottom dotplot. Although both brands have the same mean (24.0), the Chips Ahoy cookies vary from 19 chocolate chips to 30, but the Triola cookies only vary from 23 chocolate chips to 25. This characteristic of spread, or variation, or dispersion, is so important that we measure it with numbers. (*Note:* The Chips Ahoy data are real and they are listed in Table 3-1, but the Triola cookie data are fabricated. The author doesn't actually produce chocolate chip cookies, but if the writing job doesn't work out, you never know.)

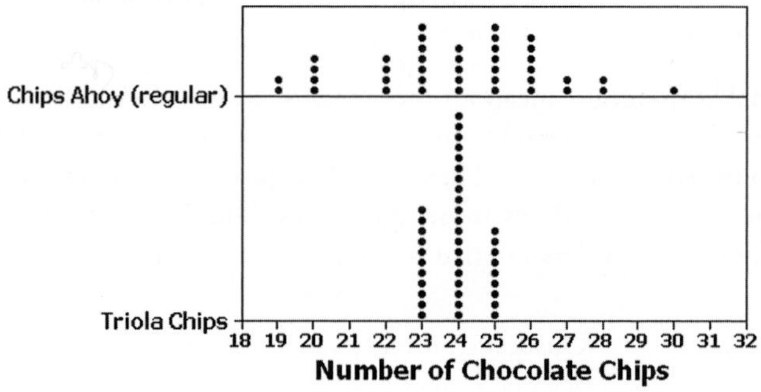

Figure 3-2 Dotplots of Numbers of Chocolate Chips in Cookies

Range

We noted that this section includes these measures of variation: (1) range; (2) standard deviation; (3) variance. We begin with the range because it is so easy to compute, even though it is not as important as the other measures of variation.

> **DEFINITION** The **range** of a set of data values is the difference between the maximum data value and the minimum data value.
>
> **Range = (maximum data value) − (minimum data value)**

Because the range uses only the maximum and the minimum data values, it is very sensitive to extreme values and isn't as useful as other measures of variation that use every data value, such as the standard deviation. To keep our round-off rules as consistent and as simple as possible, we will round the range using the same round-off rule for all measures of variation discussed in this section.

Round-Off Rule for Measures of Variation

When rounding the value of a measure of variation, carry one more decimal place than is present in the original set of data.

Example 1 Range

Find the range of these numbers of chocolate chips: 22, 22, 26, 24. (These are the first four chip counts for the Chips Ahoy cookies.)

Solution

The range is found by subtracting the lowest value from the largest value, so we get

$$\text{range} = (\text{maximum value}) - (\text{minimum value}) = 26 - 22 = 4.0$$

The result is shown with one more decimal place than is present in the original data values. We conclude that the range is 4.0 chocolate chips.

Standard Deviation of a Sample

The *standard deviation* is the measure of variation most commonly used in statistics.

DEFINITION The **standard deviation** of a set of sample values, denoted by s, is a measure of how much data values deviate away from the mean. It is calculated by using Formula 3-4 or 3-5. Formula 3-5 is just a different version of Formula 3-4, so both formulas are algebraically the same.

Formula 3-4

$$s = \sqrt{\frac{\Sigma(x - \bar{x})^2}{n - 1}}$$ sample standard deviation

Formula 3-5

$$s = \sqrt{\frac{n(\Sigma x^2) - (\Sigma x)^2}{n(n - 1)}}$$ shortcut formula for sample standard deviation (formula used by calculators and computer programs)

Later in this section we describe the reasoning behind these formulas, but for now we recommend that you use Formula 3-4 for an example or two, and then learn how to find standard deviation values using your calculator and by using a software program. (Most scientific calculators are designed so that you can enter a list of values and automatically get the standard deviation.) The following properties are consequences of the way in which the standard deviation is defined.

Important Properties of Standard Deviation

- The standard deviation is a measure of how much data values deviate away from the *mean*.

What the Median Is Not

Harvard biologist Stephen Jay Gould wrote, "The Median Isn't the Message." In it, he describes how he learned that he had abdominal mesothelioma, a form of cancer. He went to the library to learn more, and he was shocked to find that mesothelioma was incurable, with a median survival time of only *eight months* after it was discovered. Gould wrote this: "I suspect that most people, without training in statistics, would read such a statement as 'I will probably be dead in eight months' the very conclusion that must be avoided, since it isn't so, and since attitude (in fighting the cancer) matters so much." Gould went on to carefully interpret the value of the median. He knew that his chance of living longer than the median was good because he was young, his cancer was diagnosed early, and he would get the best medical treatment. He also reasoned that some could live much longer than eight months, and he saw no reason why he could not be in that group. Armed with this thoughtful interpretation of the median and a strong positive attitude, Gould lived for *20 years* after his diagnosis. He died of another cancer not related to the mesothelioma.

Note to Instructor

This is the same round-off rule used for measures of center. You might ask: "What is wrong with collecting heights of people (such as 65 in., 72 in., 60 in., and so on) and reporting that the mean is 66.234572 in. while the standard deviation is 2.5642217 in.?" That degree of precision is not justified by the original data set, and it is misleading.

Rounding Error Changes World Record

Rounding errors can often have disastrous results. Justin Gatlin was elated when he set the world record as the person to run 100 meters in the fastest time of 9.76 seconds. His record time lasted only 5 days, when it was revised to 9.77 seconds, so Gatlin then tied the world record instead of breaking it. His actual time was 9.766 seconds, and it should have been rounded up to 9.77 seconds, but the person doing the timing didn't know that a button had to be pressed for proper rounding. Gatlin's agent said that he (Gatlin) was very distraught and that the incident is "a total embarrassment to the IAAF (International Association of Athletics Federations) and our sport."

Note to Instructor

Variation is an *extremely* important topic in statistics, so this is one of the most important sections in the book. Emphasize standard deviation and don't emphasize variance. Begin with Figure 3-2 by pointing out that the two data sets have the same mean but very different amounts of variation.

Recommendation: This section is partitioned into two parts: Part 1 (Basic Concepts of Variation) and Part 2 (Beyond the Basics of Variation). If pressed for time, cover Part 1 only.

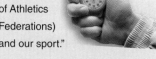

- The value of the standard deviation s is usually positive. It is zero only when all of the data values are the same number. (It is never negative.) Also, larger values of s indicate greater amounts of variation.
- The value of the standard deviation s can increase dramatically with the inclusion of one or more outliers (data values that are very far away from all of the others).
- The units of the standard deviation s (such as minutes, feet, pounds, and so on) are the same as the units of the original data values.
- The sample standard deviation s is a **biased estimator** of the population standard deviation σ, as described in Part 2 of this section.

If our goal was to develop skills for manually calculating values of standard deviations, we would focus on Formula 3-5, which simplifies the calculations. However, we prefer to show a calculation using Formula 3-4, because that formula better illustrates that the standard deviation is based on deviations of sample values away from the mean.

Example 2 Calculating Standard Deviation with Formula 3-4

Use Formula 3-4 to find the standard deviation of these numbers of chocolate chips: 22, 22, 26, 24. (These are the first four chip counts for the Chips Ahoy cookies. Here we use only four values so that we can illustrate calculations with a relatively simple example.)

Solution

The left column of Table 3-3 summarizes the general procedure for finding the standard deviation using Formula 3-4, and the right column illustrates that procedure for the sample values 22, 22, 26, and 24. The result shown in Table 3-3 is 1.9 chips, which is rounded to one more decimal place than is present in the original list of sample values (22, 22, 26, 24). Also, the units for the standard deviation are the same as the units of the original data. Because the original data are actually 22 chips, 22 chips, 26 chips, and 24 chips, the standard deviation is 1.9 chips.

Table 3-3

General Procedure for Finding Standard Deviation with Formula 3-4	Specific Example Using These Sample Values: 22, 22, 26, 24
Step 1: Compute the mean $\bar{x}$.	The sum of 22, 22, 26, and 24 is 94, so $\bar{x} = \frac{\Sigma x}{n} = \frac{22 + 22 + 26 + 24}{4} = \frac{94}{4} = 23.5$
Step 2: Subtract the mean from each individual sample value. (The result is a list of deviations of the form $(x - \bar{x})$.)	Subtract the mean of 23.5 from each sample value to get these deviations away from the mean: −1.5, −1.5, 2.5, 0.5.
Step 3: Square each of the deviations obtained from Step 2. (This produces numbers of the form $(x - \bar{x})^2$.)	The squares of the deviations from Step 2 are: 2.25, 2.25, 6.25, 0.25.
Step 4: Add all of the squares obtained from Step 3. The result is $\Sigma(x - \bar{x})^2$.	The sum of the squares from Step 3 is $2.25 + 2.25 + 6.25 + 0.25 = 11$.
Step 5: Divide the total from Step 4 by the number $n - 1$, which is 1 less than the total number of sample values present.	With $n = 4$ data values, $n - 1 = 3$, so we divide 11 by 3 to get this result: $\frac{11}{3} = 3.6667$.
Step 6: Find the square root of the result of Step 5. The result is the standard deviation, denoted by s.	The standard deviation is $\sqrt{3.6667} = 1.9149$. Rounding the result, we get $s = 1.9$ chips.

Example 3 Calculating Standard Deviation with Formula 3-5

Use Formula 3-5 to find the standard deviation of the sample values 22, 22, 26, and 24 from Example 1.

Solution

Shown below is the computation of the standard deviation of 22, 22, 26, and 24 using Formula 3-5.

$n = 4$ (because there are 4 values in the sample)

$\Sigma x = 94$ (found by adding the sample values: $22 + 22 + 26 + 24 = 94$)

$\Sigma x^2 = 2220$ (found by adding the squares of the sample values, as in $22^2 + 22^2 + 26^2 + 24^2 = 2220$)

Using Formula 3-5, we get

$$s = \sqrt{\frac{n(\Sigma x^2) - (\Sigma x)^2}{n(n-1)}} = \sqrt{\frac{4(2220) - (94)^2}{4(4-1)}} = \sqrt{\frac{44}{12}} = 1.9 \text{ chips}$$

The result of $s = 1.9$ chips is the same as the result in Example 2.

Where Are the 0.400 Hitters?

The last baseball player to hit above 0.400 was Ted Williams, who hit 0.406 in 1941. There were averages above 0.400 in 1876, 1879, 1887, 1894, 1895, 1896, 1897, 1899, 1901, 1911, 1920, 1922, 1924, 1925, and 1930, but none since 1941. Are there no longer great hitters? The late Stephen Jay Gould of Harvard University noted that the mean batting average has been steady at 0.260 for about 100 years, but the standard deviation has been decreasing from 0.049 in the 1870s to 0.031, where it is now. He argued that today's stars are as good as those from the past, but consistently better pitchers now keep averages below 0.400.

Comparing Variation in Different Samples Examples 1, 2, and 3 used only four values (22, 22, 26, 24) so that calculations are relatively simple. If we use all 80 chocolate chip counts depicted in Figure 3-2, we get the measures of center and measures of variation listed in Table 3-4. From the table we see that the range for Chips Ahoy cookies is much larger than the range for Triola cookies. Table 3-4 also shows that the standard deviation for Chips Ahoy is much larger than the standard deviation for Triola cookies, but *it's a good practice to compare two sample standard deviations only when the sample means are approximately the same.* When comparing variation in samples with very different means, it is better to use the coefficient of variation, which is defined later in this section. We also use the coefficient of variation when we want to compare variation from two samples with different scales or units of values, such as the comparison of variation of heights of men and weights of men. (See Example 8, which involves numbers of chocolate chips and weights of cola.)

Table 3-4 Comparison of Chocolate Chips in Cookies from Figure 3-2

	Chips Ahoy (regular)	Triola Chips
Number	40	40
Mean	24.0	24.0
Range	11.0	2.0
Standard Deviation	2.6	0.7

CAUTION *Compare two sample standard deviations only when the sample means are approximately the same.* When comparing variation in samples with very different means, it is better to use the coefficient of variation, which is defined later in this section.

Note to Instructor

Few instructors require that their students show all steps for manually calculating a standard deviation. Most require that students be able to compute values of standard deviations on their calculators or computers. Fully develop the concept of standard deviation in class and show how it is calculated, but then move on and allow students to obtain values from their calculators or computers. Strongly emphasize the *interpretation* of those values, as in this subsection of Range Rule of Thumb for Understanding Standard Deviation.

In class, randomly select a student and ask him or her to estimate the mean height of a student at your college; they usually do this well. Now randomly select another student and ask him or her to estimate the standard deviation of heights; comment that it's perfectly natural to have no idea. Then ask for estimates of the heights of the shortest and tallest students, and proceed to show how the range rule of thumb can be used to develop an estimate of the value of the standard deviation. Women have heights with a standard deviation around 2.5 in., and men have heights with a standard deviation around 2.8 in.

Range Rule of Thumb for Understanding Standard Deviation

Three different concepts that can help us understand and interpret values of standard deviations are (1) the range rule of thumb, (2) the empirical rule, and (3) Chebyshev's theorem. Here we discuss the range rule of thumb. The empirical rule and Chebyshev's theorem will be discussed in Part 2 of this section.

The ***range rule of thumb*** is a crude but simple tool for understanding and interpreting standard deviation. It is based on the principle that for many data sets, the vast majority (such as 95%) of sample values lie within 2 standard deviations of the mean. We could improve the accuracy of this rule by taking into account such factors as the size of the sample and the distribution, but here we sacrifice accuracy for the sake of simplicity.

Range Rule of Thumb

Interpreting a Known Value of the Standard Deviation If the standard deviation of a collection of data is a known value, use it to find rough estimates of the minimum and maximum *usual* sample values as follows:

$$\text{minimum "usual" value} = (\text{mean}) - 2 \times (\text{standard deviation})$$

$$\text{maximum "usual" value} = (\text{mean}) + 2 \times (\text{standard deviation})$$

Estimating a Value of the Standard Deviation s To roughly estimate the standard deviation from a collection of known sample data, use

$$s \approx \frac{\text{range}}{4}$$

where range = (maximum data value) − (minimum data value).

Example 4 Range Rule of Thumb for Interpreting s

Using the 40 chocolate chip counts for the Chips Ahoy (regular) cookies in Table 3-1, the mean is 24.0 chocolate chips and the standard deviation is 2.6 chocolate chips. Use the range rule of thumb to find the minimum and maximum "usual" numbers of chocolate chips; then determine whether the cookie with 30 chocolate chips is "unusual."

Solution

With a mean of 24.0 and a standard deviation of 2.6, we use the range rule of thumb to find the minimum and maximum usual numbers of chocolate chips as follows:

$$\begin{aligned}\text{minimum "usual" value} &= (\text{mean}) - 2 \times (\text{standard deviation}) \\ &= 24.0 - 2(2.6) = 18.8 \\ \text{maximum "usual" value} &= (\text{mean}) + 2 \times (\text{standard deviation}) \\ &= 24.0 + 2(2.6) = 29.2\end{aligned}$$

Interpretation

Based on these results, we expect that typical Chips Ahoy (regular) cookies have between 18.8 chocolate chips and 29.2 chocolate chips. Because 30 falls above the maximum "usual" value, we can consider it to be a cookie with an unusually high number of chocolate chips.

Example 5 Range Rule of Thumb for Estimating s

Use the range rule of thumb to estimate the standard deviation of the sample of 40 chocolate chip counts for the Chips Ahoy (regular) cookies as listed in Table 3-1. Those 40 values have a minimum of 19 and a maximum of 30.

Solution

The range rule of thumb indicates that we can estimate the standard deviation by finding the range and dividing it by 4. With a minimum of 19 and a maximum of 30, the range rule of thumb can be used to estimate the standard deviation s as follows:

$$s \approx \frac{\text{range}}{4} = \frac{30 - 19}{4} = 2.75 \text{ chips}$$

Interpretation

The actual value of the standard deviation is $s = 2.6$ chips, so the estimate of 2.75 chips is quite close. Because this estimate is based on only the minimum and maximum values, it is generally a rough estimate that might be off by a considerable amount.

Standard Deviation of a Population

The definition of standard deviation and Formulas 3-4 and 3-5 apply to the standard deviation of *sample* data. A slightly different formula is used to calculate the standard deviation σ (lowercase sigma) of a *population:* Instead of dividing by $n - 1$, we divide by the population size N, as shown here:

$$\text{population standard deviation} \quad \sigma = \sqrt{\frac{\Sigma(x - \mu)^2}{N}}$$

Because we generally deal with sample data, we will usually use Formula 3-4, in which we divide by $n - 1$. Many calculators give both the sample standard deviation and the population standard deviation, but they use a variety of different notations.

CAUTION When using a calculator to find standard deviation, identify the notation used by your particular calculator so that you get the *sample* standard deviation, not the population standard deviation.

Variance of a Sample and a Population

So far, we have used the term *variation* as a general description of the amount that values vary among themselves. (The terms *dispersion* and *spread* are sometimes used instead of *variation.*) The term *variance* has a specific meaning.

DEFINITIONS

The **variance** of a set of values is a measure of variation equal to the square of the standard deviation.

Sample variance: s^2 = square of the standard deviation s.

Population variance: σ^2 = square of the population standard deviation σ.

More Stocks, Less Risk

In their book *Investments*, authors Zvi Bodie, Alex Kane, and Alan Marcus state that "the average standard deviation for returns of portfolios composed of only one stock was 0.554. The average portfolio risk fell rapidly as the number of stocks included in the portfolio increased." They note that with 32 stocks, the standard deviation is 0.325, indicating much less variation and risk. They make the point that with only a few stocks, a portfolio has a high degree of "firm-specific" risk, meaning that the risk is attributable to the few stocks involved. With more than 30 stocks, there is very little firm-specific risk; instead, almost all of the risk is "market risk," attributable to the stock market as a whole. They note that these principles are "just an application of the well-known law of averages."

Note to Instructor

Two comments about variance are important: (1) Some later procedures use variance (such as the *F* test in Section 9-5 or all of Chapter 12). (2) The units of variance are different than the original units. For example, if the original values are expressed in dollars, then the variance will be in square dollars, which is an abstract concept that seriously impedes any chance of really understanding values of variances.

Notation Here is a summary of notation for the standard deviation and variance:

s = *sample* standard deviation
s^2 = *sample* variance
σ = *population* standard deviation
σ^2 = *population* variance

Note: Articles in professional journals and reports often use SD for standard deviation and VAR for variance.

Important Properties of Variance

- The units of the variance are the *squares* of the units of the original data values. (If the original data values are in feet, the variance will have units of ft^2; if the original data values are in seconds, the variance will have units of sec^2.)
- The value of the variance can increase dramatically with the inclusion of one or more outliers (data values that are very far away from all of the others).
- The value of the variance is usually positive. It is zero only when all of the data values are the same number. (It is never negative.)
- The sample variance s^2 is an **unbiased estimator** of the population variance σ^2, as described in Part 2 of this section.

The variance is a statistic used in some statistical methods, but for our present purposes, the variance has the serious disadvantage of using units that are *different than the units of the original data set.* This makes it difficult to understand variance as it relates to the original data set. Because of this property, it is better to focus on the standard deviation when trying to develop an understanding of variation, as we do in this section.

Note to Instructor
If pressed for time, Chebyshev's theorem can be omitted because its results are too imprecise, and the empirical rule can be omitted because it will be covered in Chapter 6.

Part 2: Beyond the Basics of Variation

In this subsection we focus on making sense of the standard deviation so that it is not some mysterious number devoid of any practical significance. We begin by addressing common questions that relate to the standard deviation.

Why Is Standard Deviation Defined as in Formula 3-4?

Why do we measure variation using Formula 3-4? In measuring variation in a set of sample data, it makes sense to begin with the individual amounts by which values deviate from the mean. For a particular data value x, the amount of **deviation** is $x - \bar{x}$, which is the difference between the individual x value and the mean. It makes sense to somehow combine those deviations into one number that can serve as a measure of the variation. Simply adding the deviations doesn't work, because the sum will always be zero. To get a statistic that measures variation (instead of always being zero), we need to avoid the canceling out of negative and positive numbers. One simple and natural approach is to add absolute values, as in $\Sigma|x - \bar{x}|$. If we find the mean of that sum, we get the **mean absolute deviation** (or **MAD**), which is the mean distance of the data from the mean:

$$\text{mean absolute deviation} = \frac{\Sigma|x - \bar{x}|}{n}$$

Why Not Use the Mean Absolute Deviation Instead of the Standard Deviation? Computation of the mean absolute deviation uses absolute values, so it uses an operation that is not "algebraic." (The algebraic operations include addition, multiplication, extracting roots, and raising to powers that are integers or fractions, but absolute value is not included among the algebraic operations.) Although the use of absolute values would be simple and easy, it would create algebraic difficulties in inferential methods of statistics discussed in later chapters. For example, Section 9-3 presents a method for making inferences about the means of two populations, and that method is built around an additive property of variances, but the mean absolute deviation has no such additive property. (Here is a simplified version of the additive property of variances: If you have two independent populations and you randomly select one value from each population and add them, such sums will have a variance equal to the sum of the variances of the two populations.) Also, the mean absolute deviation is a *biased* estimator, meaning that when you find mean absolute deviations of samples, you do not tend to target the mean absolute deviation of the population. The standard deviation has the advantage of using only algebraic operations. Because it is based on the square root of a sum of squares, the standard deviation closely parallels distance formulas found in algebra. There are many instances where a statistical procedure is based on a similar sum of squares. Therefore, instead of using absolute values, we square all deviations $(x - \bar{x})$ so that they are nonnegative. This approach leads to the standard deviation. For these reasons, scientific calculators typically include a standard deviation function, but they almost never include the mean absolute deviation.

Why Divide by $n - 1$? After finding all of the individual values of $(x - \bar{x})^2$, we combine them by finding their sum. We then divide by $n - 1$ because there are only $n - 1$ independent values. With a given mean, only $n - 1$ values can be freely assigned any number before the last value is determined. Exercise 45 illustrates that division by $n - 1$ yields a better result than division by n. That exercise shows how division by $n - 1$ causes the sample variance s^2 to target the value of the population variance σ^2, whereas division by n causes the sample variance s^2 to underestimate the value of the population variance σ^2.

How Do We Make Sense of a Value of Standard Deviation? Part 1 of this section included the range rule of thumb for interpreting a known value of a standard deviation or estimating a value of a standard deviation. (See Examples 4 and 5.) We now discuss two other approaches for interpreting standard deviation: the empirical rule and Chebyshev's theorem.

Empirical (or 68–95–99.7) Rule for Data with a Bell-Shaped Distribution

A concept helpful in interpreting the value of a standard deviation is the **empirical rule.** This rule states that *for data sets having a distribution that is approximately bell-shaped,* the following properties apply. (See Figure 3-3 on the next page.)

- About 68% of all values fall within 1 standard deviation of the mean.
- About 95% of all values fall within 2 standard deviations of the mean.
- About 99.7% of all values fall within 3 standard deviations of the mean.

Example 6 The Empirical Rule

IQ scores have a bell-shaped distribution with a mean of 100 and a standard deviation of 15. What percentage of IQ scores are between 70 and 130?

Solution

The key to solving this problem is to recognize that 70 and 130 are each exactly 2 standard deviations away from the mean of 100, as shown below:

$$2 \text{ standard deviations} = 2s = 2(15) = 30$$

Therefore, 2 standard deviations from the mean is

$$100 - 30 = 70$$

$$\text{or} \quad 100 + 30 = 130$$

The empirical rule tells us that about 95% of all values are within 2 standard deviations of the mean, so about 95% of all IQ scores are between 70 and 130.

Figure 3-3
The Empirical Rule

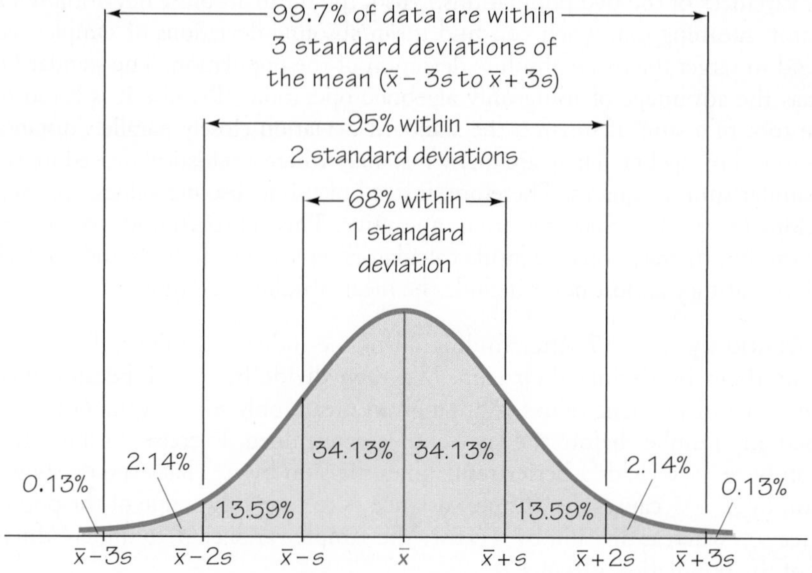

Another concept helpful in understanding or interpreting a value of a standard deviation is **Chebyshev's theorem.** The empirical rule applies only to data sets with bell-shaped distributions, but Chebyshev's theorem applies to *any* data set. Unfortunately, results from Chebyshev's theorem are only approximate. Because the results are lower limits ("at least"), Chebyshev's theorem has limited usefulness.

Chebyshev's Theorem

The proportion (or fraction) of any set of data lying within K standard deviations of the mean is always *at least* $1 - 1/K^2$, where K is any positive number greater than 1. For $K = 2$ and $K = 3$, we get the following statements:

- At least $3/4$ (or 75%) of all values lie within 2 standard deviations of the mean.
- At least $8/9$ (or 89%) of all values lie within 3 standard deviations of the mean.

Example 7 **Chebyshev's Theorem**

IQ scores have a mean of 100 and a standard deviation of 15. What can we conclude from Chebyshev's theorem?

Solution

Applying Chebyshev's theorem with a mean of 100 and a standard deviation of 15, we can reach the following conclusions:

- At least $3/4$ (or 75%) of IQ scores are within 2 standard deviations of the mean (between 70 and 130).
- At least $8/9$ (or 89%) of all IQ scores are within 3 standard deviations of the mean (between 55 and 145).

Comparing Variation in Different Populations

In Part 1 of this section, we noted that when comparing variation in two different sets of data, the standard deviations should be compared only if the two sets of data use the same scale and units and have approximately the same mean. If the means are substantially different, or if the samples use different scales or measurement units, we can use the *coefficient of variation,* defined as follows.

DEFINITION The **coefficient of variation** (or **CV**) for a set of nonnegative sample or population data, expressed as a percent, describes the standard deviation relative to the mean, and is given by the following:

Sample	Population
$CV = \frac{s}{\bar{x}} \cdot 100\%$	$CV = \frac{\sigma}{\mu} \cdot 100\%$

Round-Off Rule for the Coefficient of Variation

Round the coefficient of variation to one decimal place (such as 18.3%).

Example 8 Chocolate Chip Cookies and Coke

Compare the variation in the numbers of chocolate chips in Chips Ahoy (regular) cookies (listed in Table 3-1) and the weights of regular Coke listed in Data Set 19 of Appendix B. Using the sample data, we have these results: Cookies have $\bar{x} = 24.0$ chips and $s = 2.6$ chips; Coke has $\bar{x} = 0.81682$ lb and $s = 0.00751$ lb. Note that we want to compare variation among *numbers of chocolate chips* to variation among *weights of Coke.*

Solution

We can compare the standard deviations if the same scales and units are used and the two means are approximately equal, but here we have different scales (numbers of chocolate chips and weights of Coke) and different units of measurement (numbers and pounds), so we use the coefficients of variation:

Numbers of chocolate chips: $$CV = \frac{s}{\bar{x}} \cdot 100\% = \frac{2.6 \text{ chocolate chips}}{24.0 \text{ chocolate chips}} \cdot 100\% = 10.8\%$$

Weights of Coke: $$CV = \frac{s}{\bar{x}} \cdot 100\% = \frac{0.00751 \text{ lb}}{0.81682 \text{ lb}} \cdot 100\% = 0.9\%$$

Although the standard deviation of 2.6 chocolate chips cannot be compared to the standard deviation of 0.00751 lb, we can compare the coefficients of variation, which have no units. We can see that the numbers of chocolate chips (with $CV = 10.8\%$) vary considerably more than weights of Coke (with $CV = 0.9\%$). This makes intuitive sense, because variation among the numbers of chocolate chips is not a big deal, but if some cans of Coke were underfilled or overfilled by large amounts, the result would be angry consumers.

Biased and Unbiased Estimators

The sample standard deviation s is a **biased estimator** of the population standard deviation σ. This means that values of the sample standard deviation s do *not* target the value of the population standard deviation σ. Although individual values of s can equal or exceed the value of σ, values of s generally tend to *underestimate* the value of σ. For example, consider an IQ test designed so that the population standard deviation is 15. If you repeat the process of randomly selecting 100 subjects, giving them IQ tests, and calculating the sample standard deviation s in each case, the sample standard deviations that you obtain will tend to be less than 15, which is the population standard deviation. There is no correction that allows us to fix the bias for all distributions of data. There is a correction that allows us to fix the bias for normally distributed populations, but it is rarely used because it is too complex and makes relatively minor corrections.

The sample variance s^2 is an **unbiased estimator** of the population variance σ^2, which means that values of s^2 tend to target the value of σ^2 instead of systematically tending to overestimate or underestimate σ^2. Consider an IQ test designed so that the population variance is 225. If you repeat the process of randomly selecting 100 subjects, giving them IQ tests, and calculating the sample variance s^2 in each case, the sample variances that you obtain will tend to center around 225, which is the population variance.

The concepts of biased estimators and unbiased estimators will be discussed more in Section 6-4.

using TECHNOLOGY

STATDISK, Minitab, Excel, the TI-83/84 Plus calculator, and StatCrunch can be used for the important calculations of this section. Use the same procedures given at the end of Section 3-2.

Recommended Assignment
Exercises 1–4 and even-numbered Exercises 6–20 and 33–36. (If the coefficient of variation was covered, also include Exercises 21–24. If frequency distributions were covered, also include Exercises 37–40.)

1. The IQ scores of a class of statistics students should have less variation, because those students are a much more homogeneous group with IQ scores that are likely to be closer together.
2. Parts (a), (b), and (d) are true.
3. Variation is a general descriptive term that refers to the amount of dispersion or spread among the data values, but the variance refers specifically to the square of the standard deviation.
4. s, σ, s^2, σ^2

3-3 Basic Skills and Concepts

Statistical Literacy and Critical Thinking

1. Comparing Variation Which do you think has less variation: the IQ scores of students in your statistics class or the IQ scores of a simple random sample taken from the general population? Why?

2. Correct Statements? Which of the following statements are true?

a. If each of 25 sample values is equal to 20 min, the standard deviation of the sample is 0 min.

b. For any set of sample values, the standard deviation can *never* be a negative value.

c. If the standard deviation of a sample is 3 kg, then the variance is 9 kg.

d. If the variance of a sample is 16 sec^2, then the standard deviation is 4 sec.

e. If the standard deviation of a sample is 25 cm, then the variance is 5 cm^2.

3. Variation and Variance In statistics, how do the terms *variation* and *variance* differ?

4. Symbols Identify the symbols used for each of the following: (a) sample standard deviation; (b) population standard deviation; (c) sample variance; (d) population variance.

In Exercises 5–20, find the* range, variance, *and* standard deviation *for the given sample data. Include appropriate units (such as "minutes") in your results. (The same data were used in Section 3-2 where we found measures of center. Here we find measures of variation.) Then answer the given questions.

5. Top 10 Celebrity Incomes Listed below are the earnings (in millions of dollars) of the celebrities with the 10 highest incomes in a recent year. The celebrities in order are Steven Spielberg, Howard Stern, George Lucas, Oprah Winfrey, Jerry Seinfeld, Tiger Woods, Dan Brown, Jerry Bruckheimer, J. K. Rowling, and Tom Cruise. Can this "Top 10" list be used to learn anything about the standard deviation of the annual earnings of all celebrities?

332 302 235 225 100 90 88 84 75 67

5. Range = \$265.0 million; s^2 = 10,548.0 (the units are the square of "million dollars"); s = \$102.7 million. Because the data values are the 10 highest from the population, nothing meaningful can be known about the standard deviation of the population.

6. Top 10 Most Expensive Colleges Listed below are the annual tuition amounts of the 10 most expensive colleges in the United States for a recent year. The colleges listed in order are Sarah Lawrence, NYU, George Washington, Bates, Skidmore, Johns Hopkins, Georgetown, Connecticut College, Harvey Mudd, and Vassar. Can this "Top 10" list tell us anything about the standard deviation of the population of all U.S. college tuitions?

\$54,410 \$51,991 \$51,730 \$51,300 \$51,196
\$51,190 \$51,122 \$51,115 \$51,037 \$50,875

6. Range = \$3535.0; s^2 = 1,088,153.8 square dollars; s = \$1043.1. Because the data values are the 10 highest from the population, nothing meaningful can be known about the standard deviation of the population.

7. Car Crash Test Measurements Listed below are head injury measurements from small cars that were tested in crashes. The measurements are in "hic," which is a measurement of a standard "head injury criterion." The data are from Data Set 13 in Appendix B, which is based on data from the National Highway Traffic Safety Administration. The listed values correspond to these cars: Chevrolet Aveo, Honda Civic, Volvo S40, VW Jetta, Hyundai Elantra, Kia Rio, and Kia Spectra. Given that all of the cars are small, does it appear that the head injury measurements are about the same for all small cars?

371 356 393 544 326 520 501

7. Range = 218.0 hic; s^2 = 7879.8 hic^2; s = 88.8 hic. Although all of the cars are small, the range from 326 hic to 544 hic appears to be relatively large, so the head injury measurements are not about the same.

8. Tests of Child Booster Seats The National Highway Traffic Safety Administration conducted crash tests of child booster seats for cars. Listed below are results from those tests, with the measurements given in "hic," which is a measurement of a standard "head injury criterion." According to the safety requirement, the hic measurement should be less than 1000 hic. Do the different child booster seats have much variation among their crash test measurements?

774 649 1210 546 431 612

8. Range = 779.0 hic; s^2 = 74,383.5 hic^2; s = 272.7 hic. Yes, there appears to be much variation. The largest value is more than twice the smallest value.

9. Harry Potter Listed below are the gross amounts (in millions of dollars) earned in box office receipts for the movie *Harry Potter and the Half-Blood Prince.* The movie opened on a Wednesday, and the amounts are listed in order for the first 14 days of the movie's release. If you invested in this movie, what characteristic of the data set would you care about most, and is it a measure of center or variation?

58 22 27 29 21 10 10 8 7 9 11 9 4 4

9. Range = \$54.0 million; s^2 = 210.9 (the units are the square of "million dollars"); s = \$14.5 million. An investor would care about the gross from opening day and the rate of decline after that, but the measures of center and variation are less important.

10. Florida Manatee Deaths Listed below are the numbers of manatee deaths caused each year by collisions with watercraft. Manatees, also called "sea cows," are large mammals that live underwater, often in or near waterways. The data are listed in order for each year of the past decade. The data are from the Florida Fish and Wildlife Conservation Commission. What important feature of the data is not revealed through the different measures of variation?

78 81 95 73 69 79 92 73 90 97

10. Range = 28.0 manatees; s^2 = 101.1 manatee squared; s = 10.1 manatees. The measures of variation reveal nothing about the pattern over time.

11. Ghost Prices Listed below are the prices listed for Norton Ghost 14.0 software from these vendors: Newegg, Dell, Buycheapsoftware.com, PC Connection, Walmart, and Overstock.com. When trying to find the best deal, how helpful are the measures of variation?

\$55.99 \$69.99 \$48.95 \$48.92 \$71.77 \$59.68

11. Range = \$22.850; s^2 = 99.141 dollars squared; s = \$9.957. The measures of variation are not very helpful in trying to find the best deal.

12. Range = \$19,628,584.0; s^2 = 59,583,269,405,325.1 dollars squared (most calculators or computers will provide only the first several digits, so a result such as 59,583,300,000.0 is OK); s = \$7,719,020.0. The amount of \$1 for Jobs is an outlier, and it has a great effect on the measures of variation.

12. CEO Compensation Listed below are the recent annual compensation amounts for these chief executive officers: Mulally (Ford), Jobs (Apple), Kent (Coca-Cola), Otellini (Intel), and McNerney (Boeing). The data are from the Associated Press. What is particularly notable about these compensation amounts?

\$17,688,241 \$1 \$19,628,585 \$12,407,800 \$14,765,410

13. Range = 17.50 μg/g; s^2 = 41.75 $(\mu g/g)^2$; s = 6.46 μg/g. If the medicines contained no lead, all of the measures would be 0 μg/g, and the measures of variation would all be 0 as well.

13. Lead in Medicine Listed below are the lead concentrations (in μg/g) measured in different Ayurveda medicines. Ayurveda is a traditional medical system commonly used in India. The lead concentrations listed here are from medicines manufactured in the United States. The data are based on the article "Lead, Mercury, and Arsenic in US and Indian Manufactured Ayurvedic Medicines Sold via the Internet," by Saper et al., *Journal of the American Medical Association,* Vol. 300, No. 8. What would be the values of the measures of variation if the medicines contained no lead?

3.0 6.5 6.0 5.5 20.5 7.5 12.0 20.5 11.5 17.5

14. Range = 1.150 ppm; s^2 = 0.134 ppm^2; s = 0.366 ppm. If the tuna sushi contained no mercury, all of the measures would be 0 ppm, and the measures of variation would all be 0 as well.

14. Mercury in Sushi Listed below are the amounts of mercury (in parts per million, or ppm) found in tuna sushi sampled at different stores in New York City. The study was sponsored by the *New York Times,* and the stores (in order) are D'Agostino, Eli's Manhattan, Fairway, Food Emporium, Gourmet Garage, Grace's Marketplace, and Whole Foods. What would be the values of the measures of variation if the tuna sushi contained no mercury?

0.56 0.75 0.10 0.95 1.25 0.54 0.88

15. Range = 11.0 years; s^2 = 12.3 years^2; s = 3.5 years. No, because 12 years is within 2 standard deviations of the mean.

15. Years to Earn Bachelor's Degree Listed below are the lengths of time (in years) it took for a random sample of college students to earn bachelor's degrees (based on data from the National Center for Education Statistics). Based on these results, is it *unusual* for someone to earn a bachelor's degree in 12 years?

4 4 4 4 4 4 4.5 4.5 4.5 4.5 4.5 4.5 6 6 8 9 9 13 13 15

16. Range = 1.170 W/kg; s^2 = 0.179 $(W/kg)^2$; s = 0.423 W/kg. No. Some models of cell phones are sold much more than others, so the measures from the different models should be weighted according to their size in the population.

16. Cell Phone Radiation Listed below are the measured radiation emissions (in W/kg) corresponding to these cell phones: Samsung SGH-tss9, Blackberry Storm, Blackberry Curve, Motorola Moto, T-Mobile Sidekick, Sanyo Katana Eclipse, Palm Pre, Sony Ericsson, Nokia 6085, Apple iPhone 3G S, and Kyocera Neo E1100. The data are from the Environmental Working Group. If one of each model of cell phone is measured for radiation and the results are used to find the standard deviation, is that standard deviation equal to the standard deviation of the population of all cell phones that are in use? Why or why not?

0.38 0.55 1.54 1.55 0.50 0.60 0.92 0.96 1.00 0.86 1.46

17. Range = 43.0 min; s^2 = 231.4 min^2; s = 15.2 min. The standard deviation can never be negative.

17. JFK to LAX Flight Delays Listed below are the arrival delay times (in minutes) of randomly selected American Airline flights from New York's JFK airport to Los Angeles (LAX). Negative values correspond to flights that arrived early before the scheduled arrival time, and positive values represent lengths of delays. (The data are from the Bureau of Transportation, and more data are listed in Data Set 15 in Appendix B.) Some of the sample values are negative, but can the standard deviation ever be negative?

−15 −18 −32 −21 −9 −32 11 2

18. Range = 16 kg; s^2 = 16.5 kg^2; s = 4.1 kg. The weight gain of 6.8 kg is not unusual because it is within 2 standard deviations of the mean. Although a gain of 6.8 kg is not unusual, the mean weight gain of 1.9 kg is not close to the legendary 6.8 kg, so an individual weight gain of 6.8 kg does not support the legend.

18. Freshman 15 According to the "freshman 15" legend, college freshmen gain 15 pounds (or 6.8 kilograms) during their freshman year. Listed below are the amounts of weight change (in kilograms) for a simple random sample of freshmen included in a study ("Changes in Body Weight and Fat Mass of Men and Women in the First Year of College: A Study of the 'Freshman 15'" by Hoffman, Policastro, Quick, and Lee, *Journal of American College Health,* Vol. 55, No. 1). Positive values correspond to students who gained weight and negative values correspond to students who lost weight. Is a weight gain of 15 pounds (or 6.8 kg) *unusual*? Why or why not? If 15 pounds (or 6.8 kg) is not unusual, does that support the legend of the "freshman 15"?

11 3 0 −2 3 −2 −2 5 −2 7 2 4 1 8 1 0 −5 2

19. Saints in Super Bowl Listed below are the numbers on the jerseys of the starting lineup for the New Orleans Saints when they recently won their first Super Bowl football game. What do the measures of variation tell us about the team? Does it make sense to compute the measures of variation for these data?

9 23 25 88 12 19 74 77 76 73 78

19. Range = 79.0; $s^2 = 1017.7$; $s = 31.9$. The data are at the nominal level of measurement and it makes no sense to compute the measures of variation for these data.

20. Phenotypes of Peas Biologists conducted experiments to determine whether a deficiency of carbon dioxide in the soil affects the phenotypes of peas. Listed below are the phenotype codes, where 1 = smooth-yellow, 2 = smooth-green, 3 = wrinkled-yellow, and 4 = wrinkled-green. Can the measures of center be obtained for these values? Do the results make sense?

2 1 1 1 1 1 1 4 1 2 2 1 2 3 3 2 3 1 3 1 3 1 3 2 2

20. Range = 3.0; $s^2 = 0.9$; $s = 0.9$. Because the data are at the nominal level of measurement, these results make no sense.

In Exercises 21–24, find the* coefficient of variation *for each of the two samples; then compare the variation. (The same data were used in Section 3-2.)

21. Speeding and Race Listed below are speeds (in mi/h) of cars on the New Jersey Turnpike by race of the driver. All cars are going in the same direction, and all of the cars are from New Jersey. The data are from Statlib and the authors are Joseph Kadane and John Lamberth.

White drivers:	74	77	69	71	77	69	72	75	74	72
African American drivers:	79	70	71	76	76	74	71	75	74	74

21. White drivers: 4.0%. African American drivers: 3.7%. The variation is about the same.

22. Parking Meter Theft Listed below are amounts (in millions of dollars) collected from parking meters by Brinks and others in New York City during similar time periods. A larger data set was used to convict five Brinks employees of grand larceny. The data were provided by the attorney for New York City, and they are listed on the DASL Web site. Do the limited data listed here show evidence of stealing by Brinks employees?

Collection contractor was Brinks:	1.3	1.5	1.3	1.5	1.4	1.7	1.8	1.7	1.7	1.6
Collection contractor was not Brinks:	2.2	1.9	1.5	1.6	1.5	1.7	1.9	1.6	1.6	1.8

22. Collection contractor was Brinks: 11.5%. Collection contractor was not Brinks: 12.8%. The variation is about the same.

23. Political Contributions Listed below are contributions (in dollars) made to the two presidential candidates in the most recent election. All contributions are from the same zip code as the author, and the data are from the *Huffington Post.*

Obama: $275 $452 $300 $1000 $1000 $500 $100 $1061
$1200 $235 $875 $2000 $350 $210 $250

McCain: $50 $75 $240 $302 $250 $700 $350 $500
$1250 $1500 $500 $500 $40 $221 $400

23. Obama: 80.0%. McCain: 91.3%. The variation among the Obama contributions is a little less than the variation among the McCain contributions.

24. Customer Waiting Times Waiting times (in minutes) of customers at the Jefferson Valley Bank (where all customers enter a single waiting line) and the Bank of Providence (where customers wait in individual lines at three different teller windows) are listed below.

Jefferson Valley (single line):	6.5	6.6	6.7	6.8	7.1	7.3	7.4	7.7	7.7	7.7
Providence (individual lines):	4.2	5.4	5.8	6.2	6.7	7.7	7.7	8.5	9.3	10.0

24. Jefferson Valley: 6.7%. Providence: 25.5%. The variation among the Jefferson Valley waiting times is much less than among the Providence waiting times.

Large Data Sets from Appendix B. ***In Exercises 25–28, refer to the indicated data set in Appendix B. Use computer software or a calculator to find the* range, variance, *and* standard deviation. *Express answers using appropriate units, such as "minutes."***

25. Earthquakes Use the magnitudes (Richter scale) of the earthquakes listed in Data Set 16 in Appendix B.

25. Range = 2.950; $s^2 = 0.345$; $s = 0.587$.

26. Flight Data Refer to Data Set 15 in Appendix B and use the times required to taxi out for takeoff.

26. Range = 37.0 min; $s^2 = 85.5\ \text{min}^2$; $s = 9.2$ min.

27. Range = 36.0 years; s^2 = 94.5 years^2; s = 9.7 years.

28. Range = 42.0; s^2 = 174.5; s = 13.2.

29. 0.738, which is not substantially different from s = 0.587.

30. 9.3 min, which is very close to s = 9.2 min.

31. 9.0 years, which is reasonably close to s = 9.7 years.

32. 10.5, which is not substantially different from s = 13.2.

33. No. The pulse rate of 99 beats per minute is between the minimum usual value of 54.3 beats per minute and the maximum usual value of 100.7 beats per minute.

34. Yes. The pulse rate of 45 beats per minute is not between the minimum usual value of 46.7 beats per minute and the maximum usual value of 87.9 beats per minute.

35. Yes. The volume of 11.9 oz is not between the minimum usual value of 11.97 oz and the maximum usual value of 12.41 oz.

36. No. The weight of 0.8133 lb is between the minimum usual value of 0.8127 lb and the maximum usual value of 0.8355 lb.

37. s = 12. 3 years. The result is not substantially different from the standard deviation of 11.1 years found from the original list of sample values.

38. s = 9.7 years. The result is not substantially different from the standard deviation of 9.0 years found from the original list of sample values.

27. Presidential Longevity Refer to Data Set 12 in Appendix B and use the numbers of years that U.S. presidents have lived after their first inauguration.

28. IQ Scores Refer to Data Set 6 in Appendix B and use the listed IQ scores.

Estimating Standard Deviation with the Range Rule of Thumb. ***In Exercises 29–32, refer to the data in the indicated exercise. After finding the range of the data, use the range rule of thumb to estimate the value of the standard deviation. Compare the result to the standard deviation computed with all of the data.***

29. Exercise 25 **30.** Exercise 26

31. Exercise 27 **32.** Exercise 28

Identifying Unusual Values with the Range Rule of Thumb. ***In Exercises 33–36, use the range rule of thumb to determine whether a value is*** **unusual.**

33. Pulse Rates of Females Based on Data Set 1 in Appendix B, females have pulse rates with a mean of 77.5 beats per minute and a standard deviation of 11.6 beats per minute. Is it unusual for a female to have a pulse rate of 99 beats per minute? (All of these pulse rates are measured at rest.) Explain.

34. Pulse Rates of Males Based on Data Set 1 in Appendix B, males have pulse rates with a mean of 67.3 beats per minute and a standard deviation of 10.3 beats per minute. Is it unusual for a male to have a pulse rate of 45 beats per minute? (All of these pulse rates are measured at rest.) Explain.

35. Volumes of Coke Based on Data Set 19 in Appendix B, cans of regular Coke have volumes with a mean of 12.19 oz and a standard deviation of 0.11 oz. Is it unusual for a can to contain 11.9 oz of Coke? Explain.

36. Weights of Pepsi Based on Data Set 19 in Appendix B, cans of regular Pepsi have weights with a mean of 0.82410 lb and a standard deviation of 0.00570 lb. Is it unusual for a can to contain 0.8133 lb of Pepsi? Explain.

Finding Standard Deviation from a Frequency Distribution. ***In Exercises 37–40, find the standard deviation of sample data summarized in a frequency distribution table by using the formula below, where x represents the class midpoint, f represents the class frequency, and n represents the total number of sample values. Also, compare the computed standard deviations to these standard deviations obtained by using Formula 3-4 with the original list of data values: (Exercise 37) 11.1 years; (Exercise 38) 9.0 years; (Exercise 39) 13.4; (Exercise 40) 9.7 years.***

$$s = \sqrt{\frac{n[\Sigma(f \cdot x^2)] - [\Sigma(f \cdot x)]^2}{n(n-1)}}$$ Standard deviation for frequency distribution

37.

Age of Best Actress When Oscar Was Won	Frequency
20–29	27
30–39	34
40–49	13
50–59	2
60–69	4
70–79	1
80–89	1

38.

Age of Best Actor When Oscar Was Won	Frequency
20–29	1
30–39	26
40–49	35
50–59	13
60–69	6
70–79	1

39.

Verbal IQ Score of Subject Exposed to Lead	Frequency
50–59	4
60–69	10
70–79	25
80–89	43
90–99	26
100–109	8
110–119	3
120–129	2

40.

Years President Lived after First Inauguration	Frequency
0–4	8
5–9	2
10–14	5
15–19	7
20–24	4
25–29	6
30–34	0
35–39	1

39. $s = 13.5$. The result is very close to the standard deviation of 13.4 found from the original list of sample values.

40. $s = 9.8$ years. The result is very close to the standard deviation of 9.7 years found from the original list of sample values.

41. The Empirical Rule Based on Data Set 1 in Appendix B, blood platelet counts of women have a bell-shaped distribution with a mean of 280 and a standard deviation of 65. (All units are 1000 cells/μL.) Using the empirical rule, what is the approximate percentage of women with platelet counts

a. within 2 standard deviations of the mean, or between 150 and 410?

b. between 215 and 345?

41. a. 95% b. 68%

42. The Empirical Rule Based on Data Set 3 in Appendix B, body temperatures of healthy adults have a bell-shaped distribution with a mean of 98.20°F and a standard deviation of 0.62°F. Using the empirical rule, what is the approximate percentage of healthy adults with body temperatures

a. within 1 standard deviation of the mean, or between 97.58°F and 98.82°F?

b. between 96.34°F and 100.06°F?

42. a. 68% b. 99.7%

43. Chebyshev's Theorem Based on Data Set 1 in Appendix B, blood platelet counts of women have a bell-shaped distribution with a mean of 280 and a standard deviation of 65. (All units are 1000 cells/μL.) Using Chebyshev's theorem, what do we know about the percentage of women with platelet counts that are within 2 standard deviations of the mean? What are the minimum and maximum platelet counts that are within 2 standard deviations of the mean?

43. At least 75% of women have platelet counts within 2 standard deviations of the mean. The minimum is 150 and the maximum is 410.

44. Chebyshev's Theorem Based on Data Set 3 in Appendix B, body temperatures of healthy adults have a bell-shaped distribution with a mean of 98.20°F and a standard deviation of 0.62°F. Using Chebyshev's theorem, what do we know about the percentage of healthy adults with body temperatures that are within 3 standard deviations of the mean? What are the minimum and maximum body temperatures that are within 3 standard deviations of the mean?

44. At least 89% of healthy adults have body temperatures within 3 standard deviations of the mean. The minimum is 96.34°F and the maximum is 100.06°F.

3-3 Beyond the Basics

45. Why Divide by $n - 1$? Let a *population* consist of the values 2 min, 3 min, 8 min. (These are departure delay times taken from American Airlines flights from New York's JFK airport to Los Angeles. See Data Set 15 in Appendix B.) Assume that samples of two values are randomly selected *with replacement* from this population. (That is, a selected value is replaced before the second selection is made.)

a. Find the variance σ^2 of the population {2 min, 3 min, 8 min}.

b. After listing the nine different possible samples of two values selected with replacement, find the sample variance s^2 (which includes division by $n - 1$) for each of them; then find the mean of the nine sample variances s^2.

45. a. 6.9 min^2 b. 6.9 min^2
c. 3.4 min^2
d. Part (b), because repeated samples result in variances that target the same value (6.9 min^2) as the population variance. Use division by $n - 1$.
e. No. The mean of the sample variances (6.9 min^2) equals the population variance (6.9 min^2), but the mean of the sample standard deviations (1.9 min) does not equal the mean of the population standard deviation (2.6 min).

c. For each of the nine different possible samples of two values selected with replacement, find the variance by treating each sample as if it is a population (using the formula for population variance, which includes division by n), then find the mean of those nine population variances.

d. Which approach results in values that are better estimates of σ^2: part (b) or part (c)? Why? When computing variances of samples, should you use division by n or $n - 1$?

e. The preceding parts show that s^2 is an unbiased estimator of σ^2. Is s an unbiased estimator of σ? Explain.

46. The mean absolute deviation of the population is 2.4 min. With repeated samplings of size 2, the nine different possible samples have mean absolute deviations of 0, 0, 0, 0.5, 0.5, 2.5, 2.5, 3, 3. With many such samples, the mean of those nine results is 1.3 min, showing that the sample mean absolute deviations tend to center about the value of 1.3 min instead of the mean absolute deviation of the population, which is 2.4 min. The sample mean deviations do not target the mean deviation of the population. This is not good. This indicates that a sample mean absolute deviation is not a good estimator of the mean absolute deviation of a population.

46. Mean Absolute Deviation Use the same population of {2 min, 3 min, 8 min} from Exercise 45. Show that when samples of size 2 are randomly selected with replacement, the samples have mean absolute deviations that do not center about the value of the mean absolute deviation of the population. What does this indicate about a sample mean absolute deviation being used as an estimator of the mean absolute deviation of a population?

3-4 Measures of Relative Standing and Boxplots

Key Concept This section introduces measures of relative standing, which are numbers showing the location of data values relative to the other values within the same data set. The most important concept in this section is the z score, which will be used often in following chapters. We also discuss percentiles and quartiles, which are common statistics, as well as a new statistical graph called a boxplot.

Part 1: Basics of z Scores, Percentiles, Quartiles, and Boxplots

z Scores

A z score (or standardized value) is found by converting a value to a standardized scale, as given in the following definition. This definition shows that a z score is the number of standard deviations that a data value is away from the mean. We will use z scores extensively in Chapter 6 and later chapters.

DEFINITION A **z score** (or **standardized value**) is the number of standard deviations that a given value x is above or below the mean. The z score is calculated by using one of the following:

Sample		Population
$z = \frac{x - \bar{x}}{s}$	or	$z = \frac{x - \mu}{\sigma}$

Round-Off Rule for z Scores

Round z scores to two decimal places (such as 2.31).

This round-off rule is motivated by the format of standard tables in which z scores are expressed with two decimal places. See Table A-2 in Appendix A, which is a typical table of z scores, and notice that Table A-2 has z scores expressed with two decimal places. Example 1 illustrates how z scores can be used to compare values, even if they come from different populations.

Note to Instructor

Recommendation: Because students can usually understand z scores on their own, discuss them briefly, but spend a little more time on percentiles, quartiles, and boxplots.

Example 1 Comparing a Count and a Weight

Example 8 in Section 3-3 used the coefficient of variation to compare the variation among numbers of chocolate chips in cookies to the variation among weights of regular Coke in cans. We now consider a comparison of two *individual* data values with this question: Which of the following two data values is more extreme?

- The Chips Ahoy (regular) cookie with 30 chocolate chips (among 40 cookies with a mean of 24.0 chocolate chips and a standard deviation of 2.6 chocolate chips)
- The can of regular Coke with a weight of 0.8295 lb (among 36 cans of regular Coke with a mean weight of 0.81682 lb and a standard deviation of 0.00751 lb)

Both of the above data values are the largest values in their respective data sets, but which of them is more extreme relative to the data sets from which they came?

Solution

The two given data values are measured on different scales with different units of measurement, but we can standardize the data values by converting them to z scores. Note that in the following calculations, the individual scores are substituted for the variable x.

Chips Ahoy cookie with 30 chocolate chips:

$$z = \frac{x - \bar{x}}{s} = \frac{30 \text{ chocolate chips} - 24.0 \text{ chocolate chips}}{2.6 \text{ chocolate chips}} = 2.31$$

can of Coke with weight of 0.8295 lb:

$$z = \frac{x - \bar{x}}{s} = \frac{0.8295 \text{ lb} - 0.81682 \text{ lb}}{0.00751 \text{ lb}} = 1.69$$

Interpretation

The results show that the cookie is 2.31 standard deviations above the mean, and the can of Coke is 1.69 standard deviations above the mean. Because the cookie is more standard deviations above the mean, it is the more extreme value. A cookie with 30 chocolate chips is more extreme than a can of Coke weighing 0.8295 lb.

Changing Populations

Included among the five important data set characteristics listed in Chapter 2 is the changing pattern of data over time. Some populations change, and their important statistics change as well. Car seat belt standards haven't changed in 40 years, even though the weights of Americans have increased considerably since then. In 1960, 12.8% of adult Americans were considered obese, compared to 22.6% in 1994.

According to the National Highway Traffic Safety Administration, seat belts must fit a standard crash dummy (designed according to 1960 data) placed in the most forward position, with 4 in. to spare. In theory, 95% of men and 99% of women should fit into seat belts, but those percentages are now lower because of the increases in weight over the last half-century. Some car companies provide seat belt extenders, but some do not.

Using z Scores to Identify Unusual Values In Section 3-3 we used the range rule of thumb to conclude that a value is "unusual" if it is more than 2 standard deviations away from the mean. It follows that unusual values have z scores less than -2 or greater than $+2$, as illustrated in Figure 3-4. Using this criterion with the two individual values used in Example 1 above, we see that the cookie with 30 chocolate chips is unusual (because its z score is 2.31, which is greater than 2), but the can of Coke weighing 0.8295 lb is not unusual (because its z score is 1.69, which is between -2 and $+2$).

Note to Instructor
The concept of "unusual values" is a good preparation for some of the concepts involved with hypothesis testing introduced in Chapter 8, so be sure to emphasize the role of unusual values as described here.

Usual values: $-2 \le z \text{ score} \le 2$

Unusual values: $z \text{ score} < -2$ *or* $z \text{ score} > 2$

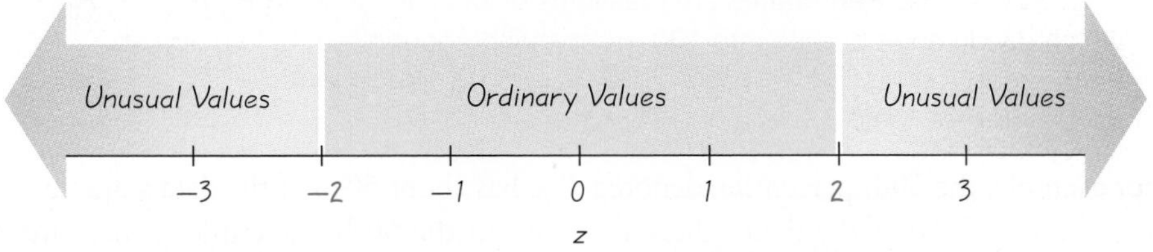

Figure 3-4
Interpreting z Scores
Unusual values are those with z scores less than -2.00 or greater than 2.00.

Americans Are Relatively Shorter

Paul Krugman writes in the *New York Times* that Americans were generally the tallest people in the world until the middle of the last century, but they are now the shortest when compared to others in advanced industrial countries. He also notes that there is a strong association between the per capita income of a country and the mean height of the country's people, so Americans should be taller than Europeans, but they no longer are. One possible explanation for this discrepancy is that Americans mistreat children by feeding them too much fast food and not giving them enough healthy food and exercise. Krugman writes that "Whatever the full explanation for America's stature deficit, our relative shortness, like our low life expectancy, suggests that something is amiss with our way of life."

Example 2 **Is a Pulse Rate of 48 Unusual?**

As the author was creating this example, he measured his pulse rate to be 48 beats per minute. (The author has too much time on his hands.) Is that pulse rate unusual? (Based on the pulse rates of males from Data Set 1 in Appendix B, assume that a large sample of adult males has pulse rates with a mean of 67.3 beats per minute and a standard deviation of 10.3 beats per minute.)

Solution

The author's pulse rate of 48 beats per minute is converted to a z score as shown below:

Pulse rate of 48:

$$z = \frac{x - \bar{x}}{s} = \frac{48 \text{ beats per minute} - 67.3 \text{ beats per minute}}{10.3 \text{ beats per minute}} = -1.87$$

Interpretation

The result shows that the author's pulse rate of 48 beats per minute is converted to the z score of -1.87. Refer to Figure 3-4 to see that $z = -1.87$ is between -2 and $+2$, so it is not unusual.

In Example 2, the pulse rate of 48 beats per minute resulted in a *negative* z score because that pulse rate is *less than* the mean of 67.3 beats per minute. Also, the units of "beats per minute" all canceled out, and the resulting z score has no units associated with it. These observations are included in the following summary describing z scores.

Properties of z Scores

1. A z score is the number of standard deviations that a given value x is above or below the mean.
2. z scores are expressed as numbers with no units of measurement.
3. A data value is unusual if its z score is less than -2 or greater than $+2$.
4. If an individual data value is less than the mean, its z score is a negative number.

A z score is a measure of position, in the sense that it describes the location of a value (in terms of standard deviations) relative to the mean. Quartiles and percentiles are other measures of position useful for comparing values within the same data set or between different sets of data.

Percentiles

Percentiles are one type of *quantiles*—or *fractiles*—which partition data into groups with roughly the same number of values in each group.

DEFINITION **Percentiles** are measures of location, denoted $P_1, P_2, \ldots, P_{99}$, which divide a set of data into 100 groups with about 1% of the values in each group.

For example, the 50th percentile, denoted P_{50}, has about 50% of the data values below it and about 50% of the data values above it. So the 50th percentile is the same as the median. There is not universal agreement on a single procedure for calculating percentiles, but we will describe two relatively simple procedures for (1) finding the percentile of a data value and (2) converting a percentile to its corresponding data value. We begin with the first procedure.

Finding the Percentile of a Data Value The process of finding the percentile that corresponds to a particular data value x is given by the following (round the result to the nearest whole number):

$$\text{Percentile of value } x = \frac{\text{number of values less than } x}{\text{total number of values}} \cdot 100$$

Example 3 Finding a Percentile

Table 3-5 lists the same counts of chocolate chips in 40 Chips Ahoy regular cookies listed in Table 3-1, but in Table 3-5 those counts are arranged in increasing order. Find the percentile for a cookie with 23 chocolate chips.

Table 3-5 *Sorted* Counts of Chocolate Chips in Chips Ahoy (Regular) Cookies

19	19	20	20	20	20	22	22	22	22
23	23	23	23	23	23	23	24	24	24
24	24	25	25	25	25	25	25	25	26
26	26	26	26	26	27	27	28	28	30

Solution

From the sorted list of chocolate chip counts in Table 3-5, we see that there are 10 cookies with fewer than 23 chocolate chips, so

$$\text{Percentile of } 23 = \frac{10}{40} \cdot 100 = 25$$

Interpretation

A cookie with 23 chocolate chips is in the 25th percentile. This can be interpreted loosely as: A cookie with 23 chocolate chips separates the lowest 25% of cookies from the highest 75%.

Example 3 shows how to convert from a given sample value to the corresponding percentile. There are several different methods for the reverse procedure of converting a given percentile to the corresponding value in the data set. The procedure we will use is summarized in Figure 3-5, which uses the following notation.

Notation

n total number of values in the data set

k percentile being used (Example: For the 25th percentile, $k = 25$.)

L locator that gives the *position* of a value (Example: For the 12th value in the sorted list, $L = 12$.)

P_k kth percentile (Example: P_{25} is the 25th percentile.)

Example 4 Converting a Percentile to a Data Value

Refer to the sorted chocolate chip counts in Table 3-5 and use the procedure in Figure 3-5 to find the value of the 18th percentile, P_{18}.

Cost of Laughing Index

There really is a Cost of Laughing Index (CLI), which tracks costs of such items as rubber chickens, Groucho Marx glasses, admissions to comedy clubs, and 13 other leading humor indicators. This is the same basic approach used in developing the Consumer Price Index (CPI), which is based on a weighted average of goods and services purchased by typical consumers. While standard scores and percentiles allow us to compare different values, they ignore any element of time. Index numbers, such as the CLI and CPI, allow us to compare the value of some variable to its value at some base time period. The value of an index number is the current value, divided by the base value, multiplied by 100.

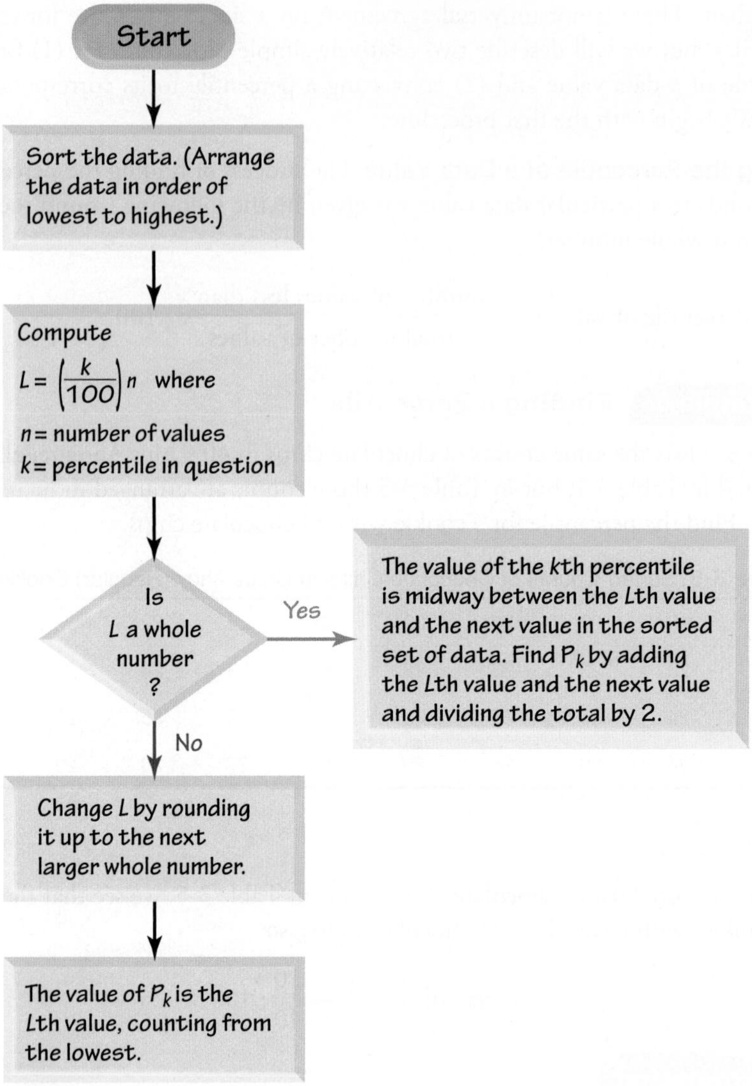
Start
Sort the data. (Arrange the data in order of lowest to highest.)
Compute
$L = \left(\frac{k}{100}\right) n$ where
n = number of values
k = percentile in question
Is L a whole number?
Yes
The value of the kth percentile is midway between the Lth value and the next value in the sorted set of data. Find P_k by adding the Lth value and the next value and dividing the total by 2.
No
Change L by rounding it up to the next larger whole number.
The value of P_k is the Lth value, counting from the lowest.

Figure 3-5
Converting from the kth percentile to the corresponding data value.

It is helpful to demonstrate use of Figure 3-5 with two examples, one that results in an integer value of L and one that does not, as in Examples 4 and 5.

Solution

From Figure 3-5, we see that the sample data are already sorted, so we can proceed to find the value of the locator L. In this computation we use $k = 18$ because we are trying to find the value of the 18th percentile. We use $n = 40$ because there are 40 data values.

$$L = \frac{k}{100} \cdot n = \frac{18}{100} \cdot 40 = 7.2$$

Since $L = 7.2$ is not a whole number, we proceed to the next lower box where we change L by rounding it up from 7.2 to 8. (In this book we typically round off the usual way, but this is one of two cases where we round *up* instead of rounding *off*.) From the last box we see that the value of P_{18} is the 8th value, counting from the lowest. In Table 3-5, the 8th value is 22. That is, $P_{18} = 22$ chocolate chips. Roughly speaking, about 18% of the cookies have fewer than 22 chocolate chips and 82% of the cookies have more than 22 chocolate chips.

Example 5 Converting a Percentile to a Data Value

Refer to the sorted chocolate chip counts in Table 3-5. Use Figure 3-5 to find the 25th percentile, denoted by P_{25}.

Solution

Referring to Figure 3-5, we see that the sample data are already sorted, so we can proceed to compute the value of the locator L. In this computation, we use $k = 25$ because we are attempting to find the value of the 25th percentile, and we use $n = 40$ because there are 40 data values.

$$L = \frac{k}{100} \cdot n = \frac{25}{100} \cdot 40 = 10$$

Since $L = 10$ is a whole number, we proceed to the box located at the right. We now see that the value of the 10th percentile is midway between the Lth (10th) value and the next value in the original set of data. That is, the value of the 10th percentile is midway between the 10th value and the 11th value. The 10th value in Table 3-5 is 22 and the 11th value is 23, so the value midway between them is 22.5 chocolate chips. We conclude that the 25th percentile is $P_{25} = 22.5$ chocolate chips.

Quartiles

Just as there are 99 percentiles that divide the data into 100 groups, there are three quartiles that divide the data into four groups.

DEFINITION Quartiles are measures of location, denoted Q_1, Q_2, and Q_3, which divide a set of data into four groups with about 25% of the values in each group.

Here are descriptions of quartiles that are more accurate than those given in the preceding definition:

Q_1 (First quartile): Same value as P_{25}. It separates the bottom 25% of the sorted values from the top 75%. (To be more precise, at least 25% of the sorted values are less than or equal to Q_1, and at least 75% of the values are greater than or equal to Q_1.)

Q_2 (Second quartile): Same as P_{50} and same as the median. It separates the bottom 50% of the sorted values from the top 50%.

Q_3 (Third quartile): Same as P_{75}. It separates the bottom 75% of the sorted values from the top 25%. (To be more precise, at least 75% of the sorted values are less than or equal to Q_3, and at least 25% of the values are greater than or equal to Q_3.)

Finding values of quartiles can be accomplished with the same procedure used for finding percentiles. Simply use the relationships shown in the margin.

$Q_1 = P_{25}$

$Q_2 = P_{50}$

$Q_3 = P_{75}$

Example 6 Finding a Quartile

Refer to the chocolate chip counts listed in Table 3-5. Find the value of the first quartile Q_1.

Solution

Finding Q_1 is really the same as finding P_{25}. See Example 5 above for the procedure used to find this result: $P_{25} = 22.5$ chocolate chips.

Nielsen Ratings for College Students

The Nielsen ratings are one of the most important measures of television viewing, and they affect billions of dollars in television advertising. In the past, the television viewing habits of college students were ignored, with the result that a large segment of the important young viewing audience was ignored. Nielsen Media Research is now including college students who do not live at home.

Some television shows have large appeal to viewers in the 18–24 age bracket, and the ratings of such shows have increased substantially with the inclusion of college students. For males, NBC's *Sunday Night Football* broadcast had an increase of 20% after male college students were included. For females, the TV show *Grey's Anatomy* had an increase of 54% after female college students were included. Those increased ratings ultimately translate into greater profits from charges to commercial sponsors. These ratings also give college students recognition that affects the programming they receive.

Just as there is not universal agreement on a procedure for finding percentiles, there is not universal agreement on a single procedure for calculating quartiles, and different computer programs often yield different results. If you use a calculator or computer software for exercises involving quartiles, you may get results that differ somewhat from the answers obtained by using the procedures described here.

CAUTION There is not universal agreement on procedures for finding quartiles, and different technologies may yield different quartile values.

In earlier sections of this chapter we described several statistics, including the mean, median, mode, range, and standard deviation. Some other statistics are defined using quartiles and percentiles, as in the following:

$$\text{Interquartile range (or IQR)} = Q_3 - Q_1$$

$$\text{Semi-interquartile range} = \frac{Q_3 - Q_1}{2}$$

$$\text{Midquartile} = \frac{Q_3 + Q_1}{2}$$

$$\text{10–90 percentile range} = P_{90} - P_{10}$$

5-Number Summary and Boxplot

The values of the three quartiles (Q_1, Q_2, Q_3) are used for the 5-number summary and the construction of boxplot graphs.

DEFINITION For a set of data, the **5-number summary** consists of these five values:

1. Minimum
2. First quartile, Q_1
3. Second quartile, Q_2 (same as the median)
4. Third quartile, Q_3
5. Maximum

A **boxplot** (or **box-and-whisker diagram**) is a graph of a data set that consists of a line extending from the minimum value to the maximum value, and a box with lines drawn at the first quartile Q_1, the median, and the third quartile Q_3. (See Figure 3-6 on the next page.)

Example 7 Finding a 5-Number Summary

Use the chocolate chip counts listed in Table 3-5 to find the 5-number summary.

Solution

Because the chocolate chip counts in Table 3-5 are sorted, it is easy to see that the minimum is 19 and the maximum is 30. The value of the first quartile is $Q_1 = 22.5$, as was found in Examples 5 and 6. Also, $Q_3 = 26.0$ can be found by using the same procedure for finding P_{75} (as summarized in Figure 3-5). The 5-number summary is 19, 22.5, 24.0, 26.0, and 30.

The 5-number summary is used to construct a boxplot, as in the following procedure.

Procedure for Constructing a Boxplot

1. Find the 5-number summary (minimum value, Q_1, Q_2, Q_3, maximum value).
2. Construct a scale with values that include the minimum and maximum data values.
3. Construct a box (rectangle) extending from Q_1 to Q_3, and draw a line in the box at the value of Q_2 (median).
4. Draw lines extending outward from the box to the minimum and maximum data values.

CAUTION Because there is not universal agreement on procedures for finding quartiles, and because boxplots are based on quartiles, different technologies may yield different boxplots.

Example 8 Constructing a Boxplot

Use the chocolate chip counts listed in Table 3-5 to construct a boxplot.

Solution

The boxplot uses the 5-number summary found in Example 7: 19, 22.5, 24.0, 26.0, and 30. Figure 3-6 is the boxplot representing the chocolate chip counts listed in Table 3-5.

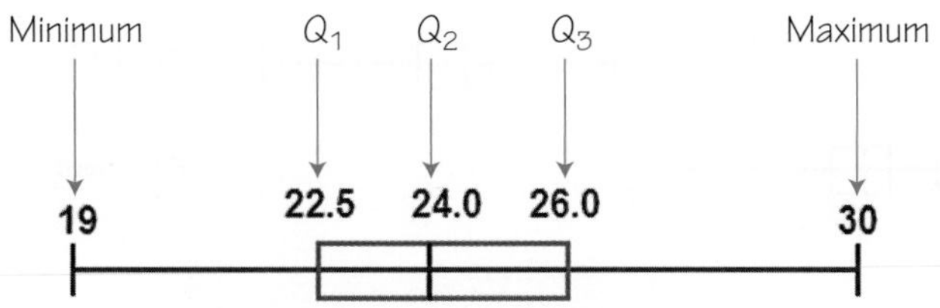

Figure 3-6 Boxplot of Chocolate Chip Counts in Regular Chips Ahoy Cookies

Boxplots give us some information about the spread of the data. Shown below is a boxplot from a data set with a normal (bell-shaped) distribution and a boxplot from a data set with a distribution that is skewed to the right (based on data from *USA Today*).

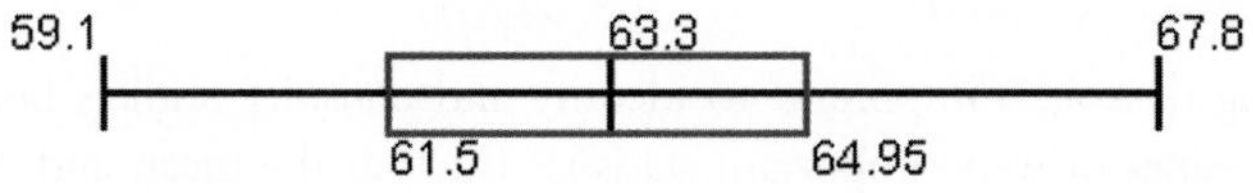

Normal Distribution: Heights from a simple random sample of women

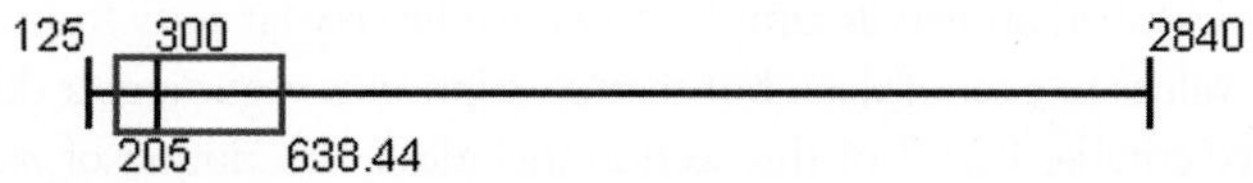

Skewed Distribution: Salaries (in thousands of dollars) of NCAA football coaches

Because the shape of a boxplot is determined by the five values from the 5-number summary, a boxplot is not a graph of the distribution of the data, and it doesn't show as much detailed information as a histogram or stemplot. However, boxplots are often great for comparing two or more data sets. When using two or more boxplots for comparing different data sets, graph the boxplots on the same scale so that comparisons can be easily made.

Example 9 Do the Different Brands of Cookies Have Different Chocolate Chip Counts?

The Chapter Problem includes Table 3-1, which lists counts of chocolate chips in cookies from different brands. Use the same scale to construct the five corresponding boxplots; then compare the results.

Solution

The STATDISK-generated boxplots shown in Figure 3-7 suggest that the numbers of chocolate chips in the different brands of cookies are very different. In particular, the counts from the Keebler and Hannaford cookies appear to be very different; the boxplots show that there isn't any overlap, and all of the Hannaford cookies have lower counts than any of the Keebler cookies. It might seem that the Hannaford brand is stingy with its chocolate chips, but the Hannaford brand had many chips that were substantially larger than those in any of the other brands.

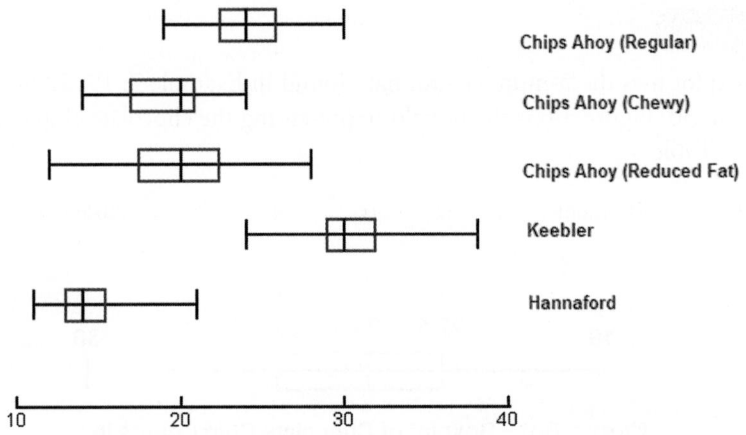

Figure 3-7 Boxplots of Counts of Chocolate Chips in Cookies

Methods discussed later in this book allow us to analyze this issue more formally. It is always wise to construct suitable graphs, such as histograms, dotplots, and boxplots, but we should not rely solely on subjective judgments based on graphs.

Outliers

When analyzing data, it is important to identify and consider outliers because they can strongly affect values of some important statistics (such as the mean and standard deviation), and they can also strongly affect important methods discussed later in this book. In Section 2-1 we described outliers as sample values that lie very far away from the vast majority of the other values in a set of data, but that description is vague and it does not provide specific objective criteria. Part 2 of this section includes a description of *modified boxplots* along with a specific definition of outliers used in the context of creating modified boxplots.

CAUTION When analyzing data, always identify outliers and consider their effects, which can be substantial.

Part 2: Outliers and Modified Boxplots

Outliers

We noted that the description of outliers is somewhat vague, but for the purposes of constructing *modified boxplots*, we can consider outliers to be data values meeting specific criteria based on quartiles and the interquartile range. (The interquartile range is often denoted by IQR, and IQR = $Q_3 - Q_1$.)

1. Find the quartiles Q_1, Q_2, and Q_3.
2. Find the interquartile range (IQR), where IQR = $Q_3 - Q_1$.
3. Evaluate 1.5 × IQR.
4. **In a modified boxplot, a data value is an *outlier* if it is**

above Q_3, by an amount greater than 1.5 × IQR

or below Q_1, by an amount greater than 1.5 × IQR

Note to Instructor
If pressed for time, skip Part 2, but it might be helpful to briefly discuss modified boxplots and how to interpret them.

Modified Boxplots

The boxplots described earlier are called **skeletal** (or **regular**) **boxplots,** but some statistical software packages provide modified boxplots, which represent outliers as special points. A **modified boxplot** is a regular boxplot constructed with these modifications: (1) A special symbol (such as an asterisk or point) is used to identify outliers as defined above, and (2) the solid horizontal line extends only as far as the minimum data value that is not an outlier and the maximum data value that is not an outlier. (*Note: Exercises involving modified boxplots are found in the "Beyond the Basics" exercises only.*)

Example 10 Modified Boxplot

Use the Hannaford chocolate chip counts from Table 3-1 to construct a modified boxplot.

Solution

Let's begin with the above four steps for identifying outliers in a modified boxplot.

1. Using the Hannaford chocolate chip counts from Table 3-1, the three quartiles are $Q_1 = 13.0$, $Q_2 = 14.0$, and $Q_3 = 15.5$.
2. The interquartile range is IQR = $Q_3 - Q_1 = 15.5 - 13.0 = 2.5$.
3. $1.5 \times \text{IQR} = 1.5 \times 2.5 = 3.75$.
4. Any outliers are above $Q_3 = 15.5$ by more than 3.75, or below $Q_1 = 13.0$ by more than 3.75. This means that any outliers are greater than 19.25 or less than 9.25.

We can now examine the original Hannaford chocolate chip counts to see that 21 is the only value greater than 19.25, and there are no values less than 9.25. Therefore, 21 is the only outlier.

We can now construct the modified boxplot shown in Figure 3-8. In Figure 3-8, the outlier is identified as a special point, the three quartiles are shown as in a regular

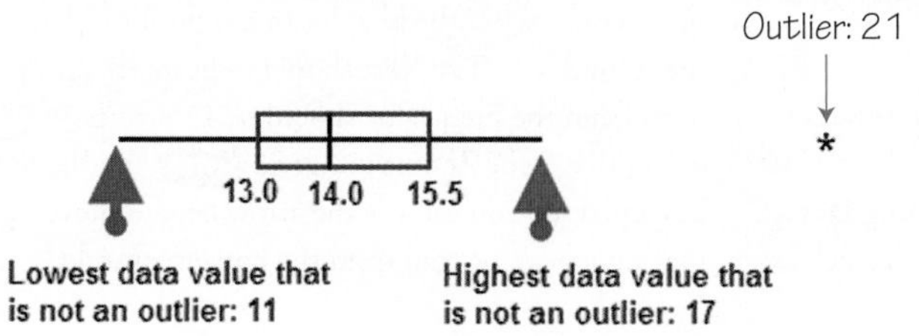

Figure 3-8
Modified Boxplots of Hannaford Chocolate Chip Counts

boxplot, and the horizontal line extends from the lowest data value that is not an outlier (11) to the highest data value that is not an outlier (17).

CAUTION Because there is not universal agreement on procedures for finding quartiles, and because modified boxplots are based on quartiles, different technologies may yield different modified boxplots.

Putting It All Together

We have discussed several basic tools commonly used in statistics. When designing an experiment, analyzing data, reading an article in a professional journal, or doing anything else with data, it is important to consider certain key factors, such as:

- Context of the data
- Source of the data
- Sampling method
- Measures of center
- Measures of variation
- Distribution
- Outliers
- Changing patterns over time
- Conclusions
- Practical implications

This is an excellent checklist, but it should not replace *thinking* about any other relevant factors. It is very possible that some application of statistics requires factors not included in the above list, and it is also possible that some of the factors in the list are not relevant for certain applications.

using TECHNOLOGY

Boxplots

STATDISK Enter or open the data in the Data Window, then click on **Data,** then **Boxplot.** Click on the columns that you want to include, then click on either **Boxplot** or **Modified Boxplot.**

MINITAB Enter or open the data in columns. With Minitab 16, click on **Assistant,** then select **Graphical Assistant,** then click on the image of a boxplot. With earlier versions of Minitab, select **Graph,** then select **Boxplot.** Select the "Simple" option for one boxplot or the "Simple" option for multiple boxplots. Enter the column names in the Variables box, then click **OK.** Minitab provides modified boxplots as described in Part 2 of this section.

EXCEL Although Excel is not designed to generate boxplots, they can be generated using XLSTAT. First enter the data in column A. Click on **XLSTAT** at the top, select **Describing Data,** then select **Descriptive Statistics.** Enter the range of cells (such as A1:A40) in the "Quantitative Data" box. Check the "Sample labels" box only if the first cell contains the name of the data set. Click on the **Charts** tab, then check the box labelled **Boxplots.** Click **OK** to continue. The result will include descriptive statistics as well as a boxplot.

TI-83/84 PLUS Enter the sample data in list L1 (or enter the data and assign them to a list with a name). Now select **STAT PLOT** by pressing 2ND Y=. Press ENTER, then select the option of ON. For a simple boxplot as described in Part 1 of this section, select the boxplot type that is positioned in the middle of the second row; for a modified boxplot as described in Part 2 of this section, select the boxplot that is positioned at the far left of the second row. The Xlist should indicate L1 (or the assigned list name) and the Freq value should be 1. Now press ZOOM and select option 9 for **ZoomStat.** Press ENTER and the boxplot should be displayed. You can use the arrow keys to move right or left so that values can be read from the horizontal scale.

STATCRUNCH Click on **Open StatCrunch,** then enter or open a data set. Click on **Graphics,** then select **Boxplot.** Select the column containing the data. Click on **Create Graph.**

5-Number Summary

STATDISK, Minitab, and the TI-83/84 Plus calculator provide the values of the 5-number summary. Use the same procedure given at the end of Section 3-2. Excel provides the minimum, maximum, and median, and the quartiles can be obtained by clicking on *fx*, selecting the function category of Statistical, and selecting QUARTILE. (In Excel 2010, select QUARTILE.INC, which is the same as QUARTILE in Excel 2003 and Excel 2007, or select the new function QUARTILE.EXC, which is supposed to be "consistent with industry best practices.")

Outliers

To identify outliers, sort the data in order from the minimum to the maximum, then examine the minimum and maximum values to determine whether they are far away from the other data values. Here are instructions for sorting data:

STATDISK Click on **Data** and select the menu item of **Sort Data.** Click on **Sort** after making the desired choices.

MINITAB Click on **Data** and select **Sort.** Enter the column in the "Sort column(s)" box and enter that same column in the "By column" box.

EXCEL In Excel 2003, click on the "sort ascending" icon, which has the letter A stacked above the letter Z and a downward arrow. In Excel 2013, 2010, or 2007, click on **Data,** then click on the "sort ascending" icon, which has the letter A stacked above the letter Z and a downward arrow.

TI-83/84 PLUS Press STAT and select **SortA** (for sort in ascending order). Press ENTER. Enter the list to be sorted, such as L1 or a named list, then press ENTER.

STATCRUNCH Click on **Open StatCrunch.** Enter or open a data set. Click on **Data,** then select the menu item of **Sort columns.** Click on the column to be sorted, then click on **Sort columns.**

3-4 Basic Skills and Concepts

Recommended Assignment
Exercises 1–16 and even-numbered Exercises 18–32.

Statistical Literacy and Critical Thinking

1. *z* Scores James Madison, the fourth President of the United States, was 163 cm tall. His height converts to the *z* score of −2.28 when included among the heights of all presidents. Is his height above or below the mean? How many standard deviations is Madison's height away from the mean?

1. Madison's height is below the mean. It is 2.28 standard deviations below the mean.

2. *z* Scores If your score on your next statistics test is converted to a *z* score, which of these *z* scores would you prefer: −2.00, −1.00, 0, 1.00, 2.00? Why?

2. 2.00 should be preferred, because it is 2.00 standard deviations above the mean and would correspond to the highest of the five different possible scores.

3. Boxplots Shown below is a STATDISK-generated boxplot of the amounts of money (in millions of dollars) that movies grossed (based on data from the Motion Picture Association of America). What do the displayed values of 5, 47, 104, 121, and 380 tell us?

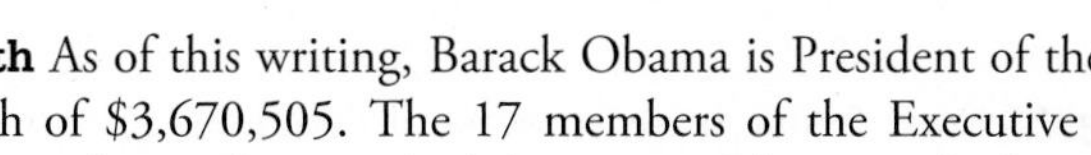

3. The lowest amount is \$5 million, the first quartile Q_1 is \$47 million, the second quartile Q_2 (or median) is \$104 million, the third quartile Q_3 is \$121 million, and the highest gross amount is \$380 million.

4. Measures of Location The values of P_{50}, Q_2, and the median are found for the net incomes reported on all individual 1040 tax forms filed last year. What do those values have in common?

4. All three values are the same.

z* Scores.** ***In Exercises 5–8, express all z scores with two decimal places.

5. Obama's Net Worth As of this writing, Barack Obama is President of the United States and he has a net worth of \$3,670,505. The 17 members of the Executive Branch have a mean net worth of \$4,939,455 with a standard deviation of \$7,775,948 (based on data from opensecrets.org).

a. What is the difference between President Obama's net worth and the mean net worth of all members of the Executive Branch?

b. How many standard deviations is that (the difference found in part (a))?

5. a. \$1,268,950
b. 0.16 standard deviations

c. $z = -0.16$
d. Usual

c. Convert President Obama's net worth to a z score.

d. If we consider "usual" amounts of net worth to be those that convert to z scores between -2 and 2, is President Obama's net worth usual or unusual?

6. a. 1.766
b. 3.01 standard deviations
c. $z = 3.01$
d. Unusual

6. Earthquakes Data Set 16 in Appendix B lists 50 magnitudes (Richter scale) of 50 earthquakes, and those earthquakes have magnitudes with a mean of 1.184 with a standard deviation of 0.587. The strongest of those earthquakes had a magnitude of 2.95.

a. What is the difference between the magnitude of the strongest earthquake and the mean magnitude?

b. How many standard deviations is that (the difference found in part (a))?

c. Convert the magnitude of the strongest earthquake to a z score.

d. If we consider "usual" magnitudes to be those that convert to z scores between -2 and 2, is the magnitude of the strongest earthquake usual or unusual?

7. a. $1,449,778
b. 2.75 standard deviations
c. $z = -2.75$
d. Unusual

7. Jobs' Job When Steve Jobs was Chief Executive Officer (CEO) of Apple, he earned an annual salary of $1. The CEOs of the 50 largest U.S. companies had a mean salary of $1,449,779 and a standard deviation of $527,651 (based on data from *USA Today*).

a. What is the difference between Jobs' salary and the mean CEO salary?

b. How many standard deviations is that (the difference found in part (a))?

c. Convert Steve Jobs' salary to a z score.

d. If we consider "usual" salaries to be those that convert to z scores between -2 and 2, is Steve Jobs' salary usual or unusual?

8. a. 15.3 beats per minute
b. 1.49 standard deviations
c. $z = -1.49$
d. Usual

8. Student's Pulse Rate A male student of the author has a measured pulse rate of 52 beats per minute. Based on Data Set 1 in Appendix B, males have a mean pulse rate of 67.3 beats per minute and a standard deviation of 10.3 beats per minute.

a. What is the difference between the student's pulse rate and the mean pulse rate of males?

b. How many standard deviations is that (the difference found in part (a))?

c. Convert the student's pulse rate to a z score.

d. If we consider "usual" pulse rates to be those that convert to z scores between -2 and 2, is the student's pulse rate usual or unusual?

Usual and Unusual Values. ***In Exercises 9–12, consider a value to be* unusual *if its z score is less than -2 or greater than 2.***

9. z scores: -2 and 2. IQ scores: 70 and 130.

9. IQ Scores The Wechsler Adult Intelligence Scale measures IQ scores with a test designed so that the mean is 100 and the standard deviation is 15. Consider the group of IQ scores that are unusual. What are the z scores that separate the unusual IQ scores from those that are usual? What are the IQ scores that separate the unusual IQ scores from those that are usual?

10. z scores: -2 and 2. Hip breadths: 31.6 cm and 41.6 cm.

10. Designing Aircraft Seats In the process of designing aircraft seats, it was found that men have hip breadths with a mean of 36.6 cm and a standard deviation of 2.5 cm. (based on anthropometric survey data from Gordon, Clauser, et al.). Consider the values of hip breadths of men that are unusual. What are the z scores that separate the unusual hip breadths from those that are usual? What are the hip breadths that separate the unusual hip breadths from those that are usual?

11. Earthquakes The Southern California Earthquake Data Center recorded magnitudes (Richter scale) of 10,594 earthquakes in a recent year. The mean is 1.240 and the standard deviation is 0.578. Consider the magnitudes that are unusual. What are the magnitudes that separate the unusual earthquakes from those that are usual?

12. Female Voice Based on data from Data Set 17 in Appendix B, the words spoken in a day by women have a mean of 16,215 words and a standard deviation of 7301 words. Consider the women with an unusual word count in a day. What are the numbers of words that separate the unusual word counts from those that are usual?

Comparing Values. *In Exercises 13–16, use z scores to compare the given values.*

13. Tallest Man and Woman As of this writing, the tallest living man is Sultan Kosen, who has a height of 247 cm. The tallest living woman is De-Fen Yao, who is 236 cm tall. Heights of men have a mean of 175 cm and a standard deviation of 7 cm. Heights of women have a mean of 162 cm and a standard deviation of 6 cm. Relative to the population of the same gender, who is taller? Explain.

14. Oscars As of this writing, Sandra Bullock was the last woman to win an Oscar for Best Actress and Jeff Bridges was the last man to win for Best Actor. At the time of the awards ceremony, Sandra Bullock was 45 years of age and Jeff Bridges was 60 years of age. Based on Data Set 11 in Appendix B, the Best Actresses have a mean age of 35.9 years and a standard deviation of 11.1 years. The Best Actors have a mean age of 44.1 years and a standard deviation of 9.0 years. (All ages are determined at the time of the awards ceremony.) Relative to their genders, who was younger when winning the Oscar: Sandra Bullock or Jeff Bridges? Explain.

15. Comparing Test Scores Scores on the SAT test have a mean of 1518 and a standard deviation of 325. Scores on the ACT test have a mean of 21.1 and a standard deviation of 4.8. Which is relatively better: a score of 1490 on the SAT test or a score of 17.0 on the ACT test? Why?

16. Red Blood Cell Counts Based on Data Set 1 in Appendix B, males have red blood cell counts with a mean of 5.072 and a standard deviation of 0.395, while females have red blood cell counts with a mean of 4.577 and a standard deviation of 0.382. Who has the higher count relative to the sample from which it came: A male with a count of 4.91 or a female with a count of 4.32? Explain.

Percentiles. ***In Exercises 17–20, use the following duration times (seconds) of 24 eruptions of the Old Faithful geyser in Yellowstone National Park. The duration times are sorted from lowest to highest. Find the percentile corresponding to the given time.***

110	120	178	213	234	234	235	237	240	243	245	245
250	250	251	252	254	255	255	259	260	266	269	273

17. 213 sec **18.** 240 sec **19.** 250 sec **20.** 260 sec

In Exercises 21–28, use the same list of 24 sorted Old Faithful eruption duration times given for Exercises 17–20. Find the indicated percentile or quartile.

21. P_{60} **22.** Q_1

23. Q_3 **24.** P_{40}

25. P_{50} **26.** P_{75}

27. P_{25} **28.** P_{85}

11. 0.084 and 2.396
12. 1613 words and 30,817 words
13. De-Fen Yao is relatively taller, because her z score is 12.33, which is greater than the z score of 10.29 for Sultan Kosen. De-Fen Yao is more standard deviations above the mean than Sultan Kosen.
14. With a z score of 0.82, Sandra Bullock was relatively younger than Jeff Bridges, who had a z score of 1.77.
15. The SAT score of 1490 has a z score of -0.09, and the ACT score of 17.0 has a z score of -0.85. The z score of -0.09 is a larger number than the z score of -0.85, so the SAT score of 1490 is relatively better.
16. The male has the higher count because his z score is -0.41, which is a higher number than the z score of -0.67 for the female.
17. 13th percentile. 18. 33rd percentile.
19. 50th percentile. 20. 83rd percentile.
21. 251 sec (Tech: Excel: 250.8 sec).
22. 234.5 sec (Tech: Minitab: 234.25 sec; Excel: 234.75 sec).
23. 255 sec.
24. 243 sec (Tech: Excel: 243.4 sec).
25. 247.5 sec. 26. 255 sec.
27. 234.5 sec (Tech: Minitab: 234.25 sec; Excel: 234.75 sec).
28. 260 sec (Tech: Excel: 259.55 sec).
29. 5-number summary: 1 sec, 8709 sec, 10,074.5 sec, 11,445 sec, 11,844 sec. (Tech: Minitab yields $Q_1 = 8338$ sec and $Q_3 = 11{,}453$ sec. Excel yields $Q_1 = 8727.75$ sec and $Q_3 = 11{,}115$ sec.)

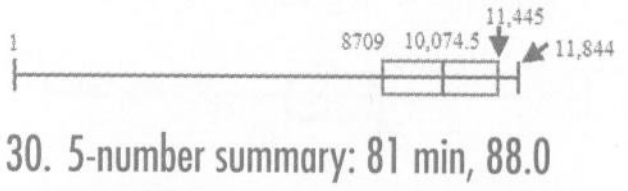

30. 5-number summary: 81 min, 88.0 min, 94.5 min, 98.0 min, 106 min. (Tech: Minitab yields $Q_1 = 87.5$ min. Excel yields $Q_1 = 88.5$ min.)

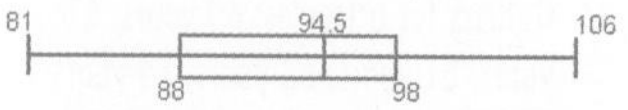

31. 5-number summary: 4 min, 14.0 min, 18.0 min, 32.0 min, 63 min. (Tech: Minitab yields $Q_1 = 12.75$ min and $Q_3 = 34.25$ min. Excel yields $Q_1 = 14.25$ min and $Q_3 = 31.5$ min.)

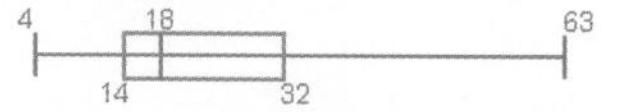

32. 5-number summary: 70 mi/h, 72.0 mi/h, 74 mi/h, 78.0 mi/h, 79 mi/h.

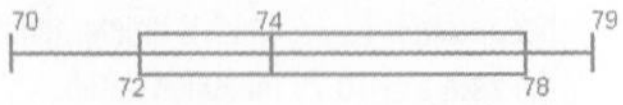

33. The top boxplot represents males. It appears that males have lower pulse rates than females.

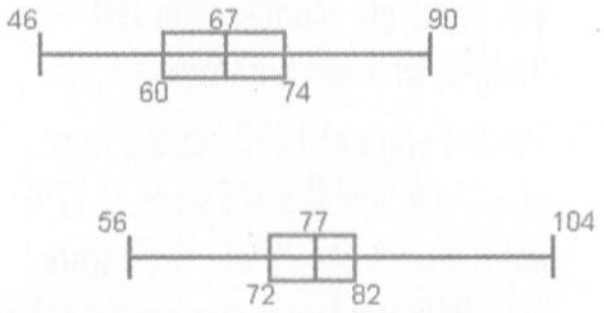

34. Although actresses include the oldest age of 80 years, the top boxplot representing actresses shows that they have ages that are generally lower than those of actors.

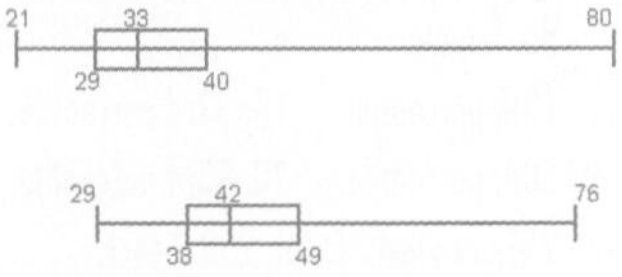

35. The weights of regular Coke represented in the top boxplot appear to be generally greater than those of diet Coke, probably due to the sugar in cans of regular Coke.

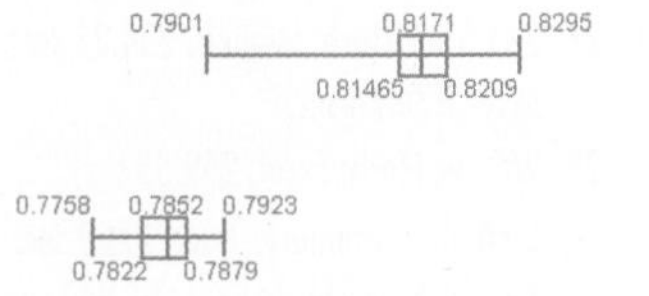

36. The low lead level group represented in the top boxplot has much more variation and the IQ scores tend to be higher than the IQ scores from the high lead level group.

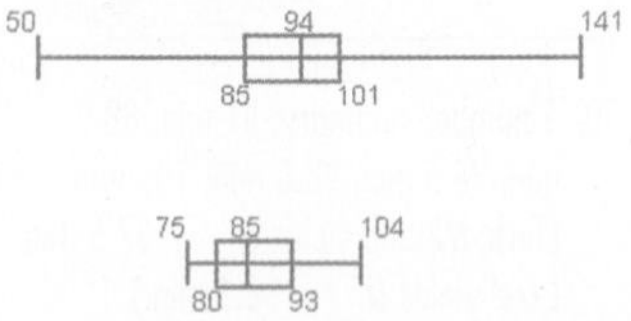

37. Outliers for actresses: 60 years, 61 years, 61 years, 63 years, 74 years, 80 years. Outliers for actors: 76 years. The modified boxplots show that only one actress has an age that is greater than any actor.

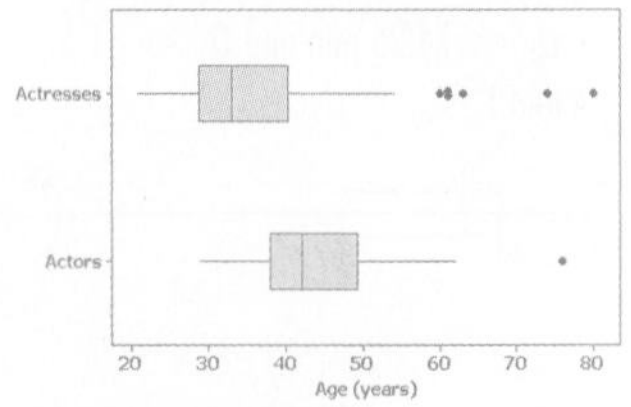

38. Using interpolation, $P_{17} = 21.6$. Using Figure 3-5, $P_{17} = 22$. In this case, the results are close, but in some other cases the results might be quite different.

Boxplots. ***In Exercises 29–32, use the given data to construct a boxplot and identify the 5-number summary.***

29. *Challenger* Flights The following are the duration times (minutes) of all missions flown by the space shuttle *Challenger*.

1 7224 8709 8784 10,060 10,089 10,125 11,445 11,476 11,844

30. Old Faithful The following are the interval times (minutes) between eruptions of the Old Faithful geyser in Yellowstone National Park (based on data from the U.S. National Park Service).

81 81 86 87 89 92 93 94 95 96 97 98 98 101 101 106

31. Grooming Times The following are amounts of time (minutes) spent on hygiene and grooming in the morning by survey respondents (based on data from an SCA survey).

4 6 7 9 14 15 15 16 18 18 25 26 30 32 41 45 55 63

32. Speeds The following are speeds (mi/h) of cars measured with a radar gun on the New Jersey Turnpike (based on data from Statlib and authors Joseph Kadane and John Lamberth).

70 70 71 72 72 73 73 74 76 77 78 78 78 79 79

Boxplots from Larger Data Sets in Appendix B. ***In Exercises 33–36, use the given data sets from Appendix B.***

33. Pulse Rates Use the same scale to construct boxplots for the pulse rates of males and females from Data Set 1 in Appendix B. Use the boxplots to compare the two data sets.

34. Ages of Oscar Winners Use the same scale to construct boxplots for the ages of the best actresses and best actors from Data Set 11 in Appendix B. Use the boxplots to compare the two data sets.

35. Weights of Regular Coke and Diet Coke Use the same scale to construct boxplots for the weights of regular Coke and diet Coke from Data Set 19 in Appendix B. Use the boxplots to compare the two data sets.

36. Lead and IQ Use the same scale to construct boxplots for the full IQ scores (IQF) for the low lead level group and the high lead level group in Data Set 5 of Appendix B.

3-4 Beyond the Basics

37. Outliers and Modified Boxplots Repeat Exercise 34 using modified boxplots. Identify any outliers as defined in Part 2 of this section. What do the modified boxplots show that the regular boxplots do not show?

38. Interpolation When finding percentiles using Figure 3-5, if the locator L is not a whole number, we round it up to the next larger whole number. An alternative to this procedure is to *interpolate*. For example, using interpolation with a locator of $L = 23.75$ leads to a value that is 0.75 (or 3/4) of the way between the 23rd and 24th values. Use this method of interpolation to find P_{17} for the chocolate chip counts in Table 3-5. How does the result compare to the value that would be found by using Figure 3-5 without interpolation?

Chapter 3 Review

This chapter presented fundamentally and critically important measures that are essential for effectively describing, exploring, and comparing data. Here are key skills that should be mastered upon completion of this chapter:

- Calculate measures of center by finding the mean and median (Section 3-2).
- Calculate measures of variation by finding the standard deviation, variance, and range (Section 3-3).
- *Understand* and *interpret* the standard deviation by using tools such as the range rule of thumb (Section 3-3).
- Compare data values by using z scores, quartiles, or percentiles (Section 3-4).
- Investigate the spread of data by constructing a boxplot.

Chapter Quick Quiz

1. Find the mean of these times that American Airlines flights used to taxi to the Los Angeles terminal after landing from a flight: 12 min, 8 min, 21 min, 17 min, 12 min. (Data are from Data Set 15 in Appendix B.)

1. 14.0 min.

2. What is the median of the sample values listed in Exercise 1?

2. 12.0 min.

3. What is the mode of the sample values listed in Exercise 1?

3. 12 min.

4. The standard deviation of the sample values in Exercise 1 is 5.0 min. What is the variance (including units)?

4. 25.0 min^2.

5. The taxi-in times for 48 flights that landed in Los Angeles have a mean of 11.4 min and a standard deviation of 7.0 min. What is the z score for a taxi-in time of 6 min?

5. −0.77.

6. You plan to investigate the variation of taxi-in times for flights that have landed in Los Angeles. Name at least two measures of variation for those data.

6. Standard deviation; variance; range; mean absolute deviation

7. Consider a sample taken from the population of all taxi-in times for all flights that land in Los Angeles. Identify the symbols used for the sample mean and the population mean.

7. $\bar{x}$; μ.

8. Consider a sample taken from the population of all taxi-in times for all flights that land in Los Angeles. Identify the symbols used for the sample standard deviation, the population standard deviation, the sample variance, and the population variance.

8. s, σ, s^2, σ^2.

9. Approximately what percentage of taxi-in times is less than the 75th percentile?

9. 75%.

10. For a sample of motorcycle speeds, name the values that constitute the 5-number summary.

10. Minimum, first quartile Q_1, second quartile Q_2 (or median), third quartile Q_3, maximum.

Review Exercises

1. Ergonomics When designing an eye-recognition security device, engineers must consider the eye heights of standing women. (It's easy for men to bend lower, but it's more difficult for women to rise higher.) Listed below are the eye heights (in millimeters) obtained from a simple random sample of standing adult women (based on anthropometric survey data from Gordon, Churchill, et al.). Use the given eye heights to find the (a) mean; (b) median; (c) mode; (d) midrange; (e) range; (f) standard deviation; (g) variance; (h) Q_1; (i) Q_3.

1550 1642 1538 1497 1571

1. a. 1559.6 mm; b. 1550.0 mm; c. none; d. 1569.5 mm; e. 145 mm; f. 53.4 mm; g. 2849.3 mm^2; h. 1538.0 mm; i. 1571.0 mm. (Tech: Minitab yields $Q_1 = 1517.5$ mm and $Q_3 = 1606.5$ mm.)

2. z Score Using the sample data from Exercise 1, find the z score corresponding to the eye height of 1642 mm. Is that eye height unusual? Why or why not?

3. Boxplot Using the same standing heights listed in Exercise 1, construct a boxplot and include the values of the 5-number summary. Does the boxplot indicate that the data are from a population with a normal (bell-shaped) distribution? Explain.

4. ZIP Codes An article in the *New York Times* noted that these new ZIP codes were created in New York City: 10065, 10021, 10075. Find the mean of these three numbers. What is fundamentally wrong with this result?

5. Comparing BMI The body mass indices (BMI) of a sample of males have a mean of 26.601 and a standard deviation of 5.359. The body mass indices of a sample of females have a mean of 28.441 and a standard deviation of 7.394 (based on Data Set 1 in Appendix B). When considered among members of the same gender, who has the relatively larger BMI: a male with a BMI of 28.00 or a female with a BMI of 29.00? Why?

6. Movies: Estimating Mean and Standard Deviation Consider the prices of regular movie tickets (not 3-D, and not discounted for children or seniors).

a. Estimate the mean price.

b. Use the range rule of thumb to make a rough estimate of the standard deviation of the prices.

7. Professors: Estimating Mean and Standard Deviation Use the range rule of thumb to estimate the standard deviation of ages of all teachers at your college.

8. Aircraft Design Engineers designing overhead bin storage in an aircraft must consider the sitting heights of male passengers. Sitting heights of adult males have a mean of 914 mm and a standard deviation of 36 mm (based on anthropometric survey data from Gordon, Churchill, et al.). Use the range rule of thumb to identify the minimum "usual" sitting height and the maximum "usual" sitting height. Which of those two values is more relevant in this situation? Why?

9. Interpreting a Boxplot Shown below is a boxplot of a sample of 20 brain volumes (cm^3). What do the numbers in the boxplot represent?

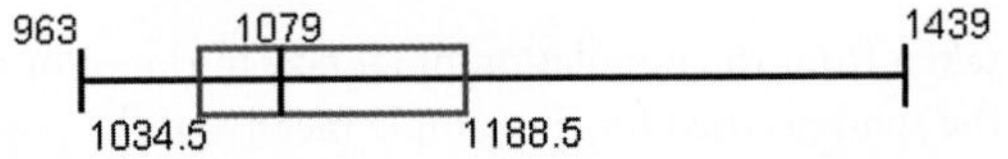

10. Mean or Median? A statistics class with 40 students consists of 30 students with no income, 10 students with small incomes from part-time jobs, and a professor with a very large income that is well deserved. Which is better for describing the income of a typical person in this class: mean or median? Explain.

Cumulative Review Exercises

Please be aware that some of the following problems may require knowledge of concepts presented in previous chapters.

1. Designing Gloves An engineer is designing a machine to manufacture gloves and she obtains the following sample of hand lengths (mm) of randomly selected adult males (based on anthropometric survey data from Gordon, Churchill, et al.):

173 179 207 158 196 195 214 199

a. Are exact hand lengths from a population that is discrete or continuous?

b. What is the level of measurement of the hand lengths? (nominal, ordinal, interval, ratio)

2. $z = 1.54$. The eye height is not unusual because its z score is between 2 and -2, so it is within 2 standard deviations of the mean.

3. Because the boxplot shows a distribution of data that is roughly symmetric, the data could be from a population with a normal distribution, but the data are not necessarily from a population with a normal distribution, because there is no way to determine whether a histogram is roughly bell-shaped.

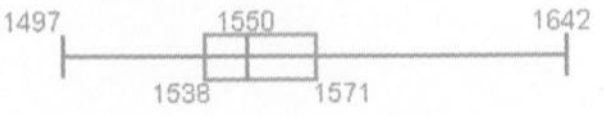

4. 10053.7. The ZIP codes do not measure or count anything. They are at the nominal level of measurement, so the mean is a meaningless statistic.

5. The male has the larger relative BMI because his z score of 0.26 is larger than the z score of 0.08 for the female.

6. The answers vary, but a mean around \$8 or \$9 is reasonable, and a standard deviation around \$1 or \$2 is a reasonable estimate.

7. Answer varies, but $s \approx 12$ years, based on a minimum of 23 years and a maximum of 70 years.

8. Minimum: 842 mm; maximum: 986 mm. The maximum usual height of 986 mm is more relevant for designing overhead bin storage.

9. The minimum volume is 963 cm^3, the first quartile Q_1 is 1034.5 cm^3, the second quartile Q_2 (or median) is 1079 cm^3, the third quartile Q_3 is 1188.5 cm^3, and the maximum volume is 1439 cm^3.

10. The median would be better because it is not affected much by the one very large income.

1. a. Continuous.
b. Ratio.

2. Frequency Distribution Use the hand lengths in Exercise 1 to construct a frequency distribution. Use a class width of 10 mm, and use 150 mm as the lower class limit of the first class.

3. Histogram Use the frequency distribution from Exercise 2 to construct a histogram.

4. Stemplot Use the hand lengths from Exercise 1 to construct a stemplot.

5. Descriptive Statistics Use the hand lengths in Exercise 1 and find the following: (a) mean; (b) median; (c) standard deviation; (d) variance; (e) range. Include the appropriate units of measurement.

6. Normal Distribution Instead of using the hand lengths in Exercise 1, a much larger sample of hand lengths is used and a frequency distribution is created. The frequencies listed in order are 1, 8, 56, 237, 382, 228, 48, 4, 1. Does it appear that the sample is from a population having a normal distribution? Explain.

7. Sampling Shortly after the World Trade Center towers were destroyed, America Online ran a poll of its Internet subscribers and asked this question: "Should the World Trade Center towers be rebuilt?" Among the 1,304,240 responses, 768,731 answered "yes," 286,756 answered "no," and 248,753 said that it was "too soon to decide." Given that this sample is extremely large, can the responses be considered to be representative of the population of the United States? Explain.

8. Histogram The accompanying histogram depicts outcomes of digits from the California Daily 4 lottery. What is the major flaw in this histogram?

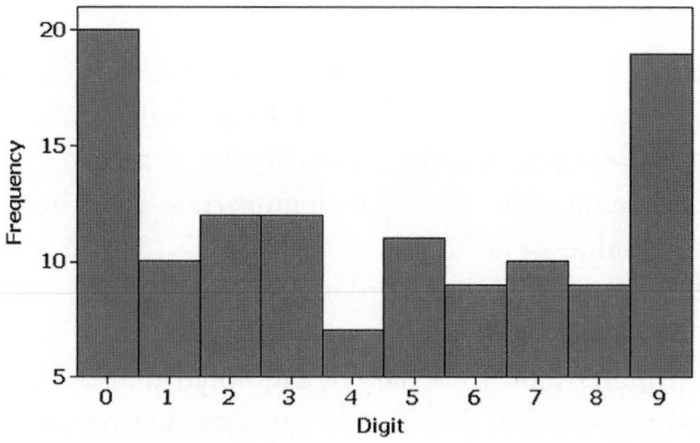

Technology Project

When dealing with large data sets, manual entry of data can become quite tedious and time-consuming. There are better things to do with your time, such as rotating the tires on your car or folding laundry. Refer to Data Set 1 in Appendix B, which includes a variety of real body measurements from randomly selected males and females. Instead of manually entering the data, use a TI-83/84 Plus calculator or STATDISK, Minitab, Excel, StatCrunch, or any other statistics software package. Load the data sets, which are available on the CD included with this book. Identify a variable to be used as a basis for comparison between the two genders. Proceed to generate histograms, any other suitable graphs, and find appropriate statistics that allow you to compare the two sets of data. Are there any outliers? Do both data sets have properties that are basically the same? Are there any significant differences? What would be a consequence of having significant differences? Write a brief report including your conclusions and supporting graphs.

2.

Hand Length (mm)	Frequency
150–159	1
160–169	0
170–179	2
180–189	0
190–199	3
200–209	1
210–219	1

3.

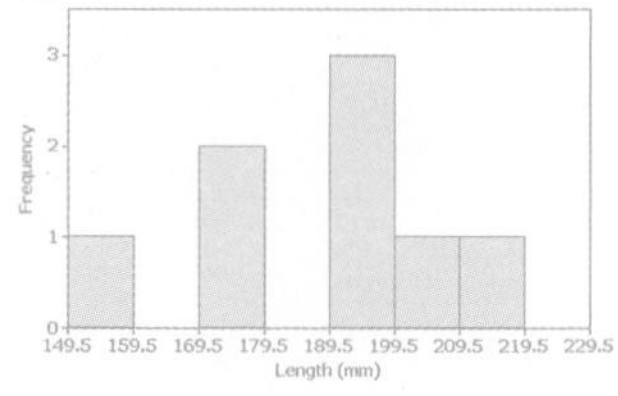

4.

15	8
16	
17	39
18	
19	569
20	7
21	4

5. a. 190.1 mm; b. 195.5 mm; c. 18.7 mm; d. 348.7 mm^2; e. 56.0 mm.

6. Yes. The frequencies increase to a maximum; then they decrease. Also, the frequencies preceding the maximum are roughly a mirror image of those that follow the maximum.

7. No. Even though the sample is large, it is a voluntary response sample, so the responses cannot be considered to be representative of the population of the United States.

8. The vertical scale does not begin at 0, so the differences among the different outcomes are exaggerated.

from data TO DECISION

Flight Planning

The *From Data to Decision* project at the end of Chapter 2 involved data from American Airline flights from New York (JFK airport) to Los Angeles (LAX airport), and the data are listed in Data Set 15 in Appendix B.

Critical Thinking

Use the methods from this chapter to explore the arrival delay times at LAX. Those times are important because they can affect passenger plans. Are there any outliers? Based on the results, should you somehow modify the scheduled arrival times? If so, how? Write a brief report of your conclusions, and provide supporting statistical evidence.

Cooperative Group Activities

1. Out-of-class activity The Chapter Problem involves counts of chocolate chips in five different brands of cookies. Obtain your own sample of chocolate chip cookies and proceed to count the number of chocolate chips in each cookie. Use the data to generate a histogram and any other suitable graphs. Find the descriptive statistics. Compare your chocolate chip counts to those given in Table 3-1. Are there any differences? Explain.

2. In-class activity In class, each student should record two pulse rates by counting the number of heartbeats in one minute. The first pulse rate should be measured while seated, and the second pulse rate should be measured while standing. Use the methods of this chapter to compare results. Do males and females appear to have different pulse rates? Do pulse rates measured while seated appear to be different from pulse rates measured while standing?

3. Out-of-class activity In the article "Weighing Anchors" in *Omni* magazine, author John Rubin observed that when people estimate a value, their estimate is often "anchored" to (or influenced by) a preceding number, even if that preceding number is totally unrelated to the quantity being estimated. To demonstrate this, he asked people to give a quick estimate of the value of $8 \times 7 \times 6 \times 5 \times 4 \times 3 \times 2 \times 1$. The mean of the answers given was 2250, but when the order of the numbers was reversed, the mean became 512. Rubin explained that when we begin calculations with larger numbers (as in $8 \times 7 \times 6$), our estimates tend to be larger. He noted that both 2250 and 512 are far below the correct product, 40,320. The article suggests that irrelevant numbers can play a role in influencing real estate appraisals, estimates of car values, and estimates of the likelihood of nuclear war.

Conduct an experiment to test this theory. Select some subjects and ask them to quickly estimate the value of

$$8 \times 7 \times 6 \times 5 \times 4 \times 3 \times 2 \times 1$$

Then select other subjects and ask them to quickly estimate the value of

$$1 \times 2 \times 3 \times 4 \times 5 \times 6 \times 7 \times 8$$

Record the estimates along with the particular order used. Carefully design the experiment so that conditions are uniform and the two sample groups are selected in a way that minimizes any bias. Don't describe the theory to subjects until after they have provided their estimates. Compare the two sets of sample results by using the methods of this chapter. Provide a printed report that includes the data collected, the detailed methods used, the method of analysis, any relevant graphs and/or statistics, and a statement of conclusions. Include a critique of the experiment, with reasons why the results might not be correct, and describe ways in which the experiment could be improved.

4. Out-of-class activity In each group of three or four students, collect an original data set of values at the interval or ratio level of measurement. Provide the following: (1) a list of sample values, (2) printed computer results of descriptive statistics and graphs, and (3) a written description of the nature of the data, the method of collection, and important characteristics.

5. Out-of-class activity Appendix B includes many real and interesting data sets. In each group of three or four students, select a data set from Appendix B and analyze it using the methods discussed so far in this book. Write a brief report summarizing key conclusions.

6. Out-of-class activity Record the service times of randomly selected customers at a drive-up window of a bank or fast-food restaurant, and describe important characteristics of those times.

7. Out-of-class activity Record the times that cars are parked at a gas pump, and describe important characteristics of those times.

4 Probability

4

chapter problem: Pre-employment drug screening

80% of companies in the United States now test employees and/or job applicants for drug use. One common approach to pre-employment drug screening is to require an initial urine test, such as the EMIT (enzyme multiplied immunoassay technique) test, which is a "five panel" test for the presence of any of five drugs: marijuana, cocaine, amphetamines, opiates, or phencyclidine. The EMIT test is one of the least expensive and most common tests used by employers. Most companies require that positive test results be confirmed by a more reliable GC-MS (gas chromotography mass spectometry) test.

Drug testing is typically a process with some degree of inaccuracy, and results are sometimes wrong. Wrong results are of two types: (1) false positive results and (2) false negative results. These terms are included among several terms commonly used in references to drug testing. A subject getting a false positive result is in the very undesirable position of appearing to be a drug user when that person is not actually a drug user.

Analyzing the Results

Table 4-1 includes results from 1000 adults in the United States. If one of the 1000 subjects from Table 4-1 is randomly selected, what is the probability that the subject will test positive if this person is not a drug user? If one of the subjects from Table 4-1 is randomly selected, what is the probability that the subject will get a correct result? We will address such questions in this chapter.

- **False positive:** *Wrong* result in which the test incorrectly indicates the presence of a condition when the subject does not actually have that condition.
- **False negative:** *Wrong* result in which the test incorrectly indicates that the subject does *not* have a condition when the subject actually does have that condition.
- **True positive:** *Correct* result in which the test correctly indicates that a condition is present when it really is present.
- **True negative:** *Correct* result in which the test correctly indicates that a condition is not present when it really is not present.
- **Test sensitivity:** The probability of a true positive, given that the subject actually has the condition being tested.
- **Test specificity:** The probability of a true negative, given that the subject does not have the condition being tested.
- **Positive predictive value:** Probability that a subject is a true positive, given that the test yields a positive result (indicating that the condition is present).
- **Negative predictive value:** Probability that the subject is a true negative, given that the test yields a negative result (indicating that the condition is not present).
- **Prevalence:** Proportion of subjects having some condition.

Table 4-1 Pre-Employment Drug Screening Results

	Positive Test Result (Drug Use Is Indicated)	Negative Test Result (Drug Use Is Not Indicated)
Subject Uses Drugs	44 (True Positive)	6 (False Negative)
Subject Is Not a Drug User	90 (False Positive)	860 (True Negative)

4-1 Review and Preview

Note to Instructor
Two important changes in this edition:

1. In the previous edition of this book, the term *unusual* was equivalent to *unlikely*, but in this new edition, we say that an event has an **unusually low number** of outcomes of a particular type or an **unusually high number** of those outcomes if that number is far from what we typically expect. Consequently, this edition modifies the definition of *unusual* so that it becomes roughly equivalent to *extreme*. In this edition, we use the term *unlikely* for an event with a low probability (such as 0.05 or less).
2. Section 4-6 (Probabilities Through Simulations) in the previous edition has been moved from the textbook to the CD included with this book.

The preceding chapters have presented some critically important concepts in the study and application of statistics. We have discussed the importance of sound sampling methods. We have discussed common measures of characteristics of data, including the mean and standard deviation. The main objective of this chapter is to develop a sound understanding of probability values, because those values constitute the underlying foundation on which methods of inferential statistics are built. Statisticians use the following *rare event rule for inferential statistics.*

Rare Event Rule for Inferential Statistics

If, under a given assumption, the probability of a particular observed event is extremely small, we conclude that the assumption is probably not correct.

Example 1 Rare Event Rule for Inferential Statistics

The Genetics & IVF Institute has developed a method of gender selection so that couples could increase the likelihood of having a baby girl. Instead of using some of their real results, let's use a more obvious example. Suppose that 100 couples use the procedure for trying to have a baby girl, and results consist of 98 girls and only 2 boys. We have two possible explanations for these results:

1. *Chance:* The gender-selection technique is not effective and the result of 98 girls and 2 boys occurred by random chance.
2. *Not Chance:* The results did not occur by chance, so it appears that the gender-selection technique is effective.

When choosing between the above two possible explanations, the *probability* of getting 98 girls and 2 boys is the deciding factor. Without calculating that probability, it is safe to say that it is *extremely* small. The probability of 98 girls and 2 boys is so small that the first explanation of chance would be rejected as being a reasonable explanation. Instead, it would be generally recognized that the results provide strong support for the claim that the gender-selection technique is effective. This is exactly how statisticians think: They reject explanations based on very low probabilities.

If you follow the thought process in Example 1, you understand a fundamental way of statistical thinking. Example 1 did not provide an actual probability value, and the main objective of this chapter is to develop a sound understanding of probability values that will be used in later chapters of this book. A secondary objective is to develop the basic skills necessary to determine probability values in a variety of important circumstances.

4-2 Basic Concepts of Probability

Key Concept Although this section presents three different approaches to finding the *probability* of an event, the most important objective of this section is to learn how to *interpret* probability values, which are expressed as values between 0 and 1. We should know that a small probability, such as 0.001, corresponds to an event that rarely occurs. In Part 2 of this section we discuss expressions of *odds* and how probability is used to determine the odds of an event occurring. The concepts related to odds are not needed for topics that follow, but odds are often used in some everyday situations, especially those that involve lotteries and gambling.

Part 1: Basics of Probability

In considering probability, we deal with procedures (such as answering a multiple-choice test question or undergoing a test for drug use) that produce outcomes.

DEFINITIONS

An **event** is any collection of results or outcomes of a procedure.

A **simple event** is an outcome or an event that cannot be further broken down into simpler components.

The **sample space** for a procedure consists of all possible *simple* events. That is, the sample space consists of all outcomes that cannot be broken down any further.

Example 1 illustrates the concepts defined above.

Example 1

In the following display, we use "b" to denote a baby boy and "g" to denote a baby girl.

Procedure	Example of Event	Sample Space (List of Simple Events)
Single birth	1 girl (simple event)	{b, g}
3 births	2 boys and 1 girl (bbg, bgb, and gbb are all simple events resulting in 2 boys and 1 girl)	{bbb, bbg, bgb, bgg, gbb, gbg, ggb, ggg}

With one birth, the result of 1 female is a *simple event* because it cannot be broken down any further. With three births, the event of "2 girls and 1 boy" is *not a simple event* because it can be broken down into simpler events, such as ggb, gbg, or bgg. With three births, the *sample space* consists of the eight simple events listed above. With three births, the outcome of ggb is considered a simple event, because it is an outcome that cannot be broken down any further. We might incorrectly think that ggb can be further broken down into the individual results of g, g, and b, but g, g, and b are not individual outcomes from three births. With three births, there are exactly eight outcomes that are simple events: bbb, bbg, bgb, bgg, gbb, gbg, ggb, and ggg.

We first list some basic notation, then we present three different approaches to finding the probability of an event.

Notation for Probabilities

P denotes a probability.

A, B, and *C* denote specific events.

$P(A)$ denotes the probability of event *A* occurring.

Probabilities That Challenge Intuition

In certain cases, our subjective estimates of probability values are dramatically different from the actual probabilities. Here is a classical example: If you take a deep breath, there is better than a 99% chance that you will inhale a molecule that was exhaled in dying Caesar's last breath. In that same morbid and unintuitive spirit, if Socrates' fatal cup of hemlock was mostly water, then the next glass of water you drink will likely contain one of those same molecules. Here's another less morbid example that can be verified: In classes of 25 students, there is better than a 50% chance that at least 2 students will share the same birthday (day and month).

Note to Instructor

This section covers basic definitions and simple concepts of probability. This section is partitioned into Part 1 (Basic Concepts of Probability) and Part 2 (Beyond the Basics of Probability: Odds). Discuss and stress the *rare event rule for inferential statistics.*

If pressed for time, skip Part 2. Some professors cover only the introductory section (Section 4-2) on probability, some cover all sections, and others prefer to cover some but not all sections. *Recommendation:* Cover Sections 4-2 (Basic Concepts of Probability), 4-3 (Addition Rule), and 4-4 (Multiplication Rule: Basics), but the sections you include depend on your individual preferences and the needs of your students. Know that students typically find probability to be the most difficult chapter in the book, so try to help them as much as possible.

Note to Instructor
Emphasize the importance of *understanding* the basic concepts of probability. Some students develop this oversimplified rule for finding probabilities: Find two numbers; then divide the smaller number by the larger one. Give at least one example where that rule doesn't work, such as this: If a quality control test shows that there are 10 defective iPhones and there are 40 that are good, the probability of randomly selecting one that is defective is 10/50, not 10/40.

Also, stress that the goal is not to simply find a probability number, but to use it for making some decision. Many exercises ask questions requiring the *interpretation* of probability values.

Activity: Give each student a thumbtack and ask them to find the probability that the tack lands with the point up. Is the answer 0.5, because the tack either lands with the point up or does not? Why or why not?

1. **Relative Frequency Approximation of Probability** Conduct (or observe) a procedure, and count the number of times that event A actually occurs. Based on these actual results, $P(A)$ is *approximated* as follows:

$$P(A) = \frac{\text{number of times } A \text{ occurred}}{\text{number of times the procedure was repeated}}$$

2. **Classical Approach to Probability (Requires Equally Likely Outcomes)** Assume that a given procedure has n different simple events and that *each of those simple events has an equal chance of occurring.* If event A can occur in s of these n ways, then

$$P(A) = \frac{\text{number of ways } A \text{ occur}}{\text{number of different simple events}} = \frac{s}{n}$$

CAUTION When using the classical approach, always verify that the outcomes are *equally likely*.

3. **Subjective Probabilities** $P(A)$, the probability of event A, is *estimated* by using knowledge of the relevant circumstances.

Note that the classical approach requires *equally likely outcomes.* If the outcomes are not equally likely, we must use the relative frequency approximation or we must rely on our knowledge of the circumstances to make an *educated guess.* Figure 4-1 illustrates the three approaches.

Figure 4-1 Three Approaches to Finding a Probability

(a) **Relative Frequency Approach:** When trying to determine the probability that an individual car crashes in a year, we must examine past results to determine the number of cars in use in a year and the number of them that crashed; then we find the ratio of the number of cars that crashed to the total number of cars. For a recent year, the result is a probability of 0.0480. (See Example 2.)
(b) **Classical Approach:** When trying to determine the probability of winning the grand prize in a lottery by selecting six numbers between 1 and 60, each combination has an equal chance of occurring. The probability of winning is 0.0000000200, which can be found by using methods presented in Section 4-6.
(c) **Subjective Probability:** When trying to estimate the probability of a passenger dying in a plane crash, we know that there are thousands of flights every day, but fatal plane crashes are quite rare, so the probability is very small. A good guess would be about 1 in 10 million, or 0.0000001.

When finding probabilities with the relative frequency approach, we obtain an *approximation* instead of an exact value. As the total number of observations increases, the corresponding approximations tend to get closer to the actual probability. This property is stated as a theorem commonly referred to as the *law of large numbers.*

Law of Large Numbers

As a procedure is repeated again and again, the relative frequency probability of an event tends to approach the actual probability.

The law of large numbers tells us that relative frequency approximations tend to get better with more observations. This law reflects a simple notion supported by common sense: A probability estimate based on only a few trials can be off by a substantial amount, but with a very large number of trials, the estimate tends to be much more accurate.

CAUTION The law of large numbers applies to behavior over a large number of trials, and it does not apply to one outcome. Don't make the foolish mistake of losing large sums of money by incorrectly thinking that a string of losses increases the chances of a win on the next bet.

Probability and Outcomes That Are Not Equally Likely One common mistake is to incorrectly assume that outcomes are equally likely just because we know nothing about the likelihood of each outcome. When we know nothing about the likelihood of different possible outcomes, we cannot necessarily assume that they are equally likely. For example, we should not conclude that the probability of passing the next statistics test is $1/2$, or 0.5 (because we either pass the test or do not). The actual probability depends on factors such as the amount of preparation and the difficulty of the test.

Example 2 Relative Frequency Probability: Smoking

A recent Harris Interactive survey of 1010 adults in the United States showed that 202 of them smoke. Find the probability that a randomly selected adult in the United States is a smoker.

Solution

We use the relative frequency approach as follows:

$$P(\text{smoker}) = \frac{\text{number of smokers}}{\text{total number of people surveyed}} = \frac{202}{1010} = 0.200$$

Note that the classical approach cannot be used since the two outcomes (smoker, not a smoker) are not equally likely.

Example 3 Classical Probability: Positive Test Result

Refer to Table 4-1 included with the Chapter Problem. Assuming that one of the 1000 subjects included in Table 4-1 is randomly selected, find the probability that the selected subject got a positive test result.

Solution

The sample space consists of results from 1000 subjects listed in Table 4-1. Among the 1000 results, 134 of them are positive test results (found from 44 + 90). Because

Winning the Lottery

In the New York State Lottery Mega Millions game, you select five numbers from 1 to 56, then you select another "Mega Ball" number from 1 to 46. To win, you must get the correct five numbers *and* the correct Mega Ball number. The chance of winning this lottery with one ticket is 1/175,711,536, even though commercials for this lottery state that "all you need is a little bit of luck." The probability of 1/175,711,536 is not easily perceived by many people, so let's consider a helpful analogy developed by Brother Donald Kelly of Marist College. A stack of 175,711,536 dimes is about 154 miles high. Commercial jets typically fly about 7 miles high, so this stack of dimes is about 22 times taller than the height of a commercial jet when it is at cruising altitude. The chance of winning the Mega Millions lottery game is equivalent to the chance of randomly selecting *one* specific dime from that pile of dimes that is 154 miles high. Any of us who spend money on this lottery should understand that the chance of winning the grand prize is very, very, very close to zero.

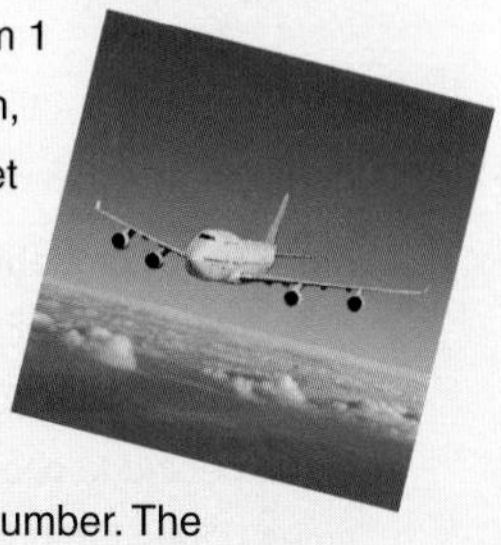

How Probable?

How do we interpret such terms as *probable, improbable,* or *extremely improbable*? The FAA interprets these terms as follows.

- *Probable:* A probability on the order of 0.00001 or greater for each hour of flight. Such events are expected to occur several times during the operational life of each airplane.
- *Improbable:* A probability on the order of 0.00001 or less. Such events are not expected to occur during the total operational life of a single airplane of a particular type, but may occur during the total operational life of all airplanes of a particular type.
- *Extremely improbable:* A probability on the order of 0.000000001 or less. Such events are so unlikely that they need not be considered to ever occur.

the subject is randomly selected, each test result is equally likely, so we can apply the classical approach as follows:

$$P(\text{positive test result from Table 4-1}) = \frac{\text{number of positive test results}}{\text{total number of results}}$$

$$= \frac{134}{1000} = 0.134$$

Example 4 Classical Probability: Three Children of the Same Gender

When three children are born, the sample space of genders is as shown in Example 1: {bbb, bbg, bgb, bgg, gbb, gbg, ggb, ggg}. If boys and girls are equally likely, then the eight simple events are equally likely. Assuming that boys and girls are equally likely, find the probability of getting three children all of the same gender when three children are born. (In reality, a boy is slightly more likely than a girl.)

Solution

The sample space {bbb, bbg, bgb, bgg, gbb, gbg, ggb, ggg} in this case includes equally likely outcomes. Among the eight outcomes, there are exactly two in which the three children are of the same gender: bbb and ggg. We can use the classical approach to get

$$P(\text{three children of the same gender}) = \frac{2}{8} = 0.25$$

Example 5 Subjective Probability: Professorial Attire

Find the probability that in your next statistics class, the professor wears a hat with a huge feather protruding from the top.

Solution

The sample space consists of two simple events: Your professor wears a hat with a huge feather protruding from the top, or does not. We can't use the relative frequency approach because we lack data on past results. We can't use the classical approach because the two possible outcomes are events that are not equally likely. We are left with making a subjective estimate. The event is possible, but highly unlikely, so we can estimate its probability as something like 0.000001.

Example 6 Subjective Probability: Stuck in an Elevator

What is the probability that you will get stuck in the next elevator that you ride?

Solution

In the absence of historical data on elevator failures, we cannot use the relative frequency approach. There are two possible outcomes (becoming stuck or not becoming stuck), but they are not equally likely, so we cannot use the classical approach. That leaves us with a subjective estimate. In this case, experience suggests that the probability is quite small. Let's estimate it to be, say, 0.0001 (equivalent to 1 chance in 10,000). That subjective estimate, based on our general knowledge, is likely to be in the general ballpark of the true probability.

CAUTION Don't make the mistake of finding a probability value by mindlessly dividing a smaller number by a larger number. Instead, think carefully about the numbers involved and what they represent. Be especially careful when determining the total number of items being considered, as in the following example.

Example 7 Tainted Currency

In a study of U.S. paper currency, bills from 17 large cities were analyzed for the presence of cocaine. Here are the results: 23 of the bills were not tainted by cocaine and 211 were tainted by cocaine. Based on these results, if a bill is randomly selected, find the probability that it is tainted by cocaine.

Solution

Hint: Instead of trying to determine an answer directly from the printed statement, begin by first summarizing the given information in a format that allows you to clearly understand the information. For example, use this format:

23	bills not tainted by cocaine
211	bills tainted by cocaine
234	total number of bills analyzed

We can now use the relative frequency approach as follows:

$$P(\text{bill tainted by cocaine}) = \frac{\text{number of bills tainted by cocaine}}{\text{total number of bills analyzed}} = \frac{211}{234}$$
$$= 0.902$$

Interpretation

There is a 0.902 probability that if a U.S. bill is randomly selected, it is tainted by cocaine.

Simulations in Baseball

Simulations are being used to identify the most effective strategies in a variety of circumstances that occur in baseball. On this topic, *New York Times* reporter Alan Schwarz quoted Harvard University statistician Carl Morris as saying that "computer simulations work pretty well in baseball for two reasons. In general, they allow you to study fairly complicated processes that you can't really get at with pure mathematics. But also, sports are great for simulations—you can play 10,000 seasons overnight." Many baseball teams now use simulations to help them decide which players should be traded.

Here are some strategies suggested by computer simulations of baseball:

- It is not wise to intentionally walk a player.
- The sacrifice bunt should be used very rarely.

Simulations The statements of the three approaches for finding probabilities and the preceding examples might seem to suggest that we should always use the classical approach when a procedure has equally likely outcomes, but many situations are so complicated that the classical approach is impractical. In the game of solitaire, for example, the outcomes (hands dealt) are all equally likely, but it is extremely difficult to try to use the classical approach to find the probability of winning. In such cases we can more easily get good estimates by using the relative frequency approach, and simulations are often helpful when using this approach. A *simulation* of a procedure is a process that behaves in the same ways as the procedure itself so that similar results are produced. (See the Technology Project near the end of this chapter.) For example, it's much easier to use the relative frequency approach for approximating the probability of winning at solitaire—that is, to play the game many times (or to run a computer simulation)—than to perform the complex calculations required with the classical approach.

Note to Instructor
Comment that instead of simply reading text, it can be really helpful to understand the problem by drawing a figure, or constructing a table, or somehow depicting the data in a format such as the one included in the solution in Example 7.

Example 8 Thanksgiving Day

If a year is selected at random, find the probability that Thanksgiving Day in the United States will be (a) on a Wednesday or (b) on a Thursday.

Solution

a. In the United States, Thanksgiving Day always falls on the fourth Thursday in November. It is therefore impossible for Thanksgiving to be on a Wednesday. When an event is impossible, we say that its probability is 0.

b. It is certain that Thanksgiving will be on a Thursday. When an event is certain to occur, we say that its probability is 1.

Because any event imaginable is impossible, certain, or somewhere in between, it follows that the mathematical probability of any event A is 0, 1, or a number between 0 and 1 (see Figure 4-2). That is, $0 \leq P(A) \leq 1$.

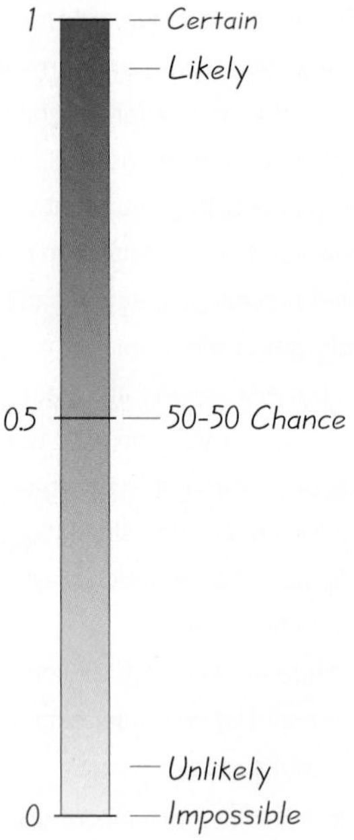

Figure 4-2 Possible Values for Probabilities

Figure 4-2 shows the possible values of probabilities and the more familiar and common expressions of likelihood.

Complementary Events

Sometimes we need to find the probability that an event A does *not* occur.

DEFINITION The **complement** of event A, denoted by $\overline{A}$, consists of all outcomes in which event A does *not* occur.

Example 9 Complement of Smoker

Results from Example 2 show that if we randomly select an adult in the United States, the probability of selecting a smoker is 0.200. Find the probability of randomly selecting an adult in the United States and getting someone who does *not* smoke.

Solution

Because 202 of the 1010 surveyed adults in the United States are smokers, it follows that the other 808 are not smokers, so

$$P(\text{not a smoker}) = \frac{808}{1010} = 0.800$$

Interpretation

The probability of randomly selecting an adult in the United States and getting someone who is *not* a smoker is 0.800.

Although it is difficult to develop a universal rule for rounding off probabilities, the following guide will apply to most problems in this text.

Rounding Off Probabilities

When expressing the value of a probability, either give the *exact* fraction or decimal or round off final decimal results to three significant digits. (*Suggestion:* When a probability is not a simple fraction such as 2/3 or 5/9, express it as a decimal so that the number can be better understood.) All digits in a number are significant except for the zeros that are included for proper placement of the decimal point.

Example 10 Rounding Probabilities

- The probability of 0.9017094017 (from Example 7) has ten significant digits (9017094017), and it can be rounded to three significant digits as 0.902.
- The probability of 1/3 can be left as a fraction or rounded to 0.333. (Do *not* round to 0.3.)
- The probability of 2/8 (from Example 4) can be expressed as 1/4 or 0.25; because 0.25 is exact, there's no need to express it with three significant digits as 0.250.

The mathematical expression of probability as a number between 0 and 1 is fundamental and common in statistical procedures, and we will use it throughout the remainder of this text. A typical computer output, for example, may include a "*P*-value" expression such as "significance less than 0.001." We will discuss the meaning of *P*-values later, but they are essentially probabilities of the type discussed in this section. For now, you should recognize that a probability of 0.001 (equivalent to 1/1000) corresponds to an event so rare that it occurs an average of only once in a thousand trials. Example 12 involves the interpretation of such a small probability value.

How Many Shuffles?

After conducting extensive research, Harvard mathematician Persi Diaconis found that it takes seven shuffles of a deck of cards to get a complete mixture. The mixture is complete in the sense that all possible arrangements are equally likely. More than seven shuffles will not have a significant effect, and fewer than seven are not enough. Casino dealers rarely shuffle as often as seven times, so the decks are not completely mixed. Some expert card players have been able to take advantage of the incomplete mixtures that result from fewer than seven shuffles.

Interpreting Probabilities: *Unlikely* Events and *Unusual* Events

We can consider an event with a small probability (such as 0.05 or less) to be *unlikely*. But we often need to determine when an event consists of an *unusually high* number of outcomes of a particular type or an *unusually low* number of such outcomes. An event has a number of particular outcomes that is unusually low or unusually high if that number is *extreme* in the sense that it is far from the number that we typically expect. The following definitions are not standard with universal use, but they are very helpful in interpreting probability values and developing concepts used extensively in later chapters.

DEFINITIONS An event is **unlikely** if its probability is very small, such as 0.05 or less. (See Figure 4-2.) An event has an **unusually low number** of outcomes of a particular type or an **unusually high number** of those outcomes if that number is far from what we typically expect.

Important change in this edition: In the previous edition of this book, the term *unusual* was equivalent to *unlikely*, but see the new definitions of *unlikely* and *unusual* as given here in this edition.

Unlikely: Small probability (such as 0.05 or less)

Unusual: Extreme result (number of outcomes of a particular type is far below or far above the typical values)

By associating "unusual" with *extreme* outcomes, we are consistent with the range rule of thumb (Section 3-3) and the use of *z* scores for identifying unusual values (Section 3-4).

Example 11 Unlikely/Unlikely

a. When a fair coin is tossed 1000 times, the result consists of exactly 500 heads. The probability of getting exactly 500 heads in 1000 tosses is 0.0252. Is this result unlikely? Is 500 heads unusually low or unusually high?

b. When a fair coin is tossed 1000 times, the result consists of 10 heads. Is this result unlikely? Is 10 heads unusually low or unusually high?

Gambling to Win

In the typical state lottery, the "house" has a 65% to 70% advantage, since only 30% to 35% of the money bet is returned as prizes. The house advantage at racetracks is usually around 15%. In casinos, the house advantage is 5.26% for roulette, 1.4% for craps, and 3% to 22% for slot machines.

The house advantage is 5.9% for blackjack, but some professional gamblers can systematically win with a 1% player advantage by using complicated card-counting techniques that require many hours of practice. If a card-counting player were to suddenly change from small bets to large bets, the dealer would recognize the card counting and the player would be ejected. Card counters try to beat this policy by working with a team. When the count is high enough, the player signals an accomplice who enters the game with large bets. A group of MIT students supposedly won millions of dollars by counting cards in blackjack.

Solution

a. Because the probability of exactly 500 heads in 1000 tosses is 0.0252, that result is *unlikely*. However, we usually get around 500 heads, so this outcome is neither unusually low nor unusually high.

b. Without actually finding the exact probability value, it is reasonable to conclude that there is a very low probability of getting 10 heads in 1000 tosses of a fair coin, so this event is unlikely. Also, the outcome of 10 heads is so far below the number of heads we typically expect (around 500), we conclude that this is an unusually low number of heads.

Example 11 considers the result of 500 heads in 1000 tosses and the result of 10 heads in 1000 tosses. Figure 4-3 is a graph of the probabilities for all possible numbers of heads in 1000 tosses. (Such a graph of outcomes and their probabilities is called a *probability histogram,* and it is formally defined in Section 5-2.) From Figure 4-3 we see that every individual number of heads is *unlikely*, because its probability is small (0.0252 or less). The red shaded regions show those outcomes that are *unusually low* or *unusually high*, because they are so far away from the outcomes that we typically expect (around 500). In Chapter 5 we will develop methods for finding the cumulative probabilities corresponding to such unusually low and unusually high numbers of outcomes. For the results shown in Figure 4-3, the probability of getting any number of heads that is unusually low or unusually high is less than 0.05.

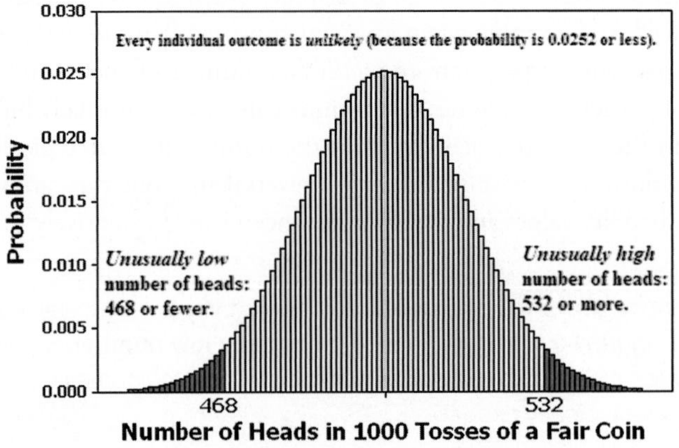

Figure 4-3 Graph of Outcomes and Their Probabilities

Section 4-1 introduced the rare event rule, and that rule is the focus of the next example.

Example 12 Rare Event Rule

In a clinical experiment of the Salk vaccine for polio, 200,745 children were given a placebo and 201,229 other children were treated with the Salk vaccine. There were 115 cases of polio among those in the placebo group and 33 cases of polio in the treatment group. If we assume that the vaccine has no effect, the probability of getting such test results is found to be "less than 0.001." What does that probability imply about the effectiveness of the vaccine?

Solution

A probability value less than 0.001 is very small. It indicates that the event will occur fewer than once in a thousand times, so it is very unlikely that the event will occur by chance, assuming that the vaccine has no effect. We have two possible explanations for the results of this clinical experiment: (1) The vaccine has no effect and the results occurred by chance; (2) the vaccine has an effect, which explains why the treatment group had a much lower incidence of polio. Because the probability is so small (less than 0.001), the second explanation is more reasonable. We conclude that the vaccine appears to be effective.

The preceding example illustrates the "rare event rule for inferential statistics" given in Section 4-1. Under the assumption of a vaccine with no effect, we find that the probability of the results is extremely small (less than 0.001), so we conclude that the assumption is probably not correct. The preceding example also illustrates the role of probability in making important conclusions about clinical experiments. For now, we should understand that when a probability is small, such as less than 0.001, it indicates that the event is very unlikely to occur.

The following list summarizes some key notation and principles discussed so far in this section.

Important Principles and Notation for Probability

- **The probability of an event is a fraction or decimal number between 0 and 1 inclusive.**
- **The probability of an impossible event is 0.**
- **The probability of an event that is certain to occur is 1.**
- **Notation: The probability of event A is denoted by $P(A)$.**
- **Notation: The probability that event A does *not* occur is denoted by $P(\overline{A})$.**

Part 2: Beyond the Basics of Probability: Odds

Expressions of likelihood are often given as *odds,* such as 50:1 (or "50 to 1"). Because the use of odds makes many calculations difficult, statisticians, mathematicians, and scientists prefer to use probabilities. The advantage of odds is that they make it easier to deal with money transfers associated with gambling, so they tend to be used in casinos, lotteries, and racetracks. Note that in the three definitions that follow, the *actual odds against* and the *actual odds in favor* are calculated with the actual likelihood of some event, but the *payoff odds* describe the relationship between the bet and the amount of the payoff. The actual odds correspond to actual probabilities of outcomes, but the payoff odds are set by racetrack and casino operators. Racetracks and casinos are in business to make a profit, so the payoff odds will not be the same as the actual odds.

DEFINITIONS

The **actual odds against** event A occurring are the ratio $P(\overline{A})/P(A)$, usually expressed in the form of $a{:}b$ (or "a to b"), where a and b are integers having no common factors.

The **actual odds in favor** of event A occurring are the ratio $P(A)/P(\overline{A})$, which is the reciprocal of the actual odds against that event. If the odds against A are $a{:}b$, then the odds in favor of A are $b{:}a$.

The **payoff odds** against event A occurring are the ratio of net profit (if you win) to the amount bet:

$$\text{payoff odds against event } A = (\text{net profit}){:}(\text{amount bet})$$

Probability of an Event That Has Never Occurred

Some events are possible, but are so unlikely that they have never occurred. Here is one such problem of great interest to political scientists: Estimate the probability that your single vote will determine the winner in a U.S. Presidential election. Andrew Gelman, Gary King, and John Boscardin write in the *Journal of the American Statistical Association* (Vol. 93, No. 441) that "the exact value of this probability is of only minor interest, but the number has important implications for understanding the optimal allocation of campaign resources, whether states and voter groups receive their fair share of attention from prospective presidents, and how formal 'rational choice' models of voter behavior might be able to explain why people vote at all." The authors show how the probability value of 1 in 10 million is obtained for close elections.

Note to Instructor
Note the difference between *actual odds* (based on the likelihood of the event) and *payoff odds* (based on the greed of the casino).

The Random Secretary

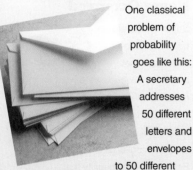

One classical problem of probability goes like this: A secretary addresses 50 different letters and envelopes to 50 different people, but the letters are randomly mixed before being put into envelopes. What is the probability that at least one letter gets into the correct envelope? Although the probability might seem like it should be small, it's actually 0.632. Even with a million letters and a million envelopes, the probability is 0.632. The solution is beyond the scope of this text—way beyond.

Example 13

If you bet \$5 on the number 13 in roulette, your probability of winning is $1/38$ and the payoff odds are given by the casino as 35:1.

a. Find the actual odds against the outcome of 13.

b. How much net profit would you make if you win by betting on 13?

c. If the casino was not operating for profit and the payoff odds were changed to match the actual odds against 13, how much would you win if the outcome were 13?

Solution

a. With $P(13) = 1/38$ and $P(\text{not } 13) = 37/38$, we get

$$\text{actual odds against 13} = \frac{P(\text{not } 13)}{P(13)} = \frac{37/38}{1/38} = \frac{37}{1} \text{ or } 37{:}1$$

b. Because the payoff odds against 13 are 35:1, we have

$$35{:}1 = (\text{net profit}){:}(\text{amount bet})$$

So there is a \$35 profit for each \$1 bet. For a \$5 bet, the net profit is \$175. The winning bettor would collect \$175 plus the original \$5 bet. After winning, the total amount collected would be \$180, for a net profit of \$175.

c. If the casino were not operating for profit, the payoff odds would be changed to 37:1, which are the actual odds against the outcome of 13. With payoff odds of 37:1, there is a net profit of \$37 for each \$1 bet. For a \$5 bet the net profit would be \$185. (The casino makes its profit by paying only \$175 instead of the \$185 that would be paid with a roulette game that is fair instead of favoring the casino.)

Recommended Assignment
Exercises 1–32.

4-2 Basic Skills and Concepts

Statistical Literacy and Critical Thinking

1. Florida Lottery Let A denote the event of placing a \$1 straight bet on the Florida Play 4 lottery and winning. The chance of event A occurring is 1 in 10,000. What is the value of $P(A)$? What is the value of $P(\bar{A})$?

1. $P(A) = 1/10{,}000$, or 0.0001. $P(\bar{A}) = 9999/10{,}000$, or 0.9999.

2. Probability Given that the following statement is incorrect, rewrite it correctly: "The probability of a baby being born a boy is 50–50."

2. The probability of a baby being born a boy is $1/2$, or 0.5.

3. Interpreting Weather While this exercise was being created, Weather.com indicated that there was a 20% chance of rain for the author's home region. Based on that report, which of the following is the most reasonable interpretation?

3. Part (c).

a. $1/5$ of the author's region will get rain today.

b. In the author's region, it will rain for $1/5$ of the day.

c. In the author's region, there is a $1/5$ probability that it will rain at some point during the day.

4. Subjective Probability Estimate the probability that the next time you ride in a car, you will *not* be delayed because of some car crash blocking the road.

4. The answer varies, but an answer in the neighborhood of 0.99 is reasonable.

5. Identifying Probability Values Which of the following values are *not* probabilities?

5:2   3/7   7/3   −0.9   0.123   123/456   456/123   0   1

5. 5:2; 7/3; −0.9; 456/123

6. U.S. President "Who was the 14th president of the United States: Taylor, Fillmore, Pierce, or Buchanan?" If you make a random guess for the answer to that question, what is the probability that your answer is the correct answer of Pierce?

6. 1/4 or 0.25

7. Digits of Pi "Which of the following is the mean of the first 100 digits of pi after the decimal point: 4.4, 4.5, 4.6, 4.7, or 4.8?" If you make a random guess for the answer to that question, what is the probability that your answer is the correct answer of 4.8?

7. 1/5 or 0.2

8. The Die Is Cast When rolling a single die, what is the probability that the outcome is 7?

8. 0

In Exercises 9–12, assume that 400 births are randomly selected. Use subjective judgment to determine whether the given outcome is unlikely, and also determine whether it is* unusual *in the sense that the result is far from what is typically expected.

9. Exactly 205 girls.

10. Exactly 380 girls.

11. Exactly 111 girls.

12. Exactly 197 girls.

9. Unlikely; neither unusually low nor unusually high.
10. Unlikely; unusually high.
11. Unlikely; unusually low.
12. Unlikely; neither unusually low nor unusually high.

In Exercises 13–20, express the indicated degree of likelihood as a probability value between 0 and 1.

13. Traffic Light When arriving at the traffic light closest to the author's home, there is a 25% chance that the light will be red.

13. 1/4, or 0.25

14. Weather While this exercise was being created, Weather.com indicated that there is a 20% chance of rain for the author's home region.

14. 0.2

15. Testing If you make a random guess for the answer to a true/false test question, there is a 50–50 chance of being correct.

15. 1/2, or 0.5

16. Pierced Ears Based on a Harris poll, there is a 50–50 chance that a randomly selected adult has pierced ears.

16. 1/2, or 0.5

17. SAT Test When guessing the answer to a multiple-choice question on an SAT test, the chance of guessing correctly is 1 in 5.

17. 1/5, or 0.2

18. Dice When rolling a pair of dice at the Bellagio Casino in Las Vegas, there is 1 chance in 36 that the outcome is a 12.

18. 1/36, or 0.0278

19. Statistics Test It is impossible to pass a statistics test without studying.

19. 0

20. Death and Taxes Benjamin Franklin said that death is a certainty of life.

20. 1

In Exercises 21–24, refer to the sample data in Table 4-1, which is included with the Chapter Problem. Assume that one of the subjects included in Table 4-1 is randomly selected.

21. Pre-Employment Drug Screening Find the probability of selecting someone who got a result that is a false negative. Who would suffer from a false negative result? Why?

22. Pre-Employment Drug Screening Find the probability of selecting someone who got a result that is a false positive. Who would suffer from a false positive result? Why?

23. Pre-Employment Drug Screening Find the probability of selecting someone who uses drugs. Is the result close to the probability of 0.134 for a positive test result?

24. Pre-Employment Drug Screening Find the probability of selecting someone who does not use drugs. Is the result close to the probability of 0.866 for a negative test result?

21. 6/1000, or 0.006. The employer would suffer because it would be at risk by hiring someone who uses drugs.
22. 90/1000, or 0.09. The person tested would suffer because he or she would be suspected of using drugs when in reality he or she does not use drugs.
23. 50/1000, or 0.05. This result is not close to the probability of 0.134 for a positive test result.
24. 950/1000, or 0.95. This result is not very close to the probability of 0.866 for a negative test result.

25. 879/945, or 0.930. Yes, the technique appears to be effective.

25. XSORT Gender Selection MicroSort's XSORT gender-selection technique is designed to increase the likelihood that a baby will be a girl. In updated results (as of this writing) of the XSORT gender-selection technique, 945 births consisted of 879 baby girls and 66 baby boys (based on data from the Genetics & IVF Institute). Based on these results, what is the probability of a girl born to a couple using MicroSort's XSORT method? Does it appear that the technique is effective in increasing the likelihood that a baby will be a girl?

26. 239/291, or 0.821. Yes, the technique appears to be effective.

26. YSORT Gender Selection MicroSort's YSORT gender-selection technique is designed to increase the likelihood that a baby will be a boy. In updated results (as of this writing) from a test of MicroSort's YSORT gender-selection technique, 291 births consisted of 239 baby boys and 52 baby girls (based on data from the Genetics & IVF Institute). Based on these results, what is the probability of a boy born to a couple using MicroSort's YSORT method? Does it appear that the technique is effective in increasing the likelihood that a baby will be a boy?

27. 0.00000101. No, the probability of being struck is much greater on an open golf course during a thunderstorm. The golfer should seek shelter.

27. Struck by Lightning In a recent year, 304 of the approximately 300,000,000 people in the United States were struck by lightning. Estimate the probability that a randomly selected person in the United States will be struck by lightning this year. Is a golfer reasoning correctly if he or she is caught out in a thunderstorm and does not seek shelter from lightning because the probability of being struck is so small?

28. $428/580 = 0.738$; yes.

28. Mendelian Genetics When Mendel conducted his famous genetics experiments with peas, one sample of offspring consisted of 428 green peas and 152 yellow peas. Based on those results, estimate the probability of getting an offspring pea that is green. Is the result reasonably close to the expected value of 3/4, as claimed by Mendel?

Using Probability to Identify Unlikely Events. ***In Exercises 29–36, consider an event to be "unlikely" if its probability is less than or equal to 0.05. (This is equivalent to the same criterion commonly used in inferential statistics, but the value of 0.05 is not absolutely rigid, and other values such as 0.01 are sometimes used instead.)***

29. a. 1/365
b. Yes
c. He already knew.
d. 0

29. Guessing Birthdays On their first date, Kelly asks Mike to guess the date of her birth, not including the year.

a. What is the probability that Mike will guess correctly? (Ignore leap years.)

b. Would it be unlikely for him to guess correctly on his first try?

c. If you were Kelly, and Mike did guess correctly on his first try, would you believe his claim that he made a lucky guess, or would you be convinced that he already knew when you were born?

d. If Kelly asks Mike to guess her age, and Mike's guess is too high by 15 years, what is the probability that Mike and Kelly will have a second date?

30. 0.670. No, it is not unlikely. Because the responses are from a voluntary response survey, the results are not likely to be very good.

30. Credit Card Purchases In a survey, 169 respondents say that they never use a credit card, 1227 say that they use it sometimes, and 2834 say that they use it frequently. What is the probability that a randomly selected person uses a credit card frequently? Is it unlikely for someone to use a credit card frequently? How are all of these results affected by the fact that the responses were obtained by those who decided to respond to the survey posted on the Internet by America OnLine?

31. 0.0767. No, a crash is not unlikely. Given that car crashes are so common, we should take precautions such as not driving after drinking and not using a cell phone or texting.

31. Car Crashes In a recent year, among 135,933,000 registered passenger cars in the United States, there were 10,427,000 crashes. Find the probability that a randomly selected passenger car in the United States will crash this year. Is it unlikely for a car to crash in a given year? What does this suggest about driving?

32. 0.000000117. Yes, it is unlikely. The air travel fatality rate is much higher than that of cars. The comparison isn't fair because car trips involve much shorter distances than trips by air.

32. Air Travel Fatalities One measure of air travel safety is this: There are 117 fatalities per billion passenger flights. Express that measure as a probability. Is it unlikely for an air passenger to be a fatality? How does air travel compare to the car fatality rate of 40 fatalities per billion trips? Is this comparison fair?

33. Texting While Driving In a *New York Times*/CBS News poll, respondents were asked if it should be legal or illegal to send a text message while driving. Eight said that it should be legal and 804 said that it should be illegal. What is the probability of randomly selecting someone who believes it should be legal to text while driving? Is it unlikely to randomly select someone with that belief?

33. 0.00985. It is unlikely.

34. Cell Phones While Driving In a *New York Times*/CBS News poll, respondents were asked if it should be legal or illegal to use hand-held cell phones while driving. One hundred forty-one said that it should be legal, and 663 said that it should be illegal. What is the probability of randomly selecting someone who believes it should be legal to use a hand-held cell phone while driving? Is it unlikely to randomly select someone with that belief?

34. 0.175. It is not unlikely.

35. Favorite Seat on a Plane Among respondents asked which is their favorite seat on a plane, 492 chose the window seat, 8 chose the middle seat, and 306 chose the aisle seat (based on data from *USA Today*). What is the probability that a passenger prefers the middle seat? Is it unlikely for a passenger to prefer the middle seat? If so, why do you think the middle seat is so unpopular?

35. 0.00993. Yes, it is unlikely. The middle seat lacks an outside view, easy access to the aisle, and a passenger in the middle seat has passengers on both sides instead of on one side only.

36. At the End of the Day In a Marist poll, respondents chose the most annoying phrases used in conversation. Nineteen chose "at the end of the day," 441 chose "whatever," 235 chose "you know," 103 chose "it is what it is," 66 chose "anyway," and 75 were unsure. Based on these results, what is the probability of selecting someone who considers "at the end of the day" to be the most annoying phrase? At the end of the day, is it unlikely to select someone with that choice?

36. 0.0202. Yes, it is unlikely.

Probability from a Sample Space. ***In Exercises 37–42, use the given sample space or construct the required sample space to find the indicated probability.***

37. Three Children Use this sample space listing the eight simple events that are possible when a couple has three children (as in Example 1): {bbb, bbg, bgb, bgg, gbb, gbg, ggb, ggg}. Assume that boys and girls are equally likely, so that the eight simple events are equally likely. Find the probability that when a couple has three children, there is exactly one girl.

37. 3/8, or 0.375

38. Three Children Using the same sample space and assumption from Exercise 37, find the probability that when a couple has three children, there are exactly two girls.

38. 3/8, or 0.375

39. Two Children Exercise 37 lists the sample space for a couple having three children. First identify the sample space for a couple having two children, then find the probability of getting one child of each gender. Again assume that boys and girls are equally likely.

39. {bb, bg, gb, gg}; 1/2, or 0.5.

40. Four Children Exercise 37 lists the sample space for a couple having three children. First identify the sample space for a couple having four children, then find the probability of getting three girls and one boy (in any order).

40. {bbbb, bbbg, bbgb, bbgg, bgbb, bgbg, bggb, bggg, gbbb, gbbg, gbgb, gbgg, ggbb, ggbg, gggb, gggg}; 4/16 or 1/4 or 0.25.

41. Genetics: Eye Color Each of two parents has the genotype brown/blue, which consists of the pair of alleles that determine eye color, and each parent contributes one of those alleles to a child. Assume that if the child has at least one brown allele, that color will dominate and the eyes will be brown. (The actual determination of eye color is somewhat more complicated.)

a. List the different possible outcomes. Assume that these outcomes are equally likely.

b. What is the probability that a child of these parents will have the blue/blue genotype?

c. What is the probability that the child will have brown eyes?

41. a. brown/brown, brown/blue, blue/brown, blue/blue
b. 1/4
c. 3/4

42. X-Linked Genetic Disease Men have XY (or YX) chromosomes and women have XX chromosomes. X-linked recessive genetic diseases (such as juvenile retinoschisis) occur when there is a defective X chromosome that occurs *without* a paired X chromosome that is good. In the following, represent a defective X chromosome with lowercase x, so a child with the xY

42. a. 0
b. 0
c. 0.5
d. 0

or Yx pair of chromosomes will have the disease and a child with XX or XY or YX or xX or Xx will not have the disease. Each parent contributes one of the chromosomes to the child.

a. If a father has the defective x chromosome and the mother has good XX chromosomes, what is the probability that a son will inherit the disease?

b. If a father has the defective x chromosome and the mother has good XX chromosomes, what is the probability that a daughter will inherit the disease?

c. If a mother has one defective x chromosome and one good X chromosome and the father has good XY chromosomes, what is the probability that a son will inherit the disease?

d. If a mother has one defective x chromosome and one good X chromosome and the father has good XY chromosomes, what is the probability that a daughter will inherit the disease?

4-2 Beyond the Basics

Odds. ***In Exercises 43–46, answer the given questions that involve odds.***

43. a. 999:1
b. 499:1
c. The description is not accurate. The odds against winning are 999:1 and the odds in favor are 1:999, not 1:1000.

43. Texas Pick 3 In the Texas Pick 3 lottery, you can bet $1 by selecting the exact order of three digits between 0 and 9 inclusive, so the probability of winning is 1/1000. If the same three numbers are drawn in the same order, you collect $500, so your net profit is $499.

a. Find the actual odds against winning.

b. Find the payoff odds.

c. The Web site www.txlottery.org indicates "Odds 1:1000" for this bet. Is that description accurate?

44. a. 18/38, or 0.474
b. 10:9
c. $18
d. $20

44. Finding Odds in Roulette A roulette wheel has 38 slots. One slot is 0, another is 00, and the others are numbered 1 through 36, respectively. You place a bet that the outcome is an odd number.

a. What is your probability of winning?

b. What are the actual odds against winning?

c. When you bet that the outcome is an odd number, the payoff odds are 1:1. How much profit do you make if you bet $18 and win?

d. How much profit would you make on the $18 bet if you could somehow convince the casino to change its payoff odds so that they are the same as the actual odds against winning? (*Recommendation:* Don't actually try to convince any casinos of this; their sense of humor is remarkably absent when it comes to things of this sort.)

45. a. $16
b. 8:1
c. About 9.75:1, which becomes 39:4
d. $21.50

46. 1/37, or 0.0270

47. Relative risk: 0.939; odds ratio: 0.938; the probability of a headache with Nasonex (0.0124) is slightly less than the probability of a headache with the placebo (0.0132), so Nasonex does not appear to pose a risk of headaches.

48. 1

49. 1/4

45. Kentucky Derby Odds When the horse Super Saver won the 136th Kentucky Derby, a $2 bet that Super Saver would win resulted in a return of $18.

a. How much net profit was made from a $2 win bet on Super Saver?

b. What were the payoff odds against a Super Saver win?

c. Based on preliminary wagering before the race, bettors collectively believed that Super Saver had a 0.093 probability of winning. Assuming that 0.093 was the true probability of a Super Saver victory, what were the actual odds against his winning?

d. If the payoff odds were the actual odds found in part (c), how much would a $2 win ticket be worth after the Super Saver win?

46. Finding Probability from Odds If the actual odds against event A are $a{:}b$, then $P(A) = b/(a + b)$. Find the probability of the horse Make Music for Me winning the 136th Kentucky Derby, given that the actual odds against his winning that race were 36:1.

47. Relative Risk and Odds Ratio In a clinical trial of 2103 subjects treated with Nasonex, 26 reported headaches. In a control group of 1671 subjects given a placebo, 22 reported headaches. Denoting the proportion of headaches in the treatment group by p_t and denoting the proportion of headaches in the control (placebo) group by p_c, the *relative risk* is p_t/p_c. The relative risk is a measure of the strength of the effect of the Nasonex treatment. Another such measure is the *odds ratio,* which is the ratio of the odds in favor of a headache for the treatment group to the odds in favor of a headache for the control (placebo) group, found by evaluating the following:

$$\frac{p_t/(1 - p_t)}{p_c/(1 - p_c)}$$

The relative risk and odds ratios are commonly used in medicine and epidemiological studies. Find the relative risk and odds ratio for the headache data. What do the results suggest about the risk of a headache from the Nasonex treatment?

48. Flies on an Orange If two flies land on an orange, find the probability that they are on points that are within the same hemisphere.

49. Points on a Stick Two points along a straight stick are randomly selected. The stick is then broken at those two points. Find the probability that the three resulting pieces can be arranged to form a triangle. (This is possibly the most difficult exercise in this book.)

Proportions of Males/Females

It is well known that when a baby is born, boys and girls are not equally likely. It is currently believed that 105 boys are born for every 100 girls, so the probability of a boy is 0.512. Kristen Navara of the University of Georgia conducted a study showing that around the world, more boys are born than girls, but the difference becomes smaller as people are located closer to the equator. She used latitudes, temperatures, unemployment rates, gross and national products from 200 countries and conducted a statistical analysis showing that the proportions of boys appear to be affected only by latitude and its related weather. So far, no one has identified a reasonable explanation for this phenomenon.

4-3 Addition Rule

Key Concept In this section we present the *addition rule* as a tool for finding $P(A \text{ or } B)$, which is the probability that either event A occurs or event B occurs (or they both occur) as the single outcome of a procedure. To find the probability of event A occurring or event B occurring, we begin by adding the number of ways that A can occur and the number of ways that B can occur, but we add without double counting. That is, we do not count any outcomes more than once. The key word in this section is *or,* which generally indicates *addition* of probabilities. Throughout this text we use the *inclusive or,* which means either one or the other or both. (Except for Exercise 41, we do not consider the *exclusive or,* which means either one or the other but not both.)

In Section 4-2 we presented the basics of probability and considered events categorized as *simple* events. In this and the following section we consider *compound events.*

DEFINITION A **compound event** is any event combining two or more simple events.

Notation for Addition Rule

$P(A \text{ or } B) = P(\text{in a single trial, event } A \text{ occurs or event } B \text{ occurs or they both occur})$

In Section 4-2 we considered simple events, such as the probability of getting a false positive when 1 test result is randomly selected from the 1000 test results listed in Table 4-1, reproduced on the next page for the convenience and pleasure of the reading audience. In Example 1 we consider P(positive test result *or* a subject uses drugs) when 1 of the 1000 test results is randomly selected. Follow the reasoning used in Example 1 and you will have a basic understanding of the addition rule that is the focus of this section.

Note to Instructor

Key points of this section: $P(A \text{ or } B)$ suggests the addition rule; associate "or" with adding. Avoid double-counting of events that are not mutually exclusive.

Recommendation: Point out that this section includes a "formal addition rule" and an "intuitive addition rule," and discourage the blind use of the formula in the formal addition rule. Encourage use of the intuitive addition rule based on an understanding of the basic concept.

Comment that a table such as Table 4-1 is called a *two-way* table or *contingency* table, and such tables are very important because they are used often in the analysis of survey results. Such tables are the focus of Section 11-3.

Coincidences?

John Adams and Thomas Jefferson (the second and third presidents) both died on July 4, 1826. President Lincoln was assassinated in Ford's Theater; President Kennedy was assassinated in a Lincoln car made by the Ford Motor Company. Lincoln and Kennedy were both succeeded by vice presidents named Johnson. Fourteen years *before* the sinking of the *Titanic,* a novel described the sinking of the *Titan,* a ship that hit an iceberg; see Martin Gardner's *The Wreck of the Titanic Foretold?* Gardner states, "In most cases of startling coincidences, it is impossible to make even a rough estimate of their probability."

Table 4-1 Pre-Employment Drug Screening Results*

	Positive Test Result (Drug Use Is Indicated)	**Negative Test Result** (Drug Use Is Not Indicated)
Subject Uses Drugs	**44** (True Positive)	**6** (False Negative)
Subject Is Not a Drug User	**90** (False Positive)	860 (True Negative)

**Numbers in bold correspond to positive test results or subjects who use drugs, and the total of those numbers is 140.*

Example 1 Pre-Employment Drug Screening

Refer to Table 4-1. If 1 subject is randomly selected from the 1000 subjects given a drug test, find the probability of selecting a subject who had a positive test result or uses drugs. This probability is denoted by P(positive test result or subject uses drugs).

Solution

Refer to Table 4-1 and carefully count the number of subjects who tested positive (first column) or use drugs (first row), but be careful to count subjects once, not twice. *When adding the frequencies from the first column and the first row, include the frequency of 44 only once.* In Table 4-1, there are 140 subjects who had positive test results or use drugs. We get this result:

$$P(\text{positive test result or subject uses drugs}) = 140/1000 = 0.140$$

In Example 1, there are several ways to count the subjects who tested positive or use drugs. Any of the following would work:

- Color the cells representing subjects who tested positive or use drugs; then add the numbers in those colored cells, being careful to add each number only once. This approach yields

$$44 + 90 + 6 = 140$$

- Add the 134 subjects who tested positive to the 50 subjects who use drugs, but the total of 184 involves double counting of 44 subjects, so compensate for the double counting by subtracting the overlap consisting of the 44 subjects who were counted twice. This approach yields a result of

$$134 + 50 - 44 = 140$$

- Start with the total of 134 subjects who tested positive, then add those subjects who use drugs and were not yet included in that total to get a result of

$$134 + 6 = 140$$

Example 1 illustrates that when finding the probability of an event A or event B, use of the word *or* suggests addition, and the addition must be done without double counting.

The preceding example suggests a general rule whereby we add the number of outcomes corresponding to each of the events in question:

When finding the probability that event *A* occurs or event *B* occurs, find the total of the number of ways *A* can occur and the number of ways *B* can occur, but *find that total in such a way that no outcome is counted more than once.*

One way to formalize the rule is to add the probability of event *A* and the probability of event *B* and, if there is any overlap, compensate by subtracting the probability of outcomes that are included twice, as in the following rule.

Formal Addition Rule

$P(A \text{ or } B) = P(A) + P(B) - P(A \text{ and } B)$

where $P(A \text{ and } B)$ denotes the probability that *A* and *B* both occur at the same time as an outcome in a trial of a procedure.

Although the formal addition rule is presented as a formula, blind use of formulas is not recommended. It is generally better to *understand* the spirit of the rule and use that understanding, as follows.

Intuitive Addition Rule

To find $P(A \text{ or } B)$, find the sum of the number of ways event *A* can occur and the number of ways event *B* can occur, *adding in such a way that every outcome is counted only once.* $P(A \text{ or } B)$ is equal to that sum, divided by the total number of outcomes in the sample space.

Note to Instructor
Emphasize use of the intuitive rule instead of the formal rule, because the intuitive rule is based on understanding of circumstances instead of blind application of a formula.

The addition rule is simplified when the events are *disjoint.*

DEFINITION Events *A* and *B* are **disjoint** (or **mutually exclusive**) if they cannot occur at the same time. (That is, disjoint events do not overlap.)

Note to Instructor
Emphasize that "disjoint" essentially means "separate" or "not overlapping." Every time you use the term *disjoint,* say "disjoint or separate events" until the meaning is well known.

Example 2 Disjoint Events

Example of disjoint events:	Randomly selecting someone who is a registered Democrat
	Randomly selecting someone who is a registered Republican
	(The selected person *cannot* be both.)
Example of events that are *not* disjoint:	Randomly selecting someone taking a statistics course
	Randomly selecting someone who is a female
	(The selected person *can* be both.)

Figure 4-4 shows a Venn diagram that provides a visual illustration of the formal addition rule. Figure 4-4 shows that the probability of *A* or *B* equals the probability of *A* (left circle) plus the probability of *B* (right circle) minus the probability of *A* and *B* (football-shaped middle region). This figure shows that the addition of the areas of the two circles will cause double counting of the football-shaped middle region. This is the basic concept that underlies the addition rule. Because of the relationship between the addition rule and the Venn diagram shown in Figure 4-4, the notation $P(A \cup B)$ is sometimes used in place of $P(A \text{ or } B)$. Similarly, the

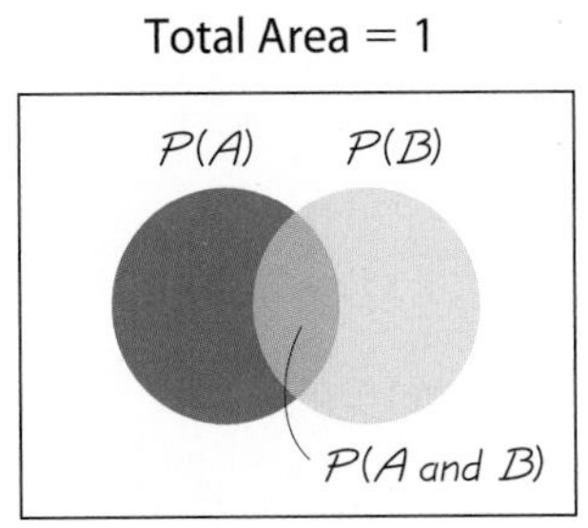

Figure 4-4 Venn Diagram for Events That Are Not Disjoint

Total Area = 1

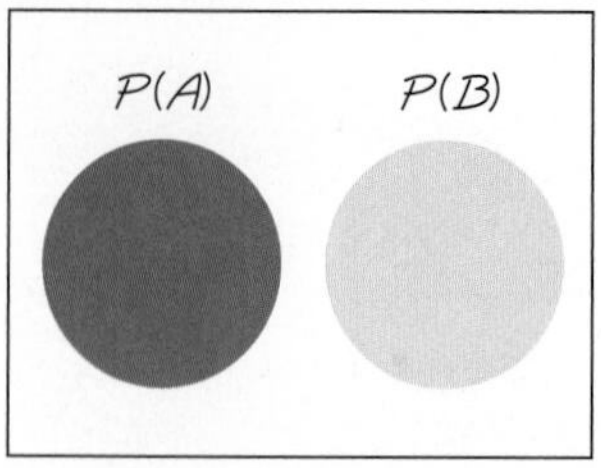

Figure 4-5 Venn Diagram for Disjoint Events

notation $P(A \cap B)$ is sometimes used in place of $P(A \text{ and } B)$, so the formal addition rule can be expressed as

$$P(A \cup B) = P(A) + P(B) - P(A \cap B)$$

Whenever A and B are disjoint, $P(A \text{ and } B)$ becomes zero in the addition rule. Figure 4-5 illustrates that when A and B are disjoint, we have $P(A \text{ or } B) = P(A) + P(B)$.

Here is a summary of the key points of this section:

1. To find $P(A \text{ or } B)$, begin by associating use of the word *or* with addition.
2. Consider whether events A and B are disjoint; that is, can they happen at the same time? If they are not disjoint (that is, they can happen at the same time), be sure to avoid (or compensate for) double counting when adding the relevant probabilities. If you understand the importance of not double counting when you find $P(A \text{ or } B)$, you don't necessarily have to calculate the value of $P(A) + P(B) - P(A \text{ and } B)$.

CAUTION Errors made when applying the addition rule often involve double counting; that is, events that are not disjoint are treated as if they were. One indication of such an error is a total probability that exceeds 1; however, errors involving the addition rule do not always cause the total probability to exceed 1.

Complementary Events and the Addition Rule

In Section 4-2 we used $\overline{A}$ to indicate that A does not occur. Common sense dictates this principle: We are certain (with probability 1) that either an event A occurs *or* it does not occur, so it follows that $P(A \text{ or } \overline{A}) = 1$. Because events A and $\overline{A}$ must be disjoint, we can use the addition rule to express that commonsense principle as follows:

$$P(A \text{ or } \overline{A}) = P(A) + P(\overline{A}) = 1$$

This result of the addition rule leads to the following three expressions that are equivalent in the sense that they are different forms of the same principle.

Rule of Complementary Events

$P(A) + P(\overline{A}) = 1$ $\quad P(\overline{A}) = 1 - P(A)$ $\quad P(A) = 1 - P(\overline{A})$

Total Area = 1

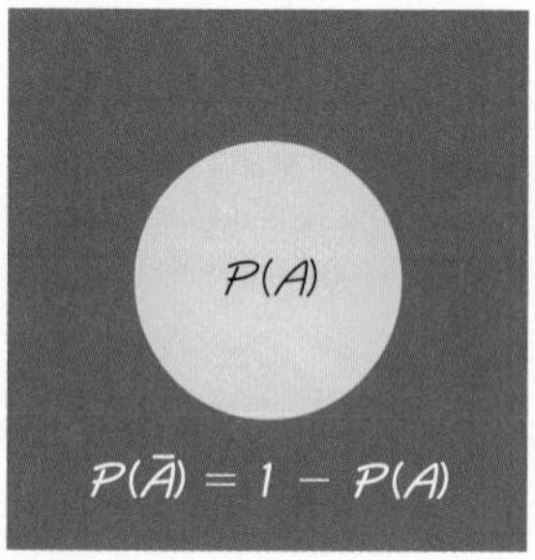

Figure 4-6 Venn Diagram for the Complement of Event *A*

Figure 4-6 visually displays this relationship between $P(A)$ and $P(\overline{A})$.

Example 3 Devilish Belief

Based on data from a Harris Interactive poll, the probability of randomly selecting someone who believes in the devil is 0.6, so $P(\text{believes in the devil}) = 0.6$. If a person is randomly selected, find the probability of getting someone who does *not* believe in the devil.

Solution

Using the rule of complementary events, we get

$P(\text{does } \textit{not} \text{ believe in the devil}) = 1 - P(\text{believes in the devil}) = 1 - 0.6 = 0.4$

The probability of randomly selecting someone who does not believe in the devil is 0.4.

A major advantage of the *rule of complementary events* is that it simplifies certain probability problems, as we illustrate in Section 4-5.

4-3 Basic Skills and Concepts

Recommended Assignment
Exercises 1–8 and 13–24.

Statistical Literacy and Critical Thinking

1. Complements What is wrong with the expression $P(A) + P(\overline{A}) = 0.5$?

1. Based on the rule of complements, the sum of $P(A)$ and $P(\overline{A})$ must always be 1, so that sum cannot be 0.5.

2. Casino Craps A gambler plans to play the casino dice game called craps, and he plans to place a bet on the "pass line." Let A be the event of winning. Based on the rules used in almost all casinos, $P(A) = 244/495$. Describe the event $\overline{A}$ and find the value of $P(\overline{A})$.

2. $\overline{A}$ is the event of betting on the pass line and not winning (or losing). $P(\overline{A}) = 251/495$, or 0.507.

3. Disjoint Events For a Gallup poll, M is the event of randomly selecting a male, and R is the event of randomly selecting a Republican. Are events M and R disjoint? Why or why not?

3. Because it is possible to select someone who is a male and a Republican, events M and R are not disjoint. Both events can occur at the same time when someone is randomly selected.

4. Rule of Complements One form of the rule of complements is this: $P(A \text{ or } \overline{A}) = 1$. Write a sentence describing the message that this rule represents.

4. It is certain that an event occurs or does not occur.

Determining Whether Events Are Disjoint. ***For Exercises 5–12, determine whether the two events are disjoint for a single trial.*** **(Hint:** ***Consider "disjoint" to be equivalent to "separate" or "not overlapping."*****)**

5. Arriving late for your next statistics class.
Arriving early for your next statistics class.

5. Disjoint.

6. Asking for a date through a Twitter post.
Asking for a date in French, the romance language.

6. Not disjoint.

7. Randomly selecting a survey respondent and getting someone who believes in UFOs.
Randomly selecting a survey respondent and getting someone who believes in the devil.

7. Not disjoint.

8. Randomly selecting a statistics student and getting one who uses a TI calculator in class.
Randomly selecting a statistics student and getting one who uses the STATDISK computer program.

8. Not disjoint.

9. Randomly selecting a drug screening result and getting one that is a false negative.
Randomly selecting a drug screening result and getting one from someone who is not a drug user.

9. Disjoint.

10. Randomly selecting a drug screening result and getting one that is a false positive.
Randomly selecting a drug screening result and getting one from someone who uses drugs.

10. Disjoint.

11. Randomly selecting a drug screening result and getting one that is a false positive.
Randomly selecting a drug screening result and getting one from someone who does not use drugs.

11. Not disjoint.

12. Randomly selecting a drug screening result and getting one that is a false positive.
Randomly selecting a drug screening result and getting one that is a false negative.

12. Disjoint.

Finding Complements. ***In Exercises 13–16, find the indicated complements.***

13. Whatever A Marist poll survey showed that 47% of respondents chose "whatever" as the most annoying phrase used in conversation. What is the probability of randomly selecting someone choosing something different from "whatever" as the most annoying phrase in conversation?

13. 0.53

14. Online Courses According to the National Association for College Admissions Counseling and *USA Today,* 19.8% of college students take at least one class online. What is the probability of randomly selecting a college student who does not take any college courses online?

14. 0.802

15. $P(\overline{D}) = 0.450$, where $P(\overline{D})$ is the probability of randomly selecting someone who does not choose a direct in-person encounter as the most fun way to flirt.

16. $P(\overline{I})$ denotes the probability of screening a driver and finding that he or she is not intoxicated, and $P(\overline{I}) = 0.99112 = 0.991$ (rounded).

15. Flirting Survey In a Microsoft Instant Messaging survey, respondents were asked to choose the most fun way to flirt, and it found that $P(D) = 0.550$, where D is directly in person. If someone is randomly selected, what does $P(\overline{D})$ represent, and what is its value?

16. Sobriety Checkpoint When the author observed a sobriety checkpoint conducted by the Dutchess County Sheriff Department, he saw that 676 drivers were screened and 6 were arrested for driving while intoxicated. Based on those results, we can estimate that $P(I) = 0.00888$, where I denotes the event of screening a driver and getting someone who is intoxicated. What does $P(\overline{I})$ denote, and what is its value?

In Exercises 17–20, use the drug screening data given in Table 4-1, which is included with the Chapter Problem.

17. 1

17. Drug Screening If one of the test subjects is randomly selected, find the probability that the subject had a positive test result or a negative test result.

18. 0.994

18. Drug Screening If one of the test subjects is randomly selected, find the probability that the subject had a positive test result or does not use drugs.

19. 0.956

19. Drug Screening If one of the subjects is randomly selected, find the probability that the subject had a negative test result or does not use drugs.

20. 0.910

20. Drug Screening If one of the subjects is randomly selected, find the probability that the subject had a negative test result or uses drugs.

Dosage Calculations. *In Exercises 21–26, use the data in the accompanying table, which lists the numbers of correct and wrong dosage amounts calculated by physicians. In a research experiment, one group of physicians was given bottles of epinephrine labeled with a* concentration *of "1 milligram in 1 milliliter solution," and another group of physicians was given bottles labeled with a* ratio *of "1 milliliter of a 1:1000 solution." The two labels describe the exact same amount, and the physicians were instructed to administer 0.12 milligrams of epinephrine. The results were reported in the* New York Times.

	Correct Dosage Calculation	Wrong Dosage Calculation
Concentration Label ("1 milligram in 1 milliliter solution")	11	3
Ratio Label ("1 milliliter of a 1:1000 solution")	2	12

21. 13/28, or 0.464. That probability is not as high as it should be.

21. Correct Dosage If one of the physicians is randomly selected, what is the probability of getting one who calculated the dose correctly? Is that probability as high as it should be?

22. 15/28, or 0.536. That probability is not as low as it should be.

22. Wrong Dosage If one of the physicians is randomly selected, what is the probability of getting one who calculated the dose incorrectly? Is that probability as low as it should be?

23. 16/28 or 4/7, or 0.571

23. Correct or Concentration If one of the physicians is randomly selected, find the probability of getting one who made a correct dosage calculation or was given the bottle with a concentration label.

24. 17/28, or 0.607

24. Wrong Dosage or Ratio If one of the physicians is randomly selected, find the probability of getting one who made a wrong dosage calculation or was given the bottle with a ratio label.

25. a. 0.786
b. 0.143
c. The physicians given the labels with concentrations appear to have done much better. The results suggest that labels described as concentrations are much better than labels described as ratios.

25. Which Group Did Better?

a. For the physicians given the bottles labeled with a concentration, find the percentage of correct dosage calculations, then express it as a probability.

b. For the physicians given the bottles labeled with a ratio, find the percentage of correct dosage calculations; then express it as a probability.

c. Does it appear that either group did better? What does the result suggest about drug labels?

26. Which Group Did Worse?

a. For the physicians given the bottles labeled with a concentration, find the percentage of wrong dosage calculations; then express it as a probability.

b. For the physicians given the bottles labeled with a ratio, find the percentage of wrong dosage calculations; then express it as a probability.

c. Does it appear that either group did worse? What does the result suggest about drug labels?

26. a. 0.214
b. 0.857
c. The physicians given the labels with ratios appear to have done much worse. The results suggest that labels described as ratios are much worse than labels described as concentrations.

Survey Refusals. ***In Exercises 27–32, refer to the following table summarizing results from a study of people who refused to answer survey questions (based on data from "I Hear You Knocking but You Can't Come In," by Fitzgerald and Fuller,* Sociological Methods and Research, *Vol. 11, No. 1). In each case, assume that one of the subjects is randomly selected.***

	Age					
	18–21	22–29	30–39	40–49	50–59	60 and over
Responded	73	255	245	136	138	202
Refused	11	20	33	16	27	49

27. Survey Refusals What is the probability that the selected person refused to answer? Does that probability value suggest that refusals are a problem for pollsters? Why or why not?

27. 156/1205 = 0.129. Yes. A high refusal rate results in a sample that is not necessarily representative of the population, because those who refuse may well constitute a particular group with opinions different from others.

28. Survey Refusals A pharmaceutical company is interested in opinions of the elderly, because they are either receiving Medicare or will receive it soon. What is the probability that the selected subject is someone 60 and over who responded?

28. 202/1205 = 0.168

29. Survey Refusals What is the probability that the selected person responded or is in the 18–21 age bracket?

29. 1060/1205 = 0.880

30. Survey Refusals What is the probability that the selected person refused to respond or is over 59 years of age?

30. 358/1205 = 0.297

31. Survey Refusals A market researcher is interested in responses, especially from those between the ages of 22 and 39, because they are the people more likely to make purchases. Find the probability that a selected subject responds or is between the ages of 22 and 39.

31. 1102/1205 = 0.915

32. Survey Refusals A market researcher is not interested in refusals or subjects below 22 years of age or over 59. Find the probability that the selected person refused to answer or is below 22 or is older than 59.

32. 431/1205 = 0.358

***In Exercises 33–38, use these results from the "1-Panel-THC" test for marijuana use, which is provided by the company Drug Test Success: Among 143 subjects with positive test results, there are 24 false positive results; among 157 negative results, there are 3 false negative results.* (Hint: *Construct a table similar to Table 4-1, which is included with the Chapter Problem.*)**

33. Screening for Marijuana Use

a. How many subjects are included in the study?

b. How many subjects did not use marijuana?

c. What is the probability that a randomly selected subject did not use marijuana?

33. a. 300
b. 178
c. 178/300 = 0.593

34. 0.487
35. 0.603
36. 122/300 = 0.407. No, the general population probably has a marijuana usage rate less than 0.407, or 40.7%.
37. 27/300 = 0.090. With an error rate of 0.090 (or 9%), the test does not appear to be highly accurate.
38. 273/300 = 0.910. Exercise 37 results in the probability of a wrong result and this exercise results in the probability of a correct result, so these exercises deal with events that are complements.

34. Screening for Marijuana Use If one of the test subjects is randomly selected, find the probability that the subject tested positive or used marijuana.

35. Screening for Marijuana Use If one of the test subjects is randomly selected, find the probability that the subject tested negative or did not use marijuana.

36. Screening for Marijuana Use If one of the test subjects is randomly selected, find the probability that the subject actually used marijuana. Do you think that the result reflects the marijuana use rate in the general population?

37. Screening for Marijuana Use Find the probability of a false positive or false negative. What does the result suggest about the test's accuracy?

38. Screening for Marijuana Use Find the probability of a correct result by finding the probability of a true positive or a true negative. How does this result relate to the result from Exercise 37?

4-3 Beyond the Basics

39. 3/4, or 0.75
40. No. Here is one example: A = event of selecting a male under 30 years of age, B = selecting a female, C = selecting a male over 18 years of age.
41. $P(A \text{ or } B) = P(A) + P(B) - 2P(A \text{ and } B)$
42. $P(A \text{ or } B \text{ or } C) = P(A) + P(B) + P(C) - P(A \text{ and } B) - P(A \text{ and } C) - P(B \text{ and } C) + P(A \text{ and } B \text{ and } C)$
43. a. $1 - P(A) - P(B) + P(A \text{ and } B)$
b. $1 - P(A \text{ and } B)$
c. No

39. Gender Selection When analyzing results from a test of the MicroSort gender-selection technique developed by the Genetics & IVF Institute, a researcher wants to compare the results to those obtained from a coin toss. Assume that boys and girls are equally likely and find $P(G \text{ or } H)$, which is the probability of getting a baby girl *or* getting heads from a coin toss.

40. Disjoint Events If events A and B are disjoint and events B and C are disjoint, must events A and C be disjoint? Give an example supporting your answer.

41. Exclusive Or The formal addition rule expressed the probability of A or B as follows: $P(A \text{ or } B) = P(A) + (B) - P(A \text{ and } B)$. The *exclusive or* means either one or the other events occurs, but not both. Rewrite the expression for $P(A \text{ or } B)$ assuming that the addition rule uses the *exclusive or* instead of the *inclusive or*. (*Hint:* Draw a Venn diagram.)

42. Extending the Addition Rule Extend the formal addition rule to develop an expression for $P(A \text{ or } B \text{ or } C)$. (*Hint:* Draw a Venn diagram.)

43. Complements and the Addition Rule

a. Develop a formula for the probability of not getting either A or B on a single trial. That is, find an expression for $P(\overline{A \text{ or } B})$.

b. Develop a formula for the probability of not getting A or not getting B on a single trial. That is, find an expression for $P(\overline{A} \text{ or } \overline{B})$.

c. Compare the results from parts (a) and (b). Does $P(\overline{A \text{ or } B}) = P(\overline{A} \text{ or } \overline{B})$?

4-4 Multiplication Rule: Basics

Note to Instructor
Key points of this section: $P(A \text{ and } B)$ suggests multiplication; adjust probabilities for *dependent* events.

Notation: The notation $P(A \text{ and } B)$ has two meanings, depending on its context. In this section, $P(A \text{ and } B)$ denotes that event A occurs in one trial and event B occurs in another trial; in Section 4-3 we used $P(A \text{ and } B)$ to denote that events A and B both occur in the same trial.

Key Concept This section presents the basic multiplication rule used for finding $P(A \text{ and } B)$, which is the probability that event A occurs and event B occurs. If the outcome of event A somehow affects the probability of event B, it is important to adjust the probability of B to reflect the occurrence of event A. The rule for finding $P(A \text{ and } B)$ is called the *multiplication rule* because it involves the multiplication of the probability of event A and the probability of event B (where, if necessary, the probability of event B is adjusted because of the outcome of event A). In Section 4-3 we associated use of the word *or* with addition; in this section we associate use of the word *and* with multiplication.

We begin with basic notation followed by the multiplication rule. We strongly suggest that you focus on the *intuitive* multiplication rule, because it is based on understanding instead of blind use of a formula.

Notation

$P(A \text{ and } B) = P(\text{event } A \text{ occurs in a first trial and event } B \text{ occurs in a second trial})$

$P(B|A)$ represents the probability of event B occurring after it is assumed that event A has already occurred. (Interpret $B|A$ as "event B occurring after event A has already occurred.")

Formal Multiplication Rule

$P(A \text{ and } B) = P(A) \cdot P(B|A)$

Intuitive Multiplication Rule

To find the probability that event A occurs in one trial and event B occurs in another trial, multiply the probability of event A by the probability of event B, but *be sure that the probability of event B takes into account the previous occurrence of event A.*

When applying the multiplication rule and considering whether event B must be adjusted to account for the previous occurrence of event A, we are focusing on whether events A and B are *independent.*

DEFINITIONS Two events *A* and *B* are **independent** if the occurrence of one does not affect the *probability* of the occurrence of the other. (Several events are similarly independent if the occurrence of any does not affect the probabilities of the occurrence of the others.) If *A* and *B* are not independent, they are said to be **dependent.**

CAUTION Don't think that *dependence* of two events means that one is the direct *cause* of the other. Having a working TV in your room and having working lights in your room are dependent events (because they have the same power source), even though neither has a direct effect on the other.

In the wonderful world of statistics, sampling methods are critically important, and the following relationships hold:

- Sampling *with replacement: Selections are independent* events.
- Sampling *without replacement: Selections are dependent* events.

Exception: Treating Dependent Events as Independent

Some cumbersome calculations can be greatly simplified by using the common practice of treating events as independent when *small samples* are drawn from *large populations.* In such cases, it is rare to select the same item twice.

Here is a common guideline routinely used with applications such as analyses of survey results:

Note to Instructor
Again, emphasize the intuitive rule that doesn't depend on blind application of a formula. Many instructors have much greater success with the intuitive rule than the formal rule.

Note to Instructor
Students sometimes believe incorrectly that dependent events must have some cause/effect relationship between them. Describe this example with two events: *A:* You walk into a kitchen and find that the lights can be turned on; *B:* You walk into a kitchen and find that the microwave oven can be turned on. These events are dependent because they have the same power source. They are dependent events even though they are not linked directly to each other. These events are dependent because the *probability* of one of them occurring is affected by the occurrence of the other. If you flip on the light switch and the lights do not turn on, it is likely that the microwave oven will not turn on.

Note to Instructor
We usually discuss independent events with two or more trials, and we usually discuss disjoint events with one trial. However, the concepts of independence and disjoint events can converge. Because two independent events must be able to occur at the same time, they cannot be disjoint (unless one of them is impossible). If two events are disjoint, they must be dependent (unless one of them is impossible).

Independent Jet Engines

Soon after departing from Miami, Eastern Airlines Flight 855 had one engine shut down because of a low oil pressure warning light.

As the L-1011 jet turned to Miami for landing, the low pressure warning lights for the other two engines also flashed. Then an engine failed, followed by the failure of the last working engine. The jet descended without power from 13,000 ft to 4000 ft when the crew was able to restart one engine, and the 172 people on board landed safely. With independent jet engines, the probability of all three failing is only 0.0001^3, or about one chance in a trillion. The FAA found that the same mechanic who replaced the oil in all three engines failed to replace the oil plug sealing rings. The use of a single mechanic caused the operation of the engines to become dependent, a situation corrected by requiring that the engines be serviced by different mechanics.

Treating Dependent Events as Independent: 5% Guideline for Cumbersome Calculations

When calculations with sampling are very cumbersome and the sample size is no more than 5% of the size of the population, treat the selections as being *independent* (even if they are actually dependent).

Example 1 illustrates the basic multiplication rule, with independent events in part (a) and dependent events in part (b). Example 2 is another illustration of the multiplication rule, and part (c) of Example 2 illustrates use of the above 5% guideline for cumbersome calculations.

Example 1 Drug Screening

Let's use only the 50 test results from the subjects who use drugs (from Table 4-1), as shown below:

Positive Test Results:	44
Negative Test Results:	6
Total Results:	50

a. If 2 of the 50 subjects are randomly selected *with replacement*, find the probability that the first selected person had a positive test result and the second selected person had a negative test result.

b. Repeat part (a) by assuming that the two subjects are selected *without* replacement.

Solution

a. *With Replacement:* First selection (with 44 positive results among 50 total results):

$$P(\text{positive test result}) = \frac{44}{50}$$

Second selection (with 6 negative test results among the same 50 total results):

$$P(\text{negative test result}) = \frac{6}{50}$$

We now apply the multiplication rule as follows:

$$P(\text{1st selection is positive and 2nd is negative}) = \frac{44}{50} \cdot \frac{6}{50} = 0.106$$

b. *Without Replacement:* Without replacement of the first subject, the calculations are the same as in part (a), except that the second probability must be adjusted to reflect the fact that the first selection was positive and is not available for the second selection. After the first positive result is selected, we have 49 test results remaining, and 6 of them are negative. The second probability is therefore 6/49, as shown in the calculation below:

$$P(\text{1st selection is positive and 2nd is negative}) = \frac{44}{50} \cdot \frac{6}{49} = 0.108$$

The key point of part (b) in Example 1 is this: *We must adjust the probability of the second event to reflect the outcome of the first event.* Because selection of the second subject is made *without* replacement of the first subject, the second probability must take into account the fact that the first selection removed a subject who tested positive, so only 49 subjects are available for the second selection, and 6 of them had a negative test result.

In Example 2, we consider three situations: (a) The items are selected *with* replacement; (b) the items are selected *without* replacement, but the required calculations are not too cumbersome; (c) the items are selected *without* replacement, but the required calculations are cumbersome and the "5% guideline for cumbersome calculations" can be used.

Convicted by Probability

A witness described a Los Angeles robber as a Caucasian woman with blond hair in a ponytail who escaped in a yellow car driven by an African-American male with a mustache and beard. Janet and Malcolm Collins fit this description, and they were convicted based on testimony that there is only about 1 chance in 12 million that any couple would have these characteristics. It was estimated that the probability of a yellow car is 1/10, and the other probabilities were estimated to be 1/10, 1/3, 1/10, and 1/1000. The convictions were later overturned when it was noted that no evidence was presented to support the estimated probabilities or the independence of the events. However, because the couple was not randomly selected, a serious error was made in not considering the probability of *other* couples being in the same region with the same characteristics.

Example 2 Airport Baggage Scales

Airport baggage scales can show that bags are overweight and high additional fees can be imposed. The New York City Department of Consumer Affairs checked all 810 scales at JFK and LaGuardia, and 102 scales were found to be defective and ordered out of use (based on data reported in the *New York Times*).

a. If 2 of the 810 scales are randomly selected *with replacement*, find the probability that they are both defective.

b. If 2 of the 810 scales are randomly selected *without replacement*, find the probability that they are both defective.

c. A larger population of 10,000 scales includes exactly 1259 defective scales. If 5 scales are randomly selected from this larger population without replacement, find the probability that all 5 are defective.

Solution

a. *With Replacement:* If the 2 scales are randomly selected *with replacement*, the two selections are independent because the second event is not affected by the first outcome. In each of the two selections there are 102 defective scales among the 810 scales available, so we get

$$\begin{aligned} P(\text{both scales are defective}) &= P(\text{first is defective } \textit{and} \text{ second is defective}) \\ &= P(\text{first is defective}) \cdot P(\text{second is defective}) \\ &= \frac{102}{810} \cdot \frac{102}{810} = 0.0159 \end{aligned}$$

b. *Without Replacement:* If the 2 scales are randomly selected *without replacement*, the two selections are dependent because the probability of the second event is affected by the first outcome. Being careful to adjust the second probability to reflect the result of a defect on the first selection, we get

$$\begin{aligned} P(\text{both scales are defective}) &= P(\text{first is defective } \textit{and} \text{ second is defective}) \\ &= P(\text{first is defective}) \cdot P(\text{second is defective}) \\ &= \frac{102}{810} \cdot \frac{101}{809} = 0.0157 \end{aligned}$$

c. With 1259 defective scales among 10,000, the exact calculation for getting 5 defective scales when 5 scales are randomly selected without replacement is cumbersome, as shown here:

$$\frac{1259}{10{,}000} \cdot \frac{1258}{9999} \cdot \frac{1257}{9998} \cdot \frac{1256}{9997} \cdot \frac{1255}{9996} = 0.0000314 \text{ (yuck!)}$$

Instead, we can use the 5% guideline for cumbersome calculations. The sample size of 5 is less than 5% of the population of 10,000, so we can treat the events as independent, even though they are actually dependent. We get this much easier (although not quite as accurate) calculation:

$$\frac{1259}{10{,}000} \cdot \frac{1259}{10{,}000} \cdot \frac{1259}{10{,}000} \cdot \frac{1259}{10{,}000} \cdot \frac{1259}{10{,}000} = \left(\frac{1259}{10{,}000}\right)^5 = 0.0000316$$

For Example 2, the following comments are important:

- In part (b) we adjust the second probability to take into account the selection of a defective scale in the first outcome.
- In part (c) we applied the 5% guideline for cumbersome calculations.
- While parts (a) and (b) apply to two events, part (c) illustrates that the multiplication rule extends quite easily to more than two events.

If you're thinking that the exact calculation in part (c) of Example 2 is not all that bad, consider a pollster who randomly selects 1000 survey subjects from the 230,118,473 adults in the United States. Pollsters sample without replacement, so the selections are dependent. Instead of doing exact calculations with some *really* messy numbers, they typically treat the selections as being independent, even though they are actually dependent. The world then becomes a much better place in which to live. Apart from trying to avoid messy calculations, in statistics we have a special interest in sampling with replacement, and this will be discussed in Section 6-4.

CAUTION In any probability calculation, it is extremely important to carefully identify the event being considered. See Example 3 where parts (a) and (b) might seem quite similar, but their solutions are very different.

Example 3 Birthdays

When two different people are randomly selected from those in your class, find the indicated probability by assuming that birthdays occur on the days of the week with equal frequencies.

a. Find the probability that the two people are born on the *same day of the week*.

b. Find the probability that the two people are both born on *Monday*.

Solution

a. Because no particular day of the week is specified, the first person can be born on any one of the seven weekdays. The probability that the second person is born on the same day as the first person is 1/7. The probability that two people are born on the same day of the week is therefore 1/7.

b. The probability that the first person is born on Monday is 1/7 and the probability that the second person is also born on Monday is 1/7. Because the two events are independent, the probability that both people are born on Monday is

$$\frac{1}{7} \cdot \frac{1}{7} = \frac{1}{49}$$

Princeton Closes Its ESP Lab

The Princeton Engineering Anomalies Research (PEAR) laboratory recently closed, after it had been in operation since 1975. The purpose of the lab was to conduct studies on extrasensory perception and telekinesis. In one of the lab's experiments, test subjects were asked to think high or think low, then a device would display a random number either above 100 or below 100. The researchers then used statistical methods to determine whether the results differed significantly from what would be expected by chance. The objective was to determine whether the test subjects could use their minds to somehow influence the behavior of the random-generating device.

Because the PEAR lab had been an embarrassment to many members of Princeton's community, they welcomed its closing. After being used for research for 28 years, and after using more than $10 million in funding, the PEAR lab failed to provide results compelling enough to convince anyone that ESP and telekinesis are real phenomena.

Important Applications of the Multiplication Rule

The following two examples illustrate practical applications of the multiplication rule. Example 4 gives us some insight into *hypothesis testing* (which is introduced in Chapter 8), and Example 5 illustrates the principle of *redundancy,* which is used to increase the reliability of many mechanical and electrical systems.

Example 4 ***Hypothesis Testing:*** **Effectiveness of Gender Selection**

MicroSort's XSORT gender-selection technique is designed to increase the likelihood that a baby will be a girl.

a. Assume that in a preliminary test of the XSORT technique, 20 couples gave birth to 20 babies, and all 20 babies were girls. If we were to assume that the XSORT method has no effect, what is the probability of getting 20 girls in 20 births by random chance? What does the result suggest?

b. Here are actual results of the XSORT gender-selection technique: Among 945 babies born, 879 were girls (based on data from the Genetics & IVF Institute). The probability of these results occurring by random chance with no effect from the XSORT method is calculated to be 0+, where 0+ denotes a positive probability that is so close to 0 that we can consider it to be 0 for all practical purposes. Does the probability of 0+ provide strong evidence to support a claim that the XSORT method is effective in increasing the likelihood that a baby will be a girl?

Solution

a. We want to find *P*(all 20 babies are girls) with the assumption that the XSORT method has no effect so that the probability of any individual offspring being a girl is 0.5. Because separate couples were used, we treat the events as being independent. We get this result:

$$
\begin{aligned}
&P(\text{all 20 babies are girls})\\
&\quad = P(\text{1st is girl and 2nd is girl and 3rd is girl} \cdots \text{and 20th is girl})\\
&\quad = P(\text{girl}) \cdot P(\text{girl}) \cdot \cdots \cdot P(\text{girl})\\
&\quad = 0.5 \cdot 0.5 \cdot \cdots \cdot 0.5\\
&\quad = 0.5^{20} = 0.000000954
\end{aligned}
$$

Because the probability of 0.000000954 is so small, it appears that random chance is a poor explanation. The more reasonable explanation is that use of the XSORT technique makes babies more likely to be girls.

b. The probability of getting 879 girls in 945 births is 0+, which is a small positive number that is almost 0. This shows that these results are nearly impossible if boys and girls are equally likely. These results do provide strong evidence to support a claim that the XSORT method is effective in increasing the likelihood that a baby will be a girl.

Redundancy

Reliability of systems can be greatly improved with redundancy of critical components. Race cars in the NASCAR Winston Cup series have two ignition systems so that if one fails, the other will keep the car running. Airplanes have two independent electrical systems, and aircraft used for instrument flight typically have two separate radios. The following is from a *Popular Science* article about stealth aircraft: "One plane built largely of carbon fiber was the Lear Fan 2100 which had to carry two radar transponders. That's because if a single transponder failed, the plane was nearly invisible to radar." Such redundancy is an application of the multiplication rule in probability theory. If one component has a 0.001 probability of failure, the probability of two independent components both failing is only 0.000001.

To Win, Bet Boldly

The New York Times published an article by Andrew Pollack in which he reported lower than expected earnings for the Mirage casino in Las Vegas. He wrote that "winnings for Mirage can be particularly volatile, because it caters to high rollers, gamblers who might bet \$100,000 or more on a hand of cards. The law of averages does not work as consistently for a few large bets as it does for thousands of smaller ones . . . ." This reflects the most fundamental principle of gambling: To win, place one big bet instead of many small bets! With the right game, such as craps, you have just under a 50% chance of doubling your money if you place one big bet. With many small bets, your chance of doubling your money drops substantially.

Example 5 Airbus 310: *Redundancy* for Better Safety

Modern aircraft are now highly reliable, and one design feature contributing to that reliability is the use of *redundancy*, whereby critical components are duplicated so that if one fails, the other will work. For example, the Airbus 310 twin-engine airliner has three independent hydraulic systems, so if any one system fails, full flight control is maintained with another functioning system. For this example, we will assume that for a typical flight, the probability of a hydraulic system failure is 0.002.

a. If the Airbus 310 were to have one hydraulic system, what is the probability that it would work for a flight?

b. Given that the Airbus 310 actually has three independent hydraulic systems, what is the probability that on a typical flight, control can be maintained with a working hydraulic system?

Solution

a. The probability of a hydraulic system failure is 0.002, so the probability that it does *not* fail is 0.998. That is, the probability that flight control can be maintained is as follows:

$$P(1 \text{ hydraulic system } \textit{does not fail})$$
$$= 1 - P(\text{failure}) = 1 - 0.002 = 0.998$$

b. With three independent hydraulic systems, flight control will be maintained provided that the three systems do not all fail. The probability of all three hydraulic systems failing is $0.002 \cdot 0.002 \cdot 0.002 = 0.000000008$. It follows that the probability of maintaining flight control is as follows:

$$P(\text{it does } \textit{not} \text{ happen that all three hydraulic systems fail})$$
$$= 1 - 0.000000008 = 0.999999992$$

Interpretation

With only one hydraulic system we have a 0.002 probability of failure, but with three independent hydraulic systems, there is only a 0.000000008 probability that flight control cannot be maintained because all three systems failed. By using three hydraulic systems instead of only one, risk of failure is decreased not by a factor of 1/3, but by a factor of 1/250,000. By using three independent hydraulic systems, risk is dramatically decreased and safety is dramatically increased.

Rationale for the Multiplication Rule

To see the reasoning that underlies the multiplication rule, consider a pop quiz consisting of (1) a true/false question and (2) a multiple-choice question with five possible answers (a, b, c, d, e). We will use the following two questions:

1. True or false: A pound of feathers is heavier than a pound of gold.
2. Who said that "By a small sample, we may judge of the whole piece"?
   (a) Judge Judy; (b) Judge Dredd; (c) Miguel de Cervantes;
   (d) George Gallup; (e) Gandhi

The answer key is T (for "true") and c. (The first answer is true, because weights of feathers are given in avoirdupois units, but weights of gold and other precious metals are given in troy units. An avoirdupois pound is 453.59 g, which is greater than the 373.24 g in a troy pound. The second answer is Cervantes, and the quote is from his famous novel *Don Quixote*.)

One way to find the probability that if someone makes random guesses for both answers, the first answer will be correct *and* the second answer will be correct, is to list the sample space as follows:

Ta Tb Tc Td Te Fa Fb Fc Fd Fe

If the two answers are random guesses, then the above 10 possible outcomes are equally likely, so

$$P(\text{both correct}) = P(T \text{ and } c) = \frac{1}{10} = 0.1$$

With $P(T \text{ and } c) = 1/10$, $P(T) = 1/2$, and $P(c) = 1/5$, we see that

$$\frac{1}{10} = \frac{1}{2} \cdot \frac{1}{5}$$

A *tree diagram* is a picture of the possible outcomes of a procedure, shown as line segments emanating from one starting point. These diagrams are sometimes helpful in determining the number of possible outcomes in a sample space, if the number of possibilities is not too large. The tree diagram shown in Figure 4-7 summarizes the outcomes of the true/false and multiple-choice questions. From Figure 4-7 we see that if both answers are random guesses, all 10 branches are equally likely and the probability of getting the correct pair (T,c) is 1/10. For each response to the first question, there are 5 responses to the second. *The total number of outcomes is 5 taken 2 times, or 10.* The tree diagram in Figure 4-7 therefore provides a visual illustration for using multiplication.

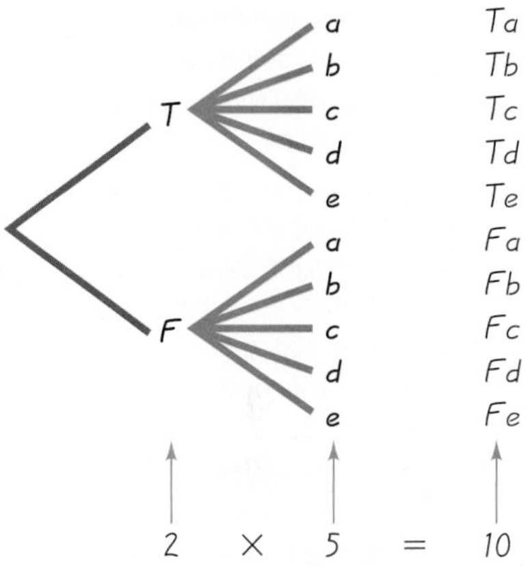

Figure 4-7 Tree Diagram of Test Answers

Probability Rules! We can summarize the addition rule from Section 4-3 and the multiplication rule from this section as follows:

Addition Rule for $P(A \text{ or } B)$: The word *or* suggests addition, and when adding $P(A)$ and $P(B)$, we must be careful to add in such a way that every outcome is counted only once.

Multiplication Rule for $P(A \text{ and } B)$: The word *and* suggests multiplication, and when multiplying $P(A)$ and $P(B)$, we must be careful to be sure that the probability of event B takes into account the previous occurrence of event A. Figure 4-8 summarizes the multiplication rule and shows the role of independence in applying it.

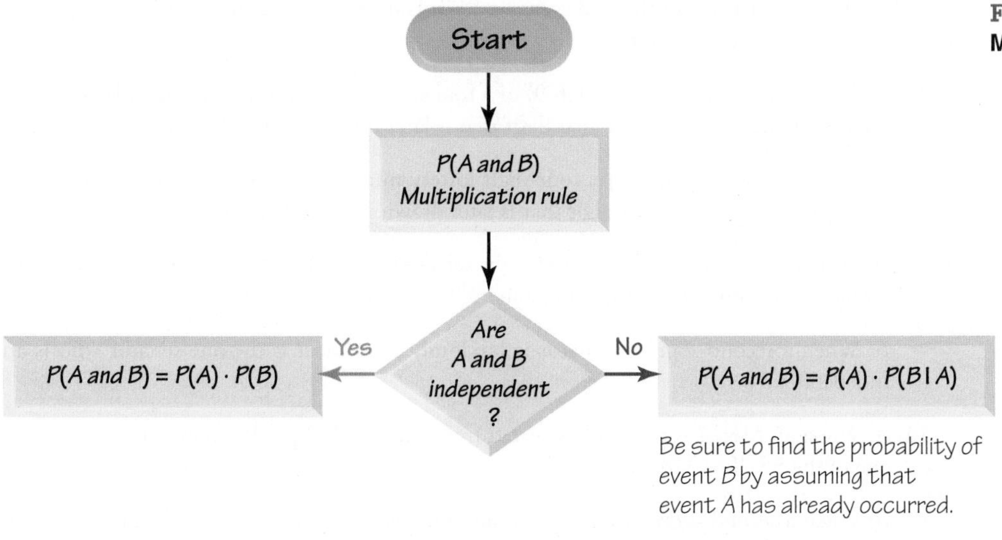

Figure 4-8 Applying the Multiplication Rule

Recommended Assignment
Exercises 1–8, 13–22.

4-4 Basic Skills and Concepts

Statistical Literacy and Critical Thinking

1. Notation Let R be the event of randomly selecting a senator and getting a Republican, and let D represent the event of randomly selecting a second *different* senator and getting a Democrat. Use words to describe what the notation $P(D \mid R)$ represents.

1. The probability that the second selected senator is a Democrat given that the first selected senator was a Republican.

2. Independent and Dependent Events Are events R and D from Exercise 1 independent or dependent? Explain.

2. *R* and *D* are dependent events, because the probability of a Democrat on the second selection is affected by the outcome of the first selection. Because it was stipulated that the second selection must be a *different* senator, the sampling is done without replacement, so only 99 senators are available for the second selection.

3. Independent and Dependent Events True or false: The event of finding that your car's radio works and the event of finding that your car's air conditioner works are independent events because they work separately from each other. Explain.

3. False. The events are dependent because the radio and air conditioner are both powered by the same electrical system. If you find that your car's radio does not work, there is a greater probability that the air conditioner will also not work.

4. Sample for a Poll There are currently 28,741,346 adults in California, and they are all included in one large numbered list. The Gallup organization uses a computer to randomly select 1068 *different* numbers between 1 and 28,741,346 and then contacts the corresponding adults for a survey. Are the events of selecting the adults actually independent or dependent? If the events are dependent, can they be treated as being independent for the purposes of calculations?

4. Because the selections are based on different numbers, the sampling is done without replacement and the events are dependent. Because the sample size of 1068 is less than 5% of the population size of 28,741,346, the events can be treated as being independent (based on the 5% guideline for cumbersome calculations).

Independent and Dependent Events. ***In Exercises 5–12, (a) determine whether events* A *and* B *are independent or dependent, and (b) find* P(A *and* B)*, the probability that events* A *and* B *both occur.***

5. *A:* When a month is randomly selected and ripped from a calendar and destroyed, it is July.

B: When a different month is randomly selected and ripped from a calendar, it is November.

5. *A:* Dependent
B: 1/132, or 0.00758

6. *A:* When a baby is born, it is a girl.

B: When a second baby is born into a different family, it is also a girl.

6. *A:* Independent
B: 1/4, or 0.25

7. *A:* When a baby is born, it is a girl.

B: When a single die is rolled, the outcome is 6.

7. *A:* Independent
B: 1/12, or 0.0833

8. *A:* When a day of the week is randomly selected, it is a Saturday.

B: When a second *different* day of the week is randomly selected, it is a Monday.

8. *A:* Dependent
B: 1/42, or 0.0238

9. *A:* When one of the 222 coins listed in Data Set 21 is randomly selected, it is one of the 5 Indian pennies.

B: When another one of the 222 coins listed in Data Set 21 is randomly selected, it is also one of the 5 Indian pennies.

9. *A:* Independent
B: 0.000507

10. *A:* When the first digit (0 through 9) of a four-digit lottery number is chosen by someone buying a ticket, it is the same first digit that is later drawn in the official lottery.

B: When the second digit of a four-digit lottery number is chosen by someone buying a ticket, it is the same second digit that is later drawn in the official lottery.

10. *A:* Independent
B: 1/100, or 0.01

11. *A:* When a survey subject is randomly selected from the 100 senators in the 111th Congress, it is one of the 58 Democrats.

B: When a second *different* senator is randomly selected, it is the one senator who is an Independent.

11. *A:* Dependent
B: 0.00586

12. *A:* When an M&M is randomly selected from the 100 M&Ms listed in Data Set 20, it is one of the 8 yellow M&Ms.

B: When a second *different* M&M is randomly selected from those listed in Data Set 20, it is also a yellow M&M.

12. *A:* Dependent
B: 0.00566

Pre-Employment Drug Screening. ***In Exercises 13–16, use the test results summarized in Table 4-1, reproduced here. Consider an event to be "unlikely" if its probability is 0.05 or less.***

Table 4-1 Pre-Employment Drug Screening Results

	Positive Test Result (Drug Use Is Indicated)	**Negative Test Result** (Drug Use Is Not Indicated)
Subject Uses Drugs	44 (True Positive)	6 (False Negative)
Subject Is Not a Drug User	90 (False Positive)	860 (True Negative)

13. Pre-Employment Drug Screening If 2 of the 1000 test subjects are randomly selected, find the probability that they both had false positive results. Is it unlikely to randomly select 2 subjects and get 2 results that are both false positive results?

a. Assume that the 2 selections are made with replacement.

b. Assume that the 2 selections are made without replacement.

13. a. 0.0081. Yes, it is unlikely.
b. 0.00802. Yes, it is unlikely.

14. Pre-Employment Drug Screening If 3 of the 1000 test subjects are randomly selected, find the probability that they all had false negative results. Is it unlikely to randomly select 3 subjects and get 3 results that are all false negative results?

a. Assume that the 3 selections are made with replacement.

b. Assume that the 3 selections are made without replacement.

14. a. 0.000000216. Yes, it is unlikely.
b. 0.000000120. Yes, it is unlikely.

15. Pre-Employment Drug Screening If 3 of the 1000 test subjects are randomly selected, find the probability that they all had correct test results (either true positive or true negative). Is such an event unlikely?

a. Assume that the 3 selections are made with replacement.

b. Assume that the 3 selections are made without replacement.

15. a. 0.739. No, it is not unlikely.
b. 0.739. No, it is not unlikely.

16. Pre-Employment Drug Screening If 4 of the 1000 test subjects are randomly selected, find the probability that they all had true negative test results. Is such an event unlikely?

a. Assume that the 4 selections are made with replacement.

b. Assume that the 4 selections are made without replacement.

16. a. 0.547. No, it is not unlikely.
b. 0.546. No, it is not unlikely.

17. Acceptance Sampling With one method of a procedure called *acceptance sampling*, a sample of items is randomly selected without replacement and the entire batch is accepted if every item in the sample is okay. Among 8834 cases of heart pacemaker malfunctions, 504 were found to be caused by firmware, which is software programmed into the device (based on data from "Pacemaker and ICD Generator Malfunctions," by Maisel et al., *Journal of the American Medical Association,* Vol. 295, No. 16). If the firmware is tested in three *different* pacemakers randomly selected from this batch of 8834, what is the probability that the firmware in the entire batch will be accepted? Does this procedure suggest that the entire batch consists of good pacemakers? Why or why not?

17. 0.838. No, the entire batch consists of malfunctioning pacemakers.

18. Acceptance Sampling With one method of a procedure called *acceptance sampling,* a sample of items is randomly selected without replacement and the entire batch is accepted if every item in the sample is okay. Among 810 airport baggage scales, 102 are defective (based on data from the New York City Department of Consumer Affairs). If four of the scales are randomly selected and tested, what is the probability that the entire batch will be accepted? Is this scheme likely to detect the large number of defects?

18. 0.583. The scheme is not likely to detect the large number of defects. With a probability of 0.583, it is more likely that the entire batch will be accepted.

19. a. 0.02
b. 0.0004
c. 0.000008
d. By using one backup drive, the probability of failure is 0.02, and with three independent disk drives, the probability drops to 0.000008. By changing from one drive to three, the likelihood of failure drops from 1 chance in 50 to only 1 chance in 125,000, and that is a very substantial improvement in reliability. Back up your data!

19. Redundancy in Computer Hard Drives It is generally recognized that it is wise to back up computer data. Assume that there is a 2% rate of disk drive failure in a year (based on data from various sources, including "Failure Trends in Large Disk Drive Population," by Pinhero et al. of Google, Inc.).

a. If you store all of your computer data on a single hard disk drive, what is the probability that the drive will fail during a year?

b. If all of your computer data are stored on a hard disk drive with a copy stored on a second hard disk drive, what is the probability that both drives will fail during a year?

c. If copies of all of your computer data are stored on three independent hard disk drives, what is the probability that all three will fail during a year?

d. Describe the improved reliability that is gained with backup drives.

20. 0.0000123. With one radio there is a 0.0035 probability of a serious problem, but with two independent radios, the probability of a serious problem drops to 0.0000123, which is substantially lower. The flight becomes much safer with two independent radios.

20. Redundancy in Aircraft Radios The FAA requires that commercial aircraft used for flying in instrument conditions must have two independent radios instead of one. Assume that for a typical flight, the probability of a radio failure is 0.0035. What is the probability that a particular flight will be threatened with the failure of both radios? Describe how the second independent radio increases safety in this case.

21. a. 1/365, or 0.00274
b. 0.00000751
c. 1/365, or 0.00274

21. Born on the 4th of July For the following, ignore leap years and assume that births on the 365 different days of the year are equally likely.

a. What is the probability that a randomly selected person was born on July 4?

b. What is the probability that two randomly selected people were both born on July 4?

c. What is the probability that two randomly selected people were born on the same day?

22. a. 1/25, or 0.04
b. 1/5, or 0.2
c. 0.000000512. Yes, it is unlikely, but perhaps there was a strong need to staff the department so that the hirings were more likely to occur on the same day.

22. Hiring Employees Assume that Google, Inc. hires employees on the different business days of the week (Monday through Friday) with equal likelihood.

a. If two different employees are randomly selected, what is the probability that they were both hired on a Monday?

b. If two different employees are randomly selected, what is the probability that they were both hired on the same day of the week?

c. What is the probability that 10 people in the same department were all hired on the same day of the week? Is such an event unlikely?

***In Exercises 23–26, use these results from the "1-Panel-THC" test for marijuana use, which is provided by the company Drug Test Success: Among 143 subjects with positive test results, there are 24 false positive results; among 157 negative results, there are 3 false negative results.* (Hint: *Construct a table similar to Table 4-1, which is included with the Chapter Problem.*)**

23. 0.828. No, it is not unlikely.

23. Screening for Marijuana Use If 2 of the subjects are randomly selected without replacement, what is the probability that they both had correct test results (either true positive or true negative)? Is such an event unlikely?

24. 0.00783. Yes, it is unlikely.

24. Screening for Marijuana Use If 2 of the subjects are randomly selected without replacement, what is the probability that they both had incorrect test results (either false positive or false negative)? Is such an event unlikely?

25. 0.000454. Yes, it is unlikely.

25. Screening for Marijuana Use If 3 of the subjects are randomly selected without replacement, what is the probability that they all had false positive test results? Is such an event unlikely?

26. Screening for Marijuana Use If 3 of the subjects are randomly selected without replacement, what is the probability that they all had true negative test results? Is such an event unlikely?

26. 0.134. No, it is not unlikely.

In Exercises 27–30, find the probabilities and indicate when the "5% guideline for cumbersome calculations" is used.

27. Road Rage In a Prince Market Research survey of 2518 motorists, 252 said that they made an obscene gesture in the previous month.

a. If 1 of the surveyed motorists is randomly selected, what is the probability that this motorist did *not* make an obscene gesture in the previous month?

b. If 50 of the surveyed motorists are randomly selected without replacement, what is the probability that none of them made an obscene gesture in the previous month?

27. a. 0.900
b. 0.00513 (using the 5% guideline for cumbersome calculations).

28. Flirting In a Microsoft Instant Messaging survey, 1021 adults were asked to identify the most fun way to flirt, and 61of them chose e-mail.

a. If 1 of the surveyed adults is randomly selected, what is the probability that this person chose something other than e-mail?

b. If 40 of the different subjects are randomly selected without replacement, what is the probability that each of them chose something other than e-mail?

28. a. 0.940
b. 0.0851 (using the 5% guideline for cumbersome calculations).

29. Online Shopping Survey 427 different adult women were randomly selected and asked what they purchase online, and 162 of the women said that they purchase clothes online (based on a survey by the *Consumer Reports* Research Center).

a. If 2 of the surveyed women are randomly selected without replacement, what is the probability that they both chose the category of clothing?

b. If 10 different surveyed women are randomly selected without replacement, what is the probability that *none* of them chose the category of clothing?

29. a. 0.143 (not 0.144)
b. 0.00848 (using the 5% guideline for cumbersome calculations).

30. Airline Survey Among respondents asked which is their favorite seat on a plane, 492 chose the window seat, 8 chose the middle seat, and 306 chose the aisle seat (based on data from *USA Today*).

a. What is the probability of randomly selecting 1 of the surveyed people and getting one who did not choose the middle seat?

b. If 2 of the surveyed people are randomly selected without replacement, what is the probability that neither of them chose the middle seat?

c. If 25 different surveyed people are randomly selected without replacement, what is the probability that none of them chose the middle seat?

30. a. 798/806, or 0.990
b. 0.980
c. 0.779 (using the 5% guideline for cumbersome calculations).

4-4 Beyond the Basics

31. System Reliability Refer to the figure at the top of the next page in which surge protectors p and q are used to protect an expensive 3D HDTV. If there is a surge in the voltage, the surge protector reduces it to a safe level. Assume that each surge protector has a 0.99 probability of working correctly when a voltage surge occurs.

a. If the two surge protectors are arranged in series, what is the probability that a voltage surge will not damage the television? (Do not round the answer.)

b. If the two surge protectors are arranged in parallel, what is the probability that a voltage surge will not damage the television? (Do not round the answer.)

31. a. 0.9999
b. 0.9801
c. The series arrangement provides better protection.

c. Which arrangement should be used for the better protection?

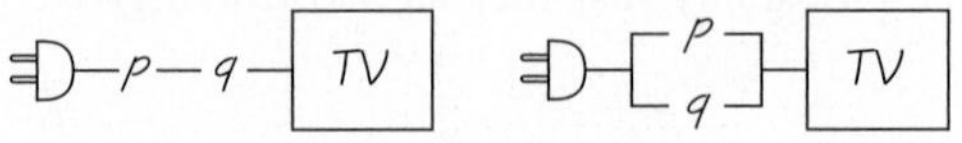

Series Configuration   Parallel Configuration

32. 0.431

32. Same Birthdays If 25 people are randomly selected, find the probability that no 2 of them have the same birthday. Ignore leap years.

4-5 Multiplication Rule: Complements and Conditional Probability

Note to Instructor
This section continues work with the multiplication rule.
Begin by clearly explaining the meaning of "at least 1." Then discuss the complement of "at least one." Then emphasize this key point: To find *P*(at least 1 of something), it's much easier to first find the probability of the complement, then subtract from 1.

Key Concept Section 4-4 presented the basic concept of the multiplication rule, and in this section we extend the use of the multiplication rule to include the following two special applications:

1. **Probability of "at least one":** Find the probability that among several trials, we get *at least one* of some specified event.
2. **Conditional probability:** Find the probability of an event occurring when we have additional information that some other event has already occurred.

We begin with situations in which we want to find the probability that among several trials, *at least one* will result in some specified outcome.

Complements: The Probability of "At Least One"

When finding the probability of some event occurring at least once, it is usually best to solve the problem "backwards" by working directly with the complementary event. In this context, the meaning of language must be clearly understood. In particular, the following principle should be well known:

In multiple trials, if *at least one* of some event occurs, then the complement (opposite) is that the event does not occur at all. If you don't get at least one occurrence of event A, then event A does not happen at all.

- "At least one" has the same meaning as "one or more."
- The *complement* of getting at least one particular event is that you get *no* occurrences of that event.

For example, not getting at least 1 defective DVD in a lot of 50 DVDs is the same as getting no defective DVDs, which is also the same as getting 50 good DVDs. The following steps describe the details of this backward method of finding the probability of getting at least one of some event:

Finding the probability of getting *at least one* of some event:

1. **Let A = getting *at least one* of some event**
2. **Then $\overline{A}$ = getting *none* of the event being considered.**
3. **Find $P(\overline{A})$ = probability that event A does not occur.** (This should be relatively easy with the multiplication rule.)
4. **Subtract the result from 1. That is, evaluate this expression:**

$$P(\textit{at least one} \text{ occurrence of event } A) = 1 - P(\textit{no} \text{ occurrences of event } A)$$

Example 1 At Least One Defective DVD

Topford Development, Ltd., supplies X-Data DVDs in lots of 50, and they have a reported defect rate of 0.5%, so the probability of an individual disk being defective is 0.005. It follows that the probability of a disk being good is 0.995. If a quality control engineer wants to carefully analyze a defective disk, what is the probability of her getting at least one defective disk in a lot of 50? Is the probability high enough that the engineer can be reasonably sure of getting a defective disk that can be used for her analysis?

Solution

Step 1: Let A = at least 1 of the 50 disks is defective.

Step 2: Identify the event that is the complement of A.

$$\overline{A} = \textit{not} \text{ getting at least 1 defective disk among 50}$$
$$= \text{all 50 disks are good}$$

Step 3: Find the probability of the complement by evaluating $P(\overline{A})$.

$$P(\overline{A}) = P(\text{all 50 disks are good})$$
$$= 0.995 \cdot 0.995 \cdot \cdots \cdot 0.995$$
$$= 0.995^{50} = 0.778$$

Step 4: Find $P(A)$ by evaluating $1 - P(\overline{A})$.

$$P(A) = 1 - P(\overline{A}) = 1 - 0.778 = 0.222$$

Interpretation

In a lot of 50 DVDs, the engineer has a 0.222 probability of getting at least 1 defective DVD. This probability is not very high, so if a defective DVD is required for analysis, the engineer should start with more than one lot consisting of 50 DVDs.

Conditional Probability

We now consider the second application of this section, which is based on the principle that the probability of an event is often affected by knowledge that some other event has occurred. For example, the probability of a golfer making a hole in one is 1/12,000 (based on past results), but if you have the additional knowledge that the selected golfer is a touring professional, the probability changes to 1/2375 (based on data from *USA Today*). In general, a *conditional probability* of an event is used when the probability is calculated with the knowledge that some other event has occurred.

DEFINITION A **conditional probability** of an event is a probability obtained with the additional information that some other event has already occurred. $P(B \mid A)$ denotes the conditional probability of event B occurring, given that event A has already occurred. $P(B \mid A)$ can be found by dividing the probability of events A and B both occurring by the probability of event A:

$$P(B \mid A) = \frac{P(A \text{ and } B)}{P(A)}$$

Prosecutor's Fallacy

The *prosecutor's fallacy* is misunderstanding or confusion of two different conditional probabilities: (1) the probability that a defendant is innocent, given that forensic evidence shows a match; (2) the probability that forensics shows a match, given that a person is innocent. The prosecutor's fallacy has led to wrong convictions and imprisonment of some innocent people.

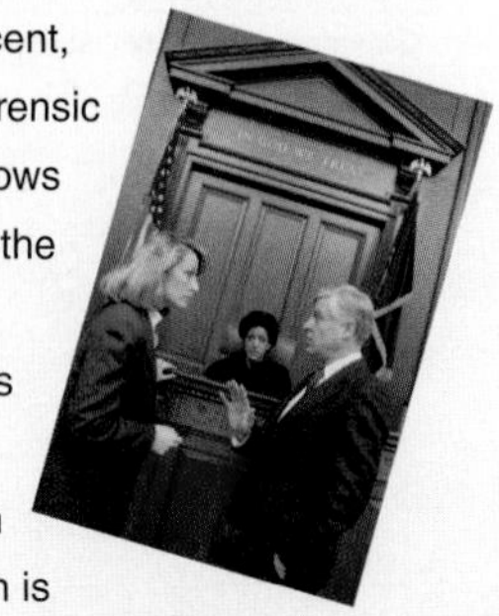

Lucia de Berk was a nurse who was convicted of murder and sentenced to prison in the Netherlands. Hospital administrators observed suspicious deaths that occurred in hospital wards where de Berk had been present. An expert testified that there was only 1 chance in 342 million that her presence was a coincidence. However, mathematician Richard Gill calculated the probability to be closer to 1/150, or possibly as low as 1/5. The court used the probability that the suspicious deaths could have occurred with de Berk present, given that she was innocent. The court should have considered the probability that de Berk is innocent, given that the suspicious deaths occurred when she was present. This error of the prosecutor's fallacy is subtle and can be very difficult to understand and recognize, yet it can lead to the imprisonment of innocent people.

The preceding formula is a formal expression of conditional probability, but blind use of formulas is not recommended. Instead, we recommend the following intuitive approach.

Note to Instructor
Students tend to find conditional probability extremely difficult when it is presented abstractly. Instead, emphasize careful reading and understanding of the available information and the probability being sought. Emphasize the intuitive approach described here.

Intuitive Approach to Conditional Probability

Finding $P(B|A)$ The conditional probability of B occurring given that A has occurred can be found by assuming that event A has occurred and then calculating the probability that event B will occur.

Example 2 Pre-Employment Drug Screening

Refer to Table 4-1 (reproduced below) to find the following:

a. If 1 of the 1000 test subjects is randomly selected, find the probability that the subject had a positive test result, given that the subject actually uses drugs. That is, find P(positive test result | subject uses drugs).

b. If 1 of the 1000 test subjects is randomly selected, find the probability that the subject actually uses drugs, given that he or she had a positive test result. That is, find P(subject uses drugs | positive test result).

Table 4-1 Pre-Employment Drug Screening Results

	Positive Test Result (Drug Use Is Indicated)	**Negative Test Result** (Drug Use Is Not Indicated)
Subject Uses Drugs	44 (True Positive)	6 (False Negative)
Subject Is Not a Drug User	90 (False Positive)	860 (True Negative)

Solution

a. *Intuitive Approach to Conditional Probability:* We want P(positive test result | subject uses drugs), the probability of getting someone with a positive test result, *given that the selected subject uses drugs*. Here is the key point: If we assume that the selected subject actually uses drugs, we are dealing only with the 50 subjects in the first row of Table 4-1. Among those 50 subjects, 44 had positive test results, so we get this result:

$$P(\text{positive test result} \mid \text{subject uses drugs}) = \frac{44}{50} = 0.88$$

Using the Formula for Conditional Probability: The same result can be found by using the formula for $P(B|A)$ given with the definition of conditional probability. We use the following notation.

$$P(B|A) = P(\text{positive test result} \mid \text{subject uses drugs})$$

where B = positive test result and A = subject uses drugs.

In the following calculation, we use P(subject uses drugs and had a positive test result) $= 44/1000$ and P(subject uses drugs) $= 50/1000$ to get the following results:

$$P(B|A) = \frac{P(A \text{ and } B)}{P(A)}$$

becomes

$$P(\text{positive test result} \mid \text{subject uses drugs})$$
$$= \frac{P(\text{subject uses drugs and had a positive test result})}{P(\text{subject uses drugs})}$$
$$= \frac{44/1000}{50/1000} = 0.88$$

By comparing the intuitive approach to the use of the formula, it should be clear that the intuitive approach is much easier to use, and it is also less likely to result in errors. The intuitive approach is based on an *understanding* of conditional probability, instead of manipulation of a formula, and understanding is so much better.

b. Here we want $P(\text{subject uses drugs} \mid \text{positive test result})$. This is the probability that the selected subject uses drugs, *given that the subject had a positive test result.* If we assume that the subject had a positive test result, we are dealing with the 134 subjects in the first column of Table 4-1. Among those 134 subjects, 44 use drugs, so

$$P(\text{subject uses drugs} \mid \text{positive test result}) = \frac{44}{134} = 0.328$$

Again, the same result can be found by applying the formula for conditional probability, but we will leave that for those with a special fondness for manipulations with formulas.

Interpretation

The first result of $P(\text{positive test result} \mid \text{subject uses drugs}) = 0.88$ indicates that a subject who uses drugs has a 0.88 probability of getting a positive test result. The second result of $P(\text{subject uses drugs} \mid \text{positive test result}) = 0.328$ indicates that for a subject who gets a positive test result, there is a 0.328 probability that this subject actually uses drugs.

Confusion of the Inverse

Note that in Example 2, $P(\text{positive test result} \mid \text{subject uses drugs}) \neq P(\text{subject uses drugs} \mid \text{positive test result})$. This example proves that in general, $P(B \mid A) \neq P(A \mid B)$. (There could be individual cases where $P(A|B)$ and $P(B|A)$ are equal, but they are generally not equal.) To incorrectly think that $P(B|A)$ and $P(A|B)$ are equal or to incorrectly use one value for the other is called *confusion of the inverse.*

Example 3 **Confusion of the Inverse**

Consider the probability that it is dark outdoors, given that it is midnight: $P(\text{dark} \mid \text{midnight}) = 1$. (We conveniently ignore the Alaskan winter and other such anomalies.) But the probability that it is midnight, given that it is dark outdoors, is almost zero. Because $P(\text{dark} \mid \text{midnight}) = 1$ but $P(\text{midnight} \mid \text{dark})$ is almost zero, we can clearly see that in this case, $P(B|A) \neq P(A|B)$. Confusion of the inverse occurs when we incorrectly switch those probability values or think that they are equal.

One study showed that physicians often give very misleading information when they confuse the inverse. They tended to confuse $P(\text{cancer} \mid \text{positive test result})$ with $P(\text{positive test result} \mid \text{cancer})$. About 95% of physicians estimated $P(\text{cancer} \mid \text{positive test result})$ to be about 10 times too high, with the result that patients were given diagnoses that were very misleading, and patients were unnecessarily distressed by the incorrect information. See Exercise 35.

DNA Evidence Misused

Micheal Bobelian wrote "DNA's Dirty Little Secret," which was published in *Washington Monthly.* He describes the use of DNA in the conviction of John Puckett for the murder of a young nurse. Thirty years after the murder, police used DNA found on the victim to identify John Puckett as the assailant, and the jury was told that there was only one chance in 1.1 million that a random person's DNA would match the DNA found on the victim. Although the DNA was degraded, this was the only physical evidence linking Puckett to the crime. Michael Bobelian wrote that when old and degraded DNA is used in a search through a large database, the "odds are exponentially higher" of getting a match with someone who is not actually guilty. He wrote that "in Puckett's case the actual chance of a false match is a staggering one in three, according to the formula endorsed by the FBI's DNA advisory board and the National Research Council." Yet Puckett's attorneys were not allowed to mention that. A serious consequence of this process is that the use of DNA evidence in similar cases might result in the conviction of innocent people. Puckett was convicted and sentenced to life in prison, but is he actually guilty?

Note to Instructor

It is helpful to briefly discuss confusion of the inverse, but let students know that they will not be required to work with it in this course. Confusion of the inverse can itself be confusing.

Recommended Assignment
Exercises 1–8 and 19–28.

4-5 Basic Skills and Concepts

Statistical Literacy and Critical Thinking

1. Subjective Probability

a. Estimate the probability that on the next test in a randomly selected statistics class, at least one student earns a grade of A.

b. Answer part (a) with the additional knowledge that the selected class is a special section for honors students with very high grade point averages.

2. Conditional Probability What is a conditional probability?

3. Notation Let event A = subject is telling the truth and event B = polygraph test indicates that the subject is lying. Use your own words to translate the notation $P(B|A)$ into a verbal statement.

4. Confusion of the Inverse Using the same events A and B described in Exercise 3, what is confusion of the inverse? (Express the answer in both words and symbols.)

Describing Complements. ***In Exercises 5–8, provide a written description of the complement of the given event, then find the probability of the complement of the given event.***

5. Five Girls When a couple has five children, all five are girls. (Assume that boys and girls are equally likely.)

6. No Girls When a couple has five children, none of the five is a girl. (Assume that boys and girls are equally likely.)

7. At Least One Zero When four digits (between 0 and 9 inclusive) are randomly selected with replacement for a lottery ticket, at least one of the digits is 0.

8. No 7s When four digits (between 0 and 9 inclusive) are randomly selected with replacement for a lottery ticket, none of the digits is a 7.

9. At Least One Correct Answer If you make random guesses for 10 multiple-choice test questions (each with five possible answers), what is the probability of getting at least 1 correct? If a very lenient instructor says that passing the test occurs if there is at least one correct answer, can you reasonably expect to pass by guessing?

10. At Least One Working Calculator A statistics student plans to use a TI-84 Plus calculator on her final exam. From past experience, she estimates that there is 0.92 probability that the calculator will work on any given day. Because the final exam is so important, she plans to use redundancy by bringing in two TI-84 Plus calculators. What is the probability that she will be able to complete her exam with a working calculator? Does she really gain much by bringing in the backup calculator? Explain.

11. Probability of a Girl Assuming that boys and girls are equally likely, find the probability of a couple having at least one boy after their fifth child is born, given that the first four children were all girls.

12. At Least One Correct Answer If you make random guesses for 10 multiple-choice test questions (each with five possible answers), what is the probability of getting at least one answer correct, given that the first nine answers are all wrong?

13. Births in the United States In the United States, the true probability of a baby being a boy is 0.512 (based on the data available at this writing). Among the next five randomly selected births in the United States, what is the probability that at least one of them is a girl?

1. a. Answer varies, but 0.98 is a reasonable estimate.
   b. Answer varies, but 0.999 is a reasonable estimate.
2. A conditional probability is a probability of an event calculated with the knowledge that some other event has occurred.
3. The probability that the polygraph indicates lying given that the subject is actually telling the truth.
4. Confusion of the inverse is to think that the following two probabilities are the same: (1) the probability of a polygraph indication of lying when the subject is telling the truth; (2) the probability of a subject telling the truth when the polygraph indicates lying. Confusion of the inverse is to think that $P(A|B) = P(B|A)$ or to use one of those probabilities in place of the other.
5. At least one of the five children is a boy. 31/32, or 0.969.
6. At least one of the five children is a girl. 31/32, or 0.969.
7. None of the digits is 0; 0.656.
8. At least one of the digits is 7; 0.344.
9. 0.893. The chance of passing is reasonably good.
10. 0.9936, or 0.994. The probability of having to complete the exam without a working calculator drops from 0.08 to 0.0064 (or 64 chances in 10,000), so she does gain a substantial increase in reliability.
11. 0.5
12. 1/5, or 0.2
13. 0.965

14. Births in China In China, the probability of a baby being a boy is 0.545. Many couples are allowed to have only one child. Among the next five randomly selected births in China, what is the probability that at least one of them is a girl? Can this system continue to work indefinitely?

14. 0.952. The system cannot continue indefinitely because eventually there would be no women to give birth.

15. Car Crashes The probability of a randomly selected car crashing during a year is 0.0423 (based on data from the *Statistical Abstract of the United States*). If a family has three cars, find the probability that at least one of them has a crash during the year. Is there any reason why this probability might be wrong?

15. 0.122. Given that the three cars are in the same family, they are not randomly selected and there is a good chance that the family members have similar driving habits, so the probability might not be accurate.

16. Cleared Burglaries According to FBI data, 12.4% of burglaries are cleared with arrests. A new detective is assigned to five different burglaries.

a. What is the probability that at least one of them is cleared with an arrest?

b. What is the probability that the detective clears five burglaries with arrests?

c. What should we conclude if the detective clears all five burglaries with arrests?

16. a. 0.484
b. 0.0000293
c. The detective is much better than average, or the detective was given five easy cases.

17. Wi-Fi Based on a poll conducted through the e-edition of *USA Today*, 67% of Internet users are more careful about personal information when using a public Wi-Fi hotspot. What is the probability that among four randomly selected Internet users, at least one is more careful about personal information when using a public Wi-Fi hotspot? How is the result affected by the additional information that the survey subjects volunteered to respond?

17. 0.988. It is very possible that the result is not valid because it is based on data from a voluntary response survey.

18. Compliments at Work Based on a poll conducted through e-mail by *USA Today*, 41% of survey respondents most liked to get compliments at work from their co-workers. Among 12 randomly selected workers, what is the probability of getting at least 1 who most likes to get compliments from co-workers? How is the result affected by the additional information that the survey subjects volunteered to respond?

18. 0.998. It is very possible that the result is not valid because it is based on data from a voluntary response survey.

In Exercises 19–24, refer to Table 4-1, included with the Chapter Problem. In each case, assume that 1 of the 1000 test subjects is randomly selected.

19. False Positive Find the probability of selecting a subject with a positive test result, given that the subject does not use drugs. Why is this particular case problematic for test subjects?

19. 90/950, or 0.0947. This is the probability of the test making it appear that the subject uses drugs when the subject is not a drug user.

20. False Negative Find the probability of selecting a subject with a negative test result, given that the subject uses drugs. Who would suffer from this type of error?

20. 6/50, or 0.12. The employer would suffer by hiring a job applicant who appears to not use drugs, but the applicant actually does use drugs.

21. Inverse Probabilities Find P(subject uses drugs | negative test result). Compare this result to the result found in Exercise 20. Are P(subject uses drugs | negative test result) *and* P(negative test result | subject uses drugs) equal?

21. 6/866, or 0.00693. This result is substantially different from the result found in Exercise 20. The probabilities P(subject uses drugs | negative test result) and P(negative test result | subject uses drugs) are not equal.

22. Inverse Probabilities

a. Find P(negative test result | subject does not use drugs).

b. Find P(subject does not use drugs | negative test result).

c. Compare the results from parts (a) and (b). Are they equal?

22. a. 860/950, or 0.905
b. 860/866, or 0.993
c. The results are different.

23. Positive Predictive Value Find the positive predictive value for the test. That is, find the probability that a subject uses drugs, given that the test yields a positive result.

23. 44/134, or 0.328

24. Negative Predictive Value Find the negative predictive value for the test. That is, find the probability that a subject does not use drugs, given that the test yields a negative result.

24. 860/866, or 0.993

Identical and Fraternal Twins. ***In Exercises 25–28, use the data in the following table. Instead of summarizing observed results, the entries reflect the actual probabilities based on births of twins (based on data from the Northern California Twin Registry and the article "Bayesians, Frequentists, and Scientists" by Bradley Efron,* Journal of the American Statistical Association, *Vol. 100, No. 469). Identical twins come from a single egg that***

splits into two embryos, and fraternal twins are from separate fertilized eggs. The table entries reflect the principle that among sets of twins, 1/3 are identical and 2/3 are fraternal. Also, identical twins must be of the same sex and the sexes are equally likely (approximately), and sexes of fraternal twins are equally likely.

Sexes of Twins

	boy/boy	boy/girl	girl/boy	girl/girl
Identical Twins	5	0	0	5
Fraternal Twins	5	5	5	5

25. a. 1/3, or 0.333
b. 0.5

25. Identical Twins

a. After having a sonogram, a pregnant woman learns that she will have twins. What is the probability that she will have identical twins?

b. After studying the sonogram more closely, the physician tells the pregnant woman that she will give birth to twin boys. What is the probability that she will have identical twins? That is, find the probability of identical twins given that the twins consist of two boys.

26. a. 2/3, or 0.667
b. 1

26. Fraternal Twins

a. After having a sonogram, a pregnant woman learns that she will have twins. What is the probability that she will have fraternal twins?

b. After studying the sonogram more closely, the physician tells the pregnant woman that she will give birth to twins consisting of one boy and one girl. What is the probability that she will have fraternal twins?

27. 0.5

27. Fraternal Twins If a pregnant woman is told that she will give birth to fraternal twins, what is the probability that she will have one child of each sex?

28. 1/4, or 0.25

28. Fraternal Twins If a pregnant woman is told that she will give birth to fraternal twins, what is the probability that she will give birth to two girls?

29. a. 0.9996 b. 0.999992

29. Redundancy in Computer Hard Drives Assume that there is a 2% rate of disk drive failure in a year (based on data from various sources, including "Failure Trends in Large Disk Drive Population," by Pinhero et al. of Google, Inc.).

a. If all of your computer data are stored on a hard disk drive with a copy stored on a second hard disk drive, what is the probability that during a year, you can avoid catastrophe with at least one working drive?

b. If copies of all of your computer data are stored on three independent hard disk drives, what is the probability that during a year, you can avoid catastrophe with at least one working drive?

30. 0.99998775. Rounding the result to three significant digits would yield 1.00, but that would be misleading because it would suggest that it is certain that both radios will work. Yes, the probability is high enough to ensure flight safety. If both radios did fail, there would also be many cell phones available for use.

31. 0.684. The probability is not low, so further testing of the individual samples will be necessary for about 68% of the combined samples.

32. 0.0248. The probability is quite low, indicating that further testing of the individual samples will be necessary for about 2% of the combined samples.

30. Redundancy in Aircraft Radios The FAA requires that commercial aircraft used for flying in instrument conditions must have two independent radios instead of one. Assume that for a typical flight, the probability of a radio failure is 0.0035. What is the probability that a particular flight will be safe with at least one working radio? Why does the usual rounding rule of three significant digits not work here? Is this probability high enough to ensure flight safety?

31. Composite Drug Screening Based on the data in Table 4-1, assume that the probability of a randomly selected person testing positive for drug use is 0.134. If drug screening samples are collected from 8 random subjects and combined, find the probability that the combined sample will reveal a positive result. Is that probability low enough so that further testing of the individual samples is rarely necessary?

32. Composite Water Samples The Fairfield County Department of Public Health tests water for contamination due to the presence of *E. coli* (*Escherichia coli*) bacteria. To reduce laboratory costs, water samples from five public swimming areas are combined for one test, and further testing is done only if the combined sample tests positive. Based on past results, there is a 0.005 probability of finding *E. coli* bacteria in a public swimming area. Find the probability that a combined sample from five public swimming areas will reveal the presence of *E. coli* bacteria. Is that probability low enough so that further testing of the individual samples is rarely necessary?

4-5 Beyond the Basics

33. Shared Birthdays Find the probability that of 25 randomly selected people,

a. no 2 share the same birthday.

b. at least 2 share the same birthday.

33. a. 0.431
b. 0.569

34. Unseen Coins A statistics professor tosses two coins that cannot be seen by any students. One student asks this question: "Did one of the coins turn up heads?" Given that the professor's response is "yes," find the probability that both coins turned up heads.

34. 1/3

35. Confusion of the Inverse In one study, physicians were asked to estimate the probability of a malignant cancer given that a test showed a positive result. They were told that the cancer had a prevalence rate of 1%, the test has a false positive rate of 10%, and the test is 80% accurate in correctly identifying a malignancy when the subject actually has the cancer. (See *Probabilistic Reasoning in Clinical Medicine* by David Eddy, Cambridge University Press.)

a. Find $P(\text{malignant} \mid \text{positive test result})$. (*Hint:* Assume that the study involves 1000 subjects and use the given information to construct a table with the same format as Table 4-1.)

b. Find $P(\text{positive test result} \mid \text{malignant})$. (*Hint:* Assume that the study involves 1000 subjects and construct a table with the same format as Table 4-1.)

c. Out of 100 physicians, 95 estimated $P(\text{malignant} \mid \text{positive test result})$ to be about 75%. Were those estimates reasonably accurate, or did they exhibit confusion of the inverse? What would be a consequence of confusion of the inverse in this situation?

35. a. 0.0748
b. 0.8
c. The estimate of 75% is dramatically greater than the actual rate of 7.48%. They exhibited confusion of the inverse. A consequence is that they would unnecessarily alarm patients who are benign, and they might start treatments that are not necessary.

4-6 Counting

Key Concept Probability problems typically require that we know the total number of simple events, but finding that number often requires one of the five rules presented in this section. With the addition rule, multiplication rule, and conditional probability, we stressed intuitive rules based on understanding and we discouraged blind use of formulas, but this section requires much greater use of formulas as we consider different methods for counting the number of possible outcomes in a variety of different situations.

Note to Instructor
Section 4-6 (Probabilities Through Simulations) in the previous edition has been moved from the textbook to the CD included with this book. Section 4-6 in this book was formerly Section 4-7 in the previous edition.

This section (Counting) can be omitted because it is not required for the following chapters. If it is omitted, consider demonstrating the combinations rule for finding the probability of winning a lottery, as in Example 5. Many instructors omit this section, but many others believe strongly that it should be included. Because of time limitations, the author recommends that this section be omitted.

Permutations and Combinations: Does *Order* Count?

When using different counting methods, it is essential to know whether different arrangements of the same items are counted only once or are counted separately. The terms *permutations* and *combinations* are standard in this context, and they are defined as follows:

DEFINITIONS

Permutations of items are arrangements in which different sequences of the same items are counted separately. For example, with the letters {a, b, c}, the arrangements of abc, acb, bac, bca, cab, and cba are all counted separately as six different permutations.

Combinations of items are arrangements in which different sequences of the same items are *not* counted separately. For example, with the letters {a, b, c}, the arrangements of abc, acb, bac, bca, cab, and cba are all considered to be same combination.

Mnemonics When trying to remember which of the two preceding terms involves order, think of **p**ermutations **p**osition, where the alliteration reminds us that with permutations, the positions of the items makes a difference. You might also use the alliteration in **c**ombinations **c**ommittee, where those words remind us that with members of a committee, rearrangements of the same members result in the same committee, so order does not count.

This section includes the following notation and counting rules. Illustrative examples follow.

Notation

The **factorial symbol (!)** denotes the product of decreasing positive whole numbers. For example, $4! = 4 \cdot 3 \cdot 2 \cdot 1 = 24$. By special definition, $0! = 1$.

Counting Rules

1. Fundamental Counting Rule
$m \cdot n =$ Number of ways that two events can occur, given that the first event can occur m ways and the second event can occur n ways. (This rule extends easily to situations with more than two events.) *Example:* For a two-character code consisting of a letter followed by a digit, the number of different possible codes is $26 \cdot 10 = 260$.

2. Factorial Rule
$n! =$ Number of different *permutations* (order counts) of n different items when all n of them are selected. (This rule reflects the fact that the first item may be selected n different ways, the second item may be selected $n - 1$ ways, and so on.) *Example:* The number of ways that the five letters {a, b, c, d, e} can be arranged is as follows:

$$5! = 5 \cdot 4 \cdot 3 \cdot 2 \cdot 1 = 120$$

3. Permutations Rule (When All of the Items Are Different)
$_nP_r = \frac{n!}{(n-r)!} =$ Number of different *permutations* (order counts) when n different items are available, but only r of them are selected *without replacement.* (Rearrangements of the same items are counted as being different.) *Example:* If the five letters {a, b, c, d, e} are available and three of them are to be selected without replacement, the number of different permutations is as follows:

$$_nP_r = \frac{n!}{(n-r)!} = \frac{5!}{(5-3)!} = 60$$

4. Permutations Rule (When Some Items Are Identical to Others)
$\frac{n!}{n_1!n_2! \cdots n_k!} =$ Number of different *permutations* (order counts) when n items are available and all n are selected *without replacement,* but some of the items are identical to others: n_1 are alike, n_2 are alike, . . . , and n_k are alike. *Example:* If the 10 letters {a, a, a, a, b, b, c, c, d, e} are available and all 10 of them are to be selected without replacement, the number of different permutations is as follows:

$$\frac{n!}{n_1!n_2! \cdots n_k!} = \frac{10!}{4!2!2!} = \frac{3{,}628{,}800}{24 \cdot 2 \cdot 2} = 37{,}800$$

5. Combinations Rule
$_nC_r = \frac{n!}{(n-r)!r!} =$ Number of different combinations (order does not count) when n different items are available, but only r of them are selected *without replacement.* (Note that rearrangements of the same items are counted as being the same.) *Example:* If the five letters {a, b, c, d, e} are available and three of them are to be selected without replacement, the number of different combinations is as follows:

$$_nC_r = \frac{n!}{(n-r)!r!} = \frac{5!}{(5-3)!3!} = \frac{120}{2 \cdot 6} = 10$$

Group Testing

During World War II, the U.S. Army tested for syphilis by giving each soldier an individual blood test that was analyzed separately. One researcher suggested mixing pairs of blood samples. After the mixed pairs were tested, those with syphilis could be identified by retesting the few blood samples that were in the pairs that tested positive. Since the total number of analyses was reduced by pairing blood specimens, why not combine them in groups of three or four or more? This technique of combining samples in groups and retesting only those groups that test positive is known as *group testing* or *pooled testing,* or *composite testing.* University of Nebraska statistician Christopher Bilder wrote an article about this topic in *Chance* magazine, and he cited some real applications. He noted that the American Red Cross uses group testing to screen for specific diseases, such as hepatitis, and group testing is used by veterinarians when cattle are tested for the bovine viral diarrhea virus.

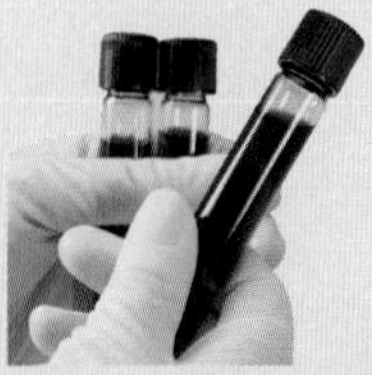

Example 1 **Fundamental Counting Rule: Computer Design**

Computers are typically designed so that the most basic unit of information is a *bit* (or binary digit), which represents either a 0 or a 1. Letters, digits, and punctuation symbols are represented as a *byte,* which is a sequence of eight bits in a particular order. For example, the ASCII coding system represents the letter *A* as 01000001 and the number 7 is represented as 00110111. How many different characters are possible if they are all to be represented as bytes?

Solution

The byte is a sequence of eight numbers, and there are only two possible numbers (0 or 1) for each of them. By applying the fundamental counting rule, the number of different possible bytes is

$$2 \cdot 2 \cdot 2 \cdot 2 \cdot 2 \cdot 2 \cdot 2 \cdot 2 = 2^8 = 256$$

Interpretation

There are 256 different characters (letters, digits, punctuations) that can be represented with different bytes. The author's keyboard has 47 keys for characters, and each of those keys is used for two different characters, so the byte system is more than adequate for such keyboards.

Example 2 **Factorial Rule: Chronological Order of Presidents**

A history pop quiz has one question in which students are asked to arrange the following presidents in chronological order: Hayes, Taft, Polk, Taylor, Grant, Pierce. If an unprepared student makes random guesses, what is the probability of selecting the correct chronological order?

Solution

The factorial rule tells us that six different items have 6! different possible rearrangements.

$$6! = 6 \cdot 5 \cdot 4 \cdot 3 \cdot 2 \cdot 1 = 720$$

Interpretation

Because only one of the 720 possible arrangements is correct, the probability of getting the correct chronological order with random guessing is 1/720, or 0.00139. With such a low probability, it is highly unlikely that a student will get the correct answer with random guessing. (The correct chronological order can be found from Data Set 12 in Appendix B.)

Example 3 **Permutations Rule (with Different Items): Exacta Bet**

In horse racing, a bet on an *exacta* in a race is won by correctly selecting the horses that finish first and second, and you must select those two horses in the correct order. The 136th running of the Kentucky Derby had a field of 20 horses. If a bettor randomly selects two of those horses for an exacta bet, what is the probability of

Choosing Personal Security Codes

All of us use personal security codes for ATM machines, computer Internet accounts, and home security systems. The safety of such codes depends on the large number of different possibilities, but hackers now have sophisticated tools that can largely overcome that obstacle. Researchers found that by using variations of the user's first and last names along with 1800 other first names, they could identify 10% to 20% of the passwords on typical computer systems. When choosing a password, *do not* use a variation of any name, a word found in a dictionary, a password shorter than seven characters, telephone numbers, or social security numbers. Do include nonalphabetic characters, such as digits or punctuation marks.

Go Figure

43,252,003,274,489,856,000: Number of possible positions on a Rubik's cube.

How to Choose Lottery Numbers

Many books and suppliers of computer programs claim to be helpful in predicting winning lottery numbers. Some use the theory that particular numbers are "due" (and should be selected) because they haven't been coming up often; others use the theory that some numbers are "cold" (and should be avoided) because they haven't been coming up often; and still others use astrology, numerology, or dreams. Because selections of winning lottery number combinations are independent events, such theories are worthless. A valid approach is to choose numbers that are "rare" in the sense that they are not selected by other people, so that if you win, you will not need to share your jackpot with many others. The combination of 1, 2, 3, 4, 5, 6 is a poor choice because many people tend to select it. In a Florida lottery with a $105 million prize, 52,000 tickets had 1, 2, 3, 4, 5, 6; if that combination had won, the prize would have been only $1000. It's wise to pick combinations not selected by many others. Avoid combinations that form a pattern on the entry card.

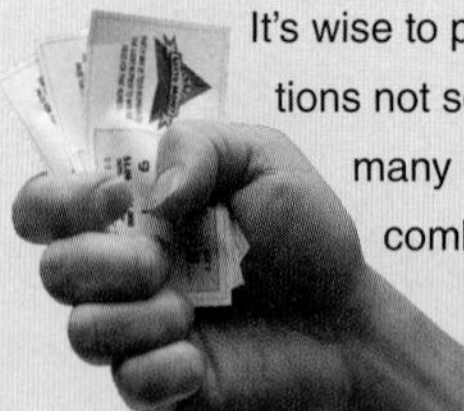

winning by selecting Super Saver to win and Ice Box to finish second (as they did)? Do all of the different possible exacta bets have the same chance of winning?

Solution

We have $n = 20$ horses available, and we must select $r = 2$ of them without replacement. The number of different sequences of arrangements is found as shown:

$$_nP_r = \frac{n!}{(n-r)!} = \frac{20!}{(20-2)!} = 380$$

There are 380 different possible arrangements of 2 horses selected from the 20 that are available. If one of those arrangements is randomly selected, there is a probability of 1/380 that the winning arrangement is selected. There are 380 different possible exacta bets, but not all of them have the same chance of winning, because some horses tend to be faster than others. (A correct $2 exacta bet in this race won $152.40.)

Example 4 **Permutations Rule (with Some Identical Items): Designing Surveys**

Here are two tricks that pollsters use when designing surveys: (1) Give different subjects rearrangements of the questions so that order does not have an effect; (2) as a check to see if a subject is thoughtlessly spewing answers just to finish the survey, repeat a question with some rewording and check to see if the answers are consistent. For one particular survey with 10 questions, 2 of the questions are the same, and 3 other questions are also identical. For this survey, how many different arrangements are possible? Is it practical to survey enough subjects so that every different possible arrangement is used?

Solution

We have 10 questions with 2 that are alike and 3 others that are alike, and we want the number of permutations. Using the rule for permutations with some items identical to others, we get

$$\frac{n!}{n_1!n_2!\cdots n_k!} = \frac{10!}{2!3!} = \frac{3{,}628{,}800}{2 \cdot 6} = 302{,}400$$

Interpretation

There are 302,400 different possible arrangements of the 10 questions. For typical surveys, the number of subjects is around 1000. For the vast majority of typical surveys, it is not practical to use 302,400 subjects; that is far too many.

Example 5 **Combinations Rule: Lottery**

In the Pennsylvania Match 6 Lotto, winning the jackpot requires that you select six different numbers from 1 to 49, and the same six numbers must be drawn in the lottery. The winning numbers can be drawn in any order, so order does not make a difference. Find the probability of winning the jackpot when one ticket is purchased.

Solution

We have 49 different numbers and we must select 6 without replacement (because the selected numbers must be different). Because order does not count, we need to find the number of different possible *combinations.* With $n = 49$ numbers available and $r = 6$ numbers selected, the number of combinations is as shown below:

$$_nC_r = \frac{n!}{(n - r)!r!} = \frac{49!}{(49 - 6)!6!} = \frac{49!}{43! \cdot 6!} = 13{,}983{,}816$$

Interpretation

If you select one 6-number combination, your probability of winning is $1/13{,}983{,}816$. Typical lotteries rely on the fact that people rarely know the value of this probability and have no realistic sense of how small that probability is. This is why the lottery is sometimes called a "tax on people who are bad at math."

Because choosing between permutations and combinations can often be tricky, we provide the following example that emphasizes the difference between them.

Example 6 Permutations and Combinations: Corporate Officials and Committees

The Teknomite Corporation must appoint a president, chief executive officer (CEO), and chief operating officer (COO). It must also appoint a Planning Committee with three different members. There are eight qualified candidates, and officers can also serve on the committee.

a. How many different ways can the officers be appointed?

b. How many different ways can the committee be appointed?

Solution

Note that in part (a), order is important because the officers have very different functions. However, in part (b), the order of selection is irrelevant because the committee members serve the same function.

a. Because order does count, we want the number of *permutations* of $r = 3$ people selected from the $n = 8$ available people. We get

$$_nP_r = \frac{n!}{(n - r)!} = \frac{8!}{(8 - 3)!} = 336$$

b. Because order does *not* count, we want the number of *combinations* of $r = 3$ people selected from the $n = 8$ available people. We get

$$_nC_r = \frac{n!}{(n - r)!r!} = \frac{8!}{(8 - 3)!3!} = 56$$

With order taken into account, there are 336 different ways that the officers can be appointed, but without order taken into account, there are 56 different possible committees.

This section presented five different counting rules summarized near the beginning of the section. Not all counting problems can be solved with these five rules, but

Bar Codes

In 1974, the first bar code was scanned on a pack of Juicy Fruit gum that cost 67¢. Now, bar codes or "Universal Product Codes" are scanned about 10 billion times each day. When used for numbers, the bar code consists of black lines that represent a sequence of 12 digits, so the total number of different bar code sequences can be found by applying the fundamental counting rule. The number of different bar code sequences is $10 \times 10 \times 10 \times 10 \times 10 \times 10 \times 10 \times 10 \times 10 \times 10 \times 10 \times 10 = 10^{12} =$ 1,000,000,000,000. The effectiveness of bar codes depends on the large number of different possible products that can be identified with unique numbers.

When a bar code is scanned, the detected number is not price; it is a number that identifies the particular product. The scanner uses that identifying number to look up the price in a central computer. Shown below is the bar code representing the author's name, so that letters are used instead of digits. There will be no price corresponding to the bar code below, because this person is priceless—at least according to most members of his immediate family.

M a r i o T r i o l a

Go Figure

10^{80} is the number of particles in the observable universe. The probability of a monkey randomly hitting keys and typing Shakespeare's *Hamlet* is $10^{-216,159}$.

they do provide a strong foundation for the most common rules that can be used in a wide variety of real applications.

Recommended Assignment
Exercises 1–24.

1. The symbol ! is the factorial symbol that represents the product of decreasing whole numbers, as in 4! = 4 · 3 · 2 · 1 = 24. Four people can stand in line 24 different ways.

2. Combinations, because order does not count and five numbers are selected (from 1 to 39) without replacement.

3. Because repetition is allowed, numbers are selected *with replacement*, so neither of the two permutation rules applies. The fundamental counting rule can be used to show that the number of possible outcomes is 10 · 10 · 10 · 10 = 10,000, so the probability of winning is 1/10,000.

4. Only the fundamental counting rule applies.

5. 1/10,000

6. 1/1,000,000,000

7. 1/362,880

8. 1/5040

9. 17,383,860. Because that number is so large, it is not practical to make a different CD for each possible combination.

10. 1/1326

11. 1/5,527,200. No, 5,527,200 is too many possibilities to list.

12. 336

4-6 Basic Skills and Concepts

Statistical Literacy and Critical Thinking

1. Notation What does the symbol ! represent? Four different people can stand in a line 4! different ways. What is the actual number of ways that four people can stand in a line?

2. California Fantasy The winning numbers for the current California Fantasy 5 lottery are 13, 18, 22, 24, and 32 in any order. Do calculations for winning this lottery involve permutations or combinations? Why?

3. California Daily 4 The winning numbers for the current California Daily 4 lottery are 5, 0, 0, and 4 in that exact order. Because order counts, do calculations for this lottery involve either of the two permutation rules presented in this section? Why or why not? If not, what rule does apply?

4. Selections with Replacement When randomly selecting items, if successive selections are made *with replacement* of previously selected items, which of the five rules of this section apply: (1) fundamental counting rule; (2) factorial rule; (3) permutations rule (when all items are different); (4) permutations rule (when some items are identical to others); (5) combinations rule?

In Exercises 5–36, express all probabilities as fractions.

5. ATM Pin Numbers A thief steals an ATM card and must randomly guess the correct PIN code that consists of four digits (0 through 9) that must be entered in the correct order. Repetition of digits is allowed. What is the probability of a correct guess on the first try?

6. Social Security Numbers A Social Security number consists of nine digits in a particular order, and repetition of digits is allowed. If randomly selecting digits for one Social Security number, what is the probability that you get the Social Security number of the president?

7. Baseball Batting Order If you know the names of the starting batters for a baseball team, what is the probability of randomly selecting a batting order and getting the order that is used in the beginning of the game?

8. Harry Potter Books There are seven books in the Harry Potter series. If the books are read in a randomly selected order, what is the probability that they are read in the order that they were written?

9. Lady Antebellum Songs A fan of Lady Antebellum music plans to make a custom CD with 12 of their 27 songs. How many different combinations of 12 songs are possible? Is it practical to make a different CD for each possible combination?

10. Blackjack In the game of blackjack played with one deck, a player is initially dealt 2 different cards from the 52 different cards in the deck. Find the probability of getting a 2-card initial hand consisting of the ace of clubs and the ace of spades in any order.

11. Scheduling Routes A presidential candidate plans to begin her campaign by visiting the capitals in 4 of the 50 states. What is the probability that she selects the route of Sacramento, Albany, Juneau, and Hartford? Is it practical to list all of the different possible routes in order to select the one that is best?

12. FedEx Deliveries With a short time remaining in the day, a FedEx driver has time to make deliveries at three locations among the eight locations remaining. How many different routes are possible?

13. Classic Counting Problem A classic counting problem is to determine the number of different ways that the letters of "Mississippi" can be arranged. Find that number.

13. 34,650

14. Statistics Count How many different ways can the letters of "statistics" be arranged?

14. 50,400

15. Connecticut Lottery Winning the jackpot in the Connecticut Classic Lotto requires that you choose six different numbers from 1 to 44, and your numbers must match the same six numbers that are later drawn. The order of the selected numbers does not matter. If you buy one ticket, what is the probability of winning the jackpot?

15. 1/7,059,052

16. Florida Lottery Winning the jackpot in the Florida Lotto requires that you choose six different numbers from 1 to 53, and your numbers must match the same six numbers that are later drawn. The order of the selected numbers does not matter. If you buy one ticket, what is the probability of winning the jackpot?

16. 1/22,957,480

17. Teed Off When four golfers are about to begin a game, they often toss a tee to randomly select the order in which they tee off. What is the probability that they tee off in alphabetical order?

17. 1/24

18. Stacking Books The author currently has seven different books in print. If those seven books are stacked in a random order, what is the probability that they are arranged in alphabetical order from top to bottom?

18. 1/5040

19. Maine Lottery

a. In the Maine Megabucks game, you win the jackpot by selecting five different whole numbers from 1 through 41 and getting the same five numbers (in any order) that are later drawn. What is the probability of winning a jackpot in this game?

b. In the Maine Pick 4 game, you win a straight bet by selecting four digits (with repetition allowed) and getting the same four digits in the exact same order they are later drawn. What is the probability of winning this game?

c. The Maine Pick 4 game returns $5000 for a winning $1 ticket. What should be the return if Maine were to run this game for no profit?

19. a. 1/749,398
b. 1/10,000
c. $10,000

20. Illinois Lottery

a. In the Illinois Little Lotto game, you win the jackpot by selecting five different whole numbers from 1 through 39 and getting the same five numbers (in any order) that are later drawn. What is the probability of winning a jackpot in this game?

b. In the Illinois Pick 3 game, you win a bet by selecting three digits (with repetition allowed) and getting the same three digits in the exact same order as they are later drawn. What is the probability of winning this game?

c. The Illinois Pick 3 game returns $500 for a winning $1 ticket. What should be the return if Illinois were to run this game for no profit?

20. a. 1/575,757
b. 1/1000
c. $1000

21. Corporate Officials and Committees The Teknomill Corporation must appoint a president, chief executive officer (CEO), chief operating officer (COO), and chief financial officer (CFO). It must also appoint a Planning Committee with four different members. There are 12 qualified candidates, and officers can also serve on the committee.

a. How many different ways can the officers be appointed?

b. How many different ways can the committee be appointed?

c. What is the probability of randomly selecting the committee members and getting the 4 youngest of the qualified candidates?

21. a. 11,880
b. 495
c. 1/495

22. Phase I of a Clinical Trial A clinical test on humans of a new drug is normally done in three phases. Phase I is conducted with a relatively small number of healthy volunteers. For

22. a. 10,461,394,944,000. (Most calculators will give a result in scientific notation, so an answer such as 10,461,395,000,000 is OK.)
b. 120
c. 1/120

example, a phase I test of bexarotene involved only 14 subjects. Assume that we want to treat 14 healthy humans with this new drug and we have 16 suitable volunteers available.

a. If the subjects are selected and treated *in sequence*, so that the trial is discontinued if anyone displays adverse effects, how many different sequential arrangements are possible if 14 people are selected from the 16 that are available?

b. If 14 subjects are selected from the 16 that are available, and the 14 selected subjects are all treated at the same time, how many different treatment groups are possible?

c. If 14 subjects are randomly selected and treated at the same time, what is the probability of selecting the 14 youngest subjects?

23. 125,000. The fundamental counting rule can be used. The different possible codes are ordered sequences of numbers, not combinations, so the name of "combination lock" is not appropriate. Given that "fundamental counting rule lock" is a bit awkward, a better name would be something like "number lock."

23. Combination Lock The typical combination lock uses three numbers between 0 and 49, and they must be selected in the correct sequence. How many different "combinations" are possible? Which of the five rules of this section is used to find that number? Is the name of "combination lock" appropriate? If not, what other name would be better?

24. 1/100,000,000. No, there are too many different possibilities.

24. Safety with Numbers The author owns a safe in which he stores all of his great ideas for the next edition of this book. The safe "combination" consists of four numbers between 0 and 99, and the safe is designed so that numbers can be repeated. If another author breaks in and tries to steal these ideas, what is the probability that he or she will get the correct combination on the first attempt? Assume that the numbers are randomly selected. Given the number of possibilities, does it seem feasible to try opening the safe by making random guesses for the combination?

25. 120; AMITY; 1/120

25. Jumble Puzzle Many newspapers carry "Jumble," a puzzle in which the reader must unscramble letters to form words. The letters MYAIT were included in newspapers on the day this exercise was written. How many ways can those letters be arranged? Identify the correct unscrambling; then determine the probability of getting that result by randomly selecting one arrangement of the given letters.

26. 360; HARROW; 1/360

26. Jumble Puzzle Repeat the preceding exercise using these letters: RAWHOR.

27. 26

27. Counting with Fingers How many different ways can you touch two or more fingers to each other on one hand?

28. a. 1/10,000,000,000,000,000
b. 1/1,000,000,000,000
c. 100,000,000. The number of possibilities is still quite large, so there is no reason to worry.

28. Identity Theft with Credit Cards Credit card numbers typically have 16 digits, but not all of them are random. Answer the following and express probabilities as fractions.

a. What is the probability of randomly generating 16 digits and getting *your* MasterCard number?

b. Receipts often show the last four digits of a credit card number. If those last four digits are known, what is the probability of randomly generating the other digits of your MasterCard number?

c. Discover cards begin with the digits 6011. If you also know the last four digits of a Discover card, what is the probability of randomly generating the other digits and getting all of them correct? Is this something to worry about?

29. 10

29. Electricity When testing for current in a cable with five color-coded wires, the author used a meter to test two wires at a time. How many different tests are required for every possible pairing of two wires?

30. a. 1/4, or 0.25
b. 3/16, or 0.188
c. Trick question. There is no finite number of attempts, because you could continue to get the wrong position every time.

30. ATM You want to obtain cash by using an ATM machine, but it's dark and you can't see your card when you insert it. The card must be inserted with the front side up and the printing configured so that the beginning of your name enters first.

a. What is the probability of selecting a random position and inserting the card with the result that the card is inserted correctly?

b. What is the probability of randomly selecting the card's position and finding that it is incorrectly inserted on the first attempt, but it is correctly inserted on the second attempt?

c. How many random selections are required to be absolutely sure that the card works because it is inserted correctly?

31. DNA Nucleotides DNA (deoxyribonucleic acid) is made of nucleotides. Each nucleotide can contain any one of these nitrogenous bases: A (adenine), G (guanine), C (cytosine), T (thymine). If one of those four bases (A, G, C, T) must be selected three times to form a linear triplet, how many different triplets are possible? All four bases can be selected for each of the three components of the triplet.

31. 64

32. World Cup Soccer Tournament Every four years, 32 soccer teams compete in a world tournament.

a. How many games are required to get one championship team from the field of 32 teams?

b. If you make random guesses for each game of the tournament, find the probability of picking the winner in every game.

32. a. 31
b. 1/2,147,483,648, or $0.5^{31} = 0.000000000466$

33. Powerball As of this writing, the Powerball lottery is run in 42 states. Winning the jackpot requires that you select the correct five numbers between 1 and 59 and, in a separate drawing, you must also select the correct single number between 1 and 39. Find the probability of winning the jackpot.

33. 1/195,249,054

34. Mega Millions As of this writing, the Mega Millions lottery is run in 42 states. Winning the jackpot requires that you select the correct five numbers between 1 and 56 and, in a separate drawing, you must also select the correct single number between 1 and 46. Find the probability of winning the jackpot.

34. 1/175,711,536

35. Designing Experiment Clinical trials of Nasonex involved a group given placebos and another group given treatments of Nasonex. Assume that a preliminary phase I trial is to be conducted with 10 subjects, including 5 men and 5 women. If 5 of the 10 subjects are randomly selected for the treatment group, find the probability of getting 5 subjects of the same sex. Would there be a problem with having members of the treatment group all of the same sex?

35. 2/252, or 1/128. Yes, if everyone treated is of one sex while everyone in the placebo group is of the opposite sex, you would not know if different reactions are due to the treatment or sex.

36. Area Codes *USA Today* reporter Paul Wiseman described the old rules for the three-digit telephone area codes by writing about "possible area codes with 1 or 0 in the second digit. (Excluded: codes ending in 00 or 11, for toll-free calls, emergency services, and other special uses.)" Codes beginning with 0 or 1 should also be excluded. How many different area codes were possible under these old rules?

36. 144

4-6 Beyond the Basics

37. Computer Variable Names A common computer programming rule is that names of variables must be between one and eight characters long. The first character can be any of the 26 letters, while successive characters can be any of the 26 letters or any of the 10 digits. For example, allowable variable names are A, BBB, and M3477K. How many different variable names are possible?

37. 2,095,681,645,538 (about 2 trillion)

38. Handshakes

a. Five managers gather for a meeting. If each manager shakes hands with each other manager exactly once, what is the total number of handshakes?

b. If n managers shake hands with each other exactly once, what is the total number of handshakes?

c. How many different ways can five managers be seated at a round table? (Assume that if everyone moves to the right, the seating arrangement is the same.)

d. How many different ways can n managers be seated at a round table?

38. a. 10
b. $n(n - 1)/2$
c. 24
d. $(n - 1)!$

39. 12

40. The probability is 0. If 9 of the letters are in the correct envelopes, the 10th letter must also be in the correct envelope, so it is impossible for the 10th letter to go into the wrong envelope.

39. Change for a Quarter How many different ways can you make change for a quarter? (Different arrangements of the same coins are not counted separately.)

40. Oldie but Goodie A secretary types 10 different letters and addresses 10 corresponding envelopes. If he is in a hurry and randomly inserts the letters into the envelopes, what is the probability that exactly 9 of the letters are in the correct envelopes? (Based on *Mathematics Magazine*, Vol. 23, No. 4, 1950)

4-7 Probabilities Through Simulations (on CD-ROM)

The CD-ROM included with this book includes another section that discusses the use of simulation methods for finding probabilities. Simulations are also discussed in the Technology Project near the end of this chapter.

4-8 Bayes' Theorem (on CD-ROM)

The CD-ROM included with this book includes another section dealing with conditional probability. This additional section discusses applications of *Bayes' theorem* (or *Bayes' rule*), which we use for revising a probability value based on additional information that is later obtained. See the CD-ROM for the discussion, examples, and exercises describing applications of Bayes' theorem.

Chapter 4 Review

The single most important concept presented in this chapter is the *rare event rule for inferential statistics*, because it forms the basis for *hypothesis testing* introduced in Chapter 8.

Rare Event Rule for Inferential Statistics

If, under a given assumption, the probability of a particular observed event is extremely small, we conclude that the assumption is probably not correct.

In Section 4-2 we presented the basic definitions and notation associated with probability. We should know that a probability value, which is expressed as a number between 0 and 1, reflects the likelihood of some event. We introduced the following important principles and notation.

Important Principles and Notation for Probability

- The probability of an event is a fraction or decimal number between 0 and 1 inclusive.
- The probability of an impossible event is 0.
- The probability of an event that is certain to occur is 1.
- Notation: The probability of event A is denoted by $P(A)$.
- Notation: The probability that event A does *not* occur is denoted by $P(\overline{A})$.

We gave three approaches to finding probabilities:

$$P(A) = \frac{\text{number of times that } A \text{ occurred}}{\text{number of times trial was repeated}} \qquad \text{(relative frequency)}$$

$$P(A) = \frac{\text{number of ways } A \text{ can occur}}{\text{number of different simple events}} = \frac{s}{n} \qquad \text{(for equally likely outcomes)}$$

$P(A)$ is *estimated* by using knowledge of the relevant circumstances. (subjective probability)

In Sections 4-3, 4-4, and 4-5 we considered compound events, which are events combining two or more simple events. We associated the word *or* with the addition rule and the word *and* with the multiplication rule.

Addition Rule for $P(A \text{ or } B)$

- $P(A \text{ or } B)$ denotes the probability that for a single trial, the outcome is event A or event B or both.
- The word *or* suggests addition, and when adding $P(A)$ and $P(B)$, we must be careful to add in such a way that every outcome is counted only once.

Multiplication Rule for $P(A \text{ and } B)$

- $P(A \text{ and } B)$ denotes the probability that event A occurs in one trial *and* event B occurs in another trial.
- The word *and* suggests multiplication, and when multiplying $P(A)$ and $P(B)$, we must be careful to ensure that the probability of event B takes into account the previous occurrence of event A.
- $P(B \mid A)$ denotes the conditional probability of event B occurring, given that event A has already occurred.
- $P($*at least one* occurrence of event $A) = 1 - P($*no* occurrences of event $A)$

Section 4-6 was devoted to the following five counting techniques, which are used to determine the total number of outcomes in probability problems:

Counting Rules

1. Fundamental Counting Rule

$m \cdot n =$ Number of ways that two events can occur, given that the first event can occur m ways and the second event can occur n ways.

2. Factorial Rule

$n! =$ Number of different *permutations* (order counts) of n different items when all n of them are selected.

3. Permutations Rule (When All of the Items Are Different)

$${}_nP_r = \frac{n!}{(n-r)!}$$ = Number of different *permutations* (order counts) when n different items are available, but only r of them are selected *without replacement.*

4. Permutations Rule (When Some Items Are Identical to Others)

$$\frac{n!}{n_1!n_2!\cdots n_k!}$$ = Number of different *permutations* (order counts) when n items are available and all n are selected *without replacement* but some of the items are identical to others: n_1 are alike, n_2 are alike, . . . , and n_k are alike.

5. Combinations Rule

$${}_nC_r = \frac{n!}{(n-r)!r!} = \text{Number of different combinations (order does not count)}$$

when n different items are available, but only r of them are selected *without replacement.*

Chapter Quick Quiz

1. 0

1. A multiple-choice question on a statistics quiz has possible correct answers of a, b, c, d, e. What is the probability that "false" is the correct answer?

2. 0.7

2. As the author is creating this exercise, a weather reporter stated that there are 3 chances in 10 of rain today. What is the probability of no rain today?

3. 1

3. If a day of the week is randomly selected, what is the probability that it is a day containing the letter *y*?

4. 0.04

4. Based on a Harris poll, 20% of adults smoke. If two adults are randomly selected, what is the probability that they both smoke?

5. Answer varies, but an answer such as 0.01 or lower is reasonable.

5. Estimate the probability that a randomly selected prime-time television show will be interrupted with a news bulletin.

In Exercises 6–10, use the following results from the 839 player challenges to referee calls in the first U.S. Open tennis tournament to use the Hawk-Eye electronic instant replay system.

	Player Challenge Was Accepted	Player Challenge Was Rejected
Male Player	201	288
Female Player	126	224

6. 512/839, or 0.610

6. If 1 of the 839 challenges is randomly selected, find the probability of getting a challenge that was rejected.

7. 713/839, or 0.850

7. If 1 of the 839 challenges is randomly selected, find the probability of getting a challenge that was made by a male player or was rejected.

8. 126/839, or 0.150

8. Find the probability of randomly selecting 1 of the 839 challenges and getting a challenge that was accepted and was made by a female player.

9. 0.0224 (not 0.0226)

9. Find the probability of randomly selecting 2 different challenges and finding that they were both accepted challenges made by female players.

10. 126/350, or 0.360

10. Find the probability of randomly selecting 1 of the 839 challenges and getting a challenge that was accepted, given that it was made by a female player.

Review Exercises

Prison and Plea. *In Exercises 1–10, use the data in the accompanying table (based on data from "Does It Pay to Plead Guilty? Differential Sentencing and the Functioning of the Criminal Courts," by Brereton and Casper,* **Law and Society Review,** *Vol. 16, No. 1). Express all probabilities as decimal numbers.*

	Guilty Plea	Plea of Not Guilty
Sentenced to Prison	392	58
Not Sentenced to Prison	564	14

1. 0.438

1. Prison and Plea If 1 of the 1028 subjects is randomly selected, find the probability of selecting someone sentenced to prison.

2. Prison and Plea Find the probability of being sentenced to prison, given that the subject entered a plea of guilty.

2. 0.410

3. Prison and Plea Find the probability of being sentenced to prison, given that the subject entered a plea of not guilty.

3. 0.806

4. Prison and Plea After comparing the results from Exercises 2 and 3, what do you conclude about the wisdom of entering a guilty plea?

4. It appears that you have a substantially better chance of avoiding prison if you enter a guilty plea.

5. Prison and Plea If 1 of the subjects is randomly selected, find the probability of selecting someone who was sentenced to prison or entered a plea of guilty.

5. 0.986

6. Prison and Plea If 2 different study subjects are randomly selected, find the probability that they both were sentenced to prison.

6. 0.191

7. Prison and Plea If 2 different study subjects are randomly selected, find the probability that they both entered pleas of not guilty.

7. 0.00484

8. Prison and Plea If 1 of the subjects is randomly selected, find the probability of selecting someone who entered a plea of not guilty or was not sentenced to prison.

8. 0.619

9. Prison and Plea If 1 of the 1028 subjects is randomly selected, find the probability of selecting someone who was sentenced to prison and entered a guilty plea.

9. 0.381

10. Prison and Plea If 1 of the subjects is randomly selected, find the probability of selecting someone who was not sentenced to prison and did not enter a plea of guilty.

10. 0.0136

11. Red Cars Use subjective probability to estimate the probability of randomly selecting a car and getting one that is red.

11. Answer varies, but DuPont data show that about 8% of cars are red, so any estimate between 0.01 and 0.2 would be reasonable.

12. Blue Eyes About 35% of the population has blue eyes (based on a study by Dr. P. Sorita Soni at Indiana University).

a. If someone is randomly selected, what is the probability that he or she does not have blue eyes?

b. If four different people are randomly selected, what is the probability that they all have blue eyes?

c. Would it be unlikely to randomly select four people and find that they all have blue eyes? Why or why not?

12. a. 0.65
b. 0.0150
c. Yes, because the probability is so small (0.0150).

13. National Statistics Day

a. If a person is randomly selected, find the probability that his or her birthday is October 18, which is National Statistics Day in Japan. Ignore leap years.

b. If a person is randomly selected, find the probability that his or her birthday is in October. Ignore leap years.

c. Estimate a subjective probability for the event of randomly selecting an adult American and getting someone who knows that October 18 is National Statistics Day in Japan.

d. Is it unlikely to randomly select an adult American and get someone who knows that October 18 is National Statistics Day in Japan?

13. a. 1/365
b. 31/365
c. Answer varies, but it is probably small, such as 0.02.
d. Yes

14. Composite Sampling for STDs Currently, the rate for sexually transmitted diseases (STDs) is 213 per 100,000 (based on data from the Centers for Disease Control and Prevention). When testing for the presence of STDs, the Acton Medical Testing Company saves money by combining blood samples for tests. The combined sample tests positive if at least one person is infected. If the combined sample tests positive, then the individual blood tests are performed. In a test for STDs, blood samples from 10 randomly selected people are combined.

14. 0.0211. No.

Find the probability that the combined sample tests positive with at least 1 of the 10 people infected. Is it likely that such combined samples test positive?

15. 1/5,245,786

15. Georgia Win for Life In the Georgia Win for Life lottery, winning the top prize of \$1000 a week for life requires that you select the correct six numbers between 1 and 42 (in any order). What is the probability of winning the top prize? Express the answer as a fraction.

16. 1/575,757

16. Georgia Fantasy 5 In the Georgia Fantasy 5 lottery, winning the top prize requires that you select the correct five numbers between 1 and 39 (in any order). What is the probability of winning the top prize? Express the answer as a fraction.

17. 1/1000

17. Numbers Game Before the proliferation of legal lotteries, illegal numbers games were commonly run by members of organized crime groups. In the typical numbers game, bettors would win by selecting the same three digits that were later drawn in the same order. A common payoff was \$600 for a winning \$1 bet. Find the probability of winning with a bet in such a numbers game.

18. 1320; 1/1320

18. Trifecta In horse racing, a trifecta is a bet that the first three finishers in a race are selected in the correct order. In a race with 12 horses, how many different trifecta bets are possible? If you randomly guess the order of the first three finishers in a race with 12 horses, what is the probability of winning?

Cumulative Review Exercises

Please be aware that some of the following problems may require knowledge of concepts presented in previous chapters.

1. a. The mean of −8.9 years is not close to the value of 0 years that would be expected with no gender discrepancy.
b. The median of −13.5 years is not close to the value of 0 years that would be expected with no gender discrepancy.
c. $s = 10.6$ years
d. $s^2 = 113.2$ years^2
e. $Q_1 = -15.0$ years
f. $Q_3 = -5.0$ years
g. The boxplot suggests that the data have a distribution that is skewed.

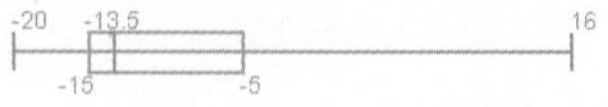

1. Oscar Winners Listed below are the differences (in years) between the ages of actresses and actors when they won Oscars. The differences are found by subtracting the age of the male winner from the age of the female winner: (actress age) − (actor age). The differences are found using results from the last 12 years (as listed in Data Set 11 in Appendix B).

$$-20 \quad -15 \quad -3 \quad -12 \quad 6 \quad -15 \quad -7 \quad -9 \quad 16 \quad -18 \quad -15 \quad -15$$

a. Find the mean. Compare the result to the value of the mean that would be expected if there is no gender discrepancy between the ages of Oscar-winning actresses and actors.

b. Find the median. Compare the result to the value of the median that would be expected if there is no gender discrepancy between the ages of Oscar-winning actresses and actors.

c. Find the standard deviation.

d. Find the variance. Be sure to include the units of measurement.

e. Find the value of the first quartile, Q_1.

f. Find the value of the third quartile, Q_3.

g. Construct a boxplot. What does the boxplot suggest about the distribution of the data?

2. a. No. The pulse rate of 100 beats per minute is within 2 standard deviations of the mean, so it is not unusual.
b. Yes. The pulse rate of 50 beats per minute is more than 2 standard deviations away from the mean, so it is unusual.
c. Yes, because the probability of 1/256 (or 0.00391) is so small.
d. No, because the probability of 1/8 (or 0.125) is not very small.

2. Unusual/Unlikely Events

a. The mean pulse rate for adult women is 77.5 beats per minute, with a standard deviation of 11.6 beats per minute (based on Data Set 1 in Appendix B). Using the range rule of thumb, would a pulse rate of 100 beats per minute be considered unusual? Explain.

b. For the pulse rates of adult women described in part (a), is a pulse rate of 50 beats per minute unusual? Explain.

c. For a couple having eight children, is it unlikely to have all girls? Explain.

d. For a couple having three children, is it unlikely to have all girls? Explain.

3. Organ Donors *USA Today* provided information about a survey of 5100 adult Internet users. Of the respondents, 2346 said they are willing to donate organs after death. In this survey conducted for Donate Life America, 100 adults were surveyed in each state and the District of Columbia, and results were weighted to account for the different state population sizes.

a. What percentage of respondents said that they are willing to donate organs after death?

b. Based on the poll results, what is the probability of randomly selecting an adult who is willing to donate organs after death?

c. What term is used to describe the sampling method of randomly selecting 100 adults from each state and the District of Columbia?

3. a. 46%
b. 0.460
c. Stratified sample

4. Driving Survey In a survey of Americans age 16 years or older, respondents were asked if they texted or e-mailed while driving during the previous 30 days. The results are summarized in the accompanying graph (based on data from AAA Foundation for Traffic Safety). What is wrong with this graph?

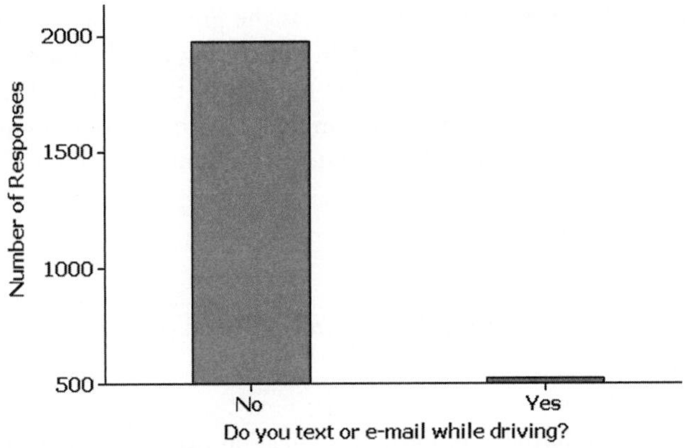

4. The graph is misleading because the vertical scale does not start at 0. The vertical scale starts at the frequency of 500 instead of 0, so the difference between the two response rates is exaggerated. The graph incorrectly makes it appear that "no" responses occurred about 60 times more often than the number of "yes" responses, but comparison of the actual frequencies shows that the "no" responses occurred about four times more often than the number of "yes" responses.

5. Sampling Eye Color Based on a study by Dr. P. Sorita Soni at Indiana University, we know that eye colors in the United States are distributed as follows: 40% brown, 35% blue, 12% green, 7% gray, 6% hazel.

a. A statistics instructor collects eye color data from her students. What is the name for this type of sample?

b. Identify one factor that might make this particular sample biased and not representative of the general population of people in the United States.

c. If one person is randomly selected, what is the probability that this person will have brown or blue eyes?

d. If two people are randomly selected, what is the probability that at least one of them has brown eyes?

5. a. Convenience sample
b. If the students at the college are mostly from a surrounding region that includes a large proportion of one ethnic group, the results will not reflect the general population of the United States.
c. 0.75
d. 0.64

6. Scatterplot of Bear Chest Sizes and Weights Given below are the chest sizes (inches) and weights (pounds) of 10 bears taken from Data Set 7 in Appendix B. Construct a scatterplot. Based on the scatterplot, does there appear to be a correlation between the chest sizes of bears and their weights?

Chest Size (in.)	26	45	54	49	35	41	41	49	38	31
Weight (lb)	80	344	416	348	166	220	262	360	204	144

6. The straight-line pattern of the points suggests that there is a correlation between chest size and weight.

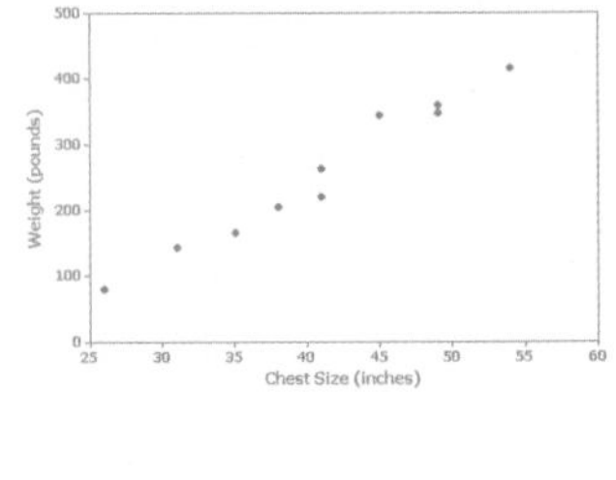

7. a. 1/575,757
b. 1/19
c. 1/10,939,383

7. Vermont Hot Lotto Winning the jackpot in the Vermont Hot Lotto game requires that you select five numbers between 1 and 39 and another number between 1 and 19. The first five numbers must match (in any order) the same five numbers that are later drawn, and the sixth number must also match the sixth number that is later drawn.

a. What is the probability that the first five selected numbers match the five numbers that are later drawn?

b. What is the probability that the sixth selected number matches the sixth number that is later drawn?

c. What is the probability of winning the jackpot?

Technology Project

Simulations

Calculating probabilities can often be painfully difficult, but *simulations* provide us with a very practical alternative to calculations based on formal rules. A **simulation** of a procedure is a process that behaves the same way as the procedure so that similar results are produced. Instead of calculating the probability of getting exactly 5 boys in 10 births, you could repeatedly toss 10 coins and count the number of times that 5 heads (or simulated "boys") occur. Better yet, you could do the simulation with a random number generator on a computer or calculator to randomly generate 1s (or simulated "boys") and 0s (or simulated "girls").

Let's consider this probability exercise:

Classic Birthday Problem

Find the probability that among 25 randomly selected people, at least 2 have the same birthday.

For the above classic birthday problem, a simulation begins by representing birthdays by integers from 1 through 365, where 1 represents a birthday of January 1, and 2 represents January 2, and so on. We can simulate 25 birthdays by using a calculator or computer to generate 25 random numbers (with repetition allowed) between 1 and 365. Those numbers can then be sorted, so it becomes easy to examine the list to determine whether any 2 of the simulated birth dates are the same. (After sorting, equal numbers are adjacent.) We can repeat the process as many times as we like, until we are satisfied that we have a good estimate of the probability. There are several ways of obtaining randomly generated numbers from 1 through 365, including the following:

STATDISK: Select **Data,** then select **Uniform Generator.** Enter a sample size of 25, a minimum of 1, and a maximum of 365, and enter 0 for the number of decimal places. The resulting STATDISK display will consist of 25 simulated birthdays. Use copy/paste to copy the data set to the **Sample Editor,** where the values can be sorted. To sort a column of data, click on **Data,** then select the first menu item of **Sort Data.** Make selections at the top and click on **Sort.**

Minitab: Click on **Calc,** select **Random Data,** then select **Integer.** In the dialog box, enter 25 for the number of rows, store the results in column C1, and enter a minimum of 1 and a maximum of 365. To sort the 25 simulated birthdays, click on **Data** and select **Sort** to arrange the data in increasing order.

Excel: Click on cell A1 (in the upper left corner), then click on the function icon ***fx*.** Select **Math & Trig,** then select **RANDBETWEEN.** In the dialog box, enter 1 for "bottom," and enter 365 for "top." After getting the random number in the first cell, click and hold down the mouse button to drag the lower right corner of this first cell, and pull it down the column until 25 cells are highlighted. When you release the mouse button, all 25 random numbers should be present.

TI-83/84 Plus: Press MATH, select **PRB,** then choose **randInt.** Enter the minimum of 1, the maximum of 365, and 25 for the number of values, all separated by commas, as in **randInt(1, 365, 25).** Press ENTER. You can store the results in list L1, then you can sort list L1 by pressing STAT and selecting **SortA** (sort in ascending order).

StatCrunch: Click on **Open StatCrunch,** then click on **Data** and select the menu item of **Simulate data.** Among the options available, select **Discrete Uniform.** In the dialog box that appears, enter 25 for the number of rows and enter 20 for the number of columns (as required for the following simulation). Enter 1 for the minimum and enter 365 for the maximum, then click on **Simulate.** You can sort columns by clicking on **Data,** then selecting the menu item of **Sort columns.** The sorted columns will appear to the right of the original columns.

Applying Simulation Methods

1. Use the above simulation method to randomly generate 20 different groups of 25 birthdays. Use the results to estimate the probability that among 25 randomly selected people, at least 2 have the same birthday.
2. One of the author's favorite class activities is to give this assignment: All students take out a coin and flip it. Students getting heads go home and actually flip a coin 200 times and record the results. Students getting tails make up their own results for 200 coin flips. In the next class, the author could select any student's results and quickly determine whether the results are real or fabricated by using this criterion: If there is a run of six heads or six tails, the results are real, but if there is no such run, the results are fabricated. This is based on the principle that when fabricating results, people almost never include a run of six or more heads or tails, but with 200 actual coin flips, there is a very high probability of getting such a run of at least six heads or tails. The calculation for the probability of getting a run of at least six heads or six tails is *extremely* difficult. Fortunately, simulations can let us know whether such runs are likely in 200 coin flips. Simulate 200 actual coin flips, repeat the simulation 20 times, then estimate the probability of getting a run of at least six heads or six tails when 200 coins are tossed. *Caution:* Do not sort generated numbers, because the probability should be based on the original sequence of coin tosses.

from data TO DECISION

Critical Thinking: Interpreting Medical Test Results

All of us undergo medical tests at various times in our lives. Some of us are tested for drug use when we apply for jobs, some of us are tested for disease when we exhibit fever or other symptoms, some of us are tested for a variety of disorders when we undergo routine physical examinations, and some of us undergo tests for pregnancy. Because such tests typically provide some wrong results, it is important that we, along with our physicians, understand how to correctly interpret results. We should know that a positive test result does not necessarily mean that a condition is present, and we should know that a negative test result does not necessarily mean that the condition is not present.

West Nile Virus (WNV) is potentially serious and is sometimes fatal. Humans commonly acquire this virus through a mosquito bite. The table below includes results based on tests given to subjects in Queens, New York (based on data from the Centers for Disease Control and Prevention).

West Nile Virus Test

	Positive Test Result (West Nile Virus Is Indicated)	**Negative Test Result** (West Nile Virus Is Not Indicated)
Subject Is Infected	17 (True Positive)	6 (False Negative)
Subject Is Not Infected	2 (False Positive)	652 (True Negative)

Analyzing the Results

1. **False Positive** Based on the results in the table, find the probability that a subject is not infected, given that the test result is positive. That is, find *P*(false positive).
2. **True Positive** Based on the results in the table, find the probability that a subject is infected, given that the test result is positive. That is, find *P*(true positive).
3. **False Negative** Based on the results in the table, find the probability that a subject is infected, given that the test result is negative. That is, find *P*(false negative).
4. **True Negative** Based on the results in the table, find the probability that a subject is not infected, given that the test result is negative. That is, find *P*(true negative).
5. **Sensitivity** Find the *sensitivity* of the test by finding the probability of a true positive, given that the subject is actually infected.
6. **Specificity** Find the *specificity* of the test by finding the probability of a true negative, given that the subject is not infected.
7. **Positive Predictive Value** Find the *positive predictive value* of the test by finding the probability of a true positive, given that the test yields a positive result.
8. **Negative Predictive Value** Find the *negative predictive value* of the test by finding the probability of a true negative, given that the test yields a negative result.
9. **Confusion of the Inverse** Find the following values, then compare them. In this case, what is confusion of the inverse?
   - *P*(subject is infected | positive test result)
   - *P*(positive test result | subject is infected)

Cooperative Group Activities

1. Out-of-class activity Divide into groups of three or four and create a new carnival game. Determine the probability of winning. Determine how much money the operator of the game can expect to take in each time the game is played.

2. In-class activity Divide into groups of three or four and use coin flipping to develop a simulation that emulates the kingdom that abides by this decree: After a mother gives birth to a son, she will not have any other children. If this decree is followed, does the proportion of girls increase?

3. In-class activity Divide into groups of three or four and use actual thumbtacks to estimate the probability that when dropped, a thumbtack will land with the point up. How many trials are necessary to get a result that appears to be reasonably accurate when rounded to the first decimal place?

4. In-class activity Divide into groups of three or four and use Hershey's Kisses candies to estimate the probability that when dropped, they land with the flat part lying on the floor. How many trials are necessary to get a result that appears to be reasonably accurate when rounded to the first decimal place?

5. In-class activity Divide into groups of three or four and use a paper cup to estimate the probability that when dropped, the cup will land upside down, with the opening on the floor. How many trials are necessary to get a result that appears to be reasonably accurate when rounded to the first decimal place?

6. Out-of-class activity Marine biologists often use the *capture-recapture method* as a way to estimate the size of a population, such as the number of fish in a lake. This method involves capturing a sample from the population, tagging each member in the sample, then returning them to the population. A second sample is later captured, and the tagged members are counted along with the total size of this second sample. The results can be used to estimate the size of the population.

Instead of capturing real fish, simulate the procedure using some uniform collection of items such as BBs, colored beads, M&Ms, Fruit Loops cereal pieces, or index cards. Start with a large collection of such items. Collect a sample of 50 and use a magic marker to "tag" each one. Replace the tagged items, mix the whole population, then select a second sample and proceed to estimate the population size. Compare the result to the actual population size obtained by counting all of the items.

7. Out-of-class activity Divide into groups of three or four. First, use subjective estimates for the probability of randomly selecting a car and getting each of these car colors: black, white, blue, red, silver, other. Then design a sampling plan for obtaining car colors through observation. Execute the sampling plan and obtain revised probabilities based on the observed results. Write a brief report of the results.

8. In-class activity The manufacturing process for a new computer integrated circuit has a yield of 1/6, meaning that 1/6 of the circuits are good and the other 5/6 are defective. Use a die to simulate this manufacturing process, and consider an outcome of 1 to be a good circuit, while outcomes of 2, 3, 4, 5, or 6 represent defective circuits. Find the mean number of circuits that must be manufactured to get one that is good.

9. In-class activity The *Monty Hall problem* is based on the old television game show *Let's Make a Deal,* hosted by Monty Hall. Suppose you are a contestant who has selected one of three doors after being told that two of them conceal nothing, but that a new red Corvette is behind one of the three. Next, the host opens one of the doors you didn't select and shows that there is nothing behind it. He then offers you the choice of *sticking* with your first selection or *switching* to the other unopened door. Should you stick with your first choice or should you switch? Divide into groups of two and simulate this game to determine whether you should stick or switch. (According to *Chance* magazine, business schools at such institutions as Harvard and Stanford use this problem to help students deal with decision making.)

5 Discrete Probability Distributions

5

chapter problem

Is the XSORT gender-selection method effective?

We live in a time with incredible advances in technology, medicine, and health care. Cloning is no longer science fiction. We have iPods, iPads, iPhones, and 3-D televisions. We carry calculators that can instantly execute many complex statistical calculations. Heart pacemakers have defibrillators capable of shocking and restarting stopped hearts. Couples use procedures that are claimed to greatly increase the chance of having a baby with a desired gender.

Some people argue that gender-selection methods should be banned, regardless of the reason, while others enthusiastically support the use of such methods. Lisa Belkin asked in the *New York Times Magazine,* "If we allow parents to choose the sex of their child today, how long will it be before they order up eye color, hair color, personality traits, and IQ?" There are some convincing arguments in favor of at least limited use of gender selection. One such argument involves couples carrying X-linked recessive genes. For some of those couples, any male children have a 50% chance of inheriting a serious disorder, but none of the female children will inherit the disorder. Those couples may want to use gender selection as a way to ensure that they have baby girls, thereby guaranteeing that a serious disorder will not be inherited by any of their children.

The Genetics & IVF Institute in Fairfax, Virginia, developed a technique called MicroSort that it claims increases the chances of a couple having a baby with a desired gender. The MicroSort XSORT method is claimed to increase the chances of a couple having a baby girl, and the MicroSort YSORT method is claimed to increase the chances of a baby boy. Current results for the XSORT method consist of 945 couples who wanted to have baby girls. After using the XSORT technique, 879 of those couples had baby girls. (See Figure 5-1 for a bar graph illustrating these results.) We usually expect that in 945 births, the number of girls should be somewhere around 472 or 473. Given that 879 out of 945 couples had girls, can we conclude that the XSORT technique is effective, or might we explain the outcome as just a chance sample result? In answering that question, we will use principles of probability to determine whether the observed birth results differ significantly from results that we would expect from random chance. This is a common goal of inferential statistics: Determine whether results can be reasonably explained by random chance or whether random chance doesn't appear to be a feasible explanation, so that other factors are influencing results. In this chapter we present methods that allow us to find the probabilities we need for determining whether the XSORT results are significant, suggesting that the method is effective.

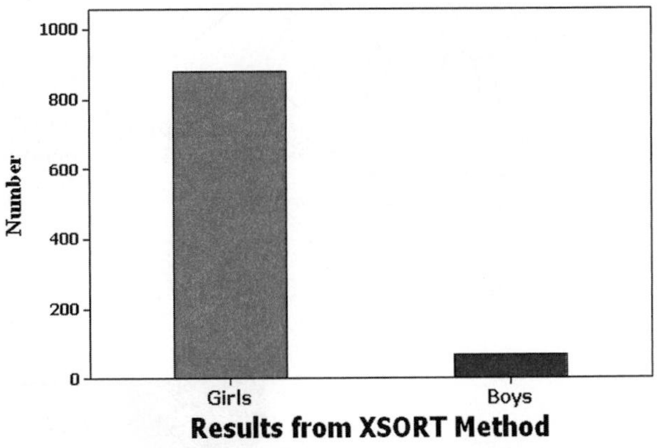

Figure 5-1 Current Results from the XSORT Method of Gender Selection

Note to Instructor

Suggestion for introducing Chapter 5: List one frequency distribution with 10 classes having the values of *x* as 0, 1, 2, . . . , 9 and corresponding frequencies of 20, 2, 3, 2, 4, 18, 5, 4, 6, 6. List a second frequency distribution with 10 classes having the values of *x* as 0, 1, 2, . . . , 9 and corresponding frequencies of 7, 6, 6, 7, 8, 6, 7, 9, 8, 6. Inform the class that both samples are the *last digits* of recorded weights of people, but one of the samples came from *measured* weights, whereas the other sample resulted from *asking* people what they weigh. Ask the class to identify the sample with digits from reported weights. Emphasize that they can make a conclusion about the nature of the data by simply examining the *distributions.* Also ask them to construct an "ideal" distribution that would result from millions of people that were actually weighed; ask them to estimate the mean and standard deviation for this distribution. (The mean should be 4.5 and the standard deviation could be estimated using the range rule of thumb; the true mean is 4.5 and the true standard deviation is around 3.)

5-1 Review and Preview

Figure 5-2 provides a visual illustration of what this chapter accomplishes. When investigating the numbers of girls in families with exactly two children, we can use the following two different approaches:

- **Use real sample data:** The approach of Chapters 2 and 3 is to collect sample data from actual families, then summarize the results in a table representing the frequency distribution, and then find statistics, such as the sample mean $\bar{x}$ and the sample standard deviation *s*.
- **Use probabilities to find the results that are expected:** Using principles of probability from Chapter 4, we can find the probability for each possible number of girls. Then we could summarize the results in a table representing a probability distribution.

In this chapter we merge the above two approaches as we create a table describing what we expect to happen, then proceed to find parameters such as the population mean μ and population standard deviation σ. (These values are population parameters because they describe the behavior of the population of all families with two children, not an actual sample of a limited number of such families.) The table at the extreme right in Figure 5-2 represents a *probability distribution,* because it describes the distribution using probabilities instead of frequency counts from actual families. The remainder of this book and the very core of inferential statistics are based on some knowledge of probability distributions. In this chapter we focus on *discrete* probability distributions. *Continuous* probability distributions are discussed in Chapter 6.

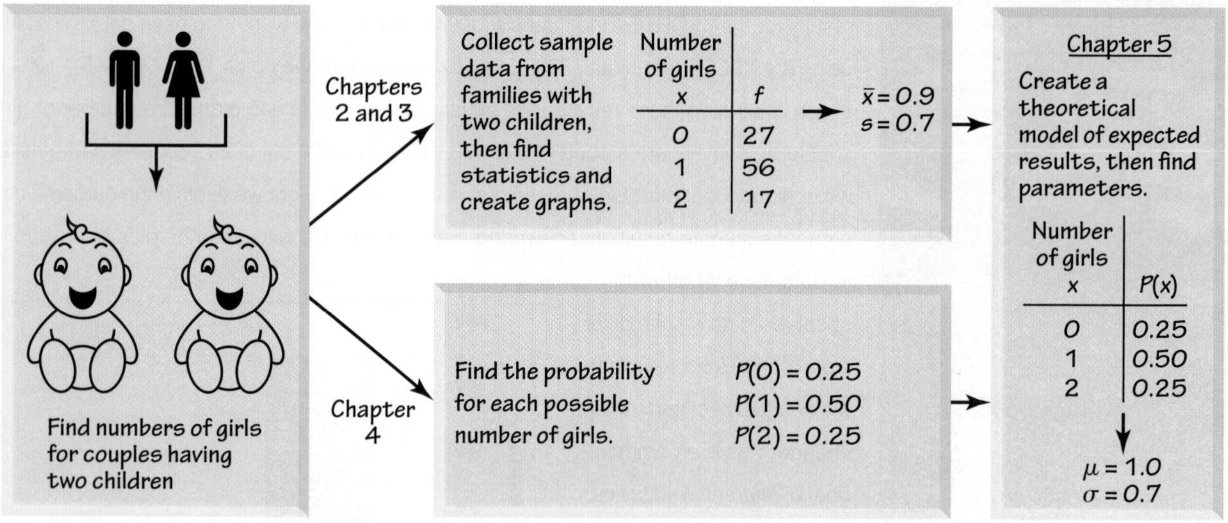

Figure 5-2

5-2 Probability Distributions

Key Concept In this section we introduce the concept of a *random variable* and the concept of a *probability distribution.* We illustrate how a *probability histogram* is a graph that visually depicts a probability distribution. We show how to find the important parameters of mean, standard deviation, and variance for a probability distribution. Most importantly, we describe how to determine whether outcomes are *unlikely* to occur by chance. We begin with the related concepts of *random variable* and *probability distribution.*

Part 1: Basic Concepts of a Probability Distribution

DEFINITIONS

A **random variable** is a variable (typically represented by x) that has a single numerical value, determined by chance, for each outcome of a procedure.

A **probability distribution** is a description that gives the probability for each value of the random variable. It is often expressed in the format of a table, formula, or graph.

In Section 1-3 we made a distinction between discrete and continuous data. Random variables may also be discrete or continuous, and the following two definitions are consistent with those given in Section 1-3.

DEFINITIONS

A **discrete random variable** has a collection of values that is finite or countable. (If there are infinitely many values, the number of values is countable if it is possible to count them individually, such as the number of tosses of a coin before getting tails.)

A **continuous random variable** has infinitely many values, and the collection of values is not countable. (That is, it is impossible to count the individual items because at least some of them are on a continuous scale.)

This chapter deals exclusively with discrete random variables, but the subsequent chapters will deal with continuous random variables.

Every probability distribution must satisfy each of the following three requirements.

Probability Distribution: Requirements

1. There is a numerical random variable x and its values are associated with corresponding probabilities.
2. $\Sigma P(x) = 1$ where x assumes all possible values. (The sum of all probabilities must be 1, but sums such as 0.999 or 1.001 are acceptable because they result from rounding errors.)
3. $0 \leq P(x) \leq 1$ for every individual value of the random variable x. (That is, each probability value must be between 0 and 1 inclusive.)

The second requirement comes from the simple fact that the random variable x represents all possible events in the entire sample space, so we are certain (with probability 1) that one of the events will occur. The third requirement comes from the basic principle that any probability value must be 0 or 1 or a value between 0 and 1.

Example 1 **Genetics**

Although the Chapter Problem involves 945 births, let's consider a simpler example that involves only two births with the following random variable:

$$x = \text{number of girls in two births}$$

The above x is a random variable because its numerical values depend on chance. With two births, the number of girls can be 0, 1, or 2, and Table 5-1 is a probability distribution because it gives the probability for each value of the random variable x and it satisfies the three requirements listed earlier:

1. The variable x is a numerical random variable and its values are associated with probabilities, as in Table 5-1.
2. $\Sigma P(x) = 0.25 + 0.50 + 0.25 = 1$
3. Each value of $P(x)$ is between 0 and 1. (Specifically, 0.25 and 0.50 and 0.25 are each between 0 and 1 inclusive.)

The random variable x in Table 5-1 is a *discrete* random variable, because it has three possible values (0, 1, 2), and 3 is a finite number, so this satisfies the requirement of being finite or countable.

Table 5-1 Probability Distribution for the Number of Girls in Two Births

Number of Girls x	$P(x)$
0	0.25
1	0.50
2	0.25

Notation

In tables such as Table 5-1 or the Binomial Probabilities in Table A-1 in Appendix A, we sometimes use 0+ to represent a probability value that is positive but very small, such as 0.000000123. (When rounding a probability value for inclusion in such a table, a rounded value of 0 would be misleading because it incorrectly suggests that the event is impossible.)

Probability Distribution: Graph

There are various ways to graph a probability distribution, but we will consider only the **probability histogram.** Figure 5-3 is a probability histogram corresponding to Table 5-1. Notice that it is similar to a relative frequency histogram (described in Section 2-3), but the vertical scale shows *probabilities* instead of relative frequencies based on actual sample results.

In Figure 5-3, we see that the values of 0, 1, 2 along the horizontal axis are located at the centers of the rectangles. This implies that the rectangles are each 1 unit wide, so the areas of the rectangles are 0.25, 0.50, and 0.25. The *areas*

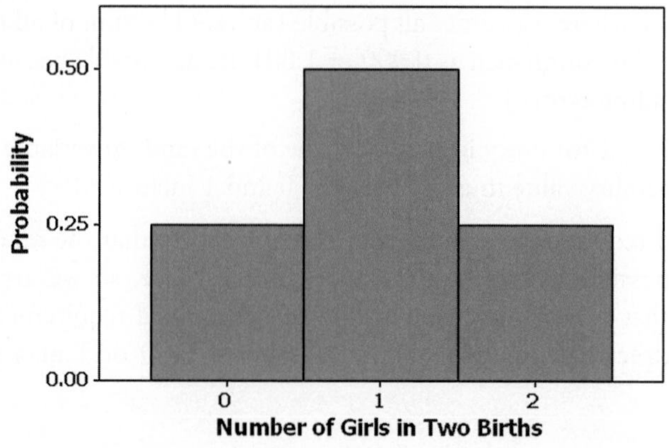

Figure 5-3 **Probability Histogram for Number of Girls in Two Births**

of these rectangles are the same as the *probabilities* in Table 5-1. We will see in Chapter 6 and future chapters that such a correspondence between area and probability is very useful.

Example 2 **Marijuana Survey**

In a Pew Research Center poll, subjects were asked if the use of marijuana should be made legal, and the results from that poll have been used to create Table 5-2. Does Table 5-2 describe a probability distribution?

Table 5-2 Should Marijuana Use Be Legal?

Response	$P(x)$
Yes	0.41
No	0.52
Don't know	0.07

Solution

Consider the three requirements listed earlier.

1. The responses (yes, no, don't know) are not numerical, so we do not have a numerical random variable. The first requirement is violated, so Table 5-2 does not describe a probability distribution.
2. The sum of the probabilities is 1.
3. Each probability is between 0 and 1 inclusive.

Because one of the three requirements is not met, Table 5-2 does not describe a probability distribution.

Example 3 **When to Discuss Salary**

Senior executives were asked when job applicants should discuss salary, and Table 5-3 is based on their responses (based on data from an Accountemps survey). Does Table 5-3 describe a probability distribution?

Table 5-3 When to Discuss Salary

Number of Interviews x	$P(x)$
1	0.30
2	0.26
3	0.10

Solution

To be a probability distribution, we must have a numerical random variable x such that $P(x)$ must satisfy the preceding three requirements.

1. The variable x is a numerical random variable and its values are associated with probabilities, as in Table 5-3.
2. $\Sigma P(x) = P(1) + P(2) + P(3)$
   $= 0.30 + 0.26 + 0.10$
   $= 0.66$ [showing that $\Sigma P(x) \neq 1$]
3. Each value of $P(x)$ is between 0 and 1 inclusive.

Because the second requirement is not satisfied, we conclude that Table 5-3 does *not* describe a probability distribution.

Example 4

Does $P(x) = \dfrac{x}{3}$ (where x can be 0, 1, or 2) determine a probability distribution?

Solution

To be a probability distribution, we must have a numerical random variable x such that $P(x)$ must satisfy the preceding three requirements. From the given formula we find that $P(0) = 0/3$ and $P(1) = 1/3$ and $P(2) = 2/3$.

continued

1. The variable x is a numerical random variable and its values (0, 1, 2) are associated with probabilities, as determined by the given formula.
2. $\Sigma P(x) = P(0) + P(1) + P(2) = \frac{0}{3} + \frac{1}{3} + \frac{2}{3} = 1$
3. Each value of $P(x)$ is between 0 and 1 inclusive.

Because the three requirements are satisfied, we conclude that the given formula does describe a probability distribution.

Life Data Analysis

Life data analysis deals with the longevity and failure rates of manufactured products. In one application, it is known that Dell computers have an "infant mortality" rate, whereby the failure rate is highest immediately after the computers are produced. Dell therefore tests or "burns-in" the computers before they are shipped. Dell can optimize profits by using an optimal burn-in time that identifies failures without wasting valuable testing time. Other products, such as cars, have failure rates that increase over time as parts wear out. If General Motors or Dell or any other company were to ignore the use of statistics and life data analysis, it would run the serious risk of going out of business because of factors such as excessive warranty repair costs or the loss of customers who experience unacceptable failure rates.

The *Weibull distribution* is a probability distribution commonly used in life data analysis applications. That distribution is beyond the scope of this book.

Parameters of a Probability Distribution: Mean, Variance, and Standard Deviation

Remember that with a probability distribution, we have a description of a *population* instead of a sample, so the values of the mean, standard deviation, and variance are *parameters* instead of statistics. The mean is the central or "average" value of the random variable. The variance and standard deviation measure the variation of the random variable. These parameters can be found with the following formulas:

Formula 5-1 $\mu = \Sigma[x \cdot P(x)]$

Mean for a probability distribution

Formula 5-2 $\sigma^2 = \Sigma[(x - \mu)^2 \cdot P(x)]$

Variance for a probability distribution (This format is easier to understand.)

Formula 5-3 $\sigma^2 = \Sigma[x^2 \cdot P(x)] - \mu^2$

Variance for a probability distribution (This format is easier for manual computations.)

Formula 5-4 $\sigma = \sqrt{\Sigma[x^2 \cdot P(x)] - \mu^2}$

Standard deviation for a probability distribution

When applying Formulas 5-1 through 5-4, use the following rule for rounding results.

Round-off Rule for μ, σ, and σ^2 from a Probability Distribution

Round results by carrying one more decimal place than the number of decimal places used for the random variable x. If the values of x are integers, round μ, σ, and σ^2 to one decimal place.

Note to Instructor
Remind your students that they should round only at the end of a computation, not at middle stages. Rounding in the middle can create errors that grow very large when used in subsequent steps.

It is sometimes necessary to use a different rounding rule because of special circumstances, such as results that require more decimal places to be meaningful. For example, with four-engine jets the mean number of jet engines working successfully throughout a flight is 3.999714286, which becomes 4.0 when rounded to one more decimal place than the original data. Here, 4.0 would be misleading because it suggests that all jet engines always work successfully. We need more precision to correctly reflect the true mean, such as the precision in the number 3.999714.

Expected Value

The mean of a discrete random variable x is the theoretical mean outcome for infinitely many trials. We can think of that mean as the *expected value* in the sense that it is the average value that we would expect to get if the trials could continue indefinitely. The uses of expected value (also called *expectation,* or *mathematical expectation*) are extensive and varied, and they play an important role in *decision theory.* (See Example 8 in Part 2 of this section.)

DEFINITION The **expected value** of a discrete random variable x is denoted by E, and it is the mean value of the outcomes, so $E = \mu$ and E can also be found by evaluating $\Sigma[x \cdot P(x)]$, as in Formula 5-1.

CAUTION An expected value need not be a whole number, even if the different possible values of x might all be whole numbers. We say that the expected number of girls in five births is 2.5, even though five specific births can never result in 2.5 girls. If we were to survey many couples with five children, we *expect* that the mean number of girls will be 2.5.

Example 5 Finding the Mean, Variance, and Standard Deviation

Table 5-1 describes the probability distribution for the number of girls in two births. Find the mean, variance, and standard deviation for the probability distribution described in Table 5-1 from Example 1.

Solution

In Table 5-4, the two columns at the left describe the probability distribution given earlier in Table 5-1; we create the three columns at the right for the purposes of the calculations required.

Using Formulas 5-1 and 5-2 and the table results, we get

Mean: $\mu = \Sigma[x \cdot P(x)] = 1.0$

Variance: $\sigma^2 = \Sigma[(x - \mu)^2 \cdot P(x)] = 0.5$

The standard deviation is the square root of the variance, so

Standard deviation: $\sigma = \sqrt{0.5} = 0.707107 = 0.7$ (rounded)

Table 5-4 Calculating μ and σ for a Probability Distribution

x	P(x)	$x \cdot P(x)$	$(x - \mu)^2 \cdot P(x)$
0	0.25	$0 \cdot 0.25 = 0.00$	$(0 - 1)^2 \cdot 0.25 = 0.25$
1	0.50	$1 \cdot 0.50 = 0.50$	$(1 - 1)^2 \cdot 0.50 = 0.00$
2	0.25	$2 \cdot 0.25 = 0.50$	$(2 - 1)^2 \cdot 0.25 = 0.25$
Total		1.00 ↑ $\mu = \Sigma[x \cdot P(x)]$	0.50 ↑ $\sigma^2 = \Sigma[(x - \mu)^2 \cdot P(x)]$

continued

Interpretation

The mean number of girls in two births is 1.0 girl, the variance is 0.50 "girl squared," and the standard deviation is 0.7 girl. Also, the expected value for the number of girls in two births is 1.0 girl, which is the same value as the mean. If we were to collect data on a large number of couples with two children, we expect to get a mean of 1.0 girl.

Making Sense of Results: Identifying Unusual Values

We present the following two different approaches for determining whether a value of a random variable x is unusually low or unusually high.

Note to Instructor
Stress that obtaining some number for an answer is not the ultimate goal. We want to *interpret* results so that they can be applied in a meaningful way. This process of determining whether events are "unusual" is extremely important in the subject of statistics, and it forms the foundation for the important methods of hypothesis testing introduced in Chapter 8.

Identifying Unusual Results with the Range Rule of Thumb

The range rule of thumb (introduced in Section 3-3) may be helpful in interpreting the value of a standard deviation. According to the range rule of thumb, the vast majority of values should lie within 2 standard deviations of the mean, so we can consider a value to be unusual if it is more than 2 standard deviations away from the mean. (The use of 2 standard deviations is not an absolutely rigid value, and other values such as 3 could be used instead.) We can therefore identify "unusual" values by determining that they lie outside of these limits:

Range Rule of Thumb

$$\textbf{maximum usual value} = \mu + 2\sigma$$

$$\textbf{minimum usual value} = \mu - 2\sigma$$

CAUTION Know that the use of the number 2 in the range rule of thumb is somewhat arbitrary, and this rule is a guideline, not an absolutely rigid rule.

Example 6 Identifying Unusual Results with the Range Rule of Thumb

In Example 5 we found that for families with two children, the mean number of girls is 1.0 and the standard deviation is 0.7 girl. Use those results and the range rule of thumb to find the maximum and minimum usual values for the number of girls. Based on the results, if a couple has two children, is 2 girls an unusually high number of girls?

Solution

Using the range rule of thumb, we can find the maximum and minimum usual values as follows:

maximum usual value: $\mu + 2\sigma = 1.0 + 2(0.7) = 2.4$

minimum usual value: $\mu - 2\sigma = 1.0 - 2(0.7) = -0.4$

Interpretation

Based on these results, we conclude that for families with two children, the number of girls should usually fall between −0.4 and 2.4, so 2 girls is not an unusually high number of girls (because 2 falls between −0.4 and 2.4). (In this case, the minimum usual value is actually 0, because that is the lowest possible number of girls, and the maximum usual value is 2 because that is the highest possible number of girls.)

Identifying Unusual Results with Probabilities: The Rare Event Rule for Inferential Statistics

If, under a given assumption (such as the assumption that boys and girls have the same chance of being born), the probability of a particular observed event (such as 879 girls in 945 births) is extremely small, we conclude that the assumption is probably not correct.

Probabilities can be used to apply the rare event rule as follows:

Using Probabilities to Determine When Results Are Unusual

- **Unusually *high* number of successes:** x successes among n trials is an *unusually high* number of successes if the probability of x or more successes is unlikely with a probability of 0.05 or less. This criterion can be expressed as follows: $P(x \text{ or more}) \leq 0.05$.*
- **Unusually *low* number of successes:** x successes among n trials is an *unusually low* number of successes if the probability of x or fewer successes is unlikely with a probability of 0.05 or less. That is, x is an unusually low number of successes if $P(x \text{ or fewer}) \leq 0.05$.*

*The value 0.05 is not absolutely rigid. Other values, such as 0.01, could be used to distinguish between results that can easily occur by chance and events that are very unlikely to occur by chance.

CAUTION The above criteria are not exactly equivalent to those used for the range rule of thumb. Chapter 8 introduces more formal and exact methods that will replace the above criteria and the range rule of thumb.

Unusually High or Unusually Low Number of Successes: Not *Exactly,* but "At Least as Extreme"

In Section 4-2 we stated that a number of successes is unusually low or unusually high if that number is far from what we typically expect. We will now refine that description by using a probability value to judge whether a particular number of successes among n trials is unusually high or unusually low. Use the probability of getting *x or any other values that are more extreme;* do *not* use only the probability of *exactly* x successes. For example, suppose you were tossing a coin to determine whether it favors heads, and suppose 1000 tosses resulted in 501 heads. We can use methods from Section 5-3 to find the following two probabilities:

- P(*exactly* 501 heads in 1000 tosses) = 0.0252 (so exactly 501 heads is *unlikely*)
- P(501 *or more* heads in 1000 tosses) = 0.487 (so 501 heads is *not unusually high*)

We use the second of the preceding two probabilities for determining whether 501 heads in 1000 tosses is an unusually high number of heads. (We do not base our

Note to Instructor
Strongly consider elaborating on this discussion, which is very important and provides a great foundation for hypothesis testing in Chapter 8. Consider using another example in class, such as this one: A question on an SAT test has the five possible answers of a, b, c, d, e. If answers from 1000 students are analyzed, about how many correct responses are expected if we assume that the question is so difficult that all students make random guesses? (Answer: About 200.) Given that P(exactly 205 correct) = 0.0289, is 205 correct answers an *unlikely* outcome? (Answer: Yes.) Is 205 correct answers an *unusually high* number of correct answers? [Answer: No, because P(205 or more correct responses) = 0.358.]

answer on the probability of *exactly* 501 heads, which is the small value of 0.0252. In this situation, any specific number of heads will have a very low probability.) Based on the 0.487 probability of getting 501 heads or more, we conclude that 501 heads in 1000 tosses of a fair coin is not an unusually high number of heads. (See Example 11 in Section 4-2.)

Example 7 Identifying Unusual Results with Probabilities

The Chapter Problem includes results consisting of 879 girls in 945 births. Is 879 girls in 945 births an unusually high number of girls? What does it suggest about the effectiveness of the XSORT method of gender selection?

Solution

The result of 879 girls in 945 births is more than we expect under normal circumstances, so we want to determine whether 879 girls is *unusually high.* Here, the relevant probability is the probability of getting *879 or more* girls in 945 births. Using methods covered later in Section 5-3, we can find that P(879 or more girls in 945 births) $= 0.000$ when rounded to three decimal places. (We can denote such a probability as 0+.) Because the probability of getting 879 or more girls is less than or equal to 0.05, we conclude that 879 girls in 945 births is an unusually high number of girls.

Interpretation

Because it is so unlikely that we would get 879 or more girls in 945 births by chance, these results suggest that the XSORT method of gender selection is effective in increasing the likelihood that a baby will be a girl. (However, this does not *prove* that the XSORT method is responsible for the large number of girls.)

Note to Instructor
If you are pressed for time, Part 2 can be omitted.

Part 2: Expected Value in Decision Theory and Rationale for Formulas 5-1 through 5-4

Expected Value in Decision Theory

In Part 1 of this section we noted that the expected value of a random variable x is equal to the mean μ. We can therefore find the expected value by computing $\Sigma[x \cdot P(x)]$, just as we do for finding the value of μ. We also noted that the concept of expected value is used in *decision theory.* In Example 8 we illustrate this use of expected value with a situation in which we must choose between two different bets. Example 8 involves a real and practical decision.

Example 8 Be a Better Bettor

You have $5 to place on a bet in the Golden Nugget casino in Las Vegas. You have narrowed your choice to one of two bets:

Roulette: Bet on the number 7 in roulette.

Craps: Bet on the "pass line" in the dice game of craps.

a. If you bet \$5 on the number 7 in roulette, the probability of losing \$5 is 37/38 and the probability of making a net gain of \$175 is 1/38. (The prize is \$180, including your \$5 bet, so the net gain is \$175.) Find your expected value if you bet \$5 on the number 7 in roulette.

b. If you bet \$5 on the pass line in the dice game of craps, the probability of losing \$5 is 251/495 and the probability of making a net gain of \$5 is 244/495. (If you bet \$5 on the pass line and win, you are given \$10 that includes your bet, so the net gain is \$5.) Find your expected value if you bet \$5 on the pass line.

Which of the preceding two bets is better? Why?

Solution

a. Roulette The probabilities and payoffs for betting \$5 on the number 7 in roulette are summarized in Table 5-5. Table 5-5 also shows that the expected value is $\Sigma[x \cdot P(x)] = -26¢$. That is, for every \$5 bet on the number 7, you can expect to *lose* an average of 26¢.

Table 5-5 Roulette

Event	x	P(x)	x · P(x)
Lose	−\$5	37/38	−\$4.868421
Win (net gain)	\$175	1/38	\$4.605263
Total			−\$0.26 (rounded) (or −26¢)

b. Dice The probabilities and payoffs for betting \$5 on the pass line in craps are summarized in Table 5-6. Table 5-6 also shows that the expected value is $\Sigma[x \cdot P(x)] = -7¢$. That is, for every \$5 bet on the pass line, you can expect to lose an average of 7¢.

Table 5-6 Dice

Event	x	P(x)	x · P(x)
Lose	−\$5	251/495	−\$2.535353
Win (net gain)	\$5	244/495	\$2.464646
Total			−\$0.07 (rounded) (or −7¢)

Interpretation

The \$5 bet in roulette results in an expected value of −26¢ and the \$5 bet in craps results in an expected value of −7¢. Because you are better off losing 7¢ instead of losing 26¢, the craps game is better in the long run, even though the roulette game provides an opportunity for a larger payoff.

Rationale for Formulas 5-1 through 5-4

Instead of blindly accepting and using formulas, it is much better to have some understanding of why they work. When computing the mean from a frequency distribution, f represents class frequency and N represents population size. In the expression below, we rewrite the formula for the mean of a frequency table so that it applies to a population. In the fraction f/N, the value of f is the frequency with which the value x occurs and N is the population size, so f/N is the probability for the value of x. When we replace f/N with $P(x)$, we make the transition from relative frequency based on a limited number of observations to probability based on infinitely many trials. This result shows why Formula 5-1 is as given earlier in this section.

$$\mu = \frac{\Sigma(f \cdot x)}{N} = \Sigma\left[\frac{f \cdot x}{N}\right] = \Sigma\left[x \cdot \frac{f}{N}\right] = \Sigma[x \cdot P(x)]$$

Similar reasoning enables us to take the variance formula from Chapter 3 and apply it to a random variable for a probability distribution; the result is Formula 5-2. Formula 5-3 is a shortcut version that will always produce the same result as Formula 5-2. Although Formula 5-3 is usually easier to work with, Formula 5-2 is easier to understand directly. Based on Formula 5-2, we can express the standard deviation as

$$\sigma = \sqrt{\Sigma[(x - \mu)^2 \cdot P(x)]}$$

or as the equivalent form given in Formula 5-4.

Recommended Assignment
Exercises 1–18

1. The random variable is x, which is the number of girls in three births. The possible values of x are 0, 1, 2, and 3. The values of the random value x are numerical.

2. The random variable is discrete because the number of possible values is 4, and 4 is a finite number. The random variable is discrete if it has a finite number of values or a countable number of values.

3. Table 5-7 does describe a probability distribution because the three requirements are satisfied. First, the variable x is a numerical random variable and its values are associated with probabilities. Second, $\Sigma P(x) = 0.125 + 0.375 + 0.375 + 0.125 = 1$ as required. Third, each of the probabilities is between 0 and 1 inclusive, as required.

4. a. Yes (because $0.0208 \leq 0.05$).

 b. No (because $0.089 > 0.05$).

5. a. Continuous random variable.

 b. Discrete random variable.

 c. Not a random variable.

 d. Discrete random variable.

 e. Continuous random variable.

 f. Discrete random variable.

5-2 Basic Skills and Concepts

Statistical Literacy and Critical Thinking

1. Random Variable Table 5-7 lists probabilities for the corresponding numbers of girls in three births. What is the random variable, what are its possible values, and are its values numerical?

2. Discrete or Continuous? Is the random variable given in Table 5-7 discrete or continuous? Explain.

3. Probability Distribution Does Table 5-7 describe a probability distribution? Show how the requirements are satisfied or are not satisfied.

4. Unusual For 200 births, the probability of exactly 90 girls is 0.0208 and the probability of 90 or fewer girls is 0.089.

a. Is exactly 90 girls in 200 births unlikely?

b. Among 200 births, is 90 girls an unusually low number of girls?

Identifying Discrete and Continuous Random Variables. *In Exercises 5 and 6, identify the given values as a* **discrete random variable, continuous random variable,** *or* **not a random variable.**

5. a. Exact weights of quarters now in circulation in the United States

b. Numbers of tosses of quarters required to get heads

c. Responses to the survey question "Did you smoke at least one cigarette in the last week?"

d. Numbers of spins of roulette wheels required to get the number 7

Table 5-7 Number of Girls in Three Births

Number of Girls x	$P(x)$
0	0.125
1	0.375
2	0.375
3	0.125

e. Exact foot lengths of humans

f. Shoe sizes (such as 8 or 8½) of humans

6. a. Eye colors of humans on commercial aircraft flights

b. Weights of humans on commercial aircraft flights

c. Numbers of passengers on commercial aircraft flights

d. Numbers of randomly generated digits before getting the digit 3

e. Political party affiliations of adults in the United States

f. Exact costs of presidential campaigns

6. a. Not a random variable.
b. Continuous random variable.
c. Discrete random variable.
d. Discrete random variable.
e. Not a random variable.
f. Discrete random variable.

Identifying Probability Distributions. ***In Exercises 7–14, determine whether a probability distribution is given. If a probability distribution is given, find its mean and standard deviation. If a probability distribution is not given, identify the requirements that are not satisfied.***

7. Genetic Disorder Four males with an X-linked genetic disorder have one child each. The random variable x is the number of children among the four who inherit the X-linked genetic disorder.

x	P(x)
0	0.0625
1	0.2500
2	0.3750
3	0.2500
4	0.0625

7. Probability distribution with $\mu = 2.0$, $\sigma = 1.0$.

8. Male Color Blindness When conducting research on color blindness in males, a researcher forms random groups with five males in each group. The random variable x is the number of males in the group who have a form of color blindness (based on data from the National Institutes of Health).

x	P(x)
0	0.659
1	0.287
2	0.050
3	0.004
4	0.001
5	0+

8. Probability distribution with $\mu = 0.4$, $\sigma = 0.6$. (The sum of the probabilities is 1.001, but that is due to rounding errors.)

9. Pickup Line Ted is not particularly creative. He uses this pickup line: "If I could rearrange the alphabet, I'd put U and I together." The random variable x is the number of girls Ted approaches before encountering one who reacts positively.

x	P(x)
1	0.001
2	0.020
3	0.105
4	0.233
5	0.242

9. Not a probability distribution because the sum of the probabilities is 0.601, which is not 1 as required. Also, Ted clearly needs a new approach.

10. Fun Ways to Flirt In a Microsoft Instant Messaging survey, respondents were asked to choose the most fun way to flirt, and the accompanying table is based on the results.

	P(x)
E-mail	0.06
In person	0.55
Instant message	0.24
Text message	0.15

10. Not a probability distribution because the responses are not values of a numerical random variable.

11. Fun Ways to Flirt A sociologist randomly selects single adults for different groups of four, and the random variable x is the number in the group who say that the most fun way to flirt is in person (based on a Microsoft Instant Messaging survey).

x	P(x)
0	0.041
1	0.200
2	0.367
3	0.299
4	0.092

11. Probability distribution with $\mu = 2.2$, $\sigma = 1.0$. (The sum of the probabilities is 0.999, but that is due to rounding errors.)

12. Probability distribution with $\mu = 4.6$, $\sigma = 1.0$. (The sum of the probabilities is 0.999, but that is due to rounding errors.)

12. Happiness Groups of people aged 15–65 are randomly selected and arranged in groups of six. The random variable x is the number in the group who say that their family and/or partner contribute most to their happiness (based on a Coca-Cola survey).

x	P(x)
0	0+
1	0.003
2	0.025
3	0.111
4	0.279
5	0.373
6	0.208

13. Not a probability distribution because the responses are not values of a numerical random variable. Also, the sum of the probabilities is 1.18 instead of being 1 as required.

13. Happiness In a survey sponsored by Coca-Cola, subjects aged 15–65 were asked what contributes most to their happiness, and the table is based on their responses.

	P(x)
Family/partner	0.77
Friends	0.15
Work/studies	0.08
Leisure	0.08
Music	0.06
Sports	0.04

14. Not a probability distribution because the sum of the probabilities is 0.967, which is not 1 as required. The discrepancy between 0.967 and 1 is too large to attribute to rounding errors.

14. Casino Games When betting on the pass line in the dice game of craps at the Mohegan Sun casino in Connecticut, the table lists the probabilities for the number of bets that must be placed in order to have a win.

x	P(x)
1	0.493
2	0.250
3	0.127
4	0.064
5	0.033

Genetics. ***In Exercises 15–18, refer to the accompanying table, which describes results from groups of 10 births from 10 different sets of parents. The random variable x represents the number of girls among 10 children.***

Number of Girls x	P(x)
0	0.001
1	0.010
2	0.044
3	0.117
4	0.205
5	0.246
6	0.205
7	0.117
8	0.044
9	0.010
10	0.001

15. $\mu = 5.0$, $\sigma = 1.6$.

15. Mean and Standard Deviation Find the mean and standard deviation for the numbers of girls in 10 births.

16. 1.8 girls to 8.2 girls; yes, 1 girl is an unusually low number of girls, because 1 girl is outside of the range of usual values.

16. Range Rule of Thumb for Unusual Events Use the range rule of thumb to identify a range of values containing the usual numbers of girls in 10 births. Based on the result, is 1 girl in 10 births an unusually low number of girls? Explain.

17. a. 0.044
b. 0.055
c. The probability from part (b).
d. No, because the probability of 8 or more girls is 0.055, which is not very low (less than or equal to 0.05).

17. Using Probabilities for Unusual Events

a. Find the probability of getting exactly 8 girls in 10 births.

b. Find the probability of getting 8 or more girls in 10 births.

c. Which probability is relevant for determining whether 8 is an unusually high number of girls in 10 births: the result from part (a) or part (b)?

d. Is 8 an unusually high number of girls in 10 births? Why or why not?

18. a. 0.010
b. 0.011
c. Part (b)
d. Yes, because the probability of 0.011 is very low (less than or equal to 0.05).

18. Using Probabilities for Unusual Events

a. Find the probability of getting exactly 1 girl in 10 births.

b. Find the probability of getting 1 or fewer girls in 10 births.

c. Which probability is relevant for determining whether 1 is an unusually low number of girls in 10 births: the result from part (a) or part (b)?

d. Is 1 an unusually low number of girls in 10 births? Why or why not?

Car Failures. *In Exercises 19–22, refer to the accompanying table, which describes results of roadworthiness tests of Ford Focus cars that are 3 years old (based on data from the Department of Transportation). The random variable x represents the number of cars that failed among six that were tested for roadworthiness.*

x	P(x)
0	0.377
1	0.399
2	0.176
3	0.041
4	0.005
5	0+
6	0+

19. Mean and Standard Deviation Find the mean and standard deviation for the numbers of cars that failed among the six cars that are tested.

19. $\mu = 0.9$ car, $\sigma = 0.9$ car

20. Range Rule of Thumb for Unusual Events Use the range rule of thumb with the results from Exercise 19 to identify a range of values containing the usual numbers of car failures among six cars tested. Based on the result, is three an unusually high number of failures among six cars tested? Explain.

20. −0.9 to 2.7; yes, 3 is above the range of usual values, so 3 is an unusually high number of failures among 6 cars tested.

21. Using Probabilities for Unusual Events

a. Find the probability of getting exactly three cars that fail among six cars tested.

b. Find the probability of getting three or more cars that fail among six cars tested.

c. Which probability is relevant for determining whether three is an unusually high number of failures among six cars tested: the result from part (a) or part (b)?

d. Is three an unusually high number of failures? Why or why not?

21. a. 0.041
b. 0.046
c. The probability from part (b).
d. Yes, because the probability of three or more failures is 0.046, which is very low (less than or equal to 0.05).

22. Using Probabilities for Unusual Events

a. Find the probability of getting exactly one car that fails among six cars tested.

b. Find the probability of getting one or fewer cars that fail among six cars tested.

c. Which probability is relevant for determining whether one is an unusually low number of cars that fail among six cars tested: the result from part (a) or part (b)?

d. Is one an unusually low number of cars that fail among six cars tested? Why or why not?

22. a. 0.399
b. 0.776
c. Part (b)
d. No, because the probability of 0.776 is not very low (less than or equal to 0.05).

5-2 Beyond the Basics

23. Expected Value for the Texas Pick 3 Game In the Texas Pick 3 lottery, you can bet \$1 by selecting three digits, each between 0 and 9 inclusive. If the same three numbers are drawn in the same order, you win and collect \$500.

a. How many different selections are possible?

b. What is the probability of winning?

c. If you win, what is your net profit?

d. Find the expected value.

e. If you bet \$1 on the pass line in the casino dice game of craps, the expected value is −1.4¢. Which bet is better: a \$1 bet in the Texas Pick 3 game or a \$1 bet on the pass line in craps? Explain.

23. a. 1000
b. 1/1000
c. \$499
d. −50¢
e. The \$1 bet on the pass line in craps is better because its expected value of −1.4¢ is much greater than the expected value of −50¢ for the Texas Pick 3 lottery.

24. Expected Value in Maine's Pick 4 Game In Maine's Pick 4 lottery game, you can pay \$1 to select a sequence of four digits, such as 1332. If you select the same sequence of four digits that are drawn, you win and collect \$5000.

a. How many different selections are possible?

b. What is the probability of winning?

24. a. 10,000
b. 1/10,000
c. \$4999
d. −50¢
e. Because both bets have the same expected value of −50¢, neither bet is better than the other.

c. If you win, what is your net profit? **d.** Find the expected value.

e. If you bet $1 in Maine's Pick 3 game, the expected value is −50¢. Which bet is better: A $1 bet in the Maine Pick 3 game or a $1 bet in the Maine Pick 4 game? Explain.

25. a. −39¢
b. The bet on the number 27 is better because its expected value of −26¢ is greater than the expected value of −39¢ for the other bet.

25. Expected Value in Roulette When playing roulette at the Venetian casino in Las Vegas, a gambler is trying to decide whether to bet $5 on the number 27 or to bet $5 that the outcome is any one of these five possibilities: 0, 00, 1, 2, 3. From Example 8, we know that the expected value of the $5 bet for a single number is −26¢. For the $5 bet that the outcome is 0, 00, 1, 2, or 3, there is a probability of 5/38 of making a net profit of $30 and a 33/38 probability of losing $5.

a. Find the expected value for the $5 bet that the outcome is 0, 00, 1, 2, or 3.

b. Which bet is better: a $5 bet on the number 27 or a $5 bet that the outcome is 0, 00, 1, 2, or 3? Why?

26. a. $131,477.54
b. $253,584.47
c. −$375,691.40 to $638,646.48
d. Yes, because the values are above the range of usual values given in part (c).

26. Expected Value for *Deal or No Deal* The television game show *Deal or No Deal* begins with individual suitcases containing the amounts of 1¢, $1, $5, $10, $25, $50, $75, $100, $200, $300, $400, $500, $750, $1000, $5000, $10,000, $25,000, $50,000, $75,000, $100,000, $200,000, $300,000, $400,000, $500,000, $750,000, and $1,000,000. If a player adopts the strategy of choosing the option of "no deal" until one suitcase remains, the payoff is one of the amounts listed, and they are all equally likely.

a. Find the expected value for this strategy. **b.** Find the value of the standard deviation.

c. Use the range rule of thumb to identify the range of usual outcomes.

d. Based on the preceding results, is a result of $750,000 or $1,000,000 unusually high? Why or why not?

TOP 20

5-3 Binomial Probability Distributions

Note to Instructor
Comment that Section 5-2 presented the concept of a discrete probability distribution, and the binomial distribution is one special category of such discrete probability distributions. In addition to the binomial distribution, there are Poisson, geometric, hypergeometric, and other discrete probability distributions that have not been given special names.
Recommendation: Consider showing the role of the binomial probability formula, but encourage the use of technology for finding binomial probabilities.

Key Concept Section 5-2 introduced the important concept of a discrete probability distribution. Among all of the different types of discrete probability distributions that exist, there are a few that are particularly important, and the focus of this section is the type that we call *binomial* probability distributions. We begin with a basic definition of a binomial probability distribution, along with notation and methods for finding probability values. As in other sections, we stress the importance of *interpreting* probability values to determine whether events are unlikely (with a low probability, such as 0.05 or less) or unusually high or low.

Binomial probability distributions allow us to deal with circumstances in which the outcomes belong to *two* relevant categories, such as acceptable/defective or survived/died. Other requirements are given in the following definition.

DEFINITION A **binomial probability distribution** results from a procedure that meets all the following requirements:

1. The procedure has a *fixed number of trials.* (A trial is a single observation.)
2. The trials must be *independent.* (The outcome of any individual trial doesn't affect the probabilities in the other trials.)
3. Each trial must have all outcomes classified into *two categories* (commonly referred to as *success* and *failure*).
4. The probability of a success remains the same in all trials.

Independence Requirement

When selecting a sample (such as survey subjects) for some statistical analysis, we usually sample without replacement. Sampling without replacement involves dependent events, which violates the second requirement in the definition on the previous page. However, we can often assume independence by applying the following 5% guideline introduced in Section 4-4:

Treating Dependent Events as Independent:
The 5% Guideline for Cumbersome Calculations

If calculations are cumbersome and if a sample size is no more than 5% of the size of the population, treat the selections as being *independent* (even if the selections are made without replacement, so that they are actually dependent).

If a procedure satisfies the four requirements on the previous page, the distribution of the random variable x (number of successes in n trials) is called a *binomial probability distribution* (or *binomial distribution*). The following notation is commonly used.

Notation for Binomial Probability Distributions

S and F (success and failure) denote the two possible categories of all outcomes.

$P(S) = p$	(p = probability of a success)
$P(F) = 1 - p = q$	(q = probability of a failure)
n	denotes the fixed number of trials.
x	denotes a specific number of successes in n trials, so x can be any whole number between 0 and n, inclusive.
p	denotes the probability of *success* in *one* of the n trials.
q	denotes the probability of *failure* in *one* of the n trials.
$P(x)$	denotes the probability of getting exactly x successes among the n trials.

The word *success* as used here is arbitrary and does not necessarily represent something good. Either of the two possible categories may be called the success S as long as its probability is identified as p. (The value of q can always be found by subtracting p from 1: If $p = 0.95$, then $q = 1 - 0.95 = 0.05$.)

CAUTION When using a binomial probability distribution, always be sure that x and p are consistent in the sense that they both refer to the *same* category being called a success.

Example 1 **Twitter**

When an adult is randomly selected (with replacement), there is a 0.85 probability that this person knows what Twitter is (based on results from a Pew Research Center survey). Suppose that we want to find the probability that exactly three of five random adults know what Twitter is.

a. Does this procedure result in a binomial distribution?

b. If this procedure does result in a binomial distribution, identify the values of n, x, p, and q.

Not at Home

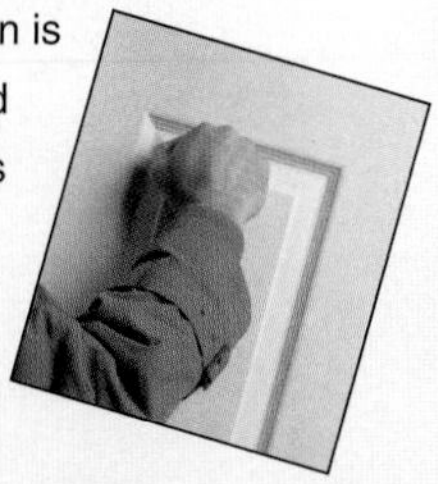

Pollsters cannot simply ignore those who were not at home when they were called the first time. One solution is to make repeated callback attempts until the person can be reached. Alfred Politz and Willard Simmons describe a way to compensate for those missing results without making repeated callbacks. They suggest weighting results based on how often people are not at home. For example, a person at home only two days out of six will have a 2/6 or 1/3 probability of being at home when called the first time. When such a person is reached the first time, his or her results are weighted to count three times as much as someone who is always home. This weighting is a compensation for the other similar people who are home two days out of six and were not at home when called the first time. This clever solution was first presented in 1949.

Note to Instructor

Strongly emphasize that x and p must be consistent in the sense that they must both refer to the same outcome. Because x counts *successes* and p is the probability of *success*, x and p must both refer to the *same* outcome. A common error is to have x count one category of outcome while p is the probability of the other category of outcome. Students also have some difficulty with the probability p; strongly emphasize that p is the probability of getting a success on just *one* individual trial.

Solution

a. This procedure does satisfy the requirements for a binomial distribution, as shown below.

1. The number of trials (5) is fixed.
2. The 5 trials are independent, because the probability of any adult knowing Twitter is not affected by results from other selected adults.
3. Each of the 5 trials has two categories of outcomes: The selected person knows what Twitter is or that person does not know what Twitter is.
4. For each randomly selected adult, there is a 0.85 probability that this person knows what Twitter is, and that probability remains the same for each of the five selected people.

b. Having concluded that the given procedure does result in a binomial distribution, we now proceed to identify the values of n, x, p, and q.

1. With five randomly selected adults, we have $n = 5$.
2. We want the probability of exactly three who know what Twitter is, so $x = 3$.
3. The probability of success (getting a person who knows what Twitter is) for one selection is 0.85, so $p = 0.85$.
4. The probability of failure (not getting someone who knows what Twitter is) is 0.15, so $q = 0.15$.

Again, it is very important to be sure that x and p both refer to the same concept of "success." In this example, we use x to count the number of people who know what Twitter is, so p must be the probability that the selected person knows what Twitter is. Therefore, x and p do use the same concept of success: knowing what Twitter is.

We now discuss three methods for finding the probabilities corresponding to the random variable x in a binomial distribution. The first method involves calculations using the *binomial probability formula* and is the basis for the other two methods. The second method involves the use of computer software or a calculator, and the third method involves the use of the Binomial Probabilities table in Table A-1 in Appendix A. (With technology so widespread, such tables are becoming obsolete.) If you are using computer software or a calculator that automatically produces binomial probabilities, we recommend that you solve one or two exercises using Method 1 to better understand the basis for the calculations. Understanding is always infinitely better than blind application of formulas.

Note to Instructor
There are three methods included here, but point out that students are not expected to know all three methods.
Recommendation: Don't require that students master calculations with the binomial probability formula. Instead, illustrate it in class, but encourage all students to find binomial probabilities using software or a calculator.

Method 1: Using the Binomial Probability Formula In a binomial probability distribution, probabilities can be calculated by using Formula 5-5.

Formula 5-5 Binomial Probability Formula

$$P(x) = \frac{n!}{(n-x)!x!} \cdot p^x \cdot q^{n-x} \qquad \text{for } x = 0, 1, 2, \ldots, n$$

where

n = number of trials
x = number of successes among n trials
p = probability of success in any one trial
q = probability of failure in any one trial $(q = 1 - p)$

The factorial symbol !, introduced in Section 4-6, denotes the product of decreasing factors. Two examples of factorials are $3! = 3 \cdot 2 \cdot 1 = 6$ and $0! = 1$ (by definition).

Example 2 Twitter

Given that there is a 0.85 probability that a randomly selected adult knows what Twitter is, use the binomial probability formula to find the probability of getting exactly three adults who know what Twitter is when five adults are randomly selected. That is, apply Formula 5-5 to find $P(3)$ given that $n = 5, x = 3, p = 0.85$, and $q = 0.15$.

Solution

Using the given values of n, x, p, and q in the binomial probability formula (Formula 5-5), we get

$$\begin{aligned} P(3) &= \frac{5!}{(5-3)!3!} \cdot 0.85^3 \cdot 0.15^{5-3} \\ &= \frac{5!}{2!3!} \cdot 0.614125 \cdot 0.0225 \\ &= (10)(0.614125)(0.0225) = 0.138178 \\ &= 0.138 \text{ (rounded)} \end{aligned}$$

The probability of getting exactly three adults who know Twitter among five randomly selected adults is 0.138.

Calculation hint: When computing a probability with the binomial probability formula, it's helpful to get a single number for $n!/[(n-x)!x!]$, a single number for p^x, and a single number for q^{n-x}, then simply multiply the three factors together as shown in the third line of the calculation in the preceding example. Don't round too much when you find those three factors; round only at the end.

Method 2: Using Technology STATDISK, Minitab, Excel, StatCrunch, SPSS, SAS, and the TI-83/84 Plus calculator are all technologies that can be used to find binomial probabilities. (Instead of directly providing probabilities for individual values of x, SPSS and SAS are more difficult to use because they provide *cumulative*

Note to Instructor
If students have a suitable calculator or computer software, they should use that technology, so Method 2 is the recommended approach. If students do not have suitable technology, stress the use of the computer software displays included in the book.

STATDISK

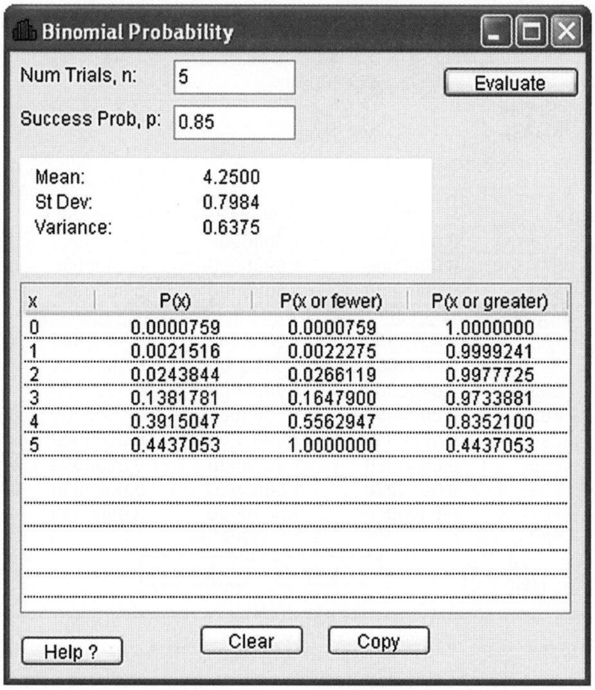

x	P(x)	P(x or fewer)	P(x or greater)
0	0.0000759	0.0000759	1.0000000
1	0.0021516	0.0022275	0.9999241
2	0.0243844	0.0266119	0.9977725
3	0.1381781	0.1647900	0.9733881
4	0.3915047	0.5562947	0.8352100
5	0.4437053	1.0000000	0.4437053

MINITAB

x	P(x)
0	0.000076
1	0.002152
2	0.024384
3	0.138178
4	0.391505
5	0.443705

EXCEL

	A	B
1	0	7.594E-05
2	1	0.0021516
3	2	0.0243844
4	3	0.1381781
5	4	0.3915047
6	5	0.4437053

TI-83/84 PLUS

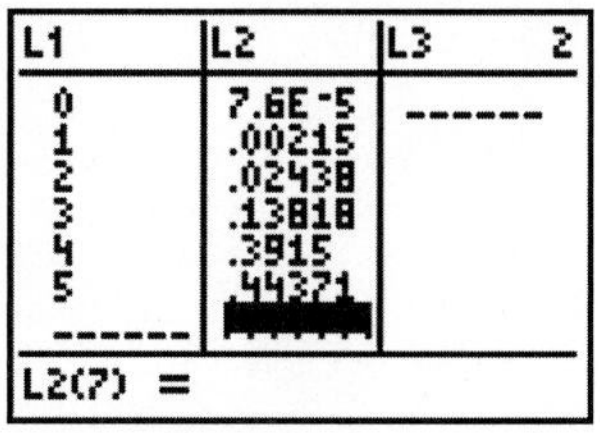

probabilities of x or fewer successes.) The screen displays list binomial probabilities for $n = 5$ and $p = 0.85$, as in Example 2. Notice that in each display, the probability distribution is given as a table. The Excel and TI-83/84 Plus display the first probability in scientific notation. (The Excel probability of 7.594E-05 is another way of expressing 7.594×10^{-5}, which can be expressed in a standard format as 0.00007594.)

Example 3 XSORT Method of Gender Selection

In the Chapter Problem, we noted that each of 945 couples gave birth to a child, and 879 of those children were girls. In order to determine whether such results can be explained by chance, we need to find the probability of getting 879 girls or a result at least as extreme as that. Find the probability of at least 879 girls in 945 births, assuming that boys and girls are equally likely.

Solution

Using the notation for binomial probabilities, we have $n = 945, p = 0.5, q = 0.5$, and we want to find the sum of all probabilities for each value of x from 879 through 945. The formula is not practical here, because we would need to apply it 67 times—that's not the way to go. Table A-1 (Binomial Probabilities) doesn't apply because $n = 945$, which is way beyond the scope of that table. Instead, we wisely choose to use technology.

The accompanying STATDISK display shows that the probability of 879 or more girls in 945 births is 0.0000000 when rounded to seven decimal places. This shows that it is *very* unlikely that we would get 879 girls in 945 births by chance. If we effectively rule out chance, we are left with the more reasonable explanation that the XSORT method appears to be effective in increasing the likelihood that a baby will be born a girl.

STATDISK

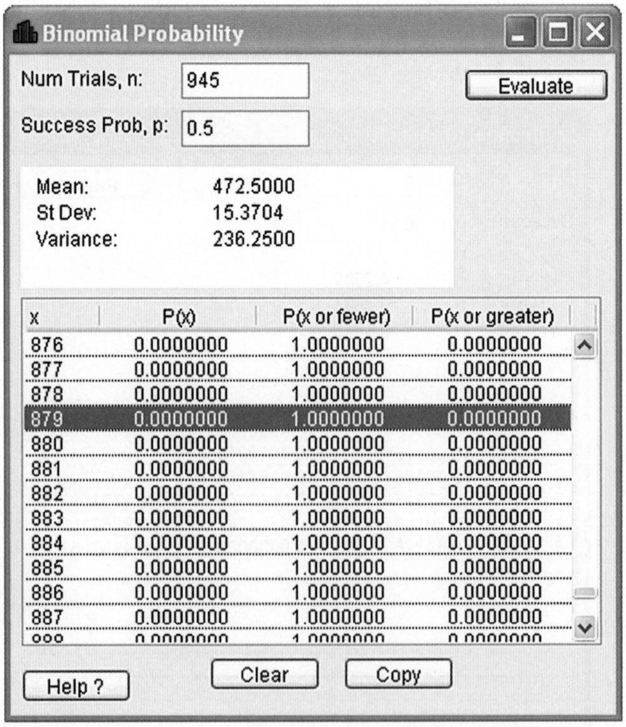

x	P(x)	P(x or fewer)	P(x or greater)
876	0.0000000	1.0000000	0.0000000
877	0.0000000	1.0000000	0.0000000
878	0.0000000	1.0000000	0.0000000
879	0.0000000	1.0000000	0.0000000
880	0.0000000	1.0000000	0.0000000
881	0.0000000	1.0000000	0.0000000
882	0.0000000	1.0000000	0.0000000
883	0.0000000	1.0000000	0.0000000
884	0.0000000	1.0000000	0.0000000
885	0.0000000	1.0000000	0.0000000
886	0.0000000	1.0000000	0.0000000
887	0.0000000	1.0000000	0.0000000

Example 3 illustrates well the power and ease of using technology. Example 3 also illustrates the rare event rule of statistical thinking: If under a given assumption (such as the assumption that the XSORT method has no effect), the probability of a particular observed event (such as 879 girls in 945 births) is extremely small (such as 0.05 or less), we conclude that the assumption is probably not correct.

Method 3: Using Table A-1 in Appendix A Table A-1 in Appendix A lists binomial probabilities for select values of n and p. It cannot be used for Example 2 because the probability of $p = 0.85$ is not one of the probabilities included. Example 3 illustrates the use of the table.

To use the table of binomial probabilities, we must first locate n and the desired corresponding value of x. At this stage, one row of numbers should be isolated. Now align that row with the desired probability of p by using the column across the top. The isolated number represents the desired probability. A very small probability, such as 0.000064, is indicated by 0+.

Note to Instructor

Change from the previous edition: In the previous edition of this book, Table A-1 (Binomial Probabilities) included values of n from $n = 2$ through $n = 16$, but this book includes values of n from $n = 2$ through $n = 8$, so Table A-1 now uses one page instead of three pages. This change is motivated by the shift to the use of technology in introductory statistics courses. If the technology is simply not available, Table A-1 can be used. Exercises and examples were designed for the values of n from 2 through 8. Also, several exercises and examples provide computer displays so students can use results from technology even if they don't have the actual technology.

Example 4 Devil of a Problem

Based on a recent Harris poll, 60% of adults believe in the devil. Assuming that we randomly select five adults, use Table A-1 to find the following:

a. The probability that exactly three of the five adults believe in the devil

b. The probability that the number of adults who believe in the devil is at least two

Solution

a. The following excerpt from the table shows that when $n = 5$ and $p = 0.6$, the probability for $x = 3$ is given by $P(3) = 0.346$.

TABLE A-1 Binomial Probabilities

n	x	.01	.50	.60	.70
5	0	.951	.031	.010	.002
	1	.048	.156	.077	.028
	2	.001	.312	.230	.132
	3	0+	.312	.346	.309
	4	0+	.156	.259	.360
	5	0+	.031	.078	.168

x	P(x)
0	0.010
1	0.077
2	0.230
3	0.346
4	0.259
5	0.078

b. The phrase "at least two" successes means that the number of successes is 2 or 3 or 4 or 5.

$$\begin{aligned} P(\text{at least 2 believe in the devil}) &= P(2 \text{ or } 3 \text{ or } 4 \text{ or } 5) \\ &= P(2) + P(3) + P(4) + P(5) \\ &= 0.230 + 0.346 + 0.259 + 0.078 \\ &= 0.913 \end{aligned}$$

If we wanted to use the binomial probability formula to find P(at least 2), as in part (b) of Example 3, we would need to apply the formula four times to compute four different probabilities, which would then be added. Given this choice between the formula and the table, it makes sense to use the table. However, given a choice that includes the formula, table, and technology, the use of technology is the way to go. Here is an effective and efficient strategy for choosing a method for finding binomial probabilities:

1. Use computer software or a T1-83/84 Plus calculator, if available.
2. If neither computer software nor a T1-83/84 Plus calculator is available, use Table A-1(Binomial Probabilities) if possible.
3. If neither computer software nor the T1-83/84 Plus calculator is available and the probabilities can't be found using Table A-1, use the binomial probability formula.

Rationale for the Binomial Probability Formula

The binomial probability formula is the basis for all three methods presented in this section. Instead of accepting and using that formula blindly, let's see why it works.

In Example 2, we used the binomial probability formula to find the probability of getting exactly three adults who know Twitter when five adults are randomly selected. With $P(\text{knows Twitter}) = 0.85$, we can use the multiplication rule from Section 4-4 to find the probability that the first three adults know Twitter while the last two adults do not know Twitter. We get the following result:

$$\begin{aligned} &P(\text{3 adults know Twitter followed by 2 adults who do not know Twitter}) \\ &\quad = 0.85 \cdot 0.85 \cdot 0.85 \cdot 0.15 \cdot 0.15 \\ &\quad = 0.85^3 \cdot 0.15^2 \\ &\quad = 0.0138 \end{aligned}$$

This result gives a probability of randomly selecting five adults and finding that the first three know Twitter and the last two do not. However, the probability of 0.0138 is not the probability of getting exactly three adults who know Twitter because it assumes a particular sequence for three people who know Twitter and two who do not. Other different sequences are possible.

In Section 4-6 we saw that with three subjects identical to each other (such as adults who know Twitter) and two other subjects identical to each other (such as adults who do not know Twitter), the total number of arrangements, or permutations, is $5!/[(5-3)!3!]$ or 10. Each of those 10 different arrangements has a probability of $0.85^3 \cdot 0.15^2$, so the total probability is as follows:

$$P(\text{3 adults know Twitter among 5}) = \frac{5!}{(5-3)!3!} \cdot 0.85^3 \cdot 0.15^2 = 0.138$$

This particular result can be generalized as the binomial probability formula (Formula 5-5). That is, the binomial probability formula is a combination of the multiplication rule of probability and the counting rule for the number of arrangements of n items when x of them are identical to each other and the other $n - x$ are identical to each other. (See Exercises 13 and 14.)

$$P(x) = \overbrace{\frac{n!}{(n-x)!x!}}^{\text{The number of outcomes with exactly } x \text{ successes among } n \text{ trials}} \cdot \overbrace{p^x \cdot q^{n-x}}^{\text{The probability of } x \text{ successes among } n \text{ trials for any one particular order}}$$

using TECHNOLOGY

STATDISK Select **Analysis** from the main menu, select **Probability Distributions,** then select the **Binomial Probabilities** option. Enter the requested values for n and p, then click on **Evaluate** and the entire probability distribution will be displayed (or you have the option of entering a single value of x to get a single probability). Other columns represent cumulative probabilities that are obtained by adding the values of $P(x)$ as you go down or up the column.

MINITAB First enter a column C1 of the x values for which you want probabilities (such as 0, 1, 2, 3, 4, 5), then select **Calc** from the main menu. Select the submenu items of **Probability Distributions** and **Binomial.** Select **Probabilities,** and enter the number of trials, the probability of success, and C1 for the input column. Click **OK.**

EXCEL List the values of x in column A (such as 0, 1, 2, 3, 4, 5). Click on cell B1, then click on ***fx*** from the toolbar. Select the function category **Statistical** and then the function name **BINOMDIST** (or **BINOM.DIST**). In the dialog box, enter A1 for the entry indicated by **Number_s** (number of successes), enter the number of trials (the value of n), enter the probability, and enter 0 for the cell indicated by **Cumulative** (instead of 1 for the cumulative binomial distribution). A value should appear in cell B1. Click and drag the lower right corner of cell B1 down the column to match the entries in column A, then release the mouse button. The probabilities should all appear in column B.

TI-83/84 PLUS Press 2ND VARS (to get **DISTR,** which denotes "distributions"), then select the option identified as **binompdf(.** Complete the entry of **binompdf(n, p, x)** with specific values for n, p, and x, then press ENTER. The result will be the probability of getting x successes among n trials.

You could also enter **binompdf(n, p)** to get a list of *all* of the probabilities corresponding to $x = 0, 1, 2, \ldots, n$. You could store this list in L2 by pressing STO› L2. You could then manually enter the values of 0, 1, 2, . . . , n in list L1, which would allow you to calculate statistics (by pressing STAT, selecting **CALC,** pressing ENTER, then entering **L1, L2**), or view the distribution in a table format (by pressing STAT, then selecting **EDIT**).

The command **binomcdf** yields *cumulative* probabilities from a binomial distribution. The command **binomcdf(n, p, x)** provides the sum of all probabilities from $x = 0$ through the specific value entered for x.

STATCRUNCH Click on **Open StatCrunch,** then click on **Stat.** Select **Calculators,** then select **Binomial.**

5-3 Basic Skills and Concepts

Recommended Assignment
Exercises 1–8, 21–30.

Statistical Literacy and Critical Thinking

1. Calculating Probabilities Based on a Saint Index survey, assume that when adults are asked to identify the most unpopular projects for their hometown, 54% include Wal-Mart among their choices. Suppose we want to find the probability that when five adults are randomly selected, exactly two of them include Wal-Mart. What is wrong with using the multiplication rule to find the probability of getting two adults who include Wal-Mart followed by three people who do not include Wal-Mart, as in this calculation: (0.54)(0.54)(0.46)(0.46)(0.46)?

2. Consistent Notation If we use the binomial probability formula (Formula 5-5) for finding the probability described in Exercise 1, what is wrong with letting p denote the probability of getting an adult who includes Wal-Mart while x counts the number of adults who do not include Wal-Mart?

3. Independent Events Based on a Saint Index survey, when 1000 adults were asked to identify the most unpopular projects for their hometown, 54% included Wal-Mart among their choices. Consider the probability that among 30 different adults randomly selected from the 1000 who were surveyed, there are at least 18 who include Wal-Mart. Given that the subjects surveyed were selected without replacement, are the 30 selections independent? Can they be treated as being independent? Can the probability be found by using the binomial probability formula? Explain.

4. Notation of 0+ Using the same survey from Exercise 3, the probability of randomly selecting 30 of the 1000 adults and getting exactly 28 who include Wal-Mart is represented as 0+. What does 0+ indicate? Does 0+ indicate that it is it impossible to get exactly 28 adults who include Wal-Mart?

1. The given calculation assumes that the first two adults include Wal-Mart and the last three adults do not include Wal-Mart, but there are other arrangements consisting of two adults who include Wal-Mart and three who do not. The probabilities corresponding to those other arrangements should also be included in the result.

2. The format of Formula 5-5 requires that the probability p and the variable x refer to the same outcome. If p is the probability of an adult including Wal-Mart, then x should count the number of people who include Wal-Mart.

3. Because the 30 selections are made without replacement, they are dependent, not independent. Based on the 5% guideline for cumbersome calculations, the 30 selections can be treated as being independent. (The 30 selections constitute 3% of the population of 1000 responses, and 3% is not more than 5% of the population.) The probability can be found by using the binomial probability formula.

4. The 0+ indicates that the probability is a very small positive value. (The actual value is 0.00000296.) The notation of 0+ does not indicate that the event is impossible; it indicates that the event is possible, but very unlikely.
5. Not binomial. Each of the weights has more than two possible outcomes.
6. Binomial.
7. Binomial.
8. Not binomial. Each of the responses has more than two possible outcomes.
9. Not binomial. Because the senators are selected without replacement, the selections are not independent. (The 5% guideline for cumbersome calculations cannot be applied because the 40 selected senators constitute 40% of the population of 100 senators, and 40% exceeds 5%.)
10. Not binomial. Because the senators are selected without replacement, the selections are not independent. (The 5% guideline for cumbersome calculations cannot be applied because the 10 selected senators constitute 10% of the population of 100 senators, and 10% exceeds 5%.) Also, the numbers of terms have more than two possible outcomes.
11. Binomial. Although the events are not independent, they can be treated as being independent by applying the 5% guideline. The sample size of 380 is not more than 5% of the population of all smartphone users.
12. Binomial. Although the events are not independent, they can be treated as being independent by applying the 5% guideline. The sample size of 427 is not more than 5% of the population of all women.
13. a. 0.128
    b. WWC, WCW, CWW; 0.128 for each
    c. 0.384
14. a. 0.0081
    b. MMXX, MXMX, MXXM, XXMM, XMXM, XMMX. Each has a probability of 0.0081.
    c. 0.0486

Identifying Binomial Distributions. ***In Exercises 5–12, determine whether the given procedure results in a binomial distribution (or a distribution that can be treated as binomial). For those that are not binomial, identify at least one requirement that is not satisfied.***

5. Clinical Trial of YSORT The YSORT method of gender selection, developed by the Genetics & IVF Institute, is designed to increase the likelihood that a baby will be a boy. When 291 couples use the YSORT method and give birth to 291 babies, the weights of the babies are recorded.

6. Clinical Trial of YSORT The YSORT method of gender selection, developed by the Genetics & IVF Institute, is designed to increase the likelihood that a baby will be a boy. When 291 couples use the YSORT method and give birth to 291 babies, the genders of the babies are recorded.

7. Veggie Survey In an Idaho Potato Commission survey of 1000 adults, subjects are asked to select their favorite vegetables, and each response was recorded as "potatoes" or "other."

8. Veggie Survey In an Idaho Potato Commission survey of 1000 adults, subjects are asked to select their favorite vegetables, and responses of potatoes, corn, broccoli, tomatoes, or "other" were recorded.

9. Surveying Senators The current Senate consists of 83 males and 17 females. Forty different senators are randomly selected without replacement, and the gender of each selected senator is recorded.

10. Surveying Senators Ten different senators are randomly selected without replacement, and the numbers of terms that they have served are recorded.

11. Smartphone Survey In a RingCentral survey, 380 different smartphone users are randomly selected without replacement. Respondents were asked to identify the only thing that they can't live without. Responses consist of whether a smartphone was identified.

12. Online Shopping In a *Consumer Reports* survey, 427 different women are randomly selected without replacement, and each woman is asked what she purchases online. Responses consist of whether clothing was identified.

13. Guessing Answers The math portion of the ACT test consists of 60 multiple-choice questions, each with five possible answers (a, b, c, d, e), one of which is correct. Assume that you guess the answers to the first three questions.

a. Use the multiplication rule to find the probability that the first two guesses are wrong and the third is correct. That is, find $P(\text{WWC})$, where C denotes a correct answer and W denotes a wrong answer.

b. Beginning with WWC, make a complete list of the different possible arrangements of two wrong answers and one correct answer, then find the probability for each entry in the list.

c. Based on the preceding results, what is the probability of getting exactly one correct answer when three guesses are made?

14. Win 4 Lottery In the New York State Win 4 lottery, you place a bet by selecting four digits. Repetition is allowed, and winning requires that your sequence of four digits matches the four digits that are later drawn. Assume that you place one bet with a sequence of four digits.

a. Use the multiplication rule to find the probability that your first two digits match those drawn and your last two digits do not match those drawn. That is, find $P(\text{MMXX})$, where M denotes a match and X denotes a digit that does not match the winning number.

b. Beginning with MMXX, make a complete list of the different possible arrangements of two matching digits and two digits that do not match, then find the probability for each entry in the list.

c. Based on the preceding results, what is the probability of getting exactly two matching digits when you select four digits for the Win 4 lottery game?

Using the Binomial Probability Table. *In Exercises 15–20, assume that random guesses are made for five multiple-choice questions on an ACT test, so that there are $n = 5$ trials, each with probability of success (correct) given by $p = 0.20$. Use the Binomial Probability table (Table A-1) to find the indicated probability for the number of correct answers.*

15. Find the probability that the number x of correct answers is exactly 3. 15. 0.051

16. Find the probability that the number x of correct answers is at least 3. 16. 0.057

17. Find the probability that the number x of correct answers is more than 2. 17. 0.057

18. Find the probability that the number x of correct answers is fewer than 3. 18. 0.943

19. Find the probability of no correct answers. 19. 0.328

20. Find the probability that all answers are correct. 20. 0+

Using Technology or the Binomial Probability Formula. *In Exercises 21–24, assume that when blood donors are randomly selected, 45% of them have blood that is Group O (based on data from the Greater New York Blood Program).*

21. If the number of blood donors is $n = 8$, find the probability that the number with Group O blood is $x = 3$. 21. 0.257

22. If the number of blood donors is $n = 16$, find the probability that the number with Group O blood is $x = 6$. 22. 0.168

23. If the number of blood donors is $n = 20$, find the probability that the number with Group O blood is $x = 16$. 23. 0.00125

24. If the number of blood donors is $n = 11$, find the probability that the number with Group O blood is $x = 9$. 24. 0.0126

Using Computer Results. *In Exercises 25–28, refer to the accompanying Excel display. (In one of Mendel's hybridization experiments with peas, the probability of offspring peas having green pods is ¾, or 0.75.) The display lists the probabilities obtained by entering the values of $n = 6$ and $p = 0.75$. Those probabilities correspond to the numbers of peas with green pods in a group of six offspring peas.*

EXCEL

Peas with Green Pods x	P(x)
0	0.000
1	0.004
2	0.033
3	0.132
4	0.297
5	0.356
6	0.178

25. Genetics Find the probability that at least two of the six offspring peas have green pods. If at least two offspring peas with green pods are required for further experimentation, is it reasonable to expect that at least two will be obtained?

26. Genetics Find the probability that at most five of the six offspring peas have green pods.

27. Genetics Find the probability that at most two of the six offspring peas have green pods. Is two an unusually low number of peas with green pods (among six)? Why or why not?

28. Genetics Find the probability that five or more of the six offspring peas will have green pods. Is five an unusually high number of offspring peas with green pods? Why or why not?

25. 0.996; yes
26. 0.822
27. 0.037; yes, because the probability of 2 or fewer peas with green pods is small (less than or equal to 0.05).
28. 0.534; no, because the probability of 0.534 is not small (such as less than or equal to 0.05).

In Exercises 29–32, use either technology or the Binomial Probability table (Table A-1).

29. See You Later Based on a Harris Interactive poll, 20% of adults believe in reincarnation. Assume that six adults are randomly selected, and find the indicated probability.

a. What is the probability that exactly five of the selected adults believe in reincarnation?

b. What is the probability that all of the selected adults believe in reincarnation?

c. What is the probability that at least five of the selected adults believe in reincarnation?

d. If six adults are randomly selected, is five an unusually high number who believe in reincarnation?

29. a. 0.002 (Tech: 0.00154)
b. 0+ (Tech: 0.000064)
c. 0.002 (Tech: 0.00160).
d. Yes, the small probability from part (c) suggests that 5 is an unusually high number.

30. a. 0.004 (Tech: 0.00430)
b. 0+ (Tech: 0.000358)
c. 0.004 (Tech: 0.00467)
d. Yes, the small probability from part (c) suggests that 2 is unusually low.

30. Live TV Based on a Comcast survey, there is a 0.8 probability that a randomly selected adult will watch prime-time TV live, instead of online, on DVR, etc. Assume that seven adults are randomly selected, and find the indicated probability.

a. Find the probability that exactly two of the selected adults watch prime-time TV live.

b. Find the probability that exactly one of the selected adults watches prime-time TV live.

c. Find the probability that fewer than three of the selected adults watch prime-time TV live.

d. If we randomly select seven adults, is two an unusually low number for those who watch prime-time TV live?

31. a. 0.328
b. 0.410
c. 0.738 (Tech: 0.737)
d. No, the probability from part (c) is not small, so 1 is not an unusually low number.

31. Too Young to Tat Based on a Harris poll, among adults who regret getting tattoos, 20% say that they were too young when they got their tattoos. Assume that five adults who regret getting tattoos are randomly selected, and find the indicated probability.

a. Find the probability that none of the selected adults say that they were too young to get tattoos.

b. Find the probability that exactly one of the selected adults says that he or she was too young to get tattoos.

c. Find the probability that the number of selected adults saying they were too young is 0 or 1.

d. If we randomly select five adults, is one an unusually low number who say that they were too young to get tattoos?

32. a. 0.430
b. 0.383
c. 0.813
d. No, the probability from part (c) is not small, so 7 is not unusually high.

32. Tainted Currency Based on the American Chemical Society, there is a 0.9 probability that in the United States, a randomly selected dollar bill is tainted with traces of cocaine. Assume that eight dollar bills are randomly selected.

a. Find the probability that all of them have traces of cocaine.

b. Find the probability that exactly seven of them have traces of cocaine.

c. Find the probability that the number of dollar bills with traces of cocaine is seven or more.

d. If we randomly select eight dollar bills, is seven an unusually high number for those with traces of cocaine?

33. 0.101. No, because the probability of exactly 12 is 0.101, the probability of 12 or more is greater than 0.101, so the probability of getting 12 or more is not very small, so 12 is not unusually high.

34. 0.185. The probability is not very small, so it is not unlikely.

35. 0.287. No, because the flights all originate from New York, they are not randomly selected flights, so the 80.5% on-time rate might not apply.

36. 0.125. No, because the probability of exactly 20 with the security concerns is 0.125, the probability of 20 or fewer having those concerns is greater than 0.125, so the probability of getting 20 or fewer is not very small, so 20 is not unusually low.

33. Death of Taxes Based on a Bellowes survey of adults, there is a 0.48 probability that a randomly selected adult uses a tax preparer to file taxes. Find the probability that among 20 randomly selected adults, exactly 12 use tax preparers to file their taxes. Among 20 random adults, is 12 an unusually high number for those who use tax preparers to file their taxes?

34. Career Choice Based on a Ridgid survey of high school students, 25% of high school students plan to choose a career in information technology. Find the probability that among 24 randomly selected high school students, exactly 6 plan to choose a career in information technology. Is it unlikely to find that among the 24 selected high school students, exactly 25% plan to choose a career in information technology?

35. On-Time Flights The U.S. Department of Transportation recently reported that 80.5% of U.S. airline flights arrived on time. Find the probability that among 12 randomly selected flights, exactly 10 arrive on time. Does that probability apply to the author, who must make 12 flights originating in New York?

36. Online Banking Based on data from a *USA Today* Snapshot, 72% of adults have security concerns about online banking. Find the probability that among 30 randomly selected adults, exactly 20 have security concerns. Among 30 random adults, is 20 an unusually low number for those who have security concerns?

37. Nielsen Rating CBS televised a recent Super Bowl football game between the New Orleans Saints and the Indianapolis Colts. That game received a rating of 45, indicating that among all U.S. households, 45% were tuned to the game (based on data from Nielsen Media Research). An advertiser wants to obtain a second opinion by conducting its own survey, and a pilot survey begins with 12 randomly selected households.

a. Find the probability that none of the households is tuned to the Saints/Colts game.

b. Find the probability that at least one household is tuned to the Saints/Colts game.

c. Find the probability that at most one household is tuned to the Saints/Colts game.

d. If at most one household is tuned to the Saints/Colts game, does it appear that the 45 share value is wrong? Why or why not?

37. a. 0.000766
b. 0.999
c. 0.00829
d. Yes, the very low probability of 0.00829 would suggest that the 45 share value is wrong.

38. Overbooking Flights When someone buys a ticket for an airline flight, there is a 0.0995 probability that the person will not show up for the flight (based on data from an IBM research paper by Lawrence, Hong, and Cherrier). The Beechcraft 1900C jet can seat 19 passengers. Is it wise to book 21 passengers for a flight on the Beechcraft 1900C? Explain.

38. With 21 booked passengers, there is a probability of 0.368 that the flight will be overbooked. It does not seem wise to schedule in such a way that the flights will be overbooked about 37% of the time.

39. XSORT Method of Gender Selection When testing a method of gender selection, we assume that the rate of female births is 50%, and we reject that assumption if we get results that are unusual in the sense that they are very unlikely to occur with the 50% rate. In a preliminary test of the XSORT method of gender selection, 14 births included 13 girls.

a. Assuming a 50% rate of female births, find the probability that in 14 births, the number of girls is 13.

b. Assuming a 50% rate of female births, find the probability that in 14 births, the number of girls is 14.

c. Assuming a 50% rate of female births, find the probability that in 14 births, the number of girls is 13 or more.

d. Do these preliminary results suggest that the XSORT method is effective in increasing the likelihood of a baby being a girl? Explain.

39. a. 0.000854
b. 0.0000610
c. 0.000916
d. Yes. The probability of getting 13 girls or a result that is more extreme is 0.000916, so chance does not appear to be a reasonable explanation for the result of 13 girls. Because 13 is an unusually high number of girls, it appears that the probability of a girl is higher with the XSORT method, and it appears that the XSORT method is effective.

40. Challenged Calls in Tennis In a recent U.S. Open tennis tournament, among 20 of the calls challenged by players, 8 were overturned after a review using the Hawk-Eye electronic system. Assume that when players challenge calls, they are successful in having them overturned 50% of the time.

a. Find the probability that among 20 challenges, exactly 8 are successfully overturned.

b. The probability that among 20 challenges, 8 or fewer are overturned is 0.252. Does this result suggest that the success rate is less than 50%? Why or why not?

40. a. 0.120
b. No. If the success rate is equal to 50%, it is likely (with probability 0.252) that we get 8 successes or a result that is more extreme (fewer than 8 successes). This indicates that with a 50% success rate, the occurrence of 8 successes in 20 challenges could be reasonably explained by chance.

Composite Sampling. ***Exercises 41 and 42 involve the method of composite sampling, whereby a medical testing laboratory saves time and money by combining blood samples for tests so that only one test is conducted for several people. A combined sample tests positive if at least one person has the disease. If a combined sample tests positive, then individual blood tests are used to identify the individual with the disease.***

41. HIV The probability of a randomly selected adult in the United States being infected with the human immunodeficiency virus (HIV) is 0.006 (based on data from the Kaiser Family Foundation). In tests for HIV, blood samples from 24 people are combined. What is the probability that the combined sample tests positive for HIV? Is it unlikely for such a combined sample to test positive?

41. 0.134. It is not unlikely for such a combined sample to test positive.

42. STD Based on data from the Centers for Disease Control, the probability of a randomly selected person having gonorrhea is 0.00114. In tests for gonorrhea, blood samples from

42. 0.0181. It is unlikely for such a combined sample to test positive.

16 people are combined. What is the probability that the combined sample tests positive for gonorrhea? Is it unlikely for such a combined sample to test positive?

Acceptance Sampling. ***Exercises 43 and 44 involve the method of acceptance sampling, whereby a shipment of a large number of items is accepted if tests of a sample of those items result in only one or none that are defective.***

43. 0.662. The probability shows that about 2/3 of all shipments will be accepted. With about 1/3 of the shipments rejected, the supplier would be wise to improve quality.

43. Aspirin The Medassist Pharmaceutical Company receives large shipments of aspirin tablets and uses this acceptance sampling plan: Randomly select and test 40 tablets, then accept the whole batch if there is only one or none that doesn't meet the required specifications. If one shipment of 5000 aspirin tablets actually has a 3% rate of defects, what is the probability that this whole shipment will be accepted? Will almost all such shipments be accepted, or will many be rejected?

44. 0.978. About 98% of all shipments will be accepted. Almost all shipments will be accepted, and only 2% of the shipments will be rejected.

44. Chocolate Chip Cookies The Killington Market chain uses this acceptance sampling plan for large shipments of packages of its generic chocolate chip cookies: Randomly select 30 packages and determine whether each is within specifications (not too many broken cookies, acceptable taste, distribution of chocolate chips, and so on). The entire shipment is accepted if at most 2 packages do not meet specifications. A shipment contains 1200 packages of chocolate chip cookies, and 2% of those packages do not meet specifications. What is the probability that this whole shipment will be accepted? Will almost all such shipments be accepted, or will many be rejected?

5-3 Beyond the Basics

45. 0.0468

45. Geometric Distribution If a procedure meets all the conditions of a binomial distribution except that the number of trials is not fixed, then the **geometric distribution** can be used. The probability of getting the first success on the xth trial is given by $P(x) = p(1 - p)^{x-1}$, where p is the probability of success on any one trial. Subjects are randomly selected for the National Health and Nutrition Examination Survey conducted by the National Center for Health Statistics, Centers for Disease Control. Find the probability that the first subject to be a universal blood donor (with group O and type Rh^- blood) is the fifth person selected. The probability that someone is a universal donor is 0.06.

46. 0.00485

46. Multinomial Distribution The binomial distribution applies only to cases involving two types of outcomes, whereas the **multinomial distribution** involves more than two categories. Suppose we have three types of mutually exclusive outcomes denoted by A, B, and C. Let $P(\text{A}) = p_1$, $P(\text{B}) = p_2$, and $P(\text{C}) = p_3$. In n independent trials, the probability of x_1 outcomes of type A, x_2 outcomes of type B, and x_3 outcomes of type C is given by

$$\frac{n!}{(x_1)!(x_2)!(x_3)!} \cdot p_1^{x_1} \cdot p_2^{x_2} \cdot p_3^{x_3}$$

A roulette wheel in the Hard Rock casino in Las Vegas has 18 red slots, 18 black slots, and 2 green slots. If roulette is played 12 times, find the probability of getting 5 red outcomes, 4 black outcomes, and 3 green outcomes.

47. a. 0.000969
b. 0.0000000715
c. 0.436

47. Hypergeometric Distribution If we sample from a small finite population without replacement, the binomial distribution should not be used because the events are not independent. If sampling is done without replacement and the outcomes belong to one of two types, we can use the **hypergeometric distribution.** If a population has A objects of one type (such as lottery numbers matching the ones you selected), while the remaining B objects are of the other type (such as lottery numbers different from the ones you selected), and if n objects are sampled without replacement (such as six drawn lottery numbers), then the probability of getting x objects of type A and $n - x$ objects of type B is

$$P(x) = \frac{A!}{(A - x)!x!} \cdot \frac{B!}{(B - n + x)!(n - x)!} \div \frac{(A + B)!}{(A + B - n)!n!}$$

In Pennsylvania's Match 6 Lotto game, a bettor selects six numbers from 1 to 49 (without repetition), and a winning six-number combination is later randomly selected. Find the probabilities of the following events and express them in decimal form.

a. You purchase one ticket with a six-number combination and you get exactly four winning numbers. (*Hint:* Use $A = 6$, $B = 43$, $n = 6$, and $x = 4$.)

b. You purchase one ticket with a six-number combination and you get all six of the winning numbers.

c. You purchase one ticket with a six-number combination and you get none of the winning numbers.

48. Poisson Distribution The **Poisson distribution** applies to occurrences of some event over a specified interval, such as time or distance. The probability of the event occurring x times over an interval is given by

$$P(x) = \frac{\mu^x \cdot e^{-\mu}}{x!}$$

where $e \approx 2.71828$ and μ is the mean number of occurrences over the interval. Over the past 100 years, the mean number of annual major earthquakes in the world is 0.93. Assuming that the Poisson distribution is a suitable model, find the probability that the number of earthquakes in a randomly selected year is

a. 0 **b.** 1 **c.** 2 **d.** 3 **e.** 4 **f.** 5 **g.** 6 **h.** 7

Here are the actual results: 47 years (0 major earthquakes); 31 years (1 major earthquake); 13 years (2 major earthquakes); 5 years (3 major earthquakes); 2 years (4 major earthquakes); 0 years (5 major earthquakes); 1 year (6 major earthquakes); 1 year (7 major earthquakes). After comparing the calculated probabilities to the actual results, is the Poisson distribution a good model?

48. a. 0.395
b. 0.367
c. 0.171
d. 0.0529
e. 0.0123
f. 0.00229
g. 0.000355
h. 0.0000471
Using the computed probabilities, the expected frequencies are 39.5, 36.7, 17.1, 5.3, 1.2, 0.2, 0.0, and 0.0, and they agree reasonably well with the actual frequencies.

5-4 Parameters for Binomial Distributions

Note to Instructor
This section can be covered quickly, and students will find that it is not too difficult. Stress the *interpretation* of results using the range rule of thumb.

Key Concept Section 5-2 introduced the general concept of a probability distribution, and Section 5-3 focused on binomial probability distributions, which constitute a specific type of discrete probability distribution. This section continues with binomial distributions as these two goals are addressed:

1. Provide an easy method for finding the parameters of the mean and standard deviation from a binomial distribution. (Because a binomial distribution describes a *population*, the mean and standard deviation are *parameters*, not statistics.)
2. Use the range rule of thumb for determining whether events are *unusual.*

Section 5-2 included Formulas 5-1, 5-3, and 5-4 for finding the mean, variance, and standard deviation from *any* discrete probability distribution. Because a binomial distribution is a particular type of discrete probability distribution, we could use those same formulas, but if we know the values of n and p, it is much easier to use these formulas:

For Binomial Distributions

Formula 5-6	Mean:	$\mu = np$
Formula 5-7	Variance:	$\sigma^2 = npq$
Formula 5-8	Standard Deviation:	$\sigma = \sqrt{npq}$

As in earlier sections, finding values for μ and σ can be great fun, but it is especially important to *interpret* and *understand* those values, so the range rule of thumb can be very helpful. Here is a brief summary of the range rule of thumb: Values are unusually high or unusually low if they differ from the mean by more than 2 standard deviations, as described by the following:

Range Rule of Thumb

maximum usual value: $\mu + 2\sigma$
minimum usual value: $\mu - 2\sigma$

Example 1 **Parameters**

The brand name of McDonald's has a 95% recognition rate (based on data from Retail Marketing Group and Harris Interactive). A special focus group consists of 12 randomly selected adults to be used for extensive market testing. For such random groups of 12 people, find the mean and standard deviation for the number of people who recognize the brand name of McDonald's.

Solution

We have $n = 12$, which is the sample size. We have $p = 0.95$ (from 95%), and we have $q = 0.05$ (from $q = 1 - p$). Using $n = 12$, $p = 0.95$, and $q = 0.05$, Formulas 5-6 and 5-8 can be applied as follows:

$$\mu = np = (12)(0.95) = 11.4$$

$$\sigma = \sqrt{npq} = \sqrt{(12)(0.95)(0.05)} = 0.754983 = 0.8 \text{ (rounded)}$$

For random groups of 12 adults, the mean number of people who recognize the brand name of McDonald's is 11.4 people, and the standard deviation is 0.8 people.

Example 2 **Unusual Outcomes**

From Example 1 we see that for groups of 12 randomly selected people, the mean number of people who recognize the brand name of McDonald's is $\mu = 11.4$ people, and the standard deviation is $\sigma = 0.8$ people.

a. For groups of 12 randomly selected people, use the range rule of thumb to find the minimum usual number and the maximum usual number of people who recognize the brand name of McDonald's.

b. In one particular randomly selected group of 12 people, is 12 an unusually high number of people who recognize the brand name of McDonald's?

Solution

a. With $\mu = 11.4$ people and $\sigma = 0.8$ people, we use the range rule of thumb as follows:

$$\text{maximum usual value: } \mu + 2\sigma = 11.4 + 2(0.8) = 13 \text{ people}$$
$$\text{minimum usual value: } \mu - 2\sigma = 11.4 - 2(0.8) = 9.8 \text{ people}$$

For groups of 12 randomly selected people, the number who recognize the brand name of McDonald's should usually fall between 9.8 and 13. (Actually, with 12 people selected, the maximum usual value cannot exceed 12, so the range of usual values is between 9.8 and 12.)

b. Because 12 falls within the range of usual values (from 9.8 to 13), we conclude that 12 is *not an unusually high number* of people who recognize the brand name of McDonald's (assuming that the recognition rate is 95%).

Example 3 XSORT Method of Gender Selection

The Chapter Problem includes results from the XSORT method of gender selection. Among 945 births, there were 879 girls born to parents using the XSORT method.

a. Assuming that boys and girls are equally likely, use Formulas 5-6 and 5-8 to find the mean and standard deviation for the numbers of girls born in groups of 945 babies (as in the Chapter Problem).

b. Use the range rule of thumb to find the minimum usual number and the maximum usual number of girls born in groups of 945 births, assuming that boys and girls are equally likely.

c. The Chapter Problem described the XSORT method of gender selection that resulted in 879 girls born in a group of 945 babies. Is 879 girls unusually high?

Solution

a. Using the values $n = 945$, $p = 0.5$, and $q = 0.5$, Formulas 5-6 and 5-8 can be applied as follows:

$$\mu = np = (945)(0.5) = 472.5 \text{ girls}$$

$$\sigma = \sqrt{npq} = \sqrt{(945)(0.5)(0.5)} = 15.4 \text{ girls (rounded)}$$

b. With $\mu = 472.5$ girls and $\sigma = 15.4$ girls, we use the range rule of thumb as follows:

$$\text{minimum usual value: } \mu - 2\sigma = 472.5 - 2(15.4) = 441.7 \text{ girls}$$

$$\text{maximum usual value: } \mu + 2\sigma = 472.5 + 2(15.4) = 503.3 \text{ girls}$$

For groups of 945 births, the number of girls should usually fall between 441.7 girls and 503.3 girls (assuming that boys and girls are equally likely).

c. Because the result of 879 girls falls above the range of usual values (from 441.7 girls to 503.3 girls), we consider the result of 879 girls to be an unusually high number of girls.

Note to Instructor

Discuss the *interpretation* of the results from Example 3. The ultimate goal is not to simply obtain numerical results but to interpret results in a practical and meaningful way. Also, determining whether a result is "unusual" is excellent preparation for the method of hypothesis testing introduced in Chapter 8.

Interpretation

The result of 879 girls in 945 births is unusually high, so it is not a result we expect to occur by chance. Because chance does not seem to be a reasonable explanation, because 879 girls is so much higher than we expect with chance, and because the 945 couples used the XSORT method of gender selection, it appears that the XSORT method is effective in increasing the likelihood that a baby is born a girl. (However, this does not *prove* that the XSORT method is the cause of the increased number of girls; such proof must be based on direct physical evidence.)

In this section we presented easy procedures for finding the parameters of μ (population mean) and σ (population standard deviation) from a binomial probability distribution. However, it is really important to be able to *interpret* those values by using such devices as the range rule of thumb for identifying a range of usual values.

Recommended Assignment
Exercises 1–14.

5-4 Basic Skills and Concepts

Statistical Literacy and Critical Thinking

1. Notation In a clinical trial of the cholesterol drug Lipitor, 270 subjects were given a placebo, and 7% of them developed headaches. For such randomly selected groups of 270 subjects given a placebo, identify the values of n, p, and q that would be used for finding the mean and standard deviation for the number of subjects who develop headaches.

2. Standard Deviation For the binomial distribution described by the conditions in Exercise 1, find the standard deviation by evaluating $\sqrt{npq}$. Some books use the expression $\sqrt{np(1-p)}$ for the standard deviation. Will that expression always yield the same result?

3. Variance An Office Team survey of 150 executives found that 93.3% of them said that they would be concerned about gaps in a résumé of a job applicant. Based on these results, such randomly selected groups of 150 executives have a mean of 140.0 executives and a standard deviation of 3.1 executives. Find the variance, and express it with the appropriate units.

4. Why a Parameter? Given the results described in Exercise 3, is the mean of 140.0 executives expressed with μ or $\bar{x}$? Explain why the mean of 140.0 executives is a parameter instead of a statistic.

Finding μ, σ, and Unusual Values. ***In Exercises 5–8, assume that a procedure yields a binomial distribution with n trials and the probability of success for one trial is p. Use the given values of n and p to find the mean μ and standard deviation σ. Also, use the range rule of thumb to find the minimum usual value $\mu - 2\sigma$ and the maximum usual value $\mu - 2\sigma$.***

5. Guessing on ACT Random guesses are made for the 60 multiple-choice questions on the math portion of the ACT test, so $n = 60$ and $p = 1/5$ (because each question has possible answers of a, b, c, d, e, and only one of them is correct).

6. Gender Selection In an analysis of preliminary test results from the XSORT gender-selection method, 14 babies are born and it is assumed that 50% of babies are girls, so $n = 14$ and $p = 0.5$.

7. Identity Theft In a Gallup poll of 1013 randomly selected adults, 66% said that they worry about identity theft, so $n = 1013$ and $p = 0.66$.

8. Clinical Trial In a clinical trial of the cholesterol drug Lipitor, 94 subjects were treated with 80 mg of Lipitor, and 6.4% of them developed headaches, so $n = 94$ and $p = 0.064$.

9. Gender Selection In a test of the YSORT method of gender selection, 291 babies are born to couples trying to have baby boys, and 239 of those babies are boys (based on data from the Genetics & IVF Institute).

a. If the gender-selection method has no effect and boys and girls are equally likely, find the mean and standard deviation for the numbers of boys born in groups of 291.

b. Is the result of 239 boys unusually high? Does it suggest that the YSORT gender-selection method appears to be effective?

10. Mendelian Genetics When Mendel conducted his famous genetics experiments with peas, one sample of offspring consisted of 580 peas, and Mendel theorized that 25% of them would be yellow peas.

a. If Mendel's theory is correct, find the mean and standard deviation for the numbers of yellow peas in such groups of 580 offspring peas.

b. The actual results consisted of 152 yellow peas. Is that result unusually high? What does this result suggest about Mendel's theory?

---

1. $n = 270, p = 0.07, q = 0.93$
2. 4.2 people. Yes, both expressions will yield the same result because they are equivalent. They are equivalent because $q = 1 - p$.
3. 9.4 executives^2 (or 9.6 executives^2 if the rounded standard deviation of 3.1 executives is used)
4. The mean of 140.0 executives is expressed with μ. The mean is calculated for the *population* of all groups of 150 executives, not just one sample group. Because the mean is calculated for a population, it is a parameter.
5. $\mu = 12.0$ correct guesses; $\sigma = 3.1$ correct guesses; minimum $= 5.8$ correct guesses; maximum $= 18.2$ correct guesses.
6. $\mu = 7.0$ girls; $\sigma = 1.9$ girls; minimum $= 3.2$ girls; maximum $=$ 10.8 girls.
7. $\mu = 668.6$ worriers; $\sigma = 15.1$ worriers; minimum $= 638.4$ worriers; maximum $= 698.8$ worriers.
8. $\mu = 6.0$ subjects with headaches; $\sigma = 2.4$ subjects with headaches; minimum $= 1.2$ subjects with headaches; maximum $= 10.8$ subjects with headaches.
9. a. $\mu = 145.5$; $\sigma = 8.5$
   b. Yes. Using the range rule of thumb, the minimum usual value is 128.5 boys and the maximum usual value is 162.5 boys. Because 239 boys is above the range of usual values, it is unusually high. Because 239 boys is unusually high, it does appear that the YSORT method of gender selection is effective.
10. a. $\mu = 145.0$, $\sigma = 10.4$
    b. No, it is within the range of usual values (124.2 to 165.8). It does not provide strong evidence against Mendel's theory.

11. Are 20% of M&M Candies Orange? Mars, Inc. claims that 20% of its M&M plain candies are orange, and a sample of 100 such candies is randomly selected.

a. Find the mean and standard deviation for the number of orange candies in such groups of 100.

b. Data Set 20 in Appendix B consists of a random sample of 100 M&Ms, including 25 that are orange. Is this result unusually high? Does it seem that the claimed rate of 20% is wrong?

11. a. $\mu = 20.0, \sigma = 4.0$
b. No, because 25 orange M&Ms is within the range of usual values (12 to 28). The claimed rate of 20% does not necessarily appear to be wrong, because that rate will usually result in 12 to 28 orange M&Ms (among 100), and the observed number of orange M&Ms is within that range.

12. Are 14% of M&M Candies Yellow? Mars, Inc. claims that 14% of its M&M plain candies are yellow, and a sample of 100 such candies is randomly selected.

a. Find the mean and standard deviation for the number of yellow candies in such groups of 100.

b. Data Set 20 in Appendix B consists of a random sample of 100 M&Ms, including 8 that are yellow. Is this result unusually low? Does it seem that the claimed rate of 14% is wrong?

12. a. $\mu = 14.0, \sigma = 3.5$
b. No, because 8 yellow M&Ms is within the range of usual values (7 to 21). The claimed rate of 14% does not necessarily appear to be wrong, because that rate will usually result in 7 to 21 yellow M&Ms (among 100), and the observed number of yellow M&Ms is within that range.

13. Cell Phones and Brain Cancer In a study of 420,095 cell phone users in Denmark, it was found that 135 developed cancer of the brain or nervous system. If we assume that the use of cell phones has no effect on developing such cancer, then the probability of a person having such a cancer is 0.000340.

a. Assuming that cell phones have no effect on developing cancer, find the mean and standard deviation for the numbers of people in groups of 420,095 that can be expected to have cancer of the brain or nervous system.

b. Based on the results from part (a), is 135 cases of cancer of the brain or nervous system unusually low or high?

c. What do these results suggest about the publicized concern that cell phones are a health danger because they increase the risk of cancer of the brain or nervous system?

13. a. $\mu = 142.8, \sigma = 11.9$
b. No, 135 is not unusually low or high because it is within the range of usual values (119.0 to 166.6).
c. Based on the given results, cell phones do not pose a health hazard that increases the likelihood of cancer of the brain or nervous system.

14. Test of Touch Therapy Nine-year-old Emily Rosa conducted this test: A professional touch therapist put both hands through a cardboard partition and Emily would use a coin flip to randomly select one of the hands. Emily would place her hand just above the hand of the therapist, who was then asked to identify the hand that Emily had selected. Touch therapists believed that they could sense the energy field and identify the hand that Emily had selected. The trial was repeated 280 times. (Based on data from "A Close Look at Therapeutic Touch," by Rosa et al., *Journal of the American Medical Association,* Vol. 279, No. 13.)

a. Assuming that the touch therapists have no special powers and made random guesses, find the mean and standard deviation for the numbers of correct responses in groups of 280 trials.

b. The professional touch therapists identified the correct hand 123 times in the 280 trials. Is that result unusually low or high? What does the result suggest about the ability of touch therapists to select the correct hand by sensing an energy field?

14. a. $\mu = 140.0, \sigma = 8.4$
b. The result of 123 correct identifications is just outside of the range of usual values (123.2 to 156.8), but this indicates that 123 is unusually low. If the touch therapists really had an ability to select the correct hand, they would have made more than 156.8 correct identifications. Therefore, they do not appear to have that ability.

15. Deciphering Messages The Central Intelligence Agency has specialists who analyze the frequencies of letters of the alphabet in an attempt to decipher intercepted messages that are sent as ciphered text. In standard English text, the letter *r* is used at a rate of 6%.

a. Find the mean and standard deviation for the number of times the letter *r* will be found on a typical page of 2600 characters.

b. In an intercepted ciphered message sent to Iran, a page of 2600 characters is found to have the letter *r* occurring 178 times. Is this unusually low or high?

15. a. $\mu = 156.0; \sigma = 12.1$
b. The minimum usual frequency is 131.8 and the maximum is 180.2. The occurrence of *r* 178 times is not unusually low or high because it is within the range of usual values (131.8 to 180.2).

16. Deciphering Messages The Central Intelligence Agency has specialists who analyze the frequencies of letters of the alphabet in an attempt to decipher intercepted messages that are sent as ciphered text. In standard English text, the letter *e* is used at a rate of 12.7%.

a. Find the mean and standard deviation for the number of times the letter *e* will be found on a typical page of 2600 characters.

b. In an intercepted ciphered message sent to France, a page of 2600 characters is found to have the letter *e* occurring 290 times. Is 290 unusually low or unusually high?

16. a. $\mu = 330.2; \sigma = 17.0$
b. The minimum usual frequency is 296.2 and the maximum is 364.2. The occurrence of *e* 290 times is unusually low because it is below the range of usual values (296.2 to 364.2).

17. a. $\mu = 74.0$; $\sigma = 7.7$
b. The minimum usual number is 58.6 and the maximum usual value is 89.4. The value of 90 is unusually high because it is above the range of usual values (58.6 to 89.4).

18. a. $\mu = 1.3$; $\sigma = 1.1$
b. The minimum usual value is -0.9 and the maximum usual value is 3.5. The result of 0 wins is not unusually low because 0 wins is within the range of usual values (from -0.9 to 3.5).

19. a. $\mu = 0.0821918$; $\sigma = 0.2862981$
b. The minimum usual number is -0.4904044 and the maximum usual number is 0.654788. The results of 2 students born on the 4th of July would be unusually high, because 2 is above the range of usual values (from -0.4904044 to 0.654788).

20. a. $\mu = 0.000013$; $\sigma = 0.003649$
b. The minimum usual number of wins is -0.007285 and the maximum usual number of wins is 0.007311. It is unusual to buy a ticket each week for 50 years and win at least once, because 1 win (or more) is outside of the range of usual values (from -0.007285 to 0.007311).

17. Too Young to Tat Based on a Harris poll of 370 adults who regret getting tattoos, 20% say that they were too young when they got their tattoos.

a. For randomly selected groups of 370 adults who regret getting tattoos, find the mean and standard deviation for the number who say that they were too young when they got their tattoos.

b. For a randomly selected group of 370 adults who regret getting tattoos, would 90 be an unusually low or high number who say that they were too young when they got their tattoos?

18. Roulette If you place a bet on the number 7 in roulette, you have a 1/38 probability of winning.

a. Find the mean and standard deviation for the number of wins for people who bet on the number 7 fifty times.

b. Would 0 wins in 50 bets be an unusually low number of wins?

19. Born on the 4th of July For the following questions, ignore leap years.

a. For classes of 30 students, find the mean and standard deviation for the number born on the 4th of July. Express results using seven decimal places.

b. For a class of 30 students, would 2 be an unusually high number who were born on the 4th of July?

20. Powerball Lottery As of this writing, the Powerball lottery is run in 42 states. If you buy one ticket, the probability of winning is 1/195,249,054. If you buy one ticket each week for 50 years, you play this lottery 2600 times.

a. Find the mean and standard deviation for the number of wins for people who buy a ticket each week for 50 years. Express the results using six decimal places.

b. Would it be unusual for someone to win this lottery at least once if they buy a ticket each week for 50 years?

5-4 Beyond the Basics

21. $n = 150$; $p = 0.4$, so that 40% of the surveyed subjects could identify at least one member of the Supreme Court; $q = 0.6$, so that 60% of surveyed subjects could not identify at least one member of the Supreme Court.

21. Finding *n, p, q* In a survey of randomly selected adults, subjects were asked if they could identify at least one current member of the Supreme Court. After obtaining the results, the range rule of thumb was used to find that for randomly selected groups of the same size, the minimum usual number who could identify at least one member of the Supreme Court is 48.0 and the maximum number is 72.0. Find the sample size *n*, the percentage of surveyed subjects who could identify at least one member of the Supreme Court, and the percentage of subjects who could not identify at least one member of the Supreme Court.

22. 170

22. Acceptable/Defective Products A new integrated circuit board is being developed for use in computers. In the early stages of development, a lack of quality control results in a 0.2 probability that a manufactured integrated circuit board has no defects. Engineers need 24 integrated circuit boards for further testing. What is the minimum number of integrated circuit boards that must be manufactured in order to be at least 98% sure that there are at least 24 that have no defects?

23. $\mu = 3.0$ and $\sigma = 1.3$ (not 1.5)

23. Hypergeometric Distribution A statistics class consists of 10 females and 30 males, and each day, 12 of the students are randomly selected without replacement. Because the sampling is from a small finite population without replacement, the hypergeometric distribution applies. (See Exercise 47 in Section 5-3.) Using the hypergeometric distribution, find the mean and standard deviation for the numbers of females that are selected on the different days.

Chapter 5 Review

This chapter introduced the important concept of a probability distribution, which describes the probability for each value of a random variable. This chapter includes only discrete probability distributions, but the following chapters will include continuous probability distributions.

In Section 5-2 we introduced probability distributions and the following definitions.

- A *random variable* has values that are determined by chance.
- A *probability distribution* consists of all values of a random variable, along with their corresponding probabilities. A probability distribution must satisfy three requirements: there is a numerical random variable x and its values are associated with corresponding probabilities, the sum of all of the probabilities for values of the random variable must be 1, and each probability value must be between 0 and 1 inclusive. The second and third requirements are expressed as $\sum P(x) = 1$ and, for each value of x, $0 \le P(x) \le 1$.
- Important characteristics of a *probability distribution* can be explored by constructing a probability histogram and by computing its mean and standard deviation using these formulas:

$$\mu = \sum [x \cdot P(x)]$$

$$\sigma = \sqrt{\sum [x^2 \cdot P(x)] - \mu^2}$$

- In Section 5-3, we introduced *binomial distributions,* which have two categories of outcomes and a fixed number of independent trials with a constant probability. The probability of x successes among n trials can be found by using the binomial probability formula, or Table A-1 (the Binomial Probability table), or computer software (such as STATDISK, Minitab, Excel, or StatCrunch), or a TI-83/84 Plus calculator.
- In Section 5-4 we noted that for a binomial distribution, the parameters of the mean and standard deviation are described by $\mu = np$ and $\sigma = \sqrt{npq}$.
- *Unusual Outcomes:* In this chapter, we saw that we could determine when an outcome has an unusually low or unusually high number of successes. We used two different criteria: (1) the range rule of thumb, and (2) the use of probabilities, described as follows.

Using the range rule of thumb to identify unusual values:

$$\textbf{maximum usual value} = \mu + 2\sigma$$

$$\textbf{minimum usual value} = \mu - 2\sigma$$

Using probabilities to identify unusual values:

Unusually high number of successes: x successes among n trials is an unusually high number of successes if $P(x \text{ or more}) \le 0.05$.*

Unusually low number of successes: x successes among n trials is an unusually low number of successes if $P(x \text{ or fewer}) \le 0.05$.*

*The value of 0.05 is commonly used but is not absolutely rigid. Other values, such as 0.01, could be used to distinguish between events that can easily occur by chance and events that are very unlikely to occur by chance.

Chapter Quick Quiz

1. Is a probability distribution defined if the only possible values of a random variable are 0, 1, 2, 3, and $P(0) = P(1) = P(2) = P(3) = 0.25$? 1. Yes

2. There are 100 questions from an SAT test, and they are all multiple choice with possible answers of a, b, c, d, e. For each question, only one answer is correct. Find the mean number of correct answers for those who make random guesses for all 100 questions. 2. 20.0

3. Using the same SAT questions described in Exercise 2, find the standard deviation for the numbers of correct answers for those who make random guesses for all 100 questions. 3. 4.0

4. Yes

5. No

6. Yes. (The sum of the probabilities is 0.999 and it can be considered to be 1 because of rounding errors.)

7. 0+ indicates that the probability is a very small positive number. It does not indicate that it is impossible for none of the five flights to arrive on time.

8. 0.945

x	P(x)
0	0+
1	0.006
2	0.048
3	0.198
4	0.409
5	0.338

9. Yes

10. No

1. 0.047 or 0.0467

2. 0.138

3. $\mu = 240.0$; $\sigma = 12.0$. Range of usual values: 216 to 264. The result of 200 with brown eyes is unusually low.

4. The probability of P(239 or fewer) = 0.484 is relevant for determining whether 239 is an unusually low number. Because that probability is not very small, it appears that 239 is not an unusually low number of people with brown eyes.

x	P(x)
0	0.674
1	0.280
2	0.044
3	0.003
4	0+

5. Yes, the three requirements are satisfied. There is a numerical random variable x and its values are associated with corresponding probabilities. $\sum P(x) = 1.001$, so the sum of the probabilities is 1 when we allow for a small discrepancy due to rounding. Also, each of the probability values is between 0 and 1 inclusive.

4. If boys and girls are equally likely, groups of 400 births have a mean of 200 girls and a standard deviation of 10 girls. Is 232 girls in 400 births an unusually high number of girls?

5. If boys and girls are equally likely, groups of 400 births have a mean of 200 girls and a standard deviation of 10 girls. Is 185 girls in 400 births an unusually low number of girls?

In Exercises 6–10, use the following: Five U.S. domestic flights are randomly selected, and the table in the margin lists the probabilities for the number that arrive on time (based on data from the Department of Transportation). Assume that five flights are randomly selected.

6. Does the table describe a probability distribution?

7. What does the probability of 0+ indicate? Does it indicate that among five randomly selected flights, it is impossible that none of them arrives on time?

8. What is the probability that at least three of the five flights arrive on time?

9. Is 0 an unusually low number of flights arriving on time?

10. Is 5 an unusually high number of flights arriving on time?

Review Exercises

In Exercises 1–4, assume that 40% of the population has brown eyes (based on data from Dr. P. Sorita at Indiana University).

1. Brown Eyes If six people are randomly selected, find the probability that none of them has brown eyes.

2. Brown Eyes Find the probability that among six randomly selected people, exactly four of them have brown eyes.

3. Brown Eyes Groups of 600 people are randomly selected. Find the mean and standard deviation for the numbers of people with brown eyes in such groups, then use the range rule of thumb to identify the range of usual values for those with brown eyes. For such a group of 600 randomly selected people, is 200 with brown eyes unusually low or high?

4. Brown Eyes When randomly selecting 600 people, the probability of exactly 239 people with brown eyes is P(239) = 0.0331. Also, P(239 or fewer) = 0.484. Which of those two probabilities is relevant for determining whether 239 is an unusually low number of people with brown eyes? Is 239 an unusually low number of people with brown eyes?

In Exercises 5 and 6, refer to the table in the margin. The random variable x is the number of males with tinnitus (ringing ears) among four randomly selected males (based on data from "Prevalence and Characteristics of Tinnitus among US Adults" by Shargorodsky et al., **American Journal of Medicine, Vol. 123, No. 8).**

5. Tinnitus Does the table describe a probability distribution? Why or why not?

6. Tinnitus Find the mean and standard deviation for the random variable x. Use the range rule of thumb to identify the range of usual values for the number of males with tinnitus among four randomly selected males. Is it unusual to get three males with tinnitus among four randomly selected males?

7. Brand Recognition In a study of brand recognition of the Kindle eReader, four consumers are interviewed. If x is the number of consumers in the group who recognize the Kindle brand name, then x can be 0, 1, 2, 3, or 4. If the corresponding probabilities are 0.026, 0.154, 0.346, 0.246, and 0.130, does the given information describe a probability distribution? Why or why not?

8. Expected Value for *Deal or No Deal* In the television game show *Deal or No Deal,* contestant Elna Hindler had to choose between acceptance of an offer of $193,000 or continuing the game. If she continued to refuse all further offers, she would have won one of these five equally likely prizes: $75, $300, $75,000, $500,000, and $1,000,000. Find her expected value if she continued the game and refused all further offers. Based on the result, should she accept the offer of $193,000, or should she continue?

9. Expected Value for a Magazine Sweepstakes *Reader's Digest* ran a sweepstakes in which prizes were listed along with the chances of winning: $1,000,000 (1 chance in 90,000,000), $100,000 (1 chance in 110,000,000), $25,000 (1 chance in 110,000,000), $5,000 (1 chance in 36,667,000), and $2,500 (1 chance in 27,500,000).

a. Assuming that there is no cost of entering the sweepstakes, find the expected value of the amount won for one entry.

b. Find the expected value if the cost of entering this sweepstakes is the cost of a postage stamp. Is it worth entering this contest?

10. Phone Calls In the month preceding the creation of this exercise, the author made no phone calls on 19 of the 30 days in the month. Assume that 10 days from that month are randomly selected with replacement. Find the following probabilities.

a. The author made no calls on any of the 10 randomly selected days.

b. The 10 randomly selected days include exactly 4 days with no calls.

c. At least 1 call was made on each of the 10 randomly selected days.

6. $\mu = 0.4$; $\sigma = 0.6$. Range of usual values: -0.8 to 1.6 (or 0 to 1.6). Yes, three is an unusually high number of males with tinnitus among four randomly selected males.
7. $\sum P(x) = 0.902$, so the sum of the probabilities is not 1 as required. Because the three requirements are not all satisfied, the given information does not describe a probability distribution.
8. $315,075. Because the offer is well below her expected value, she should continue the game (although the guaranteed prize of $193,000 had considerable appeal). (She accepted the offer of $193,000, but she would have won $500,000 if she continued the game and refused all further offers.)
9. a. 1.2¢
   b. 1.2¢ minus cost of stamp.
10. a. 0.0104
    b. 0.0821
    c. 0.0000439

Cumulative Review Exercises

Please be aware that some of the following problems may require knowledge of concepts presented in previous chapters.

1. Weekly Instruction Time The Organization for Economic Cooperation and Development provided the following mean weekly instruction times (hours) for elementary and high school students in various countries: 22.2 (United States); 24.8 (France); 24.2 (Mexico); 26.9 (China); 23.8 (Japan). Use the five given times for the following.

a. Find the mean.

b. Find the median.

c. Find the range.

d. Find the standard deviation.

e. Find the variance.

f. Use the range rule of thumb to identify the range of usual values.

g. Based on the result from part (f), are any of the times unusual? Why or why not?

h. What is the level of measurement of the data: nominal, ordinal, interval, or ratio?

i. Are the data discrete or continuous?

j. There is something fundamentally wrong with using the given times to find statistics such as the mean. What is wrong?

1. a. 24.4 hours
   b. 24.2 hours
   c. 4.7 hours
   d. 1.7 hours
   e. 2.9 hours^2
   f. Usual values: 21.0 hours to 27.8 hours.
   g. No, because none of the times are beyond the range of usual values.
   h. Ratio
   i. Continuous
   j. The given times come from countries with very different population sizes, so it does not make sense to treat the given times equally. Calculations of statistics should take the different population sizes into account. Also, the sample is very small, and there is no indication that the sample is random.

2. Ohio Pick 4 In Ohio's Pick 4 game, you pay $1 to select a sequence of four digits, such as 7709. If you buy only one ticket and win, your prize is $5000 and your net gain is $4999.

a. If you buy one ticket, what is the probability of winning?

2. a. 1/10,000 or 0.0001

b.

x	$P(x)$
−$1	0.9999
$4999	0.0001

c. 0.0365
d. 0.0352
e. −50¢
3. a. 0.282
b. 0.303
c. 0.242
d. 0.297
e. 0.0792
f. 0.738
g. 0.703

b. Construct a table describing the probability distribution corresponding to the purchase of one Pick 4 ticket.

c. If you play this game once every day, find the mean number of wins in years with exactly 365 days.

d. If you play this game once every day, find the probability of winning exactly once in 365 days.

e. Find the expected value for the purchase of one ticket.

3. Tennis Challenge In the last U.S. Open tennis tournament, there were 611 challenges made by singles players, and 172 of them resulted in referee calls that were overturned. The accompanying table lists the results by gender.

	Challenge Upheld with Overturned Call	Challenge Rejected with No Change
Challenges by Men	121	279
Challenges by Women	51	160

a. If one of the 611 challenges is randomly selected, what is the probability that it resulted in an overturned call?

b. If one of the challenges made by the men is randomly selected, what is the probability that it resulted in an overturned call?

c. If one of the challenges made by the women is randomly selected, what is the probability that it resulted in an overturned call?

d. If one of the overturned calls is randomly selected, what is the probability that the challenge was made by a woman?

e. If two different challenges are randomly selected with replacement, find the probability that they both resulted in an overturned call.

f. If one of the 611 challenges is randomly selected, find the probability that it was made by a man or was upheld with an overturned call.

g. If one of the challenged calls is randomly selected, find the probability that it was made by a man given that the call was upheld with an overturned call.

4. Because the vertical scale begins at 60 instead of 0, the difference between the two amounts is exaggerated. The graph makes it appear that men's earnings are roughly twice those of women, but men earn roughly 1.2 times the earnings of women.

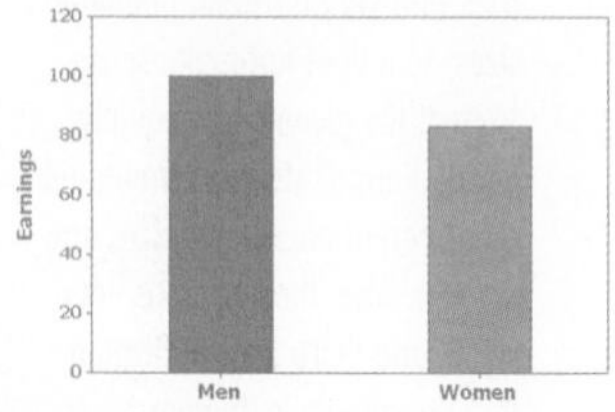

4. Gender Gap In recent years, the discrepancy between incomes of women and men has been shrinking, but it continues to exist. The accompanying graph illustrates the current gap. The graph shows that for every $100 earned by men, women earn $82.80 (based on data from the Bureau of Labor Statistics). Identify what is wrong with the graph, then redraw it so that it is not deceptive.

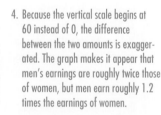

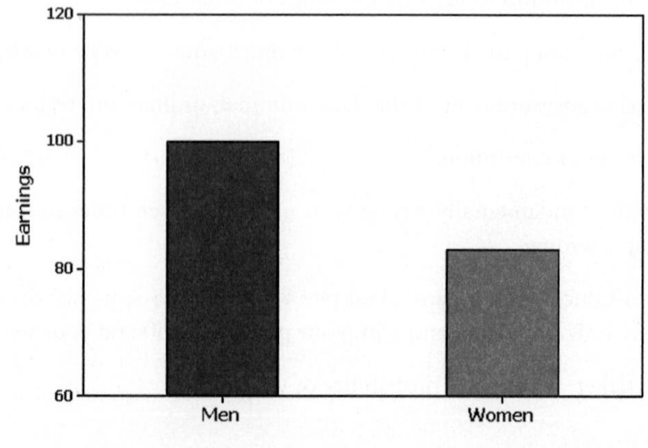

5. Random Digits The digits 0, 1, 2, 3, 4, 5, 6, 7, 8, and 9 are randomly selected for applications including the selection of lottery numbers and the selection of telephone numbers to be called as part of a survey. In the following tables, the table at the left summarizes actual results from 100 randomly selected digits, and the table at the right summarizes the probabilities of the different digits.

Digit	Frequency
0	9
1	7
2	12
3	10
4	10
5	11
6	8
7	8
8	14
9	11

Digit x	*P*(x)
0	0.1
1	0.1
2	0.1
3	0.1
4	0.1
5	0.1
6	0.1
7	0.1
8	0.1
9	0.1

a. What is the table at the left called?

b. What is the table at the right called?

c. Use the table at the left to find the mean. Is the result a statistic or a parameter?

d. Use the table at the right to find the mean. Is the result a statistic or a parameter?

e. If you were to randomly generate 1000 such digits, would you expect the mean of these 1000 digits to be close to the result from part (c) or part (d)? Why?

5. a. Frequency distribution or frequency table.
b. Probability distribution.
c. $\bar{x} = 4.7$. This value is a statistic.
d. $\mu = 4.5$. This value is a parameter.
e. The random generation of 1000 digits should have a mean close to $\mu = 4.5$ from part (d). The mean of 4.5 is the mean for the population of all random digits, so samples will have means that tend to center about 4.5.

6. Investing in College Based on a *USA Today* poll, assume that 10% of the population believes that college is no longer a good investment.

a. Find the probability that among 16 randomly selected people, exactly 4 believe that college is no longer a good investment.

b. Find the probability that among 16 randomly selected people, at least 1 believes that college is no longer a good investment.

c. The poll results were obtained by Internet users logged on to the *USA Today* Web site, and the Internet users decided whether to ignore the posted survey or respond. What type of sample is this? What does it suggest about the validity of the results?

6. a. 0.0514
b. 0.815
c. This is a voluntary response (or self-selected) sample. This suggests that the results might not be valid, because those with a strong interest in the topic are more likely to respond.

Technology Project

Overbooking Flights American Airlines Flight 201 from New York's JFK airport to LAX airport in Los Angeles uses a Boeing 767-200 with 168 seats available for passengers. Because some people with reservations don't show up, American can overbook by accepting more than 168 reservations. If the flight is not overbooked, the airline will lose revenue due to empty seats, but if too many seats are sold and some passengers are denied seats, the airline loses money from the compensation that must be given to the bumped passengers. Assume that there is a 0.0995 probability that a passenger with a reservation will not show up for the flight (based on data from the IBM research paper "Passenger-Based Predictive Modeling of Airline No-Show Rates," by Lawrence, Hong, and Cherrier). Also assume that the airline accepts 182 reservations for the 168 seats that are available.

Find the probability that when 182 reservations are accepted for American Airlines Flight 201, there are more passengers showing up than there are seats available. Table A-1 (the Binomial Probabilities table) cannot be used and calculations with the binomial probability formula would be extremely time-consuming and tedious. The best approach is to use statistics software or a TI-83/84 Plus calculator. (See Section 5-3 for instructions describing the use of STATDISK, Minitab, Excel, StatCrunch, or a TI-83/84 Plus calculator.) Is the probability of overbooking small enough so that it does not happen very often, or does it seem too high so that changes must be made to make it lower? Now use trial and error to find the maximum number of reservations that could be accepted so that the probability of having more passengers than seats is 0.05 or less.

from data TO DECISION

Critical Thinking: Did Mendel's results from plant hybridization experiments contradict his theory?

Gregor Mendel conducted original experiments to study the genetic traits of pea plants. In 1865 he wrote "Experiments in Plant Hybridization," which was published in *Proceedings of the Natural History Society.* Mendel presented a theory that when there are two inheritable traits, one of them will be dominant and the other will be recessive. Each parent contributes one gene to an offspring and, depending on the combination of genes, that offspring could inherit the dominant trait or the recessive trait. Mendel conducted an experiment using pea plants. The pods of pea plants can be green or yellow. When one pea carrying a dominant green gene and a recessive yellow gene is crossed with another pea carrying the same green/yellow genes, the offspring can inherit any one of these four combinations of genes: (1) green/green; (2) green/yellow; (3) yellow/green; (4) yellow/yellow. Because green is dominant and yellow is recessive, the offspring pod will be green if either of the two inherited genes is green. The offspring can have a yellow pod only if it inherits the yellow gene from each of the two parents. Given these conditions, we expect that 3/4 of the offspring peas should have green pods; that is, $P(\text{green pod}) = 3/4$.

When Mendel conducted his famous hybridization experiments using parent pea plants with the green/yellow combination of genes, he obtained 580 offspring. According to Mendel's theory, 3/4 of the offspring should have green pods, but the actual number of plants with green pods was 428. So the proportion of offspring with green pods to the total number of offspring is $428/580 = 0.738$. Mendel *expected* a proportion of 3/4 or 0.75, but his *actual result* is a proportion of 0.738.

a. Assuming that $P(\text{green pod}) = 3/4$, find the probability that among 580 offspring, the number of peas with green pods is *exactly* 428.

b. Assuming that $P(\text{green pod}) = 3/4$, find the probability that among 580 offspring, the number of peas with green pods is 428 *or fewer.*

c. Which of the two preceding probabilities should be used for determining whether 428 is an unusually low number of peas with green pods?

d. Use probabilities to determine whether 428 peas with green pods is an unusually low number. (*Hint:* See "Using Probabilities to Determine When Results Are Unusual" in Section 5-2.)

Cooperative Group Activities

1. In-class activity Win $1,000,000! The James Randi Educational Foundation offers a $1,000,000 prize to anyone who can show "under proper observing conditions, evidence of any paranormal, supernatural, or occult power or event." Divide into groups of three. Select one person who will be tested for extrasensory perception (ESP) by trying to correctly identify a digit randomly selected by another member of the group. Another group member should record the randomly selected digit, the digit guessed by the subject, and whether the guess was correct or wrong. Construct the table for the probability distribution of randomly generated digits, construct the relative frequency table for the random digits that were actually obtained, and construct a relative frequency table for the guesses that were made. After comparing the three tables, what do you conclude? What proportion of guesses is correct? Does it seem that the subject has the ability to select the correct digit significantly more often than would be expected by chance?

2. In-class activity See the preceding activity and *design an experiment* that would be effective in testing someone's claim that they have the ability to identify the color of a card selected from a standard deck of playing cards. Describe the experiment with great detail. Because the prize of $1,000,000 is at stake, we want to be careful to avoid the serious mistake of concluding that the person has a paranormal power when that power is not actually present. There will likely be some chance that the subject could make random guesses and be correct every time, so identify a probability that is reasonable for the event of the subject passing the test with random guesses. Be sure that the test is designed so that this probability is equal to or less than the probability value selected.

3. In-class activity Suppose we want to identify the probability distribution for the number of children born to randomly selected couples. For each student in the class, find the number of brothers and sisters and record the total number of children (including the student) in each family. Construct the relative frequency table for the result obtained. (The values of the random variable x will be 1, 2, 3, ...) What is wrong with using this relative frequency table as an estimate of the probability distribution for the number of children born to randomly selected couples?

4. Out-of-class activity The analysis of the last digits of data can sometimes reveal whether the data have been collected through actual measurements or reported by the subjects. Refer to an almanac or the Internet and find a collection of data (such as lengths of rivers in the world), then analyze the distribution of last digits to determine whether the values were obtained through actual measurements.

5. Out-of-class activity In the past, leading digits of the amounts on checks have been analyzed for fraud. For checks not involving fraud, the leading digit of 1 is expected about 30.1% of the time. Obtain a random sample of actual check amounts and record the leading digits. Compare the actual number of checks with amounts that have a leading digit of 1 to the 30.1% rate expected. Do the actual checks conform to the expected rate, or is there a substantial discrepancy? Explain.

6 Normal Probability Distributions

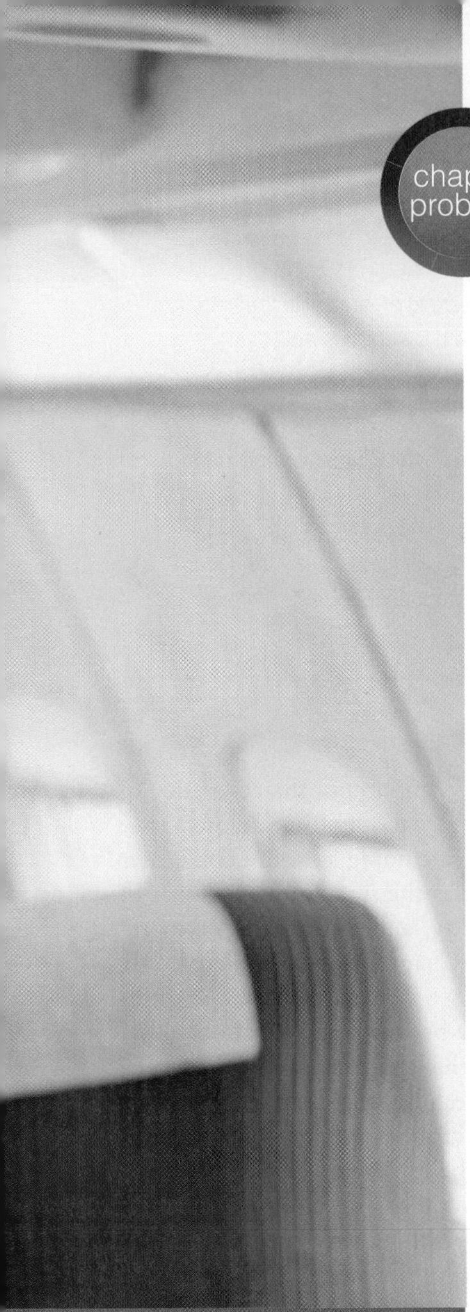

chapter problem

Reaching new heights

Ergonomics is the study of people fitting into their environments, and heights are extremely important in many applications. Section 4.4.2 of the Americans with Disabilities Act relates to height clearances with this statement: "Walks, halls, corridors, passageways, aisles, or other circulation spaces shall have 80 in. (2030 mm) minimum clear head room." A CBS News report identified many low-hanging signs in New York City subway walkways that violated that requirement with height clearances less than 80 in. Even when that 80 in. minimum height clearance is maintained, not all people can walk through without bending.

Due to aircraft cabin designs and other considerations, British Airways and many other carriers have a cabin crew height requirement between 5 ft 2 in. and 6 ft 1 in. For aesthetic reasons, Rockette dancers at New York's Radio City Music Hall must be between 66.5 in. and 71.5 in. tall. For practical reasons, the U.S. Army requires that women must be between 58 in. and 80 in. tall. For social reasons, Tall Clubs International requires that male members must be at least 6 ft 2 in. tall, and women members must be at least 5 ft 10 in. tall.

Given that heights are so important in so many different circumstances, what do we know about heights? Based on preceding chapters, we know that an investigation of heights should involve much more than simply finding a mean. We should consider the "CVDOT" elements of center, variation, distribution, outliers, and changes over time. We might use the mean as a measure of center, the standard deviation as a measure of variation, and the histogram as a tool for visualizing the distribution of the data, and we should determine whether outliers are present. We should also consider whether we are dealing with a static population or one that is changing over time. For heights of adults, we might refer to Data Set 1 in Appendix B to estimate that heights of adult males have a mean of 69.5 in. and a standard deviation of 2.4 in., while heights of adult females have a mean of 63.8 in. and a standard deviation of 2.6 in. (These values are very close to the values that would be obtained by using a much larger sample.) For distributions of heights, we might examine histograms, such as those shown here (based on Data Set 1 in Appendix B). Note that the histograms appear to be roughly bell-shaped, suggesting that the heights are from populations having normal distributions (as described in Section 2-3). For outliers, we might examine the histograms and note that there is one male with a height that is somewhat, but not dramatically, different from the others. Also, we know that heights are changing over time, so our studies will focus on current heights, not heights from past or future centuries.

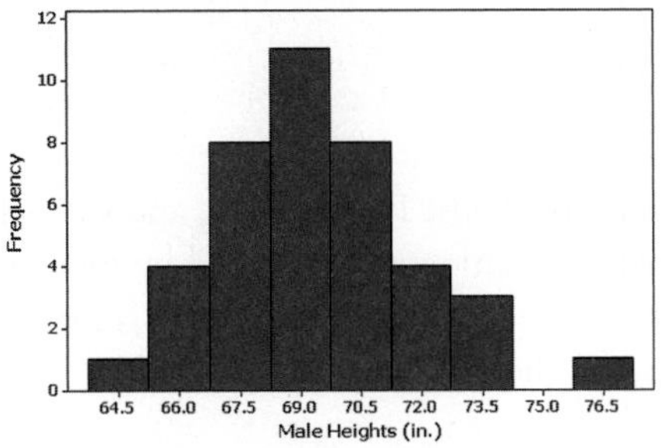

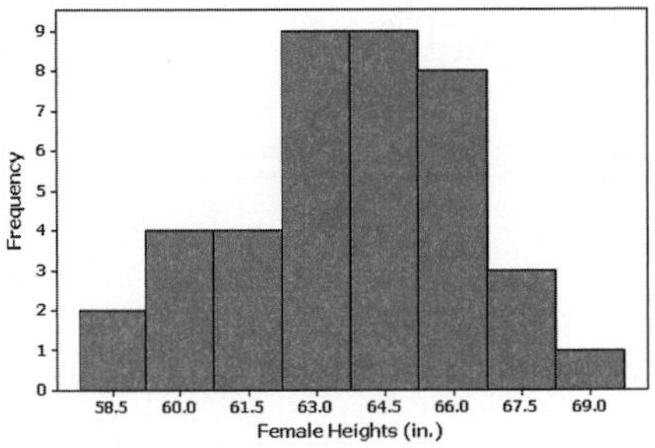

This chapter introduces the statistical tools that are basic to good ergonomic design. After completing this chapter, we will be able to solve problems in a wide variety of different disciplines as well. We will be able to answer questions such as these:

- What percentages of men and women can easily navigate in an area with the height clearance of 80 in. that is stipulated in the Americans with Disabilities Act?
- What percentages of men and women satisfy the flight cabin crew requirement of having a height between 5 ft 2 in. and 6 ft 1 in.?
- What percentage of women are eligible for membership in Tall Clubs International because they are at least 5 ft 10 in. tall?
- Current doorways are typically 6 ft 8 in. tall, but if we were to redesign doorways to accommodate 99% of the population, what should the height be?

Note to Instructor:
Organization Change: Section 6-6 ("Assessing Normality") and Section 6-7 ("Normal as Approximation to Binomial") are now switched from the order used in the previous edition. This is motivated by two factors: (1) Technology now makes it easy to generate normal quantile plots, so they become a good tool for assessing the normality of a population and "Assessing Normality" is now more accessible; (2) technology now makes it easy to find exact values of binomial probabilities, so the "Normal as Approximation to Binomial" coverage is now less important and it has been placed as the last section of Chapter 6.

Recommendation: Consider Sections 6-2, 6-3, and 6-5 to be essential. Also cover Section 6-4 if time is not a serious issue.

Recommendation: Be sure to inform your students of the materials that will be available to them on tests. Allow students to use the detachable Formula/Table card included with the book. It includes a copy of Table A-2 (Standard Normal Distribution). You might formally state that if students want to use the Formula/Table card on tests, they should not write their own notes on it.

6-1 Review and Preview

The preceding chapters introduced some extremely important characteristics of data. In Chapter 2 we considered the distribution of data, and in Chapter 3 we considered some important measures of data sets, including measures of center (such as the mean) and measures of variation (such as the standard deviation). In Chapter 4 we discussed basic principles of probability, and in Chapter 5 we presented the concept of a probability distribution. In Chapter 5 we considered only *discrete* probability distributions, but in this chapter we introduce *continuous* probability distributions. To illustrate the correspondence between area and probability, we begin with a *uniform distribution*, but most of this chapter focuses on *normal distributions.* Normal distributions occur often in real applications, and they play an important role in methods of inferential statistics. Here we present concepts of normal distributions that will be used often in the remaining chapters. Several of the statistical methods discussed in later chapters are based on concepts related to the central limit theorem discussed in Section 6-5. Many other sections require normally distributed populations, and Section 6-6 presents methods for analyzing sample data to determine whether the sample appears to be from a normally distributed population.

DEFINITION If a continuous random variable has a distribution with a graph that is symmetric and bell-shaped, as in Figure 6-1, and it can be described by the equation given as Formula 6-1, we say that it has a **normal distribution.**

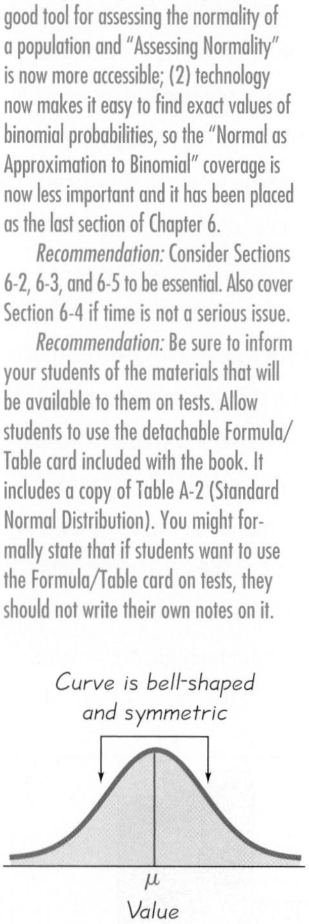

Figure 6-1
The Normal Distribution

Formula 6-1

$$y = \frac{e^{-\frac{1}{2}\left(\frac{x-\mu}{\sigma}\right)^2}}{\sigma\sqrt{2\pi}}$$

We won't actually use Formula 6-1, and we include it only to illustrate that any particular normal distribution is determined by two parameters: the mean, μ, and standard deviation, σ. In that formula, the letter π represents the constant value 3.14159 . . . and e represents the constant value 2.71828 . . . . The symbols μ and σ represent fixed values for the mean and standard deviation, respectively. Once specific values are selected for μ and σ, we can graph Formula 6-1 and the result will look like Figure 6-1. From Formula 6-1 we see that a normal distribution is determined by the fixed values of the mean μ and standard deviation σ. Fortunately, that's all we need to know about that formula.

6-2 The Standard Normal Distribution

Key Concept In this section we present the *standard normal distribution,* which has these three properties:

1. The graph of the standard normal distribution is bell-shaped (as in Figure 6-1).
2. The standard normal distribution has a mean equal to 0 (that is, $\mu = 0$).
3. The standard normal distribution has a standard deviation equal to 1 (that is, $\sigma = 1$).

In this section we develop the skill to find areas (or probabilities or relative frequencies) corresponding to various regions under the graph of the standard normal distribution. In addition, we find *z* scores that correspond to areas under the graph. These skills become important as we study nonstandard normal distributions and all of the real and important applications that they involve.

Note to Instructor
Begin Section 6-2 by asking this question in class: "If a satellite crashes at a random point on earth, what is the probability it will crash on land?" There are 54,225,000 square miles of land and 142,715,000 square miles of water, so the answer is 0.275. This demonstrates that some probability problems can be solved using areas. That's why it is important to see the relationship between area and probability, as in Examples 1 and 2 in Section 6-2. That relationship between area and probability will be used extensively in Chapter 6 and later chapters.

Comment that it is *extremely important* to master the procedures in Section 6-2 because they will be used often throughout the remainder of the course.

Uniform Distributions

The focus of this chapter is the concept of a normal probability distribution, but we begin with a *uniform distribution.* The uniform distribution allows us to see the following two very important properties:

1. The area under the graph of a probability distribution is equal to 1.
2. There is a correspondence between area and probability (or relative frequency), so some probabilities can be found by identifying the corresponding areas in the graph.

Chapter 5 considered only discrete probability distributions, but we now consider continuous probability distributions, beginning with the *uniform distribution.*

DEFINITION A continuous random variable has a **uniform distribution** if its values are spread *evenly* over the range of possibilities. The graph of a uniform distribution results in a rectangular shape.

Example 1 Subway to Mets Game

For New York City weekday late-afternoon subway travel from Times Square to the Mets stadium, you can take the #7 train that leaves Times Square every 5 minutes. Given the subway departure schedule and the arrival of a passenger, the waiting time x is between 0 min and 5 min, as described by the uniform distribution depicted in Figure 6-2. Note that in Figure 6-2, waiting times can be *any* value between 0 min and 5 min, so it is possible to have a waiting time of 2.33457 min. Note also that all of the different possible waiting times are equally likely.

The graph of a continuous probability distribution, such as in Figure 6-2, is called a **density curve.** A density curve must satisfy the following two requirements.

continued

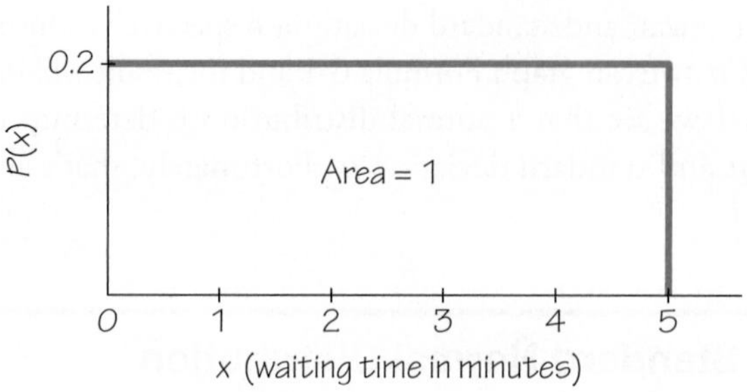

Figure 6-2 Uniform Distribution of Waiting Time

Requirements for a Density Curve

1. The total area under the curve must equal 1.
2. Every point on the curve must have a vertical height that is 0 or greater. (That is, the curve cannot fall below the *x*-axis.)

By setting the height of the rectangle in Figure 6-2 to be 0.2, we force the enclosed area to be $0.2 \times 5 = 1$, as required. (In general, the area of the rectangle becomes 1 when we make its height equal to the value of 1/range.) The requirement that the area must equal 1 simplifies probability problems, so the following statement is important:

Because the total area under the density curve is equal to 1, there is a correspondence between *area* and *probability*.

Example 2 Subway Waiting Time

Given the uniform distribution illustrated in Figure 6-2, find the probability that a randomly selected passenger has a waiting time greater than 2 minutes.

Solution

The shaded area in Figure 6-3 represents waiting times greater than 2 minutes. Because the total area under the density curve is equal to 1, there is a correspondence between area and probability. We can find the desired probability by using areas as follows:

$$\begin{aligned} P(\text{wait time greater than 2 min}) &= \text{area of shaded region in Figure 6-3} \\ &= 0.2 \times 3 \\ &= 0.6 \end{aligned}$$

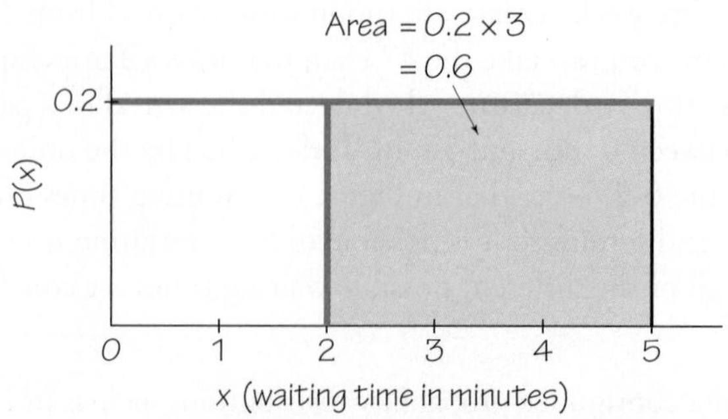

Figure 6-3 Using Area to Find Probability

Interpretation

The probability of randomly selecting a passenger with a waiting time greater than 2 minutes is 0.6.

Standard Normal Distribution

The density curve of a uniform distribution is a horizontal straight line, so we can find the area of any rectangular region by applying this formula: Area = height × width. Because the density curve of a normal distribution has a complicated bell shape as shown in Figure 6-1, it is more difficult to find areas. However, the basic principle is the same: *There is a correspondence between area and probability*. In Figure 6-4 we show that for a standard normal distribution, the area under the density curve is equal to 1.

Note to Instructor

Recommendation: Encourage students to use technology instead of Table A-2.

About Table A-2 (Standard Normal Distribution): This chapter and subsequent chapters make frequent use of Table A-2, which is based on a format of *cumulative area from the left*. This is the same format used by different technologies. Minitab, Excel, and STATDISK are all configured to work with cumulative areas from the left. (The TI-83/84 Plus calculator is configured to work with areas between any two particular values, and StatCrunch allows a variety of different formats.) Consequently, anyone using Minitab, Excel, or STATDISK will find that cumulative areas from the left are consistent with Table A-2 and their computer software. For those not using technology, Table A-2 has the advantage of being easier to work with.

DEFINITION The **standard normal distribution** is a normal distribution with the parameters of $\mu = 0$ and $\sigma = 1$. The total area under its density curve is equal to 1 (as in Figure 6-4).

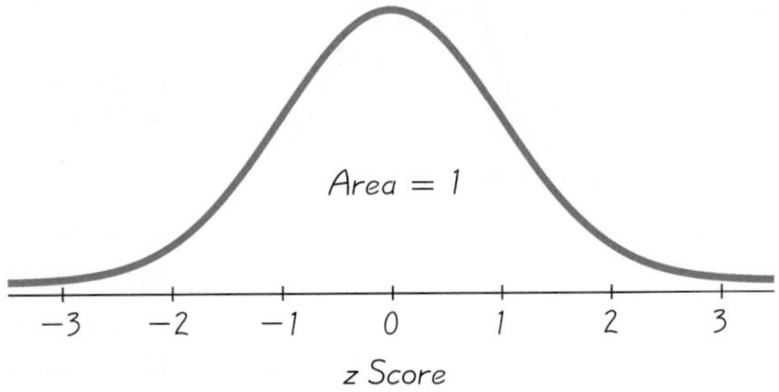

Figure 6-4 Standard Normal Distribution

It is not easy to manually find areas in Figure 6-4, but we have two other relatively easy ways of finding those areas: (1) Use technology; (2) use Table A-2 in Appendix A (the Standard Normal Distribution table in the Appendix).

Finding Probabilities when Given *z* Scores

We can find areas (or probabilities) for many different regions in Figure 6-4 by using a TI-83/84 Plus calculator or computer software such as STATDISK, Minitab, Excel, or StatCrunch, or we can also use Table A-2 (in Appendix A and the *Formulas and Tables* insert card). Key features of the different methods are summarized in Table 6-1 that follows. Because calculators or computer software generally give more accurate results than Table A-2, we strongly recommend using technology. (When there are discrepancies, answers in Appendix D will generally include results based on Table A-2 as well as answers based on technology.)

If using Table A-2, it is essential to understand these points:

1. Table A-2 is designed only for the *standard* normal distribution, which has a mean of 0 and a standard deviation of 1.

Note to Instructor

Recommendation: Require or strongly encourage the drawing of a graph for each problem solved. Point out that the graph provides a visual understanding that can be really helpful when solving the problems in this important chapter. Also, constantly stress the difference between *areas* under the curve and *z* scores that are *distances* representing the number of standard deviations that a value is away from the mean.

2. Table A-2 is on two pages, with the left page for *negative z* scores and the right page for *positive z* scores.
3. Each value in the body of the table is a *cumulative area from the left* up to a vertical boundary above a specific *z* score.
4. When working with a graph, avoid confusion between *z* scores and areas.

   z* score:** ***Distance **along the horizontal scale of the standard normal distribution; refer to the leftmost column and top row of Table A-2.**

   Area: ***Region*** **under the curve; refer to the values in the *body* of Table A-2.**

5. The part of the *z* score denoting hundredths is found across the top row of Table A-2.

CAUTION When working with a normal distribution, be careful to avoid confusion between *z* scores and areas.

Table 6-1 Methods for Finding Normal Distribution Areas

Table A-2, STATDISK, Minitab, Excel Gives the cumulative area from the left up to a vertical line above a specific value of *z*.	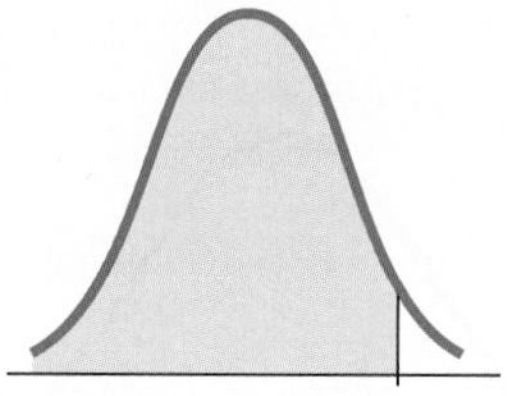	**TABLE A-2** The procedure for using Table A-2 is described in the text. **STATDISK** Select **Analysis, Probability Distributions, Normal Distribution.** Enter the *z* value, then click on **Evaluate.** **MINITAB** Select **Calc, Probability Distributions, Normal.** In the dialog box, select **Cumulative Probability, Input Constant.** **EXCEL** Select **fx, Statistical, NORMDIST.** In the dialog box, enter the value and mean, the standard deviation, and "true."
TI-83/84 Plus Calculator Gives area bounded on the left and bounded on the right by vertical lines above any specific values.	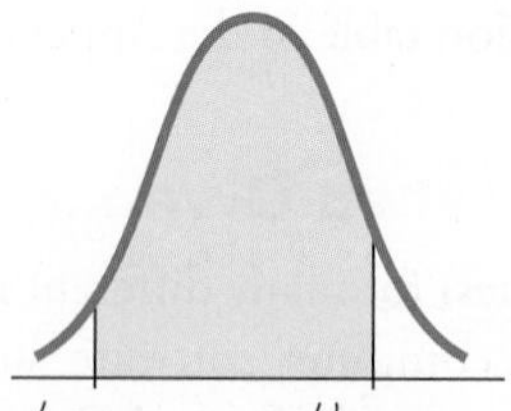	**TI-83/84** Press 2ND VARS **[2: normal cdf ( ],** then enter the two *z* scores separated by a comma, as in (left *z* score, right *z* score).

The following examples illustrate procedures that can be used with real and important applications introduced in the following sections.

Example 3 Bone Density Test

A bone mineral density test can be helpful in identifying the presence or likelihood of osteoporosis, a disease causing bones to become more fragile and more likely to break. The result of a bone density test is commonly measured as a *z* score. The population of z scores is normally distributed with a mean of 0 and a standard deviation of 1, so these test results meet the requirements of a standard normal distribution; Figure 6-4 is a graph of these test results.

A randomly selected adult undergoes a bone density test. Find the probability that the result is a reading less than 1.27.

Solution

We need to find the area in Figure 6-5 below $z = 1.27$. The *area* below $z = 1.27$ is equal to the *probability* of randomly selecting a person with a bone density test result that is less than 1.27. If using technology, see the instructions included at the end of this section. If using Table A-2, begin with the z score of 1.27 by locating 1.2 in the left column; next find the value in the adjoining row of probabilities that is directly below 0.07, as shown in the accompanying excerpt. Table A-2 shows that there is an area of 0.8980 corresponding to $z = 1.27$. We want the area *below* 1.27, and Table A-2 gives the cumulative area from the left, so the desired area is 0.8980. Because we have a correspondence between area and probability, we know that the probability of a z score below 1.27 is 0.8980.

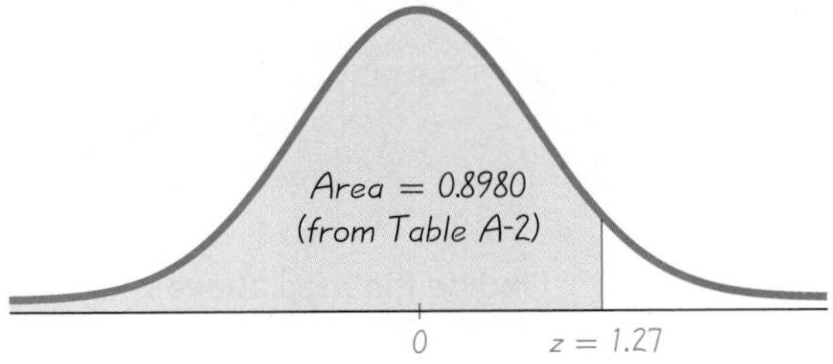

Figure 6-5
Finding Area below z = 1.27

TABLE A-2 (*continued*) Cumulative Area from the LEFT

z	.00	.01	.02	.03	.04	.05	.06	.07	.08	.09
0.0	.5000	.5040	.5080	.5120	.5160	.5199	.5239	.5279	.5319	.5359
0.1	.5398	.5438	.5478	.5517	.5557	.5596	.5636	.5675	.5714	.5753
0.2	.5793	.5832	.5871	.5910	.5948	.5987	.6026	.6064	.6103	.6141
1.0	.8413	.8438	.8461	.8485	.8508	.8531	.8554	.8577	.8599	.8621
1.1	.8643	.8665	.8686	.8708	.8729	.8749	.8770	.8790	.8810	.8830
1.2	.8849	.8869	.8888	.8907	.8925	.8944	.8962	.8980	.8997	.9015
1.3	.9032	.9049	.9066	.9082	.9099	.9115	.9131	.9147	.9162	.9177
1.4	.9192	.9207	.9222	.9236	.9251	.9265	.9279	.9292	.9306	.9319

Interpretation

The *probability* that a randomly selected person has a bone density test result below 1.27 is 0.8980, as shown as the shaded region in Figure 6-5. Another way to interpret this result is to conclude that 89.80% of people have bone density levels below 1.27.

Example 4 Bone Density Test

Using the same bone density test from Example 3, find the probability that a randomly selected person has a result above -1.00. A value above -1.00 is considered to be in the "normal" range of bone density readings.

Solution

We again find the desired *probability* by finding a corresponding *area*. We are looking for the area of the region to the right of $z = -1.00$ that is shaded in Figure 6-6. The accompanying STATDISK display shows that the area to the right of $z = -1.00$ is 0.841345.

If we use Table A-2, we should know that it is designed to apply only to cumulative areas from the *left*. Referring to the page with *negative z* scores, we find that the cumulative area from the left up to $z = -1.00$ is 0.1587 as shown in Figure 6-6. Because the total area under the curve is 1, we can find the shaded area by subtracting 0.1587 from 1. The result is 0.8413. Even though Table A-2 is designed only for cumulative areas from the left, we can use it to find cumulative areas from the right, as shown in Figure 6-6.

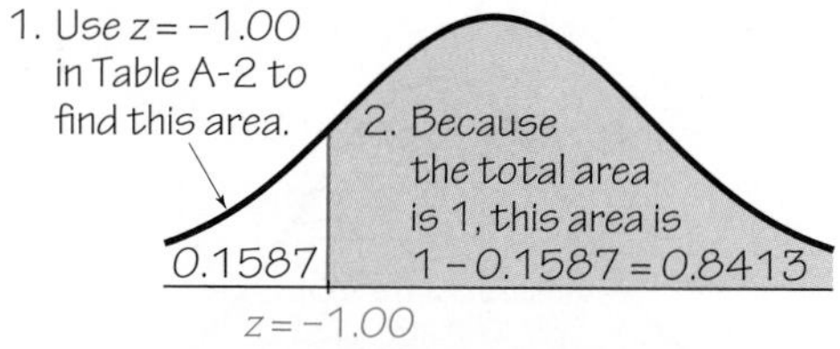

Figure 6-6 Finding the Area above z

STATDISK

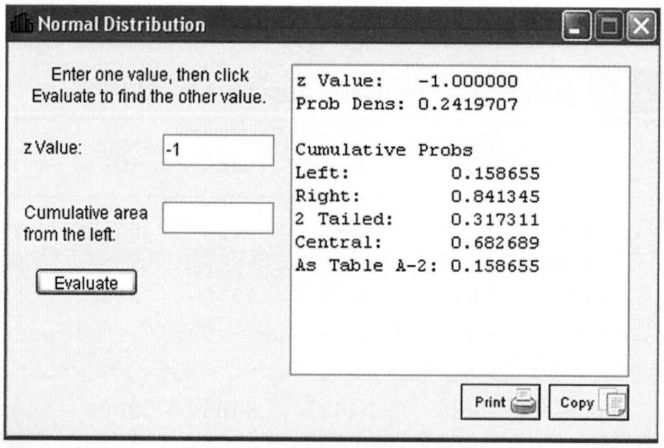

Interpretation

Because of the correspondence between probability and area, we conclude that the *probability* of randomly selecting someone with a bone density reading above -1 is 0.8413 (which is the *area* to the right of $z = -1.00$). We might also say that 84.13% of people have bone density levels above -1.00.

Example 4 illustrates a way that Table A-2 can be used indirectly to find a cumulative area from the right. The following example illustrates another way that we can find an area indirectly by using Table A-2.

Is the iPod Random Shuffle Really Random?

In *The Guardian*, Steven Levy wrote about an interview with the late Steve Jobs, former CEO of Apple, in which he presented Jobs with this dilemma: "I have a situation with my iPod. The shuffle function just doesn't seem random. Some artists come up way too much and some don't come up at all." According to Jeff Robbin, a head of the iTunes development team, "It is absolutely unequivocally random." Mathematician John Allen Paulos said that "[w]e often interpret and impose patterns on events that are random." Levy goes on to state that when we think that the iPod shuffle is not random, the problem is in our perceptions. Our minds perceive patterns and trends that don't really exist. We often hear runs of consecutive songs by the same artist and think that this is not random, but with true randomness, such consecutive runs are much more likely than we would expect.

The incorrect perception of nonrandomness caused Apple to introduce a "smart shuffle" feature in a new version of iTunes. This feature allows users to control multiple consecutive songs by the same artist. With this feature, consecutive runs by the same artist would be avoided. According to Steve Jobs, "We're making it less random to make it feel more random."

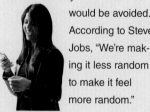

Example 5 Bone Density Test

A bone density test reading between -1.00 and -2.50 indicates that the subject has osteopenia, which is some bone loss. Find the probability that a randomly selected subject has a reading between -1.00 and -2.50.

Solution

We are again dealing with normally distributed values having a mean of 0 and a standard deviation of 1. The values between -1.00 and -2.50 correspond to the shaded region at the far right in Figure 6-7. Table A-2 cannot be used to find that area directly, but we can use it to find the following:

1. The area to the left of $z = -2.50$ is 0.0062.
2. The area to the left of $z = -1.00$ is 0.1587.
3. The area *between* $z = -2.50$ and $z = -1.00$ (the shaded area at the far right in Figure 6-7) is the difference between the areas found in the preceding two steps:

$$0.1587 - 0.0062 = 0.1525$$

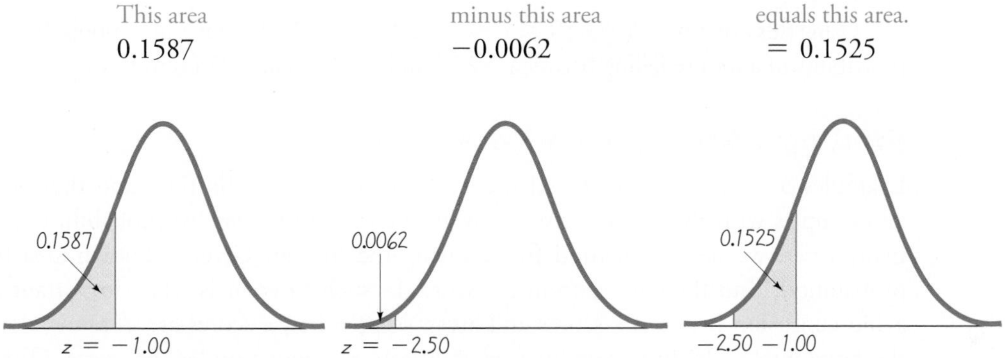

Figure 6-7 Finding the Area between Two z Scores

Interpretation

Using the correspondence between probability and area, we conclude that there is a probability of 0.1525 that a randomly selected subject has a bone density reading between -1.00 and -2.50. Another way to interpret this result is to state that 15.25% of people have osteopenia, with bone density readings between -1.00 and -2.50.

Example 5 can be generalized as the following rule: **The area corresponding to the region *between* two z scores can be found by finding the difference between the two areas found in Table A-2.** Figure 6-8 on the next page illustrates this general rule. The shaded region B can be found by calculating the *difference* between two areas found from Table A-2.

Learning hint: Don't try to memorize a rule or formula for this case. Focus on *understanding* that Table A-2 gives cumulative areas from the left only. Draw a graph, shade the desired area, then think of a way to find the desired area given the condition that Table A-2 provides only cumulative areas from the left.

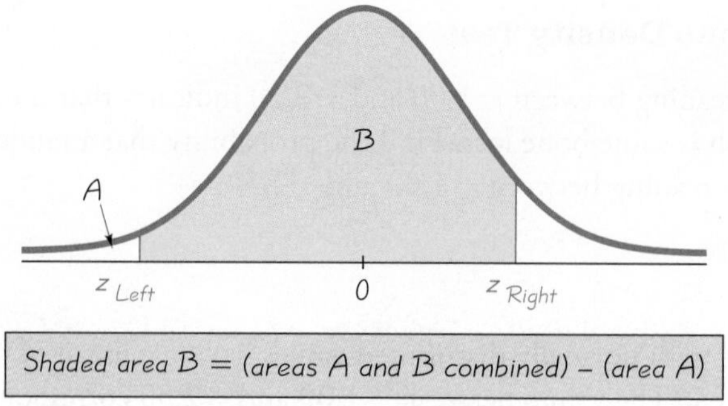

Figure 6-8 Finding the Area between Two z Scores

Note to Instructor
Many of us assume that terms such as "at most 2" or "at least 2" are clearly understood by all students, but that is an incorrect assumption. It could be very helpful to emphasize the importance of understanding and interpreting such terms. Ask them which job offer they would prefer: One that pays at most \$90,000, one that pays no more than \$90,000, or one that pays at least \$90,000.

Probabilities such as those in the preceding examples can also be expressed with the following notation.

Notation

$P(a < z < b)$ denotes the probability that the z score is between a and b.

$P(z > a)$ denotes the probability that the z score is greater than a.

$P(z < a)$ denotes the probability that the z score is less than a.

Using this notation, $P(-2.50 < z < -1.00) = 0.1525$, states in symbols that the probability of a z score falling between -2.50 and -1.00 is 0.1525 (as in Example 5).

Note to Instructor
There are three important points that should be emphasized in this subsection: (1) The numbers in the body of Table A-2 are *areas* and the numbers in the extreme left column (and across the top row) are *z* scores that are actually distances; (2) if a *z* score is to the *left* of the center line, it must be negative; (3) it's always a good idea to check answers to be sure that they are reasonable in the sense of being consistent with the locations of values in the graph.

Finding *z* Scores from Known Areas

Examples 3, 4, and 5 all involved the standard normal distribution, and they were all examples with this same format: Given *z* scores, find areas (or probabilities). In many cases, we need a method for reversing the format: Given a known area (or probability), find the corresponding *z* score. In such cases, it is really important to avoid confusion between *z* scores and areas. Remember, *z* scores are *distances* along the horizontal scale, but areas (or probabilities) are regions under the curve. (Table A-2 lists *z* scores in the left column and across the top row, but areas are found in the *body* of the table.) We should also remember that *z* scores positioned in the left half of the curve are always negative. If we already know a probability and want to find the corresponding *z* score, we use the following procedure.

Procedure for Finding a *z* Score from a Known Area

1. Draw a bell-shaped curve and identify the region under the curve that corresponds to the given probability. If that region is not a cumulative region from the left, work instead with a known region that is a cumulative region from the left.
2. Use technology or Table A-2 to find the *z* score. With Table A-2, use the cumulative area from the left, locate the closest probability in the *body* of the table, and identify the corresponding *z* score.

Example 6 **Bone Density Test**

Use the same bone density test scores described in Example 3. Those scores are normally distributed with a mean of 0 and a standard deviation of 1, so they meet the requirements of a standard normal distribution. Find the bone density score corresponding to P_{95}, the 95th percentile. That is, find the bone density score that separates the bottom 95% from the top 5%. See Figure 6-9.

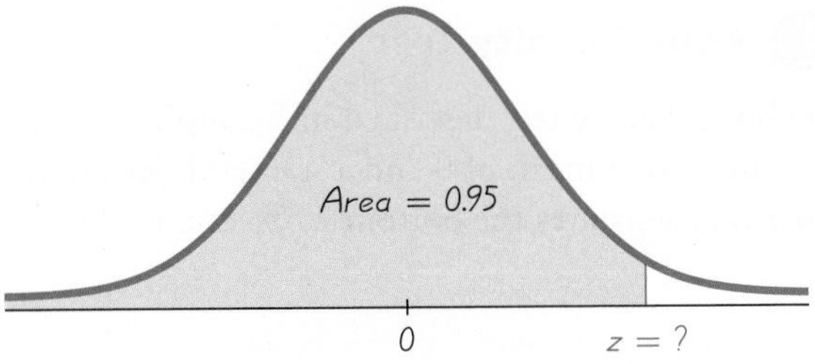

Figure 6-9 **Finding the 95th Percentile**

Solution

Figure 6-9 shows the z score that is the 95th percentile, with 95% of the area (or 0.95) below it. We could find the z score using technology. The accompanying Excel display shows that the z score with an area of 0.95 to its left is $z = 1.644853627$, or 1.645 when rounded.

If using Table A-2, search for the area of 0.95 *in the body* of the table and then find the corresponding z score. In Table A-2 we find the areas of 0.9495 and 0.9505, but there's an asterisk with a special note indicating that 0.9500 corresponds to a z score of 1.645. We can now conclude that the z score in Figure 6-9 is 1.645, so the 95th percentile is $z = 1.645$.

EXCEL

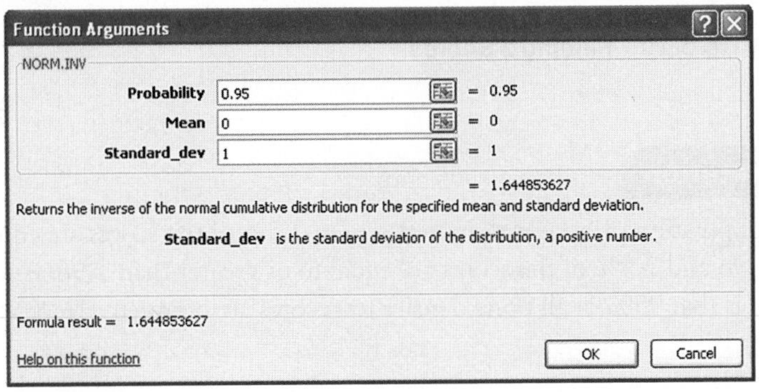

Interpretation

For bone density test scores, 95% of the scores are less than or equal to 1.645, and 5% of them are greater than or equal to 1.645.

Special Cases In the solution to Example 6, Table A-2 led to a z score of 1.645, which is midway between 1.64 and 1.65. When using Table A-2, we can usually avoid interpolation by simply selecting the closest value. The accompanying table lists special cases that are often used in a wide variety of applications. (For one of those special cases, the value of $z = 2.576$ gives an area slightly closer to the area of 0.9950, but $z = 2.575$ has the advantage of being the value midway between $z = 2.57$ and $z = 2.58$.) Except in these special cases, we can usually select the closest value in the table. (If a desired value is midway between two table values, select the larger value.) For z scores above 3.49, we can use 0.9999 as an approximation of the cumulative area from the left; for z scores below -3.49, we can use 0.0001 as an approximation of the cumulative area from the left.

Special Cases from Table A-2

z Score	Cumulative Area from the Left
1.645	0.9500
−1.645	0.0500
2.575	0.9950
−2.575	0.0050
Above 3.49	0.9999
Below −3.49	0.0001

Example 7 Bone Density Test

Using the same bone density test described in Example 3, we have a standard normal distribution with a mean of 0 and a standard deviation of 1. Find the bone density test score that separates the bottom 2.5% and find the score that separates the top 2.5%.

Solution

The required z scores are shown in Figure 6-10. Those z scores can be found using technology. If using Table A-2 to find the z score located to the left, we search the *body of the table* for an area of 0.025. The result is $z = -1.96$. To find the z score located to the right, we search *the body of Table A-2* for an area of 0.975. (Remember that Table A-2 always gives cumulative areas from the *left*.) The result is $z = 1.96$. The values of $z = -1.96$ and $z = 1.96$ separate the bottom 2.5% and the top 2.5%, as shown in Figure 6-10.

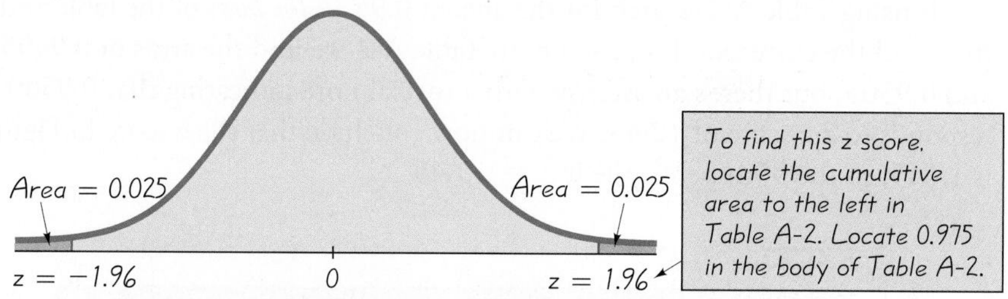

Figure 6-10 Finding z Scores

Interpretation

For the population of bone density test scores, 2.5% of the scores are equal to or less than -1.96 and 2.5% of the scores are equal to or greater than 1.96. Another interpretation is that 95% of all bone density test scores are between -1.96 and 1.96.

Critical Values For a normal distribution, a *critical value* is a z score on the borderline separating the z scores that are *likely* to occur from those that are unlikely. Common critical values are $z = -1.96$ and $z = 1.96$, and they are obtained as shown in Example 7. In Example 7, the values below $z = -1.96$ are unlikely, because only 2.5% of the population have scores below -1.96, and the values above $z = 1.96$ are unlikely because only 2.5% of the population have scores above 1.96. The reference to *critical values* is not so important in this chapter, but will become extremely important in following chapters. The following notation is used for critical z values found by using the standard normal distribution.

DEFINITION For the standard normal distribution, a **critical value** is a z score separating unlikely values from those that are likely to occur.

Notation

The expression z_α denotes the z score with an area of α to its right. (α is the Greek letter alpha.)

Example 8 **Finding z_α**

In the expression z_α, let $\alpha = 0.025$ and find the value of $z_{0.025}$.

Solution

The notation of $z_{0.025}$ is used to represent the z score with an area of 0.025 to its right. Refer to Figure 6-10 and note that the value of $z = 1.96$ has an area of 0.025 to its right, so $z_{0.025} = 1.96$.

CAUTION When using Table A-2 for finding a value of z_α for a particular value of α, note that α is the area to the *right* of z_α, but Table A-2 lists cumulative areas to the *left* of a given z score. To find the value of z_α by using the table, resolve that conflict by using the value of $1 - \alpha$. In Example 8, the value of $z_{0.025}$ can be found by locating the area of 0.9750 in the body of the table.

Examples 3 through 7 in this section are based on the real application of the bone density test, with scores that are normally distributed with a mean of 0 and standard deviation of 1, so that these scores have a standard normal distribution. Apart from the bone density test scores, it is rare to find such convenient parameters, because typical normal distributions involve means different from 0 and standard deviations different from 1. In the next section we present methods for working with such normal distributions.

using TECHNOLOGY

When working with the standard normal distribution, a technology can be used to find z scores or areas instead of Table A-2. The following instructions describe how to find such z scores or areas with technology.

STATDISK Select **Analysis, Probability Distributions, Normal Distribution.** Either enter the z score to find corresponding areas, or enter the cumulative area from the left to find the z score. After entering a value, click on the **Evaluate** button. See the STATDISK display included with Example 4.

MINITAB

- To find the cumulative area to the left of a z score (as in Table A-2), select **Calc, Probability Distributions, Normal, Cumulative probabilities.** Then enter the mean of 0 and standard deviation of 1. Click on the **Input Constant** button and enter the z score.
- To find a z score corresponding to a known probability, select **Calc, Probability Distributions, Normal.** Then select **Inverse cumulative probabilities** and the option **Input constant.** For the input constant, enter the total area to the left of the given value.

EXCEL

- To find the cumulative area to the left of a z score (as in Table A-2), click on ***fx***, then select **Statistical, NORMSDIST** (or **NORMS.DIST**). Enter the z score.
- To find a z score corresponding to a known probability, select ***fx*, Statistical, NORMSINV** (or **NORMS.INV**). Enter the total area to the left of the given value.

TI-83/84 PLUS Unlike most other technologies, the TI-83/84 Plus calculator does not base areas on cumulative regions from the left. Instead, the areas correspond to the z score that is a left boundary and another z score that is a right boundary. Press 2ND VARS and select **normalcdf.** Proceed to enter the two z scores separated by a comma, as in (left z score, right z score). Example 5 could be solved with the command of **normalcdf(− 2.50, − 1.00),** which yields a probability of 0.1524 (rounded) as shown in the accompanying screen.

TI-83/84 PLUS

```
normalcdf(-2.50,
-1.00)
             .1524455797
```

To find a z score corresponding to a known probability, press 2ND VARS and select **invNorm.** Proceed to enter the total area to the left of the z score. For example, the command of **invNorm(0.975)** yields a z score of 1.959963986, which is rounded to 1.96, as in Example 8.

STATCRUNCH Click on **Open StatCrunch,** then click on **Stat.** Select **Calculators,** then select **Normal.** You can either enter a probability or a value of x. Click on **Compute.**

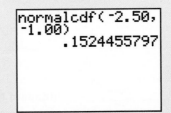

Recommended Assignment
Exercises 1–24.

6-2 Basic Skills and Concepts

Statistical Literacy and Critical Thinking

1. The word "normal" has a special meaning in statistics. It refers to a specific bell-shaped distribution that can be described by Formula 6-1.

2.

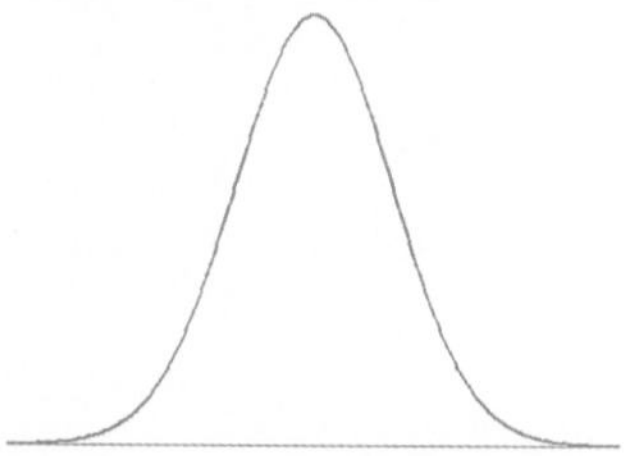

3. The mean and standard deviation have the values of $\mu = 0$ and $\sigma = 1$.

4. The notation z_α represents the z score that has an area of α to its right.

5. 0.75

6. 0.15

7. 0.4

8. 0.6

1. Normal Distribution When we refer to a "normal" distribution, does the word *normal* have the same meaning as in ordinary language, or does it have a special meaning in statistics? What exactly is a normal distribution?

2. Normal Distribution A normal distribution is informally described as a probability distribution that is "bell-shaped" when graphed. Draw a rough sketch of a curve having the bell shape that is characteristic of a normal distribution.

3. Standard Normal Distribution Identify the requirements necessary for a normal distribution to be a *standard* normal distribution.

4. Notation What does the notation z_α indicate?

Continuous Uniform Distribution. ***In Exercises 5–8, refer to the continuous uniform distribution depicted in Figure 6-2 and described in Example 1. Assume that a subway passenger is randomly selected, and find the probability that the waiting time is within the given range.***

5. Greater than 1.25 minutes

6. Less than 0.75 minutes

7. Between 1 minute and 3 minutes

8. Between 1.5 minutes and 4.5 minutes

9. 0.6700

10. 0.8508

11. 0.6992 (Tech: 0.6993)

12. 0.6063

Standard Normal Distribution. ***In Exercises 9–12, find the area of the shaded region. The graph depicts the standard normal distribution of bone density scores with mean 0 and standard deviation 1.***

9.

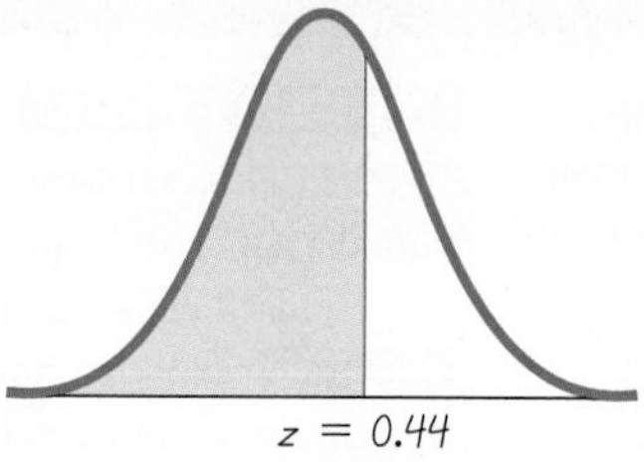

10.

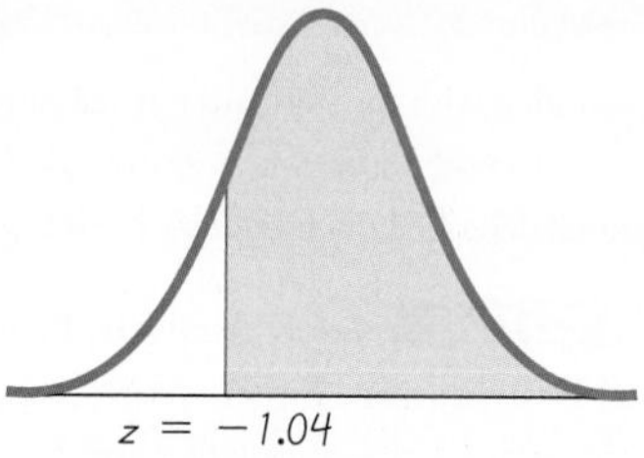

11.

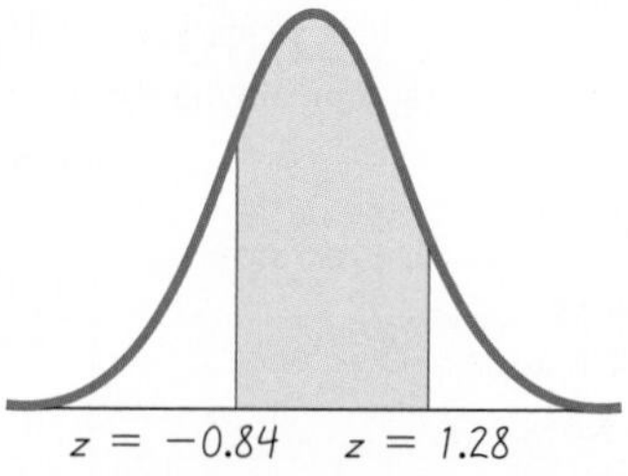

12.

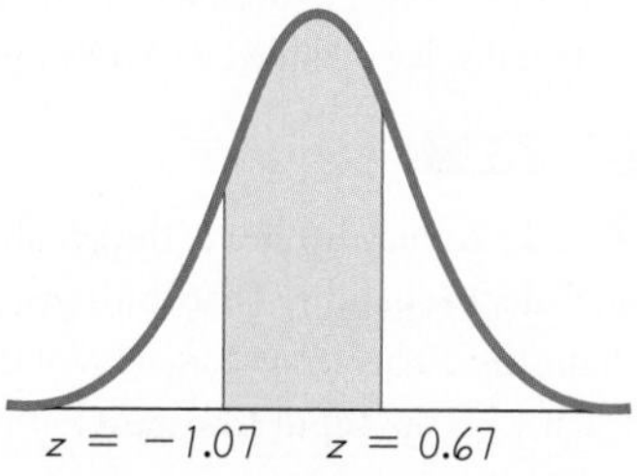

13. 1.23

14. −0.51

Standard Normal Distribution. ***In Exercises 13–16, find the indicated z score. The graph depicts the standard normal distribution of bone density scores with mean 0 and standard deviation 1.***

13.

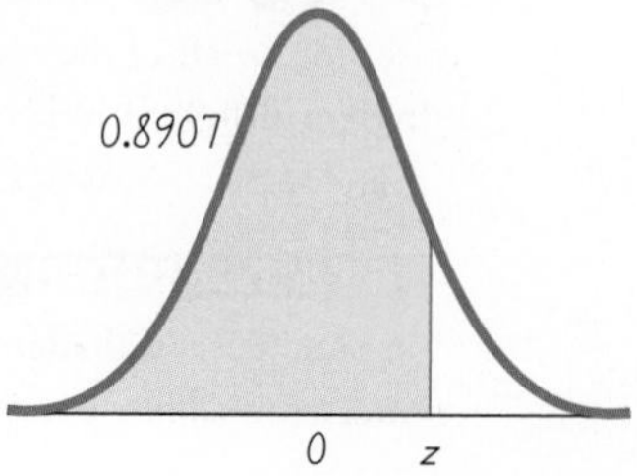

14.

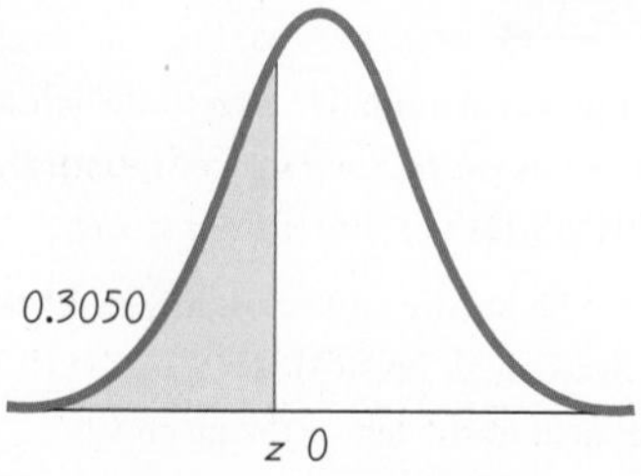

15.

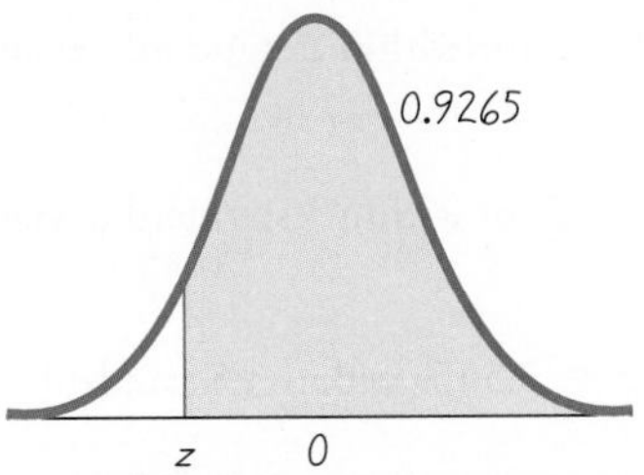

16.

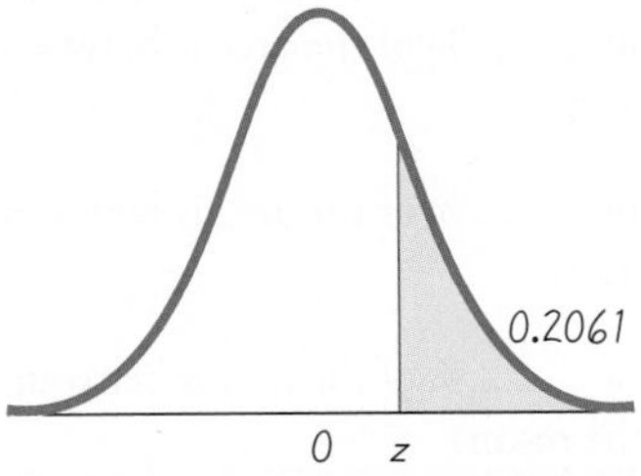

15. −1.45
16. 0.82

Standard Normal Distribution. *In Exercises 17–36, assume that a randomly selected subject is given a bone density test. Those test scores are normally distributed with a mean of 0 and a standard deviation of 1. In each case, draw a graph and find the probability of the given scores. If using technology instead of Table A-2, round answers to four decimal places.*

17. Less than −2.04

18. Less than −0.19

19. Less than 2.33

20. Less than 1.96

21. Greater than 0.82

22. Greater than 1.82

23. Greater than −1.50

24. Greater than −0.84

25. Between 0.25 and 1.25

26. Between 1.23 and 2.37

27. Between and −2.75 and −2.00

28. Between −1.93 and −0.45

29. Between −2.20 and 2.50

30. Between −0.62 and 1.78

31. Between −2.11 and 4.00

32. Between −3.90 and 2.00

33. Less than 3.65

34. Greater than −3.80

35. Greater than 0

36. Less than 0

17. 0.0207
18. 0.4247
19. 0.9901
20. 0.9750
21. 0.2061
22. 0.0344
23. 0.9332
24. 0.7995
25. 0.2957 (Tech: 0.2956)
26. 0.1004 (Tech: 0.1005)
27. 0.0198
28. 0.2996
29. 0.9799
30. 0.6949
31. 0.9825
32. 0.9771 (Tech: 0.9772)
33. 0.9999
34. 0.9999
35. 0.5000
36. 0.5000

Finding Bone Density Scores. *In Exercises 37–40 assume that a randomly selected subject is given a bone density test. Bone density test scores are normally distributed with a mean of 0 and a standard deviation of 1. In each case, draw a graph, then find the bone density test score corresponding to the given information.*

37. Find P_{90}, the 90th percentile. This is the bone density score separating the bottom 90% from the top 10%.

37. 1.28

38. Find P_5, the 5th percentile. This is the bone density score separating the bottom 5% from the top 95%.

38. −1.645

39. If bone density scores in the bottom 2.5% and the top 2.5% are used as cutoff points for levels that are too low or too high, find the two readings that are cutoff values.

39. −1.96, 1.96

40. Find the bone density scores that can be used as cutoff values separating the most extreme 1% of all scores.

40. −2.575, 2.575

Finding Critical Values. *In Exercises 41–44, find the indicated critical value.*

41. $z_{0.025}$

42. $z_{0.01}$

43. $z_{0.05}$

44. $z_{0.03}$

41. 1.96
42. 2.33
43. 1.645
44. 1.88

Basis for the Range Rule of Thumb and the Empirical Rule. *In Exercises 45–48, find the indicated area under the curve of the standard normal distribution, then convert it to a percentage and fill in the blank. The results form the basis for the range rule of thumb and the empirical rule introduced in Section 3-3.*

45. About _____% of the area is between $z = -1$ and $z = 1$ (or within 1 standard deviation of the mean).

45. 68.26% (Tech: 68.27%)

46. 95.44% (Tech: 95.45%)

46. About ____% of the area is between $z = -2$ and $z = 2$ (or within 2 standard deviations of the mean).

47. 99.74% (Tech: 99.73%)

47. About ____% of the area is between $z = -3$ and $z = 3$ (or within 3 standard deviations of the mean).

48. 99.98% (Tech: 99.95%)

48. About ____% of the area is between $z = -3.5$ and $z = 3.5$ (or within 3.5 standard deviations of the mean).

6-2 Beyond the Basics

49. a. 68.26% (Tech: 68.27%)
b. 4.56%
c. 95.00%
d. 95.44% (Tech: 95.45%)
e. 0.26% (Tech: 0.27%)

49. For bone density scores that are normally distributed with a mean of 0 and a standard deviation of 1, find the *percentage* of scores that are

a. within 1 standard deviation of the mean.

b. more than 2 standard deviations away from the mean.

c. within 1.96 standard deviations of the mean.

d. between $\mu - 2\sigma$ and $\mu + 2\sigma$.

e. more than 3 standard deviations away from the mean.

50. a. $\mu = 2.5$ min; $\sigma = 5/\sqrt{12}$ min, or 1.4 min
b. The probability is $1/\sqrt{3}$ or 0.5774, and it is very different from the probability of 0.6826 (Tech: 0.6827) that would be obtained by incorrectly using the standard normal distribution. The distribution does affect the results very much.

50. In a continuous uniform distribution,

$$\mu = \frac{\text{minimum} + \text{maximum}}{2} \quad \text{and} \quad \sigma = \frac{\text{range}}{\sqrt{12}}$$

a. Find the mean and standard deviation for the distribution of the subway waiting times represented in Figure 6-2.

b. For a continuous uniform distribution with $\mu = 0$ and $\sigma = 1$, the minimum is $-\sqrt{3}$ and the maximum is $\sqrt{3}$. For this continuous uniform distribution, find the probability of randomly selecting a value between -1 and 1, and compare it to the value that would be obtained by incorrectly treating the distribution as a standard normal distribution. Does the distribution affect the results very much?

Note to Instructor
This section involves *nonstandard* normal distributions, which do not have $\mu = 0$ and $\sigma = 1$, so students will be solving practical and applied problems that are actually used in many different circumstances. After using the simple transformation of $z = (x - \mu)/\sigma$, we can use the same basic procedures presented in Section 6-2. If students have not yet mastered the procedures from Section 6-2, they should definitely go back and master them before continuing.

TOP 20

6-3 Applications of Normal Distributions

Key Concept The objective of this section is to extend the methods of the previous section so that we can work with normal distributions having any mean and any standard deviation, instead of being limited to the standard normal distribution with mean 0 and standard deviation 1. The key element of this section is a simple conversion (Formula 6-2) that allows us to standardize any normal distribution so that the methods of the preceding section can be used with real and meaningful applications. Specifically, given some nonstandard normal distribution, we should be able to find probabilities corresponding to values of the variable x, and given some probability value, we should be able to find the corresponding value of the variable x.

When working with a normal distribution that is nonstandard (with a mean different from 0 and/or a standard deviation different from 1), we use Formula 6-2 to transform a value x to a z score, then we proceed with the same methods from Section 6-2.

Formula 6-2

$$z = \frac{x - \mu}{\sigma} \quad \text{(round } z \text{ scores to 2 decimal places)}$$

Some calculators and computer software programs do not require the above conversion to z scores because probabilities can be found directly. However, if you use Table A-2 to find probabilities, you **must** first convert values to standard z scores. Whether using technology or Table A-2, it is important to clearly understand the above conversion principle, because it is an important foundation for concepts introduced in the following chapters.

Figure 6-11 illustrates the conversion from a nonstandard to a standard normal distribution. The area in any normal distribution bounded by some score x (as in Figure 6-11(a)) is the same as the area bounded by the equivalent z score in the standard normal distribution (as in Figure 6-11(b)). This shows that when working with a nonstandard normal distribution, you can use Table A-2 the same way it was used in Section 6-2, provided that you first convert the values to z scores.

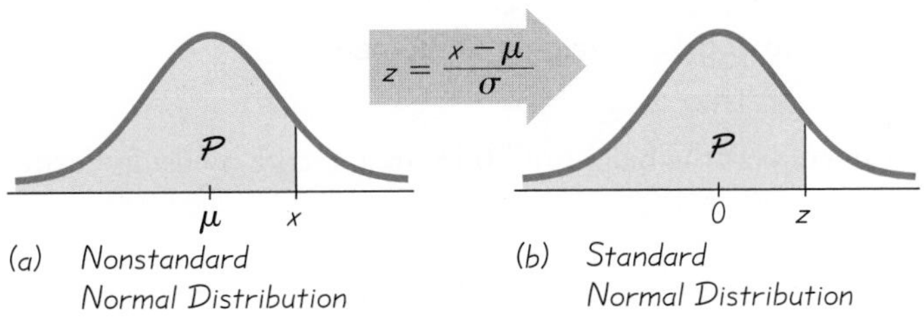

Figure 6-11 Converting Distributions

When finding areas with a nonstandard normal distribution, use the following procedure.

Procedure for Finding Areas with a Nonstandard Normal Distribution

1. Sketch a normal curve, label the mean and any specific x values, then *shade* the region representing the desired probability.
2. For each relevant value x that is a boundary for the shaded region, use Formula 6-2 to convert that value to the equivalent z score.
3. Use computer software or a calculator or Table A-2 to find the area of the shaded region. This area is the desired probability.

The following example applies these three steps to illustrate the relationship between a typical nonstandard normal distribution and the standard normal distribution.

Example 1 What Proportion of Women Are Eligible for Tall Clubs International?

The social organization Tall Clubs International has a requirement that women must be at least 70 in. tall. Given that women have normally distributed heights with a mean of 63.8 in. and a standard deviation of 2.6 in., find the percentage of women who satisfy that height requirement.

Solution

Step 1: See Figure 6-12, which incorporates this information: Women have heights that are normally distributed with a mean of 63.8 in. and a standard deviation of 2.6 in. The shaded region represents the women who satisfy the height requirement by being at least 70 in. tall.

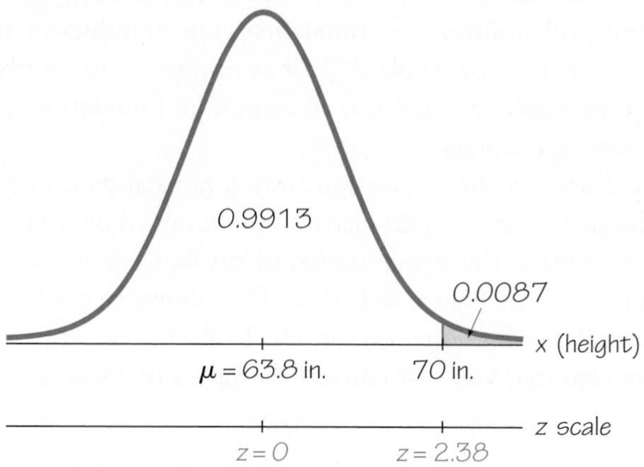

Figure 6-12 Heights of Women

Step 2: We can convert the height of 70 in. to a z score by using Formula 6-2 as follows:

$$z = \frac{x - \mu}{\sigma} = \frac{70 - 63.8}{2.6} = 2.38$$

Step 3: To use technology, refer to the instructions at the end of this section. Shown here is the STATDISK display that results from an entry of $z = 2.38$; it shows that the area to the right of $z = 2.38$ is 0.008656 (or 0.0087 rounded) and that is the shaded area in Figure 6-12.

STATDISK

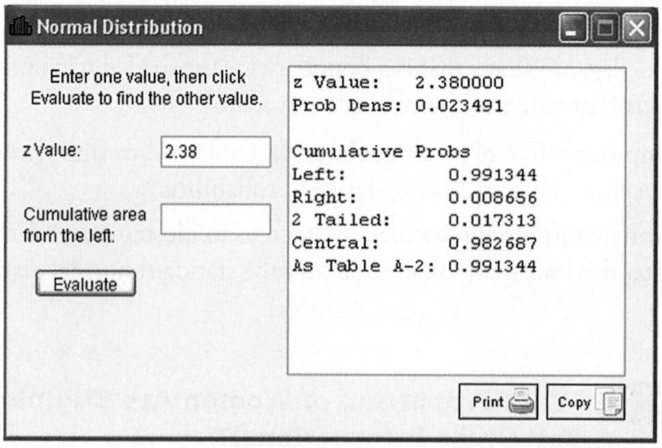

To use Table A-2, refer to that table with $z = 2.38$ and find that the cumulative area to the *left* of $z = 2.38$ is 0.9913. (Remember, Table A-2 is designed so that all areas are cumulative areas from the *left*.) Because the total area under the curve is 1, it follows that the shaded area in Figure 6-12 is $1 - 0.9913 = 0.0087$.

Interpretation

The proportion of women taller than 70 in. is 0.0087, or 0.87%. That is, just under 1% of women meet the minimum height requirement of 70 in. Tall Clubs International allows only the tallest of women.

Example 2 Airline Flight Crew Requirement

The Chapter Problem stated that British Airways and many other airlines have a requirement that a member of the cabin crew must have a height between 62 in. and 73 in. (or between 5 ft 2 in. and 6 ft 1 in.). Given that men have normally distributed heights with a mean of 69.5 in. and a standard deviation of 2.4 in., find the percentage of men who satisfy that height requirement.

Solution

Figure 6-13 shows the shaded region representing heights of men between 62 in. and 73 in.

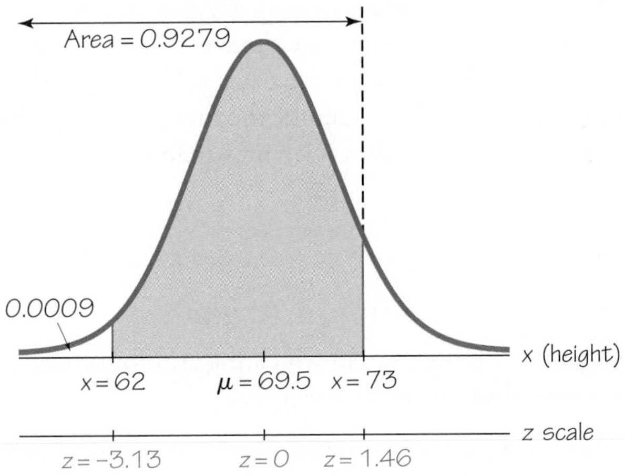

Figure 6-13 Heights of Men

Step 1: See Figure 6-13, which incorporates this information: Men have heights that are normally distributed with a mean of 69.5 in. and a standard deviation of 2.4 in. The shaded region represents the men who satisfy the height requirement by having a height between 62 in. and 73 in.

Step 2: To use technology, refer to the instructions at the end of this section. Technology will show that the shaded area in Figure 6-13 is 0.9267.

If using Table A-2, we cannot find the shaded area directly, but we can find it indirectly by using the same procedures from Section 6-2, as follows: (1) Find the cumulative area from the left up to 73 in. (or $z = 1.46$); (2) find the cumulative area from the left up to 62 in. (or $z = -3.13$); (3) find the difference between those two areas. The heights of 73 in. and 62 in. are converted to z scores by using Formula 6-2 as follows:

For $x = 73$ in.: $z = \dfrac{x - \mu}{\sigma} = \dfrac{73 - 69.5}{2.4} = 1.46$ ($z = 1.46$ yields an area of 0.9279.)

For $x = 62$ in.: $z = \dfrac{x - \mu}{\sigma} = \dfrac{62 - 69.5}{2.4} = -3.13$ ($z = -3.13$ yields an area of 0.0009.)

Step 3: If using Table A-2, refer to that table with $z = 1.46$ and find that the cumulative area to the *left* of $z = 1.46$ is 0.9279. (Remember, Table A-2 is designed so that all areas are cumulative areas from the *left*.) Table A-2 also shows that $z = -3.13$ corresponds to an area of 0.0009.

Because the areas of 0.9279 and 0.0009 are *cumulative areas from the left*, we find the shaded area in Figure 6-13 as follows:

$$\text{Shaded area in Figure 6-13} = 0.9279 - 0.0009 = 0.9270$$

There is a small discrepancy between the area of 0.9267 obtained using technology and the area of 0.9270 obtained using Table A-2. The area obtained from technology is more accurate because it is based on unrounded z scores of 1.4583333 and -3.125, whereas Table A-2 requires z scores rounded to two decimal places.

Interpretation

Expressing the result as a percentage, we conclude that about 93% of men satisfy the height requirement between 62 in. and 73 in. About 7% of men do not meet that requirement and they are not eligible to work as cabin crew members. Figure 6-13 shows that most of the ineligible men are those who are very tall, and this makes sense when we consider that some jets have low cabin heights that would cause problems for the tallest men. For example, the Embraer 145 jet has a cabin ceiling height of 71.7 in., so men taller than 71.7 in. would need to bend the entire time that they are standing.

Finding Values from Known Areas

Here are helpful hints for those cases in which the area (or probability or percentage) is known and we must find the relevant value(s):

1. *Don't confuse z scores and areas.* Remember, z scores are *distances* along the horizontal scale, but areas are *regions* under the normal curve. Table A-2 lists z scores in the left columns and across the top row, but areas are found in the body of the table.
2. *Choose the correct (right/left) side of the graph.* A value separating the *top* 10% from the others will be located on the right side of the graph, but a value separating the *bottom* 10% will be located on the left side of the graph.
3. A z score must be *negative* whenever it is located in the *left* half of the normal distribution.
4. Areas (or probabilities) are positive or zero values, but they are never negative.

Graphs are extremely helpful in visualizing, understanding, and successfully working with normal probability distributions, so they should be used whenever possible.

Procedure for Finding Values from Known Areas or Probabilities

1. Sketch a normal distribution curve, enter the given probability or percentage in the appropriate region of the graph, and identify the x value(s) being sought.
2. If using technology, refer to the instructions at the end of this section. If using Table A-2, refer to the *body* of Table A-2 to find the area to the left of x, then identify the z score corresponding to that area.

3. If you know z and must convert to the equivalent x value, use Formula 6-2 by entering the values for μ, σ, and the z score found in Step 2, then solve for x. Based on Formula 6-2, we can solve for x as follows:

$$x = \mu + (z \cdot \sigma) \qquad \text{(another form of Formula 6-2)}$$

(If z is located to the left of the mean, be sure that it is a negative number.)

4. Refer to the sketch of the curve to verify that the solution makes sense in the context of the graph and in the context of the problem.

The following example uses this procedure for finding a value from a known area.

Example 3 Designing Aircraft Cabins

When designing an environment, one common criterion is to use a design that accommodates 95% of the population. What aircraft ceiling height will allow 95% of men to stand without bumping their heads? That is, find the 95th percentile of heights of men. Assume that heights of men are normally distributed with a mean of 69.5 in. and a standard deviation of 2.4 in.

Note to Instructor
Comment that students have much greater success if they draw a graph for each problem, such as Figure 6-14.

Solution

Step 1: Figure 6-14 shows the normal distribution with the height x that we want to identify. The shaded area represents the 95% of men with heights that would allow them to stand without bumping their heads on the aircraft ceiling.

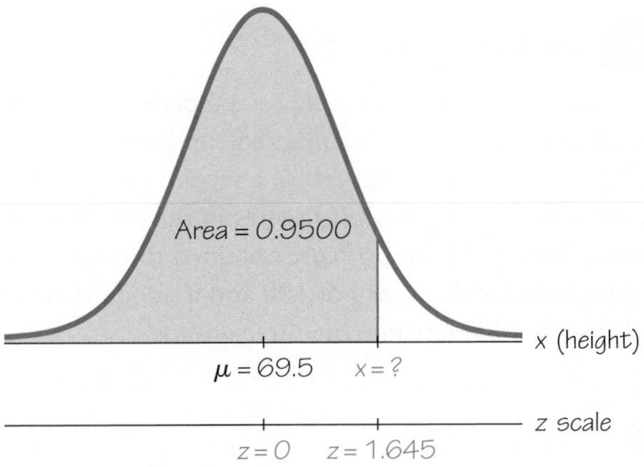

Figure 6-14 Finding the 95th Percentile

Step 2: To use technology, refer to the instructions at the end of this section. Technology will provide the value of x in Figure 6-14. For example, see the accompanying Excel display showing that $x = 73.4476487$, or 73.4 when rounded.

EXCEL

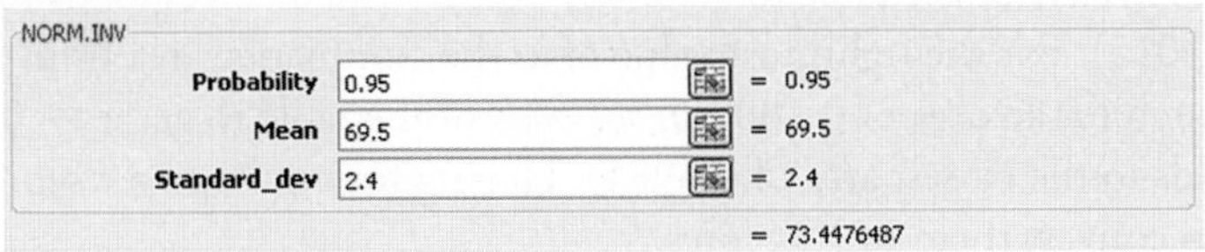

continued

If using Table A-2, search for an area of 0.9500 *in the body* of the table. (The area of 0.9500 shown in Figure 6-14 is a cumulative area from the left, and that is exactly the type of area listed in Table A-2.) The area of 0.9500 is between the Table A-2 areas of 0.9495 and 0.9505, but there is an asterisk and footnote indicating that an area of 0.9500 corresponds to $z = 1.645$.

Step 3: With $z = 1.645$, $\mu = 69.5$ in., and $\sigma = 2.4$ in., we can solve for x by using Formula 6-2:

$$z = \frac{x - \mu}{\sigma} \quad \text{becomes} \quad 1.645 = \frac{x - 69.5}{2.4}$$

The result of $x = 73.448$ in. can be found directly or by using the following version of Formula 6-2:

$$x = \mu + (z \cdot \sigma) = 69.5 + (1.645 \cdot 2.4) = 73.448 \text{ in.}$$

Step 4: The solution of $x = 73.4$ in. (rounded) in Figure 6-14 is reasonable because it is greater than the mean of 69.5 in.

Interpretation

An aircraft cabin height of 73.4 in. (or 6 ft 1.4 in.) would allow 95% of men to fit without bumping their heads. It follows that 5% of men would *not* fit without bending. The safety and comfort of passengers are factors suggesting that it might be wise to use a design with a higher ceiling.

Example 4 Ability Grouping

Some educators argue that all students are served better if they are separated into groups according to their abilities. Assume that students are to be separated into a group with IQ scores in the bottom 30%, a second group with IQ scores in the middle 40%, and a third group with IQ scores in the top 30%. The Wechsler Adult Intelligence Scale yields an IQ score obtained through a test, and the scores are normally distributed with a mean of 100 and a standard deviation of 15. Find the Wechsler IQ scores that separate the three groups.

Solution

Step 1: We begin with the graph shown in Figure 6-15. We have entered the mean of 100, and we have identified the x values separating the lowest 30% and the highest 30%.

Step 2: To use technology, refer to the instructions at the end of this section. Technology will show that the values of x in Figure 6-15 are 92.1 and 107.9 when rounded.

If using Table A-2, we must work with cumulative areas from the left. For the leftmost value of x, the cumulative area from the left is 0.3, so search for an area of 0.3000 *in the body* of the table to get $z = -0.52$ (which corresponds to the closest area of 0.3015). For the rightmost value of x, the cumulative area from the left is 0.7, so search for an area of 0.7000 *in the body* of the table to get $z = 0.52$ (which corresponds to the closest area of 0.6985). Having found the two z scores, we now proceed to convert them to IQ scores.

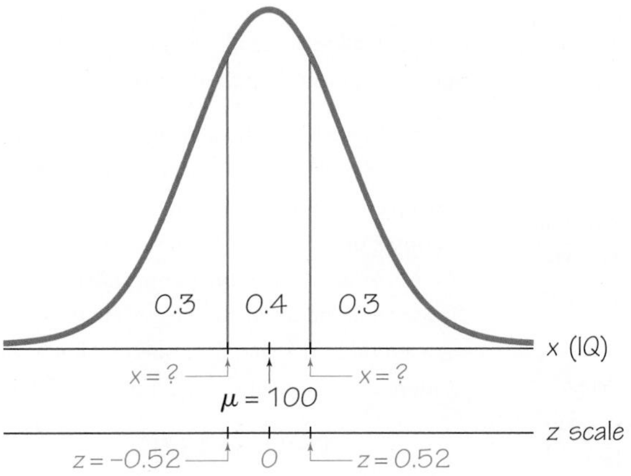

Figure 6-15 IQ Scores for Ability Grouping

Step 3: We now solve for the two values of x by using Formula 6-2 directly or by using the following version of Formula 6-2:

Leftmost value of x: $x = \mu + (z \cdot \sigma) = 100 + (-0.52 \cdot 15) = 92.2$

Rightmost value of x: $x = \mu + (z \cdot \sigma) = 100 + (0.52 \cdot 15) = 107.8$

Regarding the algebra required for solving for x in Step 3 of Example 4: Comment that this algebra need not be applied if using a suitable technology.

Step 4: Referring to Figure 6-15, we see that the leftmost value of $x = 92.2$ is reasonable because it is less than the mean of 100. Also, the rightmost value of 107.8 is reasonable because it is above the mean of 100.

There is a small discrepancy between the results from technology and the results from Table A-2, and that discrepancy is due to the fact that Table A-2 requires that we first round z scores to two decimal places. The results from technology are more accurate.

Interpretation

The Wechsler IQ scores of 92.2 and 107.8 can be used as cutoff values separating the three groups. Those in the lowest group have IQ scores below 92.2, those in the middle group have IQ scores between 92.2 and 107.8, and those in the highest group have IQ scores above 107.8.

For the methods of this section, we should carefully consider the following:

- We should *always* draw a graph to visualize the information.
- We should determine whether we want to find an area or a value of x.
- Regardless of the situation, we must usually work with a cumulative area from the *left.*
- We should know that a z score and x value are *distances* along horizontal scales, but percentages or probabilities correspond to *areas* under a curve.

using TECHNOLOGY

When working with a nonstandard normal distribution, a technology can be used to find areas or values of the relevant variable, so technology can be used instead of Table A-2. The following instructions describe the use of technology.

STATDISK Select **Analysis, Probability Distributions, Normal Distribution.** Either enter the z score to find corresponding areas or enter the cumulative area from the left to find the

continued

z score. After entering a value, click on the **Evaluate** button. (See the STATDISK display in Example 1.)

MINITAB

- *Finding Area:* To find the cumulative area to the left of a *z* score (as in Table A-2), select **Calc, Probability Distributions, Normal, Cumulative probabilities.** Enter the mean and standard deviation, then click on the **Input Constant** button and enter the value.

- *Finding x Value:* To find a value corresponding to a known area, select **Calc, Probability Distributions, Normal,** then select **Inverse cumulative probabilities.** Enter the mean and standard deviation. Select the option **Input constant,** and enter the total area to the left of the given value.

EXCEL

- *Finding Area:* To find the cumulative area to the left of a value (as in Table A-2), click on ***fx*,** then select **Statistical, NORMDIST** (or **NORM.DIST**). In the dialog box, enter the value for *x*, enter the mean and standard deviation, and enter 1 in the "cumulative" space.

- *Finding x Value:* To find a value corresponding to a known area, select ***fx*, Statistical, NORMINV** (or **NORM.INV**) and proceed to make the entries in the dialog box. When entering the probability value, enter the total area to the left of the given value. (See the Excel display in Example 3.)

TI-83/84 PLUS

- *Finding Area:* To find the area between two values, press **2ND**, then press **VARS** to get to the **DISTR** (distribution) menu. Select **normalcdf.** Enter the two values, the mean, and the standard deviation, all separated by commas, as in this format: (left value, right value, mean, standard deviation).

  Hint: If there is no left value, enter the left value as −999999, and if there is no right value, enter the right value as 999999. In Example 1 we want the area to the right of $x = 70$ in. and we have a normal distribution with mean 63.8 in. and standard deviation 2.6 in., so use the command **normalcdf (70, 999999, 63.8, 2.6)** as shown in the accompanying screen display. Because Example 1 uses a rounded *z* score, the result of 0.0085 (rounded) shown here is more accurate than the result of 0.0087 found in Example 1.

TI-83/83 PLUS

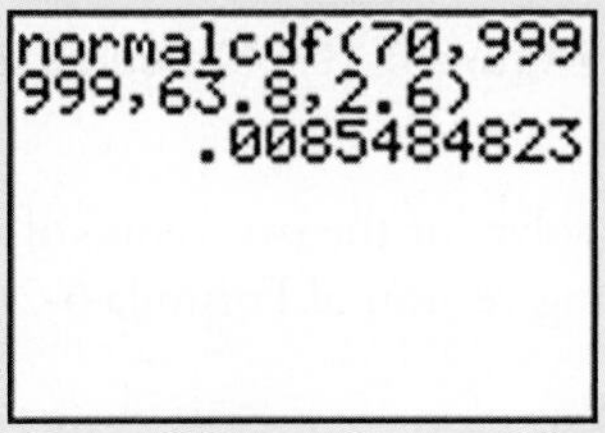

- *Finding x Value:* To find a value corresponding to a known area, press **2ND**, then press **VARS** to get to the **DISTR** (distribution) menu. Select **invNorm,** and proceed to enter the total area to the left of the value, the mean, and the standard deviation in the format of (total area to the left, mean, standard deviation) with the commas included. For Example 3, the command of **invNorm(0.95, 69.5, 2.4)** will yield the result of 73.4476487, which is 73.4 when rounded.

STATCRUNCH Click on **Open StatCrunch,** then click on **Stat.** Select **Calculators,** then select **Normal.** You can either enter a probability or a value of *x*. Click on **Compute.**

Recommended Assignment
Exercises 1–20.

6-3 Basic Skills and Concepts

Statistical Literacy and Critical Thinking

1. a. $\mu = 0$; $\sigma = 1$
b. The *z* scores are numbers without units of measurement.

1. Pulse Rates Pulse rates of women are normally distributed with a mean of 77.5 beats per minute and a standard deviation of 11.6 beats per minute (based on Data Set 1 in Appendix B).

a. What are the values of the mean and standard deviation after converting all pulse rates of women to *z* scores using $z = (x - \mu)/\sigma$?

b. The original pulse rates are measured with units of "beats per minute." What are the units of the corresponding *z* scores?

2. a. 1
b. 100
c. 100
d. 225

2. IQ Scores The Wechsler Adult Intelligence Scale is an IQ score obtained through a test, and the scores are normally distributed with a mean of 100 and a standard deviation of 15. A bell-shaped graph is drawn to represent this distribution.

a. For the bell-shaped graph, what is the area under the curve?

b. What is the value of the median?

c. What is the value of the mode?

d. What is the value of the variance?

3. Normal Distributions What is the difference between a standard normal distribution and a nonstandard normal distribution?

4. Random Digits Computers are commonly used to randomly generate digits of telephone numbers to be called when conducting a survey. Can the methods of this section be used to find the probability that when one digit is randomly generated, it is less than 3? Why or why not? What is the probability of getting a digit less than 3?

3. The standard normal distribution has a mean of 0 and a standard deviation of 1, but a nonstandard normal distribution has a different value for one or both of those parameters.

4. No. Randomly generated digits have a uniform distribution, not a normal distribution. The probability of a digit less than 3 is 0.3.

IQ Scores. ***In Exercises 5–8, find the area of the shaded region. The graphs depict IQ scores of adults, and those scores are normally distributed with a mean of 100 and a standard deviation of 15 (as on the Wechsler test).***

5.

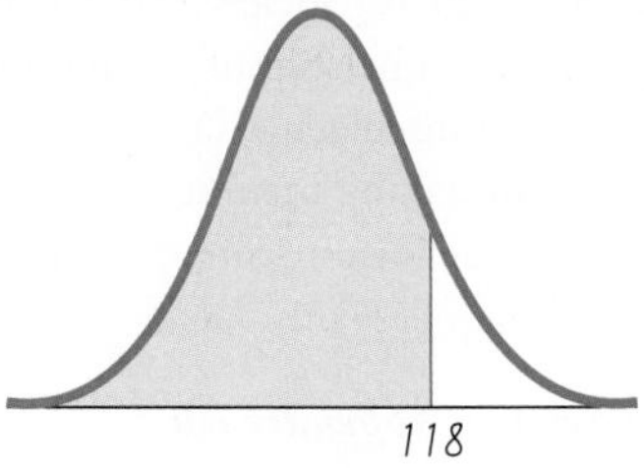

6.

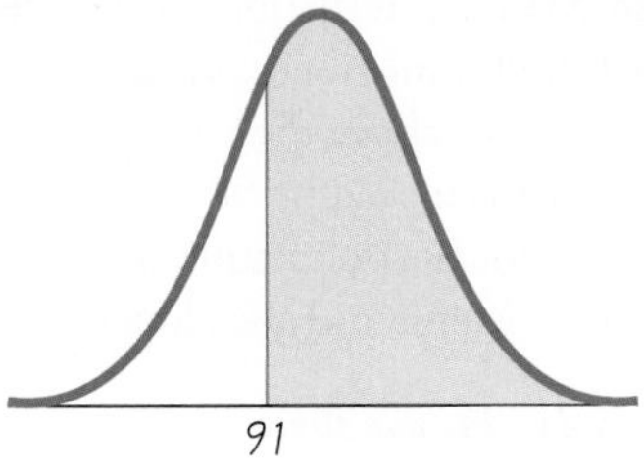

7.

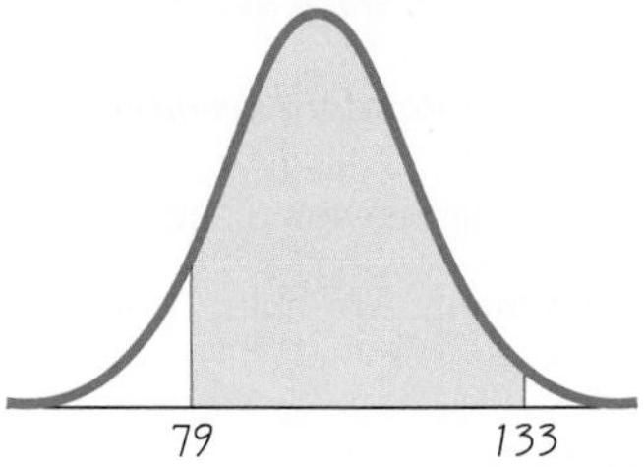

8.

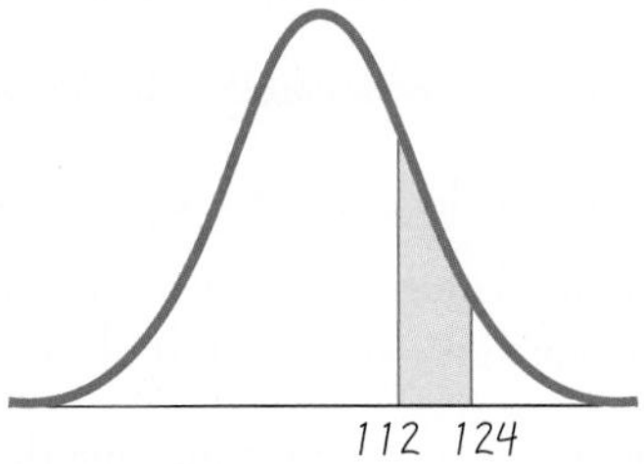

5. 0.8849
6. 0.7257
7. 0.9053
8. 0.1571

IQ Scores. ***In Exercises 9–12, find the indicated IQ score, and round to the nearest whole number. The graphs depict IQ scores of adults, and those scores are normally distributed with a mean of 100 and a standard deviation of 15 (as on the Wechsler test).***

9.

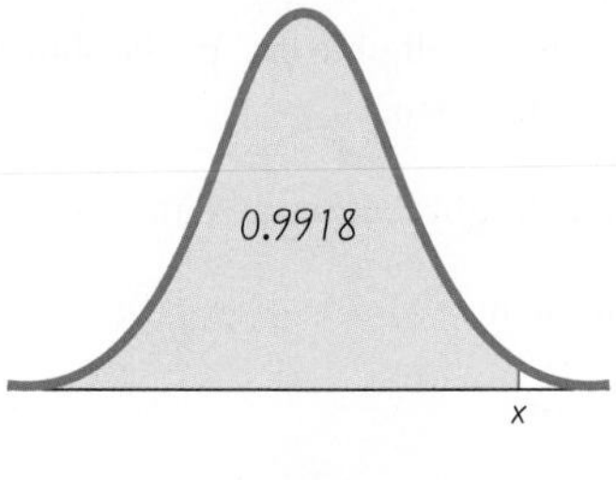

10.

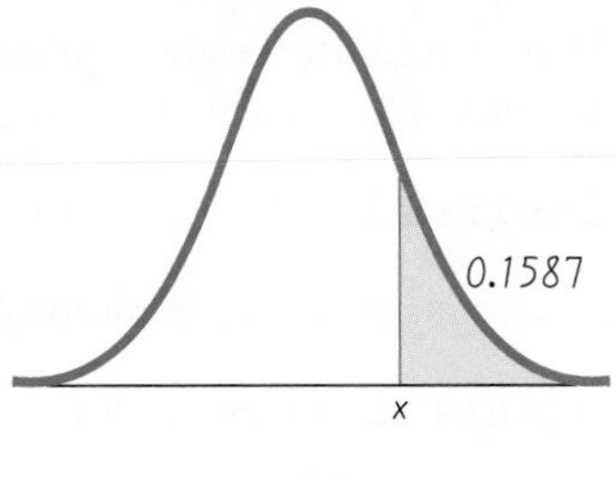

11.

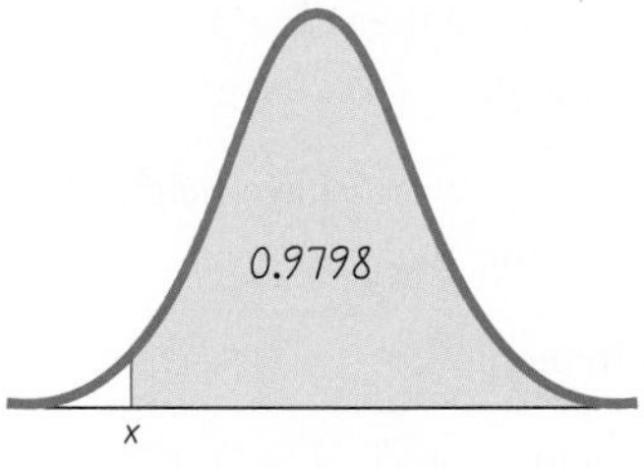

12.

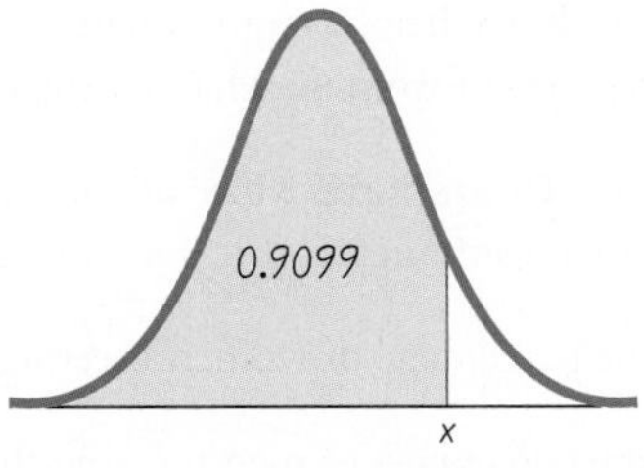

9. 136
10. 115
11. 69
12. 120

IQ Scores. ***In Exercises 13–20, assume that adults have IQ scores that are normally distributed with a mean of 100 and a standard deviation of 15 (as on the Wechsler test). For a randomly selected adult, find the indicated probability or IQ score. Round IQ scores to the nearest whole number. (Hint: Draw a graph in each case.)***

13. Find the probability of an IQ less than 85.

14. Find the probability of an IQ greater than 70 (the requirement for being a statistics textbook author).

13. 0.1587
14. 0.9772

15. 0.4972 (Tech: 0.4950)
16. 0.1596 (Tech: 0.1613)
17. 119
18. 90
19. 110
20. 131

15. Find the probability that a randomly selected adult has an IQ between 90 and 110 (referred to as the *normal* range).

16. Find the probability that a randomly selected adult has an IQ between 110 and 120 (referred to as *bright normal*).

17. Find P_{90}, which is the IQ score separating the bottom 90% from the top 10%.

18. Find the first quartile Q_1, which is the IQ score separating the bottom 25% from the top 75%.

19. Find the third quartile Q_3, which is the IQ score separating the top 25% from the others.

20. Mensa Mensa International calls itself "the international high IQ society," and it has more than 100,000 members. Mensa states that "candidates for membership of Mensa must achieve a score at or above the 98th percentile on a standard test of intelligence (a score that is greater than that achieved by 98 percent of the general population taking the test)." Find the 98th percentile for the population of Wechsler IQ scores. This is the lowest score meeting the requirement for Mensa membership.

In Exercises 21–24, use these parameters (based on Data Set 1 in Appendix B):

- ***Men's heights are normally distributed with mean 69.5 in. and standard deviation 2.4 in.***
- ***Women's heights are normally distributed with mean 63.8 in. and standard deviation 2.6 in.***

21. a. 75.48% (Tech: 75.56%). Yes, about 25% of women are not qualified because of their heights.
b. 99.90% (Tech: 99.89%). No, only about 0.1% of men are not qualified because of their heights.
c. 58.5 in. to 69.1 in.
d. 63.9 in. to 75.1 in.

21. Navy Pilots The U.S. Navy requires that fighter pilots have heights between 62 in. and 78 in.

a. Find the percentage of women meeting the height requirement. Are many women not qualified because they are too short or too tall?

b. Find the percentage of men meeting the height requirement. Are many men not qualified because they are too short or too tall?

c. If the Navy changes the height requirements so that all women are eligible except the shortest 2% and the tallest 2%, what are the new height requirements for women?

d. If the Navy changes the height requirements so that all men are eligible except the shortest 1% and the tallest 1%, what are the new height requirements for men?

22. a. 46.80% (Tech: 46.93%)
b. 98.81%
c. 58.9 in. to 74.0 in.

23. a. 99.86%
b. 98.89% (Tech: 98.90%)
c. 59.5 in. to 73.4 in.

24. a. 0.01% (Tech: 0.00%); practically no men can fit without bending.
b. 0.01% (Tech: 0.00%); practically no women can fit without bending.
c. The door design is very inadequate, but the jet is relatively small and seats only six people. A much higher door would require such major changes in the design and cost of the jet, that the greater height is not practical.
d. 70.1 in.

22. Air Force Pilots The U.S. Air Force requires that pilots have heights between 64 in. and 77 in.

a. Find the percentage of women meeting the height requirement.

b. Find the percentage of men meeting the height requirement.

c. If the Air Force height requirements are changed to exclude only the tallest 3% of men and the shortest 3% of women, what are the new height requirements?

23. Disney Characters Most of the live characters at Disney World have height requirements with a minimum of 4 ft 8 in. and a maximum of 6 ft 3 in.

a. Find the percentage of women meeting the height requirement.

b. Find the percentage of men meeting the height requirement.

c. If the height requirements are changed to exclude only the tallest 5% of men and the shortest 5% of women, what are the new height requirements?

24. Executive Jet Doorway The Gulfstream 100 is an executive jet that seats six, and it has a doorway height of 51.6 in.

a. What percentage of adult men can fit through the door without bending?

b. What percentage of adult women can fit through the door without bending?

c. Does the door design with a height of 51.6 in. appear to be adequate? Why didn't the engineers design a larger door?

d. What doorway height would allow 60% of men to fit without bending?

25. Water Taxi Safety When a water taxi sank in Baltimore's Inner Harbor, an investigation revealed that the safe passenger load for the water taxi was 3500 lb. It was also noted that the mean weight of a passenger was assumed to be 140 lb. Assume a "worst-case" scenario in which all of the passengers are adult men. (This could easily occur in a city that hosts conventions in which people of the same gender often travel in groups.) Assume that weights of men are normally distributed with a mean of 182.9 lb and a standard deviation of 40.8 lb (based on Data Set 1 in Appendix B).

a. If one man is randomly selected, find the probability that he weighs less than 174 lb (the new value suggested by the National Transportation and Safety Board).

b. With a load limit of 3500 lb, how many male passengers are allowed if we assume a mean weight of 140 lb?

c. With a load limit of 3500 lb, how many male passengers are allowed if we use a new mean weight of 182.9 lb?

d. Why is it necessary to periodically review and revise the number of passengers that are allowed to board?

25. a. 0.4129 (Tech: 0.4137)
b. 25
c. 19
d. The mean weight is increasing over time, so safety limits must be periodically updated to avoid an unsafe condition.

26. Designing a Work Station A common design requirement is that an environment must fit the range of people who fall between the 5th percentile for women and the 95th percentile for men. In designing an assembly work table, we must consider *sitting knee height,* which is the distance from the bottom of the feet to the top of the knee. Males have sitting knee heights that are normally distributed with a mean of 21.4 in. and a standard deviation of 1.2 in.; females have sitting knee heights that are normally distributed with a mean of 19.6 in. and a standard deviation of 1.1 in. (based on data from the Department of Transportation).

a. What is the minimum table clearance required to satisfy the requirement of fitting 95% of men? Why is the 5th percentile for women ignored in this case?

b. The author is writing this exercise at a table with a clearance of 23.5 in. above the floor. What percentage of men fit this table, and what percentage of women fit this table? Does the table appear to be made to fit almost everyone?

26. a. 23.4 in. If there is clearance for 95% of males, there will certainly be clearance for all women in the bottom 5%.
b. Men: 95.99%. Women: 99.99% (Tech: 99.98%). The table will fit almost everyone except about 4% of the men with the largest sitting knee heights.

27. Lengths of Pregnancies The lengths of pregnancies are normally distributed with a mean of 268 days and a standard deviation of 15 days.

a. One classical use of the normal distribution is inspired by a letter to "Dear Abby" in which a wife claimed to have given birth 308 days after a brief visit from her husband, who was serving in the Navy. Given this information, find the probability of a pregnancy lasting 308 days or longer. What does the result suggest?

b. If we stipulate that a baby is *premature* if the length of pregnancy is in the lowest 3%, find the length that separates premature babies from those who are not premature. Premature babies often require special care, and this result could be helpful to hospital administrators in planning for that care.

27. a. 0.0038; either a very rare event occurred or the husband is not the father.
b. 240 days

28. Body Temperatures Based on the sample results in Data Set 3 in Appendix B, assume that human body temperatures are normally distributed with a mean of 98.20°F and a standard deviation of 0.62°F.

a. Bellevue Hospital in New York City uses 100.6°F as the lowest temperature considered to be a fever. What percentage of normal and healthy persons would be considered to have a fever? Does this percentage suggest that a cutoff of 100.6°F is appropriate?

28. a. 0.01%; yes
b. 99.22°

b. Physicians want to select a minimum temperature for requiring further medical tests. What should that temperature be, if we want only 5.0% of healthy people to exceed it? (Such a result is a *false positive,* meaning that the test result is positive, but the subject is not really sick.)

29. a. 91.77% (Tech: 91.78%)
b. 0.01% (Tech: 0.00%)
c. 2.150. No.

29. Earthquakes Based on Data Set 16 in Appendix B, assume that Richter scale magnitudes of earthquakes are normally distributed with a mean of 1.184 and a standard deviation of 0.587.

a. Earthquakes with magnitudes less than 2.000 are considered "microearthquakes" that are not felt. What percentage of earthquakes fall into this category?

b. Earthquakes above 4.0 will cause indoor items to shake. What percentage of earthquakes fall into this category?

c. Find the 95th percentile. Will all earthquakes above the 95th percentile cause indoor items to shake?

30. 16.7 in.

30. Aircraft Seat Width Engineers want to design seats in commercial aircraft so that they are wide enough to fit 99% of all males. (Accommodating 100% of males would require very wide seats that would be much too expensive.) Men have hip breadths that are normally distributed with a mean of 14.4 in. and a standard deviation of 1.0 in. (based on anthropometric survey data from Gordon, Clauser, et al.). Find P_{99}. That is, find the hip breadth for men that separates the smallest 99% from the largest 1%.

31. $P_1 = 17.9$ chocolate chips (Tech: 18.0 chocolate chips); $P_{99} = 30.1$ chocolate chips (Tech: 30.0 chocolate chips). The values can be used to identify cookies with an unusually low number of chocolate chips or an unusually high number of chocolate chips, so those numbers can be used to monitor the production process to ensure that the numbers of chocolate chips stay within reasonable limits.

31. Chocolate Chip Cookies The Chapter Problem for Chapter 3 includes Table 3-1, which lists the numbers of chocolate chips in Chips Ahoy regular cookies. Those numbers have a distribution that is approximately normal with a mean of 24.0 chocolate chips and a standard deviation of 2.6 chocolate chips. Find P_1 and P_{99}. How might those values be helpful to the producer of Chips Ahoy regular cookies?

32. a. 61.70% (Tech: 61.71%) of legal quarters are rejected. That percentage is too high because most quarters will be rejected.
b. Accept quarters with weights between 5.55 g and 5.79 g.

32. Quarters After 1964, quarters were manufactured so that the weights had a mean of 5.67 g and a standard deviation of 0.06 g. Some vending machines are designed so that you can adjust the weights of quarters that are accepted. If many counterfeit coins are found, you can narrow the range of acceptable weights with the effect that most counterfeit coins are rejected along with some legitimate quarters.

a. If you adjust vending machines to accept weights between 5.64 g and 5.70 g, what percentage of legal quarters are rejected? Is that percentage too high?

b. If you adjust vending machines to accept all legal quarters except those with weights in the top 2.5% and the bottom 2.5%, what are the limits of the weights that are accepted?

Large Data Sets. ***In Exercises 33 and 34, refer to the data sets in Appendix B and use computer software or a calculator.***

33. a. The mean is 67.25 (67.3 rounded) beats per minute and the standard deviation is 10.334781 (10.3 rounded) beats per minute. A histogram confirms that the distribution is roughly normal.
b. 47.0 beats per minute; 87.5 beats per minute

33. Appendix B Data Set: Pulse Rates of Males Refer to Data Set 1 in Appendix B and use the pulse rates of males.

a. Find the mean and standard deviation, and verify that the pulse rates have a distribution that is roughly normal.

b. Treating the unrounded values of the mean and standard deviation as parameters, and assuming that male pulse rates are normally distributed, find the pulse rate separating the lowest 2.5% and the pulse rate separating the highest 2.5%. These values could be helpful when physicians try to determine whether pulse rates are unusually low or unusually high.

34. Appendix B Data Set: Weights of Diet Pepsi Refer to Data Set 19 in Appendix B and use the weights (pounds) of Diet Pepsi.

a. Find the mean and standard deviation, and verify that the data have a distribution that is roughly normal.

b. Treating the unrounded values of the mean and standard deviation as parameters, and assuming that the weights are normally distributed, find the weight separating the lowest 0.5% and the weight separating the highest 0.5%. These values could be helpful when quality control specialists try to control the manufacturing process so that underweight or overweight cans are rejected.

34. a. The mean is 0.783858 (0.78386 rounded) lb and the standard deviation is 0.004362 (0.00436 rounded) lb. A histogram confirms that the distribution of weights is approximately normal.
b. 0.7726 lb; 0.7951 lb

6-3 Beyond the Basics

35. Curving Test Scores A statistics professor gives a test and finds that the scores are normally distributed with a mean of 40 and a standard deviation of 10. She plans to curve the scores.

a. If she curves by adding 35 to each grade, what is the new mean? What is the new standard deviation?

b. Is it fair to curve by adding 35 to each grade? Why or why not?

c. If the grades are curved so that grades of B are given to scores above the bottom 70% and below the top 10%, find the numerical limits for a grade of B.

d. Which method of curving the grades is fairer: Adding 35 to each original score or using a scheme like the one given in part (c)? Explain.

35. a. 75; 10
b. No, the conversion should also account for variation.
c. B grade: 45.2 to 52.8
d. Use a scheme like the one given in part (c), because variation is included in the curving process.

36. Using Continuity Correction There are many situations in which a normal distribution can be used as a good approximation to a random variable that has only *discrete* values. In such cases, we can use this *continuity correction:* Represent each whole number by the interval extending from 0.5 below the number to 0.5 above it. The Chapter Problem for Chapter 3 includes Table 3-1, which lists the numbers of chocolate chips in Chips Ahoy regular cookies. Those numbers have a distribution that is approximately normal with a mean of 24.0 chocolate chips and a standard deviation of 2.6 chocolate chips. Find the percentage of such cookies having between 20 and 30 chocolate chips inclusive.

a. Find the result without using the continuity correction.

b. Find the result using the continuity correction.

c. Does the use of the continuity correction make much of a difference in this case?

36. a. 92.78% (Tech: 92.75%)
b. 95.20%
c. The use of the continuity correction changes the result by a relatively small but not insignificant amount.

37. Outliers For the purposes of constructing modified boxplots as described in Section 3-4, outliers are defined as data values that are above Q_3 by an amount greater than 1.5 × IQR or below Q_1 by an amount greater than 1.5 × IQR, where IQR is the interquartile range. Using this definition of outliers, find the probability that when a value is randomly selected from a normal distribution, it is an outlier.

37. 0.0444 (Tech: 0.0430).

38. SAT and ACT Tests Based on recent results, scores on the SAT test are normally distributed with a mean of 1511 and a standard deviation of 312. Scores on the ACT test are normally distributed with a mean of 21.1 and a standard deviation of 5.1. Assume that the two tests use different scales to measure the same aptitude.

a. If someone gets an SAT score that is in the 95th percentile, find the actual SAT score and the equivalent ACT score.

b. If someone gets an SAT score of 2100, find the equivalent ACT score.

38. a. SAT: 2024; ACT: 29.5
b. 30.7

Note to Instructor
This section is designed to introduce the general concept of a sampling distribution of a statistic and to demonstrate that some statistics (mean, variance, proportion) tend to target a population parameter while others do not.

Recommendation: Use a ***class activity*** to make the point that sample means tend to be normally distributed, provided that the sample size is large enough. For example, ask students to write the last four digits of their Social Security numbers, then find the mean of those four digits. In class, show that although the distribution of the original digits tends to be approximately uniform, the sample means tend to be approximately normal.

TOP 20

6.4 Sampling Distributions and Estimators

Key Concept In this section we consider the concept of a *sampling distribution of a statistic*. Instead of focusing on the original population, we want to focus on the values of *statistics* (such as sample means or sample proportions) obtained from the population. (The population is like parents, and sample statistics are like children; we all know how parents behave, and now we want to study the behavior of their children.) For example, if pollsters each randomly selected 50 people from the population, obtained their incomes, then computed the mean of each sample of 50 values, what do we know about the sample means that are obtained? How are those sample means distributed? What is the mean of those sample means? Table 6-2 tells us the key points that we need to know, so try really, really hard to understand the story that Table 6-2 tells.

The Story of Table 6-2 To understand Table 6-2, let's start with the top section that describes means. Beginning at the left, imagine a population having a mean μ,

Table 6-2 General Behavior of Sampling Distributions

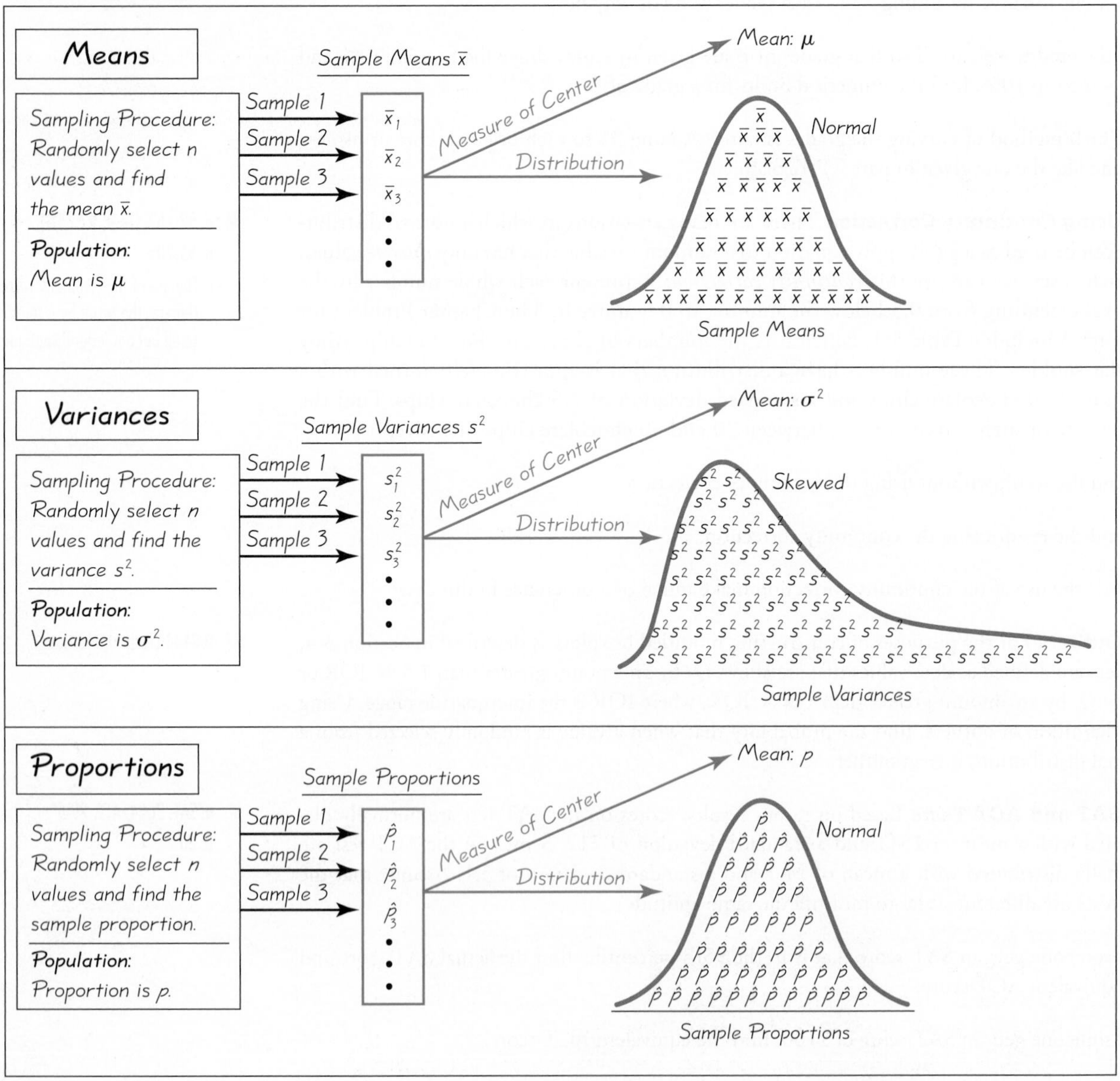

such as the population of adult incomes in the United States. Now imagine pollsters, each randomly selecting 50 people. Each pollster computes the mean of the 50 incomes, so we get a collection of sample means, which we denote as $\bar{x}_1$, $\bar{x}_2$, $\bar{x}_3$, and so on. These sample means are shown in the green box near the top of Table 6-2. What do we know about this collection of sample means? As we move farther to the right in Table 6-2, we see that the sample means tend to have a *normal* distribution (even if the original population is not normally distributed). We might write an ending to this story by saying that "the sampling distribution of the sample mean is a normal distribution." Now there's a happy ending. Let's now formally define a general definition of a sampling distribution of a statistic.

Push Polling

Push polling is the practice of political campaigning under the guise of a poll. Its name is derived from its objective of pushing voters away from opposition candidates by asking loaded questions designed to discredit them.

This survey question was used in one campaign: "Please tell me if you would be more likely or less likely to vote for Roy Romer if you knew that Gov. Romer appoints a parole board which has granted early release to an average of four convicted felons per day every day since Romer took office." The National Council on Public Polls says that push polls are unethical, and reputable pollsters do not approve of push polling.

DEFINITION The **sampling distribution of a statistic** (such as a sample mean or sample proportion) is the distribution of all values of the statistic when all possible samples of the same size *n* are taken from the same population. (The sampling distribution of a statistic is typically represented as a probability distribution in the format of a table, probability histogram, or formula.)

Sampling Distribution of the Sample Mean

The preceding definition is general, so let's consider the specific sampling distribution of the sample mean.

DEFINITION The **sampling distribution of the sample mean** is the distribution of all possible sample means (or the distribution of the variable $\bar{x}$), with all samples having the same sample size *n* taken from the same population. (The sampling distribution of the sample mean is typically represented as a probability distribution in the format of a table, probability histogram, or formula.)

These concepts are very abstract, so let's deal with a concrete application so that we can gain better understanding.

Example 1 Sampling Distribution of the Sample Mean

A friend of the author has three children with ages of 4, 5, and 9. Let's consider the population consisting of {4, 5, 9}. (We don't usually know all values in a population, and we don't usually work with such a small population, but it works well for the purposes of this example.) If two ages are randomly selected with replacement from the population {4, 5, 9}, identify the sampling distribution of the sample mean by creating a table representing the probability distribution of the sample mean. Do the values of the sample mean target the value of the population mean?

Solution

If two values are randomly selected with replacement from the population {4, 5, 9}, the leftmost column of Table 6-3 lists the nine different possible samples. Because we want to identify the sampling distribution of the sample mean, we compute the mean of each of the nine samples, and those sample means are listed in the second column of Table 6-3. The nine samples are equally likely with a probability of 1/9 (as shown in the third column of Table 6-3). We saw in Section 5-2 that a probability distribution is a description that gives the probability for each value of a random variable, as in the second and third columns of Table 6-3. The second and third columns of Table 6-3 constitute a probability distribution for the random variable representing sample means, so those two columns represent the sampling distribution of the sample mean. In Table 6-3, some of the sample mean values are repeated, so we combined equal sample mean values in Table 6-4.

Table 6-3 Sampling Distribution of Mean

Sample	Sample Mean $\bar{x}$	Probability
4, 4	4.0	1/9
4, 5	4.5	1/9
4, 9	6.5	1/9
5, 4	4.5	1/9
5, 5	5.0	1/9
5, 9	7.0	1/9
9, 4	6.5	1/9
9, 5	7.0	1/9
9, 9	9.0	1/9

$\leftrightarrow$

Table 6-4 Sampling Distribution of Mean (Condensed)

Sample Mean $\bar{x}$	Probability
4.0	1/9
4.5	2/9
5.0	1/9
6.5	2/9
7.0	2/9
9.0	1/9

Interpretation

Because Table 6-4 lists the possible values of the sample mean along with their corresponding probabilities, Table 6-4 is an example of a sampling distribution of a sample mean.

The value of the mean of the population {4, 5, 9} is $\mu = 6.0$. Using either Table 6-3 or 6-4, we could calculate the mean of the sample values; we get 6.0. Because the mean of the sample means (6.0) is equal to the mean of the population (6.0), we conclude that the values of the sample mean do *target* the value of the population mean. (The top of Table 6-2 also shows that the values of the sample mean have a measure of center equal to μ.) It's unfortunate that this sounds so much like doublespeak, but this illustrates that *the mean of the sample means is equal to the population mean* μ. Read that last sentence a few times until it makes sense.

If we were to create a probability histogram from Table 6-4, it would not have the bell shape that is characteristic of a normal distribution (as suggested by the top of Table 6-2), but that is because we are working with such small samples. If the population of {4, 5, 9} were much larger and if we were selecting samples much larger than $n = 2$ as in this example, we would get a probability histogram that is much closer to being bell-shaped, indicating a normal distribution, as in Example 2.

Example 2 Sampling Distribution of the Sample Mean

Consider repeating this process: Roll a die 5 times to randomly select 5 values from the population {1, 2, 3, 4, 5, 6}, then find the mean $\bar{x}$ of the results. What do we know about the behavior of all sample means that are generated as this process continues indefinitely?

Solution

The top portion of Table 6-5 illustrates a process of rolling a die 5 times and finding the mean of the results. Table 6-5 shows results from repeating this process 10,000 times, but the true sampling distribution of the mean involves repeating the process indefinitely. Because the values of 1, 2, 3, 4, 5, 6 are all equally likely, the population has a mean of $\mu = 3.5$, and Table 6-5 shows that the 10,000 sample means have a mean of 3.49. If the process is continued indefinitely, the mean of the sample means will be 3.5. Also, Table 6-5 shows that the distribution of the sample means is approximately a normal distribution.

Interpretation

Based on the actual sample results shown in the top portion of Table 6-5, we can describe the sampling distribution of the mean by the histogram at the top of Table 6-5. The actual sampling distribution would be described by a histogram based on all possible samples, not only the 10,000 samples included in the histogram, but the number of trials is large enough to suggest that the true sampling distribution of means is approximately a normal distribution with its characteristic bell shape.

Table 6-5 Specific Results from 10,000 Trials

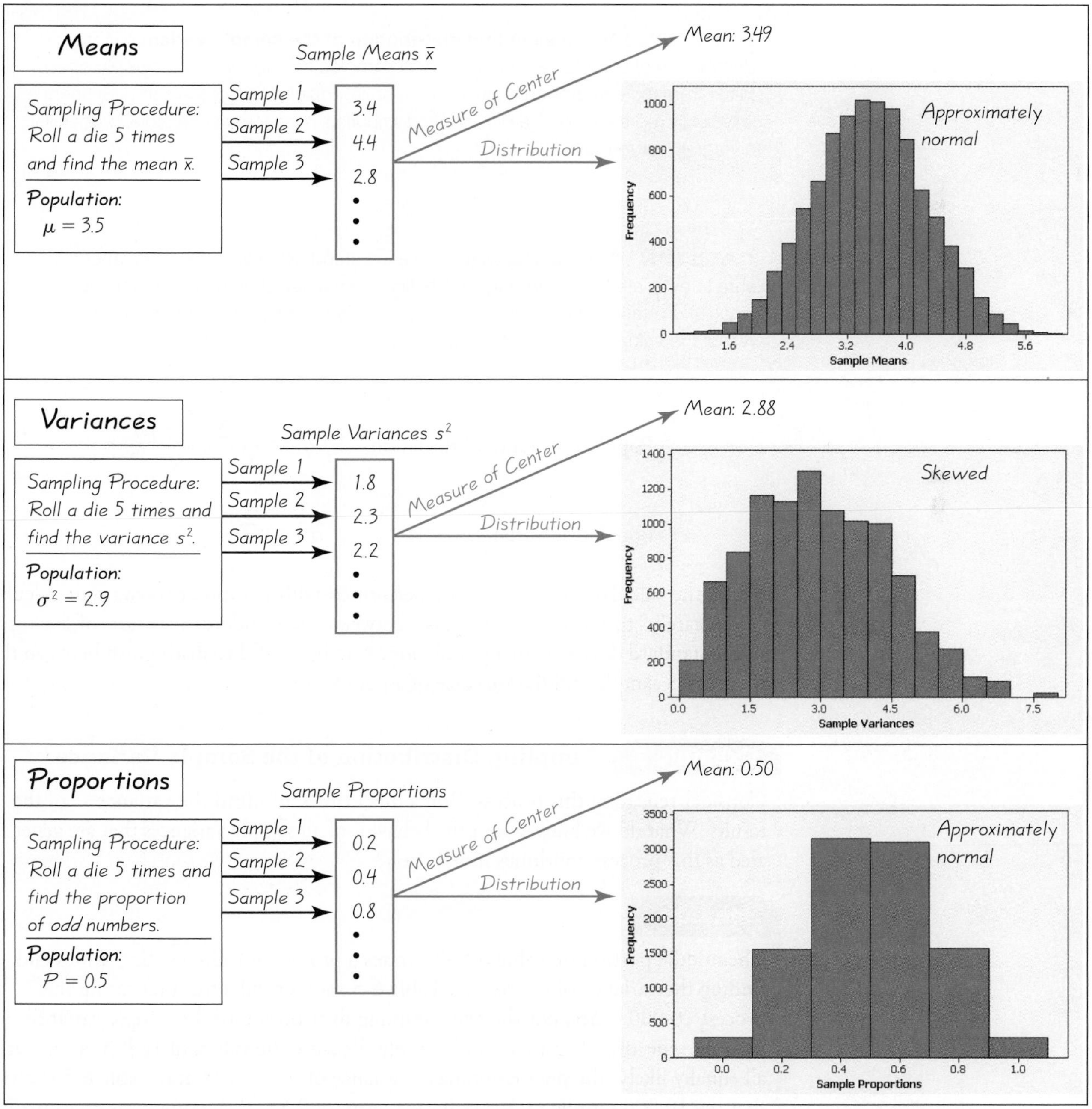

The results of Example 2 allow us to observe the following two important properties of the sampling distribution of the sample mean.

Behavior of Sample Means

1. The sample means *target* the value of the population mean. (That is, the mean of the sample means is the population mean. The expected value of the sample mean is equal to the population mean.)
2. The distribution of sample means tends to be a normal distribution. (This will be discussed further in the following section, but the distribution tends to become closer to a normal distribution as the sample size increases.)

Sampling Distribution of the Sample Variance

Let's now consider the sampling distribution of sample variances.

DEFINITION The **sampling distribution of the sample variance** is the distribution of sample variances, with all samples having the same sample size n taken from the same population. (The sampling distribution of the sample variance is typically represented as a probability distribution in the format of a table, probability histogram, or formula.)

CAUTION When working with population standard deviations or variances, be sure to evaluate them correctly. In Section 3-3 we saw that the computations for *population* standard deviations or variances involve division by the population size N (not the value of $n - 1$), as shown here.

Population standard deviation: $$\sigma = \sqrt{\frac{\Sigma(x - \mu)^2}{N}}$$

Population variance: $$\sigma^2 = \frac{\Sigma(x - \mu)^2}{N}$$

Because the calculations are typically performed with computer software or calculators, be careful to correctly distinguish between the standard deviation of a sample and the standard deviation of a population. Also be careful to distinguish between the variance of a sample and the variance of a population.

Example 3 Sampling Distribution of the Sample Variance

Consider repeating this process: Roll a die 5 times and find the variance s^2 of the results. What do we know about the behavior of all sample variances that are generated as this process continues indefinitely?

Solution

The middle portion of Table 6-5 illustrates a process of rolling a die 5 times and finding the variance of the results. Table 6-5 shows results from repeating this process 10,000 times, but the true sampling distribution of the sample variance involves repeating the process indefinitely. Because the values of 1, 2, 3, 4, 5, 6 are all equally likely, the population has a variance of $\sigma^2 = 2.9$, and Table 6-5 shows that the 10,000 sample variances have a mean of 2.88. If the process is continued indefinitely, the mean of the sample variances will be 2.9. Also, the middle portion of Table 6-5 shows that the distribution of the sample variances is a skewed distribution, not a normal distribution with its characteristic bell shape.

Interpretation

Based on the actual sample results shown in the middle portion of Table 6-5, we can describe the sampling distribution of the sample variance by the histogram in the middle of Table 6-5. The actual sampling distribution would be described by a histogram based on all possible samples, not the 10,000 samples included in the histogram, but the number of trials is large enough to suggest that the true sampling distribution of sample variances is a distribution skewed to the right.

The results of Example 3 allow us to observe the following two important properties of the sampling distribution of the sample variance.

Behavior of Sample Variances

1. The sample variances *target* the value of the population variance. (That is, the mean of the sample variances is the population variance. The expected value of the sample variance is equal to the population variance.)
2. The distribution of sample variances tends to be a distribution skewed to the right.

Sampling Distribution of Sample Proportion

We now consider the sampling distribution of a sample proportion.

DEFINITION The **sampling distribution of the sample proportion** is the distribution of sample proportions, with all samples having the same sample size n taken from the same population.

We need to distinguish between a population proportion p and some sample proportion, so the following notation is commonly used.

Notation for Proportions

p = *population* proportion

$\hat{p}$ = *sample* proportion

Example 4 Sampling Distribution of the Sample Proportion

Consider repeating this process: Roll a die 5 times and find the proportion of *odd* numbers. What do we know about the behavior of all sample proportions that are generated as this process continues indefinitely?

Solution

The bottom portion of Table 6-5 illustrates a process of rolling a die 5 times and finding the proportion of odd numbers. Table 6-5 shows results from repeating this process 10,000 times, but the true sampling distribution of the sample proportion involves repeating the process indefinitely. Because the values of 1, 2, 3, 4, 5, 6 are all equally likely, the proportion of odd numbers in the population is 0.5, and Table 6-5 shows that the 10,000 sample proportions have a mean of 0.50.

continued

If the process is continued indefinitely, the mean of the sample proportions will be 0.5. Also, the bottom portion of Table 6-5 shows that the distribution of the sample proportions is approximately a normal distribution.

Interpretation

Based on the actual sample results shown in the bottom portion of Table 6-5, we can describe the sampling distribution of the sample proportion by the histogram at the bottom of Table 6-5. The actual sampling distribution would be described by a histogram based on all possible samples, not the 10,000 samples included in the histogram, but the number of trials is large enough to suggest that the true sampling distribution of sample proportions is approximately a normal distribution.

The results of Example 4 allow us to observe the following two important properties of the sampling distribution of the proportion.

Behavior of Sample Proportions

1. Sample proportions *target* the value of the population proportion. (That is, the mean of the sample proportions is the population proportion. The expected value of the sample proportion is equal to the population proportion.)
2. The distribution of sample proportions tends to approximate a normal distribution.

Estimators: Unbiased and Biased

Note to Instructor
In Chapter 7 we build upon the properties of the unbiased estimators as we use them to construct confidence interval estimates of population parameters.

The preceding examples show that sample means, variances, and proportions tend to *target* the corresponding population parameters. More formally, we say that sample means, variances, and proportions are *unbiased estimators*. See the following two definitions.

DEFINITIONS

An **estimator** is a statistic used to infer (estimate) the value of a population parameter.

An **unbiased estimator** is a statistic that targets the value of the population parameter in the sense that the sampling distribution of the statistic has a mean that is equal to the mean of the corresponding parameter.

Unbiased Estimators These statistics are unbiased estimators. That is, they each target the value of the corresponding population parameter:

- Mean $\bar{x}$
- Variance s^2
- Proportion $\hat{p}$

Biased Estimators These statistics are biased estimators. That is, they do *not* target the value of the corresponding population parameter:

- Median
- Range
- Standard deviation s*

Important Note:* The sample standard deviations do not target the population standard deviation σ, but the bias is relatively small in large samples, so **s is often used to estimate σ even though s is a biased estimator of σ.

Example 5 Sampling Distribution of the Sample Range

As in Example 1, consider samples of size $n = 2$ randomly selected from the population {4, 5, 9}.

a. List the different possible samples along with the probability of each sample, then find the range for each sample.

b. Describe the sampling distribution of the sample range in the format of a table summarizing the probability distribution.

c. Based on the results, do the sample ranges target the population range, which is $9 - 4 = 5$?

d. What do these results indicate about the sample range as an estimator of the population range?

Solution

a. In Table 6-6 we list the nine different possible samples of size $n = 2$ selected with replacement from the population {4, 5, 9}. The nine samples are equally likely, so each has probability 1/9. Table 6-6 also shows the range for each of the nine samples.

Table 6-6 Sampling Distribution of Range

Sample	Sample Range	Probability
4, 4	0	1/9
4, 5	1	1/9
4, 9	5	1/9
5, 4	1	1/9
5, 5	0	1/9
5, 9	4	1/9
9, 4	5	1/9
9, 5	4	1/9
9, 9	0	1/9

b. The last two columns of Table 6-6 list the values of the range along with the corresponding probabilities, so the last two columns constitute a table summarizing the probability distribution. Table 6-6 therefore describes the *sampling distribution* of the sample range.

c. The mean of the sample ranges in Table 6-6 is 20/9, or 2.2. The range of the population {4, 5, 9} is $9 - 4 = 5$. Because the mean of the sample ranges (2.2) is not equal to the population range (5), the sample ranges do *not* target the value of the population range.

d. Because the sample ranges do not target the population range, the sample range is a *biased estimator* of the population range.

Interpretation

We conclude that the sample range is a biased estimator of the population range. This shows that, in general, a sample range should not be used to estimate the value of the population range.

Note to Instructor
Students should know why sampling *with replacement* is so important, given that sampling *without replacement* is used most often. Comment on the content of this paragraph.

Why Sample *with* Replacement? All of the examples in this section involved sampling *with replacement*. Sampling *without replacement* would have the very practical advantage of avoiding wasteful duplication whenever the same item is selected more than once. However, we are particularly interested in sampling *with replacement* for these two very important reasons:

1. When selecting a relatively small sample from a large population, it makes no significant difference whether we sample with replacement or without replacement.
2. Sampling with replacement results in independent events that are unaffected by previous outcomes, and independent events are easier to analyze and result in simpler calculations and formulas.

For the reasons above, we focus on the behavior of samples that are randomly selected *with replacement*. Many of the statistical procedures discussed in the following chapters are based on the assumption that sampling is conducted with replacement.

The key point of this section is to introduce the concept of a sampling distribution of a statistic. Consider the goal of trying to find the mean body temperature of all adults. Because that population is so large, it is not practical to measure the temperature of every adult. Instead, we obtain a sample of body temperatures and use it to estimate the population mean. Data Set 3 in Appendix B includes a sample of 106 such body temperatures. The mean for that sample is $\bar{x} = 98.20°F$. Conclusions that we make about the population mean temperature of all adults require that we understand the behavior of the sampling distribution of all such sample means. Even though it is not practical to obtain every possible sample and we are stuck with just one sample, we can form some very meaningful conclusions about the population of all body temperatures. A major goal of the following sections and chapters is to learn how we can effectively use a sample to form inferences or conclusions about a population. In Section 6-5 we consider more details about the sampling distribution of sample means, and in Section 6-7 we consider more details about the sampling distribution of sample proportions.

CAUTION Many methods of statistics require a *simple random sample*. Some samples, such as voluntary response samples or convenience samples, could easily result in very wrong results.

Recommended Assignment
Exercises 1–10.

1. a. The sample means will tend to center about the population parameter of 5.67 g.
   b. The sample means will tend to have a distribution that is approximately normal.
   c. The sample proportions will tend to have a distribution that is approximately normal.
2. a. Without replacement.
   b. (1) When selecting a relatively small sample from a large population, it makes no significant difference whether we sample with replacement or without replacement.
   (2) Sampling with replacement results in independent events that are unaffected by previous outcomes, and independent events are easier to analyze and they result in simpler calculations and formulas.
3. Sample mean; sample variance; sample proportion
4. No. The data set is only one sample, but the sampling distribution of the mean is the distribution of the means from all samples, not the one sample mean obtained from this single sample.
5. No. The sample is not a simple random sample from the population of all college statistics students. It is very possible that the students at Broward College do not accurately reflect the behavior of all college statistics students.
6. a. Normal b. 4.5 c. 0.5
7. a. 4.7
   b.

Sample Variance s^2	Probability
0.0	3/9
0.5	2/9
8.0	2/9
12.5	2/9

   c. 4.7
   d. Yes. The mean of the sampling distribution of the sample variances (4.7) is equal to the value of the population variance (4.7), so the sample variances target the value of the population variance.

6-4 Basic Skills and Concepts

Statistical Literacy and Critical Thinking

1. Minting Quarters In a recent year, the U.S. Mint in Denver manufactured 270 million quarters. Assume that on each day of production, a sample of 50 quarters is randomly selected, and the mean weight is obtained.

a. Given that the population of quarters has a mean weight of 5.67 g, what do you know about the mean of the sample means?

b. What do you know about the shape of the distribution of the sample means?

c. The population of quarters has a mean of 5.67 g, but the weights of individual quarters vary. For each sample of 50 quarters, consider the proportion of quarters that weigh less than 5.67 g. What do you know about the shape of the distribution of the sample proportions?

2. Sampling with Replacement In a recent year, the U.S. Mint in Denver manufactured 270 million quarters. As part of the mint's quality control program, samples of quarters are randomly selected each day for detailed inspection to confirm that they meet all required specifications.

a. Do you think the quarters are randomly selected with replacement or without replacement?

b. Give two reasons why statistical methods tend to be based on the assumption that sampling is conducted *with* replacement, instead of without replacement.

3. Unbiased Estimators Data Set 1 in Appendix B includes a sample of 40 pulse rates of women. If we compute the values of sample statistics from that sample, which of the following statistics are *unbiased* estimators of the corresponding population parameters: sample mean; sample median; sample range; sample variance; sample standard deviation; sample proportion?

4. Sampling Distribution Data Set 20 in Appendix B includes a sample of weights of 100 M&M candies. If we explore this sample of 100 weights by constructing a histogram and finding the mean and standard deviation, do those results describe the sampling distribution of the mean? Why or why not?

5. Good Sample? For the population of all college students currently taking a statistics course, you want to estimate the proportion who are women. You obtain a simple random sample of statistics students at Broward College in Florida. Is the resulting sample proportion a good estimator of the population proportion of all college statistics students? Why or why not?

6. Lottery Results Many states have a Pick 3 lottery in which three digits are randomly selected each day. Winning requires that you select the same three digits in the same order that they are drawn. Assume that you compute the mean of each set of three selected digits.

a. What is the approximate shape of the distribution of the sample means (uniform, normal, skewed, other)?

b. What value do the sample means target? That is, what is the mean of all such sample means?

c. For each set of three digits that is selected, if you find the proportion of odd numbers, what is the mean of those proportions?

In Exercises 7–10, use the same population of {4, 5, 9} that was used in Examples 1 and 5. As in Examples 1 and 5, assume that samples of size n = 2 are randomly selected with replacement.

7. Sampling Distribution of the Sample Variance

a. Find the value of the population variance σ^2.

b. Table 6-3 describes the sampling distribution of the sample mean. Construct a similar table representing the sampling distribution of the sample variance s^2. Then combine values of s^2 that are the same, as in Table 6-4. (*Hint:* See Example 1 for Tables 6-3 and 6-4 that describe the sampling distribution of the sample mean.)

c. Find the mean of the sampling distribution of the sample variance.

d. Based on the preceding results, is the sample variance an unbiased estimator of the population variance? Why or why not?

8. Sampling Distribution of the Sample Standard Deviation For the following, round results to three decimal places.

a. Find the value of the population standard deviation σ.

8. a. 2.160

b.

Sample St. Dev. s	Probability
0.000	3/9
0.707	2/9
2.828	2/9
3.536	2/9

c. 1.571

d. No. The mean of the sampling distribution of the sample standard deviations is 1.571, and it is not equal to the value of the population standard deviation (2.160), so the sample standard deviations do not target the value of the population standard deviation.

9. a. 5

b.

Sample Median	Probability
4.0	1/9
4.5	2/9
5.0	1/9
6.5	2/9
7.0	2/9
9.0	1/9

c. 6.0

d. No. The mean of the sampling distribution of the sample medians is 6.0, and it is not equal to the value of the population median (5), so the sample medians do not target the value of the population median.

10. a. 2/3, or 0.7

b.

Sample Proportion	Probability
0	1/9
0.5	4/9
1	4/9

c. 2/3, or 0.7

d. Yes. The mean of the sampling distribution of the sample proportion of odd numbers is 2/3, and it is equal to the value of the population proportion of odd numbers (2/3), so the sample proportions target the value of the population proportion.

11. a.

$\bar{x}$	Probability
46	1/16
47.5	2/16
49	1/16
51	2/16
52	2/16
52.5	2/16
53.5	2/16
56	1/16
57	2/16
58	1/16

b. The mean of the population is 52.25 and the mean of the sample means is also 52.25.

c. The sample means target the population mean. Sample means make good estimators of population means because they target the value of the population mean instead of systematically underestimating or overestimating it.

12. a. Same as Exercise 11 part (a).

b. The median of the population is 52.5, but the mean of the sample medians is 52.25, so those values are not equal.

c. The sample medians do not target the population median of 52.5, so sample medians do not make good estimators of population medians.

13. a.

Range	Probability
0	4/16
2	2/16
3	2/16
7	2/16
9	2/16
10	2/16
12	2/16

b. The range of the population is 12, but the mean of the sample ranges is 5.375. Those values are not equal.

c. The sample ranges do not target the population range of 12, so sample ranges do not make good estimators of population ranges.

14. a.

s^2	Probability
0	4/16
2	2/16
4.5	2/16
24.5	2/16
40.5	2/16
50	2/16
72	2/16

b. The population variance is 24.1875, and the mean of the sample variances is also 24.1875. Those values are equal.

c. The sample variances do target the population variance, so sample variances do make good estimators of the population variance.

15.

Proportion of Girls	Probability
0	0.25
0.5	0.50
1	0.25

Yes. The proportion of girls in 2 births is 0.5, and the mean of the sample proportions is 0.5. The result suggests that a sample proportion is an unbiased estimator of a population proportion.

b. Table 6-3 describes the sampling distribution of the sample mean. Construct a similar table representing the sampling distribution of the sample standard deviation s. Then combine values of s that are the same, as in Table 6-4. (*Hint:* See Example 1 for Tables 6-3 and 6-4 that describe the sampling distribution of the sample mean.)

c. Find the mean of the sampling distribution of the sample standard deviation.

d. Based on the preceding results, is the sample standard deviation an unbiased estimator of the population standard deviation? Why or why not?

9. Sampling Distribution of the Sample Median

a. Find the value of the population median.

b. Table 6-3 describes the sampling distribution of the sample mean. Construct a similar table representing the sampling distribution of the sample median. Then combine values of the median that are the same, as in Table 6-4. (*Hint:* See Example 1 for Tables 6-3 and 6-4 that describe the sampling distribution of the sample mean.)

c. Find the mean of the sampling distribution of the sample median.

d. Based on the preceding results, is the sample median an unbiased estimator of the population median? Why or why not?

10. Sampling Distribution of the Sample Proportion

a. For the population, find the proportion of odd numbers.

b. Table 6-3 describes the sampling distribution of the sample mean. Construct a similar table representing the sampling distribution of the sample proportion of odd numbers. Then combine values of the sample proportion that are the same, as in Table 6-4. (*Hint:* See Example 1 for Tables 6-3 and 6-4 that describe the sampling distribution of the sample mean.)

c. Find the mean of the sampling distribution of the sample proportion of odd numbers.

d. Based on the preceding results, is the sample proportion an unbiased estimator of the population proportion? Why or why not?

In Exercises 11–14, use the population of ages {56, 49, 58, 46} of the four U.S. presidents (Lincoln, Garfield, McKinley, Kennedy) when they were assassinated in office. Assume that random samples of size $n = 2$ are selected with replacement.

11. Sampling Distribution of the Sample Mean

a. After identifying the 16 different possible samples, find the mean of each sample, then construct a table representing the sampling distribution of the sample mean. In the table, combine values of the sample mean that are the same. (*Hint:* See Table 6-4 in Example 1.)

b. Compare the mean of the population {56, 49, 58, 46} to the mean of the sampling distribution of the sample mean.

c. Do the sample means target the value of the population mean? In general, do sample means make good estimators of population means? Why or why not?

12. Sampling Distribution of the Median Repeat Exercise 11using medians instead of means.

13. Sampling Distribution of the Range Repeat Exercise 11 using ranges instead of means.

14. Sampling Distribution of the Variance Repeat Exercise 11 using variances instead of means.

15. Births: Sampling Distribution of Sample Proportion When two births are randomly selected, the sample space for genders is bb, bg, gb, and gg. Assume that those four outcomes are equally likely. Construct a table that describes the sampling distribution of the sample proportion of girls from two births. Does the mean of the sample proportions equal the proportion of girls in two births? Does the result suggest that a sample proportion is an unbiased estimator of a population proportion?

16. Births: Sampling Distribution of Sample Proportion When three births are randomly selected, the sample space for genders is bbb, bbg, bgb, bgg, gbb, gbg, ggb, and ggg. Assume that those eight outcomes are equally likely. Construct a table that describes the sampling distribution of the sample proportion of girls from three births. Does the mean of the sample proportions equal the proportion of girls in three births?

17. SAT and ACT Tests Because they enable efficient procedures for evaluating answers, multiple-choice questions are commonly used on standardized tests, such as the SAT or ACT. Such questions typically have five choices, one of which is correct. Assume that you must make random guesses for two such questions. Assume that both questions have correct answers of "a."

a. After listing the 25 different possible samples, find the proportion of correct answers in each sample, then construct a table that describes the sampling distribution of the sample proportions of correct responses.

b. Find the mean of the sampling distribution of the sample proportion.

c. Is the mean of the sampling distribution (from part (b)) equal to the population proportion of correct responses? Does the mean of the sampling distribution of proportions *always* equal the population proportion?

18. Quality Control After constructing a new manufacturing machine, five prototype integrated circuit chips are produced and it is found that two are defective (D) and three are acceptable (A). Assume that two of the chips are randomly selected *with replacement* from this population.

a. After identifying the 25 different possible samples, find the proportion of defects in each of them, then use a table to describe the sampling distribution of the proportions of defects.

b. Find the mean of the sampling distribution.

c. Is the mean of the sampling distribution (from part (b)) equal to the population proportion of defects? Does the mean of the sampling distribution of proportions *always* equal the population proportion?

6-4 Beyond the Basics

19. Using a Formula to Describe a Sampling Distribution Exercise 15 requires the construction of a table that describes the sampling distribution of the proportions of girls from two births. Consider the formula shown here, and evaluate that formula using sample proportions x of 0, 0.5, and 1. Based on the results, does the formula describe the sampling distribution? Why or why not?

$$P(x) = \frac{1}{2(2 - 2x)!(2x)!} \quad \text{where } x = 0, 0.5, 1$$

20. Mean Absolute Deviation Is the mean absolute deviation of a sample a good statistic for estimating the mean absolute deviation of the population? Why or why not? (*Hint:* See Example 5.)

16.

Proportion of Girls	Probability
0	1/8
1/3	3/8
2/3	3/8
1	1/8

Yes. The proportion of girls in 3 births is 0.5, and the mean of the sample proportions is 0.5. The result suggests that a sample proportion is an unbiased estimator of a population proportion.

17. a.

Proportion Correct	Probability
0	16/25
0.5	8/25
1	1/25

b. 0.2

c. Yes. The sampling distribution of the sample proportions has a mean of 0.2 and the population proportion is also 0.2 (because there is 1 correct answer among 5 choices). Yes, the mean of the sampling distribution of the sample proportions is always equal to the population proportion.

18. a. The proportions of 0, 0.5, 1 have the corresponding probabilities of 9/25, 12/25, 4/25.

Proportion Defective	Probability
0	9/25
0.5	12/25
1	4/25

b. 0.4

c. Yes; yes

19. The formula yields $P(0) = 0.25$, $P(0.5) = 0.5$, and $P(1) = 0.25$, which does describe the sampling distribution of the sample proportions. The formula is just a different way of presenting the same information in the table that describes the sampling distribution.

20. Sample values of the mean absolute deviation (MAD) do not usually target the value of the population MAD, so a MAD statistic is not good for estimating a population MAD. If the population of {4, 5, 9} from Example 5 is used, the sample MAD values of 0.0, 0.5, 2.0, and 2.5 have the corresponding probabilities of 3/9, 2/9, 2/9, and 2/9. For these values, the population MAD is 2.0, but the sample MAD values have a mean of 1.1, so the mean of the sample MAD values (1.1) is not equal to the population MAD of 2.0.

Note to Instructor
This is an extremely important section with concepts used often in the following chapters. Strongly consider a class activity such as this: Ask each student to announce the last four digits of his or her Social Security number. As they respond, construct a dotplot and show that it depicts a distribution that is approximately uniform. Then ask each student to compute the mean of the same four digits, and proceed to show that the sample means tend to have a bell-shaped distribution. Also, the mean of the sample means should be the same as the mean of the original list of digits. Finally, question the class and try to draw out the fact that the amount of *variation* among the sample means is *less than* the amount of variation present in the original list of digits.

TOP 20

6-5 The Central Limit Theorem

Key Concept In Section 6-4 we saw that the sampling distribution of sample means tends to be a normal distribution as the sample size increases. In this section we use the sampling distribution of sample means as we introduce and apply the *central limit theorem* that allows us to use a normal distribution for some very meaningful applications.

Central Limit Theorem

For all samples of the same size n with $n > 30$, the sampling distribution of $\bar{x}$ can be approximated by a normal distribution with mean μ and standard deviation $\sigma/\sqrt{n}$.

According to the central limit theorem, the original population can have *any* distribution (uniform, skewed, and so on), but the distribution of sample means $\bar{x}$ can be approximated by a normal distribution when $n > 30$. (There are some special cases of very nonnormal distributions for which the requirement of $n > 30$ isn't quite enough, so the number 30 must be higher, but those cases are relatively rare.)

Example 1 Normal, Uniform, and U-Shaped Distributions

Table 6-7 illustrates the central limit theorem at work. The top dotplots in Table 6-7 show an approximately normal distribution, a uniform distribution, and a distribution with a shape resembling the letter *U*. In each column, the second dotplot shows the distribution of sample means for samples of size $n = 10$, and the bottom dotplot shows the distribution of sample means for samples of size $n = 50$. As we proceed down each column of Table 6-7, we can see that the distribution of sample means is approaching the shape of a normal distribution. That characteristic is included among the following observations that we can make from Table 6-7.

- As the sample size increases, the sampling distribution of sample means tends to approach a normal distribution.

Table 6-7 Sampling Distributions

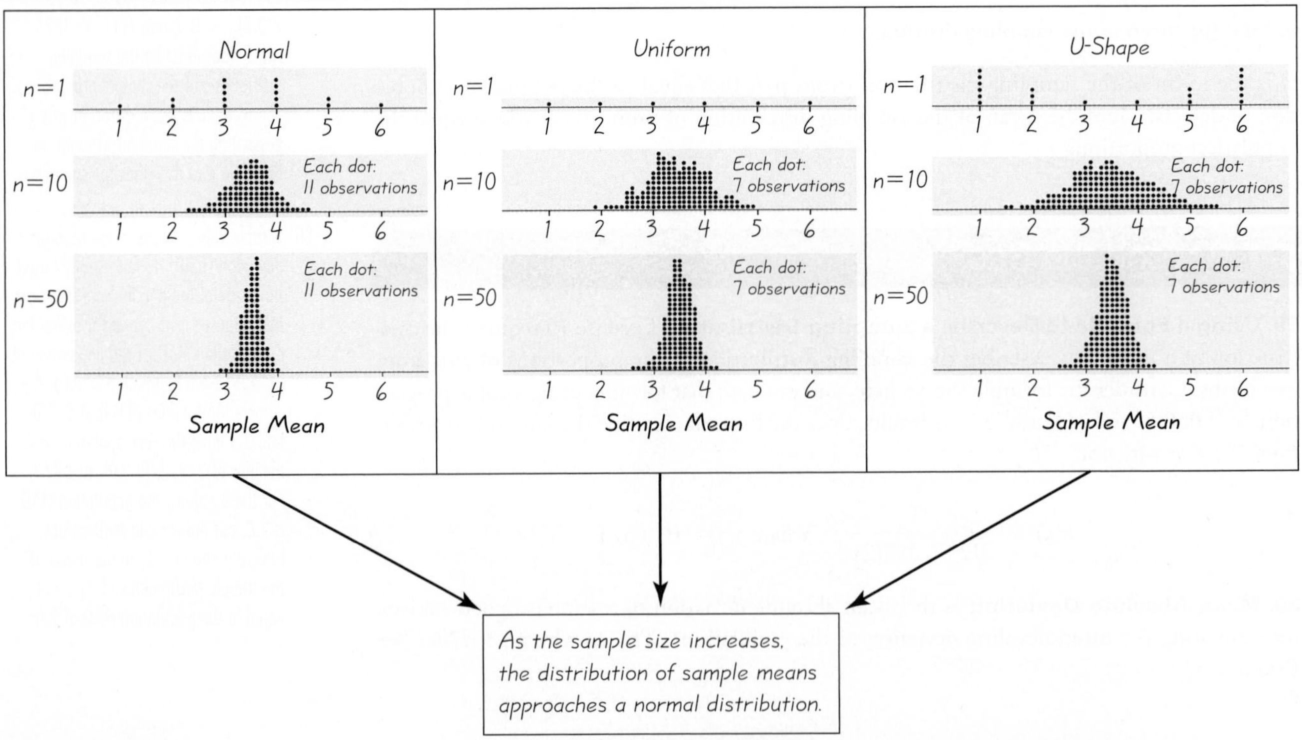

- The mean of all sample means is equal to the mean of the original population.
- As the sample size increases, the spans of the dotplots become narrower, showing that the standard deviations of sample means become smaller as the sample size increases.

The following key points form the foundation for estimating population parameters and hypothesis testing—topics discussed at length in the following chapters.

The Central Limit Theorem and the Sampling Distribution of $\bar{x}$

Given

1. The original population has mean μ and standard deviation σ.
2. Simple random samples of the same size n are selected from the population.

Practical Rules for Real Applications Involving a Sample Mean $\bar{x}$

Case 1: Original population *is* normally distributed.

For *any* sample size n: The distribution of $\bar{x}$ is a normal distribution with these parameters:

Mean of all values of $\bar{x}$: $\mu_{\bar{x}} = \mu$

Standard deviation of all values of $\bar{x}$: $\sigma_{\bar{x}} = \dfrac{\sigma}{\sqrt{n}}$

z score conversion of $\bar{x}$: $z = \dfrac{\bar{x} - \mu}{\dfrac{\sigma}{\sqrt{n}}}$

Case 2: Original population is *not* normally distributed.

For $n > 30$: The distribution of $\bar{x}$ is approximately a normal distribution with these parameters:

Mean of all values of $\bar{x}$: $\mu_{\bar{x}} = \mu$

Standard deviation of all values of $\bar{x}$: $\sigma_{\bar{x}} = \dfrac{\sigma}{\sqrt{n}}$

z score conversion of $\bar{x}$: $z = \dfrac{\bar{x} - \mu}{\dfrac{\sigma}{\sqrt{n}}}$

Distribution of Original Population	Distribution of Sample Means
Normal	Normal (for *any* sample size n)
Not normal and $n > 30$	Normal (approximately)
Not normal and $n \le 30$	*Not normal*

For $n \le 30$: The distribution of $\bar{x}$ cannot be approximated well by a normal distribution and the methods of this section do not apply. Use other methods, such as nonparametric methods or bootstrapping methods. (See the Technology Project for Chapter 7.)

Considerations for Practical Problem Solving

1. **Check Requirements:** When working with the mean from a sample, verify that the normal distribution can be used by confirming that the original population has a normal distribution or $n > 30$.
2. **Individual Value or Mean from a Sample?** Determine whether you are using a normal distribution with a single value x or the mean $\bar{x}$ from a sample of n values. See the following.

- **Individual value:** When working with an *individual* value from a normally distributed population, use the methods of Section 6-3 with $z = \dfrac{x - \mu}{\sigma}$.
- **Mean from a sample of values:** When working with a mean for some *sample* of n values, be sure to use the value of $\sigma/\sqrt{n}$ for the standard deviation of the sample means, so use $z = \dfrac{\bar{x} - \mu}{\sigma/\sqrt{n}}$.

The central limit theorem involves two different distributions: (1) the distribution of the *original population* and (2) the distribution of values of $\bar{x}$. As in previous chapters, we use the symbols μ and σ to denote the mean and standard deviation of the original population, but we use the following new notation for the mean and standard deviation of the distribution of $\bar{x}$.

Notation for the Sampling Distribution of $\bar{x}$

If all possible simple random samples of size n are selected from a population with mean μ and standard deviation σ, the mean of the sample means is denoted by $\mu_{\bar{x}}$ and the standard deviation of all sample means is denoted by $\sigma_{\bar{x}}$. ($\sigma_{\bar{x}}$ is called the standard error of the mean.)

Mean of all values of $\bar{x}$: $\mu_{\bar{x}} = \mu$

Standard deviation of all values of $\bar{x}$: $\sigma_{\bar{x}} = \dfrac{\sigma}{\sqrt{n}}$

It's very important to know when the normal distribution can be used and when it cannot be used. See Case 1 and Case 2 in the box on the preceding page.

Applying the Central Limit Theorem

Many practical problems can be solved with the central limit theorem. Example 2 is a good illustration of the central limit theorem because we can see the difference between working with an *individual* value in part (a) and working with the *mean* for a sample in part (b). Study Example 2 carefully to understand the fundamental difference between the procedures used in parts (a) and (b). In particular, note that when working with an *individual* value, we use $z = \frac{x - \mu}{\sigma}$, but when working with the mean $\bar{x}$ for a group of *sample* values, we use $z = \frac{\bar{x} - \mu}{\sigma/\sqrt{n}}$.

Note to Instructor

Instead of using an example illustrating the central limit theorem alone, it is helpful for students to see how such an example is different from the examples of earlier sections. Example 2 consists of part (a) for 1 individual male, but part (b) involves a sample of 16 males. Stress that when dealing with the mean for a *group* of values, it is essential to modify the standard deviation—divide it by the square root of the sample size *n*.

Example 2 Designing Elevators

When designing elevators, an obviously important consideration is the weight capacity. An Ohio college student died when he tried to escape from a dormitory elevator that was overloaded with 24 passengers. The elevator was rated for a capacity of 16 passengers with a total weight of 2500 lb. Weights of adults are changing over time, and Table 6-8 shows values of recent parameters (based on Data Set 1 in Appendix B). For the following, we assume a worst-case scenario in which all of the passengers are males (which could easily happen in a dormitory setting). If an elevator is loaded to a capacity of 2500 lb with 16 males, the mean weight of a passenger is 156.25 lb.

Table 6-8 Weights of Adults

	Males	Females
μ	182.9 lb	165.0 lb
σ	40.8 lb	45.6 lb
Distribution	Normal	Normal

a. Find the probability that 1 randomly selected adult male has a weight greater than 156.25 lb.

b. Find the probability that a sample of 16 randomly selected adult males has a mean weight greater than 156.25 lb (so that the total weight exceeds the maximum capacity of 2500 lb).

Solution

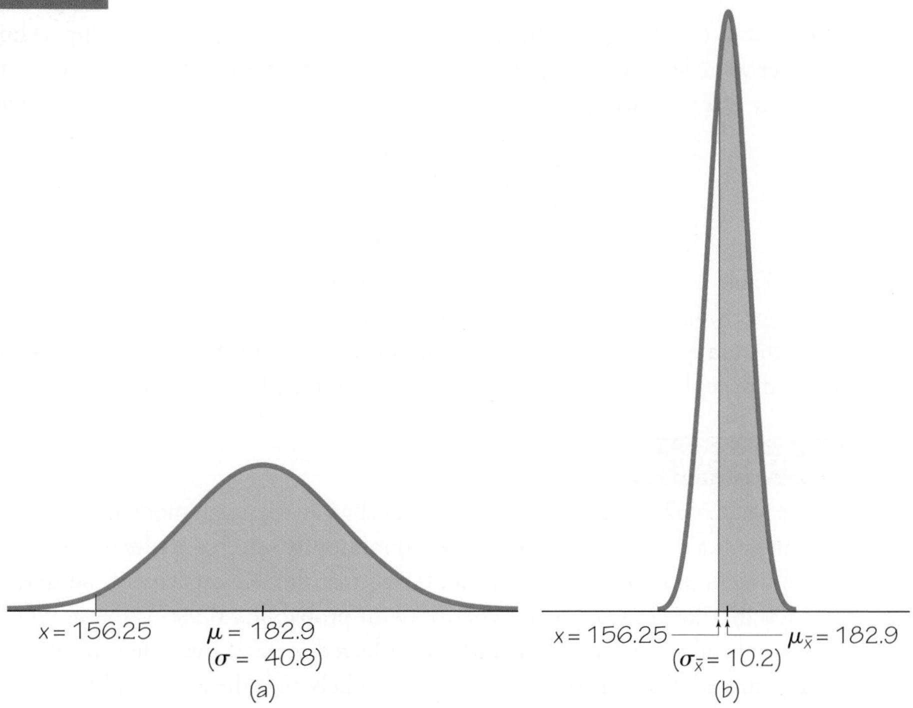

Figure 6-16 Elevator Weights

a. *Approach Used for an Individual Value: Use the methods presented in Section 6-3* (because we are dealing with an *individual* value from a normally distributed population). We seek the area of the green-shaded region in Figure 6-16(a). If using technology (as described at the end of Section 6-3), we find that the green-shaded area is 0.7432. If using Table A-2, we convert the weight of $x = 156.25$ lb to the corresponding z score of $z = -0.65$ as shown here:

$$z = \frac{x - \mu}{\sigma} = \frac{156.25 - 182.9}{40.8} = -0.65$$

We refer to Table A-2 to find that the cumulative area to the *left* of $z = -0.65$ is 0.2578, so the green-shaded area in Figure 6-16 (a) is $1 - 0.2578 = 0.7422$. (The result of 0.7432 from technology is more accurate than the result found from Table A-2.)

b. *Approach Used for the Mean of Sample Values: Use the central limit theorem* (because we are dealing with the *mean of a sample* of 16 males, not an individual male).

Requirement check We can use the normal distribution if the original population is normally distributed or $n > 30$. The sample size is not greater than 30, but the original population of weights of males has a normal distribution, so samples of *any* size will yield means that are normally distributed. ✓

Because we are now dealing with a distribution of sample means, we must use the parameters $\mu_{\bar{x}}$ and $\sigma_{\bar{x}}$, which are evaluated as follows:

$$\mu_{\bar{x}} = \mu = 182.9$$

$$\sigma_{\bar{x}} = \frac{\sigma}{\sqrt{n}} = \frac{40.8}{\sqrt{16}} = 10.2$$

We want to find the green-shaded area shown in Figure 6-16(b). If using technology, the green-shaded area in Figure 6-16(b) is 0.9955. If using Table A-2, we convert the value of $\bar{x} = 156.25$ lb to the corresponding z score of $z = -2.61$ as shown here:

$$z = \frac{\bar{x} - \mu_{\bar{x}}}{\sigma_{\bar{x}}} = \frac{156.25 - 182.9}{\frac{40.8}{\sqrt{16}}} = \frac{-26.65}{10.2} = -2.61$$

From Table A-2 we find that the cumulative area to the *left* of $z = -2.61$ is 0.0045, so the green-shaded area of Figure 6-16(b) is $1 - 0.0045 = 0.9955$ (which happens to be the same result obtained using technology). The probability that 16 randomly selected males have a mean weight greater than 156.25 lb is 0.9955.

Interpretation

There is a 0.7432 probability that an individual male will weigh more than 156.25 lb, and there is a 0.9955 probability that 16 randomly selected males will have a mean weight of more than 156.25 lb. Given that the safe capacity of the elevator is 2500 lb, there is a very good chance (with probability 0.9955) that it will be overweight if is filled with 16 randomly selected males. Given that the elevator was crammed with 24 passengers, it is very likely that the safe weight capacity was exceeded.

The calculations used here are exactly the type of calculations used by engineers when they design ski lifts, escalators, airplanes, boats, amusement park rides, and other devices that carry people.

Example 3 Designing Desks for Kindergarten Children

You need to obtain new desks for an incoming class of 25 kindergarten students who are all 5 years of age. An important characteristic of the desks is that they must accommodate the sitting heights of those students. (The sitting height is the height of a seated student from the bottom of the feet to the top of the knee.) Table 6-9 lists the parameters for sitting heights of 5-year-old children (based on data from "Nationwide Age References for Sitting Height, Leg Length, and Sitting Height/Height Ratio, and Their Diagnostic Value for Disproportionate Growth Disorders," by Fredriks et al., *Archives of Disease in Childhood,* Vol. 90, No. 8).

a. What sitting height will accommodate 95% of the boys?

b. What sitting height is greater than 95% of the means of sitting heights from random samples of 25 boys?

c. Based on the preceding results, what single value should be the minimum sitting height accommodated by the desks? Why are the sitting heights of girls not included in the calculations?

Table 6-9 Sitting Heights of 5-Year-Old Children

	Boys	Girls
μ	61.8 cm	61.2 cm
σ	2.9 cm	3.1 cm
Distribution	Normal	Normal

Solution

Requirement check We can use the normal distribution if the original population is normally distributed or $n > 30$. In parts (a) and (b), the original population is normally distributed, so the normal distribution can be used. ✓

a. If the sitting height accommodates 95% of the boys, it accommodates the *lowest* 95%. Only the boys with sitting heights in the top 5% would not fit. We therefore want the sitting height that separates the lowest 95% from the top 5%, and this corresponds to the sitting height with a cumulative left area of 0.95 under the normal distribution curve. See Figure 6-17(a). Because we are working with individual students, we use the methods of Section 6-3. If using technology, we find that the shaded area in Figure 6-17(a) is bounded by the sitting height of 66.6 cm. If using Table A-2, we refer to that table to find that $z = 1.645$ corresponds to a cumulative left area of 0.95. Using $z = (x - \mu)/\sigma$, we substitute $z = 1.645$, $\mu = 61.8$, and $\sigma = 2.9$, then we solve to get $x = 66.6$ cm.

b. Because we are now working with means from samples of 25 boys, we must use these parameters:

$$\mu_{\bar{x}} = \mu = 61.8$$

$$\sigma_{\bar{x}} = \frac{\sigma}{\sqrt{n}} = \frac{2.9}{\sqrt{25}} = 0.58$$

Figure 6-17(b) shows the shaded area corresponding to the lowest 95% of the means of sitting heights. We can find the sitting height that is at the rightmost boundary for that shaded area by using the same procedures developed in Section 6-3. If using technology, we find that the sitting height is 62.8 cm. If using Table A-2, we refer to the table to find that $z = 1.645$ corresponds to a cumulative left area of 0.95, and we get the following:

$$z = \frac{\bar{x} - \mu_{\bar{x}}}{\sigma_{\bar{x}}} \text{ becomes } 1.645 = \frac{\bar{x} - 61.8}{0.58}$$

Solving for $\bar{x}$ in the equation above results in $\bar{x} = 62.8$ cm.

c. See the following interpretation of the results.

Interpretation

Parts (a) and (b) provide us with two different sitting heights: 66.6 cm and 62.8 cm. We now temporarily leave the world of statistical calculations and enter the world of common sense. A critical consideration is that we really don't care about the means from part (b). Each desk will be occupied by one individual child, not a group of 25 children, so we are concerned with the distribution of sitting heights of individuals, as in part (a). Next, Table 6-9 shows that boys have greater sitting heights than girls, so if we accommodate 95% of the boys, more than 95% of the girls will also be accommodated. We should therefore use desks designed to accommodate a sitting height of 66.6 cm.

The Fuzzy Central Limit Theorem

In *The Cartoon Guide to Statistics,* by Gonick and Smith, the authors describe the Fuzzy Central Limit Theorem as follows: "Data that are influenced by many small and unrelated random effects are approximately normally distributed. This explains why the normal is everywhere: stock market fluctuations, student weights, yearly temperature averages, SAT scores: All are the result of many different effects." People's heights, for example, are the results of hereditary factors, environmental factors, nutrition, health care, geographic region, and other influences which, when combined, produce normally distributed values.

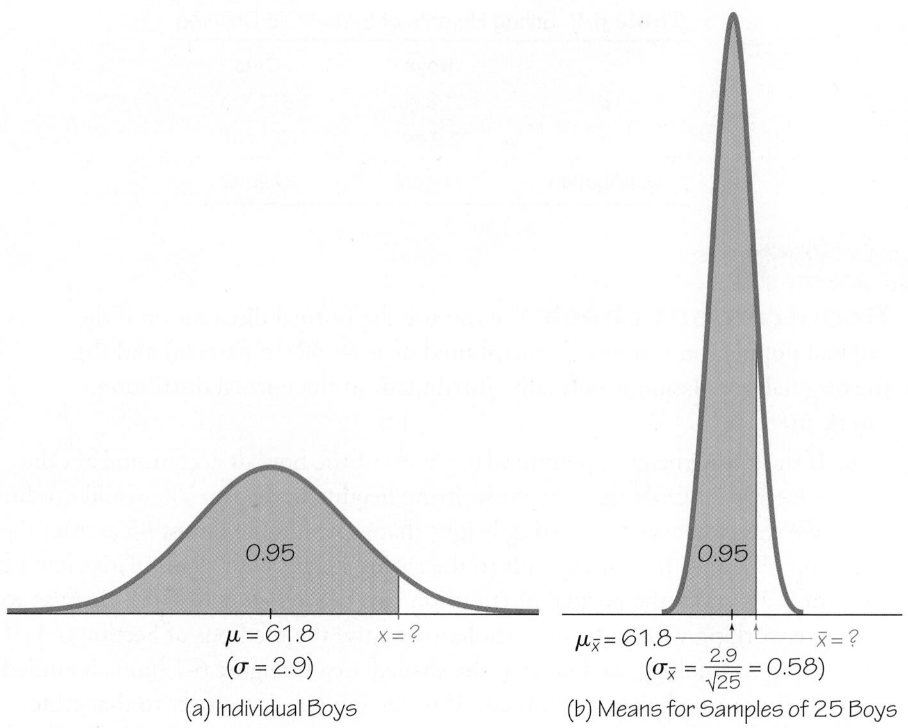

Figure 6-17 Sitting Heights

In Example 2, we were concerned with the *mean* weight of elevator passengers, but in Example 3, the most relevant consideration involves the distribution of *individual* sitting heights. It is important to use sound *critical thinking* in determining what is most relevant.

Introduction to Hypothesis Testing

Carefully examine the conclusions that are reached in the next example illustrating the type of thinking that is the basis for the important procedure of hypothesis testing (discussed in Chapter 8). Example 4 uses the rare event rule for inferential statistics, first presented in Section 4-1.

Rare Event Rule for Inferential Statistics

If, under a given assumption, the probability of a particular observed event is extremely small (such as less than 0.05), we conclude that the assumption is probably not correct.

Note to Instructor
Review the "rare event rule" because it is a good lead-in to hypothesis testing introduced in Chapter 8. Carefully explain Example 4 (or a similar example), and stress the interpretation of results. Again, obtaining numerical answers is nice, but the *interpretation* of those results is vital.

Example 4 Filling Coke Cans

Cans of regular Coke are labeled to indicate that they contain 12 oz. Data Set 19 in Appendix B lists measured amounts for a sample of Coke cans. The corresponding sample statistics are $n = 36$ and $\bar{x} = 12.19$ oz. Assuming that the Coke cans are filled so that $\mu = 12$ oz (as labeled) and the population standard deviation is $\sigma = 0.11$ oz (based on the sample results), find the probability that a sample of 36 cans will have a mean of 12.19 oz or greater. Do these results suggest that the Coke cans are filled with an amount greater than 12.00 oz?

Solution

Requirement check We can use the normal distribution if the original population is normally distributed or $n > 30$. The sample size of $n = 36$ is greater than 30, so we can approximate the sampling distribution of $\bar{x}$ with a normal distribution. ✓

The parameters of this normal distribution are as follows:

$$\mu_{\bar{x}} = \mu = 12.00 \text{ (by assumption)}$$

$$\sigma_{\bar{x}} = \frac{\sigma}{\sqrt{n}} = \frac{0.11}{\sqrt{36}} = 0.018333$$

Figure 6-18 shows the shaded area (see the small region in the right tail of the graph) corresponding to the probability we seek. Having already found the parameters that apply to the distribution shown in Figure 6-18, we can now find the shaded area by using the same procedures developed in Section 6-3.

If using technology, the shaded area in Figure 6-18 is 0.0000 when rounded to four decimal places. If using Table A-2, we first find the z score as shown here:

$$z = \frac{\bar{x} - \mu_{\bar{x}}}{\sigma_{\bar{x}}} = \frac{12.19 - 12.00}{0.018333} = 10.36$$

Referring to Table A-2, we find that $z = 10.36$ is off the chart, but for values of z above 3.49, we use 0.9999 for the cumulative left area. We therefore conclude that the shaded region in Figure 6-18 is 0.0001.

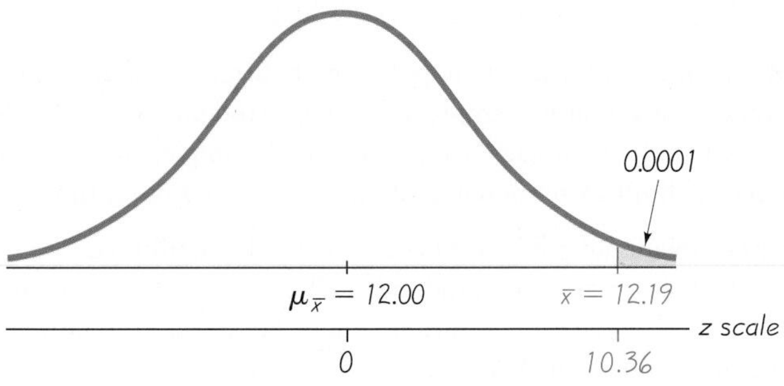

Figure 6-18 Distribution of Amounts of Coke

Interpretation

The result shows that if the mean amount in Coke cans is really 12.00 oz, then there is an extremely small probability of getting a sample mean of 12.19 oz or greater when 36 cans are randomly selected. Because we did obtain such a sample mean, there are two possible explanations: (1) Either the population mean really is 12.00 oz and the sample represents a chance event that is extremely rare; or (2) the population mean is actually greater than 12.00 oz. Because the probability is so close to 0, it seems more reasonable to conclude that the population mean is greater than 12.00 oz. It appears that Coke cans are being filled with more than 12.00 oz. However, the sample mean of 12.19 oz suggests that the mean amount of overfill is very small. It appears that the Coca-Cola company has found a way to ensure that very few cans have less than 12 oz while not wasting very much of their product.

Note to Instructor
Because of time limitations, many instructors choose to omit coverage of the finite population correction factor. If the finite population correction factor is omitted, do not assign Exercises 23, 24, and 25 (located in the group of exercises labeled "Beyond the Basics").

Correction for a Finite Population

In applying the central limit theorem, our use of $\sigma_{\bar{x}} = \sigma/\sqrt{n}$ assumes that the population has infinitely many members. When we sample with replacement (that is, put back each selected item before making the next selection), the population is effectively infinite. Yet many realistic applications involve sampling without replacement, so successive samples depend on previous outcomes. In manufacturing, quality control inspectors typically sample items from a finite production run without replacing them. For such a finite population, we may need to adjust $\sigma_{\bar{x}}$. Here is a common rule of thumb:

When sampling without replacement and the sample size n is greater than 5% of the finite population size N (that is, $n > 0.05N$), adjust the standard deviation of sample means $\sigma_{\bar{x}}$ by multiplying it by the *finite population correction factor:*

$$\sqrt{\frac{N-n}{N-1}}$$

Except for Exercises 23, 24, and 25 the examples and exercises in this section assume that the finite population correction factor does *not* apply, because we are sampling with replacement, or the population is infinite, or the sample size doesn't exceed 5% of the population size.

Recommended Assignment
Exercises 1–8, 11–15. (Exercises 23, 24, and 25 require coverage of the finite population correction factor.)

6-5 Basic Skills and Concepts

Statistical Literacy and Critical Thinking

1. Because $n > 30$, the sampling distribution of the mean ages can be approximated by a normal distribution with mean μ and standard deviation $\sigma/\sqrt{40}$.

2. No. Because the original population is normally distributed, the sample means will be normally distributed for any sample size, not just $n > 30$.

3. $\mu_{\bar{x}} = 60.5$ cm and it represents the mean of the population consisting of all sample means. $\sigma_{\bar{x}} = 1.1$ cm, and it represents the standard deviation of the population consisting of all sample means.

4. Because the digits are equally likely, they have a uniform distribution. Because the sample means are based on samples of size $n = 3$ drawn from a population that does not have a normal distribution, we should not treat the sample means as having a normal distribution.

5. a. 0.9772
b. 0.8888 (Tech: 0.8889)

1. Standard Error of the Mean The population of current statistics students has ages with mean μ and standard deviation σ. Samples of statistics students are randomly selected so that there are exactly 40 students in each sample. For each sample, the mean age is computed. What does the central limit theorem tell us about the distribution of those mean ages?

2. Small Sample Heights of adult females are normally distributed. Samples of heights of adult females, each of size $n = 3$, are randomly collected and the sample means are found. Is it correct to conclude that the sample means cannot be treated as a normal distribution because the sample size is too small? Explain.

3. Notation The population of distances that adult females can reach forward is normally distributed with a mean of 60.5 cm and a standard deviation of 6.6 cm (from the Federal Aviation Administration). If samples of 36 adult females are randomly selected, what do $\mu_{\bar{x}}$ and $\sigma_{\bar{x}}$ represent, and what are their values?

4. Lottery Numbers In each drawing for the Texas Pick 3 lottery, three digits between 0 and 9 inclusive are randomly selected. What is the distribution of the selected digits? If the mean is calculated for each drawing, can the distribution of the sample means be treated as a normal distribution?

Using the Central Limit Theorem. ***In Exercises 5–10, use this information about the overhead reach distances of adult females: $\mu = 205.5$ cm, $\sigma = 8.6$ cm, and overhead reach distances are normally distributed (based on data from the Federal Aviation Administration). The overhead reach distances are used in planning assembly work stations.***

5. a. If 1 adult female is randomly selected, find the probability that her overhead reach is less than 222.7 cm.

b. If 49 adult females are randomly selected, find the probability that they have a mean overhead reach less than 207.0 cm.

6. a. If 1 adult female is randomly selected, find the probability that her overhead reach is less than 196.9 cm.

b. If 36 adult females are randomly selected, find the probability that they have a mean overhead reach less than 205.0 cm.

6. a. 0.1587
b. 0.3632 (Tech: 0.3636)

7. a. If 1 adult female is randomly selected, find the probability that her overhead reach is greater than 218.4 cm.

b. If 9 adult females are randomly selected, find the probability that they have a mean overhead reach greater than 204.0 cm.

c. Why can the normal distribution be used in part (b), even though the sample size does not exceed 30?

7. a. 0.0668
b. 0.6985 (Tech: 0.6996)
c. Because the original population has a normal distribution, the distribution of sample means is normal for any sample size.

8. a. If 1 adult female is randomly selected, find the probability that her overhead reach is greater than 195.0 cm.

b. If 25 adult females are randomly selected, find the probability that they have a mean overhead reach greater than 203.0 cm.

c. Why can the normal distribution be used in part (b), even though the sample size does not exceed 30?

8. a. 0.8888 (Tech: 0.8889)
b. 0.9265 (Tech: 0.9270)
c. Because the original population has a normal distribution, the distribution of sample means is normal for any sample size.

9. a. If 1 adult female is randomly selected, find the probability that her overhead reach is between 179.7 cm and 231.3 cm.

b. If 40 adult females are randomly selected, find the probability that they have a mean overhead reach between 204.0 cm and 206.0 cm.

9. a. 0.9974 (Tech: 0.9973)
b. 0.5086 (Tech: 0.5085)

10. a. If 1 adult female is randomly selected, find the probability that her overhead reach is between 180.0 cm and 200.0 cm.

b. If 50 adult females are randomly selected, find the probability that they have a mean overhead reach between 198.0 cm and 206.0 cm.

10. a. 0.2596 (Tech: 0.2597)
b. 0.6590 (Tech: 0.6595)

11. Elevator Safety Example 2 referred to an Ohio elevator with a maximum capacity of 2500 lb. When rating elevators, it is common to use a 25% safety factor, so the elevator should actually be able to carry a load that is 25% greater than the stated limit. The maximum capacity of 2500 lb becomes 3125 lb after it is increased by 25%, so 16 male passengers can have a mean weight of up to 195.3 lb. If the elevator is loaded with 16 male passengers, find the probability that it is overloaded because they have a mean weight greater than 195.3 lb. (As in Example 2, assume that weights of males are normally distributed with a mean of 182.9 lb and a standard deviation of 40.8 lb.) Does this elevator appear to be safe?

11. 0.1112 (Tech: 0.1121). The elevator does not appear to be safe because there is a reasonable chance (0.1112) that it will be overloaded with 16 male passengers.

12. Elevator Safety Exercise 11 uses $\mu = 182.9$ lb, which is based on Data Set 1 in Appendix B. Repeat Exercise 11 using $\mu = 174$ lb (instead of 182.9 lb), which is the assumed mean weight that was commonly used just a few years ago. Assuming that the mean weight of males is now 182.9 lb, not the value of 174 lb that was used just a few years ago, what do you conclude about the effect of using an outdated mean that is substantially lower than it should be?

12. 0.0183 (Tech: 0.0184). The elevator appears to be relatively safe because there is a very small chance (0.0183) of overloading. Using an outdated mean that is too low has the effect of making the elevator appear to be much safer than it really is.

13. Designing Hats Women have head circumferences that are normally distributed with a mean of 22.65 in. and a standard deviation of 0.80 in. (based on data from the National Health and Nutrition Examination Survey).

a. If the Hats by Leko company produces women's hats so that they fit head circumferences between 21.00 in. and 25.00 in., what percentage of women can fit into these hats?

b. If the company wants to produce hats to fit all women except for those with the smallest 2.5% and the largest 2.5% head circumferences, what head circumferences should be accommodated?

continued

13. a. 0.9787 (Tech: 0.9788)
b. 21.08 in. to 24.22 in.
c. 0.9998. No, the hats must fit individual women, not the mean from 64 women. If all hats are made to fit head circumferences between 22.00 in. and 23.00 in., the hats won't fit about half of those women.

c. If 64 women are randomly selected, what is the probability that their mean head circumference is between 22.00 in. and 23.00 in.? If this probability is high, does it suggest that an order for 64 hats will very likely fit each of 64 randomly selected women? Why or why not?

14. a. 99.99%
b. 0.9641. No, when considering the diameters of manholes, we should use a design based on individual men, not samples of 36 men.

14. Designing Manholes According to the Web site www.torchmate.com, "manhole covers must be a minimum of 22 in. in diameter, but can be as much as 60 in. in diameter." Assume that a manhole is constructed to have a circular opening with a diameter of 22 in. Men have shoulder breadths that are normally distributed with a mean of 18.2 in. and a standard deviation of 1.0 in. (based on data from the National Health and Nutrition Examination Survey).

a. What percentage of men will fit into the manhole?

b. Assume that the Connecticut Light and Power company employs 36 men who work in manholes. If 36 men are randomly selected, what is the probability that their mean shoulder breadth is less than 18.5 in.? Does this result suggest that money can be saved by making smaller manholes with a diameter of 18.5 in.? Why or why not?

15. a. 140 lb
b. 0.9999 (Tech: 0.99999993, or 1.0000 when rounded to four decimal places)
c. 0.8078 (Tech: 0.8067)
d. Given that there is a 0.8078 probability of exceeding the 3500 lb limit when the water taxi is loaded with 20 random men, the new capacity of 20 passengers does not appear to be safe enough because the probability of overloading is too high.

15. Water Taxi Safety Passengers died when a water taxi sank in Baltimore's Inner Harbor. Men are typically heavier than women and children, so when loading a water taxi, assume a worst-case scenario in which all passengers are men. Assume that weights of men are normally distributed with a mean of 182.9 lb and a standard deviation of 40.8 lb (based on Data Set 1 in Appendix B). The water taxi that sank had a stated capacity of 25 passengers, and the boat was rated for a load limit of 3500 lb.

a. Given that the water taxi that sank was rated for a load limit of 3500 lb, what is the mean weight of the passengers if the boat is filled to the stated capacity of 25 passengers?

b. If the water taxi is filled with 25 randomly selected men, what is the probability that their mean weight exceeds the value from part (a)?

c. After the water taxi sank, the weight assumptions were revised so that the new capacity became 20 passengers. If the water taxi is filled with 20 randomly selected men, what is the probability that their mean weight exceeds 175 lb, which is the maximum mean weight that does not cause the total load to exceed 3500 lb?

d. Is the new capacity of 20 passengers safe?

16. a. 0.5239 (Tech: 0.5231)
b. 0.8944 (Tech: 0.8941)
c. Instead of filling each bag with exactly 465 M&Ms, the company probably fills the bags so that the weight is as stated. In any event, the company appears to be doing a good job of filling the bags.

16. Loading M&M Packages M&M plain candies have a mean weight of 0.8565 g and a standard deviation of 0.0518 g (based on Data Set 20 in Appendix B). The M&M candies used in Data Set 20 came from a package containing 465 candies, and the package label stated that the net weight is 396.9 g. (If every package has 465 candies, the mean weight of the candies must exceed $396.9/465 = 0.8535$ g for the net contents to weigh at least 396.9 g.)

a. If 1 M&M plain candy is randomly selected, find the probability that it weighs more than 0.8535 g.

b. If 465 M&M plain candies are randomly selected, find the probability that their mean weight is at least 0.8535 g.

c. Given these results, does it seem that the Mars Company is providing M&M consumers with the amount claimed on the label?

17. a. 0.6517 (Tech: 0.6516)
b. 0.9115
c. There is a high probability (0.9115) that the gondola will be overloaded if it is occupied by 12 men, so it appears that the number of allowed passengers should be reduced.

17. Gondola Safety A ski gondola in Vail, Colorado, carries skiers to the top of a mountain. It bears a plaque stating that the maximum capacity is 12 people or 2004 lb. That capacity will be exceeded if 12 people have weights with a mean greater than $2004/12 = 167$ lb. Because men tend to weigh more than women, a worst-case scenario involves 12 passengers who are all men. Assume that weights of men are normally distributed with a mean of 182.9 lb and a standard deviation of 40.8 lb (based on Data Set 1 in Appendix B).

a. Find the probability that if an individual man is randomly selected, his weight will be greater than 167 lb.

b. Find the probability that 12 randomly selected men will have a mean weight that is greater than 167 lb (so that their total weight is greater than the gondola maximum capacity of 2004 lb).

c. Does the gondola appear to have the correct weight limit? Why or why not?

18. Pulse Rates of Women Women have pulse rates that are normally distributed with a mean of 77.5 beats per minute and a standard deviation of 11.6 beats per minute (based on Data Set 1 in Appendix B).

a. Find the percentiles P_1 and P_{99}.

b. Dr. Puretz sees exactly 25 female patients each day. Find the probability that 25 randomly selected women have a mean pulse rate between 70 beats per minute and 85 beats per minute.

c. If Dr. Puretz wants to select pulse rates to be used as cutoff values for determining when further tests should be required, which pulse rates are better to use: the results from part (a) or the pulse rates of 70 beats per minute and 85 beats per minute from part (b)? Why?

18. a. $P_1 = 50.5$ beats per minute and $P_{99} = 104.5$ beats per minute.
b. 0.9988
c. Instead of the mean pulse rate from the patients in a day, the cutoff values should be based on individual patients, so it would be better to use the pulse rates of 50.5 beats per minute and 104.5 beats per minute.

19. Redesign of Ejection Seats When women were allowed to become pilots of fighter jets, engineers needed to redesign the ejection seats because they had been originally designed for men only. The ACES-II ejection seats were designed for men weighing between 140 lb and 211 lb. Weights of women are now normally distributed with a mean of 165.0 lb and a standard deviation of 45.6 lb (based on Data Set 1 in Appendix B).

a. If 1 woman is randomly selected, find the probability that her weight is between 140 lb and 211 lb.

b. If 36 different women are randomly selected, find the probability that their mean weight is between 140 lb and 211 lb.

c. When redesigning the fighter jet ejection seats to better accommodate women, which probability is more relevant: the result from part (a) or the result from part (b)? Why?

19. a. 0.5526 (Tech: 0.5517)
b. 0.9994 (Tech: 0.9995)
c. Part (a) because the ejection seats will be occupied by individual women, not groups of women.

20. Loading a Tour Boat The Ethan Allen tour boat capsized and sank in Lake George, New York, and 20 of the 47 passengers drowned. Based on a 1960 assumption of a mean weight of 140 lb for passengers, the boat was rated to carry 50 passengers. After the boat sank, New York State changed the assumed mean weight from 140 lb to 174 lb.

a. Given that the boat was rated for 50 passengers with an assumed mean of 140 lb, the boat had a passenger load limit of 7000 lb. Assume that the boat is loaded with 50 male passengers, and assume that weights of men are normally distributed with a mean of 182.9 lb and a standard deviation of 40.8 lb (based on Data Set 1 in Appendix B). Find the probability that the boat is overloaded because the 50 male passengers have a mean weight greater than 140 lb.

b. The boat was later rated to carry only 14 passengers, and the load limit was changed to 2436 lb. If 14 passengers are all males, find the probability that the boat is overloaded because their mean weight is greater than 174 lb (so that their total weight is greater than the maximum capacity of 2436 lb). Do the new ratings appear to be safe when the boat is loaded with 14 male passengers?

20. a. 0.9999 (Tech: 1.0000 when rounded to four decimal places)
b. 0.7939 (Tech: 0.7928). Because there is a high probability (0.7939) of overloading, the new ratings do not appear to be safe when the boat is loaded with 14 male passengers.

21. Doorway Height The Boeing 757-200 ER airliner carries 200 passengers and has doors with a height of 72 in. Heights of men are normally distributed with a mean of 69.5 in. and a standard deviation of 2.4 in. (based on Data Set 1 in Appendix B).

a. If a male passenger is randomly selected, find the probability that he can fit through the doorway without bending.

b. If half of the 200 passengers are men, find the probability that the mean height of the 100 men is less than 72 in.

c. When considering the comfort and safety of passengers, which result is more relevant: the probability from part (a) or the probability from part (b)? Why?

d. When considering the comfort and safety of passengers, why are women ignored in this case?

21. a. 0.8508 (Tech: 0.8512)
b. 0.9999 (Tech: 1.0000 when rounded to four decimal places)
c. The probability from part (a) is more relevant because it shows that 85.08% of male passengers will not need to bend. The result from part (b) gives us information about the mean for a group of 100 men, but it doesn't give us useful information about the comfort and safety of individual male passengers.
d. Because men are generally taller than women, a design that accommodates a suitable proportion of men will necessarily accommodate a greater proportion of women.

22. There is a 0.9887 probability that the aircraft is overloaded. Because that probability is so high, the pilot should take action, such as removing excess fuel and/or requiring that some passengers disembark and take a later flight.

22. Loading Aircraft Before every flight, the pilot must verify that the total weight of the load is less than the maximum allowable load for the aircraft. The Bombardier Dash 8 aircraft can carry 37 passengers, and a flight has fuel and baggage that allows for a total passenger load of 6200 lb. The pilot sees that the plane is full and all passengers are men. The aircraft will be overloaded if the mean weight of the passengers is greater than 6200 lb/37 = 167.6 lb. What is the probability that the aircraft is overloaded? Should the pilot take any action to correct for an overloaded aircraft? Assume that weights of men are normally distributed with a mean of 182.9 lb and a standard deviation of 40.8 lb (based on Data Set 1 in Appendix B).

6-5 Beyond the Basics

23. a. Yes. The sampling is without replacement and the sample size of $n = 50$ is greater than 5% of the finite population size of 275. $\sigma_{\bar{x}} = 2.0504584$.
b. 0.5947 (Tech: 0.5963)

23. Correcting for a Finite Population In a study of babies born with very low birth weights, 275 children were given IQ tests at age 8, and their scores approximated a normal distribution with $\mu = 95.5$ and $\sigma = 16.0$ (based on data from "Neurobehavioral Outcomes of School-age Children Born Extremely Low Birth Weight or Very Preterm," by Anderson et al., *Journal of the American Medical Association,* Vol. 289, No. 24). Fifty of those children are to be randomly selected for a follow-up study.

a. When considering the distribution of the mean IQ scores for samples of 50 children, should $\sigma_{\bar{x}}$ be corrected by using the finite population correction factor? Why or why not? What is the value of $\sigma_{\bar{x}}$?

b. Find the probability that the mean IQ score of the follow-up sample is between 95 and 105.

24. a. Yes. Sampling is without replacement (because each sample of 16 elevator passengers consists of 16 different people) and the sample size of $n = 16$ is greater than 5% of the finite population size of 300. $\sigma_{\bar{x}} = 9.7459365$.
b. 187.5 lb
c. 0.1401 (Tech: 0.1407). The probability is not as low as it should be.
d. 14 passengers

24. Correcting for a Finite Population The Orange County Spa began with 300 members. Those members had weights with a distribution that is approximately normal with a mean of 177 lb and a standard deviation of 40 lb. The facility includes an elevator that can hold up to 16 passengers.

a. When considering the distribution of sample means from weights of samples of 16 passengers, should $\sigma_{\bar{x}}$ be corrected by using the finite population correction factor? Why or why not? What is the value of $\sigma_{\bar{x}}$?

b. If the elevator is designed to safely carry a load of up to 3000 lb, what is the maximum safe mean weight of passengers when the elevator is loaded with 16 passengers?

c. If the elevator is filled with 16 randomly selected club members, what is the probability that the total load exceeds the safe limit of 3000 lb? Is this probability low enough?

d. What is the maximum number of passengers that should be allowed if we want at least a 0.999 probability that the elevator will not be overloaded when it is filled with randomly selected club members?

25. a. $\mu = 6.0$ and $\sigma = 2.1602469$
b. 4.5, 4.5, 6.5, 6.5, 7.0, 7.0
c. $\mu_{\bar{x}} = 6.0$ and $\sigma_{\bar{x}} = 1.0801235$
d. $\mu_{\bar{x}} = \mu = 6.0$ and $\sigma_{\bar{x}} = \frac{2.1602469}{\sqrt{2}}\sqrt{\frac{3-2}{3-1}} =$ 1.0801235, which is the same result from part (c).

25. Population Parameters Use the same population of {4, 5, 9} from Example 1 in Section 6-4. Assume that samples of size $n = 2$ are randomly selected *without* replacement.

a. Find μ and σ for the population.

b. After finding all samples of size $n = 2$ that can be obtained *without* replacement, find the population of all values of $\bar{x}$ by finding the mean of each sample of size $n = 2$.

c. Find the mean $\mu_{\bar{x}}$ and standard deviation $\sigma_{\bar{x}}$ for the population of sample means found in part (b).

d. Verify that

$$\mu_{\bar{x}} = \mu \quad \text{and} \quad \sigma_{\bar{x}} = \frac{\sigma}{\sqrt{n}}\sqrt{\frac{N-n}{N-1}}$$

6-6 Assessing Normality

Key Concept The following chapters include several important statistical methods requiring that sample data are a simple random sample from a population having a *normal* distribution. In this section we present criteria for determining whether the requirement of a normal distribution is satisfied. The criteria involve (1) visual inspection of a histogram to see if it is roughly bell-shaped; (2) identifying any outliers; and (3) constructing a graph called a *normal quantile plot*.

Part 1: Basic Concepts of Assessing Normality

When trying to determine whether a collection of data has a distribution that is approximately normal, we can visually inspect a histogram to see if it is approximately bell-shaped (as discussed in Section 2-3), we can identify outliers, and we can also use a *normal quantile plot* (discussed briefly in Section 2-3).

DEFINITION A **normal quantile plot** (or **normal probability plot**) is a graph of points (x, y) where each x value is from the original set of sample data, and each y value is the corresponding z score that is a quantile value expected from the standard normal distribution.

Procedure for Determining Whether It Is Reasonable to Assume That Sample Data Are from a Population Having a Normal Distribution

1. *Histogram:* Construct a histogram. If the histogram departs dramatically from a bell shape, conclude that the data do not have a normal distribution.
2. *Outliers:* Identify outliers. If there is more than one outlier present, conclude that the data do not have a normal distribution. (Just one outlier could be an error or the result of chance variation, but be careful, because even a single outlier can have a dramatic effect on results.)
3. *Normal quantile plot:* If the histogram is basically symmetric and the number of outliers is 0 or 1, use technology to generate a *normal quantile plot*. Apply the following criteria to determine whether or not the distribution is normal. (These criteria can be used loosely for small samples, but they should be used more strictly for large samples.)

   Normal Distribution: The population distribution is normal if the pattern of the points is reasonably close to a straight line and the points do not show some systematic pattern that is not a straight-line pattern.

   Not a Normal Distribution: The population distribution is *not* normal if either or both of these two conditions applies:

   - The points do not lie reasonably close to a straight line.
   - The points show some *systematic pattern* that is not a straight-line pattern.

Later in this section we will describe the actual process of constructing a normal quantile plot, but for now we focus on interpreting such a plot.

Note to Instructor
Organization Change: Section 6-6 ("Assessing Normality") and Section 6-7 ("Normal as Approximation to Binomial") are now switched from the order used in the previous edition. This is motivated by two factors: (1) Technology now makes it easy to generate normal quantile plots, so they become a good tool for assessing the normality of a population and "Assessing Normality" is now more accessible; (2) technology now makes it easy to find exact values of binomial probabilities, so the "Normal as Approximation to Binomial" coverage is now less important and it has been positioned as the last section of Chapter 6.

Comment that some statistical methods (such as the *t* test in Section 8-4) have a loose requirement of a normally distributed population, and some other methods (such as the chi-square test in Section 8-5) have a much stricter requirement of a normally distributed population. This is why it is important to determine whether we have sample data that are from a normally distributed population.

Example 1 Determining Normality

The following displays show histograms of data along with the corresponding normal quantile plots.

continued

Normal: The first case shows a histogram of IQ scores that is close to being bell-shaped, so the histogram suggests that the IQ scores are from a normal distribution. The corresponding normal quantile plot shows points that are reasonably close to a straight-line pattern, and the points do not show any other systematic pattern that is not a straight line. It is safe to assume that these IQ scores are from a population that has a normal distribution.

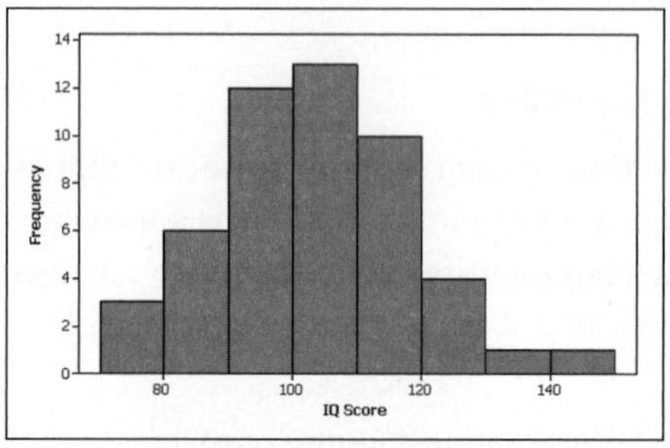

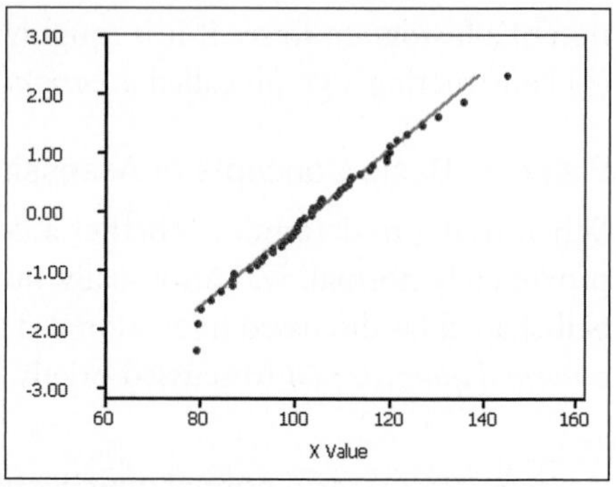

Uniform: The second case shows a histogram of data having a uniform distribution. The corresponding normal quantile plot suggests that the points are not normally distributed. Although the pattern of points is reasonably close to a straight-line pattern, *there is another systematic pattern that is not a straight-line pattern.* We conclude that these sample values are from a population having a distribution that is not normal.

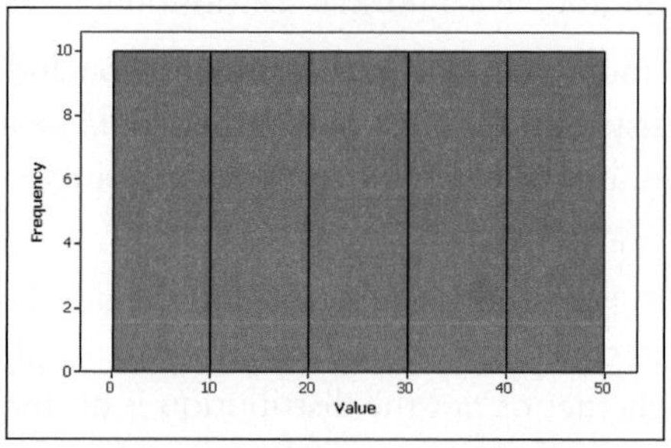

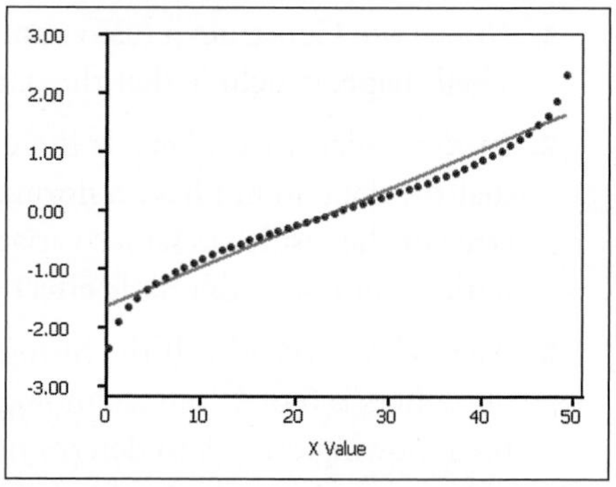

Skewed: The third case shows a histogram of the amounts of rainfall (in inches) in Boston for every Monday during one year. The shape of the histogram is skewed, not bell-shaped. The corresponding normal quantile plot shows points that are not at all close to a straight-line pattern. These rainfall amounts are from a population having a distribution that is not normal.

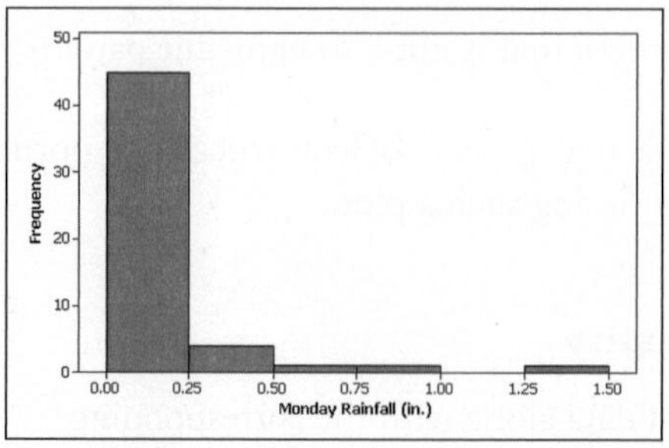

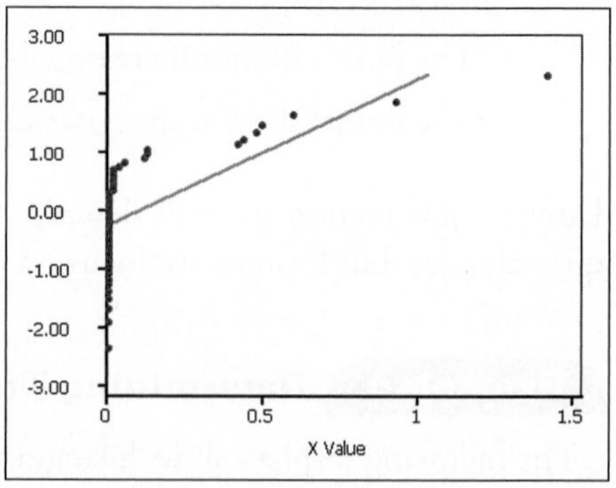

Here are some important comments about procedures for determining whether data are from a normally distributed population:

- If the requirement of a normal distribution is not too strict, simply look at a histogram and find the number of outliers. If the histogram is roughly bell-shaped and the number of outliers is 0 or 1, treat the population as if it has a normal distribution.
- Normal quantile plots can be difficult to construct on your own, but they can be generated with a TI-83/84 Plus calculator or suitable computer software, such as STATDISK, SPSS, SAS, Minitab, Excel, or StatCrunch.
- In addition to the procedures discussed in this section, there are other, more advanced procedures for assessing normality, such as the chi-square goodness-of-fit test, the Kolmogorov-Smirnov test, the Lilliefors test, the Anderson-Darling test, and the Ryan-Joiner test (discussed briefly in Part 2).

Part 2: Beyond the Basics of Assessing Normality

The following is a relatively simple procedure for manually constructing a normal quantile plot, and it is the same procedure used by STATDISK and the TI-83/84 Plus calculator. Some statistical packages use various other approaches, but the interpretation of the graph is basically the same.

Recommendation: Cover only Part 1 and stress the *interpretation* of a normal quantile plot, but do not stress the mechanics of actually creating one (as in Part 2). On a test, provide displayed results from a technology such as STATDISK, Minitab, Excel, StatCrunch, or a TI-83/84 Plus calculator, and ask whether the sample data are from a normally distributed population.

Recommendation: Regardless of the technology being used throughout the course, use STATDISK for its feature of Normality Assessment. It is very easy to copy or enter or open a set of sample data and obtain one screen displaying a histogram, normal quantile plot, and the numbers of possible outliers.

Manual Construction of a Normal Quantile Plot

Step 1: First sort the data by arranging the values in order from lowest to highest.

Step 2: With a sample of size n, each value represents a proportion of $1/n$ of the sample. Using the known sample size n, identify the areas of $\frac{1}{2n}, \frac{3}{2n}, \frac{5}{2n}$, and so on. These are the cumulative areas to the left of the corresponding sample values.

Step 3: Use the standard normal distribution (software or a calculator or Table A-2) to find the z scores corresponding to the cumulative left areas found in Step 2. (These are the z scores that are expected from a normally distributed sample.)

Step 4: Match the original sorted data values with their corresponding z scores found in Step 3, then plot the points (x, y), where each x is an original sample value and y is the corresponding z score.

Step 5: Examine the normal quantile plot and conclude that the population has a normal distribution if the pattern of the points is reasonably close to a straight line and the points do not show some systematic pattern that is not a straight-line pattern.

Example 2 Earthquake Magnitudes

Data Set 16 in Appendix B includes Richter-scale magnitudes of earthquakes. Let's consider this sample of five magnitudes: 0.70, 2.20, 1.64, 1.01, 1.62. With only five values, a histogram will not be very helpful in revealing the distribution of the data. Instead, construct a normal quantile plot for these five values and determine whether they appear to come from a population that is normally distributed.

Solution

The following steps correspond to those listed in the procedure above for constructing a normal quantile plot.

continued

Step 1: First, sort the data by arranging them in order. We get 0.70, 1.01, 1.62, 1.64, 2.20.

Step 2: With a sample of size $n = 5$, each value represents a proportion of $1/5$ of the sample, so we proceed to identify the cumulative areas to the left of the corresponding sample values. The cumulative left areas, which are expressed in general as $\frac{1}{2n}, \frac{3}{2n}, \frac{5}{2n}$, and so on, become these specific areas for this example with $n = 5$: $\frac{1}{10}, \frac{3}{10}, \frac{5}{10}, \frac{7}{10}$, and $\frac{9}{10}$. The cumulative left areas expressed in decimal form are 0.1, 0.3, 0.5, 0.7, and 0.9.

Step 3: We now use technology (or Table A-2) with the cumulative left areas of 0.1000, 0.3000, 0.5000, 0.7000, and 0.9000 to find these corresponding z scores: -1.28, -0.52, 0, 0.52, and 1.28.

Step 4: We now pair the original sorted earthquake magnitudes with their corresponding z scores. We get these (x, y) coordinates, which are plotted in the accompanying STATDISK display: $(0.70, -1.28)$, $(1.01, -0.52)$, $(1.62, 0)$, $(1.64, 0.52)$, $(2.20, 1.28)$.

STATDISK

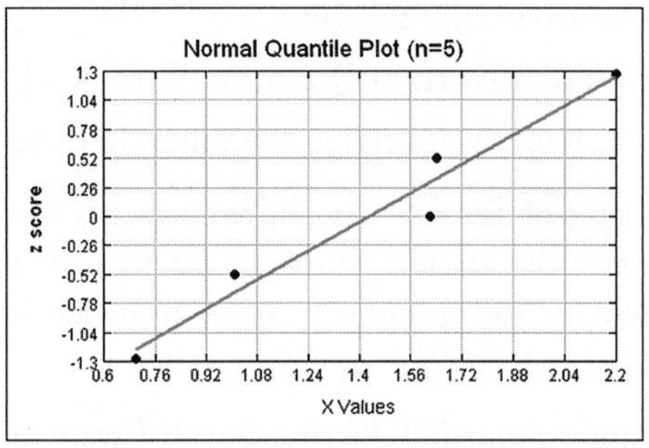

Interpretation

We examine the normal quantile plot in the STATDISK display. Because the points appear to lie reasonably close to a straight line and there does not appear to be a systematic pattern that is not a straight-line pattern, we conclude that the sample of five earthquake magnitudes appears to come from a normally distributed population.

Ryan-Joiner Test The Ryan-Joiner test is one of several formal tests of normality, each having its own advantages and disadvantages. STATDISK has a feature of **Normality Assessment** that displays a histogram, normal quantile plot, the number of potential outliers, and results from the Ryan-Joiner test. Information about the Ryan-Joiner test is readily available on the Internet.

Example 3 Earthquake Magnitudes

Example 2 used only the first five earthquake magnitudes listed in Data Set 16 from Appendix B. If we include all 50 magnitudes, we can use the **Normality Assessment** feature of STATDISK to get the accompanying display.

STATDISK

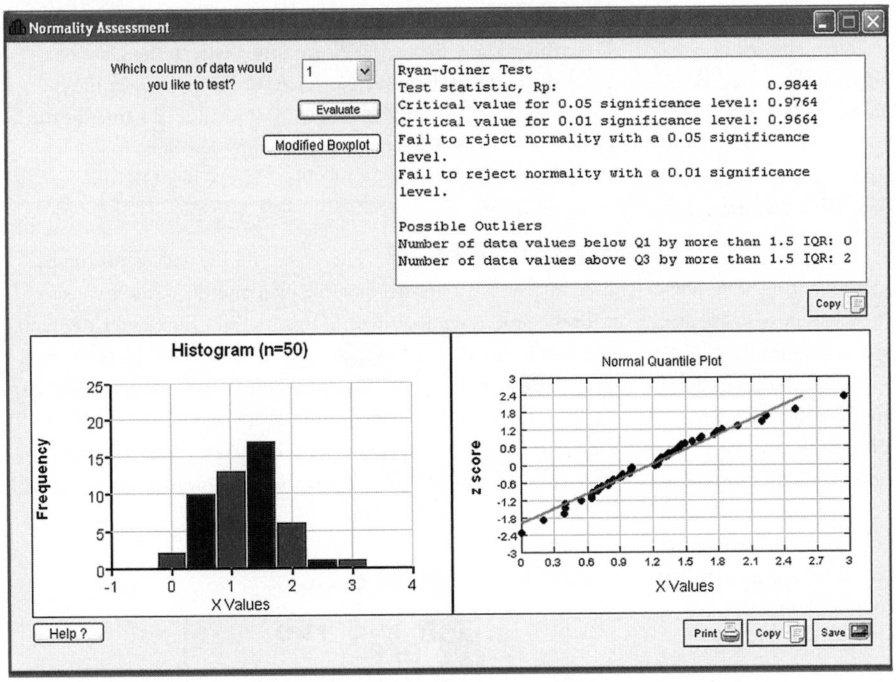

Let's use the display with the three criteria for assessing normality.

1. *Histogram:* We can see that the histogram is approximately bell-shaped.
2. *Outliers:* There are two possible outliers, but after sorting the data, we see that the five highest magnitudes are 1.98, 2.20, 2.24, 2.50, 2.95, so the possible outliers of 2.50 and 2.95 aren't too far away from the other data values.
3. *Normal quantile plot:* The points in the normal quantile plot are not far from a straight-line pattern, and there is no other pattern that is not a straight-line pattern. It is reasonable to conclude that the 50 earthquake magnitudes are from a population with a normal distribution.

Data Transformations Many data sets have a distribution that is not normal, but we can *transform* the data so that the modified values have a normal distribution. One common transformation is to transform each value of x by taking its logarithm. (You can use natural logarithms or logarithms with base 10. If any original values are 0, take logarithms of values of $x + 1$.) If the distribution of the logarithms of the values is a normal distribution, the distribution of the original values is called a **lognormal distribution.** (See Exercises 22 and 23.) In addition to transformations with logarithms, there are other transformations, such as replacing each x value with $\sqrt{x}$, or $1/x$, or x^2. In addition to getting a required normal distribution when the original data values are not normally distributed, such transformations can be used to correct deficiencies, such as a requirement (found in later chapters) that different data sets have the same variance.

using TECHNOLOGY

STATDISK STATDISK can be used to generate a normal quantile plot, and the result is consistent with the procedure described in this section. Enter the data in a column of the Sample Editor window or open a data set. Next, select **Data** from the main menu bar at the top. Select **Normal Quantile Plot** to generate the graph. Better yet, select **Normality Assessment** to obtain the normal quantile plot included in the same display with other results helpful in assessing normality. Proceed to enter the column number for the data, then click **Evaluate.**

continued

MINITAB Minitab can generate a graph similar to the normal quantile plot described in this section. Minitab's procedure is somewhat different, but the graph can be interpreted by using the same criteria given in this section. That is, normally distributed data should lie reasonably close to a straight line, and points should not reveal a pattern that is not a straight-line pattern. First enter the values in column C1 or open a data set, then select **Stat, Basic Statistics,** and **Normality Test.** Enter **C1** for the variable, then click on **OK.**

Minitab can also generate a graph that includes boundaries. If the points all lie within the boundaries, conclude that the values are normally distributed. If points lie beyond the boundaries, conclude that the values are not normally distributed. To generate the graph that includes the boundaries, first enter the values in column C1, select the main menu item of **Graph,** select **Probability Plot,** then select the option of **Simple.** Proceed to enter **C1** for the variable, then click **OK.** The accompanying Minitab display is based on the 50 earthquake *depths* (km) listed in Data Set 16 from Appendix B, and it includes the boundaries. See that there are points lying outside of the boundaries, suggesting that the data are not from a population having a normal distribution.

MINITAB

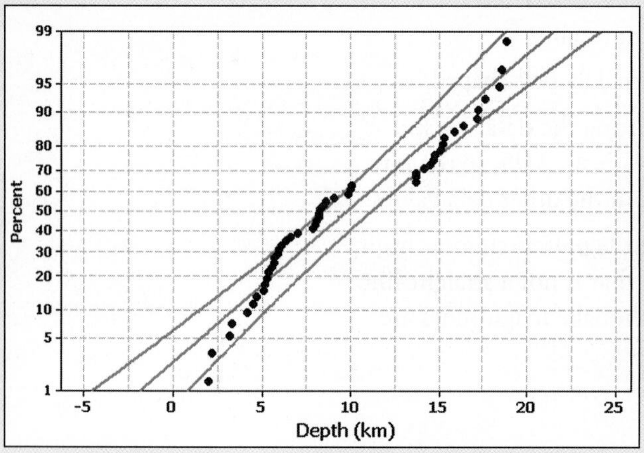

EXCEL Use XLSTAT. Click on **XLSTAT,** then click on **Describing Data,** then select **Normality Tests.** In the Data box, enter the range of data, such as A1:A50. Put a check in the "Sample labels" box only if the first cell includes a name for the data. Click on the **Charts** tab and be sure that there is a checkmark next to the "Normal Q-Q Plot" box. Click **OK** and proceed to get the graph.

TI-83/84 PLUS The TI-83/84 Plus calculator can be used to generate a normal quantile plot, and the result is consistent with the procedure described in this section. First enter the sample data in list L1. Press 2ND Y= (for **STAT PLOT**), then press ENTER. Select **ON,** select the "type" item, which is the last item in the second row of options, and enter **L1** for the data list. The screen should appear as shown here. After making all selections, press ZOOM, then **9,** and the points in the normal quantile plot will be displayed.

TI-83/84 PLUS

STATCRUNCH Click on **Open StatCrunch,** then enter or open a column of sample data. Click on **Graphics,** then select the menu item of **QQPlot.** In the dialog box that appears, select the column to be used, then click on **Create Graph.** The result will be similar to the normal quantile plot described in this section, but the axes will be switched. The interpretation will be the same as described in this section.

Recommended Assignment: Exercises 1–8. If technology can be used for generating normal quantile plots, also assign Exercises 13–16.

6-6 Basic Skills and Concepts

Statistical Literacy and Critical Thinking

1. Normal Quantile Plot Data Set 1 in Appendix B includes the heights of 40 randomly selected women. If you were to construct a histogram of those heights, what shape do you expect the histogram to have? If you were to construct a normal quantile plot of those heights, what pattern would you expect to see in the graph?

1. The histogram should be approximately bell-shaped, and the normal quantile plot should have points that approximate a straight-line pattern.

2. Normal Quantile Plot After constructing a histogram of the ages of the 40 women included in Data Set 1 in Appendix B, you see that the histogram is far from being bell-shaped. What do you now know about the normal quantile plot?

2. Either the points are not reasonably close to a straight-line pattern, or there is some systematic pattern that is not a straight-line pattern.

3. Small Sample An article includes elapsed times (hours) to lumbar puncture for 19 patients who entered emergency rooms with sudden and severe "thunderclap" headaches (based on data from "Thunderclap Headache and Normal Computed Tomographic Results: Value of Cerebrospinal Fluid Analysis," by DuPont et al., *Mayo Clinic Proceedings,* Vol. 83, No. 12). Given that the sample size is less than 30, what requirement must be met in order to treat the sample mean as a value from a normally distributed population? Identify three ways of verifying that requirement.

3. We must verify that the sample is from a population having a normal distribution. We can check for normality using a histogram, identifying the number of outliers, and constructing a normal quantile plot.

4. Assessing Normality The accompanying histogram is constructed from the platelet counts of the 40 women included in Data Set 1 in Appendix B. If you plan to conduct further statistical tests and there is a loose requirement of a normally distributed population, what do you conclude about the population distribution based on this histogram?

MINITAB

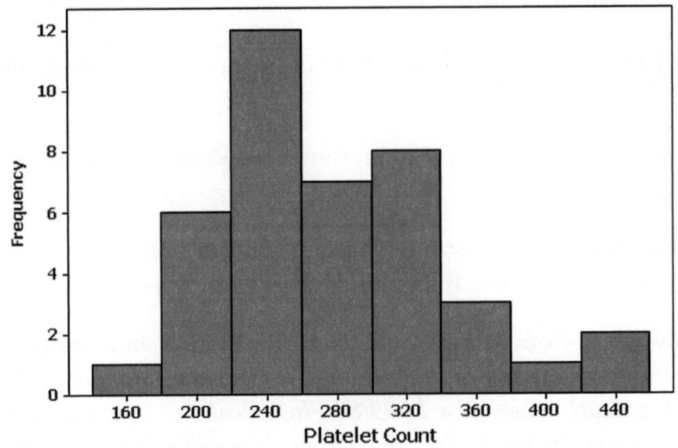

Interpreting Normal Quantile Plots. ***In Exercises 5–8, examine the normal quantile plot and determine whether the sample data appear to be from a population with a normal distribution.***

5. Ages of Oscar-Winning Actresses The normal quantile plot represents the ages of actresses when they won Oscars. The data are from Data Set 11 in Appendix B.

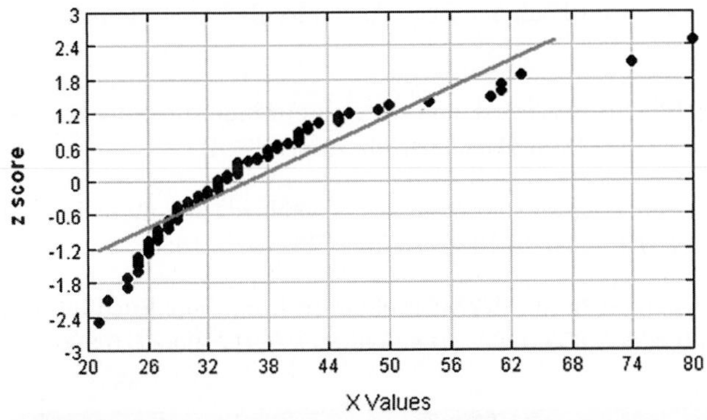

6. Body Temperatures The normal quantile plot represents body temperatures of adults from Data Set 3 in Appendix B.

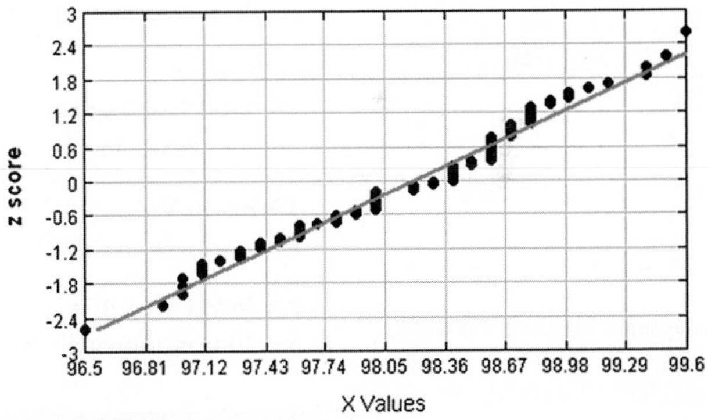

7. White Blood Cell Counts The normal quantile plot represents white blood cell counts of males from Data Set 1 in Appendix B.

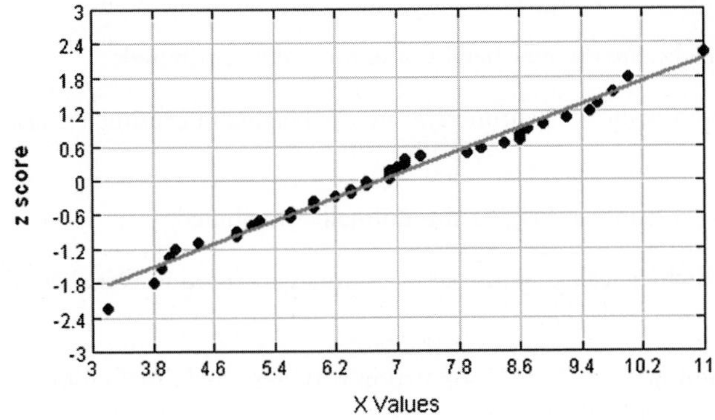

8. Flight Departure Delays The normal quantile plot represents flight departure delay times (minutes) from Data Set 15 in Appendix B.

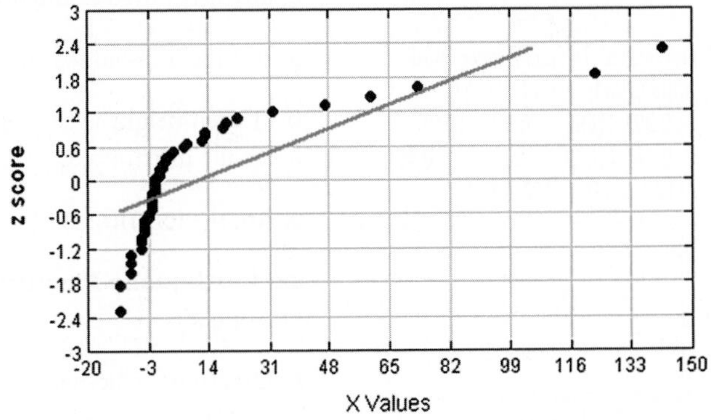

4. Because the histogram is roughly bell-shaped, conclude that the data are from a population having a normal distribution.
5. Not normal. The points show a systematic pattern that is not a straight-line pattern.
6. Normal. The points are reasonably close to a straight-line pattern, and there is no other pattern that is not a straight-line pattern.
7. Normal. The points are reasonably close to a straight-line pattern, and there is no other pattern that is not a straight-line pattern.
8. Not normal. The points are not reasonably close to a straight-line pattern, and there appears to be a pattern that is not a straight-line pattern.
9. Not normal
10. Normal
11. Normal
12. Not normal

Determining Normality. ***In Exercises 9–12, refer to the indicated sample data and determine whether they appear to be from a population with a normal distribution. Assume that this requirement is loose in the sense that the population distribution need not be exactly normal, but it must be a distribution that is roughly bell-shaped.***

9. Flight Arrival Delays The arrival delay times (minutes) as listed in Data Set 15 in Appendix B.

10. Heights of Presidents The heights (cm) of the presidents of the United States as listed in Data Set 12 in Appendix B.

11. Blood Pressure The systolic blood pressure measurements of males as listed in Data Set 1 in Appendix B.

12. Secondhand Smoke The cotinine measurements of nonsmokers who were exposed to tobacco smoke at home or work, as listed in Data Set 9 in Appendix B.

Using Technology to Generate Normal Quantile Plots. ***In Exercises 13–16, use the data from the indicated exercise in this section. Use computer software (such as STATDISK, Minitab, Excel, or StatCrunch) or a TI-83/84 Plus calculator to generate a normal quantile plot. Then determine whether the data come from a normally distributed population.***

13. Exercise 9

14. Exercise 10

15. Exercise 11

16. Exercise 12

Constructing Normal Quantile Plots. ***In Exercises 17–20, use the given data values to identify the corresponding z scores that are used for a normal quantile plot, then identify the coordinates of each point in the normal quantile plot. Construct the normal quantile plot, then determine whether the data appear to be from a population with a normal distribution.***

17. Braking Distances A sample of braking distances (in feet) measured under standard conditions for a sample of large cars from Data Set 14 in Appendix B: 139, 134, 145, 143, 131.

18. Taxi Out Times A sample of flights is selected, and the times (minutes) required to taxi out for takeoff are 37, 13, 14, 15, 31, 15 (from Data Set 15 in Appendix B).

19. Brain Volumes A sample of human brain volumes (cm^3) is obtained from those listed in Data Set 6 from Appendix B: 1272, 1051, 1079, 1034, 1070, 1173, 1079, 1067.

20. M&M Weights A sample of weights (g) of M&Ms is obtained from those listed in Data Set 20 from Appendix B: 0.864, 0.825, 0.855, 0.942, 0.825, 0.869, 0.912, 0.887, 0.886.

6-6 Beyond the Basics

21. Transformations The heights (in inches) of men listed in Data Set 1 in Appendix B have a distribution that is approximately normal, so it appears that those heights are from a normally distributed population.

a. If 2 inches is added to each height, are the new heights also normally distributed?

b. If each height is converted from inches to centimeters, are the heights in centimeters also normally distributed?

c. Are the logarithms of normally distributed heights also normally distributed?

22. Earthquake Magnitudes Richter scale earthquake magnitudes are listed in Data Set 16 of Appendix B.

a. Determine whether those magnitudes are from a population with a normal distribution.

13. Not normal

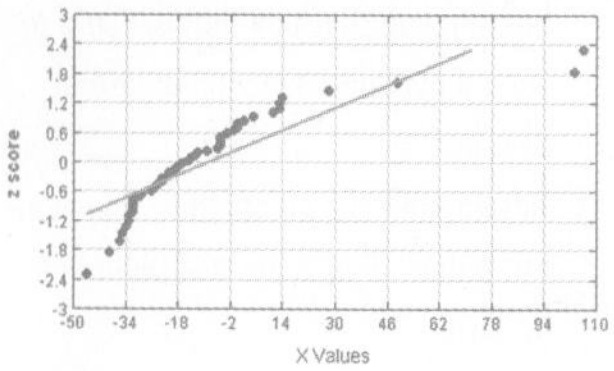

14. Normal

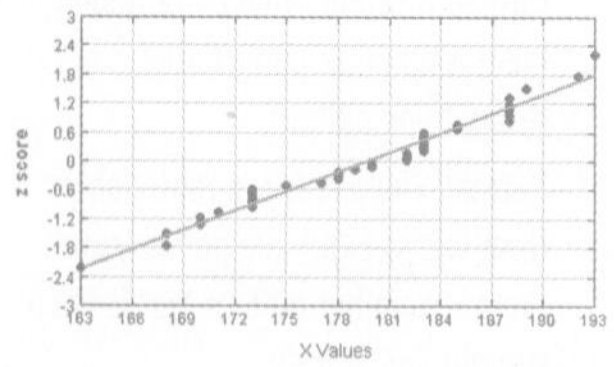

15. Normal

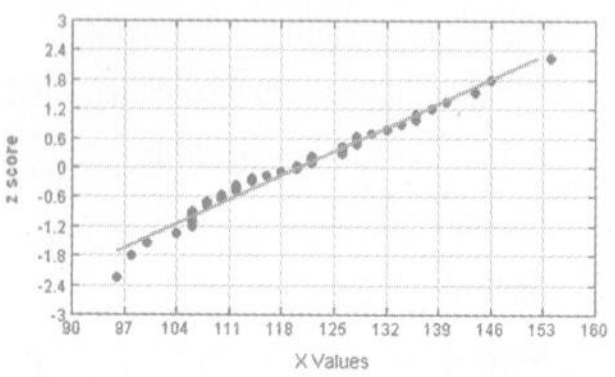

16. Not normal

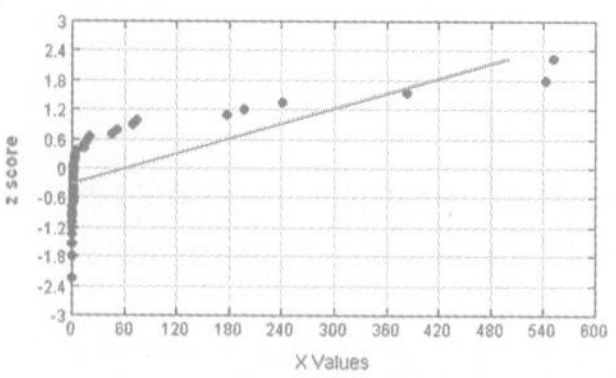

17. Normal. The points have coordinates (131, −1.28), (134, −0.52), (139, 0), (143, 0.52), (145, 1.28).

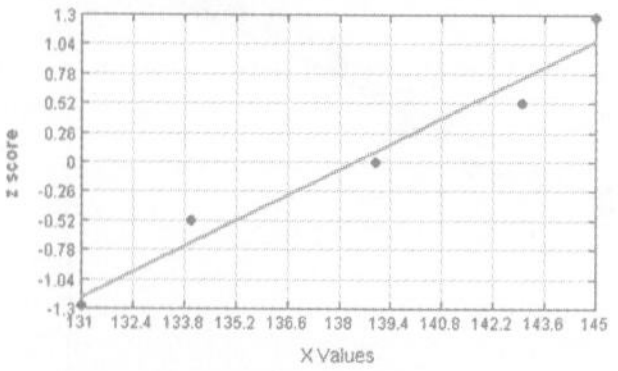

18. Not normal. The points have coordinates (13, −1.38), (14, −0.67), (15, −0.21), (15, 0.21), (31, 0.67), (37, 1.38).

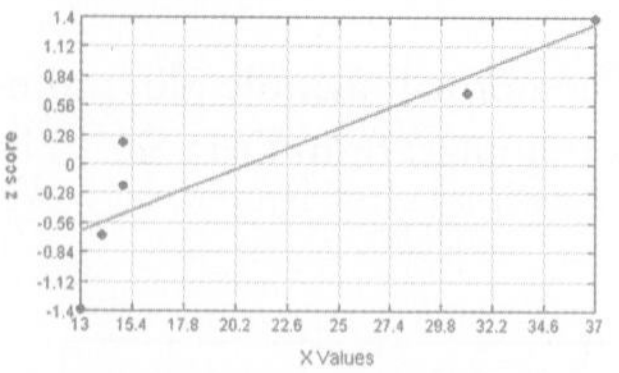

19. Not normal. The points have coordinates (1034, −1.53), (1051, −0.89), (1067, −0.49), (1070, −0.16), (1079, 0.16), (1079, 0.49), (1173, 0.89), (1272, 1.53).

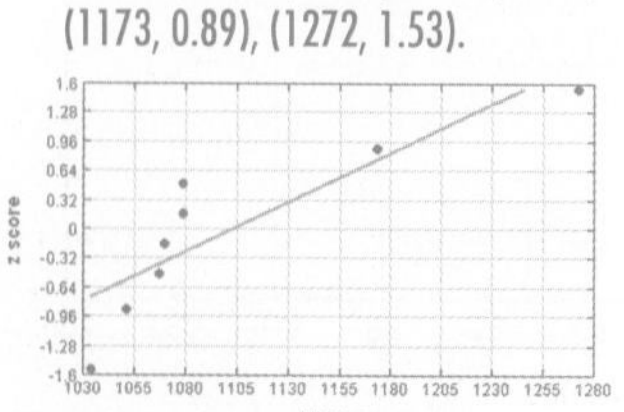

b. For the Richter scale, assume that the magnitude of the energy from an earthquake is first measured, then the logarithm (base 10) of the value is computed. Based on this description of the Richter scale and the result from part (a), what is the distribution of the original measurements before their logarithms are computed?

c. Given that the magnitudes in Data Set 16 result from computing logarithms (base 10), construct the normal quantile plot of the original magnitudes before logarithms are applied. Do the original values appear to be from a population with a normal distribution?

23. Lognormal Distribution The following are the values of net worth (in thousands of dollars) of the members of the executive branch of the current U.S. government. Test these values for normality, then take the logarithm of each value and test for normality. What do you conclude?

82490	27650	26652	21454	11494	10463	7291	5613	3784	3671	3466	3435	3395
3044	2287	1332	1305	878	872	783	556	463	397	145	27	

20. Normal. The points have coordinates (0.825, −1.59), (0.825, −0.97), (0.855, −0.59), (0.864, −0.28), (0.869, 0), (0.886, 0.28), (0.887, 0.59), (0.912, 0.97), (0.942, 1.59).

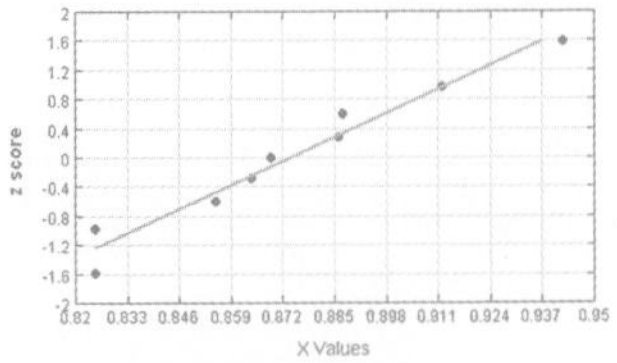

21. a. Yes b. Yes c. No

22. a. The magnitudes are normally distributed.
b. The original measurements have a lognormal distribution.
c. We can reverse the process of taking logarithms by letting each of the values be an exponent of 10. The normal quantile plot indicates that the original values are not from a population with a normal distribution.

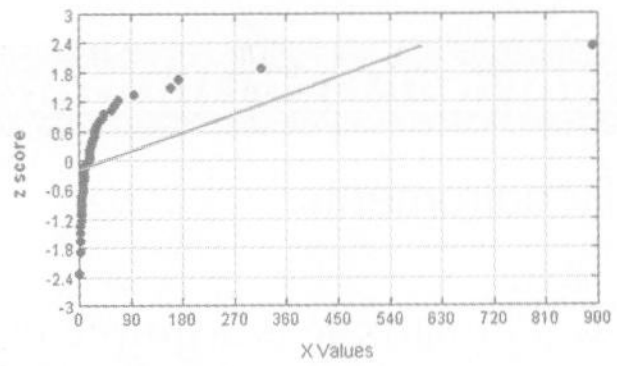

23. The original values are not from a normally distributed population. After taking the logarithm of each value, the values appear to be from a normally distributed population. The original values are from a population with a lognormal distribution.

6-7 Normal as Approximation to Binomial

Key Concept This section presents a method for using a normal distribution as an approximation to a binomial probability distribution, so that some problems involving proportions can be solved by using a normal distribution. Here are the two main points of this section:

- For a proportion p, if the conditions $np \geq 5$ and $nq \geq 5$ are both satisfied, then probabilities from a binomial probability distribution can be approximated reasonably well by using a normal distribution with these parameters: $\mu = np$ and $\sigma = \sqrt{npq}$.
- Because a binomial probability distribution typically uses only whole numbers for the random variable x, but the normal approximation is continuous, we use a "continuity correction" with a whole number x represented by the interval from $x - 0.5$ to $x + 0.5$.

Brief Review In Section 5-3 we noted that a *binomial probability distribution* has (1) a fixed number of trials; (2) trials that are independent; (3) trials that are each classified into two categories commonly referred to as *success* and *failure;* (4) trials with the property that the probability of success remains constant. Section 5-3 also introduced the following notation.

Notation

n = the fixed number of trials

x = the specific number of successes in n trials

p = probability of *success* in *one* of the n trials

q = probability of *failure* in *one* of the n trials (so $q = 1 - p$)

Rationale for Using Normal Approximation Instead of providing a theoretical derivation proving that a normal distribution can approximate a binomial distribution, we simply demonstrate it with an illustration involving this situation: In 431 professional football games that went to overtime, the teams that won the coin toss went on to win 235 games (based on data from "The Overtime Rule in the National Football League: Fair or Unfair?" by Gorgievski et al., *MathAMATYC Educator,* Vol. 2, No. 1). If we assume that the coin toss is fair, the probability of winning the game after winning the coin toss should be 0.5. With $p = 0.5$ and $n = 431$ games, we get a binomial probability

This section describes another interesting application of the normal distribution, and it follows up on the earlier work done with the binomial distribution. However, using suitable computer software or a TI-83/84 Plus calculator, we can now solve many more binomial distribution problems directly without using a normal approximation, so the methods of this section are not as necessary and important as they once were. This section is not required for future chapters and may be omitted if time is an issue.

distribution that is graphed in the accompanying Minitab display. The graph shows the probability for each number of wins from 180 to 250. (The other possible numbers of wins have probabilities that are very close to 0.) See how the graph of the binomial probabilities is close to being a normal distribution. This graph suggests that we can use a normal distribution to approximate the binomial distribution.

MINITAB

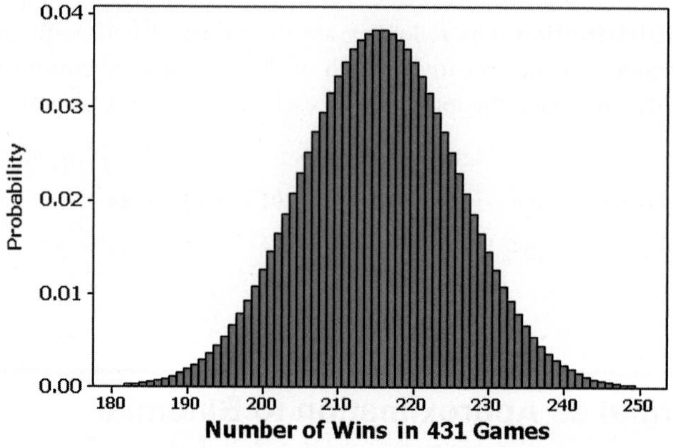

Normal Distribution as an Approximation to the Binomial Distribution

Requirements

1. The sample is a simple random sample of size n from a population in which the proportion of successes is p, or the sample is the result of conducting n independent trials of a binomial experiment in which the probability of success is p.

2. $np \geq 5$ and $nq \geq 5$.

Normal Approximation

If the above requirements are satisfied, then the probability distribution of the random variable x can be approximated by a normal distribution with these parameters:

- $\mu = np$
- $\sigma = \sqrt{npq}$

Continuity Correction

When using the normal approximation, adjust the discrete whole number x by using a *continuity correction* so that x is represented by the interval from $x - 0.5$ to $x + 0.5$.

Note that the requirements include verification of $np \geq 5$ and $nq \geq 5$. The minimum value of 5 is common, but it isn't an absolutely rigid value, and a few textbooks use 10 instead. This requirement is included in the following procedure for using a normal approximation to a binomial distribution.

Procedure for Using a Normal Distribution to Approximate a Binomial Distribution

1. Verify the two preceding requirements. (If these requirements are not both satisfied, then you must use computer software, or a calculator, or Table A-1, or calculations using the binomial probability formula.)

2. The normal distribution requires values for μ and σ, so find those values by calculating $\mu = np$ and $\sigma = \sqrt{npq}$.

3. Identify the discrete whole number x that is relevant to the binomial probability problem being considered.

   Example: If you're trying to find the probability of getting at least 235 successes among 431 trials (as in Example 1), the discrete whole number of concern is $x = 235$. First focus on the value of 235 itself, and temporarily ignore whether you want at least 235, more than 235, fewer than 235, at most 235, or exactly 235.

4. Draw a normal distribution centered about the value of μ, then draw a *vertical strip area* centered over x. Mark the left side of the strip at $x - 0.5$, and mark the right side at $x + 0.5$.

   Example: For $x = 235$, draw a strip from 234.5 to 235.5 and *consider the entire area of the entire strip to represent the probability of the discrete whole number 235.*

5. Determine whether the value of x itself should be included in the probability you want. Shade the area to the right or left of the strip from Step 4, as appropriate; shade the *interior* of the strip *if and only if x itself* is to be included. This total shaded region corresponds to the probability being sought.

   Example: The phrase "at least 235" *does* include 235 itself, but "more than 235" *does not* include 235 itself.

6. Using either $x - 0.5$ or $x + 0.5$ in place of x itself, find the area of the shaded region from Step 5 as follows:

   i. Find the z score: $z = (x - \mu)/\sigma$ (replacing x with either $x + 0.5$ or $x - 0.5$).

   ii. Use that z score to find the cumulative area to the left of the adjusted value of x.

   iii. The cumulative left area can now be used to identify the shaded area corresponding to the desired probability.

Example 1 Is the NFL Coin Toss Fair?

In 431 NFL football games that went to overtime, the teams that won the coin toss went on to win 235 of those games. If the coin-toss method is fair, we expect that the teams winning the coin toss would win about 50% of the games, so we expect about 215.5 wins in 431 overtime games. Assuming that there is a 0.5 probability of winning a game after winning the coin toss, find the probability of getting at least 235 winning games among the 431 teams that won the coin toss. That is, given $n = 431$ and $p = 0.5$, find P(at least 235 wins).

Solution

The given problem involves a binomial distribution with $n = 431$ independent trials. When a team wins the coin toss in overtime, there are two categories for each outcome: The team goes on to win the game or does not win. The assumed probability of winning ($p = 0.5$) presumably remains constant from game to game. Calculations with the binomial probability formula are not practical, because we would have to apply it 197 times, once for each value of x from 235 to 431 inclusive. Software and calculators can be used to find that the probability is 0.0335, but we will proceed to show how the normal approximation method can be used. We use the preceding six-step procedure.

Step 1: Requirement check: The 431 overtime NFL games are from recent years, but we will proceed under the assumption that we have a simple random sample. We must also verify that $np \geq 5$ and $nq \geq 5$. With $n = 431$ and $p = 0.5$ and $q = 0.5$, those requirements are both satisfied with $np = 215.5$ and $nq = 215.5$.

Step 2: We now proceed to find μ and σ needed for the normal distribution. We get the following:

$$\mu = np = 431 \cdot 0.5 = 215.5$$

$$\sigma = \sqrt{npq} = \sqrt{431 \cdot 0.5 \cdot 0.5} = 10.380270$$

Step 3: We want the probability of at least 235 wins, so the discrete whole number relevant to this example is $x = 235$.

Step 4: See Figure 6-19, which shows the normal distribution and the vertical strip from 234.5 to 235.5.

Step 5: We want to find the probability of getting *at least* 235 *wins,* so we want to shade the vertical strip representing 235 as well the area to its right. The desired area is shaded in Figure 6-19.

Step 6: We want the area to the right of 234.5 in Figure 6-19. If using technology, we find that the shaded area in Figure 6-19 is 0.0336. If using Table A-2, we must first find the z score using $x = 234.5$, $\mu = 215.5$, and $\sigma = 10.380270$ as follows:

$$z = \frac{x - \mu}{\sigma} = \frac{234.5 - 215.5}{10.380270} = 1.83$$

Using Table A-2, we find that $z = 1.83$ corresponds to cumulative left area of 0.9664, so the shaded region in Figure 6-19 is $1 - 0.9664 = 0.0336$.

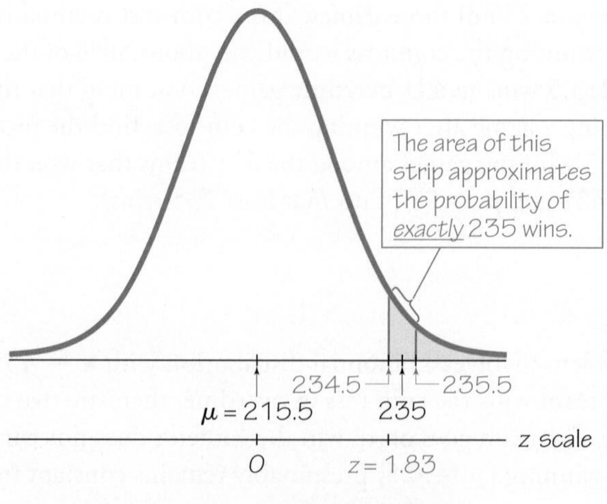

Figure 6-19 Probability of 235 Wins

Interpretation

If we assume that the overtime coin toss does not favor either team, then among 431 teams that won the overtime coin toss, there is a 0.0336 probability of at least 235 wins. This probability is low enough to suggest that the team winning the coin toss has an advantage over the team losing the coin toss. Given that winning the overtime coin toss provides an unfair advantage, it appears that overtime rules should be modified to reduce or eliminate that unfair advantage.

Continuity Correction

The procedure for using a normal distribution to approximate a binomial distribution includes a *continuity correction,* defined as follows.

Note to Instructor
Students often have difficulty applying the continuity correction. Carefully describe the three-step procedure given here. Students generally find that this procedure is quite helpful.

DEFINITION When we use the normal distribution (which is a *continuous* probability distribution) as an approximation to the binomial distribution (which is *discrete*), a **continuity correction** is made to a discrete whole number x in the binomial distribution by representing the discrete whole number x by the *interval* from $x - 0.5$ to $x + 0.5$ (that is, adding and subtracting 0.5).

In the preceding six-step procedure for using a normal distribution to approximate a binomial distribution, Steps 3 and 4 incorporate the continuity correction. (See Steps 3 and 4 in the solutions to Examples 1 and 2.)

To see examples of continuity corrections, examine the common cases illustrated in Figure 6-20. Those cases correspond to the statements in the following list:

Statement	Area
At least 235 (includes 235 and above)	To the *right* of 234.5
More than 235 (doesn't include 235)	To the *right* of 235.5
At most 235 (includes 235 and below)	To the *left* of 235.5
Fewer than 235 (doesn't include 235)	To the *left* of 234.5
Exactly 235	Between 234.5 and 235.5

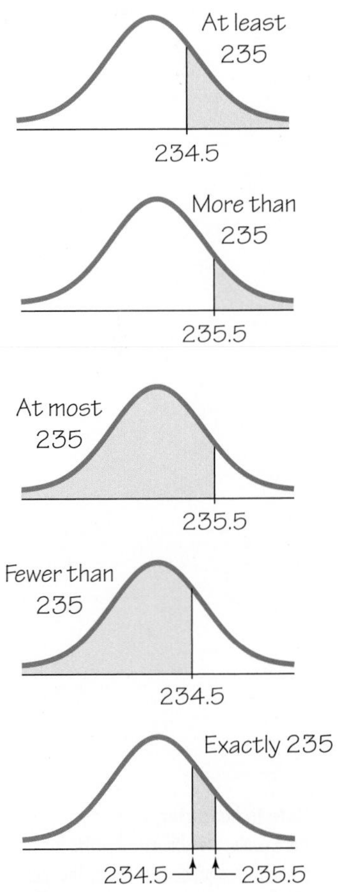

Figure 6-20
Using Continuity Corrections

Example 2 Exactly 235 Wins

Using the same information from Example 1, find the probability of *exactly* 235 wins by the 431 teams that won the coin toss in overtime. That is, given $n = 431$ and assuming that $p = 0.5$, find P(exactly 235 wins). Is this result useful for determining whether the overtime coin toss is fair?

Solution

Using the same six-step procedure given earlier, Steps 1, 2, and 3 are the same as in Example 1.

Step 4: See Figure 6-21, which shows the normal distribution with $\mu = 215.5$ and $\sigma = 10.380270$. Also, the shaded area represents the probability of *exactly* 235 wins. That region is the vertical strip between 234.5 and 235.5, as shown.

Step 5: Because we want the probability of *exactly* 235 wins, we want only the shaded area shown in Figure 6-21.

continued

Step 6: If using technology to find the shaded region in Figure 6-21, we get 0.0066. If using Table A-2 to find the shaded area in Figure 6-21, we find the two z scores corresponding to 234.5 and 235.5. We use those two z scores to find the cumulative area to the left of 234.5 and the cumulative area to the left of 235.5, then we find the *difference* between those two areas. Let's begin with the total area to the left of 235.5. If using Table A-2, we must first find the z score corresponding to 235.5. We get

$$z = \frac{235.5 - 215.5}{10.380270} = 1.93$$

We use Table A-2 to find that $z = 1.93$ has a cumulative left area of 0.9732. Now we find the area to the left of 234.5 by first finding its corresponding z score:

$$z = \frac{234.5 - 215.5}{10.380270} = 1.83$$

We use Table A-2 to find that $z = 1.83$ corresponds to a cumulative left area of 0.9664. Using Table A-2, the shaded area of Figure 6-21 is $0.9732 - 0.9664 = 0.0068$.

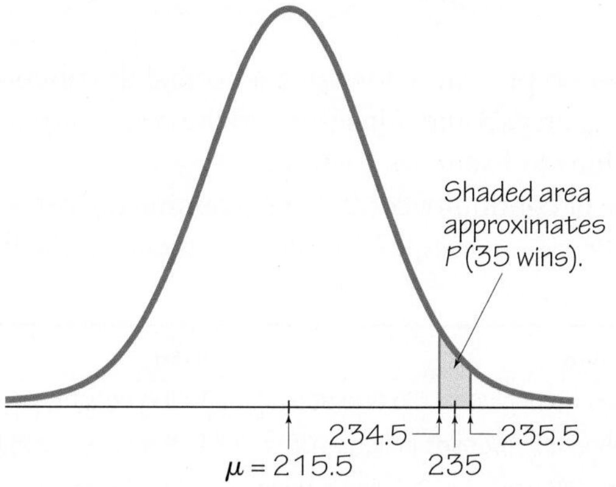

Figure 6-21 Probability of Exactly 235 Wins

Interpretation

If we assume that the overtime coin toss does not favor either team, then among 431 teams that won the overtime coin toss, the probability of *exactly* 235 wins is 0.0068. This result is not very helpful in determining whether the overtime coin toss is fair. If the coin toss gives no advantage, we expect that in 431 overtime games, the team winning the coin toss would win about 215.5 times (about half of the time). But we see that there are 235 wins, which is more than 215.5. The relevant result is the probability of getting a result *at least as extreme* as the one obtained, so the relevant result is the probability of *at least* 235 wins (as in Example 1), not the probability of exactly 235 wins (as in this example). See the following comments about interpreting results, and see Section 5-2 for the comments under the heading of "Unusually High or Unusually Low Number of Successes: Not *Exactly,* but At Least as Extreme."

Note to Instructor

Once again, the ultimate objective is not finding some probability value. The real objective is to make practical sense of the result. Carefully review the given criteria for determining whether a given number of successes among n trials is an unusually high number of successes or an unusually low number of successes.

Interpreting Results

When we use a normal distribution as an approximation to a binomial distribution, our ultimate goal is not simply to find a probability number. We often need to make some *judgment* based on the probability value. The following criterion (from Section 5-2) describes the use of probabilities to determine whether results could easily occur by chance.

Using Probabilities to Determine When Results Are Unusual

- **Unusually *high* number of successes:** x successes among n trials is an *unusually high* number of successes if the probability of x or more successes is unlikely with a probability of 0.05 or less. This criterion can be expressed as follows: $P(x \text{ or more}) \leq 0.05$.*
- **Unusually *low* number of successes:** x successes among n trials is an *unusually low* number of successes if the probability of x or fewer successes is unlikely with a probability of 0.05 or less. This criterion can be expressed as follows: $P(x \text{ or fewer}) \leq 0.05$.*

* The value 0.05 is not absolutely rigid. Other values, such as 0.01, could be used to distinguish between results that can easily occur by chance and events that are very unlikely to occur by chance.

6-7 Basic Skills and Concepts

Recommended Assignment
Exercises 1–16.

Statistical Literacy and Critical Thinking

1. Exact Value and Approximation Refer to Figure 6-21 in Example 2, and refer to the Minitab display near the beginning of this section. Figure 6-21 shows the shaded region representing exactly 235 wins among 431 games. What is the shape of the corresponding region in the Minitab display? If we use technology without an approximation, the probability of exactly 235 wins is found to be 0.0066. If we use the normal approximation to the binomial distribution, the probability of exactly 235 wins is found from Table A-2 to be 0.0068. Which result is better? Is the approximation off by much?

2. Continuity Correction In a preliminary test of the MicroSort method of gender selection, 14 couples were treated and 13 of them had baby girls. If we plan to use the normal approximation to the binomial distribution for finding the probability of exactly 13 girls (assuming that the probability of a girl is 0.5), what is the continuity correction, and how would it be applied in finding that probability?

3. Notation The SAT test uses multiple-choice test questions, each with possible answers of a, b, c, d, e, and each question has only one correct answer. For people who make random guesses for answers to a block of 25 questions, identify the values of p, q, μ, and σ. What do μ and σ measure?

4. Checking Requirement The SAT test uses multiple-choice test questions, each with possible answers of a, b, c, d, e, and each question has only one correct answer. We want to find the probability of getting exactly 10 correct answers for someone who makes random guesses for answers to a block of 25 questions. If we plan to use the methods of this section with a normal distribution used to approximate a binomial distribution, are the necessary requirements satisfied? Explain.

Using Normal Approximation. ***In Exercises 5–8, do the following: If the requirements of $np \geq 5$ and $nq \geq 5$ are both satisfied, estimate the indicated probability by using the normal distribution as an approximation to the binomial distribution; if $np < 5$ or $nq < 5$, then state that the normal approximation should not be used.***

5. With $n = 13$ and $p = 0.4$, find P(fewer than 3).

6. With $n = 12$ and $p = 0.7$, find P(fewer than 8).

7. With $n = 20$ and $p = 0.8$, find P(more than 11).

8. With $n = 25$ and $p = 0.4$, find P(more than 9).

1. The Minitab display shows that the region representing 235 wins is a rectangle. The result of 0.0068 is an approximation, but the result of 0.0066 is better because it is based on an exact calculation. The approximation differs from the exact result by a very small amount.
2. The continuity correction is used to compensate for the fact that a continuous distribution (normal) is used to approximate a discrete distribution (binomial). The discrete number of 13 is represented by the interval from 12.5 to 13.5.
3. $p = 0.2$; $q = 0.8$; $\mu = 5$; $\sigma = 2$. The value of $\mu = 5$ shows that for many people who make random guesses for the 25 questions, the mean number of correct answers is 5. For many people who make random guesses, the standard deviation of $\sigma = 2$ is a measure of how much the numbers of correct responses vary.
4. Yes. The circumstances correspond to 25 independent trials of a binomial experiment in which the probability of success is 0.2. Also, with $n = 25$, $p = 0.2$, and $q = 0.8$, the requirements of $np \geq 5$ and $nq \geq 5$ are both satisfied, because $5 \geq 5$ and $20 \geq 5$ are both true.
5. 0.0630 (Tech: 0.0632)
6. Normal approximation should not be used.
7. Normal approximation should not be used.
8. 0.5793 (Tech: 0.5809)

9. 0.2743 (Tech: 0.2731)
10. 0.2743 (Tech: 0.2731)
11. 0.0928 (Tech: 0.0933)
12. 0.0738 (Tech: 0.0740)
13. a. 0.0219 (Tech using normal approximation: 0.0214; Tech using binomial: 0.0217)
 b. 0.1711 (Tech using normal approximation: 0.1702; Tech using binomial: 0.1703). The result of 172 overturned calls is not unusually low.
 c. The result from part (b) is useful. We want the probability of getting a result that is at least as extreme as the one obtained.
 d. If the 30% rate is correct, there is a good chance (0.1711) of getting 172 or fewer calls overturned, so there is not strong evidence against the 30% rate.
14. a. 0.0012 (Tech using normal approximation: 0.0013; Tech using binomial: 0.0013)
 b. 0.0060 (Tech using normal approximation: 0.0061; Tech using binomial: 0.0056). The result of 172 overturned calls is unusually low.
 c. The result from part (b) is useful. We want the probability of getting a result that is at least as extreme as the one obtained.
 d. If the 33% rate is correct, there is a very small chance (0.0060) of getting 172 or fewer calls overturned, so there is strong evidence against the 33% rate.
15. a. 0.0318 (Tech using normal approximation: 0.0305; Tech using binomial: 0.0301)
 b. 0.2676 (Tech using normal approximation: 0.2665; Tech using binomial: 0.2650). The result of 428 peas with green pods is not unusually low.
 c. The result from part (b) is useful. We want the probability of getting a result that is at least as extreme as the one obtained.
 d. No. Assuming that Mendel's probability of 3/4 is correct, there is a good chance (0.2676) of getting the results that were obtained. The obtained results do not provide strong evidence against the claim that the probability of a pea having a green pod is 3/4.
16. a. 0.0020 (Tech using normal approximation: 0.0020; Tech using binomial: 0.0023)

Voters. ***In Exercises 9–12, use a normal approximation to find the probability of the indicated number of voters. In each case, assume that 100 eligible voters aged 18–24 are randomly selected. The most recent Census Bureau results show that among eligible voters aged 18–24, 22% of them voted.***

9. Probability that fewer than 20 voted

10. Probability that at least 25 voted

11. Probability of exactly 23 voters

12. Probability of exactly 19 voters

13. Tennis Replay In the year that this exercise was written, there were 611 challenges made to referee calls in professional tennis singles play. Among those challenges, 172 were upheld with the call overturned. Assume that 30% of the challenges are successfully upheld with the call overturned.

a. Find the probability that among the 611 challenges, the number of overturned calls is exactly 172.

b. Of the 611 challenges, the number of 172 overturned calls is fewer than 30%. Find the probability that among the 611 challenges, the number of overturned calls is 172 or fewer. If the 30% rate is correct, is 172 overturned calls for 611 challenges an unusually low number?

c. Which result is useful for determining whether the 30% rate is correct: part (a) or part (b)? Explain.

d. Is there strong evidence to suggest that the rate of overturned calls is not 30%?

14. Tennis Replay Repeat the preceding exercise after changing the assumed rate of overturned calls from 30% to 33%.

15. Mendelian Genetics When Mendel conducted his famous genetics experiments with peas, one sample of offspring consisted of 580 peas, with 428 of them having green pods. If we assume, as Mendel did, that under these circumstances, there is a 3/4 probability that a pea will have a green pod, we would expect that 435 of the peas would have green pods, so the result of 428 peas with green pods is fewer than expected. For the following, assume that the probability of a pea having a green pod is 3/4.

a. Find the probability that among the 580 offspring peas, exactly 428 of them have green pods.

b. The result of 428 peas with green pods is fewer than 3/4 of 580. Find the probability that among the 580 offspring peas, 428 or fewer have green pods. Is the result of 428 peas with green pods unusually low?

c. Which result is useful for determining whether Mendel's claimed rate of 75% is incorrect: part (a) or part (b)? Explain.

d. Is there strong evidence to suggest that Mendel's probability of 3/4 is wrong?

16. Dream Job In a Marist College poll of 1004 adults, 291 chose professional athlete as their dream job. Assume that 25% of adults consider being a professional athlete their dream job.

a. The result of 291 is more than 25% of 1004, so find the probability that among 1004 random adults, 291 or more consider being a professional athlete their dream job.

b. If the value of 25% is correct, is the result of 291 unusually high?

c. Does the result suggest that the rate is greater than 25%?

17. XSORT Gender Selection MicroSort's XSORT gender-selection technique is designed to increase the likelihood that a baby will be a girl. In updated results (as of this writing) of the XSORT gender-selection technique, 945 births consisted of 879 baby girls and 66 baby boys (based on data from the Genetics & IVF Institute). In analyzing these results, assume that the XSORT method has no effect so that boys and girls are equally likely.

a. Find the probability of getting exactly 879 girls in 945 births.

b. Find the probability of getting 879 or more girls in 945 births. If boys and girls are equally likely, is 879 girls in 945 births unusually high?

c. Which probability is relevant for trying to determine whether the XSORT method is effective: the result from part (a) or the result from part (b)?

d. Based on the results, does it appear that the XSORT method is effective? Why or why not?

18. Washing Hands Based on *observed* males using public restrooms, 85% of adult males wash their hands in a public restroom (based on data from the American Society for Microbiology and the American Cleaning Institute). In a survey of 523 adult males, 518 *reported* that they wash their hands in a public restroom. Assuming that the 85% observed rate is correct, find the probability that among 523 randomly selected adult males, 518 or more wash their hands in a public restroom. What do you conclude?

19. Voters Lying? In a survey of 1002 people, 701 said that they voted in a recent presidential election (based on data from ICR Research Group). Voting records show that 61% of eligible voters actually did vote. Given that 61% of eligible voters actually did vote, find the probability that among 1002 randomly selected eligible voters, at least 701 actually did vote. What does the result suggest?

20. Cell Phones and Brain Cancer In a study of 420,095 cell phone users in Denmark, it was found that 135 developed cancer of the brain or nervous system. Assuming that the use of cell phones has no effect on developing such cancers, there is a 0.000340 probability of a person developing cancer of the brain or nervous system. We therefore expect about 143 cases of such cancers in a group of 420,095 randomly selected people. Estimate the probability of 135 or fewer cases of such cancers in a group of 420,095 people. What do these results suggest about media reports that cell phones cause cancer of the brain or nervous system?

21. Smoking Based on a recent Harris Interactive survey, 20% of adults in the United States smoke. In a survey of 50 statistics students, it is found that 6 of them smoke. Find the probability that should be used for determining whether the 20% rate is correct for statistics students. What do you conclude?

22. Smoking Repeat the preceding exercise after changing the results so that among 50 statistics students, it is found that 3 smoke.

23. Online TV In a Comcast survey of 1000 adults, 17% said that they watch prime-time TV online. If we assume that 20% of adults watch prime-time TV online, find the probability that should be used to determine whether the 20% rate is correct or whether it should be lower than 20%? What do you conclude?

24. Internet Access Of U.S. households, 67% have Internet access (based on data from the Census Bureau). In a random sample of 250 households, 70% are found to have Internet access. Find the probability that should be used to determine whether the 67% rate is too low. What do you conclude?

6-7 Beyond the Basics

25. Decision Theory Marc Taylor plans to place 200 bets of $5 each on a game at the Mirage casino in Las Vegas.

a. One strategy is to bet on the number 7 at roulette. A win pays off with odds of 35:1 and, on any one spin, there is a probability of 1/38 that 7 will be the winning number. Among the

b. Because the probability of getting 291 or more with the value of 25% is so small, the result of 291 is unusually high.

c. The results do suggest that the rate is greater than 25%.

17. a. 0.0000 or 0+ (a very small positive probability that is extremely close to 0)

b. 0.0001 (Tech: 0.0000 or 0+, which is a very small positive probability that is extremely close to 0). If boys and girls are equally likely, 879 girls in 945 births is unusually high.

c. The result from part (b) is more relevant, because we want the probability of a result that is *at least as extreme* as the one obtained.

d. Yes. It is very highly unlikely that we would get a result as extreme as 879 girls in 945 births by chance. Given that the 945 couples were treated with the XSORT method, it appears that this method is effective in increasing the likelihood that a baby will be a girl.

18. 0.0001 (Tech using normal approximation: 0.0000 or 0+; Tech using binomial: 0.0000 or 0+). It appears that many adult males say that they wash their hands in a public restroom when they actually do not.

19. 0.0001 (Tech: 0.0000). The results suggest that the surveyed people did not respond accurately.

20. 0.2709 (Tech using normal approximation: 0.2697; Tech using binomial: 0.2726). Media reports appear to be wrong.

21. Probability of six or fewer: 0.1075 (Tech using normal approximation: 0.1080; Tech using binomial: 0.1034). Because that probability is not very small, the evidence against the rate of 20% is not very strong.

22. Probability of three or fewer: 0.0107 (Tech using normal approximation: 0.0108; Tech using binomial: 0.0057). Because that probability is so small, the evidence against the rate of 20% is very strong. It appears that the rate of smoking among statistics students is lower than the 20% rate for the general population.

23. Probability of 170 or fewer: 0.0099 (Tech using normal approximation: 0.0098; Tech using binomial: 0.0089). Because the probability of 170 or fewer is so small with the assumed 20% rate, it appears that the rate is actually less than 20%.

24. Probability of 175 or more households with Internet access: 0.1736 (Tech using normal approximation: 0.1732; Tech using binomial: 0.1734). If the Internet access rate is 67%, there is a relatively high probability (0.1732) of getting 175 or more households with Internet access when 250 households are surveyed. It does not appear that the 67% rate is too low.

25. a. 6; 0.4602 (Tech using normal approximation: 0.4583; tech using binomial: 0.4307)

b. 101; 0.3936 (Tech using normal approximation: 0.3933; tech using binomial: 0.3932)

c. The roulette game provides a better likelihood of making a profit.

26. 229 (Tech: 230)

200 bets, what is the minimum number of wins needed for Marc to make a profit? Find the probability that Marc will make a profit.

b. Another strategy is to bet on the pass line in the dice game of craps. A win pays off with odds of 1:1 and, on any one game, there is a probability of 244/495 that he will win. Among the 200 bets, what is the minimum number of wins needed for Marc to make a profit? Find the probability that Marc will make a profit.

c. Based on the preceding results, which game is the better "investment": the roulette game from part (a) or the craps game from part (b)? Why?

26. Overbooking a Boeing 767-300 A Boeing 767-300 aircraft has 213 seats. When someone buys a ticket for a flight, there is a 0.0995 probability that the person will not show up for the flight (based on data from an IBM research paper by Lawrence, Hong, and Cherrier). How many reservations could be accepted for a Boeing 767-300 for there to be at least a 0.95 probability that all reservation holders who show will be accommodated?

Chapter 6 Review

In this chapter we introduced the normal probability distribution—the most important distribution in the study of statistics.

Section 6-2 Section 6-2 introduced the standard normal distribution, which is a normal distribution with these parameters: $\mu = 0$ and $\sigma = 1$. Two extremely important basic procedures are presented in Section 6-2:

1. Given some z score, find an area under the curve representing the graph of the standard normal distribution.

2. Given some area under the curve representing the standard normal distribution, find the corresponding z score.

Section 6-3 In Section 6-3 we extended the methods from Section 6-2 so that we could work with any normal distribution, not just the standard normal distribution. A key element of Section 6-3 is the formula $z = (x - \mu)/\sigma$ that allows us to convert from a nonstandard normal distribution to the standard normal distribution so that we can solve problems such as these:

1. Given that IQ scores are normally distributed with $\mu = 100$ and $\sigma = 15$, find the probability of randomly selecting someone with an IQ above 90.

2. Given that IQ scores are normally distributed with $\mu = 100$ and $\sigma = 15$, find the IQ score separating the bottom 85% from the top 15%.

Section 6-4 In Section 6-4 we introduced the concept of a sampling distribution of a statistic. The sampling distribution of the sample mean is the probability distribution of sample means, with all samples having the same sample size n. The sampling distribution of the sample proportion is the probability distribution of sample proportions, with all samples having the same sample size n. In general, the sampling distribution of any statistic is the probability distribution of that statistic.

Section 6-5 In Section 6-5 we presented the following so that we could address many important problems involving sample means:

1. The distribution of sample means $\bar{x}$ will, as the sample size n increases, approach a normal distribution.

2. The mean of the sample means is the population mean μ.

3. The standard deviation of the sample means is $\sigma/\sqrt{n}$.

Section 6-6 In Section 6-6 we presented procedures for determining whether sample data appear to come from a population that has a normal distribution. Some of the statistical methods covered later in this book have a loose requirement of a normally distributed population. In such cases, examination of a histogram and outliers might be all that is needed. In other cases, normal quantile plots might be necessary because of factors such as a small sample or a very strict requirement that the population must have a normal distribution.

Section 6-7 In Section 6-7 we noted that a normal distribution can sometimes approximate a binomial probability distribution. Consequently, when working with probabilities or proportions or percentages, we can often use a normal distribution. If both $np \geq 5$ and $nq \geq 5$, the binomial random variable x is approximately normally distributed with the mean and standard deviation given as $\mu = np$ and $\sigma = \sqrt{npq}$. Because the binomial probability distribution deals with discrete data and the normal distribution deals with continuous data, we apply the continuity correction, with the interval from $x - 0.5$ to $x + 0.5$ representing the discrete value of x.

Chapter Quick Quiz

1. Identify the values of μ and σ for the standard normal distribution.

1. $\mu = 0$ and $\sigma = 1$

Bone Density Test. ***In Exercises 2–5, assume that scores on a bone mineral density test are normally distributed with a mean of 0 and a standard deviation of 1.***

2. Sketch a graph showing the shape of the distribution of bone density test scores.

2.

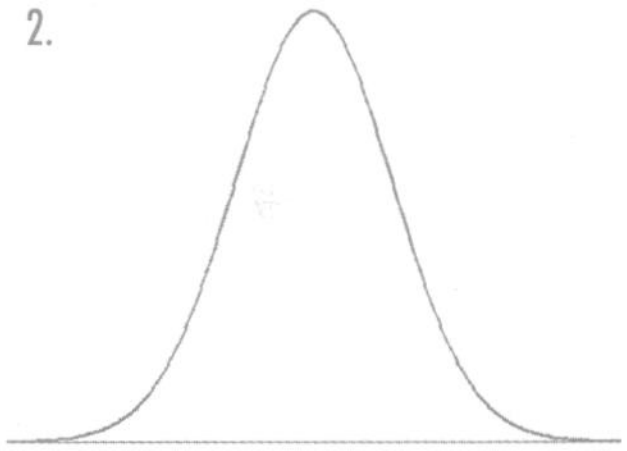

3. Find the score separating the lowest 98% of scores from the highest 2%.

4. For a randomly selected subject, find the probability of a score greater than -1.

5. For a randomly selected subject, find the probability of a score between 1.37 and 2.42.

3. $z = 2.05$ (Tech: 2.05375)
4. 0.8413
5. 0.0775 (Tech: 0.0776)

In Exercises 6–10, assume that red blood cell counts of women are normally distributed with a mean of 4.577 and a standard deviation of 0.382.

6. Find the probability that a randomly selected woman has a red blood cell count below the normal range of 4.2 to 5.4.

6. 0.1611 (Tech: 0.1618)

7. Find the probability that a randomly selected woman has a red blood cell count above the normal range of 4.2 to 5.4.

7. 0.0158 (Tech: 0.0156)

8. Find P_{80}, the 80th percentile for the red blood cell counts of women.

8. 4.898

9. If 25 women are randomly selected, find the probability that the mean of their red blood cell counts is less than 4.444.

9. 0.0409

10. What percentage of women have red blood cell counts in the normal range from 4.2 to 5.4?

10. 82.31% (Tech: 82.26%)

Review Exercises

1. Bone Density Test A bone mineral density test is used to identify a bone disease. The result of a bone density test is commonly measured as a z score, and the population of z scores is normally distributed with a mean of 0 and a standard deviation of 1.

a. For a randomly selected subject, find the probability of a bone density test score less than 2.93.

b. For a randomly selected subject, find the probability of a bone density test score greater than -1.53.

c. For a randomly selected subject, find the probability of a bone density test score between -1.07 and 2.07.

1. a. 0.9983
b. 0.9370
c. 0.8385
d. -0.52
e. 0.1401

d. Find P_{30}, the bone density test score separating the bottom 30% from the top 70%.

e. If the mean bone density test score is found for 16 randomly selected subjects, find the probability that the mean is greater than 0.27.

2. a. 7.93% (Tech: 7.89%)
b. 1369.2 mm (Tech: 1369.4 mm)

2. Door Peephole Standing eye heights of women are normally distributed with a mean of 1516 mm and a standard deviation of 63 mm (based on anthropometric survey data from Gordon, Churchill, et al.).

a. A door peephole is placed at a height that is uncomfortable for women with standing eye heights greater than 1605 mm. What percentage of women will find that height uncomfortable?

b. In selecting the height of a door peephole, the architect wants its height to be suitable for the highest 99% of standing eye heights of women. What standing eye height of women separates the highest 99% of standing eye heights from the lowest 1%?

3. a. 97.88%
b. 1742.6 mm

3. Window Placement Standing eye heights of men are normally distributed with a mean of 1634 mm and a standard deviation of 66 mm (based on anthropometric survey data from Gordon, Churchill, et al.).

a. If a window is positioned so that it is comfortable for men with standing eye heights greater than 1500 mm, what percentage of men will find that height comfortable?

b. A window is positioned to be comfortable for the lowest 95% of eye heights of men. What standing eye height of men separates the lowest 95% from the highest 5%?

4. a. Normal
b. 21.1
c. 0.57

4. Sampling Distributions Scores on the ACT test have a distribution that is approximately normal with mean 21.1 and standard deviation 5.1. A sample of 80 ACT scores is randomly selected and the sample mean is computed.

a. Describe the distribution of such sample means.

b. What is the mean of all such sample means?

c. What is the standard deviation of all such sample means?

5. a. An unbiased estimator is a statistic that targets the value of the population parameter in the sense that the sampling distribution of the statistic has a mean that is equal to the mean of the corresponding parameter.
b. Mean; variance; proportion
c. True

5. Unbiased Estimators

a. What is an unbiased estimator?

b. For the following statistics, identify those that are unbiased estimators: mean, median, range, variance, proportion.

c. Determine whether the following statement is true or false: "The sample standard deviation is a biased estimator, but the bias is relatively small in large samples, so s is often used to estimate σ."

6. a. 85.08% (Tech: 85.12%). With about 15% of all men needing to bend, the design does not appear to be adequate, but the Mark VI monorail appears to be working quite well in practice.
b. 75.1 in.

6. Monorail and Airliner Doors The Mark VI monorail used at Disney World has doors with a height of 72 in. Heights of men are normally distributed with a mean of 69.5 in. and a standard deviation of 2.4 in. (based on Data Set 1 in Appendix B).

a. What percentage of adult men can fit through the doors without bending? Does the door design with a height of 72 in. appear to be adequate? Explain.

b. What doorway height would allow 99% of adult men to fit without bending?

7. a. 0.5753 (Tech: 0.5766)
b. 0.9976. Yes, if the plane is full of male passengers, it is highly likely that it is overweight.

7. Aircraft Safety Standards Under older Federal Aviation Administration rules, airlines were required to estimate the weight of a passenger as 185 lb. (That amount is for an adult traveling in winter, and it includes 20 lb of carry-on baggage.) Rules were revised to use an estimate of 195 lb. Men now have weights that are normally distributed with a mean of 182.9 lb and a standard deviation of 40.9 lb (based on Data Set 1 in Appendix B).

a. If 1 adult male is randomly selected and is assumed to have 20 lb of carry-on baggage, find the probability that his total weight is greater than 195 lb.

b. If a Boeing 767-300 aircraft is full of 213 adult male passengers and each is assumed to have 20 lb of carry-on baggage, find the probability that the mean passenger weight (including carry-on baggage) is greater than 195 lb. Based on that probability, does a pilot have to be concerned about exceeding this weight limit?

8. Assessing Normality Listed below are the current salaries (in thousands of dollars) of players on the New York Yankees baseball team.

a. Do these salaries appear to come from a population that has a normal distribution? Why or why not?

b. Can the mean of this sample be treated as a value from a population having a normal distribution? Why or why not?

403	1250	16500	1400	6000	433	403	13000	414	21600	3750	13000	2125
6550	400	5500	13100	422	455	15000	33000	15286	5400	20625	433	5000

8. a. No. A histogram is far from bell-shaped. A normal quantile plot reveals a pattern of points that is far from a straight-line pattern.
b. No. The sample size of $n = 26$ does not satisfy the condition of $n > 30$, and the values do not appear to be from a population having a normal distribution.

9. Genetics Experiment In one of Mendel's experiments with plants, 1064 offspring consisted of 787 plants with long stems. According to Mendel's theory, 3/4 of the offspring plants should have long stems. Assuming that Mendel's proportion of 3/4 is correct, find the probability of getting 787or fewer plants with long stems among 1064 offspring plants. Based on the result, is 787 offspring plants with long stems unusually low? What does the result imply about Mendel's claimed proportion of 3/4?

9. 0.2296 (Tech using normal approximation: 0.2286; Tech using binomial: 0.2278). The occurrence of 787 offspring plants with long stems is not unusually low because its probability is not small. The results are consistent with Mendel's claimed proportion of 3/4.

10. Job Applicant Background Check There is an 80% chance that a prospective employer will check the educational background of a job applicant (based on data from the Bureau of National Affairs, Inc.). Sixty-four job applications are randomly selected.

a. Find the probability that at least 50 of the applicants have their educational backgrounds checked.

b. Find the probability that exactly 50 of the applicants have their educational backgrounds checked.

10. a. 0.7019 (Tech using normal approximation: 0.7024; Tech using binomial: 0.7100)
b. 0.1148 (Tech using normal approximation: 0.1158; Tech using binomial: 0.1119)

Cumulative Review Exercises

Please be aware that some of the following problems may require knowledge of concepts presented in previous chapters.

1. Miami Heat The following are current annual salaries (in thousands of dollars) for the starting players on the Miami Heat basketball team: 14,500, 14,500, 14,000, 5000, 3500.

a. Find the mean $\bar{x}$ and express the result in dollars instead of thousands of dollars.

b. Find the median and express the result in dollars instead of thousands of dollars.

c. Find the standard deviation s and express the result in dollars instead of thousands of dollars.

d. Find the variance s^2 and express the result in appropriate units.

e. Convert the first salary of $14,500,000 to a z score.

f. What level of measurement (nominal, ordinal, interval, ratio) describes this data set?

g. Are the salaries discrete data or continuous data?

h. Do the given salaries appear to be representative of all players for the Miami Heat team? Why or why not?

1. a. $10,300,000
b. $14,000,000
c. $5,552,027
d. 30,825,003,810,000 square dollars
e. $z = 0.76$
f. Ratio
g. Discrete
h. No, the starting players are likely to be the best players who receive the highest salaries.

2. College as an Investment Assume that 10% of us believe that college is not a good investment (based on a survey in *USA Today*).

a. Let A denote the event of selecting someone who believes that college is not a good investment. What does the event $\overline{A}$ denote?

b. Find the value of $P(\overline{A})$.

c. Find the probability of randomly selecting three different people and finding that each of them has the belief that college is not a good investment.

d. The given rate of 10% is based on results from Internet users who chose to respond to a question posted on the *USA Today* Web site. What is this type of sample called? What does this suggest about the accuracy of the 10% rate?

2. a. $\overline{A}$ is the event of selecting someone who does not have the belief that college is not a good investment. (This is not the same as selecting someone who believes that college is a good investment.)
b. 0.9
c. 0.001
d. The sample is a voluntary response (or self-selected) sample. This suggests that the 10% rate might not be very accurate, because people with strong feelings or interest about the topic are more likely to respond.

3. Birth Weights Birth weights in the United States have a distribution that is approximately normal with a mean of 3369 g and a standard deviation of 567 g (based on data from "Comparison of Birth Weight Distributions between Chinese and Caucasian Infants," by Wen, Kramer, Usher, *American Journal of Epidemiology,* Vol. 172, No. 10).

a. One definition of a premature baby is that the birth weight is below 2500 g. If a baby is randomly selected, find the probability of a birth weight below 2500 g.

b. Another definition of a premature baby is that the birth weight is in the bottom 10%. Find the birth weight that is the cutoff between the bottom 10% and the top 90%.

c. A definition of a "very low birth weight" is one that is less than 1500 g. If a baby is randomly selected, find the probability of a "very low birth weight."

d. If 25 babies are randomly selected, find the probability that their mean birth weight is greater than 3400 g.

3. a. 0.0630 (Tech: 0.0627)
b. 2643 g (Tech: 2642 g)
c. 0.0005
d. 0.3936 (Tech: 0.3923)

4. POTUS The accompanying graph is a histogram of ages of U.S. presidents at the time they were inaugurated (from Data Set 12 in Appendix B).

a. Identify two features of the vertical scale that cause the graph to be somewhat misleading.

b. Does the histogram appear to show data from a population having a normal distribution? Explain.

c. From the graph, identify the lowest and highest possible ages, then use the range rule of thumb to estimate the standard deviation of the ages. How does the result compare to the standard deviation of 6.6 years calculated from the original list of sample values?

4. a. The vertical scale does not start at 0, so differences are somewhat distorted. By using a scale ranging from 1 to 29 for frequencies that range from 2 to 14, the graph is flattened, so differences are not shown as they should be.
b. The graph depicts a distribution that is not exactly normal, but it is approximately normal because it is roughly bell-shaped.
c. Minimum: 42 years; maximum: 70 years. Using the range rule of thumb, the standard deviation is estimated to be $(70 - 42)/4 = 7$ years. The estimate of $s = 7$ years is very close to the actual standard deviation of $s = 6.6$ years, so the range rule of thumb works quite well here.

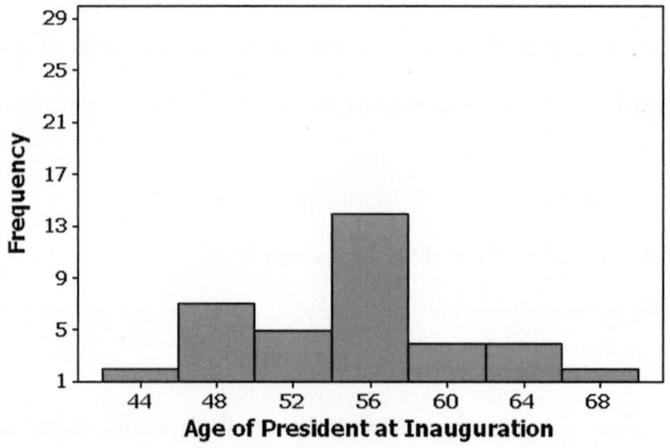

5. Left-Handedness According to data from the American Medical Association, 10% of us are left-handed.

a. If three people are randomly selected, find the probability that they are all left-handed.

b. If three people are randomly selected, find the probability that at least one of them is left-handed.

c. Why can't we solve the problem in part (b) by using the normal approximation to the binomial distribution?

d. If groups of 50 people are randomly selected, what is the mean number of left-handed people in such groups?

e. If groups of 50 people are randomly selected, what is the standard deviation for the numbers of left-handed people in such groups?

f. Use the range rule of thumb to determine whether it would be unusual to get 8 left-handed people in a randomly selected group of 50 people.

5. a. 0.001
b. 0.271
c. The requirement that $np \geq 5$ is not satisfied, indicating that the normal approximation would result in errors that are too large.
d. 5.0
e. 2.1
f. No, 8 is within two standard deviations of the mean and is within the range of values that could easily occur by chance.

Technology Projects

Some methods in this chapter are easy with technology but very difficult without it. The two projects that follow illustrate how easy it is to use technology for assessing normality and finding binomial probabilities.

1. Assessing Normality It is often necessary to determine whether sample data appear to be from a normally distributed population, and that determination is helped with the construction of a histogram and normal quantile plot. Construction of a histogram is relatively simple but time-consuming, and construction of a normal quantile plot can be a real challenge without technology. This first project shows how easy it is to get those graphs by using STATDISK. With just a few clicks of a mouse, you can painlessly and quickly generate a histogram and normal quantile plot together. If STATDISK has not yet been used, it can be downloaded from www.statdisk.org.

The data sets in Appendix B are available by clicking on **Datasets** on the top menu bar, then selecting the textbook you are using. STATDISK can be used to assess normality of a sample by clicking on **Data** and selecting the menu item of **Normality Assessment.** Use the Normality Assessment feature with the female body measurements in Data Set 1 of Appendix B. For each of the 14 columns of data, determine whether the data appear to be from a normally distributed population.

2. Binomial Probabilities Section 6-7 described a method for using a normal distribution to approximate a binomial distribution. STATDISK, Minitab, Excel, StatCrunch, and the TI-83/84 Plus calculator are all capable of generating probabilities for a binomial distribution. Instructions for these different technologies are found in the Using Technology subsection at the end of Section 5-3. Instead of using a normal approximation to a binomial distribution, use technology to find the exact binomial probabilities in Exercises 9–12 of Section 6-7.

from data TO DECISION

Critical Thinking: Designing aircraft seats

When designing seats for aircraft, we want to have sufficient room so that passengers are comfortable and safe, but we don't want too much room, because fewer seats could be installed and profits would drop. It has been estimated that removing one row of seats would cost around $8 million over the life of an aircraft.

Figure 6-22(a) shows an important human consideration: The buttock-to-knee length. The accompanying table includes relevant buttock-to-knee length parameters obtained from studies of large numbers of people. Figure 6-22(b) shows a traditional aircraft seat, and Figure 6-22(c) shows the new SkyRider seat design by the Italian company Aviointeriors. The SkyRider seat is dramatically different from traditional aircraft seats. The seats are like saddles, and they are higher so that passenger legs slant downward with weight on the legs. The most dramatic difference is that SkyRider seats have much less legroom. The distance of 23 in. shown in Figure 6-22(c) is a distance of 30 in. to 32 in. for most current economy seats. As of this writing, the SkyRider seats have not yet been approved by the Federal Aviation Administration, but approval would allow a new class of seating with very low fares.

When designing aircraft seats, we must make some hard choices. If we are to accommodate *everyone* in the population, we will have a sitting distance that is so costly in terms of reduced seating that it might not be economically feasible. Some questions we must address are: (1) What percentage of the population are we willing to exclude? (2) How much extra room do we want to provide for passenger comfort and safety?

1. A common design criterion is to accommodate people falling between the 5th percentile and the 95th percentile. Find those values for the buttock-to-knee lengths of men and women.

2. Of the four values found in Exercise 1, which single value is most important? Explain.

3. Why do the values found in Exercise 1 not apply to the SkyRider seat?

4. Apart from the drastically reduced leg room, identify at least one other disadvantage of the SkyRider seat.

5. Seat pitch is defined to be the distance between the same point on two successive seats. In addition to buttock-to-knee length, what other important factor affects the choice of seat pitch?

6. Based on the preceding results, what would you tell an engineer who is designing and configuring seats for an aircraft?

Buttock-to-Knee Length (inches)

	Mean	St. Dev.	Distribution
Males	23.5 in.	1.1 in.	Normal
Females	22.7 in.	1.0 in.	Normal

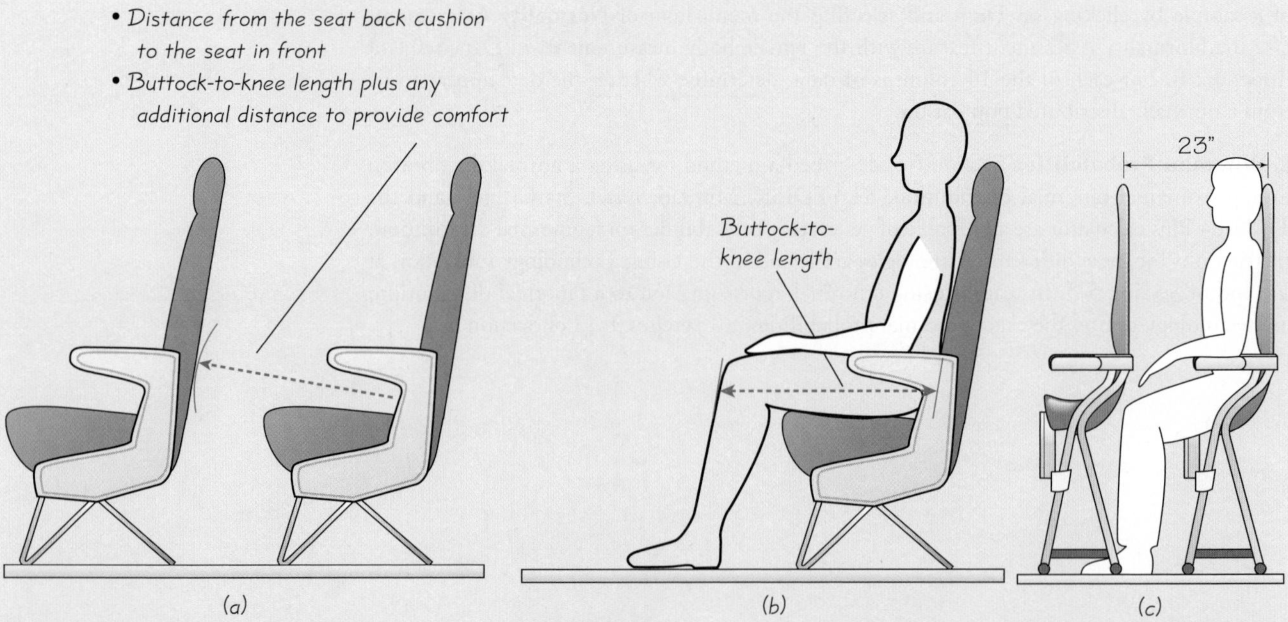

Figure 6-22 **Distances Used in the Design of Aircraft Seats**

Cooperative Group Activities

1. Out-of-class activity Bring a tape measure to a movie theater, car, bus, train, or subway, and measure the distance between the front of your vertical seat cushion and the back of the seat in front. See the Data to Decision Project for the relevant lengths and parameters. Analyze the seating design and write a brief report of the results.

2. Out-of-class activity Use the Internet to find Pick 4 lottery results for 50 different drawings. Find the 50 different means. Graph a histogram of the original 400 digits that were selected, and graph a histogram of the 50 sample means. What important principle do you observe?

3. In-class activity Divide into groups of three or four students and address these issues affecting the design of manhole covers.

- Which of the following is most relevant for determining whether a manhole cover diameter of 24 in. is large enough: weights of men, weights of women, heights of men, heights of women, hip breadths of men, hip breadths of women, shoulder breadths of men, shoulder breadths of women?

- Why are manhole covers usually round? (This was once a popular interview question asked of applicants at IBM, and there are at least three good answers. One good answer is sufficient here.)

4. Out-of-class activity Divide into groups of three or four students. In each group, develop an original procedure to illustrate the central limit theorem. The main objective is to show that when you randomly select samples from a population, the means of those samples tend to be *normally* distributed, regardless of the nature of the population distribution. For this illustration, begin with some population of values that does not have a normal distribution.

5. In-class activity Divide into groups of three or four students. Using a coin to simulate births, each individual group member should simulate 25 births and record the number of simulated girls. Combine all results from the group and record n = total number of births and x = number of girls. Given batches of n births, compute the mean and standard deviation for the number of girls. Is the simulated result unusual? Why or why not?

6. In-class activity Divide into groups of three or four students. Select a set of data from Appendix B (excluding Data Sets 1, 3, 6, 9, 10, 12, 14, 15, 16, and 20, which were used in examples or exercises in Section 6-6). Use the methods of Section 6-6 to construct a histogram and normal quantile plot, then determine whether the data set appears to come from a normally distributed population.

7 Estimates and Sample Sizes

7

chapter problem

Poll: Do you know what Twitter is?

The Pew Research Center recently conducted a survey of 1007 U.S. adults and found that 85% of those surveyed know what Twitter is. Based on the results of that poll, media reports suggested that 85% of the adult population know what Twitter is. How do they know that? Did they somehow obtain survey results from the population of *all* adults in the United States? No. Those media reports are *estimates* based on the results from the poll.

Currently, there are 241,472,385 adults in the United States. Because the Twitter poll involved only 1007 respondents, we see that only 0.0004% of the population was polled. It is natural to question the accuracy of results based on a sample that is so small when compared to the ginormous size of the population. Statistics to the rescue! Using methods presented in this chapter, we can answer questions such as these:

- How do we actually use sample results to estimate values of population parameters? Specifically, if a sample of 1007 adults is polled and it is found that 85% of them know what Twitter is, what does that sample percentage tell us about the percentage for the whole population of adults?
- How accurate is the sample result of 85% likely to be?
- Given that only 1007 people were polled in the population of 241,472,385 adults, is the sample size too small to be meaningful?
- Does the *method* of selecting the sample units have much of an effect on the results?

We can answer the last question based on the sound sampling methods discussed in Chapter 1. The method of selecting the people or households to be surveyed most definitely has an effect on the results. The results are likely to be poor if a convenience sample or some other nonrandom sampling method is used. If the sample is a simple random sample, the results are likely to be good. Because the Pew Research Center uses sophisticated sampling methods that are very carefully planned, its results are generally quite good. The Pew Research Center is a reputable polling company that uses sound sampling methods, so it is reasonable to treat the sample of 1007 adults as a simple random sample.

Our ability to understand polls and surveys and to interpret the results is crucial for our role as citizens. As we consider the topics of this chapter, we will learn much about polls and surveys and how to correctly interpret and present results.

Note to Instructor

Organization Change: In this new edition, Section 7-3 ("Estimating a Population Mean") is partitioned into Part 1 (for σ not known) and Part 2 (for σ known). The main focus in Section 7-3 is using the Student *t* distribution for constructing confidence interval estimates of a population mean. Exercises 35–38 are the only exercises in which σ is assumed to be known. The previous edition of this book had Section 7-3 ("Estimating a Population Mean: σ Known") and Section 7-4 ("Estimating a Population Mean: σ Not Known"), but this edition includes one section (Section 7-3) for these two topics.

Here is the main motivation for this change: In the real world of professional statisticians and professional journals and reports, it is extremely rare that we want to estimate an unknown value of a population mean but we somehow know the value of the population standard deviation σ. In the past, there were sound reasons for starting with the "known σ" situation, including the fact that finding critical values of *t* was often difficult without technology. Technology has now empowered us to find critical *t* values for any sample size *n*, so the situation of a known σ is now much less important. Given the time constraints in the typical introductory statistics course, it is now wise to devote little time and effort to the unrealistic "known σ" case.

7-1 Review and Preview

In Chapters 2 and 3 we used *descriptive statistics* when we summarized data using tools such as graphs and statistics such as the mean and standard deviation. This and following chapters present methods of *inferential statistics* that involve the use of sample data to form generalizations or inferences about a population. See the following two major activities.

Major Activities of Inferential Statistics

1. Use sample data to estimate values of population parameters (such as a population proportion or population mean).
2. Test hypotheses (or claims) made about population parameters.

In this chapter we begin working with the true core of inferential statistics as we use sample data to estimate values of population parameters. For example, the Chapter Problem refers to a survey of 1007 adults, and we see that 85% of them know what Twitter is. Based on the sample statistic of 85%, we will estimate the percentage of *all* U.S. adults who know what Twitter is. In so doing, we are using sample results to make an inference about the population.

This chapter focuses on the use of sample data to estimate a population parameter, and Chapter 8 will introduce the basic methods for testing claims (or hypotheses) that have been made about a population parameter.

Brief Review Because Sections 7-2 and 7-3 use *critical values,* it is helpful to review this notation introduced in Section 6-2: z_α denotes the *z* score with an area of α to its right. (α is the Greek letter alpha.) See Example 8 in Section 6-2, where it is shown that if $\alpha = 0.025$, the critical value is $z_{0.025} = 1.96$. That is, for the standard normal distribution, the critical value of $z_{0.025} = 1.96$ has an area of 0.025 to its right.

7-2 Estimating a Population Proportion

Key Concept This section presents methods for using a sample proportion to make an inference about the value of the corresponding population proportion. Here are the three main concepts included in this section:

- **Point Estimate:** The sample proportion (denoted by $\hat{p}$) is the best *point estimate* (or single value estimate) of the population proportion *p*.
- **Confidence Interval:** We can use a sample proportion to construct a *confidence interval* estimate of the true value of a population proportion, and we should know how to construct and interpret such confidence intervals.
- **Sample Size:** We should know how to find the sample size necessary to estimate a population proportion.

The concepts presented in this section are used in the following sections and chapters, so it is important to understand this section quite well.

Proportion, Probability, and Percent Although this section focuses on the population proportion *p*, we can also work with probabilities or percentages. When working with percentages, we will perform calculations with the equivalent proportion value.

In the Chapter Problem, for example, it was noted that 85% of 1007 surveyed adults know what Twitter is. The sample statistic of 85% can be expressed in decimal form as 0.85, so we will work with the sample proportion of $\hat{p} = 0.85$.

Point Estimate If we want to estimate a population proportion with a single value, the best estimate is the sample proportion $\hat{p}$. Because $\hat{p}$ consists of a single value, it is called a *point estimate.*

DEFINITION A **point estimate** is a single value (or point) used to approximate a population parameter.

The sample proportion $\hat{p}$ is the best *point estimate* of the population proportion p.

We use $\hat{p}$ as the point estimate of p because it is unbiased and it is the most consistent of the estimators that could be used. (Unbiased estimators are discussed in Section 6-4.) The sample proportion $\hat{p}$ is the most consistent estimator of p in the sense that the standard deviation of sample proportions tends to be smaller than the standard deviation of other unbiased estimators of p.

Example 1 Twitter Poll

The Pew Research Center conducted a survey of 1007 adults and found that 85% of them know what Twitter is. Based on that result, find the best point estimate of the proportion of *all* adults who know what Twitter is.

Solution

Because the sample proportion is the best point estimate of the population proportion, we conclude that the best point estimate of p is 0.85. (If using the sample results to estimate the *percentage* of all adults who know what Twitter is, the best estimate is 85%.)

Why Do We Need Confidence Intervals?

In Example 1 we saw that 0.85 was our *best* point estimate of the population proportion p, but a point estimate is a single value that gives us no indication of how *good* that best estimate is. Statisticians have cleverly developed the *confidence interval* or *interval estimate,* which consists of a range (or an interval) of values instead of just a single value. A confidence interval gives us a much better sense of how good an estimate is.

DEFINITION A **confidence interval** (or **interval estimate**) is a range (or an interval) of values used to estimate the true value of a population parameter. A confidence interval is sometimes abbreviated as CI.

DEFINITION The **confidence level** is the probability $1 - \alpha$ (such as 0.95, or 95%) that the confidence interval actually does contain the population parameter, assuming that the estimation process is repeated a large number of times. (The confidence level is also called the **degree of confidence,** or the **confidence coefficient.**)

Note to Instructor

We noted earlier that the two major activities of inferential statistics are (1) estimating values of population parameters and (2) testing claims made about population parameters. By presenting methods of estimating population parameters, this section and this chapter become the first major introduction to inferential statistics. Every introductory statistics course should include at least Sections 7-2 and 7-3. Section 7-4 can be omitted if there is not sufficient time to include it.

This section introduces two major topics: (1) confidence interval estimates of a population proportion p and (2) determining the sample size required to estimate p.

Some textbooks begin with estimates of μ, followed by estimates of p, but this book begins with estimates of p. Here are good reasons for beginning with p: Students generally see proportions much more often in the media than they see means. They are very aware of surveys and they have all heard of "margins of error" as they relate to percentages. Students also tend to be more interested in statistics expressed as proportions or percentages. Finally, methods of inferential statistics have fewer complications when proportions are involved than when means are involved. When introducing confidence interval estimates of a parameter for the first time, it is better to focus on the concepts and methods without being too concerned with complicating factors such as choosing between normal and t distributions.

Curbstoning

The glossary for the Census defines *curbstoning* as "the practice by which a census enumerator fabricates a questionnaire for a residence without actually visiting it." Curbstoning occurs when a census enumerator sits on a curbstone (or anywhere else) and fills out survey forms by making up responses. Because data from curbstoning are not real, they can affect the validity of the Census. The extent of curbstoning has been investigated in several studies, and one study showed that about 4% of Census enumerators practiced curbstoning at least some of the time.

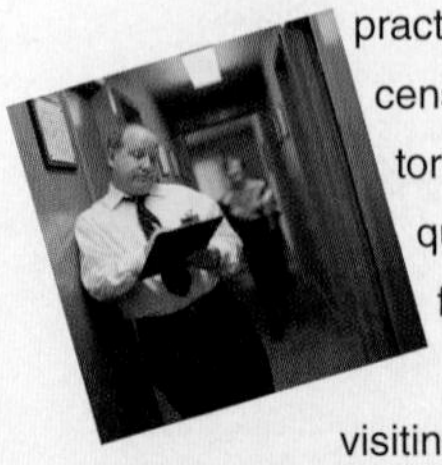

The methods of Section 7-2 assume that the sample data have been collected in an appropriate way, so if much of the sample data have been obtained through curbstoning, then the resulting confidence interval estimates might be very flawed.

A confidence interval is associated with a specific confidence level, such as 0.95 (or 95%). The confidence level gives us the *success rate of the procedure* used to construct the confidence interval. See the following and notice the relationship between the confidence level and the corresponding value of α.

Most Common Confidence Levels	Corresponding Values of α
90% (or 0.90) confidence level:	$\alpha = 0.10$
95% (or 0.95) confidence level:	$\alpha = 0.05$
99% (or 0.99) confidence level:	$\alpha = 0.01$

The confidence level of 95% is most common because it provides a good balance between precision (as reflected in the width of the confidence interval) and reliability (as expressed by the confidence level).

Here's an example of a confidence interval found later (in Example 3), which is based on the sample data of 1007 adults polled, with 85% of them knowing what Twitter is:

The 0.95 (or 95%) confidence interval estimate of the population proportion p is $0.828 < p < 0.872$.

Interpreting a Confidence Interval

We must be careful to interpret confidence intervals correctly. There is a correct interpretation and many different and creative incorrect interpretations of the confidence interval $0.828 < p < 0.872$.

Correct: "We are 95% confident that the interval from 0.828 to 0.872 actually does contain the true value of the population proportion p." This means that if we were to select many different samples of size 1007 and construct the corresponding confidence intervals, 95% of them would actually contain the value of the population proportion p. (In this correct interpretation, the confidence level of 95% refers to the *success rate of the process* used to estimate the population proportion.)

Incorrect: "There is a 95% chance that the true value of p will fall between 0.828 and 0.872."

Incorrect: "95% of sample proportions will fall between 0.828 and 0.872."

Note to Instructor
Comment that students should know the correct interpretation of a confidence interval. Discuss the correct interpretation along with the incorrect interpretations.

CAUTION Know the correct interpretation of a confidence interval, as given above.

At any specific time, a population has a fixed and constant value of the proportion p, and a confidence interval constructed from a sample either includes p or does not. Similarly, if a baby has just been born and the doctor is about to announce its gender, it's incorrect to say that there is a probability of 0.5 that the baby is a girl; the baby is a girl or is not, and there's no probability involved. A population proportion p is like the baby that has been born—the value of p is fixed, so the confidence interval limits either contain p or do not, and that is why it's incorrect to say that there is a 95% chance that p will fall between values such as 0.828 and 0.872.

A confidence level of 95% tells us that the *process* we are using will, in the long run, result in confidence interval limits that contain the true population proportion 95% of the time. Suppose that the proportion of $p = 0.90$ is the true population proportion of all adults who know what Twitter is. Then the confidence interval obtained from the Pew Research Center poll does not contain the population proportion, because $p = 0.90$ is not between 0.828 and 0.872. Figure 7-1 illustrates this, and it also shows that 19 out of 20 (or 95%) different confidence intervals contain the assumed value of $p = 0.90$. Figure 7-1 is trying to tell the story that with a 95% confidence level, we expect about 19 out of 20 confidence intervals (or 95%) to contain the true value of p.

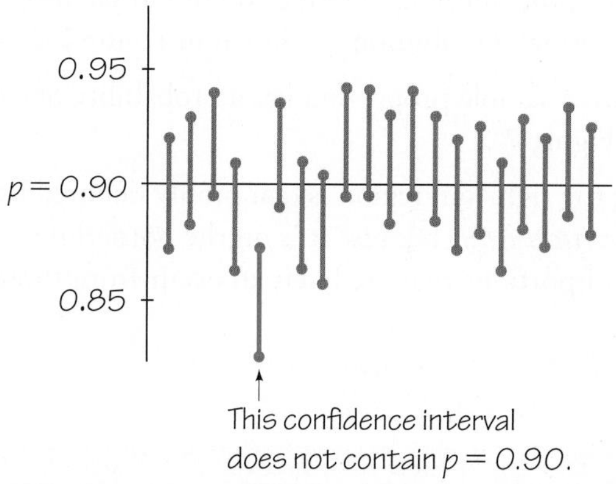

Figure 7-1 Confidence Intervals from 20 Different Samples

CAUTION Confidence intervals can be used informally to compare different data sets, but *the overlapping of confidence intervals should not be used for making formal and final conclusions about equality of proportions.* (See "On Judging the Significance of Differences by Examining the Overlap Between Confidence Intervals," by Schenker and Gentleman, *American Statistician,* Vol. 55, No. 3.)

BMI a Bad Measure?

The body mass index (BMI) of a person is calculated by finding their weight in kilograms and dividing it by the square of their height in meters. The Centers for Disease Control (CDC) publishes recommendations, including one stating that someone is obese if their BMI is 30.0 or higher. Keith Devlin wrote "Top 10 Reasons Why the BMI is Bogus" for *Your Health*, a podcast by NPR. He argues that, although it is commonly used, the BMI is not a valid measure of fat or obesity. He says that it makes no sense to square a person's height. Also, the BMI makes no adjustment for people with strong bones or good muscle tone. He says that a person with high body fat will have a high BMI, but a person with a high BMI will not necessarily have high body fat. He also says that using the BMI causes us to neglect using more reliable measures of obesity.

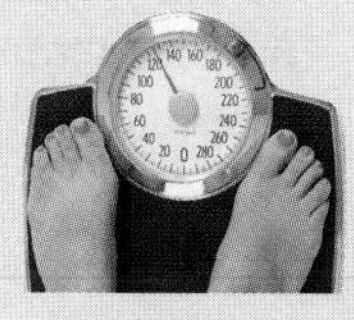

Using Confidence Intervals for Hypothesis Tests

A confidence interval can be used to *test some claim* made about a population proportion p. Formal methods of hypothesis testing are introduced in Chapter 8, and those methods might require adjustments to confidence intervals that are not described in this chapter.

Some examples and exercises in this chapter require that we address some claim, but in this chapter we do not yet use a formal method of hypothesis testing, so we simply generate a confidence interval and make an *informal judgment* based on the result. For example, if sample results consist of 70 heads in 100 tosses of a coin, the resulting 95% confidence interval of $0.610 < p < 0.790$ can be used to *informally* support a claim that the proportion of heads is *greater than* 50%. (A one-sided claim is a statement that p is *greater than* some value or is *less than* some value, and a formal hypothesis test might require that we construct a one-sided confidence interval, as in Exercise 43, or adjust the confidence level by using 90% instead of 95%.)

CAUTION Know that in this chapter, when we use a confidence interval to address a claim about a population proportion p, we simply make an *informal judgment* (that may or may not be consistent with formal methods of hypothesis testing introduced in Chapter 8).

Note to Instructor
Comment that the following chapters will use the same *critical values* that are introduced here. Critical values are used throughout the remainder of the course, so it is really helpful to understand them as soon as possible.

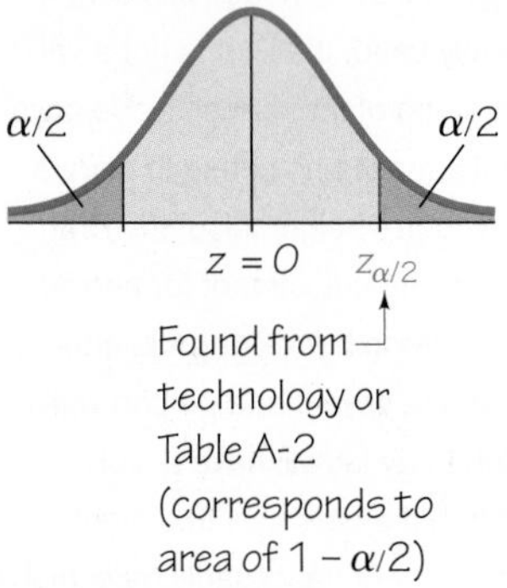

Figure 7-2
Critical Value $z_{\alpha/2}$ in the Standard Normal Distribution

Critical Values

The methods of this section (and many of the other statistical methods found in the following chapters) include reference to a standard z score that can be used to distinguish between sample statistics that are likely to occur and those that are unlikely. Such a z score is called a *critical value.* (Critical values were first presented in Section 6-2, and they are formally defined below.) Critical values are based on the following observations:

1. Under certain conditions, the sampling distribution of sample proportions can be approximated by a normal distribution, as shown in Figure 7-2.
2. A z score associated with a sample proportion has a probability of $\alpha/2$ of falling in the right tail of Figure 7-2.
3. The z score separating the right-tail region is commonly denoted by $z_{\alpha/2}$ and is referred to as a *critical value* because it is on the borderline separating z scores from sample proportions that are likely to occur from those that are unlikely.

DEFINITION A **critical value** is the number on the borderline separating sample statistics that are likely to occur from those that are unlikely. The number $z_{\alpha/2}$ is a critical value that is a z score with the property that it separates an area of $\alpha/2$ in the right tail of the standard normal distribution (as in Figure 7-2).

Example 2 Finding a Critical Value

Find the critical value $z_{\alpha/2}$ corresponding to a 95% confidence level.

Solution

A 95% confidence level corresponds to $\alpha/2 = 0.025$. Figure 7-3 shows that the area in each of the red-shaded tails is $\alpha/2 = 0.025$. We find $z_{\alpha/2} = 1.96$ by noting that the cumulative area to its left must be $1 - 0.025$, or 0.975. We can use technology or refer to Table A-2 to find that the cumulative left area of 0.9750 (found *in the body* of the table) corresponds to $z = 1.96$. For a 95% confidence level, the critical value is therefore $z_{\alpha/2} = 1.96$. To find the critical z score for a 95% confidence level, use a cumulative left area of 0.9750 (*not* 0.95).

Note: Many technologies can be used to find critical values. STATDISK, Excel, Minitab, StatCrunch, and the TI-83/84 Plus calculator all provide critical values for the normal distribution.

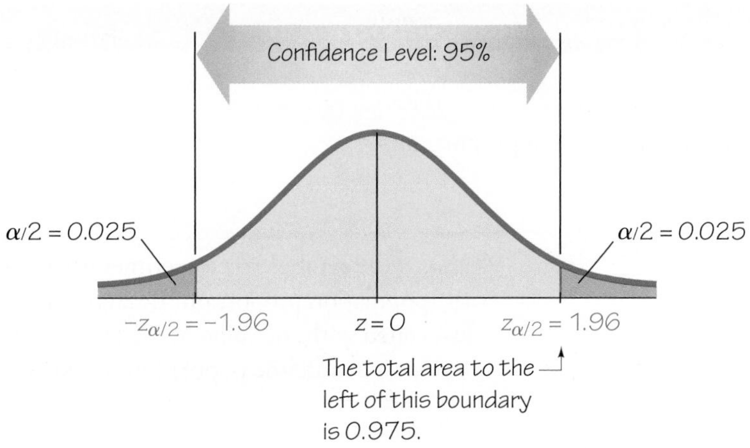

Figure 7-3 Finding $z_{\alpha/2}$ for a 95% Confidence Level

Example 2 showed that a 95% confidence level results in a critical value of $z_{\alpha/2} = 1.96$. This is the most common critical value, and it is listed with two other common values in the table that follows.

Confidence Level	α	Critical Value, $z_{\alpha/2}$
90%	0.10	1.645
95%	0.05	1.96
99%	0.01	2.575

Margin of Error

When we collect sample data that result in a sample proportion, such as the Pew Research Center poll given in Example 1, we can identify the sample proportion $\hat{p}$. Because of random variation in samples, the sample proportion $\hat{p}$ is typically different from the population proportion p. The difference between the sample proportion and the population proportion can be thought of as an error. We now define the *margin of error E* as follows.

DEFINITION When data from a simple random sample are used to estimate a population proportion p, the **margin of error,** denoted by ***E,*** is the maximum likely difference (with probability $1 - \alpha$, such as 0.95) between the observed sample proportion $\hat{p}$ and the true value of the population proportion p. The margin of error E is also called the *maximum error of the estimate* and can be found by multiplying the critical value and the standard deviation of sample proportions, as shown in Formula 7-1.

Formula 7-1 $$E = z_{\alpha/2}\sqrt{\frac{\hat{p}\hat{q}}{n}}$$ margin of error for proportions

For a 95% confidence level, $\alpha = 0.05$, so there is a probability of 0.05 that the sample proportion will be in error by more than E. This property is generalized in the following box.

Confidence Interval for Estimating a Population Proportion p

Objective

Construct a confidence interval used to estimate a population proportion p.

Notation

p = population proportion

$\hat{p}$ = sample proportion

n = number of sample values

E = margin of error

$z_{\alpha/2}$ = z score separating an area of $\alpha/2$ in the right tail of the standard normal distribution

(*Note:* The symbol π is sometimes used to denote the population proportion. Because π is so closely associated with the value of 3.14159..., this text uses p to denote the population proportion.)

Requirements

1. The sample is a simple random sample. (*Caution:* If the sample data have been obtained in a way that is not suitable, the estimate of the population proportion may be very wrong.)
2. The conditions for the binomial distribution are satisfied. That is, there is a fixed number of trials, the trials are independent, there are two categories of outcomes, and the probabilities remain constant for each trial. (See Section 5-3.)
3. There are at least 5 successes and at least 5 failures. (With the population proportions p and q unknown, we estimate their values using the sample proportion, so this requirement is a way of verifying that $np \geq 5$ and $nq \geq 5$ are both satisfied, so the normal distribution is a suitable approximation to the binomial distribution. There are procedures for dealing with situations in which the normal distribution is not a suitable approximation, as in Exercise 40.)

Confidence Interval

$$\hat{p} - E < p < \hat{p} + E \quad \text{where} \quad E = z_{\alpha/2}\sqrt{\frac{\hat{p}\hat{q}}{n}}$$

The confidence interval is often expressed in the following equivalent formats:

$$\hat{p} \pm E$$

or

$$(\hat{p} - E, \hat{p} + E)$$

Round-Off Rule for Confidence Interval Estimates of p

Round the confidence interval limits for p to three significant digits.

Confidence intervals can be easily created by using technology or Table A-2 with the following procedure:

Procedure for Constructing a Confidence Interval for p

1. Verify that the requirements in the preceding box are satisfied.
2. Use technology or Table A-2 to find the critical value $z_{\alpha/2}$ that corresponds to the desired confidence level.

3. Evaluate the margin of error $E = z_{\alpha/2}\sqrt{\hat{p}\hat{q}/n}$.
4. Using the value of the calculated margin of error E and the value of the sample proportion $\hat{p}$, find the values of the *confidence interval limits* $\hat{p} - E$ and $\hat{p} + E$. Substitute those values in the general format for the confidence interval:

$$\hat{p} - E < p < \hat{p} + E$$

or

$$\hat{p} \pm E$$

or

$$(\hat{p} - E, \hat{p} + E)$$

5. Round the resulting confidence interval limits to three significant digits.

Example 3 Constructing a Confidence Interval: Poll Results

In the Chapter Problem we noted that a Pew Research Center poll of 1007 randomly selected U.S. adults showed that 85% of the respondents know what Twitter is. The sample results are $n = 1007$ and $\hat{p} = 0.85$.

a. Find the margin of error E that corresponds to a 95% confidence level.

b. Find the 95% confidence interval estimate of the population proportion p.

c. Based on the results, can we safely conclude that more than 75% of adults know what Twitter is?

d. Assuming that you are a newspaper reporter, write a brief statement that accurately describes the results and includes all of the relevant information.

Solution

Requirement check We first verify that the necessary requirements are satisfied. (1) The polling methods used by the Pew Research Center result in samples that can be considered to be simple random samples. (2) The conditions for a binomial experiment are satisfied, because there is a fixed number of trials (1007), the trials are independent (because the response from one person doesn't affect the probability of the response from another person), there are two categories of outcome (subject knows what Twitter is or does not), and the probability remains constant. (3) With 85% of the respondents knowing what Twitter is, the number who know is 856 (or 85% of 1007). If 856 of the 1007 subjects know what Twitter is, the other 151 do not know, so the number of successes (856) and the number of failures (151) are both at least 5. The check of requirements has been successfully completed. ✓

Note to Instructor
In this chapter and throughout the remainder of this book, a formal REQUIREMENT CHECK is included in solutions whenever such a check is appropriate. See the solution to Example 3, and see the format used for this formal check of requirements.

Technology The confidence interval and margin of error can be easily found using technology. From the STATDISK display on the next page we can see the required entries on the left and the results displayed on the right. Like most technologies, STATDISK requires a value for the number of successes, so we simply find 85% of 1007 and round the result of 855.95 to the whole number of 856. The results show that the margin of error is $E = 0.022$ (rounded) and the confidence interval is $0.828 < p < 0.872$ (rounded). (The Wilson Score confidence interval is discussed near the end of this section.)

STATDISK

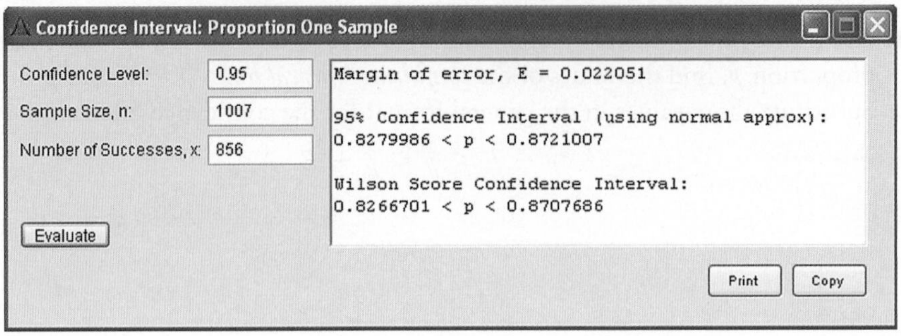

We will now proceed to show how the results for parts (a) and (b) can be obtained with manual calculations.

a. The margin of error is found by using Formula 7-1 with $z_{\alpha/2} = 1.96$ (as found in Example 2), $\hat{p} = 0.85$, $\hat{q} = 0.15$, and $n = 1007$.

$$E = z_{\alpha/2}\sqrt{\frac{\hat{p}\hat{q}}{n}} = 1.96\sqrt{\frac{(0.85)(0.15)}{1007}} = 0.0220545$$

b. Constructing the confidence interval is quite easy now that we know the values of $\hat{p}$ and E. We simply substitute those values to obtain this result:

$$\hat{p} - E < p < \hat{p} + E$$

$$0.85 - 0.0220545 < p < 0.85 + 0.0220545$$

$$0.828 < p < 0.872 \quad \text{(rounded to three significant digits)}$$

This same result could be expressed in the format of 0.85 ± 0.022 or (0.828, 0.872). If we want the 95% confidence interval for the true population *percentage,* we could express the result as $82.8\% < p < 87.2\%$.

c. Based on the confidence interval obtained in part (b), it does appear that more than 75% of adults know what Twitter is. Because the limits of 0.828 and 0.872 are likely to contain the true population proportion, it appears that the population proportion is a value greater than 0.75. (See also Exercise 43.)

d. Here is one statement that summarizes the results: 85% of U.S. adults know what Twitter is. That percentage is based on a Pew Research Center poll of 1007 randomly selected adults in the United States. In theory, in 95% of such polls, the percentage should differ by no more than 2.2 percentage points in either direction from the percentage that would be found by interviewing all adults in the United States.

Analyzing Polls Example 3 deals with a typical poll. When analyzing results from polls, we should consider the following.

1. The sample should be a simple random sample, not an inappropriate sample (such as a voluntary response sample).
2. The confidence level should be provided. (It is often 95%, but media reports often neglect to identify it.)
3. The sample size should be provided. (It is often provided by the media, but not always.)

4. Except for relatively rare cases, the quality of the poll results depends on the sampling method and the size of the sample, but the size of the population is usually not a factor.

CAUTION Never have the common misconception that poll results are unreliable if the sample size is a small percentage of the population size. The population size is usually not a factor in determining the reliability of a poll.

Note to Instructor
If you discuss the reasons underlying the format of Formula 7-3, consider presenting a geometric demonstration of why the values of $\hat{p} = 0.5$ and $\hat{q} = 0.5$ result in the largest possible product of $\hat{p}\hat{q} = 0.25$. Use a rectangle with perimeter 2, so that $L + W = 1$, as in $\hat{p} + \hat{q} = 1$. Show that given a rectangle with perimeter 2, the area is maximized by using a *square*, which has an area of 0.25, so the largest possible product of $\hat{p} \cdot \hat{q}$ is also 0.25. It's an interesting link between geometry and statistics.

Determining Sample Size

If we plan to collect sample data in order to estimate some population proportion, how do we know *how many* sample units must be obtained? If we solve the formula for the margin of error E (Formula 7-1) for the sample size n, we get Formula 7-2 below. Formula 7-2 requires $\hat{p}$ as an estimate of the population proportion p, but if no such estimate is known (as is often the case), we replace $\hat{p}$ by 0.5 and replace $\hat{q}$ by 0.5, with the result given in Formula 7-3.

Finding the Sample Size Required to Estimate a Population Proportion

Objective

Determine how large the sample n should be in order to estimate the population proportion p.

Notation

p = population proportion

$\hat{p}$ = sample proportion

n = number of sample values

E = desired margin of error

$z_{\alpha/2}$ = z score separating an area of $\alpha/2$ in the right tail of the standard normal distribution

Requirements

The sample must be a simple random sample of independent sample units.

When an estimate $\hat{p}$ is known: **Formula 7-2** $$n = \frac{[z_{\alpha/2}]^2 \hat{p}\hat{q}}{E^2}$$

When no estimate $\hat{p}$ is known: **Formula 7-3** $$n = \frac{[z_{\alpha/2}]^2 0.25}{E^2}$$

Round-Off Rule for Determining Sample Size

If the computed sample size n is not a whole number, round the value of n up to the next *larger* whole number.

If reasonable estimates of $\hat{p}$ can be made by using previous samples, a pilot study, or someone's expert knowledge, use Formula 7-2. If nothing is known about the value of $\hat{p}$, use Formula 7-3.

Role of the Population Size N Formulas 7-2 and 7-3 are remarkable because they show that the sample size does not depend on the size (N) of the population; the sample size depends on the desired confidence level, the desired margin of error, and sometimes the known estimate of $\hat{p}$. (See Exercise 39 for dealing with cases in which a relatively large sample is selected without replacement from a finite population, so the sample size n does depend on the population size N.)

Example 4 What Percentage of Adults Buy Clothing Online?

Gap, Banana Republic, J. Crew, Yahoo, and America Online are just a few of the many companies interested in knowing the percentage of adults who buy clothing online. How many adults must be surveyed in order to be 95% confident that the sample percentage is in error by no more than three percentage points?

a. Use this recent result from the Census Bureau: 66% of adults buy clothing online.

b. Assume that we have no prior information suggesting a possible value of the proportion.

Solution

a. The prior study suggests that $\hat{p} = 0.66$, so $\hat{q} = 0.34$ (found from $\hat{q} = 1 - 0.66$). With a 95% confidence level, we have $\alpha = 0.05$, so $z_{\alpha/2} = 1.96$. Also, the margin of error is $E = 0.03$ (the decimal equivalent of "three percentage points"). Because we have an estimated value of $\hat{p}$, we use Formula 7-2 as follows:

$$n = \frac{[z_{\alpha/2}]^2 \hat{p}\hat{q}}{E^2} = \frac{[1.96]^2(0.66)(0.34)}{0.03^2}$$

$$= 957.839 = 958 \qquad \text{(rounded up)}$$

We must obtain a simple random sample that includes at least 958 adults.

b. As in part (a), we again use $z_{\alpha/2} = 1.96$ and $E = 0.03$, but with no prior knowledge of $\hat{p}$ (or $\hat{q}$), we use Formula 7-3 as follows:

$$n = \frac{[z_{\alpha/2}]^2 \cdot 0.25}{E^2} = \frac{[1.96]^2 \cdot 0.25}{0.03^2}$$

$$= 1067.11 = 1068 \qquad \text{(rounded up)}$$

Interpretation

To be 95% confident that our sample percentage is within three percentage points of the true percentage for all adults, we should obtain a simple random sample of 1068 adults. By comparing this result to the sample size of 958 found in part (a), we can see that if we have no knowledge of a prior study, a larger sample is required to achieve the same results as when the value of $\hat{p}$ can be estimated.

CAUTION Try to avoid these two common errors when calculating sample size:

1. Don't make the mistake of using $E = 3$ as the margin of error corresponding to "three percentage points." If the margin of error is three percentage points, use $E = 0.03$.
2. Be sure to substitute the critical z score for $z_{\alpha/2}$. For example, if you are working with 95% confidence, be sure to replace $z_{\alpha/2}$ with 1.96. Don't make the mistake of replacing $z_{\alpha/2}$ with 0.95 or 0.05.

Finding the Point Estimate and E from a Confidence Interval

Sometimes we want to better understand a confidence interval that might have been obtained from a journal article or technology. If we already know the confidence interval limits, the sample proportion (or the best point estimate) $\hat{p}$ and the margin of error E can be found as follows:

Point estimate of p:

$$\hat{p} = \frac{(\text{upper confidence interval limit}) + (\text{lower confidence interval limit})}{2}$$

Margin of error:

$$E = \frac{(\text{upper confidence interval limit}) - (\text{lower confidence interval limit})}{2}$$

Example 5

The article "High-Dose Nicotine Patch Therapy," by Dale, Hurt, et al. (*Journal of the American Medical Association,* Vol. 274, No. 17) includes this statement: "Of the 71 subjects, 70% were abstinent from smoking at 8 weeks (95% confidence interval [CI], 58% to 81%)." Use that statement to find the point estimate $\hat{p}$ and the margin of error E.

Solution

We get the 95% confidence interval of $0.58 < p < 0.81$ from the given statement. The point estimate $\hat{p}$ is the value midway between the upper and lower confidence interval limits, so we get

$$\hat{p} = \frac{(\text{upper confidence limit}) + (\text{lower confidence limit})}{2}$$

$$= \frac{0.81 + 0.58}{2} = 0.695$$

The margin of error can be found as follows:

$$E = \frac{(\text{upper confidence limit}) - (\text{lower confidence limit})}{2}$$

$$= \frac{0.81 - 0.58}{2} = 0.115$$

Better-Performing Confidence Intervals

Important note: The exercises for this section are based on the method for constructing a confidence interval as described above, not the confidence intervals described in the following discussion.

Note to Instructor
The confidence intervals discussed in this subsection are for general information only. The exercises are based on the confidence intervals already discussed, not those cited in this subsection.

Adjusted Wald CI The confidence interval described in this section has the format typically presented in introductory statistics courses, but it does not perform as well as some other confidence intervals. The *adjusted Wald confidence interval* performs better in the sense that its probability of containing the true population proportion p is closer to the confidence level that is used. The adjusted Wald confidence interval uses this simple procedure: Add 2 to the number of successes x, add 2 to the number of failures (so that the number of trials n is increased by 4), then find the confidence interval as described in this section. For example, if we use the methods of this section with $x = 10$ and $n = 20$, we get this 95% confidence interval: $0.281 < p < 0.719$. With $x = 10$ and $n = 20$ we use the adjusted Wald confidence interval by letting $x = 12$ and $n = 24$ to get this confidence interval: $0.300 < p < 0.700$. The chance that the confidence interval $0.300 < p < 0.700$ contains p is closer to 95% than the chance that $0.281 < p < 0.719$ contains p.

Wilson Score CI Another confidence interval that performs better than the one described in this section and the adjusted Wald confidence interval is the *Wilson score confidence interval:*

$$\frac{\hat{p} + \dfrac{z_{\alpha/2}^2}{2n} \pm z_{\alpha/2}\sqrt{\dfrac{\hat{p}\hat{q} + \dfrac{z_{\alpha/2}^2}{4n}}{n}}}{1 + \dfrac{z_{\alpha/2}^2}{n}}$$

It is easy to see why this approach is not used much in introductory courses. The complexity of the above expression can be overcome by using some technologies, such as STATDISK, that provide Wilson score confidence interval results. Using $x = 10$ and $n = 20$, the 95% Wilson score confidence interval is as follows: $0.299 < p < 0.701$.

For a discussion of these and other confidence intervals for p, see "Approximation Is Better than 'Exact' for Interval Estimation of Binomial Proportions," by Agresti and Coull, *American Statistician,* Vol. 52, No. 2.

using TECHNOLOGY

Confidence Intervals

STATDISK Select **Analysis,** then **Confidence Intervals,** then **Proportion One Sample,** and proceed to enter the requested items. The confidence interval will be displayed, as in the STATDISK results included with Example 3.

MINITAB Select **Stat, Basic Statistics,** then **1 Proportion.** In the dialog box, click on the button for **Summarized Data.** Also click on the **Options** button, and enter the desired confidence level (the default is 95%). Instead of using a normal approximation, Minitab's default procedure is to determine the confidence interval limits by using an exact method. To use the normal approximation method presented in this section, click on the **Options** button and then click on the box with this statement: "Use test and interval based on normal distribution."

EXCEL Use XLSTAT. Click on **XLSTAT** at the top, click on **Parametric Tests,** then select **Tests for one proportion.** Start in the lower left corner of the dialog box by selecting **Frequency** (if you know the number of successes x) or **Proportion** (if you know the sample proportion $\hat{p}$). Enter the frequency (number of successes) or the sample proportion, enter the sample size in the "Sample size" box, and enter 0.5 in the **Test Proportion** box. Be sure that the box next to "z test" is checked. For the "Range" box, enter A1 so that the results will start at cell A1. Click on the **Options** tab, to enter the desired "Significance level (%)." Enter 5 for a 95% confidence interval. For the "Variance (confidence interval)" options, select **Sample** (so that the sample proportion is used in the computation of the confidence interval). There are four options for the type of confidence interval; accept the default of **Wald.** Click **OK.** After the results are displayed, look for "confidence interval on the proportion (Wald)."

TI-83/84 PLUS Press STAT, select **TESTS,** then select **1-PropZInt** and enter the required items. The accompanying display shows the result for Example 3. Like many technologies, the TI-83/84 Plus calculator requires entry of the number of successes, so 856 (which is 85% of the 1007 people polled) was entered for the value of x. Also like many technologies, the confidence interval limits are expressed in the format shown on the second line of the display.

TI-83/84 PLUS

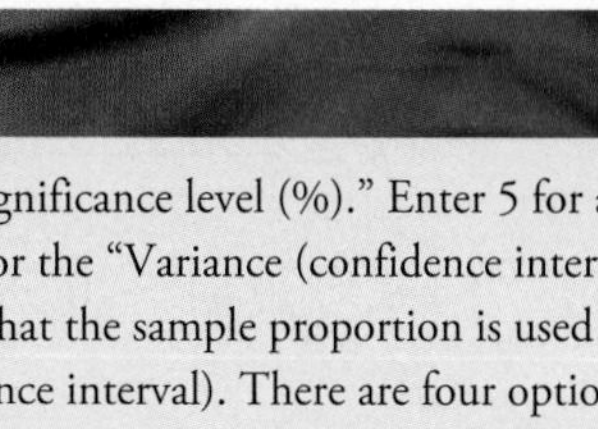

STATCRUNCH Click on **Open StatCrunch,** then enter or open a data set. Click on **Stat,** then select **Proportion.** Select **One sample,** then select **with summary.** Proceed to enter the number of successes and the number of observations, click on **Next,** then select **Confidence Interval** and use the **Standard-Wald** option.

Sample Size Determination

STATDISK Select **Analysis,** then **Sample Size Determination,** then **Estimate Proportion.** Enter the required items in the dialog box.

MINITAB 16 Minitab Release 16 has a procedure for determining sample size using the binomial distribution, not the normal approximation to the binomial distribution, so results will not agree with those found by using the methods of this section. For Minitab's procedure, select **Stat, Power and Sample Size,** then select **Sample Size for Estimation** and choose the menu item of **Proportion (Binomial).** Complete the dialog box and click **OK.**

Sample size determination is not available as a built-in function with Excel or StatCrunch or the TI-83/84 Plus calculator.

7-2 Basic Skills and Concepts

Statistical Literacy and Critical Thinking

1. Poll Results in the Media *USA Today* provided a "snapshot" illustrating poll results from 1910 professionals who interview job applicants. The illustration showed that 26% of them said the biggest interview turnoff is that the applicant did not make an effort to learn about the job or the company. The margin of error was given as ± 3 percentage points. What important feature of the poll was omitted?

2. Margin of Error For the poll described in Exercise 1, describe what is meant by the statement that "the margin of error is ± 3 percentage points."

3. Notation For the poll described in Exercise 1, what do $\hat{p}$, $\hat{q}$, n, E, and p represent? If the confidence level is 95%, what is the value of α?

4. Confidence Levels Given specific sample data, such as the data given in Exercise 1, which confidence interval is wider: the 95% confidence interval or the 80% confidence interval? Why is it wider?

Finding Critical Values. ***In Exercises 5–8, find the indicated critical z value.***

5. Find the critical value $z_{\alpha/2}$ that corresponds to a confidence level of 80%.

6. Find the critical value $z_{\alpha/2}$ that corresponds to a 99% confidence level.

7. Find $z_{\alpha/2}$ for $\alpha = 0.10$.

8. Find $z_{\alpha/2}$ for $\alpha = 0.04$.

Formats of Confidence Intervals. ***In Exercises 9–12, express the confidence interval using the indicated format. (The confidence intervals are based on the proportions of red, orange, yellow, and blue M&Ms in Data Set 20 from Appendix B.)***

9. Express the confidence interval $0.0641 < p < 0.186$ in the form of $\hat{p} \pm E$.

10. Express the confidence interval $0.165 < p < 0.335$ in the form of $\hat{p} \pm E$.

11. Express the confidence interval (0.0268, 0.133) in the form of $\hat{p} - E < p < \hat{p} + E$.

12. Express the confidence interval 0.270 ± 0.087 in the form of $\hat{p} - E < p < \hat{p} + E$.

Constructing and Interpreting Confidence Intervals. ***In Exercises 13–16, use the given sample data and confidence level. In each case, (a) find the best point estimate of the population proportion p; (b) identify the value of the margin of error E; (c) construct the confidence interval; (d) write a statement that correctly interprets the confidence interval.***

13. From a KRC Research poll in which respondents were asked if they felt vulnerable to identify theft: $n = 1002$, $x = 531$ who said "yes." Use a 95% confidence level.

Recommended Assignment
Exercises 1–18.

1. The confidence level (such as 95%) was not provided.
2. When using 26% to estimate the value of the population percentage, the maximum likely difference between 26% and the true population percentage is three percentage points, so the interval from 23% to 29% is likely to contain the true population percentage.
3. $\hat{p} = 0.26$ is the sample proportion; $\hat{q} = 0.74$ (found from evaluating $1 - \hat{p}$); $n = 1910$ is the sample size; $E = 0.03$ is the margin of error; p is the population proportion, which is unknown. The value of α is 0.05.
4. The 95% confidence interval will be wider than the 80% confidence interval. A confidence interval must be wider in order for us to be more confident that it captures the true value of the population proportion. (Think of estimating the age of a classmate. You might be 90% confident that she is between 20 and 30, but you might be 99.9% confident that she is between 10 and 40.)
5. 1.28
6. 2.575 (Tech: 2.576)
7. 1.645
8. 2.05
9. 0.125 ± 0.061
10. 0.250 ± 0.085
11. $0.0268 < p < 0.133$
12. $0.183 < p < 0.357$
13. a. 0.530
    b. $E = 0.0309$
    c. $0.499 < p < 0.561$
    d. We have 95% confidence that the interval from 0.499 to 0.561 actually does contain the true value of the population proportion.

14. From a 3M Privacy Filters poll in which respondents were asked to identify their favorite seat when they fly: $n = 806$, $x = 492$ who chose the window seat. Use a 99% confidence level.

15. From a Prince Market Research poll in which respondents were asked if they acted to annoy a bad driver: $n = 2518$, $x = 1083$ who said that they honked. Use a 90% confidence level.

16. From an Angus Reid Public Opinion poll in which respondents were asked if they felt that U.S. nuclear weapons made them feel safer: $n = 1005$, $x = 543$ who said "yes." Use a confidence level of 80%.

17. Gender Selection The Genetics & IVF Institute conducted a clinical trial of the XSORT method designed to increase the probability of conceiving a girl. As of this writing, 945 babies were born to parents using the XSORT method, and 879 of them were girls.

a. What is the best point estimate of the population proportion of girls born to parents using the XSORT method?

b. Use the sample data to construct a 95% confidence interval estimate of the proportion of girls born to parents using the XSORT method.

c. Based on the results, does the XSORT method appear to be effective? Why or why not?

18. Gender Selection The Genetics & IVF Institute conducted a clinical trial of the YSORT method designed to increase the probability of conceiving a boy. As of this writing, 291 babies were born to parents using the YSORT method, and 239 of them were boys.

a. What is the best point estimate of the population proportion of boys born to parents using the YSORT method?

b. Use the sample data to construct a 99% confidence interval estimate of the proportion of boys born to parents using the YSORT method.

c. Based on the results, does the YSORT method appear to be effective? Why or why not?

19. Touch Therapy When she was 9 years of age, Emily Rosa did a science fair experiment in which she tested professional touch therapists to see if they could sense her energy field. She flipped a coin to select either her right hand or her left hand, then she asked the therapists to identify the selected hand by placing their hand just under Emily's hand without seeing it and without touching it. Among 280 trials, the touch therapists were correct 123 times (based on data in "A Close Look at Therapeutic Touch," *Journal of the American Medical Association,* Vol. 279, No. 13).

a. Given that Emily used a coin toss to select either her right hand or her left hand, what proportion of correct responses would be expected if the touch therapists made random guesses?

b. Using Emily's sample results, what is the best point estimate of the therapist's success rate?

c. Using Emily's sample results, construct a 99% confidence interval estimate of the proportion of correct responses made by touch therapists.

d. What do the results suggest about the ability of touch therapists to select the correct hand by sensing an energy field?

14. a. 0.610
b. $E = 0.0442$
c. $0.566 < p < 0.655$
d. We have 99% confidence that the interval from 0.566 to 0.655 actually does contain the true value of the population proportion.

15. a. 0.430
b. $E = 0.0162$
c. $0.414 < p < 0.446$
d. We have 90% confidence that the interval from 0.414 to 0.446 actually does contain the true value of the population proportion.

16. a. 0.540
b. $E = 0.0201$
c. $0.520 < p < 0.560$
d. We have 80% confidence that the interval from 0.520 to 0.560 actually does contain the true value of the population proportion.

17. a. 0.930
b. $0.914 < p < 0.946$
c. Yes. The true proportion of girls with the XSORT method is substantially greater than the proportion of (about) 0.5 that is expected when no method of gender selection is used.

18. a. 0.821
b. $0.763 < p < 0.879$
c. Yes. The true proportion of boys with the YSORT method is substantially greater than the proportion of (about) 0.5 that is expected when no method of gender selection is used.

19. a. 0.5
b. 0.439
c. $0.363 < p < 0.516$
d. If the touch therapists really had an ability to select the correct hand by sensing an energy field, their success rate would be significantly greater than 0.5, but the sample success rate of 0.439 and the confidence interval suggest that they do not have the ability to select the correct hand by sensing an energy field.

20. Mendelian Genetics When Mendel conducted his famous genetics experiments with peas, one sample of offspring consisted of 428 green peas and 152 yellow peas.

a. Find a 95% confidence interval estimate of the percentage of yellow peas.

b. Based on his theory of genetics, Mendel expected that 25% of the offspring peas would be yellow. Given that the percentage of offspring yellow peas is not 25%, do the results contradict Mendel's theory? Why or why not?

20. a. $0.226 < p < 0.298$
b. No, the confidence interval includes 0.25, so the true percentage could easily equal 25%.

21. Online Books A *Consumer Reports* Research Center survey of 427 women showed that 29.0% of them purchase books online.

a. Among the 427 women who were surveyed, what is the number of women who said that they purchase books online?

b. Find a 95% confidence interval estimate of the *percentage* of all women who purchase books online.

c. Can we safely conclude that less than 50% of all women purchase books online? Why or why not?

d. Can we safely conclude that at least 25% of all women purchase books online? Why or why not?

e. What do the results tell us about the percentage of *men* who purchase books online?

21. a. 124
b. $24.7\% < p < 33.3\%$
c. Yes. Because all values of the confidence interval are less than 0.5, the confidence interval shows that the percentage of women who purchase books online is very likely less than 50%.
d. No. The confidence interval shows that it is possible that the percentage of women who purchase books online could be less than 25%.
e. Nothing.

22. Want Boss's Job? In a *USA Today* survey, 20.8% of 144 respondents said that they aspired to have their boss's job. Construct a 95% confidence interval estimate of the *percentage* of all workers who aspire to have their boss's job. How are the results affected by the additional knowledge that the respondents chose to answer an online question posted on the *USA Today* Web site?

22. $14.2\% < p < 27.4\%$ (using $x = 30$: $14.2\% < p < 27.5\%$). If the subjects chose to respond to the posted question, the sample is a voluntary response sample, so the confidence interval could be very misleading.

23. Job Interviews In a Harris poll of 514 human resource professionals, 45.9% said that body piercings and tattoos were big grooming red flags.

a. Among the 514 human resource professionals who were surveyed, how many of them said that body piercings and tattoos were big grooming red flags?

b. Construct a 99% confidence interval estimate of the proportion of all human resource professionals believing that body piercings and tattoos are big grooming red flags.

c. Repeat part (b) using a confidence level of 80%.

d. Compare the confidence intervals from parts (b) and (c) and identify the interval that is wider. Why is it wider?

23. a. 236
b. $0.402 < p < 0.516$ (using $x = 236$: $0.403 < p < 0.516$).
c. $0.431 < p < 0.487$
d. The 95% confidence interval is wider than the 80% confidence interval. A 95% confidence interval must be wider than an 80% confidence interval in order to be more confident that it captures the true value of the population proportion. (See Exercise 4.)

24. Job Interviews In a Harris poll of 514 human resource professionals, 90% said that the appearance of a job applicant is most important for a good first impression.

a. Among the 514 human resource professionals who were surveyed, how many of them said that the appearance of a job applicant is most important for a good first impression?

b. Construct a 99% confidence interval estimate of the proportion of all human resource professionals believing that the appearance of a job applicant is most important for a good first impression.

c. Repeat part (b) using a confidence level of 80%.

d. Compare the confidence intervals from parts (b) and (c) and identify the interval that is wider. Why is it wider?

24. a. 463
b. $0.866 < p < 0.934$ (using $x = 463$: $0.867 < p < 0.935$).
c. $0.883 < p < 0.917$ (using $x = 463$: $0.884 < p < 0.918$).
d. The 95% confidence interval is wider than the 80% confidence interval. A confidence interval must be wider in order to be more confident that it captures the true value of the population proportion. (See Exercise 4.)

25. $0.0168 < p < 0.143$ (Tech: $0.0169 < p < 0.143$). No, the confidence interval limits contain the value of 0.13, so the claimed rate of 13% could be the true percentage for the population of brown M&Ms.

26. a. $0.666 < p < 0.734$ (Tech: $0.666 < p < 0.733$)
b. No. Because 0.61 is not included in the confidence interval, it does not appear that the responses are consistent with the actual voter turnout.

27. a. $0.0276\% < p < 0.0366\%$ (using $x = 135$: $0.0276\% < p < 0.0367\%$).
b. No, because 0.0340% is included in the confidence interval.

28. a. 2455
b. $80.5\% < p < 82.9\%$
c. Nothing.

29. 752

30. 256 (Tech: 257)

31. 339

32. 770 (Tech: 767)

25. M&Ms The Mars candy company claims that 13% of its M&M candies are brown, but Data Set 20 in Appendix B lists data from 100 M&Ms, and 8% of them are brown. Use the sample data to construct a 98% confidence interval estimate of the proportion of brown M&Ms. Does it appear that the claimed rate of 13% rate is wrong? Why or why not?

26. Misleading Survey Responses In a survey of 1002 people, 70% said that they voted in a recent presidential election (based on data from ICR Research Group). Voting records show that 61% of eligible voters actually did vote.

a. Find a 98% confidence interval estimate of the proportion of people who say that they voted.

b. Are the survey results consistent with the actual voter turnout of 61%? Why or why not?

27. Cell Phones and Cancer A study of 420,095 Danish cell phone users found that 0.0321% of them developed cancer of the brain or nervous system. Prior to this study of cell phone use, the rate of such cancer was found to be 0.0340% for those not using cell phones. The data are from the *Journal of the National Cancer Institute*.

a. Use the sample data to construct a 90% confidence interval estimate of the *percentage* of cell phone users who develop cancer of the brain or nervous system.

b. Do cell phone users appear to have a rate of cancer of the brain or nervous system that is different from the rate of such cancer among those not using cell phones? Why or why not?

28. Medication Usage In a survey of 3005 adults aged 57 through 85 years, it was found that 81.7% of them used at least one prescription medication (based on data from "Use of Prescription and Over-the-Counter Medications and Dietary Supplements Among Older Adults in the United States," by Qato et al., *Journal of the American Medical Association,* Vol. 300, No. 24).

a. How many of the 3005 subjects used at least one prescription medication?

b. Construct a 90% confidence interval estimate of the *percentage* of adults aged 57 through 85 years who use at least one prescription medication.

c. What do the results tell us about the proportion of college students who use at least one prescription medication?

Determining Sample Size. ***In Exercises 29–36, use the given data to find the minimum sample size required to estimate a population proportion or percentage.***

29. Republicans Find the sample size needed to estimate the percentage of Republicans among registered voters in California. Use a 0.03 margin of error, use a confidence level of 90%, and assume that $\hat{p}$ and $\hat{q}$ are unknown.

30. Robberies Find the sample size needed to estimate the percentage of robberies in Texas that result in arrests. Use a 0.04 margin of error, use a confidence level of 80%, and assume that $\hat{p}$ and $\hat{q}$ are unknown.

31. Tattooed Democrats Find the sample size needed to estimate the percentage of Democrats who have tattoos. Use a 0.05 margin of error, use a confidence level of 99%, and use results from a prior Harris poll suggesting that 15% of Democrats have tattoos.

32. Fortune Tellers Find the sample size needed to estimate the percentage of adults who have consulted fortune tellers. Use a 0.03 margin of error, use a confidence level of 98%, and use results from a prior Pew Research Center poll suggesting that 15% of adults have consulted fortune tellers.

33. Airline Seating You are the operations manager for American Airlines and you are considering a higher fare level for passengers in aisle seats. You want to estimate the percentage of passengers who now prefer aisle seats. How many randomly selected air passengers must you survey? Assume that you want to be 95% confident that the sample percentage is within 2.5 percentage points of the true population percentage.

a. Assume that nothing is known about the percentage of passengers who prefer aisle seats.

b. Assume that a recent survey suggests that about 38% of air passengers prefer an aisle seat (based on a 3M Privacy Filters survey).

33. a. 1537
b. 1449

34. Windows Penetration You plan to develop a new software system that you believe will surpass the success of Google and Facebook combined. In planning for the operating system that you will use, you need to estimate the percentage of computers that use Windows. How many computers must be surveyed in order to be 99% confident that your estimate is in error by no more than one percentage point?

a. Assume that nothing is known about the percentage of computers with Windows operating systems.

b. Assume that a recent survey suggests that about 90% of computers use Windows operating systems (based on data from Net Applications).

c. Does the additional survey information from part (b) have much of an effect on the sample size that is required?

34. a. 16,577 (Tech: 16,588)
b. 5968 (Tech: 5972)
c. Yes. Using the additional survey information from part (b) dramatically reduces the sample size.

35. Twitter As manager for an advertising company, you must plan a campaign designed to increase Twitter usage. You want to first determine the percentage of adults who know what Twitter is. How many adults must you survey in order to be 90% confident that your estimate is within five percentage points of the true population percentage?

a. Assume that nothing is known about the percentage of adults who know what Twitter is.

b. Assume that a recent survey suggests that about 85% of adults know what Twitter is (based on a Pew Research Center survey).

c. Given that the required sample size is relatively small, could you simply survey the adults at the nearest college?

35. a. 271
b. 139 (Tech: 138)
c. No. A sample of students at the nearest college is a convenience sample, not a simple random sample, so it is very possible that the results would not be representative of the population of adults.

36. On-Time Rate You have been given the task of estimating the percentage of Southwest flights that arrive on time, which is no later than 15 minutes after the scheduled arrival time. How many flights must you survey in order to be 80% confident that your estimate is within three percentage points of the true population percentage?

a. Assume that nothing is known about the percentage of on-time Southwest flights.

b. Assume that for a recent year, 84% of Southwest flights were on time (based on data from the Bureau of Transportation Statistics).

c. Given that the sample size is relatively small, can you select Southwest flights between New York (LaGuardia) and San Francisco?

36. a. 456 (Tech: 457)
b. 245 (Tech: 246)
c. No. Flights between New York and San Francisco might not be representative of the population of all Southwest flights.

Using Appendix B Data Sets. ***In Exercises 37 and 38, use the indicated data set from Appendix B.***

37. Heights of Presidents Refer to Data Set 12 in Appendix B. Treat the data as a sample and find the proportion of presidents who were taller than their opponents. Use that result to construct a 95% confidence interval estimate of the population percentage. Based on the result, does it appear that greater height is an advantage for presidential candidates? Why or why not?

37. $\hat{p} = 18/34$, or 0.529. CI: $36.2\% < p < 69.7\%$. Greater height does not appear to be an advantage for presidential candidates. If greater height is an advantage, then taller candidates should win substantially more than 50% of the elections, but the confidence interval shows that the percentage of elections won by taller candidates is likely to be anywhere between 36.2% and 69.7%.

38. $\hat{p} = 44/48$, or 0.917. CI: $85.1\% < p < 98.2\%$. No, the confidence interval is based on sample data consisting of flights from New York (JFK) to Los Angeles, and arrival delays for that route might be very different from arrival delays for the population that includes all routes.

38. Arrival Delays Refer to Data Set 15 in Appendix B. A flight is considered on time if it arrives no later than 15 minutes after the scheduled arrival time. The last column of Data Set 15 lists arrival delays, so on-time flights correspond to values in that column that are 15 or less. Use the data to construct a 90% confidence interval estimate of the population percentage of on-time flights. Does the confidence interval describe the percentage of on-time flights for all American Airlines flights? Why or why not?

7-2 Beyond the Basics

39. a. 178
b. 176

39. Finite Population Correction Factor For Formulas 7-2 and 7-3 we assume that the population is infinite or very large and that we are sampling with replacement. When we have a relatively small population with size N and sample without replacement, we modify E to include the *finite population correction factor* shown here, and we can solve for n to obtain the result given here. Use this result to repeat Exercise 33, assuming that we limit our population to 200 particular passengers on a Boeing 757-200 ER aircraft.

$$E = z_{\alpha/2}\sqrt{\frac{\hat{p}\hat{q}}{n}}\sqrt{\frac{N-n}{N-1}} \qquad n = \frac{N\hat{p}\hat{q}[z_{\alpha/2}]^2}{\hat{p}\hat{q}[z_{\alpha/2}]^2 + (N-1)E^2}$$

40. $0.0395 < p < 0.710$; no

40. Confidence Interval from Small Sample Special tables are available for finding confidence intervals for proportions involving small numbers of cases, where the normal distribution approximation cannot be used. For example, given $x = 3$ successes among $n = 8$ trials, the 95% confidence interval found in *Standard Probability and Statistics Tables and Formulae* (CRC Press) is $0.085 < p < 0.755$. Find the confidence interval that would result if you were to incorrectly use the normal distribution as an approximation to the binomial distribution. Are the results reasonably close?

41. $81.4\% < p < 101.9\%$. The upper confidence interval limit is greater than 100%. Given that the percentage cannot exceed 100%, change the upper limit to 100%.

41. Interpreting Confidence Interval Limits Repeat Exercise 38 using a 99% confidence level. What is unusual about the result? Does common sense suggest a modification of the result?

42. a. The requirement of at least 5 successes and at least 5 failures is not satisfied, so the normal distribution cannot be used.
b. 0.075

42. Coping with No Success According to the *Rule of Three,* when we have a sample size n with $x = 0$ successes, we have 95% confidence that the true population proportion has an upper bound of $3/n$. (See "A Look at the Rule of Three," by Jovanovic and Levy, *American Statistician,* Vol. 51, No. 2.)

a. If n independent trials result in no successes, why can't we find confidence interval limits by using the methods described in this section?

b. If 40 couples use a method of gender selection and each couple has a baby girl, what is the 95% upper bound for p, the proportion of all babies who are boys?

43. $p > 0.831$ (Tech: $p > 0.832$). Because we have 95% confidence that p is greater than 0.831, we can safely conclude that more than 75% of adults know what Twitter is.

43. One-Sided Confidence Interval A one-sided claim about a population proportion is a claim that the proportion is less than (or greater than) some specific value. Such a claim can be formally addressed using a *one-sided confidence interval* for p, which can be expressed as $p < \hat{p} + E$ or $p > \hat{p} - E$, where the margin of error E is modified by replacing $z_{\alpha/2}$ with z_{α}. (Instead of dividing α between two tails of the standard normal distribution, put all of it in one tail.) Repeat part (c) of Example 3 by constructing an appropriate one-sided 95% confidence interval.

7-3 Estimating a Population Mean

Key Concept This section presents methods for using a sample mean $\bar{x}$ to make an inference about the value of the corresponding population mean μ. Here are the three main concepts included in this section:

- **Point Estimate:** The sample mean $\bar{x}$ is the best *point estimate* (or single value estimate) of the population mean μ.
- **Confidence Interval:** We can use a sample mean to construct a *confidence interval* estimate of the true value of a population mean, and we should know how to construct and interpret such confidence intervals.
- **Sample Size:** We should know how to find the sample size necessary to estimate a population mean.

Part 1 of this section deals with the very realistic and commonly used case in which the population standard deviation σ is not known. Part 2 includes a brief discussion of the procedure used when σ is known, which is very rare.

Note to Instructor
Organization Change: In this new edition, Section 7-3 is partitioned into Part 1 (for σ not known) and Part 2 (for σ known). The main focus in Section 7-3 is using the Student *t* distribution for constructing confidence interval estimates of a population mean. Exercises 35–38 are the only exercises in which σ is assumed to be known. The previous edition of this book had Section 7-3 ("Estimating a Population Mean: σ Known") and Section 7-4 ("Estimating a Population Mean: σ Not Known"), but this edition includes one section (Section 7-3) for these two topics.
This section introduces the *t* distribution. A good way to begin this section is to actively involve students. Have each student record his or her pulse rate as the number of beats in 1 minute. Then proceed to use the sample to construct a 95% confidence interval estimate of the mean pulse rate of all such students.

Part 1: Estimating a Population Mean When σ Is Not Known

It's rare that we want to estimate the unknown value of a population mean but we somehow know the value of the population standard deviation σ. The realistic situation is that σ is not known. When σ is not known, we construct the confidence interval by using the Student t distribution instead of the standard normal distribution.

Point Estimate The sample mean $\bar{x}$ is an *unbiased estimator* of the population mean μ, and for many populations, sample means tend to vary less than other measures of center, so the sample mean $\bar{x}$ is usually the best point estimate of the population mean μ.

The sample mean $\bar{x}$ is the best *point estimate* of the population mean μ.

Although the sample mean $\bar{x}$ is usually the *best* point estimate of the population mean μ, it does not give us any indication of just how *good* our best estimate is, so we construct a *confidence interval* (or *interval estimate*), which consists of a range (or an interval) of values instead of just a single value.

Requirement of Normality or $n > 30$ The procedure we use has a requirement that the population is normally distributed or the sample size is greater than 30. This procedure is *robust* against a departure from normality, meaning that it works reasonably well if the departure from normality is not too extreme. Verify that there are no outliers and that the histogram or dotplot has a shape that is not very far from a normal distribution.

If the original population is not itself normally distributed, we use the condition $n > 30$ for justifying use of the normal distribution, but there is no exact specific minimum sample size that works for all cases. Sample sizes of 15 to 30 are sufficient if the population has a distribution that is not far from normal, but some other populations have distributions that are extremely far from normal and sample sizes greater than 30 might be necessary. We use the simplified criterion of $n > 30$ as justification for treating the distribution of sample means as a normal distribution, regardless of how far the distribution departs from a normal distribution.

Note to Instructor
Although the normality requirement is loose, it should not be ignored. Confirmation of normality should involve verification that there are no outliers and visual inspection of a histogram and/or normal quantile plot. When assessing normality, only one outlier might be OK, but even a single outlier can have a substantial effect on a confidence interval.

Confidence Level The confidence interval is associated with a confidence level, such as 0.95 (or 95%). The confidence level gives us the *success rate of the procedure*

Confidence Interval for Estimating a Population Mean with σ Not Known

Objective

Construct a confidence interval used to estimate a population mean.

Notation

μ = population mean

$\bar{x}$ = sample mean

n = number of sample values

E = margin of error

Requirements

1. The sample is a simple random sample.

2. Either or both of these conditions are satisfied: The population is normally distributed or $n > 30$.

Confidence Interval

$$\bar{x} - E < \mu < \bar{x} + E$$

or

$$\bar{x} \pm E$$

or

$$(\bar{x} - E, \bar{x} + E)$$

where the margin of error E is found from the following:

$$E = t_{\alpha/2} \cdot \frac{s}{\sqrt{n}} \quad \text{(Use df} = n - 1.)$$

where $t_{\alpha/2}$ = critical t value separating an area of $\alpha/2$ in the right tail of the Student t distribution, and df $= n - 1$ is the number of degrees of freedom. (Find $t_{\alpha/2}$ using technology or Table A-3.)

Round-Off Rule

1. When constructing a confidence interval from the *original set of data* values, round the confidence interval limits to one more decimal place than is used for the original set of data.

2. When constructing a confidence interval from *summary statistics* (n and $\bar{x}$ and s), round the confidence interval limits to the same number of decimal places used for the sample mean.

used to construct the confidence interval. As in Section 7-2, α is the complement of the confidence level. For a 0.95 (or 95%) confidence level, $\alpha = 0.05$.

Student *t* Distribution

If σ is not known, but the relevant requirements are satisfied, we use a *Student t distribution* (instead of a normal distribution), as developed by William Gosset (1876–1937). Gosset was a Guinness Brewery employee who needed a distribution that could be used with small samples. The Irish brewery where he worked did not allow the publication of research results, so Gosset published under the pseudonym "Student." (In the interest of research and better serving his readers, the author visited the Guinness Brewery and sampled some of the product. This is true author dedication.) Gosset first referred to his distribution as a z distribution, but it was modified and he later called it a t distribution. Here are some key points about the Student t distribution:

- **Student *t* Distribution** If a population has a normal distribution, then the distribution of

$$t = \frac{\bar{x} - \mu}{\frac{s}{\sqrt{n}}}$$

is a **Student *t* distribution** for all samples of size n. A Student t distribution is commonly referred to simply as a ***t* distribution.**

- **Degrees of Freedom** Finding a critical value $t_{\alpha/2}$ requires a value for the **degrees of freedom** (or **df**). In general, the number of degrees of freedom for a collection of sample data is the number of sample values that can vary after certain restrictions have been imposed on all data values. (*Example:* If 10 test scores have the restriction that their mean is 80, then their sum must be 800, and we can freely assign values to the first 9 scores, but the 10th score would then be determined, so in this case there are 9 degrees of freedom.) For the methods of this section, the number of degrees of freedom is the sample size minus 1.

$$\textbf{degrees of freedom} = n - 1$$

- **Finding Critical Value $t_{\alpha/2}$** A critical value $t_{\alpha/2}$ can be found using technology or Table A-3. Technology can be used with any number of degrees of freedom, but a t distribution table can be used for select numbers of degrees of freedom only.
- **Using Table A-3 when a Number of Degrees of Freedom Is Not Included** If using Table A-3 to find a critical value of $t_{\alpha/2}$, but the table does not include the number of degrees of freedom, you could use the closest value, or you could be conservative by using the next lower number of degrees of freedom found in the table, or you could interpolate.

Largest Inauguration Ever?

When Barack Obama was inaugurated as the first African American president, a huge crowd filled the region from the Capitol building to the Lincoln Memorial. The National Park Service estimated that 1.8 million people attended the inauguration. Instead of making somewhat casual estimates of the size of a crowd, aerial photographs can be used to develop reasonably good estimates based on crowd density. Allison Puccioni, an analyst for IHS Jane's, studied a satellite photograph, and she estimated that the crowd size was between 1.031 million and 1.411 million people—about 22% less than the estimate made by the National Park Service.

In another case, a celebration parade for the Boston Red Sox was attended by 3.2 million fans (according to Boston city officials), or 1 million fans (according to Boston police), or at most 400,000 fans (according to Boston University Professor Farouk El-Baz, who analyzed photos). MIT physicist Bill Donnelly said that "it's a serious thing if people are just putting out any number. It means other things aren't being vetted that carefully."

Example 1 Finding a Critical *t* Value

A sample of size $n = 12$ is a simple random sample selected from a normally distributed population (as in Example 2 that follows). Find the critical value $t_{\alpha/2}$ corresponding to a 95% confidence level.

Solution

Because $n = 12$, the number of degrees of freedom is given by $n - 1 = 11$. The 95% confidence level corresponds to $\alpha = 0.05$, so there is an area of 0.025 in each of the two tails of the t distribution, as shown in Figure 7-4 on the next page. Technology or a t distribution table can be used to find that for 11 degrees of freedom and an area of 0.025 in each tail, the critical value is $t_{\alpha/2} = 2.201$. We could also express this as $t_{0.025} = 2.201$.

To find $t_{\alpha/2} = 2.201$ using Table A-3, locate the 11th row (because df = 11) by referring to the column at the extreme left, then use the column with 0.05 for the "Area in Two Tails" (or use the same column with 0.025 for the "Area in One Tail").

Shakespeare's Vocabulary

According to Bradley Efron and Ronald Thisted, Shakespeare's writings included 31,534 different words. They used probability theory to conclude that Shakespeare probably knew at least another 35,000 words that he didn't use in his writings. The problem of estimating the size of a population is an important problem often encountered in ecology studies, but the result given here is another interesting application. (See "Estimating the Number of Unseen Species: How Many Words Did Shakespeare Know?" in *Biometrika,* Vol. 63, No. 3.)

Figure 7-4 Critical Value $t_{\alpha/2}$

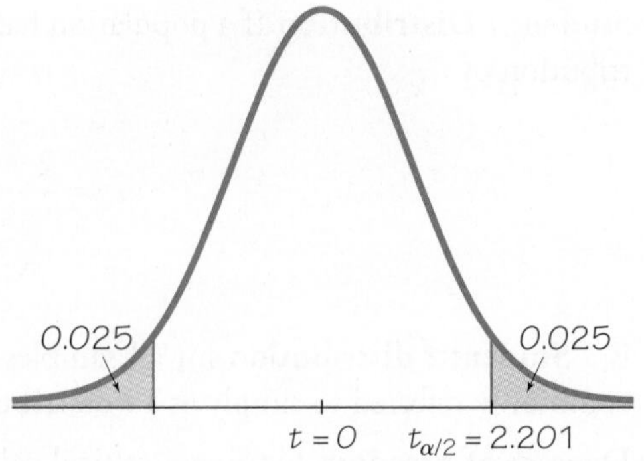

Confidence intervals can be easily constructed with technology or they can be manually constructed by using the following procedure.

Procedure for Constructing a Confidence Interval for μ

1. Verify that the two requirements in the preceding box are satisfied.
2. With σ unknown (as is usually the case), use $n - 1$ degrees of freedom and use technology or a t distribution table to find the critical value $t_{\alpha/2}$ that corresponds to the desired confidence level (as in the preceding Example 1).
3. Evaluate the margin of error using $E = t_{\alpha/2} \cdot s/\sqrt{n}$.
4. Using the value of the calculated margin of error E and the value of the sample mean $\bar{x}$, find the values of the confidence interval limits: $\bar{x} - E$ and $\bar{x} + E$. Substitute those values in one of these three general formats for the confidence interval:

$$\bar{x} - E < \mu < \bar{x} + E$$

or

$$\bar{x} \pm E$$

or

$$(\bar{x} - E, \bar{x} + E)$$

5. Round the resulting confidence interval limits using the round-off rule given in the preceding box.

Note to Instructor
Emphasize that obtaining a confidence interval is not the ultimate goal. The ultimate goal is to correctly interpret the results.

Interpreting a Confidence Interval Be careful to interpret confidence intervals correctly. If we obtain a 95% confidence interval such as $58.1 < \mu < 63.3$, there is a correct interpretation and many incorrect interpretations.

Correct: "We are 95% confident that the interval from 58.1 to 63.3 actually does contain the true value of μ." This means that if we were to select many different samples of the same size and construct the corresponding confidence intervals, in the long run 95% of them would actually contain the value of μ. (This correct interpretation refers to the *success rate of the process* being used to estimate the population mean.)

Incorrect: Because μ is a fixed constant, it would be incorrect to say "there is a 95% chance that μ will fall between 58.1 and 63.3." It would also be incorrect to say that "95% of all data values are between 58.1 and 63.3," or that "95% of sample means fall between 58.1 and 63.3." Other possible incorrect interpretations are limited only by the imagination of the reader.

Using Confidence Intervals for Hypothesis Tests

A confidence interval can be used to *test some claim* made about a population mean μ. Formal methods of hypothesis testing are introduced in Chapter 8, and those methods might require adjustments to confidence intervals that are not described in this chapter. (We might need to construct a one-sided confidence interval or adjust the confidence level by using 90% instead of 95%.)

CAUTION Know that in this chapter, when we use a confidence interval to address a claim about a population mean μ, we are making an *informal judgment* (that may or may not be consistent with formal methods of hypothesis testing introduced in Chapter 8).

Example 2 Constructing a Confidence Interval: Highway Speeds

Listed below are speeds (mi/h) measured from southbound traffic on I-280 near Cupertino, California (based on data from SigAlert). This simple random sample was obtained at 3:30 P.M. on a weekday. The speed limit for this road is 65 mi/h. Use the sample data to construct a 95% confidence interval for the mean speed. What does the confidence interval suggest about the speed limit?

62 61 61 57 61 54 59 58 59 69 60 67

Solution

Requirement check We must first verify that the requirements are satisfied. (1) The sample is a simple random sample. (2) The accompanying dotplot shows that the speeds have a distribution that is not dramatically different from a normal distribution, so the requirement that "the population is normally distributed or $n > 30$" is satisfied. ✓

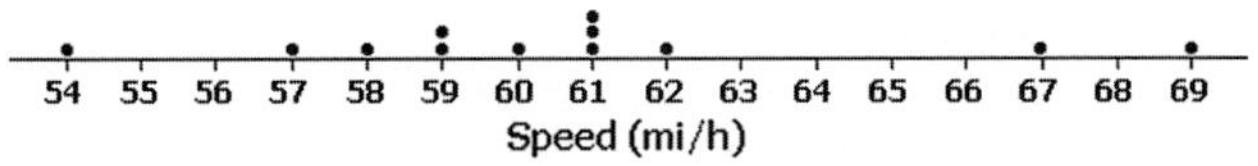

Using Technology Technology can be used to automatically construct the confidence interval. (See instructions near the end of this section.) Shown here is the Minitab display resulting from the list of 12 sample highway speeds. The confidence interval is expressed in the format of (58.08, 63.26).

MINITAB

One-Sample T: Speeds

Variable	N	Mean	StDev	SE Mean	95% CI
Speeds	12	60.67	4.08	1.18	(58.08, 63.26)

Using a *t* Distribution Table Using the listed sample values, we find that $n = 12$, $\bar{x} = 60.7$ mi/h, and $s = 4.1$ mi/h. For a confidence level of 95% and 11 degrees of freedom, the critical value is $t_{0.025} = 2.201$ as shown in Example 1. We now find the margin of error E as shown here:

$$E = t_{\alpha/2}\frac{s}{\sqrt{n}} = 2.201 \cdot \frac{4.1}{\sqrt{12}} = 2.60503$$

Estimating Wildlife Population Sizes

The National Forest Management Act protects endangered species, including the northern spotted owl, with the result that the forestry industry was not allowed to cut vast regions of trees in the Pacific Northwest.

Biologists and statisticians were asked to analyze the problem, and they concluded that survival rates and population sizes were decreasing for the female owls, known to play an important role in species survival. Biologists and statisticians also studied salmon in the Snake and Columbia Rivers in Washington State, and penguins in New Zealand. In the article "Sampling Wildlife Populations" (*Chance*, Vol. 9, No. 2), authors Bryan Manly and Lyman McDonald comment that in such studies, "biologists gain through the use of modeling skills that are the hallmark of good statistics. Statisticians gain by being introduced to the reality of problems by biologists who know what the crucial issues are."

With $\bar{x} = 60.7$ and $E = 2.60503$, we construct the confidence interval as follows:

$$\bar{x} - E < \mu < \bar{x} + E$$

$$60.7 - 2.60503 < \mu < 60.7 + 2.60503$$

$$58.1 < \mu < 63.3$$ (rounded to one decimal place more than the original sample values)

Interpretation

This result could also be expressed in the format of (58.1, 63.3) or the format of 60.7 ± 2.6. We are 95% confident that the limits of 58.1 mi/h and 63.3 mi/h actually do contain the value of the population mean μ. It appears that the mean speed is below the speed limit of 65 mi/h.

Estimating Sugar in Oranges

In Florida, members of the citrus industry make extensive use of statistical methods. One particular application involves the way in which growers are paid for oranges used to make orange juice. An arriving truckload of oranges is first weighed at the receiving plant, then a sample of about a dozen oranges is randomly selected. The sample is weighed and then squeezed, and the amount of sugar in the juice is measured. Based on the sample results, an estimate is made of the total amount of sugar in the entire truckload. Payment for the load of oranges is based on the estimate of the amount of sugar because sweeter oranges are more valuable than those less sweet, even though the amounts of juice may be the same.

Because σ is rarely known, confidence interval estimates of a population mean μ almost always use the Student t distribution instead of the standard normal distribution. Here are some important properties of the Student t distribution.

Important Properties of the Student t Distribution

1. The Student t distribution is different for different sample sizes. (See Figure 7-5 for the cases $n = 3$ and $n = 12$.)
2. The Student t distribution has the same general symmetric bell shape as the standard normal distribution, but has more variability (with wider distributions) as we expect with small samples.
3. The Student t distribution has a mean of $t = 0$ (just as the standard normal distribution has a mean of $z = 0$).
4. The standard deviation of the Student t distribution varies with the sample size, but it is greater than 1 (unlike the standard normal distribution, which has $\sigma = 1$).
5. As the sample size n gets larger, the Student t distribution gets closer to the standard normal distribution.

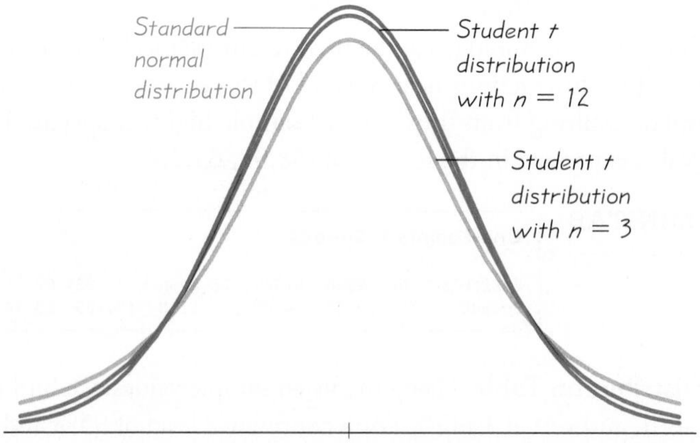

Figure 7-5 Student t Distributions for $n = 3$ and $n = 12$
The Student t distribution has the same general shape and symmetry as the standard normal distribution, but it reflects the greater variability that is expected with small samples.

Example 3 Confidence Interval for Alcohol Use in Video Games

Twelve different video games showing substance use were observed. The duration times (in seconds) of alcohol use were recorded, with the times listed below (based on data from "Content and Ratings of Teen-Rated Video Games," by Haninger and Thompson, *Journal of the American Medical Association,* Vol. 291, No. 7). The design of the study justifies the assumption that the sample can be treated as a simple random sample. Use the sample data to construct a 95% confidence interval estimate of μ, the mean duration time that the video showed the use of alcohol.

84 14 583 50 0 57 207 43 178 0 2 57

Solution

Requirement check We must first verify that the requirements are satisfied. (1) We can consider the sample to be a simple random sample. (2) When checking the requirement that "the population is normally distributed or $n > 30$," we see that the sample size is $n = 12$, so we must determine whether the data appear to be from a population with a normal distribution. Shown below are a Minitab-generated histogram and a STATDISK-generated normal quantile plot. The histogram does not appear to be bell-shaped, and the points in the normal quantile plot are not reasonably close to a straight-line pattern, so it appears that the times are not from a population having a normal distribution. The requirements are not satisfied. If we were to proceed with the construction of the confidence interval, we would get 1.8 sec $< \mu <$ 210.7 sec, but this result is questionable because it assumes incorrectly that the requirements are satisfied. ✓

MINITAB

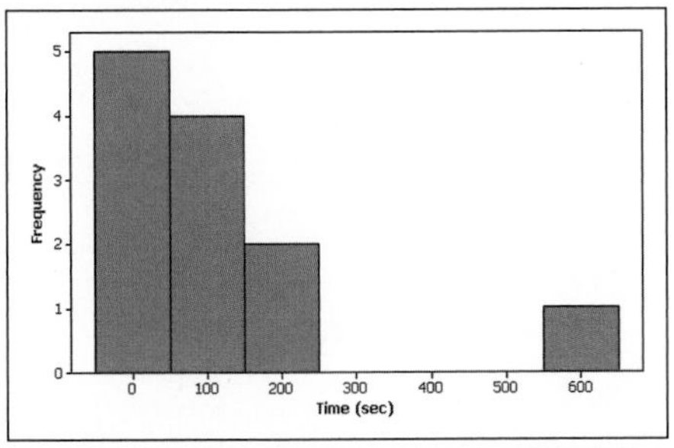

STATDISK

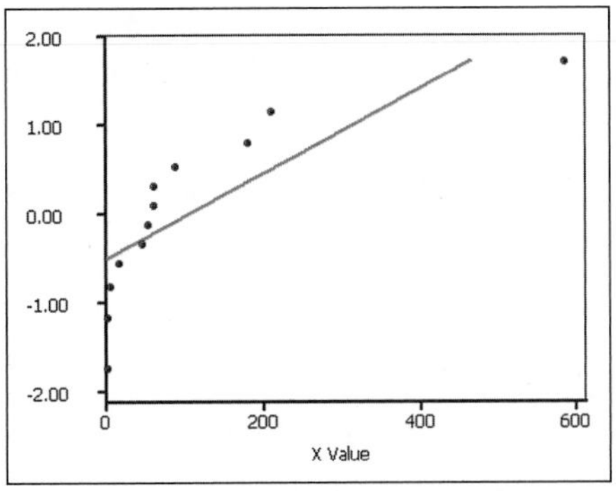

Interpretation

Because the requirement that "the population is normally distributed or $n > 30$" is not satisfied, we do not have 95% confidence that the limits of 1.8 sec and 210.7 sec actually do contain the value of the population mean. We should use some other approach for finding the confidence interval limits. For example, we could use the bootstrap resampling method described in the Technology Project at the end of this chapter.

Captured Tank Serial Numbers Reveal Population Size

During World War II, Allied intelligence specialists wanted to determine the number of tanks Germany was producing. Traditional spy techniques provided unreliable results, but statisticians obtained accurate estimates by analyzing serial numbers on captured tanks. As one example, records show that Germany actually produced 271 tanks in June 1941. The estimate based on serial numbers was 244, but traditional intelligence methods resulted in the extreme estimate of 1550. (See "An Empirical Approach to Economic Intelligence in World War II," by Ruggles and Brodie, *Journal of the American Statistical Association,* Vol. 42.)

Finding a Point Estimate and E from a Confidence Interval

When using technology to create a confidence interval, the result is often expressed in a format such as (18.128, 20.060). The sample mean $\bar{x}$ is the value midway between those limits, and the margin of error E is one-half the difference between those limits (because the upper limit is $\bar{x} + E$ and the lower limit is $\bar{x} - E$, the distance separating them is $2E$).

$$\text{Point estimate of } \mu\text{:}\quad \bar{x} = \frac{(\text{upper confidence limit}) + (\text{lower confidence limit})}{2}$$

$$\text{Margin of error:}\quad E = \frac{(\text{upper confidence limit}) - (\text{lower confidence limit})}{2}$$

Example 4 Chocolate Chips

The accompanying TI-83/84 Plus calculator screen displays results from counts of chocolate chips in a sample of 32 Chips Ahoy chewy cookies. The display shows the confidence interval limits for a 95% confidence level. Use the displayed confidence interval to find the values of the best point estimate $\bar{x}$ and the margin of error E.

TI-83/84 PLUS

```
TInterval
 (18.128,20.06)
 x̄=19.094
 Sx=2.68
 n=32
```

Solution

The following results show that for the population of Chips Ahoy chewy cookies, the mean number of chocolate chips per cookie is estimated to be 19.1 (rounded) and the margin of error is 1.0 (rounded).

$$\bar{x} = \frac{(\text{upper confidence limit}) + (\text{lower confidence limit})}{2}$$

$$= \frac{20.06 + 18.128}{2} = 19.094 \text{ chocolate chips}$$

$$E = \frac{(\text{upper confidence limit}) - (\text{lower confidence limit})}{2}$$

$$= \frac{20.06 - 18.128}{2} = 0.966 \text{ chocolate chips}$$

Using Confidence Intervals to Describe, Explore, or Compare Data

In some cases, we might use a confidence interval to achieve an ultimate goal of estimating the value of a population parameter. In other cases, confidence intervals might be among the different tools used to describe, explore, or compare data sets.

Example 5 Second-Hand Smoke

Figure 7-6 shows graphs of confidence interval estimates of the mean cotinine level in each of three samples: (1) people who smoke; (2) people who don't smoke but are exposed to tobacco smoke at home or work; (3) people who don't smoke and are not exposed to smoke. (The sample data are listed in Data Set 9 in Appendix B.) Because cotinine is produced by the body when nicotine is absorbed, cotinine is a good indication of nicotine intake. Figure 7-6 helps us see the effects of second-hand smoke. In Figure 7-6, we see that the confidence interval for smokers does not overlap the other confidence intervals, so it appears that the mean cotinine level of smokers is different from that of the other two groups. The two nonsmoking groups have confidence intervals that do overlap, so it is possible that they have the same mean cotinine level. It is helpful to compare confidence intervals or their graphs, but such comparisons should not be used for making formal and final conclusions about equality of means. Chapters 9 and 12 introduce better methods for formal comparisons of means.

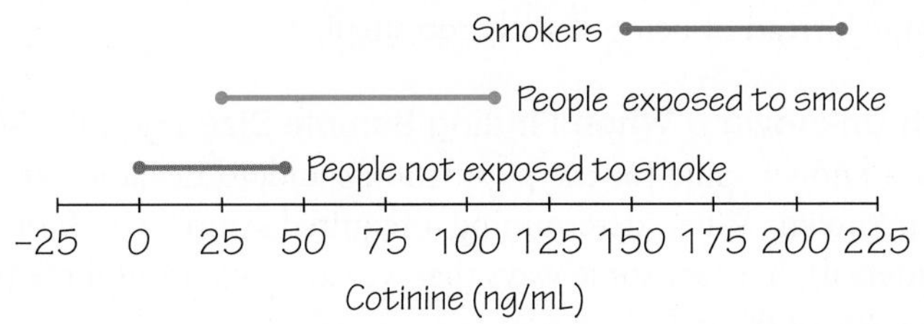

Figure 7-6 Comparing Confidence Intervals

CAUTION As in Sections 7-2 and 7-3, confidence intervals can be used *informally* to compare different data sets, but *the overlapping of confidence intervals should not be used for making formal and final conclusions about equality of means.*

Determining Sample Size If we want to collect a sample to be used for estimating a population mean μ, *how many* sample values do we need? When determining the sample size needed to estimate a population mean, we must have an estimated or known value of the population standard deviation σ, so that we can use Formula 7-4 shown in the accompanying box.

Finding the Sample Size Required to Estimate a Population Mean

Objective

Determine the sample size n required to estimate the value of a population mean μ.

Notation

μ = population mean

σ = population standard deviation

$\bar{x}$ = sample mean

E = desired margin of error

$z_{\alpha/2}$ = z score separating an area of $\alpha/2$ in the right tail of the standard normal distribution

Requirement

The sample must be a simple random sample.

Sample Size

The required sample size is found by using Formula 7-4.

Formula 7-4

$$n = \left[\frac{z_{\alpha/2}\sigma}{E}\right]^2$$

Round-Off Rule

If the computed sample size n is not a whole number, round the value of n up to the next *larger* whole number.

Population Size Formula 7-4 does not depend on the size (N) of the population (except for cases in which a relatively large sample is selected without replacement from a finite population).

Rounding The sample size must be a whole number because it is the number of sample values that must be found, but Formula 7-4 usually gives a result that is not a whole number. The round-off rule is based on the principle that when rounding is necessary, the required sample size should be rounded *upward* so that it is at least adequately large instead of being slightly too small.

Dealing with Unknown σ When Finding Sample Size Formula 7-4 requires that we substitute a known value for the population standard deviation σ, but in reality, it is usually unknown. When determining a required sample size (not constructing a confidence interval), here are some ways that we can work around the problem of not knowing the value of σ:

1. Use the range rule of thumb (see Section 3-3) to estimate the standard deviation as follows: $\sigma \approx \text{range}/4$. (With a sample of 87 or more values randomly selected from a normally distributed population, range/4 will yield a value that is greater than or equal to σ at least 95% of the time. See "Using the Sample Range as a Basis for Calculating Sample Size in Power Calculations," by Richard Browne, *American Statistician,* Vol. 55, No. 4.)
2. Start the sample collection process without knowing σ and, using the first several values, calculate the sample standard deviation s and use it in place of σ. The estimated value of σ can then be improved as more sample data are obtained, and the required sample size can be adjusted as you collect more sample data.
3. Estimate the value of σ by using the results of some other earlier study.

In addition, we can sometimes be creative in our use of other known results. For example, IQ tests are typically designed so that the mean is 100 and the standard deviation is 15. Statistics students have IQ scores with a mean greater than 100 and a standard deviation less than 15 (because they are a more homogeneous group than people randomly selected from the general population). We do not know the specific value of σ for statistics students, but we can play it safe by using $\sigma = 15$. Using a value for σ that is larger than the true value will make the sample size larger than necessary, but using a value for σ that is too small would result in a sample size that is inadequate. *When calculating the sample size n, any errors should always be conservative in the sense that they make n too large instead of too small.*

Example 6 **IQ Scores of Statistics Students**

Assume that we want to estimate the mean IQ score for the population of statistics students. How many statistics students must be randomly selected for IQ tests if we want 95% confidence that the sample mean is within 3 IQ points of the population mean?

Solution

For a 95% confidence interval, we have $\alpha = 0.05$, so $z_{\alpha/2} = 1.96$. Because we want the sample mean to be within 3 IQ points of μ, the margin of error is $E = 3$.

Also, we can assume that $\sigma = 15$ (see the discussion that immediately precedes this example). Using Formula 7-4, we get

$$n = \left[\frac{z_{\alpha/2}\sigma}{E}\right]^2 = \left[\frac{1.96 \cdot 15}{3}\right]^2 = 96.04 = 97 \quad (\text{rounded } up)$$

Interpretation

Among the thousands of statistics students, we need to obtain a simple random sample of at least 97 of their IQ scores. With a simple random sample of only 97 statistics students, we will be 95% confident that the sample mean $\bar{x}$ is within 3 IQ points of the true population mean μ.

Part 2 of this section can be omitted. It involves the unrealistic case in which we are trying to estimate an unknown population mean, but we somehow know the value of the population standard deviation. The material in Part 2 was much more important when we did not have the technology that allows us to find critical values of t for many sample sizes.

Recommendation: Skip Part 2, but draw attention to Table 7-1 and comment on the importance of using the correction distribution. Exercises 35–38 include known values of σ, so assign these exercises only if Part 2 is covered.

Part 2: Estimating a Population Mean When σ Is Known

In the real world of professional statisticians and professional journals and reports, it is extremely rare that we want to estimate an unknown value of a population mean but we somehow know the value of the population standard deviation σ. If we somehow do know the value of σ, the confidence interval is constructed using the standard normal distribution instead of the Student t distribution, as shown in the following box.

Confidence Interval for Estimating a Population Mean with σ Known

Requirements

1. The sample is a simple random sample.
2. Either or both of these conditions is satisfied: The population is normally distributed or $n > 30$.

Confidence Interval

$$\bar{x} - E < \mu < \bar{x} + E$$

or

$$\bar{x} \pm E$$

or

$$(\bar{x} - E, \bar{x} + E)$$

where the margin of error E is found from the following:

$$E = z_{\alpha/2} \cdot \frac{\sigma}{\sqrt{n}}$$

where $z_{\alpha/2}$ = critical z score separating an area of $\alpha/2$ in the right tail of the standard normal distribution. (Find $z_{\alpha/2}$ using technology or Table A-2.)

Example 7 **Confidence Interval Estimate of μ with Known σ**

Use the same sample of 12 highway speeds given in Example 2 and construct a 95% confidence interval estimate of the population mean by assuming that σ is known to be 4.1.

Solution

Requirement check The requirements were checked with the help of a dotplot in Example 2. The requirements are satisfied. ✔

continued

Power of Small Samples

The Environmental Protection Agency (EPA) had discovered that Chryslers had malfunctioning carburetors, with the result that carbon monoxide emissions were too high. Chryslers with 360- and 400-cubic-inch displacements and two-barrel carburetors were involved. The EPA ordered Chrysler to fix the problem, but Chrysler refused, and the case of *Chrysler Corporation vs. The Environmental Protection Agency* followed. That case led to the conclusion that there was "substantial evidence" that the Chryslers produced excessive levels of carbon monoxide. The EPA won the case and Chrysler was forced to recall and repair 208,000 vehicles. In discussing this case in an article in *AMSTAT News,* Chief Statistician for the EPA Barry Nussbaum wrote this: "Sampling is expensive, and environmental sampling is usually quite expensive. At the EPA, we have to do the best we can with small samples or develop models. . . . What was the sample size required to affect such a recall (of the 208,000 Chryslers)? The answer is a mere 10. It is both an affirmation of the power of inferential statistics and a challenge to explain how such a (small) sample could possibly suffice."

We can proceed with the construction of the confidence interval by first finding the critical value $z_{\alpha/2}$. With a 95% confidence level, we have $\alpha = 0.5$, and we get $z_{\alpha/2} = 1.96$ (as shown in Example 2 from Section 7-2). Using $z_{\alpha/2} = 1.96$, $\sigma = 4.1$, and $n = 12$, we find the value of the margin of error E:

$$E = z_{\alpha/2} \cdot \frac{\sigma}{\sqrt{n}}$$

$$= 1.96 \cdot \frac{4.1}{\sqrt{12}} = 2.31979$$

Using the 12 sample values listed in Example 2, we find that $\bar{x} = 60.7$. With $\bar{x} = 60.7$ and $E = 2.31979$, we find the 95% confidence interval as follows:

$$\bar{x} - E < \mu < \bar{x} + E$$

$$60.7 - 2.31979 < \mu < 60.7 + 2.31979$$

$$58.4 < \mu < 63.0$$ (rounded to one decimal place more than the original sample values)

Remember, this example illustrates the situation in which the population standard deviation σ is known, which is rare. The more realistic situation with σ unknown is considered in Part 1 of this section.

Choosing the Appropriate Distribution

When constructing a confidence interval estimate of the population mean μ, it is important to use the correct distribution. Table 7-1 summarizes the key points to consider. Table 7-1 shows that when we have a small sample ($n < 30$) drawn from a distribution that differs dramatically from a normal distribution, we can't use the methods of this chapter. In such cases, we might use nonparametric methods (see Chapter 13) or bootstrap resampling methods. (The bootstrap method is described in the Technology Project at the end of this chapter.) Remember that in reality, σ is rarely known, so estimates of μ typically involve the Student *t* distribution, provided that its requirements are met.

Table 7-1 Choosing between Student *t* and *z* (Normal) Distributions

Conditions	Method
σ not known and normally distributed population *or* σ not known and $n > 30$	Use Student *t* distribution.
σ known and normally distributed population *or* σ known and $n > 30$ (In reality, σ is rarely known.)	Use normal (*z*) distribution.
Population is not normally distributed and $n \le 30$.	Use a nonparametric method or the bootstrapping method.

Notes:

1. **Criteria for deciding whether the population is normally distributed:** The normality requirement is loose, so the distribution should appear to be somewhat symmetric with one mode and no outliers.
2. **Sample size $n > 30$:** This is a common guideline, but sample sizes of 15 to 30 are adequate if the population appears to have a distribution that is not far from being normal and there are no outliers. For some population distributions that are extremely far from normal, the sample size might need to be larger than 30.

using TECHNOLOGY

Confidence Intervals

STATDISK You must first find the sample size n, the sample mean $\bar{x}$, and the sample standard deviation s. (Find those statistics using the STATDISK procedure described in Section 3-2.) Select **Analysis** from the main menu bar, select **Confidence Intervals,** then select **Population Mean.** Enter the items in the dialog box, then click the **Evaluate** button. The confidence interval will be displayed. STATDISK will automatically choose between the normal and t distributions, depending on whether a value for the population standard deviation is entered.

MINITAB Minitab allows you to use either the summary statistics n, $\bar{x}$, and s or a list of the original sample values. Select **Stat** and **Basic Statistics.** If σ is not known, select **1-sample t** and enter the summary statistics or enter **C1** in the box located at the top right. (If σ is known, select **1-sample Z** and enter the summary statistics or enter **C1** in the box located at the top right. Also enter the value of σ in the "Standard Deviation" or "Sigma" box.) Use the **Options** button to enter the confidence level, such as 95.0.

EXCEL Use XLSTAT. Click on **XLSTAT** at the top, click on **Parametric tests,** then select **One sample t test and z test.** In the screen that appears, for the "Data" box enter the range of data, such as A1:A12 for 12 data values in column A. For "Data Format" select **One sample.** Click on the "Student's t test" box (or click on the "z test" box if σ is known). Click on the **Options** tab to enter the desired "Significance level (%)." Enter 5 for a 95% confidence interval. Select the format of "Mean 1 $\neq$ Theoretical mean" and enter any value for the theoretical mean. Click **OK.** After the results are displayed, look for "confidence interval on the mean." (The use of Excel's **CONFIDENCE** tool is not recommended, for a variety of reasons.)

TI-83/84 PLUS The TI-83/84 Plus calculator can be used to generate confidence intervals for original sample values stored in a list, or you can use the summary statistics n, $\bar{x}$, and s. Either enter the data in list L1 or have the summary statistics available, then press STAT. Now select **TESTS** and choose **TInterval** if σ is not known. (Choose **ZInterval** if σ is known.) After making the required entries, the calculator display will include the confidence interval in the format of $(\bar{x} - E, \bar{x} + E)$. For example, see the TI-83/84 Plus display that accompanies Example 4 in this section.

STATCRUNCH Click on **Open StatCrunch.** Click on **Stat,** then select **T statistics.** Select **One sample,** then select **with data** (for a list of sample data) or **with summary** (for summary statistics). Click on **Next,** then select **Confidence Interval** and click on **Calculate.**

Sample Size Determination

STATDISK Select **Analysis** from the main menu bar at the top, then select **Sample Size Determination,** followed by **Estimate Mean.** Enter the confidence level (such as 0.95) and the margin of error E. You can also enter the population standard deviation σ if it is known. There is also an option that allows you to enter a known population size N, assuming that you are sampling without replacement from a finite population.

MINITAB Using Minitab 16 or later, select **Stat** from the main menu, select **Power and Sample Size,** then select **Sample Size for Estimation.** Choose the parameter of **Mean (Normal).** Complete the dialog box and click **OK.**

Sample size determination is not available as a built-in function with Excel or StatCrunch or the TI-83/84 Plus calculator.

7-3 Basic Skills and Concepts

Recommended Assignment
Exercises 1–18.

Statistical Literacy and Critical Thinking

In Exercises 1–3, refer to the accompanying screen display that results from a sample of 40 duration times (seconds) of eruptions of the Old Faithful geyser in Yellowstone National Park.

1. Confidence Interval Refer to the accompanying screen display.

a. Express the confidence interval in the format that uses the "less than" symbol.

b. Identify the best point estimate of μ and the margin of error.

1. a. 233.4 sec $< \mu <$ 256.65 sec
   b. Best point estimate of μ is 245.025 sec. The margin of error is E = 11.625 sec.

TI-83/84 Plus

```
TInterval
 (233.4,256.65)
 x̄=245.025
 Sx=36.35754604
 n=40
```

2. a. df = 39
 b. 2.023
 c. In general, the number of degrees of freedom for a collection of sample data is the number of sample values that can vary after certain restrictions have been imposed on all data values.

2. Degrees of Freedom For the accompanying screen display, a simple random sample of size $n = 40$ was obtained from the population of duration times (seconds) of eruptions of the Old Faithful geyser.

a. What is the number of degrees of freedom that should be used for finding the critical value $t_{\alpha/2}$?

b. Find the critical value $t_{\alpha/2}$ corresponding to $n = 40$ and a 95% confidence level.

c. Give a brief general description of the number of degrees of freedom.

3. We have 95% confidence that the limits of 233.4 sec and 256.65 sec contain the true value of the mean of the population of all duration times.

4. When we say that the confidence interval methods of this section are *robust* against departures from normality, we mean that these methods work reasonably well with distributions that are not normal, provided that departures from normality are not too extreme. The given dotplot does appear to satisfy the loose normality requirement. Also, there are 40 dots, so the sample size of 40 satisfies the condition of $n > 30$.

3. Interpreting a Confidence Interval The results in the screen display are based on a 95% confidence level. Write a statement that correctly interprets the confidence interval.

4. Normality Requirement What does it mean when we say that the confidence interval methods of this section are *robust* against departures from normality? Does the dotplot below appear to satisfy the requirement of this section? Why or why not?

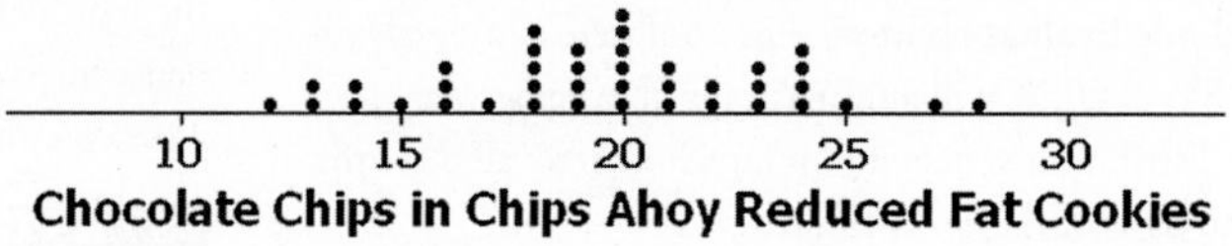

Using Correct Distribution. ***In Exercises 5–8, assume that we want to construct a confidence interval. Do one of the following, as appropriate: (a) Find the critical value $t_{\alpha/2}$, (b) find the critical value $z_{\alpha/2}$, (c) state that neither the normal distribution nor the t distribution applies.***

5. Neither the normal nor the *t* distribution applies.

5. Confidence level is 95%, σ is known to be \$4,385,000, and the dotplot of a sample of Red Sox baseball player salaries is as shown below.

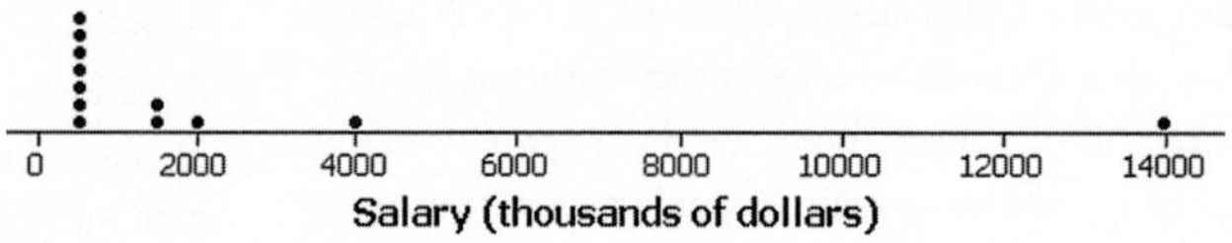

6. $t_{\alpha/2} = 1.729$

6. Confidence level is 90%, σ is not known, and the dotplot of IQ scores of 20 randomly selected statistics instructors is as shown below.

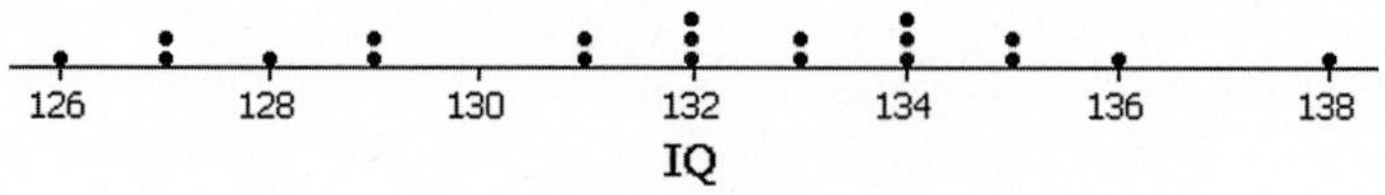

7. $t_{\alpha/2} = 2.708$

7. Confidence level is 99%, σ is not known, and the dotplot of a sample of 40 values of professional baseball player salaries is as shown below.

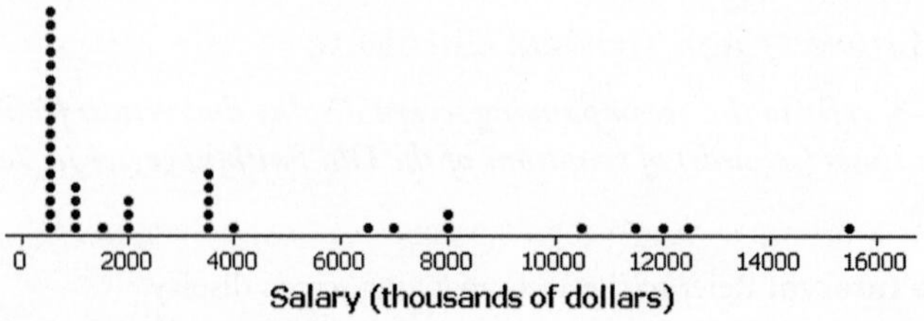

8. $z_{\alpha/2} = 2.575$ (Tech: 2.576)

8. Confidence level is 99%, σ is known to be \$4,385,000, and the dotplot of 40 sample values of professional baseball players is as shown in Exercise 7.

Confidence Intervals. *In Exercises 9–24, construct the confidence interval.*

9. Earthquake Depths Data Set 16 in Appendix B lists the depths (km) of a sample of 50 earthquakes. Those 50 depths have a mean of 9.808 km and a standard deviation of 5.013 km. Construct a 98% confidence interval estimate of the mean of all such depths. Is the confidence interval affected by the fact that the data appear to be from a population that is not normally distributed? (*Hint:* If using Table A-3 to find the critical value $t_{\alpha/2}$, that table does not include 49 degrees of freedom, so use the closest value for 50 degrees of freedom.)

9. $8.104 \text{ km} < \mu < 11.512 \text{ km}$ (Tech: $8.103 \text{ km} < \mu < 11.513 \text{ km}$). Because the sample size is greater than 30, the confidence interval yields a reasonable estimate of μ, even though the data appear to be from a population that is not normally distributed.

10. Mercury in Sushi Listed below are the amounts of mercury (in parts per million, or ppm) found in tuna sushi sampled at different stores in New York City. The study was sponsored by the *New York Times,* and the stores (in order) are D'Agostino, Eli's Manhattan, Fairway, Food Emporium, Gourmet Garage, Grace's Marketplace, and Whole Foods. The sample mean is 0.719 ppm and the standard deviation is 0.366 ppm. Construct a 90% confidence interval estimate of the mean amount of mercury in the population.

0.56 0.75 0.10 0.95 1.25 0.54 0.88

10. $0.450 \text{ ppm} < \mu < 0.988 \text{ ppm}$. (If the original values are used, the upper limit is 0.987 ppm.)

11. CEO Compensation Listed below are the recent annual compensation amounts (in thousands of dollars) for a random sample of chief executive officers (Mulally from Ford, Jobs from Apple, Kent from Coca-Cola, Otellini from Intel, and McNerney from Boeing). The mean of the sample is 12,898 (thousand dollars) and the standard deviation is 7719 (thousand dollars). Construct a 95% confidence interval estimate of the mean of the population of all such chief executive officers. Is there anything notable about the sample data that might have an effect on the result?

17,688 0.001 19,629 12,408 14,765

11. $3315.1 \text{ thousand dollars} < \mu < 22{,}480.9 \text{ thousand dollars}$ (Tech: $3313.5 \text{ thousand dollars} < \mu < 22{,}482.5 \text{ thousand dollars}$). The $1 salary of Jobs is an outlier that is very far away from the other values, and that outlier has a dramatic effect on the confidence interval.

12. Chocolate Chip Cookies The Chapter Problem for Chapter 3 includes the numbers of chocolate chips in a sample of 40 Chips Ahoy regular cookies. The mean is 23.95 chocolate chips and the standard deviation is 2.55 chocolate chips. Construct a 99% confidence interval estimate of the mean number of chocolate chips in all such cookies. How does the confidence interval not contradict the fact that most of the original values do not fall between the confidence interval limits?

12. $22.86 \text{ chocolate chips} < \mu < 25.04 \text{ chocolate chips}$. The confidence interval is an estimate of the population *mean* and it does not apply to *individual* sample values.

13. Mean Body Temperature Data Set 3 in Appendix B includes a sample of 106 body temperatures having a mean of 98.20°F and a standard deviation of 0.62°F. Construct a 95% confidence interval estimate of the mean body temperature for the entire population. What does the result suggest about the common belief that 98.6°F is the mean body temperature?

13. $98.08°\text{F} < \mu < 98.32°\text{F}$. Because the confidence interval does not contain 98.6°F, it appears that the mean body temperature is not 98.6°F, as is commonly believed.

14. Atkins Weight Loss Program In a test of weight loss programs, 40 adults used the Atkins weight loss program. After 12 months, their mean weight *loss* was found to be 2.1 lb, with a standard deviation of 4.8 lb. Construct a 90% confidence interval estimate of the mean weight loss for all such subjects. Does the Atkins program appear to be effective? Does it appear to be practical?

14. $0.8 \text{ lb} < \mu < 3.4 \text{ lb}$. Because the confidence interval does not include 0 or negative values, it does appear that the weight loss program is effective with a positive loss of weight. Because the amount of weight lost is relatively small, the weight loss program does not appear to be very practical.

15. Garlic for Reducing Cholesterol In a test of the effectiveness of garlic for lowering cholesterol, 49 subjects were treated with raw garlic. Cholesterol levels were measured before and after the treatment. The changes (before minus after) in their levels of LDL cholesterol (in mg/dL) had a mean of 0.4 and a standard deviation of 21.0 (based on data from "Effect of Raw Garlic vs Commercial Garlic Supplements on Plasma Lipid Concentrations in Adults with Moderate Hypercholesterolemia," by Gardner et al., *Archives of Internal Medicine,* Vol. 167). Construct a 98% confidence interval estimate of the mean net change in LDL cholesterol after the garlic treatment. What does the confidence interval suggest about the effectiveness of garlic in reducing LDL cholesterol?

15. $-6.8 \text{ mg/dL} < \mu < 7.6 \text{ mg/dL}$. Because the confidence interval includes the value of 0, it is very possible that the mean of the changes in LDL cholesterol is equal to 0, suggesting that the garlic treatment did not affect LDL cholesterol levels. It does not appear that garlic is effective in reducing LDL cholesterol.

16. Insomnia Treatment A clinical trial was conducted to test the effectiveness of the drug zopiclone for treating insomnia in older subjects. Before treatment with zopiclone, 16 subjects had a mean wake time of 102.8 min. After treatment with zopiclone, the

16 subjects had a mean wake time of 98.9 min and a standard deviation of 42.3 min (based on data from "Cognitive Behavioral Therapy vs Zopiclone for Treatment of Chronic Primary Insomnia in Older Adults," by Siversten et al., *Journal of the American Medical Association,* Vol. 295, No. 24). Assume that the 16 sample values appear to be from a normally distributed population and construct a 98% confidence interval estimate of the mean wake time for a population with zopiclone treatments. What does the result suggest about the mean wake time of 102.8 min before the treatment? Does zopiclone appear to be effective?

17. Harry Potter Listed below are the gross amounts (in millions of dollars) earned from box office receipts for the movie *Harry Potter and the Half-Blood Prince.* The movie opened on a Wednesday, and the amounts are listed in order for the first 14 days of the movie's release. Use the sample values to construct a 99% confidence interval estimate of the population mean. What is the population? Identify at least one major problem with this data set.

58 22 27 29 21 10 10 8 7 9 11 9 4 4

18. Years in College Listed below are the numbers of years it took for a random sample of college students to earn bachelor's degrees (based on data from the National Center for Education Statistics). Construct a 90% confidence interval estimate of the mean time required for all college students to earn bachelor's degrees. Does the confidence interval contain the value of 4 years? Is there anything about the data that would suggest that the confidence interval might not be a good result?

4 4 4 4 4 4 4.5 4.5 4.5 4.5 4.5 4.5 6 6 8 9 9 13 13 15

19. Cell Phone Radiation Listed below are the measured radiation emissions (in W/kg) corresponding to these cell phones: Samsung SGH-tss9, Blackberry Storm, Blackberry Curve, Motorola Moto, T-Mobile Sidekick, Sanyo Katana Eclipse, Palm Pre, Sony Ericsson, Nokia 6085, Apple iPhone 3G S, Kyocera Neo E1100. The data are from the Environmental Working Group. The media often present reports about the dangers of cell phone radiation as a cause of cancer. Construct a 90% confidence interval estimate of the population mean. What does the result suggest about the Federal Communications Commission standard that cell phone radiation must be 1.6 W/kg or less?

0.38 0.55 1.54 1.55 0.50 0.60 0.92 0.96 1.00 0.86 1.46

20. Ages of Race Car Drivers Listed below are the ages (years) of randomly selected race car drivers (based on data reported in *USA Today*). Construct a 98% confidence interval estimate of the mean age of all race car drivers.

32 32 33 33 41 29 38 32 33 23 27 45 52 29 25

21. Lead in Medicine Listed below are the lead concentrations (in μg/g) measured in different Ayurveda medicines. Ayurveda is a traditional medical system commonly used in India. The lead concentrations listed here are from medicines manufactured in the United States. The data are based on the article "Lead, Mercury, and Arsenic in US and Indian Manufactured Ayurvedic Medicines Sold via the Internet," by Saper et al., *Journal of the American Medical Association,* Vol. 300, No. 8. Use the sample data to construct a 95% confidence interval estimate of the mean of the lead concentrations for the population of all such medicines. If a safety standard requires lead concentrations less than 7 μg/g, does it appear that the population mean is less than that level?

3.0 6.5 6.0 5.5 20.5 7.5 12.0 20.5 11.5 17.5

22. Brain Volume Listed below are brain volumes (cm^3) of unrelated subjects used in a study. (See Data Set 6 in Appendix B.) Use the sample data to construct a 99% confidence interval estimate of the mean of the brain volume of the population. Given that typical brain volumes are between 950 cm^3 and 1800 cm^3, do these values appear to be typical?

963 1027 1272 1079 1070 1173 1067 1347 1100 1204

16. 71.4 min $< \mu <$ 126.4 min. The confidence interval includes the mean of 102.8 min that was measured before the treatment, so the mean could be the same after the treatment. This result suggests that the zopiclone treatment has no effect.

17. 4.7 million dollars $< \mu <$ 28.0 million dollars. The data appear to have a distribution that is far from normal, so the confidence interval might not be a good estimate of the population mean. The population is likely to be the list of box office receipts for each day of the movie's release. Because the values are from the first 14 days of release, the sample values are not a simple random sample, and they are likely to be the largest of all such values, so the confidence interval is not a good estimate of the population mean.

18. 5.1 years $< \mu <$ 7.9 years. The confidence interval does not contain the value of 4 years. The data appear to have a distribution that is far from normal, so the confidence interval might not be a good estimate of the population mean.

19. The sample data meet the loose requirement of having a normal distribution. CI: 0.707 W/kg $< \mu <$ 1.169 W/kg. Because the confidence interval is entirely below the standard of 1.6 W/kg, it appears that the mean amount of cell phone radiation is less than the FCC standard, but there could be individual cell phones that exceed the standard.

20. The sample data meet the loose requirement of having a normal distribution. CI: 28.4 years $< \mu <$ 38.8 years.

21. The sample data meet the loose requirement of having a normal distribution. CI: 6.43 $< \mu <$ 15.67. We cannot conclude that the population mean is less than 7 μg/g, because the confidence interval shows that the mean might be greater than that level.

22. The sample data meet the loose requirement of having a normal distribution. CI: 1009.5 cm^3 $< \mu <$ 1250.9 cm^3. The values are typical because they are between 950 cm^3 and 1800 cm^3.

23. Age Discrimination? The accompanying stemplots depict ages of applicants who were unsuccessful in winning promotion and ages of applicants who were successful in winning promotion (based on data from "Debating the Use of Statistical Evidence in Allegations of Age Discrimination" by Barry and Boland, *American Statistician*, Vol. 58, No. 2). Assume that the samples are simple random samples and use a 95% confidence level to construct the two confidence interval estimates of the two population means. Compare the results. What do you conclude?

Ages of Unsuccessful Applicants

3	4778
4	12344555689
5	3344567
6	0

Ages of Successful Applicants

3	367889
4	2233444555566778899
5	1124

23. CI for ages of unsuccessful applicants: 43.8 years $< \mu <$ 50.1 years. CI for ages of successful applicants: 42.6 years $< \mu <$ 46.4 years. Although final conclusions about means of populations should not be based on the overlapping of confidence intervals, the confidence intervals do overlap, so it appears that both populations could have the same mean, and there is not clear evidence of discrimination based on age.

24. Interbreeding of Cultures Changes in head sizes over time suggest interbreeding with people from other regions. Use the data depicted in the accompanying dotplots and construct 95% confidence intervals to determine whether skull breadths (mm) appear to have changed from 4000 B.C. to 150 A.D. Explain your conclusion.

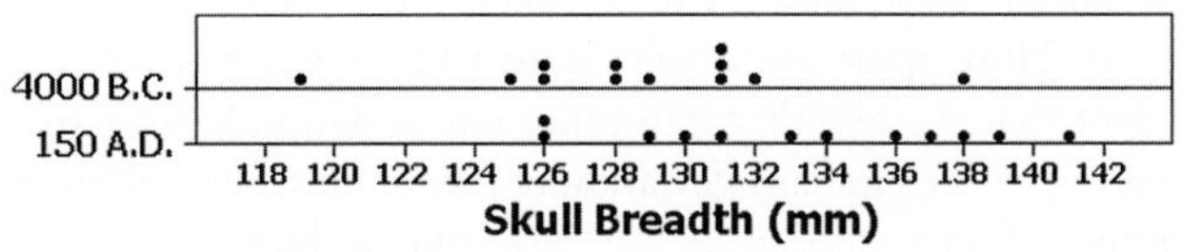

24. CI for 4000 B.C.: 125.7 mm $< \mu <$ 131.6 mm. CI for 150 A.D.: 130.1 mm $< \mu <$ 136.5 mm. Although final conclusions about means of populations should not be based on the overlapping of confidence intervals, the confidence intervals do overlap, so it appears that both populations could have the same mean, and there is not clear evidence that skull breadths changed from 4000 B.C. to 150 A.D.

Sample Size. ***In Exercises 25–32, find the sample size required to estimate the population mean.***

25. Mean IQ of Professional Pilots The Wechsler IQ test is designed so that the mean is 100 and the standard deviation is 15 for the population of normal adults. Find the sample size necessary to estimate the mean IQ score of professional pilots. We want to be 90% confident that our sample mean is within 3 IQ points of the true mean. The mean for this population is clearly greater than 100. The standard deviation for this population is probably less than 15 because it is a group with less variation than a group randomly selected from the general population; therefore, if we use $\sigma = 15$ we are being conservative by using a value that will make the sample size at least as large as necessary. Assume then that $\sigma = 15$ and determine the required sample size. Does the sample size appear to be reasonable?

25. The sample size is 68, and it does appear to be very reasonable.

26. Mean Grade Point Average As part of a study of grade inflation, you want to estimate the mean grade point average of all current college students in the United States. All grade point averages are to be standardized for a scale between 0 and 4. How many grade point averages must be obtained so that the sample mean is within 0.2 of the population mean? Assume that a 99% confidence level is desired. Also assume that a pilot study showed that the population standard deviation is estimated to be 0.79. Does it make sense to collect the entire sample at your college?

26. The required sample size is 104. Limiting the sample to students at your college would result in a convenience sample that might not be representative of the population of all college students, so it does not make sense to collect the entire sample at your college.

27. Buying a Corvette Research of selling prices for used two-year-old Corvettes reveals that they have a standard deviation of \$2157 (based on data from Edmunds.com). How many selling prices must you obtain in order to estimate the mean selling price of these cars? Assume that you want 98% confidence that your sample mean is within \$250 of the population mean. Is it likely that you will find that many two-year-old used Corvettes in your region?

27. 405 (Tech: 403). It is not likely that you would find that many two-year-old used Corvettes in your region.

28. The required sample size is 753. A major obstacle to getting a good estimate of the population mean is that it would be very difficult to actually measure times spent on Facebook, so you must rely on reported times that can be very inaccurate.

28. Facebook Time You want to estimate the mean amount of time Internet users spend on Facebook each month. How many Internet users must be surveyed in order to be 95% confident that your sample mean is within 15 minutes of the population mean? Based on results from a prior Nielsen survey, assume that the standard deviation of the population of monthly times spent on Facebook is 210 min. What is a major obstacle to getting a good estimate of the population mean?

29. Use $\sigma = 450$ to get a sample size of 110. The margin of error of 100 points seems too high to provide a good estimate of the mean SAT score.

29. Sample Size Using Range Rule of Thumb You want to estimate the mean SAT score of all college applicants. First use the range rule of thumb to make a rough estimate of the standard deviation of those scores. Possible SAT scores range from 600 to 2400. Use the estimated standard deviation to determine the sample size corresponding to a 98% confidence level and a margin of error of 100 points. What isn't quite right with this exercise?

30. Use $\sigma = \$11,250$ to get a sample size of 83,919 (Tech: 83,973). The sample size seems too large to be practical.

30. Sample Size Using Range Rule of Thumb You want to estimate the mean amount of annual tuition being paid by current full-time college students in the United States. First use the range rule of thumb to make a rough estimate of the standard deviation of the amounts spent. It is reasonable to assume that tuition amounts range from $0 to about $45,000. Then use that estimated standard deviation to determine the sample size corresponding to 99% confidence and a $100 margin of error. Does the resulting sample size seem practical?

31. With the range rule of thumb, use $\sigma = 11$ to get a required sample size of 117. With $\sigma = 10.3$, the required sample size is 102. The better estimate of σ is the standard deviation of the sample, so the correct sample size is likely to be closer to 102 than 117.

31. Sample Size Using Sample Data Refer to Data Set 1 in Appendix B and find the maximum and minimum pulse rates for males, then use those values with the range rule of thumb to estimate σ. How many adult males must you randomly select and test if you want to be 95% confident that the sample mean pulse rate is within 2 beats (per minute) of the true population mean μ. The standard deviation of the 40 pulse rates of males in Data Set 1 is 10.3 beats per minute. If, instead of using the range rule of thumb, the standard deviation of the sample is used as an estimate of σ, is the required sample size very different? Which result is likely to be closer to the correct sample size?

32. With the range rule of thumb, use $\sigma = 0.7375$ to get a required sample size of 53. With $\sigma = 0.587$, the required sample size is 34. The better estimate of σ is the standard deviation of the sample, so the sample size of 34 is the better result.

32. Sample Size Using Sample Data Refer to Data Set 16 in Appendix B and find the maximum and minimum earthquake magnitudes, then use those values with the range rule of thumb to estimate σ. How many earthquakes must you randomly select if you want to be 95% confident that the sample mean magnitude is within 0.2 of the true population mean μ? The standard deviation of the 50 earthquake magnitudes in Data Set 16 is 0.587. Find the required sample size by using the standard deviation of the sample as an estimate of σ. Which of the two results is likely to be better?

Appendix B Data Sets. ***In Exercises 33 and 34, use the data sets from Appendix B.***

33. $0.963 < \mu < 1.407$

33. Earthquake Magnitudes Use the earthquake magnitudes listed in Data Set 16 from Appendix B and construct a 99% confidence interval estimate of the mean of all such magnitudes.

34. $134.3 \text{ ng/mL} < \mu < 210.7 \text{ ng/mL}$

34. Second-Hand Smoke Figure 7-6 from Example 5 shows a graph of three different confidence intervals. Use the cotinine levels of smokers listed in Data Set 9 from Appendix B and construct the 95% confidence interval estimate of the population mean μ.

7-3 Beyond the Basics

35. $8.156 \text{ km} < \mu < 11.460 \text{ km}$ (Tech: $8.159 \text{ km} < \mu < 11.457 \text{ km}$)

36. $0.491 \text{ ppm} < \mu < 0.947 \text{ ppm}$. (If the original values are used, the upper limit is 0.946 ppm.)

37. $6131.8 \text{ thousand dollars} < \mu < 19{,}663.4 \text{ thousand dollars}$ (Tech: $6131.9 \text{ thousand dollars} < \mu < 19{,}663.3 \text{ thousand dollars}$)

Confidence Interval with Known σ. ***In Exercises 35–38, find the confidence interval using the known value of σ.***

35. Construct the confidence interval for Exercise 9 assuming that σ is known to be 5.013 km.

36. Construct the confidence interval for Exercise 10 assuming that σ is known to be 0.366 ppm.

37. Construct the confidence interval for Exercise 11 assuming that σ is known to be 7718.8 thousand dollars.

38. Construct the confidence interval for Exercise 12 assuming that σ is known to be 2.55 chocolate chips.

39. Outlier Effect If the first value of 3.0 in Exercise 21 is changed to 300, it becomes an outlier. Does this outlier have much of an effect on the confidence interval? How should you handle outliers when they are found in sample data sets that will be used for the construction of confidence intervals?

40. Finite Population Correction Factor If a simple random sample of size n is selected without replacement from a finite population of size N, and the sample size is more than 5% of the population size ($n > 0.05N$), better results can be obtained by using the finite population correction factor, which involves multiplying the margin of error E by $\sqrt{(N - n)/(N - 1)}$. For the sample of 100 weights of M&M candies in Data Set 20 from Appendix B, we get $\bar{x} = 0.8565$ g and $s = 0.0518$ g. First construct a 95% confidence interval estimate of μ assuming that the population is large, then construct a 95% confidence interval estimate of the mean weight of M&Ms in the full bag from which the sample was taken. The full bag has 465 M&Ms. Compare the results.

41. Confidence Interval for Sample of Size $n = 1$ Based on the article "An Effective Confidence Interval for the Mean with Samples of Size One and Two," by Wall, Boen, and Tweedie (*American Statistician,* Vol. 55, No. 2), a 95% confidence interval for μ can be found (using methods not discussed in this book) for a sample of size $n = 1$ randomly selected from a normally distributed population, and it can be expressed as $x \pm 9.68|x|$. Use this result to construct a 95% confidence interval using only the first sample value listed in Exercise 21. How does it compare to the result found in Exercise 21 when all of the sample values are used?

38. 22.91 chocolate chips $< \mu <$ 24.99 chocolate chips

39. The sample data do not appear to meet the loose requirement of having a normal distribution. CI: $-24.54 < \mu < 106.04$ (Tech: $-24.55 < \mu < 106.05$). The effect of the outlier on the confidence interval is very substantial. Outliers should be discarded if they are known to be errors. If an outlier is a correct value, it might be very helpful to see its effects by constructing the confidence interval with and without the outlier included.

40. 0.8462 g $< \mu <$ 0.8668 g; 0.8474 $< \mu <$ 0.8656; the second confidence interval is narrower, indicating that we have a more accurate estimate when the relatively large sample is from a relatively small finite population.

41. $-26.0 < \mu < 32.0$. The confidence interval based on the first sample value is much wider than the confidence interval based on all 10 sample values.

7-4 Estimating a Population Standard Deviation or Variance

Key Concept This section presents methods for using a sample standard deviation s (or a sample variance s^2) to estimate the value of the corresponding population standard deviation σ (or population variance σ^2). The methods of this section require that we use a *chi-square distribution.* (The Greek letter chi is pronounced "kigh.") Here are the main concepts included in this section:

- **Point Estimate:** The sample variance s^2 is the best *point estimate* (or single value estimate) of the population variance σ^2. The sample standard deviation s is commonly used as a point estimate of σ (even though it is a biased estimator). (Section 6-4 showed that s^2 is an unbiased estimator of σ^2, but s is a biased estimator of σ.)
- **Confidence Interval:** When constructing a *confidence interval* estimate of a population standard deviation (or population variance), we construct the confidence interval using the χ^2 *(chi-square) distribution.*
- **Chi-Square Distribution:** We should know how to find critical values of χ^2.

Here are some key points about the chi-square distribution:

Chi-Square Distribution: In a normally distributed population with variance σ^2, if we randomly select independent samples of size n and, for each sample, compute the sample variance s^2 (which is the square of the sample standard deviation s), the sample statistic $\chi^2 = (n - 1)s^2/\sigma^2$ has a sampling distribution called the **chi-square distribution,** so Formula 7-5 shows the format of the sample statistic.

Formula 7-5

$$\chi^2 = \frac{(n - 1)s^2}{\sigma^2}$$

Note to Instructor

Many instructors omit this section because of time limitations. The chi-square distribution is introduced here, but it can be introduced in Chapter 8. (Section 8-5 is written so that the chi-square distribution can be introduced there.) Many other instructors feel strongly that this section should be included because variation is so important.

Note to Instructor
Discuss the fact that when a sample has variance s^2 that is close to the population variance σ^2, the value of X^2 will be close to the number of degrees of freedom $n - 1$ (because the ratio of s^2/σ^2 will be close to 1). Also discuss the fact that s^2 is positive, σ^2 is positive, and $n - 1$ is positive, so X^2 will be positive. This explains why the X^2 graph begins at 0.

- **Finding Critical Values of X^2** We denote a right-tailed critical value by X_R^2 and we denote a left-tailed critical value by X_L^2. Those critical values can be found by using technology or Table A-4, and they require that we first determine a value for the number of *degrees of freedom*.
- **Degrees of Freedom** In general, the number of **degrees of freedom** (or **df**) for a collection of sample data is the number of sample values that can vary after certain restrictions have been imposed on all data values. For the methods of this section, the number of degrees of freedom is the sample size minus 1.

$$\textbf{degrees of freedom: df} = n - 1$$

CAUTION In later chapters we will encounter situations in which the degrees of freedom are not $n - 1$, so it is wrong to make the incorrect generalization that the number of degrees of freedom is always $n - 1$.

Properties of the Chi-Square Distribution

1. The chi-square distribution is not symmetric, unlike the normal and Student t distributions (see Figure 7-7). (As the number of degrees of freedom increases, the distribution becomes more symmetric, as Figure 7-8 illustrates.)
2. The values of chi-square can be zero or positive, but they cannot be negative (as shown in Figure 7-7).
3. The chi-square distribution is different for each number of degrees of freedom (as illustrated in Figure 7-8). As the number of degrees of freedom increases, the chi-square distribution approaches a normal distribution.

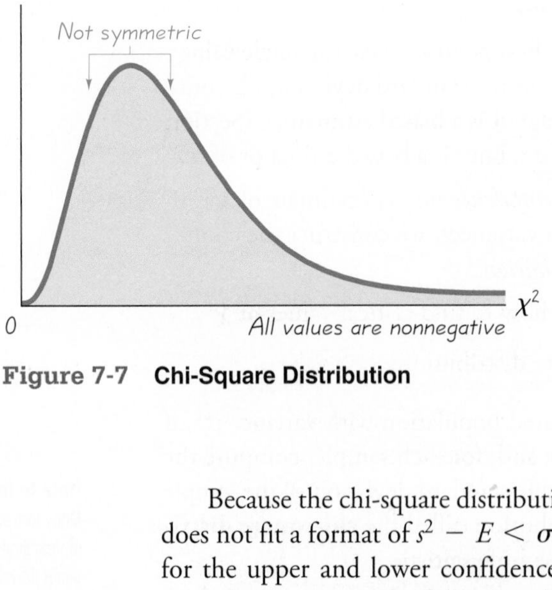

Figure 7-7 Chi-Square Distribution

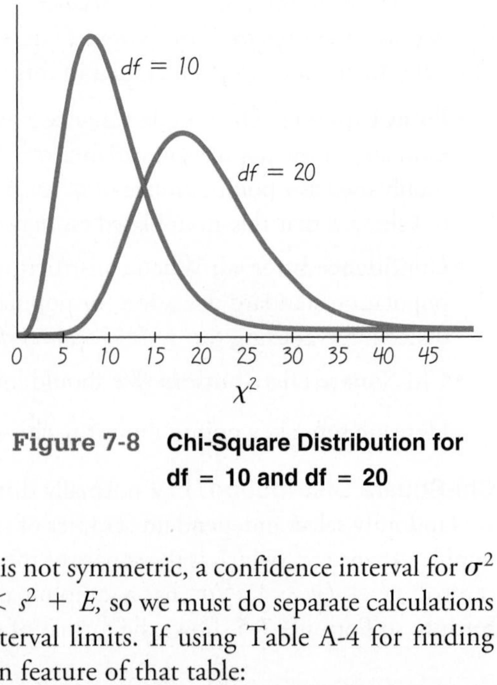

Figure 7-8 Chi-Square Distribution for df = 10 and df = 20

Because the chi-square distribution is not symmetric, a confidence interval for σ^2 does not fit a format of $s^2 - E < \sigma^2 < s^2 + E$, so we must do separate calculations for the upper and lower confidence interval limits. If using Table A-4 for finding critical values, note the following design feature of that table:

In Table A-4, each critical value of X^2 in the body of the table corresponds to an area given in the top row of the table, and each area in that top row is a *cumulative area to the right* of the critical value.

CAUTION Table A-2 for the standard normal distribution provides cumulative areas from the *left,* but Table A-4 for the chi-square distribution uses cumulative areas from the *right.*

Example 1 Finding Critical Values of χ^2

A simple random sample of 22 IQ scores is obtained (as in Example 2 which follows). Construction of a confidence interval for the population standard deviation σ requires the left and right critical values of χ^2 corresponding to a confidence level of 95% and a sample size of $n = 22$. Find the critical value of χ^2 separating an area of 0.025 in the left tail, and find the critical value of χ^2 separating an area of 0.025 in the right tail.

Solution

With a sample size of $n = 22$, the number of degrees of freedom is df $= n - 1 = 21$. See Figure 7-9.

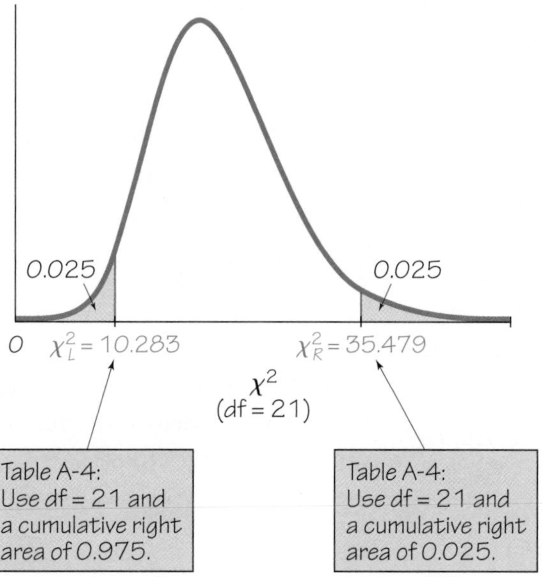

Figure 7-9 Finding Critical Values of χ^2

The critical value to the right ($\chi^2_R = 35.479$) is obtained in a straightforward manner by locating 21 in the degrees-of-freedom column at the left and 0.025 across the top row. The leftmost critical value of $\chi^2_L = 10.283$ also corresponds to 21 in the degrees-of-freedom column, but we must locate 0.975 (found by subtracting 0.025 from 1) across the top row because the values in the top row are always *areas to the right* of the critical value. Refer to Figure 7-9 and see that the total area to the right of $\chi^2_L = 10.283$ is 0.975. Figure 7-9 shows that, for a sample of 22 values taken from a normally distributed population, the chi-square statistic $(n - 1)s^2/\sigma^2$ has a 0.95 probability of falling between the chi-square critical values of 10.283 and 35.479.

Instead of using Table A-4, technology (such as STATDISK, Excel, Minitab, and StatCrunch) can be used to find critical values of χ^2. A major advantage of technology is that it can be used for any number of degrees of freedom and any confidence level, not just the limited choices included in Table A-4. Another major advantage of technology is that you can often generate the confidence interval without going through the steps of finding the critical values and manually calculating the confidence interval limits.

How Many People Do You Know?

Although the typical person cannot identify the number of people that he or she knows, methods of statistics can be used to estimate that number. The simple approach of simply asking someone how many people are known has been found to work poorly, due in large part to the same problems associated with voluntary response samples. Some other past approaches are based on the use of diaries and phonebooks. With one method, subjects are asked how many people they know in specific subpopulations. For example, subjects might be asked how many people they know who are named Adam, or how many people they know who have a twin brother or sister. Responses to such questions could be used to project the total number of people that are known. According to one estimate, the mean number of people known is 611, and the median is 472 (based on "How Many People Do You Know?: Efficiently Estimating Personal Network Size," by McCormick, Salganik, and Zheng, *Journal of the American Statistical Association,* Vol. 105, No. 489).

When obtaining critical values of X^2 from Table A-4, the numbers of degrees of freedom are consecutive integers from 1 to 30, followed by 40, 50, 60, 70, 80, 90, and 100. If a number of degrees of freedom (such as 52) is not found in the table, you can be conservative by using the next lower number of degrees of freedom, or you can use the closest critical value in the table, or you can get an approximate result with interpolation. For numbers of degrees of freedom greater than 100, use the equation given in Exercise 23, or use a more extensive table, or use technology.

Although s^2 is the best point estimate of σ^2, there is no indication of how good it actually is. To compensate for that deficiency, we develop an interval estimate (or confidence interval) that gives us a range of values associated with a confidence level.

Confidence Interval for Estimating a Population Standard Deviation or Variance

Objective

Construct a confidence interval used to estimate a population standard deviation or variance.

Notation

σ = population standard deviation

s = sample standard deviation

n = number of sample values

X_L^2 = left-tailed critical value of X^2

σ^2 = population variance

s^2 = sample variance

E = margin of error

X_R^2 = right-tailed critical value of X^2

Requirements

1. The sample is a simple random sample.

2. The population must have normally distributed values (even if the sample is large). The requirement of a normal distribution is much stricter here than in earlier sections, so departures from normal distributions can result in large errors.

Confidence Interval for the Population Variance σ^2

$$\frac{(n-1)s^2}{X_R^2} < \sigma^2 < \frac{(n-1)s^2}{X_L^2}$$

Confidence Interval for the Population Standard Deviation σ

$$\sqrt{\frac{(n-1)s^2}{X_R^2}} < \sigma < \sqrt{\frac{(n-1)s^2}{X_L^2}}$$

Round-Off Rule

1. When using the *original set of data* values to construct a confidence interval, round the confidence interval limits to one more decimal place than is used for the original set of data.

2. When using the *summary statistics* (n, s) for constructing a confidence interval, round the confidence interval limits to the same number of decimal places used for the sample standard deviation.

Confidence intervals can be easily constructed with technology or they can be constructed by using Table A-4 with the following procedure.

Procedure for Constructing a Confidence Interval for σ or σ^2

1. Verify that the two requirements in the preceding box are satisfied.
2. Using $n - 1$ degrees of freedom, use technology or refer to Table A-4 to find the critical values χ_R^2 and χ_L^2 that correspond to the desired confidence level (as in Example 1).
3. Evaluate the upper and lower confidence interval limits using this format of the confidence interval:

$$\frac{(n-1)s^2}{\chi_R^2} < \sigma^2 < \frac{(n-1)s^2}{\chi_L^2}$$

4. If a confidence interval estimate of σ is desired, take the square root of the upper and lower confidence interval limits and change σ^2 to σ.
5. Round the resulting confidence interval limits using the round-off rule given in the preceding box.

CAUTION Confidence intervals can be used *informally* to compare the variation in different data sets, but *the overlapping of confidence intervals should not be used for making formal and final conclusions about equality of variances or standard deviations.*

Using Confidence Intervals for Hypothesis Tests

A confidence interval can be used to *test some claim* made about σ or σ^2. Formal methods of hypothesis testing are introduced in Chapter 8, and those methods might require adjustments to confidence intervals that are not described in this chapter. (We might need to construct a one-sided confidence interval or adjust the confidence level by using 90% instead of 95%.)

CAUTION Know that in this chapter, when we use a confidence interval to address a claim about σ or σ^2, we are making an *informal judgment* (that may or may not be consistent with formal methods of hypothesis testing introduced in Chapter 8).

Example 2 Confidence Interval for Estimating σ of IQ Scores

Data Set 5 in Appendix B lists IQ scores for subjects in three different lead exposure groups. The 22 full IQ scores for the group with medium exposure to lead (Group 2) have a standard deviation of 14.3. Consider the sample to be a simple random sample and construct a 95% confidence interval estimate of σ, the standard deviation of the population from which the sample was obtained.

Solution

Requirement check

Step 1: We first verify that the requirements are satisfied. (1) The sample can be treated as a simple random sample. (2) The following display shows a Minitab-generated histogram. Except for one low score, the shape of the histogram is very close to the bell shape of a normal distribution, so the requirement of normality is satisfied. (It would be wise to find the confidence interval with the low score excluded so we can see its effect.) This check of requirements is Step 1 in the process of finding a confidence interval of σ, so we proceed with Step 2.

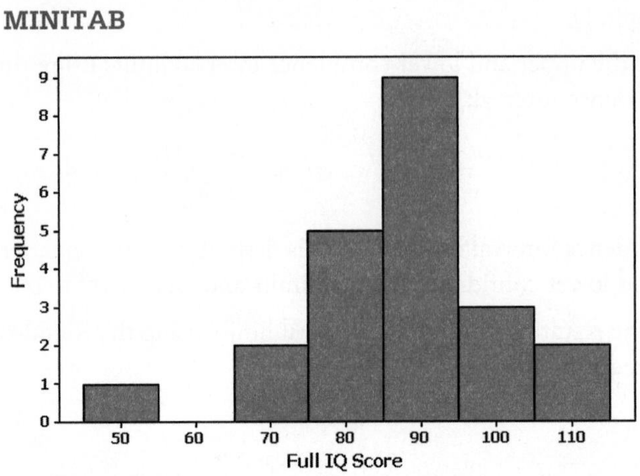

Step 2: The confidence interval can be found using technology. If using Table A-4, we first use the sample size of $n = 22$ to determine that the number of degrees of freedom is given by df $= n - 1 = 21$. If we use Table A-4, we refer to the row corresponding to 21 degrees of freedom, and we refer to the columns with areas of 0.975 and 0.025. (For a 95% confidence level, we divide $\alpha = 0.05$ equally between the two tails of the chi-square distribution, and we refer to the values of 0.975 and 0.025 across the top row of Table A-4.) The critical values are $\chi^2_L = 10.283$ and $\chi^2_R = 35.479$ (as shown in Example 1).

Step 3: Using the critical values of 10.283 and 35.479, the sample standard deviation of $s = 14.3$, and the sample size of $n = 22$, we construct the 95% confidence interval by evaluating the following:

$$\frac{(n-1)s^2}{\chi^2_R} < \sigma^2 < \frac{(n-1)s^2}{\chi^2_L}$$

$$\frac{(22-1)(14.3)^2}{35.479} < \sigma^2 < \frac{(22-1)(14.3)^2}{10.283}$$

Step 4: Evaluating the expression above results in $121.0 < \sigma^2 < 417.6$. Finding the square root of each part (before rounding), then rounding to one decimal place, yields this 95% confidence interval estimate of the population standard deviation: $11.0 < \sigma < 20.4$.

Interpretation

Based on this result, we have 95% confidence that the limits of 11.0 and 20.4 contain the true value of σ. The confidence interval can also be expressed as (11.0, 20.4).

The IQ test was designed so that the population of IQ scores would have a standard deviation of 16, and the value of 16 is contained within the confidence interval, so the variation of the IQ scores does not appear to be unusual.

Rationale for the Confidence Interval If we obtain simple random samples of size n from a normally distributed population with variance σ^2, there is a probability of $1 - \alpha$ that the statistic $(n - 1)s^2/\sigma^2$ will fall between the critical values of χ^2_L and χ^2_R. (In Figure 7-9, the confidence level of 95% corresponds to $\alpha = 0.05$, and there is a 0.95 probability that the χ^2 test statistic falls between χ^2_L and χ^2_R.) It follows that there is a $1 - \alpha$ probability that both of the following are true:

$$\frac{(n-1)s^2}{\sigma^2} < \chi^2_R \quad \text{and} \quad \frac{(n-1)s^2}{\sigma^2} > \chi^2_L$$

If we multiply both of the preceding inequalities by σ^2 and divide each inequality by the appropriate critical value of χ^2, the two preceding inequalities can be expressed in these equivalent forms:

$$\frac{(n-1)s^2}{\chi^2_R} < \sigma^2 \quad \text{and} \quad \frac{(n-1)s^2}{\chi^2_L} > \sigma^2$$

The two preceding inequalities can be combined into one inequality to get the format of the confidence interval used in this section:

$$\frac{(n-1)s^2}{\chi^2_R} < \sigma^2 < \frac{(n-1)s^2}{\chi^2_L}$$

Determining Sample Size The procedures for finding the sample size necessary to estimate σ^2 are much more complex than the procedures given earlier for means and proportions. Instead of using very complicated procedures, we will use Table 7-2.

Note to Instructor
Almost every other textbook ignores the topic of determining sample sizes required to estimate σ or σ^2, even after covering sample sizes required for estimating means and proportions. In many cases, the standard deviation is the most important parameter, so its estimation is critically important. That is why sample size determination is included in this section.

Table 7-2 Finding Sample Size

σ		σ^2	
To be 95% confident that s is within	**of the value of σ, the sample size n should be at least**	**To be 95% confident that s^2 is within**	**of the value of σ^2, the sample size n should be at least**
1%	19,205	1%	77,208
5%	768	5%	3,149
10%	192	10%	806
20%	48	20%	211
30%	21	30%	98
40%	12	40%	57
50%	8	50%	38
To be 99% confident that s is within	**of the value of σ, the sample size n should be at least**	**To be 99% confident that s^2 is within**	**of the value of σ^2, the sample size n should be at least**
1%	33,218	1%	133,449
5%	1,336	5%	5,458
10%	336	10%	1,402
20%	85	20%	369
30%	38	30%	172
40%	22	40%	101
50%	14	50%	68

STATDISK also provides sample sizes. With STATDISK, select **Analysis, Sample Size Determination,** and then **Estimate St Dev.** If using Minitab Release 16 or later, select **Stat,** then **Power and Sample Size.** Excel, StatCrunch, and the TI-83/84 Plus calculator do not provide such sample sizes.

Example 3 Finding Sample Size for Estimating σ

We want to estimate the standard deviation σ of all IQ scores of people with exposure to lead. We want to be 95% confident that our estimate is within 10% of the true value of σ. How large should the sample be? Assume that the population is normally distributed.

Solution

From Table 7-2, we can see that 95% confidence and an error of 10% for σ correspond to a sample of size 192. We should obtain a simple random sample of 192 IQ scores from the population of subjects exposed to lead.

using TECHNOLOGY

For Confidence Intervals

STATDISK First obtain the descriptive statistics and verify that the distribution is normal by using a histogram or normal quantile plot. Next, select **Analysis** from the main menu, then select **Confidence Intervals,** and **Population StDev.** Enter the required data.

MINITAB Click on **Stat,** click on **Basic Statistics,** and select 1 **Variance.** In the Data box, select the option of using a column containing the list of sample data or enter the value of the sample standard deviation or sample variance. Click on the **Options** button and enter the confidence level, such as 95.0. Click **OK** twice. The results will include a confidence interval for the standard deviation and a confidence interval for the variance.

EXCEL Neither Excel nor XLSTAT has a function for generating a confidence interval estimate of standard deviation or variance.

TI-83/84 PLUS The TI-83/84 Plus calculator does not provide confidence intervals for σ or σ^2 directly, but the program **S2INT** can be used. That program was written by Michael Lloyd of Henderson State University, and it is on the CD included with this book, or it can be downloaded from www.aw.com/triola. The program S2INT uses the program ZZINEWT, so that program must also be installed. After storing the programs on the calculator, press PRGM, select **S2INT,** and enter the sample variance s^2, the sample size n, and the confidence level (such as 0.95). Press ENTER, and wait a while for the display of the confidence interval limits for σ^2. Find the square root of the confidence interval limits if an estimate of σ is desired.

STATCRUNCH Click on **Open StatCrunch.** Click on **Stat,** then select **Variance.** Select **One sample,** then select **with data** (for a list of sample data) or **with summary** (for summary statistics). Click on **Next,** then select **Confidence Interval** and click on **Calculate.**

Recommended Assignment
Exercises 1–14.

7-4 Basic Skills and Concepts

Statistical Literacy and Critical Thinking

1. $30.3 \text{ mg/dL} < \sigma < 47.5 \text{ mg/dL}$. We have 95% confidence that the limits of 30.3 mg/dL and 47.5 mg/dL contain the true value of the standard deviation of the LDL cholesterol levels of all women.

1. LDL Cholesterol Using the 40 LDL cholesterol levels of women listed in Data Set 1, we get this 95% confidence interval estimate: $916.591 < \sigma^2 < 2252.119$, and the units of measurement are $(\text{mg/dL})^2$. Identify the corresponding confidence interval estimate of σ and include the appropriate units. Given that the original values are whole numbers, round the limits using the round-off rule given in this section. Write a statement that correctly interprets the confidence interval estimate of σ.

2. Expressing Confidence Intervals Example 2 showed how the statistics of $n = 22$ and $s = 14.3$ result in this 95% confidence interval estimate of σ: $11.0 < \sigma < 20.4$. That confidence interval can also be expressed as (11.0, 20.4), but it cannot be expressed as 15.7 ± 4.7. Given that 15.7 ± 4.7 results in values of 11.0 and 20.4, why can't we express the confidence interval as 15.7 ± 4.7?

2. The format implies that $s = 15.7$, but s is given as 14.3. In general, a confidence interval for σ does not have s at the center.

3. Pick 4 Lottery The dotplot below depicts individual digits selected in the Pick 4 lottery from different states. Can the original list of individual sample digits be identified? Can the sample data be used to construct a 95% confidence interval estimate of the population standard deviation for all such digits? Why or why not?

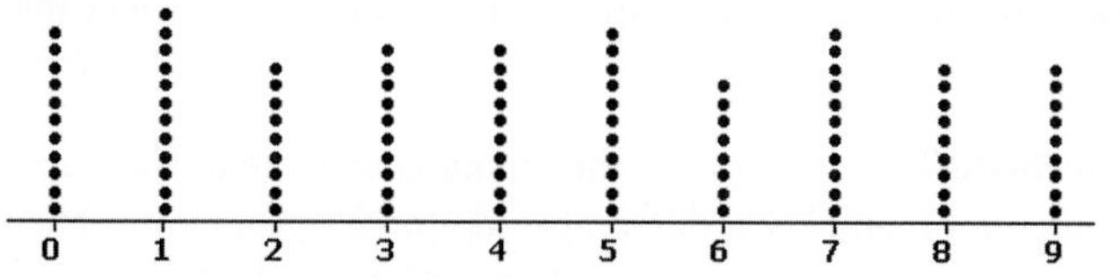

3. The original sample values can be identified, but the dotplot shows that the sample appears to be from a population having a uniform distribution, not a normal distribution as required. Because the normality requirement is not satisfied, the confidence interval estimate of σ should not be constructed using the methods of this section.

4. Normality Requirement What is different about the normality requirement for a confidence interval estimate of σ and the normality requirement for a confidence interval estimate of μ?

4. The normality requirement for a confidence interval estimate of σ has a much stricter normality requirement than the loose normality requirement for a confidence interval estimate of μ. Departures from normality have a much greater effect on confidence interval estimates of σ than on confidence interval estimates of μ.

Finding Critical Values and Confidence Intervals. ***In Exercises 5–8, use the given information to find the number of degrees of freedom, the critical values χ^2_L and χ^2_R, and the confidence interval estimate of σ. The samples are from Appendix B and it is reasonable to assume that a simple random sample has been selected from a population with a normal distribution.***

5. Nicotine in Menthol Cigarettes 99% confidence; $n = 25$, $s = 0.24$ mg.

6. Weights of Dollar Coins 99% confidence; $n = 20$, $s = 0.04111$ g.

7. Platelet Counts of Women 95% confidence; $n = 40$, $s = 65.2$.

8. Earthquake Magnitudes 95% confidence; $n = 50$, $s = 0.587$.

5. df = 24. $\chi^2_L = 9.886$ and $\chi^2_R = 45.559$. CI: 0.17 mg $< \sigma <$ 0.37 mg.

6. df = 19. $\chi^2_L = 6.844$ and $\chi^2_R = 38.582$. CI: 0.02885 g $< \sigma <$ 0.06850.

7. df = 39. $\chi^2_L = 24.433$ (Tech: 23.654) and $\chi^2_R = 59.342$ (Tech: 58.120). CI: $52.9 < \sigma < 82.4$ (Tech: $53.4 < \sigma < 83.7$).

8. df = 49. $\chi^2_L = 32.357$ (Tech: 31.555) and $\chi^2_R = 71.420$ (Tech: 70.222). CI: $0.486 < \sigma < 0.722$ (Tech: $0.490 < \sigma < 0.731$).

Finding Confidence Intervals. ***In Exercises 9–16, assume that each sample is a simple random sample obtained from a population with a normal distribution.***

9. Mean Body Temperature Data Set 3 in Appendix B includes a sample of 106 body temperatures having a mean of 98.20°F and a standard deviation of 0.62°F. Construct a 90% confidence interval estimate of the standard deviation of the body temperatures for the entire population.

9. 0.579°F $< \sigma <$ 0.720°F (Tech: 0.557°F $< \sigma <$ 0.700°F)

10. Chocolate Chip Cookies The Chapter Problem for Chapter 3 includes the numbers of chocolate chips in a sample of 40 Chips Ahoy regular cookies. The mean is 23.95 and the standard deviation is 2.55. Construct a 90% confidence interval estimate of the standard deviation of the numbers of chocolate chips in all such cookies.

10. 2.13 chocolate chips $< \sigma <$ 3.09 chocolate chips (Tech: 2.16 chocolate chips $< \sigma <$ 3.14 chocolate chips)

11. Antifreeze A container of car antifreeze is supposed to hold 3785 mL of the liquid. Realizing that fluctuations are inevitable, the quality-control manager of the Taconic Chemical Company wants to be quite sure that the standard deviation is less than 30 mL. Otherwise, some containers would overflow while others would not have enough of the coolant. She selects a simple random sample of 24 containers and finds that the mean is 3789 mL and the standard deviation is 42.8 mL. Use these sample results to construct the 99% confidence interval for the true value of σ. Does this confidence interval suggest that the variation is at an acceptable level?

11. 30.9 mL $< \sigma <$ 67.45 mL. The confidence interval shows that the standard deviation is not likely to be less than 30 mL, so the variation is too high instead of being at an acceptable level below 30 mL. (Such one-sided claims should be tested using the formal methods presented in Chapter 8.)

12. a. 7.9 beats per minute $< \sigma <$ 14.1 beats per minute (Tech: 7.9 beats per minute $< \sigma <$ 14.4 beats per minute)
b. 8.9 beats per minute $< \sigma <$ 15.9 beats per minute (Tech: 9.0 beats per minute $< \sigma <$ 16.2 beats per minute)
c. The confidence intervals are not dramatically different, so it appears that the populations of pulse rates of men and women have about the same standard deviation.

12. Pulse Rates of Men and Women Data Set 1 in Appendix B includes 40 pulse rates of men, and those pulse rates have a mean of 67.3 beats per minute and a standard deviation of 10.3 beats per minute. That data set also includes 40 pulse rates of women, and those pulse rates have a mean of 77.5 beats per minute and a standard deviation of 11.6 beats per minute.

a. Construct a 99% confidence interval estimate of the standard deviation of the pulse rates of men.

b. Construct a 99% confidence interval estimate of the standard deviation of the pulse rates of women.

c. Compare the variation of the pulse rates of men and women. Does there appear to be a difference?

13. 0.252 ppm $< \sigma <$ 0.701 ppm

13. Mercury in Sushi Listed below are the amounts of mercury (in parts per million, or ppm) found in tuna sushi sampled at different stores in New York City. The study was sponsored by the *New York Times,* and the stores (in order) are D'Agostino, Eli's Manhattan, Fairway, Food Emporium, Gourmet Garage, Grace's Marketplace, and Whole Foods. Construct a 90% confidence interval estimate of the standard deviation of the amounts of mercury in the population.

0.56 0.75 0.10 0.95 1.25 0.54 0.88

14. 2.9 mi/h $< \sigma <$ 6.9 mi/h. Because traffic conditions vary considerably at different times during the day, the confidence interval is an estimate of the standard deviation of the population of speeds at 3:30 on a weekday, not other times.

14. Highway Speeds Listed below are speeds (mi/h) measured from southbound traffic on I-280 near Cupertino, California (based on data from SigAlert). This simple random sample was obtained at 3:30 P.M. on a weekday. Use the sample data to construct a 95% confidence interval estimate of the population standard deviation. Does the confidence interval describe the standard deviation for all times during the week?

62 61 61 57 61 54 59 58 59 69 60 67

15. CI for ages of unsuccessful applicants: 5.2 years $< \sigma <$ 11.5 years. CI for ages of successful applicants: 3.7 years $< \sigma <$ 7.5 years. Although final conclusions about means of populations should not be based on the overlapping of confidence intervals, the confidence intervals do overlap, so it appears that the two populations have standard deviations that are not dramatically different.

15. Promotion and Age Listed below are ages of applicants who were unsuccessful in winning promotion and ages of applicants who were successful in winning promotion (based on data from "Debating the Use of Statistical Evidence in Allegations of Age Discrimination" by Barry and Boland, *American Statistician,* Vol. 58, No. 2). Construct 99% confidence interval estimates of the standard deviations of the two populations from which the samples were obtained. Compare the results. What do you conclude?

Ages of Unsuccessful Applicants:	34	37	37	38	41	42	43	44	44	
	45	45	45	46	48	49	53	53	54	
	54	55	56	57	60					
Ages of Successful Applicants:	33	36	37	38	38	39	42	42	43	43
	44	44	44	45	45	45	45	46	46	47
	47	48	48	49	49	51	51	52	54	

16. a. 0.33 min $< \sigma <$ 0.87 min
b. 1.25 min $< \sigma <$ 3.33 min
c. The variation appears to be significantly lower with a single line. The single line appears to be better.

16. a. Comparing Waiting Lines The values listed below are waiting times (in minutes) of customers at the Jefferson Valley Bank, where customers enter a single waiting line that feeds three teller windows. Construct a 95% confidence interval for the population standard deviation σ.

6.5 6.6 6.7 6.8 7.1 7.3 7.4 7.7 7.7 7.7

b. The values listed below are waiting times (in minutes) of customers at the Bank of Providence, where customers may enter any one of three different lines that have formed at three teller windows. Construct a 95% confidence interval for the population standard deviation σ.

4.2 5.4 5.8 6.2 6.7 7.7 7.7 8.5 9.3 10.0

c. Interpret the results found in parts (a) and (b). Do the confidence intervals suggest a difference in the variation among waiting times? Which arrangement seems better: the single-line system or the multiple-line system?

Using Large Data Sets from Appendix B. ***In Exercises 17 and 18, use the data set from Appendix B. Assume that each sample is a simple random sample obtained from a population with a normal distribution.***

17. Penny Weights Refer to Data Set 21 in Appendix B and use the weights of the post-1983 pennies to construct a 98% confidence interval estimate of the standard deviation of the weights of all post-1983 pennies.

17. 0.01239 g $< \sigma <$ 0.02100 g (Tech: 0.01291 g $< \sigma <$ 0.02255 g)

18. Ages of Presidents Refer to Data Set 12 in Appendix B and use the ages (years) of the presidents at the times of their inaugurations. Treating the data as a sample, construct a 98% confidence interval estimate of the standard deviation of the population of all such ages.

18. 5.2 years $< \sigma <$ 8.9 years

Determining Sample Size. ***In Exercises 19–22, assume that each sample is a simple random sample obtained from a normally distributed population. Use Table 7-2 to find the indicated sample size.***

19. IQ of Statistics Professors You want to estimate σ for the population of IQ scores of statistics professors. Find the minimum sample size needed to be 99% confident that the sample standard deviation s is within 1% of σ. Is this sample size practical?

19. 33,218 is too large. There aren't 33,218 statistics professors in the population, and even if there were, that sample size is too large to be practical.

20. McDonald's Waiting Times You want to estimate σ for the population of waiting times at McDonald's drive-up windows, and you want to be 95% confident that the sample standard deviation is within 20% of σ. Find the minimum sample size. Is this sample size practical?

20. The sample size of 48 is very practical, although the sample should be selected from the population of all McDonald's restaurants with drive-up windows.

21. Flight Delays You want to estimate the standard deviation of arrival delays for American Airlines flights from Chicago to Miami. Find the minimum sample size needed to be 95% confident that the sample standard deviation is within 5% of the population standard deviation. A histogram of a sample of those arrival delays suggests that the distribution is skewed, not normal. How does the distribution affect the sample size?

21. The sample size is 768. Because the population does not have a normal distribution, the computed minimum sample size is not likely to be correct.

22. U.S. Incomes You want to estimate the standard deviation of the population of current annual incomes of adults in the United States. Find the minimum sample size needed to be 99% confident that the sample standard deviation is within 5% of the population standard deviation. Does the population of incomes have a normal distribution, and how is the sample size affected by the distribution?

22. The sample size is 1336. The population of incomes does not have a normal distribution, so the computed sample size is not likely to be correct.

7-4 Beyond the Basics

23. Finding Critical Values In constructing confidence intervals for σ or σ^2, Table A-4 can be used to find the critical values χ_L^2 and χ_R^2 only for select values of n up to 101, so the number of degrees of freedom is 100 or smaller. For larger numbers of degrees of freedom, we can approximate χ_L^2 and χ_R^2 by using

$$\chi^2 = \frac{1}{2}[\pm z_{\alpha/2} + \sqrt{2k - 1}]^2$$

where k is the number of degrees of freedom and $z_{\alpha/2}$ is the critical z score described in Section 7-2. Use this approximation to find the critical values χ_L^2 and χ_R^2 for Exercise 9. How do the results compare to the actual critical values of 82.354 and 129.918?

23. $\chi_L^2 = 82.072$ and $\chi_R^2 = 129.635$ (Tech using $z_{\alpha/2} = 1.644853626$: $\chi_L^2 = 82.073$ and $\chi_R^2 = 129.632$). The approximate values are quite close to the actual critical values.

Chapter 7 Review

This chapter begins the presentation of methods for using sample data to make inferences about a population. Specifically, we use sample data to find *estimates* of population proportions, population means, and population variances (or standard deviations). This chapter included procedures for finding each of the following:

- point estimate of a population proportion, mean, standard deviation, or variance
- confidence interval estimate of a population proportion, mean, standard deviation, or variance
- sample size required to estimate a population proportion, mean, standard deviation, or variance

Because point estimates consist of single values, they have the serious disadvantage of not revealing how close to the population parameter that they are likely to be, so confidence intervals (or interval estimates) are commonly used as more informative and useful estimates. We also considered ways of determining the sample sizes necessary to estimate parameters to within given margins of error. This chapter also introduced the Student t and chi-square distributions. We must be careful to use the correct probability distribution for each set of circumstances. The following table summarizes some key elements of this chapter.

Parameter	Point Estimate	Confidence Interval
Proportion p	$\hat{p}$	$\hat{p} - E < p < \hat{p} + E$ where $E = z_{\alpha/2}\sqrt{\frac{\hat{p}\hat{q}}{n}}$
Mean μ	$\bar{x}$	$\bar{x} - E < \mu < \bar{x} + E$ where $E = t_{\alpha/2}\frac{s}{\sqrt{n}}$ or $E = z_{\alpha/2} \cdot \frac{\sigma}{\sqrt{n}}$
Standard deviation σ	s (commonly used)	$\sqrt{\frac{(n-1)s^2}{\chi_R^2}} < \sigma < \sqrt{\frac{(n-1)s^2}{\chi_L^2}}$
Variance σ^2	s^2	$\frac{(n-1)s^2}{\chi_R^2} < \sigma^2 < \frac{(n-1)s^2}{\chi_L^2}$

For the confidence interval and sample size procedures in this chapter, it is very important to verify that the requirements are satisfied. If they are not, then we cannot use the methods of this chapter and we may need to use other methods, such as the bootstrap method described in the Technology Project at the end of this chapter, or nonparametric methods, such as those discussed in Chapter 13.

Chapter Quick Quiz

1\. $36.9\% < p < 43.1\%$

1. *USA Today* reported that 40% of people surveyed planned to use accumulated loose change for paying bills. The margin of error was given as ± 3.1 percentage points. Identify the confidence interval that corresponds to that information.

2\. 0.480

3\. We have 95% confidence that the limits of 0.449 and 0.511 contain the true value of the proportion of females in the population of medical school students.

2. Here is a 95% confidence interval estimate of the proportion of female medical school students: $0.449 < p < 0.511$ (based on data from the *Journal of the American Medical Association*). What is the best point estimate of the proportion of females in the population of medical school students?

3. Write a brief statement that correctly interprets the confidence interval given in Exercise 2.

4. In a survey of 1023 high school students, 102 chose education for their career choice (based on results from a KeyStat Marketing survey). Find the critical value that would be used for constructing a 90% confidence interval estimate of the population proportion.

4. $z = 1.645$

5. Find the sample size required to estimate the percentage of college students who own a car. Assume that we want 90% confidence that the proportion from the sample is within three percentage points of the true population percentage.

5. 752

6. Find the sample size required to estimate the mean IQ of students currently taking a statistics course. Assume that we want 99% confidence that the mean from the sample is within two IQ points of the true population mean. Also assume that $\sigma = 15$.

6. 373 (Tech: 374)

7. Six human skulls from around 4000 B.C. were measured, and the lengths have a mean of 94.2 mm and a standard deviation of 4.9 mm. If you want to construct a 95% confidence interval estimate of the mean length of all such skulls, what requirements must be satisfied?

7. The sample must be a simple random sample and there is a loose requirement that the sample values appear to be from a normally distributed population.

8. In general, what does "degrees of freedom" refer to? For the sample data described in Exercise 7, find the number of degrees of freedom, assuming that you want to construct a confidence interval estimate of μ.

8. The degrees of freedom is the number of sample values that can vary after restrictions have been imposed on all of the values. For the sample data in Exercise 7, df = 5.

9. Refer to Exercise 7 and assume that the requirements are satisfied. Find the critical value that would be used for constructing a 95% confidence interval estimate of μ.

9. $t = 2.571$

10. Refer to Exercise 7 and assume that the requirements are satisfied. Find the critical values that would be used to construct a 95% confidence interval estimate of σ.

10. $\chi^2_L = 0.831$ and $\chi^2_R = 12.833$

Review Exercises

1. Underpaid In a Gallup poll of 557 randomly selected adults, 284 said that they were underpaid.

a. Identify the best point estimate of the *percentage* of all adults who say that they are underpaid.

b. Construct a 95% confidence interval estimate of the *percentage* of all adults who say that they are underpaid.

c. Can we safely conclude that the majority of adults say that they are underpaid?

1. a. 51.0%
   b. $46.8\% < p < 55.1\%$
   c. No, the confidence interval shows that the population percentage might be 50% or less, so we cannot safely conclude that the majority of adults say that they are underpaid.

2. Lefties The author had difficulty finding the percentage of people who write with their left hand. If we want to estimate that percentage based on survey results, how many people must we survey in order to be 99% confident that we are within two percentage points of the population percentage? Assume that we know nothing about the percentage of the population that writes with the left hand.

2. 4145 (Tech: 4147)

3. Lefties Yet Again There have been several studies conducted in an attempt to identify ways in which left-handed people are different from those who are right handed. Assume that you want to estimate the mean IQ of all left-handed adults. How many random left-handed adults must be tested in order to be 98% confident that the mean IQ of the sample group is within three IQ points of the mean IQ of all left-handed adults? Assume that σ is known to be 16.

3. 155 (Tech: 154)

4. Distributions Identify the distribution (normal, Student *t*, chi-square) that applies to each of the following situations. (If none of the three distributions is appropriate, then so state.)

a. In constructing a confidence interval of μ, you have 50 sample values and they appear to be from a population with a skewed distribution. The population standard deviation is not known.

4. a. Student *t* distribution
   b. Normal distribution
   c. None of the three distributions is appropriate.
   d. χ^2 (chi-square distribution)
   e. Normal distribution

b. In constructing a confidence interval estimate of μ, you have 50 sample values and they appear to be from a population with a skewed distribution. The population standard deviation is known to be 18.2 cm.

c. In constructing a confidence interval estimate of σ, you have 50 sample values and they appear to be from a population with a skewed distribution.

d. In constructing a confidence interval estimate of σ, you have 50 sample values and they appear to be from a population with a normal distribution.

e. In constructing a confidence interval estimate of p, you have 850 survey responses and 10% of them answered "yes" to the first question.

5. a. 543 (Tech: 542)
b. 247 (Tech: 246)
c. 543

5. Sample Size You have been hired by a college foundation to conduct a survey of graduates.

a. If you want to estimate the percentage of graduates who have made a donation to the college after graduation, how many graduates must you survey if you want 98% confidence that your percentage has a margin of error of five percentage points?

b. If you want to estimate the mean amount of all charitable contributions made by graduates, how many graduates must you survey if you want 98% confidence that your sample mean is in error by no more than \$50? (Based on results from a pilot study, assume that the standard deviation of donations by graduates is \$337.)

c. If you plan to obtain the estimates described in parts (a) and (b) with a single survey having several questions, how many graduates must be surveyed?

6. $61.5\% < p < 66.5\%$. Because the entire confidence interval is above 50%, we can safely conclude that the majority of adults consume alcoholic beverages (although such one-sided claims should be addressed using the formal methods presented in Chapter 8).

6. Alcohol Consumption In a Gallup poll, 1011 adults were asked if they consume alcoholic beverages, and 64% of them said that they did. Construct a 90% confidence interval estimate of the proportion of all adults who consume alcoholic beverages. Can we safely conclude that the majority of adults consume alcoholic beverages?

7. $-22.1 \text{ sec} < \mu < 308.1 \text{ sec}$

7. Wristwatch Accuracy Students of the author collected data measuring the accuracy of wristwatches. The times (sec) below show the discrepancy between the real time and the time indicated on the wristwatch. Negative values correspond to watches that are running ahead of the actual time. The data satisfy a loose requirement of appearing to come from a normally distributed population. Construct a 95% confidence interval estimate of the mean discrepancy for the population of wristwatches.

$$-85 \quad 325 \quad 20 \quad 305 \quad -93 \quad 15 \quad 282 \quad 27 \quad 555 \quad 570 \quad -241 \quad 36$$

8. $6.54 < \mu < 7.76$. Because women and men have some notable physiological differences, the confidence interval does not necessarily serve as an estimate of the mean white blood cell count of men.

8. White Blood Cell Counts Data Set 1 in Appendix B lists the white blood cell counts (1000 cells/μL) of 40 randomly selected women. The mean of those 40 values is 7.15 and the standard deviation is 2.28. Construct a 90% confidence interval estimate of the mean white blood cell count for the population of women. Does the confidence interval also serve as an estimate of the mean white blood cell count of men?

9. $37.5 \text{ g} < \mu < 47.9 \text{ g}$. There is 95% confidence that the limits of 37.5 g and 47.9 g contain the true mean deceleration measurement for all small cars.

9. Car Crash Tests Data Set 13 in Appendix B includes crash test measurements for small cars. The seven chest deceleration measurements have a mean of 42.7 g and a standard deviation of 5.6 g, where g is a force of gravity. Use the sample data to construct a 95% confidence interval estimate of the mean chest deceleration measurement for the population of all small cars. (Assume that the sample is a simple random sample and the measurements satisfy the loose requirement of being from a normally distributed population.) Write a brief statement that interprets the confidence interval.

10. $3.6 \text{ g} < \sigma < 12.3 \text{ g}$

10. Car Crash Tests Refer to the sample data described in Exercise 9 and construct a 95% confidence interval estimate of the population standard deviation.

Cumulative Review Exercises

Please be aware that some of the following problems may require knowledge of concepts presented in previous chapters.

Campus Aggravated Assaults. ***Colleges sometimes downplay occurrences of crimes on campus so that future applications are not affected. Listed below are the numbers of aggravated assaults on six college campuses. The listed values are randomly selected from the 56 largest urban campuses with residence halls, and the counts are for a recent year. In Exercises 1–5, use these values.***

6 4 2 7 2 12

1. Find the mean, median, and standard deviation.

1. $\bar{x} = 5.5$; median $= 5.0$; $s = 3.8$

2. Use the results from Exercise 1 with the range rule of thumb to find the range of usual values.

2. The range of usual values is from -2.1 to 13.1 (or from 0 to 13.1).

3. What is the level of measurement of these data (nominal, ordinal, interval, ratio)? Are the data continuous or discrete?

3. Ratio level of measurement; discrete data.

4. Find the sample size necessary to estimate the mean number of aggravated assaults on all college campuses so that there is 95% confidence that the sample mean is in error by no more than 2. Assume that a pilot study suggests that the numbers of aggravated assaults at all colleges have a standard deviation of 5.8.

4. 33 campuses

5. Campus Assaults The numbers of aggravated assaults on 40 large urban campuses with residence halls have a mean of 5.5 and a standard deviation of 5.8. Construct a 95% confidence interval estimate of the population mean. In this situation, what is the population?

5. $3.6 < \mu < 7.4$. The population should include only colleges of the same type as the sample, so the population consists of all large urban campuses with residence halls.

6. Normality Assessment Data Set 15 in Appendix B includes the times required for flights to taxi out for takeoff. All of the flights are American Airlines flights from New York (JFK) to Los Angeles and they all occurred in January of a recent year. The 48 taxi-out times are depicted in the histogram and normal quantile plot shown below. Based on those graphs, does it appear that the taxi-out times are from a population having a normal distribution? Give an explanation for the distribution shown. Do the taxi-out times appear to satisfy the requirements necessary for construction of a confidence interval estimate of the standard deviation of the population of all such times?

6. The graphs suggest that the population has a distribution that is skewed (to the right) instead of being normal. The histogram shows that some taxi-out times can be very long, and that can occur with heavy traffic, but little or no traffic cannot make the taxi-out time very low. There is a minimum time required, regardless of traffic conditions. Construction of a confidence interval estimate of a population standard deviation has a strict requirement that the sample data are from a normally distributed population, and the graphs show that this strict normality requirement is not satisfied.

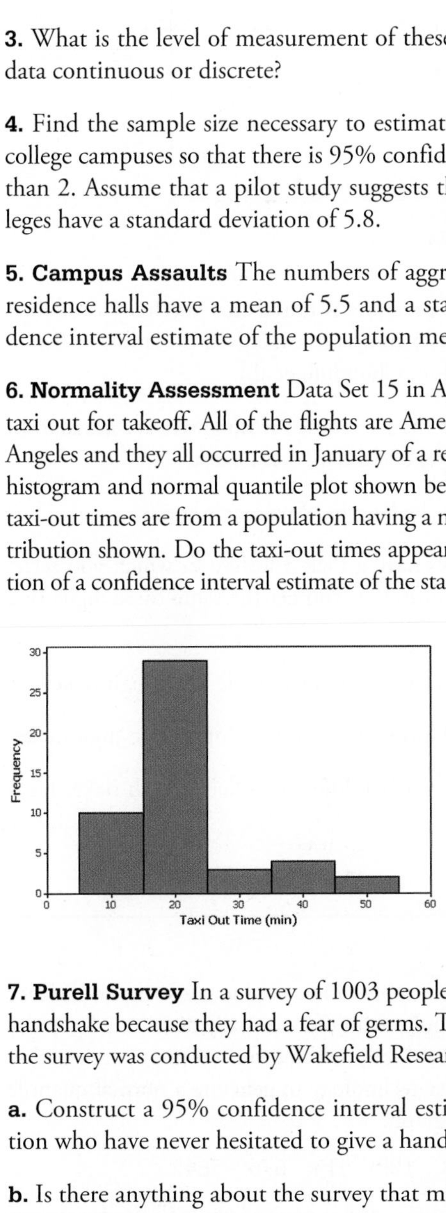

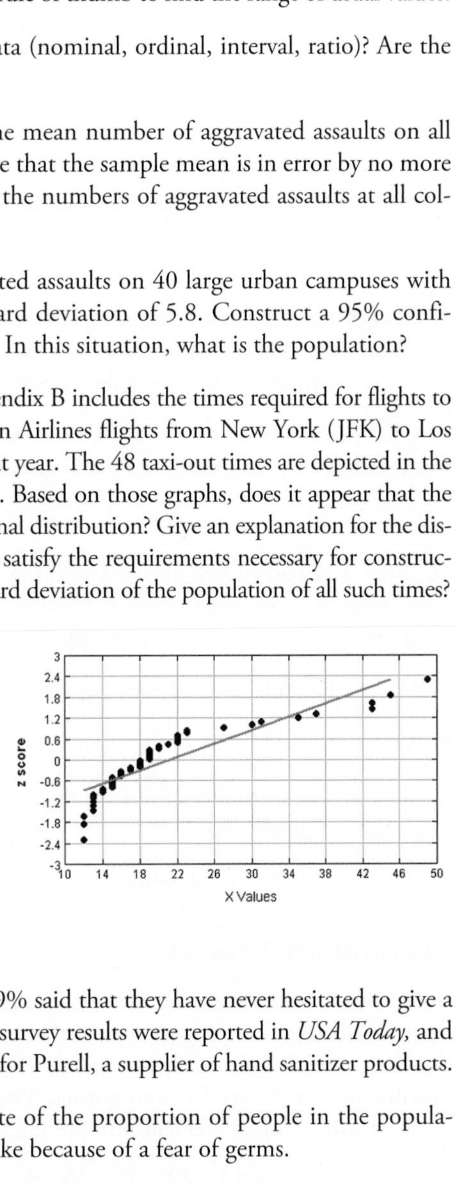

7. Purell Survey In a survey of 1003 people, 59% said that they have never hesitated to give a handshake because they had a fear of germs. The survey results were reported in *USA Today*, and the survey was conducted by Wakefield Research for Purell, a supplier of hand sanitizer products.

a. Construct a 95% confidence interval estimate of the proportion of people in the population who have never hesitated to give a handshake because of a fear of germs.

b. Is there anything about the survey that might make the results questionable?

c. If an independent pollster wanted to conduct another survey to confirm or refute the results, how many people must be surveyed? Assume that we want 90% confidence that the sample percentage is within 2.5 percentage points of the true population proportion.

7. a. $0.560 < p < 0.620$ (or $0.560 < p < 0.621$ if using $x = 592$)
b. Because the survey was about shaking hands and because it was sponsored by a supplier of hand sanitizer products, the sponsor could potentially benefit from the results, so there might be some pressure to obtain results favorable to the sponsor.
c. 1083

8. There does not appear to be a correlation between HDL and LDL cholesterol levels.

8. Cholesterol Listed below are HDL and LDL cholesterol measurements (all in mg/dL) from 10 randomly selected women (based on Data Set 1 in Appendix B). Also shown is a scatterplot of the paired data. Based on the scatterplot does there appear to be a correlation between HDL cholesterol levels and LDL cholesterol levels in women?

HDL	42	81	44	60	52	50	69	41	55	23
LDL	111	59	192	107	122	130	123	104	158	115

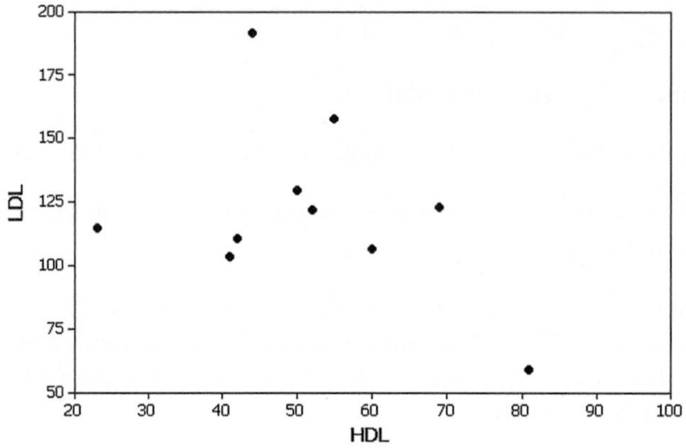

9. a. 13.35% (Tech: 13.32%). Yes, losing about 13% of the market would be a big loss.
b. 160.2 mm; 189.8 mm

9. Designing Gloves In designing gloves for women, it is found that the lengths of their hands are normally distributed with a mean of 175 mm and a standard deviation of 9 mm (based on anthropometric survey data from Gordon, Churchill, et al.).

a. What percentage of women have hands longer than 185 mm? If no gloves are made to fit hands longer than 185 mm, would a large percentage of the market be lost?

b. Find the 5th percentile and 95th percentile for the lengths of women's hands.

10. a. 1/1000
b. 999/1000
c. 0.990

10. Pick 3 Lottery Ohio and several other states have a Pick 3 lottery in which you select three digits, each between 0 and 9. Winning requires that you get the same three digits that are drawn, and they must be in the same order.

a. What is the probability of winning a Pick 3 lottery if you buy a single three-digit ticket?

b. What is the probability of *not* winning a Pick 3 lottery if you buy a single three-digit ticket?

c. What is the probability of not winning on each of 10 different tickets if each ticket has a single three-digit selection?

Technology Project

Bootstrap Resampling The *bootstrap resampling method* can be used to construct confidence intervals when the sample data suggest that the requirement of a normally distributed population is not satisfied. Listed below are 10 values randomly selected from a population with a distribution that is very far from normal. First, use technology to generate a normal quantile plot and write a brief statement interpreting that graph.

14 29 36 36 47 48 180 513 676 5642

Because methods requiring a normal distribution cannot be used, we will use **bootstrap resampling,** which has no requirements about the distribution of the population. This method typically requires a computer to repeatedly sample with replacement from the original sample. In this way, we pull the original sample up "by its own bootstraps" to simulate the original population. Using the sample data given above, construct a 95% confidence interval estimate of the population mean μ by using the following procedure.

Procedure Various technologies can be used for this bootstrap procedure. The STATDISK statistical software program is very easy to use. Enter the listed sample values in column 1 of the Data Window, then select the main menu item of **Analysis,** and select the menu item of **Bootstrap Resampling.**

a. In STATDISK, enter 500 for the number of resamplings and click on **Resample.** This will create 500 new samples, each of size 10, by selecting 10 values with replacement from the original 10 sample values given above.

b. Find the mean of each of the 500 bootstrap samples generated in part (a). In STATDISK, the means will be automatically listed in the second column of the Data Window.

c. Sort the 500 means (arrange them in order). In STATDISK, click on the main menu item of **Data,** then select **Sort Data.** Proceed to sort the means in column 2.

d. Find the percentiles $P_{2.5}$ and $P_{97.5}$ for the sorted means that result from the preceding step. ($P_{2.5}$ is the mean of the 12th and 13th values in the sorted list of means; $P_{97.5}$ is the mean of the 487th and 488th values in the sorted list of means.) Identify the resulting confidence interval by substituting the values for $P_{2.5}$ and $P_{97.5}$ in $P_{2.5} < \mu < P_{97.5}$.

Now use the same bootstrap resampling method to find a 95% confidence interval estimate of the population standard deviation σ. Use the same steps given above, but use the standard deviations of the samples. (In STATDISK, the standard deviations are automatically listed in the third column of the data window, so sort that third column and then find $P_{2.5}$ and $P_{97.5}$ for the sorted standard deviations.)

Interpretation Does the confidence interval for μ contain 1480, which is the true value of the population mean? Does the confidence interval for σ contain 2321, which is the true value of σ?

from data TO DECISION

Critical Thinking: What does the survey tell us?

Surveys have become an integral part of our lives. They directly affect us in so many ways, including public policy, the television shows we watch, the products we buy, and the political leaders we elect. Because it is so important that every citizen has the ability to interpret survey results, surveys are the focus of this project.

In a recent Pew Research Center poll, 1501 adults were randomly selected and asked this question: "From what you've read and heard, is there solid evidence that the average temperature on earth has been increasing over the past few decades, or not?" Seventy percent of the 1501 respondents answered "yes."

Analyzing the Data

1. Use the survey results to construct a 95% confidence interval estimate of the *percentage* of all adults who believe that there is solid evidence of increasing temperatures.

2. Identify the margin of error for this survey.

3. Explain why it would or would not be okay for a newspaper to make this statement: "Based on results from a recent survey, the majority of adults believe that there is solid evidence of global warming."

4. Assume that you are a newspaper reporter. Write a description of the survey results for your newspaper.

5. A common criticism of surveys is that they poll only a very small percentage of the population and therefore cannot be accurate. Is a sample of only 1501 adults taken from a population of 241,472,385 adults a sample size that is too small? Write a brief explanation of why the sample size of 1501 is or is not too small.

6. In reference to another survey, the president of a company wrote to the Associated Press about a nationwide survey of 1223 subjects. Here is what he wrote:

> When you or anyone else attempts to tell me and my associates that 1223 persons account for our opinions and tastes here in America, I get mad as hell! How dare you! When you or anyone else tells me that 1223 people represent America, it is astounding and unfair and should be outlawed.

The writer of that letter then proceeds to claim that because the sample size of 1223 people represents 120 million people, his single letter represents 98,000 (120 million divided by 1223) who share the same views. Do you agree or disagree with this claim? Write a response that either supports or refutes this claim.

Cooperative Group Activities

1. Out-of-class activity Collect sample data, and use the methods of this chapter to construct confidence interval estimates of population parameters. Here are some suggestions for parameters:

- Proportion of students at your college who can raise one eyebrow without raising the other eyebrow.
- Mean age of cars driven by statistics students and/or the mean age of cars driven by faculty.
- Mean length of words in *New York Times* editorials and mean length of words in editorials found in your local newspaper.
- Mean lengths of words in a major magazine, such as *Time*.
- Proportion of students at your college who can correctly identify the president, vice president, and secretary of state.
- Proportion of students at your college who are over the age of 18 and are registered to vote.
- Mean age of full-time students at your college.
- Proportion of motor vehicles in your region that are cars.
- Mean number of hours that students at your college study each week.
- Proportion of student cars that are painted white.

2. In-class activity Without using any measuring device, each student should draw a line believed to be 3 in. long and another line believed to be 3 cm long. Then use rulers to measure and record the lengths of the lines drawn. Find the means and standard deviations of the two sets of lengths. Use the sample data to construct a confidence interval for the length of the line estimated to be 3 in., then do the same for the length of the line estimated to be 3 cm. Do the confidence interval limits actually contain the correct length? Compare the results. Do the estimates of the 3-in. line appear to be more accurate than those for the 3-cm line?

3. In-class activity Assume that a method of gender selection can affect the probability of a baby being a girl, so that the probability becomes 1/4. Each student should simulate 20 births by drawing 20 cards from a shuffled deck. Replace each card after it has been drawn, then reshuffle. Consider the hearts to be girls and consider all other cards to be boys. After making 20 selections and recording the "genders" of the babies, construct a confidence interval estimate of the proportion of girls. Does the result appear to be effective in identifying the true value of the population proportion? (If decks of cards are not available, use some other way to simulate the births, such as using the random number generator on a calculator or using digits from phone numbers or Social Security numbers.)

4. Out-of-class activity Groups of three or four students should go to the library and collect a sample consisting of the ages of books (based on copyright dates). Plan and describe the sampling procedure, execute the sampling procedure, then use the results to construct a confidence interval estimate of the mean age of all books in the library.

5. In-class activity Each student should write an estimate of the age of the current president of the United States. All estimates should be collected and the sample mean and standard deviation should be calculated. Then use the sample results to construct a confidence interval. Do the confidence interval limits contain the correct age of the president?

6. In-class activity A class project should be designed to conduct a test in which each student is given a taste of Coke and a taste of Pepsi. The student is then asked to identify which sample is Coke. After all of the results are collected, analyze the claim that the success rate is better than the rate that would be expected with random guesses.

7. In-class activity Each student should estimate the length of the classroom. The values should be based on visual estimates, with no actual measurements being taken. After the estimates have been collected, construct a confidence interval, then measure the length of the room. Does the confidence interval contain the actual length of the classroom? Is there a "collective wisdom," whereby the class mean is approximately equal to the actual room length?

8. In-class activity Divide into groups of three or four. Examine a sample of different issues from a current magazine and find the proportion of pages that include advertising. Based on the results, construct a 95% confidence interval estimate of the percentage of all such pages that have advertising. Compare results with other groups.

9. In-class activity Divide into groups of two. First find the sample size required to estimate the proportion of times that a coin turns up heads when tossed, assuming that you want 80% confidence that the sample proportion is within 0.08 of the true population proportion. Then toss a coin the required number of times and record your results. What percentage of such confidence intervals should actually contain the true value of the population proportion, which we know is $p = 0.5$? Verify this last result by comparing your confidence interval with the confidence intervals found in other groups.

10. Out-of-class activity Identify a topic of general interest and coordinate with all members of the class to conduct a survey. Instead of conducting a "scientific" survey using sound principles of random selection, use a convenience sample consisting of respondents who are readily available, such as friends, relatives, and other students. Analyze and interpret the results. Identify the population. Identify the shortcomings of using a convenience sample, and try to identify how a sample of subjects randomly selected from the population might be different.

11. Out-of-class activity Each student should find an article in a professional journal that includes a confidence interval of the type discussed in this chapter. Write a brief report describing the confidence interval and its role in the context of the article.

8 Hypothesis Testing

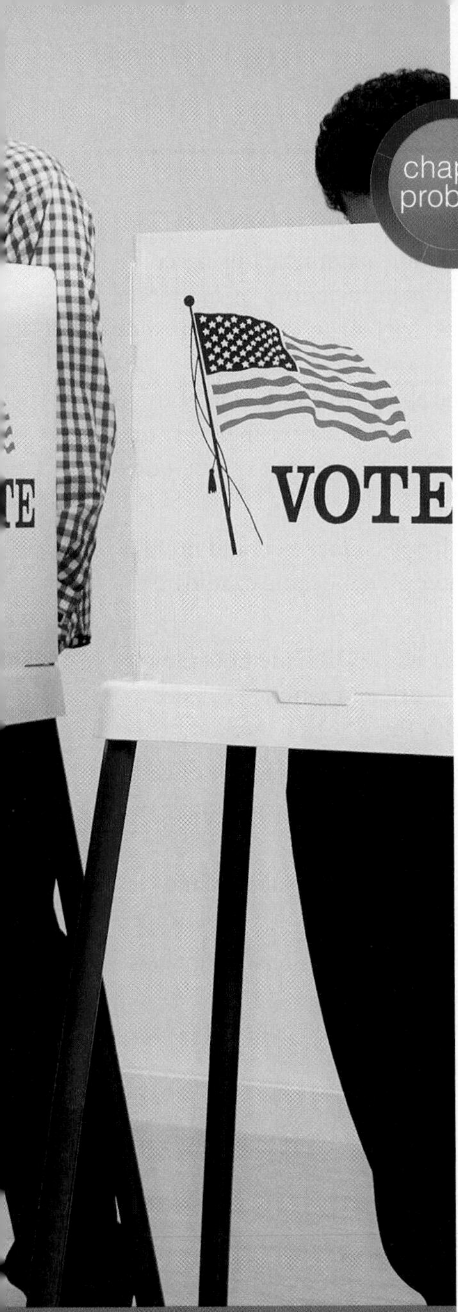

8

chapter problem

Forensic Statistics: Did the county clerk cheat on election ballots?

Some people wrongly believe that the legal profession is among those that do not require any knowledge of statistical methods. In fact, statistics often plays a central role in resolving legal disputes. Consider the case of Nicholas Caputo, the county clerk in Essex County, New Jersey. The clerk was responsible for arranging names for ballot lines used in elections. He was supposed to use a method of random selection for determining the order of the names. This was important, because a candidate can have an advantage if listed first on the ballot. Caputo made his selections using a procedure that was not observed by witnesses. Among 41 different ballots, Democrats won the first line 40 times. Republicans filed a lawsuit and made the claim that instead of using a method of random selection, Caputo was using a method that favored Democrats. Although the results appear to be lopsided and nonrandom, Caputo denied that he was rigging the results to favor Democratic candidates.

A central and key question is this: How likely is it that anyone would get results as extreme as 40 Democrats in 41 selections, assuming that each election involves the random selection of a Democrat or a Republican? This is the same likelihood as getting results as extreme as 40 heads when tossing a coin 41 times.

We must take into account this subtle but important interpretation: Instead of finding the probability of getting *exactly* 40 Democrats in 41 random selections, we need the probability of getting 40 Democrats *or any result that is more extreme*—so we need to find the probability of getting *at least* 40 Democrats in 41 random selections. (See "Identifying *Unusual* Results with Probabilities" in Section 5-2.)

This situation can be addressed by using the method of *hypothesis testing* that is presented in this chapter. We have the Republican claim that $p > 0.5$, which is the symbolic form of the verbal claim that the clerk used a selection process that favored Democrats so that Democrats have more than a 0.5 probability of being selected. We have the clerk's defense that $p = 0.5$, so that his probability of selecting a Democrat is 0.5. This chapter will present the standard methods for testing such claims.

Note to Instructor

Organization Change: In this new edition, Section 8-4 ("Testing a Claim about a Mean") is partitioned into Part 1 (for σ not known) and Part 2 (for σ known). The main focus in Section 8-4 is using the Student *t* distribution for testing a claim about a population mean. Exercises 29–32 are the only exercises in which σ is assumed to be known. The previous edition of this book included a Section 8-4 ("Testing a Claim about a Mean: σ Known") and a Section 8-5 ("Testing a Claim about a Mean: σ Not Known"), but this edition combines these two topics into one section (Section 8-4). It follows that the section "Testing a Claim about a Standard Deviation or Variance" is now Section 8-5 (instead of Section 8-6, as it was in the preceding edition).

Here is the main motivation for this change: In the real world of professional statisticians and professional journals and reports, it is extremely rare that we want to test a claim about a population mean but we somehow know the value of the population standard deviation σ. In the past, there were sound reasons for starting with the "known σ" situation, including the fact that finding critical values of *t* was often difficult without technology. Technology has now empowered us to find critical *t* values for any sample size *n*, so the situation of a known σ is now much less important. Given the time constraints in the typical introductory statistics course, it is now wise to devote little time and effort to the unrealistic "known σ" case.

Note to Instructor

Every introductory statistics course should include at least Sections 8-2, 8-3, and 8-4. Section 8-5 can be omitted if necessary because of time considerations.

Note that proportions are covered first. Proportions are very common in the media, they typically involve interesting applications, and they are easier to work with so that basic concepts can be presented with fewer complications.

Section 8-2 is partitioned into two parts: Part 1 ("Basic Concepts of Hypothesis Testing") and Part 2 ("Beyond the Basics of Hypothesis Testing: The *Power* of a Test"). It is somewhat difficult to find the power of a test, so include Part 2 only if you have sufficient time and the topic is suitable for your students.

8-1 Review and Preview

By providing methods for estimating values of population parameters using confidence intervals, Chapter 7 introduced one of the two main activities of inferential statistics. This chapter introduces the second major activity of inferential statistics: Testing claims made about population parameters.

The main objective of this chapter is to develop the ability to conduct hypothesis tests for claims made about a population proportion p, a population mean μ, or a population standard deviation σ. Here are examples of hypotheses that can be tested by the procedures in this chapter:

- **Law:** Republicans claimed in a lawsuit that a New Jersey county clerk did not use a required method of random selection when he chose a Democratic candidate to be first on the ballot in 40 out of 41 elections.
- **Genetics:** The Genetics & IVF Institute claims that its XSORT method allows couples to increase the probability of having a baby girl, and sample evidence consists of 879 girls among 945 couples treated with the XSORT method.
- **Health:** It is often claimed that the mean body temperature is 98.6°F, and we can test that claim using Data Set 3 in Appendix B, which includes a sample of 106 body temperatures with a mean of 98.2°F.
- **Business:** A newspaper cites a PriceGrabber.com survey of 1631 subjects and claims that the majority of consumers have heard of the Kindle as an e-book reader.
- **Quality Control:** When new equipment is used to manufacture aircraft altimeters, the new altimeters are better because the variation in the errors is reduced so that the readings are more consistent. (In many industries, the quality of goods and services can often be improved by reducing variation.)

8-2 Basics of Hypothesis Testing

Key Concept In this section we present the general components of a formal hypothesis test. In Part 1 we discuss the basic concepts of hypothesis testing. Because these concepts are used in the following sections and chapters, we should be able to do the following:

- Identify the null hypothesis and alternative hypothesis from a given claim, and express both in symbolic form.
- Calculate the value of the test statistic, given a claim and sample data.
- Choose the sampling distribution that is relevant.
- Either find the *P*-value or identify the critical value(s).
- State the conclusion about a claim in simple and nontechnical terms.

In Part 2 we describe the *power* of a hypothesis test.

Part 1: Basic Concepts of Hypothesis Testing

Because hypothesis testing is the main focus of this chapter, we begin with two very basic definitions.

DEFINITIONS

In statistics, a **hypothesis** is a claim or statement about a property of a population.

A **hypothesis test** (or **test of significance**) is a procedure for testing a claim about a property of a population.

The "property" referred to in the preceding definitions is often the value of a population parameter, so here are some examples of typical hypotheses (or claims):

- $\mu < 98.6°F$ The mean body temperature of humans is less than 98.6°F.
- $p > 0.5$ The XSORT method of gender selection increases the probability that a baby will be born a girl, so the probability of a girl is greater than 0.5.
- $\sigma = 15$ The population of college students has IQ scores with a standard deviation equal to 15.

Example 1 Testing the Claim That the XSORT Gender-Selection Method Is Effective

Assume that 100 babies are born to 100 couples treated with the XSORT method of gender selection that is claimed to make girls more likely. If 58 of the 100 babies are girls, test the claim that "with the XSORT method, the proportion of girls is greater than the proportion of 0.5 that occurs without any treatment." Using p to denote the proportion of girls born with the XSORT method, the claim is that $p > 0.5$.

The big picture In Example 1 we see that getting 58 girls in 100 births is more than the 50 girls that we would expect with no treatment or an ineffective treatment. But is 58 girls high enough to justify the conclusion that the XSORT method is effective? The method of hypothesis testing allows us to answer that question. We will see that without an effective treatment, there is a 0.0548 probability of getting 58 *or more* girls. Because that probability is not small, such as 0.05 or less, we will conclude that it is easy to get 58 girls in 100 births by random chance, so 58 girls is not quite high enough to justify the conclusion that the XSORT method is effective. (The actual XSORT results are more extreme, so in reality it does appear that the XSORT method is effective.)

Using technology It is easy to obtain hypothesis-testing results using technology. The accompanying screen displays show results from four different technologies, so we can use computers or calculators to do all of the computational heavy lifting. Examining the four screen displays, we see some common elements. They all display a "test statistic" of $z = 1.60$, and they all include a "P-value" of 0.055 (rounded). These two results are important, but *understanding* the hypothesis-testing procedure is critically important. Focus on *understanding* how the hypothesis-testing procedure

STATDISK

```
Claim: p > p(hyp)

Sample proportion: 0.58
Test Statistic, z: 1.6000
Critical z:        1.6449
P-Value:           0.0548
```

MINITAB

```
Test of p = 0.5 vs p > 0.5

                              95% Lower
Sample  X    N  Sample p      Bound  Z-Value  P-Value
1      58  100  0.580000   0.498817     1.60    0.055
```

TI-83/84 PLUS

```
1-PropZTest
 prop>.5
 z=1.6
 p=.0547992894
 p̂=.58
 n=100
```

STATCRUNCH

Hypothesis test results:
p : proportion of successes for population
H_0 : p = 0.5
H_A : p > 0.5

Proportion	Count	Total	Sample Prop.	Std. Err.	Z-Stat	P-value
p	58	100	0.58	0.05	1.6	0.0548

Aspirin Not Helpful for Geminis and Libras

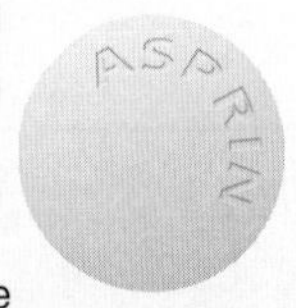

Physician Richard Peto submitted an article to *Lancet,* a British medical journal. The article showed that patients had a better chance of surviving a heart attack if they were treated with aspirin within a few hours of their heart attacks. *Lancet* editors asked Peto to break down his results into subgroups to see if recovery worked better or worse for different groups, such as males or females. Peto believed that he was being asked to use too many subgroups, but the editors insisted. Peto then agreed, but he supported his objections by showing that when his patients were categorized by signs of the zodiac, aspirin was useless for Gemini and Libra heart-attack patients, but aspirin is a lifesaver for those born under any other sign. This shows that when conducting multiple hypothesis tests with many different subgroups, there is a very large chance of getting some wrong results.

Note to Instructor
It is worth taking a few minutes to discuss Example 1 and its implications. Comment that the relevant probability is P(58 or more girls), not P(exactly 58 girls). Review the subsection of "Using Probabilities to Determine When Results Are Unusual" in Section 5-2.

works and learn the associated terminology. Only then will results from technology make sense.

The basic idea underlying the hypothesis-testing procedure is based on the rare event rule first presented in Section 4-1. Let's review that rule before proceeding.

Rare Event Rule for Inferential Statistics

If, under a given assumption, the probability of a particular observed event is extremely small, we conclude that the assumption is probably not correct.

Following this rule, we test a claim by analyzing sample data in an attempt to choose between the following two explanations:

1. The sample results could easily occur by chance.
   Example: In testing the XSORT gender-selection method that is supposed to make babies more likely to be girls, the result of 52 girls in 100 births is greater than 50%, but 52 girls could easily occur by chance, so there is not sufficient evidence to conclude that the XSORT method is effective.
2. The sample results are not likely to occur by chance.
   Example: In testing the XSORT gender-selection method that is supposed to make babies more likely to be girls, the result of 95 girls in 100 births is greater than 50%, and 95 girls is so extreme that it could *not* easily occur by chance, so there is sufficient evidence to conclude that the XSORT method is effective.

Figures 8-1 and 8-2 summarize the procedures used in two slightly different methods for conducting a formal hypothesis test. We will proceed to conduct a formal test of the claim from Example 1 that $p > 0.5$. In testing that claim, we will use the sample data consisting of 58 girls in 100 births.

Steps 1, 2, 3: Use the Claim to Create a Null Hypothesis and an Alternative Hypothesis

Objective

Identify the *null hypothesis* and *alternative hypothesis* so that the formal hypothesis test includes these standard components that are used often in many different disciplines.

Null Hypothesis (denoted by H_0)

Statement that the value of a population parameter (such as proportion, mean, or standard deviation) is *equal to* some claimed value. (The term *null* is used to indicate *no* change or no effect or no difference.) We test the null hypothesis directly in the sense that we assume (or pretend) it is true and reach a conclusion to either reject it or fail to reject it.

Example: Here is an example of a null hypothesis involving a proportion: H_0: $p = 0.5$.

Alternative Hypothesis (denoted by H_1 or H_a or H_A)

Statement that the parameter has a value that somehow differs from the null hypothesis. For the methods of this chapter, the symbolic form of the alternative hypothesis must use one of these symbols: $<, >, \neq$.

Example: Here are different examples of alternative hypotheses involving proportions:

$$H_1: p > 0.5 \quad H_1: p < 0.5 \quad H_1: p \neq 0.5$$

The *original claim* could become the null hypothesis (as in a claim that $p = 0.5$), it could become the alternative hypothesis (as in the claim that $p > 0.5$), or it might not be either the null hypothesis or the alternative hypothesis (as in the claim that $p \geq 0.5$).

Figure 8-1 ***P*-Value Method**

P-Value Method

Start

1. Identify the specific claim or hypothesis to be tested, and put in symbolic form.
2. Give the symbolic form that must be true when the original claim is false.
3. Of the two symbolic expressions obtained so far, let the alternative hypothesis H_1 be the one not containing equality, so that H_1 uses the symbol > or < or ≠. Let the null hypothesis H_0 be the symbolic expression that the parameter equals the fixed value being considered.
4. Select the significance level α based on the seriousness of a type 1 error. Make α small if the consequences of rejecting a true H_0 are severe. The values of 0.05 and 0.01 are very common.
5. Identify the statistic that is relevant to this test and determine its sampling distribution (such as normal, t, chi-square).
6. Find the test statistic and find the *P*-value (see Figure 8-5). Draw a graph and show the test statistic and *P*-value.
7. Reject H_0 if the *P*-value is less than or equal to the significance level α. Fail to reject H_0 if the *P*-value is greater than α.
8. Restate this previous decision in simple, nontechnical terms, and address the original claim.

Stop

Figure 8-2 **Critical Value Method**

Critical Value Method

Start

1. Identify the specific claim or hypothesis to be tested, and put it in symbolic form.
2. Give the symbolic form that must be true when the original claim is false.
3. Of the two symbolic expressions obtained so far, let the alternative hypothesis H_1 be the one not containing equality, so that H_1 uses the symbol > or < or ≠. Let the null hypothesis H_0 be the symbolic expression that the parameter equals the fixed value being considered.
4. Select the significance level α based on the seriousness of a type 1 error. Make α small if the consequences of rejecting a true H_0 are severe. The values of 0.05 and 0.01 are very common.
5. Identify the statistic that is relevant to this test and determine its sampling distribution (such as normal, t, chi-square).
6. Find the test statistic, the critical values, and the critical region. Draw a graph and include the test statistic, critical value(s), and critical region.
7. Reject H_0 if the test statistic is in the critical region. Fail to reject H_0 if the test statistic is not in the critical region.
8. Restate this previous decision in simple, nontechnical terms, and address the original claim.

Stop

Confidence Interval Method

Construct a confidence interval with a confidence level selected as in Table 8-1.

Because a confidence interval estimate of a population parameter contains the likely values of that parameter, reject a claim that the population parameter has a value that is not included in the confidence interval.

Table 8-1 Confidence Level for Confidence Interval

		Two-Tailed Test	One-Tailed Test
Significance Level for Hypothesis Test	0.01	99%	98%
	0.05	95%	90%
	0.10	90%	80%

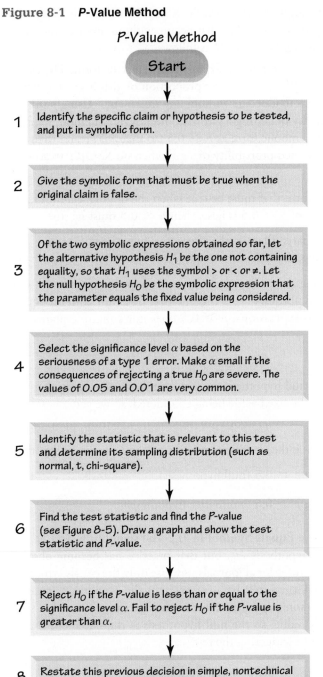

Note to Instructor
For notation, note that the null hypothesis will be expressed in terms of equality only, so expressions such as $p \leq 0.5$ or $p \geq 0.5$ will not be used for the null hypothesis. (Almost all professional journals use *only equality* for expressions of null hypotheses.) The notation H_0 is used almost universally, but it is common to see alternative hypotheses expressed as H_1 or H_a. This book uses H_1 for alternative hypotheses.

Comment that students will encounter several new terms in this section. These special terms are not unique to this book, or statistics books in general. Instead, these terms are commonly used by medical researchers, manufacturers, psychologists, educators, and many other people who use methods of statistics in their professions. When students learn these terms, they are becoming familiar with language used in many different disciplines.

Given the claim from Example 1 that "with the XSORT method, the proportion of girls is greater than the proportion of 0.5 that occurs without any treatment," we apply Steps 1, 2, and 3 in Figures 8-1 and 8-2 as follows.

Step 1: Identify the claim to be tested and express it in symbolic form. The claim is that "with the XSORT method, the proportion of girls is greater than the proportion of 0.5 that occurs without any treatment." Using p to denote the probability of getting a girl with the XSORT method, we express that claim in symbolic form as $p > 0.5$ (so that p is greater than 0.5, where 0.5 is the assumed probability of a girl when no XSORT treatment is used).

Step 2: Give the symbolic form that must be true when the original claim is false. If the original claim of $p > 0.5$ is false, then $p \leq 0.5$ must be true.

Step 3: This step is in two parts: Identify the alternative hypothesis H_1 and identify the null hypothesis H_0.

- Identify H_1: Using the two symbolic expressions $p > 0.5$ and $p \leq 0.5$, the alternative hypothesis H_1 is the one that does not contain equality. Of those two expressions, $p > 0.5$ does not contain equality, so we get

$$H_1: p > 0.5$$

- Identify H_0: The null hypothesis H_0 is the symbolic expression that the parameter *equals* the fixed value being considered, so we get

$$H_0: p = 0.5$$

The result of the first three steps is the identification of the null and alternative hypothesis:

$$H_0: p = 0.5 \text{ (null hypothesis)}$$

$$H_1: p > 0.5 \text{ (alternative hypothesis)}$$

Note to Instructor
Some mathematicians acknowledge only the research role of hypothesis testing and insist that *every* claim must be stated as an alternative hypothesis. However, many textbooks and many real applications reflect the comments made here about testing the validity of someone else's claim. Specifically, some of the claims we consider will become null hypotheses, some become alternative hypotheses, and some claims are neither null hypotheses nor alternative hypotheses. Some textbooks completely dodge the issue by wording all claims so that they all become alternative hypotheses, but that approach is not used in this book because it is far too contrived.

Note about Always Using the Equals Symbol in H_0: The symbols $\leq$ and $\geq$ were sometimes used in the null hypothesis H_0, but that practice has become obsolete. Professional statisticians and professional journals now use only the equals symbol for equality. We conduct the hypothesis test by assuming that the proportion, mean, or standard deviation is *equal to* some specified value so that we can work with a single distribution having a specific value, so the null hypothesis is the statement that we assume true for the purpose of conducting the test.

Note about Forming Your Own Claims (Hypotheses): If you are conducting a study and want to use a hypothesis test to *support* your claim, the claim must be worded so that it becomes the alternative hypothesis (and can be expressed using only the symbols $<$, $>$, or $\neq$). You can never support a claim that some parameter is *equal to* a specified value.

Step 4: Select the Significance Level α

Objective

Identify the value of the significance level α so that we have a specific criterion for distinguishing between sample results that could easily occur by chance and sample results that are unlikely to occur by chance.

The **significance level** α is the probability of making the mistake of rejecting the null hypothesis when it is true. This is the same α introduced in Section 7-2, where we defined the confidence level for a confidence interval to be the probability $1 - \alpha$. Common choices for α are 0.05, 0.01, and 0.10, with 0.05 being most common.

Step 5: Identify the Statistic Relevant to the Test and Determine Its Sampling Distribution (such as normal, *t*, or χ^2)

Objective

Based on the sample statistic that is relevant to the claim being tested, identify the sampling distribution of the statistic (such as normal, Student *t*, or χ^2) so that the correct sampling distribution can be used in Step 6.

For this chapter, verify that any requirements are satisfied, then use Table 8-2 to choose the correct sampling distribution.

Example: The claim $p > 0.5$ is a claim about the population proportion p, so use the normal distribution provided that the requirements are satisfied. (With $n = 100$, $p = 0.5$, and $q = 0.5$ as in Example 1, $np \geq 5$ and $nq \geq 5$ are both true.)

Table 8-2

Parameter	Sampling Distribution	Requirements	Test Statistic
Proportion p	Normal (z)	$np \geq 5$ and $nq \geq 5$	$z = \frac{\hat{p} - p}{\sqrt{\frac{pq}{n}}}$
Mean μ	t	σ not known and normally distributed population *or* σ not known and $n > 30$	$t = \frac{\bar{x} - \mu}{\frac{s}{\sqrt{n}}}$
Mean μ	Normal (z)	σ known and normally distributed population *or* σ known and $n > 30$	$z = \frac{\bar{x} - \mu}{\frac{\sigma}{\sqrt{n}}}$
St. dev. σ or variance σ^2	χ^2	Strict requirement: normally distributed population	$\chi^2 = \frac{(n-1)s^2}{\sigma^2}$

Note to Instructor

When conducting a test of some hypothesis or claim, it is important to use the correct distribution and the correct expression of the test statistic. Point out the four different test statistics in Table 8-2, and comment that they are on the detachable Formula/Table card included with this book. If students are allowed to use that Formula/Table card on tests, there is no need to memorize formulas. They can simply refer to the card to determine which test statistic is suitable.

Step 6: Find the Value of the Test Statistic, Then Find Either the *P*-Value or the Critical Value(s)

Objective

First transform the relevant sample statistic to a standardized score called the *test statistic*. Then find the *P-value* that can be used to make a decision about the null hypothesis, or find *critical values* that can be used with the test statistic in making a decision about the null hypothesis.

The **test statistic** is a value used in making a decision about the null hypothesis. It is found by converting the sample statistic (such as the sample proportion $\hat{p}$, the sample mean $\bar{x}$, or the sample standard deviation s) to a score (such as z, t, or χ^2) with the assumption that the null hypothesis is true. In this chapter we use the test statistics in the last column of Table 8-2.

Example: From Example 1 we have a claim made about the population proportion p, we have $n = 100$ and $x = 58$, so $\hat{p} = x/n = 0.58$. Because we also have the null hypothesis of H_0: $p = 0.5$, we are working with the assumption that $p = 0.5$, and it follows that $q = 0.5$. Using $n = 100$, $\hat{p} = 0.58$, $p = 0.5$, and $q = 0.5$, we can evaluate the test statistic as shown below. (See that the result of $z = 1.60$ is included in each of the previous displays from technology, so technology can do this calculation for us.)

$$z = \frac{\hat{p} - p}{\sqrt{\frac{pq}{n}}} = \frac{0.58 - 0.5}{\sqrt{\frac{(0.5)(0.5)}{100}}} = 1.60$$

Note to Instructor
Students usually like the hint that the symbol $<$ suggests a left-tailed test and the symbol $>$ suggests a right-tailed test. Many will want to use only that criterion, but they should also *understand* the reasoning used to determine whether a test is left-tailed, right-tailed, or two-tailed. They should not rely blindly on a rote process that is devoid of logical reasoning or common sense.

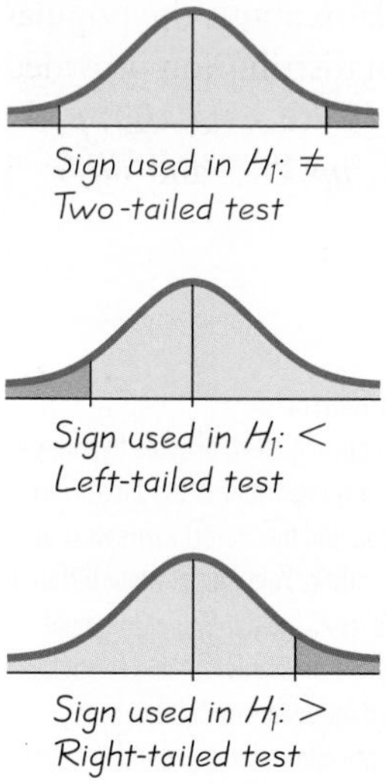

Figure 8-3 Critical Region in Two-Tailed, Left-Tailed, and Right-Tailed Tests

Types of Hypothesis Tests: Two-Tailed, Left-Tailed, Right-Tailed

The test statistic alone usually does not give us enough information to make a decision about the claim being tested. For that decision, we can use either the *P*-value approach summarized in Figure 8-1 or the critical value approach summarized in Figure 8-2. Both approaches require that we first determine whether our hypothesis test is two-tailed, left-tailed, or right-tailed.

The **critical region** (or **rejection region**) corresponds to the values of the test statistic that cause us to reject the null hypothesis. Depending on the claim being tested, the critical region could be in the two extreme tails, it could be in the left tail, or it could be in the right tail.

- **Two-tailed test:** The critical region is in the two extreme regions (tails) under the curve (as in the top graph in Figure 8-3).
- **Left-tailed test:** The critical region is in the extreme left region (tail) under the curve (as in the middle graph in Figure 8-3).
- **Right-tailed test:** The critical region is in the extreme right region (tail) under the curve (as in the bottom graph in Figure 8-3).

HINT To determine whether a test is two-tailed, left-tailed, or right-tailed, look at the alternative hypothesis and identify the region that supports that alternative hypothesis and conflicts with the null hypothesis. A useful check is summarized in Figure 8-3. *See that the inequality sign in H_1 points in the direction of the critical region.* The symbol $\neq$ is sometimes expressed in programming languages as <>, and this reminds us that an alternative hypothesis such as $p \neq 0.5$ corresponds to a two-tailed test.

Example: With H_0: $p = 0.5$ and H_1: $p > 0.5$, we reject the null hypothesis and support the alternative hypothesis only if the sample proportion is greater than 0.5 by a significant amount, so the hypothesis test in this case is *right-tailed.*

Interpreting the Test Statistic: Using the *P*-Value or Critical Value

After determining whether the hypothesis test is two-tailed, left-tailed, or right-tailed, we can proceed with either the *P*-value approach (summarized in Figure 8-1) or the critical value approach (summarized in Figure 8-2). Because technology typically provides a *P*-value in a hypothesis test, the *P*-value method is now much more common than the method based on critical values.

***P*-Value (or *p*-Value or Probability Value) Method** Find the *P*-value, which is the probability of getting a value of the test statistic that is *at least as extreme* as the one representing the sample data, assuming that the null hypothesis is true. To find the *P*-value, first find the area beyond the test statistic, then use the procedure given in Figure 8-4. That procedure can be summarized as follows:

Critical region in the left tail:	*P*-value =	area to the *left* of the test statistic
Critical region in the right tail:	*P*-value =	area to the *right* of the test statistic
Critical region in two tails:	*P*-value =	*twice* the area in the tail beyond the test statistic

EXAMPLE The test statistic of $z = 1.60$ has an area of 0.0548 to its right, so a right-tailed test with test statistic $z = 1.60$ has a *P*-value of 0.0548. (See the different technology

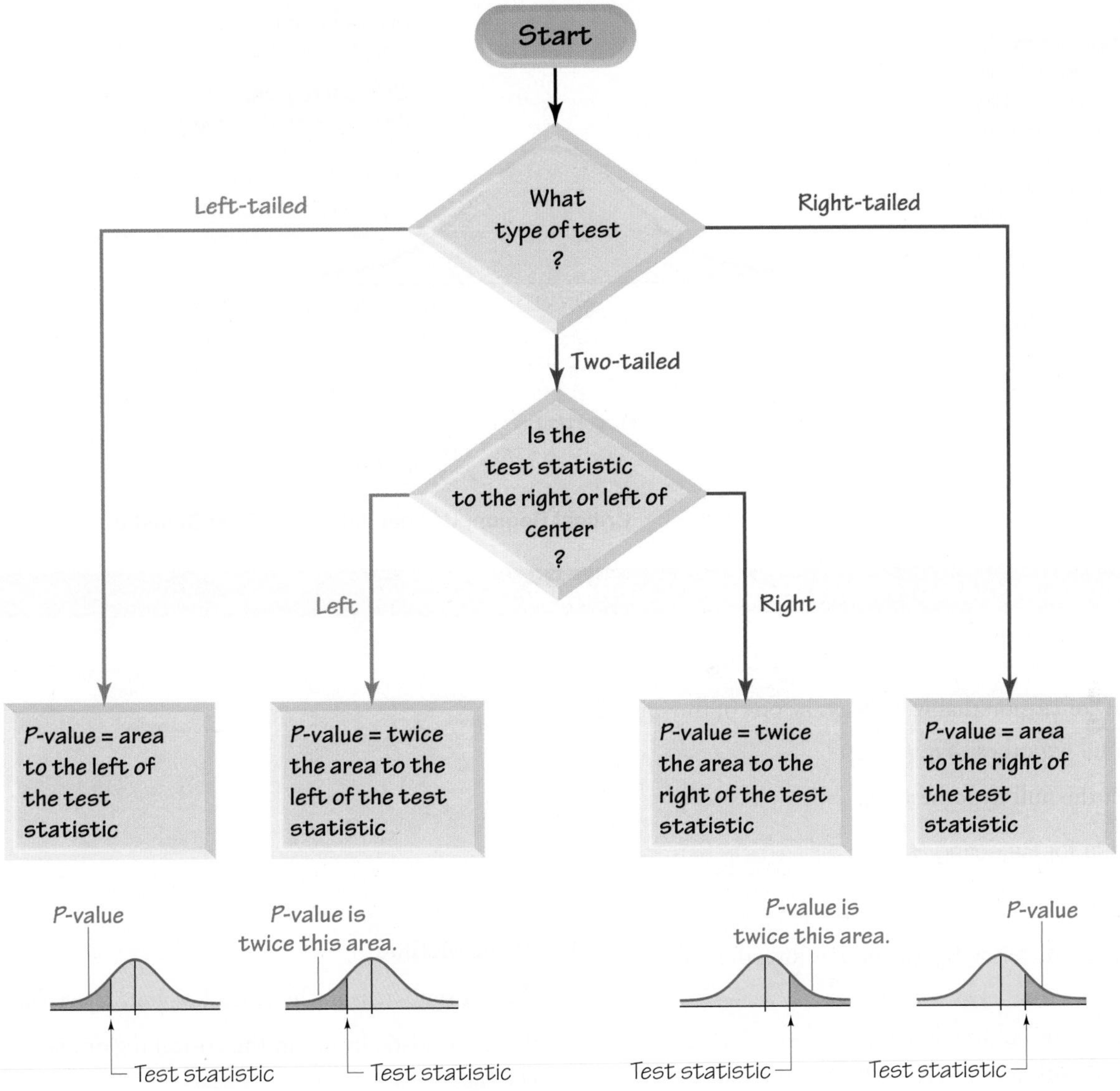

Figure 8-4 Procedure for Finding *P*-Values

displays given earlier, and note that each of them provides the same *P*-value of 0.055 after rounding.)

> **CAUTION** Don't confuse a *P*-value with a proportion *p*. Know this distinction:
>
> ***P*-value** = probability of a test statistic at least as extreme as the one obtained
>
> ***p*** = population proportion

Critical Value Method With the critical value method (or **traditional method**), we find the **critical value(s),** which separates the critical region (where we reject the null hypothesis) from the values of the test statistic that do not lead to rejection of the null hypothesis. Critical values depend on the nature of the null hypothesis, the sampling distribution, and the significance level α.

EXAMPLE The critical region in Figure 8-5 is shaded in red. Figure 8-5 shows that with $\alpha = 0.05$, the critical value is $z = 1.645$.

Note to Instructor

Change in Terminology:

The *critical value method* described in this section was referred to as the *traditional method* in the previous edition. The "critical value method" is easier to understand and remember than the "traditional" method. It's too easy to forget tradition.

Note to Instructor
Finding *P*-values can be tricky, especially with two-tailed tests. Comment that Figure 8-4 summarizes the procedure for finding *P*-values, and that figure is on the detachable Formula/Table card included with this book. It will be helpful to demonstrate the use of Figure 8-4 with a few examples in class.

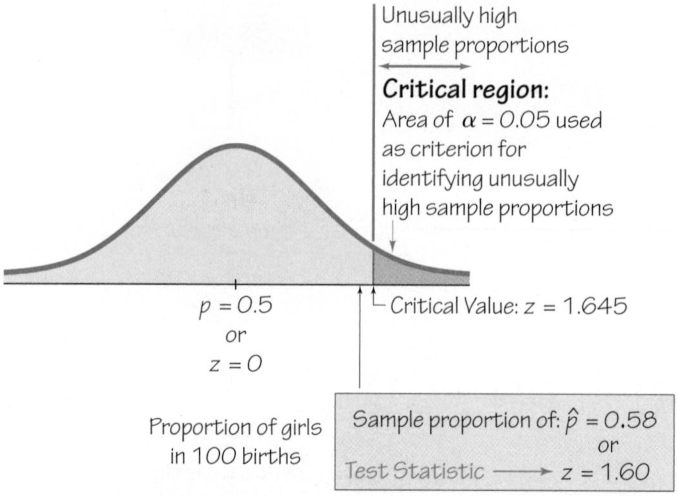

Figure 8-5 Critical Region, Critical Value, and Test Statistic

Step 7: Make a Decision: Reject H_0 or Fail to Reject H_0

Objective

Form an initial conclusion that will always be one of the following:

1. Reject the null hypothesis H_0.
2. Fail to reject the null hypothesis H_0.

Decision Criteria for Rejecting H_0

***P*-value Method:**

- If *P*-value $\leq \alpha$, reject H_0. (If the *P* is low, the null must go.)
- If *P*-value $> \alpha$, fail to reject H_0.

Example: With significance level $\alpha = 0.05$ and *P*-value $= 0.0548$, we have *P*-value $> \alpha$, so fail to reject H_0.

Critical Value Method:

- If the test statistic is in the critical region, reject H_0.
- If the test statistic in not in the critical region, fail to reject H_0.

Example: With test statistic $z = 1.60$ and the critical region from $z = 1.645$ to infinity, the test statistic does not fall within the critical region, so fail to reject H_0.

Step 8: Restate the Decision Using Simple and Nontechnical Terms

Objective

Without using technical terms not understood by most people, state a final conclusion that addresses the original claim with wording that can be understood by those without knowledge of statistical procedures.

Example: There is not sufficient evidence to support the claim that the XSORT method is effective in increasing the probability that a baby will be born a girl.

Wording the Final Conclusion For help in wording the final conclusion, refer to Table 8-3, which lists the four possible circumstances and their corresponding conclusions. Note that only the first case leads to wording indicating *support* for the original conclusion. If you want to support some claim, state it in such a way that it becomes the alternative hypothesis, and then hope that the null hypothesis gets rejected.

Table 8-3 Wording of the Final Conclusion

Condition	Conclusion
Original claim does not include equality, and you reject H_0.	"There is sufficient evidence to *support* the claim that . . . (original claim)."
Original claim does not include equality, and you fail to reject H_0.	"There is not sufficient evidence to support the claim that . . . (original claim)."
Original claim includes equality, and you reject H_0.	"There is sufficient evidence to warrant *rejection* of the claim that . . . (original claim)."
Original claim includes equality, and you fail to reject H_0.	"There is not sufficient evidence to warrant rejection of the claim that . . . (original claim)."

Note to Instructor
Clearly state that *final* conclusions of "reject the null hypothesis" or "fail to reject the null hypothesis" are not acceptable because they mean nothing to most people. The final statement should address the original claim, and it should not involve technical terms, such as "null hypothesis."

Students typically have some degree of difficulty with the correct statement of final conclusions. Stress the importance of precise wording of the final conclusion. Differences between terms such as "support" and "fail to reject" are very important. Show how Table 8-3 can be used to form the wording of the final conclusion.

Also, some students have trouble clearly understanding the meaning of "fail to reject the null hypothesis." You might also use "don't reject" instead of "fail to reject."

Accept/Fail to Reject A few textbooks continue to say "accept the null hypothesis" instead of "fail to reject the null hypothesis." The term *accept* is misleading, because it implies incorrectly that the null hypothesis has been proved, but we can never prove a null hypothesis. The phrase *fail to reject* says more correctly that the available evidence isn't strong enough to warrant rejection of the null hypothesis. In this text we use the terminology *fail to reject the null hypothesis,* instead of *accept the null hypothesis.*

Multiple Negatives When stating the final conclusion in nontechnical terms, it is possible to get correct statements with up to three negative terms. (*Example:* "There is *not* sufficient evidence to warrant *rejection* of the claim of *no* difference between 0.5 and the population proportion.") Such conclusions are confusing, so it is good to restate them in a way that makes them understandable, but be careful to not change the meaning. For example, instead of saying that "there is not sufficient evidence to warrant rejection of the claim of no difference between 0.5 and the population proportion," better statements would be these:

- Fail to reject the claim that the population proportion is equal to 0.5.
- Unless stronger evidence is obtained, continue to assume that the population proportion is equal to 0.5.

CAUTION Never conclude a hypothesis test with a statement of "reject the null hypothesis" or "fail to reject the null hypothesis." Always make sense of the conclusion with a statement that uses simple nontechnical wording that addresses the original claim.

The Big Picture Revisited Example 1 describes a clinical trial that resulted in 58 girls in 100 births. Try to follow this line of reasoning:

- If we assume that $p = 0.5$ (as in the null hypothesis), the probability of getting 58 or more girls is 0.0548.
- Because the probability of 0.0548 is not small (such as 0.05 or less), we see that random chance is a reasonable explanation for getting 58 (or more) girls in 100 births.
- Because 58 girls in 100 births can easily occur by chance, we don't have sufficient evidence to conclude that the XSORT method is effective.

The hypothesis test procedure basically formalizes the reasoning process above, and it provides a common structure and terminology recognizable by professionals in a wide variety of disciplines, not just statistics.

Errors in Hypothesis Tests

When testing a null hypothesis, we arrive at a conclusion of rejecting it or failing to reject it. Such conclusions are sometimes correct and sometimes wrong (even if we apply all procedures correctly). Table 8-4 summarizes the two different types of errors that can be made, along with the two different types of correct decisions. We distinguish between the two types of errors by calling them type I and type II errors.

- **Type I error:** The mistake of rejecting the null hypothesis when it is actually true. The symbol α (alpha) is used to represent the probability of a type I error.
- **Type II error:** The mistake of failing to reject the null hypothesis when it is actually false. The symbol β (beta) is used to represent the probability of a type II error.

MEMORY HINT Because it is difficult to remember which error is type I and which is type II, we recommend this mnemonic device: Remember "routine for fun," and using only the consonants from those words (**R**ou**T**i**N**e **F**o**R** **F**u**N**), we can easily remember that a type I error is RTN: Reject True Null (hypothesis), and a type II error is FRFN: Fail to Reject a False Null (hypothesis).

Notation

α (alpha) = probability of a type I error (the probability of rejecting the null hypothesis when it is true)

β (beta) = probability of a type II error (the probability of failing to reject a null hypothesis when it is false)

Table 8-4 Type I and Type II Errors

		True State of Nature	
		The null hypothesis is true	**The null hypothesis is false**
Decision	We decide to reject the null hypothesis	Type I error (rejecting a true null hypothesis) *P*(type I error) = α	Correct decision
	We fail to reject the null hypothesis	Correct decision	Type II error (failing to reject a false null hypothesis) *P*(type II error) = β

Example 2 Identifying Type I and Type II Errors

Consider the claim that a method of gender selection increases the likelihood of a baby girl, so that the probability of a baby girl is $p > 0.5$. Here are the null and alternative hypotheses:

H_0: $p = 0.5$

H_1: $p > 0.5$ (original claim that is addressed in the final conclusion)

Identify the following: **a.** Type I error; **b.** Type II error.

Process of Drug Approval

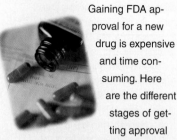

Gaining FDA approval for a new drug is expensive and time consuming. Here are the different stages of getting approval for a new drug:

- **Phase I study:** The safety of the drug is tested with a small (20–100) group of volunteers.
- **Phase II:** The drug is tested for effectiveness in randomized trials involving a larger (100–300) group of subjects. This phase often has subjects randomly assigned to either a treatment group or a placebo group.
- **Phase III:** The goal is to better understand the effectiveness of the drug as well as its adverse reactions. This phase typically involves 1,000–3,000 subjects, and it might require several years of testing.

Lisa Gibbs wrote in *Money* magazine that "the (drug) industry points out that for every 5,000 treatments tested, only 5 make it to clinical trials and only 1 ends up in drugstores." Total cost estimates vary from a low of $40 million to as much as $1.5 billion.

Note to Instructor

With a hypothesis test, it is not simply a matter of being right or wrong. Different types of errors can have dramatically different consequences, and that is why we distinguish between type I errors and type II errors. To make this point, ask the class the difference between these two errors: (1) Rejecting a perfectly good parachute and refusing to jump and (2) failing to reject a defective parachute and jumping out of a plane with it.

Also, the terms "type I error" and "type II error" reveal nothing about their meaning, so see the mnemonic of "**R**ou**T**i**N**e **F**o**R** **F**u**N**" for remembering them. Using only the consonants, the first word yields RTN for "reject true null" and the last two words yield FRFN for "fail to reject false null."

Solution

HINT Descriptions of a type I error and a type II error refer to the *null hypothesis* being true or false, but when wording a statement representing a type I error or a type II error, be sure that the conclusion addresses the *original claim* (which may or may not be the null hypothesis). In this example, the claim is that $p > 0.5$, so the following interpretations address the claim of $p > 0.5$.

a. Type I Error: A type I error is the mistake of rejecting a true null hypothesis, so the following is a type I error: In reality $p = 0.5$, but sample evidence leads us to conclude that $p > 0.5$. (In this case, a type I error is to conclude that the gender-selection method is effective when in reality it has no effect.)

b. Type II Error: A type II error is the mistake of failing to reject the null hypothesis when it is false, so the following is a type II error: In reality $p > 0.5$, but we fail to support that conclusion. (In this case, a type II error is to conclude that the gender-selection method has no effect, when it really is effective in increasing the likelihood of a baby girl.)

Controlling Type I and Type II Errors Step 4 in our standard procedure for testing hypotheses is to select a significance level α (such as 0.05), which is the probability of a type I error. The values of α, β, and the sample size n are all related, so if you choose any two of them, the third is automatically determined (although β can't be determined until an alternative value of the population parameter has been specified along with α and n). One common practice is to select the significance level α, then select a sample size that is practical, so the value of β is determined. Generally, try to use the largest α that you can tolerate, but for type I errors with more serious consequences, select smaller values of α. Then choose a sample size n as large as is reasonable, based on considerations of time, cost, and other relevant factors. Another common practice is to select α and β so the required sample size n is automatically determined. (See Example 4 in Part 2 of this section.)

Comprehensive Hypothesis Test

In this section we described the individual components used in a hypothesis test, but the following sections will combine those components in comprehensive procedures. We can test claims about population parameters by using the *P*-value method summarized in Figure 8-1 or the critical value method summarized in Figure 8-2, or we can use confidence intervals, as follows.

Confidence Interval Method for Hypothesis Testing A confidence interval estimate of a population parameter contains the likely values of that parameter. If a confidence interval does not include a claimed value of a population parameter, reject that claim. For two-tailed hypothesis tests, construct a confidence interval with a confidence level of $1 - \alpha$, but for a one-tailed hypothesis test with significance level α, construct a confidence interval with a confidence level of $1 - 2\alpha$. (See Table 8-1 for common cases.) (For a left-tailed test or a right-tailed test, we could also use a one-sided confidence interval. See Exercise 43 in Section 7-2.) After constructing the confidence interval, use this criterion:

A confidence interval estimate of a population parameter contains the likely values of that parameter. We should therefore reject a claim that the population parameter has a value that is not included in the confidence interval.

Commercials

Television networks have their own clearance departments for screening commercials and verifying claims. The National Advertising Division, a branch of the Council of Better Business Bureaus, investigates advertising claims. The Federal Trade Commission and local district attorneys also become involved. In the past, Firestone had to drop a claim that its tires resulted in 25% faster stops, and Warner Lambert had to spend $10 million informing customers that Listerine doesn't prevent or cure colds. Many deceptive ads are voluntarily dropped, and many others escape scrutiny simply because the regulatory mechanisms can't keep up with the flood of commercials.

Cheating Detected

Methods for cheating on tests include the use of cell phones to text answers to friends and the use of calculators to store notes or share answers.

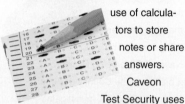

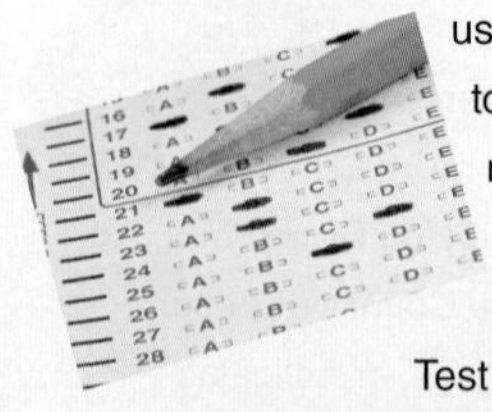

Caveon Test Security uses statistical methods to identify when cheating has occurred. That company searches for unlikely patterns, such as identical wrong answers by different students. The company even analyzes erasures on answer sheets to determine whether tests have been compromised by dishonest teachers or administrators who can gain from the appearance of higher test scores. The company also searches the Internet looking for questions or answers that were posted or discussed by previous test subjects.

Statistics plays an important role in the analyses conducted by Caveon Test Security. Unusual patterns are identified and the probabilities of those patterns are computed. If the probability of a pattern is very small, such as less than one chance in a million, the answer sheets are turned over to school administrators.

Note to Instructor

The calculations required to determine the power of a test are usually quite difficult. Exercise 35 requires interpretation of the value of power, Exercise 36 requires a computation of power, and the procedure is given in that exercise. Also, Exercise 37 uses the concept of power to determine sample size, and that exercise is very difficult.

CAUTION In some cases, a conclusion based on a confidence interval may be different from a conclusion based on a hypothesis test. The *P*-value method and critical value method are equivalent in the sense that they always lead to the same conclusion. The following table shows that for the methods included in this chapter, a confidence interval estimate of a proportion might lead to a conclusion different from that of a hypothesis test.

Parameter	Is a confidence interval equivalent to a hypothesis test in the sense that they always lead to the same conclusion?
Proportion	No
Mean	Yes
Standard Deviation or Variance	Yes

Part 2: Beyond the Basics of Hypothesis Testing: The Power of a Test

We use β to denote the probability of failing to reject a false null hypothesis, so $P(\text{type II error}) = \beta$. It follows that $1 - \beta$ is the probability of rejecting a false null hypothesis; statisticians refer to this probability as the *power* of a test, and they often use it to gauge the effectiveness of a hypothesis test in allowing us to recognize that a null hypothesis is false.

DEFINITION The **power** of a hypothesis test is the probability $1 - \beta$ of rejecting a false null hypothesis. The value of the power is computed by using a particular significance level α and a *particular* value of the population parameter that is an alternative to the value assumed true in the null hypothesis.

Note that in the definition above, determination of power requires a particular value that is an alternative to the value assumed in the null hypothesis. Consequently, a hypothesis test can have many different values of power, depending on the particular values of the population parameter chosen as alternatives to the null hypothesis.

Example 3 Power of a Hypothesis Test

Consider these preliminary results from the XSORT method of gender selection: There were 13 girls among the 14 babies born to couples using the XSORT method. If we want to test the claim that girls are more likely ($p > 0.5$) with the XSORT method, we have the following null and alternative hypotheses:

$$H_0: p = 0.5 \qquad H_1: p > 0.5$$

Let's use a significance level of $\alpha = 0.05$. In addition to all of the given test components, finding power requires that we select a particular value of p that is an alternative to the value assumed in the null hypothesis $H_0: p = 0.5$. Find the values of power corresponding to these alternative values of p: 0.6, 0.7, 0.8, and 0.9.

Solution

The values of power in the following table were found by using Minitab, and exact calculations are used instead of a normal approximation to the binomial distribution.

Specific Alternative Value of p	β	Power of Test $(1 - \beta)$
0.6	0.820	0.180
0.7	0.564	0.436
0.8	0.227	0.773
0.9	0.012	0.988

Interpretation

Based on the list of power values above, we see that this hypothesis test has power of 0.180 (or 18.0%) of rejecting H_0: $p = 0.5$ when the population proportion p is actually 0.6. That is, if the true population proportion is actually equal to 0.6, there is an 18.0% chance of making the correct conclusion of rejecting the false null hypothesis that $p = 0.5$. That low power of 18.0% is not good.

There is a 0.564 probability of rejecting $p = 0.5$ when the true value of p is actually 0.7. It makes sense that this test is more effective in rejecting the claim of $p = 0.5$ when the population proportion is actually 0.7 than when the population proportion is actually 0.6. (When identifying animals assumed to be horses, there's a better chance of rejecting an elephant as a horse—because of the greater difference—than rejecting a mule as a horse.) In general, increasing the difference between the assumed parameter value and the actual parameter value results in an increase in power, as shown in the table above.

Because the calculations of power are quite complicated, the use of technology is strongly recommended. (In this section, only Exercises 35 and 36 involve power.)

Power and the Design of Experiments

Just as 0.05 is a common choice for a significance level, a power of at least 0.80 is a common requirement for determining that a hypothesis test is effective. (Some statisticians argue that the power should be higher, such as 0.85 or 0.90.) When designing an experiment, we might consider how much of a difference between the claimed value of a parameter and its true value is an important amount of difference. If testing the effectiveness of the XSORT gender-selection method, a change in the proportion of girls from 0.5 to 0.501 is not very important. A change in the proportion of girls from 0.5 to 0.9 would be very important. Such magnitudes of differences affect power. When designing an experiment, a goal of having a power value of at least 0.80 can often be used to determine the minimum required sample size, as in the following example.

Example 4 **Finding the Sample Size Required to Achieve 80% Power**

Here is a statement similar to one in an article from the *Journal of the American Medical Association:* "The trial design assumed that with a 0.05 significance level, 153 randomly selected subjects would be needed to achieve 80% power to detect a reduction in the coronary heart disease rate from 0.5 to 0.4." Before conducting the experiment, the researchers selected a significance level of 0.05 and a power of at least 0.80. They also decided that a reduction in the proportion of coronary heart disease from 0.5 to 0.4 is an important difference that they wanted to detect (by correctly rejecting the false null hypothesis). Using a significance level of 0.05, power of 0.80, and the alternative proportion of 0.4, technology such as Minitab is used to find that the required minimum sample size is 153. The researchers can then proceed by obtaining a sample of at least 153 randomly selected subjects. Due to factors such as dropout rates, the researchers are likely to need somewhat more than 153 subjects. (See Exercise 37.)

8-2 Basic Skills and Concepts

Recommended Assignment
Exercises 1–20, 25–28, 31–34.

1. Rejection of the aspirin claim is more serious because the aspirin is a drug treatment. The wrong aspirin dosage can cause adverse reactions. M&Ms do not have those same adverse reactions. It would be wise to use a smaller significance level for testing the aspirin claim.
2. Estimates and hypothesis tests are both methods of inferential statistics, but they have different objectives. We could use the sample weights to construct a confidence interval estimate of the mean weight of all M&Ms, but hypothesis testing is used to test some claim made about the mean weight of all M&Ms.
3. a. H_0: $\mu = 98.6°F$
   b. H_1: $\mu \neq 98.6°F$
   c. Reject the null hypothesis or fail to reject the null hypothesis.
   d. No. In this case, the original claim becomes the null hypothesis. For the claim that the mean body temperature is equal to 98.6°F, we can either reject that claim or fail to reject it, but we cannot state that there is sufficient evidence to *support* that claim.
4. The *P*-value of 0.001 is preferred because it corresponds to the sample evidence that most strongly supports the alternative hypothesis that the XSORT method is effective.
5. a. $p = 0.2$
   b. H_0: $p = 0.2$ and H_1: $p \neq 0.2$
6. a. $p > 0.5$
   b. H_0: $p = 0.5$ and H_1: $p > 0.5$
7. a. $\mu \leq 76$
   b. H_0: $\mu = 76$ and H_1: $\mu > 76$

Statistical Literacy and Critical Thinking

1. M&Ms and Aspirin A package label includes a claim that the mean weight of the M&Ms is 0.8535 g, and another package label includes the claim that the mean amount of aspirin in Bayer tablets is 325 mg. Which has more serious implications: rejection of the M&M claim or rejection of the aspirin claim? Is it wise to use the same significance level for hypothesis tests of both claims?

2. Estimates and Hypothesis Tests Data Set 20 in Appendix B includes sample weights of the M&Ms referenced in Exercise 1. We could use methods of Chapter 7 for making an estimate, or we could use those values to test some claim. What is the difference between estimating and hypothesis testing?

3. Mean Body Temperature A formal hypothesis test is to be conducted using the claim that the mean body temperature is equal to 98.6°F.

a. What is the null hypothesis, and how is it denoted?

b. What is the alternative hypothesis, and how is it denoted?

c. What are the possible conclusions that can be made about the null hypothesis?

d. Is it possible to conclude that "there is sufficient evidence to support the claim that the mean body temperature is equal to 98.6°F"?

4. Interpreting *P*-value When the clinical trial of the XSORT method of gender selection is completed, a formal hypothesis test will be conducted with the alternative hypothesis of $p > 0.5$, which corresponds to the claim that the XSORT method increases the likelihood of having a girl, so that the proportion of girls is greater than 0.5. If you are responsible for developing the XSORT method and you want to show its effectiveness, which of the following *P*-values would you prefer: 0.999, 0.5, 0.95, 0.05, 0.01, 0.001? Why?

Stating Conclusions about Claims. ***In Exercises 5–8, do the following:***

a. Express the original claim in symbolic form.

b. Identify the null and alternative hypotheses.

5. Claim: 20% of adults smoke. A recent Gallup survey of 1016 randomly selected adults showed that 21% of the respondents smoke.

6. Claim: When parents use the XSORT method of gender selection, the proportion of baby girls is greater than 0.5. The latest actual results show that among 945 babies born to couples using the XSORT method of gender selection, 879 were girls.

7. Claim: The mean pulse rate (in beats per minute) of adult females is 76 or lower. For the random sample of adult females in Data Set 1 from Appendix B, the mean pulse rate is 77.5.

8. Claim: The standard deviation of pulse rates of adult women is at least 50. For the random sample of adult females in Data Set 1 from Appendix B, the pulse rates have a standard deviation of 11.6.

Forming Conclusions. ***In Exercises 9–12, refer to the exercise identified. Using only the rare event rule, make subjective estimates to determine whether results are likely, then state a conclusion about the original claim. For example, if the claim is that a coin favors heads and sample results consist of 11 heads in 20 flips, conclude that there is not sufficient evidence to support the claim that the coin favors heads (because it is easy to get 11 heads in 20 flips by chance with a fair coin).***

9. Exercise 5

10. Exercise 6

11. Exercise 7

12. Exercise 8

Finding Test Statistics. *In Exercises 13–16, find the value of the test statistic. (Refer to Table 8-2 to select the correct expression for evaluating the test statistic.)*

13. Community Involvement Claim: Three-fourths of all adults believe that it is important to be involved in their communities. Based on a *USA Today*/Gallup poll of 1021 randomly selected adults, 89% believe that it is important to be involved in their communities.

14. Tax Returns Claim: Among those who file tax returns, less than one-half file them through an accountant or other tax professional. A Fellowes survey of 1002 people who file tax returns showed that 48% of them file through an accountant or other tax professional.

15. White Blood Cell Count Claim: For adult females, the standard deviation of their white blood cell counts is equal to 5.00. The random sample of 40 adult females in Data Set 1 from Appendix B has white blood cell counts with a standard deviation of 2.28.

16. White Blood Cell Count Claim: For adult females, the mean of their white blood cell counts is equal to 8.00. The random sample of 40 adult females in Data Set 1 from Appendix B has white blood cell counts with a mean of 7.15 and a standard deviation of 2.28. (The population standard deviation σ is not known.)

Finding *P*-Values and Critical Values. *In Exercises 17–24, assume that the significance level is $\alpha = 0.05$; use the given statement and find the P-value and critical values. (See Figure 8-4.)*

17. The test statistic of $z = 2.00$ is obtained when testing the claim that $p > 0.5$.

18. The test statistic of $z = -2.00$ is obtained when testing the claim that $p < 0.5$.

19. The test statistic of $z = -1.75$ is obtained when testing the claim that $p = 1/3$.

20. The test statistic of $z = 1.50$ is obtained when testing the claim that $p \neq 0.25$.

21. With H_1: $p \neq 0.25$, the test statistic is $z = -1.23$.

22. With H_1: $p \neq 2/3$, the test statistic is $z = 2.50$.

23. With H_1: $p < 0.6$, the test statistic is $z = -3.00$.

24. With H_1: $p > 7/8$, the test statistic is $z = 2.88$.

8. a. $\sigma \geq 50$
   b. H_0: $\sigma = 50$ and H_1: $\sigma < 50$
9. There is not sufficient evidence to warrant rejection of the claim that 20% of adults smoke.
10. There is sufficient evidence to support the claim that when parents use the XSORT method of gender selection, the proportion of baby girls is greater than 0.5.
11. There is not sufficient evidence to warrant rejection of the claim that the mean pulse rate of adult females is 76 or lower.
12. There is sufficient evidence to reject the claim that pulse rates of adult females have a standard deviation of at least 50.
13. $z = 10.33$ (or $z = 10.35$ if using $x = 909$)
14. $z = -1.27$ (or $z = -1.26$ if using $x = 481$)
15. $\chi^2 = 8.110$
16. $t = -2.358$
17. *P*-value $= 0.0228$. Critical value: $z = 1.645$.
18. *P*-value $= 0.0228$. Critical value: $z = -1.645$.
19. *P*-value $= 0.0802$ (Tech: 0.0801). Critical values: $z = -1.96$, $z = 1.96$.
20. *P*-value $= 0.1336$. Critical values: $z = -1.96$, $z = 1.96$.
21. *P*-value $= 0.2186$ (Tech: 0.2187). Critical values: $z = -1.96$, $z = 1.96$.
22. *P*-value $= 0.0124$. Critical values: $z = -1.96$, $z = 1.96$.
23. *P*-value $= 0.0013$. Critical value: $z = -1.645$.
24. *P*-value $= 0.0020$. Critical value: $z = 1.645$.

Stating Conclusions. ***In Exercises 25–28, assume a significance level of $\alpha = 0.05$ and use the given information for the following:***

a. State a conclusion about the null hypothesis. (Reject H_0 or fail to reject H_0.)

b. Without using technical terms, state a final conclusion that addresses the original claim.

25. Original claim: The percentage of blue M&Ms is greater than 5%. The hypothesis test results in a *P*-value of 0.0010.

26. Original claim: Fewer than 20% of M&M candies are green. The hypothesis test results in a *P*-value of 0.0721.

27. Original claim: Women have heights with a mean equal to 160.00 cm. The hypothesis test results in a *P*-value of 0.0614.

28. Original claim: Women have heights with a standard deviation equal to 5.00 cm. The hypothesis test results in a *P*-value of 0.0055.

Terminology. ***In Exercises 29 and 30, use the given information to answer the following:***

a. Identify the null hypothesis and the alternative hypothesis.

b. What is the value of α?

c. What is the sampling distribution of the sample statistic?

d. Is the test two-tailed, left-tailed, or right-tailed?

e. What is the value of the test statistic?

f. What is the *P*-value?

g. What is the critical value?

h. What is the area of the critical region?

29. Gender Selection A 0.01 significance level is used for a hypothesis test of the claim that when parents use the XSORT method of gender selection, the proportion of baby girls is *greater* than 0.5. Assume that sample data consist of 55 girls born in 100 births, so the sample statistic of 0.55 results in a *z* score that is 1.00 standard deviation above 0.

30. Gender Selection A 0.05 significance level is used for a hypothesis test of the claim that when parents use the XSORT method of gender selection, the proportion of baby girls is *different* from 0.5. Assume that sample data consist of 55 girls born in 100 births, so the sample statistic of 0.55 results in a *z* score that is 1.00 standard deviation above 0.

Type I and Type II Errors. ***In Exercises 31–34, identify expressions that identify the type I error and the type II error that correspond to the given claim. (Although conclusions are usually expressed in verbal form, the answers here can be expressed with statements that include symbolic expressions such as $p = 0.1$.)***

31. The proportion of people who write with their left hand is equal to 0.1.

32. The proportion of gamblers who consistently win in casinos is equal to 0.001.

33. The proportion of female statistics students is greater than 0.5.

34. The proportion of husbands taller than their wives is less than 0.9.

25. a. Reject H_0.
b. There is sufficient evidence to support the claim that the percentage of blue M&Ms is greater than 5%.

26. a. Fail to reject H_0.
b. There is not sufficient evidence to support the claim that fewer than 20% of M&M candies are green.

27. a. Fail to reject H_0.
b. There is not sufficient evidence to warrant rejection of the claim that women have heights with a mean equal to 160.00 cm.

28. a. Reject H_0.
b. There is sufficient evidence to warrant rejection of the claim that women have heights with a standard deviation equal to 5.00 cm.

29. a. H_0: $p = 0.5$ and H_1: $p > 0.5$
b. $\alpha = 0.01$
c. Normal distribution.
d. Right-tailed.
e. $z = 1.00$
f. *P*-value: 0.1587
g. $z = 2.33$
h. 0.01

30. a. H_0: $p = 0.5$ and H_1: $p \neq 0.5$
b. $\alpha = 0.05$
c. Normal distribution.
d. Two-tailed.
e. $z = 1.00$
f. *P*-value: $2(0.1587) = 0.3174$ (Tech: 0.3173)
g. $z = -1.96$ and $z = 1.96$
h. 0.05

31. Type I error: In reality $p = 0.1$, but we reject the claim that $p = 0.1$. Type II error: In reality $p \neq 0.1$, but we fail to reject the claim that $p = 0.1$.

32. Type I error: In reality $p = 0.001$, but we reject the claim that $p = 0.001$. Type II error: In reality $p \neq 0.001$, but we fail to reject the claim that $p = 0.001$.

33. Type I error: In reality $p = 0.5$, but we support the claim that $p > 0.5$. Type II error: In reality $p > 0.5$, but we fail to support that conclusion.

34. Type I error: In reality $p = 0.9$, but we support the claim that $p < 0.9$. Type II error: In reality $p < 0.9$, but we fail to support that conclusion.

8-2 Beyond the Basics

35. Interpreting Power Chantix tablets are used as an aid to help people stop smoking. In a clinical trial, 129 subjects were treated with Chantix twice a day for 12 weeks, and 16 subjects experienced abdominal pain (based on data from Pfizer, Inc.). If someone claims that more than 8% of Chantix users experience abdominal pain, that claim is supported with a hypothesis test conducted with a 0.05 significance level. Using 0.18 as an alternative value of p, the power of the test is 0.96. Interpret this value of the power of the test.

36. Calculating Power Consider a hypothesis test of the claim that the MicroSort method of gender selection is effective in increasing the likelihood of having a baby girl, so that the claim is $p > 0.5$. Assume that a significance level of $\alpha = 0.05$ is used, and the sample is a simple random sample of size $n = 64$.

a. Assuming that the true population proportion is 0.65, find the power of the test, which is the probability of rejecting the null hypothesis when it is false. (*Hint:* With a 0.05 significance level, the critical value is $z = 1.645$, so any test statistic in the right tail of the accompanying top graph is in the rejection region where the claim is supported. Find the sample proportion $\hat{p}$ in the top graph, and use it to find the power shown in the bottom graph.)

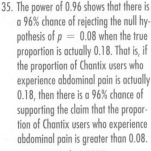

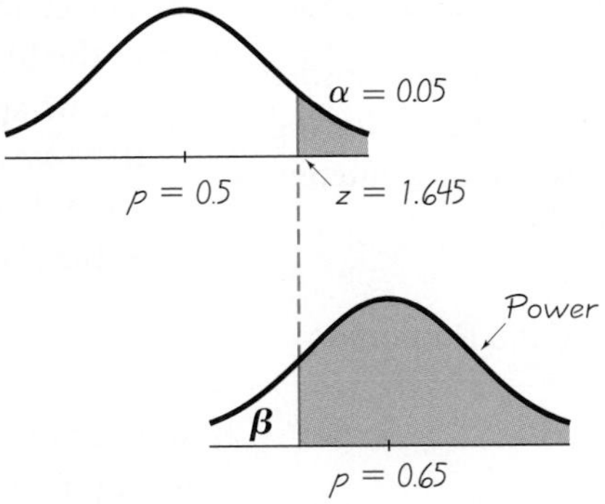

b. Explain why the red-shaded region of the bottom graph represents the power of the test.

37. Finding Sample Size to Achieve Power Researchers plan to conduct a test of a gender-selection method. They plan to use the alternative hypothesis of $H_1: p > 0.5$ and a significance level of $\alpha = 0.05$. Find the sample size required to achieve at least 80% power in detecting an increase in p from 0.5 to 0.55. (This is a very difficult exercise. *Hint:* See Exercise 36.)

35. The power of 0.96 shows that there is a 96% chance of rejecting the null hypothesis of $p = 0.08$ when the true proportion is actually 0.18. That is, if the proportion of Chantix users who experience abdominal pain is actually 0.18, then there is a 96% chance of supporting the claim that the proportion of Chantix users who experience abdominal pain is greater than 0.08.

36. a. 0.7852 (Tech: 0.7857)

b. Assuming that $p = 0.5$, as in the null hypothesis, the critical value of $z = 1.645$ corresponds to $\hat{p} = 0.6028125$, so any sample proportion greater than 0.6028125 causes us to reject the null hypothesis, as shown in the shaded critical region of the top graph. If p is actually 0.65, then the null hypothesis of $p = 0.5$ is false, and the actual probability of rejecting the null hypothesis is found by finding the area greater than $\hat{p} = 0.6028125$ in the bottom graph, which is the shaded area. That is, the shaded area in the bottom graph represents the probability of rejecting the false null hypothesis.

37. 617

TOP 20 8-3 Testing a Claim about a Proportion

Key Concept Section 8-2 presented individual components of a hypothesis test, and this section presents complete procedures for testing a hypothesis (or claim) made about a population proportion p. We illustrate hypothesis testing with the P-value method, the critical value method, and the use of confidence intervals. The methods of this section can be used with claims about population proportions, and the same methods can be used for testing claims about probabilities or the decimal equivalents of percents.

Two different approaches for testing a claim about a population proportion are (1) to use a normal distribution as an approximation to the binomial distribution, and (2) to use an exact method based on the binomial probability distribution. Part 1 of this section uses the approximation method with the normal distribution, and Part 2 of this section briefly describes the exact method.

Part 1: Basic Methods of Testing Claims about a Population Proportion p

The following box includes the key elements used for testing a claim about a population proportion.

Note to Instructor

Section 8-2 describes *components* of a hypothesis test. This section describes a *complete hypothesis-testing procedure* for testing claims made about a population proportion. This section presents three methods for testing hypotheses: (1) P-value method; (2) critical value method; (3) confidence intervals.

Recommendation: If you are making extensive use of TI-83/84 Plus calculators or computer software, emphasize the P-value approach instead of the critical value approach. The P-value approach is used most often in real applications.

This section is partitioned into Part 1 (basics) and Part 2, which is based on an exact procedure instead of using a normal approximation. If omitting Part 2, consider at least a brief discussion summarizing its main points, especially if students have technology capable of generating binomial probabilities.

Testing a Claim about a Population Proportion

Objective

Conduct a formal hypothesis test of a claim about a population proportion p.

Notation

n = sample size or number of trials

$\hat{p} = \frac{x}{n}$ (*sample* proportion)

p = population proportion (p is the value used in the statement of the null hypothesis)

$q = 1 - p$

Requirements

1. The sample observations are a simple random sample.

2. The conditions for a *binomial distribution* are satisfied. (There is a fixed number of independent trials having constant probabilities, and each trial has two outcome categories of "success" and "failure.")

3. The conditions $np \geq 5$ and $nq \geq 5$ are both satisfied, so **the binomial distribution of sample proportions can be approximated by a normal distribution with** $\mu = np$ and $\sigma = \sqrt{npq}$ (as described in Section 6-7). Note that p used here is the *assumed* proportion used in the claim, not the sample proportion $\hat{p}$.

Test Statistic for Testing a Claim about a Proportion

$$z = \frac{\hat{p} - p}{\sqrt{\frac{pq}{n}}}$$

***P*-values:** *P*-values are automatically provided by technology. If technology is not available, use the standard normal distribution (Table A-2) and refer to Figure 8-1.

Critical values: Use the standard normal distribution (Table A-2).

Note to Instructor

If Section 6-7 was omitted, state that under certain circumstances (such as those satisfying the three requirements), a binomial distribution can be approximated by a normal distribution. Refer to the Minitab display in Section 6-7 and point out that it is a graph of the probabilities for the number of heads when 431 coins are tossed. The graph appears to be a normal distribution, even though the plotted points are from a binomial distribution. If Section 6-7 was covered, comment that we use a test statistic that does not include the continuity correction from Section 6-7. The continuity correction is not included because its effects tend to be very small with large samples.

CAUTION

Reminder: Don't confuse a *P*-value with a proportion p.

- *P*-value = probability of getting a test statistic at least as extreme as the one representing the sample data, assuming that the null hypothesis H_0 is true
- p = population proportion

The test statistic above does not include a correction for continuity (as described in Section 6-7), because its effect tends to be very small with large samples.

Example 1 Reality Check

Based on information from the National Cyber Security Alliance, 93% of computer owners believe that they have antivirus programs installed on their computers. In a random sample of 400 scanned computers, it is found that 380 of them (or 95%) actually have antivirus programs. Use the sample data from the scanned computers to test the claim that 93% of computers have antivirus programs.

Requirement check We first check the three requirements.

1. The 400 computers are randomly selected.
2. There is a fixed number (400) of independent trials with two categories (computer has an antivirus program or does not).

3. The requirements $np \geq 5$ and $nq \geq 5$ are both satisfied with $n = 400$, $p = 0.93$, and $q = 0.07$. (The value of $p = 0.93$ comes from the claim. We get $np = (400)(0.93) = 372$, which is greater than or equal to 5, and we get $nq = (400)(0.07) = 28$, which is also greater than or equal to 5.)

The three requirements are satisfied.

P-Value Method

Technology Computer programs and calculators usually provide a *P*-value, so the *P*-value method is used. See the accompanying TI-83/84 Plus calculator results showing the alternative hypothesis of "prop $\neq$ 0.93," the test statistic of $z = 1.57$ (rounded), and the *P*-value of 0.1169 (rounded).

TI-83/84 PLUS

```
1-PropZTest
 prop≠.93
 z=1.567723603
 p=.1169456657
 p̂=.95
 n=400
```

If technology is not available, Figure 8-1 in the preceding section lists the steps for using the *P*-value method. Using those steps from Figure 8-1, we can test the claim in Example 1 as follows.

Step 1: The original claim is that 93% of computers have antivirus programs, and that claim can be expressed in symbolic form as $p = 0.93$.

Step 2: The opposite of the original claim is $p \neq 0.93$.

Step 3: Of the preceding two symbolic expressions, the expression $p \neq 0.93$ does not contain equality, so it becomes the alternative hypothesis. The null hypothesis is the statement that p equals the fixed value of 0.93. We can therefore express H_0 and H_1 as follows:

$$H_0: p = 0.93 \quad \text{(original claim)}$$

$$H_1: p \neq 0.93$$

Step 4: For the significance level, we select $\alpha = 0.05$, which is a very common choice.

Step 5: Because we are testing a claim about a population proportion p, the sample statistic $\hat{p}$ is relevant to this test. The sampling distribution of sample proportions $\hat{p}$ can be approximated by a normal distribution in this case.

Step 6: The test statistic $z = 1.57$ can be calculated by using $\hat{p} = 380/400$ (sample proportion), $n = 400$ (sample size), $p = 0.93$ (assumed in the null hypothesis), and $q = 1 - 0.93 = 0.07$.

$$z = \frac{\hat{p} - p}{\sqrt{\frac{pq}{n}}} = \frac{\frac{380}{400} - 0.93}{\sqrt{\frac{(0.93)(0.07)}{400}}} = 1.57$$

Lie Detectors and the Law

Why not simply require all criminal suspects to take polygraph (lie detector) tests and eliminate trials by jury? According to the Council of Scientific Affairs of the American Medical Association, when lie detectors are used to determine guilt, accuracy can range from 75% to 97%. However, a high accuracy rate of 97% can still result in a high percentage of false positives, so it is possible that 50% of innocent subjects incorrectly appear to be guilty. Such a high chance of false positives rules out the use of polygraph tests as the single criterion for determining guilt.

Note to Instructor

The *P*-value method is now so important because TI-83/84 Plus calculators and so many statistics programs generate *P*-values but not critical values. Also, many articles in professional journals include *P*-values but not critical values. When defining *P*-value, reinforce the importance of getting a value *at least as extreme* as the one found. Refer to Figure 8-1 in Section 8-2 and suggest that students use it regularly as a basic guide for including all of the steps of a complete hypothesis test. Describe exactly what you expect from them when they take tests or submit homework.

Recommendation: Require these components in all hypothesis tests: (1) statements of H_0 and H_1; (2) a graph showing the normal distribution with the test statistic and either the *P*-value or the critical value(s) and critical region; (3) a statement of either "reject H_0" or "fail to reject H_0"; and (4) a summary statement of the conclusion in nontechnical terms. Emphasize that the final conclusion should address the original claim.

We now find the *P*-value by using the following procedure, which is shown in Figure 8-4:

Left-tailed test: *P*-value = area to left of test statistic z

Right-tailed test: *P*-value = area to right of test statistic z

Two-tailed test: *P*-value = *twice* the area of the extreme region bounded by the test statistic z

Because the hypothesis test we are considering is two-tailed with a test statistic of $z = 1.57$, the *P*-value is twice the area to the right of $z = 1.57$. Referring to Table A-2, we see that the cumulative area to the *left* of $z = 1.57$ is 0.9418, so the area to the right of that test statistic is $1 - 0.9418 = 0.0582$. The *P*-value is twice 0.0582, so we get *P*-value = 0.1164. (If using technology with the unrounded z score, the *P*-value is found to be 0.1169, as shown in the TI-83/84 Plus calculator display.) Figure 8-6 shows the test statistic and *P*-value for this example.

Step 7: Because the *P*-value of 0.1164 is greater than the significance level of $\alpha = 0.05$, we fail to reject the null hypothesis. (Remember, if the *P*-value is low, the null must go, but here the *P*-value is high.)

Step 8: Because we fail to reject H_0: $p = 0.93$, we fail to reject the claim that 93% of computers have antivirus programs. We conclude that there is not sufficient sample evidence to warrant rejection of the claim that 93% of computers have antivirus programs. (See Table 8-3 for help with wording this final conclusion.)

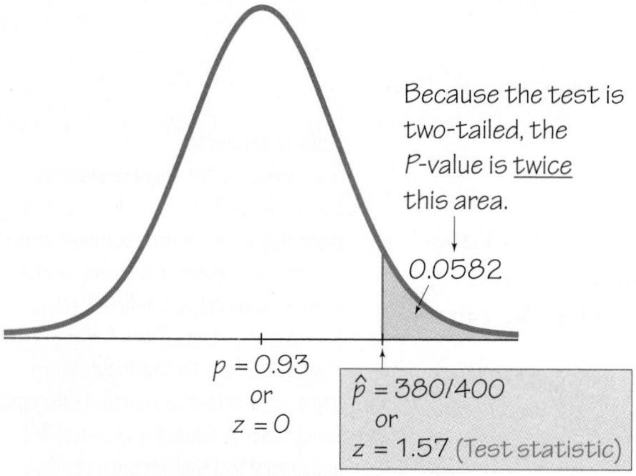

Figure 8-6 ***P*-Value Method**

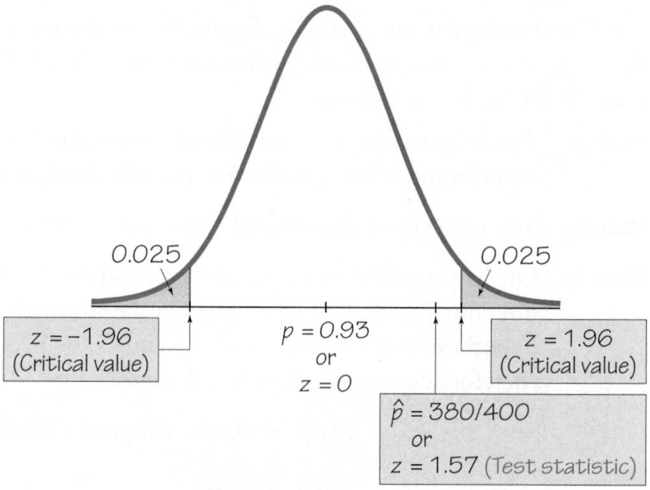

Figure 8-7 Critical Value Method

Note to Instructor
Change in Terminology: This edition uses the terminology "Critical Value Method," but the preceding edition used the term "Traditional Method." Referring to the "Critical Value Method" has the obvious advantage that students will know that this method is based on the use of critical values. It requires a decision based on a comparison of the test statistic and critical value. Comment that the TI-83/84 Plus calculator and many software packages (such as Minitab) do not include critical values with their results; they typically do provide *P*-values, so the *P*-value approach may be more suitable than the critical value method.

Critical Value Method

The critical value method of testing hypotheses is summarized in Figure 8-2 in Section 8-2. When using the critical value method with the claim given in Example 1, Steps 1 through 5 are the same as in Steps 1 through 5 for the *P*-value method, as shown above. We continue with Step 6 of the critical value method.

Step 6: The test statistic is computed to be $z = 1.57$ as shown for the preceding *P*-value method. With the critical value method, we now find the critical values (instead of the *P*-value). This is a two-tailed test, so the area of the critical region is an area of $\alpha = 0.05$, which is divided equally between the two tails. Referring to Table A-2 and applying the methods of Section 6-2, we find that the critical values of $z = -1.96$ and $z = 1.96$ are at the boundaries of the critical region, as shown in Figure 8-7.

Step 7: Because the test statistic does not fall within the critical region, we fail to reject the null hypothesis.

Step 8: Because we fail to reject H_0: $p = 0.93$, we fail to reject the claim that 93% of computers have antivirus programs. We conclude that there is not sufficient sample evidence to warrant rejection of the claim that 93% of computers have antivirus programs. (See Table 8-3 for help with wording this final conclusion.)

Confidence Interval Method

The claim of $p = 0.93$ can be tested with a 0.05 significance level by constructing a 95% confidence interval (as shown in Table 8-1 in Section 8-2). (In general, for *two-tailed* hypothesis tests construct a confidence interval with a confidence level corresponding to the significance level, as in Table 8-1.)

The 95% confidence interval estimate of the population proportion p is found using the sample data consisting of $n = 400$ and $\hat{p} = 380/400$. Using the methods of Section 7-2 we get: $0.929 < p < 0.971$. That interval contains the claimed value of 0.93. Because we are 95% confident that the limits of 0.929 and 0.971 contain the true value of p, we do not have sufficient evidence to warrant rejection of the claim that 93% of computers have antivirus programs. In this case, the conclusion is the same as with the P-value method and the critical value method.

CAUTION When testing claims about a population proportion, the critical value method and the P-value method are equivalent in the sense that they always yield the same results, but the confidence interval method is not equivalent to them and may result in a different conclusion. (Both the critical value method and P-value method use the same standard deviation based on the *claimed proportion p*, but the confidence interval uses an estimated standard deviation based on the *sample proportion.*) Here is a good strategy: Use a confidence interval to *estimate* a population proportion, but use the P-value method or critical value method for *testing a claim* about a proportion. See Exercise 36.

Finding the Number of Successes x

Computer software and calculators designed for hypothesis tests of proportions usually require input consisting of the sample size n and the number of successes x, but the sample proportion is often given instead of x. The number of successes x can be found as illustrated in Example 2. Note that in Example 2, the result of 462.6 people must be rounded to the nearest whole number.

Example 2 Finding the Number of Successes x

A Harris Interactive survey conducted for Gillette showed that among 514 human resource professionals polled, 90% said that appearance of a job applicant is most important for a good first impression. What is the actual number of respondents who said that appearance of a job applicant is most important for a good first impression?

Solution

The number of respondents who said that appearance of a job applicant is most important for a good first impression is 90% of 514, or $0.90 \times 514 = 462.6$, but the result must be a whole number, so we round the product to the nearest whole number of 463.

Win $1,000,000 for ESP

Magician James Randi instituted an educational foundation that offers a prize of $1 million to anyone who can demonstrate paranormal, supernatural, or occult powers. Anyone possessing power such as fortune telling, ESP (extrasensory perception), or the ability to contact the dead can win the prize by passing testing procedures. A preliminary test is followed by a formal test, but so far, no one has passed the preliminary test. The formal test would be designed with sound statistical methods, and it would likely involve analysis with a formal hypothesis test. According to the foundation, "We consult competent statisticians when an evaluation of the results, or experiment design, is required."

Note to Instructor

This subsection requires Chapter 7 as a prerequisite. This subsection discusses the correspondence between confidence intervals and hypothesis testing, but this approach is not emphasized in the following chapters. Greater emphasis is placed on the P-value and critical value methods.

Here is one item that students find a bit tricky: If the hypothesis test is left-tailed or right-tailed, the confidence interval should be constructed using a confidence level that is *twice* the value of the corresponding significance level. For example, if testing $p > 0.3$ with a 0.05 significance level, construct a 90% confidence interval. If testing $p \neq 0.3$ with a 0.05 significance level, construct a 95% confidence interval.

Although a media report about this study used "90%," the more precise percentage of 90.0778% is obtained by using the actual number of respondents who said that appearance of a job applicant is most important for a good first impression ($x = 463$) and the sample size ($n = 514$). When conducting the hypothesis test, better results can be obtained by using the sample proportion of 0.900778 (instead of 0.90).

Example 3 Did the County Clerk Cheat?

The Chapter Problem includes data showing that when the county clerk in Essex County, New Jersey, selected candidates for positions on election ballots, Democrats were selected first in 40 of 41 ballots. Because he was supposed to use a method of random selection, Republicans claimed that instead of using randomness, he used a method that favored Democrats. Use a 0.05 significance level to test the claim that the method favored Democrats.

Solution

Requirement check (1) For the purpose of conducting the hypothesis test, we assume that the county clerk used a process of random selection. (2) There is a fixed number (41) of independent trials with two categories (the Democrat won the top line on the ballot or did not). (3) The requirements $np \geq 5$ and $nq \geq 5$ are both satisfied with $n = 41$ and $p = 0.5$. (We get $np = (41)(0.5) = 20.5$, which is greater than or equal to 5, and we also get $nq = (41)(0.5) = 20.5$, which is greater than or equal to 5.) The three requirements are all satisfied. ✓

If using technology, the test statistic and the P-value will be provided. See the accompanying results from StatCrunch showing that the test statistic is $z = 6.09$ (rounded) and the indication that the P-value is less than 0.0001. The low P-value suggests that we reject the null hypothesis and support the alternative hypothesis that $p > 0.5$, so it appears that the county clerk did use a procedure that favored Democrats for the top position on the election ballots.

STATCRUNCH

Hypothesis test results:
p : proportion of successes for population
$H_0 : p = 0.5$
$H_A : p > 0.5$

Proportion	Count	Total	Sample Prop.	Std. Err.	Z-Stat	P-value
p	40	41	0.9756098	0.07808688	6.090777	<0.0001

If technology is not available, proceed as follows to conduct the hypothesis test using the P-value method summarized in Figure 8-1 from Section 8-2.

Step 1: The original claim is that the clerk used a method that favored Democrats. We express this in symbolic form as $p > 0.5$.

Step 2: The opposite of the original claim is $p \leq 0.5$.

Step 3: Because $p > 0.5$ does not contain equality, it becomes H_1. We get

$$H_0: p = 0.50 \quad \text{(null hypothesis)}$$

$$H_1: p > 0.50 \quad \text{(alternative hypothesis and original claim)}$$

Step 4: The significance level is $\alpha = 0.05$.

Step 5: Because the claim involves the proportion p, the statistic relevant to this test is the sample proportion $\hat{p}$ and the sampling distribution of sample proportions can be approximated by the normal distribution.

Step 6: The test statistic $z = 6.09$ is calculated as follows:

$$z = \frac{\hat{p} - p}{\sqrt{\frac{pq}{n}}} = \frac{\frac{40}{41} - 0.50}{\sqrt{\frac{(0.5)(0.5)}{41}}} = 6.09$$

Refer to Figure 8-4 for the procedure for finding the *P*-value. Figure 8-4 shows that for this right-tailed test, the *P*-value is the area to the right of the test statistic. Using Table A-2, we see that $z = 6.09$ is off the chart, so the area to the right of $z = 6.09$ is 0.0001. (Technology provides a more accurate *P*-value of 0.000000000564.)

Step 7: Because the *P*-value is less than or equal to the significance level of 0.05, we reject the null hypothesis.

Interpretation

Because we reject the null hypothesis, we support the alternative hypothesis. We therefore conclude that there is sufficient evidence to support the Republican claim that the county clerk used a method that favored Democrats. The county clerk lost his job.

Critical Value Method

If we were to repeat Example 3 using the critical value method of testing hypotheses, we would see that in Step 6 the critical value is $z = 1.645$. In Step 7 we would reject the null hypothesis because the test statistic of $z = 6.09$ would fall within the critical region bounded by $z = 1.645$. We would then reach the same conclusion given in Example 3.

Confidence Interval method

If we were to repeat Example 3 using the confidence interval method, we would use a 90% confidence level because we have a right-tailed test. (See Table 8-1.) We get this 90% confidence interval: $0.936 < p < 1.015$. Because the confidence interval limits do not contain the value of 0.5, it is very unlikely that the success rate is actually 50%, so there is sufficient evidence to reject the 50% rate. In this case, the *P*-value method, critical value method, and confidence interval method all lead to the same conclusion.

Part 2: Exact Methods for Testing Claims about a Population Proportion p

Instead of using the normal distribution as an *approximation* to the binomial distribution, we can get *exact* results by using the binomial probability distribution itself. Binomial probabilities are a nuisance to calculate manually, but technology makes this approach quite simple. Also, this exact approach does not require that $np \geq 5$ and $nq \geq 5$, so we have a method that applies when that requirement is not satisfied. To test hypotheses using the exact binomial distribution, use the binomial probability distribution with the *P*-value method, use the value of p assumed in the null hypothesis, and find *P*-values as follows:

Lefties Die Sooner?

A study by psychologists Diane Halpern and Stanley Coren received considerable media attention and generated considerable interest when it concluded that left-handed people don't live as long as right-handed people. Based on their study, it appeared that left-handed people live an average of nine years less than righties. The Halpern/Coren study has been criticized for using flawed data. They used second-hand data by surveying relatives about people who had recently died. The myth of lefties dying younger became folklore that has survived many years. However, more recent studies show that left-handed people do *not* have shorter lives than those who are right-handed.

Note to Instructor
If you have time and your students are using technology, consider including Part 2, which uses *exact* probabilities instead of a normal distribution as an approximation to a binomial distribution.

Left-tailed test: *P*-value is the probability of x or fewer successes among n trials.

Right-tailed test: *P*-value is the probability of x or more successes among n trials.

Two-tailed test: The two-tailed test can be treated with different approaches, some of which are quite complex. For example, Minitab uses a "likelihood ratio test" that is different from the following relatively simple approach:

If $\hat{p} > p$, the *P*-value is twice the probability of x or more successes.

If $\hat{p} < p$, the *P*-value is twice the probability of x or fewer successes.

Example 4 Using the Exact Method

In testing a method of gender selection, 10 randomly selected couples are treated with the method, they each have a baby, and 9 of the babies are girls. Use a 0.05 significance level to test the claim that with this method, the probability of a baby being a girl is greater than 0.75.

Solution

Requirement check The normal approximation method described in Part 1 of this section requires that $np \geq 5$ and $nq \geq 5$, but $nq = (10)(0.25) = 2.5$, so the requirement is violated. The exact method has only the first two requirements listed near the beginning of this section, and those two requirements are satisfied. Here are the null and alternative hypotheses:

$$H_0: p = 0.75 \quad \text{(null hypothesis)}$$

$$H_1: p > 0.75 \quad \text{(alternative hypothesis and original claim)}$$

Instead of using the normal distribution, we use technology to find probabilities in a binomial distribution with $p = 0.75$. Because this is a right-tailed test, the *P*-value is the probability of 9 or more successes among 10 trials, assuming that $p = 0.75$. See the accompanying STATDISK display of exact probabilities from the binomial distribution. This STATDISK display shows that the probability of 9 or more successes is 0.2440252 when rounded to seven decimal places, so the *P*-value is 0.2440252. The *P*-value is high (greater than 0.05), so we fail to reject the null hypothesis. There is not sufficient evidence to support the claim that with the gender selection method, the probability of a girl is greater than 0.75.

STATDISK

Num Trials, n: 10 Evaluate

Success Prob, p: 0.75

Mean: 7.5000
St Dev: 1.3693
Variance: 1.8750

x	P(x)	P(x or fewer)	P(x or greater)
0	0.0000010	0.0000010	1.0000000
1	0.0000286	0.0000296	0.9999990
2	0.0003862	0.0004158	0.9999704
3	0.0030899	0.0035057	0.9995842
4	0.0162220	0.0197277	0.9964943
5	0.0583992	0.0781269	0.9802723
6	0.1459980	0.2241249	0.9218731
7	0.2502823	0.4744072	0.7758751
8	0.2815676	0.7559748	0.5255928
9	0.1877117	0.9436865	0.2440252
10	0.0563135	1.0000000	0.0563135

using TECHNOLOGY

STATDISK Select **Analysis, Hypothesis Testing, Proportion-One Sample,** then enter the data in the dialog box.

MINITAB Select **Stat, Basic Statistics, 1 Proportion,** then click on the button for "Summarized data." Enter the sample size and number of successes, then click on **Options** and enter the data in the dialog box. For the confidence level, enter the complement of the significance level. (Enter 95.0 for a significance level of 0.05.) For the "test proportion" value, enter the proportion used in the null hypothesis. For "alternative," select the format used for the alternative hypothesis. Instead of using a normal approximation, Minitab's default procedure is to determine the *P*-value by using an exact method that is often the same as the one described in Part 2 of this section. (If the test is two-tailed and the assumed value of *p* is not 0.5, Minitab's exact method is different from the one described in Part 2 of this section.) To use the normal approximation method presented in Part 1 of this section, click on the **Options** button and then click on the box with this statement: "Use test and interval based on normal distribution."

EXCEL Use XLSTAT. Click on **XLSTAT** at the top, click on **Parametric Tests,** then select **Tests for one proportion.** In the screen that appears, enter the sample proportion in the "Proportion" box, enter the sample size in the "Sample size" box, and enter the claimed value of the population proportion in the "Test proportion" box. (This is the same proportion used in the null hypothesis.) Select the "Data Format" of **Proportion,** and be sure that the box next to "z test" is checked. For the "Range" box, enter A1 so that the results will be displayed in a position starting at cell A1. Click on the **Options** tab to select the type of test; select the option including $\neq$ for a two-tailed test, select the option including $<$ for a left-tailed test, or select the option including $>$ for a right-tailed test. Enter the desired "Significance level (%)." For example, enter 5 for a 0.05 significance level. Click **OK.** After the results are displayed, look for the test statistic identified as "*z* (Observed value)" and the *P*-value. Critical values will also be displayed.

T1-83/84 PLUS Press STAT, select **TESTS,** and then select **1-PropZTest.** Enter the claimed value of the population proportion for p0, then enter the values for *x* and *n*, and then select the type of test. Highlight **Calculate,** then press ENTER.

STATCRUNCH Click on **Open StatCrunch.** Click on **Stat,** then select **Proportion.** Select **One sample,** then select **with summary.** Proceed to enter the number of successes and the number of observations, click on **Next,** then select **Hypothesis Test.** Enter the claimed value of the population proportion and select the form of the test, then click on **Calculate.**

8-3 Basic Skills and Concepts

Due to limited space, answers for odd numbered exercises are not included here, but they are included in Appendix D.

Recommended Assignment
Exercises 1–20 and 36. Note that the directions for Exercises 9–32 stipulate that the *P*-value method should be used "unless your instructor specifies otherwise." The *P*-value method is recommended, but instructors can stipulate that the critical value method be used for some or all of the assigned exercises. Most Appendix D answers include both *P*-values and critical values.

2. $\hat{p} = 411/1003$ or 0.410. The symbol $\hat{p}$ is used to represent a sample proportion.

4. a. The symbol *p* represents the population proportion, but the *P*-value is a probability of getting sample results that are at least as extreme as those obtained (assuming that the null hypothesis is true).

Statistical Literacy and Critical Thinking

1. Hypothesis Tests and Confidence Intervals We can test a claim about a population proportion using the *P*-value method of hypothesis testing or the critical value method of hypothesis testing, or we could base our conclusion on a confidence interval. Assuming that all three methods are based on the same significance level, which two of the three methods always yield the same conclusion?

2. Sample Proportion In a Wakefield Research survey, respondents were asked if they ever hesitated to give a handshake because of a fear of germs. Of the respondents, 411 answered "yes" and 592 said "no." What is the sample proportion of *yes* responses, and what notation is used to represent it?

3. *P*-Value Using the sample data from Exercise 2, we can test the claim that $p < 0.5$, where *p* denotes the proportion of "yes" responses for the population. Some technologies provide results that include a *P*-value of 5.50E—9. Write that number using ordinary notation. What does the *P*-value suggest about the claim?

4. Notation and *P*-Value

a. Refer to Exercise 2 and distinguish between the value of *p* and the *P*-value.

b. In Section 8-2 we noted that we can remember how to interpret *P*-values with this: "If the *P* is low, the null must go." What does this mean?

c. Another memory trick commonly used is this: "If the *P* is high, the null will fly." Given that a hypothesis test never results in a conclusion of proving or supporting a null hypothesis, how is this memory trick misleading?

b. If the *P*-value is very low (such as less than or equal to 0.05), "the null must go" means that we should reject the null hypothesis.

c. The statement that "if the *P* is high, the null will fly" suggests that with a high *P*-value, the null hypothesis has been proved or is supported, but we should never make such a conclusion.

In Exercises 5–8, identify the indicated values or interpret the given display. Use the normal distribution as an approximation to the binomial distribution (as described in Part 1 of this section). Assume a 0.05 significance level and answer the following:

a. Is the test two-tailed, left-tailed, or right-tailed?

b. What is the test statistic?

c. What is the *P*-value?

d. What is the null hypothesis, and what do you conclude about it?

e. What is the final conclusion?

5. Adverse Reactions to Drug The drug Symbicort is used to treat asthma. In a clinical trial of Symbicort, 18 of 277 treated subjects experienced headaches (based on data from AstraZeneca). The accompanying TI-83/84 Plus calculator display shows results from a test of the claim that less than 10% of treated subjects experienced headaches.

TI-83/84 PLUS

```
1-PropZTest
 prop<.1
 z=-1.942721717
 p=.0260248259
 p̂=.0649819495
 n=277
```

6. Guns in the Home In a Gallup poll of 1003 randomly selected subjects, 373 said that they have a gun in their home. The accompanying Minitab display shows results from a test of the claim that 35% of homes have guns in them.

MINITAB

```
Test of p = 0.35 vs p not = 0.35

Variable    X     N  Sample p         95% CI              Z-Value  P-Value
Guns      373  1003  0.371884  (0.341974, 0.401795)      1.45    0.146
```

6. a. Two-tailed.
b. $z = 1.45$
c. *P*-value: 0.146
d. H_0: $p = 0.35$. Fail to reject the null hypothesis.
e. There is not sufficient evidence to warrant rejection of the claim that 35% of homes have guns in them.

7. Brand Recognition A PriceGrabber.com survey of 1631 randomly selected adults showed that 555 of them have heard of the Sony Reader. The following STATDISK display results from a test of the claim that 35% of adults have heard of the Sony Reader.

STATDISK

Sample proportion:	0.340282
Test Statistic,	z: -0.8228
Critical z:	±1.9600
P-Value:	0.4106

8. Public Speaking A TNS poll of 1000 randomly selected adults showed that 460 of them say that public speaking is an activity that they dread most. Shown next is the StatCrunch display resulting from a test of the claim that fewer than half of adults say that public speaking is the activity that they dread most.

STATCRUNCH

Hypothesis test results:
p : proportion of successes for population
H_0 : p = 0.5
H_A : p < 0.5

Proportion	Count	Total	Sample Prop.	Std. Err.	Z-Stat	P-value
p	460	1000	0.46	0.015811387	-2.529822	0.0057

8. a. Left-tailed.
b. $z = -2.53$
c. *P*-value: 0.0057
d. H_0: $p = 0.5$. Reject the null hypothesis.
e. There is sufficient evidence to support the claim that fewer than half of adults say that public speaking is the activity that they dread most.

Testing Claims about Proportions. ***In Exercises 9–32, test the given claim. Identify the null hypothesis, alternative hypothesis, test statistic, P-value or critical value(s), conclusion about the null hypothesis, and final conclusion that addresses the original claim. Use the P-value method unless your instructor specifies otherwise. Use the normal distribution as an approximation to the binomial distribution (as described in Part 1 of this section).***

9. Mendelian Genetics When Mendel conducted his famous genetics experiments with peas, one sample of offspring consisted of 428 green peas and 152 yellow peas. Use a 0.01

significance level to test Mendel's claim that under the same circumstances, 25% of offspring peas will be yellow.

10. M&Ms Data Set 20 in Appendix B lists data from 100 M&Ms, and 8% of them are brown. Use a 0.05 significance level to test the claim of the Mars candy company that the percentage of brown M&Ms is equal to 13%.

10. H_0: $p = 0.13$. H_1: $p \neq 0.13$. Test statistic: $z = -1.49$. Critical values: $z = \pm 1.96$. *P*-value: 0.1362 (Tech: 0.1371). Fail to reject H_0. There is not sufficient evidence to warrant rejection of the claim that 13% of M&Ms are brown.

11. Identify Theft In a KRC Research poll, 1002 adults were asked if they felt vulnerable to identify theft, and 531 of them said "yes." Use a 0.05 significance level to test the claim that the majority of adults feel vulnerable to identity theft.

12. Plane Seats In a 3M Privacy Filters poll, 806 adults were asked to identify their favorite seat when they fly, and 492 of them chose a window seat. Use a 0.01 significance level to test the claim that the majority of adults prefer window seats when they fly.

12. H_0: $p = 0.5$. H_1: $p > 0.5$. Test statistic: $z = 6.27$. Critical value: $z = 2.33$. *P*-value: 0.0001 (Tech: 0.000000000182). Reject H_0. There is sufficient evidence to support the claim that the majority of adults prefer window seats when they fly.

13. Gender Selection The Genetics & IVF Institute conducted a clinical trial of the XSORT method designed to increase the probability of conceiving a girl. As of this writing, 945 babies were born to parents using the XSORT method, and 879 of them were girls. Use a 0.01 significance level to test the claim that the XSORT method is effective in increasing the likelihood that a baby will be a girl.

14. Gender Selection The Genetics & IVF Institute conducted a clinical trial of the YSORT method designed to increase the probability of conceiving a boy. As of this writing, 291 babies were born to parents using the YSORT method, and 239 of them were boys. Use a 0.01 significance level to test the claim that the YSORT method is effective in increasing the likelihood that a baby will be a boy.

14. H_0: $p = 0.5$. H_1: $p > 0.5$. Test statistic: $z = 10.96$. Critical value: $z = 2.33$. *P*-value: 0.0001 (Tech: 0.0000). Reject H_0. There is sufficient evidence to support the claim that the YSORT method is effective in increasing the likelihood that a baby will be a boy.

15. Touch Therapy When she was 9 years of age, Emily Rosa did a science fair experiment in which she tested professional touch therapists to see if they could sense her energy field. She flipped a coin to select either her right hand or her left hand, then she asked the therapists to identify the selected hand by placing their hand just under Emily's hand without seeing it and without touching it. Among 280 trials, the touch therapists were correct 123 times (based on data in "A Close Look at Therapeutic Touch," *Journal of the American Medical Association,* Vol. 279, No. 13). Use a 0.10 significance level to test the claim that touch therapists use a method equivalent to random guesses. Do the results suggest that touch therapists are effective?

16. Touch Therapy Repeat the preceding exercise using a 0.01 significance level. Does the conclusion change?

16. H_0: $p = 0.5$. H_1: $p \neq 0.5$. Test statistic: $z = -2.03$. Critical values: $z = \pm 2.575$ (Tech: ± 2.576). *P*-value: 0.0424 (Tech: 0.0422). Fail to reject H_0. There is not sufficient evidence to warrant rejection of the claim that touch therapists use a method equivalent to random guesses. After changing the significance level from 0.10 to 0.01, the conclusion does change.

17. Tennis Instant Replay The Hawk-Eye electronic system is used in tennis for displaying an instant replay that shows whether a ball is in bounds or out of bounds so players can challenge calls made by referees. In the most recent U.S. Open (as of this writing), singles players made 611 challenges and 172 of them were successful with the call overturned. Use a 0.01 significance level to test the claim that fewer than 1/3 of the challenges are successful. What do the results suggest about the ability of players to see calls better than referees?

18. Perception and Reality In a presidential election, 308 out of 611 voters surveyed said that they voted for the candidate who won (based on data from ICR Survey Research Group). Use a 0.10 significance level to test the claim that among all voters, the percentage who believe that they voted for the winning candidate is equal to 43%, which is the actual percentage of votes for the winning candidate. What does the result suggest about voter perceptions?

18. H_0: $p = 0.43$. H_1: $p \neq 0.43$. Test statistic: $z = 3.70$. Critical values: $z = \pm 1.645$. *P*-value: 0.0002. Reject H_0. There is sufficient evidence to warrant rejection of the claim that the percentage who believe that they voted for the winning candidate is equal to 43%. There appears to be a substantial discrepancy between how people said that they voted and how they actually did vote.

19. Cell Phones and Cancer In a study of 420,095 Danish cell phone users, 135 subjects developed cancer of the brain or nervous system (based on data from the *Journal of the National Cancer Institute* as reported in *USA Today*). Test the claim of a somewhat common belief that such cancers are affected by cell phone use. That is, test the claim that cell phone users develop cancer of the brain or nervous system at a rate that is different from the rate of 0.0340% for people who do not use cell phones. Because this issue has such great importance, use a 0.005 significance level. Based on these results, should cell phone users be concerned about cancer of the brain or nervous system?

20. H_0: $p = 0.75$. H_1: $p > 0.75$. Test statistic: $z = 7.33$. Critical value: $z = 2.33$. P-value: 0.0001 (Tech: 0.0000). Reject H_0. There is sufficient evidence to support the claim that more than 75% of adults know what Twitter is.

22. H_0: $p = 0.5$. H_1: $p > 0.5$. Test statistic: $z = 0.83$. Critical value: $z = 1.645$. P-value: 0.2033 (Tech: 0.2031). Fail to reject H_0. There is not sufficient evidence to support the claim that among smokers who try to quit with nicotine patch therapy, the majority are smoking a year after the treatment. The results show that about half of those who use nicotine patch therapy are successful in quitting smoking.

24. H_0: $p = 0.5$. H_1: $p < 0.5$. Test statistic: $z = 1.13$. Critical value: $z = -1.645$. P-value: 0.8708 (Tech: 0.8712). Fail to reject H_0: $p = 0.5$. There is not sufficient evidence to support the claim that less than 0.5 of the deaths occur the week before Thanksgiving. Based on these results, there is no indication that people can temporarily postpone their death to survive Thanksgiving.

26. H_0: $p = 0.5$. H_1: $p < 0.5$. Test statistic: $z = -1.86$ (using $\hat{p} = 0.459$) or $z = -1.85$ (using $x = 236$). Critical value: $z = -2.33$. P-value: 0.0314 (using $\hat{p} = 0.459$) or 0.0322 (using $x = 236$) (Tech P-value: 0.0320). Fail to reject H_0. There is not sufficient evidence to support the claim that less than half of all human resource professionals say that body piercings are big grooming red flags.

28. H_0: $p = 0.61$. H_1: $p \neq 0.61$. Test statistic: $z = 5.84$ (using $\hat{p} = 0.70$) or $z = 5.81$ (using $x = 701$). Critical values: $z = \pm 1.96$ (assuming a 0.05 significance level). P-value: 0.0002 (Tech: 0.0000). Reject H_0. There is sufficient evidence to warrant rejection of the claim that the percentage of all voters who say that they voted is equal to 61%. The results suggest that either survey respondents are not being truthful or they have an incorrect perception of reality.

20. Twitter Poll The Pew Research Center conducted a survey of 1007 adults and found that 856 of them know what Twitter is. Use a 0.01 significance level to test the claim that more than 75% of adults know what Twitter is.

21. Overtime Rule in Football In "The Overtime Rule in the National Football League: Fair or Unfair?" by Gorgievski et al., *MathAMATYC Educator*, Vol. 2, No. 1, the authors report that among 414 football games won in overtime, 235 were won by the team that won the coin toss at the beginning of overtime. Using a 0.05 significance level, test the claim that the coin toss is fair in the sense that neither team has an advantage by winning it.

22. Testing Effectiveness of Nicotine Patches In one study of smokers who tried to quit smoking with nicotine patch therapy, 39 were smoking one year after the treatment and 32 were not smoking one year after the treatment (based on data from "High-Dose Nicotine Patch Therapy," by Dale et al., *Journal of the American Medical Association,* Vol. 274, No. 17). Use a 0.05 significance level to test the claim that among smokers who try to quit with nicotine patch therapy, the majority are smoking a year after the treatment. What do these results suggest about the effectiveness of nicotine patch therapy for those trying to quit smoking?

23. Smartphone Users A RingCentral survey of 380 smartphone users showed that 152 of them said that their smartphone is the only thing they could not live without. Use a 0.01 significance level to test the claim that fewer than half of smartphone users identify the smartphone as the only thing they could not live without. Do these results apply to the general population?

24. Postponing Death An interesting and popular hypothesis is that individuals can temporarily postpone death to survive a major holiday or important event such as a birthday. In a study of this phenomenon, it was found that there were 6062 deaths in the week before Thanksgiving, and 5938 deaths the week after Thanksgiving (based on data from "Holidays, Birthdays, and Postponement of Cancer Death," by Young and Hade, *Journal of the American Medical Association,* Vol. 292, No. 24). If people can postpone death until after Thanksgiving, then the proportion of deaths in the week before should be less than 0.5. Use a 0.05 significance level to test the claim that the proportion of deaths in the week before Thanksgiving is less than 0.5. Based on the result, does there appear to be any indication that people can temporarily postpone death to survive the Thanksgiving holiday?

25. Online Books A *Consumer Reports* Research Center survey of 427 women showed that 29.0% of them purchase books online. Test the claim that more than 25% of women purchase books online.

26. Job Interviews In a Harris poll of 514 human resource professionals, 45.9% said that body piercings and tattoos were big grooming red flags. Use a 0.01 significance level to test the claim that less than half of all human resource professionals say that body piercings are big grooming red flags.

27. Job Interviews In a Harris poll of 514 human resource professionals, 90% said that the appearance of a job applicant is most important for a good first impression. Use a 0.01 significance level to test the claim that more than 3/4 of all human resource professionals say that the appearance of a job applicant is most important for a good first impression.

28. Misleading Survey Responses Voting records show that 61% of eligible voters actually did vote in a recent presidential election. In a survey of 1002 people, 70% said that they voted in that election (based on data from ICR Research Group). Use the survey results to test the claim that the percentage of all voters who say that they voted is equal to 61%. What do the results suggest?

29. Bias in Jury Selection In the case of *Casteneda v. Partida,* it was found that during a period of 11 years in Hidalgo County, Texas, 870 people were selected for grand jury duty and 39% of them were Americans of Mexican ancestry. Among the people eligible for grand jury duty, 79.1% were Americans of Mexican ancestry. Use a 0.01 significance level to test the claim that the selection process is biased against Americans of Mexican ancestry. Does the jury selection system appear to be fair?

30. Finding a Job through Networking In a survey of 703 randomly selected workers, 61% got their jobs through networking (based on data from Taylor Nelson Sofres Research). Use the sample data with a 0.05 significance level to test the claim that most (more than 50%) workers get their jobs through networking. What does the result suggest about the strategy for finding a job after graduation?

31. Nielsen Ratings A recent Super Bowl football game between the New Orleans Saints and the Indianapolis Colts set a record for the number of television viewers. According to the Nielsen company, that game had a share of 77%, meaning that among the television sets in use at the time of the game, 77% were tuned to the game. Nielsen's sample size is 25,000 households. Use a 0.01 significance level to test the claim of a media report that more than 75% of television sets in use were tuned to the Super Bowl.

32. HDTV Penetration In a survey of 1500 households, it is found that 47% of them have a high-definition television (based on data from the Consumer Electronics Association). Use a 0.01 significance level to test the claim that fewer than half of all households have a high-definition television. Is the result from a few years ago likely to be valid today?

Large Data Sets. ***In Exercises 33 and 34, use the data set from Appendix B to test the given claim.***

33. M&Ms Refer to Data Set 20 in Appendix B and find the sample proportion of M&Ms that are green. Use that result to test the claim of Mars, Inc., that 16% of its plain M&M candies are green.

34. On-Time Flights Consider a flight to be on time if it arrives no later than 15 minutes after the scheduled arrival time. Refer to Data Set 15 in Appendix B, and use the sample data to test the claim made by CNN that 79.5% of flights are on time. Use a 0.05 significance level.

30. H_0: $p = 0.5$. H_1: $p > 0.5$. Test statistic: $z = 5.83$ (using $\hat{p} = 0.61$) or $z = 5.85$ (using $x = 429$). Critical value: $z = 1.645$. P-value: 0.0001 (Tech: 0.0000). Reject H_0. There is sufficient evidence to support the claim that most workers get their jobs through networking.

32. H_0: $p = 0.5$. H_1: $p < 0.5$. Test statistic: $z = -2.32$. Critical value: $z = -2.33$. P-value: 0.0102 (Tech: 0.0101). Fail to reject H_0. There is not sufficient evidence to support the claim that fewer than half of all households have a high-definition television. Because the use of high-definition televisions is growing rapidly, these results are not likely to be valid today.

34. Among 48 flights, 44 are on time. H_0: $p = 0.795$. H_1: $p \neq 0.795$. Test statistic: $z = 2.09$. Critical values: $z = \pm 1.96$. P-value: 0.0366 (Tech: 0.0368). Reject H_0. There is sufficient evidence to warrant rejection of the claim that 79.5% of flights are on time. With 91.7% of the 48 flights arriving on time, American Airlines appears to have a better on-time performance.

8-3 Beyond the Basics

35. Binomial Distribution We want to use a 0.05 significance level to test the claim that a coin favors heads, and sample data consist of 7 heads in 8 tosses. We can't use the normal approximation method because we violate the requirement that $np \geq 5$ and $nq \geq 5$. Use the binomial probability distribution to test the claim.

36. Using Confidence Intervals to Test Hypotheses When analyzing the last digits of telephone numbers in Port Jefferson, it is found that among 1000 randomly selected digits, 119 are zeros. If the digits are randomly selected, the proportion of zeros should be 0.1.

a. Use the critical value method with a 0.05 significance level to test the claim that the proportion of zeros equals 0.1.

b. Use the P-value method with a 0.05 significance level to test the claim that the proportion of zeros equals 0.1.

c. Use the sample data to construct a 95% confidence interval estimate of the proportion of zeros. What does the confidence interval suggest about the claim that the proportion of zeros equals 0.1?

d. Compare the results from the critical value method, the P-value method, and the confidence interval method. Do they all lead to the same conclusion?

37. Power For a hypothesis test with a specified significance level α, the probability of a type I error is α, whereas the probability β of a type II error depends on the particular value of p that is used as an alternative to the null hypothesis.

continued

36. a. H_0: $p = 0.10$. H_1: $p \neq 0.10$. Test statistic: $z = 2.00$. Critical values: $z = \pm 1.96$. Reject H_0. There is sufficient evidence to warrant rejection of the claim that the proportion of zeros is 0.1.

b. H_0: $p = 0.10$. H_1: $p \neq 0.10$. Test statistic: $z = 2.00$. P-value: 0.0456 (Tech: 0.0452). There is sufficient evidence to warrant rejection of the claim that the proportion of zeros is 0.1.

c. $0.0989 < p < 0.139$; because 0.1 is contained within the confidence interval, fail to reject H_0: $p = 0.10$. There is not sufficient evidence to warrant rejection of the claim that the proportion of zeros is 0.1.

d. The traditional and P-value methods both lead to rejection of the claim, but the confidence interval method does not lead to rejection of the claim.

a. Using an alternative hypothesis of $p < 0.4$, a sample size of $n = 50$, and assuming that the true value of p is 0.25, find the power of the test. See Exercise 36 in Section 8-2. (*Hint:* Use the values $p = 0.25$ and $pq/n = (0.25)(0.75)/50$.)

b. Find the value of β, the probability of making a type II error.

c. Given the conditions cited in part (a), what do the results indicate about the effectiveness of the hypothesis test?

Note to Instructor
Organization Change: In this new edition, Section 8-4 ("Testing a Claim about a Mean") is partitioned into Part 1 (for σ not known) and Part 2 (for σ known). The main focus in Section 8-4 is using the Student *t* distribution for testing a claim about a population mean. Exercises 29–32 are the only exercises in which σ is assumed to be known. The previous edition of this book had two sections, Section 8-4 ("Testing a Claim about a Mean: σ Known") and Section 8-5 ("Testing a Claim about a Mean: σ Not Known"), but this edition combines those two sections into one section (Section 8-4).

8-4 Testing a Claim about a Mean

Key Concept Because this section presents methods for testing a claim about a population mean, it is one of the most important sections in this book. Part 1 of this section deals with the very realistic and commonly used case in which the population standard deviation σ is not known. Part 2 includes a brief discussion of the procedure used when σ is known, which is very rare.

Part 1: Testing a Claim about a Population Mean When σ Is Not Known

In reality, it is very rare that we test a claim about an unknown value of a population mean but we somehow know the value of the population standard deviation σ. The realistic situation is that we test a claim about a population mean and the value of the population standard deviation σ is not known. When σ is not known, we use a "t test" that incorporates the Student t distribution. The requirements, test statistic, *P*-value, and critical values are summarized as follows.

Testing Claims about a Population Mean (with σ Not Known)

Use the formal method of hypothesis testing to test a claim about a population mean.

Notation

n = sample size

$\bar{x}$ = *sample* mean

$\mu_{\bar{x}}$ = *population* mean (this value is taken from the claim and is used in the statement of the null hypothesis)

Requirements

1. The sample is a simple random sample.

2. Either or both of these conditions are satisfied: The population is normally distributed or $n > 30$.

Test Statistic for Testing a Claim about a Mean

$$t = \frac{\bar{x} - \mu_{\bar{x}}}{\frac{s}{\sqrt{n}}}$$ (Round t to three decimal places, as in Table A-3.)

***P*-values:** Use technology or use the Student t distribution (Table A-3) with degrees of freedom given by df $= n - 1$. (Figure 8-4 in Section 8-2 summarizes the procedure for finding *P*-values.)

Critical values: Use the Student t distribution (Table A-3) with degrees of freedom given by df $= n - 1$. (If using Table A-3 to find a critical value of t, but the table does not include the number of degrees of freedom, you could be conservative by using the next lower number of degrees of freedom found in the table, or you could use the closest number of degrees of freedom in the table, or you could interpolate.)

Requirement of Normality or $n > 30$ This t test is *robust* against a departure from normality, meaning that the test works reasonably well if the departure from normality is not too extreme. Verify that there are no outliers and that the histogram or dotplot has a shape that is not very far from a normal distribution.

If the original population is not itself normally distributed, we use the condition $n > 30$ for justifying use of the normal distribution, but there is no exact specific minimum sample size that works for all cases. Sample sizes of 15 to 30 are sufficient if the population has a distribution that is not far from normal, but some other populations have distributions that are extremely far from normal, and sample sizes greater than 30 might be necessary. In this text we use the simplified criterion of $n > 30$ as justification for treating the distribution of sample means as a normal distribution, regardless of how far the distribution departs from a normal distribution.

Here is a brief review of important properties of the Student t distribution first presented in Section 7-3:

Important Properties of the Student t Distribution

1. The Student t distribution is different for different sample sizes (see Figure 7-5 in Section 7-3).
2. The Student t distribution has the same general bell shape as the standard normal distribution; its wider shape reflects the greater variability that is expected when s is used to estimate σ.
3. The Student t distribution has a mean of $t = 0$ (just as the standard normal distribution has a mean of $z = 0$).
4. The standard deviation of the Student t distribution varies with the sample size and is greater than 1 (unlike the standard normal distribution, which has $\sigma = 1$).
5. As the sample size n gets larger, the Student t distribution gets closer to the standard normal distribution.

Note to Instructor

This section uses the same t distribution first presented in Section 7-3. Also, the criteria for choosing between a normal distribution and a Student t distribution are the same in this chapter as they are in Chapter 7.

Recommendation: Conduct an actual t test with data collected from students in the class. Before class, measure your own pulse rate. In class, ask students to measure their pulse rates as the number of beats in 1 minute. Proceed to test the claim that the class has a mean pulse rate different from yours. The active involvement will provide a better learning experience.

Example 1 **Cell Phone Radiation: *P*-Value Method**

Listed below are the measured radiation emissions (in W/kg) corresponding to a sample of these cell phones: Samsung SGH-tss9, Blackberry Storm, Blackberry Curve, Motorola Moto, T-Mobile Sidekick, Sanyo Katana Eclipse, Palm Pre, Sony Ericsson, Nokia 6085, Apple iPhone 3G S, and Kyocera Neo E1100 (based on data from the Environmental Working Group). Use a 0.05 significance level to test the claim that cell phones have a mean radiation level that is less than 1.00 W/kg.

0.38 0.55 1.54 1.55 0.50 0.60 0.92 0.96 1.00 0.86 1.46

Solution

Requirement check (1) For the purposes of this test, we will assume that the sample is a simple random sample, but it appears that one cell phone of each brand was measured, so this is not a simple random sample of the cell phones in use. This will be discussed further in the interpretation that follows. (2) The sample size is $n = 11$, which is not greater than 30, so we must determine whether the sample appears to be from a population having a normal distribution. With only 11 sample values, a histogram isn't very helpful, but there are no outliers and the accompanying normal quantile plot shows that the points are reasonably close to a straight-line pattern and there is no other pattern that is not a straight line. We conclude that the sample data appear to be from a normally distributed population. The requirements are satisfied.

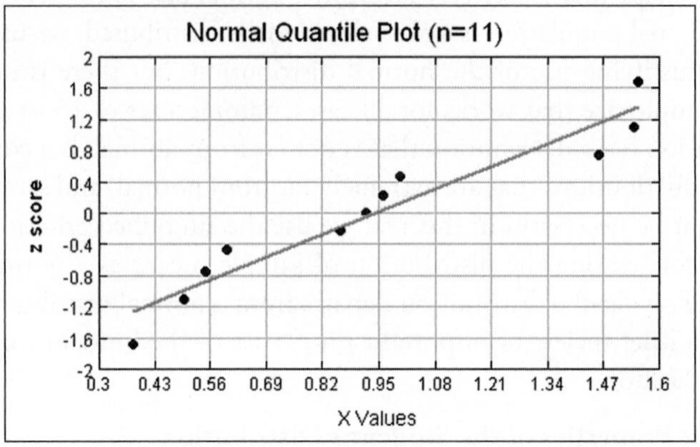

Technology We could use technology to obtain the *P*-value. Shown here are the TI-83/84 Plus calculator results for this hypothesis test, and we can see that the *P*-value is 0.3191 (rounded). Because the *P*-value is high (greater than the significance level of $\alpha = 0.05$), we fail to reject the null hypothesis and conclude that there is not sufficient evidence to support the claim that cell phones have a mean radiation level less than 1.00 W/kg.

TI-83/84 PLUS

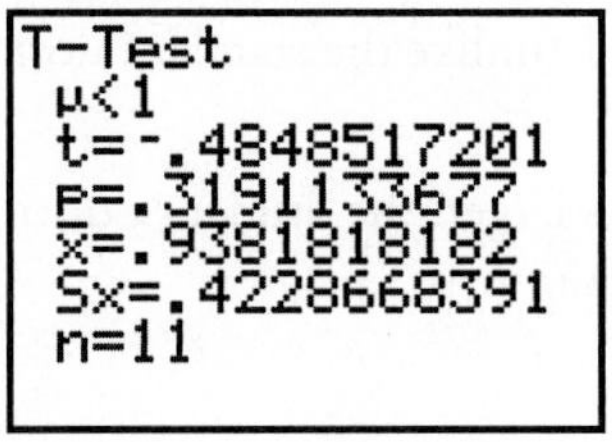
T-Test
μ<1
t=-.4848517201
p=.3191133677
x̄=.9381818182
Sx=.4228668391
n=11

If technology is not available, we can proceed to use the *P*-value approach summarized in Figure 8-1 or the critical value approach summarized in Figure 8-2. Because of the format of Table A-3, *P*-values can be somewhat difficult to find using that table, so the critical value approach is easier. We will proceed using the critical value approach.

Step 1: The claim that cell phones have a mean radiation level less than 1.00 W/kg is expressed in symbolic form as $\mu < 1.00$ W/kg.

Step 2: The alternative (in symbolic form) to the original claim is $\mu \geq 1.00$ W/kg.

Step 3: Because the statement $\mu < 1.00$ W/kg does not contain the condition of equality, it becomes the alternative hypothesis H_1. The null hypothesis H_0 is the statement that $\mu = 1.00$ W/kg.

H_0: $\mu = 1.00$ W/kg (null hypothesis)

H_1: $\mu < 1.00$ W/kg (alternative hypothesis and original claim)

Step 4: As specified in the statement of the problem, the significance level is $\alpha = 0.05$.

Step 5: Because the claim is made about the *population mean* μ, the sample statistic most relevant to this test is the *sample mean* $\bar{x}$. We use the *t* distribution as indicated in the preceding summary box.

Step 6: The sample statistics of $n = 11$ and $\bar{x} = 0.938$ W/kg and the value of $s = 0.423$ W/kg are used to calculate the test statistic as follows:

$$t = \frac{\bar{x} - \mu_{\bar{x}}}{\frac{s}{\sqrt{n}}} = \frac{0.938 - 1.00}{\frac{0.423}{\sqrt{11}}} = -0.486$$

Using this test statistic of $t = -0.486$, we now proceed to find the critical value from Table A-3. With df $= n - 1 = 10$, refer to Table A-3 and use the column corresponding to an area of 0.05 in one tail to find that the critical value is $t = -1.812$, which is shown in Figure 8-8.

Step 7: Because the test statistic of $t = -0.486$ does not fall in the critical region bounded by the critical value of $t = -1.812$ as shown in Figure 8-8, fail to reject the null hypothesis.

Interpretation

Because we fail to reject the null hypothesis, we conclude that there is not sufficient evidence to support the claim that cell phones have a mean radiation level that is less than 1.00 W/kg.

The validity of this conclusion depends on a sound sampling method. We can see from the given list of cell phone models that one phone was measured for each of 11 different models, so it is possible that we do not have a simple random sample of cell phones selected from the population of cell phones in use. It would be wise to further investigate whether this affects our results.

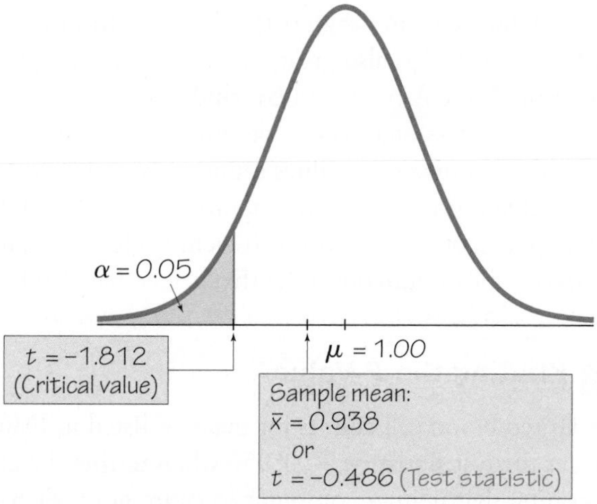

Figure 8-8 ***t*-Test: Critical Value Method**

Finding *P*-Values with the Student *t* Distribution

Example 1 used the critical value approach to hypothesis testing, but STATDISK, Minitab, XLSTAT, StatCrunch, the TI-83/84 Plus calculator, and many articles in professional journals will display *P*-values. For the preceding example, STATDISK, Minitab, XLSTAT, StatCrunch, and the TI-83/84 Plus calculator display a *P*-value of 0.3191. (Minitab displays the rounded value of 0.319.) With a significance level of 0.05 and a *P*-value greater than 0.05, we fail to reject the null hypothesis, as we did using the critical value method in Example 1. If computer software or a TI-83/84

Note to Instructor
The *P*-value approach is more difficult in this section than it was in Section 8-3. The difficulty arises from the format of Table A-3, which usually does not allow you to get exact *P*-values. If your students are not using technology, carefully explain the use of Table A-3 for finding a range of values for the *P*-value.

Plus calculator is not available, we can use Table A-3 to identify a *range of values* containing the P-value. We recommend this strategy for finding P-values using the t distribution:

1. Use software or a TI-83/84 Plus calculator. (STATDISK, Minitab, XLSTAT, StatCrunch, the TI-83/84 Plus calculator, SPSS, and SAS all provide P-values for t tests.)
2. If technology is not available, use Table A-3 to identify a range of P-values as follows: Use the number of degrees of freedom to locate the relevant row of Table A-3, then determine where the test statistic lies relative to the t values in that row. Based on a comparison of the t test statistic and the t values in the row of Table A-3, identify a range of values by referring to the area values given at the top of Table A-3.

Example 2 Finding the P-Value

Assuming that neither computer software nor a TI-83/84 Plus calculator is available, use Table A-3 to find a range of values for the P-value corresponding to the test statistic of $t = -0.486$ from Example 1.

Solution

Requirement check The requirements have already been verified in Example 1. ✓

Example 1 involves a left-tailed test, so the P-value is the area to the left of the test statistic $t = -0.486$. Because the sample size is $n = 11$, refer to Table A-3 and locate the row corresponding to 10 degrees of freedom (df $= 11 - 1 = 10$). The test statistic of $t = 0.486$ is less than every value in the row for df $= 10$, so the "area in one tail" (to the right of the test statistic) is greater than 0.10. If the area to the right of $t = 0.486$ is greater than 0.10, it follows from symmetry that the area to the left of $t = -0.486$ is also greater than 0.10. Although we can't find the exact P-value from Table A-3, we can conclude that the P-value is greater than 0.10. When creating a professional report, definitely use technology to find an exact P-value instead of a range of P-values found from Table A-3.

Because the P-value is greater than the significance level of 0.05, we again fail to reject the null hypothesis. There is not sufficient evidence to support the claim that cell phones have a mean radiation level that is less than 1.00 W/kg.

Example 3 Finding the P-Value

When using the 40 red blood cell counts for males as listed in Data Set 1 in Appendix B, we get the test statistic $t = 1.956$ when testing the claim that $\mu = 4.950$. Assuming that neither computer software nor a TI-83/84 Plus calculator is available, use Table A-3 to find a range of values for the P-value.

Solution

Because the sample size is 40, refer to Table A-3 and locate the row corresponding to 39 degrees of freedom (df $= n - 1$). In the 39th row, the test statistic $t = 1.956$ falls between 2.023 and 1.685, so the area in two tails is between 0.05 and 0.10. (Be sure to use the "Area in Two Tails" whenever the test is two-tailed.) Although we can't find the exact P-value, we can conclude that the P-value is greater than 0.05 and less than 0.10. (Computer software and the TI-83/84 Plus calculator provide the exact P-value of 0.0576.)

"How Statistics Can Help Save Failing Hearts"

A *New York Times* article by David Leonhardt featured the headline of "How Statistics Can Help Save Failing Hearts." Leonhardt writes that patients have the best chance of recovery if their clogged arteries are opened within two hours of a heart attack. In 2005, the U.S. Department of Health and Human Services began posting hospital data on its Web site www.hospitalcompare.hhs.gov, and it included the percentage of heart attack patients who received treatment for blocked arteries within two hours of arrival at the hospital. Not wanting to be embarrassed by poor data, doctors and hospitals are reducing the time it takes to unblock those clogged arteries. Leonhardt writes about the University of California, San Francisco Medical Center, which cut its time in half from almost three hours to about 90 minutes. Effective use of simple statistics can save lives.

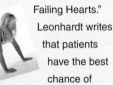

Remember, *P*-values can be easily found using computer software or a TI-83/84 Plus calculator. Also, this difficulty in finding *P*-values can be avoided by using the critical value method of testing hypotheses instead of the *P*-value method.

Confidence Interval Method

We can use a confidence interval for testing a claim about μ. For a two-tailed hypothesis test with a 0.05 significance level, we construct a 95% confidence interval, but for a one-tailed hypothesis test with a 0.05 significance level, we construct a 90% confidence interval (as described in Table 8-1 in Section 8-2).

In Section 8-3 we saw that when testing a claim about a population *proportion*, the critical value method and *P*-value method are equivalent, but the confidence interval method is somewhat different. When testing a claim about a population mean, there is no such difference, and all three methods are equivalent.

Example 4 Confidence Interval Method

Use the sample data from Example 1 to construct a confidence interval that can be used to test the claim that $\mu < 1.00$ W/kg, assuming a 0.05 significance level.

Solution

Requirement check The requirements have already been verified in Example 1. ✓

We note that a left-tailed hypothesis test conducted with a 0.05 significance level corresponds to a confidence interval with a 90% confidence level. (See Table 8-1 in Section 8-2.) Using the methods described in Section 7-4, we construct this 90% confidence interval estimate of the population mean:

$$0.707 \text{ W/kg} < \mu < 1.169 \text{ W/kg}$$

Because the assumed value of $\mu = 1.00$ W/kg is contained within the confidence interval, we cannot reject the null hypothesis that $\mu = 1.00$ W/kg. Based on the 11 sample values given in Example 1, we do not have sufficient evidence to support the claim that the mean radiation level is less than 1.00 W/kg.

Example 5 Fatal Overloading

When a plane crashed in Charlotte, North Carolina, 21 passengers were killed. In Lake George, New York, 20 passengers died when the *Ethan Allen* tour boat capsized. In Baltimore's Inner Harbor, 5 passengers died when a water taxi sank. In all of these fatal incidents, overloading was thought to be a contributing factor. Loading capacities were based on old estimates of the mean weight of men, but that mean has increased in recent years with the effect that some boats and airplanes have been overloaded and unsafe. Using the weights of the simple random sample of men from Data Set 1 in Appendix B, we obtain these sample statistics: $n = 40$, $\bar{x} = 182.9$ lb, and $s = 40.8$ lb. (The original weights are in kilograms, and they are converted to pounds before computing these statistics.) Use these results to test the claim that men have a mean weight greater than 166.3 lb, which was the weight in the National Transportation and Safety Board's recommendation M-04-04. Use a 0.05 significance level.

Solution

Requirement check (1) The sample is a simple random sample. (2) Because the sample size of $n = 40$ is greater than 30, we satisfy the requirement that "the population is normally distributed or $n > 30$." The necessary requirements are satisfied.

The t test results are shown in the accompanying screen displays. STATDISK and the TI-83/84 Plus calculator both provide a P-value of 0.0070 (rounded). SPSS provides only a two-tailed P-value of 0.014, so that value should be halved because this test is right-tailed. Using the rule that "if the P-value is low, the null must go," we reject the null hypothesis because the P-value of 0.0070 is less than the significance level of 0.05. We conclude that there is sufficient evidence to support the claim that men have a mean weight greater than the mean of 166.3 lb as was assumed by the National Transportation and Safety Board.

The use of technology makes the t test quite easy, but it is essential to understand the procedure along with the requirements and interpretations. Blind and thoughtless use of technology could easily lead to serious errors.

STATDISK

```
t Test
Test Statistic, t: 2.5747
Critical t:        1.6849
P-Value:           0.0070

90% Confidence interval:
172.0467 < μ < 193.8104
```

TI-83/84 PLUS

```
T-Test
μ>166.3
t=2.574655673
p=.0069720329
x̄=182.92856
Sx=40.8475
n=40
```

SPSS

	Test Value = 166.3					
					95% Confidence Interval of the Difference	
	t	df	Sig. (2-tailed)	Mean Difference	Lower	Upper
Weight	2.575	39	.014	16.62862	3.5650	29.6923

Note to Instructor
Part 2 can be omitted, and the author recommends that it be omitted. If Part 2 is included, these exercises apply: Exercises 29–32.

Part 2: Testing a Claim about a Population Mean When σ Is Known

In reality, it is very rare to test a claim about an unknown population mean while the population standard deviation is somehow known. For this case, the procedure is essentially the same as in Part 1 of this section, but the test statistic, P-value, and critical values are found as follows:

Test Statistic for Testing a Claim about a Mean (When σ Is Known)

$$z = \frac{\bar{x} - \mu_{\bar{x}}}{\dfrac{\sigma}{\sqrt{n}}}$$

P-Value

Provided by technology, or use the standard normal distribution (Table A-2) with the procedure summarized in Figure 8-4 from Section 8-2.

Critical Values

Use the standard normal distribution (Table A-2).

If we repeat Example 1 with the assumption that the value of $\sigma = 0.480$ W/kg is known, the test statistic is

$$z = \frac{\bar{x} - \mu_{\bar{x}}}{\frac{\sigma}{\sqrt{n}}} = \frac{0.938 - 1.00}{\frac{0.480}{\sqrt{11}}} = -0.43$$

Using this test statistic of $z = -0.43$, we can proceed to find the *P*-value. (See Figure 8-4 for the flowchart summarizing the procedure for finding *P*-values.) Example 1 refers to a left-tailed test, so the *P*-value is the area to the *left* of $z = -0.43$, which is 0.3336 (found from Table A-2). Because the *P*-value of 0.3336 is greater than the significance level of $\alpha = 0.05$, we fail to reject the null hypothesis, as we did in Example 1. As in Example 1, we conclude that there is not sufficient evidence to support a claim that the population mean is less than 1.00 W/kg.

using TECHNOLOGY

STATDISK If working with a list of the original sample values, first find the sample size, sample mean, and sample standard deviation by using the STATDISK procedure described in Section 3-2. After finding the values of n, $\bar{x}$, and s, select **Analysis** from the main menu, then select **Hypothesis Testing,** followed by **Mean-One Sample.**

MINITAB Minitab allows you to use either the summary statistics or a list of the original sample values. Select the menu items **Stat, Basic Statistics,** and **1-Sample t.** (For σ known, select 1-Sample *z*.) Enter the summary statistics or enter the column containing the list of original sample values. Use the **Options** button to change the format of the alternative hypothesis or the significance level from the default.

EXCEL Use XLSTAT. First enter the list of original sample values in a list. Click on **XLSTAT** at the top, click on **Parametric tests,** then select **One sample t test and z test.** In the screen that appears, for the "Data" box enter the range of data, such as A1:A11 for 11 data values in column A. For "Data Format" select **One sample.** Click on the "Student's t test" box (or click on the "z test" box if σ is known). Click on the **Options** tab to select the type of test; select the option including $\neq$ for a two-tailed test, select the option including $<$ for a left-tailed test, or select the option including $>$ for a right-tailed test. For the "Theoretical mean" box, enter the claimed value of the population mean, which is the same value used in the statement of the null hypothesis. Enter the desired "Significance level (%)." For example, enter 5 for a 0.05 significance level. Click **OK.** After the results are displayed, look for the test statistic identified as "*t* (Observed value)" or "*z* (Observed value)." The *P*-value and critical value(s) will also be displayed.

T1-83/84 PLUS If using a TI-83/84 Plus calculator, press STAT, then select **TESTS** and choose the menu item of **T-Test.** (For σ known, select Z-Test.) You can use the original data **(Data)** or the summary statistics **(Stats)** by providing the entries indicated in the window display. The first three items of the TI-83/84 Plus calculator results will include the alternative hypothesis, the test statistic, and the *P*-value.

STATCRUNCH Click on **Open StatCrunch.** Click on **Stat,** then select **T statistics.** Select **One sample,** then select **with data** (for a list of sample data) or **with summary** (for summary statistics). Click on **Next,** then select **Hypothesis Test.** Enter the claimed value of the population mean and enter the form of the alternative hypothesis. Click on **Calculate.**

8-4 Basic Skills and Concepts

Recommended Assignment
Exercises 1–16, 19, 20. If Part 2 was included, assign Exercises 29–32. The directions for Exercises 9–24 state that unless stipulated by the instructor, either the traditional method or P-value method can be used, so be sure to indicate a preference if you have one.

Due to limited space, answers for odd numbered exercises are not included here, but they are included in Appendix D.

Statistical Literacy and Critical Thinking

1. Requirements Twelve different video games showing alcohol use were observed. The duration times of alcohol use were recorded, with the times (seconds) listed below (based on data from "Content and Ratings of Teen-Rated Video Games," by Haninger and Thompson, *Journal of the American Medical Association,* Vol. 291, No. 7). What requirements must be satisfied to test the claim that the sample is from a population with a mean greater than 90 sec? Are the requirements all satisfied?

84 14 583 50 0 57 207 43 178 0 2 57

2. df If we are using the sample data from Exercise 1 in a t test of the claim that the population mean is greater than 90 sec, what does df denote, and what is its value?

2. df denotes the number of degrees of freedom. For the sample of 12 times, df = 11.

3. *t* Test Exercise 2 refers to a t test. What is a t test? Why are the t test methods of Part 1 in this section so much more likely to be used than the z test methods in Part 2?

4. Confidence Interval Assume that we will use the sample data from Exercise 1 with a 0.05 significance level in a test of the claim that the population mean is greater than 90 sec. If we want to construct a confidence interval to be used for testing that claim, what confidence level should be used for the confidence interval? If the confidence interval is found to be 21.1 sec $< \mu <$ 191.4 sec, what should we conclude about the claim?

4. Use a 90% confidence level. The given confidence interval does contain the value of 90 sec, so it is possible that the value of μ is equal to 90 sec or some lower value, so there is not sufficient evidence to support the claim that the mean is greater than 90 sec.

Finding *P*-values. ***In Exercises 5–8, either use technology to find the P-value or use Table A-3 to find a range of values for the P-value.***

5. Cigarette Nicotine The claim is that for the nicotine amounts in king-size cigarettes, $\mu > 1.10$ mg. The sample size is $n = 25$ and the test statistic is $t = 3.349$.

6. Cigarette Tar The claim is that for the tar amounts in king-size cigarettes, $\mu > 20.0$ mg. The sample size is $n = 25$ and the test statistic is $t = 1.733$.

6. $0.025 < P\text{-value} < 0.05$ (Tech: 0.0480).

7. Car Crash Tests The claim is that for measurements of standard head injury criteria in car crash tests, $\mu = 475$ HIC (standard *head injury condition* units). The sample size is $n = 21$ and the test statistic is $t = -2.242$.

8. Pulse Rates The claim is that for pulse rates of women, $\mu = 73$. The sample size is $n = 40$ and the test statistic is $t = 2.463$.

8. $0.01 < P\text{-value} < 0.02$ (Tech: 0.0183).

Testing Hypotheses. ***In Exercises 9–24, assume that a simple random sample has been selected and test the given claim. Unless specified by your instructor, use either the P-value method or the critical value method for testing hypotheses. Identify the null and alternative hypotheses, test statistic, P-value (or range of P-values), critical value(s), and state the final conclusion that addresses the original claim.***

9. Chocolate Chip Cookies The Chapter Problem for Chapter 3 includes the sample mean of the numbers of chocolate chips in 40 Chips Ahoy reduced fat cookies. The sample mean is $\bar{x} = 19.6$ chocolate chips and the sample standard deviation is 3.8 chocolate chips. Use a 0.05 significance level to test the claim that Chips Ahoy reduced-fat cookies have less fat because they have a mean number of chocolate chips that is less than the mean of 24 for regular Chips Ahoy cookies. See the accompanying TI-83/84 Plus display that results from this hypothesis test.

TI-83/84 PLUS

```
T-Test
 μ<24
 t=-7.323169318
 p=3.87325E-9
 x̄=19.6
 Sx=3.8
 n=40
```

10. Earthquake Depths Data Set 16 in Appendix B lists earthquake depths, and the summary statistics are $n = 50$, $\bar{x} = 9.81$ km, and $s = 5.01$ km. Use a 0.01 significance level to test the claim of a seismologist that these earthquakes are from a population with a mean depth equal to 10 km. See the accompanying Minitab display that results from this hypothesis test.

10. H_0: $\mu = 10$ km. H_1: $\mu \neq 10$ km. Test statistic: $t = -0.27$. Critical values: $t = \pm 2.678$ (approximately). P-value > 0.20. (Minitab shows a P-value of 0.790.) Fail to reject H_0. There is not sufficient evidence to warrant rejection of the claim that the earthquakes are from a population with a mean depth equal to 10 km.

MINITAB

Test of mu = 10 vs not = 10

N	Mean	StDev	SE Mean	95% CI	T	P
50	9.810	5.010	0.709	(8.386, 11.234)	-0.27	0.790

11. Oscar-Winning Actresses Data Set 11 in Appendix B lists ages of actresses when they won Oscars, and the summary statistics are $n = 82$, $\bar{x} = 35.9$ years, and $s = 11.1$ years. Use a 0.01 significance level to test the claim that the mean age of actresses when they win Oscars is 33 years.

12. Discarded Plastic The weights (lb) of discarded plastic from a sample of households is listed in Data Set 23 in Appendix B, and the summary statistics are $n = 62$, $\bar{x} = 1.911$ lb, and $s = 1.065$ lb. Use a 0.05 significance level to test the claim that the mean weight of discarded plastic from the population of households is greater than 1.800 lb.

12. H_0: $\mu = 1.800$ lb. H_1: $\mu > 1.800$ lb. Test statistic: $t = 0.821$. Critical value: $t = 1.671$ (approximately). *P*-value > 0.10 (Tech: 0.2075). Fail to reject H_0. There is not sufficient evidence to support the claim that the mean weight of discarded plastic from the population of households is greater than 1.800 lb.

13. M&M Weights A simple random sample of the weights of 19 green M&Ms has a mean of 0.8635 g and a standard deviation of 0.0570 g (as in Data Set 20 in Appendix B). Use a 0.05 significance level to test the claim that the mean weight of all green M&Ms is equal to 0.8535 g, which is the mean weight required so that M&Ms have the weight printed on the package label. Do green M&Ms appear to have weights consistent with the package label?

14. Human Body Temperature Data Set 3 in Appendix B includes a sample of 106 body temperatures with a mean of 98.20°F and a standard deviation of 0.62°F. Use a 0.05 significance level to test the claim that the mean body temperature of the population is equal to 98.6°F, as is commonly believed. Is there sufficient evidence to conclude that the common belief is wrong?

14. H_0: $\mu = 98.6$°F. H_1: $\mu \neq 98.6$°F. Test statistic: $t = -6.642$. Critical values: $t = \pm 1.984$ (approximately). *P*-value < 0.01 (Tech: 0.0000). Reject H_0. There is sufficient evidence to warrant rejection of the claim that the mean body temperature of the population is equal to 98.6°F. There is sufficient evidence to conclude that the common belief is wrong.

15. Is the Diet Practical? When 40 people used the Weight Watchers diet for one year, their mean weight loss was 3.0 lb and the standard deviation was 4.9 lb (based on data from "Comparison of the Atkins, Ornish, Weight Watchers, and Zone Diets for Weight Loss and Heart Disease Reduction," by Dansinger et al., *Journal of the American Medical Association,* Vol. 293, No. 1). Use a 0.01 significance level to test the claim that the mean weight loss is greater than 0. Based on these results, does the diet appear to be effective? Does the diet appear to have practical significance?

16. Flight Delays Data Set 15 in Appendix B lists 48 different departure delay times (minutes) for American Airlines flights from New York (JFK) to Los Angeles. Negative departure delay times correspond to flights that departed early. The mean of the 48 times is 10.5 min and the standard deviation is 30.8 min. Use a 0.01 significance level to test the claim that the mean departure delay time for all such flights is less than 12.0 min. Is a flight operations manager justified in reporting that the mean departure time is less than 12.0 min?

16. H_0: $\mu = 12.0$ min. H_1: $\mu < 12.0$ min. Test statistic: $t = -0.337$. Critical value: $t = -2.412$ (approximately). *P*-value > 0.10 (Tech: 0.3687). Fail to reject H_0. There is not sufficient evidence to support the claim that the mean departure delay time for all such flights is less than 12.0 min. A flight operations manager is not justified in reporting that the mean departure time is less than 12.0 min.

17. Garlic for Reducing Cholesterol In a test of the effectiveness of garlic for lowering cholesterol, 49 subjects were treated with raw garlic. Cholesterol levels were measured before and after the treatment. The changes (before minus after) in their levels of LDL cholesterol (in mg/dL) have a mean of 0.4 and a standard deviation of 21.0 (based on data from "Effect of Raw Garlic vs Commercial Garlic Supplements on Plasma Lipid Concentrations in Adults with Moderate Hypercholesterolemia," by Gardner et al., *Archives of Internal Medicine,* Vol. 167). Test the claim that with garlic treatment, the mean change in LDL cholesterol is greater than 0. What do the results suggest about the effectiveness of the garlic treatment?

18. Insomnia Treatment A clinical trial was conducted to test the effectiveness of the drug zopiclone for treating insomnia in older subjects. Before treatment with zopiclone, 16 subjects had a mean wake time of 102.8 min. After treatment with zopiclone, the 16 subjects had a mean wake time of 98.9 min and a standard deviation of 42.3 min (based on data from "Cognitive Behavioral Therapy vs Zopiclone for Treatment of Chronic Primary Insomnia in Older Adults," by Siversten et al., *Journal of the American Medical Association,* Vol. 295, No. 24). Assume that the 16 sample values appear to be from a normally distributed population, and test the claim that after treatment with zopiclone, subjects have a mean wake time of less than 102.8 min. Does zopiclone appear to be effective?

18. H_0: $\mu = 102.8$ min. H_1: $\mu < 102.8$ min. Test statistic: $t = -0.369$. Critical value: $t = -1.753$ (assuming a 0.05 significance level). *P*-value > 0.10 (Tech: 0.3587). Fail to reject H_0. There is not sufficient evidence to support the claim that after treatment with zopiclone, subjects have a mean wake time less than 102.8 min. This result suggests that the zopiclone treatment is not effective.

19. Years in College Listed below are the numbers of years it took for a random sample of college students to earn bachelor's degrees (based on data from the National Center for Education Statistics). Use a 0.01 significance level to test the claim that for all college students, the mean time required to earn a bachelor's degree is greater than 4.0 years. Is there anything about the data that would suggest that the conclusion might not be valid?

4 4 4 4 4 4 4.5 4.5 4.5 4.5 4.5 4.5 6 6 8 9 9 13 13 15

20. The sample data meet the loose requirement of having a normal distribution. H_0: $\mu = 30$ years. H_1: $\mu > 30$ years. Test statistic: $t = 1.818$. Critical value: $t = 1.761$. P-value < 0.05 (Tech: 0.0453). Reject H_0. There is sufficient evidence to support the claim that the mean age of all race car drivers is greater than 30 years.

20. Ages of Race Car Drivers Listed below are the ages (years) of randomly selected race car drivers (based on data reported in *USA Today*). Use a 0.05 significance level to test the claim that the mean age of all race car drivers is greater than 30 years.

32 32 33 33 41 29 38 32 33 23 27 45 52 29 25

21. Lead in Medicine Listed below are the lead concentrations (in μg/g) measured in different Ayurveda medicines. Ayurveda is a traditional medical system commonly used in India. The lead concentrations listed here are from medicines manufactured in the United States. The data are based on the article "Lead, Mercury, and Arsenic in US and Indian Manufactured Ayurvedic Medicines Sold via the Internet," by Saper et al., *Journal of the American Medical Association,* Vol. 300, No. 8. Use a 0.05 significance level to test the claim that the mean lead concentration for all such medicines is less than 14 μg/g.

3.0 6.5 6.0 5.5 20.5 7.5 12.0 20.5 11.5 17.5

22. The sample data meet the loose requirement of having a normal distribution. H_0: $\mu = 1100.0$ cm^3. H_1: $\mu \neq 1100.0$ cm^3. Test statistic: $t = 0.813$. Critical values: $t = \pm 3.250$. P-value > 0.20 (Tech: 0.4371). Fail to reject H_0. There is not sufficient evidence to warrant rejection of the claim that the population of brain volumes has a mean equal to 1100.0 cm^3.

22. Brain Volume Listed below are brain volumes (cm^3) of unrelated subjects used in a study. (See Data Set 6 in Appendix B.) Use a 0.01 significance level to test the claim that the population of brain volumes has a mean equal to 1100.0 cm^3.

963 1027 1272 1079 1070 1173 1067 1347 1100 1204

23. Heights of Supermodels Listed below are the heights (inches) for the simple random sample of supermodels Lima, Bundchen, Ambrosio, Ebanks, Iman, Rubik, Kurkova, Kerr, Kroes, and Swanepoel. Use a 0.01 significance level to test the claim that supermodels have heights with a mean that is greater than the mean height of 63.8 in. for women in the general population. Given that there are only 10 heights represented, can we really conclude that supermodels are taller than the typical woman?

70 71 69.25 68.5 69 70 71 70 70 69.5

24. H_0: $\mu = 65$ mi/h. H_1: $\mu < 65$ mi/h. Test statistic: $t = -3.684$. Critical value: $t = -1.796$. P-value < 0.005 (Tech: 0.0018). Reject H_0. There is sufficient evidence to support the claim that the sample is from a population with a mean that is less than the speed limit of 65 mi/h.

24. Highway Speeds Listed below are speeds (mi/h) measured from southbound traffic on I-280 near Cupertino, California (based on data from SigAlert). This simple random sample was obtained at 3:30 P.M. on a weekday. Use a 0.05 significance level to test the claim that the sample is from a population with a mean that is less than the speed limit of 65 mi/h.

62 61 61 57 61 54 59 58 59 69 60 67

Large Data Sets from Appendix B. ***In Exercises 25–28, use the data set from Appendix B to test the given claim. Identify the null hypothesis, alternative hypothesis, test statistic, P-value or critical value(s), conclusion about the null hypothesis, and final conclusion that addresses the original claim. Use the P-value method unless your instructor specifies otherwise.***

26. H_0: $\mu = 120.0$ mm Hg. H_1: $\mu < 120$ mm Hg. Test statistic: $t = -0.424$. Critical value: $t = -1.685$. P-value > 0.10 (Tech: 0.3370). Fail to reject H_0. There is not sufficient evidence to support the claim that the female population has a mean systolic blood pressure level less than 120.0 mm Hg.

25. Earthquake Magnitudes Use the earthquake magnitudes listed in Data Set 16 in Appendix B and test the claim that the population of earthquakes has a mean magnitude greater than 1.00. Use a 0.05 significance level.

26. Blood Pressure Use the systolic blood pressure measurements for females listed in Data Set 1 in Appendix B and test the claim that the female population has a mean systolic blood pressure level less than 120.0 mm Hg. Use a 0.05 significance level.

27. College Weights Use the September weights of males in Data Set 4 from Appendix B and test the claim that male college students have a mean weight that is less than the 83 kg mean weight of males in the general population. Use a 0.01 significance level.

28. Power Supply Data Set 18 in Appendix B lists measured voltage amounts supplied directly to the author's home. The Central Hudson power supply company states that it has a target power supply of 120 volts. Using those home voltage amounts, test the claim that the mean is 120 volts. Use a 0.01 significance level.

8-4 Beyond the Basics

Hypothesis Tests with Known σ. ***In Exercises 29–32, conduct the hypothesis test using a known value of the population standard deviation σ.***

29. Repeat Exercise 9 assuming that the population standard deviation σ is known to be 3.8 chocolate chips.

30. Repeat Exercise 10 assuming that the population standard deviation σ is known to be 5.01 km.

31. Repeat Exercise 11 assuming that the population standard deviation σ is known to be 11.1 years.

32. Repeat Exercise 12 assuming that the population standard deviation σ is known to be 1.065 lb.

33. Finding Critical *t* Values When finding critical values, we sometimes need significance levels other than those available in Table A-3. Some computer programs approximate critical *t* values by calculating

$$t = \sqrt{\text{df} \cdot (e^{A^2/\text{df}} - 1)}$$

where df $= n - 1$, $e = 2.718$, $A = z(8 \cdot \text{df} + 3)/(8 \cdot \text{df} + 1)$, and z is the critical z score. Use this approximation to find the critical t score corresponding to $n = 150$ and a significance level of 0.05 in a right-tailed case. Compare the results to the critical t value of 1.655 found from STATDISK, Minitab, or a TI-83/84 Plus calculator.

34. Using the Wrong Distribution When testing a claim about a population mean with a simple random sample selected from a normally distributed population with unknown σ, the Student t distribution should be used for finding critical values and/or a P-value. If the standard normal distribution is incorrectly used instead, does that mistake make you more or less likely to reject the null hypothesis, or does it not make a difference? Explain.

35. Interpreting Power For Example 1 in this section, the hypothesis test has power of 0.4274 of supporting the claim that $\mu < 1.00$ W/kg when the actual population mean is 0.80 W/kg.

a. Interpret the given value of the power.

b. Identify the value of β and interpret that value.

28. H_0: $\mu = 120$ V. H_1: $\mu \neq 120$ V. Test statistic: $t = 96.358$. Critical values: $t = \pm 2.708$. *P*-value $<$ 0.01 (Tech: 0.0000). Reject H_0. There is sufficient evidence to warrant rejection of the claim that the mean voltage amount is 120 volts.

30. H_0: $\mu = 10$ km. H_1: $\mu \neq 10$ km. Test statistic: $z = -0.27$. Critical values: $z = \pm 2.575$. *P*-value: 0.7872 (Tech: 0.7886). Fail to reject H_0. There is not sufficient evidence to warrant rejection of the claim that the earthquakes are from a population with a mean depth equal to 10 km.

32. H_0: $\mu = 1.800$ lb. H_1: $\mu > 1.800$ lb. Test statistic: $z = 0.82$. Critical value: $z = 1.645$. *P*-value: 0.2061 (Tech: 0.2059). Fail to reject H_0. There is not sufficient evidence to support the claim that the mean weight of discarded plastic from the population of households is greater than 1.800 lb.

34. Using the normal distribution makes you more likely to reject the null hypothesis because the critical z values are not as extreme as the corresponding critical t values.

8-5 Testing a Claim about a Standard Deviation or Variance

Key Concept This section presents methods for conducting a formal hypothesis test of a claim made about a population standard deviation σ or population variance σ^2. The methods of this section use the chi-square distribution that was first introduced in Section 7-4. The assumptions, test statistic, *P*-value, and critical values are summarized as follows.

Note to Instructor

Organization Change: Section 8-5 in this edition was Section 8-6 in the previous edition. This section can be omitted if time is an issue. If this section is included but Section 7-4 was omitted, be sure to describe the χ^2 distribution, because this would be the first time that students see it.

After covering this section, students will have studied three different parameters (p, μ, σ) along with three different distributions (normal, t, χ^2). Point out that the detachable Formula/Table card includes test statistics, and the form of the test statistic often reveals the distribution that should be used. For example, the Formula/Table card shows that the test statistic for a claim involving the standard deviation or variance is $\chi^2 = (n - 1)s^2/\sigma^2$, and that test statistic indicates that the χ^2 distribution is used. There is no need to memorize test statistic formulas, so students can focus on more important concepts instead.

Testing Claims about σ or σ^2

Objective

Conduct a hypothesis test of a claim made about a population standard deviation σ or population variance σ^2.

Notation

$n =$ sample size

$s =$ *sample* standard deviation

$s^2 =$ *sample* variance

$\sigma =$ *claimed* value of the *population* standard deviation

$\sigma^2 =$ claimed value of the *population* variance

Requirements

1. The sample is a simple random sample.
2. The population has a normal distribution. (Instead of being a loose requirement, this test has a fairly strict requirement of a normal distribution.)

Test Statistic for Testing a Claim about σ or σ^2

$$\chi^2 = \frac{(n-1)s^2}{\sigma^2}$$ (round to three decimal places, as in Table A-4)

***P*-values:** Use technology or use Table A-4 with degrees of freedom given by df $= n - 1$. (Table A-4 is based on *cumulative areas from the right.*)

Critical values: Use Table A-4 with degrees of freedom given by df $= n - 1$. (Table A-4 is based on *cumulative areas from the right.*)

CAUTION The χ^2 (chi-square) test of this section is not *robust* against a departure from normality, meaning that the test does not work well if the population has a distribution that is far from normal. The condition of a normally distributed population is therefore a much stricter requirement when testing claims about σ or σ^2 than tests of claims about a population mean μ (Section 8-4).

The chi-square distribution was introduced in Section 7-4, where we noted the following important properties.

Properties of the Chi-Square Distribution

1. All values of χ^2 are nonnegative, and the distribution is not symmetric (see Figure 8-9).
2. There is a different χ^2 distribution for each number of degrees of freedom (see Figure 8-10).
3. The critical values are found in Table A-4 using

$$\textbf{degrees of freedom} = n - 1$$

If using Table A-4 for finding critical values, note the following design feature of that table:

In Table A-4, each critical value of χ^2 in the body of the table corresponds to an area given in the top row of the table, and each area in that top row is a *cumulative area to the right* of the critical value.

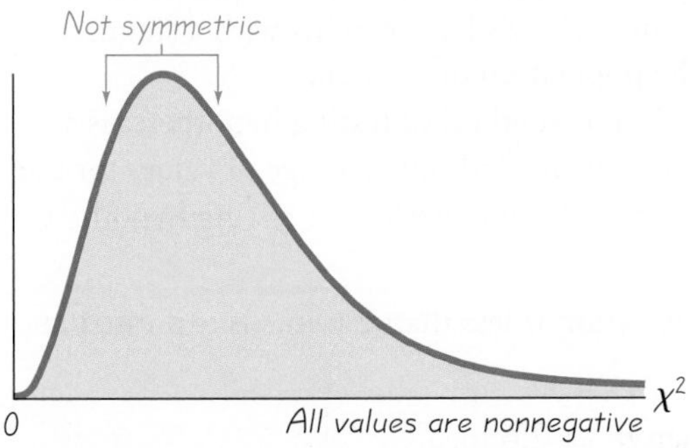

Figure 8-9 Properties of the Chi-Square Distribution

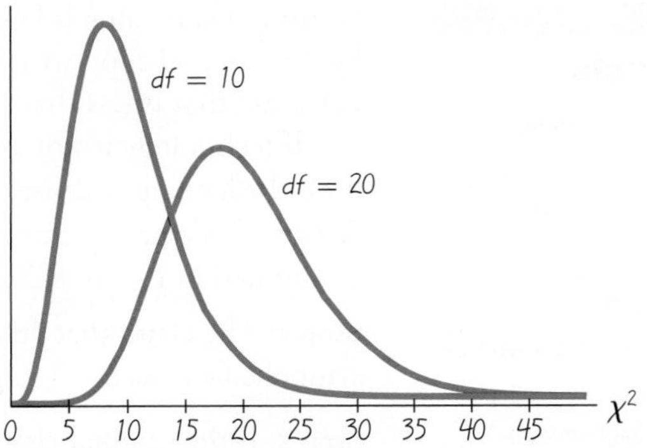

Figure 8-10 Chi-Square Distribution for df = 10 and df = 20

CAUTION Table A-2 for the standard normal distribution provides cumulative areas from the *left*, but Table A-4 for the chi-square distribution uses cumulative areas from the *right*. See Example 1 in Section 7-4.

Example 1 Supermodel Heights

Listed below are the heights (inches) for the simple random sample of supermodels Lima, Bundchen, Ambrosio, Ebanks, Iman, Rubik, Kurkova, Kerr, Kroes, and Swanepoel. Consider the claim that supermodels have heights that have much less variation than heights of women in the general population. We will use a 0.01 significance level to test the claim that supermodels have heights with a standard deviation that is less than 2.6 in. for the population of women.

70 71 69.25 68.5 69 70 71 70 70 69.5

Solution

Requirement check (1) The sample is a simple random sample. (2) In checking for normality, we see that the sample has no outliers, the accompanying normal quantile plot shows points that are reasonably close to a straight-line pattern, and there is no other pattern that is not a straight line. Both requirements are satisfied. ✓

Technology A test of the claim that $\sigma < 2.6$ in. is equivalent to a test of the claim that $\sigma^2 < 6.76$ in.^{2} (because $2.6^2 = 6.76$). Technology capable of conducting this test will typically display the *P*-value. The TI-83/84 Plus calculator can be used as described at the end of this section, and the result will be as shown in the accompanying display. The third line of the display shows that the test statistic is $\chi^2 = 0.851516185$, the last line shows that the *P*-value is 0.0002897435436 when expressed in standard form, and after rounding we have *P*-value = 0.0003.

STATDISK

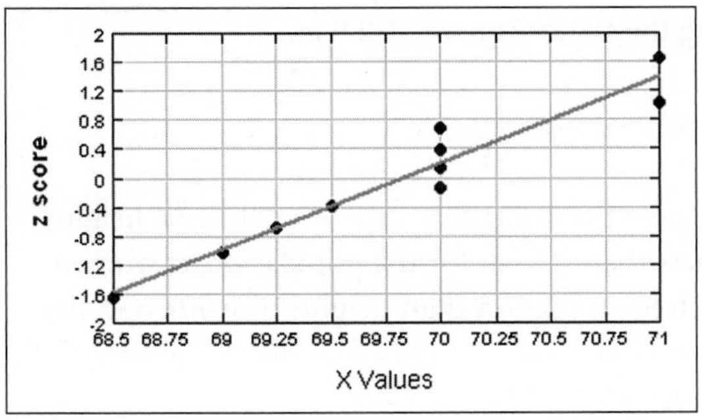

TI-83/84 PLUS

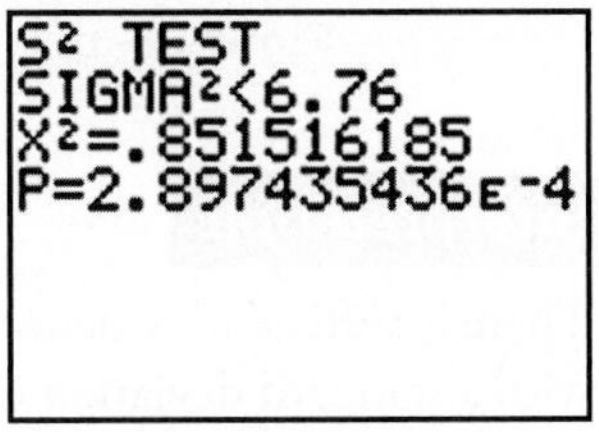

Meta-Analysis

The term *meta-analysis* refers to a technique of doing a study that essentially combines results of other studies. It has the advantage that separate smaller samples can be combined into one big sample, making the collective results more meaningful. It also has the advantage of using work that has already been done. Meta-analysis has the disadvantage of being only as good as the studies that are used. If the previous studies are flawed, the "garbage in, garbage out" phenomenon can occur. The use of meta-analysis is currently popular in medical research and psychological research. As an example, a study of migraine headache treatments was based on data from 46 other studies. (See "Meta-Analysis of Migraine Headache Treatments: Combining Information from Heterogeneous Designs," by Dominici et al., *Journal of the American Statistical Association,* Vol. 94, No. 445.)

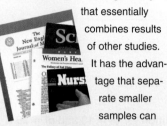

Because this *P*-value is less than the significance level of 0.01, we reject the null hypothesis and support the claim that supermodels have heights with a standard deviation that is less than 2.6 in. for the population of women.

If technology is not available, the *P*-value method of testing hypotheses is a little challenging because Table A-4 allows us to find only a range of values for the *P*-value. Let's proceed instead with the critical value method of testing hypotheses as outlined in Figure 8-2 from Section 8-2.

Step 1: The claim that "the standard deviation is less than 2.6 in." is expressed in symbolic form as $\sigma < 2.6$ in.

Step 2: If the original claim is false, then $\sigma \geq 2.6$ in.

Step 3: The expression $\sigma < 2.6$ in. does not contain equality, so it becomes the alternative hypothesis. The null hypothesis is the statement that $\sigma = 2.6$ in.

$$H_0: \sigma = 2.6 \text{ in.}$$
$$H_1: \sigma < 2.6 \text{ in.} \quad \text{(original claim)}$$

Step 4: The significance level is $\alpha = 0.01$.

Step 5: Because the claim is made about σ, we use the χ^2 (chi-square) distribution.

Step 6: The test statistic is calculated by using $\sigma = 2.6$ in. (as assumed in the above null hypothesis), $n = 10$, and $s = 0.7997395$ in., which is the unrounded standard deviation computed from the original list of 10 heights. We get this result:

$$\chi^2 = \frac{(n-1)s^2}{\sigma^2} = \frac{(10-1)(0.7997395)^2}{2.6^2} = 0.852$$

The critical value of $\chi^2 = 2.088$ is found from Table A-4, and it corresponds to 9 degrees of freedom and an "area to the right" of 0.99 (based on the significance level of 0.01 for a left-tailed test). See Figure 8-11.

Step 7: Because the test statistic is in the critical region, we reject the null hypothesis.

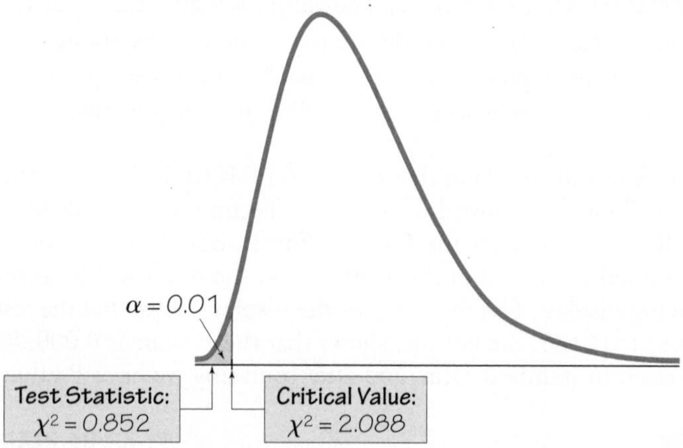

Figure 8-11 Testing the Claim That $\sigma < 2.6$ in.

Interpretation

There is sufficient evidence to support the claim that supermodels have heights with a standard deviation that is less than 2.6 in. for the population of women. Heights of supermodels have much less variation than heights of women in the general population.

P-Value Method

Example 1 can also be solved with the *P*-value method of testing hypotheses as outlined in Figure 8-1. *P*-values are typically provided by technology, but Table A-4 must be used if technology is not available. When using Table A-4, we usually cannot find *exact P*-values because the chi-square distribution table includes only selected values of α and selected numbers of degrees of freedom. See Example 2.

Example 2 Supermodel Heights: *P*-Value Method

Repeat Example 1 using the *P*-value method of testing hypotheses.

Solution

In Example 1 we noted that technology provides a *P*-value of 0.0003. Because this *P*-value is less than the significance level of 0.01, we reject the null hypothesis and support the claim that supermodels have heights with a standard deviation that is less than 2.6 in. for the population of women.

If technology is not available, refer to Table A-4 and use it as follows:

1. Locate the row for 9 degrees of freedom. (Because $n = 10$, we have df $= n - 1 = 9$.)
2. Use the test statistic of $\chi^2 = 0.852$ and see that in the 9th row of Table A-4, the test statistic of 0.852 is less than the lowest table entry of 1.735, so the area to the *right* of the test statistic is greater than 0.995, which means that the area to the *left* of the test statistic is less than $1 - 0.995 = 0.005$.
3. In this left-tailed test, the *P*-value is the area to the left of the test statistic, so we know that "*P*-value < 0.005." (This agrees with our *P*-value of 0.0003 found from technology.)

Using technology we find that *P*-value $= 0.0003$, and using Table A-4 we find that *P*-value < 0.005.

Confidence Interval Method

When testing claims about σ or σ^2, the *P*-value method, critical value method, and the confidence interval method are all equivalent in the sense that they will always lead to the same conclusion. See Example 3.

Example 3 Supermodel Heights: Confidence Interval Method

Repeat the hypothesis test in Example 1 by constructing a suitable confidence interval.

Solution

First, we should be careful to select the correct confidence level. Because the hypothesis test is left-tailed and the significance level is 0.01, we should use a confidence level of 98%, or 0.98. (See Table 8-1 from Section 8-2 for help in selecting the correct confidence level.)

Using the methods described in Section 7-4, we can use the sample data listed in Example 1 to construct a 98% confidence interval estimate of σ. We use $n = 10$, $s = 0.7997395$ in., $\chi_L^2 = 2.088$, and $\chi_R^2 = 21.666$. (The critical values χ_L^2 and χ_R^2 are found in Table A-4. Use the row with df $= n - 1 = 9$.

Human Lie Detectors

Researchers tested 13,000 people for their ability to determine when someone is lying. They found 31 people with exceptional skills at identifying lies. These human lie detectors had accuracy rates around 90%. They also found that federal officers and sheriffs were quite good at detecting lies, with accuracy rates around 80%. Psychology Professor Maureen O'Sullivan questioned those who were adept at identifying lies, and she said that "all of them pay attention to nonverbal cues and the nuances of word usages and apply them differently to different people. They could tell you eight things about someone after watching a two-second tape. It's scary, the things these people notice." Methods of statistics can be used to distinguish between people unable to detect lying and those with that ability.

Note to Instructor

In this section, the *P*-value method is more difficult than the critical value method. The difficulty arises from the format of Table A-4, which usually does not allow you to get exact *P*-values. If your students are not using a technology that automatically generates *P*-values, carefully explain the use of Table A-4 for finding a range of values for the *P*-value. Also emphasize that the format of Table A-4 is based on cumulative areas from the *right*, whereas the format of Table A-2 for the standard normal distribution is based on cumulative areas from the *left*.

The 0.98 confidence level indicates that $\alpha = 0.02$, and we divide that area of 0.02 equally between the two tails so that the areas to the *right* of the critical values are 0.99 and 0.01. Refer to Table A-4 and use the columns with areas of 0.99 and 0.01 and use the 9th row.)

$$\sqrt{\frac{(n-1)s^2}{\chi_R^2}} < \sigma < \sqrt{\frac{(n-1)s^2}{\chi_L^2}}$$

$$\sqrt{\frac{(10-1)(0.7997395^2)}{21.666}} < \sigma < \sqrt{\frac{(10-1)(0.7997395^2)}{2.088}}$$

$$0.5 \text{ in.} < \sigma < 1.7 \text{ in.}$$

Based on this confidence interval, we can support the claim that $\sigma < 2.6$ in. (because all values of the confidence interval are less than 2.6 in.). We reach the same conclusion found with the *P*-value method and the critical value method.

using TECHNOLOGY

STATDISK Select **Analysis,** then **Hypothesis Testing,** then **StDev-One Sample.** Provide the required entries in the dialog box, then click on **Evaluate.** STATDISK will display the test statistic, critical values, *P*-value, and confidence interval.

MINITAB For Minitab Release 15 and later, select **Stat,** then **Basic Statistics,** then select the menu item of σ^2 **1 Variance.** Click on the **Summarized Data** box and enter the sample size and sample standard deviation. Click on the box labeled **Perform hypothesis test** and enter the assumed value of σ from the null hypothesis. Click on the **Options** button and select the correct form of the alternative hypothesis. Click on the **OK** button twice and the *P*-value will be displayed.

In **Minitab 16,** you can use the **Stat** menu as described above, or you could also click on **Assistant,** then **Hypothesis Tests,** then select the case for **1-Sample Standard Deviation.** Fill out the dialog box, then click **OK** to get three windows of results that include the *P*-value and much other helpful information.

EXCEL Neither Excel nor XLSTAT have a function for conducting a hypothesis test of a claim about a standard deviation or variance. Excel does have a function (CHISQ.INV or CHIINV or CHISQ.INV.RT) that could be used to find critical values.

T1-83/84 PLUS The TI-83/84 Plus calculator does not test hypotheses about σ or σ^2 directly, but the program **S2TEST** can be used. That program was written by Michael Lloyd of Henderson State University, and it can be downloaded from the CD included with this book, or www.aw.com/Triola. The program S2TEST uses the program ZZINEWT, so that program must also be installed. After storing the programs on the calculator, press PRGM, select **S2TEST,** and enter the claimed variance σ^2, the sample variance s^2, and the sample size *n*. Select the format used for the alternative hypothesis and press ENTER. The *P*-value will be displayed.

STATCRUNCH Click on **Open StatCrunch.** Click on **Stat,** then select **Variance.** Select **One sample,** then select **with data** (for a list of sample data) or **with summary** (for summary statistics). Click on **Next,** then select **Hypothesis Test** and enter the claimed value of the population variance and select the form of the alternative hypothesis. Click on **Calculate.**

Recommended Assignment
Exercises 1–8, 15, 16. Because *P*-values are usually difficult to find, the directions for Exercises 5–16 stipulate that students should find either *P*-values or critical values.

Due to limited space, answers for odd numbered exercises are not included here, but they are included in Appendix D.

8-5 Basic Skills and Concepts

Statistical Literacy and Critical Thinking

1. Waiting in Line The Jefferson Valley Bank once had a separate customer waiting line at each teller window, but it now has a single waiting line that feeds the teller windows as vacancies occur. The standard deviation of customer waiting times with the old multiple-line configuration was 1.8 min. Listed below is a simple random sample of waiting times (minutes) with the single waiting line. The 10 sample values have a standard deviation of 0.5 min.

6.5 6.6 6.7 6.8 7.1 7.3 7.4 7.7 7.7 7.7

a. When the bank changed from multiple waiting lines to a single line, how was the mean waiting time affected?

b. When the bank changed from multiple waiting lines to a single line, how was the variation among waiting times affected?

c. What improvement occurred with the change from multiple waiting lines to a single line?

d. What procedure can be used to determine that the single waiting line is better?

2. Requirements If we want to use the sample data from Exercise 1 to test the claim that the sample is from a population with a standard deviation less than 1.8 min, we must satisfy the requirements of having a simple random sample and a normally distributed population. In Exercise 1 it was stated that the sample is a simple random sample.

a. In general, how does the normality requirement for a hypothesis test of a claim about a standard deviation differ from the normality requirement for a hypothesis test of a claim about a mean?

b. What methods can be used to determine whether the normality requirement is satisfied?

2. a. The normality requirement for a hypothesis test of a claim about a standard deviation is much more strict, meaning that the distribution of the population must be much closer to a normal distribution.
b. With only 10 sample values, a histogram doesn't really give us a good picture of the distribution, so a normal quantile plot would be better. Also, we should determine that there are no outliers.

3. Confidence Interval Method of Hypothesis Testing Assume that we want to use the sample data from Exercise 1 to test the claim that the sample is from a population with a standard deviation less than 1.8 min; we will use a 0.05 significance level to test that claim. If we want to use the confidence interval method of testing hypotheses, what level of confidence should be used for the confidence interval? Will the conclusion based on the confidence interval be the same as the conclusion based on a hypothesis test that uses the *P*-value method or the critical value method?

4. Hypothesis Test For the sample data from Exercise 1, we have $s = 0.5$ and we want to use that sample to test the claim that the sample is from a population with a standard deviation less than 1.8 min (as it was in the past with multiple waiting lines).

a. Identify the null and alternative hypotheses.

b. Find the value of the test statistic.

c. Using technology, the *P*-value is found to be 0.0001. What should we conclude about the null hypothesis?

d. Given that the *P*-value is 0.0001, what should we conclude about the original claim?

e. What does the conclusion suggest about the effectiveness of the change from multiple waiting lines to a single line?

4. a. H_0: $\sigma = 1.8$ min. H_1: $\sigma < 1.8$ min.
b. $\chi^2 = 0.694$
c. Reject the null hypothesis.
d. There is sufficient evidence to support the claim that the standard deviation of waiting times of all customers is less than 1.8 min.
e. The change to a single waiting line is effective because the variation among waiting times is less than it was with multiple lines.

Testing Claims about Variation. ***In Exercises 5–16, test the given claim. Identify the null hypothesis, alternative hypothesis, test statistic, P-value or critical value(s), conclusion about the null hypothesis, and final conclusion that addresses the original claim. Assume that a simple random sample is selected from a normally distributed population.***

5. Cans of Coke Data Set 19 in Appendix B includes volumes (oz) of a simple random sample of 36 cans of regular Coke. Those volumes have a mean of 12.19 oz and a standard deviation of 0.11 oz, and they appear to be from a normally distributed population. If we want the filling process to work so that almost all cans have volumes between 11.8 oz and 12.4 oz, the range rule of thumb can be used to estimate that the standard deviation should be less than 0.15 oz. Use the sample data to test the claim that the population of volumes has a standard deviation less than 0.15 oz. Use a 0.05 significance level.

6. Cans of Pepsi Repeat the preceding exercise using these statistics from a simple random sample of cans of regular Pepsi: $n = 36$, $\bar{x} = 12.29$ oz, $s = 0.09$ oz.

6. H_0: $\sigma = 0.15$ oz. H_1: $\sigma < 0.15$ oz. Test statistic: $\chi^2 = 12.600$. Critical value of χ^2 is between 18.493 and 26.509, so it is estimated to be 22.501 (Tech: 22.465). *P*-value $<$ 0.05 (Tech: 0.0002). Reject H_0. There is sufficient evidence to support the claim that the population of volumes has a standard deviation less than 0.15 oz.

7. Weights of Pennies Data Set 21 in Appendix B includes a simple random sample of 37 weights of post-1983 pennies. Those 37 weights have a mean of 2.49910 g and a standard deviation of 0.01648 g. U.S. Mint specifications require that pennies be manufactured so that the mean weight is 2.500 g. A hypothesis test will verify that the sample appears to come from a population with a mean of 2.500 g as required, but use a 0.05 significance level to test the claim that the population of weights has a *standard deviation* less than the specification of 0.0230 g.

8. H_0: $\sigma = 0.0230$ g. H_1: $\sigma >$ 0.0230 g. Test statistic: $\chi^2 =$ 98.260. Critical value of χ^2 is between 43.773 and 55.758. *P*-value $<$ 0.005 (Tech: 0.0000). Reject H_0. There is sufficient evidence to support the claim that pre-1983 pennies have a standard deviation greater than 0.0230 g. Weights of pre-1983 pennies appear to vary more than those of post-1983 pennies.

8. Weights of Pennies The preceding exercise involved the claim that *post*-1983 pennies have weights with a standard deviation less than 0.0230 g. Data Set 21 in Appendix B includes the weights of a simple random sample of 35 *pre*-1983 pennies, and that sample has a standard deviation of 0.03910 g. Use a 0.05 significance level to test the claim that pre-1983 pennies have weights with a standard deviation greater than 0.0230 g. Does it appear that weights of pre-1983 pennies vary more than those of post-1983 pennies?

9. Pulse Rates of Men A simple random sample of 40 men results in a standard deviation of 10.3 beats per minute (based on Data Set 1 in Appendix B). The normal range of pulse rates of adults is typically given as 60 to 100 beats per minute. If the range rule of thumb is applied to that normal range, the result is a standard deviation of 10 beats per minute. Use the sample results with a 0.05 significance level to test the claim that pulse rates of men have a standard deviation equal to 10 beats per minute.

10. The data appear to be from a normally distributed population. H_0: $\sigma = 10$. H_1: $\sigma \neq 10$. Test statistic: $\chi^2 = 52.478$. Critical values of χ^2: 24.433 and 59.342 (approximately). *P*-value $>$ 0.10 (Tech: 0.1463). Fail to reject H_0. There is not sufficient evidence to warrant rejection of the claim that pulse rates of women have a standard deviation equal to 10 beats per minute.

10. Pulse Rates of Women Repeat the preceding exercise using the pulse rates of women listed in Data Set 1 of Appendix B. For this sample, $n = 40$ and $s = 11.6$ beats per minute.

11. Cigarette Tar A simple random sample of 25 filtered 100-mm cigarettes is obtained, and the tar content of each cigarette is measured. The sample has a standard deviation of 3.7 mg (based on Data Set 10 in Appendix B). Use a 0.05 significance level to test the claim that the tar content of filtered 100-mm cigarettes has a standard deviation different from 3.2 mg, which is the standard deviation for unfiltered king-size cigarettes.

12. H_0: $\sigma = 28.866$ cents. H_1: $\sigma \neq$ 28.866 cents. Test statistic: $\chi^2 =$ 133.337. Critical values: $\chi^2 =$ 67.328 and 140.169 (approximately). *P*-value $>$ 0.02 (Tech: 0.0244). Fail to reject H_0. There is not sufficient evidence to warrant rejection of the claim that the standard deviation is 28.866 cents. Because the amounts from 0 cents to 99 cents are all equally likely, the requirement of a normal distribution is violated, so the results are highly questionable.

12. Analysis of Pennies In an analysis investigating the usefulness of pennies, the cents portions of 100 randomly selected credit card charges from the author are recorded, and they have a mean of 47.6 cents and a standard deviation of 33.5 cents. If the amounts from 0 cents to 99 cents are all equally likely, the mean is expected to be 49.5 cents and the population standard deviation is expected to be 28.866 cents. Use a 0.01 significance level to test the claim that the sample is from a population with a standard deviation equal to 28.866 cents. If the amounts from 0 cents to 99 cents are all equally likely, is the requirement of a normal distribution satisfied? If not, how does that affect the conclusion?

13. Ages of Race Car Drivers Listed below are the ages (years) of randomly selected race car drivers (based on data reported in *USA Today*). Most people in the general population have ages that vary between 0 and 90 years, so use of the range rule of thumb suggests that ages in the general population have a standard deviation of 22.5 years. Use a 0.01 significance level to test the claim that the standard deviation of ages of all race car drivers is less than 22.5 years.

32 32 33 33 41 29 38 32 33 23 27 45 52 29 25

14. The data appear to be from a normally distributed population. H_0: $\sigma = 5.0$ mi/h. H_1: $\sigma \neq 5.0$ mi/h. Test statistic: $\chi^2 = 7.307$. Critical values of χ^2: 3.816 and 21.920. *P*-value $>$ 0.20 (Tech: 0.4525). Fail to reject H_0. There is not sufficient evidence to warrant rejection of the claim that the standard deviation of speeds is equal to 5.0 mi/h.

14. Highway Speeds Listed below are speeds (mi/h) measured from southbound traffic on I-280 near Cupertino, California (based on data from SigAlert). This simple random sample was obtained at 3:30 P.M. on a weekday. Use a 0.05 significance level to test the claim of the highway engineer that the standard deviation of speeds is equal to 5.0 mi/h.

62 61 61 57 61 54 59 58 59 69 60 67

15. Aircraft Altimeters The Skytek Avionics company uses a new production method to manufacture aircraft altimeters. A simple random sample of new altimeters resulted in the errors listed below. Use a 0.05 level of significance to test the claim that the new production method has errors with a standard deviation greater than 32.2 ft, which was the standard deviation for the old production method. If it appears that the standard deviation is greater, does the new production method appear to be better or worse than the old method? Should the company take any action?

−42 78 −22 −72 −45 15 17 51 −5 −53 −9 −109

16. IQ of Professional Pilots The Wechsler IQ test is designed so that the mean is 100 and the standard deviation is 15 for the population of normal adults. Listed below are IQ scores of randomly selected professional pilots. It is claimed that because professional pilots are a more homogeneous group than the general population, they have IQ scores with a standard deviation less than 15. Test that claim using a 0.05 significance level.

121 116 115 121 116 107 127 98 116 101 130 114

16. The data appear to be from a normally distributed population. H_0: $\sigma = 15$. H_1: $\sigma < 15$. Test statistic: $\chi^2 = 4.416$. Critical value: $\chi^2 = 4.575$. P-value < 0.05 (Tech: 0.0439). Reject H_0. There is sufficient evidence to support the claim that IQ scores of professional pilots have a standard deviation less than 15.

Large Data Sets from Appendix B. ***In Exercises 17 and 18, use the data set from Appendix B to test the given claim. Identify the null hypothesis, alternative hypothesis, test statistic, P-value or critical value(s), conclusion about the null hypothesis, and final conclusion that addresses the original claim.***

17. Cans of Diet Coke Repeat Exercise 5 using the volumes (oz) of cans of diet Coke.

18. Weights of Pennies Data Set 21 in Appendix B includes a simple random sample of "wheat" pennies. U.S. Mint specifications now use a standard deviation of 0.0230 g for weights of pennies. Use a 0.01 significance level to test the claim that wheat pennies were manufactured so that their weights have a standard deviation equal to 0.0230 g.

18. H_0: $\sigma = 0.0230$ g. H_1: $\sigma \neq 0.0230$ g. Test statistic: $\chi^2 = 156.155$. Critical values of χ^2: 13.787 and 53.672 (approximately). P-value < 0.01 (Tech: 0.0000). Reject H_0. There is sufficient evidence to warrant rejection of the claim that the population of weights has a standard deviation equal to 0.0230 g.

8-5 Beyond the Basics

19. Finding Critical Values of χ^2 For large numbers of degrees of freedom, we can approximate critical values of χ^2 as follows:

$$\chi^2 = \frac{1}{2}(z + \sqrt{2k - 1})^2$$

Here k is the number of degrees of freedom and z is the critical value(s) found from technology or Table A-2. In Exercise 5 we have df = 35, so Table A-4 does not list an exact critical value. If we want to approximate a critical value of χ^2 in the left-tailed hypothesis test with $\alpha = 0.05$ and a sample size of 36, we let $k = 35$ with $z = -1.645$. Use this approximation to estimate the critical value of χ^2 for Exercise 5. How close is it to the value of $\chi^2 = 22.465$ obtained by using STATDISK and Minitab?

20. Finding Critical Values of χ^2 Repeat Exercise 19 using this approximation (with k and z as described in Exercise 19):

$$\chi^2 = k\left(1 - \frac{2}{9k} + z\sqrt{\frac{2}{9k}}\right)^3$$

20. Critical $\chi^2 = 22.462$, which is very close to the value of 22.465 obtained from STATDISK and Minitab.

Chapter 8 Review

The major topics in an introductory statistics course include methods of descriptive statistics (mean, standard deviation, histogram, etc.) and the two major activities of inferential statistics: estimating population parameters (as with confidence intervals) and hypothesis testing. In this chapter we introduced basic methods for testing claims about a population proportion, population mean, or population standard deviation (or variance).

In Section 8-2 we presented the basic components of a hypothesis test: null hypothesis, alternative hypothesis, test statistic, critical region, significance level, critical value, *P*-value, type I error, and type II error. We also discussed two-tailed tests, left-tailed tests, right-tailed tests, and the statement of conclusions. We used those components in identifying three different methods for testing hypotheses:

1. The *P*-value method (summarized in Figure 8-1)

2. The critical value method (or traditional method) summarized in Figure 8-2

3. Confidence intervals (Chapter 7)

In Sections 8-3 through 8-5 we discussed specific methods for dealing with different parameters. Because it is so important to select the correct distribution and test statistic, we provide Table 8-5, which summarizes some key elements of the hypothesis-testing procedures of this chapter.

Table 8-5 Hypothesis Tests

Parameter	Requirements: Simple Random Sample and . . .	Distribution and Test Statistic	Critical and *P*-values
Proportion	$np \geq 5$ and $nq \geq 5$	Normal: $z = \dfrac{\hat{p} - p}{\sqrt{\dfrac{pq}{n}}}$	Table A-2
Mean	σ not known and normally distributed population or σ not known and $n > 30$	Student *t*: $t = \dfrac{\bar{x} - \mu_{\bar{x}}}{\dfrac{s}{\sqrt{n}}}$	Table A-3
	σ known and normally distributed population or σ known and $n > 30$	Normal: $z = \dfrac{\bar{x} - \mu_{\bar{x}}}{\dfrac{\sigma}{\sqrt{n}}}$	Table A-2
	Population not normally distributed and $n \leq 30$	Use a nonparametric method or bootstrapping.	
Standard Deviation or Variance	Population normally distributed	Chi- square: $\chi^2 = \dfrac{(n-1)s^2}{\sigma^2}$	Table A-4

Chapter Quick Quiz

1. Wristwatch Accuracy Students of the author collected a simple random sample of times (sec) of wristwatch errors, and a few of those times are listed below. Negative values correspond to watches that are running ahead of the actual time. Assuming that we want to use a 0.05 significance level to test the claim that the sample is from a population with a mean equal to 0 sec, identify the null hypothesis and the alternative hypothesis.

140 −85 325 20 305 205 20 −93

2. Type of Test Refer to the hypothesis test described in Exercise 1.

a. Is the hypothesis test left-tailed, right-tailed, or two-tailed?

b. If the requirements are satisfied, what distribution should be used for the hypothesis test: normal, Student *t*, chi-square, binomial?

3. *P*-Value If we use technology to conduct the hypothesis test described in Exercise 1, we get a *P*-value of 0.1150.

a. What should we conclude about the null hypothesis?

b. What is the final conclusion that addresses the original claim?

1. H_0: $\mu = 0$ sec. H_1: $\mu \neq 0$ sec.
2. a. Two-tailed.
   b. Student *t*.
3. a. Fail to reject H_0.
   b. There is not sufficient evidence to warrant rejection of the claim that the sample is from a population with a mean equal to 0 sec.
4. There is a loose requirement of a normally distributed population in the sense that the test works reasonably well if the departure from normality is not too extreme.

4. Normality For the hypothesis test in Exercise 1, what does it mean when we say that the test is *robust* against departures from normality?

5. Death Penalty Poll In a recent Gallup poll of 511 adults, 64% said that they were in favor of the death penalty for a person convicted of murder. We want to use a 0.01 significance level to test the claim that the majority of adults are in favor of the death penalty for a person convicted of murder.

a. Identify the null and alternative hypotheses.

b. Find the value of the test statistic.

c. Technology is used to find this *P*-value: 1.263996E-10. Express the *P*-value in ordinary notation, then determine what we should conclude about the original claim.

6. *P*-Value Find the *P*-value in a test of the claim that the mean IQ score of acupuncturists is equal to 100, given that the test statistic is $z = -2.00$.

7. Equivalent Methods Which of the following statements are true?

a. When testing a claim about μ, the *P*-value method, critical value method, and confidence interval method are all equivalent.

b. When testing a claim about a population proportion p, the *P*-value method, critical value method, and confidence interval method are all equivalent.

c. When testing a claim about any population parameter, the *P*-value method, critical value method, and confidence interval method are all equivalent.

8. Chi-Square Test In a test of the claim that $\sigma = 15$ for the population of IQ scores of Facebook friends, we find the the rightmost critical value is $\chi^2 = 71.420$. Is the leftmost critical χ^2 value equal to -71.420?

9. Conclusions True or false: In hypothesis testing, it is *never* valid to form a conclusion of supporting the null hypothesis.

10. Reliability True or false: If correct methods of hypothesis testing are used with a large simple random sample, the conclusion will always be true.

Review Exercises

1. True/False Characterize each of the following statements as being true or false.

a. In a hypothesis test, a very high *P*-value indicates strong support of the alternative hypothesis.

b. The Student t distribution can be used to test a claim about a population mean whenever the sample data are randomly selected from a normally distributed population.

c. When using a χ^2 distribution to test a claim about a population standard deviation, there is a very loose requirement that the sample data be from a population having a normal distribution.

d. When conducting a hypothesis test about the claimed proportion of adults who have current passports, the problems with a convenience sample can be overcome by using a larger sample size.

e. When repeating the same hypothesis test with different random samples of the same size, the conclusions will all be the same.

2. Leisure Time In a Gallup poll, 1010 adults were randomly selected and asked if they were satisfied or dissatisfied with the amount of leisure time that they had. Of this sample 657 said that they were satisfied and 353 said that they were dissatisfied. Use a 0.01 significance level to test the claim that 2/3 of adults are satisfied with the amount of leisure time that they have.

5. a. H_0: $p = 0.5$. H_1: $p > 0.5$.
b. $z = 6.33$
c. *P*-value: 0.0000000001263996. There is sufficient evidence to support the claim that the majority of adults are in favor of the death penalty for a person convicted of murder.

6. 0.0456 (Tech: 0.0455)

7. The only true statement is the one given in part (a).

8. No. All critical values of χ^2 are greater than zero.

9. True.

10. False.

1. a. False. b. True. c. False. d. False. e. False.

2. H_0: $p = 2/3$. H_1: $p \neq 2/3$. Test statistic: $z = -1.09$. Critical values: $z = \pm 2.575$ (Tech: ± 2.576). *P*-value: 0.2758 (Tech: 0.2756). Fail to reject H_0. There is not sufficient evidence to warrant rejection of the claim that 2/3 of adults are satisfied with the amount of leisure time that they have.

3. H_0: $p = 0.75$. H_1: $p > 0.75$. Test statistic: $z = 10.65$ (if using $x = 678$) or $z = 10.66$ (if using $\hat{p} = 0.92$). Critical value: $z = \pm 2.33$. *P*-value: 0.0001 (Tech: 0.0000). Reject H_0. There is sufficient evidence to support the claim that more than 75% of us do not open unfamiliar e-mail and instant-message links. Given that the results are based on a voluntary response sample, the results are not necessarily valid.

4. H_0: $\mu = 3369$ g. H_1: $\mu < 3369$ g. Test statistic: $t = -19.962$. Critical value: $t = -2.328$ (approximately). *P*-value < 0.005 (Tech: 0.0000). Reject H_0. There is sufficient evidence to support the claim that the mean birth weight of Chinese babies is less than the mean birth weight of 3369 g for Caucasian babies.

5. H_0: $\sigma = 567$ g. H_1: $\sigma \neq 567$ g. Test statistic: $\chi^2 = 54.038$. Critical values of χ^2: 51.172 and 116.321. *P*-value is between 0.02 and 0.05 (Tech: 0.0229). Fail to reject H_0. There is not sufficient evidence to warrant rejection of the claim that the standard deviation of birth weights of Chinese babies is equal to 567 g.

3. Risky Behavior In a *USA Today* poll of 737 respondents, 92% said that they do not open unfamiliar e-mail and instant-message links. Use a 0.01 significance level to test the claim that more than 75% of us do not open unfamiliar e-mail and instant-message links. How is the validity of the results affected by the knowledge that *USA Today* posted a question on its Web site, and 737 people chose to respond?

4. Birth Weight A simple random sample of 1862 births of Chinese babies resulted in a mean birth weight of 3171 g and a standard deviation of 428 g (based on "Comparison of Birth Weight Distributions between Chinese and Caucasian Infants," by Wen et al., *American Journal of Epidemiology,* Vol. 172, No 10). Use a 0.01 significance level to test the claim that the mean birth weight of Chinese babies is less than the mean birth weight of 3369 g for Caucasian babies.

5. Birth Weights A simple random sample of 81 births of Chinese babies resulted in a mean birth weight of 3245 g and a standard deviation of 466 g. Test the claim that the standard deviation of birth weights of Chinese babies is equal to 567 g, which is the standard deviation of birth weights of Caucasian babies. Use a 0.01 significance level and assume that the birth weights are normally distributed.

6. Monitoring Lead in Air Listed below are measured amounts of lead (in micrograms per cubic meter, or $\mu g/m^3$, in the air. The Environmental Protection Agency has established an air quality standard for lead of 1.5 $\mu g/m^3$. The measurements shown below constitute a simple random sample of measurements recorded at Building 5 of the World Trade Center site on different days immediately following the destruction caused by the terrorist attacks of September 11, 2001. After the collapse of the two World Trade Center buildings, there was considerable concern about the quality of the air. Use a 0.05 significance level to test the claim that the sample is from a population with a mean greater than the EPA standard of 1.5 $\mu g/m^3$.

5.40 1.10 0.42 0.73 0.48 1.10

7. Pennsylvania Lottery In the Pennsylvania Match 6 lottery, six numbers between 1 and 49 are randomly drawn. To simulate the number selection process, a TI-83/84 Plus calculator was used to randomly generate 100 numbers between 1 and 49 inclusive. The sample has a mean of 24.2 and a standard deviation of 14.1. Use a 0.01 significance level to test the claim that the sample is selected from a population with a mean equal to 25, which is the mean of the population of all drawn numbers.

8. Type I Error and Type II Error

a. In general, what is a type I error? In general, what is a type II error?

b. For the hypothesis test in Exercise 7, write a statement that would constitute a type I error, and write another statement that would be a type II error.

9. Pennsylvania Lottery and χ^2 Test Assume that we want to use the same sample data from Exercise 7 to test the claim that the standard deviation of all drawn numbers is less than 15.0. Why can't we use a χ^2 test with the methods described in Section 8-4?

10. Tests of Child Booster Seats The National Highway Traffic Safety Administration conducted crash tests of child booster seats for cars. Listed below are results from those tests, with the measurements given in HIC (standard *head injury condition* units). The safety requirement is that the HIC measurement should be less than 1000 HIC. Use a 0.01 significance level to test the claim that the sample is from a population with a mean less than 1000 HIC. Do the results suggest that all of the child booster seats meet the specified requirement?

602 696 762 572 637

6. H_0: $\mu = 1.5\ \mu g/m^3$. H_1: $\mu > 1.5\ \mu g/m^3$. Test statistic: $t = 0.049$. Critical value: $t = 2.015$. P-value > 0.10 (Tech: 0.4814). Fail to reject H_0. There is not sufficient evidence to support the claim that the sample is from a population with a mean greater than the EPA standard of 1.5 $\mu g/m^3$. Because the sample value of 5.40 $\mu g/m^3$ appears to be an outlier and because a normal quantile plot suggests that the sample data are not from a normally distributed population, the requirements of the hypothesis test are not satisfied, and the results of the hypothesis test are therefore questionable.

7. H_0: $\mu = 25$. H_1: $\neq 25$. Test statistic: $t = -0.567$. Critical values: $t = \pm 1.984$ (approximately). P-value > 0.20 (Tech: 0.5717). Fail to reject H_0. There is not sufficient evidence to warrant rejection of the claim that the sample is selected from a population with a mean equal to 25.

8. a. A type I error is the mistake of rejecting a null hypothesis when it is actually true. A type II error is the mistake of failing to reject a null hypothesis when in reality it is false.

   b. Type I error: Reject the null hypothesis that the mean of the population is equal to 25 when in reality, the mean is actually equal to 25. Type II error: Fail to reject the null hypothesis that the population mean is equal to 25 when in reality, the mean is actually different from 25.

9. The χ^2 test has a reasonably strict requirement that the sample data must be randomly selected from a population with a normal distribution, but the numbers are selected in such a way that they are all equally likely, so the population has a uniform distribution instead of the required normal distribution. Because the requirements are not all satisfied, the χ^2 test should not be used.

10. H_0: $\mu = 1000$ HIC. H_1: $\mu < 1000$ HIC. Test statistic: $t = -10.177$. Critical value: $t = -3.747$. P-value < 0.005 (Tech: 0.0003). Reject H_0. There is sufficient evidence to support the claim that the population mean is less than 1000 HIC. The results suggest that the population mean is less than 1000 HIC, so they appear to satisfy the specified requirement.

Cumulative Review Exercises

Please be aware that some of the following problems may require knowledge of concepts presented in previous chapters.

1. Dictionary Words A simple random sample of pages from *Merriam-Webster's Collegiate Dictionary,* 11th edition, is obtained. Listed below are the numbers of words defined on those pages. Find the values of the indicated statistics.

51 63 36 43 34 62 73 39 53 79

a. Mean **b.** Median **c.** Standard deviation **d.** Variance **e.** Range

1. a. 53.3 words
b. 52.0 words
c. 15.7 words
d. 245.1 words^2
e. 45 words

2. Dictionary Words Refer to the sample data in Exercise 1.

a. What is the level of measurement of the data (nominal, ordinal, interval, ratio)?

b. Are the values discrete or continuous?

c. What does it mean to state that the sample is a simple random sample?

2. a. Ratio.
b. Discrete.
c. The sample is a simple random sample if it was selected in such a way that all possible samples of the same size have the same chance of being selected.

3. Confidence Interval for Dictionary Words Use the sample values given in Exercise 1 to construct a 95% confidence interval estimate of the population mean. Assume that the population has a normal distribution.

3. 42.1 words $< \mu <$ 64.5 words

4. Hypothesis Test for Dictionary Words Refer to the sample data given in Exercise 1. Given that the dictionary has 1459 pages with defined words, the claim that there are more than 70,000 defined words is the same as the claim that the mean number of defined words on a page is greater than 48.0. Use a 0.05 significance level to test the claim that the mean number of defined words on a page is greater than 48.0. What does the result suggest about the claim that there are more than 70,000 defined words in the dictionary?

4. H_0: μ = 48.0 words. H_1: $\mu >$ 48.0 words. Test statistic: t = 1.070. Critical value: t = 1.833. *P*-value $>$ 0.10 (Tech: 0.1561). Fail to reject H_0. There is not sufficient evidence to support the claim that the mean number of words on a page is greater than 48.0. There is not enough evidence to support the claim that there are more than 70,000 words in the dictionary.

5. Designing Cars The sitting height (from seat to top of head) of drivers must be considered in the design of a new car model. Men have sitting heights that are normally distributed with a mean of 36.0 in. and a standard deviation of 1.4 in. (based on anthropometric survey data from Gordon, Churchill, et al.).

a. One car is designed to accommodate sitting heights of 38.8 in. or less. Find the percentage of men with sitting heights greater than 38.8 in.

b. If the sitting height is to be changed so that 98% of men will be accommodated, what is the new sitting height?

c. If a car is occupied by four men, find the probability that they have a mean sitting height less than 37.0 in.

5. a. 2.28%
b. 38.9 in.
c. 0.9236 (Tech: 0.9234)

6. Left-Handedness Among Americans, 9.7% of males are left-handed and 12.5% of females are left-handed.

a. If three females are randomly selected, find the probability that they are all left-handed. When randomly selecting three females, is it unlikely that all of them are left-handed? Why or why not?

b. If one male is randomly selected and one female is randomly selected, find the probability that they are both left-handed.

c. If five females are randomly selected, find the probability that at least one of them is left-handed.

6. a. 0.00195. It is unlikely because the probability of the event occurring is so small.
b. 0.0121
c. 0.487

7. Normal Rainfall The following histogram is obtained from the daily rainfall amounts (in.) in Boston for a year. If we want to conduct a hypothesis test of a claim about the mean, there is a requirement that the population of rainfall amounts must have a normal distribution.

7. No. The distribution is very skewed. A normal distribution would be approximately bell-shaped, but the displayed distribution is very far from being bell-shaped.

Based on the histogram, does it appear that the population satisfies the requirement of a normal distribution? Why or why not?

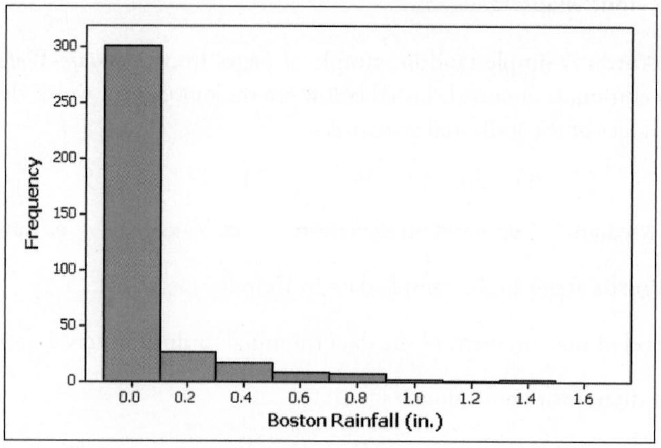

8. Because the vertical scale starts at 7000 and not at 0, the difference between the number of males and the number of females is exaggerated, so the graph is deceptive by creating the wrong impression that there are many more male graduates than female graduates.

8. Analyzing Graph The accompanying graph depicts results from a recent year in which there were 8878 male graduates and 8203 female graduates from medical schools in the United States. Does the graph depict the data in a way that is fair and objective, or is it somehow deceptive? Explain.

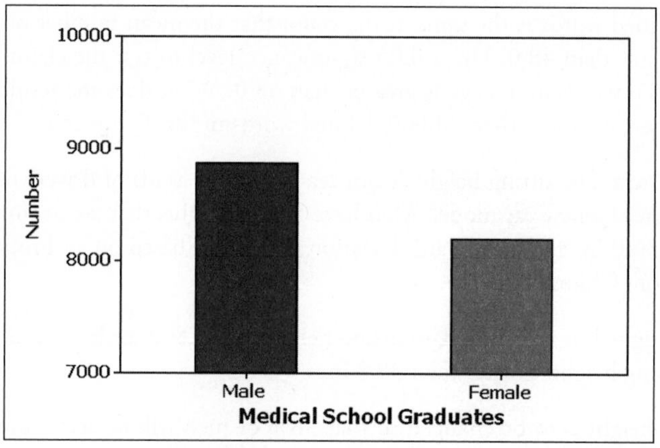

9. a. 373
b. $34.2\% < p < 40.2\%$
c. Yes. With test statistic $z = -8.11$ and with a *P*-value close to 0, there is sufficient evidence to support the claim that less than 50% of adults answer "yes."
d. The required sample size depends on the confidence level and the sample proportion, not the population size.

10. H_0: $p = 0.5$. H_1: $p < 0.5$. Test statistic: $z = -8.11$. Critical value: $z = -2.33$. *P*-value: 0.0001 (Tech: 0.0000). Reject H_0. There is sufficient evidence to support the claim that fewer than 50% of Americans say that they have a gun in their home.

9. Gun Survey In a recent Gallup poll, 1003 randomly selected adults in the United States were asked if they have a gun in their home, and 37.2% of them answered "yes."

a. What is the number of respondents who answered "yes"?

b. Construct a 95% confidence interval estimate of the percentage of all adults who would answer "yes" when asked if they have a gun in their home.

c. Based on a hypothesis test, can we safely conclude that less than 50% of adults answer "yes" when asked if they have a gun in their home? Why or why not?

d. What is a sensible response to the criticism that the Gallup poll cannot provide good results because the sample size is only 1003 adults selected from a large population of 241,472,385 adults in the United States?

10. Hypothesis Test for Gun Survey Refer to the survey results from Exercise 9 and use a 0.01 significance level to test the claim that fewer than 50% of Americans say that they have a gun in their home.

Technology Project

Simulation for the Chapter Problem The Chapter Problem describes a situation in which a county clerk was supposed to use a method of random selection for determining whether the Democrat or Republican would get the top line on each of 41 different election ballots. His procedure resulted in the Democrat being selected 40 times in the 41 trials. Section 8-3 discussed the formal hypothesis test of his claim that he used a random method. This project involves a different approach consisting of simulations. The basic idea is to assume that randomness is used with a 0.5 probability of selecting a Democrat, then making the random selection 41 times. After repeating that simulation 100 times, we will see how often the results are as extreme as 40 or 41 Democrats selected in the 41 trials. We will then understand how unlikely it is to get 40 Democrats in 41 trials. The simulation is basically the same as repeatedly tossing 41 coins and seeing how often the result is 40 heads or 41 heads.

Use STATDISK, Minitab, Excel, StatCrunch, the TI-83/84 Plus calculator, or any other technology that can randomly generate data from a binomial distribution. Conduct 100 simulations of the process of randomly selecting a Democrat or Republican for the top line on each of 41 different election ballots. Examine the results and write a brief statement explaining how they either support or refute the claim that the county clerk did not really use a method of random selection.

Here are instructions for different technologies.

STATDISK: Click on **Data,** then click on **Binomial Generator.** In the dialog box, enter 100 for the sample size (because we want to repeat the selection process 100 times), enter 0.5 for the success probability (so that there is a 0.5 probability of selecting a Democrat), and enter 41 for the number of trials (to simulate the 41 election ballots). Click on **Generate.** Scroll through the results and count the number of times that you got 40 Democrats or 41 Democrats using a process of random selection.

Minitab: Click on the main menu item of **Calc,** then click on **Random Data,** then **Binomial.** In the dialog box, enter 100 for the sample size (because we want to repeat the selection process 100 times), enter C1 as the column to store the results, enter 41 for the number of trials (to simulate 41 election ballots), and enter 0.5 for the event probability (so that there is a 0.5 probability of selecting a Democrat). Click on **OK.** Scroll through the list of results and count the number of times that you got 40 Democrats or 41 Democrats using a process of random selection.

Excel: Click on *fx* just below the tool bar. Select the category of **Math & Trig,** then select **RANDBETWEEN.** Click on **OK.** Proceed to enter 0 for "bottom" and enter 1 for "top." Click on **OK.** Excel will randomly select either 0 or 1 and the result will appear in cell A1. Click on the lower right corner of cell A1 and, while holding the mouse button down, drag the mouse downward until 41 rows are highlighted. When the mouse button is released, column A should include 41 values, each of which is 0 or 1. Now click on cell A41 and, while holding the mouse button down, drag the mouse to the right until reaching column DV, so that 100 columns are highlighted. When the mouse is released, you will have 100 columns of simulated election ballots. In cell A42, enter the expression **=SUM(A1:A41)** so that the sum of the entries in column A will be shown. This is the number of times a Democrat was selected in 41 trials. Click on the lower right corner of cell A42 and, while holding down the mouse button, drag it to the right until reaching column DV. Row 42 will show the numbers of times a Democrat was selected in 41 ballots. Scroll across row 42 and see how often you get 40 Democrats or 41 Democrats when using a process of random selection.

TI-83/84 Plus: Press MATH, then select **PRB** and select the menu item of **randBin.** Make the entries to get **randBin(41, 0.5, 100).** That command causes 100 simulations of the process of randomly selecting Democrats in each of 41 election ballots. Store the results in list L1 by pressing STO› L1. Press ENTER. Now press STAT and select **Edit,** then scroll through list L1 to see how often you get 40 Democrats or 41 Democrats when using a process of random selection.

StatCrunch: Click on **Data,** then click on **Simulate Data,** then select the menu item of **Binomial.** In the dialog box, enter 100 for the number of rows, enter 1 for the number of columns, enter 41 for *n,* and enter 0.5 for *p.* Click on **Simulate.** Scroll through the list of results and count the number of times that you got 40 Democrats or 41 Democrats using a process of random selection.

from data TO DECISION

Critical Thinking: Designing an aircraft cockpit

In designing a cockpit for a Boeing aircraft, the overhead grip reach of a seated pilot is being considered as an important factor for placement of landing light switches to be located directly above the pilot. Listed below are the measured overhead grip reaches (cm) of a simple random sample of women (based on anthropometric data from Gordon, Churchill, et al.). Use a 0.01 significance level to test the claim that the mean overhead grip reach of women is less than the value of 123 cm that is being planned for an aircraft.

120	115	130	123	118	118	116
121	119	131	125	119	124	122
121	129	125	126	115	122	

Analyzing the Results

a. It is not too difficult to conduct the hypothesis test, but is that hypothesis test the best tool for determining whether the value of 123 cm is suitable? If not, determine whether 123 cm is suitable, and if it is not suitable, find a value that is suitable.

b. In this application, why does it make sense to ignore the overhead grip reach of men?

Cooperative Group Activities

1. In-class activity Without using any measuring device, each student should draw a line believed to be 3 in. long and another line believed to be 3 cm long. Then use rulers to measure and record the lengths of the lines drawn. Find the means and standard deviations of the two sets of lengths. Test the claim that the lines estimated to be 3 in. have a mean length that is equal to 3 in. Test the claim that the lines estimated to be 3 cm have a mean length that is equal to 3 cm. Compare the results. Do the estimates of the 3-in. line appear to be more accurate than those for the 3-cm line? Do an Internet search to identify the countries that do not yet use the metric system. What do these results suggest?

2. In-class activity Assume that a method of gender selection can affect the probability of a baby being a girl so that the probability becomes 1/4. Each student should simulate 20 births by drawing 20 cards from a shuffled deck. Replace each card after it has been drawn, then reshuffle. Consider the hearts to be girls and consider all other cards to be boys. After making 20 selections and recording the "genders" of the babies, use a 0.10 significance level to test the claim that the proportion of girls is equal to 1/4. How many students are expected to get results leading to the wrong conclusion that the proportion is not 1/4? How does that relate to the probability of a type I error? Does this procedure appear to be effective in identifying the effectiveness of the gender-selection method? (If decks of cards are not available, use some other way to simulate the births, such as using the random number generator on a calculator or using digits from phone numbers or Social Security numbers.)

3. Out-of-class activity Groups of three or four students should go to the library and collect a sample consisting of the ages of books (based on copyright dates). Plan and describe the sampling plan, execute the sampling procedure, then use the results to test the claim that the mean age of books in the library is greater than 15 years.

4. In-class activity Each student should write an estimate of the age of the current president of the United States. All estimates should be collected and the sample mean and standard deviation should be calculated. Then test the hypothesis that the mean of all such estimates is equal to the actual current age of the president.

5. In-class activity A class project should be designed to conduct a test in which each student is given a taste of Coke and a taste of Pepsi. The student is then asked to identify which sample is Coke. After all of the results are collected, test the claim that the success rate is better than the rate that would be expected with random guesses.

6. In-class activity Each student should estimate the length of the classroom. The values should be based on visual estimates, with no actual measurements being taken. After the estimates have been collected, measure the length of the room, then test the claim that the sample mean is equal to the actual length of the classroom. Is there a "collective wisdom," whereby the class mean is approximately equal to the actual room length?

7. Out-of-class activity Using a wristwatch that is reasonably accurate, set the time to be exact. Use a radio station or telephone time report which states that "at the tone, the time is . . . ." If you cannot set the time to the nearest second, record the error for the watch you are using. Now compare the time on your watch to the time on others. Record the errors with negative signs for watches that are ahead of the actual time and positive signs for those watches that are behind the actual time. Use the data to test the claim that the mean error of all wristwatches is equal to 0. Do we collectively run on time, or are we early or late? Also test the claim that the standard deviation of errors is less than 1 min. What are the practical implications of a standard deviation that is excessively large?

8. In-class activity In a group of three or four people, conduct an ESP experiment by selecting one of the group members as the subject. Draw a circle on one small piece of paper and draw a square on another sheet of the same size. Repeat this experiment 20 times: Randomly select the circle or the square and place it in the subject's hand behind his or her back so that it cannot be seen, then ask the subject to identify the shape (without seeing it); record whether the response is correct. Test the claim that the subject has ESP because the proportion of correct responses is greater than 0.5.

9. In-class activity After dividing into groups of between 10 and 20 people, each group member should record the number of heartbeats in a minute. After calculating the sample mean and standard deviation, each group should proceed to test the claim that the mean is greater than 48, which is the author's result. (When people exercise, they tend to have lower pulse rates, and the author runs 5 miles a few times each week. What a guy!)

10. Out-of-class activity In groups of three or four, collect data to determine whether subjects have a Facebook page, then combine the results and test the claim that more than 1/4 of students have a Facebook page.

11. Out-of-class activity Each student should find an article in a professional journal that includes a hypothesis test of the type discussed in this chapter. Write a brief report describing the hypothesis test and its role in the context of the article.

9 Inferences from Two Samples

9

chapter problem

Using colors for creativity and accuracy

Some studies are quite intriguing as they reveal various facets of human behavior. One such study is reported in the article "Blue or Red? Exploring the Effect of Color on Cognitive Task Performances," by Mehta and Zhu, *Science Express,* DOI: 10.1126. The researchers from the University of British Columbia conducted studies to investigate the effects of color on creativity and cognitive tasks.

To investigate the effects of color on creativity, subjects with a red background were asked to think of creative uses for a brick, while other subjects with a blue background were given the same task. Both groups were allowed one minute. Responses were scored by a panel of judges and results from scores of creativity were as follows:

Creativity Scores

Red Background:	$n = 35, \bar{x} = 3.39, s = 0.97$
Blue Background:	$n = 36, \bar{x} = 3.97, s = 0.63$

In other trials, subjects were given detail-oriented tasks consisting of words displayed on a computer screen with background colors of red and blue. The subjects studied 36 words for 2 minutes, then they were asked to recall as many of the words as they could after waiting 20 minutes. Results from scores on the word recall test are given below.

Accuracy Scores

Red Background:	$n = 35, \bar{x} = 15.89, s = 5.90$
Blue Background:	$n = 36, \bar{x} = 12.31, s = 5.48$

Examining the above results, we can see that creativity scores were higher for those having a blue background. It's obvious that the mean of 3.97 is greater than the mean of 3.39, but is the difference *significant*? We can also see that the accuracy scores were higher for those having a red background. The mean of 15.89 is greater than the mean of 12.31, but is the difference *significant*? This is an ideal situation for the use of hypothesis tests. Unlike the hypothesis tests of Chapter 8, this situation involves *two* populations instead of just one. Section 9-3 will present methods for testing claims about *two* population means, and we will consider the test results above in that section. We will then be able to determine whether it appears that the color blue fosters creativity while red fosters accuracy in detail-oriented tasks.

Note to Instructor
Some instructors cover this entire chapter while many others omit it because they don't have enough time for it. If you don't have class time for this chapter, strongly consider assigning some exercises from it. Once students understand the basic concepts of hypothesis testing and confidence intervals, they can use computer software or a TI-83/84 Plus calculator to do the number crunching, then they can focus on the interpretation of the results. Also, there is a real pedagogical advantage in teaching the topic of hypothesis testing in Chapter 8, then having students *independently* apply the same general concepts to the different circumstances included in this chapter.

Sections 9-2, 9-3, and 9-4 are especially important, so consider covering those sections or assigning exercises from them.

9-1 Review and Preview

Inferential statistics involves forming conclusions (or inferences) about a population parameter. Two major activities of inferential statistics are (1) estimating values of population parameters using confidence intervals, and (2) testing claims made about population parameters. Chapter 7 introduced methods for constructing confidence interval estimates of a population proportion, population mean, or a population standard deviation or variance. Chapter 8 introduced methods for testing claims about a population proportion, population mean, or population standard deviation or variance. Chapters 7 and 8 both involved methods for dealing with a sample from *one* population. The objective of this chapter is to extend the methods for *estimating* values of population parameters and the methods for *testing hypotheses* to situations involving *two* populations. The following are examples typical of those found in this chapter.

- Test the claim that when people in one group are each given four quarters while people in another group are each given a $1 bill, the group given the $1 bill is less likely to spend the money.
- Test the claim that the proportion of children who contract polio is less for children given the Salk vaccine than for children given a placebo.
- Test the claim that the mean body temperature of men is different from the mean body temperature of women.

Because there are so many studies involving a comparison of *two* samples, the methods of this chapter are very important because they apply to a wide variety of real applications.

Note to Instructor
Because so many real applications involve a claim about proportions from two different populations (such as a treatment group and a placebo group), this is one of the most important sections in the book.

TOP 20

9-2 Two Proportions

Key Concept In this section we present methods for (1) testing a claim made about two population proportions and (2) constructing a confidence interval estimate of the difference between two population proportions. Although the focus of this section is proportions, we can use the same methods for dealing with probabilities or the decimal equivalents of percentages.

Objectives

(1) Test a claim about two population proportions or (2) construct a confidence interval estimate of the difference between two population proportions.

Notation for Two Proportions

For population 1 we let

$p_1 =$ *population* proportion

$n_1 =$ size of the sample

$x_1 =$ number of successes in the sample

$\hat{p}_1 = \dfrac{x_1}{n_1}$ (*sample* proportion)

$\hat{q}_1 = 1 - \hat{p}_1$ (complement of $\hat{p}_1$)

The corresponding notations $p_2, n_2, x_2, \hat{p}_2$, and $\hat{q}_2$ apply to population 2.

Pooled Sample Proportion

The **pooled sample proportion** is denoted by $\bar{p}$ and is given by

$$\bar{p} = \frac{x_1 + x_2}{n_1 + n_2}$$

$$\bar{q} = 1 - \bar{p}$$

Requirements

1. The sample proportions are from two simple random samples that are *independent.* (Samples are *independent* if the sample values selected from one population are not related to or somehow naturally paired or matched with the sample values selected from the other population.)
2. For each of the two samples, there are at least 5 successes and at least 5 failures. (That is, $n\hat{p} \geq 5$ and $n\hat{q} \geq 5$ for each of the two samples.)

Test Statistic for Two Proportions (with H_0: $p_1 = p_2$)

$$z = \frac{(\hat{p}_1 - \hat{p}_2) - (p_1 - p_2)}{\sqrt{\dfrac{\bar{p}\bar{q}}{n_1} + \dfrac{\bar{p}\bar{q}}{n_2}}} \quad \text{where } p_1 - p_2 = 0 \text{ (assumed in the null hypothesis)}$$

$$\hat{p}_1 = \frac{x_1}{n_1} \quad \text{and} \quad \hat{p}_2 = \frac{x_2}{n_2} \quad \text{(sample proportions)}$$

$$\bar{p} = \frac{x_1 + x_2}{n_1 + n_2} \quad (\textit{pooled} \text{ sample proportion}) \quad \text{and} \quad \bar{q} = 1 - \bar{p}$$

***P*-values:** *P*-values are automatically provided by technology. If technology is not available, use the computed value of the test statistic with the standard normal distribution (Table A-2) and find the *P*-value by following the procedure summarized in Figure 8-4 in Section 8-2.

Critical values: Use Table A-2. (Based on the significance level α, find critical values by using the same procedures introduced in Section 8-2.)

Confidence Interval Estimate of $p_1 - p_2$

The confidence interval estimate of the difference $p_1 - p_2$ is

$$(\hat{p}_1 - \hat{p}_2) - E < (p_1 - p_2) < (\hat{p}_1 - \hat{p}_2) + E$$

where the margin of error E is given by $E = z_{\alpha/2}\sqrt{\dfrac{\hat{p}_1\hat{q}_1}{n_1} + \dfrac{\hat{p}_2\hat{q}_2}{n_2}}$

Rounding: Round the confidence interval limits to three significant digits.

Hypothesis Tests

For tests of hypotheses made about two population proportions, we consider only tests having a null hypothesis of $p_1 = p_2$ (so the null hypothesis is given as H_0: $p_1 = p_2$). The following example will help clarify the roles of x_1, n_1, $\hat{p}_1$, $\bar{p}$, and so on. Note that with the assumption of equal proportions, the best estimate of the common proportion is obtained by pooling both samples into one big sample, so that $\bar{p}$ is the estimator of the common population proportion.

The Lead Margin of Error

Authors Stephen Ansolabehere and Thomas Belin wrote in their article "Poll Faulting" (*Chance* magazine) that "our greatest criticism of the reporting of poll results is with the margin of error of a single proportion (usually ±3%) when media attention is clearly drawn to the *lead* of one candidate." They point out that the lead is really the *difference* between two proportions ($p_1 - p_2$) and go on to explain how they developed the following rule of thumb: The lead is approximately $\sqrt{3}$ times larger than the margin of error for any one proportion. For a typical pre-election poll, a reported ±3% margin of error translates to about ±5% for the lead of one candidate over the other. They write that the margin of error for the lead should be reported.

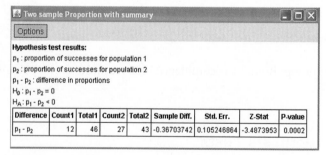

Example 1 Large Denominations Less Likely to Be Spent?

In the article "The Denomination Effect" by Priya Raghubir and Joydeep Srivastava, *Journal of Consumer Research,* Vol. 36, researchers reported results from studies conducted to determine whether people have different spending characteristics when they have larger bills, such as a $20 bill, instead of smaller bills, such as twenty $1 bills. In one of the trials that they conducted, 89 undergraduate business students from two different colleges were randomly assigned to two different groups. In the "dollar bill" group, 46 subjects were given dollar bills; the "quarter" group consisted of 43 subjects given quarters. All subjects from both groups were given a choice of keeping the money or buying gum or mints. The article includes the claim that "money in a large denomination is less likely to be spent relative to an equivalent amount in many smaller denominations." Let's test that claim using a 0.05 significance level with the following sample data from the study.

	Group 1	Group 2
	Subjects Given $1 Bill	**Subjects Given 4 Quarters**
Spent the money	$x_1 = 12$	$x_2 = 27$
Subjects in group	$n_1 = 46$	$n_2 = 43$

Solution

Requirement check We first verify that the two necessary requirements are satisfied. (1) Because the 89 subjects were randomly assigned to the two groups, we will consider the two samples to be simple random samples (but see the interpretation at the end of this example for more comments). The two samples are independent because subjects are not matched or paired in any way. (2) The subjects given $1 bills include 12 who spent the money and 34 who did not, so the number of successes is at least 5 and the number of failures is at least 5. The subjects given four quarters include 27 who spent the money and 16 who did not, so the number of successes is at least 5 and the number of failures is at least 5. The requirements are satisfied. ✓

Technology Computer programs and calculators usually provide a *P*-value, so the *P*-value method is typically used. See the accompanying StatCrunch results showing the null and alternative hypotheses, the test statistic of $z = -3.49$ (rounded), and the *P*-value of 0.0002.

STATCRUNCH

Two sample Proportion with summary

Options

Hypothesis test results:
p_1 : proportion of successes for population 1
p_2 : proportion of successes for population 2
$p_1 - p_2$: difference in proportions
$H_0 : p_1 - p_2 = 0$
$H_A : p_1 - p_2 < 0$

Difference	Count1	Total1	Count2	Total2	Sample Diff.	Std. Err.	Z-Stat	P-value
$p_1 - p_2$	12	46	27	43	-0.36703742	0.105246864	-3.4873953	0.0002

If technology is not available, see Figure 8-1 in Section 8-2 for the *P*-value method; we will now proceed with that method.

Step 1: The claim that "money in a large denomination is less likely to be spent relative to an equivalent amount in many smaller denominations" can be expressed as $p_1 < p_2$.

Step 2: If $p_1 < p_2$ is false, then $p_1 \geq p_2$.

Step 3: Because the claim of $p_1 < p_2$ does not contain equality, it becomes the alternative hypothesis. The null hypothesis is the statement of equality, so we have

$$H_0: p_1 = p_2 \quad H_1: p_1 < p_2 \text{ (original claim)}$$

Step 4: The significance level is $\alpha = 0.05$.

Step 5: We will use the normal distribution (with the test statistic given earlier in this section) as an approximation to the binomial distribution. We estimate the common value of p_1 and p_2 with the pooled sample estimate $\bar{p}$ calculated as shown below, with extra decimal places used to minimize rounding errors in later calculations.

$$\bar{p} = \frac{x_1 + x_2}{n_1 + n_2} = \frac{12 + 27}{46 + 43} = 0.438202$$

With $\bar{p} = 0.438202$, it follows that $\bar{q} = 1 - 0.438202 = 0.561798$.

Step 6: Because we assume in the null hypothesis that $p_1 = p_2$, the value of $p_1 - p_2$ is 0 in the following calculation of the test statistic:

$$z = \frac{(\hat{p}_1 - \hat{p}_2) - (p_1 - p_2)}{\sqrt{\frac{\bar{p}\bar{q}}{n_1} + \frac{\bar{p}\bar{q}}{n_2}}}$$

$$= \frac{\left(\frac{12}{46} - \frac{27}{43}\right) - 0}{\sqrt{\frac{(0.438202)(0.561798)}{46} + \frac{(0.438202)(0.561798)}{43}}}$$

$$= -3.49$$

This is a left-tailed test, so the P-value is the area to the left of the test statistic $z = -3.49$ (as indicated by Figure 8-4). Refer to Table A-2 and find that the area to the left of the test statistic $z = -3.49$ is 0.0002, so the P-value is 0.0002. The test statistic and P-value are shown in Figure 9-1(a).

Step 7: Because the P-value of 0.0002 is less than the significance level of $\alpha = 0.05$, we reject the null hypothesis of $p_1 = p_2$.

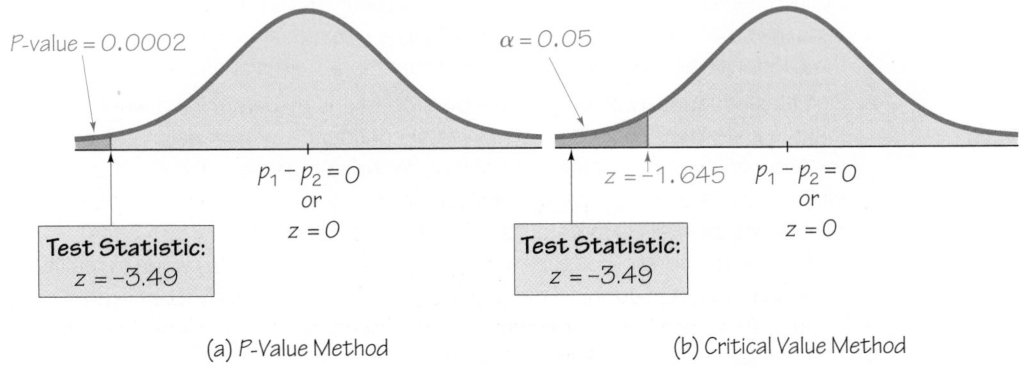

Figure 9-1 Hypothesis Test with Two Proportions

continued

Author as a Witness

The author was asked to testify in New York State Supreme Court by a former student who was contesting a lost reelection to the office of Dutchess County Clerk. The author testified by using statistics to show that the voting behavior in one contested district was significantly different from the behavior in all other districts. When the opposing attorney asked about results of a confidence interval, he asked if the 5% error (from a 95% confidence level) could be added to the three percentage point margin of error to get a total error of 8%, thereby indicating that he did not understand the basic concept of a confidence interval. The judge cited the author's testimony, upheld the claim of the former student, and ordered a new election in the contested district. That judgment was later overturned by the appellate court on the grounds that the ballot irregularities should have been contested before the election, not after.

Polio Experiment

In 1954 an experiment was conducted to test the effectiveness of the Salk vaccine as protection against the devastating effects of polio.

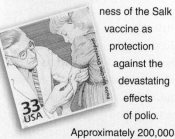

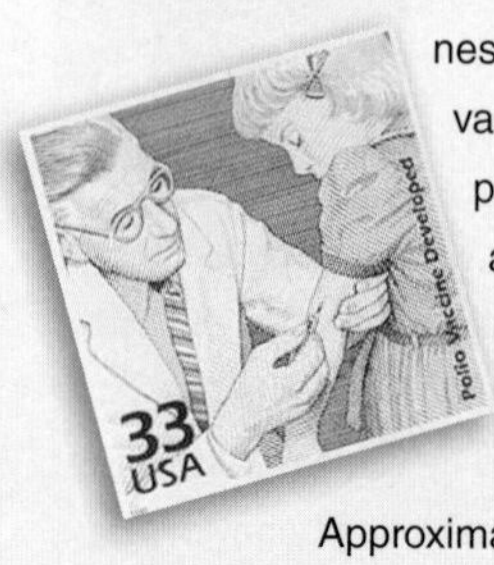

Approximately 200,000 children were injected with an ineffective salt solution, and 200,000 other children were injected with the vaccine. The experiment was "double blind" because the children being injected didn't know whether they were given the real vaccine or the placebo, and the doctors giving the injections and evaluating the results didn't know either. Only 33 of the 200,000 vaccinated children later developed paralytic polio, whereas 115 of the 200,000 injected with the salt solution later developed paralytic polio. Statistical analysis of these and other results led to the conclusion that the Salk vaccine was indeed effective against paralytic polio.

Note to Instructor
Reminder about Terminology: The "critical value method" was called the "traditional method" in the previous edition. Using the terminology of "critical value method" makes it easier for students to remember that this method uses critical values instead of *P*-values.

Interpretation

We must address the original claim that "money in a large denomination is less likely to be spent relative to an equivalent amount in many smaller denominations." Because we reject the null hypothesis, we conclude that there is sufficient evidence to support the claim that $p_1 < p_2$. That is, there is sufficient evidence to support the claim that people with money in large denominations are less likely to spend relative to people with an equivalent amount of money in smaller denominations. (See Table 8-3 in Section 8-2 for help in wording the final conclusion.) Based on these results, it appears that the original claim is supported.

It should be noted that the study subjects were 89 undergraduate business students from two different colleges. It would be wise to qualify the preceding conclusions by saying that the results do not necessarily apply to the general population.

Critical Value Method of Testing Hypotheses

The critical value method of testing hypotheses can also be used for Example 1. In Step 6, instead of finding the *P*-value, find the critical value. With a significance level of $\alpha = 0.05$ in a left-tailed test based on the normal distribution, we refer to Table A-2 and find that an area of $\alpha = 0.05$ in the left tail corresponds to the critical value of $z = -1.645$. In Figure 9-1(b) we can see that the test statistic of $z = -3.49$ falls within the critical region bounded by the critical value of $z = -1.645$. We again reject the null hypothesis. The conclusions are the same as in Example 1.

Confidence Intervals

Using the format given earlier in this section, we can construct a confidence interval estimate of the difference between population proportions $(p_1 - p_2)$. If a confidence interval estimate of $p_1 - p_2$ does not include 0, we have evidence suggesting that p_1 and p_2 have different values. The confidence interval uses a standard deviation based on estimated values of the population proportions, whereas a hypothesis test uses a standard deviation based on the assumption that the two population proportions are equal. Consequently, a conclusion based on a confidence interval might be different from a conclusion based on a hypothesis test. See the following cautions.

CAUTIONS

1. When testing a claim about two population proportions, the *P*-value method and the critical value method are equivalent, but they are *not* equivalent to the confidence interval method. If you want to test a claim about two population proportions, use the *P*-value method or critical value method; if you want to estimate the difference between two population proportions, use a confidence interval.
2. Don't test for equality of two population proportions by determining whether there is an overlap between two individual confidence interval estimates of the two individual population proportions. When compared to the confidence interval estimate of $p_1 - p_2$, the analysis of overlap between two individual confidence intervals is more conservative (by rejecting equality less often), and it has less power (because it is less likely to reject $p_1 = p_2$ when in reality $p_1 \neq p_2$). (See "On Judging the Significance of Differences by Examining the Overlap Between Confidence Intervals," by Schenker and Gentleman, *American Statistician,* Vol. 55, No. 3.) See Exercise19.

Example 2 Confidence Interval for Claim about Denominations

Use the sample data given in Example 1 to construct a 90% confidence interval estimate of the difference between the two population proportions. (As shown in Table 8-1 in Section 8-2, a one-tailed hypothesis test with significance level $\alpha = 0.05$ requires a confidence level of 90%.) What does the result suggest about the claim that "money in a large denomination is less likely to be spent relative to an equivalent amount in many smaller denominations"?

Solution

Requirement check We are using the same data from Example 1, and the same requirement check applies here, so the requirements are satisfied. ✔

With a 90% confidence level, $z_{\alpha/2} = 1.645$ (from Table A-2). We first calculate the value of the margin of error E as shown here.

$$E = z_{\alpha/2}\sqrt{\frac{\hat{p}_1\hat{q}_1}{n_1} + \frac{\hat{p}_2\hat{q}_2}{n_2}} = 1.645\sqrt{\frac{\left(\frac{12}{46}\right)\left(\frac{34}{46}\right)}{46} + \frac{\left(\frac{27}{43}\right)\left(\frac{16}{43}\right)}{43}}$$
$$= 0.161387$$

With $\hat{p}_1 = 12/46 = 0.260870$ and $\hat{p}_2 = 27/43 = 0.627907$, $\hat{p}_1 - \hat{p}_2 = -0.367037$. With $\hat{p}_1 - \hat{p}_2 = -0.367037$ and $E = 0.161387$, the confidence interval is evaluated as follows, with the confidence interval limits rounded to three significant digits:

$$(\hat{p}_1 - \hat{p}_2) - E < (p_1 - p_2) < (\hat{p}_1 - \hat{p}_2) + E$$
$$-0.367037 - 0.161387 < (p_1 - p_2) < -0.367037 + 0.161387$$
$$-0.528 < (p_1 - p_2) < -0.206$$

Interpretation

The confidence interval limits do not contain 0, implying that there is a significant difference between the two proportions. The confidence interval suggests that the value of p_1 is less than the value of p_2, so there does appear to be sufficient evidence to support the claim that "money in a large denomination is less likely to be spent relative to an equivalent amount in many smaller denominations."

Rationale: Why Do the Procedures of This Section Work? The test statistic given for hypothesis tests is justified by the following reasoning.

With $n_1\hat{p}_1 \geq 5$ and $n_1\hat{q}_1 \geq 5$, the distribution of $\hat{p}_1$ can be approximated by a normal distribution with mean p_1, standard deviation $\sqrt{p_1q_1/n_1}$, and variance p_1q_1/n_1 (based on Sections 6-7 and 7-2). They also apply to the second sample. Because the distributions of $\hat{p}_1$ and $\hat{p}_2$ are each approximated by a normal distribution, the difference $\hat{p}_1 - \hat{p}_2$ will also be approximated by a normal distribution with mean $p_1 - p_2$ and variance

$$\sigma^2_{(\hat{p}_1-\hat{p}_2)} = \sigma^2_{\hat{p}_1} + \sigma^2_{\hat{p}_2} = \frac{p_1q_1}{n_1} + \frac{p_2q_2}{n_2}$$

(The result above is based on this property: The variance of the *differences* between two independent random variables is the *sum* of their individual variances.) The pooled estimate of the common value of p_1 and p_2 is $\bar{p} = (x_1 + x_2)/(n_1 + n_2)$. If we

New Technology, New Data, New Insight

Residents of New York City believed that taxi cabs became scarce around rush hour in the late afternoon. Complaints could not be addressed, because there were no data to support that alleged shortage. However, as GPS units were installed on cabs, officials became able to track their locations. It was found that 20% fewer cabs were in service between 4:00 P.M. and 5:00 P.M. than in the preceding hour. Two factors were found to be responsible for this decrease. First, the 12-hour shifts were scheduled to change at 5:00 P.M. so that drivers on both shifts would get an equal share at a rush hour. Second, rising rents in Manhattan forced many cab companies to house their cabs in Queens, so drivers had to start returning around 4:00 P.M. so that they could make it back in time and avoid fines for being late. As of this writing, officials are considering regulations designed to eliminate or reduce the late-afternoon shortage. Any changes will result from planners' turning to the new GPS technology for data that could be objectively analyzed, instead of relying on subjective beliefs and anecdotal stories.

replace p_1 and p_2 by $\bar{p}$ and replace q_1 and q_2 by $\bar{q} = 1 - \bar{p}$, the variance above leads to the following standard deviation:

$$\sigma_{(\hat{p}_1 - \hat{p}_2)} = \sqrt{\frac{\bar{p}\bar{q}}{n_1} + \frac{\bar{p}\bar{q}}{n_2}}$$

We now know that the distribution of $p_1 - p_2$ is approximately normal, with mean $p_1 - p_2$ and standard deviation as shown above, so the z test statistic has the form given earlier.

The form of the confidence interval requires an expression for the variance different from the one given above. When constructing a confidence interval estimate of the difference between two proportions, we don't assume that the two proportions are equal, and we estimate the standard deviation as

$$\sqrt{\frac{\hat{p}_1\hat{q}_1}{n_1} + \frac{\hat{p}_2\hat{q}_2}{n_2}}$$

In the test statistic

$$z = \frac{(\hat{p}_1 - \hat{p}_2) - (p_1 - p_2)}{\sqrt{\dfrac{\hat{p}_1\hat{q}_1}{n_1} + \dfrac{\hat{p}_2\hat{q}_2}{n_2}}}$$

use the positive and negative values of z (for two tails) and solve for $p_1 - p_2$. The results are the limits of the confidence interval given in the box near the beginning of this section.

using TECHNOLOGY

STATDISK Select **Analysis** from the main menu bar, then select either **Hypothesis Testing** or **Confidence Intervals.** Select the menu item of **Proportion-Two Samples.** Enter the required items in the dialog box, then click on the **Evaluate** button. The accompanying display is from Example 1 in this section.

```
Claim:  p1 < p2

Pooled proportion: 0.4382022
Test Statistic,       z: -3.4874
Critical z:           -1.6449
P-Value:              0.0002

90% Confidence interval:
-0.5284103 < p1-p2 < -0.2056645
```

MINITAB Select **Stat** from the main menu bar, then select **Basic Statistics,** then **2 Proportions.** Click on the button for **Summarized data** and enter the sample values. Click on the **Options** bar and enter the desired confidence level. (Enter 95 for a hypothesis test with a 0.05 significance level.) If testing a hypothesis, enter 0 for the claimed value of $p_1 - p_2$, then select the format for the alternative hypothesis, and click on the box to use the pooled estimate of p for the test. Click **OK** twice.

In **Minitab 16,** you can also click on **Assistant,** then **Hypothesis Tests,** then select the case for **2-Sample % Defective.** Fill out the dialog box, then click **OK** to get three windows of results that include the P-value and much other helpful information.

EXCEL **Hypothesis Test:** Use XLSTAT. Click on **XLSTAT** at the top. Click on **Parametric tests,** then select **Test for two proportions.** In the dialog box that appears, enter the frequency and sample size for each of the two samples. For the "Data format" options, select **Frequencies.** Be sure that there is a checkmark in the box next to "z test." Click on the **Options** tab and select the type of test; for a two-tailed test, select the case including the symbol $\neq$; for a left-tailed test, select the case including $<$; and for a right-tailed test, select the case including $>$. Enter a value in the "Significance level (%)" box. For example, enter 5 for a 0.05 significance level. For the options listed under "Variance," select pq(1/n1 + 1/n2). Click **OK** to get results that include the test statistic and P-value.

Confidence Interval: Use XLSTAT. Click on **XLSTAT** at the top. Click on **Parametric tests,** then select **Test for two proportions.** In the dialog box that appears, enter the frequency and sample size for each of the two samples. For the "Data format" options, select **Frequencies.** Be sure that there is a checkmark in the box next to "z test." Click on the **Options** tab. For the alternative hypothesis, select the format of a two-tailed test. Enter a value in the "Significance level (%)" box. For example, enter 5 for a 95% confidence interval. For the options listed under "Variance," select p1q1/n1 + p2q2/n2. Click **OK** to get results that include the confidence interval.

TI-83/84 PLUS The TI-83/84 Plus calculator can be used for hypothesis tests and confidence intervals. Press **STAT** and select **TESTS.** Then choose the option of **2-PropZTest** (for a hypothesis test) or **2-PropZInt** (for a confidence interval). When testing hypotheses, the TI-83/84 Plus calculator will display a *P*-value instead of critical values, so the *P*-value method of testing hypotheses is used.

STATCRUNCH Click on **Open StatCrunch.** Click on **Stat,** then select **Proportions.** Select **Two sample,** then select **with summary.** Proceed to enter the numbers of successes and the numbers of observations, click on **Next,** then select either **Hypothesis Test** or **Confidence Interval.** Enter the required values, then click on **Calculate.**

9-2 Basic Skills and Concepts

Recommended Assignment
Exercises 1–10, 15, 16.

Due to limited space, answers for odd numbered exercises are not included here, but they are included in Appendix D.

Statistical Literacy and Critical Thinking

1. Verifying Requirements In the largest clinical trial ever conducted, 401,974 children were randomly assigned to two groups. The treatment group consisted of 201,229 children given the Salk vaccine for polio, and the other 200,745 children were given a placebo. Among those in the treatment group, 33 developed polio, and among those in the placebo group, 115 developed polio. If we want to use the methods of this section to test the claim that the rate of polio is less for children given the Salk vaccine, are the requirements for a hypothesis test satisfied? Explain.

2. Notation For the sample data given in Exercise 1, consider the Salk vaccine treatment group to be the first sample. Identify the values of $n_1, \hat{p}_1, \hat{q}_1, n_2, \hat{p}_2, \hat{q}_2, \bar{p}$, and $\bar{q}$. Round all values so that they have six significant digits.

2. $n_1 = 201{,}229$, $\hat{p}_1 = 0.000163992$, $\hat{q}_1 = 0.999836$, $n_2 = 200{,}745$, $\hat{p}_2 = 0.000572866$, $\hat{q}_2 = 0.999427$, $\bar{p} = 0.000368183$, and $\bar{q} = 0.999632$.

3. Hypotheses and Conclusions Refer to the hypothesis test described in Exercise 1.

a. Identify the null hypothesis and the alternative hypothesis.

b. If the *P*-value for the test is reported as "less than 0.001," what should we conclude about the original claim?

4. Using Confidence Intervals

a. Assume that we want to use a 0.05 significance level to test the claim that $p_1 < p_2$. If we want to test that claim by using a confidence interval, what confidence level should we use?

b. If we test the claim in part (a) using the sample data in Exercise 1, we get this confidence interval: $-0.000508 < p_1 - p_2 < -0.000309$. What does this confidence interval suggest about the claim?

c. In general, when dealing with inferences for two population proportions, which two of the following are equivalent: confidence interval method; *P*-value method; critical value method?

4. a. 0.90, or 90%
b. Because the confidence interval limits do not contain 0, there appears to be a significant difference between the two proportions. Because the confidence interval consists of negative values only, it appears that the first proportion is less than the second proportion. There is sufficient evidence to support the claim that the rate of polio is less for children given the Salk vaccine than it is for children given a placebo.
c. The *P*-value method and the critical value method are equivalent in the sense that they will always lead to the same conclusion, but the confidence interval method is not equivalent to them.

Interpreting Displays. ***In Exercises 5 and 6, use the results from the given displays.***

5. Flu Vaccine A *USA Today* article about an experimental nasal spray vaccine for children stated, "In a trial involving 1602 children only 14 (1%) of the 1070 who received the vaccine developed the flu, compared with 95 (18%) of the 532 who got a placebo." The accompanying TI-83/84 Plus calculator display results from a test of the claim that the vaccine had no effect. Identify the test statistic and *P*-value. What should you conclude about the claim?

TI-83/84 PLUS

```
2-PropZTest
 p1≠p2
 z=-12.38798074
 p=3.137085E-35
 p̂1=.0130841121
 p̂2=.1785714286
↓p̂=.0680399501
```

6. E-mail Privacy Among 436 workers surveyed in a Gallup poll, 192 said that it was seriously unethical to monitor employee e-mail. Among 121 senior-level managers, 40 said that it was seriously unethical to monitor employee e-mail. Consider the claim that for those saying that monitoring e-mail is seriously unethical, the proportion of workers is the same as

6. Test statistic: $z = 2.17$. P-value: 0.030. Because the P-value is greater than the significance level of 0.01, conclude that there is not sufficient evidence to warrant rejection of the claim that for those saying that monitoring e-mail is seriously unethical, the proportion of workers is the same as the proportion of managers.

8. a. H_0: $p_1 = p_2$. H_1: $p_1 < p_2$. Test statistic: $z = -1.66$. Critical value: $z = -2.33$. P-value: 0.0485 (Tech: 0.0484). Fail to reject H_0. There is not sufficient evidence to support the claim that the rate of dementia among those who use ginkgo is less than the rate of dementia among those who use a placebo. There is not sufficient evidence to support the claim that ginkgo is effective in preventing dementia.

b. 98% CI: $-0.0542 < p_1 - p_2 < 0.00909$ (Tech: $-0.0541 < p_1 - p_2 < 0.00904$). Because the confidence interval limits include 0, there does not appear to be a significant difference between dementia rates for those treated with ginkgo and those given a placebo. There is not sufficient evidence to support the claim that the rate of dementia among those who use ginkgo is less than the rate of dementia among those who use a placebo. There is not sufficient evidence to support the claim that ginkgo is effective in preventing dementia.

c. The sample results suggest that ginkgo is not effective in preventing dementia.

10. a. H_0: $p_1 = p_2$. H_1: $p_1 \neq p_2$. Test statistic: $z = 18.26$. Critical values: $z = \pm 2.575$ (Tech: ± 2.576). P-value: 0.0002 (Tech: 0.0000). Reject H_0. There is sufficient evidence to warrant rejection of the claim that the survival rates are the same for day and night.

the proportion of managers. Identify the test statistic and P-value. If using a 0.01 significance level, what should you conclude about the claim?

MINITAB

```
Difference = p (1) - p (2)
Estimate for difference:  0.109788
95% CI for difference:  (0.0138876, 0.205689)
Test for difference = 0 (vs not = 0):  Z = 2.17  P-Value = 0.030
```

Testing Claims about Proportions. ***In Exercises 7–18, test the given claim. Identify the null hypothesis, alternative hypothesis, test statistic, P-value or critical value(s), conclusion about the null hypothesis, and final conclusion that addresses the original claim.***

7. Dreaming in Black and White A study was conducted to determine the proportion of people who dream in black and white instead of color. Among 306 people over the age of 55, 68 dream in black and white, and among 298 people under the age of 25, 13 dream in black and white (based on data from "Do We Dream in Color?" by Eva Murzyn, *Consciousness and Cognition*, Vol. 17, No. 4). We want to use a 0.01 significance level to test the claim that the proportion of people over 55 who dream in black and white is greater than the proportion for those under 25.

a. Test the claim using a hypothesis test.

b. Test the claim by constructing an appropriate confidence interval.

c. An explanation given for the results is that those over the age of 55 grew up exposed to media that was mostly displayed in black and white. Can the results from parts (a) and (b) be used to verify that explanation?

8. Ginkgo for Dementia The herb ginkgo biloba is commonly used as a treatment to prevent dementia. In a study of the effectiveness of this treatment, 1545 elderly subjects were given ginkgo and 1524 elderly subjects were given a placebo. Among those in the ginkgo treatment group, 246 later developed dementia, and among those in the placebo group, 277 later developed dementia (based on data from "Ginkgo Biloba for Prevention of Dementia," by DeKosky et al., *Journal of the American Medical Association,* Vol. 300, No. 19). We want to use a 0.01 significance level to test the claim that ginkgo is effective in preventing dementia.

a. Test the claim using a hypothesis test.

b. Test the claim by constructing an appropriate confidence interval.

c. Based on the results, is ginkgo effective in preventing dementia?

9. Are Seat Belts Effective? A simple random sample of front-seat occupants involved in car crashes is obtained. Among 2823 occupants not wearing seat belts, 31 were killed. Among 7765 occupants wearing seat belts, 16 were killed (based on data from "Who Wants Airbags?" by Meyer and Finney, *Chance,* Vol. 18, No. 2). We want to use a 0.05 significance level to test the claim that seat belts are effective in reducing fatalities.

a. Test the claim using a hypothesis test.

b. Test the claim by constructing an appropriate confidence interval.

c. What does the result suggest about the effectiveness of seat belts?

10. Cardiac Arrest at Day and Night A study investigated survival rates for in-hospital patients who suffered cardiac arrest. Among 58,593 patients who had cardiac arrest during the day, 11,604 survived and were discharged. Among 28,155 patients who suffered cardiac arrest at night, 4139 survived and were discharged (based on data from "Survival from

In-Hospital Cardiac Arrest During Nights and Weekends," by Peberdy et al., *Journal of the American Medical Association,* Vol. 299, No. 7). We want to use a 0.01 significance level to test the claim that the survival rates are the same for day and night.

a. Test the claim using a hypothesis test.

b. Test the claim by constructing an appropriate confidence interval.

c. Based on the results, does it appear that for in-hospital patients who suffer cardiac arrest, the survival rate is the same for day and night?

11. Is Echinacea Effective for Colds? Rhinoviruses typically cause common colds. In a test of the effectiveness of echinacea, 40 of the 45 subjects treated with echinacea developed rhinovirus infections. In a placebo group, 88 of the 103 subjects developed rhinovirus infections (based on data from "An Evaluation of Echinacea Angustifolia in Experimental Rhinovirus Infections," by Turner et al., *New England Journal of Medicine,* Vol. 353, No. 4). We want to use a 0.05 significance level to test the claim that echinacea has an effect on rhinovirus infections.

a. Test the claim using a hypothesis test.

b. Test the claim by constructing an appropriate confidence interval.

c. Based on the results, does echinacea appear to have any effect on the infection rate?

12. Bednets to Reduce Malaria In a randomized controlled trial in Kenya, insecticide-treated bednets were tested as a way to reduce malaria. Among 343 infants using bednets, 15 developed malaria. Among 294 infants not using bednets, 27 developed malaria (based on data from "Sustainability of Reductions in Malaria Transmission and Infant Mortality in Western Kenya with Use of Insecticide-Treated Bednets," by Lindblade et al., *Journal of the American Medical Association,* Vol. 291, No. 21). We want to use a 0.01 significance level to test the claim that the incidence of malaria is lower for infants using bednets.

a. Test the claim using a hypothesis test.

b. Test the claim by constructing an appropriate confidence interval.

c. Based on the results, do the bednets appear to be effective?

13. Tennis Challenges Since the Hawk-Eye instant replay system for tennis was introduced at the U.S. Open in 2006, men challenged 1412 referee calls, with the result that 421 of the calls were overturned. Women challenged 759 referee calls, and 220 of the calls were overturned. We want to use a 0.05 significance level to test the claim that men and women have equal success in challenging calls.

a. Test the claim using a hypothesis test.

b. Test the claim by constructing an appropriate confidence interval.

c. Based on the results, does it appear that men and women have equal success in challenging calls?

14. Police Gunfire In a study of police gunfire reports during a recent year, it was found that among 540 shots fired by New York City police, 182 hit their targets; and among 283 shots fired by Los Angeles police, 77 hit their targets (based on data from the *New York Times*). We want to use a 0.05 significance level to test the claim that New York City police and Los Angeles police have the same proportion of hits.

a. Test the claim using a hypothesis test.

b. Test the claim by constructing an appropriate confidence interval. *continued*

b. 99% CI: $0.0441 < p_1 - p_2 < 0.0579$. Because the confidence interval limits do not contain 0, there appears to be a significant difference between the two proportions. There is sufficient evidence to warrant rejection of the claim that the survival rates are the same for day and night.

c. The data suggest that for in-hospital patients who suffer cardiac arrest, the survival rate is not the same for day and night. It appears that the survival rate is higher for in-hospital patients who suffer cardiac arrest during the day.

12. a. H_0: $p_1 = p_2$. H_1: $p_1 < p_2$. Test statistic: $z = -2.44$. Critical value: $z = -2.33$. *P*-value: 0.0074. Reject H_0. There is sufficient evidence to support the claim that the incidence of malaria is lower for infants who use the bednets.

b. 98% CI: $0.0950 < p_1 - p_2 < -0.00118$ (Tech: $-0.0950 < p_1 - p_2 < -0.00125$). Because the confidence interval does not include 0 and it includes only negative values, it appears that the rate of malaria is lower for infants who use the bednets.

c. The bednets appear to be effective.

14. a. H_0: $p_1 = p_2$. H_1: $p_1 \neq p_2$. Test statistic: $z = 1.91$. Critical values: $z = \pm 1.96$. *P*-value: 0.0562 (Tech: 0.0567). Fail to reject H_0. There is not sufficient evidence to warrant rejection of the claim that New York City police and Los Angeles police have the same proportion of hits.

b. 95% CI: $-0.000455 < p_1 - p_2 < 0.130$ (Tech: $-0.000454 < p_1 - p_2 < 0.130$). Because the confidence interval limits contain 0, there does not appear to be a significant difference between the two proportions. There is not sufficient evidence to warrant rejection of the claim that New York City police and Los Angeles police have the same proportion of hits.

c. There does not appear to be a difference between the hit rates of New York City police and Los Angeles police.

c. Based on the results, does there appear to be a difference between the hit rates of New York City police and Los Angeles police?

15. Headache Treatment In a study of treatments for very painful "cluster" headaches, 150 patients were treated with oxygen and 148 other patients were given a placebo consisting of ordinary air. Among the 150 patients in the oxygen treatment group, 116 were free from headaches 15 minutes after treatment. Among the 148 patients given the placebo, 29 were free from headaches 15 minutes after treatment (based on data from "High-Flow Oxygen for Treatment of Cluster Headache," by Cohen, Burns, and Goadsby, *Journal of the American Medical Association,* Vol. 302, No. 22). We want to use a 0.01 significance level to test the claim that the oxygen treatment is effective.

a. Test the claim using a hypothesis test.

b. Test the claim by constructing an appropriate confidence interval.

c. Based on the results, is the oxygen treatment effective?

16. Spending Large and Small Bills In the same study cited in Example 1, another trial was conducted with 75 women in China given a 100-Yuan bill, while another 75 women in China were given 100 Yuan in the form of smaller bills (a 50-Yuan bill plus two 20-Yuan bills plus two 5-Yuan bills). Among those given the single bill, 60 spent some or all of the money. Among those given the smaller bills, 68 spent some or all of the money. We want to use a 0.05 significance level to test the claim that when given a single large bill, a smaller proportion of women in China spend some or all of the money when compared to the proportion of women in China given the same amount in smaller bills.

a. Test the claim using a hypothesis test.

b. Test the claim by constructing an appropriate confidence interval.

c. If the significance level is changed to 0.01, does the conclusion change?

17. Lefties In a random sample of males, it was found that 23 write with their left hands and 217 do not. In a random sample of females, it was found that 65 write with their left hands and 455 do not (based on data from "The Left-Handed: Their Sinister History," by Elaine Fowler Costas, Education Resources Information Center, Paper 399519). We want to use a 0.01 significance level to test the claim that the rate of left-handedness among males is less than that among females.

a. Test the claim using a hypothesis test.

b. Test the claim by constructing an appropriate confidence interval.

c. Based on the results, is the rate of left-handedness among males less than the rate of left-handedness among females?

18. Marathon Finishers In a recent New York City marathon, 25,221 men finished and 253 dropped out. Also, 12,883 women finished and 163 dropped out (based on data from the *New York Times*). We want to use a 0.01 significance level to test the claim that the rate of those who finish is the same for men and women.

a. Test the claim using a hypothesis test.

b. Test the claim by constructing an appropriate confidence interval.

c. Based on the results, do men and women finish the New York City marathon at the same rate?

16. a. H_0: $p_1 = p_2$. H_1: $p_1 < p_2$. Test statistic: $z = -1.85$. Critical value: $z = -1.645$. *P*-value: 0.0322 (Tech: 0.0324). Reject H_0. There is sufficient evidence to support the claim that when given a single large bill, a smaller proportion of women in China spend some or all of the money when compared to the proportion of women in China given the same amount in smaller bills.

b. 90% CI: $-0.201 < p_1 - p_2 < -0.0127$. Because the confidence interval does not include 0 and it includes only negative values, it appears that the first proportion is less than the second proportion. There is sufficient evidence to support the claim that when given a single large bill, a smaller proportion of women in China spend some or all of the money when compared to the proportion of women in China given the same amount in smaller bills.

c. Because the *P*-value is 0.0322 (Tech: 0.0324), the difference is significant at the 0.05 significance level, but not at the 0.01 significance level. The conclusion does change.

18. a. H_0: $p_1 = p_2$. H_1: $p_1 \neq p_2$. Test statistic: $z = 2.30$. Critical values: $z = \pm 2.575$ (Tech: ± 2.576). *P*-value: 0.0214 (Tech: 0.0213). Fail to reject H_0. There is not sufficient evidence to warrant rejection of the claim that the rate of those who finish is the same for men and women.

b. 99% CI: $-0.000409 < p_1 - p_2 < 0.00553$ (Tech: $-0.000409 < p_1 - p_2 < 0.00554$). Because the confidence interval limits contain 0, there does not appear to be a significant difference between the two proportions. There is not sufficient evidence to warrant rejection of the claim that the rate of those who finish is the same for men and women.

c. It appears that men and women finish the New York City marathon at the same rate.

9-2 Beyond the Basics

19. Interpreting Overlap of Confidence Intervals In the article "On Judging the Significance of Differences by Examining the Overlap Between Confidence Intervals," by Schenker and Gentleman (*American Statistician,* Vol. 55, No. 3), the authors consider sample data in this statement: "Independent simple random samples, each of size 200, have been drawn, and 112 people in the first sample have the attribute, whereas 88 people in the second sample have the attribute."

a. Use the methods of this section to construct a 95% confidence interval estimate of the difference $p_1 - p_2$. What does the result suggest about the equality of p_1 and p_2?

b. Use the methods of Section 7-2 to construct individual 95% confidence interval estimates for each of the two population proportions. After comparing the overlap between the two confidence intervals, what do you conclude about the equality of p_1 and p_2?

c. Use a 0.05 significance level to test the claim that the two population proportions are equal. What do you conclude?

d. Based on the preceding results, what should you conclude about the equality of p_1 and p_2? Which of the three preceding methods is least effective in testing for the equality of p_1 and p_2?

20. Equivalence of Hypothesis Test and Confidence Interval Two different simple random samples are drawn from two different populations. The first sample consists of 20 people with 10 having a common attribute. The second sample consists of 2000 people with 1404 of them having the same common attribute. Compare the results from a hypothesis test of $p_1 = p_2$ (with a 0.05 significance level) and a 95% confidence interval estimate of $p_1 - p_2$.

20. Hypothesis test: With a test statistic of $z = -1.9615$, P-value $= 0.05$ (Tech: 0.0498), reject $p_1 = p_2$. Confidence interval: $-0.422 < p_1 - p_2 < 0.0180$, which suggests that we should not reject $p_1 = p_2$ (because 0 is included). The hypothesis test and confidence interval lead to different conclusions about the equality of $p_1 = p_2$.

21. Determining Sample Size The sample size needed to estimate the difference between two population proportions to within a margin of error E with a confidence level of $1 - \alpha$ can be found by using the following expression:

$$E = z_{\alpha/2}\sqrt{\frac{p_1 q_1}{n_1} + \frac{p_2 q_2}{n_2}}$$

Replace n_1 and n_2 by n in the formula above (assuming that both samples have the same size) and replace each of p_1, q_1, p_2, and q_2 by 0.5 (because their values are not known). Solving for n results in this expression:

$$n = \frac{z_{\alpha/2}^2}{2E^2}$$

Use this expression to find the size of each sample if you want to estimate the difference between the proportions of adult men and women who are college graduates. Assume that you want 90% confidence that your error is no more than 0.02.

9-3 Two Means: Independent Samples

Note to Instructor

This section has been partitioned into Parts 1 and 2. Part 1 involves situations in which the standard deviations of the two populations are unknown and are not assumed to be equal. Part 2 involves two other situations: (1) The two population standard deviations are both known; (2) the two population standard deviations are unknown but are assumed to be equal.

Recommendation: Cover Part 1 and skip Part 2. In class, use the measured pulse rates from the male students and female students to test for a difference between the mean pulse rate of men and the mean pulse rate of women. The active involvement will enhance the learning experience. It will also ensure that all students are awake and alive.

Key Concept This section presents methods for using sample data from two independent samples to test hypotheses made about two population means or to construct confidence interval estimates of the difference between two population means. In Part 1 we discuss situations in which the standard deviations of the two populations are unknown and are not assumed to be equal. In Part 2 we discuss two other situations: (1) The two population standard deviations are both known; (2) the two population standard deviations are unknown but are assumed to be equal.

Using Statistics to Identify Thieves

Methods of statistics can be used to determine that an employee is stealing, and they can also be used to estimate the amount stolen. The following are some of the indicators that have been used. For comparable time periods, samples of sales have means that are significantly different. The mean sale amount decreases significantly. There is a significant increase in the proportion of "no sale" register openings. There is a significant decrease in the ratio of cash receipts to checks. Methods of hypothesis testing can be used to identify such indicators. (See "How to Catch a Thief," by Manly and Thomson, *Chance,* Vol. 11, No. 4.)

Part 1: Independent Samples with σ_1 and σ_2 Unknown and Not Assumed Equal

This section involves two *independent* samples, and the following section deals with samples that are *dependent*. It is important to know the difference between independent samples and dependent samples.

DEFINITIONS

Two samples are **independent** if the sample values from one population are not related to or somehow naturally paired or matched with the sample values from the other population.

Two samples are **dependent** (or consist of **matched pairs**) if the sample values are somehow matched, where the matching is based on some inherent relationship. (That is, each pair of sample values consists of two measurements from the same subject—such as before/after data—or each pair of sample values consists of matched pairs—such as husband/wife data—where the matching is based on some meaningful relationship.)

Example 1

***Independent Samples:* Proctored Tests and Nonproctored Tests in Online Courses** Researchers Michael Flesch and Elliot Ostler investigated the reliability of test assessment. One group consisted of 30 students who took proctored tests. A second group consisted of 32 students who took tests online without a proctor. The two samples are independent, because the subjects were not paired or matched in any way.

***Independent Samples:* Weights of M&Ms** Data Set 20 in Appendix B includes the following weights (grams) of a sample of yellow M&Ms and a sample of brown M&Ms. The yellow and brown weights might appear to be paired because of the way that they are listed, but they are not matched according to some inherent relationship. They are actually two independent samples that just happen to be listed in a way that might cause us to incorrectly think that they are matched.

Yellow	0.883	0.769	0.859	0.784	0.824	0.858	0.848	0.851
Brown	0.696	0.876	0.855	0.806	0.840	0.868	0.859	0.982

***Dependent Samples:* Heights of Husbands and Wives** Students of the author collected data consisting of the heights (cm) of husbands and the heights (cm) of their wives. Five of those pairs of heights are listed below. These two samples are dependent, because the height of each husband is matched with the height of his wife.

Height of Husband	175	180	173	176	178
Height of Wife	160	165	163	162	166

For inferences about means from two independent populations, the following box summarizes key elements of a hypothesis test and a confidence interval estimate of the difference between the population means.

Objectives

1. Conduct a hypothesis test of a claim about two independent population means, or
2. Construct a confidence interval estimate of the difference between two independent population means.

Notation

For population 1 we let

μ_1 = *population* mean

$\bar{x}_1$ = *sample* mean

σ_1 = *population* standard deviation

s_1 = *sample* standard deviation

n_1 = size of the first sample

The corresponding notations μ_2, σ_2, $\bar{x}_2$, s_2, and n_2 apply to population 2.

Requirements

1. The values of σ_1 and σ_2 are unknown and we do not assume that they are equal.
2. The two samples are *independent.*
3. Both samples are *simple random samples.*
4. Either or both of these conditions is satisfied: The two sample sizes are both *large* (with $n_1 > 30$ and $n_2 > 30$) or both samples come from populations having normal distributions. (The methods used here are *robust* against departures from normality, so for small samples, the normality requirement is loose in the sense that the procedures perform well as long as there are no outliers and departures from normality are not too extreme.)

Hypothesis Test Statistic for Two Means: Independent Samples

$$t = \frac{(\bar{x}_1 - \bar{x}_2) - (\mu_1 - \mu_2)}{\sqrt{\dfrac{s_1^2}{n_1} + \dfrac{s_2^2}{n_2}}} \quad \text{(where } \mu_1 - \mu_2 \text{ is often assumed to be 0)}$$

Degrees of freedom: When finding critical values or *P*-values, use the following for determining the number of degrees of freedom, denoted by df. (Although these two methods typically result in different numbers of degrees of freedom, the conclusion of a hypothesis test is rarely affected by the choice.)

1. In this book we use this simple and conservative estimate:

$$\text{df} = \textbf{smaller of } n_1 - 1 \textbf{ and } n_2 - 1$$

2. Statistical software packages typically use the more accurate but more difficult estimate given in Formula 9-1. (We will not use Formula 9-1 for the examples and exercises in this book.)

Formula 9-1

$$\text{df} = \frac{(A + B)^2}{\dfrac{A^2}{n_1 - 1} + \dfrac{B^2}{n_2 - 1}}$$

$$\text{where } A = \frac{s_1^2}{n_1} \text{ and } B = \frac{s_2^2}{n_2}$$

***P*-values:** *P*-values are automatically provided by technology. If technology is not available, refer to the *t* distribution in Table A-3. Use the procedure summarized in Figure 8-4 from Section 8-2.

Critical values: Refer to the *t* distribution in Table A-3.

continued

Confidence Interval Estimate of $\mu_1 - \mu_2$: Independent Samples

The confidence interval estimate of the difference $\mu_1 - \mu_2$ is

$$(\bar{x}_1 - \bar{x}_2) - E < (\mu_1 - \mu_2) < (\bar{x}_1 - \bar{x}_2) + E$$

where

$$E = t_{\alpha/2}\sqrt{\frac{s_1^2}{n_1} + \frac{s_2^2}{n_2}}$$

and the number of degrees of freedom df is as described above for hypothesis tests. (In this book, we use df = smaller of $n_1 - 1$ and $n_2 - 1$.)

Note to Instructor
Very observant students might wonder why the test statistic uses a standard deviation that includes a *sum* of two terms, whereas the numerator uses a *difference*. Explain that the variance of the differences between two independent random variables equals the variance of the first random variable *plus* the variance of the second random variable.

CAUTION Before conducting a hypothesis test, consider the context of the data, the source of the data, and the sampling method, and explore the data with graphs and descriptive statistics. Be sure to verify that the requirements are satisfied.

Equivalent Methods

The *P*-value method of hypothesis testing, the critical value method of hypothesis testing, and confidence intervals all use the same distribution and standard error, so they are all equivalent in the sense that they result in the same conclusions.

Example 2 Are We More Creative with Blue?

In the Chapter Problem, we noted that researchers from the University of British Columbia conducted trials to investigate the effects of color on creativity. Subjects with a red background were asked to think of creative uses for a brick; other subjects with a blue background were given the same task. Responses were scored by a panel of judges and results from scores of creativity are given below. The researchers make the claim that "blue enhances performance on a creative task." Test that claim using a 0.01 significance level.

Creativity Scores

Red Background:	$n = 35, \bar{x} = 3.39, s = 0.97$
Blue Background:	$n = 36, \bar{x} = 3.97, s = 0.63$

Solution

Requirement check (1) The values of the two population standard deviations are not known and we are not making an assumption that they are equal. (2) The two samples are independent because different subjects were used for the two different color groups. (3) The samples are simple random samples because subjects were randomly assigned to each of the two different color groups. (4) Both samples are large (greater than 30), so we satisfy the requirement that "the two sample sizes are both *large* (with $n_1 > 30$ and $n_2 > 30$) or both samples come from populations having normal distributions." The requirements are all satisfied. ✓

TI-83/84 PLUS

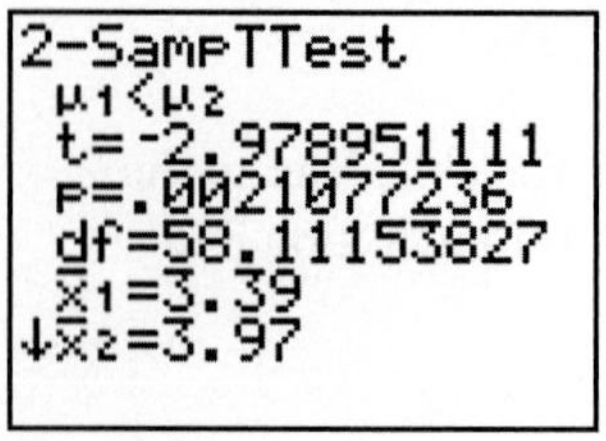

Technology Computer programs and calculators usually provide a *P*-value, so the *P*-value method is used. See the accompanying TI-83/84 Plus calculator results showing the alternative hypothesis of $\mu_1 < \mu_2$ (from the claim that the second group of blue has a higher mean), the test statistic of $t = -2.979$ (rounded), and the P-value of 0.0021 (rounded). Because the P-value of 0.0021 is less than the significance level of $\alpha = 0.01$, we reject the null hypothesis. See the interpretation at the end of this example.

We can test the claim using the critical value method as follows.

Step 1: The claim that "blue enhances performance on a creative task" can be restated as the claim that people with a blue background have a higher mean creativity score than those in the first group with a red background. The first group of red background subjects therefore has a lower mean creativity score than the second blue group, and this can be expressed as $\mu_1 < \mu_2$.

Step 2: If the original claim is false, then $\mu_1 \geq \mu_2$.

Step 3: The alternative hypothesis is the expression not containing equality, and the null hypothesis is an expression of equality, so we have

$$H_0: \mu_1 = \mu_2 \qquad H_1: \mu_1 < \mu_2 \text{ (original claim)}$$

We now proceed with the assumption that $\mu_1 = \mu_2$, or $\mu_1 - \mu_2 = 0$.

Step 4: The significance level is $\alpha = 0.01$.

Step 5: Because we have two independent samples and we are testing a claim about the two population means, we use a t distribution with the test statistic given earlier in this section.

Step 6: The test statistic is calculated as follows:

$$t = \frac{(\bar{x}_1 - \bar{x}_2) - (\mu_1 - \mu_2)}{\sqrt{\dfrac{s_1^2}{n_1} + \dfrac{s_2^2}{n_2}}} = \frac{(3.39 - 3.97) - 0}{\sqrt{\dfrac{0.97^2}{35} + \dfrac{0.63^2}{36}}} = -2.979$$

Critical Values Because we are using a t distribution, the critical value of $t = -2.441$ is found from Table A-3. With an area of 0.01 in the left tail, we want the t value corresponding to 34 degrees of freedom, which is the smaller of $n_1 - 1$ and $n_2 - 1$ (or the smaller of 34 and 35). The test statistic, critical values, and critical region are shown in Figure 9-2.

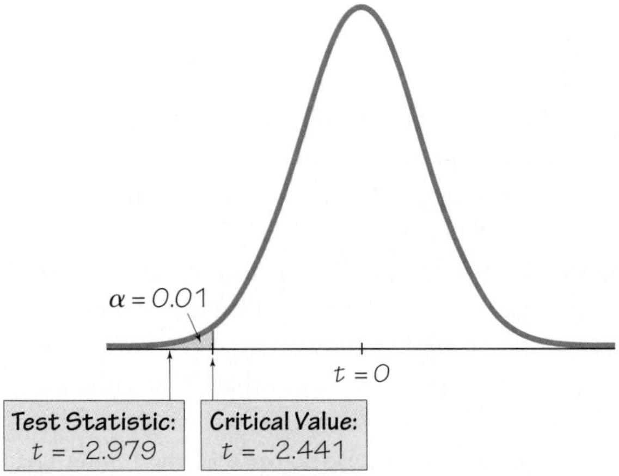

Figure 9-2 Hypothesis Test of Means from Two Independent Populations

Step 7. Because the test statistic does fall within the critical region, we reject the null hypothesis $\mu_1 = \mu_2$ (or $\mu_1 - \mu_2 = 0$).

Do Real Estate Agents Get You the Best Price?

When a real estate agent sells a home, does he or she get the best price for the seller? This question was addressed by Steven Levitt and Stephen Dubner in *Freakonomics*. They collected data from thousands of homes near Chicago, including homes owned by the agents themselves. Here is what they write: "There's one way to find out: measure the difference between the sales data for houses that belong to real-estate agents themselves and the houses they sold on behalf of clients. Using the data from the sales of those 100,000 Chicago homes, and controlling for any number of variables—location, age and quality of the house, aesthetics, and so on—it turns out that a real-estate agent keeps her own home on the market an average of ten days longer and sells it for an extra 3-plus percent, or $10,000 on a $300,000 house." A conclusion such as this can be obtained by using the methods of this section.

Interpretation

There is sufficient evidence to support the claim that the red background group has a lower mean creativity score than the blue background group. This supports the conclusion that higher creativity scores were achieved by the subjects with the blue background, but the results do not justify the conclusion that the blue background is the *cause* of the higher creativity scores.

Example 3 Confidence Interval for Creativity Scores

Using the sample data given in Example 2, construct a 98% confidence interval estimate of the difference between the mean creativity score for those with a red background and the mean creativity score for those with a blue background.

Solution

Requirement check Because we are using the same data from Example 2, the same requirement check applies here, so the requirements are satisfied. ✓

We first find the value of the margin of error E. We use the same critical value of $t_{\alpha/2} = 2.441$ found in Example 2. (A more accurate critical value is 2.392.)

$$E = t_{\alpha/2}\sqrt{\frac{s_1^2}{n_1} + \frac{s_2^2}{n_2}} = 2.441\sqrt{\frac{0.97^2}{35} + \frac{0.63^2}{36}} = 0.475261$$

Using $E = 0.475261$, $\bar{x}_1 = 3.39$, and $\bar{x}_2 = 3.97$, we can now find the confidence interval as follows:

$$(\bar{x}_1 - \bar{x}_2) - E < (\mu_1 - \mu_2) < (\bar{x}_1 - \bar{x}_2) + E$$
$$-1.06 < (\mu_1 - \mu_2) < -0.10$$

If we use technology to obtain more accurate results, we get the confidence interval of $-1.05 < (\mu_1 - \mu_2) < -0.11$, so we can see that the confidence interval above is quite good, even though we used a simplified method for finding the number of degrees of freedom (instead of getting more accurate results by using Formula 9-1 to compute the number of degrees of freedom).

Interpretation

We are 98% confident that the limits of -1.05 and -0.11 actually do contain the difference between the two population means. Because those limits do not contain 0, this confidence interval suggests that there is a significant difference between the two means. (We used a 98% confidence level, so the one-sided hypothesis test has a corresponding significance level of 0.01.) Also, because the confidence interval consists entirely of negative values, there is sufficient evidence to support the claim that the red background group has a lower mean creativity score than the blue background group. This supports the claim that "blue enhances performance on a creative task" as claimed by the researchers, but we should be careful to avoid a statement that the blue background is the *cause* of the higher creativity scores.

Rationale for the Test Statistic and Confidence Interval

If the given assumptions are satisfied, the sampling distribution of $\bar{x}_1 - \bar{x}_2$ can be approximated by a t distribution with mean equal to $\mu_1 - \mu_2$ and standard deviation

Expensive Diet Pill

There are many past examples in which ineffective treatments were marketed for substantial profits. Capsules of "Fat Trapper" and "Exercise in a Bottle," manufactured by the Enforma Natural Products company, were advertised as being effective treatments for weight reduction. Advertisements claimed that after taking the capsules, fat would be blocked and calories would be burned, even without exercise. Because the Federal Trade Commission identified claims that appeared to be unsubstantiated, the company was fined $10 million for deceptive advertising.

The effectiveness of such treatments can be determined with experiments in which one group of randomly selected subjects is given the treatment, while another group of randomly selected subjects is given a placebo. The resulting weight losses can be compared using statistical methods, such as those described in this section.

Note to Instructor

Point out that a common objective in constructing a confidence interval estimate of the difference $\mu_1 - \mu_2$ is to determine whether the confidence interval limits contain 0. If those limits do contain 0, then there is not a significant difference between the two sample means, which suggests that the two population means are equal. If those limits do not contain 0, then it appears that the two population means are different. Because the hypothesis test and confidence interval use the same distribution and standard error, they are equivalent in the sense that they result in the same conclusions.

equal to $\sqrt{s_1^2/n_1 + s_2^2/n_2}$. This last expression for the standard deviation is based on the property that the variance of the *differences* between two independent random variables equals the variance of the first random variable *plus* the variance of the second random variable.

Note to Instructor
Part 2 can be omitted. Part 2 includes two alternative methods for these two situations: (1) The two population variances are somehow known; (2) the two population variances are not known, but they are assumed to be equal. The first of these two situations is not likely to occur in reality, but the second situation does arise. For example, if we use randomization to assign subjects to two different treatment groups (such as a group given a drug and another group given a placebo), we know that the experiment has subjects randomly selected from the same population, so it is reasonable to assume that they have the same variance.

Part 2: Alternative Methods

Part 1 of this section dealt with situations in which the two population standard deviations are unknown and are not assumed to be equal. In Part 2 we address two other situations:

1. The two population standard deviations are both known.
2. The two population standard deviations are unknown but are assumed to be equal.

Alternative Method Used When σ_1 and σ_2 Are Known

In reality, the population standard deviations σ_1 and σ_2 are almost never known, but if they are somehow known, the test statistic and confidence interval are based on the normal distribution instead of the t distribution. See the summary box below.

Inferences about Means of Two Independent Populations, with σ_1 and σ_2 Known

Requirements

1. The two population standard deviations σ_1 and σ_2 are both known.
2. The two samples are *independent.*
3. Both samples are *simple random samples.*
4. Either or both of these conditions is satisfied: The two sample sizes are both *large* (with $n_1 > 30$ and $n_2 > 30$) or both samples come from populations having normal distributions. (For small samples, the normality requirement is loose in the sense that the procedures perform well as long as there are no outliers and departures from normality are not too extreme.)

Hypothesis Test

Test statistic:

$$z = \frac{(\bar{x}_1 - \bar{x}_2) - (\mu_1 - \mu_2)}{\sqrt{\dfrac{\sigma_1^2}{n_1} + \dfrac{\sigma_2^2}{n_2}}}$$

***P*-values:** *P*-values are automatically provided by technology. If technology is not available, refer to the normal distribution in Table A-2. Use the procedure summarized in Figure 8-4 from Section 8-2.

Critical values: Refer to Table A-2.

Confidence Interval Estimate of $\mu_1 - \mu_2$

$$(\bar{x}_1 - \bar{x}_2) - E < (\mu_1 - \mu_2) < (\bar{x}_1 - \bar{x}_2) + E$$

where

$$E = z_{\alpha/2}\sqrt{\frac{\sigma_1^2}{n_1} + \frac{\sigma_2^2}{n_2}}$$

Alternative Method: Assume That $\sigma_1 = \sigma_2$ and *Pool* the Sample Variances

Even when the specific values of σ_1 and σ_2 are not known, if it can be assumed that they have the *same* value, the sample variances s_1^2 and s_2^2 can be *pooled* to obtain an estimate of the common population variance σ^2. The **pooled estimate of σ^2** is denoted by s_p^2 and is a weighted average of s_1^2 and s_2^2, which is described in the following box.

Inferences about Means of Two Independent Populations, Assuming That $\sigma_1 = \sigma_2$

Requirements

1. The two population standard deviations are not known, but they are assumed to be equal. That is, $\sigma_1 = \sigma_2$.
2. The two samples are *independent.*
3. Both samples are *simple random samples.*
4. Either or both of these conditions is satisfied: The two sample sizes are both *large* (with $n_1 > 30$ and $n_2 > 30$) or both samples come from populations having normal distributions. (For small samples, the normality requirement is loose in the sense that the procedures perform well as long as there are no outliers and departures from normality are not too extreme.)

Hypothesis Test

Test statistic:

$$t = \frac{(\bar{x}_1 - \bar{x}_2) - (\mu_1 - \mu_2)}{\sqrt{\dfrac{s_p^2}{n_1} + \dfrac{s_p^2}{n_2}}}$$

where

$$s_p^2 = \frac{(n_1 - 1)s_1^2 + (n_2 - 1)s_2^2}{(n_1 - 1) + (n_2 - 1)} \quad \text{(pooled sample variance)}$$

and the number of degrees of freedom is $\text{df} = n_1 + n_2 - 2$.

Confidence Interval Estimate of $\mu_1 - \mu_2$

$$(\bar{x}_1 - \bar{x}_2) - E < (\mu_1 - \mu_2) < (\bar{x}_1 - \bar{x}_2) + E$$

where

$$E = t_{\alpha/2}\sqrt{\frac{s_p^2}{n_1} + \frac{s_p^2}{n_2}}$$

and s_p^2 is as given in the test statistic above, and $\text{df} = n_1 + n_2 - 2$.

If we want to use this method based on equal population standard deviations, how do we determine that $\sigma_1 = \sigma_2$? One approach is to use a hypothesis test of the null hypothesis $\sigma_1 = \sigma_2$, but that approach is not recommended and we will not use the preliminary test of $\sigma_1 = \sigma_2$. In the article "Homogeneity of Variance in the Two-Sample Means Test" (by Moser and Stevens, *American Statistician,* Vol. 46, No. 1), the authors note that we rarely know that $\sigma_1 = \sigma_2$. They analyze the performance of the different tests by considering sample sizes and powers of the tests.

They conclude that more effort should be spent learning the method given in Part 1, and less emphasis should be placed on the method based on the assumption of $\sigma_1 = \sigma_2$. Unless instructed otherwise, we use the following strategy, which is consistent with the recommendations in the article by Moser and Stevens:

> **Assume that σ_1 and σ_2 are unknown, do *not* assume that $\sigma_1 = \sigma_2$, and use the test statistic and confidence interval given in Part 1 of this section.**

Figure 9-3 summarizes the methods for inferences about two independent population means.

Why Not Eliminate the Method of Pooling Sample Variances?

If we use randomness to assign subjects to treatment and placebo groups, we know that the samples are drawn from the same population. So if we conduct a hypothesis test assuming that two population means are equal, it is not unreasonable to also assume that the samples are from populations with the same standard deviations (but we should still check that assumption). The advantage of this alternative method of

Inferences about Two Independent Means

Start

Are σ_1 and σ_2 known? — Yes → Use normal distribution with standard error. $\sqrt{\dfrac{\sigma_1^2}{n_1} + \dfrac{\sigma_2^2}{n_2}}$

This case almost never occurs in reality.

No ↓

Can it be assumed that $\sigma_1 = \sigma_2$? — Yes → Use *t* distribution with POOLED standard error.

Some statisticians recommend against this approach.

No ↓

Approximate method: Use *t* distribution with standard error. $\sqrt{\dfrac{s_1^2}{n_1} + \dfrac{s_2^2}{n_2}}$

Use this method unless instructed otherwise.

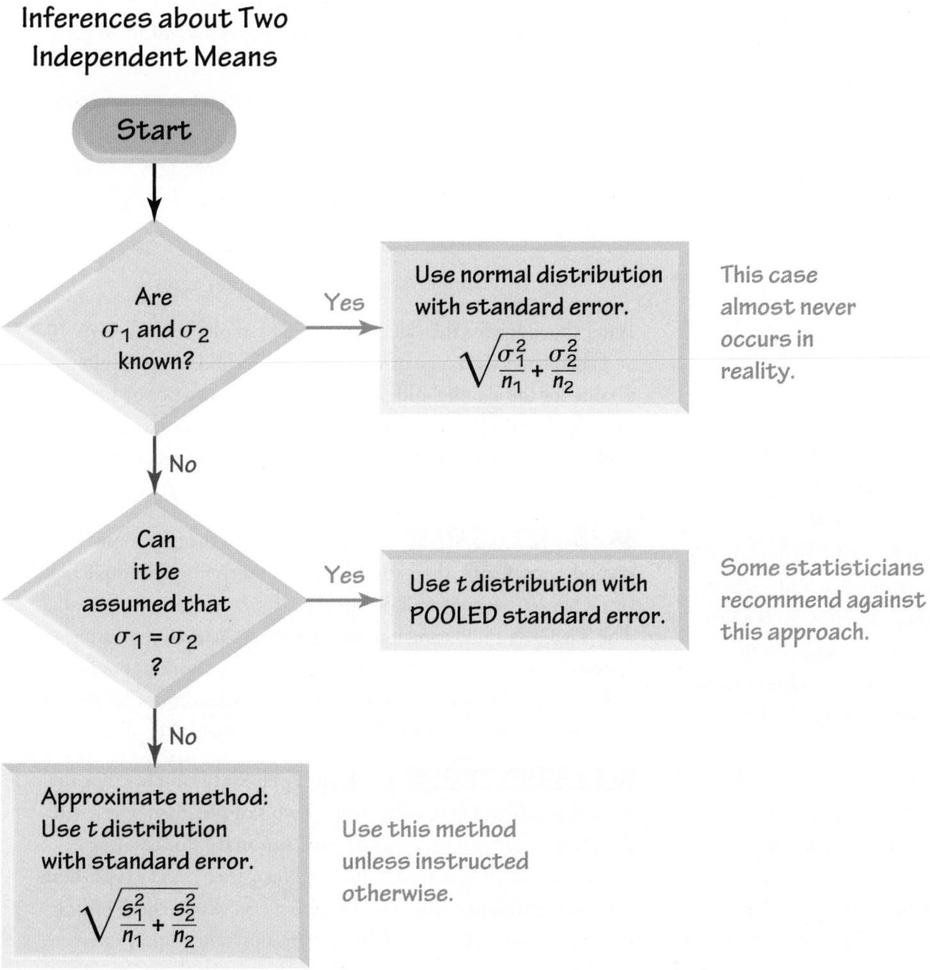

Figure 9-3 **Methods for Inferences about Two Independent Means**

Super Bowls

Students were invited to a Super Bowl game and half of them were given large 4-liter snack bowls while the other half were given smaller 2-liter bowls. Those using the large bowls consumed 56% more than those using the smaller bowls. (See "Super Bowls: Serving Bowl Size and Food Consumption," by Wansink and Cheney, *Journal of the American Medical Association,* Vol. 293, No. 14.)

A separate study showed that there is "a significant increase in fatal motor vehicle crashes during the hours following the Super Bowl telecast in the United States." Researchers analyzed 20,377 deaths on 27 Super Bowl Sundays and 54 other Sundays used as controls. They found a 41% increase in fatalities after Super Bowl games. (See "Do Fatal Crashes Increase Following a Super Bowl Telecast?" by Redelmeier and Stewart, *Chance,* Vol. 18, No. 1.)

pooling sample variances is that the number of degrees of freedom is a little higher, so hypothesis tests have more power and confidence intervals are a little narrower. Consequently, statisticians sometimes use this method of pooling, and that is why we include it in this subsection.

using TECHNOLOGY

STATDISK Select the menu item of **Analysis.** Select either **Hypothesis Testing** or **Confidence Intervals,** then select **Mean-Two Independent Samples.** Enter the required values in the dialog box. You have the options of "Not Eq vars: NO POOL," "Eq vars: POOL," or "Prelim F Test." The option of **Not Eq vars: NO POOL** is consistent with the methods described in Part 1 of this section, and this method is recommended.

MINITAB Minitab allows the use of summary statistics or original lists of sample data. If the original sample values are known, enter them in columns C1 and C2. Select the options **Stat, Basic Statistics,** and **2-Sample t.** Make the required entries in the window that pops up. Use the **Options** button to select a confidence level, enter a claimed value of the difference, or select a format for the alternative hypothesis. The Minitab display also includes the confidence interval limits.

If the two population variances appear to be equal, Minitab does allow use of a pooled estimate of the common variance. There will be a box next to **Assume equal variances,** so click on that box only if you want to assume that the two populations have equal variances, but this approach is not recommended.

In **Minitab 16,** you can also click on **Assistant,** then **Hypothesis Tests,** then select the case for **2-Sample t.** Fill out the dialog box, then click **OK** to get three windows of results that include the *P*-value and much other helpful information.

EXCEL Excel requires entry of the original lists of sample data. Enter the data for the two samples in columns A and B and use either XLSTAT or Excel's Data Analysis add-in.

XLSTAT for Hypothesis Test: Click on **XLSTAT** at the top. Click on **Parametric tests,** then select **Two sample t test and z test.** In the dialog box that appears, enter the range of values for the first sample (such as A1:A13) and enter the range of values for the second sample. For "Data format," select the option of **One column per sample.** For the "Column labels" box, include a checkmark only if the first row of the sample data consists of names or labels. Put a checkmark next to "Student's t test" (use z test if σ is known). Click on the **Options** tab and select the type of test; for a two-tailed test, select the case including the symbol $\neq$, for a left-tailed test select the case including $<$, and for a right-tailed test, select the case including $>$. Enter a value in the "Significance level (%)" box. For example, enter 5 for a 0.05 significance level. Put a checkmark next to "Cochran-Cox" so that the variances are not assumed to be equal. Click **OK** to get results that include the test statistic and *P*-value.

XLSTAT for Confidence Interval: Click on **XLSTAT** at the top. Click on **Parametric tests,** then select **Two sample t test and z test.** In the dialog box that appears, enter the range of values for the first sample (such as A1:A13) and enter the range of values for the second sample. For "Data format," select the option of **One column per sample.** For the "Column labels" box, include a checkmark only if the first row of the sample data consists of names or labels. Put a checkmark next to "Student's t test" (use z test if σ is known). Click on the **Options** tab. For the alternative hypothesis, select the format of a two-tailed test. Enter a value in the "Significance level (%)" box. For example, enter 5 for a 95% confidence interval. Put a checkmark next to "Cochran-Cox" so that the variances are not assumed to be equal. Click **OK** to get results that include the confidence interval.

Data Analysis Add-in: If using Excel 2013, 2010, or 2007, click on **Data,** then **Data Analysis;** if using Excel 2003, click on **Tools** and select **Data Analysis.** Select one of the following two items (we recommend the assumption of *unequal variances*):

t-test: Two-Sample Assuming Equal Variances

t-test: Two-Sample Assuming Unequal Variances

Enter the range for the values of the first sample (such as A1:A50) and then enter the range of values for the second sample. Enter a value for the claimed difference between the two population means, which will often be 0. Enter the significance level in the Alpha box and click on **OK.** (Excel does not provide a confidence interval.)

TI-83/84 PLUS To conduct tests of the type found in this section, press **STAT**, then select **TESTS** and choose **2-SampTTest** (for a hypothesis test) or **2-SampTInt** (for a confidence interval). The TI-83/84 Plus calculator does give you the option of using "pooled" variances (if you believe that $\sigma_1^2 = \sigma_2^2$) or not pooling the variances, but we recommend that the variances not be pooled, so we recommend selecting **No** for the line labeled "Pooled."

STATCRUNCH Click on **Open StatCrunch.** Click on **Stat,** then select **T statistics.** Select **Two sample,** then select **with data** (for lists of sample data) or **with summary** (for entering known summary statistics). Click on **Next,** then select **Hypothesis Test** or **Confidence Interval.** There is a box with a check indicating that variances be pooled, but it is recommended that variances should not be pooled. Click on **Next,** then make the required entries. Click on **Calculate.**

9-3 Basic Skills and Concepts

Statistical Literacy and Critical Thinking

1. Independent and Dependent Samples Which of the following involve independent samples?

a. To test the effectiveness of the Atkins diet, 36 randomly selected subjects are weighed before the diet and six months after treatment with the diet. The two samples consist of the before/after weights.

b. To determine whether smoking affects memory, 50 randomly selected smokers are given a test of word recall and 50 randomly selected nonsmokers are given the same test. Sample data consist of the scores from the two groups.

c. IQ scores are obtained from a random sample of 75 wives and IQ scores are obtained from their husbands.

d. Annual incomes are obtained from a random sample of 1200 residents of Alaska and from another random sample of 1200 residents of Hawaii.

e. Scores from a standard test of mathematical reasoning are obtained from a random sample of statistics students and another random sample of sociology students.

2. Interpreting Confidence Intervals If the heights of men and women from Data Set 1 in Appendix B are used to construct a 95% confidence interval for the difference between the two population means, the result is 11.61 cm $< \mu_1 - \mu_2 <$ 17.32 cm, where heights of men correspond to population 1 and heights of women correspond to population 2. Express the confidence interval with heights of women being population 1 and heights of men being population 2.

3. Interpreting Confidence Intervals What does the confidence interval in Exercise 2 suggest about the heights of men and women?

4. Hypothesis Tests and Confidence Intervals

a. In general, if you conduct a hypothesis test using the methods of Part 1 of this section, will the *P*-value method, the critical value method, and the confidence interval method result in the same conclusion?

b. Assume that you want to use a 0.05 significance level to test the claim that the mean height of men is greater than the mean height of women. What *confidence level* should be used if you want to test that claim using a confidence interval?

In Exercises 5–20, assume that the two samples are independent simple random samples selected from normally distributed populations, and do not assume that the population standard deviations are equal. Do the following:

a. ***Test the given claim using the P-value method or critical value method.***

b. ***Construct a confidence interval suitable for testing the given claim.***

5. Color and Creativity Repeat Example 2 with these changes: Use a 0.05 significance level, and test the claim that the two samples are from populations with the same mean.

6. Color and Cognition In the Chapter Problem, it was noted that researchers conducted a study to investigate the effects of color on cognitive tasks. Words were displayed on a computer screen with background colors of red and blue. Results from scores on a test of word recall are given below. Use a 0.05 significance level to test the claim that the samples are from populations with the same mean. Does the background color appear to have an effect

Recommended Assignment
Exercises 1–6, 12–14, 16, 17.

Due to limited space, answers for some odd numbered exercises are not included here, but they are included in Appendix D.

1. Independent: b, d, e
2. -17.32 cm $< \mu_1 - \mu_2 <$ -11.61 cm
3. Because the confidence interval does not contain 0, it appears that there is a significant difference between the mean height of women and the mean height of men. Based on the confidence interval, it appears that the mean height of men is greater than the mean height of women.
4. a. Yes.
   b. 90%
5. a. H_0: $\mu_1 = \mu_2$. H_1: $\mu_1 \neq \mu_2$. Test statistic: $t = -2.979$. Critical values: $t = \pm 2.032$ (Tech: ± 2.002). *P*-value < 0.01 (Tech: 0.0042). Reject H_0. There is sufficient evidence to warrant rejection of the claim that the samples are from populations with the same mean. Color does appear to have an effect on creativity scores. Blue appears to be associated with higher creativity scores.
   b. 95% CI: $-0.98 < \mu_1 - \mu_2 < -0.18$ (Tech: $-0.97 < \mu_1 - \mu_2 < -0.19$)
6. a. H_0: $\mu_1 = \mu_2$. H_1: $\mu_1 \neq \mu_2$. Test statistic: $t = 2.647$. Critical values: $t = \pm 2.032$ (Tech: ± 1.995). *P*-value < 0.02 (Tech: 0.0101). Reject H_0. There is sufficient evidence to warrant rejection of the claim that the samples are from populations with the same mean. Color does appear to have an effect on word recall scores. Red appears to be associated with higher word memory recall scores.
   b. 95% CI: $0.83 < \mu_1 - \mu_2 < 6.33$ (Tech: $0.88 < \mu_1 - \mu_2 < 6.28$)

on word recall scores? If so, which color appears to be associated with higher word memory recall scores?

Red Background: $n = 35, \bar{x} = 15.89, s = 5.90$

Blue Background: $n = 36, \bar{x} = 12.31, s = 5.48$

7. Magnet Treatment of Pain People spend around \$5 billion annually for the purchase of magnets used to treat a wide variety of pains. Researchers conducted a study to determine whether magnets are effective in treating back pain. Pain was measured using the visual analog scale, and the results given below are among the results obtained in the study (based on data from "Bipolar Permanent Magnets for the Treatment of Chronic Lower Back Pain: A Pilot Study," by Collacott, Zimmerman, White, and Rindone, *Journal of the American Medical Association,* Vol. 283, No. 10). Use a 0.05 significance level to test the claim that those treated with magnets have a greater mean reduction in pain than those given a sham treatment (similar to a placebo). Does it appear that magnets are effective in treating back pain? Is it valid to argue that magnets might appear to be effective if the sample sizes are larger?

Reduction in Pain Level after Magnet Treatment: $n = 20, \bar{x} = 0.49, s = 0.96$

Reduction in Pain Level after Sham Treatment: $n = 20, \bar{x} = 0.44, s = 1.4$

8. Do Men Talk Less Than Women? The accompanying table gives results from a study of the words spoken in a day by men and women, and the original data are in Data Set 17 in Appendix B (based on "Are Women Really More Talkative Than Men?" by Mehl et al., *Science,* Vol. 317, No. 5834). Use a 0.01 significance level to test the claim that the mean number of words spoken in a day by men is less than that for women.

9. Do Women Have a Higher Mean Body Temperature? If we use the body temperatures from 8 A.M. on Day 2 as listed in Data Set 3 in Appendix B, we get the statistics given in the accompanying table. Use these data with a 0.01 significance level to test the claim that women have a higher mean body temperature than men.

10. Do Men and Women Have the Same Mean Body Temperature? Consider the sample of body temperatures (°F) listed in the last column of Data Set 3 in Appendix B. The summary statistics are given in the accompanying table. Use a 0.01 significance level to test the claim that men and women have different mean body temperatures.

11. Skull Measurements from Different Times Researchers measured skulls from different time periods in an attempt to determine whether interbreeding of cultures occurred. Results are given below (based on data from *Ancient Races of the Thebaid,* by Thomson and Randall-Maciver, Oxford University Press). Use a 0.01 significance level to test the claim that the mean maximal skull breadth in 4000 B.C. is less than the mean in A.D. 150.

4000 B.C. (Maximal Skull Breadth): $n = 30, \bar{x} = 131.37$ mm, $s = 5.13$ mm

A.D. 150 (Maximal Skull Breadth): $n = 30, \bar{x} = 136.17$ mm, $s = 5.35$ mm

12. Flight Arrival Delays Data Set 15 in Appendix B lists arrival delay times (min) for randomly selected flights from New York (JFK) to Los Angeles (LAX). Statistics for times are given below. Use a 0.05 significance level to test the claim that Flight 1 and Flight 3 have the same mean arrival delay time.

Flight 1 $n = 12, \bar{x} = -20.5$ min, $s = 12.38401$ min

Flight 3 $n = 12, \bar{x} = -15.08333$ min, $s = 15.62317$ min

13. Proctored and Nonproctored Tests In a study of proctored and nonproctored tests in an online Intermediate Algebra course, researchers obtained the data for test results given below (based on "Analysis of Proctored versus Non-proctored Tests in Online Algebra Courses,"

8. a. $H_0: \mu_1 = \mu_2$. $H_1: \mu_1 < \mu_2$. Test statistic: $t = -0.676$. Critical value: $t = -2.345$ (Tech: -2.337). P-value > 0.10 (Tech: 0.2499). Fail to reject H_0. There is not sufficient evidence to support the claim the mean number of words spoken in a day by men is less than that for women.
   b. 98% CI: -2443.6 words $< (\mu_1 - \mu_2) < 1350.6$ words (Tech: -2436.8 words $< (\mu_1 - \mu_2) < 1343.8$ words)

10. a. $H_0: \mu_1 = \mu_2$. $H_1: \mu_1 \neq \mu_2$. Test statistic: $t = 1.559$. Critical values: $t = \pm 2.977$ (Tech: ± 2.789). P-value > 0.10 (Tech: 0.1316). Fail to reject H_0. There is not sufficient evidence to support the claim that men and women have different mean body temperatures.

Table for Exercise 8

Men	Women
$n_1 = 186$	$n_2 = 210$
$\bar{x}_1 = 15{,}668.5$	$\bar{x}_2 = 16{,}215.0$
$s_1 = 8632.5$	$s_2 = 7301.2$

Table for Exercise 9

Women	Men
$n_1 = 11$	$n_2 = 59$
$\bar{x}_1 = 97.69°F$	$\bar{x}_2 = 97.45°F$
$s_1 = 0.89°F$	$s_2 = 0.66°F$

Table for Exercise 10

Women	Men
$n_1 = 15$	$n_2 = 91$
$\bar{x}_1 = 98.38°F$	$\bar{x}_2 = 98.17°F$
$s_1 = 0.45°F$	$s_2 = 0.65°F$

   b. 99% CI: $-0.19°F < (\mu_1 - \mu_2) < 0.61°F$ (Tech: $-0.17 < (\mu_1 - \mu_2) < 0.59°F$)

12. a. $H_0: \mu_1 = \mu_2$. $H_1: \mu_1 \neq \mu_2$. Test statistic: $t = -0.941$. Critical value: $t = \pm 2.201$ (Tech: 2.080). P-value > 0.20 (Tech: 0.3573). Fail to reject H_0. There is not sufficient evidence to warrant rejection of the claim that Flight 1 and Flight 3 have the same mean arrival delay time.
   b. 95% CI: -18.1 min $< (\mu_1 - \mu_2) < 7.3$ min (Tech: -17.4 min $< (\mu_1 - \mu_2) < 6.6$ min)

by Flesch and Ostler, *MathAMATYC Educator,* Vol. 2, No. 1). Use a 0.01 significance level to test the claim that students taking nonproctored tests get a higher mean than those taking proctored tests.

Group 1 (Proctored): $n = 30, \bar{x} = 74.30, s = 12.87$

Group 2 (Nonproctored): $n = 32, \bar{x} = 88.62, s = 22.09$

14. Proctored and Nonproctored Tests In the same study described in the preceding exercise, the same groups of students took a nonproctored test; the results are given below. Use a 0.01 significance level to test the claim that the samples are from populations with the same mean.

Group 1 (Nonproctored): $n = 30, \bar{x} = 70.29, s = 22.09$

Group 2 (Nonproctored): $n = 32, \bar{x} = 74.26, s = 18.15$

15. BMI We know that the mean weight of men is greater than the mean weight of women, and the mean height of men is greater than the mean height of women. A person's body mass index (BMI) is computed by dividing weight (kg) by the square of height (m). Given below are the BMI statistics for random samples of males and females from Data Set 1 in Appendix B. Use a 0.05 significance level to test the claim that males and females have the same mean BMI.

Male BMI $n = 40, \bar{x} = 28.44075, s = 7.394076$

Female BMI $n = 40, \bar{x} = 26.6005, s = 5.359442$

16. IQ and Lead Exposure Data Set 5 in Appendix B lists full IQ scores for a random sample of subjects with low lead levels in their blood and another random sample of subjects with high lead levels in their blood. The statistics are summarized below. Use a 0.05 significance level to test the claim that the mean IQ score of people with low lead levels is higher than the mean IQ score of people with high lead levels.

Low Lead Level: $n = 78, \bar{x} = 92.88462, s = 15.34451$

High Lead Level: $n = 21, \bar{x} = 86.90476, s = 8.988352$

17. IQ and Lead Exposure Repeat Exercise 16 after replacing the low lead level group with the following full IQ scores from the medium lead level group.

72	90	92	71	86	79	83	114	100	93	91
98	91	46	85	82	97	91	92	77	111	78

18. Heights of Supermodels Listed below are the heights (inches) for the simple random sample of supermodels Lima, Bundchen, Ambrosio, Ebanks, Iman, Rubik, Kurkova, Kerr, Kroes, and Swanepoel. Data Set 1 in Appendix B includes the heights of a simple random sample of 40 women from the general population, and here are the statistics for those heights: $n = 40, \bar{x} = 63.7815$ in., and $s = 2.59665$ in. Use a 0.01 significance level to test the claim that the supermodels have heights with a mean that is greater than the mean height of women in the general population.

70	71	69.25	68.5	69	70	71	70	70	69.5

19. Longevity Listed below are the numbers of years that popes and British monarchs (since 1690) lived after their election or coronation (based on data from *Computer-Interactive Data Analysis,* by Lunn and McNeil, John Wiley & Sons). Treat the values as simple random samples from a larger population. Use a 0.01 significance level to test the claim that the mean longevity for popes is less than the mean for British monarchs after coronation.

Popes:	2	9	21	3	6	10	18	11	6	25	23	6	
	2	15	32	25	11	8	17	19	5	15	0	26	
Kings and Queens:	17	6	13	12	13	33	59	10	7	63	9	25	36	15

14. a. H_0: $\mu_1 = \mu_2$. H_1: $\mu_1 \neq \mu_2$. Test statistic: $t = -0.770$. Critical values: $t = \pm 2.756$ (Tech: ± 2.666). P-value > 0.20 (Tech: 0.4443). Fail to reject H_0. There is not sufficient evidence to warrant rejection of the claim that the samples are from populations with the same mean.
b. 99% CI: $-18.17 < \mu_1 - \mu_2 < 10.23$ (Tech: $-17.71 < (\mu_1 - \mu_2) < 9.77$)

16. a. H_0: $\mu_1 = \mu_2$. H_1: $\mu_1 > \mu_2$. Test statistic: $t = 2.282$. Critical value: $t = 1.725$ (Tech: 2.004). P-value < 0.05 (Tech: 0.0132). Reject H_0. There is sufficient evidence to support the claim that the mean IQ score of people with low lead levels is higher than the mean IQ score of people with high lead levels.
b. 90% CI: $1.5 < \mu_1 - \mu_2 < 10.5$ (Tech: $1.6 < \mu_1 - \mu_2 < 10.4$)

18. a. The sample data meet the loose requirement of having a normal distribution. H_0: $\mu_1 = \mu_2$. H_1: $\mu_1 > \mu_2$. Test statistic: $t = 12.533$. Critical value: $t = 2.821$ (Tech: 2.411). P-value < 0.005 (Tech: 0.0000). Reject H_0. There is sufficient evidence to support the claim that supermodels have heights with a mean that is greater than the mean height of women in the general population. We can conclude that supermodels are taller than typical women.
b. 98% CI: 4.7 in $< \mu_1 - \mu_2 <$ 7.4 in. (Tech: 4.9 in. $< \mu_1 - \mu_2 <$ 7.2 in.)

19. a. H_0: $\mu_1 = \mu_2$. H_1: $\mu_1 < \mu_2$. Test statistic: $t = -1.810$. Critical value: $t = -2.650$ (Tech: -2.574). P-value > 0.025 (Tech: 0.0442). Fail to reject H_0. There is not sufficient evidence to support the claim that the mean longevity for popes is less than the mean for British monarchs after coronation.
b. 98% CI: -23.6 years $< \mu_1 - \mu_2 <$ 4.4 years (Tech: -23.2 years $< \mu_1 - \mu_2 <$ 4.0 years)

20. a. H_0: $\mu_1 = \mu_2$. H_1: $\mu_1 > \mu_2$. Test statistic: $t = 3.265$. Critical value: $t = 1.796$ (Tech: $t = 1.746$). *P*-value < 0.005 (Tech: 0.0024). Reject H_0. There is sufficient evidence to support the claim that the mean amount of strontium-90 from Pennsylvania residents is greater than the mean from New York residents.
b. 90% CI: 5.0 mBq $< (\mu_1 - \mu_2) <$ 17.3 mBq (Tech: 5.2 mBq $< \mu_1 - \mu_2 <$ 17.1 mBq)
22. -9.1 years $< (\mu_1 - \mu_2) <$ 5.4 years (Tech: -9.0 years $< (\mu_1 - \mu_2) <$ 5.3 years). Because the confidence interval includes 0, there is not a significant difference between the two population means. It appears that the sample of men and the sample of women are from populations with the same mean.
24. H_0: $\mu_1 = \mu_2$. H_1: $\mu_1 \neq \mu_2$. Test statistic: $t = 22.095$. Critical values: $t = \pm 2.023$ (Tech: ± 2.003). *P*-value < 0.01 (Tech: 0.0000). Reject H_0. There is sufficient evidence to warrant rejection of the claim that the two populations have equal means. The difference is due to the sugar in regular Coke that is not in diet Coke.
26. a. H_0: $\mu_1 = \mu_2$. H_1: $\mu_1 < \mu_2$. Test statistic: $t = -0.682$. Critical value: $t = -2.336$. *P*-value $>$ 0.10 (Tech: 0.2477). Fail to reject H_0. There is not sufficient evidence to support the claim that the mean number of words spoken in a day by men is less than that for women.
b. -2417.4 words $< (\mu_1 - \mu_2) <$ 1324.4 words (Tech: -2417.2 words $< (\mu_1 - \mu_2) <$ 1324.3 words). The test statistic became larger, the *P*-value became smaller, and the confidence interval became narrower, so pooling had the effect of attributing more significance to the results.
27. H_0: $\mu_1 = \mu_2$. H_1: $\mu_1 \neq \mu_2$. Test statistic: $t = 15.322$. Critical values: $t = \pm 2.080$. *P*-value < 0.01 (Tech: 0.0000). Reject H_0. There is sufficient evidence to warrant rejection of the claim that the two populations have the same mean.

20. Radiation in Baby Teeth Listed below are amounts of strontium-90 (in millibecquerels, or mBq, per gram of calcium) in a simple random sample of baby teeth obtained from Pennsylvania residents and New York residents born after 1979 (based on data from "An Unexpected Rise in Strontium-90 in U.S. Deciduous Teeth in the 1990s," by Mangano et al., *Science of the Total Environment,* Vol. 317). Use a 0.05 significance level to test the claim that the mean amount of strontium-90 from Pennsylvania residents is greater than the mean amount from New York residents.

Pennsylvania:	155	142	149	130	151	163	151	142	156	133	138	161
New York:	133	140	142	131	134	129	128	140	140	140	137	143

Large Data Sets. ***In Exercises 21–24, use the indicated Data Sets from Appendix B. Assume that the two samples are independent simple random samples selected from normally distributed populations. Do not assume that the population standard deviations are equal.***

21. Weights of Quarters Vending machines reject coins based on weight. Refer to Data Set 21 in Appendix B and use a 0.05 significance level to test the claim that the mean weight of pre-1964 quarters is equal to the mean weight of post-1964 quarters. Given the relatively small sample sizes from the large populations of millions of quarters, can we really conclude that the mean weights are different?

22. Baseline Characteristics Reports of results from clinical trials often include statistics about "baseline characteristics," so we can see that different groups have the same basic characteristics. Refer to Data Set 1 in Appendix B and construct a 95% confidence interval estimate of the difference between the mean age of men and the mean age of women. Based on the result, does it appear that the sample of men and the sample of women are from populations with the same mean?

23. Weights of Pepsi Refer to Data Set 19 in Appendix B and construct a 95% confidence interval estimate of the difference between the mean weight of the cola in cans of regular Pepsi and the mean weight of cola in cans of Diet Pepsi. Does there appear to be a difference between those two means? If there is a difference in the mean weights, identify the most likely explanation for that difference.

24. Weights of Coke Refer to Data Set 19 in Appendix B and use a 0.05 significance level to test the claim that because they contain the same amount of cola, the mean weight of cola in cans of regular Coke is the same as the mean weight of cola in cans of Diet Coke. If there is a difference in the mean weights, identify the most likely explanation for that difference.

9-3 Beyond the Basics

Pooling. ***In Exercises 25 and 26, assume that the two samples are independent simple random samples selected from normally distributed populations. Also assume that the population standard deviations are equal ($\sigma_1 = \sigma_2$), so that the standard error of the differences between means is obtained by pooling the sample variances as described in Part 2 of this section.***

25. Do Men Have a Higher Mean Body Temperature? Repeat Exercise 9 with the additional assumption that $\sigma_1 = \sigma_2$. How are the results affected by this additional assumption?

26. Do Men Talk Less Than Women? Repeat Exercise 8 with the additional assumption that $\sigma_1 = \sigma_2$. How are the results affected by this additional assumption?

27. No Variation in a Sample An experiment was conducted to test the effects of alcohol. Researchers measured the breath alcohol levels for a treatment group of people who drank ethanol and another group given a placebo. The results are given below. Use a 0.05 significance level to test the claim that the two sample groups come from populations with the same mean. The given results are based on data from "Effects of Alcohol Intoxication on Risk Taking, Strategy, and Error Rate in Visuomotor Performance," by Streufert et al., *Journal of Applied Psychology,* Vol. 77, No. 4.

Treatment Group: $n_1 = 22, \bar{x}_1 = 0.049, s_1 = 0.015$

Placebo Group: $n_2 = 22, \bar{x}_2 = 0.000, s_2 = 0.000$

28. Calculating Degrees of Freedom The confidence interval given in Exercise 2 is based on df = 39, which is the "smaller of $n_1 - 1$ and $n_2 - 1$." Use Formula 9-1 to find the number of degrees of freedom. Using the number of degrees of freedom from Formula 9-1 results in this confidence interval: 11.65 cm $< \mu_1 - \mu_2 <$ 17.28 cm. In what sense is "df = smaller of $n_1 - 1$ and $n_2 - 1$" a more conservative estimate of the number of degrees of freedom than the estimate obtained with Formula 9-1?

29. One Standard Deviation Known We sometimes know the value of one population standard deviation from an extensive history of data, but a new procedure or treatment results in sample values with an unknown standard deviation. If σ_1 is known but σ_2 is unknown, use the procedures in Part 1 of this section with these changes: Replace s_1 with the known value of σ_1 and find the number of degrees of freedom using the expression below. (See "The Two-Sample *t* Test with One Variance Unknown," by Maity and Sherman, *The American Statistician,* Vol. 60, No. 2.) Repeat Exercise 13 by assuming that $\sigma = 15$ for proctored tests.

$$df = \frac{\left(\frac{\sigma_1^2}{n_1} + \frac{s_2^2}{n_2}\right)^2}{\frac{(s_2^2/n_2)^2}{n_2 - 1}}$$

28. df = 77.3502249. Using "df = smaller of $n_1 - 1$ and $n_2 - 1$" is a more conservative estimate of the number of degrees of freedom (than the estimate obtained with Formula 9-1) in the sense that the confidence interval is wider, so the difference between the sample means needs to be more extreme to be considered a significant difference.

29. a. H_0: $\mu_1 = \mu_2$. H_1: $\mu_1 < \mu_2$. Test statistic: $t = -3.002$. Critical value based on 68.9927614 degrees of freedom: $t = -2.381$ (Tech: -2.382). *P*-value < 0.005 (Tech: 0.0019). Reject H_0. There is sufficient evidence to support the claim that students taking the nonproctored test get a higher mean than those taking the proctored test.
b. $-25.68 < \mu_1 - \mu_2 < -2.96$ (Tech: $-25.69 < (\mu_1 - \mu_2) < -2.95$)

9-4 Two Dependent Samples (Matched Pairs)

Key Concept In this section we present methods for testing hypotheses and constructing confidence intervals involving the mean of the differences of the values from two populations which are dependent in the sense that the data consist of matched pairs. The pairs must be matched according to some relationship, such as before/after measurements from the same subjects or IQ scores of husbands and wives. The following two sets of sample data look similar because they both consist of two samples with five heights each, but the first case has dependent samples because there is a relationship that is a basis for matching the pairs of data. The second case consists of independent samples because there is no relationship that serves as a basis for matching the pairs of data.

Note to Instructor
Warn students against blindly using the methods of this section whenever they have paired data. For example, suppose we have pulse rates of students matched with their heights. Even though the data are paired, it would make no sense to apply the methods of this section. Such data might be analyzed using methods of correlation and regression, but we should not conduct any analysis based on differences between pulse rates and heights. We should constantly warn students against blind use of formulas or procedures.

Dependent Samples: Matched pairs of heights of U.S. presidents and heights of their main opponents.

Height (cm) of President	189	173	183	180	179
Height (cm) of Main Opponent	170	185	175	180	178

Independent Samples: Heights of males and females that are listed together, but there is no relationship between the males and females.

Height (cm) of Male	178.8	177.5	187.8	172.4	181.7
Height (cm) of Female	163.7	165.5	163.1	166.3	163.6

Note to Instructor
Comment on the point made here about experimental design. It might be helpful to present these two scenarios in testing the effectiveness of a training program designed to increase SAT scores: (1) 100 subjects take the SAT test before the training program and again after the training program; (2) the SAT test is given to 100 subjects who have completed the training program and it is also given to another 100 subjects who have not had the training program. Ask students to identify which of the two approaches is better, and ask them to identify problems with the second approach.

Good Experimental Design

Suppose we want to conduct an experiment to compare the effectiveness of two different types of fertilizer (one organic and one chemical). The fertilizers are to be used on 20 plots of land with equal area but varying soil quality. To make a fair comparison, we should divide each of the 20 plots in half so that one half is treated with organic fertilizer and the other half is treated with chemical fertilizer. The yields can then be matched by the plots they share, resulting in dependent data. The advantage of using matched pairs is that we reduce extraneous variation, which could occur if each plot were treated with one type of fertilizer rather than both—that is, if the samples were independent. This strategy for designing an experiment can be generalized by the following design principle:

When designing an experiment or planning an observational study, using dependent samples with paired data is generally better than using two independent samples.

***t* Test** There are no exact procedures for dealing with dependent samples, but the following approximation methods are commonly used.

Objectives

1. Use the differences from two dependent samples (matched pairs) to test a claim about the mean of the population of all such differences.
2. Use the differences from two dependent samples (matched pairs) to construct a confidence interval estimate of the mean of the population of all such differences.

Notation for Dependent Samples

d = individual difference between the two values in a single matched pair

μ_d = mean value of the differences d for the *population* of all matched pairs of data

$\bar{d}$ = mean value of the differences d for the paired *sample* data

s_d = standard deviation of the differences d for the paired *sample* data

n = number of *pairs* of sample data

Requirements

1. The sample data are dependent (matched pairs).
2. The samples are simple random samples.
3. Either or both of these conditions is satisfied: The number of pairs of sample data is large ($n > 30$) or the pairs of values have differences that are from a population having a distribution that is approximately normal. (These methods are robust against departures for normality, so for small samples, the normality requirement is loose in the sense that the procedures perform well as long as there are no outliers and departures from normality are not too extreme.)

Hypothesis Test Statistic for Dependent Samples

$$t = \frac{\bar{d} - \mu_d}{\frac{s_d}{\sqrt{n}}} \quad \text{(For degrees of freedom use df} = n - 1.)$$

***P*-values:** *P*-values are automatically provided by technology. If technology is not available, refer to the t distribution in Table A-3. Use the procedure summarized in Figure 8-4 from Section 8-2.

Critical values: Use Table A-3 (t distribution) with degrees of freedom found by df $= n - 1$.

Confidence Intervals for Dependent Samples

$$\bar{d} - E < \mu_d < \bar{d} + E$$

where

$$E = t_{\alpha/2}\frac{s_d}{\sqrt{n}}$$

Critical values of $t_{\alpha/2}$: Use Table A-3 (t distribution) with degrees of freedom df $= n - 1$.

If we compare the components of the preceding box to those used in Part 1 of Section 8-4 for inferences about the mean of a population (with σ not known), we can see that they are very similar. The methods of this section are actually the same as those in Section 8-4, except that here we use the differences.

Procedures for Inferences with Dependent Samples

1. Verify that the sample data consist of dependent samples (or matched pairs), and verify that the preceding requirements are satisfied.
2. Find the difference d for each pair of sample values. (*Caution:* Be sure to subtract in a consistent manner.)
3. Find the mean of the differences (denoted by $\bar{d}$), and find the standard deviation of the differences (denoted by s_d).
4. For hypothesis tests and confidence intervals, use the same t test procedures described in Part 1 of Section 8-5.

Equivalent Methods Because the hypothesis test and confidence interval use the same distribution and standard error, they are *equivalent* in the sense that they result in the same conclusions. Consequently, the null hypothesis that the mean difference equals 0 can be tested by determining whether the confidence interval includes 0.

Crest and Dependent Samples

In the late 1950s, Procter & Gamble introduced Crest toothpaste as the first such product with fluoride. To test the effectiveness of Crest in reducing cavities, researchers conducted experiments with several sets of twins. One of the twins in each set was given Crest with fluoride, while the other twin continued to use ordinary toothpaste without fluoride. It was believed that each pair of twins would have similar eating, brushing, and genetic characteristics. Results showed that the twins who used Crest had significantly fewer cavities than those who did not. This use of twins as dependent samples allowed the researchers to control many of the different variables affecting cavities.

Example 1 **Are Presidents Taller Than Their Main Campaign Opponents?**

Data Set 12 in Appendix B lists heights of U.S. presidents and their main opponents in the presidential campaigns. We will use only the sample data included in Table 9-1. (We use only a small random selection of the available data so that we can better illustrate the method of hypothesis testing.) Use the sample data in Table 9-1 with a 0.05 significance level to test the claim that for the population of heights of presidents and their main opponents, the differences have a mean greater than 0 cm (so presidents tend to be taller than their opponents).

Table 9-1 Heights (cm) of Presidents and Their Main Opponents

Height (cm) of President	189	173	183	180	179
Height (cm) of Main Opponent	170	185	175	180	178
Difference d	19	−12	8	0	1

Twins in Twinsburg

During the first weekend in August of each year, Twinsburg, Ohio, celebrates its annual "Twins Days in Twinsburg" festival. Thousands of twins from around the world have attended this festival in the past. Scientists saw the festival as an opportunity to study identical twins. Because they have the same basic genetic structure, identical twins are ideal for studying the different effects of heredity and environment on a variety of traits, such as male baldness, heart disease, and deafness—traits that were recently studied at one Twinsburg festival. A study of twins showed that myopia (near-sightedness) is strongly affected by hereditary factors, not by environmental factors such as watching television, surfing the Internet, or playing computer or video games.

Solution

Requirement check We address the three requirements listed earlier in this section. (1) The samples are dependent because the values are paired. Each pair consists of the height of the winning president and the height of the main opponent in the same election. (2) The pairs of data are randomly selected from the data available as of this writing. We will consider the data to be a simple random sample. (3) The number of pairs of data in Table 9-1 is $n = 5$, which is not large, so we should check for normality of the differences and we should check for outliers. Inspection of the differences shows that there are no outliers. A histogram isn't too helpful with only five data values, but the accompanying STATDISK display shows the normal quantile plot. Because the points approximate a straight-line pattern with no other pattern, we conclude that the differences satisfy the loose requirement of being from a normally distributed population. All requirements are satisfied. ✓

STATDISK

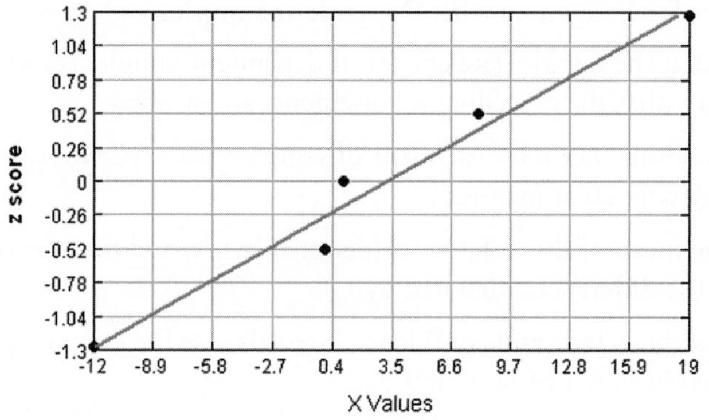

If we use μ_d (where the subscript d denotes "difference") to denote the mean of the differences in height, the claim is that $\mu_d > 0$ cm.

Technology Computer programs and calculators usually provide a P-value, so the P-value method is typically used. See the accompanying SPSS results showing the test statistic of $t = 0.628$. SPSS provides the two-tailed P-value of 0.564, so we halve that value for this right-tailed test to get a P-value of 0.282. Because the P-value of 0.282 is greater than the significance level of 0.05, we fail to reject the null hypothesis and we conclude that there is not sufficient evidence to support the claim that presidents tend to be taller than their opponents.

SPSS

Paired Samples Test

		Paired Differences					t	df	Sig. (2-tailed)
					90% Confidence Interval of the Difference				
		Mean	Std. Deviation	Std. Error Mean	Lower	Upper			
Pair 1	President - Opponent	3.20000	11.38859	5.09313	-7.65778	14.05778	.628	4	.564

If technology is not available, the P-value method, the critical value method, or a confidence interval are all described in Section 8-2. We can test the claim using the critical value method as follows. We will follow the same basic method of hypothesis testing that was introduced in Chapter 8, but we use the test statistic for dependent samples that was given earlier in this section.

Step 1: The claim is that $\mu_d > 0$ cm. (That is, the mean difference is greater than 0 cm.)

Step 2: If the original claim is not true, we have $\mu_d \le 0$ cm.

Step 3: The null hypothesis must express equality and the alternative hypothesis cannot include equality, so we have

$$H_0: \mu_d = 0 \text{ cm} \quad H_1: \mu_d > 0 \text{ cm (original claim)}$$

Step 4: The significance level is $\alpha = 0.05$.

Step 5: We use the Student t distribution.

Step 6: Before finding the value of the test statistic, we must first find the values of $\bar{d}$ and s_d. Refer to Table 9-1 and use the differences of 19, −12, 8, 0, and 1 to find these sample statistics: $\bar{d} = 3.2$ cm and $s_d = 11.4$ cm. Using these sample statistics and the assumption of the hypothesis test that $\mu_d = 0$ cm, we can now find the value of the test statistic.

$$t = \frac{\bar{d} - \mu_d}{\frac{s_d}{\sqrt{n}}} = \frac{3.2 - 0}{\frac{11.4}{\sqrt{5}}} = 0.628$$

Because we are using a t distribution, we refer to Table A-3 to find the critical value of $t = 2.132$ as follows: Use the column for 0.05 (Area in One Tail), and use the row with degrees of freedom of $n - 1 = 4$. Figure 9-4 shows the test statistic, critical value, and critical region.

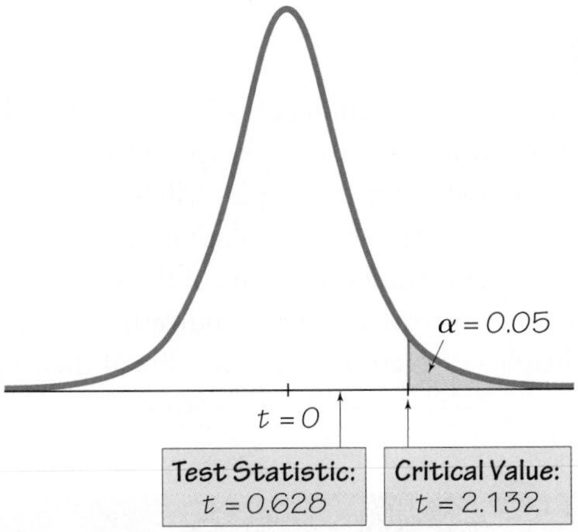

Figure 9-4 Hypothesis Test with Dependent Samples

Step 7: Because the test statistic does not fall in the critical region, we fail to reject the null hypothesis.

Interpretation

We conclude that there is not sufficient evidence to support $\mu_d > 0$ cm. That is, there is not sufficient evidence to support the claim that for the population of heights of presidents and their main opponents, the differences have a mean greater than 0 cm. That is, presidents do not appear to be taller than their opponents.

Example 2 Confidence Interval for Estimating the Mean of the Height Differences

Using the same sample data in Table 9-1, construct a 90% confidence interval estimate of μ_d, which is the mean of the differences in height. By using a confidence level of 90%, we get a result that could be used for the hypothesis test in Example 1.

Gender Gap in Drug Testing

A study of the relationship between heart attacks and doses of aspirin involved 22,000 male physicians. This study, like many others, excluded women. The General Accounting Office recently criticized the National Institutes of Health for not including both sexes in many studies because results of medical tests on males do not necessarily apply to females. For example, women's hearts are different from men's in many important ways. When forming conclusions based on sample results, we should be wary of an inference that extends to a population larger than the one from which the sample was drawn.

(Because the hypothesis test is one-tailed with a significance level of $\alpha = 0.05$, the confidence level should be 90%. See Table 8-1.)

Solution

Requirement check The solution for Example 1 includes verification that the requirements are satisfied. ✓

The preceding SPSS display shows the lower and upper confidence interval limits for a 90% confidence interval, but we will show how those values are obtained. We use the values of $\bar{d} = 3.2$ cm, $s_d = 11.4$ cm, and $t_{\alpha/2} = 2.132$ (found from Table A-3 with $n - 1 = 4$ degrees of freedom and an area of 0.10 in two tails). We first find the value of the margin of error E.

$$E = t_{\alpha/2}\frac{s_d}{\sqrt{n}} = 2.132 \cdot \frac{11.4}{\sqrt{5}} = 10.8694$$

We now find the confidence interval.

$$\bar{d} - E < \mu_d < \bar{d} + E$$
$$3.2 - 10.8694 < \mu_d < 3.2 + 10.8694$$
$$-7.7 \text{ cm} < \mu_d < 14.1 \text{ cm}$$

Interpretation

We have 90% confidence that the limits of −7.7 cm and 14.1 cm contain the true value of the mean of the differences in height (president's height − opponent's height). In the long run, 90% of such samples will lead to confidence interval limits that actually do contain the true population mean of the differences. See that the confidence interval includes the value of 0 cm, so it is very possible that the mean of the differences is equal to 0 cm, indicating that there is no significant difference between heights of presidents and heights of their opponents.

using TECHNOLOGY

General Comment: If it appears that your technology does not have a procedure designed specifically for dependent samples, find the differences d and then use technology for conducting a t test of a claim about one population mean, or use the differences to construct a confidence interval estimate of the mean of one population.

STATDISK First enter the matched data in columns of the STATDISK Data Window, then select **Analysis** from the main menu. Select either **Hypothesis Testing** or **Confidence Intervals,** then select **Mean-Matched Pairs.** Complete the entries and make any selections in the dialog box, then click on **Evaluate.**

MINITAB Enter the paired sample data in columns C1 and C2. Click on **Stat,** select **Basic Statistics,** then select **Paired t.** Enter C1 for the first sample, enter C2 for the second sample, then click on the **Options** box so that you can change the confidence level or the form of the alternative hypothesis or to use a value of μ_d different from zero.

EXCEL Enter the paired sample data in columns A and B and proceed to use either XLSTAT or Excel's Data Analysis add-in.

XLSTAT for Hypothesis Test: Click on **XLSTAT** at the top. Click on **Parametric tests,** then select **Two sample t test and z test.** In the dialog box that appears, enter the range of values for the first sample (such as A1:A13) and enter the range of values for the second sample. For "Data format," select the option of **Paired Samples.** For the "Column labels" box, include a checkmark only if the first row of the sample data consists of names or labels. Put a checkmark next to "Student's t test" (use z test if σ_1 and σ_2 are known). Click on the **Options** tab and select the type of test. For a two-tailed test, select the case including the symbol ≠; for a left-tailed test, select the case including <; and for a right-tailed test, select the case including >. Enter a value in the "Significance level (%)" box. For example, enter 5 for a 0.05 significance level. Click **OK** to get results that include the test statistic and P-value.

XLSTAT for Confidence Interval: Click on **XLSTAT** at the top. Click on **Parametric tests,** then select **Two sample t test and z test.** In the dialog box that appears, enter the range of values for the first sample (such as A1:A13) and enter the range of values for the second sample. For "Data format," select the option of **Paired samples.** For the "Column labels" box, include a checkmark only if the first row of the sample data consists of names or labels. Put a checkmark next to "Student's t test." Click on the **Options** tab. For the alternative hypothesis, select the format of a two-tailed test. Enter a value in the "Significance level (%)" box. For example, enter 5 for a 95% confidence interval. Click **OK** to get results that include the confidence interval.

Data Analysis Add-in: If using Excel 2013, 2010, or 2007, click on **Data,** then **Data Analysis;** if using Excel 2003, click on **Tools,** found on the main menu bar, then select **Data Analysis,** and proceed to select **t-test Paired Two Sample for Means.** In the dialog box, enter the range of values for each of the two samples, enter the assumed value of the population mean difference (typically 0), and enter the significance level. The displayed results will include the test statistic, the *P*-values for a one-tailed test and a two-tailed test, and the critical values for a one-tailed test and a two-tailed test.

TI-83/84 PLUS *Caution:* Do not use the menu item **2-SampTTest** because it applies to *independent* samples. Instead, enter the data for the first variable in list L1, enter the data for the second variable in list L2, then clear the screen and enter **L1 – L2** STO› **L3** so that list L3 will contain the individual differences *d*. Now press STAT, then select **TESTS,** and choose the option of **T-Test** (for a hypothesis test) or **TInterval** (for a confidence interval). Use the input option of **Data.** For the list, enter L3. If using **T-Test,** also enter the assumed value of the population mean difference (typically 0) for μ_0. Press ENTER when finished.

STATCRUNCH Click on **Open StatCrunch.** Enter the sample data in columns, or open a data set. Click on **Stat,** then select **T statistics.** Select **Paired.** Click on **Next,** then select **Hypothesis Test** or **Confidence Interval.** Make the required entries. Click on **Calculate.**

9-4 Basic Skills and Concepts

Recommended Assignment
Exercises 1–14.

Due to limited space, answers for some odd numbered exercises are not included here, but they are included in Appendix D.

1. Parts (c) and (e) are true.
2. $\bar{d} = 2.4$ mi/gal and $s_d = 1.1$ mi/gal. μ_d represents the mean of the differences from the population of paired data.
3. The test statistic will remain the same. The confidence interval limits will be expressed in the equivalent values of km/L.
4. The first confidence interval shows that we have 95% confidence that the limits of 1.0 mi/gal and 3.8 mi/gal contain the mean of the population of differences, but the second confidence interval shows that we have 95% confidence that the limits of −7.8 mi/gal and 12.6 mi/gal contain the difference between the two population means. Because the first confidence interval does not include 0 mi/gal and consists of positive values only, it appears that the old ratings are higher than the new ratings. Because the second confidence interval does include 0 mi/gal, there does not appear to be a significant different between the mean of the old ratings and the mean of the new ratings.

Statistical Literacy and Critical Thinking

1. True Statements? For the methods of this section, which of the following statements are true?

a. The requirement of a simple random sample is satisfied if we have matched pairs of voluntary response data.

b. If we have more than 10 matched pairs of sample data, we can consider the sample to be large and there is no need to check for normality.

c. If we have five matched pairs of sample data, there is a loose requirement that the five differences appear to be from a normally distributed population.

d. The methods of this section apply if we have 50 heights of women from Texas and 50 heights of women from Utah.

e. If we want to use a confidence interval to test the claim that $\mu_d > 0$ with a 0.05 significance level, the confidence interval should have a confidence level of 90%.

2. Notation Listed below are combined city–highway fuel consumption ratings (in mi/gal) for five different cars measured under both the old rating system and a new rating system introduced in 2008 (based on data from *USA Today*). The new ratings were implemented in response to complaints that the old ratings were too high. Find the values of $\bar{d}$ and s_d. In general, what does μ_d represent?

Old Rating (mi/gal)	16	18	27	17	33
New Rating (mi/gal)	15	16	24	15	29

3. Units of Measure If the values listed in Exercise 2 are changed so that they are expressed in kilometers per liter (km/L) instead of mi/gal, how is a test statistic affected? How are the confidence interval limits affected?

4. Confidence Intervals If we use the sample data in Exercise 2, we get this 95% confidence interval estimate: 1.0 mi/gal $< \mu_d <$ 3.8 mi/gal. Treating the same data as *independent samples* yields -7.8 mi/gal $< \mu_1 - \mu_2 <$ 12.6 mi/gal for a 95% confidence level. What is the difference between interpretations of these two confidence intervals?

TI-83/84 PLUS

```
T-Test
 μ>0
 t=.035713579
 p=.4858629522
 x̄=.0588235
 Sx=9.604106
 n=34
```

5. POTUS Hypothesis Test Example 1 in this section used only five pairs of data from Data Set 12 in Appendix B for a 95% confidence level. Repeat Example 1 using all of the cases with heights for both the president and the main opponent. Results are shown in the accompanying TI-83/84 Plus calculator display.

TI-83/84 PLUS

```
TInterval
 (-2.729,2.8463)
 x̄=.0588235
 Sx=9.604106
 n=34
```

6. POTUS Confidence Interval Example 2 in this section used only five pairs of data from Data Set 12 in Appendix B. Repeat Example 2 using all of the cases with heights for both the president and the main opponent. The accompanying TI-83/84 Plus calculator display shows results for a 90% confidence interval constructed from the list of differences in height. In this display, $\bar{x}$ is used instead of $\bar{d}$, and Sx is used instead of s_d. What feature of the confidence interval causes us to reach the same conclusion as in Exercise 4?

Calculations with Paired Sample Data. ***In Exercises 7 and 8, assume that you want to use a 0.05 significance level to test the claim that paired sample data come from a population for which the mean difference is $\mu_d = 0$. Find (a) $\bar{d}$, (b) s_d, (c) the t test statistic, and (d) the critical values.***

7. Oscars Listed below are ages of actresses and actors at the time that they won Oscars for the categories of Best Actress and Best Actor. The data are from Data Set 11 in Appendix B.

Actress	22	37	28	63	32
Actor	44	41	62	52	41

8. Body Temperatures Listed below are body temperatures of four subjects measured at two different times in a day (from Data Set 3 in Appendix B).

Body Temperature (°F) at 8 A.M. on Day 1	98	97.0	98.6	97.4
Body Temperature (°F) at 12 A.M. on Day 1	98	97.6	98.8	98.0

In Exercises 9–20, assume that the paired sample data are simple random samples and that the differences have a distribution that is approximately normal.

9. Oscars Use the sample data from Exercise 7 to test for a difference between the ages of actresses and actors when they win Oscars. Use a significance level of $\alpha = 0.05$.

10. Body Temperatures Use the sample data from Exercise 8 to test the claim that there is no difference between body temperatures measured at 8 A.M. and at 12 A.M. Use a 0.05 significance level.

11. Flight Operations The table below lists the times (min) required for randomly selected flights to taxi out for takeoff and the corresponding times (min) required to taxi in after landing. (See Data Set 15 in Appendix B.) All flights are Flight 1 of American Airlines from New York (JFK) to Los Angeles (LAX). Construct a 90% confidence interval estimate of the difference between taxi-out times and taxi-in times. What does the confidence interval suggest about the claim of the flight operations manager that for flight delays, more of the blame is attributable to taxi-out times at JFK than taxi-in times at LAX?

Taxi-Out Time	30	19	12	19	18	22	37	13	14	15	31	15
Taxi-In Time	12	13	8	21	17	11	12	12	15	26	9	11

12. Brain Volumes of Twins Listed below are brain volumes (cm^3) of twins listed in Data Set 6 of Appendix B. Construct a 99% confidence interval estimate of the mean of the

6. -2.7 cm $< \mu_d <$ 2.8 cm. The confidence interval includes 0 cm, so it is very possible that the mean of the differences is equal to 0 cm, indicating that there is no significant difference between heights of presidents and heights of their opponents.

8. a. $\bar{d} = -0.35$°F
   b. $s_d = 0.30$°F
   c. $t = -2.333$
   d. $t = \pm 3.182$

9. H_0: $\mu_d = 0$. H_1: $\mu_d \neq 0$. Test statistic: $t = -1.507$. Critical values: $t = \pm 2.776$. *P*-value > 0.20 (Tech: 0.2063). Fail to reject H_0. There is not sufficient evidence to support the claim that there is a difference between the ages of actresses and actors when they win Oscars.

10. H_0: $\mu_d = 0$. H_1: $\mu_d \neq 0$. Test statistic: $t = -2.333$. Critical values: $t = \pm 3.182$. *P*-value > 0.10 (Tech: 0.1018). Fail to reject H_0. There is not sufficient evidence to warrant rejection of the claim that there is no difference between body temperatures measured at 8 A.M. and at 12 A.M.

12. -66.8 $cm^3 < \mu_d <$ 49.8 cm^3 (Tech: -66.7 $cm^3 < \mu_d <$ 49.7 cm^3). Because the confidence interval includes 0 cm^3, the mean of the differences could be equal to 0 cm^3, so there does not appear to be a significant difference.

differences between volumes for the first-born and the second-born twins. What does the confidence interval suggest?

First Born	1005	1035	1281	1051	1034	1079	1104	1439	1029	1160
Second Born	963	1027	1272	1079	1070	1173	1067	1347	1100	1204

13. Speaking Couples Listed below are the numbers of words spoken in a day by each member of six different couples. The data are randomly selected from the first two columns in Data Set 17 from Appendix B. Use a 0.05 significance level to test the claim that among couples, males speak more words in a day than females.

Male	5638	21,319	17,572	26,429	46,978	25,835
Female	5198	11,661	19,624	13,397	31,553	18,667

13. H_0: $\mu_d = 0$. H_1: $\mu_d > 0$. Test statistic: $t = 2.579$. Critical value: $t = 2.015$. P-value < 0.025 (Tech: 0.0247). Reject H_0. There is sufficient evidence to support the claim that among couples, males speak more words in a day than females.

14. Is Blood Pressure the Same for Both Arms? Listed below are systolic blood pressure measurements (mm Hg) taken from the right and left arms of the same woman (based on data from "Consistency of Blood Pressure Differences Between the Left and Right Arms," by Eguchi et al., *Archives of Internal Medicine,* Vol. 167). Use a 0.01 significance level to test for a difference between the measurements from the two arms. What do you conclude?

Right Arm	102	101	94	79	79
Left Arm	175	169	182	146	144

14. H_0: $\mu_d = 0$. H_1: $\mu_d \neq 0$. Test statistic: $t = -17.339$. Critical values: $t = \pm 4.604$. P-value > 0.01 (Tech: 0.0001). Reject H_0. There is sufficient evidence to support the claim of a difference in measurements between the two arms. The right and left arms should yield the same measurements, but the given data show that this is not happening.

15. Is Friday the 13th Unlucky? Researchers collected data on the numbers of hospital admissions resulting from motor vehicle crashes, and results are given below for Fridays on the 6th of a month and Fridays on the following 13th of the same month (based on data from "Is Friday the 13th Bad for Your Health?" by Scanlon et al., *British Medical Journal,* Vol. 307, as listed in the *Data and Story Line* online resource of data sets). Construct a 95% confidence interval estimate of the mean of the population of differences between hospital admissions on days that are Friday the 6th of a month and days that are Friday the 13th of a month. Use the confidence interval to test the claim that when the 13th day of a month falls on a Friday, the numbers of hospital admissions from motor vehicle crashes are not affected.

Friday the 6th	9	6	11	11	3	5
Friday the 13th	13	12	14	10	4	12

15. $-6.5 < \mu_d < -0.2$. Because the confidence interval does not include 0, it appears that there is sufficient evidence to warrant rejection of the claim that when the 13th day of a month falls on a Friday, the numbers of hospital admissions from motor vehicle crashes are not affected. Hospital admissions do appear to be affected.

16. Self-Reported and Measured Male Heights As part of the National Health and Nutrition Examination Survey, the Department of Health and Human Services obtained self-reported heights and measured heights for males aged 12–16. All measurement are in inches. Listed below are sample results. Construct a 99% confidence interval estimate of the mean difference between reported heights and measured heights. Interpret the resulting confidence interval, and comment on the implications of whether the confidence interval limits contain 0.

Reported Height	68	71	63	70	71	60	65	64	54	63	66	72
Measured Height	67.9	69.9	64.9	68.3	70.3	60.6	64.5	67.0	55.6	74.2	65.0	70.8

16. $-4.2 < \mu_d < 2.2$. Because the confidence interval limits contain 0, there is not sufficient evidence to support a claim that there is a difference between self-reported heights and measured heights. We might believe that males would tend to exaggerate their heights, but the given data do not provide enough evidence to support that belief.

17. Harry Potter The Harry Potter books and movies grossed huge sums of money. The table below lists the amounts grossed (in millions of dollars) during the first few days of release of the movies *Harry Potter and the Half-Blood Prince* and *Harry Potter and the Order of the Phoenix.* Use a 0.05 significance level to test the claim that *Harry Potter and the Half-Blood Prince* did better at the box office. Apart from this hypothesis test, what is a better way to judge the validity of the claim?

Day of Release	1	2	3	4	5	6	7	8	9	10
Phoenix	44.2	18.4	25.8	28.3	23.0	10.4	9.1	8.4	7.6	10.2
Prince	58.2	22.0	26.8	29.2	21.8	9.9	9.5	7.5	6.9	9.3

17. H_0: $\mu_d = 0$. H_1: $\mu_d < 0$. Test statistic: $t = -1.080$. Critical value: $t = -1.833$. P-value > 0.10 (Tech: 0.1540). Fail to reject H_0. There is not sufficient evidence to support the claim that *Harry Potter and the Half-Blood Prince* did better at the box office. After a few years, the gross amounts from both movies can be identified, and the conclusion can then be judged objectively without using a hypothesis test.

18. H_0: $\mu_d = 0$. H_1: $\mu_d > 0$. Test statistic: $t = 6.371$. Critical value: $t = 2.718$. *P*-value < 0.005 (Tech: 0.00003). Reject H_0. There is sufficient evidence to support the claim that Captopril is effective in lowering systolic blood pressure.

19. $0.69 < \mu_d < 5.56$. Because the confidence interval limits do not contain 0 and they consist of positive values only, it appears that the "before" measurements are greater than the "after" measurements, so hypnotism does appear to be effective in reducing pain.

20. $-7.3°\text{F} < \mu_d < 6.3°\text{F}$. Because the confidence interval limits do contain 0°F, there is not a significant difference between the actual high temperatures and those that were forecast five days earlier. This suggests that the forecast temperatures are reasonably accurate.

21. H_0: $\mu_d = 0$. H_1: $\mu_d \neq 0$. Test statistic: $t = -5.553$. Critical values: $t = \pm 1.990$. *P*-value < 0.01 (Tech: 0.0000). Reject H_0. There is sufficient evidence to support the claim that there is a difference between the ages of actresses and actors when they win Oscars.

22. H_0: $\mu_d = 0$. H_1: $\mu_d \neq 0$. Test statistic: $t = 0.124$. Critical values: $t = \pm 2.028$. *P*-value > 0.20 (Tech: 0.9023). Fail to reject H_0. There is not sufficient evidence to warrant rejection of the claim that there is no difference between body temperatures measured at 8 A.M. and at 12 A.M.

23. H_0: $\mu_d = 0$. H_1: $\mu_d < 0$. Test statistic: $t = -1.560$. Critical value of t is between -1.671 and -1.676 (Tech: -1.673). *P*-value > 0.05 (Tech: 0.0622). Fail to reject H_0. There is not sufficient evidence to support the claim that among couples, males speak fewer words in a day than females.

24. H_0: $\mu_d = 0$ sec. H_1: $\mu_d > 0$ sec. Test statistic: $t = 0.938$. Critical value: $t = 1.694$. *P*-value > 0.10 (Tech: 0.1776). Fail to reject H_0. There is not sufficient evidence to support the claim that the mean of the differences is greater than 0 sec. There is not sufficient evidence to support the claim that more time is devoted to showing tobacco than alcohol. For animated children's movies, *no* time should be spent showing the use of tobacco or alcohol.

18. Before/After Treatment Results Captopril is a drug designed to lower systolic blood pressure. When subjects were treated with this drug, their systolic blood pressure readings (in mm Hg) were measured before and after the drug was taken. Results are given in the accompanying table (based on data from "Essential Hypertension: Effect of an Oral Inhibitor of Angiotensin-Converting Enzyme," by MacGregor et al., *British Medical Journal,* Vol. 2). Using a 0.01 significance level, is there sufficient evidence to support the claim that Captopril is effective in lowering systolic blood pressure?

Subject	A	B	C	D	E	F	G	H	I	J	K	L
Before	200	174	198	170	179	182	193	209	185	155	169	210
After	191	170	177	167	159	151	176	183	159	145	146	177

19. Hypnotism for Reducing Pain A study was conducted to investigate the effectiveness of hypnotism in reducing pain. Results for randomly selected subjects are given in the accompanying table (based on "An Analysis of Factors That Contribute to the Efficacy of Hypnotic Analgesia," by Price and Barber, *Journal of Abnormal Psychology,* Vol. 96, No. 1). The values are before and after hypnosis; the measurements are in centimeters on a pain scale. Construct a 95% confidence interval for the mean of the "before/after" differences. Does hypnotism appear to be effective in reducing pain?

Subject	A	B	C	D	E	F	G	H
Before	6.6	6.5	9.0	10.3	11.3	8.1	6.3	11.6
After	6.8	2.4	7.4	8.5	8.1	6.1	3.4	2.0

20. Forecast and Actual Temperatures The author recorded actual temperatures (°F) along with the temperatures (°F) that were predicted five days earlier. Results are listed below. Construct a 99% confidence interval estimate of the mean of the population of all "actual/forecast" differences. What does the result suggest about the accuracy of the forecast temperatures?

Date	9/1	9/5	9/12	9/15	9/22	9/23	9/27	9/30
Actual High	80	73	78	73	82	81	74	62
High Forecast Five Days Earlier	80	79	79	78	73	79	70	69

Large Data Sets. ***In Exercises 21–24, use the indicated Data Sets from Appendix B. Assume that the paired sample data are simple random samples and the differences have a distribution that is approximately normal.***

21. Oscars Use the sample data from Data Set 11 in Appendix B to test for a difference between the ages of actresses and actors when they win Oscars. Use a significance level of $\alpha = 0.05$.

22. Body Temperatures Use the sample data from 8 A.M. and 12 A.M. on Day 1 as listed in Data Set 3 in Appendix B. Test the claim that there is no difference between body temperatures measured at 8 A.M. and at 12 A.M. Use a 0.05 significance level.

23. Speaking Couples Use the data in the first two columns of Data Set 17 in Appendix B. Those columns list the numbers of words spoken in a day by each member of 56 different couples. Use a 0.05 significance level.

24. Tobacco and Alcohol in Children's Movies Refer to Data Set 8 in Appendix B and use the times (seconds) that animated Disney movies showed the use of tobacco and the times that they showed the use of alcohol. Use a 0.05 significance level to test the claim that the mean of the differences is greater than 0 sec so that more time is devoted to showing tobacco than alcohol. For animated children's movies, how much time should be spent showing the use of tobacco and alcohol?

9-4 Beyond the Basics

25. Freshman 15 The "Freshman 15" refers to the urban legend that is the common belief that students gain an average of 15 lb (or 6.8 kg) during their freshman year. Refer to Data Set 4 in Appendix B and consider the sample values in this format: (April weight) − (September weight). In this format, positive differences represent *gains* in weight, and negative differences represent *losses* of weight. If we use μ_d to denote the mean of the "April − September" differences in weights of college students during their freshman year, the "Freshman 15" is the claim that $\mu_d = 15$ lb or $\mu_d = 6.8$ kg. Test the claim that $\mu_d = 6.8$ kg using a 0.05 significance level with the 67 subjects from Data Set 4 in Appendix B. What do these results suggest about the "Freshman 15"?

25. H_0: $\mu_d = 6.8$ kg. H_1: $\mu_d \neq 6.8$ kg. Test statistic: $t = -11.833$. Critical values: $t = \pm 1.994$ (Tech: ± 1.997). *P*-value < 0.01 (Tech: 0.0000). Reject H_0. There is sufficient evidence to warrant rejection of the claim that $\mu_d = 6.8$ kg. It appears that the "Freshman 15" is a myth, and college freshman might gain some weight, but they do not gain as much as 15 pounds.

Chapter 9 Review

In this chapter we considered methods for using two samples for making inferences about two populations. Specifically, we presented methods for testing claims or constructing confidence interval estimates based on the following:

Section 9-2: Proportions from two samples selected from two populations

Section 9-3: Means from samples selected from two independent populations

Section 9-4: Mean of the differences from two dependent samples consisting of matched pairs

Two main activities of inferential statistics are (1) constructing confidence interval estimates of population parameters (as in Chapter 7), and (2) using methods of hypothesis testing to test claims about population parameters (as in Chapter 8). Chapters 7 and 8 considered only cases involving a single population, but in this chapter we considered two samples drawn from two populations.

When using the methods of this chapter, it is important to check the given requirements, including the requirement that the sample data have been collected in an appropriate way. Using data collected with methods that are not appropriate, such as voluntary response samples, could easily provide results that are dramatically wrong.

Chapter Quick Quiz

In Exercises 1–4, use the following survey results: Randomly selected subjects were asked if they agreed with the statement "It is morally wrong for married people to have an affair." Among the 386 women surveyed, 347 agreed with the statement. Among the 359 men surveyed, 305 agreed with the statement (based on data from a Pew Research poll).

1. Identify the null and alternative hypotheses resulting from the claim that the proportion of women who agree with the given statement is equal to the proportion of men who agree.

1. H_0: $p_1 = p_2$. H_1: $p_1 \neq p_2$.

2. Find the value of the pooled proportion $\bar{p}$ obtained when testing the claim given in Exercise 1.

2. 0.875

3. When testing the claim that $p_1 = p_2$, a test statistic of $z = 2.04$ is obtained. Find the *P*-value obtained from this test statistic.

3. 0.0414

4. When using the given sample data to construct a 95% confidence interval estimate of the difference between the two population proportions, the result of (0.00172, 0.0970) is obtained from technology. Express that confidence interval in a format that uses the symbol <.

4. $0.00172 < p_1 - p_2 < 0.0970$

5. Because the data consist of matched pairs, they are dependent.
6. H_0: $\mu_d = 0$. H_1: $\mu_d > 0$.
7. There is not sufficient evidence to support the claim that front repair costs are greater than the corresponding rear repair costs.
8. True
9. False.
10. True.

5. Listed below are the costs (in dollars) of repairing the front ends and rear ends of different cars when they were damaged in controlled low-speed crash tests (based on data from the Insurance Institute for Highway Safety). The cars are Toyota, Mazda, Volvo, Saturn, Subaru, Hyundai, Honda, Volkswagen, and Nissan. Are the data independent or dependent?

Front repair cost	936	978	2252	1032	3911	4312	3469	2598	4535
Rear repair cost	1480	1202	802	3191	1122	739	2769	3375	1787

6. Refer to the sample data given in Exercise 5 and identify the null and alternative hypotheses resulting from the claim that the mean of the differences is greater than zero so that front repair costs are greater than the corresponding rear repair costs. Express those hypotheses in symbolic form.

7. When testing the hypotheses from Exercise 6, we get the test statistic $t = 1.302$ and the P-value of 0.2293. What should we conclude?

8. Determine whether the following statement is true or false: When testing a claim about the means of two independent populations, the alternative hypothesis can never contain the condition of equality.

9. Determine whether the following statement is true or false: When testing the claim that the mean annual income of statistics professors in New York is equal to the mean annual income of statistics professors in Illinois, either the two population standard deviations must be known or both samples must include more than 30 values.

10. Determine whether the following statement is true or false: When using the methods of this chapter to test a claim that two population proportions are equal, each of the two samples must satisfy the requirement that $np \geq 5$ and $nq \geq 5$.

1. H_0: $p_1 = p_2$. H_1: $p_1 > p_2$. Test statistic: $z = 3.12$. Critical value: $z = 2.33$. P-value: 0.0009. Reject H_0. There is sufficient evidence to support a claim that the proportion of successes with surgery is greater than the proportion of successes with splinting. When treating carpal tunnel syndrome, surgery should generally be recommended instead of splinting.
2. 98% CI: $0.0581 < p_1 - p_2 < 0.332$ (Tech: $0.0583 < p_1 - p_2 < 0.331$). The confidence interval limits do not contain 0; the interval consists of positive values only. This suggests that the success rate with surgery is greater than the success rate with splints.
3. H_0: $p_1 = p_2$. H_1: $p_1 < p_2$. Test statistic: $z = -1.91$. Critical value: $z = -1.645$. P-value: 0.0281 (Tech: 0.0280). Reject H_0. There is sufficient evidence to support the claim that the fatality rate of occupants is lower for those in cars equipped with airbags.
4. H_0: $\mu_d = 0$. H_1: $\mu_d > 0$. Test statistic: $t = 4.712$. Critical value: $t = 3.143$. P-value < 0.005 (Tech: 0.0016). Reject H_0. There is sufficient evidence to support the claim that flights scheduled 1 day in advance cost more than flights scheduled 30 days in advance. Save money by scheduling flights 30 days in advance.

Review Exercises

1. Carpal Tunnel Syndrome Carpal tunnel syndrome is a common wrist complaint resulting from a compressed nerve, and it is often caused by repetitive wrist movements. In a randomized controlled trial, among 73 patients treated with surgery and evaluated one year later, 67 were found to have successful treatments. Among 83 patients treated with splints and evaluated one year later, 60 were found to have successful treatments (based on data from "Splinting vs Surgery in the Treatment of Carpal Tunnel Syndrome," by Gerritsen et al., *Journal of the American Medical Association,* Vol. 288, No. 10). In a journal article about the trial, authors claimed that "treatment with open carpal tunnel release surgery resulted in better outcomes than treatment with wrist splinting for patients with CTS (carpal tunnel syndrome)." Use a 0.01 significance level to test the stated claim. What treatment strategy is suggested by the results?

2. Carpal Tunnel Syndrome Construct a confidence interval suitable for testing the claim from Exercise 1. What feature of the resulting confidence interval leads to the same conclusion from Exercise 1?

3. Airbags Save Lives In a study of the effectiveness of airbags in cars, 11,541 occupants were observed in car crashes with airbags available, and 41 of them were fatalities. Among 9853 occupants in crashes with airbags not available, 52 were fatalities (based on data from "Who Wants Airbags?" by Meyer and Finney, *Chance,* Vol. 18, No. 2). Use a 0.05 significance level to test the claim that the fatality rate of occupants is lower for those in cars equipped with airbags.

4. Are Flights Cheaper When Scheduled Earlier? Listed below are the costs (in dollars) of flights from New York (JFK) to San Francisco for US Air, Continental, Delta, United,

American, Alaska Airlines, and Northwest. Use a 0.01 significance level to test the claim that flights scheduled one day in advance cost more than flights scheduled 30 days in advance. What strategy appears to be effective in saving money when flying?

Flight scheduled one day in advance	456	614	628	1088	943	567	536
Flight scheduled 30 days in advance	244	260	264	264	278	318	280

5. Self-Reported and Measured Female Heights As part of the National Health and Nutrition Examination Survey conducted by the Department of Health and Human Services, self-reported heights and measured heights were obtained for females aged 12 to 16. Listed below are sample results. Is there sufficient evidence to support the claim that there is a difference between self-reported heights and measured heights of females aged 12 to 16? Use a 0.05 significance level.

Reported height	53	64	61	66	64	65	68	63
Measured height	58.1	62.7	61.1	64.8	63.2	66.4	67.6	63.5

6. Eyewitness Accuracy of Police Does stress affect the recall ability of police eyewitnesses? This issue was studied in an experiment that tested eyewitness memory a week after a nonstressful interrogation of a cooperative suspect and a stressful interrogation of an uncooperative and belligerent suspect. The numbers of details recalled a week after the incident were recorded, and the summary statistics are given below (based on data from "Eyewitness Memory of Police Trainees for Realistic Role Plays," by Yuille et al., *Journal of Applied Psychology,* Vol. 79, No. 6). Use a 0.01 significance level to test the claim in the article that "stress decreases the amount recalled."

Nonstress: $n = 40, \bar{x} = 53.3, s = 11.6$

Stress: $n = 40, \bar{x} = 45.3, s = 13.2$

7. Eyewitness Accuracy of Police Construct a confidence interval suitable for testing the claim from Exercise 6. What feature of the resulting confidence interval leads to the same conclusion from Exercise 6?

8. Effect of Blinding Among 13,200 submitted abstracts that were blindly evaluated (with authors and institutions not identified), 26.7% were accepted for publication. Among 13,433 abstracts that were not blindly evaluated, 29.0% were accepted (based on data from "Effect of Blinded Peer Review on Abstract Acceptance," by Ross et al., *Journal of the American Medical Association,* Vol. 295, No. 14). Use a 0.01 significance level to test the claim that the acceptance rate is the same with or without blinding. How might the results be explained?

9. Comparing Means The *baseline characteristics* of different treatment groups are often included in journal articles. In one study, 49 subjects treated with raw garlic had LDL cholesterol measurements with a mean of 151 and a standard deviation of 15, while 48 subjects given placebos had LDL cholesterol measurements with a mean of 149 and a standard deviation of 14 (based on data from "Effect of Raw Garlic vs Commercial Garlic Supplements on Plasma Lipid Concentrations in Adults with Moderate Hypercholesterolemia," by Gardner et al., *Archives of Internal Medicine,* Vol. 167). Use a 0.05 significance level to test the claim that there is no difference between the mean LDL cholesterol levels of subjects treated with raw garlic and subjects given placebos. Do both groups appear to be about the same?

10. Comparing Means Repeat Exercise 9 by using a confidence interval instead of a hypothesis test.

5. H_0: $\mu_d = 0$. H_1: $\mu_d \neq 0$. Test statistic: $t = -0.574$. Critical values: $t = \pm 2.365$. P-value > 0.20 (Tech: 0.5840). Fail to reject H_0. There is not sufficient evidence to support the claim that there is a difference between self-reported heights and measured heights of females aged 12–16.
6. H_0: $\mu_1 = \mu_2$. H_1: $\mu_1 > \mu_2$. Test statistic: $t = 2.879$. Critical value: $t = 2.426$ (Tech: 2.376). P-value < 0.005 (Tech: 0.0026). Reject H_0. There is sufficient evidence to support the claim that "stress decreases the amount recalled."
7. 98% CI: $1.3 < (\mu_1 - \mu_2) < 14.7$ Tech: $1.4 < (\mu_1 - \mu_2) < 14.6$). The confidence interval limits do not contain 0; the interval consists of positive values only. This suggests that the numbers of details recalled are lower for those in the stress population.
8. H_0: $p_1 = p_2$. H_1: $p_1 \neq p_2$. Test statistic: $z = -4.20$. Critical values: $z = \pm 2.575$. P-value: 0.0002 (Tech: 0.0000). Reject H_0. There is sufficient evidence to warrant rejection of the claim that the acceptance rate is the same with or without blinding. Without blinding, reviewers know the names and institutions of the abstract authors, and they might be influenced by that knowledge.
9. H_0: $\mu_1 = \mu_2$. H_1: $\mu_1 \neq \mu_2$. Test statistic: $t = 0.679$. Critical values: $t = \pm 2.014$ approximately (Tech: ± 1.985). P-value > 0.20 (Tech: 0.4988). Fail to reject H_0. There is not sufficient evidence to warrant rejection of the claim of no difference between the mean LDL cholesterol levels of subjects treated with raw garlic and subjects given placebos. Both groups appear to be about the same.
10. $-3.9 < (\mu_1 - \mu_2) < 7.9$ (Tech: $-3.8 < (\mu_1 - \mu_2) < 7.8$) Because the confidence interval contains zero, it is possible that the two population means are equal, so there is not sufficient evidence to warrant rejection of the claim of no difference between the mean LDL cholesterol levels of subjects treated with raw garlic and subjects given placebos.

Cumulative Review Exercises

Heights of Mothers and Daughters. *In Exercises 1–5, use the following heights (in.) of mothers, fathers, and their adult daughters. The data are matched so that each column consists of heights from the same family.*

Mother	63	67	62	69	63	64	63	64	60	65
Father	69	70	69	62	66	76	69	68	66	68
Daughter	62.2	67.2	63.4	68.4	62.2	64.7	59.6	61.0	64.0	65.4

1. a. Are the three samples independent or dependent? Why?

b. Find the mean, median, mode, range, standard deviation, and variance of the heights of the daughters. Express results with the appropriate units.

c. What is the level of measurement of the sample data (nominal, ordinal, interval, ratio)?

2. Scatterplot Construct a scatterplot of the paired mother/daughter heights. What does the result suggest?

3. Confidence Interval Construct a 95% confidence interval estimate of the mean height of daughters. Write a brief statement that interprets the confidence interval.

4. Hypothesis Test Use a 0.05 significance level to test the claim that there is no significant difference between the heights of mothers and the heights of their daughters.

5. Assessing Normality Refer to the accompanying normal quantile plot and determine whether the sample of heights of daughters appears to be from a normally distributed population.

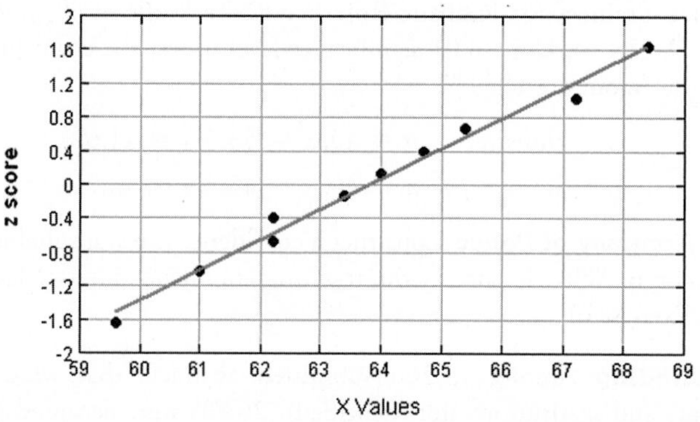

6. Dark Survey In a survey of 1032 Americans, respondents were asked what they miss most when electrical power is lost. Of the respondents 134 indicated their computer and/or Internet access (based on results from a Utility Pulse survey). Use the results to construct a 95% confidence interval estimate of the proportion of all Americans who feel that their computer and/or Internet access is missed most when they lose power. Do the results justify the statement that "fewer than 20% of Americans choose their computer and/or Internet access when identifying what they miss most when electrical power is lost"? Explain.

7. Backup Generator The *USA Today* Web site posted this question: "Have you considered getting a home generator for backup power?" Among 928 Internet users who chose to respond, 41% answered "yes." If we use those results to construct a 95% confidence interval estimate of the proportion of the population who have considered getting a generator for backup power, we get this: $0.378 < p < 0.441$. Are we 95% confident that the limits of 0.378 and 0.441 contain the true proportion of the population who have considered getting a home generator for backup power? Explain.

8. Juke Survey Late-night talk show host David Letterman made jokes about the name "Juke" for a Nissan motor vehicle. (It is kind of funny.) If you have been commissioned to

1. a. Because the sample data are matched with each column consisting of heights from the same family, the data are dependent.
   b. Mean: 63.81 in.; median: 63.70 in.; mode: 62.2 in.; range: 8.80 in.; standard deviation: 2.73 in.; variance: 7.43 in^2
   c. Ratio
2. There does not appear to be a correlation or association between the heights of mothers and the heights of their daughters.

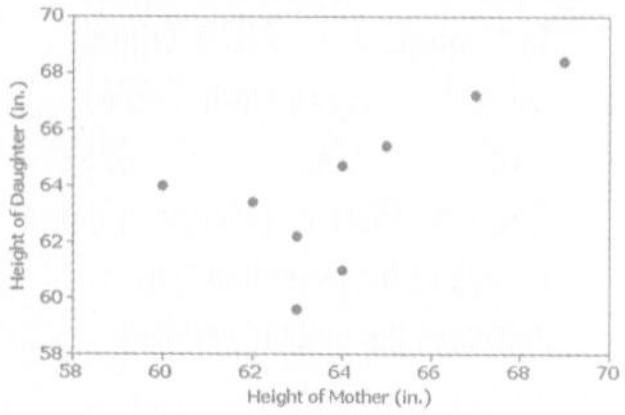

3. 61.86 in. $< \mu <$ 65.76 in. We have 95% confidence that the limits of 61.86 in. and 65.76 in. actually contain the true value of the mean height of all adult daughters.
4. H_0: $\mu_d = 0$. H_1: $\mu_d \neq 0$. Test statistic: $t = 0.283$. Critical values: $t = \pm 2.262$. P-value > 0.20 (Tech: 0.7834). Fail to reject H_0. There is not sufficient evidence to warrant rejection of the claim of no significant difference between the heights of mothers and the heights of their daughters.
5. Because the points lie reasonably close to a straight-line pattern and there is no other pattern that is not a straight-line pattern and there are no outliers, the sample data appear to be from a population with a normal distribution.
6. $0.109 < p < 0.150$. Because the entire range of values in the confidence interval lies below 0.20, the results do justify the statement that "fewer than 20% of Americans choose their computer and/or Internet access when identifying what they miss most when electrical power is lost."
7. No. Because the Internet users chose to respond, we have a voluntary response sample, so the results are not necessarily valid.

conduct a survey of American drivers who see humor in the Juke name, how many American drivers must you survey? Assume that you want 97% confidence that your sample percentage is in error by no more than two percentage points. Is it OK to save money by conducting a telephone survey using only local phone numbers? Why or why not?

9. Normal Distribution Based on the measurements in Data Set 1 of Appendix B, assume that heights of women are normally distributed with a mean of 162.0 cm and a standard deviation of 6.6 cm.

a. If a woman is randomly selected, find the probability that she is taller than 152.1 cm.

b. If four women are randomly selected, find the probability that their mean height is greater than 152.1 cm.

c. Find P_{80}.

10. Mean Income Another (clearly inferior) statistics textbook includes an exercise in which students are given the mean income for each of the 50 states. The students are then asked to compute the mean of those 50 values. Is the result the mean income for the U.S. population? Why or why not?

8. 2944. The survey should not be conducted using only local phone numbers. Such a convenience sample could easily lead to results that are dramatically different from results that would be obtained by randomly selecting respondents from the entire population, not just those having local phone numbers.
9. a. 0.9332
   b. 0.9987
   c. 167.5 cm (Tech: 167.6 cm)
10. No. Because the states have different population sizes, the mean cannot be found by adding the 50 state means and dividing the total by 50. The mean income for the U.S. population can be found by using a weighted mean that incorporates the population size of each state.

Technology Project

STATDISK, Minitab, Excel, StatCrunch, the TI 83-84 Plus calculator, and many other statistical software packages are all capable of generating normally distributed data drawn from a population with a specified mean and standard deviation. In Example 3 of Section 6-2, we noted that bone density test scores are measured as z scores having a normal distribution with a mean of 0 and a standard deviation of 1. Generate two sets of sample data that represent simulated bone density scores, as shown below.

Bone Density Scores of Treatment Group: Generate 10 sample values from a normally distributed population with mean 0 and standard deviation 1.

Bone Density Scores of Placebo Group: Generate 15 sample values from a normally distributed population with mean 0 and standard deviation 1.

STATDISK: Select **Data,** then select **Normal Generator.**

Minitab: Select **Calc, Random Data, Normal.**

Excel: If using Excel 2013, 2010, or 2007, select **Data;** if using Excel 2003, select **Tools.** Select **Data Analysis, Random Number Generator,** and be sure to select **Normal** for the distribution.

TI-83/84 Plus: Press MATH, select **PRB,** then select **randNorm** and enter the mean, standard deviation, and the number of data values in the format of $(\bar{x}, s, n)$.

StatCrunch: Click on **Data,** then click on **Simulate Data,** then select the menu item of **Normal.** In the dialog box, enter 10 for the number of rows, and enter 1 for the number of columns. Click on **Simulate.** Repeat this process using 15 for the number of rows.

Because each of the two samples consists of random selections from a normally distributed population with a mean of 0 and a standard deviation of 1, the data are generated so that both data sets really come from the same population, so there should be no difference between the two sample means.

a. After generating the two data sets, use a 0.10 significance level to test the claim that the two samples come from populations with the same mean.

b. If this experiment is repeated many times, what is the expected percentage of trials leading to the conclusion that the two population means are different? How does this relate to a type I error?

c. If your generated data should lead to the conclusion that the two population means are different, would this conclusion be correct or incorrect in reality? How do you know?

d. If part (a) is repeated 20 times, what is the probability that none of the hypothesis tests leads to rejection of the null hypothesis?

e. Repeat part (a) 20 times. How often was the null hypothesis of equal means rejected? Is this the result you expected?

from data TO DECISION

Critical Thinking: Ages of workers killed in the Triangle Factory fire

Listed below are the ages (years) of the 146 employees who perished in the Triangle Factory fire that occurred on March 25, 1911 in Manhattan (based on data from the Kheel Center and the *New York Times*). One factor contributing to the large number of deaths is that almost all exits were locked so that employees could be checked for theft when they finished work at the end of the day. That fire revealed grossly poor and unsafe working conditions that led to changes in building codes and labor laws.

Analyzing the Results

1. First *explore* the combined male and female ages using suitable statistics and graphs. What is the mean age? What are the minimum and maximum ages? What is the standard deviation of the ages? Are there any outliers? Describe the distribution of the ages.

2. Examination of the two lists shows that relatively few men perished in the fire. Treat the ages as sample data and determine whether there is sufficient evidence to support the claim that among the workers who perish in such circumstances, the majority are women.

3. Construct a 95% confidence interval estimate of the mean age of males and construct another 95% confidence interval estimate of the mean age of females. Compare the results.

4. Treat the ages as sample data and determine whether there is sufficient evidence to support the claim that female workers have a mean age that is less than that of male workers.

5. Treat the ages as sample data and determine whether there is sufficient evidence to support the claim that ages of males and females have different standard deviations.

6. Based on the preceding results, identify any particularly notable features of the data.

Males

38	19	30	24	23	23	19	18	19	33	17	22	33	25	20	23	22

Females

24	16	25	31	22	18	19	22	16	23	17	15	21	18	17	17	17	31	20	36
18	25	30	16	25	25	21	19	17	18	20	18	26	26	16	18	18	17	22	17
20	22	18	20	16	25	18	40	21	18	19	19	18	18	19	16	19	16	16	21
33	21	14	22	19	19	23	19	18	21	39	20	14	27	22	15	19	16	16	19
18	21	18	19	19	20	18	43	16	20	18	30	21	22	18	21	35	22	21	22
21	22	17	24	25	20	18	32	20	21	19	24	17	18	30	18	16	22	22	17
22	20	15	20	17	21	21	18	17											

Cooperative Group Activities

1. Out-of-class activity Survey married couples and record the number of credit cards each person has. Analyze the paired data to determine whether husbands have more credit cards, wives have more credit cards, or they both have about the same number of credit cards. Try to identify reasons for any discrepancy.

2. Out-of-class activity Measure and record the height of the husband and the height of the wife from each of several different married couples. Estimate the mean of the differences between heights of husbands and the heights of their wives. Compare the result to the difference between the mean height of men and the mean height of women included in Data Set 1 in Appendix B. Do the results suggest that height is a factor when people select marriage partners?

3. Out-of-class activity Are estimates influenced by anchoring numbers? Refer to the related Chapter 3 Cooperative Group Activity. In Chapter 3 we noted that, according to author John Rubin, when people must estimate a value, their estimate is often "anchored" to (or influenced by) a preceding number. In that Chapter 3 activity, some subjects were asked to quickly estimate the value of $8 \times 7 \times 6 \times 5 \times 4 \times 3 \times 2 \times 1$, and others were asked to quickly estimate the value of $1 \times 2 \times 3 \times 4 \times 5 \times 6 \times 7 \times 8$. In Chapter 3, we could compare the two sets of results by using statistics (such as the mean) and graphs (such as boxplots). The methods of Chapter 9 now allow us to compare the results with a formal hypothesis test. Specifically, collect your own sample data and test the claim that when we begin with larger numbers (as in $8 \times 7 \times 6$), our estimates tend to be larger.

4. In-class activity Divide into groups according to gender, with about 10 or 12 students in each group. Each group member should record his or her pulse rate by counting the number of heartbeats in 1 minute, and calculate the group statistics ($n, \bar{x}, s$). The groups should test the null hypothesis of no difference between their mean pulse rate and the mean of the pulse rates for the population from which subjects of the same gender were selected for Data Set 1 in Appendix B.

5. Out-of-class activity Randomly select a sample of male students and a sample of female students and ask each selected person a yes/no question, such as whether they support a death penalty for people convicted of murder, or whether they believe that the federal government should fund stem cell research. Record the response, the gender of the respondent, and the gender of the person asking the question. Use a formal hypothesis test to determine whether there is a difference between the proportions of *yes* responses from males and females. Also, determine whether the responses appear to be influenced by the gender of the interviewer.

6. Out-of-class activity Use a watch to record the waiting times of a sample of McDonald's customers and the waiting times of a sample of Burger King customers. Use a hypothesis test to determine whether there is a significant difference.

7. Out-of-class activity Construct a short survey of just a few questions, including a question asking the subject to report his or her height. After the subject has completed the survey, measure the subject's height (without shoes) using an accurate measuring system. Record the gender, reported height, and measured height of each subject. Do male subjects appear to exaggerate their heights? Do female subjects appear to exaggerate their heights? Do the errors for males appear to have the same mean as the errors for females?

8. In-class activity Without using any measuring device, ask each student to draw a line believed to be 3 in. long and another line believed to be 3 cm long. Then use rulers to measure and record the lengths of the lines drawn. Record the errors along with the genders of the students making the estimates. Test the claim that when estimating the length of a 3-in. line, the mean error from males is equal to the mean error from females. Also, do the results show that we have a better understanding of the British system of measurement (inches) than the SI system (centimeters)?

9. Out-of-class activity Obtain simple random samples of cars in the student and faculty parking lots, and test the claim that students and faculty have the same proportions of foreign cars.

10. Out-of-class activity Obtain simple random samples of cars in parking lots of a discount store and an upscale department store, and test the claim that cars are newer in the parking lot of the upscale department store.

11. Out-of-class activity Obtain sample data and test the claim that husbands are older than their wives.

12. Out-of-class activity Obtain sample data to test the claim that in the college library, science books have a mean age that is less than the mean age of novels.

13. Out-of-class activity Conduct experiments and collect data to test the claim that there are no differences in taste between ordinary tap water and different brands of bottled water.

14. Out-of-class activity Collect sample data and test the claim that people who exercise tend to have pulse rates that are lower than those who do not exercise.

15. Out-of-class activity Collect sample data and test the claim that the proportion of female students who smoke is equal to the proportion of male students who smoke.

16. Out-of-class activity Collect sample data to test the claim that women carry more pocket change than men.

10 Correlation and Regression

10

chapter problem

CSI Statistics: Can we use footprint evidence to estimate a suspect's height?

Police sometimes use footprint evidence to estimate the height of a suspect, and the height is included in a description that becomes part of a BOLO ("be on the lookout"). (Aren't acronyms fun? Wait until you get to ANOVA.) Around 1877, anthropologist Paul Topinard collected foot/height measurements and used them to develop this rule: Estimate a person's height by dividing their foot length by 0.15. (An equivalent calculation is to estimate height by multiplying foot length by 6.67.) Try it yourself by measuring the length of your foot, then divide by 0.15 (or multiply by 6.67) to get your estimated height. Is the result reasonably accurate?

Table 10-1 includes some measurements taken from Data Set 2 in Appendix B. Table 10-1 includes shoe length and height measurements from five males, but Data Set 2 includes more measurements from larger samples of males and females. (The data are from "Estimation of Stature from Foot and Shoe Length: Applications in Forensic Science," by Brenda Rohren M.A., MFS, LIMHP, LADC, MAC. Brenda Rohren was a graduate student at Nebraska Wesleyan University when she conducted the research and wrote the report in Fall 2006. The data are used with her permission.)

Using shoe print measurements, such as those in Table 10-1, can we confirm that there is a relationship between lengths of our shoe prints and our heights? If we do conclude that there is a relationship, how do we use it to develop a formula for estimating height based on the length of a shoe print? Such questions are critically important for many applications, and this chapter provides the tools for answering them.

Table 10-1 Shoe Print Lengths and Heights of Males

Shoe Print (cm)	29.7	29.7	31.4	31.8	27.6
Height (cm)	175.3	177.8	185.4	175.3	172.7

Note to Instructor
Recommendation: Cover at least Sections 10-2 (correlation) and 10-3 (regression). Manual calculations would be awful, so make effective use of technology for finding the linear correlation coefficient *r* and the slope b_1 and *y*-intercept b_0 for the regression line.

Some instructors prefer to cover the basics of correlation/regression early in their course. Sections 10-2 and 10-3 are organized so that they can follow Chapter 3. In Section 10-2, simply exclude Part 2 (Formal Hypothesis Test).

10-1 Review and Preview

A major focus of this chapter is to analyze *paired* sample data. In Section 9-4 we considered two dependent samples, with each value of one sample somehow paired with a value from the other sample. The goal in Section 9-4 was to describe methods for testing hypotheses and constructing confidence intervals for the *mean of the differences* from the matched pairs. In this chapter we again consider paired sample data, but the objective is fundamentally different from that of Section 9-4. In this chapter we introduce methods for determining whether a *correlation,* or association, between two variables exists and whether the correlation is linear. For linear correlations, we can identify an equation of a straight line that best fits the data, and we can use that equation to predict the value of one variable given the value of the other variable.

Note to Instructor
Consider beginning this chapter by making the important point that the *context* of data is extremely important. The context can dramatically affect the methods we use. Present an unidentified table of values, such as this one:

x	78	85	92	100	85
y	89	93	99	100	84

(1) What is the key issue if these data are test grades of subjects before and after formal instruction? ("Is the instruction effective, as indicated by higher *y* scores?") This issue would be addressed by using the methods for dependent samples in Section 9-4.) (2) What is the key issue if these data are reasoning tests for a sample of men (*x*) and a separate sample of women (*y*)? ("Does the population of men have the same mean as the population of women?" This issue would be addressed by using the methods for two independent samples in Section 9-3.) (3) What is the key issue if each pair represents a math reasoning score *x* and a starting salary *y* (in thousands of dollars) for the same person? ("Is there an association between starting salary and math reasoning?" This issue would be addressed by using the methods for correlation presented in Section 10-2.)

TOP 20

10-2 Correlation

Key Concept In Part 1 of this section we introduce the *linear correlation coefficient r*, which is a number that measures how well paired sample data fit a straight-line pattern when graphed. We use the sample of paired data (sometimes called **bivariate data**) to find the value of *r* (usually using technology), then we use that value to decide whether there is (or is not) a linear correlation between the two variables. In this section we consider only *linear* relationships, which means that when graphed in a scatterplot, the points approximate a straight-line pattern. In Part 2, we discuss methods for conducting a formal hypothesis test that can be used to decide whether there is a linear correlation between two variables.

Part 1: Basic Concepts of Correlation

We begin with the basic definition of *correlation,* a term commonly used in the context of an association between two variables.

DEFINITIONS

A **correlation** exists between two variables when the values of one variable are somehow associated with the values of the other variable.

A **linear correlation** exists between two variables when there is a correlation and the plotted points of paired data result in a pattern that can be approximated by a straight line.

Table 10-1, for example, includes paired sample data consisting of lengths of shoe prints and the corresponding heights of five different males. We will determine whether there is a linear correlation between the variable *x* (length of shoe print) and the variable *y* (height). Instead of blindly jumping into the calculation of the linear correlation coefficient *r,* it is wise to first *explore* the data.

Explore!

Because it is always wise to explore sample data before applying a formal statistical procedure, we should use a scatterplot to explore the paired data visually. Figure 10-1(a) shows the scatterplot of the shoe/height data from Table 10-1. The scatterplot suggests that there might be a pattern, but it isn't very strong. Also, there are no outliers, which are points far away from all the other points. Figure 10-1(b) shows the scatterplot of all 40 pairs of shoe print and height measurements from Data Set 2 in Appendix B. Because Figure 10-1(b) includes more sample data, it gives us a better picture of the relationship between the two variables. Figure 10-1(b) shows that there is a pattern of points tending to rise as they move farther to the right, but the pattern does not appear to be very distinct. Again, there are no outliers.

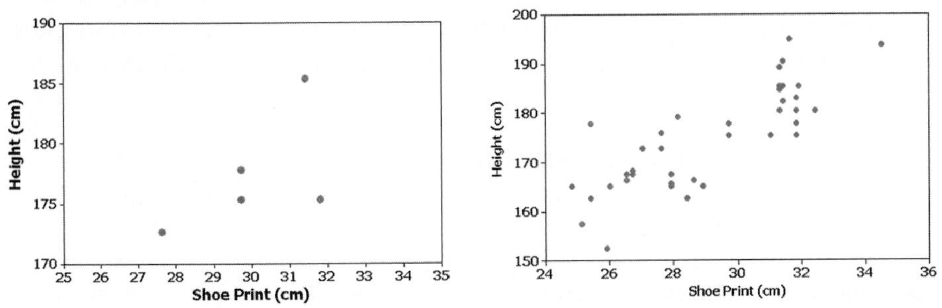

(a) (b)

Figure 10-1 **Minitab Scatterplots**

Interpreting Scatterplots

Figure 10-2 shows four scatterplots with different characteristics. The scatterplot in Figure 10-2(a) shows a distinct straight-line, or linear, pattern. We say that there is a *positive* linear correlation between x and y, since as the x-values increase, the

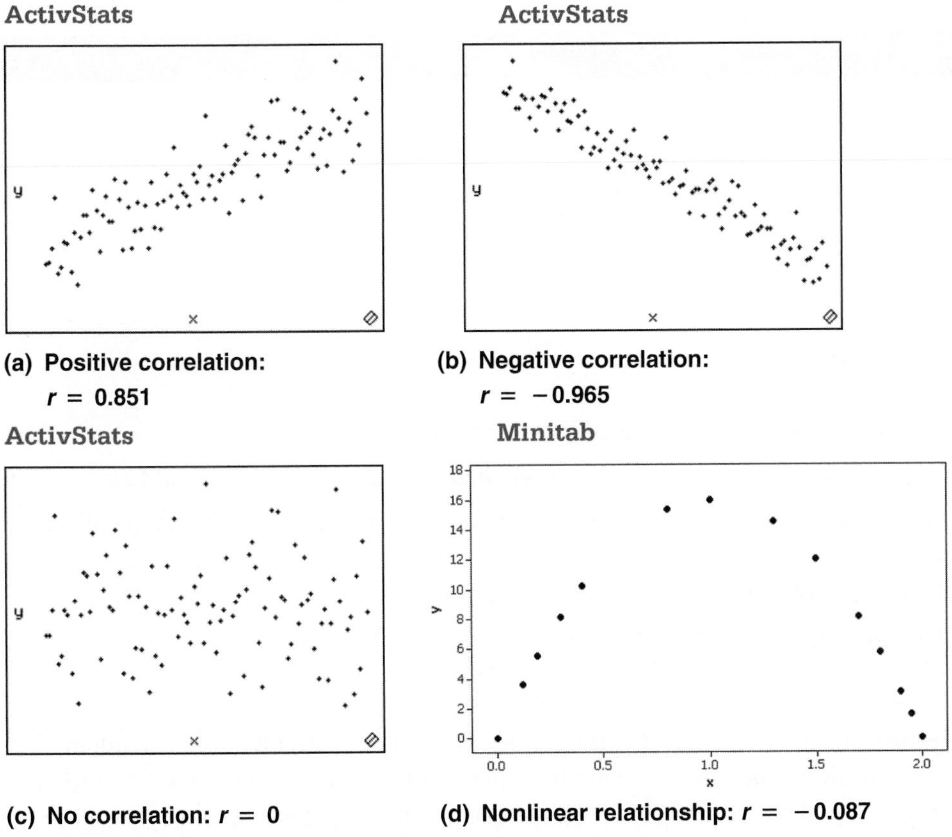

(a) Positive correlation: $r = 0.851$

(b) Negative correlation: $r = -0.965$

(c) No correlation: $r = 0$

(d) Nonlinear relationship: $r = -0.087$

Figure 10-2 **Scatterplots**

Speeding Out-of-Towners Ticketed More?

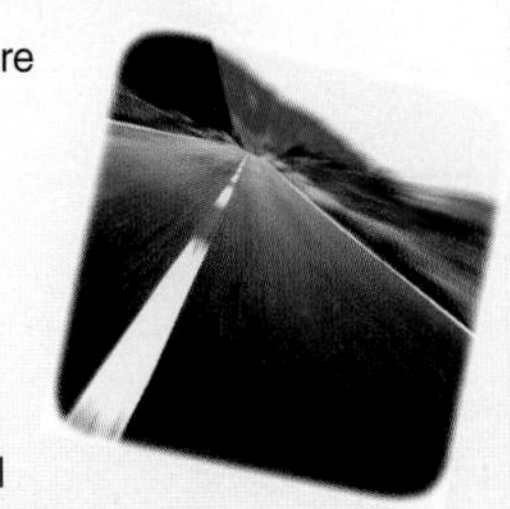

Are police more likely to issue a ticket to a speeding driver who is out-of-town than to a local driver? George Mason University researchers Michael Makowsky and Thomas Stratmann addressed this question by examining more than 60,000 warnings and tickets issued by Massachusetts police in one year. They found that out-of-town drivers from Massachusetts were 10% more likely to be ticketed than local drivers, and the 10% figure rose to 20% for out-of-state drivers. They also found a statistical association between a town's finances and speeding tickets. When compared to local drivers, out-of-town drivers had a 37% greater chance of being ticketed when speeding in a town in which voters had rejected a proposition to raise taxes more than the 2.5% amount allowed by state law. Such analyses can be conducted using methods of correlation and regression.

Note to Instructor

Scatterplots were first introduced in Section 2-4. They should be discussed here. Because of their subjective nature, students might feel that scatterplots are effectively worthless. Point out that although analysis of scatterplots is largely subjective, they are extremely helpful in detecting patterns that are not straight-line patterns. Exercises 9 and 10 involve nonlinear data, and those exercises are ideal for making the point that you can make a big mistake if you ignore scatterplots and proceed directly to the calculation of r.

corresponding y-values also increase. The scatterplot in Figure 10-2(b) shows a distinct linear pattern. We say that there is a *negative* linear correlation between x and y, since as the x-values increase, the corresponding y-values decrease. The scatterplot in Figure 10-2(c) shows no distinct pattern and suggests that there is no linear correlation between x and y. The scatterplot in Figure 10-2(d) shows a distinct pattern suggesting a correlation between x and y, but the pattern is not that of a straight line.

Measure the Strength of the Linear Correlation

Because conclusions based on visual examinations of scatterplots are largely subjective, we need more objective measures. We use the linear correlation coefficient r, which is a number that measures the strength of the (linear) association between the two variables.

DEFINITION The **linear correlation coefficient r** measures the strength of the linear correlation between the paired quantitative x- and y-values in a *sample.* (Its value is computed by using Formula 10-1 or Formula 10-2, included in the following box. [The linear correlation coefficient is sometimes referred to as the **Pearson product moment correlation coefficient** in honor of Karl Pearson (1857–1936), who originally developed it.]

Because the linear correlation coefficient r is calculated using sample data, it is a sample statistic used to measure the strength of the linear correlation between x and y. If we had every pair of x and y values from an entire population, the result of Formula 10-1 or Formula 10-2 would be a population parameter, represented by ρ (Greek letter rho).

Objective

Determine whether there is a linear correlation between two variables.

Notation for the Linear Correlation Coefficient

- n number of pairs of sample data.
- Σ denotes addition of the items indicated.
- Σx sum of all x-values.
- Σx^2 indicates that each x-value should be squared and then those squares added.
- $(\Sigma x)^2$ indicates that the x-values should be added and the total then squared. Avoid confusing Σx^2 and $(\Sigma x)^2$.
- Σxy indicates that each x-value should first be multiplied by its corresponding y-value. After obtaining all such products, find their sum.
- r linear correlation coefficient for *sample* data.
- ρ linear correlation coefficient for a *population* of paired data.

Requirements

Given any collection of sample paired quantitative data, the linear correlation coefficient r can always be computed, but the following requirements should be satisfied when using the sample data to make a conclusion about correlation in the population.

1. The sample of paired (x, y) data is a simple random sample of quantitative data. (It is important that the sample data have not been collected using some inappropriate method, such as using a voluntary response sample.)

2. Visual examination of the scatterplot must confirm that the points approximate a straight-line pattern.
3. Because results can be strongly affected by the presence of outliers, any outliers must be removed if they are known to be errors. The effects of any other outliers should be considered by calculating r with and without the outliers included.

Note: Requirements 2 and 3 above are simplified attempts at checking this formal requirement: The pairs of (x, y) data must have a **bivariate normal distribution.** Normal distributions are discussed in Chapter 6, but this assumption basically requires that for any fixed value of x, the corresponding values of y have a distribution that is approximately normal, and for any fixed value of y, the values of x have a distribution that is approximately normal. This requirement is usually difficult to check, so for now, we will use Requirements 2 and 3 as listed above.

Formulas for Calculating r

Formula 10-1

$$r = \frac{n(\Sigma xy) - (\Sigma x)(\Sigma y)}{\sqrt{n(\Sigma x^2) - (\Sigma x)^2}\sqrt{n(\Sigma y^2) - (\Sigma y)^2}}$$

Formula 10-1 simplifies manual calculations, but r is usually calculated with computer software or a calculator. Formula 10-2 is better for understanding the reasoning behind r; see the rationale discussed later in this section.

Formula 10-2

$$r = \frac{\Sigma(z_x z_y)}{n - 1}$$

where z_x denotes the z score for an individual sample value x and z_y is the z score for the corresponding sample value y.

Interpreting the Linear Correlation Coefficient r

- ***Using Computer Software to Interpret r:*** If the P-value computed from r is less than or equal to the significance level, conclude that there is sufficient evidence to support a claim of a linear correlation. Otherwise, there is not sufficient evidence to support a claim of a linear correlation.
- ***Using Table A-5 to Interpret r:*** Consider critical values from Table A-5 as being both positive and negative, and draw a graph similar to Figure 10-3 that accompanies Example 4.

Correlation If the computed linear correlation coefficient r lies in the left tail beyond the leftmost critical value or if it lies in the right tail beyond the rightmost critical value, conclude that there is sufficient evidence to support the claim of a linear correlation.

No Correlation If the computed linear correlation coefficient lies *between* the two critical values, conclude that there is not sufficient evidence to support the claim of a linear correlation.

(Here are equivalent criteria: If the absolute value of r, denoted $|r|$, exceeds the value in Table A-5, conclude that there is a linear correlation. Otherwise, there is not sufficient evidence to support the conclusion of a linear correlation.)

CAUTION Know that the methods of this section apply to a *linear* correlation. If you conclude that there does not appear to be linear correlation, it is possible that there might be some other association that is not linear, as in Figure 10-2(d).

Rounding the Linear Correlation Coefficient r

Round the linear correlation coefficient r to three decimal places (so that its value can be directly compared to critical values in Table A-5). If manually calculating r and other statistics in this chapter, rounding in the middle of a calculation often creates substantial errors, so try to round only the final result.

Note to Instructor

Formula 10-1 is the "shortcut" version of the formula for r and it is intended to simplify calculations. Formula 10-2 is better when trying to gain some understanding of what r represents. There are other versions of r given later in this section when the "Rationale" is described. Some instructors prefer introducing r with one of these other versions. In any case, it is important to go beyond the formula and focus on using technology to find r, then correctly interpret the result.

Properties of the Linear Correlation Coefficient *r*

1. The value of r is always between -1 and 1 inclusive. That is, $-1 \leq r \leq 1$.
2. *If all values of either variable are converted to a different scale, the value of r does not change.*
3. *The value of r is not affected by the choice of x or y.* Interchange all x- and y-values and the value of r will not change.
4. *r measures the strength of a linear relationship.* It is not designed to measure the strength of a relationship that is not linear (as in Figure 10-2(d)).
5. r is very sensitive to outliers in the sense that a single outlier can dramatically affect its value.

Palm Reading

Some people believe that the length of their palm's lifeline can be used to predict longevity. In a letter published in the *Journal of the American Medical Association,* authors M. E. Wilson and L. E. Mather refuted that belief with a study of cadavers. Ages at death were recorded, along with the lengths of palm lifelines. The authors concluded that there is no correlation between age at death and length of lifeline. Palmistry lost, hands down.

Calculating the Linear Correlation Coefficient *r*

The following three examples illustrate three different methods for finding the value of the linear correlation coefficient r, but you need to use only one method. *The use of computer software or a calculator (as in Example 1) is strongly recommended.* If manual calculations are absolutely necessary, Formula 10-1 is recommended (as in Example 2). If a better understanding of r is desired, Formula 10-2 is recommended (as in Example 3).

Note to Instructor
Example 1 shows how technology is used to find r. Example 2 illustrates the use of Formula 10-1, and Example 3 illustrates the use of Formula 10-2. Decide how you want your students to evaluate r, and be sure to let them know of your choice.

Recommendation: Announce that students can use their technology to find values of r, as in Example 1.

Example 1 Finding *r* Using Computer Software

The paired shoe/height data from five males are listed in Table 10-1. Use computer software or a calculator with these paired sample values to find the value of the linear correlation coefficient r for the paired sample data.

Solution

Requirement check We can always calculate the linear correlation coefficient r from paired quantitative data, but we should check the requirements if we want to use that value for making a conclusion about correlation. (1) The data are a simple random sample of quantitative data. (2) The plotted points in the scatterplot in Figure 10-1(a) appear to roughly approximate a straight-line pattern. (3) The scatterplot in Figure 10-1(a) also shows that there are no outliers. The requirements are satisfied. ✓

If using computer software or a calculator, the value of r will be automatically calculated. For example, see the following Minitab, TI-83/84 Plus calculator, and STATDISK displays showing that $r = 0.591$. Excel and many other computer software packages and calculators provide the same value of $r = 0.591$.

MINITAB

```
Correlations: Shoe Print, Height

Pearson correlation of Shoe Print and Height = 0.591
P-Value = 0.294
```

TI-83/84 PLUS

```
LinRegTTest
 y=a+bx
 β≠0 and ρ≠0
↑b=1.72745239
 s=4.538762333
 r²=.3495991312
 r=.5912690853
```

STATDISK

```
Correlation Results:
Correlation coeff, r: 0.5912691
Critical r:           ±0.8783393
P-value (two-tailed): 0.29369
```

Example 2 Finding r Using Formula 10-1

Use Formula 10-1 to find the value of the linear correlation coefficient r for the paired shoe print and height data given in Table 10-1.

Solution

Requirement check See the discussion of the requirement check in Example 1. The same comments apply here. ✓

Using Formula 10-1, the value of r is calculated as shown below. Here, the variable x is used for the shoe print lengths, and the variable y is used for the heights. Because there are five pairs of data, $n = 5$. Other required values are computed in Table 10-2.

Table 10-2 Calculating r with Formula 10-1

x (Shoe Print)	y (Height)	x^2	y^2	xy
29.7	175.3	882.09	30730.09	5206.41
29.7	177.8	882.09	31612.84	5280.66
31.4	185.4	985.96	34373.16	5821.56
31.8	175.3	1011.24	30730.09	5574.54
27.6	172.7	761.76	29825.29	4766.52
$\Sigma x = 150.2$	$\Sigma y = 886.5$	$\Sigma x^2 = 4523.14$	$\Sigma y^2 = 157271.47$	$\Sigma xy = 26649.69$

Using Formula 10-1 with the results from Table 10-2, r is calculated as follows:

$$r = \frac{n\Sigma xy - (\Sigma x)(\Sigma y)}{\sqrt{n(\Sigma x^2) - (\Sigma x)^2}\sqrt{n(\Sigma y^2) - (\Sigma y)^2}}$$

$$= \frac{5(26649.69) - (150.2)(886.5)}{\sqrt{5(4523.14) - (150.2)^2}\sqrt{5(157271.47) - (886.5)^2}}$$

$$= \frac{96.15}{\sqrt{55.66}\sqrt{475.10}} = 0.591$$

Example 3 Finding r Using Formula 10-2

Use Formula 10-2 to find the value of the linear correlation coefficient r for the paired shoe print and height data given in Table 10-1.

Solution

Requirement check See the discussion of the requirement check in Example 1. The same comments apply here. ✓

If manual calculations are absolutely necessary, Formula 10-1 is much easier than Formula 10-2, but Formula 10-2 has the advantage of making it easier to *understand* how r works. (See the rationale for r discussed later in this section.) As in Example 2, the variable x is used for the shoe print lengths, and the variable y is used for the heights. In Formula 10-2, each sample value is replaced by its corresponding

Teacher Evaluations Correlate with Grades

Student evaluations of faculty are often used to measure teaching effectiveness. Many studies reveal a correlation with higher student grades being associated with higher faculty evaluations. One study at Duke University involved student evaluations collected before and after final grades were assigned. The study showed that "grade expectations or received grades caused a change in the way students perceived their teacher and the quality of instruction." It was noted that with student evaluations, "the incentives for faculty to manipulate their grading policies in order to enhance their evaluations increase." It was concluded that "the ultimate consequence of such manipulations is the degradation of the quality of education in the United States." (See "Teacher Course Evaluations and Student Grades: An Academic Tango," by Valen Johnson, *Chance*, Vol. 15, No. 3.)

z score. For example, the shoe print lengths have a mean of $\bar{x} = 30.04$ cm and a standard deviation of $s_x = 1.66823$ cm, so the first shoe print length of 29.7 cm is converted to a z score of -0.20381 as shown here:

$$z_x = \frac{x - \bar{x}}{s_x} = \frac{29.7 - 30.04}{1.66823} = -0.20381$$

Table 10-3 lists the z scores for all of the shoe print lengths (see the third column) and the z scores for all of the heights (see the fourth column). The last column of Table 10-3 lists the products $z_x \cdot z_y$.

Table 10-3 Calculating r with Formula 10-2

x (Shoe Print)	y (Height)	z_x	z_y	$z_x \cdot z_y$
29.7	175.3	−0.20381	−0.41035	0.08363
29.7	177.8	−0.20381	0.10259	−0.02091
31.4	185.4	0.81524	1.66191	1.35485
31.8	175.3	1.05501	−0.41035	−0.43292
27.6	172.7	−1.46263	−0.94380	1.38043
				$\Sigma(z_x \cdot z_y) = 2.36508$

Using $\Sigma(z_x \cdot z_y) = 2.36508$ from Table 10-3, the value of r is calculated by using Formula 10-2 as shown below.

$$r = \frac{\Sigma(z_x \cdot z_y)}{n - 1} = \frac{2.36508}{4} = 0.591$$

Is There a Linear Correlation?

We know from the preceding three examples that the value of the linear correlation coefficient is $r = 0.591$ for the sample data in Table 10-1. We now proceed to interpret its meaning, and our goal is to decide whether there appears to be a linear correlation between shoe print lengths and heights of people. Using the criteria given in the preceding box, we can base our interpretation on a P-value or a critical value from Table A-5. See the criteria for "Interpreting the Linear Correlation Coefficient r" given in the preceding box.

Example 4 Is There a Linear Correlation?

In Examples 1, 2, and 3, we used the sample data from Table 10-1 to find that $r = 0.591$. If we use a significance level of 0.05, is there sufficient evidence to support a claim that there is a linear correlation between shoe print lengths and heights of people?

Solution

Requirement check The requirement check in Example 1 also applies here. ✔

We can base our conclusion about correlation on either the P-value obtained from computer software or the critical value found in Table A-5. (See the criteria for "Interpreting the Linear Correlation Coefficient r"given in the preceding box.)

- ***Using Computer Software to Interpret r:*** If the computed P-value is less than or equal to the significance level, conclude that there is a linear correlation. Otherwise, there is not sufficient evidence to support the conclusion of a linear correlation. Example 1 includes Minitab, TI-83/84 Plus calculator, and STATDISK displays. Those technologies provide the P-value of 0.294. Because that P-value is *not* less than or equal to the significance level of 0.05, we conclude that *there is not sufficient evidence to support the conclusion that there is a linear correlation between shoe print lengths and heights of people.*
- ***Using Table A-5 to Interpret r:*** Consider critical values from Table A-5 as being both positive and negative, and draw a graph like Figure 10-3. For the sample data in Table 10-1, Table A-5 yields $r = 0.878$ (for five pairs of data and a 0.05 significance level). We can now compare the computed value of $r = 0.591$ to the critical values of $r = \pm 0.878$ as shown in Figure 10-3.

  Correlation If the computed linear correlation coefficient r lies in the left or right tail region beyond the critical value for that tail, conclude that there is sufficient evidence to support the claim of a linear correlation.

  No Correlation If the computed linear correlation coefficient lies between the two critical values, conclude that there is not sufficient evidence to support the claim of a linear correlation.

Because Figure 10-3 shows that the computed value of $r = 0.591$ lies between the two critical values, we conclude that *there is not sufficient evidence to support the claim of a linear correlation between shoe print lengths and heights.*

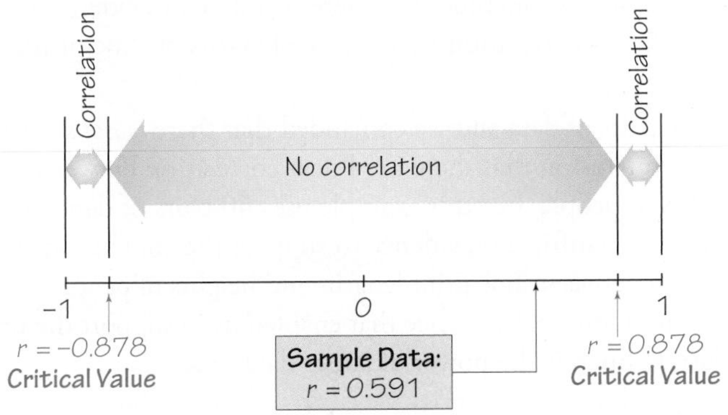

Figure 10-3 Critical Values from Table A-5 and the Computed Value of *r*

Example 5 Larger Data Set

Example 4 used only the shoe lengths and heights from five males. Let's now use the shoe lengths and heights from the 40 subjects listed in Data Set 2 in Appendix B. The scatterplot is shown in Figure 10-1(b). Refer to the following results obtained by using XLSTAT with Excel. See that the value of the linear correlation coefficient is $r = 0.813$ and the P-value is less than 0.0001. Is there sufficient evidence to support a claim of a linear correlation between the lengths of shoe prints and heights?

XLSTAT

Correlation matrix (Pearson):		
Variables	Shoe Print	Height
Shoe Print	1	**0.8129**
Height	**0.8129**	1
Values in bold are different from 0 with a significance level alpha=0.05		

p-values:		
Variables	Shoe Print	Height
Shoe Print	0	**< 0.0001**
Height	**< 0.0001**	0
Values in bold are different from 0 with a significance level alpha=0.05		

Solution

Requirement check (1) The sample is a simple random sample of quantitative data. (2) The points in the scatterplot of Figure 10-1(b) roughly approximate a straight-line pattern. (3) There are no outliers that are far away from almost all of the other pairs of data. ✔

Using Software: The *P*-value obtained from XLSTAT is less than 0.0001. Because the *P*-value is less than or equal to 0.05, we conclude that there is sufficient evidence to support a claim of a linear correlation between the lengths of shoe prints and heights.

Using Table A-5: If we refer to Table A-5 with $n = 40$ pairs of sample data, we obtain the critical values of -0.312 and 0.312 for $\alpha = 0.05$. Because the computed value of $r = 0.813$ *does* exceed the critical value of 0.312 from Table A-5, the computed value of r lies in the right tail beyond the rightmost critical value of 0.312, so we conclude that there is sufficient evidence to support a claim of a linear correlation between the lengths of shoe prints and heights.

Example 4 used only five pairs of data and we concluded that there is *not* sufficient evidence to support the conclusion that there is a linear correlation between shoe print lengths and heights of people, but this example uses 40 pairs of data and here we conclude that there *is* sufficient evidence to support the conclusion that there is a linear correlation between shoe print lengths and heights of people. This larger data set provided the additional evidence that enabled us to support the presence of a linear correlation. Such is the power of larger data sets.

Interpreting *r*: Explained Variation

If we conclude that there is a linear correlation between x and y, we can find a linear equation that expresses y in terms of x, and that equation can be used to predict values of y for given values of x. In Section 10-3 we will describe a procedure for finding such equations and show how to predict values of y when given values of x. But a predicted value of y will not necessarily be the exact result that occurs because in addition to x, there are other factors affecting y, such as random variation and other characteristics not included in the study. The following principle describes the relationship between the linear correlation coefficient r and the explained variation in y.

The value of r^2 is the proportion of the variation in y that is explained by the linear relationship between x and y.

Example 6 Explained Variation

In Example 5 we noted that for the 40 pairs of shoe print lengths and heights listed in Data Set 2 from Appendix B, the linear correlation coefficient is $r = 0.813$. What proportion of the variation in height can be explained by the variation in the length of shoe print?

Solution

With $r = 0.813$, we get $r^2 = 0.661$.

Interpretation

We conclude that 0.661 (or about 66%) of the variation in height can be explained by the linear relationship between lengths of shoe prints and heights. This implies that about 34% of the variation in heights cannot be explained by lengths of shoe prints.

Common Errors Involving Correlation

Here are three of the most common errors made in interpreting results involving correlation:

1. *A common error is to assume that correlation implies causality.* One classic example involves paired data consisting of the stork population in Copenhagen and the number of human births. For several years, the data suggested a linear correlation. *Bulletin:* Storks do not actually cause births, and births do not cause storks. Both variables were affected by some other variable lurking in the background. (A **lurking variable** is one that affects the variables being studied but is not included in the study.)
2. *Another error arises with data based on averages.* Averages suppress individual variation and may inflate the correlation coefficient. One study produced a 0.4 linear correlation coefficient for paired data relating income and education among individuals, but the linear correlation coefficient became 0.7 when regional averages were used.
3. *A third error involves the property of linearity.* If there is no linear correlation, there might be some other correlation that is not linear, as in Figure 10-2(d). (Figure 10-2(d) is a scatterplot that depicts the relationship between distance above ground and time elapsed for an object thrown upward.)

Note to Instructor
The first common error should be strongly emphasized in class: "Correlation does not imply causality." Here's a classic example: There is a correlation between per capita beer consumption and teachers' salaries, but (most) teachers don't use salary increases to buy more beer.

CAUTION Know that *correlation does not imply causality.*

Part 2: Formal Hypothesis Test (Requires Coverage of Chapter 8)

Here in Part 2 we describe formal procedures for conducting hypothesis tests to determine whether there is a significant linear correlation between two variables. The following box contains key elements of the hypothesis test.

Note to Instructor
As the title of this subsection indicates, this material requires prior coverage of Chapter 8. If you are including Sections 10-2 and 10-3 early (such as following Chapter 3), skip this subsection. If you do include this subsection, be sure to inform the class of the method that you prefer: the critical value method based on the test statistic r and critical values found from Table A-5, or the *P*-value method based on a t test statistic.

Hypothesis Test for Correlation (Using Test Statistic r)

Notation

n = number of pairs of sample data

r = linear correlation coefficient for a *sample* of paired data

ρ = linear correlation coefficient for a *population* of paired data

Requirements

The requirements are the same as those given in the preceding box from Part 1 of this section.

Hypotheses

$H_0: \rho = 0$ (There is no linear correlation.)
$H_1: \rho \neq 0$ (There is a linear correlation.)

Test Statistic: r

Critical values: Refer to Table A-5.

Conclusion

Consider critical values from Table A-5 as being both positive and negative, and draw a graph similar to Figure 10-3.

- **Correlation** If the computed linear correlation coefficient r lies in the left tail beyond the leftmost critical value or if it lies in the right tail beyond the rightmost critical value, reject H_0 and conclude that there is sufficient evidence to support the claim of a linear correlation.
- **No Correlation** If the computed linear correlation coefficient lies *between* the two critical values, fail to reject H_0 and conclude that there is not sufficient evidence to support the claim of a linear correlation.

The following criteria are equivalent to those given above.

- **Correlation** If $|r| >$ critical value from Table A-5, reject H_0 and conclude that there is sufficient evidence to support the claim of a linear correlation.
- **No Correlation** If $|r| \leq$ critical value, fail to reject H_0 and conclude that there is not sufficient evidence to support the claim of a linear correlation.

Example 7 Hypothesis Test Based on r

Use the paired shoe print lengths and heights in Table 10-1 to conduct a formal hypothesis test of the claim that there is a linear correlation between the two variables. Use a 0.05 significance level.

Solution

Requirement check The solution in Example 1 already includes verification that the requirements are satisfied. ✓

To claim that there is a linear correlation is to claim that the population linear correlation coefficient ρ is different from 0. We therefore have the following hypotheses:

$H_0: \rho = 0$ (There is no linear correlation.)
$H_1: \rho \neq 0$ (There is a linear correlation.)

The test statistic is $r = 0.591$ (from Examples 1, 2, and 3). The critical values of $r = \pm 0.878$ are found in Table A-5 with $n = 5$ and $\alpha = 0.05$. See Example 4 and Figure 10-3. Because the test statistic $r = 0.591$ is between the critical values of -0.878 and 0.878, we fail to reject H_0: $\rho = 0$.

Interpretation

We conclude that there is not sufficient evidence to support the claim of a linear correlation between shoe print lengths and heights.

P-Value Method for a Hypothesis Test for Linear Correlation

The preceding method of hypothesis testing using the test statistic r involves relatively simple calculations. Software packages and the TI-83/84 Plus calculator typically use a *P*-value method based on a t test. The key components of the t test are as follows.

Hypothesis Test for Correlation (Using *P*-Value from a *t* Test)

Hypotheses

H_0: $\rho = 0$ (There is no linear correlation.)

H_1: $\rho \neq 0$ (There is a linear correlation.)

Test Statistic

$$t = \frac{r}{\sqrt{\frac{1 - r^2}{n - 2}}}$$

***P*-value:** Use software or a TI-83/84 Plus calculator or use Table A-3 with $n - 2$ degrees of freedom to find the *P*-value corresponding to the test statistic t.

Conclusion

- **Correlation** If the *P*-value is less than or equal to the significance level, reject H_0 and conclude that there is sufficient evidence to support the claim of a linear correlation.
- **No Correlation** If the *P*-value is greater than the significance level, fail to reject H_0 and conclude that there is not sufficient evidence to support the claim of a linear correlation.

Example 8 Hypothesis Test Based on *P*-Value from *t* Test

Use the paired shoe print lengths and heights in Table 10-1 to conduct a formal hypothesis test of the claim that there is a linear correlation between the two variables. Base the conclusion on a *P*-value and use a 0.05 significance level.

Solution

Requirement check The solution in Example 1 already includes verification that the requirements are satisfied. ✓

To claim that there is a linear correlation is to claim that the population linear correlation coefficient ρ is different from 0. We therefore have the following hypotheses:

H_0: $\rho = 0$ (There is no linear correlation.)

H_1: $\rho \neq 0$ (There is a linear correlation.)

The linear correlation coefficient is $r = 0.591$ (from Examples 1, 2, and 3) and $n = 5$ (because there are five pairs of sample data), so the test statistic is

$$t = \frac{r}{\sqrt{\frac{1 - r^2}{n - 2}}} = \frac{0.591}{\sqrt{\frac{1 - 0.591^2}{5 - 2}}} = 1.269$$

Computer software packages use more precision to obtain the more accurate test statistic of $t = 1.270$. With $n - 2 = 3$ degrees of freedom, Table A-3 shows that the test statistic of $t = 1.269$ yields a *P*-value that is greater than 0.20. Software packages and the TI-83/84 Plus calculator show that the *P*-value is 0.2937. Because the *P*-value of 0.2937 is greater than the significance level of 0.05, we fail to reject H_0. ("If the *P* is low, the null must go." The *P*-value of 0.2937 is not low.)

Interpretation

We conclude that there is not sufficient evidence to support the claim of a linear correlation between shoe print lengths and heights.

One-Tailed Tests Examples 7 and 8 illustrate a two-tailed hypothesis test. The examples and exercises in this section will generally involve only two-tailed tests, but one-tailed tests can occur with a claim of a positive linear correlation or a claim of a negative linear correlation. In such cases, the hypotheses will be as shown here.

Claim of Negative Correlation (Left-tailed test)	Claim of Positive Correlation (Right-tailed test)
H_0: $\rho = 0$	H_0: $\rho = 0$
H_1: $\rho < 0$	H_1: $\rho > 0$

For these one-tailed tests, the *P*-value method can be used as in earlier chapters.

Rationale for Methods of This Section We have presented Formulas 10-1 and 10-2 for calculating *r* and have illustrated their use. Those formulas are given below along with some other formulas that are "equivalent," in the sense that they all produce the same values.

Formula 10-1

$$r = \frac{n\Sigma xy - (\Sigma x)(\Sigma y)}{\sqrt{n(\Sigma x^2) - (\Sigma x)^2}\sqrt{n(\Sigma y^2) - (\Sigma y)^2}}$$

Formula 10-2

$$r = \frac{\Sigma(z_x z_y)}{n - 1}$$

$$r = \frac{\Sigma(x - \bar{x})(y - \bar{y})}{(n - 1)s_x s_y} \qquad r = \frac{\Sigma\left[\frac{(x - \bar{x})}{s_x}\frac{(y - \bar{y})}{s_y}\right]}{n - 1} \qquad r = \frac{s_{xy}}{\sqrt{s_{xx}}\sqrt{s_{yy}}}$$

We will use Formula 10-2 to help us understand the reasoning that underlies the development of the linear correlation coefficient. Because Formula 10-2 uses z scores, the value of $\Sigma(z_x z_y)$ does not depend on the scale that is used. Figure 10-1(a) shows the scatterplot of the shoe print and height data from Table 10-1, and Figure 10-4 shows the scatterplot of the z scores from the same sample data. Compare Figure 10-1(a) to Figure 10-4 and see that they are essentially the same scatterplots with different scales. The red lines in Figure 10-4 form the same coordinate axes that we have all come to know and love from earlier mathematics courses. The red lines partition Figure 10-4 into four quadrants.

If the points of the scatterplot approximate an uphill line (as in the figure), individual values of the product $z_x \cdot z_y$ tend to be positive (because most of the points are found in the first and third quadrants, where the values of z_x and z_y are either both positive or both negative), so $\Sigma(z_x z_y)$ tends to be positive. If the points of the scatterplot approximate a downhill line, most of the points are in the second and fourth quadrants, where z_x and z_y are opposite in sign, so $\Sigma(z_x z_y)$ tends to be negative. Points that follow no linear pattern tend to be scattered among the four quadrants, so the value of $\Sigma(z_x z_y)$ tends to be close to 0.

We can therefore use $\Sigma(z_x z_y)$ as a measure of how the points are configured among the four quadrants. A large positive sum suggests that the points are predominantly in the first and third quadrants (corresponding to a positive linear correlation), a large negative sum suggests that the points are predominantly in the second and fourth quadrants (corresponding to a negative linear correlation), and a sum near 0 suggests that the points are scattered among the four quadrants (with no linear correlation). We divide $\Sigma(z_x z_y)$ by $n - 1$ to get a type of average instead of a statistic that becomes larger simply because there are more data values. (The reasons for dividing by $n - 1$ instead of n are essentially the same reasons that relate to the standard deviation.) The end result is Formula 10-2, which can be algebraically manipulated into any of the other expressions for r.

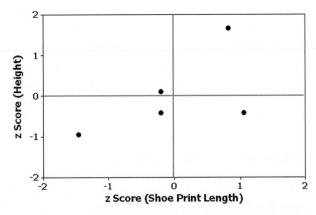

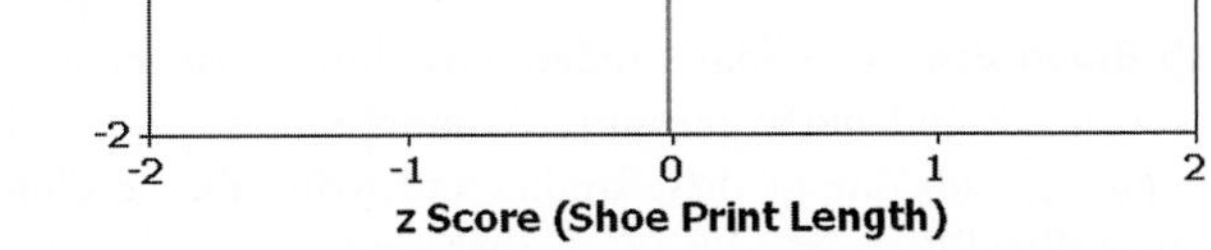

Figure 10-4 **Scatterplot of z Scores from Shoe Print Lengths and Heights in Table 10-1**

using TECHNOLOGY

STATDISK Enter the paired data in columns of the Statdisk Data Window. Select **Analysis** from the main menu bar, then use the option **Correlation and Regression.** Enter a value for the significance level. Select the columns of data to be used, then click on the **Evaluate** button. The STATDISK display will include the value of the linear correlation coefficient along with the critical value of r, the P-value, and other results to be discussed in later sections. A scatterplot can also be obtained by clicking on the **Scatterplot** button. See part of a STATDISK display in Example 1.

continued

MINITAB Enter the paired data in columns C1 and C2, then select **Stat** from the main menu bar, choose **Basic Statistics,** followed by **Correlation,** and enter C1 and C2 for the columns to be used. Minitab will provide the value of the linear correlation coefficient *r* as well as a *P*-value. To obtain a scatterplot, select **Graph, Scatterplot,** then enter C1 and C2 for *X* and *Y*, and click **OK.**

With Minitab 16, you can also click on **Assistant,** then **Regression.** Click on the box with the message **Click to perform analysis.** Fill out the dialog box and select **Linear** for the methods of this section. Click **OK** to get four windows of results that include the value of the linear correlation coefficient *r*, the *P*-value, a scatterplot, and much more helpful information.

EXCEL Excel has a function that calculates the value of the linear correlation coefficient. First enter the paired sample data in columns A and B. Click on the ***fx*** function key located on the main menu bar. Select the function category **Statistical** and the function name **CORREL,** then click **OK.** In the dialog box, enter the cell range of values for *x*, such as A1:A6. Also enter the cell range of values for *y* such as B1:B6. To obtain a scatterplot, click on the **Insert** tab, then use the chart type identified as **Scatter.** The style of the scatterplot can be edited by right-clicking on the desired feature.

XLSTAT can also be used. The instructions are included near the end of the following section.

TI-83/84 PLUS Enter the paired data in lists L1 and L2, then press STAT and select **TESTS.** Using the option of **LinRegT-Test** will result in several displayed values, including the value of the linear correlation coefficient *r* and a *P*-value.

To obtain a scatterplot, press 2ND, then Y= (for STAT PLOT). Press ENTER ENTER to turn Plot 1 on, then select the first graph type, which resembles a scatterplot. Set the *X* list and *Y* list labels to L1 and L2 and press ZOOM, then select **ZoomStat** and press ENTER.

STATCRUNCH Click on **Open StatCrunch.** Enter the columns of data or open a data set. Click on **Stat,** then select **Regression,** then select **Simple Linear.** Enter the columns to be used, then click on **Calculate.**

Recommended Assignment
Exercises 1–18.

10-2 Basic Skills and Concepts

Statistical Literacy and Critical Thinking

1. Notation For each of several randomly selected years, the sunspsot number and the high value of the Dow Jones Industrial Average are recorded. The sunspot number is a measure of sunspot activity on the sun, and the Dow Jones Industrial Average is one measure of stock market value. For this sample of paired data, what does *r* represent? What does ρ represent? Without doing any research or calculations, estimate the value of *r*.

2. Physics Experiment A physics experiment consists of recording paired data consisting of the time (seconds) elapsed since the beginning of the experiment and the distance (cm) of a robot from its point of origin. Using the paired time/distance data, the value of *r* is calculated to be 0. Is it correct to conclude that there is no relationship between time and distance? Explain.

3. Cause of High Blood Pressure Some studies have shown that there is a correlation between consumption of salt and blood pressure. As more salt is consumed, blood pressure tends to rise. A reporter reads one of these studies and writes the headline "Increased Salt Consumption Causes Blood Pressure to Rise." Is that headline justified? If not, help this reporter by rewriting the headline so that it is correct.

4. Weight Loss and Correlation In a test of the Weight Watchers weight loss program, weights of 40 subjects are recorded before and after the program. Assume that the before/after weights result in $r = 0.876$. Is there sufficient evidence to support a claim of a linear correlation between the before/after weights? Does the value of *r* indicate that the program is effective in reducing weight? Why or why not?

Interpreting r. ***In Exercises 5–8, use a significance level of $\alpha = 0.05$.***

5. Old Faithful For 40 eruptions of the Old Faithful geyser in Yellowstone National Park, duration times (sec) were recorded along with the time intervals (min) to the next eruption. The paired durations and interval times were used to obtain the results shown in the accompanying

1. *r* represents the value of the linear correlation computed by using the paired sample data. ρ represents the value of the linear correlation coefficient that would be computed by using all of the paired data in the population. The value of *r* is estimated to be 0 (because there is no correlation between sunspot numbers and the Dow Jones Industrial Average).
2. No. The value of $r = 0$ suggests that there is no *linear* relationship, but there might be some other relationship that is not linear in the sense that the pattern of points in the scatterplot is not a straight-line pattern.
3. The headline is not justified because it states that increased salt consumption is the *cause* of higher blood pressure levels, but the presence of a correlation between two variables does not necessarily imply that one is the *cause* of the other. Correlation does not imply causality. A correct headline would be this: "Study Shows That Increased Salt Consumption Is Associated with Higher Blood Pressure."

STATDISK display. Is there sufficient evidence to support the claim that there is a linear correlation between the durations of eruptions and the time intervals to the next eruptions? Explain.

STATDISK

```
Correlation Results:
Correlation coeff, r:  0.6869123
Critical r:           ±0.3120061
P-value (two-tailed):  0.000
```

6. Old Faithful For 40 eruptions of the Old Faithful geyser in Yellowstone National Park, duration times (sec) were recorded along with the heights (ft) of the eruptions. The paired duration and height measurements were used to obtain the results shown in the accompanying Excel display. Is there sufficient evidence to support the claim that there is a linear correlation between the durations of eruptions and the heights of the eruptions? Explain.

EXCEL

A	B	C
	Column 1	*Column 2*
Column 1	1	
Column 2	0.091548	1

7. Heights of Fathers and Sons The heights (in inches) of a sample of 10 father/son pairs of subjects were measured. Minitab results are shown below (based on data from the National Health Examination Survey). Is there sufficient evidence to support the claim that there is a linear correlation between the heights of fathers and the heights of their sons? Explain.

MINITAB

```
Pearson correlation of FATHER and SON = 0.149
P-Value = 0.681
```

8. Cereal Killers The amounts of sugar (grams of sugar per gram of cereal) and calories (per gram of cereal) were recorded for a sample of 16 different cereals. TI-83/84 Plus calculator results are shown here. Is there sufficient evidence to support the claim that there is a linear correlation between sugar and calories in a gram of cereal? Explain.

TI-83/84 PLUS

```
LinRegTTest
 y=a+bx
 β≠0 and ρ≠0
↑b=.5789830508
 s=.1117312583
 r²=.5858430396
 r=.7654038409
```

Importance of Graphing. *Exercises 9 and 10 provide two data sets from "Graphs in Statistical Analysis," by F. J. Anscombe,* The American Statistician, *Vol. 27. For each exercise,*

a. Construct a scatterplot.

b. Find the value of the linear correlation coefficient r, then determine whether there is sufficient evidence to support the claim of a linear correlation between the two variables.

c. Identify the feature of the data that would be missed if part (b) was completed without constructing the scatterplot.

4. Table A-5 shows that the critical values of r are ± 0.312 (assuming a 0.05 significance level), so there is sufficient evidence to support a claim of a linear correlation between the before and after weights. The value of r does *not* indicate that the diet is effective in reducing weight. While the diet might be effective in reducing weight, there could be a linear correlation if the diet has no effect so that the before and after weights are about the same, or there could be a linear correlation if the diet causes people to gain weight.
5. Yes. With $r = 0.687$ and critical values of ± 0.312, there is sufficient evidence to support the claim that there is a linear correlation between the durations of eruptions and the time intervals to the next eruptions.
6. No. With $r = 0.091$ and critical values of ± 0.312, there is not sufficient evidence to support the claim that there is a linear correlation between the durations of eruptions and the heights of eruptions.
7. No. With $r = 0.149$ and a *P*-value of 0.681 (or critical values of ± 0.632), there is not sufficient evidence to support the claim that there is a linear correlation between the heights of fathers and the heights of their sons.
8. Yes. With $r = 0.765$ and critical values of ± 0.497, there is sufficient evidence to support the claim that there is a linear correlation between calories and sugar in a gram of cereal.

9.

x	10	8	13	9	11	14	6	4	12	7	5
y	9.14	8.14	8.74	8.77	9.26	8.10	6.13	3.10	9.13	7.26	4.74

10.

x	10	8	13	9	11	14	6	4	12	7	5
y	7.46	6.77	12.74	7.11	7.81	8.84	6.08	5.39	8.15	6.42	5.73

11. Effects of an Outlier Refer to the accompanying Minitab-generated scatterplot.

a. Examine the pattern of all 10 points and subjectively determine whether there appears to be a correlation between *x* and *y*.

b. After identifying the 10 pairs of coordinates corresponding to the 10 points, find the value of the correlation coefficient *r* and determine whether there is a linear correlation.

c. Now remove the point with coordinates (10, 10) and repeat parts (a) and (b).

d. What do you conclude about the possible effect from a single pair of values?

MINITAB

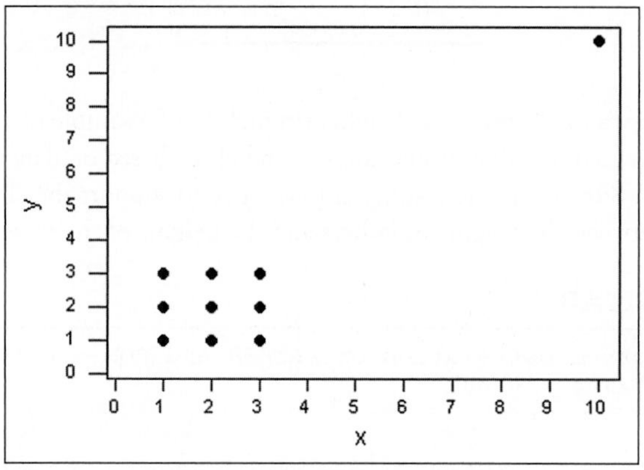

12. Effects of Clusters Refer to the following Minitab-generated scatterplot. The four points in the lower left corner are measurements from women, and the four points in the upper right corner are from men.

MINITAB

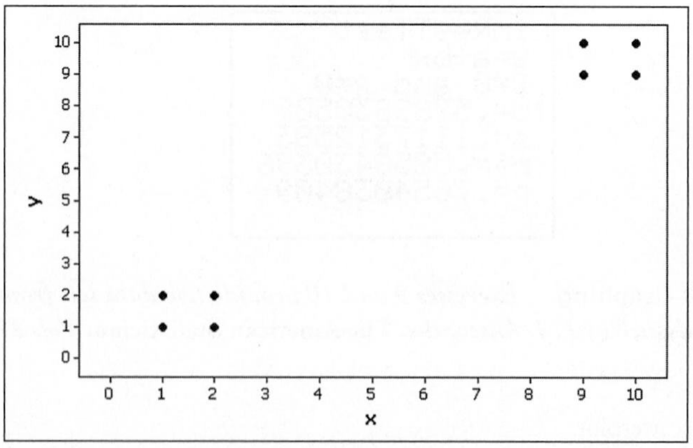

a. Examine the pattern of the four points in the lower left corner (from women) only, and subjectively determine whether there appears to be a correlation between *x* and *y* for women.

b. Examine the pattern of the four points in the upper right corner (from men) only, and subjectively determine whether there appears to be a correlation between *x* and *y* for men.

9. a.

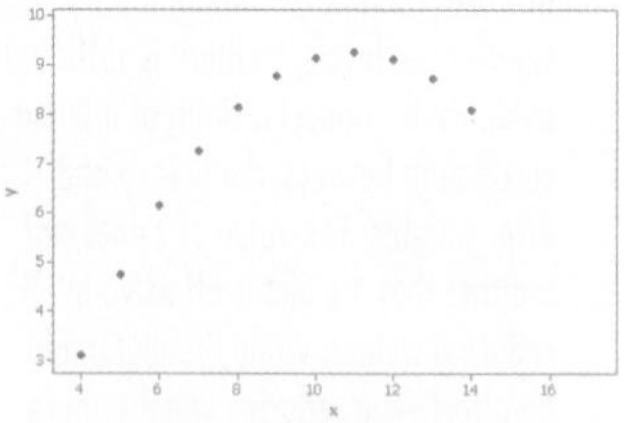

b. $r = 0.816$. Critical values: $r = \pm 0.602$. *P*-value $= 0.002$. There is sufficient evidence to support the claim of a linear correlation between the two variables.

c. The scatterplot reveals a distinct pattern that is not a straight-line pattern.

10. a.

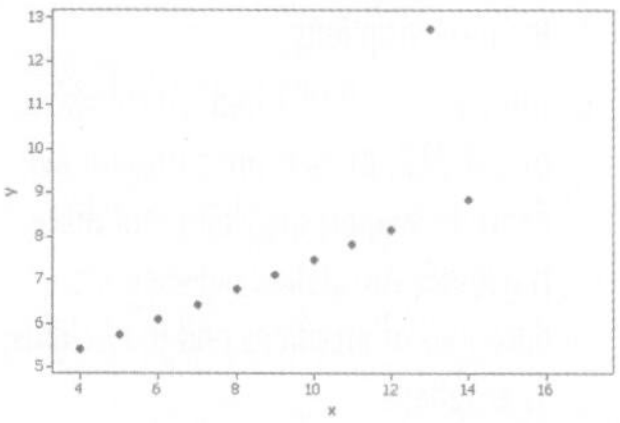

b. $r = 0.816$. Critical values: $r = \pm 0.602$. *P*-value $= 0.002$. There is sufficient evidence to support the claim of a linear correlation between the two variables.

c. The scatterplot reveals a perfect straight-line pattern, except for the presence of one outlier.

11. a. There appears to be a linear correlation.

b. $r = 0.906$. Critical values: $r = \pm 0.632$ (for a 0.05 significance level). There is a linear correlation.

c. $r = 0$. Critical values: $r = \pm 0.666$ (for a 0.05 significance level). There does not appear to be a linear correlation.

d. The effect from a single pair of values can be very substantial, and it can change the conclusion.

12. a. There does not appear to be a linear correlation.

b. There does not appear to be a linear correlation.

c. $r = 0$. Critical values: $r = \pm 0.950$ (for a 0.05 significance level). There does not appear to be a linear correlation. The same results are obtained with the four points in the upper right corner.

d. $r = 0.985$. Critical values: $r = \pm 0.707$ (for a 0.05 significance level). There is a linear correlation.

c. Find the linear correlation coefficient using only the four points in the lower left corner (for women). Will the four points in the upper left corner (for men) have the same linear correlation coefficient?

d. Find the value of the linear correlation coefficient using all eight points. What does that value suggest about the relationship between *x* and *y*?

e. Based on the preceding results, what do you conclude? Should the data from women and the data from men be considered together, or do they appear to represent two different and distinct populations that should be analyzed separately?

e. There are two different populations that should be considered separately. It is misleading to use the combined data from women and men and conclude that there is a relationship between *x* and *y*.

Testing for a Linear Correlation. ***In Exercises 13–28, construct a scatterplot, and find the value of the linear correlation coefficient r. Also find the P-value or the critical values of r from Table A-5 using $\alpha = 0.05$. Determine whether there is sufficient evidence to support a claim of a linear correlation between the two variables. (Save your work because the same data sets will be used in Section 10-3 exercises.)***

13. Lemons and Car Crashes Listed below are annual data for various years. The data are weights (metric tons) of lemons imported from Mexico and U.S. car crash fatality rates per 100,000 population [based on data from "The Trouble with QSAR (or How I Learned to Stop Worrying and Embrace Fallacy)" by Stephen Johnson, *Journal of Chemical Information and Modeling*, Vol. 48, No. 1]. Is there sufficient evidence to conclude that there is a linear correlation between weights of lemon imports from Mexico and U.S. car fatality rates? Do the results suggest that imported lemons cause car fatalities?

Lemon Imports	230	265	358	480	530
Crash Fatality Rate	15.9	15.7	15.4	15.3	14.9

13. $r = -0.959$. Critical values: $r = \pm 0.878$. *P*-value = 0.010. There is sufficient evidence to support the claim that there is a linear correlation between weights of lemon imports from Mexico and U.S. car fatality rates. The results do not suggest any cause-effect relationship between the two variables.

14. PSAT and SAT Scores Listed below are PSAT scores and SAT scores from prospective college applicants. The scores were reported by subjects who responded to a request posted by the Web site talk.collegeconfidential.com. Is there sufficient evidence to conclude that there is a linear correlation between PSAT scores and SAT scores? Is there anything about the data that might make the results questionable?

PSAT	183	207	167	206	197	142	193	176
SAT	2200	2040	1890	2380	2290	2070	2370	1980

14. $r = 0.543$. Critical values: $r = \pm 0.707$. *P*-value = 0.164. There is not sufficient evidence to support the claim that there is a linear correlation between PSAT scores and SAT scores. Because the data are from a voluntary response sample, the results are very questionable.

15. Campus Crime Listed below are numbers of enrolled students (in thousands) and numbers of burglaries for randomly selected large colleges in a recent year (based on data from the *New York Times*). Is there sufficient evidence to conclude that there is a linear correlation between enrollment and burglaries? Do the results change if the actual enrollments are listed as 32,000, 31,000, 53,000, and so on?

Enrollment	32	31	53	28	27	36	42	30	34	46
Burglaries	103	103	86	57	32	131	157	20	27	161

15. $r = 0.561$. Critical values: $r = \pm 0.632$. *P*-value = 0.091. There is not sufficient evidence to support the claim that there is a linear correlation between enrollment and burglaries. The results do not change if the actual enrollments are listed as 32,000, 31,000, 53,000, and so on.

16. Altitude and Temperature Listed below are altitudes (thousands of feet) and outside air temperatures (degrees Fahrenheit) recorded by the author during Delta Flight 1053 from New Orleans to Atlanta. Is there sufficient evidence to conclude that there is a linear correlation between altitude and outside air temperature? Do the results change if the altitudes are reported in meters and the temperatures are converted to the Celsius scale?

Altitude	3	10	14	22	28	31	33
Temperature	57	37	24	−5	−30	−41	−54

16. $r = -0.997$. Critical values: $r = \pm 0.754$. *P*-value = 0.000. There is sufficient evidence to support the claim that there is a linear correlation between altitude and outside air temperature. The results do not change if the altitudes are converted to meters and the temperatures are converted to the Celsius scale.

17. $r = 0.864$. Critical values: $r = \pm 0.666$. *P*-value $= 0.003$. There is sufficient evidence to support the claim that there is a linear correlation between court incomes and justice salaries. The correlation does not imply that court incomes directly affect justice salaries, but it does appear that justices might profit by levying larger fines, or perhaps justices with higher salaries impose larger fines.

18. $r = 0.947$. Critical values: $r = \pm 0.878$. *P*-value $= 0.015$. There is sufficient evidence to support the claim that there is a linear correlation between the opening bids suggested by the auctioneer and the final winning bids.

19. $r = 1.000$. Critical values: $r = \pm 0.811$. *P*-value $= 0.000$. There is sufficient evidence to support the claim that there is a linear correlation between amounts of redshift and distances to clusters of galaxies. Because the linear correlation coefficient is 1.000, it appears that the distances can be directly computed from the amounts of redshift.

20. $r = 0.968$. Critical values: $r = \pm 0.811$. *P*-value $= 0.002$. There is sufficient evidence to support the claim that there is a linear correlation between weights and prices. The results do not necessarily apply to other populations of diamonds, such as those with different color and clarity ratings.

21. $r = 0.948$. Critical values: $r = \pm 0.811$. *P*-value $= 0.004$. There is sufficient evidence to support the claim of a linear correlation between the overhead width of a seal in a photograph and the weight of a seal.

17. Town Courts Listed below are amounts of court fine revenue and salaries paid to the town justices (based on data from the *Poughkeepsie Journal*). All amounts are in thousands of dollars, and all of the towns are in Dutchess County, New York. Is there sufficient evidence to conclude that there is a linear correlation between court incomes and justice salaries? Based on the results, does it appear that justices might profit by levying larger fines?

Court Income	65	404	1567	1131	272	252	111	154	32
Justice Salary	30	44	92	56	46	61	25	26	18

18. Auction Bids The author is a member of the board of directors of a nonprofit organization that held a fund-raising auction. He recorded the opening bids suggested by the auctioneer and the final winning bids for several items. The amounts are listed below. Is there sufficient evidence to conclude that there is a linear correlation between the opening bids suggested by the auctioneer and the final winning bids?

Opening Bid	1500	500	500	400	300
Winning Bid	650	175	125	275	125

19. Galaxy Distances The table below lists measured amounts of redshift and the distances (billions of light-years) to randomly selected clusters of galaxies (based on data from *The Cosmic Perspective* by Bennett et al., Benjamin Cummings). Is there sufficient evidence to conclude that there is a linear correlation between amounts of redshift and distances to clusters of galaxies? What do the results suggest about the relationship between the two variables?

Redshift	0.0233	0.0539	0.0718	0.0395	0.0438	0.0103
Distance	0.32	0.75	1.00	0.55	0.61	0.14

20. Diamond Prices The table below lists weights (carats) and prices (dollars) of randomly selected diamonds. All of the diamonds are round with ratings of "very good" cut, they all have a color rating of F ("slight color"), and a clarity rating of VSI ("very slightly included"). The values are based on data from the retailer Blue Nile. For diamonds of the type described, is there sufficient evidence to conclude that there is a linear correlation between weights and prices? Do the results also apply to other types of diamonds, such as those with different color and clarity ratings?

Weight	0.3	0.4	0.5	0.5	1.0	0.7
Price	510	1151	1343	1410	5669	2277

21. Measuring Seals from Photos Listed below are the overhead widths (in cm) of seals measured from photographs and the weights (in kg) of the seals (based on "Mass Estimation of Weddell Seals Using Techniques of Photogrammetry," by R. Garrott of Montana State University). The purpose of the study was to determine if weights of seals could be determined from overhead photographs. Is there sufficient evidence to conclude that there is a linear correlation between overhead widths of seals from photographs and the weights of the seals?

Overhead Width	7.2	7.4	9.8	9.4	8.8	8.4
Weight	116	154	245	202	200	191

22. Car Repair Costs Listed below are repair costs (in dollars) for cars crashed at 6 mi/h in full-front crash tests and the same cars crashed at 6 mi/h in full-rear crash tests (based on data from the Insurance Institute for Highway Safety). The cars are the Toyota Camry, Mazda 6, Volvo S40, Saturn Aura, Subaru Legacy, Hyundai Sonata, and Honda Accord. Is there sufficient evidence to conclude that there is a linear correlation between the repair costs from full-front crashes and full-rear crashes?

Front	936	978	2252	1032	3911	4312	3469
Rear	1480	1202	802	3191	1122	739	2767

22. $r = -0.283$. Critical values: $r = \pm 0.754$. P-value $= 0.539$. There is not sufficient evidence to support the claim of a linear correlation between the repair costs from full-front crashes and full-rear crashes.

23. Blood Pressure Measurements Listed below are systolic blood pressure measurements (in mm Hg) obtained from the same woman (based on data from "Consistency of Blood Pressure Differences Between the Left and Right Arms," by Eguchi et al., *Archives of Internal Medicine,* Vol. 167). Is there sufficient evidence to conclude that there is a linear correlation between right and left arm systolic blood pressure measurements?

Right Arm	102	101	94	79	79
Left Arm	175	169	182	146	144

23. $r = 0.867$. Critical values: $r = \pm 0.878$. P-value $= 0.057$. There is not sufficient evidence to support the claim of a linear correlation between the systolic blood pressure measurements of the right and left arm.

24. Crickets and Temperature One classic application of correlation involves the association between the temperature and the number of times a cricket chirps in a minute. Listed below are the numbers of chirps in 1 min and the corresponding temperatures in °F (based on data from *The Song of Insects* by George W. Pierce, Harvard University Press). Is there sufficient evidence to conclude that there is a linear correlation between the number of chirps in 1 min and the temperature?

Chirps in 1 min	882	1188	1104	864	1200	1032	960	900
Temperature (°F)	69.7	93.3	84.3	76.3	88.6	82.6	71.6	79.6

24. $r = 0.874$. Critical values: $r = \pm 0.707$. P-value $= 0.005$. There is sufficient evidence to support the claim of a linear correlation between the number of cricket chirps and the temperature.

25. Gas Prices Gas prices have been very volatile in recent years. The table below lists gas prices (dollars per gallon) at randomly selected Connecticut stations at the time of this writing (based on data from AOL Autos). Is there sufficient evidence to conclude that there is a linear correlation between prices of regular gas and prices of premium gas? For the gas stations that sometimes post only the price of regular gas, can you use those prices to get a good sense of the price of premium gas?

Regular	2.77	2.77	2.79	2.81	2.78	2.86	2.75	2.77
Mid-Grade	3.00	2.77	2.89	2.93	2.93	2.96	2.86	2.91
Premium	3.07	3.09	3.00	3.06	3.03	3.06	3.02	3.03

25. $r = 0.197$. Critical values: $r = \pm 0.707$. P-value $= 0.640$. There is not sufficient evidence to support the claim that there is a linear correlation between prices of regular gas and prices of premium gas. Because there does not appear to be a linear correlation between prices of regular and premium gas, knowing the price of regular gas is not very helpful in getting a good sense for the price of premium gas.

26. Gas Prices Repeat the preceding exercise using the prices of regular gas and the prices of mid-grade gas.

26. $r = 0.399$. Critical values: $r = \pm 0.707$. P-value $= 0.327$. There is not sufficient evidence to support the claim that there is a linear correlation between prices of regular gas and prices of mid-grade gas. Because there does not appear to be a linear correlation between prices of regular and mid-grade gas, knowing the price of regular gas is not very helpful in getting a good sense for the price of mid-grade gas.

27. Sports Diameters (cm), circumferences (cm), and volumes (cm^3) from balls used in different sports are listed in the table below. Is there sufficient evidence to conclude that there is a linear correlation between diameters and circumferences? Does the scatterplot confirm a *linear* association?

	Baseball	Basketball	Golf	Soccer	Tennis	Ping-Pong	Volleyball	Softball
Diameter	7.4	23.9	4.3	21.8	7.0	4.0	20.9	9.7
Circumference	23.2	75.1	13.5	68.5	22.0	12.6	65.7	30.5
Volume	212.2	7148.1	41.6	5424.6	179.6	33.5	4780.1	477.9

27. $r = 1.000$. Critical values: $r = \pm 0.707$. P-value $= 0.000$. There is sufficient evidence to support the claim that there is a linear correlation between diameters and circumferences. The scatterplot confirms a linear association.

28. $r = 0.978$. Critical values: $r = \pm 0.707$. P-value $= 0.000$. There is sufficient evidence to support the claim that there is a linear correlation between diameters and volumes. Although the results suggest that there is a linear correlation between diameters and volumes, the scatterplot suggests that there is a very strong correlation that is not linear.

29. $r = -0.063$. Critical values: $r = \pm 0.444$. P-value $= 0.791$. There is not sufficient evidence to support the claim of a linear correlation between IQ and brain volume.

30. $r = 0.917$. Critical values: $r = \pm 0.279$ (approximately) (Tech: ± 0.285). P-value $= 0.000$. There is sufficient evidence to support the claim of a linear correlation between departure delay times and arrival delay times.

31. $r = 0.319$. Critical values: $r = \pm 0.254$ (approximately) (Tech: ± 0.263). P-value $= 0.017$. There is sufficient evidence to support the claim of a linear correlation between the numbers of words spoken by men and women who are in couple relationships.

32. $r = 0.027$. Critical values: $r = \pm 0.279$. P-value $= 0.852$. There is not sufficient evidence to support the claim of a linear correlation between magnitudes of earthquakes and their depths.

33. a. 0.911
b. 0.787
c. 0.9999 (largest)
d. 0.976
e. -0.948

34. ± 0.445

28. Sports Repeat the preceding exercise using diameters and volumes.

Large Data Sets. ***In Exercises 29–32, use the data from Appendix B to construct a scatterplot, find the value of the linear correlation coefficient r, and find either the P-value or the critical values of r from Table A-5 using $\alpha = 0.05$. Determine whether there is sufficient evidence to support the claim of a linear correlation between the two variables. (Save your work because the same data sets will be used in Section 10-3 exercises.)***

29. IQ and Brain Volume Refer to Data Set 6 in Appendix B and use the paired data consisting of brain volume (cm^3) and IQ score.

30. Flight Delays Refer to Data Set 15 in Appendix B and use the departure delay times and the arrival delay times.

31. Word Counts of Men and Women Refer to Data Set 17 in Appendix B and use the word counts measured from men and women in couple relationships listed in the first two columns of Data Set 17.

32. Earthquakes Refer to Data Set 16 in Appendix B and use the magnitudes and depths from the earthquakes.

10-2 Beyond the Basics

33. Transformed Data In addition to testing for a linear correlation between x and y, we can often use *transformations* of data to explore other relationships. For example, we might replace each x value by x^2 and use the methods of this section to determine whether there is a linear correlation between y and x^2. Given the paired data in the accompanying table, construct the scatterplot and then test for a linear correlation between y and each of the following. Which case results in the largest value of r?

a. x **b.** x^2 **c.** $\log x$ **d.** $\sqrt{x}$ **e.** $1/x$

x	2	3	20	50	95
y	0.3	0.5	1.3	1.7	2.0

34. Finding Critical *r* Values Table A-5 lists critical values of r for selected values of n and α. More generally, critical r values can be found by using the formula

$$r = \frac{t}{\sqrt{t^2 + n - 2}}$$

where the t value is found from the table of critical t values (Table A-3) assuming a two-tailed case with $n - 2$ degrees of freedom. Use the formula for r given here and Table A-3 (with $n - 2$ degrees of freedom) to find the critical r values corresponding to H_1: $\rho \neq 0$, $\alpha = 0.02$, and $n = 27$.

10-3 Regression

Key Concept Suppose that we have a collection of paired data and we use the methods of Section 10-2 to conclude that there is a linear correlation between two variables. This section presents methods for finding the equation of the straight line that best fits a scatterplot of the sample data. That best-fitting straight line is called the *regression line*, and its equation is called the *regression equation*. We can use the regression equation to make predictions for the value of one of the variables given some specific value of the other variable. In Part 2 of this section we discuss marginal change, influential points, and residual plots as tools for analyzing correlation and regression results.

Part 1: Basic Concepts of Regression

In some cases, two variables are related in a *deterministic* way, meaning that given a value for one variable, the value of the other variable is exactly determined without any error, as in the equation $y = 2.54x$ for converting a distance x from inches to centimeters. Such equations are considered in algebra courses, but statistics courses focus on *probabilistic* models, which are equations with a variable that is not determined completely by the other variable. For example, the height of a child cannot be determined completely by the height of the father and/or mother. Sir Francis Galton (1822–1911) studied the phenomenon of heredity and showed that when tall or short couples have children, the heights of those children tend to *regress,* or revert to the more typical mean height for people of the same gender. We continue to use Galton's "regression" terminology, even though our data do not involve the same height phenomena studied by Galton.

DEFINITIONS

Given a collection of paired sample data, the **regression line** (or *line of best fit,* or *least-squares line*) is the straight line that "best" fits the scatterplot of the data. (The specific criterion for the "best-fitting" straight line is the "least-squares" property described later.)

The **regression equation**

$$\hat{y} = b_0 + b_1x$$

algebraically describes the regression line.

The regression equation expresses a relationship between x (called the **explanatory variable,** or **predictor variable,** or **independent variable**) and $\hat{y}$ (called the **response variable,** or **dependent variable**). The preceding definition shows that in statistics, the typical equation of a straight line $y = mx + b$ is expressed in the form $\hat{y} = b_0 + b_1x$, where b_0 is the y-intercept and b_1 is the slope.

The values of the slope b_1 and y-intercept b_0 can be easily found by using any one of the many computer programs and calculators designed to provide those values. (See "Using Technology" at the end of this section.)

Prediction Worth $1 Million

Netflix is a large movie rental company that recently sponsored a contest with a $1 million prize for developing a system for predicting whether someone will like a movie based on how much they liked other movies. Netflix had developed its own system it called Cinematch, but the contest required a new system with a substantial improvement over Cinematch. The $1 million prize was won by a team called BellKor's Pragmatic Chaos, which included members of the Statistics Research Department at AT&T and others.

Contestants used data from 100 million movie ratings. The objective was to use the past movie ratings to predict which movies people would prefer, and the predictions were compared to movies that the people later viewed and rated. The winning prediction system is quite complex and is beyond the scope of this text.

Note to Instructor

Section 10-3 is partitioned into Part 1 ("Basic Concepts of Regression") and Part 2 ("Beyond the Basics of Regression").

Recommendation: Cover Part 1, and cover Part 2 only if time is not an issue. It is helpful to briefly review the $y = mx + b$ format of the equation of a straight line. What does m represent? What does b represent? What if the equation is changed to a format of $y = b_0 + b_1x$? What is the slope? What is the y-intercept? You might comment that we don't use the format of $y = mx + b$ for this reason: The format of $y = b_0 + b_1x_1$ can be easily and naturally expanded to include more variables, as in $y = b_0 + b_1x_1 + b_2x_2 + b_3x_3$, which is a typical expression used for multiple regression.

Objective

Find the equation of a regression line.

Notation for the Equation of a Regression Line

	Population Parameter	Sample Statistic
y-Intercept of regression equation	β_0	b_0
Slope of regression equation	β_1	b_1
Equation of the regression line	$y = \beta_0 + \beta_1$	$\hat{y} = b_0 + b_1x$

Requirements

1. The sample of paired (x, y) data is a *random* sample of quantitative data.
2. Visual examination of the scatterplot shows that the points approximate a straight-line pattern.
3. Outliers can have a strong effect on the regression equation, so remove any outliers if they are known to be errors. Consider the effects of any outliers that are not known errors.

Note: Requirements 2 and 3 above are simplified attempts at checking these formal requirements for regression analysis:

- For each fixed value of x, the corresponding values of y have a normal distribution.
- For the different fixed values of x, the distributions of the corresponding y-values all have the same standard deviation. (This is violated if part of the scatterplot shows points very close to the regression line while another portion of the scatterplot shows points that are much farther away from the regression line. See the discussion of residual plots in Part 2 of this section.)
- For the different fixed values of x, the distributions of the corresponding y-values have means that lie along the same straight line.

The methods of this section are not seriously affected if departures from normal distributions and equal standard deviations are not too extreme.

Formulas for Finding the Slope b_1 and y-Intercept b_0 in the Regression Equation $\hat{y} = b_0 + b_1x$

Formula 10-3 **Slope:** $b_1 = r\dfrac{s_y}{s_x}$ where r is the linear correlation coefficient, s_y is the standard deviation of the y values, and s_x is the standard deviation of the x values.

Formula 10-4 ***y*-Intercept:** $b_0 = \bar{y} - b_1\bar{x}$

The slope b_1 and y-intercept b_0 can also be found using the following formulas that are useful for manual calculations or writing computer programs:

$$b_1 = \frac{n(\Sigma xy) - (\Sigma x)(\Sigma y)}{n(\Sigma x^2) - (\Sigma x)^2} \qquad b_0 = \frac{(\Sigma y)(\Sigma x^2) - (\Sigma x)(\Sigma xy)}{n(\Sigma x^2) - (\Sigma x)^2}$$

Note to Instructor

Recommendation: Illustrate the use of Formulas 10-3 and 10-4 once, but then go beyond the formula. Allow students to find the slope and intercept from their calculators or computer software. This is consistent with the trend of making the statistics course much more meaningful than laboriously cranking out values by applying formulas.

Rounding the Slope b_1 and the y-Intercept b_0

Round b_1 and b_0 to three significant digits. It's difficult to provide a simple universal rule for rounding values of b_1 and b_0, but this rule will work for most situations in this book. (Depending on how you round, this book's answers to examples and exercises may be slightly different from your answers.)

Example 1 Using Technology to Find the Regression Equation

Refer to the sample data given in Table 10-1 in the Chapter Problem. Use technology to find the equation of the regression line in which the explanatory variable (or x variable) is shoe print length and the response variable (or y variable) is the corresponding height of a person.

Solution

Requirement check (1) The data are assumed to be a simple random sample. (2) Figure 10-1(a) is a scatterplot showing a pattern of points. This pattern is very roughly a straight-line pattern. (3) There are no outliers. The requirements are satisfied. ✓

Using technology: The use of computer software or a calculator is recommended for finding the equation of a regression line. Shown below are the results from STATDISK, Minitab, XLSTAT, the TI-83/84 Plus calculator, SPSS, and JMP. Minitab and XLSTAT provide the actual equation; the other technologies list the values of the y-intercept and the slope. All of these technologies show that the regression equation can be expressed as $\hat{y} = 125 + 1.73x$, where $\hat{y}$ is the predicted height of a person and x is the length of the shoe print.

STATDISK

```
Regression Results:
Y= b0 + b1x:
Y Intercept, b0:        125.4073
Slope, b1:              1.727452
```

EXCEL (XLSTAT)

Equation of the model:

Height = 125.40733+1.72745*Shoe Print

MINITAB

Regression Analysis: Height versus Shoe Print

```
The regression equation is
Height = 125 + 1.73 Shoe Print
```

TI-83/84 PLUS

```
LinRegTTest
 y=a+bx
 β≠0 and ρ≠0
↑a=125.4073302
 b=1.72745239
 s=4.538762333
↓r²=.3495991312
```

SPSS

Model		Unstandardized Coefficients		Standardized Coefficients		
		B	Std. Error	Beta	t	Sig.
1	(Constant)	125.407	40.915		3.065	.055
	Shoeprint	1.727	1.360	.591	1.270	.294

JMP

Term	Estimate	Std Error	t Ratio	Prob>|t|
Intercept	125.40733	40.91531	3.07	0.0548
Shoeprint	1.7274524	1.360351	1.27	0.2937

We should know that the regression equation is an *estimate* of the true regression equation for the population of paired data. This estimate is based on one particular set of sample data, but another sample drawn from the same population would probably lead to a slightly different equation.

Example 2 Using Manual Calculations to Find the Regression Equation

Refer to the sample data given in Table 10-1 in the Chapter Problem. Use Formulas 10-3 and 10-4 to find the equation of the regression line in which the explanatory variable (or x variable) is the shoe print length and the response variable (or y variable) is the corresponding height of the person.

Solution

Requirement check The requirements are verified in Example 1. ✓

We begin by finding the slope b_1 with Formula 10-3 as follows (with extra digits included for greater accuracy). Remember, r is the linear correlation coefficient,

Postponing Death

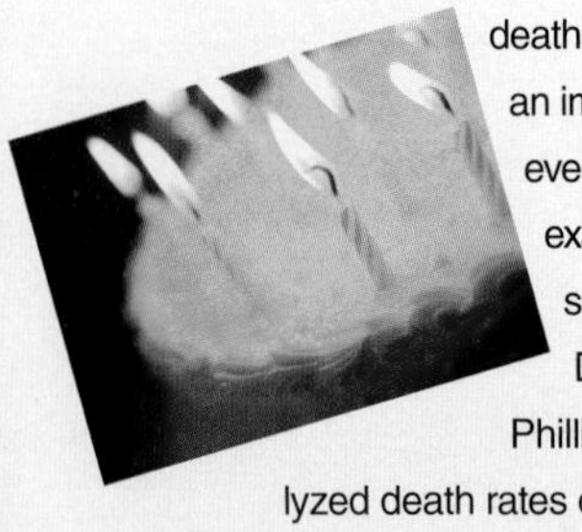

Several studies have addressed the ability of people to postpone their death until after an important event. For example, sociologist David Phillips analyzed death rates of Jewish men who died near Passover, and he found that the death rate dropped dramatically in the week before Passover, but rose the week after. Other researchers of cancer patients concluded that there is "no pattern to support the concept that 'death takes a holiday.'" (See "Holidays, Birthdays, and Postponement of Cancer Death," by Young and Hade, *Journal of the American Medical Association,* Vol. 292, No. 24.) Based on records of 1.3 million deaths, this more recent study found no relationship between the time of death and Christmas, Thanksgiving, or the person's birthday. The findings were disputed by David Phillips, who said that the study focused on cancer patients, but they are least likely to have psychosomatic effects.

Note to Instructor
Students typically experience some difficulty in determining the best predicted value of a variable. Begin by randomly selecting a student and asking him or her to predict the IQ score of a male who is 6 ft tall, then explain why the answer of 100 makes sense. (There is no correlation between IQ and height, and the answer of 100 is not based on any regression equation.) Now randomly select another student and ask him or her to predict the time it would take to drive 100 miles. Answers such as 2 or 3 hours are good, and they involve a calculation using an estimated regression equation.

s_y is the standard deviation of the sample y values, and s_x is the standard deviation of the sample x values.

$$b_1 = r\frac{s_y}{s_x} = 0.591269 \cdot \frac{4.87391}{1.66823} = 1.72745$$

After finding the slope b_1, we can now use Formula 10-4 to find the y-intercept as follows:

$$b_0 = \bar{y} - b_1\bar{x} = 177.3 - (1.72745)(30.04) = 125.40740$$

After rounding, the slope and y-intercept are $b_1 = 1.73$ and $b_0 = 125$. We can now express the regression equation as $\hat{y} = 125 + 1.73x$, where $\hat{y}$ is the predicted height of a person and x is the length of the shoe print.

Example 3 Graphing the Regression Line

Graph the regression equation $\hat{y} = 125 + 1.73x$ (found in Examples 1 and 2) on the scatterplot of the shoe print and height data from Table 10-1 and examine the graph to subjectively determine how well the regression line fits the data.

Solution

Shown below is the Minitab display of the scatterplot with the graph of the regression line included. We can see that the regression line doesn't fit the data very well.

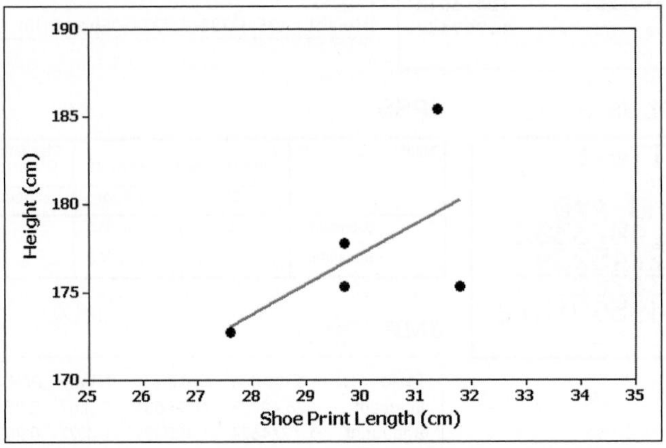

Using the Regression Equation for Predictions

Regression equations are often useful for *predicting* the value of one variable, given some specific value of the other variable. When making predictions, we should consider the following:

1. Use the regression equation for predictions only if the graph of the regression line on the scatterplot confirms that the regression line fits the points reasonably well.
2. Use the regression equation for predictions only if the linear correlation coefficient r indicates that there is a linear correlation between the two variables (as described in Section 10-2).
3. Use the regression line for predictions only if the data do not go much beyond the scope of the available sample data. (Predicting too far beyond the scope of the available sample data is called *extrapolation,* and it could result in bad predictions.)
4. If the regression equation does not appear to be useful for making predictions, the best predicted value of a variable is its sample mean.

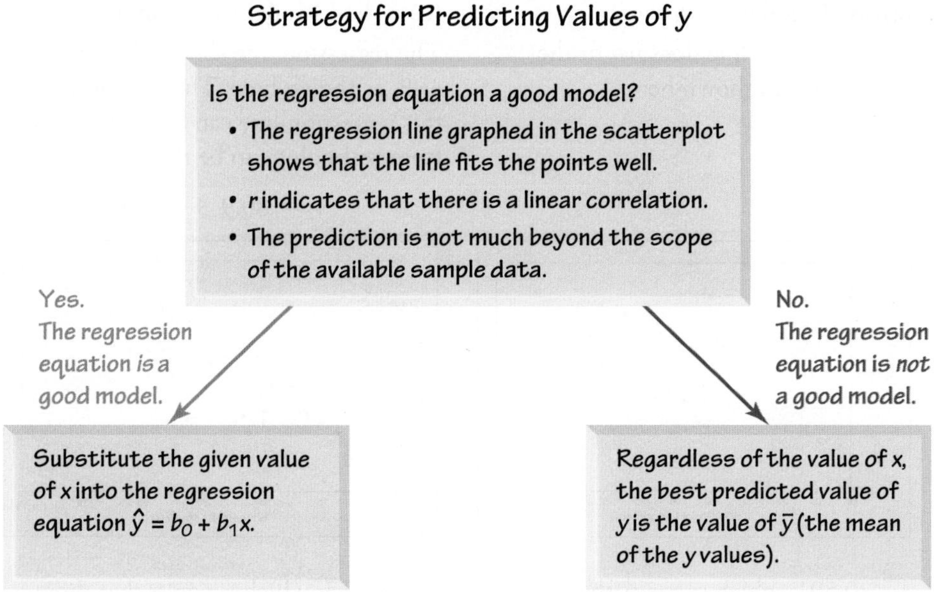
Strategy for Predicting Values of y
Is the regression equation a good model?
• The regression line graphed in the scatterplot shows that the line fits the points well.
• r indicates that there is a linear correlation.
• The prediction is not much beyond the scope of the available sample data.
Yes. The regression equation is a good model.
No. The regression equation is not a good model.
Substitute the given value of x into the regression equation ŷ = b0 + b1x.
Regardless of the value of x, the best predicted value of y is the value of ȳ (the mean of the y values).

Figure 10-5 Recommended Strategy for Predicting Values of *y*

Figure 10-5 summarizes a strategy for predicting values of a variable y when given some value of x. Figure 10-5 shows that if the regression equation is a good model, then we substitute the value of x into the regression equation to find the predicted value of y. However, if the regression equation is not a good model, the best predicted value of y is simply $\bar{y}$, the mean of the y values. Remember, this strategy applies to *linear* patterns of points in a scatterplot. If the scatterplot shows a pattern that is not a straight-line pattern, we may be able to use other methods not included in this book.

Example 4 Predicting Height

Use the given data to predict the height of someone with a shoe print length of 29 cm.

a. Use the 5 pairs of shoe print lengths and heights from Table 10-1 to predict the height of a person with a shoe print length of 29 cm.

b. Use the 40 pairs of shoe print lengths and heights from Data Set 2 in Appendix B and predict the height of a person with a shoe print length of 29 cm.

Solution

Shown below are key points in the solutions for parts (a) and (b). Note that in part (a), the paired data *do not* result in a good regression model, so the predicted height is $\bar{y}$, the mean of the five heights. However, part (b) shows that the shoe print and height data *do* result in a good regression model, so the predicted height is found by substituting the value of $x = 29$ cm into the regression equation.

(a) Bad Model: Use $\bar{y}$ for Predictions

Use the 5 pairs of sample data from Table 10-1 to predict the height of someone with a shoe print length of 29 cm:

(b) Good Model: Use the Regression Equation for Predictions

Use the 40 pairs of sample data from Data Set 2 in Appendix B to predict the height of someone with a shoe print length of 29 cm:

1° Forecast Error = $1 Billion

5 Day Forecast
83
83
85

The prediction of forecast temperatures might seem to be an inexact science, but many companies are working feverishly to obtain more accurate predictions. *USA Today* reporter Del Jones wrote that we could save $1 billion in a year if we could forecast temperatures more accurately by just 1 degree Fahrenheit. Jones reported that for the region served by the Tennessee Valley Authority, forecast temperatures have been off by about 2.35 degrees, and that error is common for forecasts in the United States. He states that reducing the 2.35 degree error to 1.35 degrees would save the TVA an estimated $100,000 every day. Forecast temperatures are used to determine the allocation of power from generators, nuclear plants, hydroelectric plants, coal, natural gas, and wind. Statistical forecasting techniques are now being refined so that we can all benefit from savings of money and natural resources.

(a) Bad Model *(continued)*

The regression line does *not* fit the points well, as shown here.

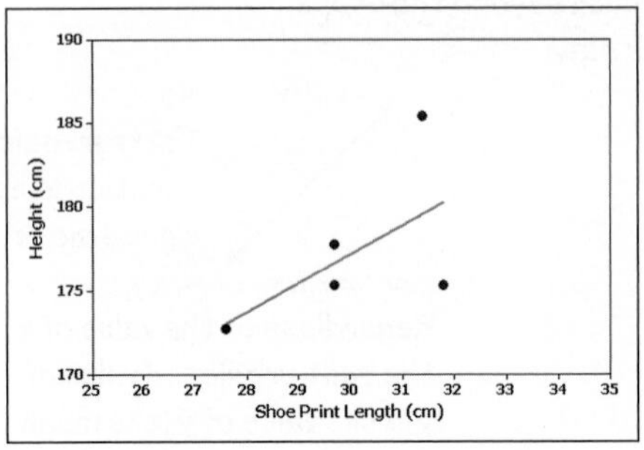

$r = 0.591$, which suggests that there is *not* a linear correlation between shoe print length and height. (The *P*-value is 0.294.)

The given shoe print length of 29 cm is not beyond the scope of the available data.

↓

Because the regression equation $\hat{y} = 125 + 1.73x$ is *not* a good model, the best predicted height is simply the mean of the sample heights: $\bar{y} = 177.3$ cm (or 69.8 in.).

(b) Good Model *(continued)*

The regression line *does* fit the points well, as shown here. The equation of this regression line can be easily found from technology to be this:

$$\hat{y} = 80.9 + 3.22x$$

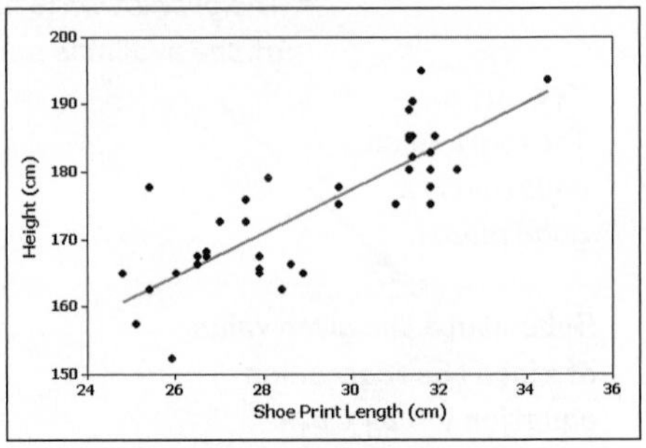

$r = 0.813$, which suggests that there *is* a linear correlation between shoe print length and height. (The *P*-value is 0.000.)

The given shoe print length of 29 cm is not beyond the scope of the available data.

↓

Because the regression equation $\hat{y} = 80.9 + 3.22x$ *is* a good model, substitute $x = 29$ cm into this regression equation to get a predicted height of 174.3 cm (or 68.6 in.).

Interpretation

Key point: Use the regression equation for predictions only if it is a good model. If the regression equation is not a good model, use the predicted value of $\bar{y}$.

Note to Instructor
Include Part 2 only if you have an abundance of time available and/or your students have suitable mathematics backgrounds.

Part 2: Beyond the Basics of Regression

In Part 2 we consider the concept of marginal change, which is helpful in interpreting a regression equation; then we consider the effects of outliers and special points called *influential points*. We also consider residual plots.

Interpreting the Regression Equation: Marginal Change

We can use the regression equation to see the effect on one variable when the other variable changes by some specific amount.

DEFINITION In working with two variables related by a regression equation, the **marginal change** in a variable is the amount that it changes when the other variable changes by exactly one unit. The slope b_1 in the regression equation represents the marginal change in y that occurs when x changes by one unit.

Let's consider the 40 pairs of shoe print lengths and heights included in Data Set 2 of Appendix B. Those 40 pairs of values result in this regression equation: $\hat{y} = 80.9 + 3.22x$. The slope of 3.22 tells us that if we increase x (the length of the shoe print) by 1 cm, the predicted height of the person will increase by 3.22 cm. That is, for every additional 1 cm increase in the length of a shoe print, we expect the height to increase by 3.22 cm.

Outliers and Influential Points

A correlation/regression analysis of bivariate (paired) data should include an investigation of *outliers* and *influential points,* defined as follows.

DEFINITIONS

In a scatterplot, an **outlier** is a point lying far away from the other data points.

Paired sample data may include one or more **influential points,** which are points that strongly affect the graph of the regression line.

To determine whether a point is an outlier, examine the scatterplot to see if the point is far away from the others. Here's how to determine whether a point is an influential point: First graph the regression line resulting from the data with the point included, then graph the regression line resulting from the data with the point excluded. If the regression line changes by a considerable amount, the point is influential.

Example 5 Influential Point

Consider the 40 pairs of shoe print lengths and heights from Data Set 2 in Appendix B. The scatterplot located to the left below shows the regression line. If we include this additional pair of data: $x = 35$ cm, $y = 25$ cm (shoe length is 35 cm and height is 25 cm), we get the regression line shown to the right below. The additional point (35 cm, 25 cm) is an influential point because the graph of the regression line did change considerably, as shown by the regression line located to the right below. Compare the two graphs and you will see clearly that the addition of that one pair of values has a very dramatic effect on the regression line, so that additional point is an influential point. The additional point is also an outlier because it is far from the other points.

Original Paired Shoe Print and Height Data

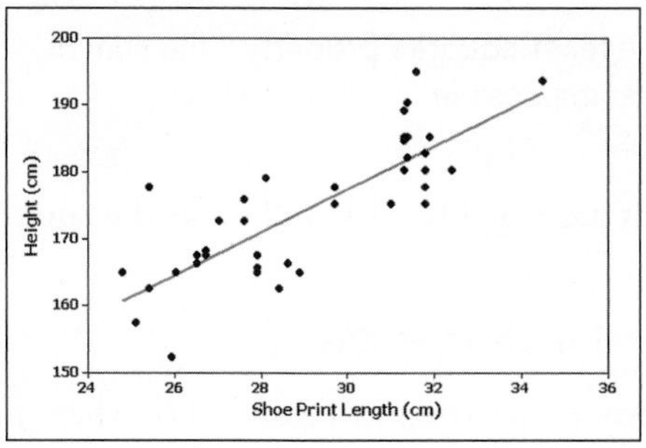

Shoe Print and Height Data with an Additional Point (35 cm, 25 cm)

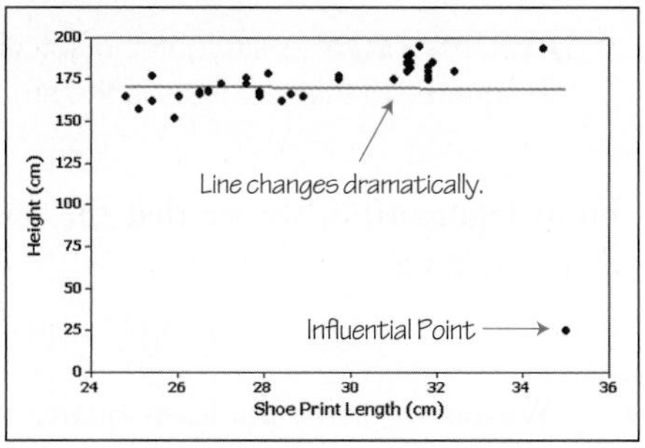

Residuals and the Least-Squares Property

We stated that the regression equation represents the straight line that "best" fits the data. The criterion to determine the line that is better than all others is based on the vertical distances between the original data points and the regression line. Such distances are called *residuals.*

DEFINITION For a pair of sample *x* and *y* values, the **residual** is the difference between the *observed* sample value of *y* and the *y* value that is *predicted* by using the regression equation. That is,

$$\text{residual} = \text{observed } y - \text{predicted } y = y - \hat{y}$$

x	1	2	4	5
y	4	24	8	32

The paired data are plotted as blue points in Figure 10-6.

So far, this definition hasn't yet won any prizes for simplicity, but you can easily understand residuals by referring to Figure 10-6, which corresponds to the paired sample data shown in the margin. In Figure 10-6, the residuals are represented by the dashed lines.

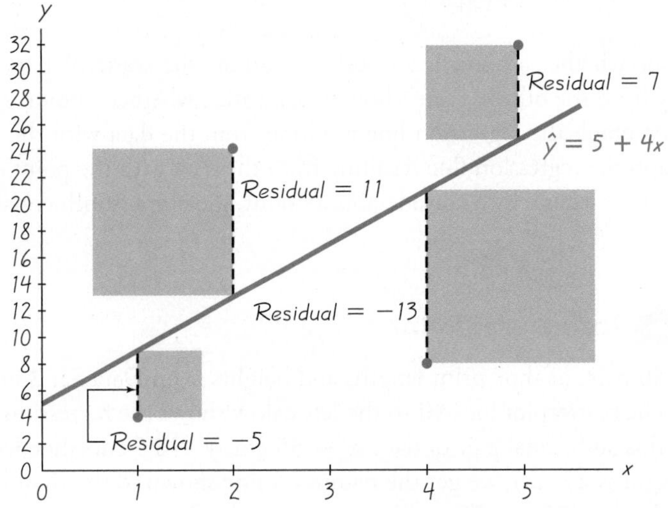

Figure 10-6 Residuals and Squares of Residuals

Consider the sample point with coordinates of (5, 32). If we substitute $x = 5$ into the regression equation $\hat{y} = 5 + 4x$, we get a predicted value of $\hat{y} = 25$. But the actual observed sample value is $y = 32$. The difference $y - \hat{y} = 32 - 25 = 7$ is a residual.

The regression equation represents the line that "best" fits the points according to the following least-squares property.

DEFINITION A straight line satisfies the **least-squares property** if the sum of the squares of the residuals is the smallest sum possible.

From Figure 10-6, we see that the residuals are −5, 11, −13, and 7, so the sum of their squares is

$$(-5)^2 + 11^2 + (-13)^2 + 7^2 = 364$$

We can visualize the least-squares property by referring to Figure 10-6, where the squares of the residuals are represented by the red-square areas. The sum of the red-square areas is 364, which is the smallest sum possible. Use any other straight line, and the red squares will combine to produce an area larger than the combined red area of 364.

Fortunately, we need not deal directly with the least-squares property when we want to find the equation of the regression line. Calculus has been used to build the least-squares property into Formulas 10-3 and 10-4. Because the derivations of these formulas require calculus, we don't include the derivations in this text, and for that, we should be very thankful.

Residual Plots

In this section and the preceding section we listed simplified requirements for the effective analyses of correlation and regression results. We noted that we should always begin with a scatterplot, and we should verify that the pattern of points is approximately a straight-line pattern. We should also consider outliers. A *residual plot* can be another helpful tool for analyzing correlation and regression results and for checking the requirements necessary for making inferences about correlation and regression.

DEFINITION A **residual plot** is a scatterplot of the (x, y) values after each of the y-coordinate values has been replaced by the residual value $y - \hat{y}$ (where $\hat{y}$ denotes the predicted value of y). That is, a residual plot is a graph of the points $(x, y - \hat{y})$.

To construct a residual plot, draw a horizontal reference line through the residual value of 0, then plot the paired values of $(x, y - \hat{y})$. Because the manual construction of residual plots can be tedious, the use of computer software is strongly recommended. When analyzing a residual plot, look for a pattern in the way the points are configured, and use these criteria:

- The residual plot should not have any obvious pattern (not even a straight-line pattern). (This confirms that a scatterplot of the sample data is a straight-line pattern and not some other pattern that is not a straight line.)
- The residual plot should not become much wider (or thinner) when viewed from left to right. (This confirms the requirement that for the different fixed values of x, the distributions of the corresponding y values all have the same standard deviation.)

Example 6 Residual Plot

The shoe print and height data from Table 10-1 are used to obtain the accompanying Minitab-generated residual plot. The first sample x value of 29.7 cm is substituted into the regression equation of $\hat{y} = 125 + 1.73x$ (found in Examples 1 and 2). The result is the predicted value of $\hat{y} = 176.4$ cm. For the first x value of 29.7 cm, the actual corresponding y value is 175.3 cm, so the value of the residual is

$$\text{observed } y - \text{predicted } y = y - \hat{y} = 175.3 - 176.4 = -1.1$$

(The result is -1.4 if we use greater precision in the calculations.) Using the x value of 29.7 cm and the residual of -1.1, we get the coordinates of the point $(29.7, -1.1)$, which is one of the points in the residual plot shown on the following page. This residual plot becomes thicker, suggesting that the regression equation might not be a good model.

MINITAB

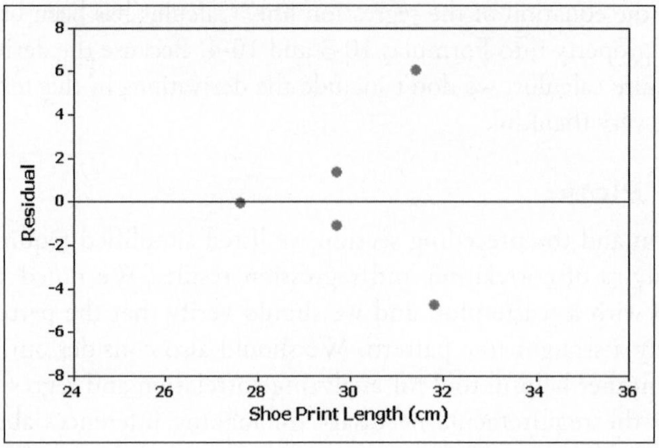

See the three residual plots below. The leftmost residual plot suggests that the regression equation is a good model. The middle residual plot shows a distinct pattern, suggesting that the sample data do not follow a straight-line pattern as required. The rightmost residual plot becomes thicker, which suggests that the requirement of equal standard deviations is violated.

Residual Plot Suggesting That the Regression Equation Is a Good Model

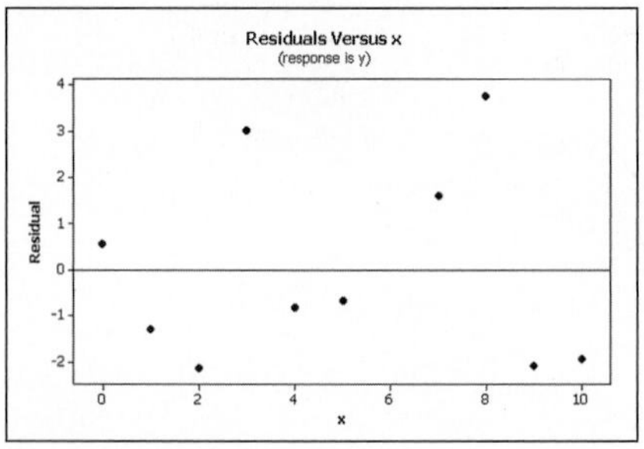

Residual Plot with an Obvious Pattern, Suggesting That the Regression Equation Is Not a Good Model

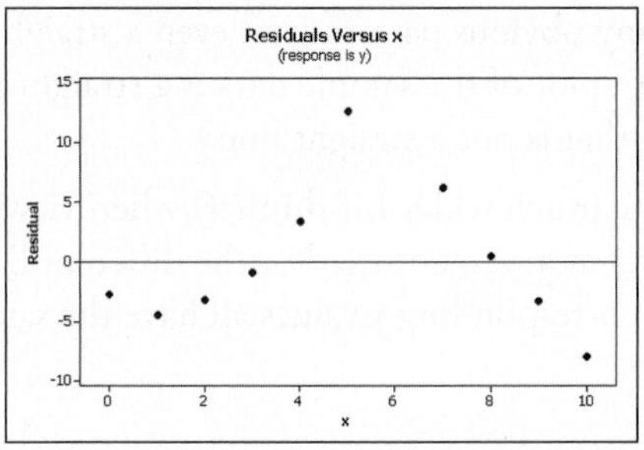

Residual Plot That Becomes Thicker, Suggesting That the Regression Equation Is Not a Good Model

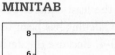

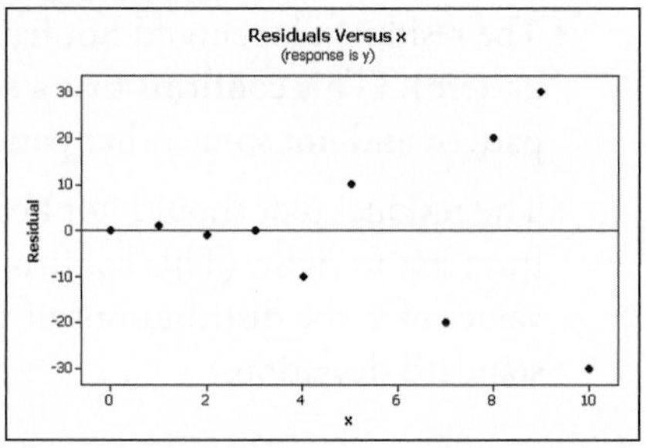

Complete Regression Analysis

In Part 1 of this section, we identified simplified criteria for determining whether a regression equation is a good model. A more complete and thorough analysis can be implemented with the following steps:

1. Construct a scatterplot and verify that the pattern of the points is approximately a straight-line pattern without outliers. (If there are outliers, consider their effects by comparing results that include the outliers to results that exclude the outliers.)
2. Construct a residual plot and verify that there is no pattern and also verify that the residual plot does not become thicker (or thinner).
3. Use a histogram and/or normal quantile plot to confirm that the values of the residuals have a distribution that is approximately normal.
4. Consider any effects of a pattern over time.

using TECHNOLOGY

Because of the messy calculations involved, the linear correlation coefficient r and the slope and y-intercept of the regression line are usually found using a calculator or computer software.

STATDISK First enter the paired data in columns of the Statdisk Data Window. Select **Analysis** from the main menu bar, then use the option **Correlation and Regression.** Enter a value for the significance level and select the columns of data. Click on the **Evaluate** button. The display will include the value of the linear correlation coefficient along with the critical value of r and the P-value, the conclusion about correlation, and the intercept and slope of the regression equation, as well as some other results. Click on **Plot** to get a graph of the scatterplot with the regression line included.

MINITAB First enter the x values in column C1 and enter the y values in column C2 (or use any other columns). In Section 10-2 we saw that we could find the value of the linear correlation coefficient r by selecting **Stat/Basic Statistics/Correlation.** To get the equation of the regression line, select **Stat/Regression/Regression,** and enter C2 for "response" and C1 for "predictor." To get the graph of the scatterplot with the regression line, select **Stat/Regression/Fitted Line Plot,** then enter C2 for the response variable and C1 for the predictor variable. Select the "linear" model.

With Minitab Release 16 or later, you can also click on **Assistant,** then select **Regression.** Click on the icon to perform the analysis, then complete the dialog box that appears. There will be much helpful information displayed.

EXCEL Enter the paired data in columns A and B. Use either XLSTAT or Excel's Data Analysis add-in.

XLSTAT Click on **XLSTAT,** then select **Modeling Data,** then **Linear Regression.** In the dialog box that appears, first enter the range of the sample values for the dependent y variable in the box identified as "Y/Dependent variable." Next, enter the range of quantitative values for the independent x variable in the box identified as "X/Explanatory variable." Put a checkmark next to the "Variable labels" box only if the first row of each column consists of names or labels. Click **OK.** After the results are displayed, look for the value of the linear correlation coefficient r in the "Correlation matrix" table. The P-value can be found in the last column of the "Analysis of Variance" table. The equation of the regression line is identified as "Equation of the model."

Data Analysis add-in: If using Excel 2103, 2010, or 2007, click on **Data,** then click on **Data Analysis;** if using Excel 2003, click on **Tools,** then click on **Data Analysis.** Select **Regression,** then click on **OK.** Enter the range for the y values, such as B1:B10. Enter the range for the x values, such as A1:A10. Click on the box adjacent to Line Fit Plots, then click **OK.** Among all of the information provided by Excel, the slope and intercept of the regression equation can be found under the table heading "Coefficient." The displayed graph will include a scatterplot of the original sample points along with the points that would be predicted by the regression equation. You can easily get the regression line by connecting the "predicted y" points.

TI-83/84 PLUS Enter the paired data in lists L1 and L2, then press STAT and select **TESTS,** then choose the option **LinRegTTest.** The displayed results will include the y-intercept and slope of the regression equation. Instead of b_0 and b_1, the TI-83/84 display represents these values as a and b.

STATCRUNCH Click on **Open StatCrunch.** Enter the columns of data or open a data set. Click on **Stat,** then select **Regression,** then select **Simple Linear.** Enter the columns to be used, then click on **Calculate.** The equation of the regression line will be found in the fourth line of the display.

10-3 Basic Skills and Concepts

Recommended Assignment
Exercises 1–18.

Statistical Literacy and Critical Thinking

1. Notation and Terminology If we use the paired height/pulse data for females from Data Set 1 in Appendix B, we get this regression equation: $\hat{y} = 73.9 + 0.0223x$, where x represents height (cm) and the pulse rate is in beats per minute. What does the symbol $\hat{y}$ represent? In this case, what does the predictor variable represent? What does the response variable represent?

2. Best-Fit Line In what sense is the regression line the straight line that "best" fits the points in a scatterplot?

3. Correlation and Slope Formula 10-3 shows that the slope of a regression line can be found by evaluating $r \cdot s_y/s_x$. What do we know about the graph of the regression line if r is a positive value? What do we know about the graph of the regression line if r is a negative value?

4. Notation What is the difference between the regression equation $\hat{y} = b_0 + b_1x$ and the regression equation $y = \beta_0 + \beta_1x$?

1. The symbol $\hat{y}$ represents the predicted pulse rate. The predictor variable represents height. The response variable represents pulse rate.
2. The regression line has the property that the sum of squares of the residuals is the lowest possible sum (where a residual is the difference between an observed value of y and a predicted value of y).
3. If r is positive, the regression line has a positive slope and rises from left to right. If r is negative, the slope of the regression line is negative and it falls from left to right.

4. The first equation represents the regression line that best fits *sample* data, whereas the second equation represents the regression line that best fits all paired data in a *population.*
5. The best predicted time for an interval after the eruption is 69.0 min.
6. The best predicted height is $\bar{y} = 127.2$ ft.
7. The best predicted height is $\bar{y} = 68.0$ in.
8. The best predicted value is 3.86 calories.
9. $\hat{y} = 3.00 + 0.500x$. The data have a pattern that is not a straight line.
10. $\hat{y} = 3.00 + 0.500x$. There is an outlier.
11. a. $\hat{y} = 0.264 + 0.906x$
    b. $\hat{y} = 2 + 0x$ (or $\hat{y} = 2$)
    c. The results are very different, indicating that one point can dramatically affect the regression equation.
12. a. $\hat{y} = 0.0846 + 0.985x$
    b. $\hat{y} = 1.5 + 0x$ (or $\hat{y} = 1.5$)
    c. $\hat{y} = 9.5 + 0x$ (or $\hat{y} = 9.5$)
    d. The results are very different, indicating that combinations of clusters can produce results that differ dramatically from results within each cluster alone.

Making Predictions. ***In Exercises 5–8, let the predictor variable x be the first variable given. Use the given data to find the regression equation and the best predicted value of the response variable. Be sure to follow the prediction procedure summarized in Figure 10-5.***

5. Old Faithful For 40 eruptions of the Old Faithful geyser in Yellowstone National Park, duration times (sec) were recorded along with the time intervals (min) after the eruptions. The linear correlation coefficient is $r = 0.687$ and the regression equation is $\hat{y} = 47.4 + 0.180x$, where x represents duration time. The mean of the 40 duration times is 245.0 sec and the mean of the 40 interval times is 91.4 min. What is the best predicted interval time following an eruption with a duration time of 120 min?

6. Old Faithful For 40 eruptions of the Old Faithful geyser in Yellowstone National Park, duration times (sec) were recorded along with the heights (ft) of the eruptions. The linear correlation coefficient is $r = 0.0915$ and the regression equation is $\hat{y} = 119 + 0.0331x$, where x represents duration time. The mean of the 40 duration times is 245.0 sec and the mean of the 40 heights is 127.2 ft. What is the best predicted height of an eruption with a duration time of 120 min?

7. Heights of Fathers and Sons The heights (in inches) of a sample of 10 father/son pairs of subjects were measured. The linear correlation coefficient is $r = 0.149$ and the regression equation is $\hat{y} = 66.8 + 0.016x$, where x represents the father's height. The mean of the 10 father heights is 69.7 in. and the mean of the 10 son heights is 68.0 in. What is the best predicted height of a son who has a father with a height of 72.0 in.?

8. Cereal Killers The amounts of sugar (grams of sugar per gram of cereal) and calories (per gram of cereal) were recorded for a sample of 16 different cereals. The linear correlation coefficient is $r = 0.765$ and the regression equation is $\hat{y} = 3.46 + 1.01x$, where x represents the amount of sugar. The mean of the 16 amounts of sugar is 0.295 grams and the mean of the 16 calorie counts is 3.76. What is the best predicted calorie count for a cereal with a measured sugar amount of 0.40 g?

Finding the Equation of the Regression Line. ***In Exercises 9 and 10, use the given data to find the equation of the regression line. Examine the scatterplot and identify a characteristic of the data that is ignored by the regression line.***

9.

x	10	8	13	9	11	14	6	4	12	7	5
y	9.14	8.14	8.74	8.77	9.26	8.10	6.13	3.10	9.13	7.26	4.74

10.

x	10	8	13	9	11	14	6	4	12	7	5
y	7.46	6.77	12.74	7.11	7.81	8.84	6.08	5.39	8.15	6.42	5.73

11. Effects of an Outlier Refer to the Minitab-generated scatterplot given in Exercise 11 of Section 10-2.

a. Using the pairs of values for all 10 points, find the equation of the regression line.

b. After removing the point with coordinates (10, 10), use the pairs of values for the remaining 9 points and find the equation of the regression line.

c. Compare the results from parts (a) and (b).

12. Effects of Clusters Refer to the Minitab-generated scatterplot given in Exercise 12 of Section 10-2.

a. Using the pairs of values for all 8 points, find the equation of the regression line.

b. Using only the pairs of values for the 4 points in the lower left corner, find the equation of the regression line.

c. Using only the pairs of values for the 4 points in the upper right corner, find the equation of the regression line.

d. Compare the results from parts (a), (b), and (c).

Regression and Predictions. ***Exercises 13–28 use the same data sets as Exercises 13–28 in Section 10-2. In each case, find the regression equation, letting the first variable be the predictor (x) variable. Find the indicated predicted value by following the prediction procedure summarized in Figure 10-5.***

13. Lemons and Car Crashes Find the best predicted crash fatality rate for a year in which there are 500 metric tons of lemon imports.

Lemon Imports	230	265	358	480	530
Crash Fatality Rate	15.9	15.7	15.4	15.3	14.9

13. $\hat{y} = 16.5 - 0.00282x$; best predicted value is 15.1 fatalities per 100,000 population.

14. PSAT and SAT Scores One subject not included in the given table had a PSAT score of 229. Find the best predicted SAT score for this student. Is the result close to the reported value of 2400? Given that the data are from volunteered responses, are the results valid?

PSAT	183	207	167	206	197	142	193	176
SAT	2200	2040	1890	2380	2290	2070	2370	1980

14. $\hat{y} = 1314 + 4.56x$; best predicted value is $\bar{y} = 2153$. The result is not close to the actual reported value of 2400. Because the data are from a voluntary response sample, the results have questionable validity.

15. Campus Crime Find the best predicted number of burglaries for Ohio State, which had an enrollment of 51,800 students. Is the predicted value close to 329, which was the actual number of burglaries?

Enrollment	32	31	53	28	27	36	42	30	34	46
Burglaries	103	103	86	57	32	131	157	20	27	161

15. $\hat{y} = -36.8 + 3.47x$; best predicted value is $\bar{y} = 87.7$ burglaries. The predicted value is not close to the actual value of 329 burglaries.

16. Altitude and Temperature At 6327 ft (or 6.327 thousand feet), the author recorded the temperature. Find the best predicted temperature at that altitude. How does the result compare to the actual recorded value of 48°F?

Altitude	3	10	14	22	28	31	33
Temperature	57	37	24	–5	–30	–41	–54

16. $\hat{y} = 72.5 - 3.68x$; best predicted value is 49.2°F. The predicted value is close to the actual value of 48°F.

17. Town Courts The court for the town of Beekman had income of \$83,941 (or \$83.941 thousand). Find the best predicted salary for the justice. Is the result close to the actual salary of \$26,088?

Court Income	65	404	1567	1131	272	252	111	154	32
Justice Salary	30	44	92	56	46	61	25	26	18

17. $\hat{y} = 27.7 + 0.0373x$; best predicted value is \$30,800. The predicted value is not very close to the actual salary of \$26,088.

18. Auction Bids An item not included in the table is a pair of New York Knicks tickets. The auctioneer started the bidding at \$300. Find the best predicted winning bid. Is the result close to the actual winning bid of \$250?

Opening Bid	1500	500	500	400	300
Winning Bid	650	175	125	275	125

18. $\hat{y} = -4.62 + 0.429x$; best predicted value is \$124. The predicted value is not very close to the actual winning bid of \$250.

19. $\hat{y} = -0.00440 + 14.0x$; best predicted value is 0.172 billion light-years. The predicted value is very close to the actual distance of 0.18 light-years.

19. Galaxy Distances The cluster Hydra has a measured redshift of 0.0126. Find the best predicted distance to that cluster. Is the result close to the actual distance of 0.18 billion light-years?

Redshift	0.0233	0.0539	0.0718	0.0395	0.0438	0.0103
Distance	0.32	0.75	1.00	0.55	0.61	0.14

20. $\hat{y} = -2010 + 7180x$; best predicted value is \$8760 (Tech: \$8759). The predicted value is far from the actual price of \$16,097. The weight of 1.50 carats is well beyond the scope of the available sample weights, so the extrapolation might be off by a considerable amount.

20. Diamond Prices Not included in the table below is a diamond with a weight of 1.50 carats. Find the best predicted price for this diamond. Is the result close to the actual price of \$16,097? What is wrong with predicting the price of a 1.50-carat diamond?

Weight	0.3	0.4	0.5	0.5	1.0	0.7
Price	510	1151	1343	1410	5669	2277

21. $\hat{y} = -157 + 40.2x$; best predicted weight is −76.6 kg (Tech: −76.5 kg). That prediction is a negative weight that cannot be correct. The overhead width of 2 cm is well beyond the scope of the available sample widths, so the extrapolation might be off by a considerable amount.

21. Measuring Seals from Photos Find the best predicted weight of a seal if the overhead width measured from a photograph is 2 cm. Can the prediction be correct? What is wrong with predicting the weight in this case?

Overhead Width	7.2	7.4	9.8	9.4	8.8	8.4
Weight	116	154	245	202	200	191

22. $\hat{y} = 2060 - 0.186x$; best predicted cost is $\bar{y}$ = \$1615. The predicted cost of \$1615 is very different from the actual cost of \$982.

22. Car Repair Costs Find the best predicted repair costs for a full-rear crash for a Volkswagen Passat, given that its repair cost from a full-front crash is \$4594. How does the result compare to the \$982 actual repair cost from a full-rear crash?

Front	936	978	2252	1032	3911	4312	3469
Rear	1480	1202	802	3191	1122	739	2767

23. $\hat{y} = 43.6 + 1.31x$; best predicted value is $\bar{y}$ = 163.2 mm Hg.

23. Blood Pressure Measurements Find the best predicted systolic blood pressure in the left arm given that the systolic blood pressure in the right arm is 100 mm Hg.

Right Arm	102	101	94	79	79
Left Arm	175	169	182	146	144

24. $\hat{y} = 27.6 + 0.0523x$; best predicted value is 185°F (Tech: 184°F). The value of 3000 chirps in 1 minute is well beyond the scope of the available sample data, so the extrapolation might be off by a considerable amount.

24. Crickets and Temperature Find the best predicted temperature at a time when a cricket chirps 3000 times in 1 minute. What is wrong with this predicted value?

Chirps in 1 min	882	1188	1104	864	1200	1032	960	900
Temperature (°F)	69.7	93.3	84.3	76.3	88.6	82.6	71.6	79.6

25. $\hat{y} = 2.57 + 0.172x$; best predicted value is $\bar{y}$ = \$3.05. The predicted price is not very close to the actual price of \$2.93.

25. Gas Prices One gas station not included in the table below had a listed price of \$2.78 for regular gas. Find the best predicted price of premium gas at this station. Is the result close to the actual price of \$2.93 for premium gas?

Regular	2.77	2.77	2.79	2.81	2.78	2.86	2.75	2.77
Mid-Grade	3.00	2.77	2.89	2.93	2.93	2.96	2.86	2.91
Premium	3.07	3.09	3.00	3.06	3.03	3.06	3.02	3.03

26. Gas Prices Using the data from the preceding exercise, find the best predicted price for mid-grade gas for a station that posted $2.78 as the price of regular gas. Is the result close to the actual price of $2.84 for mid-grade gas?

27. Sports Find the best predicted circumference of a marble with a diameter of 1.50 cm. How does the result compare to the actual circumference of 4.7 cm?

	Baseball	Basketball	Golf	Soccer	Tennis	Ping-Pong	Volleyball	Softball
Diameter	7.4	23.9	4.3	21.8	7.0	4.0	20.9	9.7
Circumference	23.2	75.1	13.5	68.5	22.0	12.6	65.7	30.5
Volume	212.2	7148.1	41.6	5424.6	179.6	33.5	4780.1	477.9

28. Sports Using the data from the preceding exercise, find the best predicted volume of a marble with a diameter of 1.50 cm. How does the result compare to the actual volume of 1.8 cm³?

Large Data Sets. *Exercises 29–32 use the same Appendix B data sets as Exercises 29–32 in Section 10-2. In each case, find the regression equation, letting the first variable be the predictor (x) variable. Find the indicated predicted values following the prediction procedure summarized in Figure 10-5.*

29. IQ and Brain Volume Refer to Data Set 6 in Appendix B and use the paired data consisting of IQ score and brain volume (cm³). Find the best predicted IQ score for someone with a brain volume of 1000 cm³.

30. Flight Delays Refer to Data Set 15 in Appendix B and use the departure delay times and the arrival delay times. Find the best predicted arrival delay time for a flight with no departure delay.

31. Word Counts of Men and Women Refer to Data Set 17 in Appendix B and use the word counts measured from men and women in couple relationships listed in the first two columns of Data Set 17. Find the best predicted word count for a woman who is in a couple relationship with a man having a word count of 10,000.

32. Earthquakes Refer to Data Set 16 in Appendix B and use the magnitudes and depths from the earthquakes. Find the best predicted depth of an earthquake with a magnitude of 1.50.

10-3 Beyond the Basics

33. Equivalent Hypothesis Tests Explain why a test of the null hypothesis $H_0: \rho = 0$ is equivalent to a test of the null hypothesis $H_0: \beta_1 = 0$, where ρ is the linear correlation coefficient for a population of paired data, and β_1 is the slope of the regression line for that same population.

34. Least-Squares Property According to the least-squares property, the regression line minimizes the sum of the squares of the residuals. Refer to the data in Table 10-1.

a. Find the sum of squares of the residuals.

b. Show that the regression equation $\hat{y} = 120 + 2.00x$ results in a larger sum of squares of residuals.

---

26. $\hat{y} = 0.640 + 0.813x$; best predicted value is $\bar{y} = \$2.91$. The predicted price is not too far from the actual price.

27. $\hat{y} = -0.00396 + 3.14x$; best predicted value is 4.7 cm. Even though the diameter of 1.50 cm is beyond the scope of the sample diameters, the predicted value yields the actual circumference.

28. $\hat{y} = -2010 + 347x$; best predicted value is -1489.5 cm³ (Tech: -1489.8 cm³). The predicted value is negative and is far from the actual volume of 1.8 cm³. The diameter of 1.50 cm is beyond the scope of the sample diameters, and the predicted value is way wrong. The scatterplot suggests that a nonlinear model would yield better results.

29. $\hat{y} = 109 - 0.00670x$; best predicted IQ score is $\bar{y} = 101$.

30. $\hat{y} = -18.4 + 904x$; best predicted arrival delay time is -18.4 minutes. That is, if a flight has no departure delay, we can predict that the flight will arrive 18.4 minutes early.

31. $\hat{y} = 13,400 + 0.302x$; best predicted value is 16,400 (Tech: 16,458).

32. $\hat{y} = 9.53 + 0.231x$; best predicted value is $\bar{y} = 9.81$ km.

33. With $\beta_1 = 0$, the regression line is horizontal so that different values of x result in the same y value, and there is no correlation between x and y.

34. a. 61.8

b. The sum of squares of the residuals is 101.3, which is larger than 61.8.

Note to Instructor
Because this section deals with the correlation between two variables, it could be covered along with Section 10-2. Section 10-2 has a requirement of a normal distribution, but this section does not require a normal distribution or any other particular distribution.

10-4 Rank Correlation

Key Concept In this section we describe the nonparametric method of the *rank correlation test,* which uses ranks of paired data to test for an association between two variables. In Section 10-2 we used paired sample data to compute values for the linear correlation coefficient r, but in this section we use *ranks* as the basis for computing the rank correlation coefficient r_s. As in Section 10-2, we should begin an analysis of paired data by exploring with a scatterplot so that we can identify any patterns in the data.

DEFINITION The **rank correlation test** (or **Spearman's rank correlation test**) is a nonparametric test that uses ranks of sample data consisting of matched pairs. It is used to test for an association between two variables.

We use the notation r_s for the rank correlation coefficient so that we don't confuse it with the linear correlation coefficient r. The subscript s does *not* refer to a standard deviation; it is used in honor of Charles Spearman (1863–1945), who originated the rank correlation approach. In fact, r_s is often called **Spearman's rank correlation coefficient.** Key components of the rank correlation test are given in the following box, and the rank correlation procedure is summarized in Figure 10-7.

Rank Correlation

Objective

Compute the rank correlation coefficient r_s and use it to test for an association between two variables. The null and alternative hypotheses are as follows:

H_0: $\rho_s = 0$ (There is no correlation between the two variables.)

H_1: $\rho_s \neq 0$ (There is a correlation between the two variables.)

Notation

r_s = rank correlation coefficient for sample paired data (r_s is a sample statistic)

ρ_s = rank correlation coefficient for all the population data (ρ_s is a population parameter)

n = number of pairs of sample data

d = difference between ranks for the two values within an individual pair

Requirements

The paired data are a simple random sample.
The data are ranks or can be converted to ranks.
Note: Unlike the parametric methods of Section 10-2, there is *no* requirement that the sample pairs of data have a bivariate normal distribution (as described in Section 10-2). There is *no* requirement of a normal distribution for any population.

Test Statistic

Within each sample, first convert the data to *ranks*, then find the exact value of the rank correlation coefficient r_s by using Formula 10-1:

$$r_s = \frac{n\Sigma xy - (\Sigma x)(\Sigma y)}{\sqrt{n(\Sigma x^2) - (\Sigma x)^2}\sqrt{n(\Sigma y^2) - (\Sigma y)^2}}$$

No ties: After converting the data in each sample to ranks, if there are no ties among ranks for the first variable and there are no ties among ranks for the second variable, the exact value of the test statistic can be calculated using Formula 10-1 (at the left) or with this relatively simple formula:

$$r_s = 1 - \frac{6\Sigma d^2}{n(n^2 - 1)}$$

P-Values

P-values are sometimes provided by technology. (*Caution:* When finding r_s, you can convert the sample data to ranks and then use the same technology that was used for the linear correlation coefficient described in Section 10-2, but *do not use P-values from linear correlation for the methods of rank correlation.* Use the *P*-value from technology only if the technology has a procedure designed specifically for rank correlation. See the "Using Technology" instructions given at the end of this section.)

Critical Values

1. If $n \le 30$, critical values are found in Table A-6.
2. If $n > 30$, critical values of r_s are found using Formula 10-5.

Formula 10-5

$$r_s = \frac{\pm z}{\sqrt{n - 1}} \quad \text{(critical values when } n > 30\text{)}$$

where the value of z corresponds to the significance level. (For example, if $\alpha = 0.05$, $z = 1.96$.)

CAUTION When working with data having ties among ranks, the rank correlation coefficient r_s can be calculated using Formula 10-1. Technology can be used instead of manual calculations with Formula 10-1, but the displayed *P*-values for linear correlation do not apply to the methods of rank correlation. *Do not use P-values from linear correlation for methods of rank correlation.*

Note to Instructor

An easy way to evaluate r_s is to convert the sample data to ranks, then use technology with the same procedure from Section 10-2 for the linear correlation coefficient *r*. However, enthusiastically stress that we should not use a *P*-value from that result. A *P*-value can be used for rank correlation only if the technology has a procedure designed specifically for rank correlation.

Advantages: Rank correlation has these advantages over the parametric methods discussed in Section 10-2:

1. Rank correlation can be used with paired data that are ranks or can be converted to ranks. Unlike the parametric methods of Section 10-2, the method of rank correlation does *not* require a normal distribution for any population.
2. Rank correlation can be used to detect some (not all) relationships that are not linear.

Disadvantage: A not very serious disadvantage of rank correlation is its efficiency rating of 0.91. This efficiency rating shows that with all other circumstances being equal, the nonparametric approach of rank correlation requires 100 pairs of sample data to achieve the same results as only 91 pairs of sample observations analyzed through the parametric approach, assuming that the stricter requirements of the parametric approach are met.

Direct Link Between Smoking and Cancer

When we find a statistical correlation between two variables, we must be extremely careful to avoid the mistake of concluding that there is a cause-effect link. The tobacco industry has consistently emphasized that correlation does not imply causality as they denied that tobacco products cause cancer. However, Dr. David Sidransky of Johns Hopkins University and other researchers found a direct physical link that involves mutations of a specific gene among smokers. Molecular analysis of genetic changes allows researchers to determine whether cigarette smoking is the cause of a cancer. (See "Association Between Cigarette Smoking and Mutation of the p53 Gene in Squamous-Cell Carcinoma of the Head and Neck," by Brennan, Boyle, et al, *New England Journal of Medicine*, Vol 332, No. 11.) Although statistical methods cannot prove that smoking *causes* cancer, statistical methods can be used to identify an association, and physical proof of causation can then be sought by researchers.

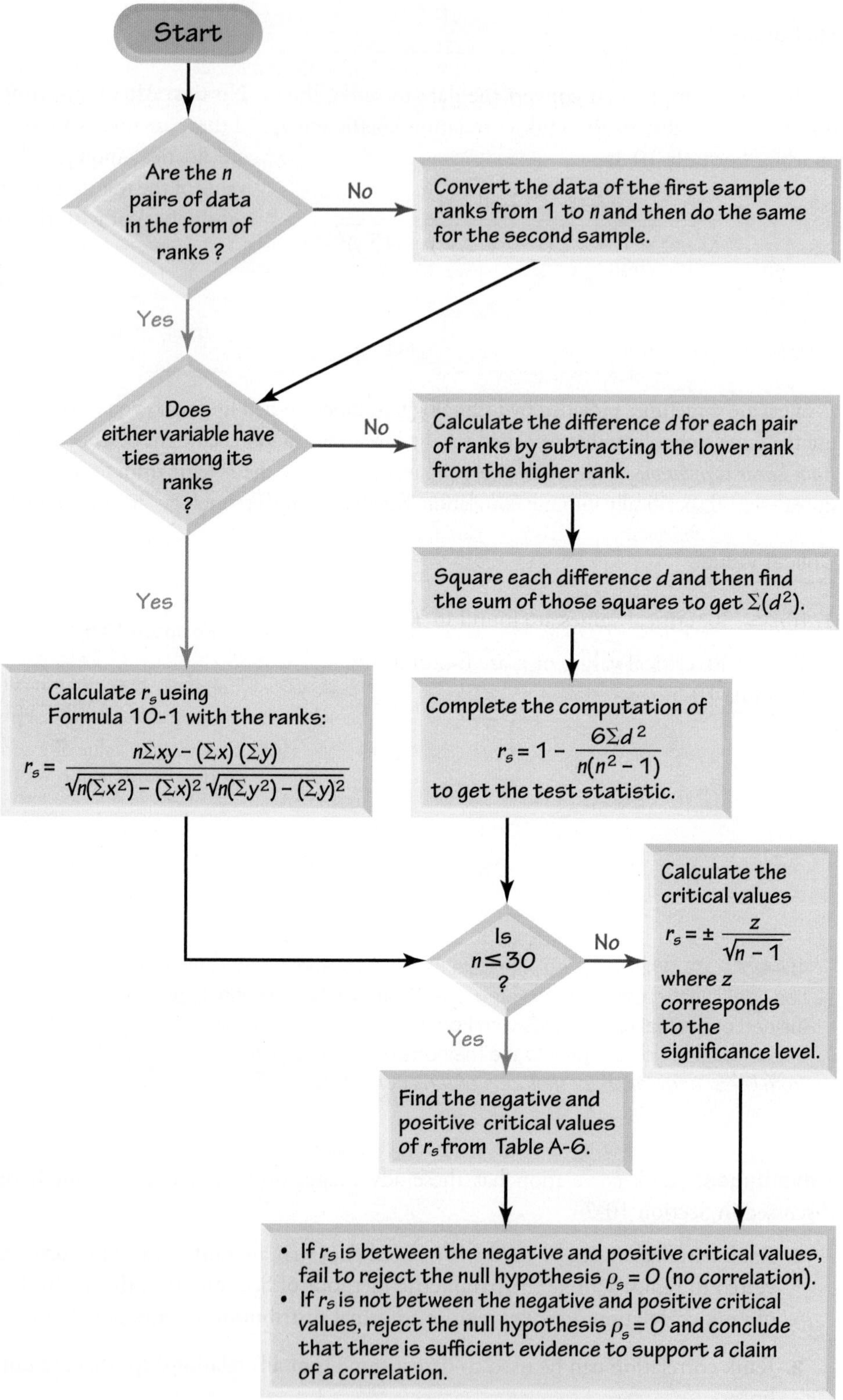

Figure 10-7 Rank Correlation Procedure for Testing H_0: $\rho_s = 0$

Example 1 Are the Best Televisions the Most Expensive?

Table 10-4 lists quality rankings and prices of 37-inch LCD televisions (based on data from *Consumer Reports*). Find the value of the rank correlation coefficient and use it to determine whether there is a correlation between quality and price. Use a 0.05 significance level. Based on the result, does it appear that you can get better quality by spending more?

Table 10-4 Overall Quality Scores and Prices of LCD Televisions

Quality rank	1	2	3	4	5	6	7
Price (dollars)	1900	1200	1300	2000	1700	1400	2700

Solution

Requirement check The only requirement is that the paired data are a simple random sample. The sample data are a simple random sample from the televisions that were tested. ✓

The quality ranks are consecutive integers and are not from a population that is normally distributed, so we use the rank correlation coefficient to test for a relationship between quality and price. The null and alternative hypotheses are as follows:

H_0: $\rho_s = 0$ (There is *no* correlation between quality and price.)

H_1: $\rho_s \neq 0$ (There is a correlation between quality and price.)

Following the procedure of Figure 10-7, we begin by converting the data in Table 10-4 into their corresponding ranks shown in Table 10-5. The lowest price of \$1200 is assigned a rank of 1, the next lowest price of \$1300 is assigned a rank of 2, and so on.

Table 10-5 Ranks of Data from Table 10-4

Quality rank	1	2	3	4	5	6	7
Price rank	5	1	2	6	4	3	7
Difference d	4	1	1	2	1	3	0
d^2	16	1	1	4	1	9	0

Neither of the two variables has ties among ranks, so the exact value of the test statistic can be calculated as shown below. We use $n = 7$ (for 7 pairs of data) and $\Sigma d^2 = 16 + 1 + 1 + 4 + 1 + 9 + 0 = 32$.

$$r_s = 1 - \frac{6\Sigma d^2}{n(n^2 - 1)} = 1 - \frac{6(32)}{7(7^2 - 1)}$$

$$= 1 - \frac{192}{336} = 0.429$$

Now we refer to Table A-6 to find the critical values of ± 0.786 (based on $\alpha = 0.05$ and $n = 7$). Because the test statistic $r_s = 0.429$ is between the critical values of -0.786 and 0.786, we fail to reject the null hypothesis. There is not sufficient evidence to support a claim of a correlation between quality and price. Based on the given sample data, it appears that you don't necessarily get better quality by paying more.

Example 2 **Large Sample Case**

Refer to the measured systolic and diastolic blood pressure measurements of 40 randomly selected females in Data Set 1 in Appendix B and use a 0.05 significance level to test the claim that among women, there is a correlation between systolic blood pressure and diastolic blood pressure.

Solution

Requirement check The data are a simple random sample. ✔

Test Statistic The value of the rank correlation coefficient is $r_s = 0.505$, which can be found by using technology.

Critical Values Because there are 40 pairs of data, we have $n = 40$. Because n exceeds 30, we find the critical values from Formula 10-5 instead of Table A-6. With $\alpha = 0.05$ in two tails, we let $z = 1.96$ to get the critical values of -0.314 and 0.314, as shown below.

$$r_s = \frac{\pm 1.96}{\sqrt{40 - 1}} = \pm 0.314$$

The test statistic of $r_s = 0.505$ is not between the critical values of -0.314 and 0.314, so we reject the null hypothesis of $r_s = 0$. There is sufficient evidence to support the claim that among women, there is a correlation between systolic blood pressure and diastolic blood pressure.

Detecting Nonlinear Patterns Rank correlation methods sometimes allow us to detect relationships that we cannot detect with the methods of Section 10-2. See the accompanying scatterplot that shows an S-shaped pattern of points suggesting that there is a correlation between x and y. The methods of Section 10-2 result in the linear correlation coefficient of $r = 0.590$ and critical values of ± 0.632, suggesting

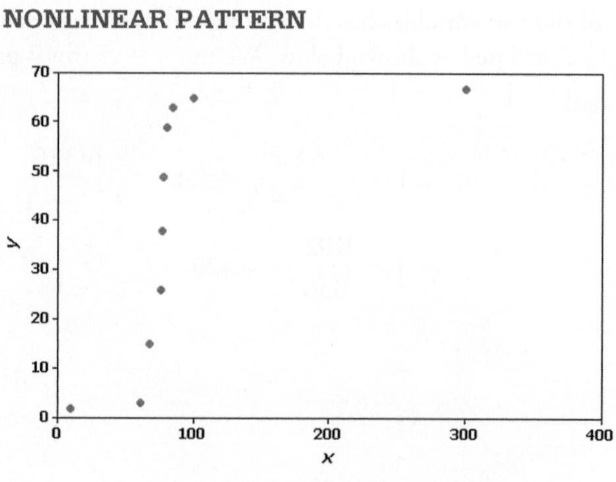

that there is not a linear correlation between x and y. But if we use the methods of this section, we get $r_s = 1$ and critical values of ± 0.648, suggesting that there is a correlation between x and y. *With rank correlation, we can sometimes detect relationships that are not linear.*

using TECHNOLOGY

STATDISK Enter the sample data in columns of the data window. Select **Analysis** from the main menu bar, then select **Rank Correlation.** Select the two columns of data to be included, then click **Evaluate.** The STATDISK results include the exact value of the test statistic r_s and the critical value.

MINITAB Enter the paired data in columns C1 and C2. If the data are not already ranks, select **Data** and **Rank** to convert the data to ranks, then select **Stat,** followed by **Basic Statistics,** followed by **Correlation.** Minitab will display the exact value of the test statistic r_s. Although Minitab identifies it as the Pearson correlation coefficient described in Section 10-2, it is actually the Spearman correlation coefficient described in this section (because it is based on ranks). *Caution:* Ignore the *P*-value, because it is calculated using the methods of Section 10-2, not the methods of this section.

EXCEL Excel does not have a function that calculates the rank correlation coefficient from original sample values, but the exact value of the test statistic r_s can be found as follows. First replace each of the original sample values by its corresponding rank. Enter those ranks in columns A and B. Click on the ***fx*** function key located on the main menu bar. Select the function category **Statistical** and the function name **CORREL,** then click **OK.** In the dialog box, enter the cell range of values for x, such as A1:A10. Also enter the cell range of values for y, such as B1:B10. Excel will display the exact value of the rank correlation coefficient r_s.

XLSTAT XLSTATcan be used by selecting **Correlation/Association tests.** For the type of test, select **Spearman**. The value of r_s will be listed in the table identified as "Correlation matrix (Spearman)," and if the value is displayed in a bold font, we can reject the claim of no correlation.

TI-83/84 PLUS If using a TI-83/84 Plus calculator or any other calculator with 2-variable statistics, you can find the exact value of r_s as follows: (1) Replace each sample value by its corresponding rank, then (2) calculate the value of the linear correlation coefficient r with the same procedures used in Section 10-2. Enter the paired ranks in lists L1 and L2, then press STAT and select **TESTS.** Using the option **LinRegTTest** will result in several displayed values, including the exact value of the rank correlation coefficient r_s. *Caution:* Ignore the *P*-value, because it is calculated using the methods of Section 10-2, not the methods of this section.

STATCRUNCH Replace each sample value by its corresponding rank, then use the same StatCrunch procedure described in Section 10-2. *Caution:* Ignore the *P*-value, because it is calculated using the methods of Section 10-2, not the methods of this section.

10-4 Basic Skills and Concepts

Statistical Literacy and Critical Thinking

1. Regression If the methods of this section are used with paired sample data, and the conclusion is that there is sufficient evidence to support the claim of a correlation between the two variables, can we use the methods of Section 10-3 to find the regression equation that can be used for predictions? Why or why not?

2. Level of Measurement Which of the levels of measurement (nominal, ordinal, interval, ratio) describe data that cannot be used with the methods of rank correlation? Explain.

3. Notation What do r, r_s, ρ, and ρ_s denote? Why is the subscript s used? Does the subscript s represent the same standard deviation s introduced in Section 3-3?

4. Efficiency The *efficiency* of the rank correlation test is 0.91. What does that value tell us about the test?

Recommended Assignment
Exercises 1–8, 11, 12.

1. The methods of Section 10-3 should not be used for predictions. The regression equation is based on a *linear* correlation between the two variables, but the methods of this section do not require a linear relationship. The methods of this section could suggest that there is a correlation with paired data associated by some nonlinear relationship, so the regression equation would not be a suitable model for making predictions.
2. Data at the nominal level of measurement have no ordering that enables them to be converted to ranks, so data at the nominal level of measurement cannot be used with the methods of rank correlation.

In Exercises 5 and 6, use the scatterplot to find the value of the rank correlation coefficient r_s and the critical values corresponding to a 0.05 significance level used to test the null hypothesis of $\rho_s = 0$. Determine whether there is a correlation.

5. Distance/Time Data for a Dropped Object

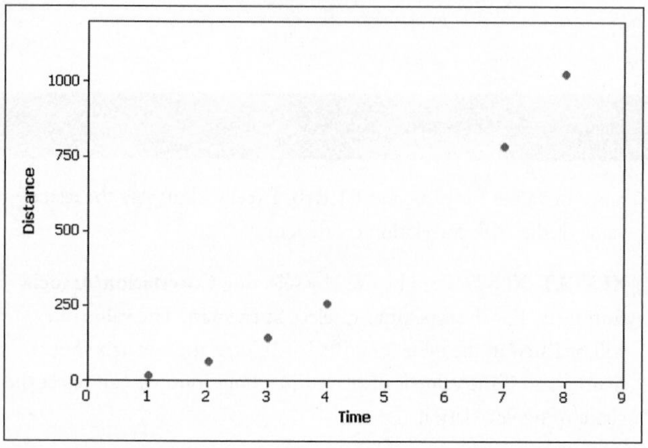

6. Altitude/Time Data for a Descending Aircraft

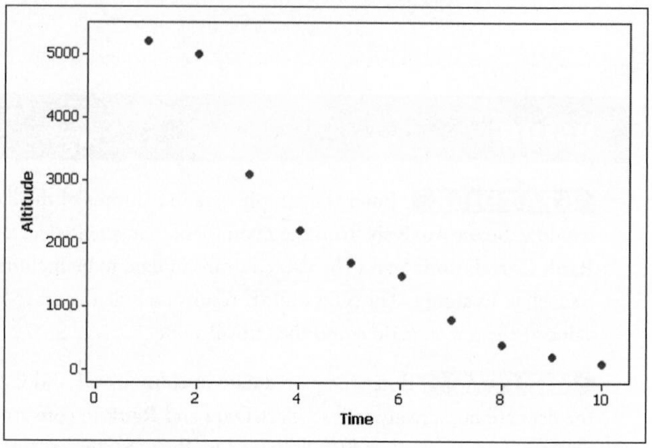

Testing for Rank Correlation. *In Exercises 7–12, use the rank correlation coefficient to test for a correlation between the two variables. Use a significance level of $\alpha = 0.05$.*

7. Are the Best Televisions the Most Expensive? The following table lists overall quality scores and prices of 40-inch LCD televisions (based on data from *Consumer Reports*). Do these data suggest that you can get better quality by spending more?

Quality score	74	71	68	65	63	62	57
Price (dollars)	2700	3000	3800	2300	2000	1300	1400

8. Are the Best Paints the Most Expensive? The following table lists overall quality scores and prices for a gallon of exterior paints (based on data from *Consumer Reports*). Do these data suggest that you can get better quality by spending more?

Quality score	90	87	82	78	62	56	23	19
Price (dollars)	27	32	19	15	39	24	15	18

9. Judges of Marching Bands Two judges ranked seven bands in the Texas state finals competition of marching bands (Coppell, Keller, Grapevine, Dickinson, Poteet, Fossil Ridge, Heritage), and their rankings are listed below (based on data from the University Interscholastic League). Test for a correlation between the two judges. Do the judges appear to rank about the same or are they very different?

Band	Cpl	Klr	Grp	Dck	Ptt	FR	Her
First judge	1	3	4	7	5	6	2
Second judge	6	4	5	1	3	2	7

10. Judges of Marching Bands In the same competition described in Exercise 9, a third judge ranked the bands with the results shown below. Test for a correlation between the first and third judges. Do the judges appear to rank about the same or are they very different?

Band	Cpl	Klr	Grp	Dck	Ptt	FR	Her
First judge	1	3	4	7	5	6	2
Third judge	3	4	1	5	7	6	2

3. r represents the linear correlation coefficient computed from sample paired data; ρ represents the parameter of the linear correlation coefficient computed from a population of paired data; r_s denotes the rank correlation coefficient computed from sample paired data; ρ_s represents the rank correlation coefficient computed from a population of paired data. The subscript s is used so that the rank correlation coefficient can be distinguished from the linear correlation coefficient r. The subscript does not represent the standard deviation s. It is used in recognition of Charles Spearman, who introduced the rank correlation method.

4. The efficiency rating of 0.91 indicates that with all other factors being the same, rank correlation requires 100 pairs of sample observations to achieve the same results as 91 pairs of observations with the parametric test using linear correlation, assuming that the stricter requirements for using linear correlation are met.

5. $r_s = 1$. Critical values are -0.886 and 0.886. Reject the null hypothesis of $\rho_s = 0$. There is sufficient evidence to support a claim of a correlation between distance and time.

6. $r_s = -1$. Critical values are -0.648 and 0.648. Reject the null hypothesis of $\rho_s = 0$. There is sufficient evidence to support a claim of a correlation between altitude and time.

7. $r_s = 0.821$. Critical values: -0.786, 0.786. Reject the null hypothesis of $\rho_s = 0$. There is sufficient evidence to support the claim of a correlation between the quality scores and prices. These results do suggest that you get better quality by spending more.

11. Measuring Seals from Photos Listed below are the overhead widths (in cm) of seals measured from photographs and the weights of the seals (in kg). The data are based on "Mass Estimation of Weddell Seals Using Techniques of Photogrammetry," by R. Garrott of Montana State University. The purpose of the study was to determine if weights of seals could be determined from overhead photographs. Is there sufficient evidence to conclude that there is a correlation between overhead widths of seals from photographs and the weights of the seals?

Overhead width	7.2	7.4	9.8	9.4	8.8	8.4
Weight	116	154	245	202	200	191

12. Crickets and Temperature The association between the temperature and the number of times a cricket chirps in 1 min was studied. Listed below are the numbers of chirps in 1 min and the corresponding temperatures in degrees Fahrenheit (based on data from *The Song of Insects* by George W. Pierce, Harvard University Press). Is there sufficient evidence to conclude that there is a relationship between the number of chirps in 1 min and the temperature?

Chirps in 1 min	882	1188	1104	864	1200	1032	960	900
Temperature (°F)	69.7	93.3	84.3	76.3	88.6	82.6	71.6	79.6

Appendix B Data Sets. ***In Exercises 13–16, use the data from Appendix B to test for rank correlation with a 0.05 significance level.***

13. Blood Pressure Refer to the measured systolic and diastolic blood pressure measurements of 40 randomly selected males in Data Set 1 in Appendix B and test the claim that among men, there is a correlation between systolic blood pressure and diastolic blood pressure.

14. IQ and Brain Volume Refer to Data Set 6 in Appendix B and use the paired data consisting of brain volume (cm^3) and IQ score.

15. Flight Delays Refer to Data Set 15 in Appendix B and use the departure delay times and the arrival delay times.

16. Earthquakes Refer to Data Set 16 in Appendix B and use the magnitudes and depths from the earthquakes.

10-4 Beyond the Basics

17. Finding Critical Values An alternative to using Table A-6 to find critical values for rank correlation is to compute them using this approximation:

$$r_s = \pm\sqrt{\frac{t^2}{t^2 + n - 2}}$$

Here t is the critical t value from Table A-3 corresponding to the desired significance level and $n - 2$ degrees of freedom. Use this approximation to find critical values of r_s for the following cases. Are the resulting approximations close to the values from Table A-6?

a. $n = 8$, $\alpha = 0.05$

b. $n = 30$, $\alpha = 0.01$

8. $r_s = 0.467$. Critical values: -0.738, 0.738. Fail to reject the null hypothesis of $\rho_s = 0$. There is not sufficient evidence to support the claim of a correlation between the quality scores and prices. These results do not suggest that you get better quality by spending more.
9. $r_s = -0.929$. Critical values: -0.786, 0.786. Reject the null hypothesis of $\rho_s = 0$. There is sufficient evidence to support the claim of a correlation between the two judges. Examination of the results shows that the first and third judges appear to have opposite rankings.
10. $r_s = 0.607$. Critical values: -0.786, 0.786. Fail to reject the null hypothesis of $\rho_s = 0$. There is not sufficient evidence to support the claim of a correlation between the two judges. The two judges appear to rank the bands very differently.
11. $r_s = 1$. Critical values: -0.886, 0.886. Reject the null hypothesis of $\rho_s = 0$. There is sufficient evidence to conclude that there is a correlation between overhead widths of seals from photographs and the weights of the seals.
12. $r_s = 0.857$. Critical values: -0.738, 0.738. Reject the null hypothesis of $\rho_s = 0$. There is sufficient evidence to conclude that there is a correlation between the number of chirps in 1 min and the temperature.
13. $r_s = 0.394$. Critical values: -0.314, 0.314. Reject the null hypothesis of $\rho_s = 0$. There is sufficient evidence to conclude that there is a correlation between the systolic and diastolic blood pressure levels in males.
14. $r_s = 0.106$. Critical values: -0.447, 0.447. Fail to reject the null hypothesis of $\rho_s = 0$. There is not sufficient evidence to conclude that there is a correlation between brain volumes and IQ scores.
15. $r_s = 0.651$. Critical values: -0.286, 0.286. Reject the null hypothesis of $\rho_s = 0$. There is sufficient evidence to conclude that there is a correlation between departure delay times and arrival delay times.
16. $r_s = 0.0428$. Critical values: -0.280, 0.280. Fail to reject the null hypothesis of $\rho_s = 0$. There is not sufficient evidence to conclude that there is a correlation between magnitudes and depths of earthquakes.
17. a. ± 0.707 is not very close to the values of ± 0.738 found in Table A-6.
    b. ± 0.463 is quite close to the values of ± 0.467 found in Table A-6.

Chapter 10 Review

The core content of this chapter focuses on a correlation between two variables, so much of the chapter deals with paired data. Section 9-4 includes methods for forming inferences from two dependent samples, so Section 9-4 also deals with paired data, but this chapter and Section 9-4 have fundamentally different objectives. Consider the table below with the two different scenarios that follow to see the basic difference between them.

x	75	83	66	90	55
y	87	83	69	92	72

Scenario 1: Dependent Samples The data in the table represent measurements of strength. The variable x is the measurement *before* a training program and y is the measurement of the same person *after* a training program. Here, the issue is whether the training program is effective, so we want to test the hypothesis that the sample differences are from a population with a mean that is less than 0, indicating that the posttraining scores are higher than the pretraining scores. The methods of Section 9-4 apply.

Scenario 2: Paired Sample Data The data in the table are measurements from a sample of five subjects. The x values are scores on a test of depth perception, and the y values are the times (sec) required to complete a particular task. Here, the issue is whether there is a correlation between the two variables, and the methods of this chapter apply. The following is a brief review of those methods.

- Section 10-2 includes methods for using scatterplots and the linear correlation coefficient r to determine whether there is sufficient evidence to support a claim of a linear correlation between two variables.

- In Section 10-3 we presented methods for finding the equation of the regression line that best fits a graph of the paired data. When the regression line fits the data reasonably well, the regression equation can be used to predict the value of a variable, given some value of the other variable.

- In Section 10-4 we introduce the concept of *rank correlation*, which illustrates one of the methods of "nonparametric statistics." Methods of nonparametric statistics do not have the stricter requirements of parametric methods, such as the requirement of a bivariate normal distribution, as described in Section 10-2.

Chapter Quick Quiz

The exercises are based on the following sample data obtained from different second-year medical students who took blood pressure measurements of the same person (based on data from Marc Triola, MD).

Systolic	138	130	135	140	120
Diastolic	82	91	100	100	80

1. If you plan to use a 0.05 significance level in a test of a correlation between the systolic and diastolic readings, what are the critical values of r?

2. The linear correlation coefficient r is found to be 0.585. What should you conclude?

3. The sample data result in a linear correlation coefficient of $r = 0.585$ and the regression equation $\hat{y} = -1.99 + 0.698x$. What is the best predicted diastolic reading given a systolic reading of 125, and how was it found?

4. Repeat the preceding exercise assuming that the linear correlation coefficient is $r = 0.989$.

1. ± 0.878

2. Based on the critical values of ± 0.878 (assuming a 0.05 significance level), conclude that there is not sufficient evidence to support the claim of a linear correlation between systolic and diastolic readings.

3. The best predicted diastolic reading is 90.6, which is the mean of the five sample diastolic readings.

4. The best predicted diastolic reading is 85.3, which is found by substituting 125 for x in the regression equation.

5. Given that the linear correlation coefficient r is found to be 0.585, what is the proportion of the variation in diastolic blood pressure that is explained by the linear relationship between systolic and diastolic blood pressure?

5. $r^2 = 0.342$

6. True or false: If there is no linear correlation between systolic and diastolic blood pressure, then those two variables are not related in any way.

6. False.

7. True or false: If the sample data lead us to the conclusion that there is sufficient evidence to support the claim of a linear correlation between systolic and diastolic blood pressure, then we could also conclude that systolic blood pressure causes diastolic blood pressure.

7. False.

8. If each systolic reading is exactly twice the diastolic reading, what is the value of the linear correlation coefficient r?

8. $r = 1$

9. If you had computed the value of the linear correlation coefficient to be 3.335, what should you conclude?

9. Because r must be between -1 and 1 inclusive, the value of 3.335 is the result of an error in the calculations.

10. If the sample data were to result in the scatterplot shown here, what is the value of the linear correlation coefficient r?

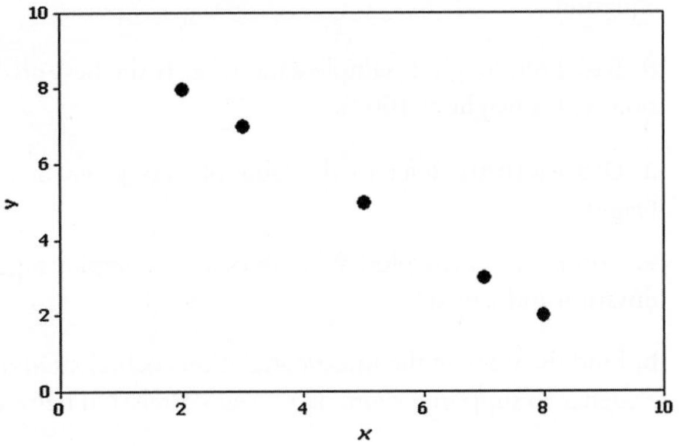

10. $r = -1$

Review Exercises

1. Old Faithful The table below lists measurements from eight different eruptions of the Old Faithful geyser in Yellowstone National Park. The data consist of the duration (sec) of the eruption, height (ft) of the eruption, time interval (min) before the eruption, and time interval (min) after the eruption. Shown below are Minitab results obtained by using the durations and interval-after times.

a. Determine whether there is sufficient evidence to support a claim of a linear correlation between duration and interval-after time for eruptions of the Old Faithful geyser.

b. What percentage of the variation in interval-after times can be explained by the linear correlation between interval-after times and durations?

c. Letting y represent interval-after time and letting x represent duration time, identify the regression equation.

d. If an eruption has a duration of 200 sec, what is the best predicted value for the time interval after the eruption to the next eruption?

1. a. $r = 0.926$. Critical values: $r = \pm 0.707$ (assuming a 0.05 significance level). P-value $= 0.001$. There is sufficient evidence to support the claim that there is a linear correlation between duration and interval-after time.
b. 85.7%
c. $\hat{y} = 34.8 + 0.234x$
d. 81.6 min

Eruptions of the Old Faithful Geyser

Duration	240	120	178	234	235	269	255	220
Height	140	110	125	120	140	120	125	150
Interval Before	98	90	92	98	93	105	81	108
Interval After	92	65	72	94	83	94	101	87

```
Pearson correlation of DURATION and AFTER = 0.926
P-Value = 0.001

The regression equation is
AFTER = 34.8 + 0.234 DURATION
```

2. Old Faithful Refer to the table of data given in Exercise 1 and use the heights and interval-after times.

a. Construct a scatterplot. What does the scatterplot suggest about a linear correlation between heights of eruptions and interval-after times?

b. Find the value of the linear correlation coefficient and determine whether there is sufficient evidence to support a claim of a linear correlation between heights of eruptions and interval-after times.

c. Letting y represent the interval-after time and letting x represent height, find the regression equation.

d. Based on the given sample data, what is the best predicted interval-after time for an eruption with a height of 100 ft?

3. Old Faithful Refer to the table of data given in Exercise 1 and use the durations and heights.

a. Construct a scatterplot. What does the scatterplot suggest about a linear correlation between duration and height?

b. Find the value of the linear correlation coefficient and determine whether there is sufficient evidence to support a claim of a linear correlation between duration and height.

c. Letting y represent height and letting x represent duration, find the regression equation.

d. If an eruption has a duration of 200 sec, what is its best predicted height?

4. Time and Motion In a physics experiment at Doane College, a soccer ball was thrown upward from the bed of a moving truck. The table below lists the time (sec) that has lapsed from the throw and the height (m) of the soccer ball. What do you conclude about the relationship between time and height? What horrible mistake would be easy to make if the analysis is conducted without a scatterplot?

Time (sec)	0.0	0.2	0.4	0.6	0.8	1.0	1.2	1.4	1.6	1.8
Height (m)	0.0	1.7	3.1	3.9	4.5	4.7	4.6	4.1	3.3	2.1

5. Student and *U.S. News and World Report* Rankings of Colleges Each year, *U.S. News and World Report* publishes rankings of colleges based on statistics such as admission rates, graduation rates, class size, faculty–student ratio, faculty salaries, and peer ratings of administrators. Economists Christopher Avery, Mark Glickman, Caroline Minter Hoxby, and Andrew Metrick took an alternative approach of analyzing the college choices of 3240 high-achieving school seniors. They examined the colleges that offered admission along with the colleges that the students chose to attend. The table below lists rankings for a small sample of colleges. Find the value of the rank correlation coefficient and use it to determine whether there is a correlation between the student rankings and the rankings of the magazine.

Student ranks	1	2	3	4	5	6	7	8
U.S. News and World Report ranks	1	2	5	4	7	6	3	8

2. a. The scatterplot suggests that there is not sufficient sample evidence to support the claim of a linear correlation between heights of eruptions and interval-after times.

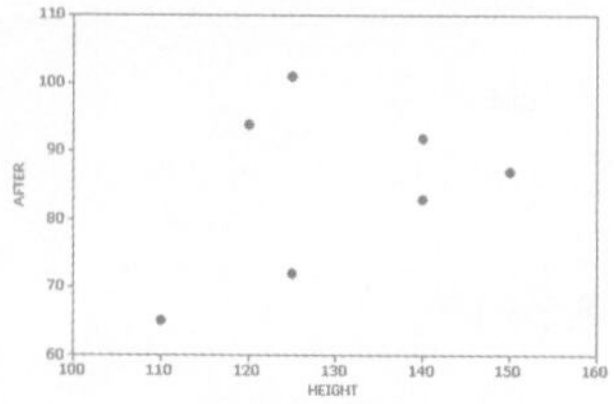

b. $r = 0.269$. Critical values: $r = \pm 0.707$ (assuming a 0.05 significance level). P-value $= 0.519$. There is not sufficient evidence to support the claim that there is a linear correlation between height and interval-after time.

c. $\hat{y} = 54.3 + 0.246x$

d. 86.0 min

3. a. The scatterplot suggests that there is not sufficient sample evidence to support the claim of a linear correlation between duration and height.

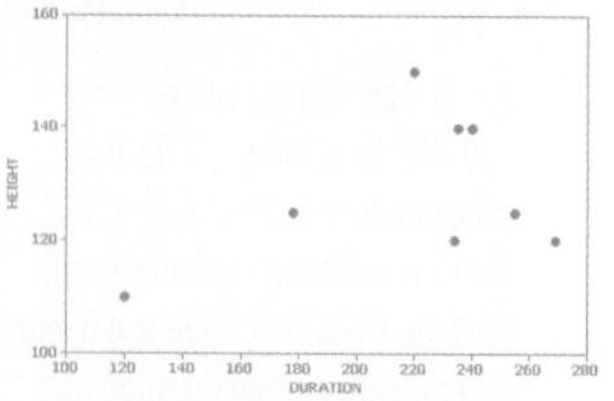

b. $r = 0.389$. Critical values: $r = \pm 0.707$ (assuming a 0.05 significance level). P-value $= 0.340$. There is not sufficient evidence to support the claim that there is a linear correlation between duration and height.

c. $\hat{y} = 105 + 0.108x$

d. 128.8 ft

4. $r = 0.450$. Critical values: $r = \pm 0.632$ (assuming a 0.05 significance level). P-value $= 0.192$. There is not sufficient evidence to support the claim that there is a linear correlation between time and height. Although there is no *linear* correlation between time and height, the scatterplot shows a very distinct pattern revealing that time and height are associated by some function that is not linear.

5. $r_s = 0.714$. Critical values: ± 0.738. Fail to reject the null hypothesis of $\rho_s = 0$. There is not sufficient evidence to support the claim that there is a correlation between the student ranks and the magazine ranks. When ranking colleges, students and the magazine do not appear to agree.

Cumulative Review Exercises

Please be aware that some of the following problems may require knowledge of concepts presented in previous chapters.

Effectiveness of Diet. ***Listed below are weights (lb) of subjects before and after the Zone diet. (Data are based on results from "Comparison of the Atkins, Ornish, Weight Watchers, and Zone Diets for Weight Loss and Heart Disease Risk Reduction," by Dansinger et al.,*** **Journal of the American Medical Association,** ***Vol. 293, No. 1.) Use the data for Exercises 1–5.***

Before	183	212	177	209	155	162	167	170
After	179	198	180	208	159	155	164	166

1. Diet Clinical Trial: Statistics Find the mean and standard deviation of the "before − after" differences.

2. Diet Clinical Trial: *z* Score Using only the weights before the diet, identify the highest weight and convert it to a *z* score. In the context of these sample data, is that highest value an "unusual" weight? Why or why not?

3. Diet Clinical Trial: Hypothesis Test Use a 0.05 significance level to test the claim that the diet is effective.

4. Diet Clinical Trial: Confidence Interval Construct a 95% confidence interval estimate of the mean weight of subjects before the diet. Write a brief statement interpreting the confidence interval.

5. Diet Clinical Trial: Correlation Use the before/after weights listed above.

a. Test for a correlation between the before and after weights.

b. If each subject were to weigh exactly the same after the diet as before, what would be the value of the linear correlation coefficient?

c. If all subjects were to lose 5% of their weight from the diet, what would be the value of the linear correlation coefficient found from the before/after weights?

d. What do the preceding results suggest about the suitability of correlation as a tool for testing the effectiveness of the diet?

6. Birth Weights Birth weights in the United States are normally distributed with a mean of 3420 g and a standard deviation of 495 g.

a. What percentage of babies are born with a weight greater than 3500 g?

b. Find P_{10}, which is the 10th percentile.

c. The Rockland Medical Center requires special treatment for babies that are less than 2450 g (unusually underweight) or more than 4390 g (unusually overweight). What is the percentage of babies who require special treatment? Under these conditions, do many babies require special treatment?

1. $\bar{x} = 3.3$ lb, $s = 5.7$ lb
2. The highest weight before the diet is 212 lb, which converts to $z = 1.55$. The highest weight is not unusual because its z score of 1.55 shows that it is within 2 standard deviations of the mean.
3. H_0: $\mu_d = 0$. H_1: $\mu_d > 0$. Test statistic: $t = 1.613$. Critical value: $t = 1.895$. P-value > 0.05 (Tech: 0.075). Fail to reject H_0. There is not sufficient evidence to support the claim that the diet is effective.
4. $161.8 \text{ lb} < \mu < 197.0$ lb. We have 95% confidence that the interval limits of 161.8 lb and 197.0 lb contain the true value of the mean of the population of all subjects before the diet.
5. a. $r = 0.965$. Critical values: $r = \pm 0.707$ (assuming a 0.05 significance level). P-value $= 0.000$. There is sufficient evidence to support the claim that there is a linear correlation between before and after weights.
   b. $r = 1$
   c. $r = 1$
   d. The effectiveness of the diet is determined by the amounts of weight lost, but the linear correlation coefficient is not sensitive to different amounts of weight loss. Correlation is not a suitable tool for testing the effectiveness of the diet.
6. a. 43.64% (Tech: 43.58%)
   b. 2786.4 g (Tech: 2785.6 g)
   c. 5.00%. Yes, many of the babies do require special treatment.

7. a. H_0: $p = 0.5$. H_1: $p > 0.5$. Test statistic: $z = 3.84$. Critical value: $z = 1.645$. *P*-value: 0.0001. Reject H_0. There is sufficient evidence to support the claim that the majority of us say that honesty is always the best policy.
b. The sample is a voluntary response (or self-selected) sample. This type of sample suggests that the results given in part (a) are not necessarily valid.

8. a. Nominal.
b. Ratio.
c. Discrete.
d. 0.575
e. Parameter.

9. a. 0.330
b. 0.870
c. 0.972
d. 7.37%

10.

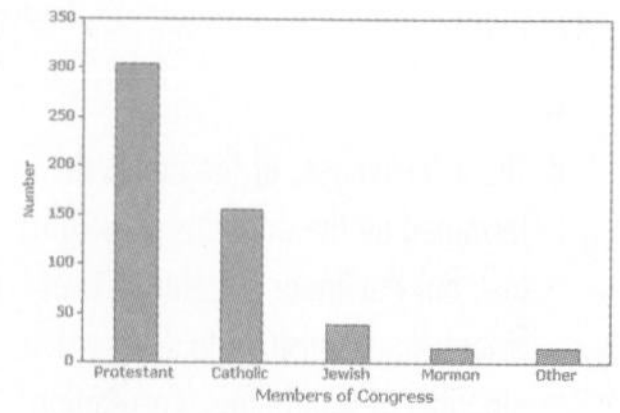

7. Honesty In a *USA Today* survey of 456 subjects, 269 answered "yes" to this question: "Is honesty always the best policy?"

a. Use a 0.05 significance level to test the claim that the majority of us say that honesty is always the best policy.

b. The survey results were obtained from Internet users who visited the *USA Today* Web site and chose to respond to the question that was posted. What is the term used to describe this type of sample? What does this sampling method suggest about the results given in part (a)?

Congress and Religion. ***Based on data from the Pew Forum on Religion and Political Life, the members of Congress have these religious affiliations: Catholic (156), Jewish (39), Protestant (304), Mormon (15), other (15). There were six members of Congress who did not specify a religious affiliation, and the data are current at the time of this writing. Ignore the six unspecified religious affiliations and use these data for Exercises 8–10.***

8. Congress and Religion

a. What is the level of measurement of the religious affiliations (Catholic, Jewish, and so on)?

b. What is the level of measurement of the frequency counts (156, 39, and so on)?

c. Are the frequency counts data that are discrete or continuous?

d. What proportion of the members of Congress are Protestant?

e. Given that the data describe the population of members of Congress, is the result from part (c) a statistic or a parameter?

9. Congress and Religion: Probability

a. If two different members of Congress are randomly selected, what is the probability that they are both Protestant?

b. If a member of Congress is randomly selected, what is the probability that he or she is Catholic or Protestant?

c. If a member of Congress is randomly selected, what is the probability that he or she is not Mormon?

d. What *percentage* of members of Congress are Jewish?

10. Congress and Religion: Graph Construct the graph that is most effective in depicting the religious affiliations of members of Congress.

Technology Project

IQ scores are commonly measured using the Wechsler Adult Intelligence Scale (WAIS). Scores are obtained from tests. WAIS scores are normally distributed with a mean of 100 and a standard deviation of 15.

Much effort is spent studying IQ scores of identical twins that were separated at birth and raised apart in different environments. Identical twins occur when a single fertilized egg splits in two, so both twins share the same genetic makeup. By obtaining IQ scores of identical twins separated at birth, researchers hope to identify the effects of heredity and environment on intelligence.

a. The table below lists IQ scores from 10 sets of twins, but technology was used to simulate the data. Use the methods of Section 10-2 to test for a linear correlation between the IQ scores of the first-born twins and the IQ scores of the second-born twins. Use a significance level of 0.10.

First-Born Twin	121	113	91	118	81	77	80	90	103	95
Second-Born Twin	108	110	79	111	138	96	100	80	102	109

b. Many technologies (including STATDISK, Minitab, Excel, StatCrunch, and the TI-83/84 Plus calculator) have a feature that allows you to randomly generate data from a normally distributed population with any mean and standard deviation. Use a technology to randomly generate IQ scores for each of 10 pairs of twins, as in the above table. Using your own simulated IQ scores, test for a linear correlation between the IQ scores of the first-born twin and the IQ scores of the second-born twin. Use a significance level of 0.10.

c. For the test conducted in part (b), a type I error is the mistake of rejecting a true null hypothesis, which, in this case, is to conclude that there is a linear correlation when in reality there is no linear correlation. What proportion of simulated sets of twins should result in a type I error? Repeat part (b) to verify that this proportion is approximately correct.

from data TO DECISION

Critical Thinking:
Is replication validation?

The Chapter Problem includes Table 10-1, which lists shoe print lengths and heights for five male subjects. The data in Table 10-1 are from Data Set 2 in Appendix B. Data Set 2 lists a few different measurements from a sample of 19 males and 21 females. Listed below are foot lengths (cm) and heights (cm) for a sample of 50 males used in an anthropometric survey conducted by the U.S. Army. The data are paired according to their corresponding positions. The first male has a foot length of 26.0 cm and a height of 173.5 cm, the second male has a foot length of 29.0 cm and a height of 183.0 cm, and so on.

Foot Length (cm)

26.0	29.0	25.4	27.1	24.0	26.3	26.5	27.2	27.3	27.1
26.8	27.2	26.5	25.1	25.5	25.3	26.9	28.2	26.5	28.1
27.7	25.9	26.0	28.5	27.9	26.9	26.0	24.4	26.4	28.5
25.2	26.0	27.3	27.1	29.7	28.9	29.7	25.7	28.8	25.2
29.0	26.7	26.3	27.8	27.7	27.2	26.7	29.0	28.6	27.1

Height (cm)

173.5	183.0	172.6	178.3	166.9	175.8	178.8	182.9	174.5	164.3
169.8	172.7	173.4	168.4	166.7	168.0	178.5	188.5	184.1	170.2
173.8	173.2	173.9	187.7	172.5	170.4	174.8	164.7	169.9	176.1
167.1	164.4	173.9	174.9	184.5	195.4	184.4	160.1	184.6	167.2
193.4	186.2	169.4	174.0	173.9	171.9	172.2	177.6	189.7	173.0

Analyzing the Results

1. Use the given data to construct a scatterplot, then use the methods of Section 10-2 to test for a linear correlation between foot length and height. Compare the results to those found using the foot lengths and heights of the 19 males listed in Data Set 2 in Appendix B.

2. Use the given data to find the equation of the regression line. Let the response (y) variable be height. Compare the results to those found using the foot lengths and heights of the 19 males listed in Data Set 2 in Appendix B.

3. It was noted in the Chapter Problem that you can estimate a person's height by dividing foot length by 0.15. (An equivalent calculation is to estimate height by multiplying foot length by 6.67.) How does this rule compare to the use of a regression equation?

Cooperative Group Activities

1. In-class activity The Chapter Problem involves the relationship between shoe print length and height. For each student in the class, measure those two variables. Test for a linear correlation and identify the equation of the regression line. Measure the shoe print length of the professor and use it to estimate his or her height. How close is the estimated height to the actual height?

2. In-class activity Divide into groups of 8 to 12 people. For each group member, measure the person's height and also measure his or her navel height, which is the height from the floor to the navel. Is there a correlation between height and navel height? If so, find the regression equation with height expressed in terms of navel height. According to one theory, the average person's ratio of height to navel height is the golden ratio: $(1 + \sqrt{5})/2 \approx 1.6$. Does this theory appear to be reasonably accurate?

3. In-class activity Divide into groups of 8 to 12 people. For each group member, measure height and arm span. For the arm span, the subject should stand with arms extended, like the wings on an airplane. Using the paired sample data, is there a correlation between height and arm span? If so, find the regression equation with height expressed in terms of arm span. Can arm span be used as a reasonably good predictor of height?

4. In-class activity Divide into groups of 8 to 12 people. For each group member, use a string and ruler to measure head circumference and forearm length. Is there a relationship between these two variables? If so, what is it?

5. In-class activity Use a ruler as a device for measuring reaction time. One person should suspend the ruler by holding it at the top while the subject holds his or her thumb and forefinger at the bottom edge ready to catch the ruler when it is released. Record the distance that the ruler falls before it is caught. Convert that distance to the time (in seconds) that it took the subject to react and catch the ruler. (If the distance is measured in inches, use $t = \sqrt{d/192}$. If the distance is measured in centimeters, use $t = \sqrt{d/487.68}$.) Test each subject once with the right hand and once with the left hand, and record the paired data. Test for a correlation. Find the equation of the regression line. Does the equation of the regression line suggest that the dominant hand has a faster reaction time?

6. In-class activity Divide into groups of 8 to 12 people. Record the pulse rate of each group member while seated. Then record the pulse rate of each group member while standing. Is there a relationship between sitting and standing pulse rate? If so, what is it?

7. In-class activity Divide into groups of three or four people. Appendix B includes many data sets not yet included in examples or exercises in this chapter. Search Appendix B for a pair of variables of interest, then investigate correlation and regression. State your conclusions and try to identify practical applications.

8. Out-of-class activity Divide into groups of three or four people. Investigate the relationship between two variables by collecting your own paired sample data and using the methods of this chapter to determine whether there is a significant linear correlation. Also identify the regression equation and describe a procedure for predicting values of one of the variables when given values of the other variable. Suggested topics:

- Is there a relationship between taste and cost of different brands of chocolate chip cookies (or colas)? Taste can be measured on some number scale, such as 1 to 10.
- Is there a relationship between salaries of professional baseball (or basketball, or football) players and their season achievements?
- Is there a relationship between the lengths of men's (or women's) feet and their heights?
- Is there a relationship between student grade point averages and the amount of television watched? If so, what is it?

11 Chi-Square and Analysis of Variance

11

chapter problem

Are four quarters the same as one dollar?

Are four quarters the same as one dollar? There may be a temptation to answer yes and move on to the next chapter. Using our extensive knowledge base of arithmetic and currency, we do know that in terms of pure purchasing power, four quarters are the same as one dollar. But this Chapter Problem deals more with psychology than basic arithmetic. It involves the *denomination effect,* which refers to the tendency of people to spend money more readily when it is in the form of lower denominations (such as four quarters) instead of higher denominations (such as a $1 bill). This is discussed in "The Denomination Effect" by Priya Raghubir and Joydeep Srivastava, *Journal of Consumer Research*, Vol. 36. In one study of this phenomenon, 43 college students were each given $1 in the form of four quarters, while 46 other college students were each given $1 in the form of a dollar bill. All of the students were then given two choices: (1) Keep the money; (2) spend the money on gum. The results are given in Table 11-1.

Table 11-1 Results from a Study of the Denomination Effect

	Purchased Gum	Kept the Money
Students Given Four Quarters	27	16
Students Given a $1 Bill	12	34

Table 11-1 is called a *two-way table* (because the data are partitioned according to *two* different variables) or a *contingency table* (because we want to determine whether there is a *dependence* between the row and column categories). Similar tables occur often in real applications, so they are extremely important in the study of statistics.

Analyzing the Results

The purpose of the gum/money study is to determine whether the form of the gift (four quarters or $1 bill) appears to affect the decision to purchase gum or keep the money. Table 11-1 has two rows corresponding to the variable of the form of the money, and it has two columns corresponding to the variable of how the subjects chose to use the money. We want to determine whether the row variable has an effect on the column variable. If there is an effect, then people behave differently with money in different denominations, so the "denomination effect" appears to be real. Analyzing the data in Table 11-1, we see that there does appear to be an effect, because 63% of those given four quarters chose to spend the money, but only 26% of those given a $1 bill chose to spend the money. But is that a *significant* difference? That determination can be made using the methods introduced in Section 11-3.

Note to Instructor
Recommendation: Cover Section 11-3. Analysis of survey results often involves the use of contingency tables, so the concepts of Section 11-3 are used often in real applications.

χ^2 notation: We usually use Greek letters for population parameters, but here we use χ^2 for a test statistic. Although not consistent, this is very common notation in this context. (Ralph Waldo Emerson said that "a foolish consistency is the hobgoblin of little minds.")

11-1 Review and Preview

By introducing basic concepts of estimating population parameters (with confidence intervals) and methods of hypothesis tests, Chapters 7 and 8 moved us into methods of inferential statistics. Chapters 9 and 10 then involved us with different configurations of data.

In Section 11-2 we consider hypothesis tests of a claim that observed frequency counts agree with some claimed distribution, so there is a "good fit" of the sample data with the claimed distribution. In Section 11-3 we analyze contingency tables (or two-way frequency tables), which consist of frequency counts arranged in a table with at least two rows and two columns. The objective is to determine whether there appears to be some dependence between the row variable and the column variable. In Section 11-4 we introduce analysis of variance as a method for testing equality of three or more population means.

Sections 11-2 and 11-3 use the same χ^2 (chi-square) distribution that was first introduced in Section 7-4. See Section 7-4 for a quick review of properties of the χ^2 distribution. Section 11-4 introduces a new distribution: the *F* distribution.

Note to Instructor
Section 11-2 requires the χ^2 distribution, so be sure to discuss it if it has not yet been introduced. Basic concepts of the χ^2 distribution are included here as well as in Section 7-4 and again in Section 8-5.

Encourage the class to develop their own reasoning process for why the tests of this section are all right-tailed. Ask the class these questions: If there are large discrepancies between the observed frequencies and those that are expected, what do we know about the $O - E$ values? The $(O - E)^2$ values? The value of the χ^2 test statistic? Where on the χ^2 distribution do large discrepancies fall?

11-2 Goodness-of-Fit

Key Concept By "goodness-of-fit" we mean that sample data consisting of observed frequency counts arranged in a single row or column (called a *one-way frequency table*) agree with some particular distribution being considered. We will use a hypothesis test for the claim that the observed frequency counts agree with some claimed distribution.

DEFINITION A **goodness-of-fit test** is used to test the hypothesis that an observed frequency distribution fits (or conforms to) some claimed distribution.

Objective

Conduct a goodness-of-fit test. That is, conduct a hypothesis test to determine whether a single row (or column) of frequency counts agrees with some specific distribution (such as uniform or normal).

Notation

O represents the *observed frequency* of an outcome, found from the sample data.
E represents the *expected frequency* of an outcome, found by assuming that the distribution is as claimed.
k represents the *number of different categories* or cells.
n represents the total *number of trials* (or observed sample values).

Requirements

1. The data have been randomly selected.
2. The sample data consist of frequency counts for each of the different categories.
3. For each category, the *expected* frequency is at least 5. (The expected frequency for a category is the frequency that would occur if the data actually have the distribution that is being claimed. There is no requirement that the *observed* frequency for each category must be at least 5.)

Null and Alternative Hypotheses

H_0: The frequency counts agree with the claimed distribution.
H_1: The frequency counts do not agree with the claimed distribution.

Test Statistic for Goodness-of-Fit Tests

$$\chi^2 = \sum \frac{(O - E)^2}{E}$$

***P*-values:** P-values are typically provided by technology, or a range of P-values can be found from Table A-4.

Critical values: **1.** Critical values are found in Table A-4 by using $k - 1$ degrees of freedom, where k is the number of categories.

2. Goodness-of-fit hypothesis tests are always *right-tailed.*

Finding Expected Frequencies

Conducting a goodness-of-fit test requires that we identify the observed frequencies, then determine the frequencies expected with the claimed distribution. Table 11-2 (on the next page) includes observed frequencies with a sum of 100, so $n = 100$. If we assume that the 100 digits were obtained from a population in which all digits are equally likely, then we *expect* that each digit should occur in 1/10 of the 100 trials, so each of the 10 expected frequencies is given by $E = 10$. In general, if we are assuming that all of the expected frequencies are equal, each expected frequency is $E = n/k$, where n is the total number of observations and k is the number of categories. In other cases in which the expected frequencies are not all equal, we can often find the expected frequency for each category by multiplying the sum of all observed frequencies and the probability p for the category, so $E = np$. We summarize these two procedures here.

- **Expected frequencies are equal: $E = n/k$.**
- **Expected frequencies are not all equal: $E = np$ for each individual category.**

As good as these two preceding formulas for E might be, it is better to use an informal approach. Just ask, "How can the observed frequencies be split up among the different categories so that there is perfect agreement with the claimed distribution?" Also, note that the *observed* frequencies are all whole numbers because they represent actual counts, but the *expected* frequencies need not be whole numbers. If Table 11-2 had 75 observations instead of 100, each expected frequency would be 7.5.

We know that sample frequencies typically differ somewhat from the values we theoretically expect, so we now present the key question: Are the differences between the actual *observed* frequencies O and the theoretically *expected* frequencies E statistically significant? We need a measure of the discrepancy between the O and E values, so we use the test statistic given in the preceding box. (Later we will explain how this test statistic was developed, but you can see that it has differences of $O - E$ as a key component.)

The χ^2 test statistic is based on differences between the observed and expected values. If the observed and expected values are *close,* the χ^2 test statistic will be small and the P-value will be large. If the observed and expected frequencies are *far apart,* the χ^2 test statistic will be large and the P-value will be small. Figure 11-1 on the next page summarizes this relationship. The hypothesis tests of this section are always

Note to Instructor
Briefly review the concept of *expected value,* which was first introduced in Section 5-2. Present a few simple and obvious examples such as this one: Find the expected number of girls born in groups of 100 babies. When students respond with the correct answer of 50, ask them to describe the exact thought process that led to the answer. They will respond that they found 1/2 of 100, which can be generalized as $p \times n$, which leads to $E = np$. Also point out that the expected number of girls among 3 babies is 1.5, so the expected value need not be a whole number.

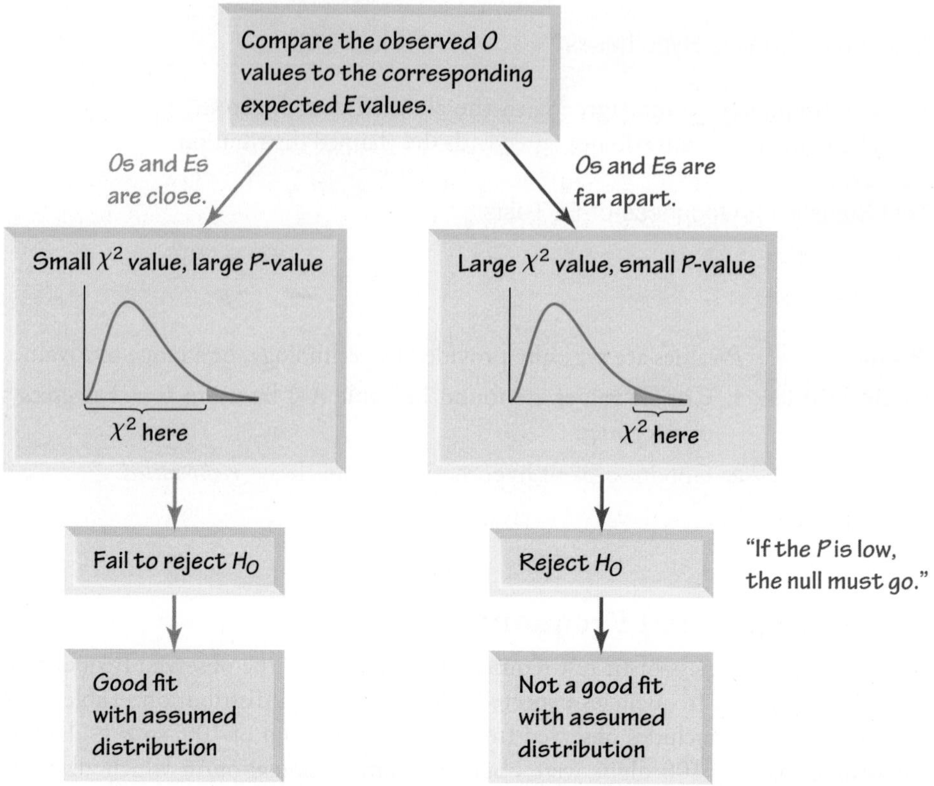

Figure 11-1 **Relationships among the χ^2 Test Statistic, *P*-Value, and Goodness-of-Fit**

right-tailed, because the critical value and critical region are located at the extreme right of the distribution. If confused, just remember this:

> **"If the *P* is low, the null must go."**
>
> **(If the *P*-value is small, reject the null hypothesis that the distribution is as claimed.)**

Table 11-2
Last Digits of Weights

Last Digit	Frequency
0	46
1	1
2	2
3	3
4	3
5	30
6	4
7	0
8	8
9	3

Note to Instructor
The analysis of the last digits of data is often used to detect data that have been reported instead of measured.

Example 1 Last Digits of Weights

A random sample of 100 weights of Californians is obtained, and the last digits of those weights are summarized in Table 11-2 (based on data from the California Department of Public Health). When obtaining weights of subjects, it is extremely important to actually measure their weights instead of asking them to report their weights. By analyzing the *last digits* of weights, researchers can verify that they were obtained through actual measurements instead of being reported. When people report weights, they tend to round, so a weight of 197 lb might be rounded and reported as a more desirable 170 lb. Reported weights tend to have many last digits consisting of 0 or 5. In contrast, if people are actually weighed, the weights tend to have last digits that are uniformly distributed, with 0, 1, 2, . . . , 9 all occurring with roughly the same frequencies.

Test the claim that the sample is from a population of weights in which the last digits do *not* occur with the same frequency. Based on the results, what can we conclude about the procedure used to obtain the weights?

Solution

Requirement check (1) The data come from randomly selected subjects. (2) The data do consist of frequency counts, as shown in Table 11-2. (3) With 100 sample values and 10 categories that are claimed to be equally likely, each expected frequency is 10, so each expected frequency does satisfy the requirement of being a value of at least 5. All of the requirements are satisfied.

The claim that the digits do not occur with the same frequency is equivalent to the claim that the relative frequencies or probabilities of the 10 cells ($p_0, p_1, \ldots, p_9$) are not all equal. (This is equivalent to testing the claim that the distribution of digits is not a uniform distribution.) We will use the critical value method for testing hypotheses (introduced in Section 8-2).

Step 1: The original claim is that the digits do not occur with the same frequency. That is, at least one of the probabilities $p_0, p_1, \ldots, p_9$ is different from the others.

Step 2: If the original claim is false, then all of the probabilities are the same. That is, $p_0 = p_1 = p_2 = p_3 = p_4 = p_5 = p_6 = p_7 = p_8 = p_9$.

Step 3: The null hypothesis must contain the condition of equality, so we have

H_0: $p_0 = p_1 = p_2 = p_3 = p_4 = p_5 = p_6 = p_7 = p_8 = p_9$

H_1: At least one of the probabilities is different from the others.

Step 4: No significance level was specified, so we select $\alpha = 0.05$.

Step 5: Because we are testing a claim about the distribution of the last digits being a uniform distribution (with all of the digits having the same probability), we use the goodness-of-fit test described in this section. The χ^2 distribution is used with the test statistic given in the preceding box.

Step 6: The observed frequencies O are listed in Table 11-2. Each corresponding expected frequency E is equal to 10 (because the 100 digits would be uniformly distributed among the 10 categories). The Excel add-in XLSTAT is used to obtain the results shown in the accompanying screen display, and Table 11-3 shows the computation of the χ^2 test statistic. The test statistic is $\chi^2 = 212.800$. The critical value is $\chi^2 = 16.919$ (found in Table A-4 with $\alpha = 0.05$ in the right tail and degrees of freedom equal to $k - 1 = 9$). The P-value is less than 0.0001. The test statistic and critical value are shown in Figure 11-2.

XLSTAT

Chi-square (Observed value)	212.8000
Chi-square (Critical value)	16.9190
DF	9
p-value	< 0.0001
alpha	0.05

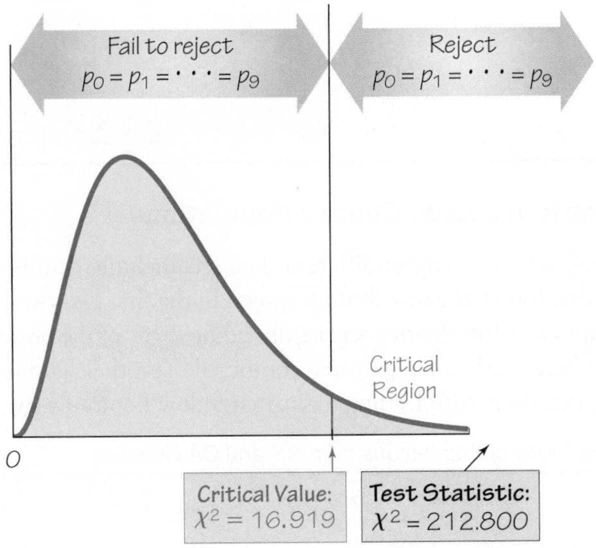

Figure 11-2 **Test of $p_0 = p_1 = p_2 = p_3 = p_4 = p_5 = p_6 = p_7 = p_8 = p_9$**

continued

Which Car Seats Are Safest?

Many people believe that the back seat of a car is the safest place to sit, but is it? University of Buffalo researchers analyzed more than 60,000 fatal car crashes and found that the middle back seat is the safest place to sit in a car. They found that sitting in that seat makes a passenger 86% more likely to survive than those who sit in the front seats, and they are 25% more likely to survive than those sitting in either of the back seats nearest the windows. An analysis of seat belt use showed that when not wearing a seat belt in the back seat, passengers are three times more likely to die in a crash than those wearing seat belts in that same seat. Passengers concerned with safety should sit in the middle back seat wearing a seat belt.

Step 7: If we use the *P*-value method of testing hypotheses, we see that the *P*-value is small (less than 0.0001), so we reject the null hypothesis. If we use the critical value method of testing hypotheses, Figure 11-2 shows that the test statistic falls in the critical region, so there is sufficient evidence to reject the null hypothesis.

Step 8: There is sufficient evidence to support the claim that the last digits do not occur with the same relative frequency.

Interpretation

This goodness-of-fit test suggests that the last digits do not provide a good fit with the claimed uniform distribution of equally likely frequencies. Instead of the subjects being weighed, it appears that they reported their weights. In fact, the weights are from the California Health Interview Survey (CHIS), and the title of that survey indicates that subjects were interviewed, not measured. Because those weights are reported, the reliability of the data is very questionable.

Example 1 involves a situation in which the expected frequencies *E* for the different categories are all equal. The methods of this section can also be used when the expected frequencies are different, as shown in Example 2.

Table 11-3 Calculating the χ^2 Test Statistic for the Last Digits of Weights

Last Digit	Observed Frequency *O*	Expected Frequency *E*	$O - E$	$(O - E)^2$	$\frac{(O-E)^2}{E}$
0	46	10	36	1296	129.6
1	1	10	−9	81	8.1
2	2	10	−8	64	6.4
3	3	10	−7	49	4.9
4	3	10	−7	49	4.9
5	30	10	20	400	40.0
6	4	10	−6	36	3.6
7	0	10	−10	100	10.0
8	8	10	−2	4	0.4
9	3	10	−7	49	4.9

$$\chi^2 = \sum \frac{(O-E)^2}{E} = 212.8$$

Note to Instructor
Benford's law is fun, so Example 2 should generate a reasonable amount of interest.

Example 2 Benford's Law: County Populations

According to *Benford's law*, a variety of different data sets includes numbers with leading (first) digits that follow the distribution shown in the first two rows of Table 11-4. The bottom row lists the frequencies of leading digits of the populations of all 120 counties from New York and California combined. Test the claim that those 120 counties have populations with leading digits that follow Benford's law.

Table 11-4 Leading Digits of Populations from NY and CA Counties

Leading Digit	1	2	3	4	5	6	7	8	9
Benford's Law: Distribution of Leading Digits	30.1%	17.6%	12.5%	9.7%	7.9%	6.7%	5.8%	5.1%	4.6%
CA and NY County Populations	33	22	10	15	10	9	5	7	9

Solution

Requirement check (1) The sample data are not randomly selected from a larger population, but we treat them as a random sample for the purpose of determining whether they are typical results that might be obtained from such a random sample. (2) The sample data do consist of frequency counts. (3) Each expected frequency is at least 5, as will be shown later in this solution. All of the requirements are satisfied. ✓

Step 1: The original claim is that the leading digits fit the distribution given as Benford's law. Using subscripts corresponding to the leading digits, we can express this claim as $p_1 = 0.301$ and $p_2 = 0.176$ and $p_3 = 0.125$ and . . . and $p_9 = 0.046$.

Step 2: If the original claim is false, then at least one of the proportions does not have the value as claimed.

Step 3: The null hypothesis must contain the condition of equality, so we have

H_0: $p_1 = 0.301$ and $p_2 = 0.176$ and $p_3 = 0.125$ and . . . and $p_9 = 0.046$.

H_1: At least one of the proportions is not equal to the given claimed value.

Step 4: The significance level is not specified, so we use the common choice of $\alpha = 0.05$.

Step 5: Because we are testing a claim that the distribution of leading digits fits the distribution given by Benford's law, we use the goodness-of-fit test described in this section. The χ^2 distribution is used with the test statistic given earlier.

Step 6: Table 11-5 shows the calculations of the components of the χ^2 test statistic for the leading digits of 1 and 2. If we include all nine leading digits, we get the test statistic of $\chi^2 = 5.958$, as shown in the accompanying TI-84 Plus calculator display. The critical value is $\chi^2 = 15.507$ (found in Table A-4 with $\alpha = 0.05$ in the right tail and degrees of freedom equal to $k - 1 = 8$). The TI-84 Plus calculator display shows the value of the test statistic as well as the P-value of 0.652. (The bottom row of the display shows the expected values, which can be viewed by scrolling to the right. CNTRB is an abbreviated form of "contribution," and the values are the individual contributions to the total value of the χ^2 test statistic.)

Table 11-5 Calculating the χ^2 Test Statistic for Leading Digits in Table 11-4

Leading Digit	Observed Frequency O	Expected Frequency $E = np$	$O - E$	$(O - E)^2$	$\frac{(O-E)^2}{E}$
1	33	$120 \cdot 0.301 = 36.12$	−3.12	9.7344	0.2695
2	22	$120 \cdot 0.176 = 21.12$	0.88	0.7744	0.0367

Step 7: The P-value of 0.652 is greater than the significance level of 0.05, so there is not sufficient evidence to reject the null hypothesis. (Also, the test statistic of $\chi^2 = 5.958$ does not fall in the critical region bounded by the critical value of 15.507, so there is not sufficient evidence to reject the null hypothesis.)

Step 8: There is not sufficient evidence to warrant rejection of the claim that the 120 counties have populations with leading digits that fit the distribution given by Benford's law.

Safest Seats in a Commercial Jet

A study by aviation writer and researcher David Noland showed that sitting farther back in a commercial jet will increase your chances of surviving in the event of a crash. The study suggests that the chance of surviving is not the same for each seat, so a goodness-of-fit test would lead to rejection of the null hypothesis that every seat has the same probability of surviving. Records from the 20 commercial jet crashes that occurred since 1971 were analyzed. It was found that if you sit in business or first class, you have a 49% chance of surviving a crash, if you sit in coach over the wing or ahead of the wing you have a 56% chance of surviving, and if you sit in the back behind the wing you have a 69% chance of surviving.

In commenting on this study, David Noland stated that he does not seek a rear seat when he flies. He says that because the chance of a crash is so small, he doesn't worry about where he sits, but he prefers a window seat.

TI-84 PLUS

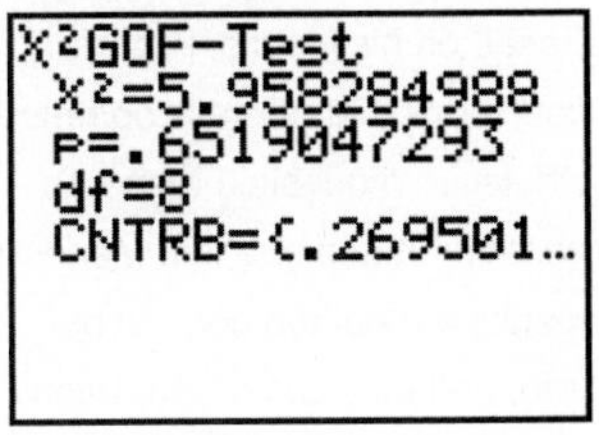

Interpretation

The sample of leading digits does not provide enough evidence to conclude that the Benford distribution is not being followed.

In Figure 11-3 we use a red line to graph the expected proportions given by Benford's law (as in Table 11-4) along with a green line for the observed proportions from Table 11-4. Figure 11-3 allows us to visualize the "goodness-of-fit" between the distribution given by Benford's law and the frequencies that were observed. In Figure 11-3, the red and green lines agree reasonably well, so it appears that the green line for the observed data fits the red line for the expected values reasonably well.

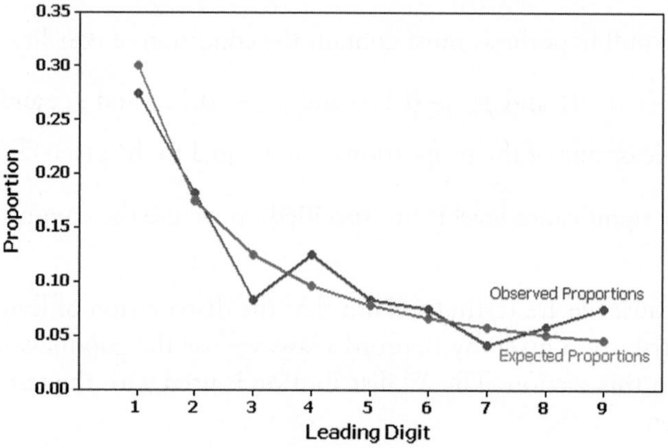

Figure 11-3 **Observed Proportions and Proportions Expected with Benford's Law**

Rationale for the Test Statistic Examples 1 and 2 show that the χ^2 test statistic is a measure of the discrepancy between observed and expected frequencies. Simply summing the differences $O - E$ between observed and expected values tells us nothing, because that sum is always 0. Squaring the $O - E$ gives us a better statistic. (The reasons for squaring the $O - E$ values are essentially the same as the reasons for squaring the $x - \bar{x}$ values in the formula for standard deviation.) The value of $\Sigma(O - E)^2$ measures only the magnitude of the differences, but we need to find the magnitude of the differences relative to what was expected. We need a type of average instead of a cumulative total. This relative magnitude is found through division by the expected frequencies, as in the test statistic.

The theoretical distribution of $\Sigma(O - E)^2/E$ is a discrete distribution because the number of possible values is finite. The distribution can be approximated by a chi-square distribution, which is continuous. This approximation is generally considered acceptable, provided that all expected values E are at least 5. (There are ways of circumventing the problem of an expected frequency that is less than 5, such as combining categories so that all expected frequencies are at least 5. Also, there are other methods that can be used when not all expected frequencies are at least 5.)

The number of degrees of freedom reflects the fact that we can freely assign frequencies to $k - 1$ categories before the frequency for every category is determined. (Although we say that we can "freely" assign frequencies to $k - 1$ categories, we cannot have negative frequencies, nor can we have frequencies so large that their sum exceeds the total of the observed frequencies for all categories combined.)

Mendel's Data Falsified?

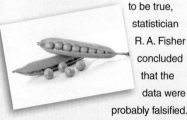

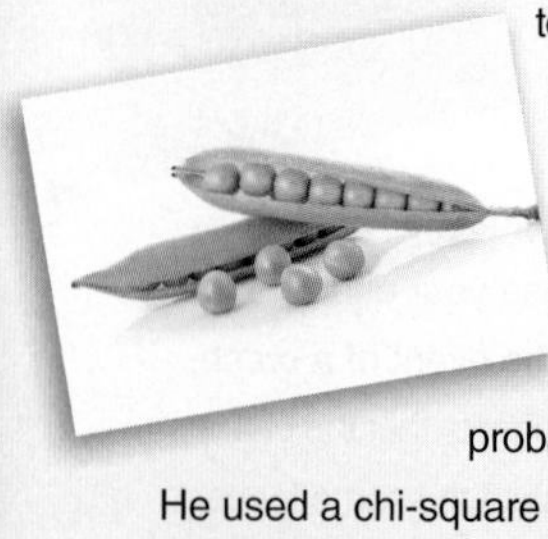

Because some of Mendel's data from his famous genetics experiments seemed too perfect to be true, statistician R. A. Fisher concluded that the data were probably falsified. He used a chi-square distribution to show that when a test statistic is extremely far to the left and results in a *P*-value very close to 1, the sample data fit the claimed distribution almost perfectly, and this is evidence that the sample data have not been randomly selected. It has been suggested that Mendel's gardener knew what results Mendel's theory predicted, and subsequently adjusted results to fit that theory.

Ira Pilgrim wrote in *The Journal of Heredity* that this use of the chi-square distribution is not appropriate. He notes that the question is not about goodness-of-fit with a particular distribution, but whether the data are from a sample that is truly random. Pilgrim used the binomial probability formula to find the probabilities of the results obtained in Mendel's experiments. Based on his results, Pilgrim concludes that "there is no reason whatever to question Mendel's honesty." It appears that Mendel's results are not too good to be true, and they could have been obtained from a truly random process.

using TECHNOLOGY

STATDISK First enter the observed frequencies in the first column of the Data Window. If the expected frequencies are not all equal, enter a second column that includes either expected proportions or actual expected frequencies. Select **Analysis** from the main menu bar, then select the option **Goodness-of-Fit.** Choose between "equal expected frequencies" and "unequal expected frequencies" and enter the data in the dialog box, then click on **Evaluate.**

MINITAB Enter observed frequencies in column C1. If the expected frequencies are not all equal, enter them as proportions in column C2. Select **Stat, Tables,** and **Chi-Square Goodness-of-Fit Test.** Make the entries in the window and click on **OK.**

EXCEL First enter the observed frequencies in one column, then compute the values of the corresponding expected frequencies and enter them in another column. Proceed by using XLSTAT. Click on **XLSTAT** at the top, then click on **Parametric tests,** then select **Multinomial goodness of fit test.** In the "Frequencies" box enter the range of cells containing the observed frequencies, such as A1: A10. In the "Expected frequencies" box, enter the range of cells containing the expected frequencies. Put a check next to the box identified as "Chi-square." Enter the significance level. For example, enter 5 for a 0.05 significance level. Click **OK** to get the results, which will include the chi-square test statistic and the *P*-value.

TI-83/84 PLUS Enter the observed frequencies in list L1, then identify the expected frequencies and enter them in list L2. With a TI-84 Plus calculator, press **STAT**, select **TESTS,** select **x^2 GOF-Test,** then enter L1 and L2 and the number of degrees of freedom when prompted. (The number of degrees of freedom is 1 less than the number of categories.) With a TI-83 Plus calculator, use the program **X2GOF.** Press **PRGM**, select **X2GOF,** then enter L1 and L2 when prompted. Results will include the test statistic and *P*-value.

STATCRUNCH Click on **Open StatCrunch.** Enter the observed frequencies in one column and enter the expected frequencies in another column. Click on **Stat,** then select **Goodness-of-fit,** then select **Chi-Square test.** Identify the columns to be used, then click on **Calculate.** Results will include the test statistic and *P*-value.

11-2 Basic Skills and Concepts

Recommended Assignment
Exercises 1–10, 15–18.

Statistical Literacy and Critical Thinking

1. Quality Family Time The table below lists days of the week selected by a random sample of 1005 subjects who were asked to identify the day of the week that is best for quality family time (based on results from a Pillsbury survey reported in *USA Today*). Consider the claim that the days of the week are selected with a uniform distribution so that all days have the same chance of being selected. If we test that claim using the goodness-of-fit test described in this section, what is it that we actually test?

Sun	Mon	Tues	Wed	Thurs	Fri	Sat
523	20	9	19	11	41	382

2. Expected Value Exercise 1 includes results from a survey of 1005 randomly selected subjects. Consider the claim that when respondents select days of the week, the seven different days have the same chance of being selected. If that claim is true for the 1005 respondents, what is the expected value for each of the seven days? Identify the values of O and E for Sunday.

3. χ^2 Value Without performing actual calculations, examine the frequencies in the table given with Exercise 1. Do you expect the value of the χ^2 test statistic to be large or small? Do you expect the *P*-value to be large or small? Explain.

4. Goodness-of-Fit Test For the goodness-of-fit test described in Exercise 1, identify the number of degrees of freedom and the critical value of χ^2, assuming a 0.05 significance level.

In Exercises 5–20, conduct the hypothesis test and provide the test statistic, critical value, and/or P-value, and state the conclusion.

5. Quality Family Time The accompanying STATDISK display shows results from the claim and data given in Exercise 1. Test that claim.

Due to limited space, answers for some odd numbered exercises are not included here, but they are included in Appendix D.

1. The test is to determine whether the observed frequency counts agree with the claimed uniform distribution so that frequencies for the different days are equally likely.
2. $E = 1005/7$, or 143.571, for each of the seven days of the week. For Sunday, $O = 523$ and $E = 143.571$.
3. Because the given frequencies differ substantially from frequencies that are all about the same, the χ^2 test statistic should be large and the *P*-value should be small.
4. $df = 6$. Critical value: $\chi^2 = 12.592$.

STATDISK

Num Categories:	7
Degrees of freedom:	6
Expected Freq:	143.5714
Test Statistic, X^2:	1934.9791
Critical X^2:	12.59157
P-Value:	0.0000

5. Test statistic: $\chi^2 = 1934.979$. Critical value: $\chi^2 = 12.592$. P-value $= 0.000$. There is sufficient evidence to warrant rejection of the claim that the days of the week are selected with a uniform distribution with all days having the same chance of being selected.
6. Test statistic: $\chi^2 = 6.6$. Critical value: $\chi^2 = 16.919$. P-value $= 0.679$. There is not sufficient evidence to support the claim that the sample is from a population of heights in which the last digits do *not* occur with the same frequency.
7. Critical value: $\chi^2 = 16.919$. P-value > 0.10 (Tech: 0.516). There is not sufficient evidence to warrant rejection of the claim that the observed outcomes agree with the expected frequencies. The slot machine appears to be functioning as expected.
8. Test statistic: $\chi^2 = 4.600$. Critical value: $\chi^2 = 7.815$. P-value > 0.10 (Tech: P-value $= 0.204$). There is not sufficient evidence to warrant rejection of the claim that the tires selected by the students are equally likely. It appears that students do not have the ability to select the same tire.
10. Test statistic: $\chi^2 = 93.072$. Critical value: $\chi^2 = 19.675$. P-value < 0.005 (Tech: 0.000). There is sufficient evidence to warrant rejection of the claim that American-born major league baseball players are born in different months with the same frequency. The sample data appear to support Gladwell's claim.
11. Test statistic: $\chi^2 = 5.860$. Critical value: $\chi^2 = 11.071$. P-value > 0.10 (Tech: P-value $= 0.320$). There is not sufficient evidence to support the claim that the outcomes are not equally likely. The outcomes appear to be equally likely, so the loaded die does not appear to behave differently from a fair die.
12. Test statistic: $\chi^2 = 16.895$. Critical value: $\chi^2 = 16.812$. P-value < 0.01 (Tech: 0.0097). There is sufficient evidence to warrant rejection of the claim that births occur on the days of the week with equal frequency. Because many births are induced or involve Caesarean section, they are scheduled for days other than Saturday or Sunday, so those two days have smaller numbers of births.

6. Last Digits of Heights Example 1 in this section involved an analysis of the last digits of weights from a random sample of 100 Californians. Using those same subjects, the last digits of their heights are listed in the table below (based on data from the California Department of Public Health). Use a 0.05 significance level to test the claim that the sample is from a population of heights in which the last digits do *not* occur with the same frequency. The accompanying Minitab display results from the data in the table.

Last Digit	0	1	2	3	4	5	6	7	8	9
Frequency	12	8	14	9	11	9	13	8	11	5

MINITAB

N	DF	Chi-Sq	P-Value
100	9	6.6	0.679

7. Testing a Slot Machine The author purchased a slot machine (Bally Model 809) and tested it by playing it 1197 times. There are 10 different categories of outcomes, including no win, win jackpot, win with three bells, and so on. When testing the claim that the observed outcomes agree with the expected frequencies, the author obtained a test statistic of $\chi^2 = 8.185$. Use a 0.05 significance level to test the claim that the actual outcomes agree with the expected frequencies. Does the slot machine appear to be functioning as expected?

8. Flat Tire and Missed Class A classic story involves four carpooling students who missed a test and gave as an excuse a flat tire. On the makeup test, the instructor asked the students to identify the particular tire that went flat. If they really didn't have a flat tire, would they be able to identify the same tire? The author asked 41 other students to identify the tire they would select. The results are listed in the following table (except for one student who selected the spare). Use a 0.05 significance level to test the author's claim that the results fit a uniform distribution. What does the result suggest about the ability of the four students to select the same tire when they really didn't have a flat?

Tire	Left Front	Right Front	Left Rear	Right Rear
Number Selected	11	15	8	6

9. NYC Homicides For a recent year, the following are the numbers of homicides that occurred each month in New York City: 38, 30, 46, 40, 46, 49, 47, 50, 50, 42, 37, 37. Use a 0.05 significance level to test the claim that homicides in New York City are equally likely for each of the 12 months. Is there sufficient evidence to support the police commissioner's claim that homicides occur more often in the summer when the weather is better?

10. Baseball Player Births In his book *Outliers,* author Malcolm Gladwell argues that more baseball players have birthdates in the months immediately following July 31, because that was the cutoff date for nonschool baseball leagues. Here is a sample of frequency counts of months of birthdates of American-born major league baseball players starting with January: 387, 329, 366, 344, 336, 313, 313, 503, 421, 434, 398, 371. Using a 0.05 significance level, is there sufficient evidence to warrant rejection of the claim that American-born major league baseball players are born in different months with the same frequency? Do the sample values appear to support Gladwell's claim?

11. Loaded Die The author drilled a hole in a die and filled it with a lead weight, then proceeded to roll it 200 times. Here are the observed frequencies for the outcomes of 1, 2, 3, 4, 5, and 6, respectively: 27, 31, 42, 40, 28, 32. Use a 0.05 significance level to test the claim that the outcomes are not equally likely. Does it appear that the loaded die behaves differently than a fair die?

12. Births Records of randomly selected births were obtained and categorized according to the day of the week that they occurred (based on data from the National Center for Health Statistics). Because babies are unfamiliar with our schedule of weekdays, a reasonable claim is that births occur on the different days with equal frequency. See the table that follows. Use a 0.01 significance level to test that claim. Can you provide an explanation for the result?

Day	Sun	Mon	Tues	Wed	Thurs	Fri	Sat
Number of Births	77	110	124	122	120	123	97

13. Kentucky Derby The table below lists the frequency of wins for different post positions in the Kentucky Derby horse race (current as of this writing). A post position of 1 is closest to the inside rail, so that horse has the shortest distance to run. (Because the number of horses varies from year to year, only the first 10 post positions are included.) Use a 0.05 significance level to test the claim that the likelihood of winning is the same for the different post positions. Based on the result, should bettors consider the post position of a horse racing in the Kentucky Derby?

Post Position	1	2	3	4	5	6	7	8	9	10
Wins	19	14	11	15	14	7	8	12	5	11

14. Win 4 Lottery The author recorded all digits selected in New York's Win 4 Lottery for two drawings held each day in a recent year. The frequencies of the digits from 0 through 9 are 280, 303, 331, 289, 285, 294, 283, 274, 297, and 284. Use a 0.05 significance level to test the claim of lottery officials that the digits are selected in a way that they are equally likely.

15. Police Calls The police department in Madison, Connecticut, released the following numbers of calls for the different days of the week during a recent February that had 28 days: Monday (114); Tuesday (152); Wednesday (160); Thursday (164); Friday (179); Saturday (196); Sunday (130). Use a 0.01 significance level to test the claim that the different days of the week have the same frequencies of police calls. Is there anything notable about the observed frequencies?

16. Police Calls Repeat the preceding exercise using these observed frequencies for police calls received during the month of March: Monday (208); Tuesday (224); Wednesday (246); Thursday (173); Friday (210); Saturday (236); Sunday (154). What is a fundamental error with this analysis?

17. World Series Games The table below lists the numbers of games played in the baseball World Series as of this writing. That table also includes the expected proportions for the numbers of games in a World Series, assuming that in each series, both teams have about the same chance of winning. Use a 0.05 significance level to test the claim that the actual numbers of games fit the distribution indicated by the expected proportions.

Games Played	4	5	6	7
Actual World Series Contests	20	23	23	37
Expected Proportion	2/16	4/16	5/16	5/16

18. American Idol The contestants on the TV show *American Idol* try to win a singing contest. At one point, the Web site WhatNotToSing.com listed the actual numbers of eliminations for different orders of singing, and the expected number of eliminations was also listed. The results are in the table below. Use a 0.05 significance level to test the claim that the actual eliminations agree with the expected numbers. Does there appear to be support for the claim that the leadoff singers appear to be at a disadvantage?

Singing Order	1	2	3	4	5	6	7 through 12
Eliminations	20	12	9	8	6	5	9
Expected Eliminations	12.9	12.9	9.9	7.9	6.4	5.5	13.5

19. M&M Candies Mars, Inc. claims that its M&M plain candies are distributed with the following color percentages: 16% green, 20% orange, 14% yellow, 24% blue, 13% red, and 13% brown. Refer to Data Set 20 in Appendix B and use the sample data to test the claim that the color distribution is as claimed by Mars, Inc. Use a 0.05 significance level.

13. Test statistic: $\chi^2 = 13.483$. Critical value: $\chi^2 = 16.919$. *P*-value > 0.10 (Tech: 0.142). There is not sufficient evidence to warrant rejection of the claim that the likelihood of winning is the same for the different post positions. Based on these results, post position should not be considered when betting on the Kentucky Derby race.

14. Test statistic: $\chi^2 = 8.021$. Critical value: $\chi^2 = 16.919$. *P*-value > 0.10 (Tech: 0.532). There is not sufficient evidence to warrant rejection of the claim that the digits are selected in a way that they are equally likely.

15. Test statistic: $\chi^2 = 29.814$. Critical value: $\chi^2 = 16.812$. *P*-value < 0.005 (Tech: 0.000). There is sufficient evidence to warrant rejection of the claim that the different days of the week have the same frequencies of police calls. The highest numbers of calls appear to fall on Friday and Saturday, and these are weekend days with disproportionately more partying and drinking.

16. Test statistic: $\chi^2 = 31.963$. Critical value: $\chi^2 = 16.812$. *P*-value < 0.005 (Tech: 0.000). There is sufficient evidence to warrant rejection of the claim that the different days of the week have the same frequencies of police calls. Because March has 31 days, three of the days of the week occur more often than the other days of the week, so the comparison does not make sense with the given data.

17. Test statistic: $\chi^2 = 7.579$. Critical value: $\chi^2 = 7.815$. *P*-value > 0.05 (Tech: 0.056). There is not sufficient evidence to warrant rejection of the claim that the actual numbers of games fit the distribution indicated by the proportions listed in the given table.

18. Test statistic: $\chi^2 = 5.624$. Critical value: $\chi^2 = 12.592$. *P*-value > 0.10 (Tech: 0.467). There is not sufficient evidence to warrant rejection of the claim that the actual eliminations agree with the expected numbers. The leadoff singers do appear to be at a disadvantage because 20 of them were eliminated compared to the expected value of 12.9 eliminations, but that result is not significant in the context of the available sample data.

20. Do World War II Bomb Hits Fit a Poisson Distribution? In analyzing hits by V-1 buzz bombs in World War II, South London was subdivided into regions, each with an area of 0.25 km^2. Shown below is a table of actual frequencies of hits and the frequencies expected with the Poisson distribution. (The Poisson distribution is described in Section 5-5.) Use the values listed and a 0.05 significance level to test the claim that the actual frequencies fit a Poisson distribution.

Number of Bomb Hits	0	1	2	3	4 or more
Actual Number of Regions	229	211	93	35	8
Expected Number of Regions (from Poisson Distribution)	227.5	211.4	97.9	30.5	8.7

Benford's Law. According to Benford's law, a variety of different data sets include numbers with leading (first) digits that follow the distribution shown in the table below. In Exercises 21–24, test for goodness-of-fit with Benford's law.

Leading Digit	1	2	3	4	5	6	7	8	9
Benford's Law: Distribution of Leading Digits	30.1%	17.6%	12.5%	9.7%	7.9%	6.7%	5.8%	5.1%	4.6%

21. Detecting Fraud When working for the Brooklyn district attorney, investigator Robert Burton analyzed the leading digits of the amounts from 784 checks issued by seven suspect companies. The frequencies were found to be 0, 15, 0, 76, 479, 183, 8, 23, and 0, and those digits correspond to the leading digits of 1, 2, 3, 4, 5, 6, 7, 8, and 9, respectively. If the observed frequencies are substantially different from the frequencies expected with Benford's law, the check amounts appear to result from fraud. Use a 0.01 significance level to test for goodness-of-fit with Benford's law. Does it appear that the checks are the result of fraud?

22. Author's Check Amounts Exercise 21 lists the observed frequencies of leading digits from amounts on checks from seven suspect companies. Here are the observed frequencies of the leading digits from the amounts on checks written by the author: 68, 40, 18, 19, 8, 20, 6, 9, 12. (Those observed frequencies correspond to the leading digits of 1, 2, 3, 4, 5, 6, 7, 8, and 9, respectively.) Using a 0.05 significance level, test the claim that these leading digits are from a population of leading digits that conform to Benford's law. Do the author's check amounts appear to be legitimate?

23. Tax Cheating? Frequencies of leading digits from IRS tax files are 152, 89, 63, 48, 39, 40, 28, 25, and 27 (corresponding to the leading digits of 1, 2, 3, 4, 5, 6, 7, 8, and 9, respectively, based on data from Mark Nigrini, who sells software for Benford data analysis). Using a 0.05 significance level, test for goodness-of-fit with Benford's law. Does it appear that the tax entries are legitimate?

24. Author's Computer Files The author recorded the leading digits of the sizes of the files stored on his computer, and the leading digits have frequencies of 45, 32, 18, 12, 9, 3, 13, 9, and 9 (corresponding to the leading digits of 1, 2, 3, 4, 5, 6, 7, 8, and 9, respectively). Using a 0.05 significance level, test for goodness-of-fit with Benford's law.

19. Test statistic: $\chi^2 = 6.682$. Critical value: $\chi^2 = 11.071$ (assuming a 0.05 significance level). *P*-value > 0.10 (Tech: 0.245). There is not sufficient evidence to warrant rejection of the claim that the color distribution is as claimed.

20. Test statistic: $\chi^2 = 0.976$. Critical value: $\chi^2 = 9.488$. *P*-value > 0.10 (Tech: 0.913). There is not sufficient evidence to warrant rejection of the claim that the actual frequencies fit a Poisson distribution.

21. Test statistic: $\chi^2 = 3650.251$. Critical value: $\chi^2 = 20.090$. *P*-value < 0.005 (Tech: 0.000). There is sufficient evidence to warrant rejection of the claim that the leading digits are from a population with a distribution that conforms to Benford's law. It does appear that the checks are the result of fraud (although the results cannot confirm that fraud is the cause of the discrepancy between the observed results and the expected results).

22. Test statistic: $\chi^2 = 14.432$. Critical value: $\chi^2 = 15.507$. *P*-value > 0.05 (Tech: 0.071). There is not sufficient evidence to warrant rejection of the claim that the leading digits are from a population with a distribution that conforms to Benford's law. The author's check amounts appear to be legitimate.

23. Test statistic: $\chi^2 = 1.762$. Critical value: $\chi^2 = 15.507$. *P*-value > 0.10 (Tech: 0.988). There is not sufficient evidence to warrant rejection of the claim that the leading digits are from a population with a distribution that conforms to Benford's law. The tax entries do appear to be legitimate.

24. Test statistic: $\chi^2 = 10.299$. Critical value: $\chi^2 = 15.507$. *P*-value > 0.10 (Tech: 0.245). There is not sufficient evidence to warrant rejection of the claim that the leading digits are from a population with a distribution that conforms to Benford's law.

11-2 Beyond the Basics

25. Testing Goodness-of-Fit with a Normal Distribution Refer to Data Set 1 in Appendix B for the 40 heights of females.

Height (cm)	Less Than 155.410	155.410–162.005	162.005–168.601	Greater Than 168.601
Frequency				

25. a. 6, 13, 15, 6
b. 0.1587, 0.3413, 0.3413, 0.1587 (Tech: 0.1587, 0.3413, 0.3414, 0.1586)
c. 6.348, 13.652, 13.652, 6.348 (Tech: 6.348, 13.652, 13.656, 6.344)

a. Enter the observed frequencies in the preceding table.

b. Assuming a normal distribution with mean and standard deviation given by the sample mean and standard deviation, use the methods of Chapter 6 to find the probability of a randomly selected height belonging to each class.

c. Using the probabilities found in part (b), find the expected frequency for each category.

d. Use a 0.01 significance level to test the claim that the heights were randomly selected from a normally distributed population. Does the goodness-of-fit test suggest that the data are from a normally distributed population?

d. Test statistic: $\chi^2 = 0.202$ (Tech: 0.201). Critical value: $\chi^2 = 11.345$. P-value > 0.10 (Tech: 0.977). There is not sufficient evidence to warrant rejection of the claim that heights were randomly selected from a normally distributed population. The test suggests that the data are from a normally distributed population.

11-3 Contingency Tables

Key Concept This section presents methods for analyzing *contingency tables* (or two-way frequency tables), which include frequency counts for categorical data arranged in a table with at least two rows and at least two columns. In Part 1 of this section, we present a method for conducting a hypothesis test of the null hypothesis that the row and column variables are independent of each other. This test of independence is used in real applications quite often. In Part 2, we will consider three variations of the basic method presented in Part 1: (1) test of homogeneity, (2) Fisher exact test, and (3) McNemar's test for matched pairs.

Part 1: Basic Concepts of Testing for Independence

In this section we use standard statistical methods to analyze frequency counts in a contingency table (or two-way frequency table). We begin with the definition of a contingency table.

DEFINITION A **contingency table** (or **two-way frequency table**) is a table consisting of frequency counts of categorical data corresponding to two different variables. (One variable is used to categorize rows, and a second variable is used to categorize columns.)

Example 1 Contingency Table for Different Treatments

Table 11-6 is a contingency table with four rows and two columns. The cells of the table contain frequency counts. The row variable identifies the treatment used for a stress fracture in a foot bone, and the column variable identifies the outcome as a success or failure (based on data from "Surgery Unfounded for Tarsal Navicular Stress Fracture," by Bruce Jancin, *Internal Medicine News*, Vol. 42, No. 14).

Table 11-6 Study of Success with Different Treatments for Stress Fracture

	Success	Failure
Surgery	54	12
Weight-Bearing Cast	41	51
Non-Weight-Bearing Cast for 6 Weeks	70	3
Non-Weight-Bearing Cast Less Than 6 Weeks	17	5

Note to Instructor
It will be obvious that the test statistic given here is identical to the χ^2 test statistic given in Section 11-2. The hypothesis tests of this section are all right-tailed, as in Section 11-2. Differences between this section and Section 11-2 are found in the method used for finding expected values E and the calculation of the number of degrees of freedom.

The word *contingent* has a few different meanings, one of which refers to a *dependence* on some other factor. We use the term *contingency table* because we test for *independence* between the row and column variables. We first define a *test of independence* and we provide key elements of the test in the box that follows.

DEFINITION In a **test of independence,** we test the null hypothesis that in a contingency table, the row and column variables are independent. (That is, there is no dependency between the row variable and the column variable.)

Objective

Conduct a hypothesis test for independence between the row variable and column variable in a contingency table.

Notation

O represents the *observed frequency* in a cell of a contingency table.

E represents the *expected frequency* in a cell, found by assuming that the row and column variables are independent.

r represents the number of rows in a contingency table (not including labels).

c represents the number of columns in a contingency table (not including labels).

Requirements

1. The sample data are randomly selected.
2. The sample data are represented as frequency counts in a two-way table.
3. For every cell in the contingency table, the expected frequency E is at least 5. (There is no requirement that every observed frequency must be at least 5. Also, there is no requirement that the population must have a normal distribution or any other specific distribution.)

Null and Alternative Hypotheses

The null and alternative hypotheses are as follows:

H_0: The row and column variables are independent.

H_1: The row and column variables are dependent.

Test Statistic for a Test of Independence

$$\chi^2 = \sum \frac{(O - E)^2}{E}$$

where O is the observed frequency in a cell and E is the expected frequency found by evaluating

$$E = \frac{(\text{row total})\,(\text{column total})}{(\text{grand total})}$$

P-Values

P-values are typically provided by technology, or a range of *P*-values can be found from Table A-4.

Critical Values

1. The critical values are found in Table A-4 using

$$\textbf{degrees of freedom} = (r - 1)(c - 1)$$

where r is the number of rows and c is the number of columns.

2. Tests of independence with a contingency table are always *right-tailed.*

The test statistic allows us to measure the amount of disagreement between the frequencies actually observed and those that we would theoretically expect when the two variables are independent. Large values of the X^2 test statistic are in the rightmost region of the chi-square distribution, and they reflect significant differences between observed and expected frequencies. The distribution of the test statistic X^2 can be approximated by the chi-square distribution, provided that all expected frequencies are at least 5. The number of degrees of freedom $(r - 1)(c - 1)$ reflects the fact that because we know the total of all frequencies in a contingency table, we can freely assign frequencies to only $r - 1$ rows and $c - 1$ columns before the frequency for every cell is determined. However, we cannot have negative frequencies or frequencies so large that any row (or column) sum exceeds the total of the observed frequencies for that row (or column).

Finding Expected Values *E*

The test statistic X^2 is found by using the values of O (observed frequencies) and the values of E (expected frequencies). An individual expected frequency E can be found for a cell by simply multiplying the total of the row frequencies by the total of the column frequencies, then dividing by the grand total of all frequencies, as shown in Example 2.

Example 2 Finding Expected Frequency

Refer to Table 11-6 and find the expected frequency for the first cell, where the observed frequency is 54. (See Table 11-6 on page 547.)

Solution

The first cell lies in the first row (with a total frequency of 66) and the first column (with total frequency of 182). The "grand total" is the sum of all frequencies in the table, which is 253. The expected frequency of the first cell is

$$E = \frac{(\text{row total})(\text{column total})}{(\text{grand total})} = \frac{(66)(182)}{253} = 47.478$$

Interpretation

We know that the first cell has an observed frequency of $O = 54$ and an expected frequency of $E = 47.478$. We can interpret the expected value by stating that if we assume that success is independent of the treatment, then we expect to find that 47.478 of the subjects would be treated with surgery and that treatment would be successful. There is a discrepancy between $O = 54$ and $E = 47.478$ and such discrepancies are key components of the test statistic that is a collective measure of the overall disagreement between the observed frequencies and the frequencies expected with independence between the row and column variables.

Rationale for Expected Frequencies To better understand expected frequencies, pretend that we know only the row and column totals in Table 11-6. Let's assume that the row and column variables are independent and that one of the 253 study subjects is randomly selected. The probability of getting someone counted in the first cell of Table 11-6 is calculated as follows:

$$P(\text{surgery treatment}) = 66/253 \text{ and } P(\text{success}) = 182/253$$

$$P(\text{surgery treatment } \textit{and} \text{ success}) = \frac{66}{253} \cdot \frac{182}{253} = 0.187661$$

Polls and Psychologists

Poll results can be dramatically affected by the wording of questions. A phrase such as "over the last few years" is interpreted differently by different people. Over the last few years (actually, since 1980), survey researchers and psychologists have been working together to improve surveys by decreasing bias and increasing accuracy. In one case, psychologists studied the finding that 10 to 15 percent of those surveyed say they voted in the last election when they did not. They experimented with theories of faulty memory, a desire to be viewed as responsible, and a tendency of those who usually vote to say that they voted in the most recent election, even if they did not. Only the last theory was actually found to be part of the problem.

With a probability of 0.187661 for the first cell, we expect that among 253 subjects, there are $253 \cdot 0.187661 = 47.478$ subjects in the first cell. If we generalize these calculations, we get the following:

$$\text{Expected frequency } E = (\text{grand total}) \cdot \frac{(\text{row total})}{(\text{grand total})} \cdot \frac{(\text{column total})}{(\text{grand total})}$$

This expression can be simplified to

$$E = \frac{(\text{row total})(\text{column total})}{(\text{grand total})}$$

We now proceed to conduct a hypothesis test of independence, as in Example 3.

Example 3 Does the Choice of Treatment Affect Success?

If we analyze the data in Table 11-6, it appears that the choice of treatment does affect success. However, we must determine whether those differences are *significant,* and that is the purpose of the test of independence. Use a 0.05 significance level to test the claim that success is independent of the treatment group. What does the result indicate about the increasing trend to use surgery?

Solution

Requirement check (1) Based on the description of the study, we will treat the subjects as being randomly selected and randomly assigned to the different treatment groups. (2) The results are expressed as frequency counts in Table 11-6. (3) The expected frequencies are all at least 5. (The lowest expected frequency is 6.174.) The requirements are satisfied. ✓

The null hypothesis and alternative hypothesis are as follows:

H_0: Success is independent of the treatment.

H_1: Success and the treatment are dependent.

The significance level is $\alpha = 0.05$.

Because the data are in the form of a contingency table, we use the χ^2 distribution with this test statistic:

$$\chi^2 = \sum \frac{(O - E)^2}{E} = \frac{(54 - 47.478)^2}{47.478} + \cdots + \frac{(5 - 6.174)^2}{6.174}$$

$$= 58.393$$

XLSTAT

Chi-square (Observed value)	58.3933
Chi-square (Critical value)	7.8147
DF	3
p-value	< 0.0001
alpha	0.05

***P*-Value** If using technology, results typically include the χ^2 test statistic and the *P*-value. For example, see the accompanying XLSTAT display showing the test statistic is $\chi^2 = 58.393$ and the *P*-value is less than 0.0001. Because the *P*-value is less than the significance level of 0.05, we reject the null hypothesis of independence between success and treatment.

Critical Value If using the critical value method of hypothesis testing, the critical value of $\chi^2 = 7.815$ is found from Table A-4 with $\alpha = 0.05$ in the right tail and the number of degrees of freedom given by $(r - 1)(c - 1) = (4 - 1)(2 - 1) = 3$. The test statistic and critical value are shown in Figure 11-4. Because the test statistic does fall within the critical region, we reject the null hypothesis of independence between success and treatment.

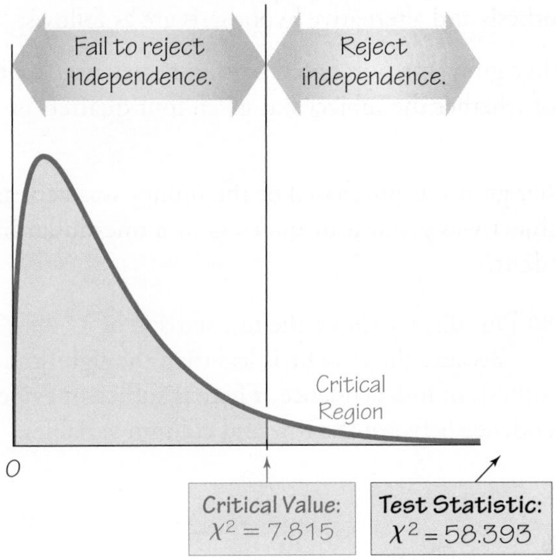

Figure 11-4

Interpretation

It appears that success is dependent on the treatment. Although the results of this test do not tell us which treatment is best, we can see that the success rates of 81.8%, 44.6%, 95.9%, and 77.3% suggest that the best treatment is to use a non-weight-bearing cast for 6 weeks. These results suggest that the increasing use of surgery is a treatment strategy that is not supported by the evidence.

Example 4 Are Four Quarters the Same as One Dollar?

Table 11-1 provided with the Chapter Problem consists of a contingency table with a row variable (whether subject was given four quarters or a one-dollar bill) and a column variable (whether the subject purchased gum or kept the money). Use a 0.05 significance level to test the claim that the row variable is independent of the column variable. What do the results of the test tell us? The table is shown below with results from a TI-83/84 Plus calculator.

Table 11-1 Results from a Study of the Denomination Effect

	Purchased Gum	Kept the Money
Students Given Four Quarters	27	16
Students Given a $1 Bill	12	34

TI-83/84 PLUS

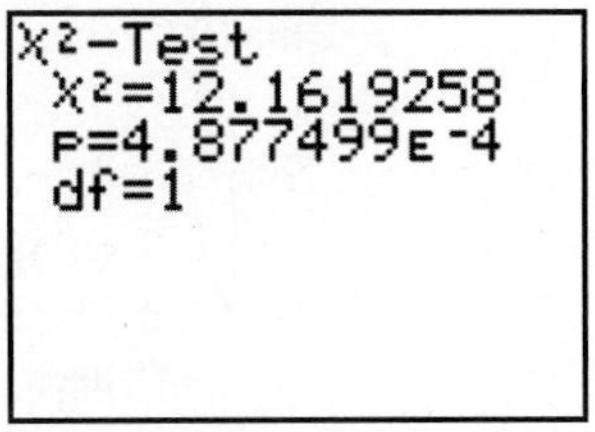

Solution

Requirement check (1) The data in Table 11-1 are from 89 undergraduate business students who were assigned at random to one of two groups. We will treat the data as being random. (2) The sample data are represented as frequency counts in a two-way table. (3) Each expected frequency is at least 5. (The expected frequencies are 18.843, 24.157, 20.157, and 25.843.) The requirements are satisfied. ✓

The null hypothesis and alternative hypothesis are as follows:

H_0: Whether gum was purchased or the money was kept is independent of whether the subject was given four quarters or a one-dollar bill.

H_1: Whether gum was purchased or the money was kept and whether the subject was given four quarters or a one-dollar bill are dependent.

The given TI-83/84 Plus display shows the test statistic of $\chi^2 = 12.162$ and the P-value of 0.000488. Because the P-value is less than the significance level of 0.05, reject the null hypothesis of independence. There is sufficient evidence to warrant rejection of independence between the row and column variables.

Interpretation

We reject independence between the row and column variables. It appears that whether the subject purchases gum or keeps the money is dependent on whether the subject is given four quarters or a one-dollar bill. The evidence therefore supports the concept of a "denomination effect."

As in Section 11-2, if observed and expected frequencies are close, the χ^2 test statistic will be small and the P-value will be large. If observed and expected frequencies are not close, the χ^2 test statistic will be large and the P-value will be small. These relationships are summarized and illustrated in Figure 11-5.

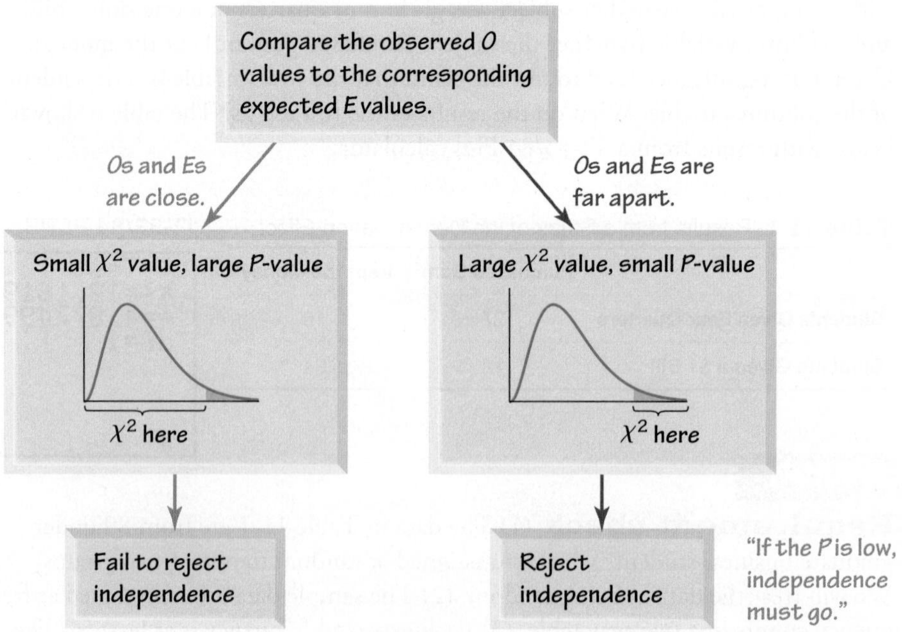

Figure 11-5 Relationships among Key Components in Test of Independence

Part 2: Test of Homogeneity, Fisher Exact Test, and McNemar's Test for Matched Pairs

Test of Homogeneity

In Part 1 of this section, we focused on the test of *independence* between the row and column variables in a contingency table. In Part 1, the sample data are from one population, and individual sample results are categorized with the row and column variables. In a *chi-square test of homogeneity,* we have samples randomly selected from different populations, and we want to determine whether those populations have the same proportions of some characteristic being considered. (The word *homogeneous* means "having the same quality," and in this context, we are testing to determine whether the proportions are the same.) Section 9-2 presented a procedure for testing a claim about *two* populations with categorical data having two possible outcomes, but a chi-square test of homogeneity allows us to use two or more populations with outcomes from several categories.

Note to Instructor
Part 2 can be omitted if there is not sufficient time to include it.

DEFINITION A **chi-square test of homogeneity** is a test of the claim that *different populations* have the same proportions of some characteristics.

Sampling Plan In a typical test of independence as described in Part 1 of this section, sample subjects are randomly selected from one population (such as people treated for stress fractures in a foot bone) and values of two different variables are observed (such as success/failure for people receiving different treatments). In a typical chi-square test of homogeneity, subjects are randomly selected from the different populations separately.

Procedure In conducting a test of homogeneity, we can use the same notation, requirements, test statistic, critical value, and procedures given in Part 1 of this section, with this exception: Instead of testing the null hypothesis of independence between the row and column variables, we test the null hypothesis that *the different populations have the same proportions of some characteristics.*

Example 5 Do the Four Treatment Populations Have the Same Success Rate?

Example 3 used the sample results from Table 11-6 to test for *independence* between treatment and success. If we want to use the same data from Table 11-6 in a test of the null hypothesis that the four populations corresponding to the four different treatment groups have the same proportion of success, we could use the chi-square test of homogeneity. The test statistic, critical value, and P-value are the same as those found in Example 3. Based on those values, we should reject the null hypothesis that the four treatment populations have the same success rate.

Fisher Exact Test

The procedures for testing hypotheses with contingency tables have the requirement that every cell must have an expected frequency of at least 5. This requirement is necessary for the χ^2 distribution to be a suitable approximation to the exact distribution of the χ^2 test statistic. The *Fisher exact test* is often used for a 2 × 2 contingency table with one or more expected frequencies that are below 5. The Fisher exact test provides an *exact* P-value and does not require an approximation technique. Because the calculations are quite complex, it's a good idea to use technology when using the Fisher exact test. STATDISK, Minitab, XLSTAT, and StatCrunch all have the ability to perform the Fisher exact test.

Note to Instructor
The discussions about the Fisher exact text and McNemar's test are provided for relevant information, but the exercises use the methods already discussed in Part 1 of this section.

McNemar's Test for Matched Pairs

The methods in Part 1 of this section are based on independent data. For 2×2 tables consisting of frequency counts that result from matched pairs, the frequency counts within each matched pair are not independent and, for such cases, we can use McNemar's test of the null hypothesis that the frequencies from the discordant (different) categories occur in the same proportion.

Table 11-7 shows a general format for summarizing results from data consisting of frequency counts from matched pairs. Table 11-7 refers to two different treatments (such as two different eye drop solutions) applied to two different parts of each subject (such as left eye and right eye). It's a bit difficult to correctly read a table such as Table 11-7. The total number of subjects is $a + b + c + d$, and each of those subjects yields results from each of two parts of a matched pair. If $a = 100$, then 100 subjects were cured with both treatments. If $b = 50$ in Table 11-7, then each of 50 subjects had no cure with treatment X but they were each cured with treatment Y. Remember, the entries in Table 11-7 are frequency counts of subjects, not the total number of individual components in the matched pairs. If 500 people have each eye treated with two different ointments, the value of $a + b + c + d$ is 500 (the number of subjects), not 1000 (the number of treated eyes).

Table 11-7 2×2 Table with Frequency Counts from Matched Pairs

		Treatment *X*	
		Cured	**Not Cured**
Treatment *Y*	**Cured**	a	b
	Not Cured	c	d

McNemar's test requires that for a table such as Table 11-7, the frequencies are such that $b + c \geq 10$. The test is a right-tailed chi-square test with the following test statistic:

$$\chi^2 = \frac{(|b - c| - 1)^2}{b + c}$$

P-values are typically provided by software, and critical values can be found in Table A-4 using 1 degree of freedom. *Caution:* When applying McNemar's test, be careful to use only the two frequency counts from *discordant* (different) pairs, such as the frequency b in Table 11-7 (with different pairs of cured/not cured) and frequency c in Table 11-7 (with different pairs of not cured/cured).

Note to Instructor
Stress that with McNemar's test, the calculations must be based on frequencies for the two *discordant* cases only. We should always identify the two discordant cases, and they might not be located in the upper right cell and the lower left cell.

Example 6 Are Hip Protectors Effective?

A randomized controlled trial was designed to test the effectiveness of hip protectors in preventing hip fractures in the elderly. Nursing home residents each wore protection on one hip, but not the other. Results are summarized in Table 11-8 (based on data from "Efficacy of Hip Protector to Prevent Hip Fracture in Nursing Home Residents," by Kiel et al., *Journal of the American Medical Association*, Vol. 298, No. 4). McNemar's test can be used to test the null hypothesis that the following two proportions are the same:

- The proportion of subjects with no hip fracture on the protected hip and a hip fracture on the unprotected hip.

- The proportion of subjects with a hip fracture on the protected hip and no hip fracture on the unprotected hip.

Using the discordant (different) pairs with the general format from Table 11-7, we have $b = 10$ and $c = 15$, so the test statistic is calculated as follows:

$$\chi^2 = \frac{(|b - c| - 1)^2}{b + c} = \frac{(|10 - 15| - 1)^2}{10 + 15} = 0.640$$

With a 0.05 significance level and degrees of freedom given by df = 1, we refer to Table A-4 to find the critical value of $\chi^2 = 3.841$ for this right-tailed test. The test statistic of $\chi^2 = 0.640$ does not exceed the critical value of $\chi^2 = 3.841$, so we fail to reject the null hypothesis. (Also, the *P*-value is 0.424, which is greater than 0.05, indicating that we fail to reject the null hypothesis.) The proportion of hip fractures with the protectors worn is not significantly different from the proportion of hip fractures without the protectors worn. The hip protectors do not appear to be effective in preventing hip fractures.

Table 11-8 Randomized Controlled Trial of Hip Protectors

		No Hip Protector Worn	
		No Hip Fracture	**Hip Fracture**
Hip Protector Worn	**No Hip Fracture**	309	10
	Hip Fracture	15	2

using TECHNOLOGY

STATDISK Enter the observed frequencies in the Data Window as they appear in the contingency table. Select **Analysis** from the main menu, then select **Contingency Tables.** Enter a significance level and proceed to identify the columns containing the frequencies. Click on **Evaluate.** The STATDISK results include the test statistic, critical value, and *P*-value, as shown in the display resulting from Table 11-1.

STATDISK

Degrees of freedom: 1	
Test Statistic, X^2:	12.1619
Critical X^2:	3.841456
P-Value:	0.0005

MINITAB First enter the observed frequencies in columns, then select **Stat** from the main menu bar. Next select the option **Tables,** then select **Chi Square Test (Two-Way Table in Worksheet)** and enter the names of the columns containing the observed frequencies, such as C1 C2 C3 C4. Minitab provides the test statistic and *P*-value, the expected frequencies, and the individual terms of the χ^2 test statistic.

EXCEL Use XLSTAT. First enter the contingency table in rows and columns. Click on **XLSTAT** at the top. Select **Correlation/Association tests,** then select **Test on contingency table.** In the "Contingency table" box, enter the range of cells containing the frequency counts of the contingency table. For example, enter A1:B4 for a contingency table with two columns (A and B) and four rows. For the "Data format," select the **Contingency Table** option. Click on the **Options** tab, leave a checkmark next to "Chi-square test," and enter a value for "Significance level (%)." For example, enter 5 for a 0.05 significance level. Click **OK** and results including the chi-square test statistic and *P*-value will be displayed.

TI-83/84 PLUS First enter the contingency table as a matrix by pressing 2ND x⁻¹ to get the MATRIX menu (or the MATRIX key on the TI-83). Select **EDIT**, and press ENTER. Enter the dimensions of the matrix (rows by columns) and proceed to enter the individual frequencies. When finished, press STAT, select **TESTS,** and then select the option **χ^2-Test.** Be sure that the observed matrix is the one you entered, such as matrix A. The expected frequencies will be automatically calculated and stored in the separate matrix identified as "Expected." Scroll down to **Calculate** and press ENTER to get the test statistic, *P*-value, and number of degrees of freedom.

STATCRUNCH Click on **Open StatCrunch.** Enter the row labels in the first column and enter the cell frequencies in the following columns. Click on **Stat,** select **Tables,** then select **Contingency.** Select the option of **with summary.** Enter the columns to be used, then click on **Calculate.** Results will include the test statistic and *P*-value.

Recommended Assignment
Exercises 1–12.

11-3 Basic Skills and Concepts

Statistical Literacy and Critical Thinking

1. Smoking Cessation The accompanying table summarizes successes and failures when subjects used different methods when trying to stop smoking. The determination of smoking or not smoking was made five months after the treatment was begun, and the data are based on results from the Centers for Disease Control and Prevention. If we test the claim that success is independent of the method used, the TI-83/84 Plus calculator provides a *P*-value of 0.216 (rounded). What does the *P*-value tell us about that claim?

	Nicotine Gum	Nicotine Patch	Nicotine Inhaler
Smoking	191	263	95
Not Smoking	59	57	27

2. Terminology The table in Exercise 1 is called a contingency table or two-way table. Why is the term *contingency* used? Why is the terminology of *two-way* table used?

3. Degrees of Freedom and Critical Value For the hypothesis test in Exercise 1, the test statistic is 3.062. Find the number of degrees of freedom used to find the critical value, then find the critical value. Assume a 0.05 significance level.

4. Right-Tailed, Left-Tailed, Two-Tailed Is the hypothesis test in Exercise 1 right-tailed, left-tailed, or two-tailed? Explain your choice.

In Exercises 5–18, test the given claim.

5. Denomination Effect In a study of the "denomination effect" described in the Chapter Problem, 150 women in China were given either a single 100 Yuan bill or a total of 100 Yuan in smaller bills. The value of 100 Yuan is about \$15. The women were given the choice of spending the money on specific items or they could keep the money. The results are summarized in the table below, and STATDISK results are provided in the screen display. Use a 0.05 significance level to test the claim that the form of the 100 Yuan is independent of whether the money was spent. What does the result suggest about a denomination effect?

	Spent the Money	Kept the Money
Women Given a Single 100-Yuan Bill	60	15
Women Given 100 Yuan in Smaller Bills	68	7

STATDISK

```
Test Statistic, X^2: 3.4091
Critical X^2:        3.841456
P-Value:             0.0648
```

6. Which Treatment Is Better? A randomized controlled trial was designed to compare the effectiveness of splinting versus surgery in the treatment of carpal tunnel syndrome. Results are given in the table below (based on data from "Splinting vs. Surgery in the Treatment of Carpal Tunnel Syndrome," by Gerritsen et al., *Journal of the American Medical Association*, Vol. 288, No. 10). The results are based on evaluations made one year after the treatment. Minitab results are given below the table. Using a 0.01 significance level, test the claim that success is independent of the type of treatment. What do the results suggest about treating carpal tunnel syndrome?

	Successful Treatment	Unsuccessful Treatment
Splint Treatment	60	23
Surgery Treatment	67	6

MINITAB

```
Chi-Sq = 9.750, DF = 1, P-Value = 0.002
```

1. Because the *P*-value of 0.216 is not small (such as 0.05 or lower), fail to reject the null hypothesis of independence between the treatment and whether the subject stops smoking. This suggests that the choice of treatment doesn't appear to make much of a difference.
2. In this context, the word *contingency* refers to a dependency of one variable on another, and we use a test of independence between the row variable and the column variable to determine whether one variable appears to be contingent on the other. We use the terminology of *two-way* table because the frequency counts are arranged in a table with two variables: the row variable and the column variable.
3. df = 2 and the critical value is $\chi^2 = 5.991$.
4. The test is right-tailed. The test statistic is based on differences between observed frequencies and the frequencies expected with the assumption of independence between the row and column variables. Only large values of the test statistic correspond to substantial differences between the observed and expected values, and such large values are located in the right tail of the distribution.
5. Test statistic: $\chi^2 = 3.409$. Critical value: $\chi^2 = 3.841$. *P*-value > 0.05 (Tech: 0.0648). There is not sufficient evidence to warrant rejection of the claim that the form of the 100-Yuan gift is independent of whether the money was spent. There is not sufficient evidence to support the claim of a denomination effect.
6. Test statistic: $\chi^2 = 9.750$. Critical value: $\chi^2 = 6.635$. *P*-value < 0.005 (Tech: 0.002). There is sufficient evidence to warrant rejection of the claim that success is independent of the type of treatment. The results suggest that the surgery treatment is better.

7. Testing a Lie Detector The table below includes results from polygraph (lie detector) experiments conducted by researchers Charles R. Honts (Boise State University) and Gordon H. Barland (Department of Defense Polygraph Institute). In each case, it was known if the subject lied or did not lie, so the table indicates when the polygraph test was correct. Use a 0.05 significance level to test the claim that whether a subject lies is independent of the polygraph test indication. Do the results suggest that polygraphs are effective in distinguishing between truths and lies?

7. Test statistic: $\chi^2 = 25.571$. Critical value: $\chi^2 = 3.841$. *P*-value < 0.005 (Tech: 0.000). There is sufficient evidence to warrant rejection of the claim that whether a subject lies is independent of the polygraph test indication. The results suggest that polygraphs are effective in distinguishing between truths and lies, but there are many false positives and false negatives, so they are not highly reliable.

	Did the Subject Actually Lie?	
	No (Did Not Lie)	Yes (Lied)
Polygraph test indicated that the subject lied.	15	42
Polygraph test indicated that the subject did not lie.	32	9

8. Discrimination The U.S. Supreme Court considered a case involving the exam for firefighter lieutenant in the city of New Haven, Connecticut. Results from the exam are shown in the table below. Is there sufficient evidence to support the claim that results from the test should be thrown out because they are discriminatory? Use a 0.01 significance level.

8. Test statistic: $\chi^2 = 4.423$. Critical value: $\chi^2 = 6.635$. *P*-value > 0.025 (Tech: 0.0355). There is not sufficient evidence to support the claim that the results are discriminatory.

	Passed	Failed
White Candidates	17	16
Minority Candidates	9	25

9. Is Sentence Independent of Plea? Many people believe that criminals who plead guilty tend to get lighter sentences than those who are convicted in trials. The accompanying table summarizes randomly selected sample data for San Francisco defendants in burglary cases (based on data from "Does It Pay to Plead Guilty? Differential Sentencing and the Functioning of the Criminal Courts," by Brereton and Casper, *Law and Society Review*, Vol. 16, No. 1). All of the subjects had prior prison sentences. Use a 0.05 significance level to test the claim that the sentence (sent to prison or not sent to prison) is independent of the plea. If you were an attorney defending a guilty defendant, would these results suggest that you should encourage a guilty plea?

9. Test statistic: $\chi^2 = 42.557$. Critical value: $\chi^2 = 3.841$. *P*-value < 0.005 (Tech: 0.000). There is sufficient evidence to warrant rejection of the claim that the sentence is independent of the plea. The results encourage pleas for guilty defendants.

	Guilty Plea	Not Guilty Plea
Sent to Prison	392	58
Not Sent to Prison	564	14

10. Nurse a Serial Killer? Alert nurses at the Veteran's Affairs Medical Center in Northampton, Massachusetts, noticed an unusually high number of deaths at times when another nurse, Kristen Gilbert, was working. Those same nurses later noticed missing supplies of the drug epinephrine, which is a synthetic adrenaline that stimulates the heart. Kristen Gilbert was arrested and charged with four counts of murder and two counts of attempted murder. When seeking a grand jury indictment, prosecutors provided a key piece of evidence consisting of the table below. Use a 0.01 significance level to test the defense claim that deaths on shifts are independent of whether Gilbert was working. What does the result suggest about the guilt or innocence of Gilbert?

10. Test statistic: $\chi^2 = 86.481$. Critical value: $\chi^2 = 6.635$. *P*-value < 0.005 (Tech: 0.000). There is sufficient evidence to warrant rejection of the claim that deaths on shifts are independent of whether Gilbert was working. The results favor the guilt of Gilbert.

	Shifts with a Death	Shifts without a Death
Gilbert Was Working	40	217
Gilbert Was Not Working	34	1350

11. Test statistic: $\chi^2 = 0.164$. Critical value: $\chi^2 = 3.841$. *P*-value > 0.10 (Tech: 0.686). There is not sufficient evidence to warrant rejection of the claim that the gender of the tennis player is independent of whether the call is overturned.

11. Tennis Challenges The table below shows results since 2006 of challenged referee calls in the U.S. Open. Use a 0.05 significance level to test the claim that the gender of the tennis player is independent of whether the call is overturned.

	Was the Challenge to the Call Successful?	
	Yes	No
Men	421	991
Women	220	539

12. Test statistic: $\chi^2 = 1.364$. Critical value: $\chi^2 = 3.841$. *P*-value > 0.10 (Tech: 0.243). There is not sufficient evidence to warrant rejection of the claim that left-handedness is independent of gender.

12. Lefties A random sample of 760 subjects was obtained, and each was tested for left-hand writing. Results are in the table below (based on data from "The Left-Handed: Their Sinister History," by Elaine Fowler Costas, Education Resources Information Center, Paper 399519). Use a 0.05 significance level to test the claim that left-handedness is independent of gender.

	Writes with Left Hand?	
	Yes	No
Male	23	217
Female	65	455

13. Test statistic: $\chi^2 = 14.589$. Critical value: $\chi^2 = 9.488$. *P*-value < 0.01 (Tech: 0.0056). There is sufficient evidence to warrant rejection of the claim that the direction of the kick is independent of the direction of the goalkeeper jump. The results do not support the theory that because the kicks are so fast, goalkeepers have no time to react, so the directions of their jumps are independent of the directions of the kicks.

13. Soccer Strategy In soccer, serious fouls result in a penalty kick with one kicker and one defending goalkeeper. The table below summarizes results from 286 kicks during games among top teams (based on data from "Action Bias Among Elite Soccer Goalkeepers: The Case of Penalty Kicks," by Bar-Eli et al., *Journal of Economic Psychology,* Vol. 28, No. 5). In the table, jump direction indicates which way the goalkeeper jumped, where the kick direction is from the perspective of the goalkeeper. Use a 0.05 significance level to test the claim that the direction of the kick is independent of the direction of the goalkeeper jump. Do the results support the theory that because the kicks are so fast, goalkeepers have no time to react, so the directions of their jumps are independent of the directions of the kicks?

	Goalkeeper Jump		
	Left	Center	Right
Kick to Left	54	1	37
Kick to Center	41	10	31
Kick to Right	46	7	59

14. Test statistic: $\chi^2 = 1.358$. Critical value: $\chi^2 = 7.815$ (assuming a 0.05 significance level). *P*-value > 0.10 (Tech: 0.715). There is not sufficient evidence to warrant rejection of the claim that the amount of smoking is independent of seat belt use. The theory is not supported by the given data.

14. Is Seat Belt Use Independent of Cigarette Smoking? A study of seat belt users and nonusers yielded the randomly selected sample data summarized in the given table (based on data from "What Kinds of People Do Not Use Seat Belts?" by Helsing and Comstock, *American Journal of Public Health,* Vol. 67, No. 11). Test the claim that the amount of smoking is independent of seat belt use. A plausible theory is that people who smoke more are less concerned about their health and safety and are therefore less inclined to wear seat belts. Is this theory supported by the sample data?

	Number of Cigarettes Smoked per Day			
	0	1–14	15–34	35 and over
Wear Seat Belts	175	20	42	6
Don't Wear Seat Belts	149	17	41	9

15. Clinical Trial of Echinacea In a clinical trial of the effectiveness of echinacea for preventing colds, the results in the table below were obtained (based on data from "An Evaluation of *Echinacea Angustifolia* in Experimental Rhinovirus Infections," by Turner et al., *New England Journal of Medicine,* Vol. 353, No. 4). Use a 0.05 significance level to test the claim that getting a cold is independent of the treatment group. What do the results suggest about the effectiveness of echinacea as a prevention against colds?

15. Test statistic: $\chi^2 = 2.925$. Critical value: $\chi^2 = 5.991$. P-value > 0.10 (Tech: 0.232). There is not sufficient evidence to warrant rejection of the claim that getting a cold is independent of the treatment group. The results suggest that echinacea is not effective for preventing colds.

	Treatment Group		
	Placebo	Echinacea: 20% extract	Echinacea: 60% extract
Got a Cold	88	48	42
Did Not Get a Cold	15	4	10

16. Injuries and Motorcycle Helmet Color A case-control (or retrospective) study was conducted to investigate a relationship between the colors of helmets worn by motorcycle drivers and whether they are injured or killed in a crash. Results are given in the table below (based on data from "Motorcycle Rider Conspicuity and Crash Related Injury: Case-Control Study," by Wells et al., *BMJ USA,* Vol. 4). Test the claim that injuries are independent of helmet color. Should motorcycle drivers choose helmets with a particular color? If so, which color appears best?

16. Test statistic: $\chi^2 = 9.971$. Critical value: $\chi^2 = 9.488$ (assuming a 0.05 significance level). P-value < 0.05 (Tech: 0.041). There is sufficient evidence to warrant rejection of the claim that injuries are independent of helmet color. It appears that motorcycle drivers should use yellow or orange helmets.

	Color of Helmet				
	Black	White	Yellow/Orange	Red	Blue
Controls (not injured)	491	377	31	170	55
Cases (injured or killed)	213	112	8	70	26

17. Survey Refusals A study of people who refused to answer survey questions provided the randomly selected sample data shown in the table below (based on data from "I Hear You Knocking But You Can't Come In," by Fitzgerald and Fuller, *Sociological Methods and Research,* Vol. 11, No. 1). At the 0.01 significance level, test the claim that the cooperation of the subject (response or refusal) is independent of the age category. Does any particular age group appear to be particularly uncooperative?

17. Test statistic: $\chi^2 = 20.271$. Critical value: $\chi^2 = 15.086$. P-value < 0.005 (Tech: 0.0011). There is sufficient evidence to warrant rejection of the claim that cooperation of the subject is independent of the age category. The age group of 60 and over appears to be particularly uncooperative.

	Age					
	18–21	22–29	30–39	40–49	50–59	60 and over
Responded	73	255	245	136	138	202
Refused	11	20	33	16	27	49

18. Baseball Player Births In his book *Outliers,* author Malcolm Gladwell argues that more American-born baseball players have birthdates in the months immediately following July 31 because that was the cutoff date for nonschool baseball leagues. The table below lists months of births for a sample of American-born baseball players and foreign-born baseball players. Using a 0.05 significance level, is there sufficient evidence to warrant rejection of the claim that months of births of baseball players are independent of whether they are born in America? Do the data appear to support Gladwell's claim?

18. Test statistic: $\chi^2 = 20.054$. Critical value: $\chi^2 = 19.675$. P-value > 0.05 (Tech: 0.0446). There is sufficient evidence to warrant rejection of the claim that months of births of baseball players are independent of whether they are born in America. The data do appear to support Gladwell's claim.

	Jan	Feb	March	April	May	June	July	Aug	Sept	Oct	Nov	Dec
Born in America	387	329	366	344	336	313	313	503	421	434	398	371
Foreign Born	101	82	85	82	94	83	59	91	70	100	103	82

19. Test statistic: $\chi^2 = 0.773$. Critical value: $\chi^2 = 11.345$. P-value > 0.10 (Tech: 0.856). There is not sufficient evidence to warrant rejection of the claim that getting an infection is independent of the treatment. The atorvastatin treatment does not appear to have an effect on infections.

19. Clinical Trial of Lipitor Lipitor is the trade name of the drug atorvastatin, which is used to reduce cholesterol in patients. (Until its patent expired in 2011, this was the largest-selling drug in the world, with annual sales of \$13 billion.) Adverse reactions have been studied in clinical trials, and the table below summarizes results for infections in patients from different treatment groups (based on data from Parke-Davis). Use a 0.01 significance level to test the claim that getting an infection is independent of the treatment. Does the atorvastatin treatment appear to have an effect on infections?

	Placebo	Atorvastatin 10 mg	Atorvastatin 40 mg	Atorvastatin 80 mg
Infection	27	89	8	7
No Infection	243	774	71	87

20. Test statistic: $\chi^2 = 784.647$. Critical value: $\chi^2 = 11.345$. P-value < 0.005 (Tech: 0.000). There is sufficient evidence to warrant rejection of the claim that left-handedness is independent of parental handedness. It appears that handedness of the parents has an effect on handedness of the offspring, so left-handedness appears to be an inherited trait.

20. Genetics and Handedness In a study of left-handedness as a possible inherited trait, the data in the table below were obtained (based on data from "Why Are Some People Left-Handed? An Evolutionary Perspective," by Laurens and Faurie, *Philosophical Transactions,* Vol. 364). Use a 0.01 significance level to test the claim that left-handedness is independent of parental handedness. What do the results suggest about the inheritability of left-handedness?

Parental Handedness	Offspring Left-Handed?	
Father/Mother	Yes	No
Right/Right	5360	50,928
Right/Left	767	2736
Left/Right	741	3667
Left/Left	94	289

11-3 Beyond the Basics

21. Test statistics: $\chi^2 = 12.1619258$ and $z = 3.487395274$, so that $z^2 = \chi^2$. Critical values: $\chi^2 = 3.841$ and $z = \pm 1.96$, so $z^2 = \chi^2$ (approximately).

22. Without Yates's correction, the test statistic is $\chi^2 = 12.162$. With Yates's correction, the test statistic is $\chi^2 = 10.717$. Yates's correction decreases the test statistic so that sample data must be more extreme in order to reject the null hypothesis of independence.

21. Equivalent Tests A χ^2 test involving a 2×2 table is equivalent to the test for the difference between two proportions, as described in Section 9-2. Using the claim and table in Example 4, verify that the χ^2 test statistic and the z test statistic (found from the test of equality of two proportions) are related as follows: $z^2 = \chi^2$. Also show that the critical values have that same relationship.

22. Using Yates's Correction for Continuity The chi-square distribution is continuous, whereas the test statistic used in this section is discrete. Some statisticians use *Yates's correction for continuity* in cells with an expected frequency of less than 10 or in all cells of a contingency table with two rows and two columns. With Yates's correction, we replace

$$\sum \frac{(O - E)^2}{E} \quad \text{with} \quad \sum \frac{(|O - E| - 0.5)^2}{E}$$

Given the contingency table in Example 4, find the value of the χ^2 test statistic using Yates's correction. What effect does Yates's correction have?

Note to Instructor
Recommendation: Given that many statistics professors don't have enough time to include analysis of variance, consider this approach: Ask students to read Section 11-4 on their own and assign Exercises 15 and 16 for extra credit.

11-4 Analysis of Variance

Key Concept In this section we introduce the method of *one-way analysis of variance,* which is used for tests of hypotheses that three or more populations have means that are all equal, as in H_0: $\mu_1 = \mu_2 = \mu_3$. Because the calculations are very complicated, we emphasize the interpretation of results obtained by using technology. Here is a recommended study strategy for this section.

1. Understand that a small P-value (such as 0.05 or less) leads to rejection of the null hypothesis of equal means. ("If the P is low, the null must go.") With a large P-value (such as greater than 0.05), fail to reject the null hypothesis of equal means.
2. Develop an understanding of the underlying rationale by studying the examples in this section.

Part 1: Basics of One-Way Analysis of Variance

When testing for equality of three or more population means, use the method of one-way analysis of variance.

Note to Instructor
Instead of choosing between the terms *factor* and *treatment*, consider referring to *factor or treatment* repeatedly so students get to know that those two terms are synonymous. Minitab and TI-83/84 Plus displays include the term *factor*, so it's important to recognize and understand that reference.

DEFINITION One-way analysis of variance (ANOVA) is a method of testing the equality of three or more population means by analyzing sample variances. One-way analysis of variance is used with data categorized with *one* **factor** (or **treatment**), so there is one characteristic used to separate the sample data into the different categories.

One-Way Analysis of Variance for Testing Equality of Three or More Population Means

Objective

Use samples from three or more different populations to test a claim that the populations all have the same mean.

Requirements

1. The populations have distributions that are approximately normal. (This is a loose requirement, because the method works well unless a population has a distribution that is very far from normal.)
2. The populations have the same variance σ^2 (or standard deviation σ). This is a loose requirement, because the method works well unless the population variances differ by large amounts. Statistician George E. P. Box showed that as long as the sample sizes are equal (or nearly equal), the variances can differ by amounts that make the largest up to nine times the smallest and the results of ANOVA will continue to be essentially reliable.
3. The samples are simple random samples of quantitative data.
4. The samples are independent of each other. (The samples are not matched or paired in any way.)
5. The different samples are from populations that are categorized in only one way.

Procedure for Testing H_0: $\mu_1 = \mu_2 = \mu_3 = \cdots = \mu_k$

1. Use STATDISK, Minitab, Excel, StatCrunch, a TI-83/84 Plus calculator, or any other technology to obtain results.
2. Identify the P-value from the display. (The ANOVA test is right-tailed because only large values of the test statistic cause us to reject equality of the population means.)
3. Form a conclusion based on these criteria that use the significance level α:
   - **Reject:** If the P-value $\leq \alpha$, reject the null hypothesis of equal means and conclude that at least one of the population means is different from the others.
   - **Fail to Reject:** If the P-value $> \alpha$, fail to reject the null hypothesis of equal means.

In the preceding definition, the term *treatment* is used because early applications of analysis of variance involved agricultural experiments in which different plots of farmland were treated with different fertilizers, seed types, insecticides, and so on. For example, Table 11-9 uses the one "treatment" (or factor) of blood lead level. That factor has three different categories: low, medium, and high blood lead levels (as defined in Data Set 5 from Appendix B).

Table 11-9 Performance IQ Scores of Children

Low Blood Lead Level															
85	90	107	85	100	97	101	64	111	100	76	136	100	90	135	104
149	99	107	99	113	104	101	111	118	99	122	87	118	113	128	121
111	104	51	100	113	82	146	107	83	108	93	114	113	94	106	92
79	129	114	99	110	90	85	94	127	101	99	113	80	115	85	112
112	92	97	97	91	105	84	95	108	118	118	86	89	100		
Medium Blood Lead Level															
78	97	107	80	90	83	101	121	108	100	110	111	97	51	94	80
101	92	100	77	108	85										
High Blood Lead Level															
93	100	97	79	97	71	111	99	85	99	97	111	104	93	90	107
108	78	95	78	86											

F Distribution

The analysis of variance (ANOVA) methods of this section require the *F* distribution, which has the following properties (see Figure 11-6):

1. The *F* distribution is not symmetric.
2. Values of the *F* distribution cannot be negative.
3. The exact shape of the *F* distribution depends on the two different degrees of freedom.

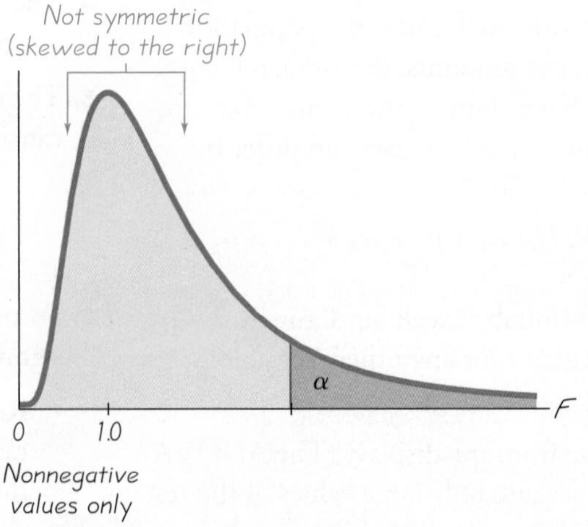

Figure 11-6 ***F* Distribution**

There is a different *F* distribution for each different pair of degrees of freedom for numerator and denominator.

Example 1 Lead and Performance IQ Scores

Table 11-9 lists samples of performance IQ scores obtained from a study involving children who lived near a large ore smelter in Texas. The children were partitioned into groups based on their measured blood lead levels. (See Data Set 5 in Appendix B for the specific blood lead level cutoff values.) Use the performance IQ scores listed in Table 11-9 and a significance level of $\alpha = 0.05$ to test the claim that the three samples come from populations with means that are all equal.

Solution

Requirement check (1) Based on the three samples listed in Table 11-9, the three populations appear to have distributions that are approximately normal, as indicated by normal quantile plots. (2) The three samples in Table 11-9 have standard deviations that are not dramatically different, so the three population variances appear to be about the same. (3) Based on the careful design of the study, we can treat the samples as simple random samples. (4) The samples are independent of each other; the performance IQ scores are not matched in any way. (5) The three samples are from populations categorized according to the single factor of lead level (low, medium, high). The requirements are satisfied.

The null hypothesis and the alternative hypothesis are as follows:

H_0: $\mu_1 = \mu_2 = \mu_3$.

H_1: At least one of the means is different from the others.

The significance level is $\alpha = 0.05$.

Step 1: Use technology to obtain ANOVA results, such as one of those shown in the accompanying displays.

STATDISK

Source:	DF:	SS:	MS:	Test Stat, F:	Critical F:	P-Value:
Treatment:	2	2022.729906	1011.364953	4.071122	3.073087	0.01951
Error:	118	29314.046953	248.424127			
Total:	120	31336.77686				

MINITAB

One-way ANOVA: Low, Medium, High

Source	DF	SS	MS	F	P
Factor	2	2023	1011	4.07	0.020
Error	118	29314	248		
Total	120	31337			

S = 15.76 R-Sq = 6.45% R-Sq(adj) = 4.87%

EXCEL

ANOVA

Source of Variation	SS	df	MS	F	P-value	F crit
Between Groups	2022.729906	2	1011.364953	4.071122103	0.019510383	3.073090341
Within Groups	29314.04695	118	248.4241267			
Total	31336.77686	120				

STATCRUNCH

ANOVA table

Source	df	SS	MS	F-Stat	P-value
Treatments	2	2022.7299	1011.3649	4.071122	0.0195
Error	118	29314.047	248.42413		
Total	120	31336.777			

TI-83/84 PLUS

```
One-way ANOVA
 F=4.071122103
 p=.0195103826
 Factor
  df=2
  SS=2022.72991
↓ MS=1011.36495
```

SPSS

	Sum of Squares	df	Mean Square	F	Sig.
Between Groups	2022.730	2	1011.365	4.071	.020
Within Groups	29314.047	118	248.424		
Total	31336.777	120			

JMP

Source	DF	Sum of Squares	Mean Square	F Ratio
Model	2	2022.730	1011.36	4.0711
Error	118	29314.047	248.42	Prob > F
C. Total	120	31336.777		0.0195*

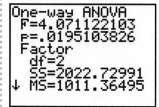

Step 2: The displays all show that the P-value is 0.020 when rounded.

Step 3: Because the P-value of 0.020 is less than the significance level of $\alpha = 0.05$, we reject the null hypothesis of equal means. (If the P is low, the null must go.)

Interpretation

There is sufficient evidence to warrant rejection of the claim that the three samples come from populations with means that are all equal. Based on the samples of measurements listed in Table 11-9, we conclude that those values come from populations having means that are not all the same. On the basis of this ANOVA test, we cannot conclude that any particular mean is different from the others, but we can informally note that the sample mean for the low blood lead group is higher than the means for the medium and high blood lead groups. It appears that greater blood lead levels are associated with lower performance IQ scores.

CAUTION When we conclude that there is sufficient evidence to reject the claim of equal population means, we cannot conclude from ANOVA that any particular mean is different from the others. (There are several other methods that can be used to identify the specific means that are different, and some of them are discussed in Part 2 of this section.)

How Is the P-Value Related to the Test Statistic? *Larger* values of the test statistic result in *smaller* P-values, so the ANOVA test is right-tailed. Figure 11-7 shows the relationship between the F test statistic and the P-value. Assuming that the populations have the same variance σ^2 (as required for the test), the F test statistic is the ratio of these two estimates of σ^2: (1) variation *between* samples (based on variation among sample means); and (2) variation *within* samples (based on the sample variances).

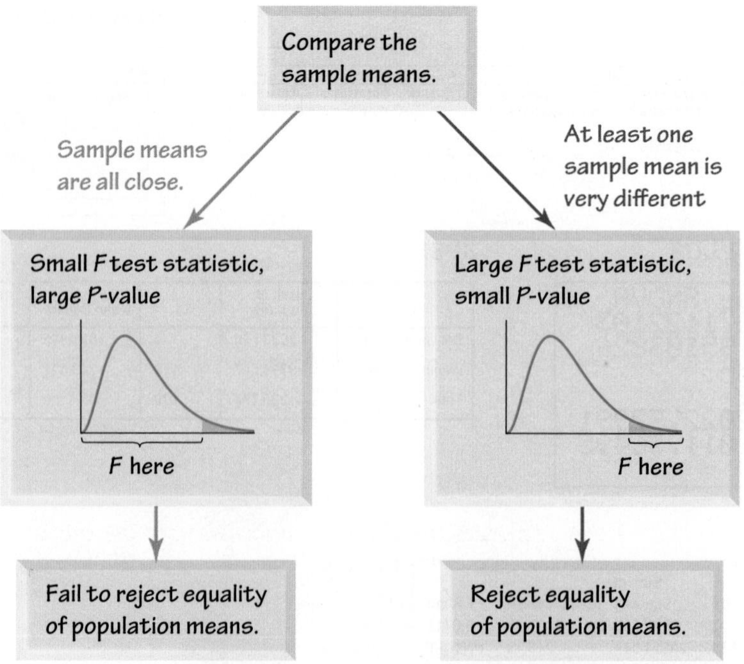

Figure 11-7 Relationship between the F Test Statistic and the P-Value

Why 0.05?

In 1925, R. A. Fisher published a book that introduced the method of analysis of variance, and he needed a table of critical values based on numerator degrees of freedom and denominator degrees of freedom. Because the table uses two different degrees of freedom, it becomes very long if many different critical values are used, so Fisher included a table using 0.05 only. In a later edition he also included the significance level of 0.01.

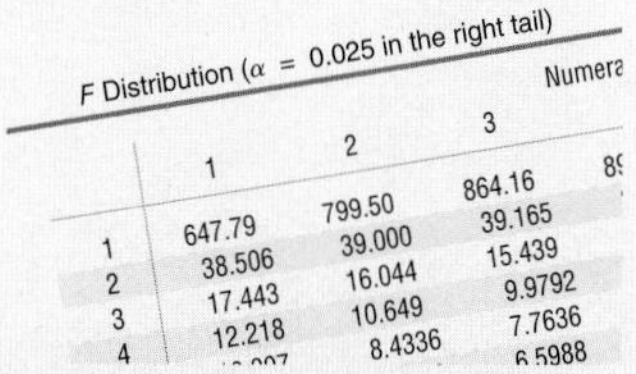

Stephen Stigler, a notable historian of statistics, wrote in *Chance* magazine that the choice of a significance level of 0.05 is a convenient round number that is somewhat arbitrary. Although it is arbitrary, the choice of 0.05 accomplishes the following important goals. (1) The value of a 0.05 significance level results in sample sizes that are reasonable and not too large. (2) The choice of 0.05 is large enough to give us a reasonable chance of identifying important effects (by correctly rejecting a null hypothesis of no effect when there really is an effect). (3) The choice of 0.05 is not so small that it forces us to miss important effects (by making the mistake of failing to reject a null hypothesis of no effect when there really is an effect).

Test Statistic for One-Way ANOVA: $F = \dfrac{\text{variance between samples}}{\text{variance within samples}}$

The numerator of the F test statistic measures variation between sample means. The estimate of variance in the denominator depends only on the sample variances and is not affected by differences among the sample means. Consequently, sample means that are close in value result in a small F test statistic and a large P-value, so we conclude that there are no significant differences among the sample means. Sample means that are very far apart in value result in a large F test statistic and a small P-value, so we reject the claim of equal means.

Why Not Just Test Two Samples at a Time? If we want to test for equality among three or more population means, why do we need a new procedure when we can test for equality of two means using the methods presented in Section 9-3? For example, if we want to use the sample data from Table 11-9 to test the claim that the three populations have the same mean, why not simply pair them off and test two at a time by testing H_0: $\mu_1 = \mu_2$, H_0: $\mu_2 = \mu_3$, and H_0: $\mu_1 = \mu_3$? For the data in Table 11-9, the approach of testing equality of two means at a time requires three different hypothesis tests. If we use a 0.05 significance level for each of those three hypothesis tests, the actual overall confidence level could be as low as 0.95^3 (or 0.857). In general, as we increase the number of individual tests of significance, we increase the risk of finding a difference by chance alone (instead of a real difference in the means). The risk of a type I error—finding a difference in one of the pairs when no such difference actually exists—is far too high. The method of analysis of variance helps us avoid that particular pitfall (rejecting a true null hypothesis) by using *one test* for equality of several means, instead of several tests that each compare two means at a time.

CAUTION When testing for equality of three or more populations, use analysis of variance. (Using multiple hypothesis tests with two samples at a time could wreak havoc with the significance level.)

Note to Instructor
Recommendation: Cover Part 2 only if you have an abundance of available class time, such as in a two-semester course. Otherwise, omit Part 2.

Part 2: Calculations and Identifying Means That Are Different

Calculations with Equal Sample Sizes *n*

Table 11-10 can be very helpful in understanding the methods of ANOVA. In Table 11-10, compare Data Set A to Data Set B to see that Data Set A is the same as Data Set B with this notable exception: the Sample 1 values each differ by 10. If the data sets all have the same sample size (as in $n = 4$ for Table 11-10), the following calculations aren't too difficult, as shown on the next page.

Variance *between* Samples Find the variance *between* samples by evaluating $ns_{\bar{x}}^2$, where $s_{\bar{x}}^2$ is the variance of the sample means and n is the size of each of the samples. That is, consider the sample means to be an ordinary set of values and calculate the variance. (From the central limit theorem, $\sigma_{\bar{x}} = \sigma/\sqrt{n}$ can be solved for σ to get $\sigma = \sqrt{n}\cdot\sigma_{\bar{x}}$, so that we can estimate σ^2 with $ns_{\bar{x}}^2$.) For example, the sample means for Data Set A in Table 11-10 are 5.5, 6.0, and 6.0; these three values have a variance of $s_{\bar{x}}^2 = 0.0833$, so that

$$\text{variance between samples} = ns_{\bar{x}}^2 = 4(0.0833) = 0.3332$$

Variance *within* Samples Estimate the variance *within* samples by calculating s_p^2, which is the pooled variance obtained by finding the mean of the sample variances. The sample variances in Table 11-10 are 3.0, 2.0, and 2.0, so that

$$\text{variance within samples} = s_p^2 = \frac{3.0 + 2.0 + 2.0}{3} = 2.3333$$

Table 11-10 Effect of a Mean on the F Test Statistic

	A			B (Sample 1 of A, add 10)		
	Sample 1	Sample 2	Sample 3	Sample 1	Sample 2	Sample 3
	7	6	4	**17**	6	4
	3	5	7	**13**	5	7
	6	5	6	**16**	5	6
	6	8	7	**16**	8	7
	$n_1 = 4$	$n_2 = 4$	$n_3 = 4$	$n_1 = 4$	$n_2 = 4$	$n_3 = 4$
	$\bar{x}_1 = 5.5$	$\bar{x}_2 = 6.0$	$\bar{x}_3 = 6.0$	$\bar{x}_1 = \mathbf{15.5}$	$\bar{x}_2 = 6.0$	$\bar{x}_3 = 6.0$
	$s_1^2 = 3.0$	$s_2^2 = 2.0$	$s_3^2 = 2.0$	$s_1^2 = 3.0$	$s_2^2 = 2.0$	$s_3^2 = 2.0$
Variance between samples	$ns_{\bar{x}}^2 = 4(0.0833) = 0.3332$			$ns_{\bar{x}}^2 = 4(30.0833) = 120.3332$		
Variance within samples	$s_p^2 = \frac{3.0 + 2.0 + 2.0}{3} = 2.3333$			$s_p^2 = \frac{3.0 + 2.0 + 2.0}{3} = 2.3333$		
F test statistic	$F = \frac{ns_{\bar{x}}^2}{s_p^2} = \frac{0.3332}{2.3333} = \mathbf{0.1428}$			$F = \frac{ns_{\bar{x}}^2}{s_p^2} = \frac{120.3332}{2.3333} = \mathbf{51.5721}$		
P-value (found from Excel)	P-value = 0.8688			P-value = 0.0000118		

Calculate the Test Statistic Evaluate the F test statistic as follows:

$$F = \frac{\text{variance between samples}}{\text{variance within samples}} = \frac{ns_{\bar{x}}^2}{s_p^2} = \frac{0.3332}{2.3333} = 0.1428$$

The critical value of F is found by assuming a right-tailed test, because large values of F correspond to significant differences among means. With k samples each having n values, the numbers of degrees of freedom are as follows.

Degrees of Freedom: k = number of samples and n = sample size

$$\text{numerator degrees of freedom} = k - 1$$

$$\text{denominator degrees of freedom} = k(n - 1)$$

For Data Set A in Table 11-10, $k = 3$ and $n = 4$, so the degrees of freedom are 2 for the numerator and $3(4 - 1) = 9$ for the denominator. With $\alpha = 0.05$, 2 degrees of freedom for the numerator, and 9 degrees of freedom for the denominator, the critical F value can be found to be 4.2565. If we were to use the critical value method of hypothesis testing with Data Set A in Table 11-10, we would see that this right-tailed test has a test statistic of $F = 0.1428$ and a critical value of $F = 4.2565$, so the test statistic is not in the critical region. We therefore fail to reject the null hypothesis of equal means.

To really see how the method of analysis of variance works, consider both collections of sample data in Table 11-10. Note that the three samples in Data Set A are identical to the three samples in Data Set B, except that each value in Sample 1 of Data Set B is 10 more than the corresponding value in Data Set A. The three sample means in A are very close, but there are substantial differences in B. However, the three sample variances in A are identical to those in B.

Adding 10 to each data value in the first sample of Table 11-10 has a dramatic effect on the test statistic, with F changing from 0.1428 to 51.5721. Adding 10 to each data value in the first sample also has a dramatic effect on the P-value, which changes from 0.8688 (not significant) to 0.0000118 (significant). Note that the variance between samples in A is 0.3332, but for B it is 120.3332 (indicating that the sample means in B are farther apart). Note also that the variance within samples is 2.3333 in both parts, because the variance *within* a sample isn't affected when we add a constant to every sample value. *The change in the F test statistic and the P-value is attributable only to the change in* $\bar{x}_1$. This illustrates the key point underlying the method of one-way analysis of variance: **The F test statistic is very sensitive to sample *means,* even though it is obtained through two different estimates of the common population *variance.*** Adding 10 to each value of the first sample causes the three sample means to grow farther apart, with the result that the F test statistic increases and the P-value decreases.

Calculations with Unequal Sample Sizes

While the calculations for cases with equal sample sizes are somewhat reasonable, they become much more complicated when the sample sizes are not all the same, but the same basic reasoning applies. We calculate an F test statistic that is the ratio of two different estimates of the common population variance σ^2. With unequal sample sizes, we must base the calculations on *weighted* measures that take the sample sizes into account. The test statistic is essentially the same as the one given earlier, and its interpretation is also the same as described earlier. The denominator depends only on the sample variances that measure variation within the treatments and is not affected by the differences among the sample means. In contrast, the numerator is affected by differences among the sample means. If the differences among the sample means are very large, those large differences will cause the numerator to be very large, so F will also be very large. Consequently, very large values of F suggest unequal means, and the ANOVA test is therefore right-tailed. Instead of providing the relevant messy formulas required for cases with unequal sample sizes, we wisely and conveniently assume that technology should be used to obtain the P-value for the analysis of variance. We become unencumbered by complex computations and we can focus on checking requirements and interpreting results.

Designing the Experiment With one-way (or single-factor) analysis of variance, we use one factor as the basis for partitioning the data into different categories. If we conclude that the differences among the means are significant, we can't be absolutely sure that the differences can be explained by the factor being used. It is possible that the variation of some other unknown factor is responsible. One way to reduce the effect of the extraneous factors is to design the experiment so that it has a **completely randomized design,** in which each sample value is given the same chance of belonging to the different factor groups. For example, you might assign subjects to two different treatment groups and a third placebo group through a process of random selection equivalent to picking slips of paper from a bowl. Another way to reduce the effect of extraneous factors is to use a **rigorously controlled design,** in which sample values are carefully chosen so that all other factors have no variability. In general, good results require that the experiment be carefully designed and executed.

Identifying Which Means Are Different

After conducting an analysis of variance test, we might conclude that there is sufficient evidence to reject a claim of equal population means, but we cannot conclude from ANOVA that any *particular* means are different from the others. There are

Note to Instructor
This subsection is intended mainly for information purposes and can be easily omitted. Exercise 17 refers to the Tukey test, and Exercise 18 refers to the Bonferroni test, but both exercises have displayed results.

several formal and informal procedures that can be used to identify the specific means that are different. Here are two *informal* methods for comparing means:

- Construct boxplots of the different samples to see if one or more of them is very different from the others.
- Construct confidence interval estimates of the means for each of the different samples, then compare those confidence intervals to see if one or more of them does not overlap with the others.

There are several formal procedures for identifying which means are different. Some of the tests, called **range tests,** allow us to identify subsets of means that are not significantly different from each other. Other tests, called **multiple comparison tests,** use pairs of means, but they make adjustments to overcome the problem of having a significance level that increases as the number of individual tests increases. There is no consensus on which test is best, but some of the more common tests are the Duncan test, Student-Newman-Keuls test (or SNK test), Tukey test (or Tukey honestly significant difference test), Scheffé test, Dunnett test, least significant difference test, and the Bonferroni test. Let's consider the Bonferroni test to see one example of a multiple comparison test. Here is the procedure.

Bonferroni Multiple Comparison Test

Step 1: Do a separate t test for each pair of samples, but make the adjustments described in the following steps.

Step 2: For an estimate of the variance σ^2 that is common to all of the involved populations, use the value of MS(error), which uses all of the available sample data. The value of MS(error) is typically obtained when conducting the analysis of variance test. Using the value of MS(error), calculate the value of the test statistic t, as shown below. The particular test statistic calculated below is based on the choice of Sample 1 and Sample 2; change the subscripts and use another pair of samples until all of the different possible pairs of samples have been tested.

$$t = \frac{\bar{x}_1 - \bar{x}_2}{\sqrt{\text{MS(error)} \cdot \left(\frac{1}{n_1} + \frac{1}{n_2}\right)}}$$

Step 3: After calculating the value of the test statistic t for a particular pair of samples, find either the critical t value or the P-value, but make the following adjustment so that the overall significance level does not increase.

***P*-value:** Use the test statistic t with df $= N - k$, where N is the total number of sample values and k is the number of samples, and find the P-value the usual way, but adjust the P-value by multiplying it by the number of different possible pairings of two samples. (For example, with three samples, there are three different possible pairings, so adjust the P-value by multiplying it by 3.)

Critical value: When finding the critical value, adjust the significance level α by dividing it by the number of different possible pairings of two samples. (For example, with three samples, there are three different possible pairings, so adjust the significance level by dividing it by 3.)

Note that in Step 3 of the preceding Bonferroni procedure, either an individual test is conducted with a much lower significance level or the P-value is greatly increased. Rejection of equality of means therefore requires differences that are much farther apart. This adjustment in Step 3 compensates for the fact that we are doing several tests instead of only one test.

Example 2 Bonferroni Test

Example 1 in this section used analysis of variance with the sample data in Table 11-9. We concluded that there is sufficient evidence to warrant rejection of the claim of equal means. Use the Bonferroni test with a 0.05 significance level to identify which mean is different from the others.

Solution

The Bonferroni test requires a separate t test for each of three different possible pair of samples. Here are the null hypotheses to be tested:

$$H_0: \mu_1 = \mu_2 \qquad H_0: \mu_1 = \mu_3 \qquad H_0: \mu_2 = \mu_3$$

We begin with $H_0: \mu_1 = \mu_2$. Using the sample data given in Table 11-9 and carrying some extra decimal places for greater accuracy in the calculations, we have $n_1 = 78$ and $\bar{x} = 102.705128$. Also, $n_2 = 22$ and $\bar{x}_2 = 94.136364$. From the technology results shown in Example 1 we also know that MS(error) = 248.424127. We now evaluate the test statistic using the unrounded sample means:

$$t = \frac{\bar{x}_1 - \bar{x}_2}{\sqrt{\text{MS(error)} \cdot \left(\frac{1}{n_1} + \frac{1}{n_2}\right)}}$$

$$= \frac{102.705128 - 94.136364}{\sqrt{248.424127 \cdot \left(\frac{1}{78} + \frac{1}{22}\right)}} = 2.252$$

The number of degrees of freedom is df $= N - k = 121 - 3 = 118$. ($N = 121$ because there are 121 different sample values in all three samples combined, and $k = 3$ because there are three different samples.) With a test statistic of $t = 2.252$ and with df $= 118$, the two-tailed P-value is 0.026172, but we adjust this P-value by multiplying it by 3 (the number of different possible pairs of samples) to get a final P-value of 0.078516, or 0.079 when rounded. Because this P-value is not small (less than 0.05), we fail to reject the null hypothesis. It appears that Samples 1 and 2 do not have significantly different means.

Instead of continuing with separate hypothesis tests for the other two pairings, see the SPSS display showing all of the Bonferroni test results. (The first row of numerical results corresponds to the results found here; see the value of 0.079, which is calculated here.) The display shows that the pairing of low/high yields a P-value of 0.090, so there is not a significant difference between the means from the low and high blood lead levels. Also, the SPSS display shows that the pairing of medium/high yields a P-value of 1.000, so there is not a significant difference between the means from the medium and high blood lead levels.

SPSS BONFERRONI RESULTS

(I) Level	(J) Level	Mean Difference (I-J)	Std. Error	Sig.	95% Confidence Interval	
					Lower Bound	Upper Bound
1.00	2.00	8.56876	3.80486	.079	-.6717	17.8092
	3.00	8.51465	3.87487	.090	-.8958	17.9251
2.00	1.00	-8.56876	3.80486	.079	-17.8092	.6717
	3.00	-.05411	4.80851	1.000	-11.7320	11.6238
3.00	1.00	-8.51465	3.87487	.090	-17.9251	.8958
	2.00	.05411	4.80851	1.000	-11.6238	11.7320

Interpretation

Although the analysis of variance test tells us that at least one of the means is different from the others, the Bonferroni test results do not identify any one particular sample mean that is significantly different from the others. In the original article discussing these results, the authors state that "our findings indicate that a chronic absorption of particulate lead . . . may result in subtle but statistically significant impairment in the non-verbal cognitive and perceptual motor skills measured by the performance scale of the Wechsler intelligence tests." That statement confirms these results: From analysis of variance we know that at least one mean is different from the others, but such differences are not strong enough so that the Bonferroni test results identify any one particular mean as being significantly different. However, the sample means of 102.7 (low blood lead level), 94.1 (medium blood lead level), and 94.2 (high blood lead level) suggest that medium and high blood lead levels are associated with lower mean performance IQ scores than the low blood level group.

using TECHNOLOGY

STATDISK Enter the data in columns of the Data Window. Select **Analysis** from the main menu bar, then select **One-Way Analysis of Variance,** and select the columns of sample data to be used. Click **Evaluate.**

MINITAB First enter the sample data in columns C1, C2, C3 . . . . Next, select **Stat, ANOVA, ONEWAY (UNSTACKED),** and enter C1 C2 C3 . . . in the box identified as "Responses" (in separate columns).

EXCEL You can use either XLSTAT or Excel's Data Analysis add-in. An advantage of the Data Analysis add-in is that you are not required to stack all of the data in one column with corresponding category names in another column.

XLSTAT First stack all of the sample data in column B with the corresponding variable names listed in column A. Click on **XLSTAT,** then select **Modeling Data,** then select **ANOVA.** In the "Quantitative" box, enter the range of cells containing the sample data, such as B1:B121. In the "Qualitative" box, enter the range of cells containing the variable names, such as A1:A121. Put a check next to the "Variable labels" box only if the first row consists of labels. Click **OK.** In the results, look for the "Analysis of Variance" table that includes the *F* test statistic and the *P*-value.

Data Analysis add-in: Enter the data in columns A, B, C, ... In Excel 2013, 2010, and 2007, click on **Data;** in Excel 2003, click on **Tools.** Now click on **Data Analysis** and select **Anova: Single Factor.** In the dialog box, enter the range containing the sample data. (For example, enter A1:C30 if the first value is in row 1 of column A and the longest column has 30 data values.)

TI-83/84 PLUS First enter the data as lists in L1, L2, L3 . . . then press STAT, select **TESTS,** and choose the option **ANOVA.** Enter the column labels. For example, if the data are in columns L1, L2, and L3, enter those columns to get **ANOVA (L1, L2, L3),** and press ENTER.

STATCRUNCH Click on **Open StatCrunch.** Enter the columns of data or open a data set. Click on **Stat,** then select **ANOVA,** then select **One Way.** Enter the columns to be used, then click on **Calculate.** Results will include the test statistic and *P*-value.

Recommended Assignment
Exercises 1–10. (None of Exercises 1–10 require the use of computer software or a TI-83/84 Plus calculator. Exercises 11–16 require the use of technology.) If students are using Minitab, their results might not be as precise as answers given in the back of the book. For example, a student might use Minitab to get a test statistic of $F = 15.81$, whereas the answer in the back of the book might be $F = 15.8142$. Consider informing students that they shouldn't be concerned about those last digits unless the test statistic is very close to the critical value—a condition that is very rare.

11-4 Basic Skills and Concepts

Statistical Literacy and Critical Thinking

In Exercises 1–4, use the following listed chest deceleration measurements (in g, where g is the force of gravity) from samples of small, midsize, and large cars. (These values are from Data Set 13 in Appendix B.) Also shown (on the next page) are the SPSS results for analysis of variance. Assume that we plan to use a 0.05 significance level to test the claim that the different size categories have the same mean chest deceleration in the standard crash test.

Chest Deceleration Measurements (*g*) from a Standard Crash Test

Small	44	39	37	54	39	44	42
Midsize	36	53	43	42	52	49	41
Large	32	45	41	38	37	38	33

SPSS

	Sum of Squares	df	Mean Square	F	Sig.
Between Groups	200.857	2	100.429	3.288	.061
Within Groups	549.714	18	30.540		
Total	750.571	20			

1. ANOVA

a. What characteristic of the data above indicates that we should use *one-way* analysis of variance?

b. If the objective is to test the claim that the three size categories have the same *mean* chest deceleration, why is the method referred to as analysis of *variance*?

2. Why Not Test Two at a Time? Refer to the sample data given in Exercise 1. If we want to test for equality of the three means, why don't we use three separate hypothesis tests for $\mu_1 = \mu_2$, $\mu_1 = \mu_3$, and $\mu_2 = \mu_3$?

3. Test Statistic What is the value of the test statistic? What distribution is used with the test statistic?

4. *P*-Value If we use a 0.05 significance level in analysis of variance with the sample data given in Exercise 1, what is the *P*-value? What should we conclude?

In Exercises 5–16, use analysis of variance for the indicated test.

5. Lead and Verbal IQ Scores Example 1 used measured *performance* IQ scores for three different blood lead levels. If we use the same three categories of blood lead levels with measured *verbal* IQ scores, we get the accompanying Minitab display. (The data are listed in Data Set 5 of Appendix B.) Using a 0.05 significance level, test the claim that the three categories of blood lead level have the same mean verbal IQ score. Does exposure to lead appear to have an effect on verbal IQ scores?

MINITAB

Source	DF	SS	MS	F	P
LEAD	2	142	71	0.39	0.677
Error	118	21441	182		
Total	120	21584			

6. Lead and Full IQ Scores Example 1 used measured *performance* IQ scores for three different blood lead levels. If we use the same three categories of blood lead levels with the *full* IQ scores, we get the accompanying Excel display. (The data are listed in Data Set 5 of Appendix B.) Using a 0.05 significance level, test the claim that the three categories of blood lead level have the same mean full IQ score. Does it appear that exposure to lead has an effect on full IQ scores?

EXCEL

ANOVA						
Source of Variation	SS	df	MS	F	P-value	F crit
Between Groups	938.3653	2	469.1827	2.303395	0.104395	3.07309
Within Groups	24035.63	118	203.6918			
Total	24974	120				

1. a. The chest deceleration measurements are categorized according to the one characteristic of size.
   b. The terminology of *analysis of variance* refers to the method used to test for equality of the three population means. That method is based on two different estimates of a common population variance.
2. As we increase the number of individual tests of significance, we increase the risk of finding a difference by chance alone (instead of a real difference in the means). The risk of a type I error—finding a difference in one of the pairs when no such difference actually exists—is too high. The method of analysis of variance helps us avoid that particular pitfall (rejecting a true null hypothesis) by using *one test* for equality of several means, instead of several tests that each compare two means at a time.
3. The test statistic is $F = 3.288$, and the *F* distribution applies.
4. The *P*-value is 0.061. Because the *P*-value is greater than the significance level of 0.05, we fail to reject the null hypothesis of equal means. There is not sufficient evidence to warrant rejection of the claim that the three different categories of car sizes have the same mean chest deceleration in the standard car crash test.
5. Test statistic: $F = 0.39$. *P*-value: 0.677. Fail to reject H_0: $\mu_1 = \mu_2 = \mu_3$. There is not sufficient evidence to warrant rejection of the claim that the three categories of blood lead level have the same mean verbal IQ score. Exposure to lead does not appear to have an effect on verbal IQ scores.
6. Test statistic: $F = 2.3034$. *P*-value: 0.1044. Fail to reject H_0: $\mu_1 = \mu_2 = \mu_3$. There is not sufficient evidence to warrant rejection of the claim that the three categories of blood lead level have the same mean full IQ score. Exposure to lead does not appear to have an effect on full IQ scores.

7. Test statistic: $F = 11.6102$. P-value: 0.000577. Reject H_0: $\mu_1 = \mu_2 = \mu_3$. There is sufficient evidence to warrant rejection of the claim that the three size categories have the same mean highway fuel consumption. The size of a car does appear to affect highway fuel consumption.

7. Highway Fuel Consumption Data Set 14 in Appendix B lists highway fuel consumption amounts (mi/gal) for cars categorized by size (small, midsize, large). If we use those highway fuel consumption amounts arranged into the three separate size categories, we get the TI-83/84 Plus calculator results shown below. Using a 0.05 significance level, test the claim that the three size categories have the same mean highway fuel consumption. Does the size of a car appear to affect highway fuel consumption?

TI-83/84 PLUS

```
One-way ANOVA
 F=11.61019737
 p=5.773554E-4
 Factor
  df=2
  SS=224.095238
↓ MS=112.047619
```

8. Test statistic: $F = 23.9457$. P-value: 0.000008. Reject H_0: $\mu_1 = \mu_2 = \mu_3$. There is sufficient evidence to warrant rejection of the claim that the three size categories have the same mean city fuel consumption. The size of a car does appear to affect city fuel consumption.

8. City Fuel Consumption Data Set 14 in Appendix B lists city fuel consumption amounts (mi/gal) for cars categorized by size (small, midsize, large). If we use those city fuel consumption amounts arranged into the three separate size categories, we get the STATDISK results shown here. Using a 0.05 significance level, test the claim that the three size categories have the same mean city fuel consumption. Does the size of a car appear to affect city fuel consumption?

STATDISK

Source:	DF:	SS:	MS:	Test Stat, F:	Critical F:	P-Value:
Treatment:	2	209.809524	104.904762	23.945652	3.554561	0.000008
Error:	18	78.857143	4.380952			
Total:	20	288.666667				

9. Test statistic: $F = 0.161$. P-value: 0.852. Fail to reject H_0: $\mu_1 = \mu_2 = \mu_3$. There is not sufficient evidence to warrant rejection of the claim that the three size categories have the same mean head injury measurement. The size of a car does not appear to affect head injuries.

9. Head Injury Crash Test Data Exercises 1–4 use chest deceleration data for three different size categories (small, midsize, large). The data are from a standard crash test and they are listed in Data Set 13 in Appendix B. If we use the head injury measurements (in HIC, which is a standard head injury criterion) with the same three size categories, we get the SPSS results shown here. Using a 0.05 significance level, test the claim that the three size categories have the same mean head injury measurement. Does the size of a car appear to affect head injuries?

SPSS

	Sum of Squares	df	Mean Square	F	Sig.
Between Groups	7366.952	2	3683.476	.161	.852
Within Groups	411540.286	18	22863.349		
Total	418907.238	20			

10. Test statistic: $F = 0.3476$. P-value: 0.7111. Fail to reject H_0: $\mu_1 = \mu_2 = \mu_3$. There is not sufficient evidence to warrant rejection of the claim that the three size categories have the same mean pelvis injury measurement. The size of a car does not appear to affect pelvis injuries.

10. Pelvis Injury Crash Test Data Exercises 1–4 use chest deceleration data for three different size categories (small, midsize, large). The data are from a standard crash test and they are listed in Data Set 13 in Appendix B. If we use the pelvis injury measurements (g) with the same three size categories, we get the XLSTAT results shown here. Using a 0.05 significance level, test the claim that the three size categories have the same mean pelvis injury measurement. Does the size of a car appear to affect pelvis injuries?

XLSTAT

Source	DF	Sum of squares	Mean squares	F	Pr > F
Model	2	79.1429	39.5714	0.3476	0.7111
Error	18	2049.4286	113.8571		
Corrected	20	2128.5714			

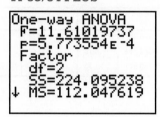

11. Triathlon Times Jeff Parent is a statistics instructor who participates in triathlons. Listed below are times (in minutes and seconds) he recorded while riding a bicycle for five laps through each mile of a 3-mile loop. Use a 0.05 significance level to test the claim that it takes the same time to ride each of the miles. Does one of the miles appear to have a hill?

Mile 1	3:15	3:24	3:23	3:22	3:21
Mile 2	3:19	3:22	3:21	3:17	3:19
Mile 3	3:34	3:31	3:29	3:31	3:29

11. Test statistic: $F = 27.2488$. P-value: 0.000. Reject H_0: $\mu_1 = \mu_2 = \mu_3$. There is sufficient evidence to warrant rejection of the claim that the three different miles have the same mean time. These data suggest that the third mile appears to take longer, and a reasonable explanation is that the third lap has a hill.

12. Clancy, Rowling, Tolstoy Readability Pages were randomly selected by the author from *The Bear and the Dragon* by Tom Clancy, *Harry Potter and the Sorcerer's Stone* by J. K. Rowling, and *War and Peace* by Leo Tolstoy. The Flesch Reading Ease scores for those pages are listed below. Use a 0.05 significance level to test the claim that the three samples are from populations with the same mean. Do the books appear to have different reading levels of difficulty?

Clancy	58.2	73.4	73.1	64.4	72.7	89.2	43.9	76.3	76.4	78.9	69.4	72.9
Rowling	85.3	84.3	79.5	82.5	80.2	84.6	79.2	70.9	78.6	86.2	74.0	83.7
Tolstoy	69.4	64.2	71.4	71.6	68.5	51.9	72.2	74.4	52.8	58.4	65.4	73.6

12. Test statistic: $F = 9.4695$. P-value: 0.000562. Reject H_0: $\mu_1 = \mu_2 = \mu_3$. There is sufficient evidence to warrant rejection of the claim that the three books have the same mean Flesch Reading Ease score. The data suggest that the books appear to have mean scores that are not all the same.

13. Poplar Tree Weights Weights (kg) of poplar trees were obtained from trees planted in a rich and moist region. The trees were given different treatments identified in the table below. The data are from a study conducted by researchers at Pennsylvania State University and were provided by Minitab, Inc. Use a 0.05 significance level to test the claim that the four treatment categories yield poplar trees with the same mean weight. Is there a treatment that appears to be most effective?

No Treatment	Fertilizer	Irrigation	Fertilizer and Irrigation
1.21	0.94	0.07	0.85
0.57	0.87	0.66	1.78
0.56	0.46	0.10	1.47
0.13	0.58	0.82	2.25
1.30	1.03	0.94	1.64

13. Test statistic: $F = 6.1413$. P-value: 0.0056. Reject H_0: $\mu_1 = \mu_2 = \mu_3 = \mu_4$. There is sufficient evidence to warrant rejection of the claim that the four treatment categories yield poplar trees with the same mean weight. Although not justified by the results from analysis of variance, the treatment of fertilizer and irrigation appears to be most effective.

14. Poplar Tree Weights Weights (kg) of poplar trees were obtained from trees planted in a sandy and dry region. The trees were given different treatments identified in the table below. The data are from a study conducted by researchers at Pennsylvania State University and were provided by Minitab, Inc. Use a 0.05 significance level to test the claim that the four treatment categories yield poplar trees with the same mean weight. Is there a treatment that appears to be most effective in the sandy and dry region?

No Treatment	Fertilizer	Irrigation	Fertilizer and Irrigation
0.24	0.92	0.96	1.07
1.69	0.07	1.43	1.63
1.23	0.56	1.26	1.39
0.99	1.74	1.57	0.49
1.80	1.13	0.72	0.95

14. Test statistic: $F = 0.3801$. P-value: 0.769. Fail to reject H_0: $\mu_1 = \mu_2 = \mu_3 = \mu_4$. There is not sufficient evidence to warrant rejection of the claim that the four treatment categories yield poplar trees with the same mean weight. In the sandy and dry region, there does not appear to be a treatment that is more effective than the others.

15. Test statistic: $F = 18.9931$. P-value: 0.000. Reject H_0: $\mu_1 = \mu_2 = \mu_3$. There is sufficient evidence to warrant rejection of the claim that the three different types of cigarettes have the same mean amount of nicotine. Given that the king-size cigarettes have the largest mean of 1.26 mg per cigarette, compared to the other means of 0.87 mg per cigarette and 0.92 mg per cigarette, it appears that the filters do make a difference (although this conclusion is not justified by the results from analysis of variance).
16. Test statistic: $F = 20.8562$. P-value: 0.000. Reject H_0: $\mu_1 = \mu_2 = \mu_3 = \mu_4$. There is sufficient evidence to warrant rejection of the claim that the three samples are from populations with the same mean. It appears that cotinine levels are greater with more exposure to tobacco smoke. (The samples do not appear to be from normally distributed populations, but ANOVA is robust against departures from normality.)
17. The Tukey test results show different P-values, but they are not dramatically different. The Tukey results suggest the same conclusions as the Bonferroni test.
18. a. In Exercise 13 we reject the null hypothesis of equal means. The displayed Bonferroni results show that with a P-value of 0.039, there is a significant difference between the mean of the no treatment group (group 1) and the mean of the group treated with both fertilizer and irrigation (group 4).
    b. The test statistic is $t = -4.007$. P-value $= 6(0.001018) = 0.00611$. Reject the null hypothesis that the mean weight from the irrigation treatment group is equal to the mean from the group treated with both fertilizer and irrigation.

In Exercises 15 and 16, use the data set from Appendix B.

15. Nicotine in Cigarettes Refer to Data Set 10 in Appendix B and use the amounts of nicotine (mg per cigarette) in the king-size cigarettes, the 100-mm menthol cigarettes, and the 100-mm nonmenthol cigarettes. The king-size cigarettes are nonfiltered, nonmenthol, and nonlight. The 100-mm menthol cigarettes are filtered and nonlight. The 100-mm nonmenthol cigarettes are filtered and nonlight. Use a 0.05 significance level to test the claim that the three categories of cigarettes yield the same mean amount of nicotine. Given that only the king-size cigarettes are not filtered, do the filters appear to make a difference?

16. Secondhand Smoke Refer to Data Set 9 in Appendix B and use the measured serum cotinine levels (in mg/mL) from the three groups of subjects (smokers, nonsmokers exposed to tobacco smoke, and nonsmokers not exposed to tobacco smoke). When nicotine is absorbed by the body, cotinine is produced. Use a 0.05 significance level to test the claim that the three samples are from populations with the same mean. What do the results suggest about the effects of secondhand smoke?

11-4 Beyond the Basics

17. Tukey Test This section included a display of the Bonferroni test results from Table 11-9. Shown here is the SPSS-generated display of results from the Tukey test using the same data. Compare the Tukey test results to those from the Bonferroni test.

SPSS

(I) Level	(J) Level	Mean Difference (I-J)	Std. Error	Sig.	95% Confidence Interval Lower Bound	95% Confidence Interval Upper Bound
1.00	2.00	8.56876	3.80486	.067	-.4626	17.6002
	3.00	8.51465	3.87487	.076	-.6830	17.7123
2.00	1.00	-8.56876	3.80486	.067	-17.6002	.4626
	3.00	-.05411	4.80851	1.000	-11.4678	11.3596
3.00	1.00	-8.51465	3.87487	.076	-17.7123	.6830
	2.00	.05411	4.80851	1.000	-11.3596	11.4678

18. Bonferroni Test Shown below are partial results from using the Bonferroni test with the sample data from Exercise 13. Assume that a 0.05 significance level is being used.

a. What do the displayed results tell us?

b. Use the Bonferroni test procedure to test for a significant difference between the mean amount of the irrigation treatment group and the group treated with both fertilizer and irrigation. Identify the test statistic and either the P-value or critical values. What do the results indicate?

SPSS

(I) Treatment	(J) Treatment	Mean Difference (I-J)	Std. Error	Sig.	95% Confidence Interval Lower Bound	95% Confidence Interval Upper Bound
1.00	2.00	-.02200	.26955	1.000	-.8329	.7889
	3.00	.23600	.26955	1.000	-.5749	1.0469
	4.00	-.84400*	.26955	.039	-1.6549	-.0331

Chapter 11 Review

Sections 11-2 and 11-3 involve applications of the χ^2 distribution to categorical data consisting of frequency counts. In Section 11-2 we described methods for using frequency counts from different categories for testing goodness-of-fit with some claimed distribution. The test statistic given below is used in a right-tailed test in which the χ^2 distribution has $k - 1$

degrees of freedom, where k is the number of categories. This test requires that each of the expected frequencies must be at least 5.

$$\text{Test statistic is } \chi^2 = \sum \frac{(O - E)^2}{E}.$$

In Section 11-3 we described methods for testing claims involving contingency tables (or two-way frequency tables), which have at least two rows and two columns. Contingency tables have two variables: One variable is used for determining the row that describes a sample value, and the second variable is used for determining the column that describes a sample value. We conduct a test of independence between the row and column variables by using the test statistic given below. This test statistic is used in a right-tailed test in which the χ^2 distribution has the number of degrees of freedom given by $(r - 1)(c - 1)$, where r is the number of rows and c is the number of columns. This test requires that each of the expected frequencies must be at least 5.

$$\text{Test statistic is } \chi^2 = \sum \frac{(O - E)^2}{E}.$$

In Section 11-4 we presented the method of one-way analysis of variance, which is used to test for equality of three or more population means. The requirements and procedures are presented in Section 11-4. Because of the complex and difficult nature of the required calculations, we focused on interpretation of P-values obtained using technology. When using one-way analysis of variance for testing equality of three or more population means, we use the following decision criteria:

- **Reject Equality of Means:** If the P-value is small, such as 0.05 or less, reject the null hypothesis of equal population means and conclude that at least one of the population means is different from the others.
- **Fail to Reject Equal Means:** If the P-value is large, such as greater than 0.05, fail to reject the null hypothesis of equal population means. Conclude that there is not sufficient evidence to warrant rejection of equal population means.

Chapter Quick Quiz

Questions 1–5 refer to the sample data in the following table (based on data from the Bureau of Labor Statistics). The table lists frequencies for randomly selected nonfatal occupation injuries arranged according to day of the week. Assume that we want to use a 0.05 significance level to test the claim that such injuries occur with equal frequency on different days of the week.

Day	Mon	Tues	Wed	Thurs	Fri
Number	23	23	21	21	19

1. What are the null and alternative hypotheses corresponding to the stated claim?

2. When testing the claim in Question 1, what are the observed and expected frequencies for Monday?

3. Is the hypothesis test left-tailed, right-tailed, or two-tailed?

4. If using a 0.05 significance level to test the stated claim, find the number of degrees of freedom and the critical value.

5. Given that the P-value for the hypothesis test is 0.971, what do you conclude?

1. H_0: $p_1 = p_2 = p_3 = p_4 = p_5$. H_1: At least one of the probabilities is different from the others.
2. $O = 23$ and $E = 21.4$.
3. Right-tailed.
4. $df = 4$ and the critical value is $\chi^2 = 9.488$.
5. There is not sufficient evidence to warrant rejection of the claim that occupation injuries occur with equal frequency on the different days of the week.

Questions 6–10 refer to the sample data in the following table (based on data from the Pew Research Center). Randomly selected subjects were asked about the use of marijuana for medical purposes. Assume that we want to use a 0.05 significance level to test the claim that response to the question is independent of gender.

	In Favor	Oppose	Don't Know
Men	538	167	29
Women	557	186	31

6. Identify the null and alternative hypotheses corresponding to the stated claim.

7. What distribution is used to test the stated claim (normal, *t*, *F*, chi-square, uniform)?

8. Is the hypothesis test left-tailed, right-tailed, or two-tailed?

9. Find the number of degrees of freedom and the critical value.

10. Given that the *P*-value for the hypothesis test is 0.836, what do you conclude?

6. H_0: Response to the question is independent of gender.
H_1: Response to the question and gender are dependent.
7. Chi-square distribution.
8. Right-tailed.
9. $df = 2$ and the critical value is $\chi^2 = 5.991$.
10. There is not sufficient evidence to warrant rejection of the claim that response is independent of gender.

Review Exercises

1. Auto Fatalities The table below lists auto fatalities by day of the week for a recent year (based on data from the Federal Highway Administration). Minitab results are also shown. Use a 0.01 significance level to test the claim that auto fatalities occur on the different days of the week with the same frequency. Can you provide an explanation for the results?

Day	Mon	Tues	Wed	Thurs	Fri	Sat	Sun
Frequency	3797	3615	3724	4004	4867	5786	5004

MINITAB

N	DF	Chi-Sq	P-Value
30797	6	931.347	0.000

2. Measuring Weights When certain quantities are measured, the last digits tend to be uniformly distributed, but if they are estimated or reported, the last digits tend to have disproportionately more 0s or 5s. If we use the last digits (decimal portion) of the 80 weights in Data Set 1 from Appendix B, we get the frequency counts in the table below. Use a 0.05 significance level to test the claim that the last digits of 0, 1, 2, . . . , 9 occur with the same frequency. Does it appear that the weights were obtained through measurements?

Last Digit	0	1	2	3	4	5	6	7	8	9
Frequency	6	7	4	11	10	8	5	9	10	10

3. Weather-Related Deaths Listed below are the numbers of weather-related deaths in a recent year (based on data from the National Weather Service). Use a 0.01 significance level to test the claim that weather-related deaths occur in the different months with the same frequency. Can you provide an explanation for the result?

Month	Jan	Feb	March	April	May	June	July	Aug	Sept	Oct	Nov	Dec
Number of Deaths	39	17	20	19	60	79	109	73	28	5	5	35

4. Bicycle Helmets A study was conducted of 531 persons injured in bicycle crashes, and randomly selected sample results are summarized in the table shown on the next page (based on results from "A Case-Control Study of the Effectiveness of Bicycle Safety Helmets in Preventing Facial Injury," by Thompson et al., *American Journal of Public Health,* Vol. 80, No. 12). The TI-83/84 Plus calculator results also are shown. At the 0.05 significance level, test the claim that wearing a helmet has no effect on whether facial injuries are received. Based on these results, does a helmet seem to be effective in helping to prevent facial injuries in a crash?

1. Test statistic: $\chi^2 = 931.347$. Critical value: $\chi^2 = 16.812$. *P*-value: 0.000. There is sufficient evidence to warrant rejection of the claim that auto fatalities occur on the different days of the week with the same frequency. Because people generally have more free time on weekends and more drinking occurs on weekends, the days of Friday, Saturday, and Sunday appear to have disproportionately more fatalities.
2. Test statistic: $\chi^2 = 6.500$. Critical value: $\chi^2 = 16.919$. *P*-value > 0.10 (Tech: 0.689). There is not sufficient evidence to warrant rejection of the claim that the last digits of 0, 1, 2, . . . , 9 occur with the same frequency. It does appear that the weights were obtained through measurements.
3. Test statistic: $\chi^2 = 288.448$. Critical value: $\chi^2 = 24.725$. *P*-value < 0.005 (Tech: 0.000). There is sufficient evidence to warrant rejection of the claim that weather-related deaths occur in the different months with the same frequency. The summer months appear to have disproportionately more weather-related deaths, and that is probably due to the fact that vacations and outdoor activities are much greater during those months.
4. Test statistic: $\chi^2 = 10.708$. Critical value: $\chi^2 = 3.841$. *P*-value: 0.00107. There is sufficient evidence to warrant rejection of the claim that wearing a helmet has no effect on whether facial injuries are received. It does appear that a helmet is helpful in preventing facial injuries in a crash.

	Helmet Worn	No Helmet
Facial Injuries Received	30	182
All Injuries Nonfacial	83	236

TI-83/84 PLUS

```
X²-Test
 X²=10.7080789
 P=.0010666873
 df=1
```

5. Flipping and Spinning Pennies Use the data in the table below with a 0.05 significance level to test the claim that when flipping or spinning a penny, the outcome is independent of whether the penny was flipped or spun. (The data are from experimental results given in *Chance News*.) Does the conclusion change if the significance level is changed to 0.01?

	Heads	Tails
Flipping	2048	1992
Spinning	953	1047

6. Home Field Advantage Winning-team data were collected for teams in different sports, with the results given in the accompanying table (based on data from "Predicting Professional Sports Game Outcomes from Intermediate Game Scores," by Copper, DeNeve, and Mosteller, *Chance*, Vol. 5, No. 3–4). Use a 0.05 level of significance to test the claim that home/visitor wins are independent of the sport.

	Basketball	Baseball	Hockey	Football
Home Team Wins	127	53	50	57
Visiting Team Wins	71	47	43	42

7. Car Weight and Fuel Consumption Data Set 14 in Appendix B includes highway fuel consumption amounts for cars with 4 cylinders, 6 cylinders, and 8 cylinders. (The 5-cylinder Volvo S60 is excluded from this exercise.) Analysis of variance is used with the highway fuel consumption amounts (mi/gal) categorized according to the number of cylinders (4, 6, 8); the results are shown in the following display. Identify the null hypothesis, test statistic, and *P*-value, and state the final conclusion in nontechnical terms.

MINITAB

```
Source  DF     SS     MS      F      P
Factor   2  215.3  107.7  10.10  0.001
Error   17  181.3   10.7
Total   19  396.6
```

5. Test statistic: $\chi^2 = 4.955$. Critical value: $\chi^2 = 3.841$. *P*-value < 0.05 (Tech: 0.0260). There is sufficient evidence to warrant rejection of the claim that when flipping or spinning a penny, the outcome is independent of whether the penny was flipped or spun. It appears that the outcome is affected by whether the penny is flipped or spun. If the significance level is changed to 0.01, the critical value changes to 6.635, and we fail to reject the given claim, so the conclusion does change.

6. Test statistic: $\chi^2 = 4.737$. Critical value: $\chi^2 = 7.815$. *P*-value > 0.10 (Tech: 0.192). There is not sufficient evidence to warrant rejection of the claim that home/visitor wins are independent of the sport.

7. H_0: $\mu_1 = \mu_2 = \mu_3$. Test statistic: $F = 10.10$. *P*-value: 0.001. Reject the null hypothesis. There is sufficient evidence to warrant rejection of the claim that 4-cylinder cars, 6-cylinder cars, and 8-cylinder cars have the same mean highway fuel consumption amount.

Cumulative Review Exercises

Please be aware that some of the following problems may require knowledge of concepts presented in previous chapters.

1. Weather-Related Deaths Review Exercise 3 involved weather-related deaths. Among the 489 deaths included in the table, 325 are males. Use a 0.05 significance level to test the claim that among those who die in weather-related deaths, the percentage of males is equal to 50%.

2. Cigarette Costs In an American Express survey of 1000 adults 18 and older, 62% said that cigarettes should cost more to help offset potential negative health effects. Use the sample data to construct a 95% confidence interval estimate of the percentage of all adults who share that same belief. What can we conclude about a claim that half of all adults share that same belief?

3. ICU Patients Listed below are the ages of randomly selected patients in intensive care units (based on data from "A Multifaceted Intervention for Quality Improvement in a Network of Intensive Care Units," by Scales et al., *Journal of the American Medical Association,*

1. H_0: $p = 0.5$. H_1: $p \neq 0.5$. Test statistic: $z = 7.28$. Critical values: $z = \pm 1.96$. *P*-value: 0.0002 (Tech: 0.0000). Reject H_0. There is sufficient evidence to warrant rejection of the claim that among those who die in weather-related deaths, the percentage of males is equal to 50%.

2. $59.0\% < p < 65.0\%$. Because the confidence interval does not include 50% (or "half"), we should reject the stated claim.

3. $\bar{x} = 53.7$ years, median $= 60.0$ years, $s = 16.1$ years. Because an age of 16 differs from the mean by more than 2 standard deviations, it is an unusual age.

Vol. 305, No. 4). Find the mean, median, and standard deviation. Based on the results, is an age of 16 years *unusual*? Why or why not?

38 64 35 67 42 29 68 62 74 58

4. 42.2 years $< \mu <$ 65.2 years. Yes, the confidence interval limits do contain the value of 65.0 years that was found from a sample of 9269 ICU patients.

4. ICU Patients Use the sample of ages from Exercise 3 to construct a 95% confidence interval estimate of the mean age of the population of ICU patients. Do the confidence interval limits contain the value of 65.0 years that was found from a sample of 9269 ICU patients?

5. a. $r = -0.0458$. Critical values: $r = \pm 0.632$. P-value $= 0.900$. There is not sufficient evidence to support the claim that there is a linear correlation between the numbers of boats and the numbers of manatee deaths.
b. $\hat{y} = 96.1 - 0.137x$
c. 83.2 manatee deaths (the value of $\bar{y}$). The predicted value is not very accurate because it is not very close to the actual value of 78 manatee deaths.

5. Boats and Manatees The table below lists the numbers of registered pleasure boats (thousands) in Florida and the numbers of watercraft-related manatee deaths for each year of the past decade.

Boats	90	92	94	95	97	99	99	97	95	90
Manatee Deaths	81	95	73	69	79	92	73	90	97	83

a. Test for a linear correlation between the numbers of boats and the numbers of manatee deaths.

b. Find the equation of the regression line. Use the numbers of boats for the independent x variable.

c. What is the best predicted number of manatee deaths for the year preceding those included in the table? For that year, there were 84 (thousand) registered pleasure boats in Florida. How accurate is that predicted value, given that there were actually 78 manatee deaths in that year?

6. a. 630 mm
b. 14.46% (Tech: 14.48%). That percentage is too high, because too many women would not be accommodated.
c. 0.7611 (Tech: 0.7599). Groups of 16 women do not occupy a cockpit; because *individual* women occupy the cockpit, this result has no effect on the design.

6. Forward Grip Reach and Ergonomics When designing cars and aircraft, we must consider the forward grip reach of women. Women have normally distributed forward grip reaches with a mean of 686 mm and a standard deviation of 34 mm (based on anthropometric survey data from Gordon, Churchill, et al.)

a. If a car dashboard is positioned so that it can be reached by 95% of women, what is the shortest forward grip reach that can access the dashboard?

b. If a car dashboard is positioned so that it can be reached by women with a grip reach greater than 650 mm, what percentage of women cannot reach the dashboard? Is that percentage too high?

c. Find the probability that 16 randomly selected women have forward grip reaches with a mean greater than 680 mm. Does this result have any effect on the design?

7. a. Statistic.
b. Quantitative.
c. Discrete.
d. The sampling is conducted so that all samples of the same size have the same chance of being selected.
e. The sample is a voluntary response sample (or self-selected sample), and those with strong feelings about the topic are more likely to respond, so it is not a valid sampling plan.

7. Honesty Is the Best Policy In a *USA Today* survey of 456 subjects, 269 agreed that honesty is always the best policy. Assume that we plan to use the survey results to estimate the proportion of the population who believe that honesty is always the best policy.

a. Is the sample proportion of 269/456 a statistic or a parameter?

b. Are frequency counts such as 269 categorical data or quantitative data?

c. Are frequency counts such as 269 discrete or continuous?

d. What does it mean for a sample, such as the 456 surveyed subjects, to be a simple random sample?

e. The *USA Today* survey was conducted as follows: A question was posted on the *USA Today* Web site and Internet users could choose to respond. In statistics, is this a valid sampling plan?

8. a. 0.130
b. 0.4

8. Probability and Honesty Based on the sample described in the preceding exercise, assume that when someone is randomly selected, there is a 0.6 probability that the selected person believes that honesty is always the best policy.

a. If four people are randomly selected, find the probability that they all believe that honesty is always the best policy.

b. If someone is randomly selected, what is the probability that he or she does *not* believe that honesty is always the best policy?

Technology Project

Use STATDISK, Minitab, Excel, StatCrunch, or a TI-83/84 Plus calculator, or any other software package or calculator capable of generating equally likely random digits between 0 and 9 inclusive. Generate 5000 digits and record the results in the accompanying table. Use a 0.05 significance level to test the claim that the sample digits come from a population with a uniform distribution (so that all digits are equally likely). Does the random number generator appear to be working as it should?

Digit	0	1	2	3	4	5	6	7	8	9
Frequency										

from data TO DECISION

Critical Thinking: Was Allstate wrong?

The Allstate insurance company once issued a press release listing revised zodiac signs along with the corresponding numbers of accidents, as shown in the first and last columns in the table below.

In the original press release, Allstate included comments such as one stating that Virgos are worried and shy, and they were involved in 211,650 accidents, making them the worst offenders. Allstate quickly issued another press release saying that the original press release was meant to be a joke.

Zodiac Sign	Dates	Length (days)	Accidents
Capricorn	Jan. 18 – Feb. 15	29	128,005
Aquarius	Feb. 16 – March 11	24	106,878
Pisces	March 12 – April 16	36	172,030
Aries	April 17 – May 13	27	112,402
Taurus	May 14 – June 19	37	177,503
Gemini	June 20 – July 20	31	136,904
Cancer	July 21 – Aug. 9	20	101,539
Leo	Aug. 10 – Sept. 15	37	179,657
Virgo	Sept. 16 – Oct. 30	45	211,650
Libra	Oct. 31 – Nov. 22	23	110,592
Scorpio	Nov. 23 – Nov. 28	6	26,833
Ophiuchus	Nov. 29 – Dec. 17	19	83,234
Sagittarius	Dec. 18 – Jan. 17	31	154,477

Analyzing the Results

The original Allstate press release did not include the lengths (days) of the different zodiac signs. The preceding table lists those lengths in the third column. A reasonable explanation for the different numbers of accidents is that they should be proportional to the lengths of the zodiac signs. For example, people are born under the Capricorn sign on 29 days out of the 365 days in the year, so they should have 29/365 of the total number of accidents. Use the methods of this chapter to determine whether this appears to explain the results in the table. Write a brief report of your findings.

Cooperative Group Activities

1. Out-of-class activity Divide into groups of four or five students. The instructions for Exercises 21–24 in Section 11-2 noted that according to Benford's law, a variety of different data sets include numbers with leading (first) digits that follow the distribution shown in the table below. Collect original data and use the methods of Section 11-2 to support or refute the claim that the data conform reasonably well to Benford's law. Here are some possibilities: (1) leading digits of amounts on the checks that you wrote; (2) leading digits of the prices of stocks; (3) leading digits of the numbers on street addresses; (5) leading digits of the lengths of rivers in the world.

Leading Digit	1	2	3	4	5	6	7	8	9
Benford's Law	30.1%	17.6%	12.5%	9.7%	7.9%	6.7%	5.8%	5.1%	4.6%

2. Out-of-class activity Divide into groups of four or five students and collect past results from a state lottery. Such results are often available on Web sites for individual state lotteries. Use the methods of Section 11-2 to test that the numbers are selected in such a way that all possible outcomes are equally likely.

3. Out-of-class activity Divide into groups of four or five students. Each group member should survey at least 15 male students and 15 female students at the same college by asking two questions: (1) Which political party does the subject favor most? (2) If the subject were to make up an absence excuse of a flat tire, which tire would he or she say went flat if the instructor asked? (See Exercise 8 in Section 11-2.) Ask the subject to write the two responses on an index card, and also record the gender of the subject and whether the subject wrote with the right or left hand. Use the methods of this chapter to analyze the data collected. Include these claims:

- The four possible choices for a flat tire are selected with equal frequency.
- The tire identified as being flat is independent of the gender of the subject.
- Political party choice is independent of the gender of the subject.
- Political party choice is independent of whether the subject is right- or left-handed.
- The tire identified as being flat is independent of whether the subject is right- or left-handed.
- Gender is independent of whether the subject is right- or left-handed.
- Political party choice is independent of the tire identified as being flat.

4. Out-of-class activity Divide into groups of four or five students. Each group member should select about 15 other students and first ask them to "randomly" select four digits each. After the four digits have been recorded, ask each subject to write the last four digits of his or her Social Security number. Take the "random" sample results and mix them into one big sample, then mix the Social Security digits into a second big sample. Using the "random" sample set, test the claim that students select digits randomly. Then use the Social Security digits to test the claim that they come from a population of random digits. Compare the results. Does it appear that students can randomly select digits? Are they likely to select any digits more often than others? Are they likely to select any digits less often than others? Do the last digits of Social Security numbers appear to be randomly selected?

5. In-class activity Divide into groups of three or four students. Each group should be given a die along with the instruction that it should be tested for "fairness." Is the die fair or is it biased? Describe the analysis and results.

6. Out-of-class activity Divide into groups of two or three students. The analysis of last digits of data can sometimes reveal whether values are the results of actual measurements or whether they are reported estimates. Refer to an almanac and find the lengths of rivers in the world, then analyze the last digits to determine whether those lengths appear to be actual measurements or whether they appear to be reported estimates.

Instead of lengths of rivers, you could use other variables, such as the following:

- Heights of mountains
- Heights of tallest buildings
- Lengths of bridges
- Heights of roller coasters

7. In-class activity Divide the class into three groups. One group should record the pulse rate of each member while remaining seated. The second group should record the pulse rate of each member while standing. The third group should record the pulse rate of each member immediately after standing and sitting 10 times. Analyze the results. What do the results indicate?

8. In-class activity Ask each student in the class to estimate the length of the classroom. Specify that the length is the distance between the chalkboard and the opposite wall. On the same sheet of paper, each student should also write his or her gender (male/female) and major. Then divide into groups of three or four, and use the data from the entire class to address these questions:

- Is there a significant difference between the mean estimate for males and the mean estimate for females?
- Is there sufficient evidence to reject equality of the mean estimates for different majors? Describe how the majors were categorized.
- Does an interaction between gender and major have an effect on the estimated length?
- Does gender appear to have an effect on estimated length?
- Does major appear to have an effect on estimated length?

Appendix A Tables

Table A-1 Binomial Probabilities

		p													
n	*x*	.01	.05	.10	.20	.30	.40	.50	.60	.70	.80	.90	.95	.99	*x*
2	0	.980	.902	.810	.640	.490	.360	.250	.160	.090	.040	.010	.002	0+	0
	1	.020	.095	.180	.320	.420	.480	.500	.480	.420	.320	.180	.095	.020	1
	2	0+	.002	.010	.040	.090	.160	.250	.360	.490	.640	.810	.902	.980	2
3	0	.970	.857	.729	.512	.343	.216	.125	.064	.027	.008	.001	0+	0+	0
	1	.029	.135	.243	.384	.441	.432	.375	.288	.189	.096	.027	.007	0+	1
	2	0+	.007	.027	.096	.189	.288	.375	.432	.441	.384	.243	.135	.029	2
	3	0+	0+	.001	.008	.027	.064	.125	.216	.343	.512	.729	.857	.970	3
4	0	.961	.815	.656	.410	.240	.130	.062	.026	.008	.002	0+	0+	0+	0
	1	.039	.171	.292	.410	.412	.346	.250	.154	.076	.026	.004	0+	0+	1
	2	.001	.014	.049	.154	.265	.346	.375	.346	.265	.154	.049	.014	.001	2
	3	0+	0+	.004	.026	.076	.154	.250	.346	.412	.410	.292	.171	.039	3
	4	0+	0+	0+	.002	.008	.026	.062	.130	.240	.410	.656	.815	.961	4
5	0	.951	.774	.590	.328	.168	.078	.031	.010	.002	0+	0+	0+	0+	0
	1	.048	.204	.328	.410	.360	.259	.156	.077	.028	.006	0+	0+	0+	1
	2	.001	.021	.073	.205	.309	.346	.312	.230	.132	.051	.008	.001	0+	2
	3	0+	.001	.008	.051	.132	.230	.312	.346	.309	.205	.073	.021	.001	3
	4	0+	0+	0+	.006	.028	.077	.156	.259	.360	.410	.328	.204	.048	4
	5	0+	0+	0+	0+	.002	.010	.031	.078	.168	.328	.590	.774	.951	5
6	0	.941	.735	.531	.262	.118	.047	.016	.004	.001	0+	0+	0+	0+	0
	1	.057	.232	.354	.393	.303	.187	.094	.037	.010	.002	0+	0+	0+	1
	2	.001	.031	.098	.246	.324	.311	.234	.138	.060	.015	.001	0+	0+	2
	3	0+	.002	.015	.082	.185	.276	.312	.276	.185	.082	.015	.002	0+	3
	4	0+	0+	.001	.015	.060	.138	.234	.311	.324	.246	.098	.031	.001	4
	5	0+	0+	0+	.002	.010	.037	.094	.187	.303	.393	.354	.232	.057	5
	6	0+	0+	0+	0+	.001	.004	.016	.047	.118	.262	.531	.735	.941	6
7	0	.932	.698	.478	.210	.082	.028	.008	.002	0+	0+	0+	0+	0+	0
	1	.066	.257	.372	.367	.247	.131	.055	.017	.004	0+	0+	0+	0+	1
	2	.002	.041	.124	.275	.318	.261	.164	.077	.025	.004	0+	0+	0+	2
	3	0+	.004	.023	.115	.227	.290	.273	.194	.097	.029	.003	0+	0+	3
	4	0+	0+	.003	.029	.097	.194	.273	.290	.227	.115	.023	.004	0+	4
	5	0+	0+	0+	.004	.025	.077	.164	.261	.318	.275	.124	.041	.002	5
	6	0+	0+	0+	0+	.004	.017	.055	.131	.247	.367	.372	.257	.066	6
	7	0+	0+	0+	0+	0+	.002	.008	.028	.082	.210	.478	.698	.932	7
8	0	.923	.663	.430	.168	.058	.017	.004	.001	0+	0+	0+	0+	0+	0
	1	.075	.279	.383	.336	.198	.090	.031	.008	.001	0+	0+	0+	0+	1
	2	.003	.051	.149	.294	.296	.209	.109	.041	.010	.001	0+	0+	0+	2
	3	0+	.005	.033	.147	.254	.279	.219	.124	.047	.009	0+	0+	0+	3
	4	0+	0+	.005	.046	.136	.232	.273	.232	.136	.046	.005	0+	0+	4
	5	0+	0+	0+	.009	.047	.124	.219	.279	.254	.147	.033	.005	0+	5
	6	0+	0+	0+	.001	.010	.041	.109	.209	.296	.294	.149	.051	.003	6
	7	0+	0+	0+	0+	.001	.008	.031	.090	.198	.336	.383	.279	.075	7
	8	0+	0+	0+	0+	0+	.001	.004	.017	.058	.168	.430	.663	.923	8

NOTE: 0+ represents a positive probability less than 0.0005.

From Frederick C. Mosteller, Robert E. K. Rourke, and George B. Thomas, Jr., *Probability with Statistical Applications,* 2nd ed., © 1970. Reprinted and electronically reproduced by permission of Pearson Education, Inc., Upper Saddle River, New Jersey.

NEGATIVE *z* Scores

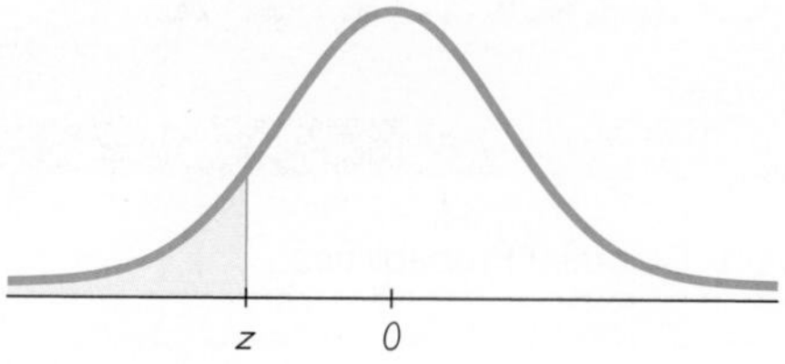

Table A-2 Standard Normal (*z*) Distribution: Cumulative Area from the LEFT

z	.00	.01	.02	.03	.04	.05	.06	.07	.08	.09
−3.50 and lower	.0001									
−3.4	.0003	.0003	.0003	.0003	.0003	.0003	.0003	.0003	.0003	.0002
−3.3	.0005	.0005	.0005	.0004	.0004	.0004	.0004	.0004	.0004	.0003
−3.2	.0007	.0007	.0006	.0006	.0006	.0006	.0006	.0005	.0005	.0005
−3.1	.0010	.0009	.0009	.0009	.0008	.0008	.0008	.0008	.0007	.0007
−3.0	.0013	.0013	.0013	.0012	.0012	.0011	.0011	.0011	.0010	.0010
−2.9	.0019	.0018	.0018	.0017	.0016	.0016	.0015	.0015	.0014	.0014
−2.8	.0026	.0025	.0024	.0023	.0023	.0022	.0021	.0021	.0020	.0019
−2.7	.0035	.0034	.0033	.0032	.0031	.0030	.0029	.0028	.0027	.0026
−2.6	.0047	.0045	.0044	.0043	.0041	.0040	.0039	.0038	.0037	.0036
−2.5	.0062	.0060	.0059	.0057	.0055	.0054	.0052	.0051 *	.0049	.0048
−2.4	.0082	.0080	.0078	.0075	.0073	.0071	.0069	.0068	.0066	.0064
−2.3	.0107	.0104	.0102	.0099	.0096	.0094	.0091	.0089	.0087	.0084
−2.2	.0139	.0136	.0132	.0129	.0125	.0122	.0119	.0116	.0113	.0110
−2.1	.0179	.0174	.0170	.0166	.0162	.0158	.0154	.0150	.0146	.0143
−2.0	.0228	.0222	.0217	.0212	.0207	.0202	.0197	.0192	.0188	.0183
−1.9	.0287	.0281	.0274	.0268	.0262	.0256	.0250	.0244	.0239	.0233
−1.8	.0359	.0351	.0344	.0336	.0329	.0322	.0314	.0307	.0301	.0294
−1.7	.0446	.0436	.0427	.0418	.0409	.0401	.0392	.0384	.0375	.0367
−1.6	.0548	.0537	.0526	.0516	.0505 *	.0495	.0485	.0475	.0465	.0455
−1.5	.0668	.0655	.0643	.0630	.0618	.0606	.0594	.0582	.0571	.0559
−1.4	.0808	.0793	.0778	.0764	.0749	.0735	.0721	.0708	.0694	.0681
−1.3	.0968	.0951	.0934	.0918	.0901	.0885	.0869	.0853	.0838	.0823
−1.2	.1151	.1131	.1112	.1093	.1075	.1056	.1038	.1020	.1003	.0985
−1.1	.1357	.1335	.1314	.1292	.1271	.1251	.1230	.1210	.1190	.1170
−1.0	.1587	.1562	.1539	.1515	.1492	.1469	.1446	.1423	.1401	.1379
−0.9	.1841	.1814	.1788	.1762	.1736	.1711	.1685	.1660	.1635	.1611
−0.8	.2119	.2090	.2061	.2033	.2005	.1977	.1949	.1922	.1894	.1867
−0.7	.2420	.2389	.2358	.2327	.2296	.2266	.2236	.2206	.2177	.2148
−0.6	.2743	.2709	.2676	.2643	.2611	.2578	.2546	.2514	.2483	.2451
−0.5	.3085	.3050	.3015	.2981	.2946	.2912	.2877	.2843	.2810	.2776
−0.4	.3446	.3409	.3372	.3336	.3300	.3264	.3228	.3192	.3156	.3121
−0.3	.3821	.3783	.3745	.3707	.3669	.3632	.3594	.3557	.3520	.3483
−0.2	.4207	.4168	.4129	.4090	.4052	.4013	.3974	.3936	.3897	.3859
−0.1	.4602	.4562	.4522	.4483	.4443	.4404	.4364	.4325	.4286	.4247
−0.0	.5000	.4960	.4920	.4880	.4840	.4801	.4761	.4721	.4681	.4641

NOTE: For values of *z* below −3.49, use 0.0001 for the area.

*Use these common values that result from interpolation:

z Score	Area
−1.645	0.0500
−2.575	0.0050

(*continued*)

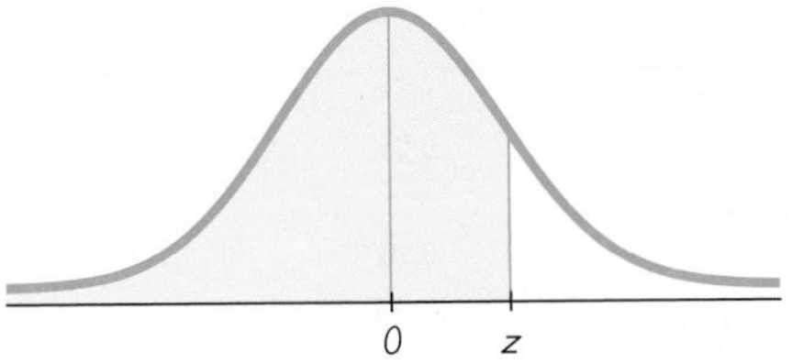

POSITIVE *z* Scores

Table A-2 *(continued)* Cumulative Area from the LEFT

z	.00	.01	.02	.03	.04	.05	.06	.07	.08	.09
0.0	.5000	.5040	.5080	.5120	.5160	.5199	.5239	.5279	.5319	.5359
0.1	.5398	.5438	.5478	.5517	.5557	.5596	.5636	.5675	.5714	.5753
0.2	.5793	.5832	.5871	.5910	.5948	.5987	.6026	.6064	.6103	.6141
0.3	.6179	.6217	.6255	.6293	.6331	.6368	.6406	.6443	.6480	.6517
0.4	.6554	.6591	.6628	.6664	.6700	.6736	.6772	.6808	.6844	.6879
0.5	.6915	.6950	.6985	.7019	.7054	.7088	.7123	.7157	.7190	.7224
0.6	.7257	.7291	.7324	.7357	.7389	.7422	.7454	.7486	.7517	.7549
0.7	.7580	.7611	.7642	.7673	.7704	.7734	.7764	.7794	.7823	.7852
0.8	.7881	.7910	.7939	.7967	.7995	.8023	.8051	.8078	.8106	.8133
0.9	.8159	.8186	.8212	.8238	.8264	.8289	.8315	.8340	.8365	.8389
1.0	.8413	.8438	.8461	.8485	.8508	.8531	.8554	.8577	.8599	.8621
1.1	.8643	.8665	.8686	.8708	.8729	.8749	.8770	.8790	.8810	.8830
1.2	.8849	.8869	.8888	.8907	.8925	.8944	.8962	.8980	.8997	.9015
1.3	.9032	.9049	.9066	.9082	.9099	.9115	.9131	.9147	.9162	.9177
1.4	.9192	.9207	.9222	.9236	.9251	.9265	.9279	.9292	.9306	.9319
1.5	.9332	.9345	.9357	.9370	.9382	.9394	.9406	.9418	.9429	.9441
1.6	.9452	.9463	.9474	.9484	.9495 *	.9505	.9515	.9525	.9535	.9545
1.7	.9554	.9564	.9573	.9582	.9591	.9599	.9608	.9616	.9625	.9633
1.8	.9641	.9649	.9656	.9664	.9671	.9678	.9686	.9693	.9699	.9706
1.9	.9713	.9719	.9726	.9732	.9738	.9744	.9750	.9756	.9761	.9767
2.0	.9772	.9778	.9783	.9788	.9793	.9798	.9803	.9808	.9812	.9817
2.1	.9821	.9826	.9830	.9834	.9838	.9842	.9846	.9850	.9854	.9857
2.2	.9861	.9864	.9868	.9871	.9875	.9878	.9881	.9884	.9887	.9890
2.3	.9893	.9896	.9898	.9901	.9904	.9906	.9909	.9911	.9913	.9916
2.4	.9918	.9920	.9922	.9925	.9927	.9929	.9931	.9932	.9934	.9936
2.5	.9938	.9940	.9941	.9943	.9945	.9946	.9948	.9949 *	.9951	.9952
2.6	.9953	.9955	.9956	.9957	.9959	.9960	.9961	.9962	.9963	.9964
2.7	.9965	.9966	.9967	.9968	.9969	.9970	.9971	.9972	.9973	.9974
2.8	.9974	.9975	.9976	.9977	.9977	.9978	.9979	.9979	.9980	.9981
2.9	.9981	.9982	.9982	.9983	.9984	.9984	.9985	.9985	.9986	.9986
3.0	.9987	.9987	.9987	.9988	.9988	.9989	.9989	.9989	.9990	.9990
3.1	.9990	.9991	.9991	.9991	.9992	.9992	.9992	.9992	.9993	.9993
3.2	.9993	.9993	.9994	.9994	.9994	.9994	.9994	.9995	.9995	.9995
3.3	.9995	.9995	.9995	.9996	.9996	.9996	.9996	.9996	.9996	.9997
3.4	.9997	.9997	.9997	.9997	.9997	.9997	.9997	.9997	.9997	.9998
3.50 and up	.9999									

NOTE: For values of *z* above 3.49, use 0.9999 for the area.
*Use these common values that result from interpolation:

z Score	Area
1.645	0.9500
2.575	0.9950

Common Critical Values

Confidence Level	Critical Value
0.90	1.645
0.95	1.96
0.99	2.575

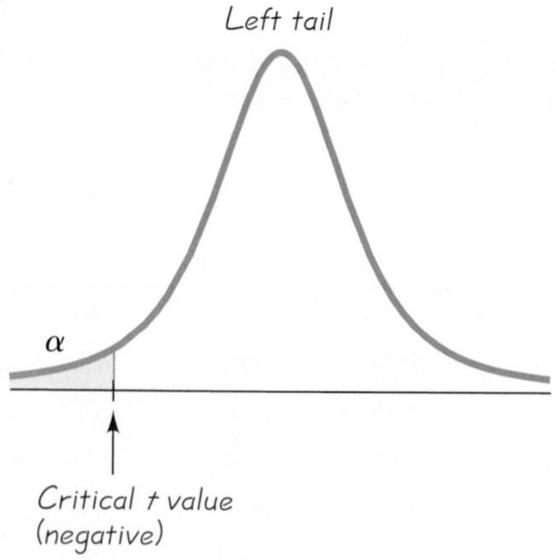

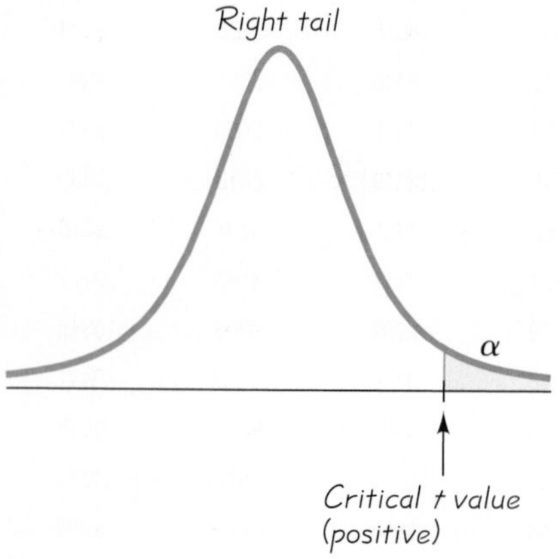

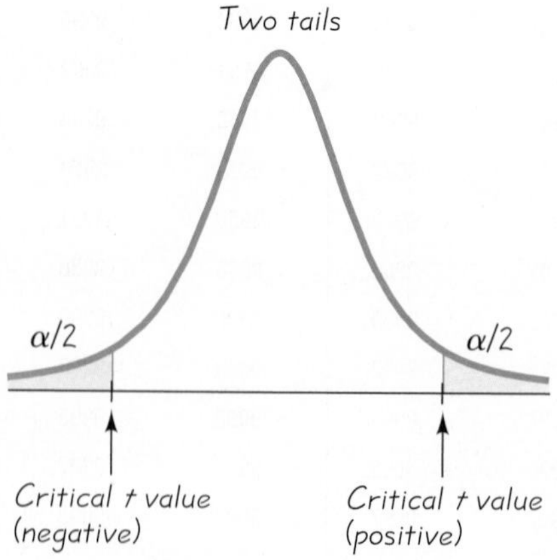

Table A-3 *t* Distribution: Critical *t* Values

	Area in One Tail				
	0.005	0.01	0.025	0.05	0.10
	Area in Two Tails				
Degrees of Freedom	0.01	0.02	0.05	0.10	0.20
1	**63.657**	31.821	**12.706**	6.314	3.078
2	**9.925**	6.965	**4.303**	2.920	1.886
3	**5.841**	4.541	**3.182**	2.353	1.638
4	**4.604**	3.747	**2.776**	2.132	1.533
5	**4.032**	3.365	**2.571**	2.015	1.476
6	**3.707**	3.143	**2.447**	1.943	1.440
7	**3.499**	2.998	**2.365**	1.895	1.415
8	**3.355**	2.896	**2.306**	1.860	1.397
9	**3.250**	2.821	**2.262**	1.833	1.383
10	**3.169**	2.764	**2.228**	1.812	1.372
11	**3.106**	2.718	**2.201**	1.796	1.363
12	**3.055**	2.681	**2.179**	1.782	1.356
13	**3.012**	2.650	**2.160**	1.771	1.350
14	**2.977**	2.624	**2.145**	1.761	1.345
15	**2.947**	2.602	**2.131**	1.753	1.341
16	**2.921**	2.583	**2.120**	1.746	1.337
17	**2.898**	2.567	**2.110**	1.740	1.333
18	**2.878**	2.552	**2.101**	1.734	1.330
19	**2.861**	2.539	**2.093**	1.729	1.328
20	**2.845**	2.528	**2.086**	1.725	1.325
21	**2.831**	2.518	**2.080**	1.721	1.323
22	**2.819**	2.508	**2.074**	1.717	1.321
23	**2.807**	2.500	**2.069**	1.714	1.319
24	**2.797**	2.492	**2.064**	1.711	1.318
25	**2.787**	2.485	**2.060**	1.708	1.316
26	**2.779**	2.479	**2.056**	1.706	1.315
27	**2.771**	2.473	**2.052**	1.703	1.314
28	**2.763**	2.467	**2.048**	1.701	1.313
29	**2.756**	2.462	**2.045**	1.699	1.311
30	**2.750**	2.457	**2.042**	1.697	1.310
31	**2.744**	2.453	**2.040**	1.696	1.309
32	**2.738**	2.449	**2.037**	1.694	1.309
33	**2.733**	2.445	**2.035**	1.692	1.308
34	**2.728**	2.441	**2.032**	1.691	1.307
35	**2.724**	2.438	**2.030**	1.690	1.306
36	**2.719**	2.434	**2.028**	1.688	1.306
37	**2.715**	2.431	**2.026**	1.687	1.305
38	**2.712**	2.429	**2.024**	1.686	1.304
39	**2.708**	2.426	**2.023**	1.685	1.304
40	**2.704**	2.423	**2.021**	1.684	1.303
45	**2.690**	2.412	**2.014**	1.679	1.301
50	**2.678**	2.403	**2.009**	1.676	1.299
60	**2.660**	2.390	**2.000**	1.671	1.296
70	**2.648**	2.381	**1.994**	1.667	1.294
80	**2.639**	2.374	**1.990**	1.664	1.292
90	**2.632**	2.368	**1.987**	1.662	1.291
100	**2.626**	2.364	**1.984**	1.660	1.290
200	**2.601**	2.345	**1.972**	1.653	1.286
300	**2.592**	2.339	**1.968**	1.650	1.284
400	**2.588**	2.336	**1.966**	1.649	1.284
500	**2.586**	2.334	**1.965**	1.648	1.283
1000	**2.581**	2.330	**1.962**	1.646	1.282
2000	**2.578**	2.328	**1.961**	1.646	1.282
Large	**2.576**	2.326	**1.960**	1.645	1.282

Table A-4 Chi-Square (χ^2) Distribution

Degrees of Freedom	Area to the Right of the Critical Value									
	0.995	0.99	0.975	0.95	0.90	0.10	0.05	0.025	0.01	0.005
1	—	—	0.001	0.004	0.016	2.706	3.841	5.024	6.635	7.879
2	0.010	0.020	0.051	0.103	0.211	4.605	5.991	7.378	9.210	10.597
3	0.072	0.115	0.216	0.352	0.584	6.251	7.815	9.348	11.345	12.838
4	0.207	0.297	0.484	0.711	1.064	7.779	9.488	11.143	13.277	14.860
5	0.412	0.554	0.831	1.145	1.610	9.236	11.071	12.833	15.086	16.750
6	0.676	0.872	1.237	1.635	2.204	10.645	12.592	14.449	16.812	18.548
7	0.989	1.239	1.690	2.167	2.833	12.017	14.067	16.013	18.475	20.278
8	1.344	1.646	2.180	2.733	3.490	13.362	15.507	17.535	20.090	21.955
9	1.735	2.088	2.700	3.325	4.168	14.684	16.919	19.023	21.666	23.589
10	2.156	2.558	3.247	3.940	4.865	15.987	18.307	20.483	23.209	25.188
11	2.603	3.053	3.816	4.575	5.578	17.275	19.675	21.920	24.725	26.757
12	3.074	3.571	4.404	5.226	6.304	18.549	21.026	23.337	26.217	28.299
13	3.565	4.107	5.009	5.892	7.042	19.812	22.362	24.736	27.688	29.819
14	4.075	4.660	5.629	6.571	7.790	21.064	23.685	26.119	29.141	31.319
15	4.601	5.229	6.262	7.261	8.547	22.307	24.996	27.488	30.578	32.801
16	5.142	5.812	6.908	7.962	9.312	23.542	26.296	28.845	32.000	34.267
17	5.697	6.408	7.564	8.672	10.085	24.769	27.587	30.191	33.409	35.718
18	6.265	7.015	8.231	9.390	10.865	25.989	28.869	31.526	34.805	37.156
19	6.844	7.633	8.907	10.117	11.651	27.204	30.144	32.852	36.191	38.582
20	7.434	8.260	9.591	10.851	12.443	28.412	31.410	34.170	37.566	39.997
21	8.034	8.897	10.283	11.591	13.240	29.615	32.671	35.479	38.932	41.401
22	8.643	9.542	10.982	12.338	14.042	30.813	33.924	36.781	40.289	42.796
23	9.260	10.196	11.689	13.091	14.848	32.007	35.172	38.076	41.638	44.181
24	9.886	10.856	12.401	13.848	15.659	33.196	36.415	39.364	42.980	45.559
25	10.520	11.524	13.120	14.611	16.473	34.382	37.652	40.646	44.314	46.928
26	11.160	12.198	13.844	15.379	17.292	35.563	38.885	41.923	45.642	48.290
27	11.808	12.879	14.573	16.151	18.114	36.741	40.113	43.194	46.963	49.645
28	12.461	13.565	15.308	16.928	18.939	37.916	41.337	44.461	48.278	50.993
29	13.121	14.257	16.047	17.708	19.768	39.087	42.557	45.722	49.588	52.336
30	13.787	14.954	16.791	18.493	20.599	40.256	43.773	46.979	50.892	53.672
40	20.707	22.164	24.433	26.509	29.051	51.805	55.758	59.342	63.691	66.766
50	27.991	29.707	32.357	34.764	37.689	63.167	67.505	71.420	76.154	79.490
60	35.534	37.485	40.482	43.188	46.459	74.397	79.082	83.298	88.379	91.952
70	43.275	45.442	48.758	51.739	55.329	85.527	90.531	95.023	100.425	104.215
80	51.172	53.540	57.153	60.391	64.278	96.578	101.879	106.629	112.329	116.321
90	59.196	61.754	65.647	69.126	73.291	107.565	113.145	118.136	124.116	128.299
100	67.328	70.065	74.222	77.929	82.358	118.498	124.342	129.561	135.807	140.169

Source: Donald B. Owen, *Handbook of Statistical Tables.*

Degrees of Freedom

$n - 1$	**Confidence interval or hypothesis test** for a standard deviation σ or variance σ^2
$k - 1$	**Goodness-of-fit** with k categories
$(r - 1)(c - 1)$	**Contingency table** with r rows and c columns
$k - 1$	**Kruskal-Wallis test** with k samples

Linear Correlation (Section 10-2)

Table A-5 Critical Values of the Pearson Correlation Coefficient *r*

n	$\alpha = .05$	$\alpha = .01$
4	.950	.990
5	.878	.959
6	.811	.917
7	.754	.875
8	.707	.834
9	.666	.798
10	.632	.765
11	.602	.735
12	.576	.708
13	.553	.684
14	.532	.661
15	.514	.641
16	.497	.623
17	.482	.606
18	.468	.590
19	.456	.575
20	.444	.561
25	.396	.505
30	.361	.463
35	.335	.430
40	.312	.402
45	.294	.378
50	.279	.361
60	.254	.330
70	.236	.305
80	.220	.286
90	.207	.269
100	.196	.256

NOTE: To test $H_0: \rho = 0$ against $H_1: \rho \neq 0$, reject H_0 if the absolute value of r is greater than the critical value in the table.

Rank Correlation (Section 10-4)

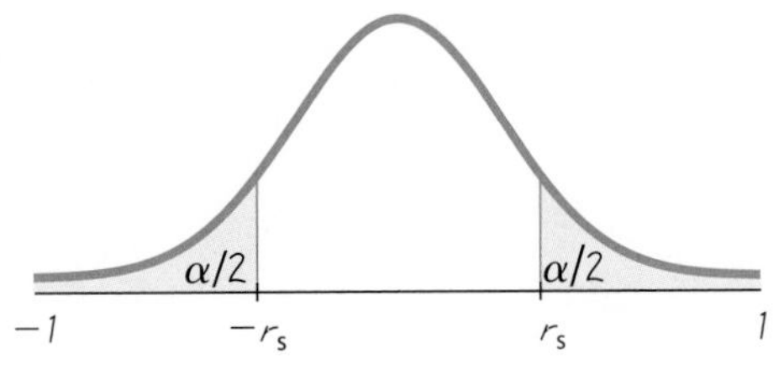

Table A-6 Critical Values of Spearman's Rank Correlation Coefficient r_s

n	$\alpha = 0.10$	$\alpha = 0.05$	$\alpha = 0.02$	$\alpha = 0.01$
5	.900	—	—	—
6	.829	.886	.943	—
7	.714	.786	.893	.929
8	.643	.738	.833	.881
9	.600	.700	.783	.833
10	.564	.648	.745	.794
11	.536	.618	.709	.755
12	.503	.587	.678	.727
13	.484	.560	.648	.703
14	.464	.538	.626	.679
15	.446	.521	.604	.654
16	.429	.503	.582	.635
17	.414	.485	.566	.615
18	.401	.472	.550	.600
19	.391	.460	.535	.584
20	.380	.447	.520	.570
21	.370	.435	.508	.556
22	.361	.425	.496	.544
23	.353	.415	.486	.532
24	.344	.406	.476	.521
25	.337	.398	.466	.511
26	.331	.390	.457	.501
27	.324	.382	.448	.491
28	.317	.375	.440	.483
29	.312	.368	.433	.475
30	.306	.362	.425	.467

NOTES:

1. For $n > 30$ use $r_s = \pm z/\sqrt{n-1}$, where z corresponds to the level of significance. For example, if $\alpha = 0.05$, then $z = 1.96$.
2. If the absolute value of the test statistic r_s exceeds the positive critical value, then reject H_0: $\rho_s = 0$ and conclude that there is a correlation.

Based on data from *Biostatistical Analysis*, 4th edition © 1999, by Jerrold Zar, Prentice Hall, Inc., Upper Saddle River, New Jersey, and "Distribution of Sums of Squares of Rank Differences to Small Numbers with Individuals," *The Annals of Mathematical Statistics*, Vol. 9, No. 2.

Appendix B Data Sets

Data Set 1: Body Measurements

Data Set 2: Foot and Height Measurements

Data Set 3: Body Temperatures of Healthy Adults

Data Set 4: Freshman 15 Data

Data Set 5: IQ and Lead Exposure

Data Set 6: IQ and Brain Size

Data Set 7: Bear Measurements

Data Set 8: Alcohol and Tobacco Use in Animated Children's Movies

Data Set 9: Passive and Active Smoke

Data Set 10: Cigarette Tar, Nicotine, and Carbon Monoxide

Data Set 11: Ages of Oscar Winners

Data Set 12: POTUS

Data Set 13: Car Crash Tests

Data Set 14: Car Measurements

Data Set 15: Flight Data

Data Set 16: Earthquake Measurements

Data Set 17: Word Counts by Males and Females

Data Set 18: Voltage Measurements from a Home

Data Set 19: Cola Weights and Volumes

Data Set 20: M&M Plain Candy Weights

Data Set 21: Coin Weights

Data Set 22: Axial Loads of Aluminum Cans

Data Set 23: Weights of Discarded Garbage for One Week

Additional data sets are available at the Web site aw.com/Triola.

Data Set 1: Body Measurements

AGE is in years, PULSE is pulse rate (beats per minute), SYS is systolic blood pressure (mm Hg), DIAS is diastolic blood pressure (mm Hg), HDL is HDL cholesterol (mg/dL), LDL is LDL cholesterol (mg/dL), WHITE is white blood cell count (1000 cells/μL), RED is red blood cell count (million cells/μL), PLATE is platelet count (1000 cells/μL), WT is weight (kg), HT is height (cm), WAIST is circumference (cm), ARMC is arm circumference (cm), and BMI is body mass index (kg/m^2). Data are from the National Center for Health Statistics.

STATDISK: Data set name for males is Body Measurements Male.
Minitab: Worksheet name for males is MBODY.MTW.
Excel: Workbook name for males is MBODY.XLS.
TI-83/84 Plus: App name for male data is MBODY and the file names are the same as for text files.
Text file names for males: MAGE, MPULS, MSYS, MDIAS, MHDL, MLDL, MWHT, MRED, MPLAT, MWT, MHT, MWAST, MARMC, MBMI.

Males

AGE	PULSE	SYS	DIAS	HDL	LDL	WHITE	RED	PLATE	WT	HT	WAIST	ARMC	BMI
18	60	132	68	44	213	8.7	4.91	409	64.4	178.8	81.4	28.4	20.14
20	74	120	68	41	88	5.9	5.59	187	61.8	177.5	74.8	26.8	19.62
43	86	106	84	71	174	7.3	4.44	250	78.5	187.8	84.1	32.3	22.26
39	54	104	60	41	121	6.2	4.80	273	86.3	172.4	95.5	39.0	29.04
60	90	122	80	57	124	5.9	5.17	278	73.1	181.7	90.1	29.4	22.14
18	80	122	62	50	99	6.4	5.24	279	58.5	169.0	69.8	28.7	20.48
57	66	126	78	60	114	3.9	4.51	237	134.3	186.9	137.8	42.8	38.45
27	68	118	60	47	122	6.4	4.77	200	79.8	183.1	94.4	32.0	23.80
20	68	110	40	44	112	9.8	5.02	209	64.8	176.4	74.2	29.2	20.82
18	56	96	56	33	67	4.0	5.10	203	58.1	183.4	69.9	26.2	17.27
63	80	114	76	48	108	3.8	4.45	206	76.1	169.6	95.2	32.0	26.46
20	62	128	78	46	94	5.6	5.41	174	118.4	185.4	117.5	39.9	34.45
24	74	108	76	62	69	6.6	5.45	219	56.2	166.1	77.0	26.2	20.37
46	60	126	78	45	77	4.9	4.27	302	73.4	169.3	90.5	33.1	25.61
29	52	116	82	43	120	9.6	4.90	285	126.9	193.5	122.0	37.9	33.89
63	60	138	76	39	84	8.6	4.60	254	84.4	173.1	104.0	32.2	28.17
21	66	110	50	41	86	5.2	4.93	195	97.8	171.6	104.1	37.9	33.21
45	64	144	84	47	77	9.2	4.28	290	66.5	180.8	76.4	29.9	20.34
40	64	112	70	46	201	6.6	4.74	240	90.5	175.3	98.6	37.0	29.45
50	46	130	74	56	196	3.2	4.98	270	83.7	178.3	89.7	36.0	26.33
48	68	136	82	38	179	6.9	5.20	264	88.7	172.8	102.1	33.7	29.71
64	58	146	70	37	140	7.1	5.17	208	80.4	165.2	104.6	32.6	29.46
18	68	106	62	61	117	6.9	4.85	271	64.0	176.6	77.6	26.1	20.52
50	70	136	76	41	117	7.0	5.77	386	65.7	174.2	87.5	30.2	21.65
20	56	122	86	56	87	8.4	5.09	203	58.9	181.7	69.1	25.1	17.84
20	66	112	80	59	78	5.9	5.24	223	82.4	176.0	83.4	36.0	26.60
47	78	128	90	40	159	5.1	5.18	261	85.1	173.7	96.1	33.2	28.21
19	68	114	56	54	113	10.0	5.34	282	87.7	177.1	97.2	35.3	27.96
55	62	100	68	40	155	4.4	4.89	184	111.9	174.5	122.4	36.5	36.75
23	70	106	78	75	85	7.1	5.59	204	70.9	180.4	81.5	31.0	21.79
21	72	134	78	55	90	7.9	5.94	287	101.8	177.3	97.5	38.8	32.38
19	74	128	68	48	104	8.1	5.39	219	99.0	172.5	106.7	36.4	33.27
64	64	108	80	50	154	5.6	5.24	257	100.5	180.2	109.6	36.9	30.95
30	50	106	68	57	114	4.1	4.72	206	78.9	172.7	91.8	33.2	26.45
43	70	112	68	38	156	9.5	5.34	342	79.2	168.6	96.7	34.0	27.86
23	58	120	68	78	103	4.9	5.23	319	86.1	178.3	86.6	38.2	27.08
64	60	140	62	45	145	8.9	4.98	251	100.8	176.7	118.1	36.0	32.28
40	88	98	56	82	135	11.0	5.06	178	76.4	174.8	81.5	34.7	25.00
23	84	126	68	44	123	8.6	5.64	220	77.4	172.1	92.1	31.9	26.13
44	76	154	86	51	170	6.9	5.47	268	89.7	173.4	103.3	37.5	29.83

(continued)

Data Set 1: Body Measurements *(continued)*

STATDISK: Data set name for females is Body Measurements Female.
Minitab: Worksheet name for females is FBODY.MTW.
Excel: Workbook name for females is FBODY.XLS.

TI-83/84 Plus: App name for female data is FBODY and the file names are the same as for text files.
Text file names for females: FAGE, FPULS, FSYS, FDIAS, FHDL, FLDL, FWHT, FRED, FPLAT, FWT, FHT, FWAST, FARMC, FBMI.

Females

AGE	PULSE	SYS	DIAS	HDL	LDL	WHITE	RED	PLATE	WT	HT	WAIST	ARMC	BMI
60	78	122	60	74	140	9.6	4.70	317	59.3	163.7	82.3	27.5	22.13
24	80	120	64	56	142	7.1	4.64	224	74.5	165.5	88.5	30.0	27.20
49	68	90	48	70	72	7.5	4.61	248	77.7	163.1	94.9	34.4	29.21
62	56	150	80	40	164	6.8	4.09	309	97.9	166.3	111.8	38.7	35.40
53	76	132	58	67	109	5.6	4.36	335	71.7	163.6	86.5	32.2	26.79
18	78	88	52	96	93	5.4	4.64	278	60.9	170.9	78.4	26.7	20.85
41	78	100	60	43	122	6.7	4.98	312	60.5	153.5	81.7	30.5	25.68
21	90	114	72	80	116	8.6	3.76	338	88.2	155.7	111.2	36.9	36.38
21	96	94	54	77	84	10.2	4.30	290	43.8	153.0	79.2	23.2	18.71
19	60	100	54	41	55	4.1	4.66	236	47.9	157.0	63.4	22.6	19.43
19	98	110	74	23	115	13.0	5.66	310	126.6	163.7	103.8	40.5	47.24
58	66	188	0	76	98	9.2	4.71	460	64.8	161.3	90.4	29.0	24.91
44	100	106	74	39	139	5.9	4.81	317	75.6	168.6	93.6	32.1	26.60
52	76	130	76	52	122	8.0	4.95	304	81.0	156.4	111.8	32.8	33.11
48	64	126	76	38	100	7.0	4.32	239	72.8	159.4	97.2	31.9	28.65
36	82	90	56	37	99	9.1	4.35	243	67.3	164.2	83.9	30.8	24.96
48	62	168	90	57	131	5.7	3.93	351	58.0	169.0	80.3	26.7	20.31
34	72	110	66	81	59	4.6	4.69	242	107.8	161.6	120.6	40.7	41.28
22	78	98	60	37	90	6.0	4.66	217	54.1	161.1	80.5	25.2	20.85
61	74	112	70	50	130	5.7	5.26	173	59.6	150.8	97.1	29.7	26.21
21	90	116	66	73	144	8.9	4.23	228	48.9	166.6	79.7	21.6	17.62
33	90	128	72	61	84	6.4	4.81	259	115.7	175.7	119.9	39.9	37.48
32	68	116	72	33	112	8.1	4.50	193	75.3	169.0	96.3	32.0	26.36
37	72	112	66	41	128	7.9	4.36	417	114.9	160.5	124.4	43.2	44.60
19	82	126	60	60	107	4.4	4.68	193	60.0	159.1	74.5	29.4	23.70
51	72	94	60	62	209	4.9	4.64	244	67.3	166.7	83.8	31.0	24.22
35	78	120	78	67	77	5.3	4.01	266	77.3	151.5	94.0	37.2	33.68
18	104	94	32	79	39	5.3	4.11	286	49.7	156.2	71.2	24.2	20.37
60	62	120	76	64	85	4.7	4.37	206	58.4	148.7	90.7	30.0	26.41
58	72	148	62	44	162	9.8	4.66	426	82.2	168.3	103.6	33.4	29.02
60	72	126	76	42	111	5.3	4.89	244	79.5	159.6	98.3	36.4	31.21
48	88	126	72	41	104	4.9	4.35	295	80.3	165.9	96.6	33.0	29.18
31	74	112	72	44	192	6.3	4.56	282	56.4	147.6	79.1	29.5	25.89
29	72	120	78	57	110	5.4	5.10	211	64.3	165.4	85.7	31.0	23.50
46	82	110	76	69	123	7.0	5.30	355	102.1	159.8	119.6	39.5	39.98
18	78	98	66	60	70	13.5	4.83	357	62.3	158.5	83.2	30.6	24.80
50	78	130	72	47	155	10.0	4.65	272	74.6	165.6	97.5	31.0	27.20
20	98	114	54	67	108	10.3	4.29	251	116.3	164.9	140.0	35.5	42.77
56	72	174	86	55	158	5.1	4.40	219	92.6	172.0	109.0	39.0	31.30
18	64	108	58	82	50	6.6	4.26	234	65.0	170.2	81.8	27.5	22.44

Data Set 2: Foot and Height Measurements

Sex is gender of subject, Age is age in years, Foot Length is foot length (cm), Shoe Print is length of shoe (cm), Shoe Size is reported shoe size, and Height is height (cm) of the subject.

Data from Rohren, Brenda, "Estimation of Stature from Foot and Shoe Length: Applications in Forensic Science." Copyright © 2006. Reprinted by permission of the author.

Brenda Rohren (M.A., MFS, LIMHP, LADC, MAC) was a graduate student at Nebraska Wesleyan University when she conducted the research and wrote the report.

STATDISK: Data set name is Foot and Height Measurements.
Minitab: Worksheet name for males is FOOT.MTW.
Excel: Workbook name for males is FOOT.XLS.
TI-83/84 Plus: App name is FOOT and the file names are the same as for text files.
Text file names: FTSEX, FTAGE, FTLN, SHOPT, SHOSZ, HT

Sex	Age	Foot Length	Shoe Print	Shoe Size	Height
M	67	27.8	31.3	11.0	180.3
M	47	25.7	29.7	9.0	175.3
M	41	26.7	31.3	11.0	184.8
M	42	25.9	31.8	10.0	177.8
M	48	26.4	31.4	10.0	182.3
M	34	29.2	31.9	13.0	185.4
M	26	26.8	31.8	11.0	180.3
M	29	28.1	31.0	10.5	175.3
M	60	25.4	29.7	9.5	177.8
M	48	27.9	31.4	11.0	185.4
M	30	27.5	31.4	11.0	190.5
M	43	28.8	31.6	12.0	195.0
M	54	26.7	31.8	10.0	175.3
M	31	26.7	32.4	10.5	180.3
M	42	25.1	27.6	9.0	172.7
M	21	28.7	31.8	12.5	182.9
M	59	29.2	31.3	11.0	189.2
M	58	27.9	31.3	11.5	185.4
M	42	28.6	34.5	14.0	193.7
F	47	23.2	24.8	7.0	165.1
F	19	24.3	28.6	9.0	166.4
F	20	26.0	25.4	10.0	177.8
F	27	23.8	26.7	8.0	167.6
F	19	25.1	26.7	9.0	168.3
F	21	25.4	27.9	8.5	165.7
F	32	21.9	27.9	8.0	165.1
F	19	26.2	28.9	11.0	165.1
F	27	23.8	27.9	8.0	165.1
F	18	22.2	25.9	9.5	152.4
F	26	24.6	25.4	8.5	162.6
F	36	24.6	28.1	9.0	179.1
F	28	23.7	27.6	9.0	175.9
F	29	25.6	26.5	8.5	166.4
F	58	24.1	26.5	7.0	167.6
F	30	23.8	28.4	9.0	162.6
F	23	23.3	26.5	8.0	167.6
F	26	23.5	26.0	8.0	165.1
F	47	25.1	27.0	10.0	172.7
F	36	24.1	25.1	7.5	157.5
F	19	23.8	27.9	10.0	167.6

Data Set 3: Body Temperatures (in degrees Fahrenheit) of Healthy Adults

Data provided by Dr. Steven Wasserman, Dr. Philip Mackowiak, and Dr. Myron Levine of the University of Maryland.

STATDISK: Data set name for the 12 A.M. temperatures on Day 2 is Body Temperatures of Healthy Adults.

Minitab: Worksheet name for the 12 A.M. temperatures on Day 2 is BODYTEMP.MTW.

Excel: Workbook name for the 12 A.M. temperatures on Day 2 is BODYTEMP.XLS.

TI-83/84 Plus: App name for 12 A.M. temperatures on Day 2 is BTEMP and the file name is BTEMP.

Text files: Text file name is BTEMP.

				Temperature Day 1		Temperature Day 2	
Subject	Age	Sex	Smoke	8 AM	12 AM	8 AM	12 AM
1	22	M	Y	98.0	98.0	98.0	98.6
2	23	M	Y	97.0	97.6	97.4	—
3	22	M	Y	98.6	98.8	97.8	98.6
4	19	M	N	97.4	98.0	97.0	98.0
5	18	M	N	98.2	98.8	97.0	98.0
6	20	M	Y	98.2	98.8	96.6	99.0
7	27	M	Y	98.2	97.6	97.0	98.4
8	19	M	Y	96.6	98.6	96.8	98.4
9	19	M	Y	97.4	98.6	96.6	98.4
10	24	M	N	97.4	98.8	96.6	98.4
11	35	M	Y	98.2	98.0	96.2	98.6
12	25	M	Y	97.4	98.2	97.6	98.6
13	25	M	N	97.8	98.0	98.6	98.8
14	35	M	Y	98.4	98.0	97.0	98.6
15	21	M	N	97.6	97.0	97.4	97.0
16	33	M	N	96.2	97.2	98.0	97.0
17	19	M	Y	98.0	98.2	97.6	98.8
18	24	M	Y	—	—	97.2	97.6
19	18	F	N	—	—	97.0	97.7
20	22	F	Y	—	—	98.0	98.8
21	20	M	Y	—	—	97.0	98.0
22	30	F	Y	—	—	96.4	98.0
23	29	M	N	—	—	96.1	98.3
24	18	M	Y	—	—	98.0	98.5
25	31	M	Y	—	98.1	96.8	97.3
26	28	F	Y	—	98.2	98.2	98.7
27	27	M	Y	—	98.5	97.8	97.4
28	21	M	Y	—	98.5	98.2	98.9
29	30	M	Y	—	99.0	97.8	98.6
30	27	M	N	—	98.0	99.0	99.5
31	32	M	Y	—	97.0	97.4	97.5
32	33	M	Y	—	97.3	97.4	97.3
33	23	M	Y	—	97.3	97.5	97.6
34	29	M	Y	—	98.1	97.8	98.2
35	25	M	Y	—	—	97.9	99.6
36	31	M	N	—	97.8	97.8	98.7
37	25	M	Y	—	99.0	98.3	99.4
38	28	M	N	—	97.6	98.0	98.2
39	30	M	Y	—	97.4	—	98.0
40	33	M	Y	—	98.0	—	98.6
41	28	M	Y	98.0	97.4	—	98.6
42	22	M	Y	98.8	98.0	—	97.2
43	21	F	Y	99.0	—	—	98.4
44	30	M	N	—	98.6	—	98.6
45	22	M	Y	—	98.6	—	98.2
46	22	F	N	98.0	98.4	—	98.0
47	20	M	Y	—	97.0	—	97.8
48	19	M	Y	—	—	—	98.0
49	33	M	N	—	98.4	—	98.4

(continued)

Data Set 3: Body Temperatures (in degrees Fahrenheit) of Healthy Adults *(continued)*

				Temperature Day 1		Temperature Day 2	
Subject	Age	Sex	Smoke	8 AM	12 AM	8 AM	12 AM
50	31	M	Y	99.0	99.0	—	98.6
51	26	M	N	—	98.0	—	98.6
52	18	M	N	—	—	—	97.8
53	23	M	N	—	99.4	—	99.0
54	28	M	Y	—	—	—	96.5
55	19	M	Y	—	97.8	—	97.6
56	21	M	N	—	—	—	98.0
57	27	M	Y	—	98.2	—	96.9
58	29	M	Y	—	99.2	—	97.6
59	38	M	N	—	99.0	—	97.1
60	29	F	Y	—	97.7	—	97.9
61	22	M	Y	—	98.2	—	98.4
62	22	M	Y	—	98.2	—	97.3
63	26	M	Y	—	98.8	—	98.0
64	32	M	N	—	98.1	—	97.5
65	25	M	Y	—	98.5	—	97.6
66	21	F	N	—	97.2	—	98.2
67	25	M	Y	—	98.5	—	98.5
68	24	M	Y	—	99.2	97.0	98.8
69	25	M	Y	—	98.3	97.6	98.7
70	35	M	Y	—	98.7	97.5	97.8
71	23	F	Y	—	98.8	98.8	98.0
72	31	M	Y	—	98.6	98.4	97.1
73	28	M	Y	—	98.0	98.2	97.4
74	29	M	Y	—	99.1	97.7	99.4
75	26	M	Y	—	97.2	97.3	98.4
76	32	M	N	—	97.6	97.5	98.6
77	32	M	Y	—	97.9	97.1	98.4
78	21	F	Y	—	98.8	98.6	98.5
79	20	M	Y	—	98.6	98.6	98.6
80	24	F	Y	—	98.6	97.8	98.3
81	21	F	Y	—	99.3	98.7	98.7
82	28	M	Y	—	97.8	97.9	98.8
83	27	F	N	98.8	98.7	97.8	99.1
84	28	M	N	99.4	99.3	97.8	98.6
85	29	M	Y	98.8	97.8	97.6	97.9
86	19	M	N	97.7	98.4	96.8	98.8
87	24	M	Y	99.0	97.7	96.0	98.0
88	29	M	N	98.1	98.3	98.0	98.7
89	25	M	Y	98.7	97.7	97.0	98.5
90	27	M	N	97.5	97.1	97.4	98.9
91	25	M	Y	98.9	98.4	97.6	98.4
92	21	M	Y	98.4	98.6	97.6	98.6
93	19	M	Y	97.2	97.4	96.2	97.1
94	27	M	Y	—	—	96.2	97.9
95	32	M	N	98.8	96.7	98.1	98.8
96	24	M	Y	97.3	96.9	97.1	98.7
97	32	M	Y	98.7	98.4	98.2	97.6
98	19	F	Y	98.9	98.2	96.4	98.2
99	18	F	Y	99.2	98.6	96.9	99.2
100	27	M	N	—	97.0	—	97.8
101	34	M	Y	—	97.4	—	98.0
102	25	M	N	—	98.4	—	98.4
103	18	M	N	—	97.4	—	97.8
104	32	M	Y	—	96.8	—	98.4
105	31	M	Y	—	98.2	—	97.4
106	26	M	N	—	97.4	—	98.0
107	23	M	N	—	98.0	—	97.0

Data Set 4: Freshman 15 Data

Weights are in kilograms, and BMI denotes measured body mass index. Measurements were made in September of freshman year and then later in April of freshman year. Results are published in Hoffman, D.J., Policastro, P., Quick, V., Lee, S.K.: "Changes in Body Weight and Fat Mass of Men and Women in the First Year of College: A Study of the 'Freshman 15.'" *Journal of American College Health*, July 1, 2006, vol. 55, no. 1, p. 41.

STATDISK:	Data set name is Freshman 15 Study Data.
Minitab:	Worksheet name is FRESH15.MTW.
Excel:	Workbook name is FRESH15.XLS.
TI-83/84 Plus:	App name is FRESH and the file names are the same as for text files.
Text file names:	WTSEP, WTAPR, BMISP, BMIAP.

Sex	Weight in September	Weight in April	BMI in September	BMI in April
M	72	59	22.02	18.14
M	97	86	19.70	17.44
M	74	69	24.09	22.43
M	93	88	26.97	25.57
F	68	64	21.51	20.10
M	59	55	18.69	17.40
F	64	60	24.24	22.88
F	56	53	21.23	20.23
F	70	68	30.26	29.24
F	58	56	21.88	21.02
F	50	47	17.63	16.89
M	71	69	24.57	23.85
M	67	66	20.68	20.15
F	56	55	20.97	20.36
F	70	68	27.30	26.73
F	61	60	23.30	22.88
F	53	52	19.48	19.24
M	92	92	24.74	24.69
F	57	58	20.69	20.79
M	67	67	20.49	20.60
F	58	58	21.09	21.24
F	49	50	18.37	18.53
M	68	68	22.40	22.61
F	69	69	28.17	28.43
M	87	88	23.60	23.81
M	81	82	26.52	26.78
M	60	61	18.89	19.27
F	52	53	19.31	19.75
M	70	71	20.96	21.32
F	63	64	21.78	22.22
F	56	57	19.78	20.23
M	68	69	22.40	22.82
M	68	69	22.76	23.19
F	54	56	20.15	20.69
M	80	82	22.14	22.57
M	64	66	20.27	20.76
F	57	59	22.15	22.93
F	63	65	23.87	24.67
F	54	56	18.61	19.34
F	56	58	21.73	22.58
M	54	56	18.93	19.72
M	73	75	25.88	26.72
M	77	79	28.59	29.53
F	63	66	21.89	22.79
F	51	54	18.31	19.28
F	59	62	19.64	20.63
F	65	68	23.02	24.10
F	53	56	20.63	21.91
F	62	65	22.61	23.81
F	55	58	22.03	23.42
M	74	77	20.31	21.34
M	74	78	20.31	21.36
M	64	68	19.59	20.77
M	64	68	21.05	22.31
F	57	61	23.47	25.11
F	64	68	22.84	24.29
F	60	64	19.50	20.90
M	64	68	18.51	19.83
M	66	71	21.40	22.97
F	52	57	17.72	19.42
M	71	77	22.26	23.87
F	55	60	21.64	23.81
M	65	71	22.51	24.45
M	75	82	23.69	25.80
F	42	49	15.08	17.74
M	74	82	22.64	25.33
M	94	105	36.57	40.86

Data Set 5: IQ and Lead Exposure

Data are measured from children in two consecutive years, and the children were living close to a lead smelter. LEAD is blood lead level group [1 = *low lead level* (blood lead levels < 40 micrograms/100 mL in both years); 2 = *medium lead level* (blood lead levels ≥ 40 micrograms/100 mL in exactly one of two years); 3 = *high lead level* (blood lead level ≥ 40 micrograms/100 mL in both years)]. Age is age in years. Sex is sex of subject (1 = male; 2 = female). YEAR1 is blood lead level in first year, and YEAR2 is blood lead level in second year. IQV is measured verbal IQ score. IQP is measured performance IQ score. IQF is measured full IQ score.

Data are from "Neuropsychological Dysfunction in Children with Chronic Low-Level Lead Absorption," by P. J. Landrigan, R. H. Whitworth, R.W. Baloh, N. W. Staehling, W. F Barthel, and B. F. Rosenblum, *Lancet,* Vol. 1, Issue 7909.

STATDISK: Data set name is IQ and Lead Exposure.
Minitab: Worksheet name for males is IQLEAD.MTW.
Excel: Workbook name for males is IQLEAD.XLS.
TI-83/84 Plus: App name is IQLEAD and the file names are the same as for text files.
Text file names: LEAD, IQAGE, IQSEX, YEAR1, YEAR2, IQV, IQP, IQF

LEAD	AGE	SEX	YEAR1	YEAR2	IQV	IQP	IQF
1	11	1	25	18	61	85	70
1	9	1	31	28	82	90	85
1	11	1	30	29	70	107	86
1	6	1	29	30	72	85	76
1	11	1	2	34	72	100	84
1	6	1	29	25	95	97	96
1	6	1	25	24	89	101	94
1	15	2	24	15	57	64	56
1	7	2	24	16	116	111	115
1	7	1	31	24	95	100	97
1	13	2	21	19	82	76	77
1	10	2	29	27	116	136	128
1	12	1	32	29	99	100	99
1	12	1	36	32	74	90	80
1	15	1	30	25	100	135	118
1	10	1	29	23	72	104	86
1	15	1	28	28	126	149	141
1	9	2	28	19	80	99	88
1	8	1	34	22	86	107	96
1	11	1	21	22	94	99	96
1	7	1	35	27	100	113	107
1	11	2	39	38	72	104	86
1	6	1	36	31	63	101	80
1	6	1	19	25	101	111	107
1	9	1	29	24	85	118	101
1	13	2	1	24	85	99	91
1	9	2	22	20	124	122	125
1	6	1	23	18	105	87	96
1	13	1	21	18	81	118	99
1	6	1	32	26	87	113	99
1	12	2	26	27	100	128	115
1	8	2	20	24	91	121	106
1	9	1	2	22	99	111	105
1	7	2	36	31	89	104	96
1	12	1	24	34	57	51	50
1	6	2	38	37	99	100	99
1	9	2	14	25	58	113	85
1	14	2	18	20	97	82	88
1	15	1	24	16	94	146	120
1	14	1	20	21	82	107	93
1	9	2	33	34	92	83	87
1	7	2	36	27	89	108	98
1	12	2	18	22	69	93	78
1	8	1	31	26	87	114	100
1	7	1	33	30	97	113	105
1	15	2	27	33	82	94	87
1	6	2	28	26	85	106	94
1	14	2	24	23	89	92	89
1	7	1	30	23	85	79	80
1	13	1	24	23	92	129	111
1	12	1	29	24	95	114	104
1	6	1	38	31	76	99	85
1	7	1	31	39	80	110	94
1	9	1	28	32	66	90	75
1	12	2	22	23	67	85	73
1	15	1	22	16	63	94	76
1	10	1	27	28	87	127	107
1	9	1	24	33	79	101	88
1	13	2	36	34	82	99	89
1	12	2	10	38	82	113	96
1	6	2	30	21	70	80	72

(continued)

LEAD	AGE	SEX	YEAR1	YEAR2	IQV	IQP	IQF
1	11	2	27	31	81	115	97
1	15	2	34	33	72	85	76
1	4	1	35	34	101	112	107
1	4	2	27	27	95	112	104
1	3	1	23	26	81	92	85
1	4	1	27	20	61	97	76
1	4	1	2	27	94	97	95
1	3	1	32	28	85	91	86
1	4	1	33	31	76	105	89
1	4	1	32	28	72	84	76
1	4	2	35	37	97	95	96
1	5	1	35	34	95	108	101
1	4	2	30	34	97	118	108
1	4	2	34	23	87	118	102
1	5	2	18	25	74	86	77
1	4	1	38	25	66	89	74
1	6	2	36	23	86	100	92
2	14	1	42	38	72	78	72
2	7	2	40	34	85	97	90
2	7	1	40	34	80	107	92
2	13	2	41	33	67	80	71
2	6	1	41	32	85	90	86
2	15	1	41	30	79	83	79
2	12	1	43	32	70	101	83
2	9	1	66	25	105	121	114
2	9	2	51	39	92	108	100
2	8	1	44	31	89	100	93
2	14	2	64	38	74	110	91
2	9	1	43	32	86	111	98
2	11	2	43	24	87	97	91
2	7	1	40	37	51	51	46
2	9	2	40	35	80	94	85
2	4	1	47	22	87	80	82
2	3	2	44	34	94	101	97
2	5	2	45	29	91	92	91
2	4	1	41	34	92	100	92
2	5	2	51	30	81	77	77
2	3	1	42	32	111	108	111
2	3	1	50	36	75	85	78
3	10	1	68	53	75	93	82
3	10	2	53	49	87	100	93
3	9	1	48	40	76	97	85
3	6	1	41	40	76	79	75
3	5	1	45	47	76	97	85
3	12	2	62	45	92	71	80
3	6	1	49	43	91	111	101
3	7	2	52	58	82	99	89
3	12	2	61	48	80	85	80
3	8	1	41	50	91	99	94
3	6	1	59	40	81	97	88
3	6	1	57	58	97	111	104
3	8	1	58	57	76	104	88
3	11	2	40	51	85	93	88
3	9	1	57	48	80	90	83
3	8	1	48	43	100	107	104
3	4	1	57	43	86	108	96
3	4	1	51	49	79	78	76
3	4	1	59	47	70	95	80
3	3	1	44	45	84	78	79
3	3	2	42	45	69	86	75

Data Set 6: IQ and Brain Size

The data are obtained from monozygotic (identical) twins. PAIR identifies the set of twins, SEX is the gender of the subject (1 = male, 2 = female), ORDER is the birth order, IQ is full-scale IQ, VOL is total brain volume (cm^3), AREA is total brain surface area (cm^2), CCSA is corpus callosum (fissure connecting left and right cerebral hemispheres) surface area (cm^2), CIRC is head circumference (cm), and WT is body weight (kg).

Data provided by M. J. Tramo, W. C. Loftus, T. A. Stukel, J. B. Weaver, M. S. Gazziniga. See "Brain Size, Head Size, and IQ in Monozygotic Twins," *Neurology,* Vol. 50.

STATDISK: Data set name is IQ and Brain Size.
Minitab: Worksheet name is IQBRAIN.MTW.
Excel: Workbook name is IQBRAIN.XLS.
TI-83/84 Plus: App name is IQBRAIN and the individual column names are the same as for text files.
Text file names: PAIR, SEX, ORDER, IQ, VOL, AREA, CCSA, CIRC, WT

PAIR	SEX	ORDER	IQ	VOL	AREA	CCSA	CIRC	WT
1	2	1	96	1005	1913.88	6.08	54.7	57.607
1	2	2	89	963	1684.89	5.73	54.2	58.968
2	2	1	87	1035	1902.36	6.22	53.0	64.184
2	2	2	87	1027	1860.24	5.80	52.9	58.514
3	2	1	101	1281	2264.25	7.99	57.8	63.958
3	2	2	103	1272	2216.40	8.42	56.9	61.690
4	2	1	103	1051	1866.99	7.44	56.6	133.358
4	2	2	96	1079	1850.64	6.84	55.3	107.503
5	2	1	127	1034	1743.04	6.48	53.1	62.143
5	2	2	126	1070	1709.30	6.43	54.8	83.009
6	1	2	101	1173	1689.60	7.99	57.2	61.236
6	1	1	96	1079	1806.31	8.76	57.2	61.236
7	1	2	93	1067	2136.37	6.32	57.2	83.916
7	1	1	88	1104	2018.92	6.32	57.2	79.380
8	1	2	94	1347	1966.81	7.60	55.8	97.524
8	1	1	85	1439	2154.67	7.62	57.2	99.792
9	1	1	97	1029	1767.56	6.03	57.2	81.648
9	1	2	114	1100	1827.92	6.59	56.5	88.452
10	1	2	113	1204	1773.83	7.52	59.2	79.380
10	1	1	124	1160	1971.63	7.67	58.5	72.576

Data Set 7: Bears (measurements from anesthetized wild bears)

AGE is in months, MONTH is the month of measurement (1 = January), SEX is coded with 1 = male and 2 = female, HEADLEN is head length (inches), HEADWTH is width of head (inches), NECK is distance around neck (in inches), LENGTH is length of body (inches), CHEST is distance around chest (inches), and WEIGHT is measured in pounds. Data are from Gary Alt and Minitab, Inc.

STATDISK:	Data set name is Bears.
Minitab:	Worksheet name is BEARS.MTW.
Excel:	Workbook name is BEARS.XLS.
TI-83/84 Plus:	App name is BEARS and the file names are the same as for text files.
Text file names:	BAGE, BMNTH, BSEX, BHDLN, BHDWD, BNECK, BLEN, BCHST, BWGHT.

Age	Month	Sex	Headlen	Headwth	Neck	Length	Chest	Weight
19	7	1	11.0	5.5	16.0	53.0	26.0	80
55	7	1	16.5	9.0	28.0	67.5	45.0	344
81	9	1	15.5	8.0	31.0	72.0	54.0	416
115	7	1	17.0	10.0	31.5	72.0	49.0	348
104	8	2	15.5	6.5	22.0	62.0	35.0	166
100	4	2	13.0	7.0	21.0	70.0	41.0	220
56	7	1	15.0	7.5	26.5	73.5	41.0	262
51	4	1	13.5	8.0	27.0	68.5	49.0	360
57	9	2	13.5	7.0	20.0	64.0	38.0	204
53	5	2	12.5	6.0	18.0	58.0	31.0	144
68	8	1	16.0	9.0	29.0	73.0	44.0	332
8	8	1	9.0	4.5	13.0	37.0	19.0	34
44	8	2	12.5	4.5	10.5	63.0	32.0	140
32	8	1	14.0	5.0	21.5	67.0	37.0	180
20	8	2	11.5	5.0	17.5	52.0	29.0	105
32	8	1	13.0	8.0	21.5	59.0	33.0	166
45	9	1	13.5	7.0	24.0	64.0	39.0	204
9	9	2	9.0	4.5	12.0	36.0	19.0	26
21	9	1	13.0	6.0	19.0	59.0	30.0	120
177	9	1	16.0	9.5	30.0	72.0	48.0	436
57	9	2	12.5	5.0	19.0	57.5	32.0	125
81	9	2	13.0	5.0	20.0	61.0	33.0	132
21	9	1	13.0	5.0	17.0	54.0	28.0	90
9	9	1	10.0	4.0	13.0	40.0	23.0	40
45	9	1	16.0	6.0	24.0	63.0	42.0	220
9	9	1	10.0	4.0	13.5	43.0	23.0	46
33	9	1	13.5	6.0	22.0	66.5	34.0	154
57	9	2	13.0	5.5	17.5	60.5	31.0	116
45	9	2	13.0	6.5	21.0	60.0	34.5	182
21	9	1	14.5	5.5	20.0	61.0	34.0	150
10	10	1	9.5	4.5	16.0	40.0	26.0	65
82	10	2	13.5	6.5	28.0	64.0	48.0	356
70	10	2	14.5	6.5	26.0	65.0	48.0	316
10	10	1	11.0	5.0	17.0	49.0	29.0	94
10	10	1	11.5	5.0	17.0	47.0	29.5	86
34	10	1	13.0	7.0	21.0	59.0	35.0	150
34	10	1	16.5	6.5	27.0	72.0	44.5	270
34	10	1	14.0	5.5	24.0	65.0	39.0	202
58	10	2	13.5	6.5	21.5	63.0	40.0	202
58	10	1	15.5	7.0	28.0	70.5	50.0	365
11	11	1	11.5	6.0	16.5	48.0	31.0	79
23	11	1	12.0	6.5	19.0	50.0	38.0	148
70	10	1	15.5	7.0	28.0	76.5	55.0	446
11	11	2	9.0	5.0	15.0	46.0	27.0	62
83	11	2	14.5	7.0	23.0	61.5	44.0	236
35	11	1	13.5	8.5	23.0	63.5	44.0	212
16	4	1	10.0	4.0	15.5	48.0	26.0	60
16	4	1	10.0	5.0	15.0	41.0	26.0	64
17	5	1	11.5	5.0	17.0	53.0	30.5	114
17	5	2	11.5	5.0	15.0	52.5	28.0	76
17	5	2	11.0	4.5	13.0	46.0	23.0	48
8	8	2	10.0	4.5	10.0	43.5	24.0	29
83	11	1	15.5	8.0	30.5	75.0	54.0	514
18	6	1	12.5	8.5	18.0	57.3	32.8	140

Data Set 8: Alcohol and Tobacco Use in Animated Children's Movies

Movie lengths are in minutes, tobacco use times are in seconds, and alcohol use times are in seconds. The data are based on Goldstein, Adam O., Sobel, Rachel A., Newman, Glen R.; "Tobacco and Alcohol Use in G-Rated Children's Animated Films." *Journal of the American Medical Association,* March 24/31, 1999, vol. 281, no. 12, p. 1132.

STATDISK: Data set name is Alcohol and Tobacco in Movies.
Minitab: Worksheet name is CHMOVIE.MTW.
Excel: Workbook name is CHMOVIE.XLS.
TI-83/84 Plus: App name is CHMOVIE and the file names are the same as for text files.
Text file names: CHLEN, CHTOB, CHALC.

Movie	Company	Length (min)	Tobacco Use (sec)	Alcohol Use (sec)
Snow White	Disney	83	0	0
Pinocchio	Disney	88	223	80
Fantasia	Disney	120	0	0
Dumbo	Disney	64	176	88
Bambi	Disney	69	0	0
Three Caballeros	Disney	71	548	8
Fun and Fancy Free	Disney	76	0	4
Cinderella	Disney	74	37	0
Alice in Wonderland	Disney	75	158	0
Peter Pan	Disney	76	51	33
Lady and the Tramp	Disney	75	0	0
Sleeping Beauty	Disney	75	0	113
101 Dalmatians	Disney	79	299	51
Sword and the Stone	Disney	80	37	20
Jungle Book	Disney	78	0	0
Aristocats	Disney	78	11	142
Robin Hood	Disney	83	0	39
Rescuers	Disney	77	0	0
Winnie the Pooh	Disney	71	0	0
Fox and the Hound	Disney	83	0	0
Black Cauldron	Disney	80	0	34
Great Mouse Detective	Disney	73	165	414
Oliver and Company	Disney	72	74	0
Little Mermaid	Disney	82	9	0
Rescuers Down Under	Disney	74	0	76
Beauty and the Beast	Disney	84	0	123
Aladdin	Disney	90	2	3
Lion King	Disney	89	0	0
Pocahontas	Disney	81	6	7
Toy Story	Disney	81	0	0
Hunchback of Notre Dame	Disney	90	23	46
James and the Giant Peach	Disney	79	206	38
Hercules	Disney	92	9	13
Secret of NIMH	MGM	82	0	0
All Dogs Go to Heaven	MGM	89	205	73
All Dogs Go to Heaven 2	MGM	82	162	72
Babes in Toyland	MGM	74	0	0
Thumbelina	Warner Bros	86	6	5
Troll in Central Park	Warner Bros	76	1	0
Space Jam	Warner Bros	81	117	0
Pippi Longstocking	Warner Bros	75	5	0
Cats Don't Dance	Warner Bros	75	91	0
An American Tail	Universal	77	155	74
Land Before Time	Universal	70	0	0
Fievel Goes West	Universal	75	24	28
We're Back: Dinosaur Story	Universal	64	55	0
Land Before Time 2	Universal	73	0	0
Balto	Universal	74	0	0
Once Upon a Forest	20th Century Fox	71	0	0
Anastasia	20th Century Fox	94	17	39

Data Set 9: Passive and Active Smoke

All values are measured levels of serum cotinine (in ng/mL), a metabolite of nicotine. (When nicotine is absorbed by the body, cotinine is produced.) Data are from the U.S. Department of Health and Human Services, National Center for Health Statistics, Third National Health and Nutrition Examination Survey.

STATDISK: Data set name is Passive and Active Smoke.
Minitab: Worksheet name is COTININE.MTW.
Excel: Workbook name is COTININE.XLS.
TI-83/84 Plus: App name is COTININE and the file names are the same as for text files.
Text file names: SMKR, ETS, NOETS.

SMKR (smokers, or subjects reported tobacco use)

1	0	131	173	265	210	44	277	32	3
35	112	477	289	227	103	222	149	313	491
130	234	164	198	17	253	87	121	266	290
123	167	250	245	48	86	284	1	208	173

ETS (nonsmokers exposed to environmental tobacco smoke at home or work)

384	0	69	19	1	0	178	2	13	1
4	0	543	17	1	0	51	0	197	3
0	3	1	45	13	3	1	1	1	0
0	551	2	1	1	1	0	74	1	241

NOETS (nonsmokers with no exposure to environmental tobacco smoke at home or work)

0	0	0	0	0	0	0	0	0	0
0	9	0	0	0	0	0	0	244	0
1	0	0	0	90	1	0	309	0	0
0	0	0	0	0	0	0	0	0	0

Data Set 10: Cigarette Tar, Nicotine, and Carbon Monoxide

All measurements are in milligrams per cigarette. CO denotes carbon monoxide. The king-size cigarettes are nonfiltered, nonmenthol, and nonlight. The menthol cigarettes are 100 mm long, filtered, and nonlight. The cigarettes in the third group are 100 mm long, filtered, nonmenthol, and nonlight. Data are from the Federal Trade Commission.

STATDISK: Data set name is Cigarette Tar, Nicotine, and Carbon Monoxide.
Minitab: Worksheet name is CIGARET.MTW.
Excel: Workbook name is CIGARET.XLS.
TI-83/84 Plus: App name is CIGARET and the file names are the same as for text files.
Text file names: KGTAR, KGNIC, KGCO, MNTAR, MNNIC, MNCO, FLTAR, FLNIC, FLCO (where KG denotes the king size cigarettes, MN denotes the menthol cigarettes, and FL denotes the filtered cigarettes that are not menthol types).

King Size

Brand	Tar	Nicotine	CO
Austin	20	1.1	16
Basic	27	1.7	16
Bristol	27	1.7	16
Cardinal	20	1.1	16
Cavalier	20	1.1	16
Chesterfield	24	1.4	17
Cimarron	20	1.1	16
Class A	23	1.4	15
Doral	20	1.0	16
GPC	22	1.2	14
Highway	20	1.1	16
Jacks	20	1.1	16
Marker	20	1.1	16
Monaco	20	1.1	16
Monarch	20	1.1	16
Old Gold	10	1.8	14
Pall Mall	24	1.6	16
Pilot	20	1.1	16
Prime	21	1.2	14
Pyramid	25	1.5	18
Raleigh Extra	23	1.3	15
Sebring	20	1.1	16
Summit	22	1.3	14
Sundance	20	1.1	16
Worth	20	1.1	16

Menthol

Brand	Tar	Nicotine	CO
Alpine	16	1.1	15
Austin	13	0.8	17
Basic	16	1.0	19
Belair	9	0.9	9
Best Value	14	0.8	17
Cavalier	13	0.8	17
Doral	12	0.8	15
Focus	14	0.8	17
GPC	14	0.9	15
Highway	13	0.8	17
Jacks	13	0.8	17
Kool	16	1.2	15
Legend	13	0.8	17
Marker	13	0.8	17
Maverick	18	1.3	18
Merit	9	0.7	11
Newport	19	1.4	18
Now	2	0.2	3
Pilot	13	0.8	17
Players	14	1.0	14
Prime	14	0.8	15
Pyramid	15	0.8	22
Salem	16	1.2	16
True	6	0.6	7
Vantage	8	0.7	9

Filtered 100-mm Nonmenthol

Brand	Tar	Nicotine	CO
Barclay	5	0.4	4
Basic	16	1.0	19
Camel	17	1.2	17
Highway	13	0.8	18
Jacks	13	0.8	18
Kent	14	1.0	13
Lark	15	1.1	17
Marlboro	15	1.1	15
Maverick	15	1.1	15
Merit	9	0.8	12
Monaco	13	0.8	18
Monarch	13	0.8	17
Mustang	13	0.8	18
Newport	15	1.0	16
Now	2	0.2	3
Old Gold	15	1.1	18
Pall Mall	15	1.0	15
Pilot	13	0.8	18
Players	14	1.0	15
Prime	15	0.9	17
Raleigh	16	1.1	15
Tareyton	15	1.1	15
True	7	0.6	7
Viceroy	17	1.3	16
Winston	15	1.1	14

Data Set 11: Ages of Oscar Winners

Ages (years) of actresses and actors at the times that they won Oscars for the categories of Best Actress and Best Actor. The ages are listed in chronological order by row, so that corresponding locations in the two tables are from the same year. (*Notes:* In 1968 there was a tie in the Best Actress category, and the mean of the two ages is used; in 1932 there was a tie in the Best Actor category, and the mean of the two ages is used. These data are suggested by the article "Ages of Oscar-winning Best Actors and Actresses" by Richard Brown and Gretchen Davis, *Mathematics Teacher* magazine. In that article, the year of birth of the award winner was subtracted from the year of the awards ceremony, but the ages listed here are calculated from the birth date of the winner and the date of the awards ceremony.) The data are complete as of this writing.

STATDISK:	Data set name is Ages of Oscar Winners.
Minitab:	Worksheet name is OSCR.MTW.
Excel:	Workbook name for males is OSCR.XLS.
TI-83/84 Plus:	App name is OSCR and the file names are the same as for text files.
Text file names:	OSCRF, OSCRM

Best Actresses

22	37	28	63	32	26	31	27	27	28
30	26	29	24	38	25	29	41	30	35
35	33	29	38	54	24	25	46	41	28
40	39	29	27	31	38	29	25	35	60
43	35	34	34	27	37	42	41	36	32
41	33	31	74	33	50	38	61	21	41
26	80	42	29	33	35	45	49	39	34
26	25	33	35	35	28	30	29	61	32
33	45								

Best Actors

44	41	62	52	41	34	34	52	41	37
38	34	32	40	43	56	41	39	49	57
41	38	42	52	51	35	30	39	41	44
49	35	47	31	47	37	57	42	45	42
44	62	43	42	48	49	56	38	60	30
40	42	36	76	39	53	45	36	62	43
51	32	42	54	52	37	38	32	45	60
46	40	36	47	29	43	37	38	45	50
48	60								

Data Set 12: POTUS (Presidents of the United States)

AGE is age in years at time of inauguration. Days is the number of days served as president. Years is the number of years lived after the first inauguration. Ht is height (cm) of the president. HtOpp is the height (cm) of the major opponent for the presidency. *Note:* Presidents who took office as the result of an assassination or resignation are not included. Data are complete as of this writing.

STATDISK:	Data set name is POTUS.
Minitab:	Worksheet name for males is POTUS.MTW.
Excel:	Workbook name for males is POTUS.XLS.
TI-83/84 Plus:	App name is POTUS and the file names are the same as for text files. *Caution:* For TI-83/84 Plus calculators, missing values are entered as 9999.
Text file names:	PRAGE, DAYS, YEARS, PRHT, HTOPP

	Age	Days	Years	Ht (cm)	HtOpp (cm)
Washington	57	2864	10	188	
J. Adams	61	1460	29	170	189
Jefferson	57	2921	26	189	170
Madison	57	2921	28	163	
Monroe	58	2921	15	183	
J. Q. Adams	57	1460	23	171	191
Jackson	61	2921	17	185	171
Van Buren	54	1460	25	168	180
W. H. Harrison	68	31	0	173	168
Polk	49	1460	4	173	185
Taylor	64	491	1	173	174
Pierce	48	1460	16	178	196
Buchanan	65	1460	12	183	175
Lincoln	52	1503	4	193	188
Grant	46	2921	17	173	
Hayes	54	1460	16	173	178
Garfield	49	199	0	183	187
Cleveland	47	2920	24	180	180
B. Harrison	55	1460	12	168	180
McKinley	54	1655	4	170	178
T. Roosevelt	42	2727	18	178	175
Taft	51	1460	21	182	178
Wilson	56	2921	11	180	182
Harding	55	881	2	183	178
Coolidge	51	2039	9	178	180
Hoover	54	1460	36	182	180
F. Roosevelt	51	4452	12	188	182
Truman	60	2810	28	175	173
Eisenhower	62	2922	16	179	178
Kennedy	43	1036	3	183	182
Johnson	55	1886	9	192	180
Nixon	56	2027	25	182	180
Carter	52	1461		177	183
Reagan	69	2922	23	185	177
G. H. W. Bush	64	1461		188	173
Clinton	46	2922		188	188
G. W. Bush	54	2920		183	185
Obama	47			188	175

Data Set 13: Car Crash Tests

The same cars are used in Data Set 14. The data are measurements from cars crashed into a fixed barrier at 35 mi/h with a crash test dummy in the driver's seat. HIC is a measurement of a standard "head injury criterion," CHEST is chest deceleration (in g, where g is a force of gravity), FEML is the measured load on the left femur (in lb), FEMR is the measured load on the right femur (in lb), TTI is a measurement of the side thoracic trauma index, and PLVS is pelvis deceleration (in g, where g is a force of gravity). Data are from the National Highway Traffic Safety Administration.

STATDISK:	Data set name is Car Crash Tests.
Minitab:	Worksheet name is CRASH.MTW.
Excel:	Workbook name is CRASH.XLS.
TI-83/84 Plus:	App name is CRASH and the individual column names are the same as for text files.
Text file names:	HIC, CHEST, FEML, FEMR, TTI, PLVS.

CAR	SIZE	HIC	CHEST	FEML	FEMR	TTI	PLVS
Chev Aveo	Small	371	44	1188	1261	62	71
Honda Civic	Small	356	39	289	324	63	71
Mitsubishi Lancer	Small	275	37	329	446	35	45
VW Jetta	Small	544	54	707	1048	44	66
Hyundai Elantra	Small	326	39	602	1474	58	71
Kia Rio	Small	520	44	245	1046	64	84
Subaru Impreza	Small	443	42	334	455	50	53
Ford Fusion	Midsize	366	36	399	844	51	78
Nissan Altima	Midsize	287	53	317	713	53	53
Nissan Maxima	Midsize	255	43	301	133	44	59
Honda Accord	Midsize	249	42	297	236	59	55
Volvo S60	Midsize	502	52	810	687	49	67
VW Passat	Midsize	502	49	280	905	43	61
Toyota Camry	Midsize	505	41	411	547	42	57
Toyota Avalon	Large	342	32	215	752	36	55
Hyundai Azera	Large	698	45	1636	1202	48	75
Cadillac DTS	Large	346	41	738	772	61	75
Lincoln Town	Large	608	38	882	554	57	61
Dodge Charger	Large	216	37	937	669	63	53
Merc Gr Marq	Large	608	38	882	554	57	61
Buick Lucerne	Large	169	33	472	290	64	76

Data Set 14: Car Measurements

The same cars are used in Data Set 13. The data are measurements from cars that have automatic transmissions and were manufactured in the same recent year. WT is weight (lb), LN is length (inches), BRK is braking distance (feet) from 60 mi/h, CYL is the number of cylinders, DISP is the engine displacement (liters), CITY is the fuel consumption (mi/gal) for city driving conditions, HWY is the fuel consumption (mi/gal) for highway driving conditions, and GHG is a measure of greenhouse gas emissions (in tons/year, expressed as CO_2 equivalents).

STATDISK: Data set name is Car Measurements.
Minitab: Worksheet name is CARS.MTW.
Excel: Workbook name is CARS.XLS.
TI-83/84 Plus: App name is CARS and the individual column names are the same as for text files.
Text file names: CWT, CLN, CBRK, CCYL, CDISP, CCITY, CHWY, CGHG.

CAR	SIZE	WT	LN	BRK	CYL	DISP	CITY	HWY	GHG
Chev Aveo	Small	2560	154	133	4	1.6	25	34	6.6
Honda Civic	Small	2740	177	136	4	1.8	25	36	6.3
Mitsubishi Lancer	Small	3610	177	126	4	2.0	22	28	7.7
VW Jetta	Small	3225	179	137	4	2.0	29	40	6.4
Hyundai Elantra	Small	2895	177	138	4	2.0	25	33	6.6
Kia Rio	Small	2615	167	132	4	1.6	27	35	6.1
Subaru Impreza	Small	3110	180	135	4	2.5	20	26	8.3
Ford Fusion	Midsize	3320	191	136	4	2.3	20	28	8.0
Nissan Altima	Midsize	3215	190	136	4	2.5	23	31	7.1
Nissan Maxima	Midsize	3555	191	128	6	3.5	19	26	8.3
Honda Accord	Midsize	3270	194	140	4	2.4	21	30	7.7
Volvo S60	Midsize	3465	180	140	5	2.5	19	28	8.0
VW Passat	Midsize	3465	188	135	4	2.0	19	29	8.0
Toyota Camry	Midsize	3530	189	137	4	2.4	21	31	7.3
Toyota Avalon	Large	3600	197	139	6	3.5	19	28	8.0
Hyundai Azera	Large	3835	193	134	6	3.3	18	26	8.7
Cadillac DTS	Large	4085	208	145	8	4.6	15	23	10.2
Lincoln Town	Large	4415	215	143	8	4.6	16	24	9.6
Dodge Charger	Large	4170	200	131	6	2.7	18	26	8.7
Merc Gr Marq	Large	4180	212	140	8	4.6	16	24	9.6
Buick Lucerne	Large	4095	203	143	8	4.6	17	25	10.2

Data Set 15: Flight Data

All flights are American Airlines flights from New York (JFK) to Los Angeles (LAX), and all flights occurred in January of a recent year. ID is the identification number on the tail of the aircraft. Dep Delay is the departure delay time (minutes), and negative numbers correspond to flights that departed early. Taxi Out is the time (minutes) that the flight used to taxi from the terminal to the runway for departure. Taxi In is the time (minutes) that the flight used to taxi to the terminal after landing. Arr Delay is the arrival delay time (minutes), and negative numbers correspond to flights that arrived early (before the scheduled arrival time). Data are from the Bureau of Transportation.

STATDISK:	Data set name is Flight Data.
Minitab:	Worksheet name is FLIGHTS.MTW.
Excel:	Workbook name is FLIGHTS.XLS.
TI-83/84 Plus:	App name is FLIGHTS and the individual column names are the same as for text files.
Text file names:	DPDLY, TXOUT, TXIN, ARDLY

Flight	ID	Dep Delay	Taxi Out	Taxi In	Arr Delay
1	N338AA	−2	30	12	−32
1	N329AA	−1	19	13	−25
1	N319AA	−2	12	8	−26
1	N319AA	2	19	21	−6
1	N329AA	−2	18	17	5
1	N320AA	0	22	11	−15
1	N321AA	−2	37	12	−17
1	N338AA	−3	13	12	−36
1	N327AA	−5	14	15	−29
1	N319AA	−4	15	26	−18
1	N320AA	2	31	9	−12
1	N320AA	−2	15	11	−35
3	N327AA	22	16	6	2
3	N332AA	−11	14	7	−33
3	N321AA	7	15	4	−5
3	N338AA	0	27	11	0
3	N328AA	−5	19	10	0
3	N319AA	3	22	7	−1
3	N338AA	−8	22	11	−33
3	N328AA	8	23	10	−5
3	N336AA	−2	16	3	−14
3	N320AA	−8	13	7	−39
3	N323AA	−3	16	7	−21
3	N338AA	−4	18	5	−32
19	N329AA	19	15	10	−5
19	N322AA	−4	12	10	−32
19	N328AA	−5	19	16	−13
19	N319AA	−1	18	13	−9
19	N335AA	−4	21	9	−19
19	N332AA	73	20	8	49
19	N329AA	0	13	4	−30
19	N338AA	1	15	3	−23
19	N328AA	13	43	8	14
19	N327AA	−1	18	16	−21
19	N322AA	−8	17	9	−32
19	N319AA	32	19	5	11
21	N336AA	18	13	13	−23
21	N336AA	60	20	4	28
21	N327AA	142	12	6	103
21	N336AA	−1	17	21	−19
21	N323AA	−11	35	29	−5
21	N324AA	−1	19	5	−46
21	N327AA	47	22	27	13
21	N320AA	13	43	9	−3
21	N322AA	12	49	12	13
21	N323AA	123	45	7	106
21	N335AA	1	13	36	−34
21	N332AA	4	23	12	−24

Data Set 16: Earthquake Measurements

Fifty matched pairs of magnitude/depth measurements randomly selected from 10,594 earthquakes recorded in one year from a location in southern California. MAG is magnitude measured on the Richter scale and DEPTH is depth in km. The magnitude and depth both describe the source of the earthquake. In the two tables below, magnitudes and depths are paired by their corresponding positions in the two tables. The data are from the Southern California Earthquake Data Center.

STATDISK:	Data set name is Earthquakes.
Minitab:	Worksheet name is QUAKE.MTW.
Excel:	Workbook name is QUAKE.XLS.
TI-83/84 Plus:	App name is QUAKE and the file names are the same as for text files.
Text file names:	MAG, DEPTH.

Mag

0.70	2.20	1.64	1.01	1.62	1.28	0.92	1.00	1.49	1.42
0.74	1.98	1.32	1.26	1.83	0.83	1.00	2.24	0.84	1.35
0.64	0.64	2.95	0.00	0.99	1.34	0.79	2.50	1.42	0.93
0.39	1.22	0.90	0.65	1.56	0.54	0.79	1.79	1.00	0.40
0.70	0.20	1.76	1.46	0.40	1.25	1.44	1.25	1.25	1.39

Depth

6.6	2.2	18.5	7.0	13.7	5.4	5.3	5.9	4.7	14.5
2.0	14.8	8.1	18.6	4.5	17.7	15.9	15.1	8.6	5.2
15.3	5.6	10.0	8.2	8.3	9.9	13.7	8.5	8.2	7.9
17.2	6.1	13.7	5.7	6.0	17.3	4.2	14.7	15.2	3.3
3.2	9.1	8.0	18.9	14.2	5.1	5.7	16.4	10.1	6.4

Data Set 17: Word Counts by Males and Females

The columns are counts of the numbers of words spoken in a day by male (M) and female (F) subjects in six different sample groups. Column M1 denotes the word counts for males in Sample 1, F1 is the count for females in Sample 1, and so on.

Sample 1: Recruited couples ranging in age from 18 to 29
Sample 2: Students recruited in introductory psychology classes, aged 17 to 23
Sample 3: Students recruited in introductory psychology classes in Mexico, aged 17 to 25
Sample 4: Students recruited in introductory psychology classes, aged 17 to 22
Sample 5: Students recruited in introductory psychology classes, aged 18 to 26
Sample 6: Students recruited in introductory psychology classes, aged 17 to 23

Results were published in "Are Women Really More Talkative Than Men?" by Mehl, Vazire, Ramirez-Esparza, Slatcher, Pennebaker, *Science,* Vol. 317, No. 5834.

STATDISK: Data set name is Word Counts from Men and Women.
Minitab: Worksheet name is WORDS.MTW.
Excel: Workbook name is WORDS.XLS.
TI-83/84 Plus: App name is WORDS, and the file names are the same as for text files.
Text file names: Text file names correspond to the columns below: M1, F1, M2, F2, M3, F3, M4, F4, M5, F5 M6, F6.

M1	F1	M2	F2	M3	F3	M4	F4	M5	F5	M6	F6
27531	20737	23871	16109	21143	6705	47016	11849	39207	15962	28408	15357
15684	24625	5180	10592	17791	21613	27308	25317	20868	16610	10084	13618
5638	5198	9951	24608	36571	11935	42709	40055	18857	22497	15931	9783
27997	18712	12460	13739	6724	15790	20565	18797	17271	5004	21688	26451
25433	12002	17155	22376	15430	17865	21034	20104		10171	37786	12151
8077	15702	10344	9351	11552	13035	24150	17225		31327	10575	8391
21319	11661	9811	7694	11748	24834	24547	14356		8758	12880	19763
17572	19624	12387	16812	12169	7747	22712	20571			11071	25246
26429	13397	29920	21066	15581	3852	20858	12240			17799	8427
21966	18776	21791	32291	23858	11648	3189	10031			13182	6998
11680	15863	9789	12320	5269	25862	10379	13260			8918	24876
10818	12549	31127	19839	12384	17183	15750	22871			6495	6272
12650	17014	8572	22018	11576	11010	4288	26533			8153	10047
21683	23511	6942	16624	17707	11156	12398	26139			7015	15569
19153	6017	2539	5139	15229	11351	25120	15204			4429	39681
1411	18338	36345	17384	18160	25693	7415	18393			10054	23079
20242	23020	6858	17740	22482	13383	7642	16363			3998	24814
10117	18602	24024	7309	18626	19992	16459	21293			12639	19287
20206	16518	5488	14699	1118	14926	19928	12562			10974	10351
16874	13770	9960	21450	5319	14128	26615	15422			5255	8866
16135	29940	11118	14602		10345	21885	29011				10827
20734	8419	4970	18360		13516	10009	17085				12584
7771	17791	10710	12345		12831	35142	13932				12764
6792	5596	15011	17379		9671	3593	2220				19086
26194	11467	1569	14394		17011	15728	5909				26852
10671	18372	23794	11907		28575	19230	10623				17639
13462	13657	23689	8319		23557	17108	20388				16616
12474	21420	11769	16046		13656	23852	13052				
13560	21261	26846	5112		8231	11276	12098				
18876	12964	17386	8034		10601	14456	19643				
13825	33789	7987	7845		8124	11067	21747				
9274	8709	25638	7796			18527	26601				

(continued)

Data Set 17: Word Counts by Males and Females *(continued)*

M1	F1	M2	F2	M3	F3	M4	F4	M5	F5	M6	F6
20547	10508	695	20910			11478	17835				
17190	11909	2366	8756			6025	14030				
10578	29730	16075	5683			12975	7990				
14821	20981	16789	8372			14124	16667				
15477	16937	9308	17191			22942	5342				
10483	19049		8380			12985	12729				
19377	20224		16741			18360	18920				
11767	15872		16417			9643	24261				
13793	18717		2363			8788	8741				
5908	12685		24349			7755	14981				
18821	17646					9725	1674				
14069	16255					11033	20635				
16072	28838					10372	16154				
16414	38154					16869	4148				
19017	25510					16248	5322				
37649	34869					9202					
17427	24480					11395					
46978	31553										
25835	18667										
10302	7059										
15686	25168										
10072	16143										
6885	14730										
20848	28117										

Data Set 18: Voltage Measurements from a Home

All measurements are from the author's home. The voltage measurements are from the electricity supplied directly to the home, an independent Generac generator (model PP 5000), and an uninterruptible power supply (APC model CS 350) connected to the home power supply.

STATDISK:	Data set name is Voltage Measurements from a Home.
Minitab:	Worksheet name is VOLTAGE.MTW.
Excel:	Workbook name is VOLTAGE.XLS.
TI-83/84 Plus:	App name is VOLTAGE and the file names are the same as for text files.
Text file names:	Text file names are VHOME, VGEN, VUPS.

Day	Home (volts)	Generator (volts)	UPS (volts)
1	123.8	124.8	123.1
2	123.9	124.3	123.1
3	123.9	125.2	123.6
4	123.3	124.5	123.6
5	123.4	125.1	123.6
6	123.3	124.8	123.7
7	123.3	125.1	123.7
8	123.6	125.0	123.6
9	123.5	124.8	123.6
10	123.5	124.7	123.8
11	123.5	124.5	123.7
12	123.7	125.2	123.8
13	123.6	124.4	123.5
14	123.7	124.7	123.7
15	123.9	124.9	123.0
16	124.0	124.5	123.8
17	124.2	124.8	123.8
18	123.9	124.8	123.1
19	123.8	124.5	123.7
20	123.8	124.6	123.7
21	124.0	125.0	123.8
22	123.9	124.7	123.8
23	123.6	124.9	123.7
24	123.5	124.9	123.8
25	123.4	124.7	123.7
26	123.4	124.2	123.8
27	123.4	124.7	123.8
28	123.4	124.8	123.8
29	123.3	124.4	123.9
30	123.3	124.6	123.8
31	123.5	124.4	123.9
32	123.6	124.0	123.9
33	123.8	124.7	123.9
34	123.9	124.4	123.9
35	123.9	124.6	123.6
36	123.8	124.6	123.2
37	123.9	124.6	123.1
38	123.7	124.8	123.0
39	123.8	124.3	122.9
40	123.8	124.0	123.0

Data Set 19: Cola Weights and Volumes

Weights are in pounds and volumes are in ounces.

STATDISK: Data set name is Weights and Volumes of Cola.
Minitab: Worksheet name is COLA.MTW.
Excel: Workbook name is COLA.XLS.
TI-83/84 Plus: App name is COLA, and the file names are the same as for text files.
Text file names: CRGWT, CRGVL, CDTWT, CDTVL, PRGWT, PRGVL, PDTWT, PDTVL.

Weight Regular Coke	Volume Regular Coke	Weight Diet Coke	Volume Diet Coke	Weight Regular Pepsi	Volume Regular Pepsi	Weight Diet Pepsi	Volume Diet Pepsi
0.8192	12.3	0.7773	12.1	0.8258	12.4	0.7925	12.3
0.8150	12.1	0.7758	12.1	0.8156	12.2	0.7868	12.2
0.8163	12.2	0.7896	12.3	0.8211	12.2	0.7846	12.2
0.8211	12.3	0.7868	12.3	0.8170	12.2	0.7938	12.3
0.8181	12.2	0.7844	12.2	0.8216	12.2	0.7861	12.2
0.8247	12.3	0.7861	12.3	0.8302	12.4	0.7844	12.2
0.8062	12.0	0.7806	12.2	0.8192	12.2	0.7795	12.2
0.8128	12.1	0.7830	12.2	0.8192	12.2	0.7883	12.3
0.8172	12.2	0.7852	12.2	0.8271	12.3	0.7879	12.2
0.8110	12.1	0.7879	12.3	0.8251	12.3	0.7850	12.3
0.8251	12.3	0.7881	12.3	0.8227	12.2	0.7899	12.3
0.8264	12.3	0.7826	12.3	0.8256	12.3	0.7877	12.2
0.7901	11.8	0.7923	12.3	0.8139	12.2	0.7852	12.2
0.8244	12.3	0.7852	12.3	0.8260	12.3	0.7756	12.1
0.8073	12.1	0.7872	12.3	0.8227	12.2	0.7837	12.2
0.8079	12.1	0.7813	12.2	0.8388	12.5	0.7879	12.2
0.8044	12.0	0.7885	12.3	0.8260	12.3	0.7839	12.2
0.8170	12.2	0.7760	12.1	0.8317	12.4	0.7817	12.2
0.8161	12.2	0.7822	12.2	0.8247	12.3	0.7822	12.2
0.8194	12.2	0.7874	12.3	0.8200	12.2	0.7742	12.1
0.8189	12.2	0.7822	12.2	0.8172	12.2	0.7833	12.2
0.8194	12.2	0.7839	12.2	0.8227	12.3	0.7835	12.2
0.8176	12.2	0.7802	12.1	0.8244	12.3	0.7855	12.2
0.8284	12.4	0.7892	12.3	0.8244	12.2	0.7859	12.2
0.8165	12.2	0.7874	12.2	0.8319	12.4	0.7775	12.1
0.8143	12.2	0.7907	12.3	0.8247	12.3	0.7833	12.2
0.8229	12.3	0.7771	12.1	0.8214	12.2	0.7835	12.2
0.8150	12.2	0.7870	12.2	0.8291	12.4	0.7826	12.2
0.8152	12.2	0.7833	12.3	0.8227	12.3	0.7815	12.2
0.8244	12.3	0.7822	12.2	0.8211	12.3	0.7791	12.1
0.8207	12.2	0.7837	12.3	0.8401	12.5	0.7866	12.3
0.8152	12.2	0.7910	12.4	0.8233	12.3	0.7855	12.2
0.8126	12.1	0.7879	12.3	0.8291	12.4	0.7848	12.2
0.8295	12.4	0.7923	12.4	0.8172	12.2	0.7806	12.2
0.8161	12.2	0.7859	12.3	0.8233	12.4	0.7773	12.1
0.8192	12.2	0.7811	12.2	0.8211	12.3	0.7775	12.1

Data Set 20: M&M Plain Candy Weights (grams)

STATDISK: Data set name is M&M Candy Weights.
Minitab: Worksheet name is M&M.MTW.
Excel: Workbook name is M&M.XLS.
TI-83/84 Plus: App name is MM, and the file names are the same as for text files.
Text file names: Text file names are RED, ORNG, YLLW, BROWN, BLUE, GREEN.

Red	Orange	Yellow	Brown	Blue	Green
0.751	0.735	0.883	0.696	0.881	0.925
0.841	0.895	0.769	0.876	0.863	0.914
0.856	0.865	0.859	0.855	0.775	0.881
0.799	0.864	0.784	0.806	0.854	0.865
0.966	0.852	0.824	0.840	0.810	0.865
0.859	0.866	0.858	0.868	0.858	1.015
0.857	0.859	0.848	0.859	0.818	0.876
0.942	0.838	0.851	0.982	0.868	0.809
0.873	0.863			0.803	0.865
0.809	0.888			0.932	0.848
0.890	0.925			0.842	0.940
0.878	0.793			0.832	0.833
0.905	0.977			0.807	0.845
	0.850			0.841	0.852
	0.830			0.932	0.778
	0.856			0.833	0.814
	0.842			0.881	0.791
	0.778			0.818	0.810
	0.786			0.864	0.881
	0.853			0.825	
	0.864			0.855	
	0.873			0.942	
	0.880			0.825	
	0.882			0.869	
	0.931			0.912	
				0.887	
				0.886	

Data Set 21: Coin Weights (grams)

The "pre-1983 pennies" were made after the Indian and wheat pennies, and they are 97% copper and 3% zinc. The "post-1983 pennies" are 3% copper and 97% zinc. The "pre-1964 silver quarters" are 90% silver and 10% copper. The "post-1964 quarters" are made with a copper-nickel alloy.

STATDISK:	Data set name is Coin Weights.
Minitab:	Worksheet name is COINS.MTW.
Excel:	Workbook name is COINS.XLS.
TI-83/84 Plus:	App name is COINS, and the file names are the same as for text files.
Text file names:	Text file names are CPIND, CPWHT, CPPRE, CPPST, CPCAN, CQPRE, CQPST, CDOL.

Indian Pennies	Wheat Pennies	Pre-1983 Pennies	Post-1983 Pennies	Canadian Pennies	Pre-1964 Quarters	Post-1964 Quarters	Dollar Coins
3.0630	3.1366	3.1582	2.5113	3.2214	6.2771	5.7027	8.1008
3.0487	3.0755	3.0406	2.4907	3.2326	6.2371	5.7495	8.1072
2.9149	3.1692	3.0762	2.5024	2.4662	6.1501	5.7050	8.0271
3.1358	3.0476	3.0398	2.5298	2.8357	6.0002	5.5941	8.0813
2.9753	3.1029	3.1043	2.4950	3.3189	6.1275	5.7247	8.0241
	3.0377	3.1274	2.5127	3.2612	6.2151	5.6114	8.0510
	3.1083	3.0775	2.4998	3.2441	6.2866	5.6160	7.9817
	3.1141	3.1038	2.4848	2.4679	6.0760	5.5999	8.0954
	3.0976	3.1086	2.4823	2.7202	6.1426	5.7790	8.0658
	3.0862	3.0586	2.5163	2.5120	6.3415	5.6841	8.1238
	3.0570	3.0603	2.5222		6.1309	5.6234	8.1281
	3.0765	3.0502	2.5004		6.2412	5.5928	8.0307
	3.1114	3.1028	2.5248		6.1442	5.6486	8.0719
	3.0965	3.0522	2.5058		6.1073	5.6661	8.0345
	3.0816	3.0546	2.4900		6.1181	5.5361	8.0775
	3.0054	3.0185	2.5068		6.1352	5.5491	8.1384
	3.1934	3.0712	2.5016		6.2821	5.7239	8.1041
	3.1461	3.0717	2.4797		6.2647	5.6555	8.0894
	3.0185	3.0546	2.5067		6.2908	5.6063	8.0538
	3.1267	3.0817	2.5139		6.1661	5.5709	8.0342
	3.1524	3.0704	2.4762		6.2674	5.5591	
	3.0786	3.0797	2.5004		6.2718	5.5864	
	3.0131	3.0713	2.5170		6.1949	5.6872	
	3.1535	3.0631	2.4925		6.2465	5.6274	
	3.0480	3.0866	2.4876		6.3172	5.6157	
	3.0050	3.0763	2.4933		6.1487	5.6668	
	3.0290	3.1299	2.4806		6.0829	5.7198	
	3.1038	3.0846	2.4907		6.1423	5.6694	
	3.0357	3.0917	2.5017		6.1970	5.5454	
	3.0064	3.0877	2.4950		6.2441	5.6646	
	3.0936	2.9593	2.4973		6.3669	5.5636	
	3.1031	3.0966	2.5252		6.0775	5.6485	
	3.0408	2.9800	2.4978		6.1095	5.6703	
	3.0561	3.0934	2.5073		6.1787	5.6848	
	3.0994	3.1340	2.4658		6.2130	5.5609	
			2.4529		6.1947	5.7344	
			2.5085		6.1940	5.6449	
					6.0257	5.5804	
					6.1719	5.6010	
					6.3278	5.6022	

Data Set 22: Axial Loads of Aluminum Cans

Axial loads are measured in pounds. Axial loads are applied when the tops are pressed into place.

STATDISK:	Data set name is Axial Loads of Cans.
Minitab:	Worksheet name is CANS.MTW.
Excel:	Workbook name is CANS.XLS.
TI-83/84 Plus:	App name is CANS, and the file names are the same as for text files.
Text file names:	CN109, CN111.

Sample	Aluminum cans 0.0109 in. thick Axial load (pounds)						
1	270	273	258	204	254	228	282
2	278	201	264	265	223	274	230
3	250	275	281	271	263	277	275
4	278	260	262	273	274	286	236
5	290	286	278	283	262	277	295
6	274	272	265	275	263	251	289
7	242	284	241	276	200	278	283
8	269	282	267	282	272	277	261
9	257	278	295	270	268	286	262
10	272	268	283	256	206	277	252
11	265	263	281	268	280	289	283
12	263	273	209	259	287	269	277
13	234	282	276	272	257	267	204
14	270	285	273	269	284	276	286
15	273	289	263	270	279	206	270
16	270	268	218	251	252	284	278
17	277	208	271	208	280	269	270
18	294	292	289	290	215	284	283
19	279	275	223	220	281	268	272
20	268	279	217	259	291	291	281
21	230	276	225	282	276	289	288
22	268	242	283	277	285	293	248
23	278	285	292	282	287	277	266
24	268	273	270	256	297	280	256
25	262	268	262	293	290	274	292

Sample	Aluminum cans 0.0111 in. thick Axial load (pounds)						
1	287	216	260	291	210	272	260
2	294	253	292	280	262	295	230
3	283	255	295	271	268	225	246
4	297	302	282	310	305	306	262
5	222	276	270	280	288	296	281
6	300	290	284	304	291	277	317
7	292	215	287	280	311	283	293
8	285	276	301	285	277	270	275
9	290	288	287	282	275	279	300
10	293	290	313	299	300	265	285
11	294	262	297	272	284	291	306
12	263	304	288	256	290	284	307
13	273	283	250	244	231	266	504
14	284	227	269	282	292	286	281
15	296	287	285	281	298	289	283
16	247	279	276	288	284	301	309
17	284	284	286	303	308	288	303
18	306	285	289	292	295	283	315
19	290	247	268	283	305	279	287
20	285	298	279	274	205	302	296
21	282	300	284	281	279	255	210
22	279	286	293	285	288	289	281
23	297	314	295	257	298	211	275
24	247	279	303	286	287	287	275
25	243	274	299	291	281	303	269

Data Set 23: Weights of Discarded Garbage for One Week

Weights are in pounds. HHSIZE is the household size. Data provided by Masakuza Tani, the Garbage Project, University of Arizona.

STATDISK: Data set name is Weights of Garbage.
Minitab: Worksheet name is GARBAGE.MTW.
Excel: Workbook name is GARBAGE.XLS.
TI-83/84 Plus: App name is GARBAGE, and the file names are the same as for text files.
Text file names: HHSIZ, METAL, PAPER, PLAS, GLASS, FOOD, YARD, TEXT, OTHER, TOTAL.

Household	HHSize	Metal	Paper	Plas	Glass	Food	Yard	Text	Other	Total
1	2	1.09	2.41	0.27	0.86	1.04	0.38	0.05	4.66	10.76
2	3	1.04	7.57	1.41	3.46	3.68	0.00	0.46	2.34	19.96
3	3	2.57	9.55	2.19	4.52	4.43	0.24	0.50	3.60	27.60
4	6	3.02	8.82	2.83	4.92	2.98	0.63	2.26	12.65	38.11
5	4	1.50	8.72	2.19	6.31	6.30	0.15	0.55	2.18	27.90
6	2	2.10	6.96	1.81	2.49	1.46	4.58	0.36	2.14	21.90
7	1	1.93	6.83	0.85	0.51	8.82	0.07	0.60	2.22	21.83
8	5	3.57	11.42	3.05	5.81	9.62	4.76	0.21	10.83	49.27
9	6	2.32	16.08	3.42	1.96	4.41	0.13	0.81	4.14	33.27
10	4	1.89	6.38	2.10	17.67	2.73	3.86	0.66	0.25	35.54
11	4	3.26	13.05	2.93	3.21	9.31	0.70	0.37	11.61	44.44
12	7	3.99	11.36	2.44	4.94	3.59	13.45	4.25	1.15	45.17
13	3	2.04	15.09	2.17	3.10	5.36	0.74	0.42	4.15	33.07
14	5	0.99	2.80	1.41	1.39	1.47	0.82	0.44	1.03	10.35
15	6	2.96	6.44	2.00	5.21	7.06	6.14	0.20	14.43	44.44
16	2	1.50	5.86	0.93	2.03	2.52	1.37	0.27	9.65	24.13
17	4	2.43	11.08	2.97	1.74	1.75	14.70	0.39	2.54	37.60
18	4	2.97	12.43	2.04	3.99	5.64	0.22	2.47	9.20	38.96
19	3	1.42	6.05	0.65	6.26	1.93	0.00	0.86	0.00	17.17
20	3	3.60	13.61	2.13	3.52	6.46	0.00	0.96	1.32	31.60
21	2	4.48	6.98	0.63	2.01	6.72	2.00	0.11	0.18	23.11
22	2	1.36	14.33	1.53	2.21	5.76	0.58	0.17	1.62	27.56
23	4	2.11	13.31	4.69	0.25	9.72	0.02	0.46	0.40	30.96
24	1	0.41	3.27	0.15	0.09	0.16	0.00	0.00	0.00	4.08
25	4	2.02	6.67	1.45	6.85	5.52	0.00	0.68	0.03	23.22
26	6	3.27	17.65	2.68	2.33	11.92	0.83	0.28	4.03	42.99
27	11	4.95	12.73	3.53	5.45	4.68	0.00	0.67	19.89	51.90
28	3	1.00	9.83	1.49	2.04	4.76	0.42	0.54	0.12	20.20
29	4	1.55	16.39	2.31	4.98	7.85	2.04	0.20	1.48	36.80
30	3	1.41	6.33	0.92	3.54	2.90	3.85	0.03	0.04	19.02
31	2	1.05	9.19	0.89	1.06	2.87	0.33	0.01	0.03	15.43
32	2	1.31	9.41	0.80	2.70	5.09	0.64	0.05	0.71	20.71
33	2	2.50	9.45	0.72	1.14	3.17	0.00	0.02	0.01	17.01
34	4	2.35	12.32	2.66	12.24	2.40	7.87	4.73	0.78	45.35
35	6	3.69	20.12	4.37	5.67	13.20	0.00	1.15	1.17	49.37
36	2	3.61	7.72	0.92	2.43	2.07	0.68	0.63	0.00	18.06
37	2	1.49	6.16	1.40	4.02	4.00	0.30	0.04	0.00	17.41
38	2	1.36	7.98	1.45	6.45	4.27	0.02	0.12	2.02	23.67
39	2	1.73	9.64	1.68	1.89	1.87	0.01	1.73	0.58	19.13
40	2	0.94	8.08	1.53	1.78	8.13	0.36	0.12	0.05	20.99
41	3	1.33	10.99	1.44	2.93	3.51	0.00	0.39	0.59	21.18
42	3	2.62	13.11	1.44	1.82	4.21	4.73	0.64	0.49	29.06

(continued)

Household	HHSize	Metal	Paper	Plas	Glass	Food	Yard	Text	Other	Total
43	2	1.25	3.26	1.36	2.89	3.34	2.69	0.00	0.16	14.95
44	2	0.26	1.65	0.38	0.99	0.77	0.34	0.04	0.00	4.43
45	3	4.41	10.00	1.74	1.93	1.14	0.92	0.08	4.60	24.82
46	6	3.22	8.96	2.35	3.61	1.45	0.00	0.09	1.12	20.80
47	4	1.86	9.46	2.30	2.53	6.54	0.00	0.65	2.45	25.79
48	4	1.76	5.88	1.14	3.76	0.92	1.12	0.00	0.04	14.62
49	3	2.83	8.26	2.88	1.32	5.14	5.60	0.35	2.03	28.41
50	3	2.74	12.45	2.13	2.64	4.59	1.07	0.41	1.14	27.17
51	10	4.63	10.58	5.28	12.33	2.94	0.12	2.94	15.65	54.47
52	3	1.70	5.87	1.48	1.79	1.42	0.00	0.27	0.59	13.12
53	6	3.29	8.78	3.36	3.99	10.44	0.90	1.71	13.30	45.77
54	5	1.22	11.03	2.83	4.44	3.00	4.30	1.95	6.02	34.79
55	4	3.20	12.29	2.87	9.25	5.91	1.32	1.87	0.55	37.26
56	7	3.09	20.58	2.96	4.02	16.81	0.47	1.52	2.13	51.58
57	5	2.58	12.56	1.61	1.38	5.01	0.00	0.21	1.46	24.81
58	4	1.67	9.92	1.58	1.59	9.96	0.13	0.20	1.13	26.18
59	2	0.85	3.45	1.15	0.85	3.89	0.00	0.02	1.04	11.25
60	4	1.52	9.09	1.28	8.87	4.83	0.00	0.95	1.61	28.15
61	2	1.37	3.69	0.58	3.64	1.78	0.08	0.00	0.00	11.14
62	2	1.32	2.61	0.74	3.03	3.37	0.17	0.00	0.46	11.70

Appendix C Bibliography of Books and Web Sites

Books

***An asterisk denotes a book recommended for reading. Other books are recommended as reference texts.**

Bennett, D. 1998. *Randomness.* Cambridge, Mass.: Harvard University Press.

*Best, J. 2001. *Damned Lies and Statistics.* Berkeley: University of California Press.

*Best, J. 2004. *More Damned Lies and Statistics.* Berkeley: University of California Press.

*Butros, M. 2013. *Student Solutions Manual to Accompany Elementary Statistics.* 12th ed. Boston: Addison-Wesley.

*Campbell, S. 2004. *Flaws and Fallacies in Statistical Thinking.* Mineola, N.Y.: Dover Publications.

*Crossen, C. 1996. *Tainted Truth: The Manipulation of Fact in America.* New York: Simon & Schuster.

*Freedman, D., R. Pisani, R. Purves, and A. Adhikari. 2007. *Statistics.* 4th ed. New York: Norton.

*Gonick, L., and W. Smith. 1993. *The Cartoon Guide to Statistics.* New York: HarperCollins.

*Heyde, C., and E. Seneta, eds. 2001. *Statisticians of the Centuries.* New York: Springer-Verlag.

*Hollander, M., and F. Proschan. 1984. *The Statistical Exorcist: Dispelling Statistics Anxiety.* New York: Marcel Dekker.

*Holmes, C. 1990. *The Honest Truth About Lying with Statistics.* Springfield, Ill.: Charles C. Thomas.

*Hooke, R. 1983. *How to Tell the Liars from the Statisticians.* New York: Marcel Dekker.

*Huff, D. 1993. *How to Lie with Statistics.* New York: Norton.

*Jaffe, A., and H. Spirer. 1998. *Misused Statistics.* New York: Marcel Dekker.

Kotz, S., and D. Stroup. 1983. *Educated Guessing—How to Cope in an Uncertain World.* New York: Marcel Dekker.

*Moore, D., and W. Notz. 2009. *Statistics: Concepts and Controversies.* 6th ed. San Francisco: Freeman.

*Paulos, J. 2001. *Innumeracy: Mathematical Illiteracy and Its Consequences.* New York: Hill and Wang.

*Reichmann, W. 1981. *Use and Abuse of Statistics.* New York: Penguin.

*Rossman, A., and B. Chance. 2008. *Workshop Statistics: Discovery with Data.* Emeryville, Calif.: Key Curriculum Press.

*Salsburg, D. 2000. *The Lady Tasting Tea: How Statistics Revolutionized the Twentieth Century.* New York: W. H. Freeman.

Sheskin, D. 2011. *Handbook of Parametric and Nonparametric Statistical Procedures.* 5th ed. Boca Raton, Fla.: CRC Press.

Simon, J. 1997. *Resampling: The New Statistics.* 2nd ed. Arlington, Va.: Resampling Stats.

*Stigler, S. 1986. *The History of Statistics.* Cambridge, Mass.: Harvard University Press.

Triola, M. 2013. *Minitab Student Laboratory Manual and Workbook.* 12th ed. Boston: Addison-Wesley.

Triola, M. 2013. *STATDISK 12 Student Laboratory Manual and Workbook.* 12th ed. Boston: Addison-Wesley.

Triola, M., and L. Franklin. 1994. *Business Statistics.* Boston: Addison-Wesley.

Triola, M., and M. Triola. 2006. *Biostatistics for the Biological and Health Sciences.* Boston: Addison-Wesley.

*Tufte, E. 2001. *The Visual Display of Quantitative Information.* 2nd ed. Cheshire, Conn.: Graphics Press.

Tukey, J. 1977. *Exploratory Data Analysis.* Boston: Addison-Wesley.

Vickers, A. 2009. *What Is a P-Value Anyway?* Boston: Pearson.

Zwillinger, D., and S. Kokoska. 2000. *CRC Standard Probability and Statistics Tables and Formulae.* Boca Raton, Fla.: CRC Press.

Web Sites

Triola Statistics Series: www.aw.com/triola

Triola Stats Blog and Resources: www.TriolaStats.com and www.facebook.com/TriolaStats

STATDISK: www.statdisk.org

StatCrunch: www.statcrunch.com

Data and Story Library: http://lib.stat.cmu.edu/DASL/

Statistical Science Web Data Sets: http://www.statsci.org/datasets.html

***Journal of Statistics Education* Data Archive:** http://www.amstat.org/publications/jse/jse_data_archive.htm

UCLA Statistics Data Sets: http://www.stat.ucla.edu/data/

Appendix D

Answers to Odd-Numbered Section Exercises, plus Answers to All Chapter Quick Quizzes, Chapter Review Exercises, and Cumulative Review Exercises

Chapter 1 Answers

Section 1-2

1. Statistical significance is indicated when methods of statistics are used to reach a conclusion that some treatment or finding is effective, but common sense might suggest that the treatment or finding does not make enough of a difference to justify its use or to be practical. Yes, it is possible for a study to have statistical significance but not practical significance.
3. A voluntary response sample is a sample in which the subjects themselves decide whether to be included in the study. A voluntary response sample is generally not suitable for a statistical study, because the sample may have a bias resulting from participation by those with a special interest in the topic being studied.
5. There does appear to be a potential to create a bias.
7. There does not appear to be a potential to create a bias.
9. The sample is a voluntary response sample and is therefore flawed.
11. The sampling method appears to be sound.
13. Because there is a 30% chance of getting such results with a diet that has no effect, it does not appear to have statistical significance, but the average loss of 45 pounds does appear to have practical significance.
15. Because there is a 23% chance of getting such results with a program that has no effect, the program does not appear to have statistical significance. Because the success rate of 23% is not much better than the 20% rate that is typically expected with random guessing, the program does not appear to have practical significance.
17. The male and female pulse rates in the same column are not matched in any meaningful way. It does not make sense to use the difference between any of the pulse rates that are in the same column.
19. The data can be used to address the issue of whether males and females have pulse rates with the same average (mean) value.
21. Yes, each IQ score is matched with the brain volume in the same column, because they are measurements obtained from the same person. It does not make sense to use the difference between each IQ score and the brain volume in the same column, because IQ scores and brain volumes use different units of measurement. It would make no sense to find the difference between an IQ score of 87 and a brain volume of 1035 cm^3.
23. Given that the researchers do not appear to benefit from the results, they are professionals at prestigious institutions, and funding is from a U.S. government agency, the source of the data appears to be unbiased.
25. It is questionable that the sponsor is the Idaho Potato Commission and the favorite vegetable is potatoes.
27. The correlation, or association, between two variables does not mean that one of the variables is the cause of the other. Correlation does not imply causation.
29. a. 397.02 adults
    b. No. Because the result is a count of people among the 1018 who were surveyed, the result must be a whole number.
    c. 397 adults d. 25%
31. a. 322.28 adults
    b. No. Because the result is a count of adults among the 2302 who were surveyed, the result must be a whole number.
    c. 322 adults d. 2%
33. Because a reduction of 100% would eliminate all of the size, it is not possible to reduce the size by 100% or more.
35. If foreign investment fell by 100%, it would be totally eliminated, so it is not possible for it to fall by more than 100%.
37. Without our knowing anything about the number of ATVs in use, or the number of ATV drivers, or the amount of ATV usage, the number of 740 fatal accidents has no context. Some information should be given so that the reader can understand the *rate* of ATV fatalities.
39. The wording of the question is biased and tends to encourage negative responses. The sample size of 20 is too small. Survey respondents are self-selected instead of being selected by the newspaper. If 20 readers respond, the percentages should be multiples of 5, so 87% and 13% are not possible results.

Section 1-3

1. A parameter is a numerical measurement describing some characteristic of a population, whereas a statistic is a numerical measurement describing some characteristic of a sample.
3. Parts a and c describe discrete data.

5. Statistic	7. Parameter	9. Parameter
11. Statistic	13. Continuous	15. Discrete
17. Discrete	19. Continuous	21. Nominal
23. Interval	25. Ratio	27. Ordinal

29. The numbers are not counts or measures of anything, so they are at the nominal level of measurement, and it makes no sense to compute the average (mean) of them.
31. The numbers are used as substitutes for the categories of low, medium, and high, so the numbers are at the ordinal level of measurement. It does not make sense to compute the average (mean) of such numbers.
33. a. Continuous, because the number of possible values is infinite and not countable
    b. Discrete, because the number of possible values is finite
    c. Discrete, because the number of possible values is finite
    d. Discrete, because the number of possible values is infinite and countable
35. With no natural starting point, temperatures are at the interval level of measurement, so ratios such as "twice" are meaningless.

Section 1-4

1. No. Not every sample of the same size has the same chance of being selected. For example, the sample with the first two names has no chance of being selected. A simple random sample of n items is selected in such a way that every sample of the same size has the same chance of being selected.
3. The population consists of the adult friends on the list. The simple random sample is selected from the population of adult friends on the list, so the results are not likely to be representative of the much larger general population of adults in the United States.
5. Because the subjects are subjected to anger and confrontation, they are given a form of treatment, so this is an experiment, not an observational study.
7. This is an observational study because the therapists were not given any treatment. Their responses were observed.
9. Cluster 11. Random 13. Convenience
15. Systematic 17. Random 19. Convenience
21. The sample is not a simple random sample. Because every 1000th pill is selected, some samples have no chance of being selected. For example, a sample consisting of two consecutive pills has no chance of being selected, and this violates the requirement of a simple random sample.
23. The sample is a simple random sample. Every sample of size 500 has the same chance of being selected.
25. The sample is not a simple random sample. Not every sample has the same chance of being selected. For example, a sample that includes people who do not appear to be approachable has no chance of being selected.
27. Prospective study 29. Cross-sectional study
31. Matched pairs design 33. Completely randomized design
35. Blinding is a method whereby a subject (or a person who evaluates results) in an experiment does not know whether the subject is treated with the DNA vaccine or the adenoviral vector vaccine. It is important to use blinding so that results are not somehow distorted by knowledge of the particular treatment used.

Chapter 1: Quick Quiz

1. No. The numbers do not measure or count anything.
2. Nominal 3. Continuous 4. Quantitative data
5. Ratio 6. False 7. No
8. Statistic 9. Observational study
10. False

Chapter 1: Review Exercises

1. a. Discrete b. Ratio c. Stratified d. Cluster
   e. The mailed responses would be a voluntary response sample, so those with strong opinions are more likely to respond. It is very possible that the results do not reflect the true opinions of the population of all customers.
2. The survey was sponsored by the American Laser Centers, and 24% said that the favorite body part is the face, which happens to be a body part often chosen for some type of laser treatment. The source is therefore questionable.
3. The sample is a voluntary response sample, so the results are questionable.
4. a. It uses a voluntary response sample, and those with special interests are more likely to respond, so it is very possible that the sample is not representative of the population.
   b. Because the statement refers to 72% of all Americans, it is a parameter (but it is probably based on a 72% rate from the sample, and the sample percentage is a statistic).
   c. Observational study
5. a. If they have no fat at all, they have 100% less than any other amount with fat, so the 125% figure cannot be correct.
   b. 686 c. 28%
6. The Gallup poll used randomly selected respondents, but the AOL poll used a voluntary response sample. Respondents in the AOL poll are more likely to participate if they have strong feelings about the candidates, and this group is not necessarily representative of the population. The results from the Gallup poll are more likely to reflect the true opinions of American voters.
7. Because there is only a 4% chance of getting the results by chance, the method appears to have statistical significance. The result of 112 girls in 200 births is above the approximately 50% rate expected by chance, but it does not appear to be high enough to have practical significance. Not many couples would bother with a procedure that raises the likelihood of a girl from 50% to 56%.
8. a. Random b. Stratified c. Nominal
   d. Statistic, because it is based on a sample.
   e. The mailed responses would be a voluntary response sample. Those with strong opinions about the topic would be more likely to respond, so it is very possible that the results would not reflect the true opinions of the population of all adults.
9. a. Systematic b. Random c. Cluster
   d. Stratified e. Convenience
10. a. 780 adults b. 23%
    c. Men: 48.5%; women: 51.5%
    d. No, although this is a subjective judgment.
    e. No, although this is a subjective judgment.

Chapter 1: Cumulative Review Exercises

1. The mean is 11. Because the flight numbers are not measures or counts of anything, the result does not have meaning.
2. The mean is 101, and it is reasonably close to the population mean of 100.
3. 11.83 is an unusually high value.
4. 0.46 5. 1067 6. 0.0037
7. 28.0 8. 5.3 9. 0.00078364164
10. 68,719,476,736 (or about 68,719,476,000)
11. 678,223,072,849 (or about 678,223,070,000)
12. 0.0000059049

Chapter 2 Answers

Section 2-2

1. No. For each class, the frequency tells us how many values fall within the given range of values, but there is no way to determine the exact IQ scores represented in the class.
3. No. The sum of the percents is 199%, not 100%, so each respondent could answer "yes" to more than one category. The table does not show the distribution of a data set among all of several different categories. Instead, it shows responses to five separate questions.

5. Class width: 10. Class midpoints: 24.5, 34.5, 44.5, 54.5, 64.5, 74.5, 84.5. Class boundaries: 19.5, 29.5, 39.5, 49.5, 59.5, 69.5, 79.5, 89.5.
7. Class width: 10. Class midpoints: 54.5, 64.5, 74.5, 84.5, 94.5, 104.5, 114.5, 124.5. Class boundaries: 49.5, 59.5, 69.5, 79.5, 89.5, 99.5, 109.5, 119.5, 129.5.
9. Class width: 2.0. Class midpoints: 3.95, 5.95, 7.95, 9.95, 11.95. Class boundaries: 2.95, 4.95, 6.95, 8.95, 10.95, 12.95.
11. No. The frequencies do not satisfy the requirement of being roughly symmetric about the maximum frequency of 34.
13. 18, 7, 4
15. The actresses appear to be younger than the actors.

Age When Oscar Was Won	Relative Frequency (Actresses)	Relative Frequency (Actors)
20–29	32.9%	1.2%
30–39	41.5%	31.7%
40–49	15.9%	42.7%
50–59	2.4%	15.9%
60–69	4.9%	7.3%
70–79	1.2%	1.2%
80–89	1.2%	0.0%

17.

Age (years) of Best Actress When Oscar Was Won	Cumulative Frequency
Less than 30	27
Less than 40	61
Less than 50	74
Less than 60	76
Less than 70	80
Less than 80	81
Less than 90	82

19. Because there are disproportionately more 0s and 5s, it appears that the heights were reported instead of measured. Consequently, it is likely that the results are not very accurate.

x	Frequency
0	9
1	2
2	1
3	3
4	1
5	15
6	2
7	0
8	3
9	1

21. Yes, the distribution appears to be a normal distribution.

Pulse Rate (Male)	Frequency
40–49	1
50–59	7
60–69	17
70–79	9
80–89	5
90–99	1

23. No, the distribution does not appear to be a normal distribution.

Magnitude	Frequency
0.00–0.49	5
0.50–0.99	15
1.00–1.49	19
1.50–1.99	7
2.00–2.49	2
2.50–2.99	2

25. Yes, the distribution appears to be roughly a normal distribution.

Red Blood Cell Count	Frequency
4.00–4.39	2
4.40–4.79	7
4.80–5.19	15
5.20–5.59	13
5.60–5.99	3

27. Yes. Among the 48 flights, 36 arrived on time or early, and 45 of the 48 flights arrived no more than 30 minutes late.

Arrival Delay (min)	Frequency
(−60)–(−31)	11
(−30)–(−1)	25
0–29	9
30–59	1
60–89	0
90–119	2

29.

Category	Relative Frequency
Male Survivors	16.2%
Males Who Died	62.8%
Female Survivors	15.5%
Females Who Died	5.5%

31. Pilot error is the most serious threat to aviation safety. Better training and stricter pilot requirements can improve aviation safety.

Cause	Relative Frequency
Pilot Error	50.5%
Other Human Error	6.1%
Weather	12.1%
Mechanical	22.2%
Sabotage	9.1%

33. An outlier can dramatically affect the frequency table.

Weight (lb)	With Outlier	Without Outlier
200–219	6	6
220–239	5	5
240–259	12	12
260–279	36	36
280–299	87	87
300–319	28	28
320–339	0	
340–359	0	
360–379	0	
380–399	0	
400–419	0	
420–439	0	
440–459	0	
460–479	0	
480–499	0	
500–519	1	

Section 2-3

1. It is easier to see the distribution of the data by examining the graph of the histogram than by examining the numbers in a frequency distribution.
3. With a data set that is so small, the true nature of the distribution cannot be seen with a histogram. The data set has an outlier of 1 min. That duration time corresponds to the last flight, which ended in an explosion that killed the seven crew members.
5. Identifying the exact value is not easy, but answers not too far from 200 are good answers.
7. The tallest person is about 108 in., or about 9 ft, tall. That tallest height is depicted in the bar that is farthest to the right in the histogram. That height is an outlier because it is very far away from all of the other heights. That height of 9 ft must be an error, because nobody is that tall.
9. The digits 0 and 5 seem to occur much more often than the other digits, so it appears that the heights were reported and not actually measured. This suggests that the results might not be very accurate.

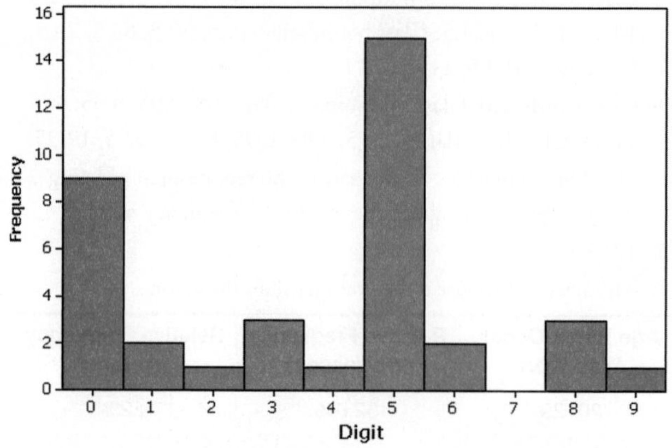

11. The histogram does appear to depict a normal distribution. The frequencies increase to a maximum and then decrease, and the histogram is symmetric with the left half being roughly a mirror image of the right half.

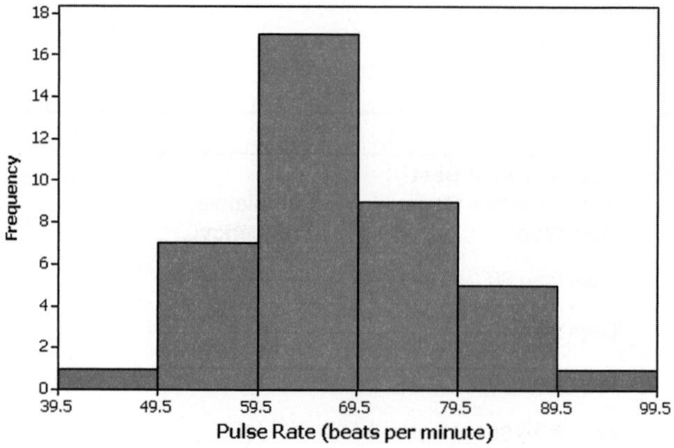

13. The histogram appears to roughly approximate a normal distribution. The frequencies increase to a maximum and then tend to decrease, and the histogram is symmetric with the left half being roughly a mirror image of the right half.

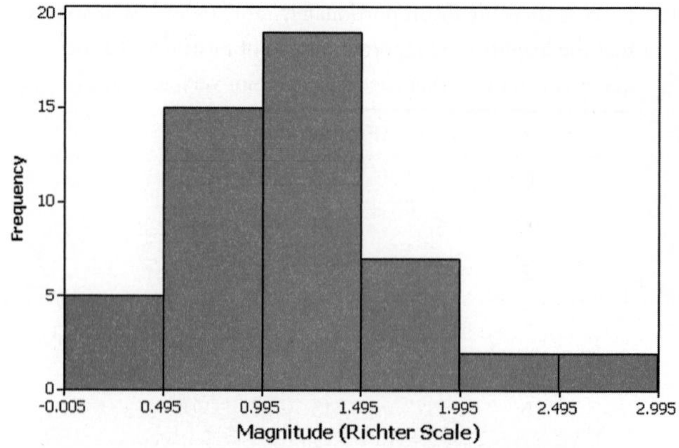

15. The histogram appears to roughly approximate a normal distribution. The frequencies increase to a maximum and then decrease, and the histogram is symmetric with the left half being roughly a mirror image of the right half.

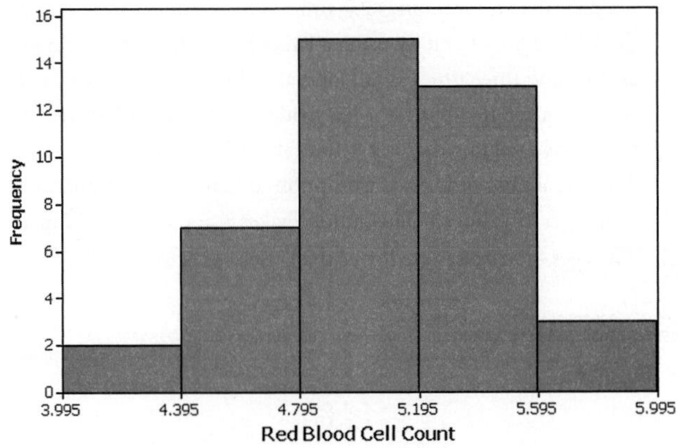

17. The two leftmost bars depict flights that arrived early, and the other bars to the right depict flights that arrived late.

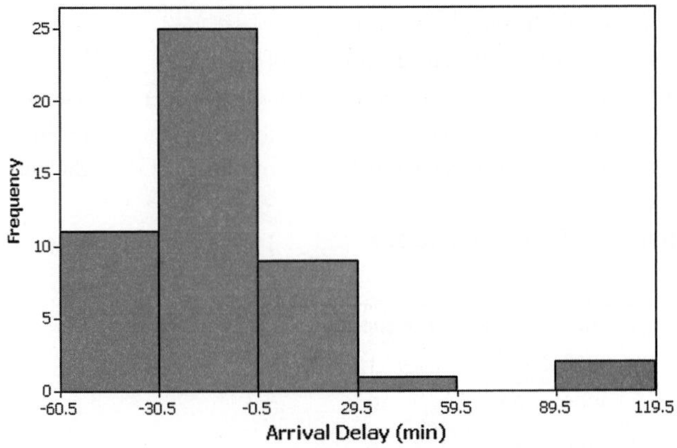

19. The ages of actresses are lower than those of actors.

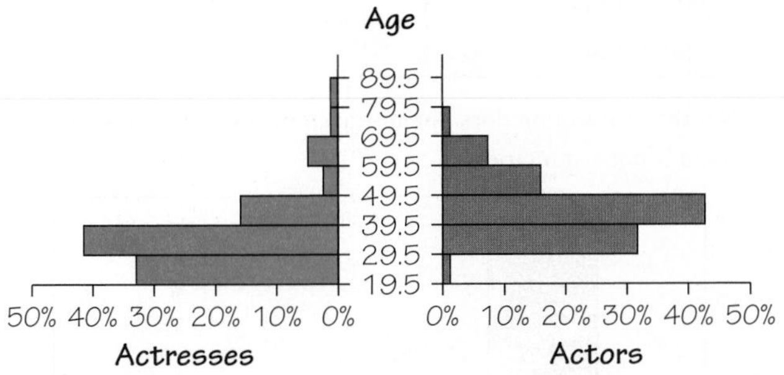

Section 2-4

1. In a Pareto chart, the bars are always arranged in descending order according to frequencies. The Pareto chart helps us understand data by drawing attention to the more important categories, which have the highest frequencies.
3. The data set is too small for a graph to reveal important characteristics of the data. With such a small data set, it would be better to simply list the data or place them in a table.
5. Because the points are scattered throughout with no obvious pattern, there does not appear to be a correlation.

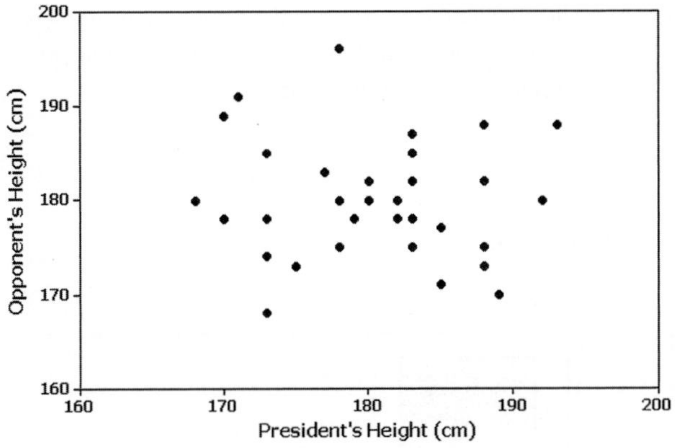

7. Yes. There is a very distinct pattern showing that bears with larger chest sizes tend to weigh more.

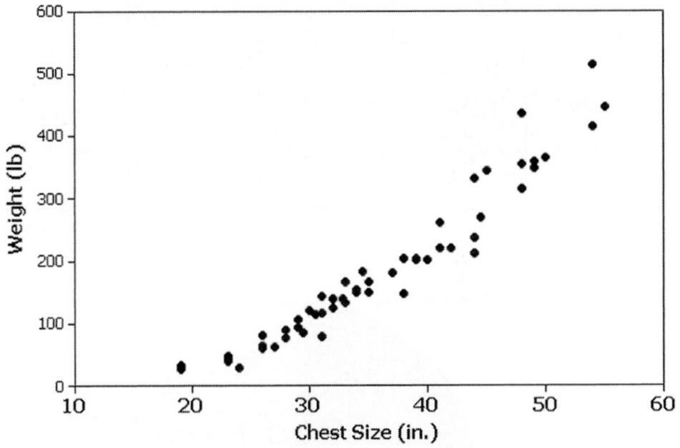

9. The first amount is highest for the opening day, when many Harry Potter fans are most eager to see the movie; the third and fourth values are from the first Friday and the first Saturday, which are the popular weekend days when movie attendance tends to spike.

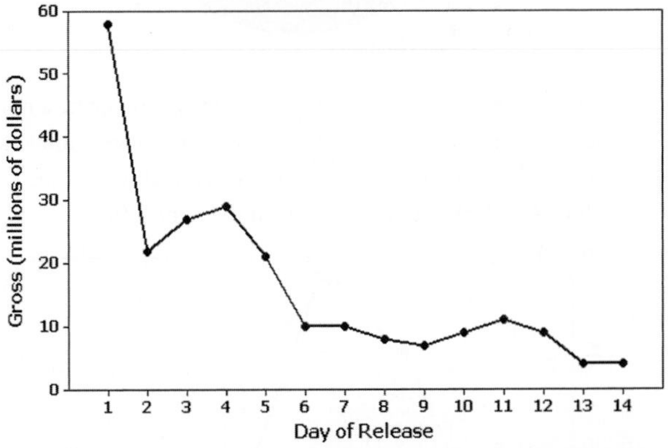

11. Yes, because the configuration of the points is roughly a bell shape, the volumes appear to be from a normally distributed population. The volume of 11.8 oz appears to be an outlier.

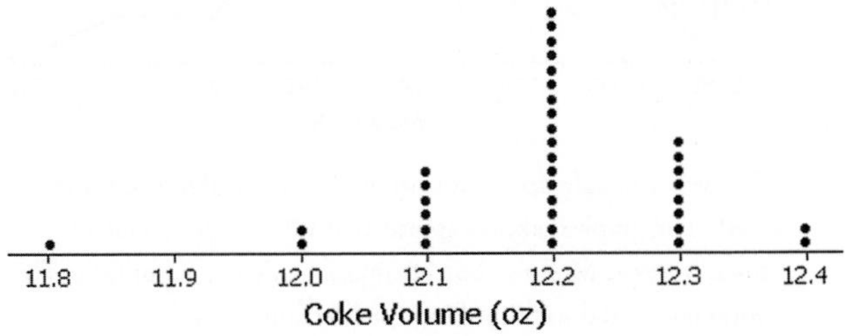

13. No. The distribution is not dramatically far from being a normal distribution with a bell shape, so there is not strong evidence against a normal distribution.

```
4 | 5
5 | 3335579
6 | 11167
7 | 1115568
8 | 4
```

15.

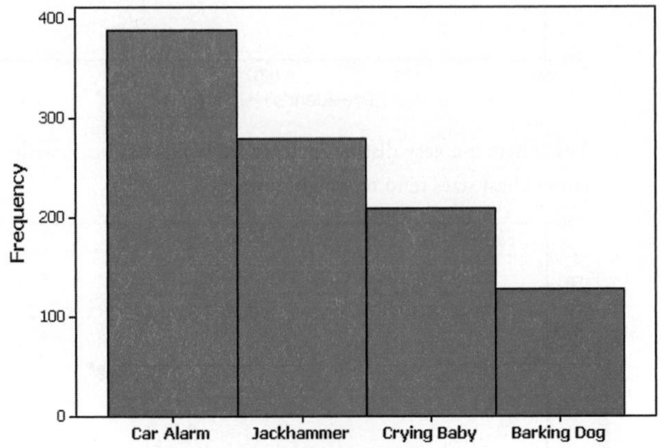

17.

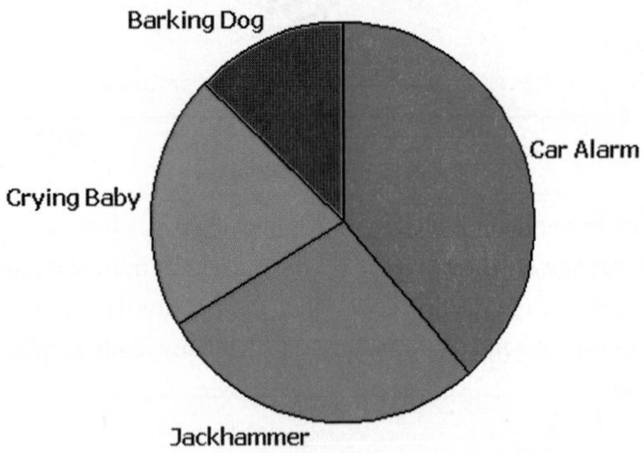

19. The frequency polygon appears to roughly approximate a normal distribution. The frequencies increase to a maximum and then tend to decrease, and the graph is symmetric with the left half being roughly a mirror image of the right half.

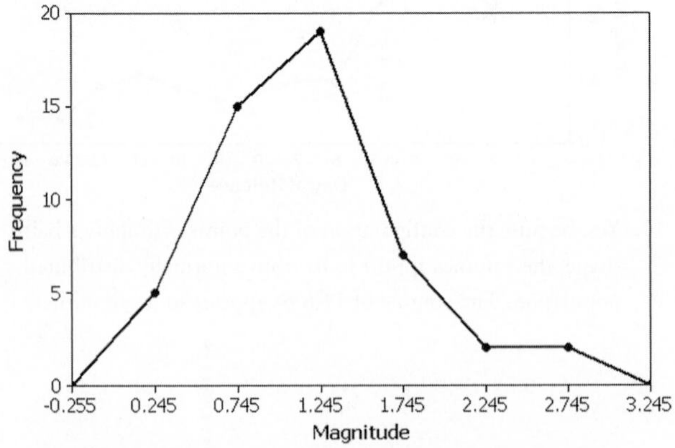

21. The vertical scale does not start at 0, so the difference is exaggerated. The graph makes it appear that Obama got about twice as many votes as McCain, but Obama actually got about 69 million votes compared to 60 million for McCain.

23. China's oil consumption is 2.7 times (or roughly 3 times) that of the United States, but by using a larger barrel that is three times as wide and three times as tall (and also three times as deep) as the smaller barrel, the illustrator has made it appear that the larger barrel has a volume that is 27 times that of the smaller barrel. The actual ratio of U.S. consumption to China's consumption is roughly 3 to 1, but the illustration makes it appear to be 27 to 1.

25. The ages of actresses are lower than those of actors.

```
                       Actresses         Actors
         999999988887777666655554421 | 2 | 9
  9988887765555555544433333222111000 | 3 | 00122244455666777788888999
                       9655322111110 | 4 | 00011111122222233334445555677788999
                                  40 | 5 | 011222234667 7
                                3110 | 6 | 000222
                                   4 | 7 | 6
                                   0 | 8 |
```

Chapter 2: Quick Quiz

1. 1.00 2. –0.005 and 0.995 3. No
4. 61 min, 62 min, 62 min, 62 min, 62 min, 67 min, 69 min
5. No 6. Bar graph 7. Scatterplot
8. Pareto chart 9. The distribution of the data
10. The bars of the histogram start relatively low, increase to some maximum, and then decrease. Also, the histogram is symmetric with the left half being roughly a mirror image of the right half.

Chapter 2: Review Exercises

1.

Volume (cm^3)	Frequency
900–999	1
1000–1099	10
1100–1199	4
1200–1299	3
1300–1399	1
1400–1499	1

2. No, the distribution does not appear to be normal because the graph is not symmetric.

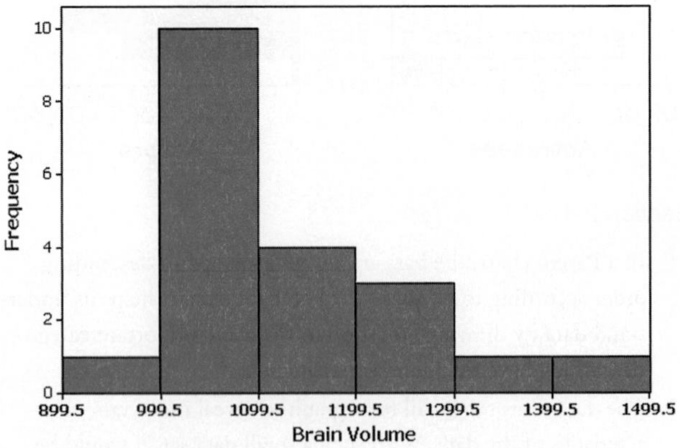

3. Although there are differences among the frequencies of the digits, the differences are not too extreme given the relatively small sample size, so the lottery appears to be fair.

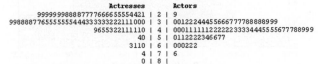

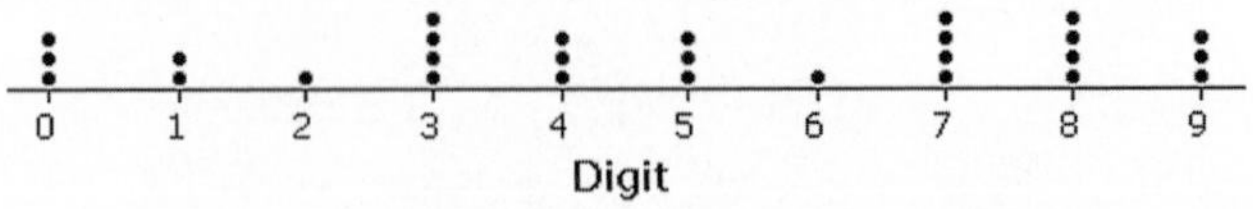

4. The sample size is not large enough to reveal the true nature of the distribution of IQ scores for the population from which the sample is obtained.

8	779
9	66
10	133

5. A time-series graph is best. It suggests that the amounts of carbon monoxide emissions in the United States are increasing.

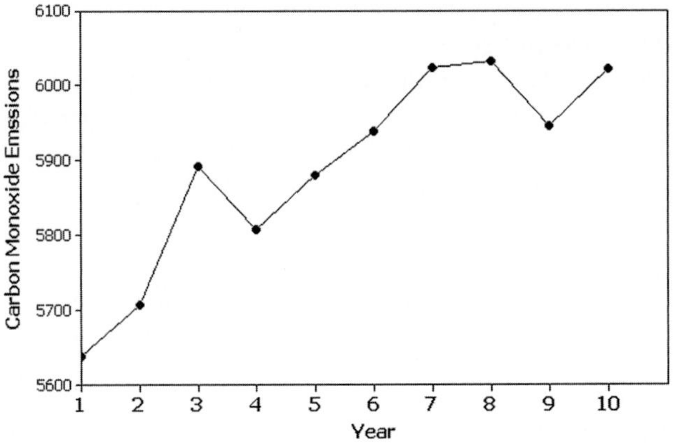

6. A scatterplot is best. The scatterplot does not suggest that there is a relationship.

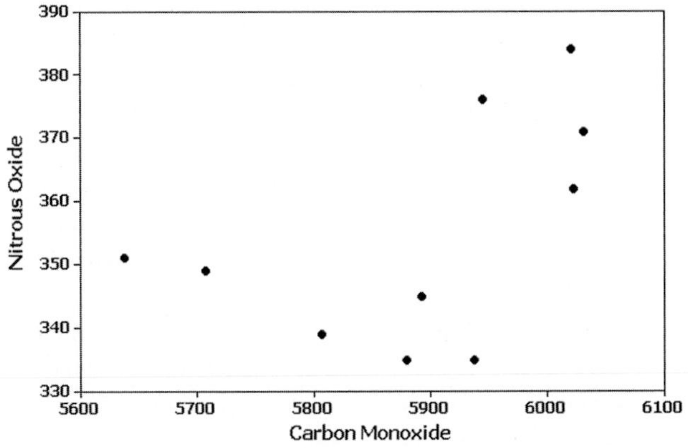

7. A Pareto chart is best.

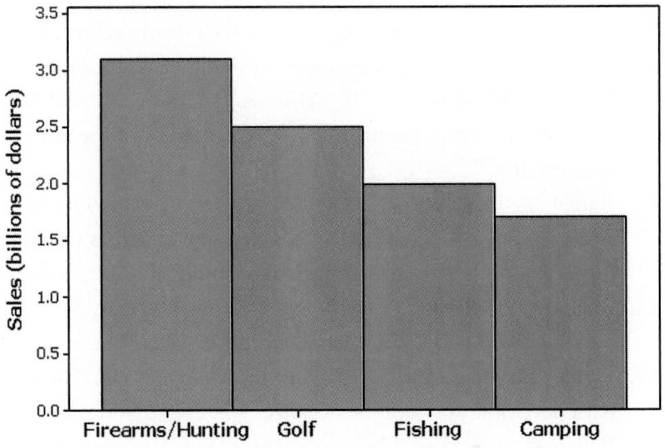

Chapter 2: Cumulative Review Exercises

1. Pareto chart
2. Nominal, because the responses consist of names only. The responses do not measure or count anything, and they cannot be arranged in order according to some quantitative scale.
3. Voluntary response sample (or self-selected sample). The voluntary response sample is not likely to be representative of the population, because those with special interests or strong feelings about the topic are more likely than others to respond, and their views might be very different from those of the general population.
4. By using a vertical scale that does not begin at 0, the graph exaggerates the differences in the numbers of responses. The graph could be modified by starting the vertical scale at 0 instead of 50.
5. 37.6% chose the category of boss. Because it is based on a sample (not on the population), that percentage is a statistic.
6.

Grooming Time (min)	Frequency
0–9	2
10–19	3
20–29	9
30–39	4
40–49	2

7. Because the frequencies increase to a maximum and then decrease, and the left half of the histogram is roughly a mirror image of the right half, the data do appear to be from a population with a normal distribution.

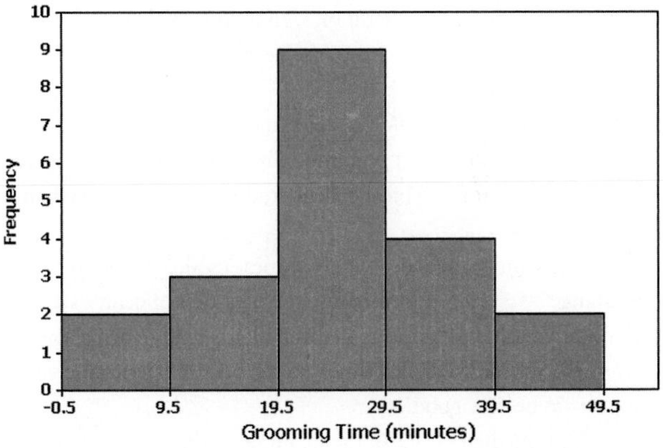

8.

0	05
1	255
2	024555778
3	0055
4	05

Chapter 3: Answers

Section 3-2

1. No. The numbers do not measure or count anything, so the mean would be a meaningless statistic.
3. No. The price exactly in between the highest and lowest is the midrange, not the median.

5. $\bar{x}$ = \$159.8 million; median = \$95.0 million; mode: none; midrange = \$199.5 million. Apart from the obvious and trivial fact that the mean annual earnings of all celebrities is less than \$332 million, nothing meaningful can be known about the mean of the population.
7. $\bar{x}$ = 430.1 hic; median = 393.0 hic; mode: none; midrange = 435.0 hic. The safest of these cars appears to be the Hyundai Elantra. Because the measurements appear to vary substantially from a low of 326 hic to a high of 544 hic, it appears that some small cars are considerably safer than others.
9. $\bar{x}$ = \$16.4 million; median = \$10.0 million; mode: \$4 million, \$9 million, and \$10 million; midrange = \$31 million. The measures of center do not reveal anything about the pattern of the data over time, and that pattern is a key component of a movie's success. The first amount is highest for the opening day when many Harry Potter fans are most eager to see the movie; the third and fourth values are from the first Friday and the first Saturday, which are the popular weekend days when movie attendance tends to spike.
11. $\bar{x}$ = \$59.217; median = \$57.835; mode: none; midrange = \$60.345. None of the measures of center are most important here. The most relevant statistic in this case is the minimum value of \$48.92, because that is the lowest price for the software. Here, we generally care about the lowest price, not the mean price or median price.
13. $\bar{x}$ = 11.05 μg/g; median = 9.50 μg/g; mode: 20.5 μg/g; midrange = 11.75 μg/g. There is not enough information given here to assess the true danger of these drugs, but ingestion of any lead is generally detrimental to good health. All of the decimal values are either 0 or 5, so it appears that the lead concentrations were rounded to the nearest one-half unit of measurement (μg/g).
15. $\bar{x}$ = 6.5 years; median = 4.5 years; mode: 4 and 4.5; midrange = 9.5 years. It is common to earn a bachelor's degree in four years, but the typical college student requires more than four years.
17. $\bar{x}$ = −14.3 min; median = −16.5 min; mode: −32 min; midrange = −10.5 min. Because the measures of center are all negative values, it appears that the flights tend to arrive early *before* the scheduled arrival times, so the on-time performance appears to be very good.
19. $\bar{x}$ = 50.4; median = 73.0; mode: none; midrange = 48.5. The numbers do not measure or count anything; they are simply replacements for names. The data are at the nominal level of measurement, and it makes no sense to compute the measures of center for these data.
21. White drivers: $\bar{x}$ = 73.0 mi/h; median = 73.0 mi/h. African American drivers: $\bar{x}$ = 74.0 mi/h; median = 74.0 mi/h. Although the African American drivers have a mean speed greater than the white drivers, the difference is very small, so it appears that drivers of both races appear to speed about the same amount.
23. Obama: $\bar{x}$ = \$653.9; median = \$452.0. McCain: $\bar{x}$ = \$458.5; median = \$350.0. The contributions appear to favor Obama because his mean and median are substantially higher. With 66 contributions to Obama and 20 contributions to McCain, Obama collected substantially more in total contributions.
25. $\bar{x}$ = 1.184; median = 1.235. Yes, it is an outlier because it is a value that is very far away from all of the other sample values.
27. $\bar{x}$ = 15.0 years; median = 16.0 years. Presidents receive Secret Service protection after they leave office, so the mean is helpful in planning for the cost and resources used for that protection.
29. $\bar{x}$ = 35.8 years. This result is quite close to the mean of 35.9 years found by using the original list of data values.
31. $\bar{x}$ = 84.7. This result is close to the mean of 84.4 found by using the original list of data values.
33. a. 0.6 parts per million b. $n - 1$
35. $\bar{x}$ = 39.070; 10% trimmed mean: 27.677; 20% trimmed mean: 27.176. By deleting the outlier of 472.4, the trimmed means are substantially different from the untrimmed mean.
37. Geometric mean: 1.036711036, or 1.0367 when rounded. Single percentage growth rate: 3.67%. The result is not exactly the same as the mean, which is 3.68%.
39. 34.0 years (rounded from 33.970588 years); the value of 33.0 years is better because it is based on the original data and does not involve interpolation.

Section 3-3

1. The IQ scores of a class of statistics students should have less variation, because those students are a much more homogeneous group with IQ scores that are likely to be closer together.
3. Variation is a general descriptive term that refers to the amount of dispersion or spread among the data values, but the variance refers specifically to the square of the standard deviation.
5. Range = \$265.0 million; s^2 = 10,548.0 (the units are the square of "million dollars"); s = \$102.7 million. Because the data values are the 10 highest from the population, nothing meaningful can be known about the standard deviation of the population.
7. Range = 218.0 hic; s^2 = 7879.8 hic^2; s = 88.8 hic. Although all of the cars are small, the range from 326 hic to 544 hic appears to be relatively large, so the head injury measurements are not about the same.
9. Range = \$54.0 million; s^2 = 210.9 (the units are the square of "million dollars"); s = \$14.5 million. An investor would care about the gross from opening day and the rate of decline after that, but the measures of center and variation are less important.
11. Range = \$22.850; s^2 = 99.141 dollars squared; s = \$9.957. The measures of variation are not very helpful in trying to find the best deal.
13. Range = 17.50 μg/g; s^2 = 41.75 (μg/g)^2; s = 6.46 μg/g. If the medicines contained no lead, all of the measures would be 0 μg/g, and the measures of variation would all be 0 as well.
15. Range = 11.0 years; s^2 = 12.3 years^2; s = 3.5 years. No, because 12 years is within 2 standard deviations of the mean.
17. Range = 43.0 min; s^2 = 231.4 min^2; s = 15.2 min. The standard deviation can never be negative.
19. Range = 79.0; s^2 = 1017.7; s = 31.9. The data are at the nominal level of measurement and it makes no sense to compute the measures of variation for these data.
21. White drivers: 4.0%. African American drivers: 3.7%. The variation is about the same.

23. Obama: 80.0%. McCain: 91.3%. The variation among the Obama contributions is a little less than the variation among the McCain contributions.
25. Range = 2.950; $s^2 = 0.345$; $s = 0.587$.
27. Range = 36.0 years; $s^2 = 94.5$ years^2; $s = 9.7$ years.
29. 0.738, which is not substantially different from $s = 0.587$.
31. 9.0 years, which is reasonably close to $s = 9.7$ years.
33. No. The pulse rate of 99 beats per minute is between the minimum usual value of 54.3 beats per minute and the maximum usual value of 100.7 beats per minute.
35. Yes. The volume of 11.9 oz is not between the minimum usual value of 11.97 oz and the maximum usual value of 12.41 oz.
37. $s = 12.3$ years. The result is not substantially different from the standard deviation of 11.1 years found from the original list of sample values.
39. $s = 13.5$. The result is very close to the standard deviation of 13.4 found from the original list of sample values.
41. a. 95% b. 68%
43. At least 75% of women have platelet counts within 2 standard deviations of the mean. The minimum is 150 and the maximum is 410.
45. a. 6.9 min^2 b. 6.9 min^2 c. 3.4 min^2
    d. Part (b), because repeated samples result in variances that target the same value (6.9 min^2) as the population variance. Use division by $n - 1$.
    e. No. The mean of the sample variances (6.9 min^2) equals the population variance (6.9 min^2), but the mean of the sample standard deviations (1.9 min) does not equal the mean of the population standard deviation (2.6 min).

Section 3-4

1. Madison's height is below the mean. It is 2.28 standard deviations below the mean.
3. The lowest amount is \$5 million, the first quartile Q_1 is \$47 million, the second quartile Q_2 (or median) is \$104 million, the third quartile Q_3 is \$121 million, and the highest gross amount is \$380 million.
5. a. \$1,268,950 b. 0.16 standard deviations
   c. $z = -0.16$ d. Usual
7. a. \$1,449,778 b. 2.75 standard deviations
   c. $z = -2.75$ d. Unusual
9. z scores: −2 and 2. IQ scores: 70 and 130.
11. 0.084 and 2.396
13. De-Fen Yao is relatively taller, because her z score is 12.33, which is greater than the z score of 10.29 for Sultan Kosen. De-Fen Yao is more standard deviations above the mean than Sultan Kosen.
15. The SAT score of 1490 has a z score of −0.09, and the ACT score of 17.0 has a z score of −0.85. The z score of −0.09 is a larger number than the z score of −0.85, so the SAT score of 1490 is relatively better.
17. 13th percentile.
19. 50th percentile.
21. 251 sec (Tech: Excel: 250.8 sec).
23. 255 sec.
25. 247.5 sec.
27. 234.5 sec (Tech: Minitab: 234.25 sec; Excel: 234.75 sec).
29. 5-number summary: 1 sec, 8709 sec, 10,074.5 sec, 11,445 sec, 11,844 sec. (Tech: Minitab yields $Q_1 = 8338$ sec and $Q_3 = 11,453$ sec. Excel yields $Q_1 = 8727.75$ sec and $Q_3 = 11,115$ sec.)

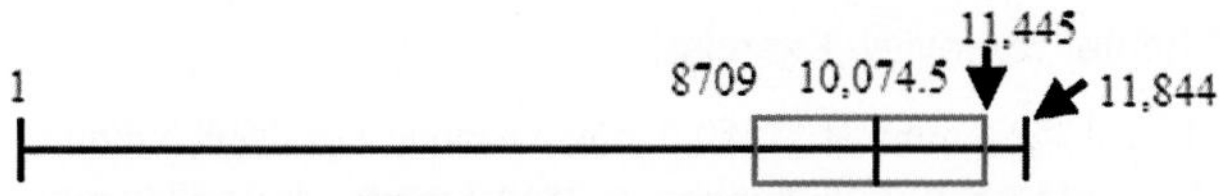

31. 5-number summary: 4 min, 14.0 min, 18.0 min, 32.0 min, 63 min. (Tech: Minitab yields $Q_1 = 12.75$ min and $Q_3 = 34.25$ min. Excel yields $Q_1 = 14.25$ min and $Q_3 = 31.5$ min.)

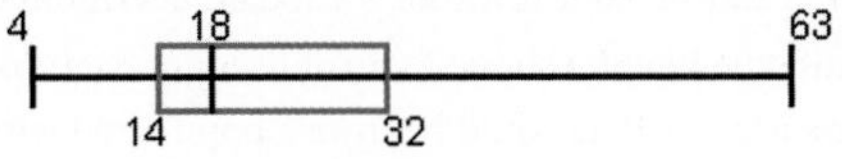

33. The top boxplot represents males. It appears that males have lower pulse rates than females.

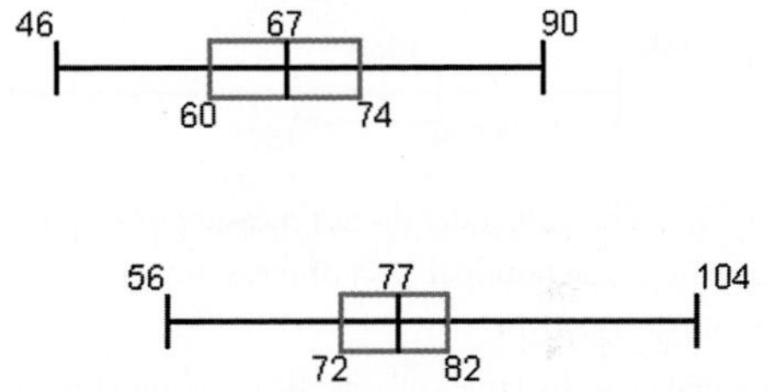

35. The weights of regular Coke represented in the top boxplot appear to be generally greater than those of diet Coke, probably due to the sugar in cans of regular Coke.

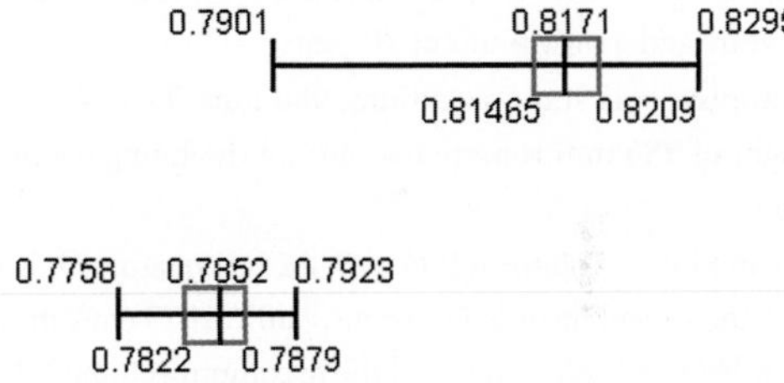

37. Outliers for actresses: 60 years, 61 years, 61 years, 63 years, 74 years, 80 years. Outliers for actors: 76 years. The modified boxplots show that only one actress has an age that is greater than any actor.

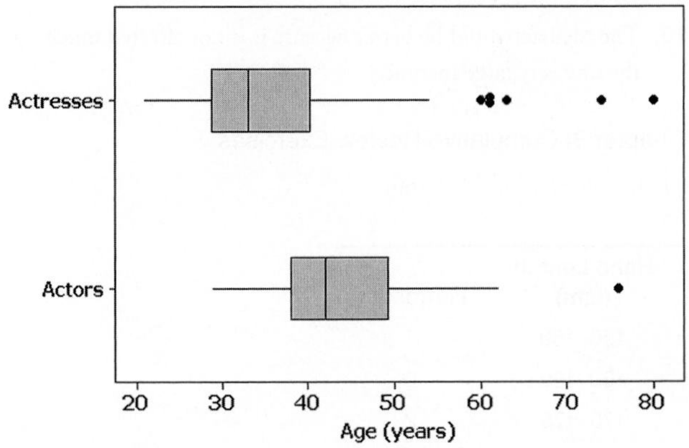

Chapter 3: Quick Quiz

1. 14.0 min.
2. 12.0 min.
3. 12 min.
4. 25.0 min^2.
5. −0.77.

6. Standard deviation; variance; range; mean absolute deviation
7. $\bar{x}$; μ. 8. s, σ, s^2, σ^2. 9. 75%.
10. Minimum, first quartile Q_1, second quartile Q_2 (or median), third quartile Q_3, maximum.

Chapter 3: Review Exercises

1. a. 1559.6 mm; b. 1550.0 mm; c. none; d. 1569.5 mm; e. 145 mm; f. 53.4 mm; g. 2849.3 mm^2; h. 1538.0 mm; i. 1571.0 mm. (Tech: Minitab yields $Q_1 = 1517.5$ mm and $Q_3 = 1606.5$ mm.)
2. $z = 1.54$. The eye height is not unusual because its z score is between 2 and −2, so it is within 2 standard deviations of the mean.
3. Because the boxplot shows a distribution of data that is roughly symmetric, the data could be from a population with a normal distribution, but the data are not necessarily from a population with a normal distribution, because there is no way to determine whether a histogram is roughly bell-shaped.

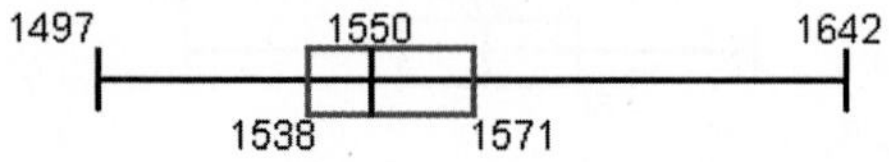

4. 10053.7. The ZIP codes do not measure or count anything. They are at the nominal level of measurement, so the mean is a meaningless statistic.
5. The male has the larger relative BMI because his z score of 0.26 is larger than the z score of 0.08 for the female.
6. The answers vary, but a mean around \$8 or \$9 is reasonable, and a standard deviation around \$1 or \$2 is a reasonable estimate.
7. Answer varies, but $s \approx 12$ years, based on a minimum of 23 years and a maximum of 70 years.
8. Minimum: 842 mm; maximum: 986 mm. The maximum usual height of 986 mm is more relevant for designing overhead bin storage.
9. The minimum volume is 963 cm^3, the first quartile Q_1 is 1034.5 cm^3, the second quartile Q_2 (or median) is 1079 cm^3, the third quartile Q_3 is 1188.5 cm^3, and the maximum volume is 1439 cm^3.

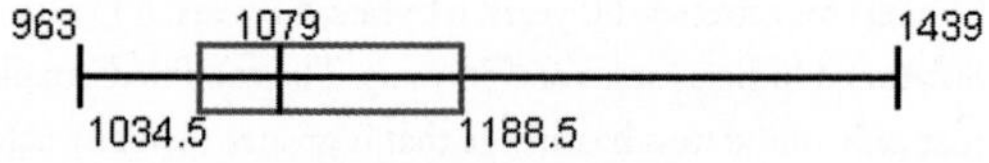

10. The median would be better because it is not affected much by the one very large income.

Chapter 3: Cumulative Review Exercises

1. a. Continuous. b. Ratio.
2.

Hand Length (mm)	Frequency
150–159	1
160–169	0
170–179	2
180–189	0
190–199	3
200–209	1
210–219	1

3.

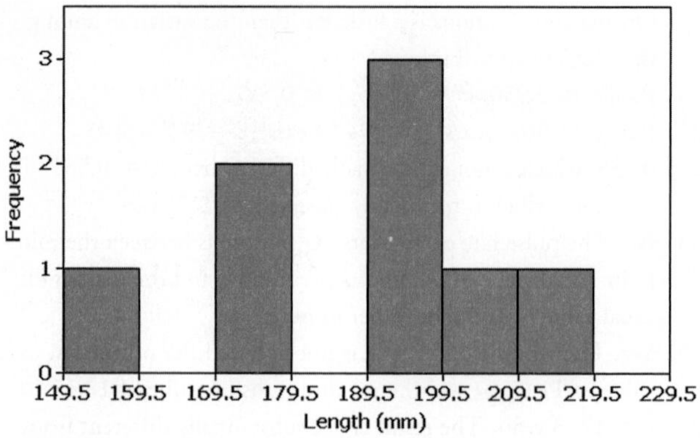

4.

```
15 | 8
16 |
17 | 39
18 |
19 | 569
20 | 7
21 | 4
```

5. a. 190.1 mm; b. 195.5 mm; c. 18.7 mm; d. 348.7 mm^2 e. 56.0 mm.
6. Yes. The frequencies increase to a maximum; then they decrease. Also, the frequencies preceding the maximum are roughly a mirror image of those that follow the maximum.
7. No. Even though the sample is large, it is a voluntary response sample, so the responses cannot be considered to be representative of the population of the United States.
8. The vertical scale does not begin at 0, so the differences among the different outcomes are exaggerated.

Chapter 4 Answers

Section 4-2

1. $P(A) = 1/10{,}000$, or 0.0001. $P(\bar{A}) = 9999/10{,}000$, or 0.9999.
3. Part (c).
5. 5:2; 7/3; −0.9; 456/123
7. 1/5 or 0.2
9. Unlikely; neither unusually low nor unusually high.
11. Unlikely; unusually low.
13. 1/4, or 0.25
15. 1/2, or 0.5 17. 1/5, or 0.2 19. 0
21. 6/1000, or 0.006. The employer would suffer because it would be at risk by hiring someone who uses drugs.
23. 50/1000, or 0.05. This result is not close to the probability of 0.134 for a positive test result.
25. 879/945, or 0.930. Yes, the technique appears to be effective.
27. 0.00000101. No, the probability of being struck is much greater on an open golf course during a thunderstorm. The golfer should seek shelter.
29. a. 1/365 b. Yes c. He already knew. d. 0
31. 0.0767. No, a crash is not unlikely. Given that car crashes are so common, we should take precautions such as not driving after drinking and not using a cell phone or texting.

33. 0.00985. It is unlikely.
35. 0.00993. Yes, it is unlikely. The middle seat lacks an outside view, easy access to the aisle, and a passenger in the middle seat has passengers on both sides instead of on one side only.
37. 3/8, or 0.375
39. {bb, bg, gb, gg}; 1/2, or 0.5.
41. a. brown/brown, brown/blue, blue/brown, blue/blue
b. 1/4 c. 3/4
43. a. 999:1 b. 499:1
c. The description is not accurate. The odds against winning are 999:1 and the odds in favor are 1:999, not 1:1000.
45. a. $16 b. 8:1 c. About 9.75:1, which becomes 39:4
d. $21.50
47. Relative risk: 0.939; odds ratio: 0.938; the probability of a headache with Nasonex (0.0124) is slightly less than the probability of a headache with the placebo (0.0132), so Nasonex does not appear to pose a risk of headaches.
49. 1/4

Section 4-3

1. Based on the rule of complements, the sum of $P(A)$ and $P(\overline{A})$ must always be 1, so that sum cannot be 0.5.
3. Because it is possible to select someone who is a male and a Republican, events M and R are not disjoint. Both events can occur at the same time when someone is randomly selected.
5. Disjoint. 7. Not disjoint. 9. Disjoint.
11. Not disjoint. 13. 0.53
15. $P(\overline{D}) = 0.450$, where $P(\overline{D})$ is the probability of randomly selecting someone who does not choose a direct in-person encounter as the most fun way to flirt.
17. 1 19. 0.956
21. 13/28, or 0.464. That probability is not as high as it should be.
23. 16/28 or 4/7 or 0.571
25. a. 0.786 b. 0.143
c. The physicians given the labels with concentrations appear to have done much better. The results suggest that labels described as concentrations are much better than labels described as ratios.
27. 156/1205 = 0.129. Yes. A high refusal rate results in a sample that is not necessarily representative of the population, because those who refuse may well constitute a particular group with opinions different from others.
29. 1060/1205 = 0.880
31. 1102/1205 = 0.915
33. a. 300 b. 178 c. 178/300 = 0.593
35. 0.603
37. 27/300 = 0.090. With an error rate of 0.090 (or 9%), the test does not appear to be highly accurate.
39. 3/4, or 0.75
41. $P(A \text{ or } B) = P(A) + P(B) - 2P(A \text{ and } B)$
43. a. $1 - P(A) - P(B) + P(A \text{ and } B)$
b. $1 - P(A \text{ and } B)$
c. No

Section 4-4

1. The probability that the second selected senator is a Democrat given that the first selected senator was a Republican.
3. False. The events are dependent because the radio and air conditioner are both powered by the same electrical system. If you find that your car's radio does not work, there is a greater probability that the air conditioner will also not work.
5. a. Dependent b. 1/132, or 0.00758
7. a. Independent b. 1/12, or 0.0833
9. a. Independent b. 0.000507
11. a. Dependent b. 0.00586
13. a. 0.0081. Yes, it is unlikely.
b. 0.00802. Yes, it is unlikely.
15. a. 0.739. No, it is not unlikely.
b. 0.739. No, it is not unlikely.
17. 0.838. No, the entire batch consists of malfunctioning pacemakers.
19. a. 0.02 b. 0.0004 c. 0.000008
d. By using one backup drive, the probability of failure is 0.02, and with three independent disk drives, the probability drops to 0.000008. By changing from one drive to three, the likelihood of failure drops from 1 chance in 50 to only 1 chance in 125,000, and that is a very substantial improvement in reliability. Back up your data!
21. a. 1/365, or 0.00274
b. 0.00000751
c. 1/365, or 0.00274
23. 0.828. No, it is not unlikely.
25. 0.000454. Yes, it is unlikely.
27. a. 0.900
b. 0.00513 (using the 5% guideline for cumbersome calculations).
29. a. 0.143 (not 0.144)
b. 0.00848 (using the 5% guideline for cumbersome calculations).
31. a. 0.9999 b. 0.9801
c. The series arrangement provides better protection.

Section 4-5

1. a. Answer varies, but 0.98 is a reasonable estimate.
b. Answer varies, but 0.999 is a reasonable estimate.
3. The probability that the polygraph indicates lying given that the subject is actually telling the truth.
5. At least one of the five children is a boy. 31/32, or 0.969.
7. None of the digits is 0; 0.656.
9. 0.893. The chance of passing is reasonably good.
11. 0.5 13. 0.965
15. 0.122. Given that the three cars are in the same family, they are not randomly selected and there is a good chance that the family members have similar driving habits, so the probability might not be accurate.
17. 0.988. It is very possible that the result is not valid because it is based on data from a voluntary response survey.
19. 90/950, or 0.0947. This is the probability of the test making it appear that the subject uses drugs when the subject is not a drug user.
21. 6/866, or 0.00693. This result is substantially different from the result found in Exercise 20. The probabilities P(subject uses drugs | negative test result) and P(negative test result | subject uses drugs) are not equal.

23. 44/134, or 0.328
25. a. 1/3, or 0.333
    b. 0.5
27. 0.5
29. a. 0.9996 b. 0.999992
31. 0.684. The probability is not low, so further testing of the individual samples will be necessary for about 68% of the combined samples.
33. a. 0.431 b. 0.569
35. a. 0.0748 b. 0.8
    c. The estimate of 75% is dramatically greater than the actual rate of 7.48%. They exhibited confusion of the inverse. A consequence is that they would unnecessarily alarm patients who are benign, and they might start treatments that are not necessary.

Section 4-6

1. The symbol ! is the factorial symbol that represents the product of decreasing whole numbers, as in $4! = 4 \cdot 3 \cdot 2 \cdot 1 = 24$. Four people can stand in line 24 different ways.
3. Because repetition is allowed, numbers are selected *with replacement*, so neither of the two permutation rules applies. The fundamental counting rule can be used to show that the number of possible outcomes is $10 \cdot 10 \cdot 10 \cdot 10 = 10{,}000$, so the probability of winning is 1/10,000.
5. 1/10,000
7. 1/362,880
9. 17,383,860. Because that number is so large, it is not practical to make a different CD for each possible combination.
11. 1/5,527,200. No, 5,527,200 is too many possibilities to list.
13. 34,650
15. 1/7,059,052
17. 1/24
19. a. 1/749,398 b. 1/10,000 c. $10,000
21. a. 11,880 b. 495 c. 1/495
23. 125,000. The fundamental counting rule can be used. The different possible codes are ordered sequences of numbers, not combinations, so the name of "combination lock" is not appropriate. Given that "fundamental counting rule lock" is a bit awkward, a better name would be something like "number lock."
25. 120; AMITY; 1/120
27. 26 29. 10 31. 64
33. 1/195,249,054
35. 2/252, or 1/128. Yes, if everyone treated is of one sex while everyone in the placebo group is of the opposite sex, you would not know if different reactions are due to the treatment or sex.
37. 2,095,681,645,538 (about 2 trillion)
39. 12

Chapter 4: Quick Quiz

1. 0 2. 0.7 3. 1 4. 0.04
5. Answer varies, but an answer such as 0.01 or lower is reasonable.
6. 512/839, or 0.610
7. 713/839, or 0.850
8. 126/839, or 0.150
9. 0.0224 (not 0.0226)
10. 126/350, or 0.360

Chapter 4: Review Exercises

1. 0.438 2. 0.410 3. 0.806
4. It appears that you have a substantially better chance of avoiding prison if you enter a guilty plea.
5. 0.986 6. 0.191 7. 0.00484
8. 0.619 9. 0.381 10. 0.0136
11. Answer varies, but DuPont data show that about 8% of cars are red, so any estimate between 0.01 and 0.2 would be reasonable.
12. a. 0.65
    b. 0.0150
    c. Yes, because the probability is so small (0.0150).
13. a. 1/365
    b. 31/365
    c. Answer varies, but it is probably small, such as 0.02.
    d. Yes
14. 0.0211. No. 15. 1/5,245,786
16. 1/575,757 17. 1/1000
18. 1320; 1/1320

Chapter 4: Cumulative Review Exercises

1. a. The mean of −8.9 years is not close to the value of 0 years that would be expected with no gender discrepancy.
   b. The median of −13.5 years is not close to the value of 0 years that would be expected with no gender discrepancy.
   c. $s = 10.6$ years
   d. $s^2 = 113.2$ years^2
   e. $Q_1 = -15.0$ years
   f. $Q_3 = -5.0$ years
   g. The boxplot suggests that the data have a distribution that is skewed.

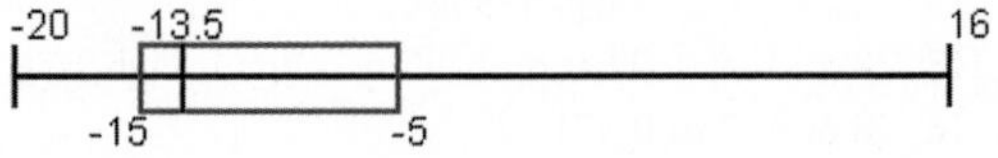

2. a. No. The pulse rate of 100 beats per minute is within 2 standard deviations of the mean, so it is not unusual.
   b. Yes. The pulse rate of 50 beats per minute is more than 2 standard deviations away from the mean, so it is unusual.
   c. Yes, because the probability of 1/256 (or 0.00391) is so small.
   d. No, because the probability of 1/8 (or 0.125) is not very small.
3. a. 46% b. 0.460 c. Stratified sample
4. The graph is misleading because the vertical scale does not start at 0. The vertical scale starts at the frequency of 500 instead of 0, so the difference between the two response rates is exaggerated. The graph incorrectly makes it appear that "no" responses occurred about 60 times more often than the number of "yes" responses, but comparison of the actual frequencies shows that the "no" responses occurred about four times more often than the number of "yes" responses.
5. a. Convenience sample
   b. If the students at the college are mostly from a surrounding region that includes a large proportion of one ethnic group, the results will not reflect the general population of the United States.
   c. 0.75 d. 0.64

6. The straight-line pattern of the points suggests that there is a correlation between chest size and weight.

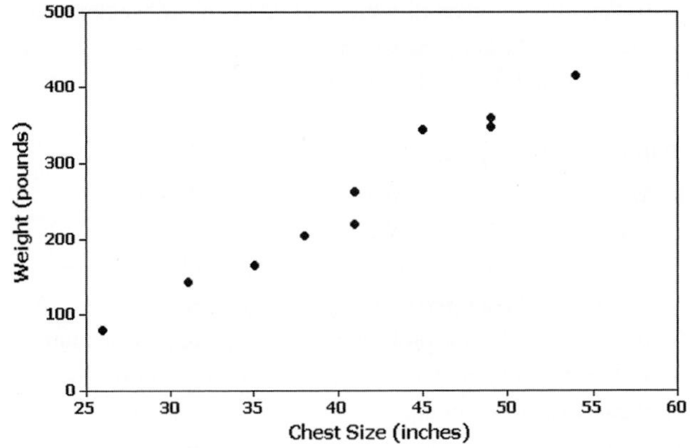

7. a. 1/575,757 b. 1/19 c. 1/10,939,383

Chapter 5: Answers

Section 5-2

1. The random variable is x, which is the number of girls in three births. The possible values of x are 0, 1, 2, and 3. The values of the random value x are numerical.
3. Table 5-7 does describe a probability distribution because the three requirements are satisfied. First, the variable x is a numerical random variable and its values are associated with probabilities. Second, $\Sigma P(x) = 0.125 + 0.375 + 0.375 + 0.125 = 1$ as required. Third, each of the probabilities is between 0 and 1 inclusive, as required.
5. a. Continuous random variable. b. Discrete random variable. c. Not a random variable. d. Discrete random variable. e. Continuous random variable. f. Discrete random variable.
7. Probability distribution with $\mu = 2.0$, $\sigma = 1.0$.
9. Not a probability distribution because the sum of the probabilities is 0.601, which is not 1 as required. Also, Ted clearly needs a new approach.
11. Probability distribution with $\mu = 2.2$, $\sigma = 1.0$. (The sum of the probabilities is 0.999, but that is due to rounding errors.)
13. Not a probability distribution because the responses are not values of a numerical random variable. Also, the sum of the probabilities is 1.18 instead of being 1 as required.
15. $\mu = 5.0$, $\sigma = 1.6$.
17. a. 0.044 b. 0.055
    c. The probability from part (b).
    d. No, because the probability of 8 or more girls is 0.055, which is not very low (less than or equal to 0.05).
19. $\mu = 0.9$ car, $\sigma = 0.9$ car
21. a. 0.041 b. 0.046 c. The probability from part (b).
    d. Yes, because the probability of three or more failures is 0.046, which is very low (less than or equal to 0.05).
23. a. 1000 b. 1/1000 c. \$499 d. −50¢
    e. The \$1 bet on the pass line in craps is better because its expected value of −1.4¢ is much greater than the expected value of −50¢ for the Texas Pick 3 lottery.
25. a. −39¢
    b. The bet on the number 27 is better because its expected value of −26¢ is greater than the expected value of −39¢ for the other bet.

Section 5-3

1. The given calculation assumes that the first two adults include Wal-Mart and the last three adults do not include Wal-Mart, but there are other arrangements consisting of two adults who include Wal-Mart and three who do not. The probabilities corresponding to those other arrangements should also be included in the result.
3. Because the 30 selections are made without replacement, they are dependent, not independent. Based on the 5% guideline for cumbersome calculations, the 30 selections can be treated as being independent. (The 30 selections constitute 3% of the population of 1000 responses, and 3% is not more than 5% of the population.) The probability can be found by using the binomial probability formula.
5. Not binomial. Each of the weights has more than two possible outcomes.
7. Binomial.
9. Not binomial. Because the senators are selected without replacement, the selections are not independent. (The 5% guideline for cumbersome calculations cannot be applied because the 40 selected senators constitute 40% of the population of 100 senators, and 40% exceeds 5%.)
11. Binomial. Although the events are not independent, they can be treated as being independent by applying the 5% guideline. The sample size of 380 is not more than 5% of the population of all smartphone users.
13. a. 0.128
    b. WWC, WCW, CWW; 0.128 for each
    c. 0.384
15. 0.051 17. 0.057 19. 0.328
21. 0.257 23. 0.00125 25. 0.996; yes
27. 0.037; yes, because the probability of 2 or fewer peas with green pods is small (less than or equal to 0.05).
29. a. 0.002 (Tech: 0.00154)
    b. 0+ (Tech: 0.000064)
    c. 0.002 (Tech: 0.00160).
    d. Yes, the small probability from part (c) suggests that 5 is an unusually high number.
31. a. 0.328 b. 0.410 c. 0.738 (Tech: 0.737)
    d. No, the probability from part (c) is not small, so 1 is not an unusually low number.
33. 0.101. No, because the probability of exactly 12 is 0.101, the probability of 12 or more is greater than 0.101, so the probability of getting 12 or more is not very small, so 12 is not unusually high.
35. 0.287. No, because the flights all originate from New York, they are not randomly selected flights, so the 80.5% on-time rate might not apply.
37. a. 0.000766 b. 0.999 c. 0.00829
    d. Yes, the very low probability of 0.00829 would suggest that the 45 share value is wrong.

39. a. 0.000854 b. 0.0000610 c. 0.000916
    d. Yes. The probability of getting 13 girls or a result that is more extreme is 0.000916, so chance does not appear to be a reasonable explanation for the result of 13 girls. Because 13 is an unusually high number of girls, it appears that the probability of a girl is higher with the XSORT method, and it appears that the XSORT method is effective.
41. 0.134. It is not unlikely for such a combined sample to test positive.
43. 0.662. The probability shows that about 2/3 of all shipments will be accepted. With about 1/3 of the shipments rejected, the supplier would be wise to improve quality.
45. 0.0468
47. a. 0.000969 b. 0.0000000715 c. 0.436

Section 5-4

1. $n = 270, p = 0.07, q = 0.93$
3. 9.4 executives^2 (or 9.6 executives^2 if the rounded standard deviation of 3.1 executives is used)
5. $\mu = 12.0$ correct guesses; $\sigma = 3.1$ correct guesses; minimum = 5.8 correct guesses; maximum = 18.2 correct guesses.
7. $\mu = 668.6$ worriers; $\sigma = 15.1$ worriers; minimum = 638.4 worriers; maximum = 698.8 worriers.
9. a. $\mu = 145.5; \sigma = 8.5$
   b. Yes. Using the range rule of thumb, the minimum usual value is 128.5 boys and the maximum usual value is 162.5 boys. Because 239 boys is above the range of usual values, it is unusually high. Because 239 boys is unusually high, it does appear that the YSORT method of gender selection is effective.
11. a. $\mu = 20.0, \sigma = 4.0$
    b. No, because 25 orange M&Ms is within the range of usual values (12 to 28). The claimed rate of 20% does not necessarily appear to be wrong, because that rate will usually result in 12 to 28 orange M&Ms (among 100), and the observed number of orange M&Ms is within that range.
13. a. $\mu = 142.8, \sigma = 11.9$
    b. No, 135 is not unusually low or high because it is within the range of usual values (119.0 to 166.6).
    c. Based on the given results, cell phones do not pose a health hazard that increases the likelihood of cancer of the brain or nervous system.
15. a. $\mu = 156.0; \sigma = 12.1$
    b. The minimum usual frequency is 131.8 and the maximum is 180.2. The occurrence of *r* 178 times is not unusually low or high because it is within the range of usual values (131.8 to 180.2).
17. a. $\mu = 74.0; \sigma = 7.7$
    b. The minimum usual number is 58.6 and the maximum usual value is 89.4. The value of 90 is unusually high because it is above the range of usual values (58.6 to 89.4).
19. a. $\mu = 0.0821918; \sigma = 0.2862981$
    b. The minimum usual number is −0.4904044 and the maximum usual number is 0.654788. The results of 2 students born on the 4th of July would be unusually high, because 2 is above of the range of usual values (from −0.4904044 to 0.654788).
21. $n = 150; p = 0.4$, so that 40% of the surveyed subjects could identify at least one member of the Supreme Court; $q = 0.6$, so that 60% of surveyed subjects could not identify at least one member of the Supreme Court.
23. $\mu = 3.0$ and $\sigma = 1.3$ (not 1.5)

Chapter 5: Quick Quiz

1. Yes 2. 20.0 3. 4.0
4. Yes 5. No
6. Yes. (The sum of the probabilities is 0.999 and it can be considered to be 1 because of rounding errors.)
7. 0+ indicates that the probability is a very small positive number. It does not indicate that it is impossible for none of the five flights to arrive on time.
8. 0.945 9. Yes 10. No

Chapter 5: Review Exercises

1. 0.047 or 0.0467 2. 0.138
3. $\mu = 240.0; \sigma = 12.0$. Range of usual values: 216 to 264. The result of 200 with brown eyes is unusually low.
4. The probability of $P(239 \text{ or fewer}) = 0.484$ is relevant for determining whether 239 is an unusually low number. Because that probability is not very small, it appears that 239 is not an unusually low number of people with brown eyes.
5. Yes, the three requirements are satisfied. There is a numerical random variable *x* and its values are associated with corresponding probabilities. $\Sigma P(x) = 1.001$, so the sum of the probabilities is 1 when we allow for a small discrepancy due to rounding. Also, each of the probability values is between 0 and 1 inclusive.
6. $\mu = 0.4; \sigma = 0.6$. Range of usual values: −0.8 to 1.6 (or 0 to 1.6). Yes, three is an unusually high number of males with tinnitus among four randomly selected males.
7. $\Sigma P(x) = 0.902$, so the sum of the probabilities is not 1 as required. Because the three requirements are not all satisfied, the given information does not describe a probability distribution.
8. $315,075. Because the offer is well below her expected value, she should continue the game (although the guaranteed prize of $193,000 had considerable appeal). (She accepted the offer of $193,000, but she would have won $500,000 if she continued the game and refused all further offers.)
9. a. 1.2¢ b. 1.2¢ minus cost of stamp.
10. a. 0.0104
    b. 0.0821
    c. 0.0000439

Chapter 5: Cumulative Review Exercises

1. a. 24.4 hours b. 24.2 hours c. 4.7 hours
   d. 1.7 hours e. 2.9 hours^2
   f. Usual values: 21.0 hours to 27.8 hours.
   g. No, because none of the times are beyond the range of usual values.
   h. Ratio i. Continuous
   j. The given times come from countries with very different population sizes, so it does not make sense to treat the given times equally. Calculations of statistics should take the different population sizes into account. Also, the

sample is very small, and there is no indication that the sample is random.

2. a. 1/10,000 or 0.0001

b.

x	P(x)
−$1	0.9999
$4999	0.0001

c. 0.0365 d. 0.0352 e. -50¢

3. a. 0.282 b. 0.303 c. 0.242 d. 0.297
e. 0.0792 f. 0.738 g. 0.703

4. Because the vertical scale begins at 60 instead of 0, the difference between the two amounts is exaggerated. The graph makes it appear that men's earnings are roughly twice those of women, but men earn roughly 1.2 times the earnings of women.

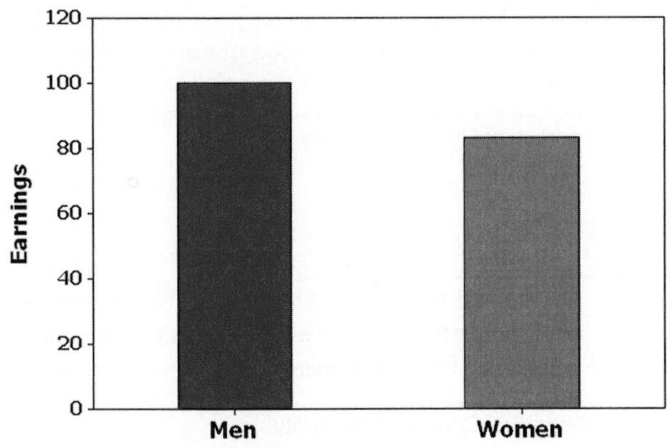

5. a. Frequency distribution or frequency table.
b. Probability distribution.
c. $\bar{x} = 4.7$. This value is a statistic.
d. $\mu = 4.5$. This value is a parameter.
e. The random generation of 1000 digits should have a mean close to $\mu = 4.5$ from part (d). The mean of 4.5 is the mean for the population of all random digits, so samples will have means that tend to center about 4.5.

6. a. 0.0514 b. 0.815
c. This is a voluntary response (or self-selected) sample. This suggests that the results might not be valid, because those with a strong interest in the topic are more likely to respond.

Chapter 6 Answers

Section 6-2

1. The word "normal" has a special meaning in statistics. It refers to a specific bell-shaped distribution that can be described by Formula 6-1.
3. The mean and standard deviation have the values of $\mu = 0$ and $\sigma = 1$.
5. 0.75 7. 0.4 9. 0.6700
11. 0.6992 (Tech: 0.6993)
13. 1.23 15. −1.45 17. 0.0207
19. 0.9901 21. 0.2061 23. 0.9332
25. 0.2957 (Tech: 0.2956)
27. 0.0198 29. 0.9799 31. 0.9825
33. 0.9999 35. 0.5000 37. 1.28
39. −1.96, 1.96 41. 1.96 43. 1.645
45. 68.26% (Tech: 68.27%)
47. 99.74% (Tech: 99.73%)
49. a. 68.26% (Tech: 68.27%)
b. 4.56% c. 95.00%
d. 95.44% (Tech: 95.45%)
e. 0.26% (Tech: 0.27%)

Section 6-3

1. a. $\mu = 0; \sigma = 1$
b. The z scores are numbers without units of measurement.
3. The standard normal distribution has a mean of 0 and a standard deviation of 1, but a nonstandard normal distribution has a different value for one or both of those parameters.
5. 0.8849 7. 0.9053 9. 136
11. 69 13. 0.1587
15. 0.4972 (Tech: 0.4950)
17. 119 19. 110
21. a. 75.48% (Tech: 75.56%). Yes, about 25% of women are not qualified because of their heights.
b. 99.90% (Tech: 99.89%). No, only about 0.1% of men are not qualified because of their heights.
c. 58.5 in. to 69.1 in.
d. 63.9 in. to 75.1 in.
23. a. 99.86%
b. 98.89% (Tech: 98.90%)
c. 59.5 in. to 73.4 in.
25. a. 0.4129 (Tech: 0.4137)
b. 25 c. 19
d. The mean weight is increasing over time, so safety limits must be periodically updated to avoid an unsafe condition.
27. a. 0.0038; either a very rare event occurred or the husband is not the father.
b. 240 days
29. a. 91.77% (Tech: 91.78%)
b. 0.01% (Tech: 0.00%)
c. 2.150. No.
31. $P_1 = 17.9$ chocolate chips (Tech: 18.0 chocolate chips); $P_{99} = 30.1$ chocolate chips (Tech: 30.0 chocolate chips). The values can be used to identify cookies with an unusually low number of chocolate chips or an unusually high number of chocolate chips, so those numbers can be used to monitor the production process to ensure that the numbers of chocolate chips stay within reasonable limits.
33. a. The mean is 67.25 (67.3 rounded) beats per minute and the standard deviation is 10.334781 (10.3 rounded) beats per minute. A histogram confirms that the distribution is roughly normal.
b. 47.0 beats per minute; 87.5 beats per minute
35. a. 75; 10
b. No, the conversion should also account for variation.
c. B grade: 45.2 to 52.8
d. Use a scheme like the one given in part (c), because variation is included in the curving process.
37. 0.0444 (Tech: 0.0430).

Section 6-4

1. a. The sample means will tend to center about the population parameter of 5.67 g.
   b. The sample means will tend to have a distribution that is approximately normal.
   c. The sample proportions will tend to have a distribution that is approximately normal.
3. Sample mean; sample variance; sample proportion
5. No. The sample is not a simple random sample from the population of all college statistics students. It is very possible that the students at Broward College do not accurately reflect the behavior of all college statistics students.
7. a. 4.7
   b.

Sample Variance s^2	Probability
0.0	3/9
0.5	2/9
8.0	2/9
12.5	2/9

   c. 4.7
   d. Yes. The mean of the sampling distribution of the sample variances (4.7) is equal to the value of the population variance (4.7), so the sample variances target the value of the population variance.
9. a. 5
   b.

Sample Median	Probability
4.0	1/9
4.5	2/9
5.0	1/9
6.5	2/9
7.0	2/9
9.0	1/9

   c. 6.0
   d. No. The mean of the sampling distribution of the sample medians is 6.0, and it is not equal to the value of the population median (5), so the sample medians do not target the value of the population median.
11. a.

$\bar{x}$	Probability
46	1/16
47.5	2/16
49	1/16
51	2/16
52	2/16
52.5	2/16
53.5	2/16
56	1/16
57	2/16
58	1/16

   b. The mean of the population is 52.25 and the mean of the sample means is also 52.25.
   c. The sample means target the population mean. Sample means make good estimators of population means because they target the value of the population mean instead of systematically underestimating or overestimating it.
13. a.

Range	Probability
0	4/16
2	2/16
3	2/16
7	2/16
9	2/16
10	2/16
12	2/16

   b. The range of the population is 12, but the mean of the sample ranges is 5.375. Those values are not equal.
   c. The sample ranges do not target the population range of 12, so sample ranges do not make good estimators of population ranges.
15.

Proportion of Girls	Probability
0	0.25
0.5	0.50
1	0.25

Yes. The proportion of girls in 2 births is 0.5, and the mean of the sample proportions is 0.5. The result suggests that a sample proportion is an unbiased estimator of a population proportion.
17. a.

Proportion Correct	Probability
0	16/25
0.5	8/25
1	1/25

   b. 0.2
   c. Yes. The sampling distribution of the sample proportions has a mean of 0.2 and the population proportion is also 0.2 (because there is 1 correct answer among 5 choices). Yes, the mean of the sampling distribution of the sample proportions is always equal to the population proportion.
19. The formula yields $P(0) = 0.25$, $P(0.5) = 0.5$, and $P(1) = 0.25$, which does describe the sampling distribution of the sample proportions. The formula is just a different way of presenting the same information in the table that describes the sampling distribution.

Section 6-5

1. Because $n > 30$, the sampling distribution of the mean ages can be approximated by a normal distribution with mean μ and standard deviation $\sigma/\sqrt{40}$.
3. $\mu_{\bar{x}} = 60.5$ cm and it represents the mean of the population consisting of all sample means. $\sigma_{\bar{x}} = 1.1$ cm, and it represents the standard deviation of the population consisting of all sample means.
5. a. 0.9772
   b. 0.8888 (Tech: 0.8889)

7. a. 0.0668
b. 0.6985 (Tech: 0.6996)
c. Because the original population has a normal distribution, the distribution of sample means is normal for any sample size.
9. a. 0.9974 (Tech: 0.9973)
b. 0.5086 (Tech: 0.5085)
11. 0.1112 (Tech: 0.1121). The elevator does not appear to be safe because there is a reasonable chance (0.1112) that it will be overloaded with 16 male passengers.
13. a. 0.9787 (Tech: 0.9788)
b. 21.08 in. to 24.22 in.
c. 0.9998. No, the hats must fit individual women, not the mean from 64 women. If all hats are made to fit head circumferences between 22.00 in. and 23.00 in., the hats won't fit about half of those women.
15. a. 140 lb
b. 0.9999 (Tech: 0.99999993, or 1.0000 when rounded to four decimal places)
c. 0.8078 (Tech: 0.8067)
d. Given that there is a 0.8078 probability of exceeding the 3500 lb limit when the water taxi is loaded with 20 random men, the new capacity of 20 passengers does not appear to be safe enough because the probability of overloading is too high.
17. a. 0.6517 (Tech: 0.6516)
b. 0.9115
c. There is a high probability (0.9115) that the gondola will be overloaded if it is occupied by 12 men, so it appears that the number of allowed passengers should be reduced.
19. a. 0.5526 (Tech: 0.5517)
b. 0.9994 (Tech: 0.9995)
c. Part (a) because the ejection seats will be occupied by individual women, not groups of women.
21. a. 0.8508 (Tech: 0.8512)
b. 0.9999 (Tech: 1.0000 when rounded to four decimal places)
c. The probability from part (a) is more relevant because it shows that 85.08% of male passengers will not need to bend. The result from part (b) gives us information about the mean for a group of 100 men, but it doesn't give us useful information about the comfort and safety of individual male passengers.
d. Because men are generally taller than women, a design that accommodates a suitable proportion of men will necessarily accommodate a greater proportion of women.
23. a. Yes. The sampling is without replacement and the sample size of $n = 50$ is greater than 5% of the finite population size of 275. $\sigma_{\bar{x}} = 2.0504584$.
b. 0.5947 (Tech: 0.5963)
25. a. $\mu = 6.0$ and $\sigma = 2.1602469$
b. 4.5, 4.5, 6.5, 6.5, 7.0, 7.0
c. $\mu_{\bar{x}} = 6.0$ and $\sigma_{\bar{x}} = 1.0801235$
d. $\mu_{\bar{x}} = \mu = 6.0$ and $\sigma_{\bar{x}} = \frac{2.1602469}{\sqrt{2}}\sqrt{\frac{3-2}{3-1}} = 1.0801235$, which is the same result from part (c).

Section 6-6

1. The histogram should be approximately bell-shaped, and the normal quantile plot should have points that approximate a straight-line pattern.
3. We must verify that the sample is from a population having a normal distribution. We can check for normality using a histogram, identifying the number of outliers, and constructing a normal quantile plot.
5. Not normal. The points show a systematic pattern that is not a straight-line pattern.
7. Normal. The points are reasonably close to a straight-line pattern, and there is no other pattern that is not a straight-line pattern.
9. Not normal
11. Normal
13. Not normal

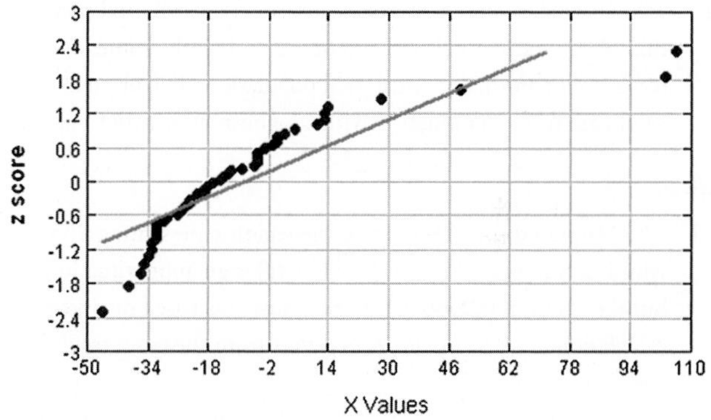

15. Normal

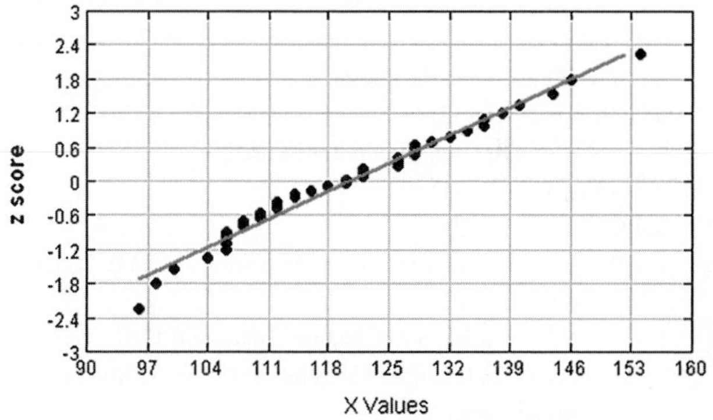

17. Normal. The points have coordinates (131, −1.28), (134, −0.52), (139, 0), (143, 0.52), (145, 1.28).

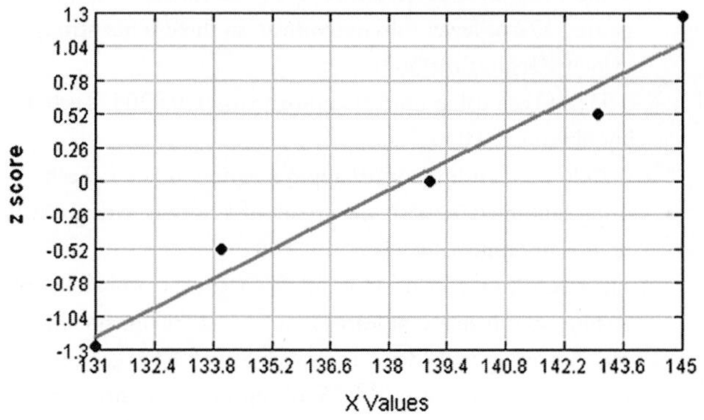

19. Not normal. The points have coordinates (1034, −1.53), (1051, −0.89), (1067, −0.49), (1070, −0.16), (1079, 0.16), (1079, 0.49), (1173, 0.89), (1272, 1.53).

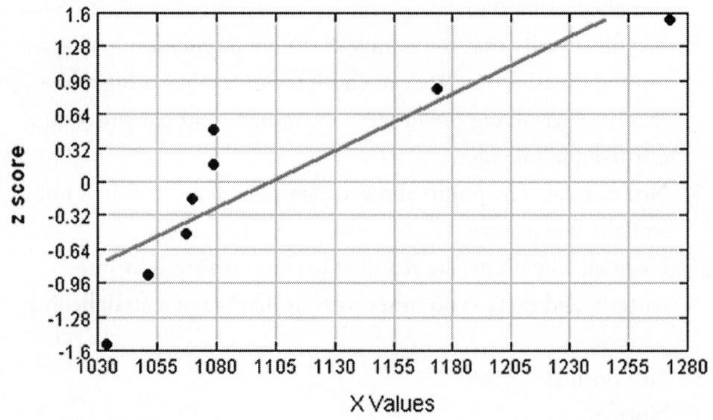

21. a. Yes b. Yes c. No
23. The original values are not from a normally distributed population. After taking the logarithm of each value, the values appear to be from a normally distributed population. The original values are from a population with a lognormal distribution.

Section 6-7

1. The Minitab display shows that the region representing 235 wins is a rectangle. The result of 0.0068 is an approximation, but the result of 0.0066 is better because it is based on an exact calculation. The approximation differs from the exact result by a very small amount.
3. $p = 0.2$; $q = 0.8$; $\mu = 5$; $\sigma = 2$. The value of $\mu = 5$ shows that for many people who make random guesses for the 25 questions, the mean number of correct answers is 5. For many people who make random guesses, the standard deviation of $\sigma = 2$ is a measure of how much the numbers of correct responses vary.
5. 0.0630 (Tech: 0.0632)
7. Normal approximation should not be used.
9. 0.2743 (Tech: 0.2731)
11. 0.0928 (Tech: 0.0933)
13. a. 0.0219 (Tech using normal approximation: 0.0214; Tech using binomial: 0.0217)
    b. 0.1711 (Tech using normal approximation: 0.1702; Tech using binomial: 0.1703). The result of 172 overturned calls is not unusually low.
    c. The result from part (b) is useful. We want the probability of getting a result that is at least as extreme as the one obtained.
    d. If the 30% rate is correct, there is a good chance (0.1711) of getting 172 or fewer calls overturned, so there is not strong evidence against the 30% rate.
15. a. 0.0318 (Tech using normal approximation: 0.0305; Tech using binomial: 0.0301)
    b. 0.2676 (Tech using normal approximation: 0.2665; Tech using binomial: 0.2650). The result of 428 peas with green pods is not unusually low.
    c. The result from part (b) is useful. We want the probability of getting a result that is at least as extreme as the one obtained.
    d. No. Assuming that Mendel's probability of 3/4 is correct, there is a good chance (0.2676) of getting the results that were obtained. The obtained results do not provide strong evidence against the claim that the probability of a pea having a green pod is 3/4.
17. a. 0.0000 or 0+ (a very small positive probability that is extremely close to 0)
    b. 0.0001 (Tech: 0.0000 or 0+, which is a very small positive probability that is extremely close to 0). If boys and girls are equally likely, 879 girls in 945 births is unusually high.
    c. The result from part (b) is more relevant, because we want the probability of a result that is *at least as extreme* as the one obtained.
    d. Yes. It is very highly unlikely that we would get a result as extreme as 879 girls in 945 births by chance. Given that the 945 couples were treated with the XSORT method, it appears that this method is effective in increasing the likelihood that a baby will be a girl.
19. 0.0001 (Tech: 0.0000). The results suggest that the surveyed people did not respond accurately.
21. Probability of six or fewer: 0.1075 (Tech using normal approximation: 0.1080; Tech using binomial: 0.1034). Because that probability is not very small, the evidence against the rate of 20% is not very strong.
23. Probability of 170 or fewer: 0.0099 (Tech using normal approximation: 0.0098; Tech using binomial: 0.0089). Because the probability of 170 or fewer is so small with the assumed 20% rate, it appears that the rate is actually less than 20%.
25. a. 6; 0.4602 (Tech using normal approximation: 0.4583; tech using binomial: 0.4307)
    b. 101; 0.3936 (Tech using normal approximation: 0.3933; tech using binomial: 0.3932)
    c. The roulette game provides a better likelihood of making a profit.

Chapter 6: Quick Quiz

1. $\mu = 0$ and $\sigma = 1$
2. 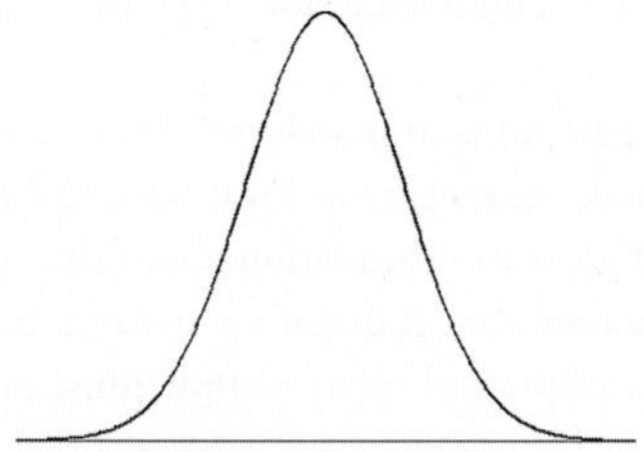
3. $z = 2.05$ (Tech: 2.05375)
4. 0.8413
5. 0.0775 (Tech: 0.0776)
6. 0.1611 (Tech: 0.1618)
7. 0.0158 (Tech: 0.0156)
8. 4.898
9. 0.0409
10. 82.31% (Tech: 82.26%)

Chapter 6: Review Exercises

1. a. 0.9983 b. 0.9370 c. 0.8385
   d. −0.52 e. 0.1401
2. a. 7.93% (Tech: 7.89%)
   b. 1369.2 mm (Tech: 1369.4 mm)
3. a. 97.88% b. 1742.6 mm

4. a. Normal b. 21.1 c. 0.57
5. a. An unbiased estimator is a statistic that targets the value of the population parameter in the sense that the sampling distribution of the statistic has a mean that is equal to the mean of the corresponding parameter.
   b. Mean; variance; proportion c. True
6. a. 85.08% (Tech: 85.12%). With about 15% of all men needing to bend, the design does not appear to be adequate, but the Mark VI monorail appears to be working quite well in practice.
   b. 75.1 in.
7. a. 0.5753 (Tech: 0.5766)
   b. 0.9976. Yes, if the plane is full of male passengers, it is highly likely that it is overweight.
8. a. No. A histogram is far from bell-shaped. A normal quantile plot reveals a pattern of points that is far from a straight-line pattern.
   b. No. The sample size of $n = 26$ does not satisfy the condition of $n > 30$, and the values do not appear to be from a population having a normal distribution.
9. 0.2296 (Tech using normal approximation: 0.2286; Tech using binomial: 0.2278). The occurrence of 787 offspring plants with long stems is not unusually low because its probability is not small. The results are consistent with Mendel's claimed proportion of 3/4.
10. a. 0.7019 (Tech using normal approximation: 0.7024; Tech using binomial: 0.7100)
    b. 0.1148 (Tech using normal approximation: 0.1158; Tech using binomial: 0.1119)

Chapter 6: Cumulative Review Exercises

1. a. $10,300,000
   b. $14,000,000
   c. $5,552,027
   d. 30,825,003,810,000 square dollars
   e. $z = 0.76$ f. Ratio g. Discrete
   h. No, the starting players are likely to be the best players who receive the highest salaries.
2. a. $\overline{A}$ is the event of selecting someone who does not have the belief that college is not a good investment. (This is not the same as selecting someone who believes that college is a good investment.)
   b. 0.9 c. 0.001
   d. The sample is a voluntary response (or self-selected) sample. This suggests that the 10% rate might not be very accurate, because people with strong feelings or interest about the topic are more likely to respond.
3. a. 0.0630 (Tech: 0.0627)
   b. 2643 g (Tech: 2642 g)
   c. 0.0005
   d. 0.3936 (Tech: 0.3923)
4. a. The vertical scale does not start at 0, so differences are somewhat distorted. By using a scale ranging from 1 to 29 for frequencies that range from 2 to 14, the graph is flattened, so differences are not shown as they should be.
   b. The graph depicts a distribution that is not exactly normal, but it is approximately normal because it is roughly bell-shaped.
   c. Minimum: 42 years; maximum: 70 years. Using the range rule of thumb, the standard deviation is estimated to be $(70 - 42)/4 = 7$ years. The estimate of $s = 7$ years is very close to the actual standard deviation of $s = 6.6$ years, so the range rule of thumb works quite well here.
5. a. 0.001 b. 0.271
   c. The requirement that $np \geq 5$ is not satisfied, indicating that the normal approximation would result in errors that are too large.
   d. 5.0 e. 2.1
   f. No, 8 is within two standard deviations of the mean and is within the range of values that could easily occur by chance.

Chapter 7: Answers

Section 7-2

1. The confidence level (such as 95%) was not provided.
3. $\hat{p} = 0.26$ is the sample proportion; $\hat{q} = 0.74$ (found from evaluating $1 - \hat{p}$); $n = 1910$ is the sample size; $E = 0.03$ is the margin of error; p is the population proportion, which is unknown. The value of α is 0.05.
5. 1.28
7. 1.645
9. 0.125 ± 0.061
11. $0.0268 < p < 0.133$
13. a. 0.530 b. $E = 0.0309$ c. $0.499 < p < 0.561$
    d. We have 95% confidence that the interval from 0.499 to 0.561 actually does contain the true value of the population proportion.
15. a. 0.430 b. $E = 0.0162$ c. $0.414 < p < 0.446$
    d. We have 90% confidence that the interval from 0.414 to 0.446 actually does contain the true value of the population proportion.
17. a. 0.930 b. $0.914 < p < 0.946$
    c. Yes. The true proportion of girls with the XSORT method is substantially greater than the proportion of (about) 0.5 that is expected when no method of gender selection is used.
19. a. 0.5 b. 0.439 c. $0.363 < p < 0.516$
    d. If the touch therapists really had an ability to select the correct hand by sensing an energy field, their success rate would be significantly greater than 0.5, but the sample success rate of 0.439 and the confidence interval suggest that they do not have the ability to select the correct hand by sensing an energy field.
21. a. 124 b. $24.7\% < p < 33.3\%$
    c. Yes. Because all values of the confidence interval are less than 0.5, the confidence interval shows that the percentage of women who purchase books online is very likely less than 50%.
    d. No. The confidence interval shows that it is possible that the percentage of women who purchase books online could be less than 25%.
    e. Nothing.
23. a. 236
    b. $0.402 < p < 0.516$ (using $x = 236$: $0.403 < p < 0.516$).
    c. $0.431 < p < 0.487$

d. The 95% confidence interval is wider than the 80% confidence interval. A confidence interval must be wider in order to be more confident that it captures the true value of the population proportion. (See Exercise 4.)

25. $0.0168 < p < 0.143$ (Tech: $0.0169 < p < 0.143$). No, the confidence interval limits contain the value of 0.13, so the claimed rate of 13% could be the true percentage for the population of brown M&Ms.

27. a. $0.0276\% < p < 0.0366\%$ (using $x = 135$: $0.0276\% < p < 0.0367\%$).
    b. No, because 0.0340% is included in the confidence interval.

29. 752

31. 339

33. a. 1537 b. 1449

35. a. 271 b. 139 (Tech: 138)
    c. No. A sample of students at the nearest college is a convenience sample, not a simple random sample, so it is very possible that the results would not be representative of the population of adults.

37. $\hat{p} = 18/34$, or 0.529. CI: $36.2\% < p < 69.7\%$. Greater height does not appear to be an advantage for presidential candidates. If greater height is an advantage, then taller candidates should win substantially more than 50% of the elections, but the confidence interval shows that the percentage of elections won by taller candidates is likely to be anywhere between 36.2% and 69.7%.

39. a. 178 b. 176

41. $81.4\% < p < 101.9\%$. The upper confidence interval limit is greater than 100%. Given that the percentage cannot exceed 100%, change the upper limit to 100%.

43. $p > 0.831$ (Tech: $p > 0.832$). Because we have 95% confidence that p is greater than 0.831, we can safely conclude that more than 75% of adults know what Twitter is.

Section 7-3

1. a. 233.4 sec $< \mu <$ 256.65 sec
   b. Best point estimate of μ is 245.025 sec. The margin of error is $E = 11.625$ sec.

3. We have 95% confidence that the limits of 233.4 sec and 256.65 sec contain the true value of the mean of the population of all duration times.

5. Neither the normal nor the t distribution applies.

7. $t_{\alpha/2} = 2.708$

9. 8.104 km $< \mu <$ 11.512 km (Tech: 8.103 km $< \mu <$ 11.513 km). Because the sample size is greater than 30, the confidence interval yields a reasonable estimate of μ, even though the data appear to be from a population that is not normally distributed.

11. 3315.1 thousand dollars $< \mu <$ 22,480.9 thousand dollars (Tech: 3313.5 thousand dollars $< \mu <$ 22,482.5 thousand dollars). The $1 salary of Jobs is an outlier that is very far away from the other values, and that outlier has a dramatic effect on the confidence interval.

13. 98.08°F $< \mu <$ 98.32°F. Because the confidence interval does not contain 98.6°F, it appears that the mean body temperature is not 98.6°F, as is commonly believed.

15. −6.8 mg/dL $< \mu <$ 7.6 mg/dL. Because the confidence interval includes the value of 0, it is very possible that the mean of the changes in LDL cholesterol is equal to 0, suggesting that the garlic treatment did not affect LDL cholesterol levels. It does not appear that garlic is effective in reducing LDL cholesterol.

17. 4.7 million dollars $< \mu <$ 28.0 million dollars. The data appear to have a distribution that is far from normal, so the confidence interval might not be a good estimate of the population mean. The population is likely to be the list of box office receipts for each day of the movie's release. Because the values are from the first 14 days of release, the sample values are not a simple random sample, and they are likely to be the largest of all such values, so the confidence interval is not a good estimate of the population mean.

19. The sample data meet the loose requirement of having a normal distribution. CI: 0.707 W/kg $< \mu <$ 1.169 W/kg. Because the confidence interval is entirely below the standard of 1.6 W/kg, it appears that the mean amount of cell phone radiation is less than the FCC standard, but there could be individual cell phones that exceed the standard.

21. The sample data meet the loose requirement of having a normal distribution. CI: $6.43 < \mu < 15.67$. We cannot conclude that the population mean is less than 7 μg/g, because the confidence interval shows that the mean might be greater than that level.

23. CI for ages of unsuccessful applicants: 43.8 years $< \mu <$ 50.1 years. CI for ages of successful applicants: 42.6 years $< \mu <$ 46.4 years. Although final conclusions about means of populations should not be based on the overlapping of confidence intervals, the confidence intervals do overlap, so it appears that both populations could have the same mean, and there is not clear evidence of discrimination based on age.

25. The sample size is 68, and it does appear to be very reasonable.

27. 405 (Tech: 403). It is not likely that you would find that many two-year-old used Corvettes in your region.

29. Use $\sigma = 450$ to get a sample size of 110. The margin of error of 100 points seems too high to provide a good estimate of the mean SAT score.

31. With the range rule of thumb, use $\sigma = 11$ to get a required sample size of 117. With $\sigma = 10.3$, the required sample size is 102. The better estimate of σ is the standard deviation of the sample, so the correct sample size is likely to be closer to 102 than 117.

33. $0.963 < \mu < 1.407$

35. 8.156 km $< \mu <$ 11.460 km (Tech: 8.159 km $< \mu <$ 11.457 km)

37. 6131.8 thousand dollars $< \mu <$ 19,663.4 thousand dollars (Tech: 6131.9 thousand dollars $< \mu <$ 19,663.3 thousand dollars)

39. The sample data do not appear to meet the loose requirement of having a normal distribution. CI: $-24.54 < \mu < 106.04$ (Tech: $-24.55 < \mu < 106.05$). The effect of the outlier on the confidence interval is very substantial. Outliers should be discarded if they are known to be errors. If an outlier is a correct value, it might be very helpful to see its effects by constructing the confidence interval with and without the outlier included.

41. $-26.0 < \mu < 32.0$. The confidence interval based on the first sample value is much wider than the confidence interval based on all 10 sample values.

Section 7-4

1. 30.3 mg/dL $< \sigma <$ 47.5 mg/dL. We have 95% confidence that the limits of 30.3 mg/dL and 47.5 mg/dL contain the true value of the standard deviation of the LDL cholesterol levels of all women.
3. The original sample values can be identified, but the dotplot shows that the sample appears to be from a population having a uniform distribution, not a normal distribution as required. Because the normality requirement is not satisfied, the confidence interval estimate of σ should not be constructed using the methods of this section.
5. df = 24. $\chi^2_L = 9.886$ and $\chi^2_R = 45.559$. CI: 0.17 mg $< \sigma <$ 0.37 mg.
7. df = 39. $\chi^2_L = 24.433$ (Tech: 23.654) and $\chi^2_R = 59.342$ (Tech: 58.120). CI: $52.9 < \sigma < 82.4$ (Tech: $53.4 < \sigma < 83.7$).
9. 0.579°F $< \sigma <$ 0.720°F (Tech: 0.557°F $< \sigma <$ 0.700°F)
11. 30.9 mL $< \sigma <$ 67.45 mL. The confidence interval shows that the standard deviation is not likely to be less than 30 mL, so the variation is too high instead of being at an acceptable level below 30 mL. (Such one-sided claims should be tested using the formal methods presented in Chapter 8.)
13. 0.252 ppm $< \sigma <$ 0.701 ppm
15. CI for ages of unsuccessful applicants: 5.2 years $< \sigma <$ 11.5 years. CI for ages of successful applicants: 3.7 years $< \sigma <$ 7.5 years. Although final conclusions about means of populations should not be based on the overlapping of confidence intervals, the confidence intervals do overlap, so it appears that the two populations have standard deviations that are not dramatically different.
17. 0.01239 g $< \sigma <$ 0.02100 g (Tech: 0.01291 g $< \sigma <$ 0.02255 g)
19. 33,218 is too large. There aren't 33,218 statistics professors in the population, and even if there were, that sample size is too large to be practical.
21. The sample size is 768. Because the population does not have a normal distribution, the computed minimum sample size is not likely to be correct.
23. $\chi^2_L = 82.072$ and $\chi^2_R = 129.635$ (Tech using $z_{\alpha/2} = 1.644853626$: $\chi^2_L = 82.073$ and $\chi^2_R = 129.632$). The approximate values are quite close to the actual critical values.

Chapter 7: Quick Quiz

1. $36.9\% < p < 43.1\%$
2. 0.480
3. We have 95% confidence that the limits of 0.449 and 0.511 contain the true value of the proportion of females in the population of medical school students.
4. $z = 1.645$
5. 752
6. 373 (Tech: 374)
7. The sample must be a simple random sample and there is a loose requirement that the sample values appear to be from a normally distributed population.
8. The degrees of freedom is the number of sample values that can vary after restrictions have been imposed on all of the values. For the sample data in Exercise 7, df = 5.
9. $t = 2.571$
10. $\chi^2_L = 0.831$ and $\chi^2_R = 12.833$

Chapter 7: Review Exercises

1. a. 51.0% b. $46.8\% < p < 55.1\%$
   c. No, the confidence interval shows that the population percentage might be 50% or less, so we cannot safely conclude that the majority of adults say that they are underpaid.
2. 4145 (Tech: 4147) 3. 155 (Tech: 154)
4. a. Student *t* distribution b. Normal distribution
   c. The distribution is not normal, Student *t*, or chi-square.
   d. χ^2 (chi-square distribution) e. Normal distribution
5. a. 543 (Tech: 542) b. 247 (Tech: 246) c. 543
6. $61.5\% < p < 66.5\%$. Because the entire confidence interval is above 50%, we can safely conclude that the majority of adults consume alcoholic beverages.
7. -22.1 sec $< \mu <$ 308.1 sec
8. $6.54 < \mu < 7.76$. Because women and men have some notable physiological differences, the confidence interval does not necessarily serve as an estimate of the mean white blood cell count of men.
9. 37.5 g $< \mu <$ 47.9 g. There is 95% confidence that the limits of 37.5 g and 47.9 g contain the true mean deceleration measurement for all small cars.
10. 3.6 g $< \sigma <$ 12.3 g

Chapter 7: Cumulative Review Exercises

1. $\bar{x} = 5.5$; median = 5.0; $s = 3.8$
2. The range of usual values is from -2.1 to 13.1 (or from 0 to 13.1).
3. Ratio level of measurement; discrete data.
4. 33 campuses
5. $3.6 < \mu < 7.4$. The population should include only colleges of the same type as the sample, so the population consists of all large urban campuses with residence halls.
6. The graphs suggest that the population has a distribution that is skewed (to the right) instead of being normal. The histogram shows that some taxi-out times can be very long, and that can occur with heavy traffic, but little or no traffic cannot make the taxi-out time very low. There is a minimum time required, regardless of traffic conditions. Construction of a confidence interval estimate of a population standard deviation has a strict requirement that the sample data are from a normally distributed population, and the graphs show that this strict normality requirement is not satisfied.
7. a. $0.560 < p < 0.620$ (or $0.560 < p < 0.621$ if using $x = 592$)
   b. Because the survey was about shaking hands and because it was sponsored by a supplier of hand sanitizer products, the sponsor could potentially benefit from the results, so there might be some pressure to obtain results favorable to the sponsor.
   c. 1083

8. There does not appear to be a correlation between HDL and LDL cholesterol levels.

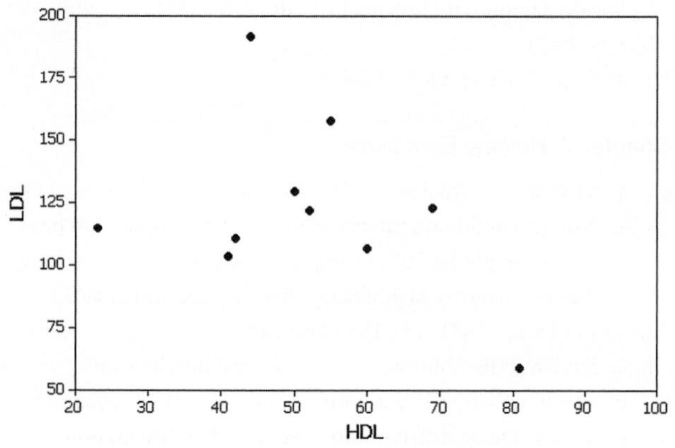

9. a. 13.35% (Tech: 13.32%). Yes, losing about 13% of the market would be a big loss.
   b. 160.2 mm; 189.8 mm
10. a. 1/1000 b. 999/1000 c. 0.990

Chapter 8

Section 8-2

1. Rejection of the aspirin claim is more serious because the aspirin is a drug treatment. The wrong aspirin dosage can cause adverse reactions. M&Ms do not have those same adverse reactions. It would be wise to use a smaller significance level for testing the aspirin claim.
3. a. H_0: $\mu = 98.6$°F b. H_1: $\mu \neq 98.6$°F
   c. Reject the null hypothesis or fail to reject the null hypothesis.
   d. No. In this case, the original claim becomes the null hypothesis. For the claim that the mean body temperature is equal to 98.6°F, we can either reject that claim or fail to reject it, but we cannot state that there is sufficient evidence to *support* that claim.
5. a. $p = 0.2$ b. H_0: $p = 0.2$ and H_1: $p \neq 0.2$
7. a. $\mu \leq 76$ b. H_0: $\mu = 76$ and H_1: $\mu > 76$
9. There is not sufficient evidence to warrant rejection of the claim that 20% of adults smoke.
11. There is not sufficient evidence to warrant rejection of the claim that the mean pulse rate of adult females is 76 or lower.
13. $z = 10.33$ (or $z = 10.35$ if using $x = 909$)
15. $\chi^2 = 8.110$
17. *P*-value = 0.0228. Critical value: $z = 1.645$.
19. *P*-value = 0.0802 (Tech: 0.0801). Critical values: $z = -1.96$, $z = 1.96$.
21. *P*-value = 0.2186 (Tech: 0.2187). Critical values: $z = -1.96$, $z = 1.96$.
23. *P*-value = 0.0013. Critical value: $z = -1.645$.
25. a. Reject H_0.
    b. There is sufficient evidence to support the claim that the percentage of blue M&Ms is greater than 5%.
27. a. Fail to reject H_0.
    b. There is not sufficient evidence to warrant rejection of the claim that women have heights with a mean equal to 160.00 cm.
29. a. H_0: $p = 0.5$ and H_1: $p > 0.5$ b. $\alpha = 0.01$
    c. Normal distribution. d. Right-tailed.
    e. $z = 1.00$ f. *P*-value: 0.1587 g. $z = 2.33$ h. 0.01
31. Type I error: In reality $p = 0.1$, but we reject the claim that $p = 0.1$. Type II error: In reality $p \neq 0.1$, but we fail to reject the claim that $p = 0.1$.
33. Type I error: In reality $p = 0.5$, but we support the claim that $p > 0.5$. Type II error: In reality $p > 0.5$, but we fail to support that conclusion.
35. The power of 0.96 shows that there is a 96% chance of rejecting the null hypothesis of $p = 0.08$ when the true proportion is actually 0.18. That is, if the proportion of Chantix users who experience abdominal pain is actually 0.18, then there is a 96% chance of supporting the claim that the proportion of Chantix users who experience abdominal pain is greater than 0.08.
37. 617

Section 8-3

1. The *P*-value method and the critical value method always yield the same conclusion. The confidence interval method might or might not yield the same conclusion obtained by using the other two methods.
3. *P*-value: 0.00000000550. Because the *P*-value is so low, we have sufficient evidence to support the claim that $p < 0.5$.
5. a. Left-tailed. b. $z = -1.94$
   c. *P*-value: 0.0260 (rounded)
   d. H_0: $p = 0.1$. Reject the null hypothesis.
   e. There is sufficient evidence to support the claim that less than 10% of treated subjects experience headaches.
7. a. Two-tailed. b. $z = -0.82$ c. *P*-value: 0.4106
   d. H_0: $p = 0.35$. Fail to reject the null hypothesis.
   e. There is not sufficient evidence to warrant rejection of the claim that 35% of adults have heard of the Sony Reader.
9. H_0: $p = 0.25$. H_1: $p \neq 0.25$. Test statistic: $z = 0.67$. Critical values: $z = \pm 2.575$ (Tech: ± 2.576). *P*-value: 0.5028 (Tech: 0.5021). Fail to reject H_0. There is not sufficient evidence to warrant rejection of the claim that 25% of offspring peas will be yellow.
11. H_0: $p = 0.5$. H_1: $p > 0.5$. Test statistic: $z = 1.90$. Critical value: $z = 1.645$. *P*-value: 0.0287 (Tech: 0.0290). Reject H_0. There is sufficient evidence to support the claim that the majority of adults feel vulnerable to identify theft.
13. H_0: $p = 0.5$. H_1: $p > 0.5$. Test statistic: $z = 26.45$. Critical value: $z = 2.33$. *P*-value: 0.0001 (Tech: 0.0000). Reject H_0. There is sufficient evidence to support the claim that the XSORT method is effective in increasing the likelihood that a baby will be a girl.
15. H_0: $p = 0.5$. H_1: $p \neq 0.5$. Test statistic: $z = -2.03$. Critical values: $z = \pm 1.645$. *P*-value: 0.0424 (Tech: 0.0422). Reject H_0. There is sufficient evidence to warrant rejection of the claim that touch therapists use a method equivalent to random guesses. However, their success rate of 123/280 (or 43.9%) indicates that they performed *worse* than random guesses, so they do not appear to be effective.
17. H_0: $p = 1/3$. H_1: $p < 1/3$. Test statistic: $z = -2.72$. Critical value: $z = -2.33$. *P*-value: 0.0033. Reject H_0. There is sufficient evidence to support the claim that fewer than 1/3 of the

challenges are successful. Players don't appear to be very good at recognizing referee errors.

19. H_0: $p = 0.000340$. H_1: $p \neq 0.000340$. Test statistic: $z = -0.66$. Critical values: $z = \pm 2.81$. P-value: 0.5092 (Tech: 0.5122). Fail to reject H_0. There is not sufficient evidence to support the claim that the rate is different from 0.0340%. Cell phone users should not be concerned about cancer of the brain or nervous system.
21. H_0: $p = 0.5$. H_1: $p \neq 0.5$. Test statistic: $z = 2.75$. Critical values: $z = \pm 1.96$. P-value: 0.0060 (Tech: 0.0059). Reject H_0. There is sufficient evidence to warrant rejection of the claim that the coin toss is fair in the sense that neither team has an advantage by winning it. The coin toss rule does not appear to be fair.
23. H_0: $p = 0.5$. H_1: $p < 0.5$. Test statistic: $z = -3.90$. Critical value: $z = -2.33$. P-value: 0.0001 (Tech: 0.0000484). Reject H_0. There is sufficient evidence to support the claim that fewer than half of smartphone users identify the smartphone as the only thing they could not live without. Because only smartphone users were surveyed, the results do not apply to the general population.
25. H_0: $p = 0.25$. H_1: $p > 0.25$. Test statistic: $z = 1.91$ (using $\hat{p} = 0.29$) or $z = 1.93$ (using $x = 124$). Critical value: $z = 1.645$ (assuming a 0.05 significance level). P-value: 0.0281 (using $\hat{p} = 0.29$) or 0.0268 (using $x = 124$) (Tech P-value: 0.0269). Reject H_0. There is sufficient evidence to support the claim that more than 25% of women purchase books online.
27. H_0: $p = 3/4$. H_1: $p > 3/4$. Test statistic: $z = 7.85$ (using $\hat{p} = 0.9$) or $z = 7.89$ (using $x = 463$). Critical value: $z = 2.33$. P-value: 0.0001 (Tech: 0.0000). Reject H_0. There is sufficient evidence to support the claim that more than 3/4 of all human resource professionals say that the appearance of a job applicant is most important for a good first impression.
29. H_0: $p = 0.791$. H_1: $p < 0.791$. Test statistic: $z = -29.09$ (using $\hat{p} = 0.39$) or $z = -29.11$ (using $x = 339$). Critical value: $z = -2.33$. P-value: 0.0001 (Tech: 0.0000). Reject H_0. There is sufficient evidence to support the claim that the percentage of selected Americans of Mexican ancestry is less than 79.1%, so the jury selection process appears to be unfair.
31. H_0: $p = 0.75$. H_1: $p > 0.75$. Test statistic: $z = 7.30$. Critical value: $z = 2.33$. P-value: 0.0001 (Tech: 0.0000). Reject H_0. There is sufficient evidence to support the claim that more than 75% of of television sets in use were tuned to the Super Bowl.
33. Among 100 M&Ms, 19 are green. H_0: $p = 0.16$. H_1: $p \neq 0.16$. Test statistic: $z = 0.82$. Critical values: $z = \pm 1.96$. P-value: 0.4122 (Tech: 0.4132). Fail to reject H_0. There is not sufficient evidence to warrant rejection of the claim that 16% of plain M&M candies are green.
35. H_0: $p = 0.5$. H_1: $p > 0.5$. Using the binomial probability distribution with an assumed proportion of $p = 0.5$, the probability of 7 or more heads is 0.0352, so the P-value is 0.0352. Reject H_0. There is sufficient evidence to support the claim that the coin favors heads.
37. a. 0.7224 (Tech: 0.7219)
    b. 0.2776 (Tech: 0.2781)
    c. The power of 0.7224 shows that there is a reasonably good chance of making the correct decision of rejecting the false null hypothesis. It would be better if the power were even higher, such as greater than 0.8 or 0.9.

Section 8-4

1. The requirements are (1) the sample must be a simple random sample, and (2) either or both of these conditions must be satisfied: The population is normally distributed or $n > 30$. There is not enough information given to determine whether the sample is a simple random sample. Because the sample size is not greater than 30, we must check for normality, but the value of 583 sec appears to be an outlier, and a normal quantile plot or histogram suggests that the sample does not appear to be from a normally distributed population.
3. A t test is a hypothesis test that uses the Student t distribution, such as the method of testing a claim about a population mean as presented in this section. The t test methods are much more likely to be used than the z test methods because the t test does not require a known value of σ, and realistic hypothesis tests of claims about μ typically involve a population with an unknown value of σ.
5. P-value < 0.005 (Tech: 0.0013).
7. $0.02 < P\text{-value} < 0.05$ (Tech: 0.0365).
9. H_0: $\mu = 24$. H_1: $\mu < 24$. Test statistic: $t = -7.323$. Critical value: $t = -1.685$. P-value < 0.005. (The display shows that the P-value is 0.00000000387325.) Reject H_0. There is sufficient evidence to support the claim that Chips Ahoy reduced-fat cookies have a mean number of chocolate chips that is less than 24 (but this does not provide conclusive evidence of reduced fat).
11. H_0: $\mu = 33$ years. H_1: $\mu \neq 33$ years. Test statistic: $t = 2.367$. Critical values: $t = \pm 2.639$ (approximately). P-value > 0.02 (Tech: 0.0204). Fail to reject H_0. There is not sufficient evidence to warrant rejection of the claim that the mean age of actresses when they win Oscars is 33 years.
13. H_0: $\mu = 0.8535$ g. H_1: $\mu \neq 0.8535$ g. Test statistic: $t = 0.765$. Critical values: $t = \pm 2.101$. P-value > 0.20 (Tech: 0.4543). Fail to reject H_0. There is not sufficient evidence to warrant rejection of the claim that the mean weight of all green M&Ms is equal to 0.8535 g. The green M&Ms do appear to have weights consistent with the package label.
15. H_0: $\mu = 0$ lb. H_1: $\mu > 0$ lb. Test statistic: $t = 3.872$. Critical value: $t = 2.426$. P-value < 0.005 (Tech: 0.0002). Reject H_0. There is sufficient evidence to support the claim that the mean weight loss is greater than 0. Although the diet appears to have statistical significance, it does not appear to have practical significance, because the mean weight loss of only 3.0 lb does not seem to be worth the effort and cost.
17. H_0: $\mu = 0$. H_1: $\mu > 0$. Test statistic: $t = 0.133$. Critical value: $t = 1.676$ (approximately, assuming a 0.05 significance level). P-value > 0.10 (Tech: 0.4472). Fail to reject H_0. There is not sufficient evidence to support the claim that with garlic treatment, the mean change in LDL cholesterol is greater than 0. The results suggest that the garlic treatment is not effective in reducing LDL cholesterol levels.
19. H_0: $\mu = 4$ years. H_1: $\mu > 4$ years. Test statistic: $t = 3.189$. Critical value: $t = 2.539$. P-value < 0.005 (Tech: 0.0024). Reject H_0. There is sufficient evidence to support the claim that the mean time required to earn a bachelor's degree is greater than 4.0 years. Because $n \leq 30$ and the data do not appear to be from a normally distributed population, the requirement that "the population is normally distributed or $n > 30$" is not satisfied, so

the conclusion from the hypothesis test might not be valid. However, some of the sample values are equal to 4 years and others are greater than 4 years, so the claim does appear to be justified.

21. The sample data meet the loose requirement of having a normal distribution. H_0: $\mu = 14$ μg/g. H_1: $\mu < 14$ μg/g. Test statistic: $t = -1.444$. Critical value: $t = -1.833$. P-value > 0.05 (Tech: 0.0913). Fail to reject H_0. There is not sufficient evidence to support the claim that the mean lead concentration for all such medicines is less than 14 μg/g.
23. The sample data meet the loose requirement of having a normal distribution. H_0: $\mu = 63.8$ in. H_1: $\mu > 63.8$ in. Test statistic: $t = 23.824$. Critical value: $t = 2.821$. P-value < 0.005 (Tech: 0.0000). Reject H_0. There is sufficient evidence to support the claim that supermodels have heights with a mean that is greater than the mean height of 63.8 in. for women in the general population. We can conclude that supermodels are taller than typical women.
25. H_0: $\mu = 1.00$. H_1: $\mu > 1.00$. Test statistic: $t = 2.218$. Critical value: $t = 1.676$ (approximately). P-value < 0.025 (Tech: 0.0156). Reject H_0. There is sufficient evidence to support the claim that the population of earthquakes has a mean magnitude greater than 1.00.
27. H_0: $\mu = 83$ kg. H_1: $\mu < 83$ kg. Test statistic: $t = -5.524$. Critical value: $t = -2.453$. P-value < 0.005 (Tech: 0.0000). Reject H_0. There is sufficient evidence to support the claim that male college students have a mean weight that is less than the 83 kg mean weight of males in the general population.
29. H_0: $\mu = 24$. H_1: $\mu < 24$. Test statistic: $z = -7.32$. Critical value: $z = -1.645$. P-value: 0.0001 (Tech: 0.0000). Reject H_0. There is sufficient evidence to support the claim that Chips Ahoy reduced-fat cookies have a mean number of chocolate chips that is less than 24 (but this does not provide conclusive evidence of reduced fat).
31. H_0: $\mu = 33$ years. H_1: $\mu \neq 33$ years. Test statistic: $z = 2.37$. Critical values: $z = \pm 2.575$. P-value: 0.0178 (Tech: 0.0180). Fail to reject H_0. There is not sufficient evidence to warrant rejection of the claim that the mean age of actresses when they win Oscars is 33 years.
33. The approximation yields a critical value of $t = 1.655$, which is the same as the result from STATDISK or a TI-83/84 Plus calculator.
35. a. The power of 0.4274 shows that there is a 42.74% chance of supporting the claim that $\mu < 1.00$ W/kg when the true mean is actually 0.80 W/kg. This value of power is not very high, and it shows that the hypothesis test is not very effective in recognizing that the mean is less than 1.00 W/kg when the actual mean is 0.80 W/kg.
    b. $\beta = 0.5726$. The probability of a type II error is 0.5726. That is, there is a 0.5726 probability of making the mistake of not supporting the claim that $\mu < 1.00$ W/kg when in reality the population mean is 0.80 W/kg.

Section 8-5

1. a. The mean waiting time remains the same.
   b. The variation among waiting times is lowered.
   c. Because customers all have waiting times that are roughly the same, they experience less stress and are generally more satisfied. Customer satisfaction is improved.
   d. The single line is better because it results in lower variation among waiting times, so a hypothesis test of a claim of a lower standard deviation is a good way to verify that the variation is lower with a single waiting line.
3. Use a 90% confidence interval. The conclusion based on the 90% confidence interval will be the same as the conclusion from a hypothesis test using the P-value method or the critical value method.
5. H_0: $\sigma = 0.15$ oz. H_1: $\sigma < 0.15$ oz. Test statistic: $\chi^2 = 18.822$. Critical value of χ^2 is between 18.493 and 26.509, so it is estimated to be 22.501 (Tech: 22.465). P-value < 0.05 (Tech: 0.0116). Reject H_0. There is sufficient evidence to support the claim that the population of volumes has a standard deviation less than 0.15 oz.
7. H_0: $\sigma = 0.0230$ g. H_1: $\sigma < 0.0230$ g. Test statistic: $\chi^2 = 18.483$. Critical value of χ^2 is between 18.493 and 26.509, so it is estimated to be 22.501 (Tech: 23.269). P-value < 0.05 (Tech: 0.0069). Reject H_0. There is sufficient evidence to support the claim that the population of weights has a standard deviation less than the specification of 0.0230 g.
9. The data appear to be from a normally distributed population. H_0: $\sigma = 10$. H_1: $\sigma \neq 10$. Test statistic: $\chi^2 = 41.375$. Critical values of χ^2 : 24.433 and 59.342 (approximately). P-value $>$ 0.20 (Tech: 0.7347). Fail to reject H_0. There is not sufficient evidence to warrant rejection of the claim that pulse rates of men have a standard deviation equal to 10 beats per minute.
11. H_0: $\sigma = 3.2$ mg. H_1: $\sigma \neq 3.2$ mg. Test statistic: $\chi^2 = 32.086$. Critical values: $\chi^2 = 12.401$ and 39.364. P-value > 0.20 (Tech: 0.2498). Fail to reject H_0. There is not sufficient evidence to support the claim that filtered 100-mm cigarettes have tar amounts with a standard deviation different from 3.2 mg. There is not enough evidence to conclude that filters have an effect.
13. The data appear to be from a normally distributed population. H_0: $\sigma = 22.5$ years. H_1: $\sigma < 22.5$ years. Test statistic: $\chi^2 = 1.627$. Critical value: $\chi^2 = 4.660$. P-value < 0.005 (Tech: 0.0000). Reject H_0. There is sufficient evidence to support the claim that the standard deviation of ages of all race car drivers is less than 22.5 years.
15. H_0: $\sigma = 32.2$ ft. H_1: $\sigma > 32.2$ ft. Test statistic: $\chi^2 = 29.176$. Critical value: $\chi^2 = 19.675$. P-value: 0.0021. Reject H_0. There is sufficient evidence to support the claim that the new production method has errors with a standard deviation greater than 32.2 ft. The variation appears to be greater than in the past, so the new method appears to be worse, because there will be more altimeters that have larger errors. The company should take immediate action to reduce the variation.
17. H_0: $\sigma = 0.15$ oz. H_1: $\sigma < 0.15$ oz. Test statistic: $\chi^2 = 10.173$. Critical value of χ^2 is between 18.493 and 26.509, so it is estimated to be 22.501 (Tech: 22.465). P-value < 0.01 (Tech: 0.0000). Reject H_0. There is sufficient evidence to support the claim that the population of volumes has a standard deviation less than 0.15 oz.
19. Critical $\chi^2 = 22.189$, which is reasonably close to the value of 22.465 obtained from STATDISK and Minitab.

Chapter 8: Quick Quiz

1. H_0: $\mu = 0$ sec. H_1: $\mu \neq 0$ sec.
2. a. Two-tailed. b. Student *t*.
3. a. Fail to reject H_0.
   b. There is not sufficient evidence to warrant rejection of the claim that the sample is from a population with a mean equal to 0 sec.
4. There is a loose requirement of a normally distributed population in the sense that the test works reasonably well if the departure from normality is not too extreme.
5. a. H_0: $p = 0.5$. H_1: $p > 0.5$. b. $z = 6.33$
   c. *P*-value: 0.0000000001263996. There is sufficient evidence to support the claim that the majority of adults are in favor of the death penalty for a person convicted of murder.
6. 0.0456 (Tech: 0.0455)
7. The only true statement is the one given in part (a).
8. No. All critical values of χ^2 are greater than zero.
9. True. 10. False.

Chapter 8: Review Exercises

1. a. False. b. True. c. False.
   d. False. e. False.
2. H_0: $p = 2/3$. H_1: $p \neq 2/3$. Test statistic: $z = -1.09$. Critical values: $z = \pm 2.575$ (Tech: ± 2.576). *P*-value: 0.2758 (Tech: 0.2756). Fail to reject H_0. There is not sufficient evidence to warrant rejection of the claim that 2/3 of adults are satisfied with the amount of leisure time that they have.
3. H_0: $p = 0.75$. H_1: $p > 0.75$. Test statistic: $z = 10.65$ (if using $x = 678$) or $z = 10.66$ (if using $\hat{p} = 0.92$). Critical value: $z = \pm 2.33$. *P*-value: 0.0001 (Tech: 0.0000). Reject H_0. There is sufficient evidence to support the claim that more than 75% of us do not open unfamiliar e-mail and instant-message links. Given that the results are based on a voluntary response sample, the results are not necessarily valid.
4. H_0: $\mu = 3369$ g. H_1: $\mu < 3369$ g. Test statistic: $t = -19.962$. Critical value: $t = -2.328$ (approximately). *P*-value < 0.005 (Tech: 0.0000). Reject H_0. There is sufficient evidence to support the claim that the mean birth weight of Chinese babies is less than the mean birth weight of 3369 g for Caucasian babies.
5. H_0: $\sigma = 567$ g. H_1: $\sigma \neq 567$ g. Test statistic: $\chi^2 = 54.038$. Critical values of χ^2: 51.172 and 116.321. *P*-value is between 0.02 and 0.05 (Tech: 0.0229). Fail to reject H_0. There is not sufficient evidence to warrant rejection of the claim that the standard deviation of birth weights of Chinese babies is equal to 567 g.
6. H_0: $\mu = 1.5\ \mu g/m^3$. H_1: $\mu > 1.5\ \mu g/m^3$. Test statistic: $t = 0.049$. Critical value: $t = 2.015$. *P*-value > 0.10 (Tech: 0.4814). Fail to reject H_0. There is not sufficient evidence to support the claim that the sample is from a population with a mean greater than the EPA standard of $1.5\ \mu g/m^3$. Because the sample value of $5.40\ \mu g/m^3$ appears to be an outlier and because a normal quantile plot suggests that the sample data are not from a normally distributed population, the requirements of the hypothesis test are not satisfied, and the results of the hypothesis test are therefore questionable.
7. H_0: $\mu = 25$. H_1: $\neq 25$. Test statistic: $t = -0.567$. Critical values: $t = \pm 1.984$ (approximately). *P*-value > 0.20 (Tech: 0.5717). Fail to reject H_0. There is not sufficient evidence to warrant rejection of the claim that the sample is selected from a population with a mean equal to 25.
8. a. A type I error is the mistake of rejecting a null hypothesis when it is actually true. A type II error is the mistake of failing to reject a null hypothesis when in reality it is false.
   b. Type I error: Reject the null hypothesis that the mean of the population is equal to 25 when in reality, the mean is actually equal to 25. Type II error: Fail to reject the null hypothesis that the population mean is equal to 25 when in reality, the mean is actually different from 25.
9. The χ^2 test has a reasonably strict requirement that the sample data must be randomly selected from a population with a normal distribution, but the numbers are selected in such a way that they are all equally likely, so the population has a uniform distribution instead of the required normal distribution. Because the requirements are not all satisfied, the χ^2 test should not be used.
10. H_0: $\mu = 1000$ HIC. H_1: $\mu < 1000$ HIC. Test statistic: $t = -10.177$. Critical value: $t = -3.747$. *P*-value < 0.005 (Tech: 0.0003). Reject H_0. There is sufficient evidence to support the claim that the population mean is less than 1000 HIC. The results suggest that the population mean is less than 1000 HIC, so they appear to satisfy the specified requirement.

Chapter 8: Cumulative Review Exercises

1. a. 53.3 words b. 52.0 words c. 15.7 words
   d. 245.1 words^2 e. 45 words
2. a. Ratio. b. Discrete.
   c. The sample is a simple random sample if it was selected in such a way that all possible samples of the same size have the same chance of being selected.
3. 42.1 words $< \mu <$ 64.5 words
4. H_0: $\mu = 48.0$ words. H_1: $\mu > 48.0$ words. Test statistic: $t = 1.070$. Critical value: $t = 1.833$. *P*-value > 0.10 (Tech: 0.1561). Fail to reject H_0. There is not sufficient evidence to support the claim that the mean number of words on a page is greater than 48.0. There is not enough evidence to support the claim that there are more than 70,000 words in the dictionary.
5. a. 2.28% b. 38.9 in. c. 0.9236 (Tech: 0.9234)
6. a. 0.00195. It is unlikely because the probability of the event occurring is so small.
   b. 0.0121 c. 0.487
7. No. The distribution is very skewed. A normal distribution would be approximately bell-shaped, but the displayed distribution is very far from being bell-shaped.
8. Because the vertical scale starts at 7000 and not at 0, the difference between the number of males and the number of females is exaggerated, so the graph is deceptive by creating the wrong impression that there are many more male graduates than female graduates.
9. a. 373 b. $34.2\% < p < 40.2\%$
   c. Yes. With test statistic $z = -8.11$ and with a *P*-value close to 0, there is sufficient evidence to support the claim that less than 50% of adults answer "yes."

d. The required sample size depends on the confidence level and the sample proportion, not the population size.

10. H_0: $p = 0.5$. H_1: $p < 0.5$. Test statistic: $z = -8.11$. Critical value: $z = -2.33$. P-value: 0.0001 (Tech: 0.0000). Reject H_0. There is sufficient evidence to support the claim that fewer than 50% of Americans say that they have a gun in their home.

Chapter 9 Answers

Section 9-2

1. The samples are simple random samples that are independent. For each of the two groups, the number of successes is at least 5 and the number of failures is at least 5. (Depending on what we call a success, the four numbers are 33, 115, 201,229, and 200,745 and all of those numbers are at least 5.) The requirements are satisfied.

3. a. H_0: $p_1 = p_2$. H_1: $p_1 < p_2$.
   b. If the P-value is less than 0.001 we should reject the null hypothesis and conclude that there is sufficient evidence to support the claim that the rate of polio is less for children given the Salk vaccine than it is for children given a placebo.

5. Test statistic: $z = -12.39$ (rounded). The P-value of 3.137085E^{-}35 is 0.0000 when rounded to four decimal places. There is sufficient evidence to warrant rejection of the claim that the vaccine has no effect.

7. a. H_0: $p_1 = p_2$. H_1: $p_1 > p_2$. Test statistic: $z = 6.44$. Critical value: $z = 2.33$. P-value: 0.0001 (Tech: 0.0000). Reject H_0. There is sufficient evidence to support the claim that the proportion of people over 55 who dream in black and white is greater than the proportion for those under 25.
   b. 98% CI: $0.117 < p_1 - p_2 < 0.240$. Because the confidence interval limits do not include 0, it appears that the two proportions are not equal. Because the confidence interval limits include only positive values, it appears that the proportion of people over 55 who dream in black and white is greater than the proportion for those under 25.
   c. The results suggest that the proportion of people over 55 who dream in black and white is greater than the proportion for those under 25, but the results cannot be used to verify the cause of that difference.

9. a. H_0: $p_1 = p_2$. H_1: $p_1 > p_2$. Test statistic: $z = 6.11$. Critical value: $z = 1.645$. P-value: 0.0001 (Tech: 0.0000). Reject H_0. There is sufficient evidence to support the claim that the fatality rate is higher for those not wearing seat belts.
   b. 90% CI: $0.00556 < p_1 - p_2 < 0.0122$. Because the confidence interval limits do not include 0, it appears that the two fatality rates are not equal. Because the confidence interval limits include only positive values, it appears that the fatality rate is higher for those not wearing seat belts.
   c. The results suggest that the use of seat belts is associated with lower fatality rates than not using seat belts.

11. a. H_0: $p_1 = p_2$. H_1: $p_1 \neq p_2$. Test statistic: $z = 0.57$. Critical values: $z = \pm 1.96$. P-value: 0.5686 (Tech: 0.5720). Fail to reject H_0. There is not sufficient evidence to support the claim that echinacea treatment has an effect.
   b. 95% CI: $-0.0798 < p_1 - p_2 < 0.149$. Because the confidence interval limits do contain 0, there is not a significant difference between the two proportions. There is not sufficient evidence to support the claim that echinacea treatment has an effect.
   c. Echinacea does not appear to have a significant effect on the infection rate. Because it does not appear to have an effect, it should not be recommended.

13. a. H_0: $p_1 = p_2$. H_1: $p_1 \neq p_2$. Test statistic: $z = 0.40$. Critical values: $z = \pm 1.96$. P-value: 0.6892 (Tech: 0.6859). Fail to reject H_0. There is not sufficient evidence to warrant rejection of the claim that men and women have equal success in challenging calls.
   b. 95% CI: $-0.0318 < p_1 - p_2 < 0.0484$. Because the confidence interval limits contain 0, there is not a significant difference between the two proportions. There is not sufficient evidence to warrant rejection of the claim that men and women have equal success in challenging calls.
   c. It appears that men and women have equal success in challenging calls.

15. a. H_0: $p_1 = p_2$. H_1: $p_1 > p_2$. Test statistic: $z = 9.97$. Critical value: $z = 2.33$. P-value: 0.0001 (Tech: 0.0000). Reject H_0. There is sufficient evidence to support the claim that the cure rate with oxygen treatment is higher than the cure rate for those given a placebo. It appears that the oxygen treatment is effective.
   b. 98% CI: $0.467 < p_1 - p_2 < 0.687$. Because the confidence interval limits do not include 0, it appears that the two cure rates are not equal. Because the confidence interval limits include only positive values, it appears that the cure rate with oxygen treatment is higher than the cure rate for those given a placebo. It appears that the oxygen treatment is effective.
   c. The results suggest that the oxygen treatment is effective in curing cluster headaches.

17. a. H_0: $p_1 = p_2$. H_1: $p_1 < p_2$. Test statistic: $z = -1.17$. Critical value: $z = -2.33$. P-value: 0.1210 (Tech: 0.1214). Fail to reject H_0. There is not sufficient evidence to support the claim that the rate of left-handedness among males is less than that among females.
   b. 98% CI: $-0.0849 < p_1 - p_2 < 0.0265$ (Tech: $-0.0848 < p_1 - p_2 < 0.0264$). Because the confidence interval limits include 0, there does not appear to be a significant difference between the rate of left-handedness among males and the rate among females. There is not sufficient evidence to support the claim that the rate of left-handedness among males is less than that among females.
   c. The rate of left-handedness among males does not appear to be less than the rate of left-handedness among females.

19. a. $0.0227 < p_1 - p_2 < 0.217$; because the confidence interval limits do not contain 0, it appears that $p_1 = p_2$ can be rejected.
   b. $0.491 < p_1 < 0.629$; $0.371 < p_2 < 0.509$; because the confidence intervals do overlap, it appears that $p_1 = p_2$ cannot be rejected.
   c. H_0: $p_1 = p_2$. H_1: $p_1 \neq p_2$. Test statistic: $z = 2.40$. P-value: 0.0164. Critical values: $z = \pm 1.96$. Reject H_0. There is sufficient evidence to reject $p_1 = p_2$.
   d. Reject $p_1 = p_2$. Least effective: Using the overlap between the individual confidence intervals.

21. 3383 (Tech: 3382)

Section 9-3

1. Independent: b, d, e
3. Because the confidence interval does not contain 0, it appears that there is a significant difference between the mean height of women and the mean height of men. Based on the confidence interval, it appears that the mean height of men is greater than the mean height of women.
5. a. H_0: $\mu_1 = \mu_2$. H_1: $\mu_1 \neq \mu_2$. Test statistic: $t = -2.979$. Critical values: $t = \pm 2.032$ (Tech: ± 2.002). P-value < 0.01 (Tech: 0.0042). Reject H_0. There is sufficient evidence to warrant rejection of the claim that the samples are from populations with the same mean. Color does appear to have an effect on creativity scores. Blue appears to be associated with higher creativity scores.
   b. 95% CI: $-0.98 < \mu_1 - \mu_2 < -0.18$ (Tech: $-0.97 < \mu_1 - \mu_2 < -0.19$)
7. a. H_0: $\mu_1 = \mu_2$. H_1: $\mu_1 > \mu_2$. Test statistic: $t = 0.132$. Critical value: $t = 1.729$. P-value > 0.10 (Tech: 0.4480). Fail to reject H_0. There is not sufficient evidence to support the claim that the magnets are effective in reducing pain. It is valid to argue that the magnets might appear to be effective if the sample sizes are larger.
   b. 90% CI: $-0.61 < \mu_1 - \mu_2 < 0.71$ (Tech: $-0.59 < \mu_1 - \mu_2 < 0.69$)
9. a. The sample data meet the loose requirement of having a normal distribution. H_0: $\mu_1 = \mu_2$. H_1: $\mu_1 > \mu_2$. Test statistic: $t = 0.852$. Critical value: $t = 2.764$ (Tech: 2.676). P-value > 0.10 (Tech: 0.2054). Fail to reject H_0. There is not sufficient evidence to support the claim that men have a higher mean body temperature than women.
   b. 98% CI: $-0.54°\text{F} < (\mu_1 - \mu_2) < 1.02°\text{F}$ (Tech: $-0.51°\text{F} < (\mu_1 - \mu_2) < 0.99°\text{F}$)
11. a. H_0: $\mu_1 = \mu_2$. H_1: $\mu_1 < \mu_2$. Test statistic: $t = -3.547$. Critical value: $t = -2.462$ (Tech: -2.392). P-value $<$ 0.005 (Tech: 0.0004). Reject H_0. There is sufficient evidence to support the claim that the mean maximal skull breadth in 4000 B.C. is less than the mean in A.D. 150.
   b. 98% CI: $-8.13 \text{ mm} < \mu_1 - \mu_2 < -1.47 \text{ mm}$ (Tech: $-8.04 \text{ mm} < (\mu_1 - \mu_2) < -1.56 \text{ mm}$)
13. a. H_0: $\mu_1 = \mu_2$. H_1: $\mu_1 < \mu_2$. Test statistic: $t = -3.142$. Critical value: $t = -2.462$ (Tech: -2.403). P-value < 0.005 (Tech: 0.0014). Reject H_0. There is sufficient evidence to support the claim that students taking the nonproctored test get a higher mean than those taking the proctored test.
   b. 98% CI: $-25.54 < \mu_1 - \mu_2 < -3.10$ (Tech: $-25.27 < (\mu_1 - \mu_2) < -3.37$)
15. a. H_0: $\mu_1 = \mu_2$. H_1: $\mu_1 \neq \mu_2$. Test statistic: $t = 1.274$. Critical values: $t = \pm 2.023$ (Tech: ± 1.994). P-value $>$ 0.20 (Tech: 0.2066). Fail to reject H_0. There is not sufficient evidence to warrant rejection of the claim that males and females have the same mean BMI.
   b. 95% CI: $-1.08 < \mu_1 - \mu_2 < 4.76$ (Tech: $-1.04 < \mu_1 - \mu_2 < 4.72$)
17. a. H_0: $\mu_1 = \mu_2$. H_1: $\mu_1 > \mu_2$. Test statistic: $t = 0.089$. Critical value: $t = 1.725$ (Tech: 2.029). P-value > 0.10 (Tech: 0.4648.) Fail to reject H_0. There is not sufficient evidence to support the claim that the mean IQ score of people with medium lead levels is higher than the mean IQ score of people with high lead levels.
   b. 90% CI: $-5.9 < \mu_1 - \mu_2 < 6.6$ (Tech: $-5.8 < (\mu_1 - \mu_2) < 6.4$)
19. a. H_0: $\mu_1 = \mu_2$. H_1: $\mu_1 < \mu_2$. Test statistic: $t = -1.810$. Critical value: $t = -2.650$ (Tech: -2.574). P-value $>$ 0.025 (Tech: 0.0442). Fail to reject H_0. There is not sufficient evidence to support the claim that the mean longevity for popes is less than the mean for British monarchs after coronation.
   b. 98% CI: $-23.6 \text{ years} < (\mu_1 - \mu_2) < 4.4 \text{ years}$ (Tech: $-23.2 \text{ years} < (\mu_1 - \mu_2) < 4.0 \text{ years}$)
21. H_0: $\mu_1 = \mu_2$. H_1: $\mu_1 \neq \mu_2$. Test statistic: $t = 32.773$. Critical values: $t = \pm 2.023$ (Tech: ± 1.994). P-value < 0.01 (Tech: 0.0000). Reject H_0. There is sufficient evidence to warrant rejection of the claim that the two populations have equal means. The difference is highly significant, even though the samples are relatively small.
23. $0.03795 \text{ lb} < (\mu_1 - \mu_2) < 0.04254 \text{ lb}$ (Tech: $0.03786 \text{ lb} < (\mu_1 - \mu_2) < 0.04263 \text{ lb}$). Because the confidence interval does not include 0, there appears to be a significant difference between the two population means. It appears that the cola in cans of regular Pepsi weighs more than the cola in cans of Diet Pepsi, and that is probably due to the sugar in regular Pepsi that is not in Diet Pepsi.
25. a. The sample data meet the loose requirement of having a normal distribution. H_0: $\mu_1 = \mu_2$. H_1: $\mu_1 > \mu_2$. Test statistic: $t = 1.046$. Critical value: $t = 2.381$ (Tech: 2.382). P-value > 0.10 (Tech: 0.1496). Fail to reject H_0. There is not sufficient evidence to support the claim that men have a higher mean body temperature than women.
   b. $-0.31°\text{F} < (\mu_1 - \mu_2) < 0.79°\text{F}$. The test statistic became larger, the P-value became smaller, and the confidence interval became narrower, so pooling had the effect of attributing more significance to the results.
27. H_0: $\mu_1 = \mu_2$. H_1: $\mu_1 \neq \mu_2$. Test statistic: $t = 15.322$. Critical values: $t = \pm 2.080$. P-value < 0.01 (Tech: 0.0000). Reject H_0. There is sufficient evidence to warrant rejection of the claim that the two populations have the same mean.
29. a. H_0: $\mu_1 = \mu_2$. H_1: $\mu_1 < \mu_2$. Test statistic: $t = -3.002$. Critical value based on 68.9927614 degrees of freedom: $t = -2.381$ (Tech: -2.382). P-value < 0.005 (Tech: 0.0019). Reject H_0. There is sufficient evidence to support the claim that students taking the nonproctored test get a higher mean than those taking the proctored test.
   b. $-25.68 < \mu_1 - \mu_2 < -2.96$ (Tech: $-25.69 < \mu_1 - \mu_2 < -2.95$)

Section 9-4

1. Parts (c) and (e) are true.
3. The test statistic will remain the same. The confidence interval limits will be expressed in the equivalent values of km/L.
5. H_0: $\mu_d = 0$ cm. H_1: $\mu_d > 0$ cm. Test statistic: $t = 0.036$ (rounded). Critical value: $t = 1.692$. P-value > 0.10 (Tech: 0.4859). Fail to reject H_0. There is not sufficient evidence to

support the claim that for the population of heights of presidents and their main opponents; the differences have a mean greater than 0 cm (with presidents tending to be taller than their opponents).

7. a. $\bar{d} = -11.6$ years b. $s_d = 17.2$ years
c. $t = -1.507$ d. $t = \pm 2.776$

9. H_0: $\mu_d = 0$. H_1: $\mu_d \neq 0$. Test statistic: $t = -1.507$. Critical values: $t = \pm 2.776$. P-value > 0.20 (Tech: 0.2063). Fail to reject H_0. There is not sufficient evidence to support the claim that there is a difference between the ages of actresses and actors when they win Oscars.

11. 1.0 min $< \mu_d <$ 12.0 min. Because the confidence interval includes only positive values and does not include 0 min, it appears that the taxi-out times are greater than the corresponding taxi-in times, so there is sufficient evidence to support the claim of the flight operations manager that for flight delays, more of the blame is attributable to taxi-out times at JFK than taxi-in times at LAX.

13. H_0: $\mu_d = 0$. H_1: $\mu_d > 0$. Test statistic: $t = 2.579$. Critical value: $t = 2.015$. P-value < 0.025 (Tech: 0.0247). Reject H_0. There is sufficient evidence to support the claim that among couples, males speak more words in a day than females.

15. $-6.5 < \mu_d < -0.2$. Because the confidence interval does not include 0, it appears that there is sufficient evidence to warrant rejection of the claim that when the 13th day of a month falls on a Friday, the numbers of hospital admissions from motor vehicle crashes are not affected. Hospital admissions do appear to be affected.

17. H_0: $\mu_d = 0$. H_1: $\mu_d < 0$. Test statistic: $t = -1.080$. Critical value: $t = -1.833$. P-value > 0.10 (Tech: 0.1540). Fail to reject H_0. There is not sufficient evidence to support the claim that *Harry Potter and the Half-Blood Prince* did better at the box office. After a few years, the gross amounts from both movies can be identified, and the conclusion can then be judged objectively without using a hypothesis test.

19. $0.69 < \mu_d < 5.56$. Because the confidence interval limits do not contain 0 and they consist of positive values only, it appears that the "before" measurements are greater than the "after" measurements, so hypnotism does appear to be effective in reducing pain.

21. H_0: $\mu_d = 0$. H_1: $\mu_d \neq 0$. Test statistic: $t = -5.553$. Critical values: $t = \pm 1.990$. P-value < 0.01 (Tech: 0.0000). Reject H_0. There is sufficient evidence to support the claim that there is a difference between the ages of actresses and actors when they win Oscars.

23. H_0: $\mu_d = 0$. H_1: $\mu_d < 0$. Test statistic: $t = -1.560$. Critical value of t is between -1.671 and -1.676 (Tech: -1.673). P-value > 0.05 (Tech: 0.0622). Fail to reject H_0. There is not sufficient evidence to support the claim that among couples, males speak fewer words in a day than females.

25. H_0: $\mu_d = 6.8$ kg. H_1: $\mu_d \neq 6.8$ kg. Test statistic: $t = -11.833$. Critical values: $t = \pm 1.994$ (Tech: ± 1.997). P-value < 0.01 (Tech: 0.0000). Reject H_0. There is sufficient evidence to warrant rejection of the claim that $\mu_d = 6.8$ kg. It appears that the "Freshman 15" is a myth, and college freshman might gain some weight, but they do not gain as much as 15 pounds.

Chapter 9: Quick Quiz

1. H_0: $p_1 = p_2$. H_1: $p_1 \neq p_2$.
2. 0.875
3. 0.0414
4. $0.00172 < p_1 - p_2 < 0.0970$
5. Because the data consist of matched pairs, they are dependent.
6. H_0: $\mu_d = 0$. H_1: $\mu_d > 0$.
7. There is not sufficient evidence to support the claim that front repair costs are greater than the corresponding rear repair costs.
8. True.
9. False.
10. True.

Chapter 9: Review Exercises

1. H_0: $p_1 = p_2$. H_1: $p_1 > p_2$. Test statistic: $z = 3.12$. Critical value: $z = 2.33$. P-value: 0.0009. Reject H_0. There is sufficient evidence to support a claim that the proportion of successes with surgery is greater than the proportion of successes with splinting. When treating carpal tunnel syndrome, surgery should generally be recommended instead of splinting.

2. 98% CI: $0.0581 < p_1 - p_2 < 0.332$ (Tech: $0.0583 < p_1 - p_2 < 0.331$). The confidence interval limits do not contain 0; the interval consists of positive values only. This suggests that the success rate with surgery is greater than the success rate with splints.

3. H_0: $p_1 = p_2$. H_1: $p_1 < p_2$. Test statistic: $z = -1.91$. Critical value: $z = -1.645$. P-value: 0.0281 (Tech: 0.0280). Reject H_0. There is sufficient evidence to support the claim that the fatality rate of occupants is lower for those in cars equipped with airbags.

4. H_0: $\mu_d = 0$. H_1: $\mu_d > 0$. Test statistic: $t = 4.712$. Critical value: $t = 3.143$. P-value < 0.005 (Tech: 0.0016). Reject H_0. There is sufficient evidence to support the claim that flights scheduled 1 day in advance cost more than flights scheduled 30 days in advance. Save money by scheduling flights 30 days in advance.

5. H_0: $\mu_d = 0$. H_1: $\mu_d \neq 0$. Test statistic: $t = -0.574$. Critical values: $t = \pm 2.426$. P-value > 0.20 (Tech: 0.5840). Fail to reject H_0. There is not sufficient evidence to support the claim that there is a difference between self-reported heights and measured heights of females aged 12–16.

6. H_0: $\mu_1 = \mu_2$. H_1: $\mu_1 > \mu_2$. Test statistic: $t = 2.879$. Critical value: $t = 2.426$ (Tech: 2.376). P-value < 0.005 (Tech: 0.0026). Reject H_0. There is sufficient evidence to support the claim that "stress decreases the amount recalled."

7. 98% CI: $1.3 < (\mu_1 - \mu_2) < 14.7$ (Tech: $1.4 < (\mu_1 - \mu_2) < 14.6$). The confidence interval limits do not contain 0; the interval consists of positive values only. This suggests that the numbers of details recalled are lower for those in the stress population.

8. H_0: $p_1 = p_2$. H_1: $p_1 \neq p_2$. Test statistic: $z = -4.20$. Critical values: $z = \pm 2.575$. P-value: 0.0002 (Tech: 0.0000). Reject H_0. There is sufficient evidence to warrant rejection of the claim that the acceptance rate is the same with or without blinding. Without blinding, reviewers know the names and institutions

of the abstract authors, and they might be influenced by that knowledge.

9. H_0: $\mu_1 = \mu_2$. H_1: $\mu_1 \neq \mu_2$. Test statistic: $t = 0.679$. Critical values: $t = \pm 2.014$ approximately (Tech: ± 1.985). P-value > 0.20 (Tech: 0.4988). Fail to reject H_0. There is not sufficient evidence to warrant rejection of the claim of no difference between the mean LDL cholesterol levels of subjects treated with raw garlic and subjects given placebos. Both groups appear to be about the same.
10. $-3.9 < (\mu_1 - \mu_2) < 7.9$ (Tech: $-3.8 < (\mu_1 - \mu_2) < 7.8$)

Chapter 9: Cumulative Review Exercises

1. a. Because the sample data are matched with each column consisting of heights from the same family, the data are dependent.
   b. Mean: 63.81 in.; median: 63.70 in.; mode: 62.2 in.; range: 8.80 in.; standard deviation: 2.73 in.; variance: 7.43 in^2
   c. Ratio
2. There does not appear to be a correlation or association between the heights of mothers and the heights of their daughters.

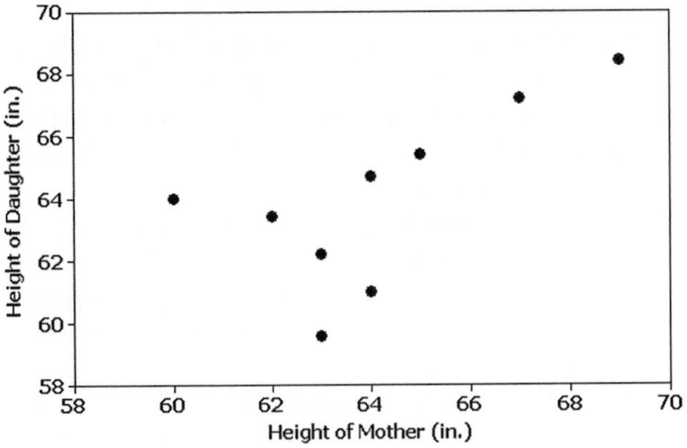

3. 61.86 in. $< \mu <$ 65.76 in. We have 95% confidence that the limits of 61.86 in. and 65.76 in. actually contain the true value of the mean height of all adult daughters.
4. H_0: $\mu_d = 0$. H_1: $\mu_d \neq 0$. Test statistic: $t = 0.283$. Critical values: $t = \pm 2.262$. P-value > 0.20 (Tech: 0.7834). Fail to reject H_0. There is not sufficient evidence to warrant rejection of the claim of no significant difference between the heights of mothers and the heights of their daughters.
5. Because the points lie reasonably close to a straight-line pattern and there is no other pattern that is not a straight-line pattern and there are no outliers, the sample data appear to be from a population with a normal distribution.
6. $0.109 < p < 0.150$. Because the entire range of values in the confidence interval lies below 0.20, the results do justify the statement that "fewer than 20% of Americans choose their computer and/or Internet access when identifying what they miss most when electrical power is lost."
7. No. Because the Internet users chose to respond, we have a voluntary response sample, so the results are not necessarily valid.
8. 2944. The survey should not be conducted using only local phone numbers. Such a convenience sample could easily lead to results that are dramatically different from results that would be obtained by randomly selecting respondents from the entire population, not just those having local phone numbers.
9. a. 0.9332 b. 0.9987
   c. 167.5 cm (Tech: 167.6 cm)
10. No. Because the states have different population sizes, the mean cannot be found by adding the 50 state means and dividing the total by 50. The mean income for the U.S. population can be found by using a weighted mean that incorporates the population size of each state.

Chapter 10

Section 10-2

1. r represents the value of the linear correlation computed by using the paired sample data. ρ represents the value of the linear correlation coefficient that would be computed by using all of the paired data in the population. The value of r is estimated to be 0 (because there is no correlation between sunspot numbers and the Dow Jones Industrial Average).
3. The headline is not justified because it states that increased salt consumption is the *cause* of higher blood pressure levels, but the presence of a correlation between two variables does not necessarily imply that one is the *cause* of the other. Correlation does not imply causality. A correct headline would be this: "Study Shows That Increased Salt Consumption Is Associated with Higher Blood Pressure."
5. Yes. With $r = 0.687$ and critical values of ± 0.312, there is sufficient evidence to support the claim that there is a linear correlation between the durations of eruptions and the time intervals to the next eruptions.
7. No. With $r = 0.149$ and a P-value of 0.681 (or critical values of ± 0.632), there is not sufficient evidence to support the claim that there is a linear correlation between the heights of fathers and the heights of their sons.
9. a.

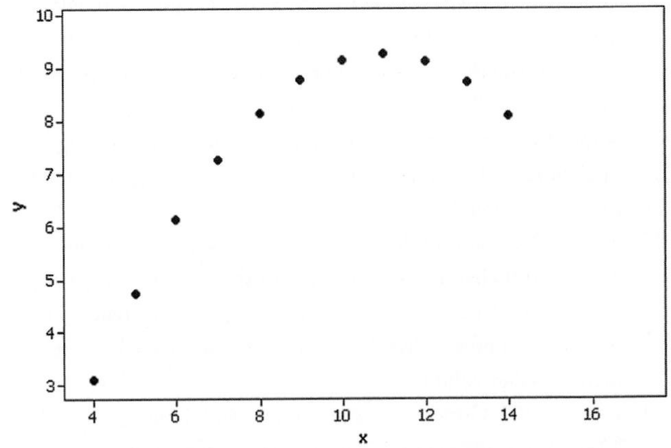

   b. $r = 0.816$. Critical values: $r = \pm 0.602$. P-value $= 0.002$. There is sufficient evidence to support the claim of a linear correlation between the two variables.
   c. The scatterplot reveals a distinct pattern that is not a straight-line pattern.
11. a. There appears to be a linear correlation.
    b. $r = 0.906$. Critical values: $r = \pm 0.632$ (for a 0.05 significance level). There is a linear correlation.

c. $r = 0$. Critical values: $r = \pm 0.666$ (for a 0.05 significance level). There does not appear to be a linear correlation.
d. The effect from a single pair of values can be very substantial, and it can change the conclusion.

13. $r = -0.959$. Critical values: $r = \pm 0.878$. P-value $= 0.010$. There is sufficient evidence to support the claim that there is a linear correlation between weights of lemon imports from Mexico and U.S. car fatality rates. The results do not suggest any cause-effect relationship between the two variables.
15. $r = 0.561$. Critical values: $r = \pm 0.632$. P-value $= 0.091$. There is not sufficient evidence to support the claim that there is a linear correlation between enrollment and burglaries. The results do not change if the actual enrollments are listed as 32,000, 31,000, 53,000, and so on.
17. $r = 0.864$. Critical values: $r = \pm 0.666$. P-value $= 0.003$. There is sufficient evidence to support the claim that there is a linear correlation between court incomes and justice salaries. The correlation does not imply that court incomes directly affect justices' salaries, but it does appear that justices might profit by levying larger fines, or perhaps justices with higher salaries impose larger fines.
19. $r = 1.000$. Critical values: $r = \pm 0.811$. P-value $= 0.000$. There is sufficient evidence to support the claim that there is a linear correlation between amounts of redshift and distances to clusters of galaxies. Because the linear correlation coefficient is 1.000, it appears that the distances can be directly computed from the amounts of redshift.
21. $r = 0.948$. Critical values: $r = \pm 0.811$. P-value $= 0.004$. There is sufficient evidence to support the claim of a linear correlation between the overhead width of a seal in a photograph and the weight of a seal.
23. $r = 0.867$. Critical values: $r = \pm 0.878$. P-value $= 0.057$. There is not sufficient evidence to support the claim of a linear correlation between the systolic blood pressure measurements of the right and left arm.
25. $r = 0.197$. Critical values: $r = \pm 0.707$. P-value $= 0.640$. There is not sufficient evidence to support the claim that there is a linear correlation between prices of regular gas and prices of premium gas. Because there does not appear to be a linear correlation between prices of regular and premium gas, knowing the price of regular gas is not very helpful in getting a good sense for the price of premium gas.
27. $r = 1.000$. Critical values: $r = \pm 0.707$. P-value $= 0.000$. There is sufficient evidence to support the claim that there is a linear correlation between diameters and circumferences. A scatterplot confirms that there is a *linear* association between diameters and volumes.
29. $r = -0.063$. Critical values: $r = \pm 0.444$. P-value $= 0.791$. There is not sufficient evidence to support the claim of a linear correlation between IQ and brain volume.
31. $r = 0.319$. Critical values: $r = \pm 0.254$ (approximately) (Tech: ± 0.263). P-value $= 0.017$. There is sufficient evidence to support the claim of a linear correlation between the numbers of words spoken by men and women who are in couple relationships.
33. a. 0.911 b. 0.787
c. 0.9999 (largest) d. 0.976 e. -0.948

Section 10-3

1. The symbol $\hat{y}$ represents the predicted pulse rate. The predictor variable represents height. The response variable represents pulse rate.
3. If r is positive, the regression line has a positive slope and rises from left to right. If r is negative, the slope of the regression line is negative and it falls from left to right.
5. The best predicted time for an interval after the eruption is 69.0 min.
7. The best predicted height is $\bar{y} = 68.0$ in.
9. $\hat{y} = 3.00 + 0.500x$. The data have a pattern that is not a straight line.
11. a. $\hat{y} = 0.264 + 0.906x$
b. $\hat{y} = 2 + 0x$ (or $\hat{y} = 2$)
c. The results are very different, indicating that one point can dramatically affect the regression equation.
13. $\hat{y} = 16.5 - 0.00282x$; best predicted value is 15.1 fatalities per 100,000 population.
15. $\hat{y} = -36.8 + 3.47x$; best predicted value is $\bar{y} = 87.7$ burglaries. The predicted value is not close to the actual value of 329 burglaries.
17. $\hat{y} = 27.7 + 0.0373x$; best predicted value is \$30,800. The predicted value is not very close to the actual salary of \$26,088.
19. $\hat{y} = -0.00440 + 14.0x$; best predicted value is 0.172 billion light-years. The predicted value is very close to the actual distance of 0.18 light-years.
21. $\hat{y} = -157 + 40.2x$; best predicted weight is -76.6 kg (Tech: -76.5 kg). That prediction is a negative weight that cannot be correct. The overhead width of 2 cm is well beyond the scope of the available sample widths, so the extrapolation might be off by a considerable amount.
23. $\hat{y} = 43.6 + 1.31x$; best predicted value is $\bar{y} = 163.2$ mm Hg.
25. $\hat{y} = 2.57 + 0.172x$; best predicted value is $\bar{y} = \$3.05$. The predicted price is not very close to the actual price of \$2.93.
27. $\hat{y} = -0.00396 + 3.14x$; best predicted value is 4.7 cm. Even though the diameter of 1.50 cm is beyond the scope of the sample diameters, the predicted value yields the actual circumference.
29. $\hat{y} = 109 - 0.00670x$; best predicted IQ score is $\bar{y} = 101$.
31. $\hat{y} = 13{,}400 + 0.302x$; best predicted value is 16,400 (Tech: 16,458).
33. With $\beta_1 = 0$, the regression line is horizontal so that different values of x result in the same y value, and there is no correlation between x and y.

Section 10-4

1. The methods of Section 10-3 should not be used for predictions. The regression equation is based on a linear correlation between the two variables, but the methods of this section do not require a linear relationship. The methods of this section could suggest that there is a correlation with paired data associated by some nonlinear relationship, so the regression equation would not be a suitable model for making predictions.
3. r represents the linear correlation coefficient computed from sample paired data; ρ represents the parameter of the linear correlation coefficient computed from a population of paired

data; r_s denotes the rank correlation coefficient computed from sample paired data; ρ_s represents the rank correlation coefficient computed from a population of paired data. The subscript s is used so that the rank correlation coefficient can be distinguished from the linear correlation coefficient r. The subscript does not represent the standard deviation s. It is used in recognition of Charles Spearman, who introduced the rank correlation method.

5. $r_s = 1$. Critical values are -0.886 and 0.886. Reject the null hypothesis of $\rho_s = 0$. There is sufficient evidence to support a claim of a correlation between distance and time.
7. $r_s = 0.821$. Critical values: -0.786, 0.786. Reject the null hypothesis of $\rho_s = 0$. There is sufficient evidence to support the claim of a correlation between the quality scores and prices. These results do suggest that you get better quality by spending more.
9. $r_s = -0.929$. Critical values: -0.786, 0.786. Reject the null hypothesis of $\rho_s = 0$. There is sufficient evidence to support the claim of a correlation between the two judges. Examination of the results shows that the first and third judges appear to have opposite rankings.
11. $r_s = 1$. Critical values: -0.886, 0.886. Reject the null hypothesis of $\rho_s = 0$. There is sufficient evidence to conclude that there is a correlation between overhead widths of seals from photographs and the weights of the seals.
13. $r_s = 0.394$. Critical values: -0.314, 0.314. Reject the null hypothesis of $\rho_s = 0$. There is sufficient evidence to conclude that there is a correlation between the systolic and diastolic blood pressure levels in males.
15. $r_s = 0.651$. Critical values: -0.286, 0.286. Reject the null hypothesis of $\rho_s = 0$. There is sufficient evidence to conclude that there is a correlation between departure delay times and arrival delay times.
17. a. ± 0.707 is not very close to the values of ± 0.738 found in Table A-9.
    b. ± 0.463 is quite close to the values of ± 0.467 found in Table A-9.

Chapter 10: Quick Quiz

1. ± 0.878
2. Based on the critical values of ± 0.878 (assuming a 0.05 significance level), conclude that there is not sufficient evidence to support the claim of a linear correlation between systolic and diastolic readings.
3. The best predicted diastolic reading is 90.6, which is the mean of the five sample diastolic readings.
4. The best predicted diastolic reading is 85.3, which is found by substituting 125 for x in the regression equation.
5. $r^2 = 0.342$
6. False. 7. False. 8. $r = 1$
9. Because r must be between -1 and 1 inclusive, the value of 3.335 is the result of an error in the calculations.
10. $r = -1$

Chapter 10: Review Exercises

1. a. $r = 0.926$. Critical values: $r = \pm 0.707$ (assuming a 0.05 significance level). P-value $= 0.001$. There is sufficient evidence to support the claim that there is a linear correlation between duration and interval-after time.
   b. 85.7% c. $\hat{y} = 34.8 + 0.234x$ d. 81.6 min
2. a. The scatterplot suggests that there is not sufficient sample evidence to support the claim of a linear correlation between heights of eruptions and interval-after times.

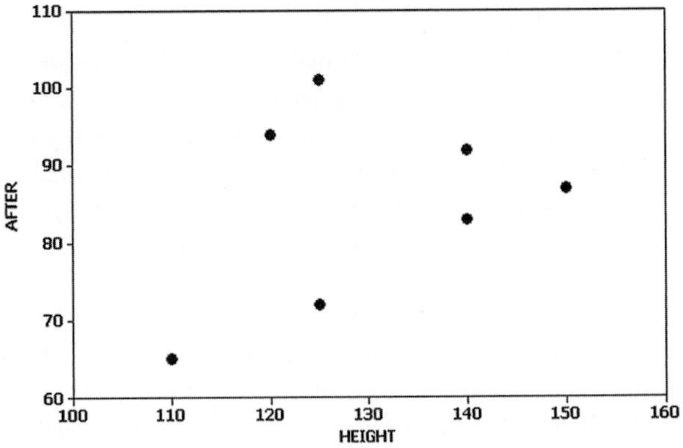

   b. $r = 0.269$. Critical values: $r = \pm 0.707$ (assuming a 0.05 significance level). P-value $= 0.519$. There is not sufficient evidence to support the claim that there is a linear correlation between height and interval-after time.
   c. $\hat{y} = 54.3 + 0.246x$
   d. 86.0 min
3. a. The scatterplot suggests that there is not sufficient sample evidence to support the claim of a linear correlation between duration and height.

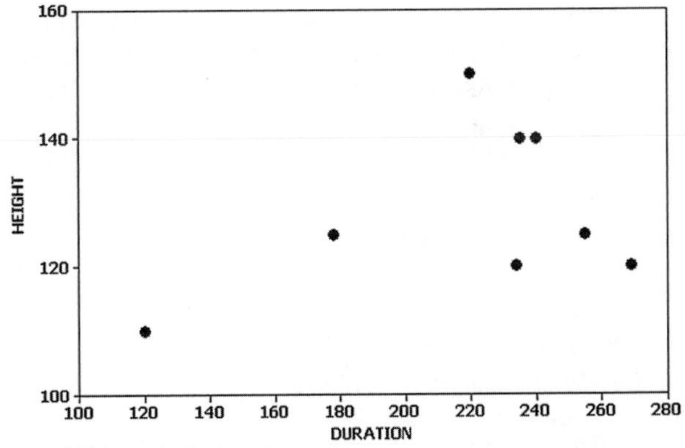

   b. $r = 0.389$. Critical values: $r = \pm 0.707$ (assuming a 0.05 significance level). P-value $= 0.340$. There is not sufficient evidence to support the claim that there is a linear correlation between duration and height.
   c. $\hat{y} = 105 + 0.108x$
   d. 128.8 ft
4. $r = 0.450$. Critical values: $r = \pm 0.632$ (assuming a 0.05 significance level). P-value $= 0.192$. There is not sufficient evidence to support the claim that there is a linear correlation between time and height. Although there is no *linear* correlation between time and height, the scatterplot shows a very distinct pattern revealing that time and height are associated by some function that is not linear.

5. $r_s = 0.714$. Critical values: ± 0.738. Fail to reject the null hypothesis of $\rho_s = 0$. There is not sufficient evidence to support the claim that there is a correlation between the student ranks and the magazine ranks. When ranking colleges, students and the magazine do not appear to agree.

Chapter 10: Cumulative Review Exercises

1. $\bar{x} = 3.3$ lb, $s = 5.7$ lb
2. The highest weight before the diet is 212 lb, which converts to $z = 1.55$. The highest weight is not unusual because its z score of 1.55 shows that it is within 2 standard deviations of the mean.
3. H_0: $\mu_d = 0$. H_1: $\mu_d > 0$. Test statistic: $t = 1.613$. Critical value: $t = 1.895$. P-value > 0.05 (Tech: 0.075). Fail to reject H_0. There is not sufficient evidence to support the claim that the diet is effective.
4. 161.8 lb $< \mu <$ 197.0 lb. We have 95% confidence that the interval limits of 161.8 lb and 197.0 lb contain the true value of the mean of the population of all subjects before the diet.
5. a. $r = 0.965$. Critical values: $r = \pm 0.707$ (assuming a 0.05 significance level). P-value $= 0.000$. There is sufficient evidence to support the claim that there is a linear correlation between before and after weights.
   b. $r = 1$ c. $r = 1$
   d. The effectiveness of the diet is determined by the amounts of weight lost, but the linear correlation coefficient is not sensitive to different amounts of weight loss. Correlation is not a suitable tool for testing the effectiveness of the diet.
6. a. 43.64% (Tech: 43.58%)
   b. 2786.4 g (Tech: 2785.6 g)
   c. 5.00%. Yes, many of the babies do require special treatment.
7. a. H_0: $p = 0.5$. H_1: $p > 0.5$. Test statistic: $z = 3.84$. Critical value: $z = 1.645$. P-value: 0.0001. Reject H_0. There is sufficient evidence to support the claim that the majority of us say that honesty is always the best policy.
   b. The sample is a voluntary response (or self-selected) sample. This type of sample suggests that the results given in part (a) are not necessarily valid.
8. a. Nominal. b. Ratio.
   c. Discrete. d. 0.575
   e. Parameter.
9. a. 0.330 b. 0.870 c. 0.972 d. 7.37%
10.

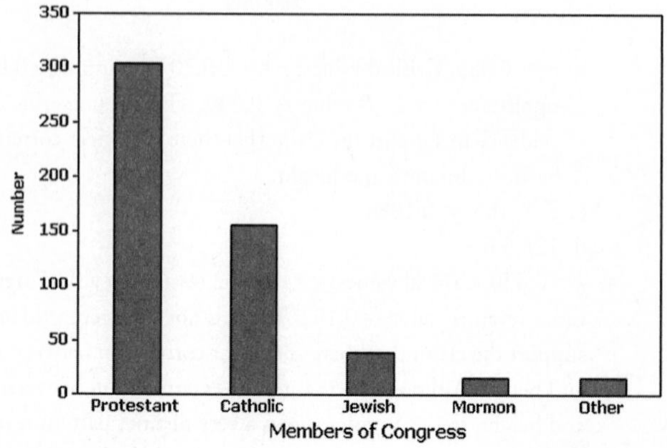

Chapter 11

Section 11-2

1. The test is to determine whether the observed frequency counts agree with the claimed uniform distribution so that frequencies for the different days are equally likely.
3. Because the given frequencies differ substantially from frequencies that are all about the same, the χ^2 test statistic should be large and the P-value should be small.
5. Test statistic: $\chi^2 = 1934.979$. Critical value: $\chi^2 = 12.592$. P-value $= 0.000$. There is sufficient evidence to warrant rejection of the claim that the days of the week are selected with a uniform distribution with all days having the same chance of being selected.
7. Critical value: $\chi^2 = 16.919$. P-value > 0.10 (Tech: 0.516). There is not sufficient evidence to warrant rejection of the claim that the observed outcomes agree with the expected frequencies. The slot machine appears to be functioning as expected.
9. Test statistic: $\chi^2 = 10.375$. Critical value: $\chi^2 = 19.675$. P-value > 0.10 (Tech: 0.497). There is not sufficient evidence to warrant rejection of the claim that homicides in New York City are equally likely for each of the 12 months. There is not sufficient evidence to support the police commissioner's claim that homicides occur more often in the summer when the weather is better.
11. Test statistic: $\chi^2 = 5.860$. Critical value: $\chi^2 = 11.071$. P-value > 0.10 (Tech: P-value $= 0.320$). There is not sufficient evidence to support the claim that the outcomes are not equally likely. The outcomes appear to be equally likely, so the loaded die does not appear to behave differently from a fair die.
13. Test statistic: $\chi^2 = 13.483$. Critical value: $\chi^2 = 16.919$. P-value > 0.10 (Tech: 0.142). There is not sufficient evidence to warrant rejection of the claim that the likelihood of winning is the same for the different post positions. Based on these results, post position should not be considered when betting on the Kentucky Derby race.
15. Test statistic: $\chi^2 = 29.814$. Critical value: $\chi^2 = 16.812$. P-value < 0.005 (Tech: 0.000). There is sufficient evidence to warrant rejection of the claim that the different days of the week have the same frequencies of police calls. The highest numbers of calls appear to fall on Friday and Saturday, and these are weekend days with disproportionately more partying and drinking.
17. Test statistic: $\chi^2 = 7.579$. Critical value: $\chi^2 = 7.815$. P-value > 0.05 (Tech: 0.056). There is not sufficient evidence to warrant rejection of the claim that the actual numbers of games fit the distribution indicated by the proportions listed in the given table.
19. Test statistic: $\chi^2 = 6.682$. Critical value: $\chi^2 = 11.071$ (assuming a 0.05 significance level). P-value > 0.10 (Tech: 0.245). There is not sufficient evidence to warrant rejection of the claim that the color distribution is as claimed.
21. Test statistic: $\chi^2 = 3650.251$. Critical value: $\chi^2 = 20.090$. P-value < 0.005 (Tech: 0.000). There is sufficient evidence to warrant rejection of the claim that the leading digits are from a population with a distribution that conforms to Benford's law. It does appear that the checks are the result of fraud

(although the results cannot confirm that fraud is the cause of the discrepancy between the observed results and the expected results).

23. Test statistic: $X^2 = 1.762$. Critical value: $X^2 = 15.507$. P-value > 0.10 (Tech: 0.988). There is not sufficient evidence to warrant rejection of the claim that the leading digits are from a population with a distribution that conforms to Benford's law. The tax entries do appear to be legitimate.
25. a. 6, 13, 15, 6
    b. 0.1587, 0.3413, 0.3413, 0.1587 (Tech: 0.1587, 0.3413, 0.3414, 0.1586)
    c. 6.348, 13.652, 13.652, 6.348 (Tech: 6.348, 13.652, 13.656, 6.344)
    d. Test statistic: $X^2 = 0.202$ (Tech: 0.201). Critical value: $X^2 = 11.345$. P-value > 0.10 (Tech: 0.977). There is not sufficient evidence to warrant rejection of the claim that heights were randomly selected from a normally distributed population. The test suggests that the data are from a normally distributed population.

Section 11-3

1. Because the P-value of 0.216 is not small (such as 0.05 or lower), fail to reject the null hypothesis of independence between the treatment and whether the subject stops smoking. This suggests that the choice of treatment doesn't appear to make much of a difference.
3. $df = 2$ and the critical value is $X^2 = 5.991$.
5. Test statistic: $X^2 = 3.409$. Critical value: $X^2 = 3.841$. P-value > 0.05 (Tech: 0.0648). There is not sufficient evidence to warrant rejection of the claim that the form of the 100-Yuan gift is independent of whether the money was spent. There is not sufficient evidence to support the claim of a denomination effect.
7. Test statistic: $X^2 = 25.571$. Critical value: $X^2 = 3.841$. P-value < 0.005 (Tech: 0.000). There is sufficient evidence to warrant rejection of the claim that whether a subject lies is independent of the polygraph test indication. The results suggest that polygraphs are effective in distinguishing between truths and lies, but there are many false positives and false negatives, so they are not highly reliable.
9. Test statistic: $X^2 = 42.557$. Critical value: $X^2 = 3.841$. P-value < 0.005 (Tech: 0.000). There is sufficient evidence to warrant rejection of the claim that the sentence is independent of the plea. The results encourage pleas for guilty defendants.
11. Test statistic: $X^2 = 0.164$. Critical value: $X^2 = 3.841$. P-value > 0.10 (Tech: 0.686). There is not sufficient evidence to warrant rejection of the claim that the gender of the tennis player is independent of whether the call is overturned.
13. Test statistic: $X^2 = 14.589$. Critical value: $X^2 = 9.488$. P-value < 0.01 (Tech: 0.0056). There is sufficient evidence to warrant rejection of the claim that the direction of the kick is independent of the direction of the goalkeeper jump. The results do not support the theory that because the kicks are so fast, goalkeepers have no time to react, so the directions of their jumps are independent of the directions of the kicks.
15. Test statistic: $X^2 = 2.925$. Critical value: $X^2 = 5.991$. P-value > 0.10 (Tech: 0.232). There is not sufficient evidence to warrant rejection of the claim that getting a cold is independent of the treatment group. The results suggest that echinacea is not effective for preventing colds.
17. Test statistic: $X^2 = 20.271$. Critical value: $X^2 = 15.086$. P-value < 0.005 (Tech: 0.0011). There is sufficient evidence to warrant rejection of the claim that cooperation of the subject is independent of the age category. The age group of 60 and over appears to be particularly uncooperative.
19. Test statistic: $X^2 = 0.773$. Critical value: $X^2 = 11.345$. P-value > 0.10 (Tech: 0.856). There is not sufficient evidence to warrant rejection of the claim that getting an infection is independent of the treatment. The atorvastatin treatment does not appear to have an effect on infections.
21. Test statistics: $X^2 = 12.1619258$ and $z = 3.487395274$, so that $z^2 = X^2$. Critical values: $X^2 = 3.841$ and $z = \pm 1.96$, so $z^2 = X^2$ (approximately).

Section 11-4

1. a. The chest deceleration measurements are categorized according to the one characteristic of size.
   b. The terminology of *analysis of variance* refers to the method used to test for equality of the three population means. That method is based on two different estimates of a common population variance.
3. The test statistic is $F = 3.288$, and the F distribution applies.
5. Test statistic: $F = 0.39$. P-value: 0.677. Fail to reject H_0: $\mu_1 = \mu_2 = \mu_3$. There is not sufficient evidence to warrant rejection of the claim that the three categories of blood lead level have the same mean verbal IQ score. Exposure to lead does not appear to have an effect on verbal IQ scores.
7. Test statistic: $F = 11.6102$. P-value: 0.000577. Reject H_0: $\mu_1 = \mu_2 = \mu_3$. There is sufficient evidence to warrant rejection of the claim that the three size categories have the same mean highway fuel consumption. The size of a car does appear to affect highway fuel consumption.
9. Test statistic: $F = 0.161$. P-value: 0.852. Fail to reject H_0: $\mu_1 = \mu_2 = \mu_3$. There is not sufficient evidence to warrant rejection of the claim that the three size categories have the same mean head injury measurement. The size of a car does not appear to affect head injuries.
11. Test statistic: $F = 27.2488$. P-value: 0.000. Reject H_0: $\mu_1 = \mu_2 = \mu_3$. There is sufficient evidence to warrant rejection of the claim that the three different miles have the same mean time. These data suggest that the third mile appears to take longer, and a reasonable explanation is that the third lap has a hill.
13. Test statistic: $F = 6.1413$. P-value: 0.0056. Reject H_0: $\mu_1 = \mu_2 = \mu_3 = \mu_4$. There is sufficient evidence to warrant rejection of the claim that the four treatment categories yield poplar trees with the same mean weight. Although not justified by the results from analysis of variance, the treatment of fertilizer and irrigation appears to be most effective.
15. Test statistic: $F = 18.9931$. P-value: 0.000. Reject H_0: $\mu_1 = \mu_2 = \mu_3$. There is sufficient evidence to warrant rejection of the claim that the three different types of cigarettes have the same mean amount of nicotine. Given that the king-size

cigarettes have the largest mean of 1.26 mg per cigarette, compared to the other means of 0.87 mg per cigarette and 0.92 mg per cigarette, it appears that the filters do make a difference (although this conclusion is not justified by the results from analysis of variance).

17. The Tukey test results show different P-values, but they are not dramatically different. The Tukey results suggest the same conclusions as the Bonferroni test.

Chapter 11: Quick Quiz

1. H_0: $p_1 = p_2 = p_3 = p_4 = p_5$. H_1: At least one of the probabilities is different from the others.
2. $O = 23$ and $E = 21.4$.
3. Right-tailed.
4. $df = 4$ and the critical value is $\chi^2 = 9.488$.
5. There is not sufficient evidence to warrant rejection of the claim that occupation injuries occur with equal frequency on the different days of the week.
6. H_0: Response to the question is independent of gender. H_1: Response to the question and gender are dependent.
7. Chi-square distribution.
8. Right-tailed.
9. $df = 2$ and the critical value is $\chi^2 = 5.991$.
10. There is not sufficient evidence to warrant rejection of the claim that response is independent of gender.

Chapter 11: Review Exercises

1. Test statistic: $\chi^2 = 931.347$. Critical value: $\chi^2 = 16.812$. P-value: 0.000. There is sufficient evidence to warrant rejection of the claim that auto fatalities occur on the different days of the week with the same frequency. Because people generally have more free time on weekends and more drinking occurs on weekends, the days of Friday, Saturday, and Sunday appear to have disproportionately more fatalities.
2. Test statistic: $\chi^2 = 6.500$. Critical value: $\chi^2 = 16.919$. P-value > 0.10 (Tech: 0.689). There is not sufficient evidence to warrant rejection of the claim that the last digits of 0, 1, 2, . . . , 9 occur with the same frequency. It does appear that the weights were obtained through measurements.
3. Test statistic: $\chi^2 = 288.448$. Critical value: $\chi^2 = 24.725$. P-value < 0.005 (Tech: 0.000). There is sufficient evidence to warrant rejection of the claim that weather-related deaths occur in the different months with the same frequency. The summer months appear to have disproportionately more weather-related deaths, and that is probably due to the fact that vacations and outdoor activities are much greater during those months.
4. Test statistic: $\chi^2 = 10.708$. Critical value: $\chi^2 = 3.841$. P-value: 0.00107. There is sufficient evidence to warrant rejection of the claim that wearing a helmet has no effect on whether facial injuries are received. It does appear that a helmet is helpful in preventing facial injuries in a crash.
5. Test statistic: $\chi^2 = 4.955$. Critical value: $\chi^2 = 3.841$. P-value < 0.05 (Tech: 0.0260). There is sufficient evidence to warrant rejection of the claim that when flipping or spinning a penny, the outcome is independent of whether the penny was flipped or spun. It appears that the outcome is affected by whether the penny is flipped or spun. If the significance level is changed to 0.01, the critical value changes to 6.635, and we fail to reject the given claim, so the conclusion does change.
6. Test statistic: $\chi^2 = 4.737$. Critical value: $\chi^2 = 7.815$. P-value > 0.10 (Tech: 0.192). There is not sufficient evidence to warrant rejection of the claim that home/visitor wins are independent of the sport.
7. H_0: $\mu_1 = \mu_2 = \mu_3$. Test statistic: $F = 10.10$. P-value: 0.001. Reject the null hypothesis. There is sufficient evidence to warrant rejection of the claim that 4-cylinder cars, 6-cylinder cars, and 8-cylinder cars have the same mean highway fuel consumption amount.

Chapter 11: Cumulative Review Exercises

1. H_0: $p = 0.5$. H_1: $p \neq 0.5$. Test statistic: $z = 7.28$. Critical values: $z = \pm 1.96$. P-value: 0.0002 (Tech: 0.0000). Reject H_0. There is sufficient evidence to warrant rejection of the claim that among those who die in weather-related deaths, the percentage of males is equal to 50%.
2. $59.0\% < p < 65.0\%$. Because the confidence interval does not include 50% (or "half"), we should reject the stated claim.
3. $\bar{x} = 53.7$ years, median $= 60.0$ years, $s = 16.1$ years. Because an age of 16 differs from the mean by more than 2 standard deviations, it is an unusual age.
4. 42.2 years $< \mu <$ 65.2 years. Yes, the confidence interval limits do contain the value of 65.0 years that was found from a sample of 9269 ICU patients.
5. a. $r = -0.0458$. Critical values: $r = \pm 0.632$. P-value $= 0.900$. There is not sufficient evidence to support the claim that there is a linear correlation between the numbers of boats and the numbers of manatee deaths.
   b. $\hat{y} = 96.1 - 0.137x$
   c. 83.2 manatee deaths (the value of $\bar{y}$). The predicted value is not very accurate because it is not very close to the actual value of 78 manatee deaths.
6. a. 630 mm
   b. 14.46% (Tech: 14.48%). That percentage is too high, because too many women would not be accommodated.
   c. 0.7611 (Tech: 0.7599). Groups of 16 women do not occupy a cockpit; because individual women occupy the cockpit, this result has no effect on the design.
7. a. Statistic.
   b. Quantitative.
   c. Discrete.
   d. The sampling is conducted so that all samples of the same size have the same chance of being selected.
   e. The sample is a voluntary response sample (or self-selected sample), and those with strong feelings about the topic are more likely to respond, so it is not a valid sampling plan.
8. a. 0.130 b. 0.4

Credits

Photographs

Title page: top left, Stephen Bonk/Fotolia; top right, Blend Images/Alamy; bottom left, Domine/Shutterstock; bottom right, AISPIX by Image Source/Shutterstock.

The pencil graphic shown on the title page and chapter openers is from Dinnyd/iStockphoto.

Chapter 1

Pages 2–3, Lightpost/Shutterstock
Page 5, Jacek Chabraszewski/Fotolia
Page 7, Sergey Kelin/Shutterstock
Page 8, Ryan McVay/Photodisc/Getty Images
Page 9, Jozsef Szasz-Fabian/Shuterstock
Page 10, William Perugini/Shutterstock
Page 16, Allstar Picture Library/Alamy
Page 17, Andriy Solovyov/Fotolia
Page 18, PVMil/Fotolia
Page 19, Olly/Shutterstock
Page 24, Andersen Ross/Stockbyte/Getty Images
Page 26, Karramba Production/Shutterstock
Page 27, Digital Vision/Getty Vision
Page 28, Tatyana Gladskih/Fotolia
Page 29, Yuri Arcurs/Shutterstock
Page 31, Mearicon/Shutterstock

Chapter 2

Pages 42–43, Guryanov Andrey Vladimirovich/Shutterstock
Page 44, Scott David Patterson/Shutterstock
Page 46, United States Mint Headquarters
Page 47, David Castillo Dominici/Shutterstock
Page 48, Lightpoet/Shutterstock
Page 50, Corbis
Page 61, Stockbyte/Getty Images
Page 62, North Wind Picture Archives/Alamy

Chapter 3

Pages 78–79, Stephen Bonk/Fotolia
Page 81, Anne Kitzman/Shutterstock
Page 82, Lightpoet/Fotolia
Page 96, Fabian Petzold/Fotolia
Page 97, Yuri Arcurs/Fotolia
Page 98, Sergej Razvodovskij/Shutterstock
Page 99, Scott Martin/AP Images
Page 101, Photodisc/Getty Images
Page 113, Andrew Wakeford/Photodisc/Getty Images
Page 114, Jozsef Szasz-Fabian/Shutterstock
Page 115, Shutterstock
Page 118, Shutterstock

Chapter 4

Pages 132–133, Kenon/Shutterstock
Page 135, Studiovespa/Fotolia
Page 136, (a) Pakhnyushchyy/Fotolia; (b) Thumb/Shutterstock; (c) Andrejs Zavadskis/Shutterstock
Page 137, Binkski/Shutterstock
Page 138, Fotolia
Page 139, Shutterstock
Page 140, Olga Lyubkina/Shutterstock
Page 141, Serg Shalimoff/Shutterstock
Page 142, Oleg Svyatoslavsky/Life File/Getty Images
Page 143, S. Meltzer/PhotoLink/Getty Images
Page 144, Photodisc/Getty Images
Page 149, Michael Pettigrew/Fotolia
Page 150, Everett Collection
Page 158, Dejan Milinkovic/Shutterstock
Page 159, L. Barnwell/Shutterstock
Page 160, Dorling Kindersley, Ltd
Page 161, Mayonaise/Fotolia
Page 162, Chepe Nicoli/Shutterstock
Page 169, Corbis
Page 171, Scott Rothstein/Shutterstock
Page 176, Don Farrall/PhotoDisc/Getty Images
Page 177, Ene/Shutterstock
Page 178, Don Farrall/PhotoDisc/Getty Images

Chapter 5

Pages 194–195, Serhiy Kobyakov/Fotolia
Page 200, Ene/Shutterstock
Page 211, Barbara Penoyar/Photodisc/Getty Images

Chapter 6

Pages 236–237, Andres Rodriguez/Fotolia
Page 267, Jacob Hamblin/Shutterstock
Page 283, Image Source/Getty Images
Page 285, Chieh Cheng/iStockphoto

Chapter 7

Pages 316–317, Lucky Business/Shutterstock
Page 320, Ryan McVay/Digital Vision/Getty Images
Page 321, Jason Stitt/Fotolia
Page 339, Paul Hakimata/Alamy
Page 340, Lebrecht/The Image Works
Page 341, Eric Isselée/Fotolia.
Page 342, Anna Kaminska/Shutterstock
Page 343, Gary Blakeley/Shutterstock
Page 348, Jürgen Fälchle/Fotolia
Page 357, Yuri Arcurs/Fotolia

Chapter 8

Pages 374–375, Blend Images/Alamy
Page 377, Bruce Rolff/Shutterstock
Page 386, Photodisc/Getty Images
Page 387, Charlie Hutton/Shutterstock
Page 388, Chad McDermott/Shutterstock
Page 395, Digital Vision/Photodisc/Getty Images
Page 397, Pete Saloutos/Shutterstock
Page 399, Photodisc/Getty Images
Page 410, Ken Hurst/Fotolia
Page 420, Pearson Education, Inc
Page 421, Mark Burnett/Alamy

Chapter 9

Pages 434–435, Helder Almeida/Shutterstock
Page 438, S. Meltzer/Photodisc/Getty Images
Page 439, Rafa Irusta/Fotolia
Page 440, Topham/The Image Works
Page 441, Beboy/Fotolia
Page 448, Keith Brofsky/Stockbyte/Getty Images
Page 451, James Steidl/Shutterstock
Page 452, AVAVA/Fotolia
Page 455, JupiterImages/Arts Corporation/Brand X
Page 463, Milos Luzanin/Shutterstock
Page 464, Barbara Penoyar/Photodisc/Getty Images
Page 465, Monkey Business Images/Shutterstock

Chapter 10

Pages 480–481, Domine/Shutterstock
Page 483, Juan Fuertes/Shutterstock
Page 486, Anneka/Shutterstock
Page 487, Alexander Raths/Fotolia
Page 503, NetPics/Alamy
Page 506, Ryan McVay/Photodisc/Getty Images
Page 507, Digital Vision/Getty Images
Page 520, Michael Matisse/Photodisc/Getty Images

Chapter 11

Pages 534–535, Haveseen/Fotolia
Page 540, Shutterstock
Page 541, Chad McDermott/Shutterstock
Page 542, Serhiy Shullye/Shutterstock
Page 549, Jack Star/PhotoLink/Photodisc/Getty Images

Technology

Index of Applications

CP = Chapter Problem
IE = In-Text Example
M = Margin Example
E = Exercise
BB = Beyond the Basics
R = Review Exercise
CR = Cumulative Review Exercise
DD = Data to Decision
CGA = Cooperative Group Activity
TP = Technology Project

Finance

Food/Drink

Games

General Interest

Health

Surveys and Opinion Polls

Technology

Transportation

War/Warfare

Index

T